CAMBRIDGE LIBRARY COLLECTION

Books of enduring scholarly value

Botany and Horticulture

Until the nineteenth century, the investigation of natural phenomena, plants
and animals was considered either the preserve of elite scholars or a pastime
for the leisured upper classes. As increasing academic rigour and systematisation
was brought to the study of 'natural history', its subdisciplines were adopted
into university curricula, and learned societies (such as the Royal Horticultural
Society, founded in 1804) were established to support research in these areas.
A related development was strong enthusiasm for exotic garden plants,
which resulted in plant collecting expeditions to every corner of the globe,
some-times with tragic consequences. This series includes accounts of some
of those expeditions, detailed reference works on the flora of different regions,
and practical advice for amateur and professional gardeners.

Flora Capensis

This seminal publication began life as a collaborative effort between the Irish
botanist William Henry Harvey (1811–66) and his German counterpart
Otto Wilhelm Sonder (1812–81). Relying on many contributors of specimens
and descriptions from colonial South Africa – and building on the foundations
laid by Carl Peter Thunberg, whose *Flora Capensis* (1823) is also reissued in
this series – they published the first three volumes between 1860 and 1865.
These were reprinted unchanged in 1894, and from 1896 the project was
supervised by William Thiselton-Dyer (1843–1928), director of the
Royal Botanic Gardens at Kew. A final supplement appeared in 1933.
Reissued now in ten parts, this significant reference work catalogues more
than 11,500 species of plant found in South Africa. Volume 5 appeared in
three parts, the second comprising sections published between 1915 and
1925, covering Thymelaeaceae to Ceratophylleae. The 1933 supplement on
Gymnospermae is also incorporated in this reissue.

Flora Capensis

*Being a Systematic Description
of the Plants of the Cape Colony,
Caffraria & Port Natal,
and Neighbouring Territories*

VOLUME 5: PART 2
THYMELAEACEAE TO CERATOPHYLLEAE;
GYMNOSPERMAE

WILLIAM H. HARVEY *ET AL.*

CAMBRIDGE
UNIVERSITY PRESS

CAMBRIDGE
UNIVERSITY PRESS

University Printing House, Cambridge, CB2 8BS, United Kingdom

Cambridge University Press is part of the University of Cambridge.

It furthers the University's mission by disseminating knowledge in the pursuit of
education, learning and research at the highest international levels of excellence.

www.cambridge.org
Information on this title: www.cambridge.org/9781108068123

© in this compilation Cambridge University Press 2014

This edition first published 1925 and 1933
This digitally printed version 2014

ISBN 978-1-108-06812-3 Paperback

FLORA CAPENSIS.

VOL. V. SECT. 2.

DATES OF PUBLICATION OF THE SEVERAL PARTS
OF THIS VOLUME.

PART I., pp. 1–192, was published *April*, 1915.

PART II., pp. 193–384, was published *October*, 1915.

PART III., pp. 385-528, was published *May*, 1920.

PART IV., pp. 529–606, was published *January*, 1925.

FLORA CAPENSIS:

BEING A

Systematic Description of the Plants

OF THE

CAPE COLONY, CAFFRARIA, & PORT NATAL

(AND NEIGHBOURING TERRITORIES)

BY

VARIOUS BOTANISTS.

EDITED BY

SIR WILLIAM T. THISELTON-DYER, K.C.M.G.,
C.I.E., LL.D., D.Sc., F.R.S.

HONORARY STUDENT OF CHRIST CHURCH, OXFORD,
LATE DIRECTOR, ROYAL BOTANIC GARDENS, KEW.

*Published under the authority of the Governments of the
Cape of Good Hope, Natal and the Transvaal.*

VOLUME V. SECTION 2.
THYMELÆACEÆ--CERATOPHYLLEÆ.

LONDON:
L. REEVE & CO., LTD.,
6, HENRIETTA STREET, COVENT GARDEN.
Publishers to the Home, Colonial and Indian Governments.
1925.

LONDON:
PRINTED BY WILLIAM CLOWES AND SONS, LIMITED,
DUKE STREET, STAMFORD STREET, S.E. 1, AND GREAT WINDMILL STREET, W. 1.

PREFACE.

THE first Section of Volume V was published in 1912. The second Section, now at last completed, fell under the shadow of the Great War. Two Parts were, however, issued in 1915 of which the contents had been slowly maturing. Of these the *Thymelæaceæ* were contributed by Mr. C. H. WRIGHT, A.L.S., the *Penæaceæ* by Miss EDITH LAYARD STEPHENS, B.A., of the Botany Department, South African College, Cape Town, the *Loranthaceæ* by Mr. T. A. SPRAGUE, F.L.S., and the *Santalaceæ* by Dr. HILL, F.R.S., the Director of the Royal Botanic Gardens, Kew. The vast order *Euphorbiaceæ*, which bulks so largely in South African vegetation, required prolonged study; the genus *Euphorbia* was undertaken by Mr. N. E. BROWN, A.L.S., who is an authority without a rival on Succulent Plants, and Sir DAVID PRAIN, Treas. R.S., divided with Mr. JOHN HUTCHINSON, F.L.S., the remaining genera. For the present part, Mr. N. E. BROWN also undertook *Urticaceæ* and Mr. HUTCHINSON *Myricaceæ* as well as the genus *Ficus*. The *Betulaceæ* and *Salicineæ* are contributed by Mr. S. A. SKAN.

During the remainder of the War, work on the Flora was in abeyance, but in 1920 a third Part was published not without some difficulty. The cost of printing had enhanced greatly. Part III contained nine sheets and these "have cost within a few shillings of the total cost of Parts I and II," which aggregate twenty-four sheets.

The preface to Section 1 commemorated the loss of many whose generous assistance and co-operation have made the preparation of the Flora possible. Two more—and, as will be seen below, both contributors of material—must be added to the obituary list. HENRY HAROLD WELCH PEARSON, F.R.S., Harry Bolus Professor of Botany in the South African College, and Hon. Director, National Botanic Gardens, Cape Town, died on 3rd November, 1916 (obituary notice and bibliography, *Kew Bulletin*, 1916, pp. 271–281). Sir ISAAC BAYLEY BALFOUR, K.B.E., F.R.S., Professor of Botany in the University of Edinburgh, died 30th November, 1922 (obituary notice and bibliography, *Kew Bulletin*, 1923, pp. 30–35).

For the limits of the regions under which the localities are cited, reference may be made to the Preface to Vol. VI; for the maps which have been used to the Preface and Section 1 of Vol. V.

For the loan or contribution of specimens used in working out the Orders included in the present Section, Kew is indebted to the following :—

Sir I. B. BALFOUR. Loan of *Euphorbiaceæ.*

ALWIN BERGER, La Mortola, loan and gift of specimens of *Euphorbia.*

Mrs. F. BOLUS. Loan of *Santalaceæ* and *Euphorbiaceæ* from the Bolus Herbarium.

Dr. J. I. BRIQUET. Loan of *Thesium* from Boissier Herbarium.

J. L. DRÈGE, Port Elizabeth. Living and dried specimens of *Euphorbia.*

D. J. W. C. GOETHART. Loan of *Euphorbiaceæ* from the Leiden Herbarium.

Dr. RUDOLPH MARLOTH, Capetown. Loan and gift of specimens of *Euphorbia.*

Dr. C. H. OSTENFELD. Loan of *Euphorbiaceæ* from the Universitatets Botaniske Museum, Copenhagen.

Prof. H. H. W. PEARSON. Living and dried plants collected on a Sladen Expedition.

Dr. A. B. RENDLE. Loan of *Thymelæaceæ* and *Euphorbiaceæ* from British Museum.

Archdeacon F. A. ROGERS. Gift of South African plants.

Prof. A. C. SEWARD. Loan of *Santalaceæ* and *Euphorbiaceæ* from Cambridge University Herbarium.

I continue to be indebted for invaluable aid to Mr. C. H. WRIGHT, A.L.S., and to Mr. N. E. BROWN, A.L.S., the former in reading the proofs and in other ways, the latter for working out the localities and distribution.

The present Part completing Vol. V., Sect. 2, completes also the enumeration and description of the Flowering Plants of South Africa belonging to DICOTYLEDONES, as far as Herbarium material at each moment has been available. Vol. V., Sect. 3, with the two succeeding volumes treats in the same way with the MONOCOTYLEDONES. The two classes taken together constitute the subdivision of the Vegetable Kingdom, ANGIOSPERMÆ, in which seeds are enclosed in a seed-vessel.

W. T. T-D.

WITCOMBE,
23rd September, 1924.

SEQUENCE OF ORDERS CONTAINED IN
VOL. V. SECT. 2, WITH BRIEF CHARACTERS.

Continuation of Series V. Daphnales. Ord. CXVIII–CXIX.

CXVIII. THYMELÆACEÆ (page 1). *Calyx-lobes* 4–5, imbricate. *Petals* 4, 5, 8 or 12, sometimes 0, membranous or thick. *Stamens* 4–5 at the throat of the calyx and opposite its lobes, sometimes with 4–5 others in the tube and alternate with the lobes. *Ovary* superior 1–2-celled ; ovule solitary, pendulous. *Fruit* dry or fleshy, indehiscent. *Seeds* albuminous or not.

CXIX. PENÆACEÆ (page 81). *Perianth-lobes* 4, valvate or reduplicate valvate. *Stamens* 4, alternate with the perianth lobes ; anthers basifixed, rarely versatile ; connective conspicuous. *Seeds* exalbuminous ; cotyledons very minute.

CXIX A. GEISSOLOMACEÆ (page 98). *Perianth-lobes* 4, imbricate. *Stamens* 8 ; anthers versatile ; connective scarcely manifest. *Seeds* albuminous ; cotyledons long.

Series vi. Achlamydosporeæ. *Ovary* usually inferior, 1-celled ; ovules 1–3, usually not evident before flowering. *Seeds* without a testa, sometimes adhering to the pericarp. *Perianth* sometimes coloured.

CXIX B. LORANTHACEÆ (page 100). Green shrubs, more rarely herbs, parasitic. *Ovule* solitary, erect.

CXX. SANTALACEÆ (page 135). Herbs, shrubs or trees, often parasitic. *Ovules* 2–4, pendulous from a free-central placenta.

CXX A. BALANOPHORACEÆ (page 212). Fleshy herbs parasitic on roots, without chlorophyll, but usually brightly coloured. *Leaves* reduced to scales.

Series vii. Unisexuales. *Flowers* unisexual. *Ovary* syncarpous or monocarpous ; styles as many as the carpels, often bipartite ; ovules solitary or 2 collateral. *Seed* albuminous or exalbuminous. *Perianth* calycine, small or none ; petal present in some *Euphorbiaceæ*. (*Herbs, shrubs or trees.*)

CXXI. EUPHORBIACEÆ (page 216). *Inflorescence*, perianth and stamens very variable. *Ovary* 2–3-(rarely many-) celled rarely of 1 carpel. *Fruit* usually breaking into 2-valved cocci (winged in *Hymenocardia*), sometimes drupaceous or nutlike. *Albumen* usually copious and fleshy, sometimes thin or none ; radicle superior. (*Herbs, shrubs or trees, often with milky juice, sometimes cactus-like.*)

CXXII. ULMACEÆ (page 516). *Flowers* hermaphrodite or unisexual. *Perianth* 4–5-merous. *Stamens* as many as the perianth-lobes and opposite to them or more ; filaments not inflexed. *Ovule* pendulous. *Fruit* fleshy with hard endocarp (in the South African genera). *Leaves* alternate. *Style* 2-partite. (*Trees or shrubs.*)

CXXII A. MORACEÆ (page 522). *Flowers* unisexual in spikes or globose heads or enclosed within a globose or pear-shaped receptacle or seated upon or immersed in a flattened receptacle. *Perianth* 4–6-lobed. *Filaments* inflexed or erect in bud. *Ovary* 1-celled ; ovule solitary, pendulous ; styles 2 or 1. *Fruit* of achenes.

CXXII B. URTICACEÆ (page 541). *Flowers* monoecious or dioecious. *Perianth* 2–5-lobed or partite or urceolate or tubular or bract-like, sometimes absent from female flowers. *Stamens* 1–5 ; filaments inflexed in bud. *Ovary* 1-celled ; ovule solitary, erect ; style solitary. *Fruit* an achene or fleshy.

CXXIII. MYRICACEÆ (page 561). *Flowers* unisexual, in spikes, without perianth. *Stamens* 2 to many ; filaments short, more or less connate ; anthers erect. *Ovary* 1-celled ; ovule solitary, erect. *Fruit* a drupe. *Seed* without albumen. (*Trees or shrubs, frequently aromatic.*)

CXXIV. BETULACEÆ (page 573). *Flowers* monoecious, in catkins or budlike heads. *Perianth* 4- or fewer-lobed, sometimes none. *Stamens* 2–12 ; filaments free, often split ; anthers erect. *Ovary* inferior, 2-celled ; styles 2 ; ovules solitary, pendulous. *Fruit* a nutlet. (*Trees or shrubs.*)

SERIES viii. ORDINES ANOMALI. Orders nearest allied to those of series vii. Unisexuales, but not sufficiently closely so as to be joined to any one of them.

CXXV. SALICINEÆ (page 574). *Flowers* dioecious, in catkins or racemes. *Perianth* 0. *Stamens* 2 to many ; filaments free connate or. *Ovary* 1-celled ; placentas 2–4 parietal ; ovules 2-many. *Capsule* 2–4-valved. *Seeds* exalbuminous, with silky hairs. (*Trees or shrubs.*)

CXXV A. CERATOPHYLLEÆ (page 580). *Flowers* unisexual, axillary. *Perianth* multipartite. *Stamens* many. *Ovary* 1-celled, 1-ovuled. *Fruit* a nutlet. *Seed* pendulous, exalbuminous. (*Submerged aquatic herb.*)

FLORA CAPENSIS.

ORDER CXVIII. THYMELÆACEÆ.

(By C. H. WRIGHT.)

Flowers hermaphrodite in the South African genera. *Calyx* superior, tubular, often swollen below; lobes 4–5, spreading, two inner sometimes smaller, imbricate. *Petals* 4, 5, 8 or 12, sometimes absent, alternating singly or in groups of 2–3 with the calyx-lobes, smaller than calyx-lobes, membranous or thick, sometimes anther-like, glabrous or surrounded by hairs (*Struthiola*). *Scales* sometimes present in the middle of the calyx-tube (*Cryptadenia*). *Stamens* 4–5 at the throat of the calyx and opposite its lobes, sometimes 4–5 others included in the tube and alternate with the lobes; filaments usually very short; anthers 2-celled, introrse, dehiscing by longitudinal slits. *Hypogynous disc* usually absent from the South African species, sometimes cupular. *Ovary* superior, 1–2-celled, entire, often compressed; style long or short, excentric or central; stigma capitate, penicillate or small; ovule solitary, lateral, pendulous, anatropous. *Fruit* dry or fleshy, indehiscent. *Seed* solitary, pendulous or laterally affixed; testa crustaceous, rarely membranous; embryo straight, cotyledons fleshy; albumen fleshy, copious or scanty, or absent.

Trees or large or small shrubs, with tough fibrous bark, rarely herbs; leaves opposite, alternate or scattered, entire, small and heath-like with inconspicuous nerves, or broader and pinnately nerved, exstipulate; flowers usually terminal in bracteate or ebracteate sessile or peduncled heads, short racemes or spikes, rarely solitary and axillary.

DISTRIB. Genera about 40; species about 400 in Tropical and South Africa, the Mediterranean region and Australia, a few in Asia and North and South America.

Tribe I. EUTHYMELÆEÆ.—*Ovary* 1-celled; ovule solitary.

*Petals 0; *calyx-tube without scales inside.*

    I. **Dais.**—*Stamens* twice as many as the calyx-lobes, included or the upper exserted. *Flowers* 5-merous. *Disc* cupular.

    II. **Arthrosolen.**—*Stamens* twice as many as the calyx-lobes, included. *Flowers* 4- (rarely 5-)merous. *Disc* 0. *Calyx-tube* slender or cylindrical. *Fruit* dry.

    III. **Passerina.**—*Stamens* twice as many as the calyx-lobes, exserted. *Flowers* 4-merous. *Disc* 0. *Calyx* urceolate; lobes as long as the tube. *Fruit* dry.

    IV. **Chymococca.**—*Stamens* twice as many as the calyx-lobes. *Flowers* 4-merous. *Disc* 0. *Calyx* urceolate. *Fruit* baccate.

**Petals 0; *calyx-tube with scales below the insertion of the stamens.*

    V. **Cryptadenia.**—*Flowers* axillary, solitary.

2 THYMELÆACEÆ (Wright).

VI. **Lachnæa.**—*Flowers* terminal, capitate, rarely solitary.

****Petals present, fleshy or membranous.*

†Stamens as many as the calyx-lobes.

VII. **Struthiola.**—Only South African genus.

††Stamens twice as many as the calyx-lobes.

VIII. **Gnidia.**—*Flowers* in bracteate heads or spikes, rarely axillary, 4-merous.

IX. **Lasiosiphon.**—*Flowers* in bracteate heads or spikes, rarely axillary, 5-merous.

X. **Englerodaphne.**—*Flowers* in ebracteate terminal fascicles, 4-merous.

XI. **Synaptolepis.**—*Flowers* axillary, 5-merous.

Tribe II. PHALERIEÆ.—*Ovary* 2-celled, cells 1-ovuled.

XII. **Peddiea.**—*Flowers* pedicelled.   *Petals* 0.

I. DAIS, Linn.

Flowers hermaphrodite.   *Calyx-tube* cylindrical, often curved, naked in the throat; lobes 5, patent.   *Stamens* 10, included, or the upper series or all shortly exserted; filaments short; anthers oblong; connective inconspicuous.   *Disc* hypogynous, cupular, membranous, truncate or toothed.   *Ovary* villous, 1-celled; style long, filiform; stigma ovoid or capitate.   *Fruit* dry, enclosed in the base of the calyx; pericarp membranous.   *Seeds* exalbuminous; testa crustaceous; cotyledons broad, fleshy.

Shrubs; leaves opposite or scattered, often collected at the ends of the branches; flowers in dense stalked heads at the tips of the branches; bracts 2-6, broad, forming an involucre; calyx silky.

DISTRIB. Species 2, one in Madagascar, the other South African.

1. **D. cotinifolia** (Linn. Sp. Pl. ed. ii. 556); a shrub up to 9 ft. high; bark fibrous; leaves opposite and alternate, obovate or oblong, 2½ in. long, 1½ in. wide, acute, cuneate at the base, glabrous, lateral nerves about 10, spreading, looping within the margin; petiole 2 lin. long, stout; peduncle 1½-2 in. long; involucral bracts 4, broadly ovate or almost rotundate, 5 lin. long, chestnut-brown when dry; calyx-tube 6 lin. long, ¼ lin. in diam., cylindrical, silky outside especially near the base, less densely so inside; lobes 3 lin. long, 1¼ lin. wide, ovate, obtuse; filaments filiform, 1 lin. long; anthers oblong, ½ lin. long; ovary ovoid; style filiform, reaching nearly to the lower anthers; stigma globose. *Lam. Ill. t.* 368, *fig.* 1; *Bot. Mag. t.* 147; *Wikstr. in Vet. Acad. Handl. Stockh.* 1818, 348; *Meisn. in Linnæa,* xiv. 388, *and in DC. Prodr.* xiv. 529, *incl. var. major; Drège, Zwei Pfl. Documente,* 148; *Harv. Gen. S. Afr. Pl. ed.* 2, 325; *Wood, Natal Pl. t.* 308; *Sim, For. Fl. Cape Col.* 302, *t.* 153, *fig.* 5.

VAR. β, **parvifolia** (Meisn. in Linnæa, xiv. 388); leaves oval or elliptic, not more than 1 in. long. *DC. Prodr.* xiv. 329; *Drège, Zwei Pfl. Documente,* 142.

VAR. γ, **laurifolia** (Meisn. in DC. Prodr. xiv. 529); leaves lanceolate-oblong, tapering towards both ends, 2-3 in. long, about 1 in. wide. *D. laurifolia, Jacq. Coll.* i. 146, *Ic. t.* 77.

COAST REGION: Komgha Div.; near Keimouth, *Flanagan*, 422! British Kaffraria; Fort Murray, 1200 ft., *Sim*, 1480! Var. β: King Williamstown Div.; Buffalo River, under 1000 ft., *Drège.*

KALAHARI REGION: Orange River Colony; Nelsons Kop, *Cooper*, 849! 3091! Transvaal; near Lydenburg, *Atherstone! Wilms*, 1292!

EASTERN REGION: Transkei; Kentani, *Miss Pegler*, 316! Tembuland; bank of the Umtata River, under 1000 ft., *Drège!* Bazeia, 2000-3000 ft., *Baur*, 22! Griqualand East; by banks of rivers near Clydesdale, 2500 ft., *Tyson in MacOwan & Bolus, Herb. Austr.-Afr.*, 1226! banks of the Umzimkulu River, *Tyson*, 1429! Natal; Nottingham, *Buchanan*, 152! Friedenau Farm, Dumisa, *Rudatis*, 580! and without precise locality, *Sutherland! Gerrard*, 1388! *Wood*, 3155!

Kaffir name "*Intozane*." "The bark of this yields the strongest fibre known to the natives of the colony of Natal." (*Sutherland*.) "The natives [of Kentani] use the bark as a thread." (*Miss Pegler*.) The variety *laurifolia* is known only in cultivation.

Imperfectly known species.

2. **D. canescens** (Bartl. ex Meisn. in Linnæa, xiv. 388); branches stout; branchlets flexuous, canescent and leafy at the apex, leafless and scarred below; leaves alternate, erect, subimbricate, oblong, attenuate at the base, rather obtuse, 2–4 lin. long, flat, 1-nerved on the back, canescent on both surfaces; involucral leaves 6–8, linear-oblong, obtuse, 6 lin. long, scarious at the margin, brownish inside, somewhat silky-canescent; heads 3–6-flowered; calyx silky-pilose; tube filiform, 1 in. long, persistently long-hairy at the base; lobes equal, a quarter the length of the tube, linear-oblong, very obtuse, glabrous inside, slightly revolute at the margin; anthers linear, 5 inserted in the calyx throat, 5 slightly lower. *Meisn. in DC. Prodr.* xiv. 529.

SOUTH AFRICA: without locality or collector's name.

3. **D. eriocephala** (Lichtenst. ex Bartl. in Linnæa,, xiv. 389); branches short, densely leafy; leaves alternate, scarcely imbricate, linear, mucronate, 3–6 lin. long, slightly concave, as well as the branchlets glaucescent and quite glabrous, with a slightly prominent keel; involucral leaves 6–8, oblong-lanceolate, acute, villous outside, fuscous, paler at the margin, the inner larger than the outer; heads many-flowered; calyx unknown.

CENTRAL REGION: Albert Div.; Leeuwefontein, *Lichtenstein*.

Meisner suggests that this may be an *Arthrosolen.*

II. ARTHROSOLEN, C. A. Meyer.

Flowers hermaphrodite. *Calyx-tube* cylindric, slender, circum-scissile above the ovary, naked at the throat; lobes 4, rarely 5, patent. *Stamens* 8, rarely 10, 2-seriate, included; filaments short; anthers oblong or linear; connective inconspicuous. *Disc* none. *Ovary* subsessile, 1-celled; style filiform, usually long; stigma small, capitate. *Fruit* dry, enclosed in the persistent calyx-base;

pericarp membranous.   *Seed* with scanty or no albumen ; testa
crustaceous ; cotyledons fleshy.

Branched shrubs, often small ; leaves scattered or opposite, flat ; flowers spicate,
capitate, or axillary in the upper part of the stem, bracteate or ebracteate.

DISTRIB.   Species 18, 8 in Tropical Africa, one of which extends into South
Africa.

Flowers in terminal heads :
   Involucral bracts conspicuous, coloured   ...   ... (1) **polycephalus.**

   Involucral bracts not conspicuous :
     Leaves linear, glabrous   ...   . (2) **sericocephalus.**

     Leaves oblong-lanceolate, silky   ...   . (3) **calocephalus.**

Flowers in the axils of the upper leaves :
   Leaves oblong-lanceolate, acute, long-pilose ...   ... (4) **ornatus.**

   Leaves ovate, obtuse, glabrous above, villous beneath (5) **spicatus.**

Flowers in terminal (at first subcapitate) spikes :
   Leaves opposite ...   . (6) **fraternus.**

   Leaves alternate, rarely subopposite :
     Leaves obtuse, ½–1 lin. wide ...   . (7) **laxus.**

     Leaves acute, 1–3 lin. wide :
      Bracts 0 ; leaves 2 lin. wide   ...   . (8) **gymnostachys.**

      Bracts present ; leaves 1–1½ lin. wide   ...   ... (9) **variabilis.**

      Bracts 0 ; leaves 3 lin. wide   ...   . (10) **phæotrichus.**

Flowers in terminal pairs or solitary and axillary ; leaves
   opposite ...   . (11) **inconspicuus.**

1. **A. polycephalus** (C. A. Meyer in Bull. Phys.-Math. Acad.
Pétersb. i. 1845, 359) ; a much-branched plant about 1½ ft. high ;
branches erect, virgate, terete, glabrous ; leaves linear-lanceolate,
acute, slightly constricted at the base, 6 lin. long, ½ lin. wide,
glabrous ; heads terminal, ovoid, 4–6-flowered ; inner bracts broadly
ovate, 7 lin. long, 5 lin. wide, outer 5 lin. long and nearly as wide,
all obtuse and silky outside, concave, coloured ; receptacle silky ; calyx
rich yellow ; tube 9 lin. long, the lower ⅓ persistent, narrowly ovoid
and clothed with silky hairs 4 lin. long, the upper part cylindric or
only slightly swollen at the insertion of the lower anthers, silky ;
lobes 3 lin. long, ¾ lin. wide, silky outside ; anthers oblong, ¾ lin.
long ; ovary oblong, 1½ lin. long, silky especially above ; style
filiform, reaching to the calyx-throat.   *Meisn. in DC. Prodr.* xiv.
560 ; *Sim, For. Fl. Cape Col.* 302, *t.* 153, *fig.* 8 ; *N. E. Br. in Kew
Bulletin,* 1909, 135.   *Passerina polycephala, E. Meyer ex Meisn. in
Linnæa,* xiv. 390 ; *Drège, Zwei Pfl. Documente,* 55, 57, 58, 59.
Dais virgata, Lichtenst. ex Meisn. in DC. Prodr. xiv. 530, 561.
Gnidia polyclada, Gilg in O. Kuntze, Rev. Gen. Pl. iii. ii. 281 ;
Dinter, Deutsch Südw. Afr. 96.   *Lasiosiphon polycephalus, Pearson
in Dyer, Fl. Trop. Afr.* vi. i. 228.

SOUTH AFRICA : without locality, *Wallich* !
COAST REGION : Tulbagh Div. ; on the Witzenberg Range near Tulbagh,
Burchell, 8696 ! Albany Div., *Bowker* !
CENTRAL REGION : Calvinia Div. ; Hantam, *Meyer* ! Tarka Div. ; Tarkastad to
Katberg, *Shaw* ! Cradock Div. ; Fish River, near Cradock, *Burke* ! Graaff Reinet

Div. ; Sneeuwberg Range, 4000–5000 ft., *Drège*, 5000–6000 ft., *Bolus*, 129 !
2500 ft., *MacOwan*, 971 ! Murraysburg Div. ; near Murraysburg, 4000 ft., *Tyson*,
196 ! Beaufort West Div. ; Nieuwveld, between Rhinoster Kop and Ganzefontein,
3500–4500 ft., *Drège* ! Fraserburg Div. ; between Zak River and Kopjes Fontein,
Burchell, 1499 ! Carnarvon Div. ; at Leeuwefontein, *Burchell*, 1522 ! near
Carnarvon, *Burchell*, 1537 ! at Buffels Bout, *Burchell*, 1595 ! near Van Wyks
Vley, *Alston* in *MacOwan*, *Herb. Austr.-Afr.*, 1787 ! Victoria West; Nieuwveld,
between Brak River and Uitvlugt, 3000–4000 ft., *Drège* ! Richmond Div. ;
Winterveld, near Limoenfontein and Groot Tafelberg, 3000–4000 ft., *Drège* !
Middleburg Div. ; near Middleburg Road, *Flanagan*, 1380 ! *Sim*, 2207 ! Colesberg
Div. ; Colesberg, *Arnot* !

WESTERN REGION : Great Namaqualand ; Lachanabis, *Dinter*, 901 !

KALAHARI REGION : Griqualand West ; diamond fields, *Nelson*, 256 ! Warrenton,
Miss Adams, 81 ! Orange Free State ; on the Orange River near Aliwal North,
Kuntze ! Bechuanaland ; Pellat Plains, *Burchell*, 2243 ! south of Takun,
Burchell, 2222 ! Transvaal ; Sterkstroom River, *Burke*, 517 ! and without precise
locality, *Holub* !

Also in Tropical Africa.

2. **A. sericocephalus** (Meisn. in DC. Prodr. xiv. 561) ; an erect
sparingly branched shrub nearly 1 ft. high ; branches terete, pilose ;
leaves linear, acutely acuminate, 9–12 lin. long, ½–1 lin. wide,
glabrous, 1-nerved; head globose, 9 lin. in diam., many-flowered ;
peduncle 2–8 in. long, pilose, bearing a few reduced leaves ; involucral bracts few, ovate, acuminate, 5 lin. long, 1½ lin. wide ; calyx
about 4 lin. long ; tube subcylindric, pilose outside, circumscissile
just below the middle and with a tuft of hairs 2½ lin. long near the
base ; lobes 5, about ½ lin. long, ovate, subacute, fleshy ; stamens 10 ;
achene ovoid, 1½ lin. long, granular-punctate. *Gnidia Pretoriæ*,
Gilg in O. Kuntze, Rev. Gen. Pl. iii. ii. 281.

SOUTH AFRICA : without locality, *Zeyher*, 1494 !
KALAHARI REGION : Orange River Colony ; Rhinoster River, *Burke* ! Bechuanaland ; between Mafeking and Ramoutsa, *Lugard* ! Transvaal ; Mooi River, *Burke* !
Aapjes River, *Burke* ! Pretoria, *Kuntze* ! Wonderboompoort, *Rehmann*, 4529 !
Bohfontein, Rustenberg, *Jenkins*, 6904 ! *Collins*, 123 ! Warmbaths, *Miss Leendertz*,
1333 ! Potgietersrust, *Miss Leendertz*, 1919 ! and without precise locality, *McLea
in Herb. Bolus*, 5798 !

3. **A. calocephalus** (C. A. Meyer in Bull. Phys.-Math. Acad.
Pétersb. i. 1845, 359) ; a branched shrub about 1½ ft. high ;
branches virgate, usually simple, slender, terete, often reddish-
brown, pilose ; leaves oblong-lanceolate, up to 1¼ in. long and 4 lin.
wide, the uppermost sometimes elliptic, 7 lin. long, 4 lin. wide, all
acute or subobtuse, with numerous long adpressed hairs on both
surfaces, silky when young ; heads terminal, many-flowered ; involucral bracts similar to the upper leaves, long-silky ; receptacle
hemispherical ; calyx white, 6 lin. long, densely villous outside,
constricted and circumscissile about 1½ lin. above the base, slightly
wider above ; lobes 5, ovate or oblong, ¾ lin. long ; upper anthers
slightly exserted ; ovary oblong ; achene oblong, finely granular
beneath the shining pellucid epidermis, enclosed in the glabrescent
calyx-base. *Meisn. in DC. Prodr.* xiv. 561. *Passerina calo-*

cephala, Meisn. in Linnæa, xiv. 393 ; *Drège, Zwei Pfl. Documente*,
158. *Gnidia calocephala, Gilg in Engl. & Prantl, Pflanzenfam.* iii.
6 A, 228 ; *O. Kuntze, Rev. Gen. Pl.* iii. ii. 280.

EASTERN REGION : Pondoland ; St. Andrews, 1000 ft., *Tyson*, 3138 ! Natal ;
near Durban, *Sanderson*, 85 ! Intschanga, *Rehmann*, 7885 ! between Umkomanzi
and Umlazi Rivers, *Drège* ! Inanda, *Wood*, 38 ! 156. Hillside near Bothas,
2500 ft., *Wood* ! *and in MacOwan, Herb. Austr.-Afr.*, 1525 ! Dumisa Station,
Rudatis, 420 ! Krantz Kloof, *Kuntze* ! and without precise locality, *Gerrard*, 282 !
Zululand, *Mrs. McKenzie* !

4. A. ornatus (Meisn. in DC. Prodr. xiv. 559) ; a much-branched
shrub ; branches at first pilose, finally glabrous and rather rough ;
leaves opposite or occasionally alternate, closely placed, oblong-
lanceolate, acute, 6 lin. long, 1½–3 lin. wide, pilose with long hairs
on the margin and undersurface ; flowers in the axils of the uppermost leaves, subcapitate ; calyx pilose outside ; tube about 7 lin.
long, slightly curved, glabrous inside ; lobes 4, two outer broadly
ovate, rounded at the base, 3½ lin. long, 2¼ lin. wide, pilose outside,
two inner oblong, about 3 lin. long, 1 lin. wide, glabrous on both
surfaces, all obtuse ; anthers oblong, nearly 1 lin. long, upper
shortly exserted ; ovary oblong, pilose at the apex ; style excentric,
filiform, clavate at the apex. *Gnidia vesiculosa, Eckl. & Zeyh. ex
Drège in Linnæa*, xx. 208 ; *Meisn. in DC. Prodr.* xiv. 559.
G. ornata, Gilg in Engl. & Prantl, Pflanzenfam. iii. 6 A, 228.

VAR. β, Gueinzii (Meisn. in DC. Prodr. xiv. 559) ; branches and leaves pilose
longer than in the type ; calyx smaller.

SOUTH AFRICA : without locality, *Ecklon & Zeyher*. Var. β, *Gueinzius*.
COAST REGION : Caledon Div. ; Kleinrivier Mountains, 1000–3000 ft., *Zeyher*,
3755 ! Bredasdorp Div. ; Rhenoster Kop, *Schlechter*, 10603 ! near Elim, *Bolus*,
7852 !

5. A. spicatus (C. A. Meyer in Bull. Phys.-Math. Acad. Pétersb.
i. 1845, 359) ; a shrub about 2 or more ft. high ; stem terete,
glabrous except when young, rough with the scars of fallen leaves ;
leaves ovate, obtuse, imbricate, 3 lin. long, 1¼ lin. wide, glabrous
and somewhat tuberculate (when dry) above, villous below when
young ; flowers solitary in the axils of the upper leaves ; bracteoles 2,
linear, 3 lin. long ; calyx white, pubescent outside ; tube 3½ lin.
long, lower 1½ lin. oval, persistent, upper part funnel-shaped ;
lobes 4, 2 lin. long, 2 outer ovate, acute, 1 lin. wide, 2 inner oblong,
obtuse, ¾ lin. wide ; anthers oblong ; ovary oblong, 1 lin. long,
pubescent at the apex ; style slender, reaching nearly to the upper
anthers ; stigma capitate, hairy. *Meisn. in DC. Prodr.* xiv. 559 ;
Bolus & Wolley-Dod in Trans. S. Afr. Phil. Soc. xiv. 315. *Passerina
spicata, Linn. f. Suppl.* 226 ; *Wikstr. in Vet. Acad. Handl. Stockh.*
1818, 346 ; *Wendl. Bot. Beobacht.* 19, *t.* 2, *fig.* 19 ; *Lodd. Cab. t.* 311 ;
Thunb. Fl. Cap. ed. Schult. 377 ; *Meisn. in Linnæa*, xiv. 398 ; *Drège,
Zwei Pfl. Documente*, 101, 122. *P. lateriflora, Hort. ex Wikstr.
l.c.* 347. *Genista spicata, Eckl. & Zeyh. ex Meisn. in DC. Prodr.*

xiv. 559.   *Gnidia spicata, Gilg in Engl. & Prantl, Pflanzenfam.* iii.
6 A, 228.

SOUTH AFRICA: without locality, *Sieber,* 63! *Forster! Burke,* 180! *Armstrong,*
189!
COAST REGION: Cape Div.; Cape Flats, *Burchell,* 8393! *Mund,* 6! near
Durban Road, *MacOwan & Bolus, Herb. Norm. Austr.-Afr.,* 250! Vygeskraal
River, *Wolley-Dod,* 331! Muizenberg Vley, *Wolley-Dod,* 923! Flats near Doorn
Hoogte, *Wolley-Dod,* 615! sand dunes, *Bolus,* 3698! Pearl Div.; Klein Draken-
stein Mountains and Dal Josaphat, under 1000 ft., *Drège!* Stellenbosch Div.;
between Stellenbosch and Cape Flats, *Burchell,* 8350! Bredasdorp Div.; foot of
Pot Berg, *Mund,* 18! limestone hills between Cape Agulhas and Pot Berg, *Drège.*
Elim, *Schlechter,* 7729! Swellendam Div.; between Buffeljagts River Drift and
Swellendam, *Burchell,* 7291!

6. **A. fraternus** (N. E. Br. in Kew Bulletin, 1901, 132); an
undershrub 6–10 in. high; branches many, slender, adpressed-
pilose; leaves opposite, subsessile, narrowly lanceolate, 4–6 lin.
long, 1–1½ lin. wide, flat, erect, glabrous; spike 6–9 lin. long, laxly
6–10-flowered, ebracteate; calyx adpressed-pilose outside; tube
3½ lin. long, very slender; lobes 4, oblong, 1½ lin. long, ½ lin. wide,
obtuse, yellowish-purple at the tip; stamens 8; achene about 2 lin.
long, narrowly ovoid, smooth, enclosed in the persistent membranous
base of the calyx.

COAST REGION: Queenstown Div.; Mountains near Queenstown, 4000–5000 ft.,
Galpin, 1771!

7. **A. laxus** (C. A. Meyer in Bull. Phys.-Math. Acad. Pétersb. i.
1845, 359); a shrub, 2 ft. or more high, very much branched;
branches terete, pubescent to pilose; leaves linear or oblong, 2–5
lin. long, ½–1 lin. wide, obtuse, pilose when young, finally glabrous
or with a terminal tuft of hairs; flowers 4–8 in a terminal very
short spike; bracts similar to the leaves or slightly wider; calyx
greenish-yellow, 3–4 lin. long, inflated below, very slender at the
centre, funnel-shaped above, pilose outside, circumscissile about the
middle; lobes 4, about ⅓ lin. long, rounded; anthers ⅓ lin. long;
ovary ovoid; achene 1½ lin. long, pilose at the apex, transversely
rugose. *Bolus & Wolley-Dod in Trans. S. Afr. Phil. Soc.* xiv. 315.
Passerina laxa, Linn. f. Suppl. 226; *Wikstr. in Vet. Acad. Handl.*
Stockh. 1818, 346; *Wendl. Bot. Beobacht.* 20, *t.* 2, *fig.* 20; *Thunb.*
Fl. Cap. ed. Schult. 376; *Lodd. Cab. t.* 755; *Drège, Zwei Pfl.*
Documente, 88; *Meisn. in Linnæa,* xiv. 396, *and in DC. Prodr.* xiv.
560. *P. tenuiflora, Willd. Enum. Hort. Berol.* i. 426; *Wikstr.*
l.c. 342; *Bartl. in Linnæa,* xiv. 403. *Gnidia laxa, Gilg in Engl.*
& Prantl, Pflanzenfam. iii. 6 A, 226. *Rhytidosolen laxus, Van*
Tiegh. in Bull. Soc. Bot. France, xl. 75, 1893.

SOUTH AFRICA: without locality, *Forster!*
COAST REGION: Cape Div.; near Capetown, 50 ft., *Bolus,* 3318! between
Cape Town and Table Mountain, *Burchell,* 19! Table Mountain, 1000–2000 ft.,
Drège! Devils Mountain, near Rondebosch, 300 ft., *Bolus in MacOwan & Bolus,*
Herb. Norm. Austr.-Afr., 380! Raapenberg Vley, *Wolley-Dod,* 2108! Signal Hill,
Wolley-Dod, 3037! Stellenbosch Div.; Lowrys Pass, 500 ft., *Schlechter,* 7799!

8. A. gymnostachys (C. A. Meyer in Bull. Phys.-Math. Acad. Pétersb. i. 1845, 359); an undershrub; stems simple or branched, terete, with yellowish adpressed hairs; leaves alternate, rarely the lower subopposite, linear-lanceolate, or oblong, acute, up to 6 lin. long and 2 lin. wide, glabrous above, with long white adpressed hairs on the margins and undersurface of the midrib; spikes terminal, rather lax, 1–2 in. long, ebracteate; calyx densely pilose outside; tube 2–3 lin. long, slender, cylindrical; lobes 4, oblong, obtuse, 1 lin. long; ovary ovoid, compressed, ¾ lin. long, hairy at the apex; style exserted; stigma capitate. *Passerina gymnostachya,* Meisn. *in Linnæa,* xiv. 397, *and in DC. Prodr.* xiv. 560; *Drège, Zw?i Pfl. Documente,* 50.

CENTRAL REGION: Wodehouse Div.; Mooi Flats, 5000–6000 ft., *Drège*! KALAHARI REGION: Orange River Colony; without precise locality, *Cooper,* 825! Transvaal; near Lydenburg, *Wilms,* 1287. Waterval River, *Wilms,* 1034. EASTERN REGION: Natal; Ladysmith, *Rehmann,* 7135!

9. A. variabilis (C. H. Wright); stem slender, terete, woody, glabrous; branches many, virgate, covered at least when young with rather long adpressed hairs; leaves alternate, more rarely subopposite, sessile, oblong or oblong-lanceolate, acute at both ends, silky beneath, glabrous or nearly so above, about 6 lin. long and 1–1½ lin. wide; flowers in a gradually elongating terminal spike; bracts shorter than the calyx; calyx 3–4 lin. long, fleshy, pink or yellowish, persistent part ovoid, about 1½ lin. long; tube cylindrical, adpressed-hairy outside; lobes 4, obtuse, about ¾ lin. long, outer ovate, ½ lin. wide, inner oblong, ⅓ lin. wide; stamens 8; ovary ovate, compressed, tipped by a tuft of straight white hairs ¼ lin. long; style excentric, reaching to the lower anthers; fruit ovoid, acuminate, 1½ lin. long.

KALAHARI REGION: Orange River Colony; Besters Vlei, near Witzies Hoek, 5300 ft., *Bolus,* 8243! Transvaal; near Ermelo, *Burtt-Davy,* 960! Lydenburg, *Wilms,* 1287! 1288! EASTERN REGION: Griqualand East; by streams near Kokstadt, 4250 ft., *Tyson,* 1214! Natal; Weenen County, 4000–5000 ft., *Wood,* 4550! grassy hill near Newcastle, 3000–4000 ft., *Wood,* 7200! near Charlestown, 5000–6000 ft., *Wood,* 4802! and without precise locality, *Gerrard,* 284!

This species varies in the length and colour of its calyx and also in the length and amount of the indumentum upon it. No satisfactory line can be drawn between the forms included here, which merit further study in the field. *A. fraternus,* N. E. Br., the nearest ally, differs in having opposite leaves and in the absence of bracts.

10. A. phæotrichus (C. H. Wright); stems slightly branched, 1 ft. or rather more high, at first hirsute, at length glabrous or nearly so, reddish-brown, 2 lin. in diam.; leaves alternate, usually oblong, acute, entire, 7 lin. long, 3. lin. wide, densely long-silky beneath, more sparingly so above; flowers in a short terminal spike; calyx urceolate, not expanded in the upper part, densely hairy outside, swollen in the lower 2 lin., cylindrical above; lobes 4,

¾ lin. long, ¼ lin. wide, oblong, rounded at the apex, all alike; petals 0; stamens 8 in 2 series; anthers dark brown, subsessile; ovary ovate, compressed, 1 lin. long, with a terminal tuft of straight hairs; style excentric, filiform, reaching to the base of anthers; stigma slightly enlarged. *Gnidia phæotricha, Gilg in O. Kuntze, Rev. Gen. Pl.* iii. ii. 281.

EASTERN REGION: Natal; Inanda, *Wood*, 241! Brakwaal, 5000 ft., *Wood*, 6583, near Durban, *Wood*, 236, near Charlestown, 5000–6000 ft., *Wood*, 5689! Drakensberg Range, *Cooper*, 3079! Van Reenans Pass, *Kuntze*! Bosch Berg, 4000 ft., *MacOwan*, 1539! and without precise locality, *Mrs. Saunders*! *Gerrard*, 283!

11. **A. inconspicuus** (Meisn. in DC. Prodr. xiv. 560); a dwarf undershrub, much branched; branches opposite or almost verticillate, terete, at first minutely puberulous, afterwards ash-coloured and uneven; leaves opposite, rarely almost in whorls of 4, sessile, spreading, ovate-oblong or lanceolate, acute, 2–3 (rarely 4) lin. long, nerveless or obscurely 1-nerved beneath, densely silky-pilose on both surfaces; flowers in terminal pairs or solitary in the axils of the uppermost leaves, about as long as the leaves, white or yellowish silky outside; calyx-tube scarcely swollen at the apex; lobes 4, half as long as the tube, oblong, obtuse, glabrous inside, two outer rather wider, lilac inside, two inner yellow; anthers 8, sessile, upper 4 exserted. *Passerina inconspicua, Meisn. in Linnæa,* xiv. 397.

SOUTH AFRICA: without locality, *Drège,* 7372.

III. PASSERINA, Linn.

Flowers hermaphrodite. *Calyx-tube* slender from an ovoid base, circumscissile above the ovary, naked in the throat; lobes 4, patent, about as long as the tube. *Stamens* 8, inserted in the calyx-throat, exserted; filaments rather shorter than the calyx-lobes; anthers ovate or oblong; connective narrow. *Disc* 0. *Ovary* subsessile, glabrous, 1-celled; style slenderly filiform; stigma globose, papillate. *Fruit* dry, included in the persistent base of the calyx; pericarp membranous. *Seed* with crustaceous testa; albumen fleshy; cotyledons fleshy, plano-convex.

Heath-like shrubs; leaves small, decussate; flowers in the axils of bracts (usually wider than the leaves) arranged in terminal spikes.

DISTRIB. Species about 10, all South African.

Leaves spreading, falcate, 4 lin. long   ...   . (1) **falcifolia.**
Leaves not spreading, nor closely adpressed to the branches:
    Leaves stout, 2½ lin. long   ...   . (2) **Galpini.**
    Leaves acicular, 1½–3 lin. long:
      Leaves glabrous   ...   . (3) **filiformis.**
      Leaves pilose when young   ...   . (4) **comosa.**

Leaves closely adpressed to the branches, not more than
1½ lin. long:
Bracts smooth when dry :
  Calyx woolly ... ... (5) **laniflora.**
  Calyx glabrous ... ... (6) **paleacea.**
Bracts sulcately marked when dry :
  Leaves shorter than the internodes ; spikes elongated (7) **rubra.**
  Leaves usually longer than the internodes ; spikes
  congested :
    Branches at first tomentose :
      Leaves linear or ovate-oblong ... ... (8) **ericoides.**
      Leaves ovate ... ... (9) **rigida.**
    Branches at first puberulous ... ... (10) **corymbosa.**

1. **P. falcifolia** (C. H. Wright) ; stem terete, much-branched, closely tomentose when young ; leaves falcate, triangular in section, slightly grooved above, acute beneath, 4 lin. long, ⅓ lin. thick, obtuse, glabrous ; flowers in dense terminal spikes about 6 lin. long ; bracts ovate, cuspidate, 3 lin. long, 2 lin. wide, glabrous outside, densely woolly within ; calyx-tube 3 lin. long, ovoid in the lower third, cylindrical above, densely woolly outside ; lobes 4, ovate, obtuse, patent, 1½ lin. long, 1 lin. wide, the 2 outer more concave than the inner ; stamens inserted in the calyx-throat, shorter than the lobes ; anthers ovoid, 1 lin. long ; ovary oblong, 1 lin. long. *P. filiformis, var. falcifolia, Meisn. in Linnæa,* xiv. 399, *and in DC. Prodr.* xiv. 562 ; *Drège, Zwei Pfl. Documente,* 118, 124.

SOUTH AFRICA : without locality, *Wallich* ! *Mund* !
COAST REGION : George Div. ; Wolf-drift, Malgaten River, *Burchell,* 6109 ! Knysna Div. ; Ruigte Valley, under 500 ft., *Drège* ! Uitenhage Div. ; Uitenhage, in the stony channel of the Zwartkops River, *Zeyher,* 277 ! Van Stadens Berg, 1000–2000 ft., *Drège* !

2. **P. Galpini** (C. H. Wright) ; branches short, glabrous ; leaves closely placed, subcylindrical, slightly incurved, 3 lin. long, ½ lin. thick, glabrous, shining ; flowers clustered at the ends of the branches ; bracts transversely oblong, 2½ lin. wide, scarious, produced above into a thick subulate obtuse lobe 1 lin. long ; bracteoles 0 ; woolly at the base inside ; calyx glabrous ; tube narrowly ovoid, 1½ lin. long, ½ lin. in diam. below ; lobes 1½ lin. long, 1 lin. wide, ovate, concave, obtuse ; longer stamens as long as the calyx-lobes ; anthers obtusely cordate, ½ lin. long ; ovary oblong, glabrous ; style slightly longer than the calyx-tube ; stigma penicillate.

COAST REGION : Riversdale Div. ; Milkwoodfontein, about 600 ft., *Galpin,* 4491 !

3. **P. filiformis** (Linn. Sp. Pl. ed. i. 559) ; stem much branched, terete ; branches puberulous ; leaves acicular, triquetrous, acute, 1½–3 lin. long, adpressed to the branches, straight or very slightly incurved ; spikes terminal, very dense, about 9 lin. long ; bracts

ovate, cuspidate, 2½ lin. long, 1½ lin. wide, glabrous outside, very woolly within, midrib prominent on the back; calyx-tube about as long as the bract, ovoid and nearly glabrous below, slender, cylindrical and clothed with patent white hairs in the upper half; lobes patent, elliptical, rounded at the apex, 1½ lin. long, not quite half as broad, glabrous; stamens exserted; anther-cells diverging below; ovary elliptic, compressed; style filiform, about as long as the stamens. *Thunb. Prodr.* 75, *and Fl. Cap. ed. Schult.* 374; *Wendl. Beobacht.* 18, *t.* 2, *fig.* 15; *Willd. Sp. Pl.* ii. 429; *Poir. Encycl.* v. 40; *Wikstr. in Vet. Acad. Handl. Stockh.* 1818, 324; *Meisn. in Linnæa,* xiv. 399, *and in DC. Prodr.* xiv. 562; *Bolus & Wolley-Dod in Trans. S. Afr. Phil. Soc.* xiv. 315, *partly.*

SOUTH AFRICA : without locality, *Sieber,* 74! *Pappe!* *Forster!* *Forsyth!* *Thom,* 553! 577! *Zeyher!* *Mund!* *Harvey,* 642! 692! *Hooker!* *Miller!* *Talbot!* *Laubner!*

COAST REGION. Paarl Div.; Paarl Mountains, *Drège!* Paarl, *Bolus,* 2924! Cape Div.; hills and flats near Cape Town, *Burchell,* 66! 276! 473! *Ecklon,* 508! *Bolus,* 2925! *Wilms,* 3590! *Wolley-Dod,* 3103! *Mund and Maire!* *Wright!* Swellendam Div.; Swellendam, *Burke,* 45! Riversdale Div.; Great Valsch River, *Burchell,* 6544! Mossel Bay Div.; between the landing place at Mossel Bay and Cape St. Blaize, *Burchell,* 6267! Attaques Kloof, *Gill!* Humansdorp Div.; Diep River, near Humansdorp, *Bolus,* 2440! Port Elizabeth Div.; Port Elizabeth, drift sands, *Sim,* 20! Bathurst Div.; Port Alfred, *Hutton,* 302! Albany Div.; Howisons Poort, near Grahamstown, *MacOwan,* 103! Zwartwater Poort, *Burchell,* 3419! and without precise locality, *Zeyher,* 72! King Williamstown Div. ; Mount Coke, 1000 ft., *Sim,* 1380!

CENTRAL REGION : Clanwilliam Div.; Wupperthal, *Baron Th. von Wurmb.*

WESTERN REGION : Little Namaqualand; Modderfontein Berg, 4000–5000 ft., *Drège.*

EASTERN REGION : Tembuland ; Equtyeni, near All Saints, *Baur,* 1162! Natal; Inanda, *Wood,* 1182! Van Reenan, 5000 ft., *Wood,* 6601! and without precise locality, *Gerrard,* 1478! *Sutherland!*

4. P. comosa (C. H. Wright); stem terete, pilose when young; leaves 4-ranked, closely adpressed, almost straight, more or less convex on the back, subobtuse, pilose (especially when young) in the upper part, about 2 lin. long and ⅓ lin. thick; spikes terminal; bracts ovate, obtuse, 1½ lin. long, densely pilose; calyx 2 lin. long, pilose outside; tube ovoid below, shortly cylindrical above; lobes 4, obovate, 1½ lin. long, ¾ lin. wide, obtuse; stamens exserted; filaments filiform, ¾ lin. long, the alternate 1¼ lin. long. *P. filiformis, var. comosa, Meisn. in Linnæa,* xiv. 399, *and in DC. Prodr.* xiv. 562.

WESTERN REGION : Little Namaqualand; Kamiesberg Range, *Drège,* 2570!

5. P. laniflora (C. H. Wright); a woody plant with numerous short branches, pilose at their tips; leaves strictly 4-ranked, oblong, slightly incurved, 2 lin. long, ¼ lin. wide, obtuse, glabrous; flowers clustered at the end of the branches; bracts ovate, 1½ lin. long; bracteoles 2, thickly triquetrous, 1½ lin. long; calyx densely woolly all over; tube subcylindrical, 1½ lin. long; lobes 4, broadly ovate,

1½ lin. long, 1 lin. wide, obtuse; filaments slender, the 4 longer half as long as the calyx-lobes; appendages cylindrical, ¼ lin. long; ovary ovoid; style filiform, 2 lin. long; stigma penicillate.

COAST REGION : Clanwilliam Div.; Cederberg Range, Sneeuw Kop, 4500 ft., *Bodkin in Herb. Bolus,* 9086!

6. P. paleacea (Wikstr. in Vet. Acad. Handl. Stockh. 1818, 323); a much-branched shrub; branches lax, with a thick covering of wool at the apex, finally glabrous; leaves linear, obtuse, trigonous, 1–1½ lin. long, adpressed, subimbricate, glabrous; spikes short, comose; bracts scaly, coloured, imbricate, broadly ovate, obtuse, 1½ lin. long and wide, convex and obtusely keeled on the back, smooth, nerveless; calyx glabrous; tube as long the bracts. *Meisn. in Linnæa,* xiv. 400, *and in DC. Prodr.* xiv. 562. *P. salsolæfolia, Lam. Encycl.* v. 41; *Meisn. in Linnæa,* xiv. 400, *partly. Lachnæa paleacea, Herb. Banks. ex Wikstr. in Vet. Acad. Handl. Stockh.* 1818, 324.

SOUTH AFRICA : without locality, *Sparrmann, Drège, Ludwig.*

7. P. rubra (C. H. Wright); branches virgate, slender, white-pubescent when young, at length glabrous, internodes slightly longer than the leaves; leaves oblong, obtuse, triquetrous, 1½ lin. long, ⅓ lin. thick, glabrous; flowers crowded near the upper part of the branches; bracts ovate, 2 lin. long, 1½ lin. wide, densely silky inside, glabrous and strongly 5- or 7-ribbed outside; bracteoles 0; calyx-tube ovoid, 1½ lin. long; lobes elliptic, concave, obtuse, 1⅓ lin. long, ⅔ lin. wide, subscarious; stamens 8, the longer as long as the calyx-lobes; ovary ovoid, glabrous; style filiform; stigma penicillate.

COAST REGION : Riversdale Div.; Muiskraal, near Garcias Pass, 1200 ft., *Galpin,* 4492! Albany Div.; mountains near Howisons Poort, 2000 ft., *MacOwan,* 103!

8. P. ericoides (Linn. Mant. 236); branches numerous, slender, short, at first tomentose, finally glabrous; leaves closely adpressed to the branches, linear or ovate-oblong, 1 lin. long, obtuse or truncate, thick, glabrous; flowers in short terminal spikes; bracts transversely oblong, 1 lin. long, 1½ lin. wide, obtuse, glabrous outside, woolly within; calyx-tube ovoid, 1½ lin. long, glabrous, submembranous; lobes elliptic, concave, obtuse, scarious, 1 lin. long, ¾ lin. wide; longer stamens two-thirds as long as the calyx-lobes; ovary ovate; style filiform, longer than the calyx-tube. *Lam. Encycl.* v. 41; *Wikstr. in Vet. Acad. Handl. Stockh.* 1818, 325; *Meisn. in Linnæa,* xiv. 401, *and in DC. Prodr.* xiv. 562; *Drège, Zwei Pfl. Documente,* 114, 129. *P. glomerata, Thunb. Prodr.* 75, *and Fl. Cap. ed. Schult.* 374. *Lachnæa conglomerata, Linn. Sp. Pl. ed.* i. 560; *Willd. Sp. Pl.* ii. 434.

COAST REGION : Cape Div. ; Simons Bay, *Wright*! Cape Flats, *Burchell,* 8389 ! near Wynberg, *Bolus,* 2926 ! Ceres Div. ; between Hex River Mountains and Bokkeveld, 3000–4000 ft., *Drège*! Riversdale Div. ; hills near Zoetemelks River, *Burchell,* 6781! Swellendam Div. ; on dry hills near Breede River, *Burchell,* 7463 ! Mossel Bay Div. ; between Great Brak River and Little Brak River, *Burchell,* 6163 ! Uitenhage Div. ; without precise locality, *Zeyher,* 156 ! Port Elizabeth Div. ; Port Elizabeth, on the sand hills and rocky shores, under 100 ft., *Drège*! Albany Div.; Howisons Poort, *Cooper,* 2301 ! Bathurst Div. ; between Kasuga River and Port Alfred, *Burchell,* 4049 !

CENTRAL REGION : Somerset Div. ; Somerset, *Bowker*! Molteno Div. ; Broughton, near Molteno, *Flanagan,* Albert Div. ; without precise locality, *Cooper,* 625 !

KALAHARI REGION : Orange Free State ; without precise locality, *Cooper,* 842 ! Basutoland ; without precise locality, *Cooper,* 702 ! 2302 ! Transvaal ; MacMac, *Mudd* !

EASTERN REGION: Natal ; near Durban, *Peddie* ! Mooi River, *Wood* ! Van Reenen, *Wood,* 11405 ! Klip River, *Sutherland* !

9. **P. rigida** (Wikstr. in Vet. Acad. Handl. Stockh. 1818, 326) ; a much-branched shrub ; branchlets tomentose in the upper part ; leaves opposite, 4-ranked, imbricate, linear, obtuse, almost triquetrous, 1–2 lin. long, ½ lin. wide ; flowers few at the ends of the branches ; bracts broadly ovate, 1 lin. long, ¾ lin. wide, obtuse or subacute ; calyx-lobes oblong, obtuse, 1 lin. long. *Meisn. in DC. Prodr.* xiv. 563, *and in Linnæa,* xiv. 402. *P. filiformis, Bolus & Wolley-Dod in Trans. S. Afr. Phil. Soc.* xiv. 315, *partly, not of Linn.*

VAR. β, **truncata** (Meisn. in Linnæa, xiv. 402) ; leaves adpressed to the branches, approximately quadrifarious, obtuse or almost præmorse, about 1½ lin. long ; spikes terminal, not comose ; calyx-lobes oval, obtuse. *DC. Prodr.* xiv. 563. *P. rigida, Drège, Zwei Pfl. Documente,* 68, 69.

VAR. γ, **tetragona** (Meisn. in Linnæa, xiv. 402) ; leaves very slightly spreading, strictly quadrifarious, obtuse ; spikes terminal, not comose ; bracts ovate or subrotund, obtuse, keeled, distinctly nerved ; calyx-lobes oblong. *Meisn. in DC. Prodr.* xiv. 563. *P. rigida, var.* β, *Drège, Zwei Pfl. Documente,* 75.

VAR. δ, **comosa** (Meisn. in Linnæa, xiv. 402) ; leaves adpressed to the branches, 1–1½ lin. long, convex, obtuse, not or very slightly keeled near the apex ; flowers in tufted slender spikes at the ends of the branches ; bracts ovate-oblong, nerves indistinct ; calyx-lobes narrowly oblong. *Meisn. in DC. Prodr.* xiv. 563. *P. rigida, Drège in Linnæa,* xx. 210.

SOUTH AFRICA : without locality, *Sparrmann.*

COAST REGION : Cape Div. ; Table Mountain, *Burchell,* 626 ! near Simonstown, *Wolley-Dod,* 1878 ! 2927 ! Houts Bay, *Wolley-Dod,* 1575 ! Riversdale Div. ; summit of Kampsche Berg, *Burchell,* 7129 ! George Div. ; on the Cradock Berg, near George, *Burchell,* 5929 ! Port Elizabeth Div. ; shore near Port Elizabeth, *Zeyher,* 405 ! Bathurst Div. ; near Port Alfred, *Burchell,* 3835 ! 3950 ! King Williamstown Div. ; Perie, 3000 ft., *Sim,* 68 ! Var. β : Tulbagh Div. ; Tulbagh, *Pappe* ! near Tulbagh waterfall, *Zeyher,* 43 ! near Saron, *Schlechter,* 10627 ! Knysna Div. ; Ruigte Vallei, under 500 ft., *Drège* ! Var. γ : Clanwilliam Div. ; Ezels Bank, Ceder Bergen, 4000–5000 ft., *Drège* ! Var. δ : Caledon Div. ; tops of the mountains of the Baviaans Kloof, near Genadendal, *Burchell,* 7761 ! Genadendal, *Bolus in Herb. Norm. Austr.-Afr.,* 687 ! Uitenhage Div. ; Zwartkops River, *Zeyher,* 7381 ! Port Elizabeth Div. ; Krakakamma, *Zeyher,* 3780 !

CENTRAL REGION : Ceres Div. ; Gydouw Mountain, *Schlechter,* 10220 !

WESTERN REGION : Var. β : Little Namaqualand ; near Spektakel, 3000 ft., *Bolus,* 9507 !

KALAHARI REGION : Orange Free State ; Harrismith, *Sankey,* 69 !

EASTERN REGION : Natal ; on sandhills near the sea beach, Durban, *Wood,* 1712 ! Var. δ : Natal ; Durban, *Sutherland* ! *Wilms,* 2277 ! and without precise locality, *Gerrard,* 95 !

10. P. corymbosa (Eckl. ex Meisn. in DC. Prodr. xiv. 562) ; a much-branched undershrub ; branchlets slender, at first puberulous ; leaves oblong, obtuse, about 1½ lin. long, smooth ; flowers collected in short thick spikes at the ends of the branches ; bracts ovate, subacute, 1½ lin. long, densely woolly inside, glabrous, ribbed and keeled outside ; calyx-tube 2 lin. long, inflated and very finely pubescent in the lower half, cylindrical, densely pubescent and ⅓ lin. in diam. above ; lobes elliptical, obtuse, concave, 1 lin. long ; stamens inserted in the calyx-throat, the outer slightly longer than the lobes ; ovary ¾ lin. long, compressed, glabrous ; style lateral.

COAST REGION : Tulbagh Div. ; Piquetberg Road, *Tyson*, 2318 ! Saron, 3000 ft., *Schlechter*, 10660 ! Swellendam Div. ; Swellendam, *Zeyher* ! Albany Div. ; Highlands, *Misses Daly & Sole*, 297 ! Bathurst Div. ; Port Alfred, *Hutton*, 1603 ! Queenstown Div. ; summit of Andriesberg, near Bailey, 6400–6800, *Galpin*, 2028 ! CENTRAL REGION : Graaff Reinet Div. ; summit of Oude Berg, 5000 ft., and Sneewberg Mountains, 5000–7500 ft., *Bolus*, 170 ! KALAHARI REGION : Basutoland ; Machacha, 9000 ft., *Bryce* ! EASTERN REGION : Natal ; Mooi River, 4000 ft., *Wood*, 4036 !

Imperfectly known species.

11. P. eriophora (Gandog. in Bull. Soc. Bot. France, lx. 418, 1913) ; a dwarf shrub, very densely branched ; branches white-woolly, arcuately deflexed or pendulous ; leaves triangular, dilated at the base, keeled, glabrous, straight, imbricate ; flowers white, in a short terminal spike, axis very woolly ; berry subglobose, reddish.

EASTERN REGION : Natal ; near Durban, *Wood*, 1702, 6592. Allied to *P. filiformis*, Linn.

12. P. hamulata (Gandog. in Bull. Soc. Bot. France, lx. 418, 1913) ; quite glabrous except the branches ; branches rigid, straight, slender, tomentose ; leaves about 1 lin. long, quadrifarious, imbricate, uncinate, pallid, dilated, mucronate and hooked at the apex ; flowers white, capitate ; berry globose.

COAST REGION : Cape Div. ; on sand dunes near Wynberg, *Bolus*. This may be the same as *P. ericoides*, Linn.

IV. CHYMOCOCCA, Meisn.

Flowers hermaphrodite. *Calyx-tube* urceolate, constricted above the ovary, naked in the throat ; lobes 4, patent. *Stamens* 8, fixed in the calyx-throat, exserted ; filaments a little shorter than the calyx-lobes ; anthers ovate-oblong, connective narrow. Hypogynous *disc* 0. *Ovary* subsessile, glabrous, 1-celled ; style filiform ; stigma globose. *Fruit* baccate, included in the base of the calyx or not, striate. *Seed* albuminous ; testa rather thick, at length subcrustaceous ; cotyledons fleshy.

A heath-like shrub with the habit of *Passerina* ; bracts not different from the stem leaves.

DISTRIB. Monotypic, endemic.

1. C. empetroides (Meisn. in DC. Prodr. xiv. 565) ; a small much-branched shrub ; branchlets whitish-tomentose ; leaves opposite or subopposite, oblong, trigonous, obtuse, rather thick, 1–1½ lin. long, smooth or rugulose ; bracts nearly 1 lin. wide, acuminate ; flowers in a short spike at the apex of the branchlets ; calyx green (*Meisner*), 1½ lin. long, deciduous ; tube oblong, puberulous outside and in the throat ; lobes roundish, equal, glabrous, ¾ lin. across ; stamens 8, the 4 opposite the calyx-lobes nearly as long as them, the others rather shorter ; ovary ovoid ; style slender, nearly as long as the ovary ; stigma capitate, papillose ; berry globose, 3 lin. in diam., shining scarlet (*Bolus*) ; seed ovoid, slightly compressed, shining black. *Bolus & Wolley-Dod in Trans. S. Afr. Phil. Soc.* xiv. 315. *Passerina filiformis, var. crassifolia, Eckl. & Zeyh. ex Meisn. l.c.*

SOUTH AFRICA : without locality, *Zeyher* ! *Harvey*, 641 !
COAST REGION : Cape Div. ; Simons Bay, *Wright* ! Kalk Bay, 50 ft., *Bolus*, 4498 !
WESTERN REGION : Vanrhynsdorp Div. ; Zout River, *Bolus*, 42 !

V. CRYPTADENIA, Meisn.

Flowers hermaphrodite. *Calyx-tube* cylindrical or funnel-shaped, constricted and at length circumscissile above the ovary ; naked inside above the filaments ; lobes 4, patent, as long as or longer than the tube. *Stamens* 8, all or the 4 upper shortly exserted ; filaments filiform ; anthers oblong ; connective narrow. *Scales* 8 to many in a single series below the stamens. *Disc* 0. *Ovary* sessile, 1-celled ; style filiform, sometimes thicker and hairy above. *Fruit* dry, included in the persistent thinly membranous base of the calyx. *Seed* with a shining crustaceous testa ; albumen fleshy ; embryo subterete.

Much-branched, heath-like undershrubs ; leaves decussate, small, acerose or obtuse ; flowers solitary at the apex of the branches or in the axils of the upper leaves, often silky outside, bibracteolate.

DISTRIB. Species 5, endemic.

Calyx-lobes 6 lin. long ... ... (1) **grandiflora.**

Calyx-lobes 3 lin. long :
    Branchlets glabrous ... ... (2) **uniflora.**

    Branches at first puberulous ... ... (3) **laxa.**

Calyx-lobes 2 lin. long, oblong ... ... (4) **breviflora.**

Calyx-lobes 1¼ lin. long, ovate ... ... (5) **filicaulis.**

1. **C. grandiflora** (Meisn. in Linnæa, xiv. 405); a small much-branched shrub; branchlets slightly pubescent; leaves opposite, sessile, adpressed, linear or oblong, concave, rounded on the back or keeled towards the apex, acute, 2–4 lin. long, nearly 1 lin. wide, the uppermost slightly wider, glabrous; calyx mauve (*Galpin*), silky outside; tube cylindrical, ¾ lin. in diam.; lobes oblong, obtuse or acute, up to 6 lin. long and 3 lin. wide; scales oblong, ⅓ lin. long; the 4 stamens opposite the calyx-lobes a third as long as them, the others shorter; ovary oblong, ¾ lin. long; style slightly oblique, slender, reaching to the level of the longer stamens; stigma subcapitate. *Meisn. in Hook. Lond. Journ. Bot.* ii. 552, *and in DC. Prodr.* xiv. 573; *Bolus & Wolley-Dod in Trans. S. Afr. Phil. Soc.* xiv. 315; *Gilg in Engl. & Prantl, Pflanzenfam.* iii. 6 A, 234, *fig.* 83 E. *Gnidia pachyphylla, Spreng. ex Meisn. in Linnæa,* xiv. 405. *Calysericos caniculata, Eckl. & Zeyh. ex Drège in Linnæa,* xx. 210.

VAR. β, **latifolia** (Meisn. in DC. Prodr. xiv. 573); leaves oval or oblong, flat or slightly incurved at the margins, 2½–4 lin. long, 1½–2½ lin. wide, acute, nerveless; calyx half as large as in the type; tube as long as the leaves; lobes ovate, obtuse, 4–4½ lin. long, 3 lin. wide.

SOUTH AFRICA : without locality, *Sieber*! *Grey*! *Wallich*! *Drège, Harvey*, 643! Var. β, without collector's name.
COAST REGION : Malmesbury Div.; Mamre, *Baur*! Cape Div.; Cape Flats, near Rondebosch, *Burchell*, 163! 194! *Ecklon*, 361! Cape Peninsula, Kommetjes, 100 ft., *Galpin*, 4494! Vygeskraal Farm, *Wolley-Dod*, 663! Muizenberg, 1000 ft., *Bolus*, 4635! Simons Bay, *Wright*! *MacGillivray*, 632! 633! *Milne*, 170! Stellenbosch Div.; between Stellenbosch and Bottelary Hill, *Burchell*, 8342! between Lowrys Pass and Jonkers Hoek, *Burchell*, 8326! Caledon Div.; Houw Hoek, 1500 ft., *Schlechter*, 9375!

2. **C. uniflora** (Meisn. in Linnæa, xiv. 406); an undershrub; branches slender, glabrous; leaves opposite, subacicular, 3–6 lin. long, ⅓ lin. wide, subpungent, keeled, glabrous; bracts 4, like the leaves; flowers solitary; calyx silky outside; tube 3 lin. long, cylindrical, hairy in the throat; lobes ovate-elliptic, acute, 3 lin. long, 1¼ lin. wide, purple, glabrous within; scales minute, oblong, inserted at about the middle of the calyx-tube; filaments ½ lin. long, stout; anthers oblong, about as long as the free part of the filament; ovary oblong, ¾ lin. long, glabrous; style filiform; stigma clavate. *Meisn. in DC. Prodr.* xiv. 573; *Drege, Zwei Pfl. Documente,* 104, 110, 111, 112, 119; *Bot. Mag. t.* 4143; *Bolus & Wolley-Dod in Trans. S. Afr. Phil. Soc.* xiv. 315. *C. ciliata, Meisn. in Linnæa,* xiv. 407, *and in DC. Prodr.* xiv. 574. *Passerina uniflora, Linn. Sp. Pl. ed.* i. 560; *Lam. Ill. t.* 291, *fig.* 1; *Wikstr. in Vet. Acad. Handl. Stockh.* 1818, 344; *var. purpurea, Berg. Descr. Pl. Cap.* 128. *Calycoseris uniflora, Eckl. & Zeyh. ex Meisn. in DC. Prodr.* xiv. 573.—*Thymelæa ramosa, linearibus, etc., Burm. Rar. Afr. Pl. Dec.* v. 134, *t.* 48, *fig.* 1.

SOUTH AFRICA: without locality, *Mund*! *Villet*! *Krebs*. 61!
COAST REGION: Clanwilliam Div.; between Pikeniers Kloof and Markus Kraal,

Drège ! between Kromrivier and Berg Vallei, under 1000 ft., *Drège*, Piquetberg Div. ; Twenty-four River, under 1000 ft., *Drège*. Malmesbury Div. ; between Groene Kloof and Dassen Berg, under 500 ft., *Drège*. Tulbagh Div. ; Roodezand, near Tulbagh, *Drège*, Ceres Road, *Schlechter*, 9089 ! Worcester Div. ; Breede River, near Bains Kloof, 800 ft., *Bolus*, 2923 ! Cape Div. ; Cape Flats, *Zeyher*, 1485 ! *Ecklon*, 362 ! *Harvey* ! *Tyson*, 1459 ! Wynberg Flats, *Wilms*, 3592 ! *Drège*, near Claremont, 100 ft., *Bolus*, 2923 B ! *Wolley-Dod*, 611 ! Vygeskraal, *Wolley-Dod*, 482 ! sand flats between Paarden Island, Tygerberg and Blueberg, *Drège* ! near Cape Town, 100 ft., *Bolus*, 4592 ! Princess Vley, *MacOwan*, *Herb. Austr.-Afr.*, 1637 !

3. **C. laxa** (C. H. Wright) ; plant not exceeding 1 ft. in height ; branches arising close to the ground, wiry, at first puberulous, finally glabrous ; internodes usually as long as the leaves ; leaves opposite, ovate-lanceolate, acute, up to 4 lin. long and nearly 1 lin. wide, concave, smooth on the back or slightly keeled above, glabrous ; flowers solitary, terminal ; calyx silky outside ; tube cylindrical, slender, 2–3 lin. long ; lobes ovate, 3 lin. long, $1\frac{1}{4}$ lin. wide, subacute ; longer stamens reaching to the middle of the calyx-lobes ; anthers oblong, $\frac{1}{2}$ lin. long ; scales oblong, slender, $\frac{1}{4}$ lin. long, at base of free part of filament ; ovary oblong, glabrous ; style lateral, filiform, thickened and hairy above ; stigma shortly clavate.

SOUTH AFRICA : without locality, *Harvey* !
COAST REGION : Caledon Div. ; Zwart Berg, near Caledon, 3200 ft., *Bolus*, 7875 ! near Houw Hoek, 2000 ft., *Bolus*, 9208 ! Bredasdorp Div. ; Elim, *Bolus*, 7876 !

4. **C. breviflora** (Meisn. in Linnæa, xiv. 406) ; branches pubescent when young, at length glabrous ; leaves decussate, patulous, linear, acute, keeled on the back, glabrous, $2\frac{1}{2}$ lin. long, $\frac{1}{2}$ lin. wide ; flowers terminal, solitary ; calyx silky outside ; tube 2 lin. long, infundibuliform ; lobes oblong, 2 lin. long, 1 lin. wide ; anthers oblong, obtuse, $\frac{3}{4}$ lin. long ; ovary ovoid ; style columnar, lateral. *Drège, Zwei Pfl. Documente*, 99 ; *Meisn. in DC. Prodr.* xiv. 573 ; *Bolus & Wolley-Dod in Trans. S. Afr. Phil. Soc.* xiv. 315. *Passerina uniflora*, β *alba, Berg. Descr. Pl. Cap.* 129. *P. campanulata*, *E. Meyer ex Meisn. in DC. Prodr.* xiv. 574. *P. grandiflora, Curt. Bot. Mag. t.* 292. *Calycoseris tabularis*, *C. subulata* and *C. parviflora, Eckl. & Zeyh. ex Meisn. in DC. Prodr.* xiv. 574.—*Thymelæa foliis triquetris, angustis, etc., Burm. Rar. Afr. Pl. Dec.* v. 132, *t.* 48, *fig.* 2. *Thymelæa fruticosa Pinastri, etc., Pluk. Mant.* 180, *and Phyt. t.* 445, *fig.* 8.

SOUTH AFRICA : without locality, *Niven, Eschscholtz, Lalande, Phillips*, 3796 !
COAST REGION : Paarl Div. ; between Paarl and Lady Grey Railway Bridge, *Drège* ! Cape Div. ; Table Mountain, *Ecklon*, 360 ! Doornhoogte, Cape Flats, *Zeyher*, 3744, near Smitswinkel Bay not far from Cape Point, *MacOwan* ! Stellenbosch Div. ; Hottentots Holland, *MacOwan*, 3182 ! Caledon Div. ; Houw Hoek, 1000 ft., *Galpin*, 4493 !

5. **C. filicaulis** (Meisn. in Linnæa, xiv. 407) ; an undershrub about 1 ft. high ; branches slender and wiry, glabrous ; leaves

decussate, oblong, acute, 3 lin. long, $\frac{1}{2}$ lin. wide, flat or slightly concave above, convex beneath, glabrous; flowers usually solitary, terminal; calyx pubescent outside; tube ovoid for about 1 lin., then suddenly dilated into a cup $\frac{1}{2}$ lin. deep and $1\frac{1}{2}$ lin. in diam.; lobes ovate, 2 acute, 2 obtuse, $1\frac{1}{4}$ lin. long, $\frac{3}{4}$ lin. wide; longer stamens reaching to about the middle of the calyx-lobes; anthers oblong, $\frac{1}{4}$ lin. long; scales yellow, smaller than the anthers; ovary ovoid; style lateral, columnar. *Meisn. in DC. Prodr.* xiv. 574; *Drège in Linnæa*, xx. 210. *Gnidia genistaefolia, Eckl. & Zeyh. ex Meisn. l.c.*

SOUTH AFRICA: without locality, *Drège*, 7367!
COAST REGION: Caledon Div. ; River Zonder Einde, near Appelskraal, *Zeyher*, 3748! Vogelgat, 100 ft., *Schlechter*, 9579! Zwarteberg, *Zeyher*, 3747!
CENTRAL REGION: Ceres Div. ; between Witzenberg and Skurfdeberg, 2000–5000 ft., *Zeyher*, 5! 1486! Cold Bokkeveld, Klyn Vlei, 4000 ft., *Schlechter*, 10067!

VI. LACHNÆA, Linn.

Flowers hermaphrodite. *Calyx-tube* cylindrical, or constricted above the ovary, circumscissile; lobes 4, patent, the outer usually rather larger; throat naked. *Petals* 0. *Stamens* 8, usually shortly exserted; filaments filiform; anthers small, connective narrow. *Scales* 8, linear or clavate, included in the calyx-tube below the stamens. *Disc* 0. *Ovary* sessile, 1-celled; style filiform; stigma globose or ovoid. *Fruit* dry, enclosed in the persistent base of the calyx; pericarp thinly membranous. *Seed* with shining crustaceous testa; albumen fleshy; embryo subterete or with slightly flattened cotyledons.

Small heath-like shrubs; leaves opposite or scattered, usually small, coriaceous; flowers in terminal bracteate or ebracteate heads, rarely solitary.
DISTRIB. Species 19, endemic.

Branches glabrous :
Calyx silky ... ... (10) **capitata.**
Calyx pubescent ... ... (11) **diosmoides.**
Heads few-flowered :
Leaves acerose, pungent ... ... (12) **ambigua.**
Leaves oblong, obtuse ... ... (13) **ericoides.**
Leaves adpressed to the stem :
Branches long, slender, finally glabrous :
Leaves linear, pungent ... ... (14) **globulifera.**
Leaves linear-lanceolate, obtuse ... ... (15) **funicaulis.**
Branches short, white-tomentose :
Leaves lanceolate, triquetrous ... ... (16) **penicillata.**
Leaves lanceolate, concave ... ... (17) **passerinoides.**
Flowers solitary, axillary :
Branches straight, erect ... ... (18) **micrantha.**
Branches flexuous, interlacing ... ... (19) **axillaris.**

1. **L. macrantha** (Meisn. in DC. Prodr. xiv. 575); an erect
subshrub about 1 ft. high; branches stout, glabrous except when
quite young; leaves concave, oval or obovate, obtuse, contracted
at the base, quite glabrous, about 6 lin. long and 4 lin. wide;
flowers collected into terminal heads, white (*Burchell*); calyx-tube
oblong, constricted above, densely hairy; lobes 4, densely silky on
both surfaces, the abaxial oblong, 6 lin. long, 3 lin. wide, rounded
at the apex, the adaxial oblong, 2 lin. long, 1 lin. wide, lateral
lobes linear, 2½ lin. long, ½ lin. wide; filaments filiform, the longer
1½ lin. long; scales in the throat of the calyx ¼ lin. long, obtuse;
ovary obovoid, 1 lin. long; style 2½ lin. long, with silky upward-
pointing hairs; stigma plumose. *L. buxifolia, Eckl. & Zeyh. ex
Meisn. l.c., not of Lam.*

SOUTH AFRICA : without locality, *Bowie*!
COAST REGION : Riversdale Div. ; summit of Kampsche Berg, *Burchell*, 7120 !

2. **L. buxifolia** (Lam. Encycl. iii. 373); a shrub; branches robust,
glabrous; leaves alternate, imbricate, oval or ovate, acute, sessile,
quite glabrous, 9 lin. long, 4–5 lin. wide; flowers congested at the
apex of the branches, without a special involucre; calyx-tube 6–8
lin. long, cylindrical or very slightly widened at the base, densely
and patently hairy; lobes pale yellow, ovate, subacute, 4 lin. long,
1½ lin. wide, densely hairy on both surfaces; scales in the calyx-
throat ½ lin. long, clavate; filaments shorter than the calyx-lobes,
very slender ; anthers elliptic, ⅓ lin. long; ovary elliptic, compressed,
1 lin. long, hairy at the apex; style lateral, very slender, densely
hairy in the uppermost line. *Illustr. t. 292, fig. 1; Meisn. in
Linnæa, xiv. 410, and in DC. Prodr. xiv. 574; var. virens, Sims in
Bot. Mag. t. 1657. L. filamentosa, Gilg in Engl. & Prantl, Pflanzenfam. iii. 6 A, 240, not of Meisn. Gnidia filamentosa, Linn. f.
Suppl. 224; Willd. Sp. Pl. ii. 425.*

c 2

SOUTH AFRICA : without locality, *Thunberg*! *Forster*!
COAST REGION : Tulbagh Div. ; Tulbagh, *Pappe*! Ceres Div. ; near Ceres,
amongst stones on the Skurfdeberg Range, 1800 ft., *Bolus*, 7349 ! and in *Herb.*
Norm. Austr.-Afr., 1091 ! Worcester Div. ; Goudinie, *Cooper*, 3149 ! Dutoits
Kloof, *Drège*! mouth of Els Kloof, Hex River, *Wolley-Dod*, 2758 !

3. L. filamentosa (Meisn. in Linnæa, xiv. 410) ; a shrub ; leaves
alternate, imbricate, elliptic-oblong, acute, 5 lin. long, $1\frac{1}{2}$ lin. wide,
sessile, glabrous ; flowers in terminal heads ; calyx-tube cylindrical
and slender from a subglobose base, 4 lin. long, densely villous ;
lobes unequal, the abaxial elliptic, subacute, 4 lin. long, $1\frac{1}{2}$ lin.
wide, the others half as large, all densely hairy on both surfaces ;
stamens shorter than the smaller calyx-lobes ; scales oblong, $\frac{1}{2}$ lin.
long, almost hidden by hair ; ovary oblong, glabrous ; style filiform,
villous in the upper part ; stigma penicillate. *Meisn. in DC. Prodr.*
xiv. 575 ; *Drège, Zwei Pfl. Documente*, 73.   *Gnidia filamentosa*,
Thunb. Fl. Cap. ed. Schult. 378, *not of Linn.*

VAR. β, **major** (Meisn. in DC. Prodr. xiv. 575) ; a more robust plant than
the type ; leaves up to 10 lin. long and $4\frac{1}{2}$ lin. wide ; calyx up to 1 in. long.
L. glauca, Eckl. & Zeyh. ex Meisn. l.c.
COAST REGION : Clanwilliam Div. ; Ceder Berg Range, near Honig Valley, 3000–
4000 ft., *Drège*! Sneeuwkop, 5000 ft., *Bodkin in Herb. Bolus*, 9085 ! Var. β :
Tulbagh Div. ; Winterhoek Mountains, near Tulbagh, *Ecklon & Zeyher*, 72, 2500–
5000 ft., *Bolus*, 5260 ! *Marloth in MacOwan, Herb. Norm. Austr.-Afr.*, 1952!

4. L. nervosa (Meisn. in Linnæa, xiv. 417) ; an erect, more or
less virgate shrub ; branches slender, pubescent ; leaves alternate,
acerose-linear, pungent, 5 lin. long, $\frac{1}{4}$ lin. wide, sessile ; heads
terminal, few-flowered ; calyx densely silky outside ; tube 2 lin.
long, inflated lower part slightly shorter than the cylindrical upper ;
lobes equal, lanceolate, acute, $2\frac{1}{2}$ lin. long, $\frac{3}{4}$ lin. wide ; longer
stamens reaching to the middle of the lobes ; scales filiform, nearly
1 lin. long ; ovary ovoid ; style lateral, filiform, hairy above ;
stigma penicillate.   *Meisn. in DC. Prodr.* xiv. 579 ; *Drège in*
Linnæa, xx. 209.   *Passerina nervosa, Thunb. Fl. Cap. ed. Schult.*
375 ; *Wikstr. in Vet. Acad. Handl. Stockh.* 1818, 328.   *Gonophylla*
stricta, Eckl. & Zeyh. ex Meisn. in DC. Prodr. xiv. 579.

SOUTH AFRICA : without locality, *Villet* ! *Niven, Drège.*
COAST REGION : Swellendam Div. ; Grootvaders Bosch and neighbourhood,
1000–4000 ft., *Zeyher*, 3767 ! *Ludwig, Ecklon.*

5. L. eriocephala (Linn. Sp. Pl. ed. i. 560) ; a branched under-
shrub, $\frac{1}{2}$–$1\frac{1}{2}$ ft. high ; branches moderately stout, at first pubescent ;
leaves quadrifarious, imbricate, linear-triquetrous, 4 lin. long, $\frac{1}{2}$ lin.
wide, glabrous ; flowers in terminal solitary heads ; involucral bracts
broadly ovate, acute, keeled, 5 lin. long, about 4 lin. wide, hairy on
both surfaces, more conspicuously so on the margin near the apex ;
calyx white, densely hairy ; tube 5 lin. long, slightly inflated in the

lowest fifth, cylindrical above ; lobes unequal, the abaxial 4 lin.
long, 1½ lin. wide, subacute, the others oblanceolate, 3 lin. long,
1 lin. wide ; glands minute, almost concealed by hairs ; filaments
slender, shorter than the calyx-lobes ; ovary oblong, glabrous ;
style excentric, filiform ; stigma penicillate. *Willd. Sp. Pl.* ii. 434 ;
Gærtn. Fruct. iii. *t.* 215 ; *Andr. Rep. t.* 104 ; *Bot. Mag. t.* 1295 ;
Lam. Encycl. iii. 374, *and Ill. t.* 292, *fig.* 2 ; *Herb. Amat.* iv. *t.* 234 ;
Drège, Zwei Pfl. Documente, 78, 102 ; *Meisn. in Linnæa,* xiv. 411,
and in DC. Prodr. xiv. 575. *L. sphærocephala, Burm. Prodr. Fl.
Cap.* 12. *Passerina cephalophora, Thunb. Fl. Cap. ed. Schult.* 375 ;
Wikstr. in Vet. Acad. Handl. Stockh. 1818, 322.

COAST REGION : Malmesbury Div. ; Dassen Berg, *Drège*! Stellenbosch Div. ;
Hottentots Holland, on sandy hills under mountains, *Niven*! *Mund,* 9! near
Lowrys Pass, *Burchell,* 8277! below Lowrys Pass, 600 ft., *Bolus,* 5553!

6. **L. purpurea** (Andr. Rep. t. 293) ; branches long, simple, from
the lower part of the stem, silky when young ; leaves 5–8 lin. long,
½ lin. wide, linear-triquetrous, acute, glabrous, slightly spreading ;
flowers in solitary terminal heads ; bracts broadly ovate, glabrous
outside, villous along the top edge and more shortly so within ;
calyx purple ; tube 5 lin. long, glabrous and slightly inflated in the
lower third, cylindrical and patently villous in the upper part ;
lobes unequal, more or less elliptic, acute, the largest 3½ lin. long,
2 lin. wide ; filaments filiform, shorter than the calyx-lobes ; scales
concealed by hairs ; style filiform, hairy above ; stigma penicillate.
Ait. Hort. Kew. ed. 2, ii. 415 ; *Bot. Mag. t.* 1594 ; *Lodd. Cab.
t.* 273. *L. eriocephala, var. purpurea, Meisn. in DC. Prodr.* xiv.
576. *Passerina purpurea, Wikstr. in Vet. Acad. Handl. Stockh.*
1818, 323.

COAST REGION : Tulbagh Div.; Roode Zand, *Niven*! on sandy mountains near
Tulbagh, *Ecklon & Zeyher,* 31, 77, New Kloof, 1200 ft., *Drège* ; *MacOwan in
Herb. Norm. Austr.-Afr.,* 541! Worcester Div. ; Worcester, *Zeyher*!

7. **L. aurea** (Eckl. & Zeyh. ex Meisn. in DC. Prodr. xiv. 576) ; a
woody plant about 1 ft. high ; branches straight, glabrous ; leaves
alternate, imbricate, linear or lanceolate, acute or submucronate,
sometimes slightly keeled on the back just below the apex ; flowers
in terminal involucrate heads ; bracts ovate or elliptic, acuminate,
4 lin. long, 2 lin. wide, quite glabrous, margins membranous ; calyx
yellow ; tube 2 lin. long, inflated, glabrous and brown in the lower
half, densely villous above ; lobes unequal, villous outside, glabrous
within, the anterior 2 lin. long, 1 lin. wide, the posterior 6 lin.
long, 1½ lin. wide, the lateral 4½ lin. long, 1 lin. wide, all obtuse ;
longer filaments 2 lin. long ; anthers oblong ; ovary oblong, compressed ; style lateral, filiform, villous above, 2 lin. long ; stigma
penicillate.

COAST REGION : Caledon Div. ; Kleinriver Mountains, near Hemel en Aarde,
Zeyher! Bredasdorp Div. ; Elim, 200 ft., *Bolus,* 9186! 400 ft., *Schlechter,* 7610!

8. L. striata (Meisn. in Linnæa, xiv. 415); a shrub, corymbosely branched; branches terete, pubescent, slender; leaves alternate, sessile, ovate-lanceolate or lanceolate, acuminate, subpungent, with 3–5 nerves prominent beneath, conspicuously ciliate on the margins, otherwise glabrous; heads terminal, 8–16-flowered, sessile or shortly stalked, not involucrate; calyx-tube 1½ lin. long, glabrous, oblong and inflated in the lower half, cylindrical above; lobes ovate-oblong, acute, 1½ lin. long, 1 lin. wide, hairy on both surfaces, red or white; scales ligulate, nearly 1 lin. long; stamens shorter than the calyx-lobes; ovary oblong, glabrous; style lateral, filiform; stigma penicillate. *Meisn. in DC. Prodr.* xiv. 577; *Drège, Zwei Pfl. Documente*, 77. *Passerina striata, Lam. Encycl.* v. 44, *and Ill. t.* 291, *fig.* 1. *P. nervosa, Wikstr. in Vet. Acad. Handl. Stockh.* 1818, 328. *Gonophylla nervosa, Eckl. & Zeyh. ex Meisn. in DC. Prodr.* xiv. 577; *Drège in Linnæa*, xx. 209.

COAST REGION: Clanwilliam Div.; Clanwilliam, *Mader in Herb. MacOwan*, 2210! Pakhuis Pass, Ceder Berg Range, 3000 ft., *Bolus*, 9087! Tulbagh Div.; Great Winter Hoek, 3000–4000 ft., *Drège*! Worcester Div.; Worcester, *Zeyher*!

9. L. densiflora (Meisn. in DC. Prodr. xiv. 578); a much-branched shrub; branches terete, at first pubescent; leaves acerose-linear, sessile, not conspicuously nerved, 3–5 lin. long, ¼ lin. wide, glabrous; heads solitary, terminal, globose, many-flowered; peduncle 3 lin. long, turbinate at the apex, tomentose; calyx white, shortly hairy outside; tube ¾ lin. long, constricted above; lobes ovate, 1 lin. long, ¾ lin. wide, acute, hairy inside as well as outside; stamens shorter than the calyx-lobes; scales filiform, ¼ lin. long; ovary oblong, glabrous; style terminal, hairy in the upper part; stigma papillate. *Bolus & Wolley-Dod in Trans. S. Afr. Phil. Soc.* xiv. 315.

SOUTH AFRICA: without locality, *Burke*! *Thom*! *Sieber*! *Grey*! *Zeyher*, 1483! COAST REGION: Cape Div.; near Cape Town, *Mund*, 8! *Bolus*, 3776, partly! Buffelsriver, *Marloth*, 480! Patrys Vley, *Wolley-Dod*, 1498! near Wynberg, *MacOwan in Herb. Norm. Austr.-Afr.*, 763! Simons Bay, *Wright*! Caledon Div.; Hermanuspetrusfontein, 20–50 ft., *Galpin*, 4499! Hawston, 50 ft., *Schlechter*, 9461!

10. L. capitata (Meisn. in Linnæa, xiv. 414, incl. vars. *pauciflora* and *multiflora*); a small shrub with slender virgate reddish glabrous branches; leaves alternate, linear, pungent, 5 lin. long, ½ lin. wide, slightly narrowed at the base; flowers in terminal heads which often elongate into spikes 4 lin. long, 8- to many-flowered; peduncles about 3 lin. long, incrassate and ribbed above, pubescent; calyx white, densely silky outside and on the inside of the lobes; tube 1½ lin. long, inflated below, cylindrical in the middle, funnel-shaped above; lobes broadly ovate, nearly as long as the tube, 1 lin. wide; stamens much shorter than the calyx-lobes; scales filiform, almost concealed in the dense hair of the throat; ovary oblong, glabrous; style filiform, densely hairy; stigma subcapitate.

Meisn. in DC. Prodr. xiv. 577 ; *Drège, Zwei Pfl. Documente*, 112 ;
Bolus & Wolley-Dod in Trans. S. Afr. Phil. Soc. xiv. 315. *L. phylicoides, Lam. Encycl.* iii. 374. *Passerina capitata, Linn. Amœn. Acad.*
vi. 88, *and Sp. Pl. ed.* ii. 513 ; *Lam. Encycl.* v. 41, *and Ill. t.* 291,
fig. 3 ; *Thunb. Prodr.* 75, *and Fl. Cap. ed. Schult.* 376 ; *Wikstr. in
Vet. Acad. Handl. Stockh.* 1818 329 ; *Wendl. Bot. Beobacht.* 18, *t.* 2,
fig. 17. *Gonophylla acuminata, G. capitata, and G. conglomerata,
Eckl. & Zeyh. ex Meisn. in DC. Prodr.* xiv. 578.—*Thymelæa foliis
linearibus, etc., J. Burm. Rar. Afr. Pl.* 133, *t.* 48, *fig.* 3.

SOUTH AFRICA : without locality, *Burke* ! *Grey* ! *Forster* ! *Zeyher*, 1484 !
Harvey, 685 !
COAST REGION : Clanwilliam Div. ; Zeekoe Vley, 800 ft., *Schlechter*, 8508 !
Cape Div. ; Cape Flats, near Rondebosch, *Burchell*, 712 ! *Mund*, 7 ! *Wolley-Dod*,
503 ! Claremont Flats, *Dümmer*, 297 ! Tokay Flat, *Wolley-Dod*, 2565 ! near
Doornhoogte, *Wolley-Dod*, 653 ! near Cape Town, *Bolus*, 3776, partly ! *MacOwan*,
Herb. Austr.-Afr., 1786 ! Kommetjes, 100 ft., *Galpin*, 4500 ; Muizenberg, near
Kalkberg, *Wilms*, 3594 ! Stellenbosch Div. ; between Tygerberg and Simons Berg,
Drège ! between Lowrys Pass and Jonkers Hoek, *Burchell*, 8298 ! Caledon Div. ;
near Grabouw, on the Palmiet River, *Bolus*, 4195 !

Meisner in founding the varieties *pauciflora* and *multiflora* expresses doubt as to
their validity.   The subsequent accumulation of material shows there is every
gradation from few-flowered to many-flowered states, with corresponding changes
in the length of the receptacle, so that the varieties cannot stand.

11. **L. diosmoides** (Meisn. in Linnæa, xiv. 418, var. *elatior*) ; a
shrub ; branches virgate, slender, glabrous ; leaves alternate, linear,
acute, flat, 5 lin. long, ½ lin. wide, the uppermost very slightly
wider ; heads terminal, about 10-flowered, sometimes clustered ;
calyx pubescent outside ; tube 1 lin. long, slightly inflated below ;
lobes ovate, obtuse, about as long as the tube, the two outer slightly
narrower than the inner ; stamens two-thirds as long as the calyx-
lobes ; scales filiform, ¼ lin. long ; ovary glabrous ; style excentric,
filiform and glabrous below, thicker and hairy above ; stigma
penicillate. *Meisn. in DC. Prodr.* xiv. 578, *incl. var. elatior. L.
phylicoides, Lam. Ill. t.* 292, *fig.* 3, *not Encycl.* iii. 374 ; *Drège, Zwei
Pfl. Documente*, 124.

VAR. β, **tenella** (Meisn. in Linnæa, xiv. 419, and in DC. Prodr. xiv. 579) ;
branches more slender ; leaves more distant ; heads 2-4-flowered. *L. phylicoides,
Drège, Zwei Pfl. Documente*, 117.

SOUTH AFRICA : without locality, *Wallich* ! *Mund & Maire*, 208 ! 234 !
COAST REGION : George Div. ; George, *Zeyher*, 1475 ! *Pappe* ! *Bolus*, 8690 !
Cradock Berg, 900 ft., *Galpin*, 4501 ! *Burchell*, 5953 ! Hoogekraal River, on
mountains under 1000 ft., *Galpin*, 4502 ! Uitenhage Div. ; Van Stadensberg Range, 800 ft,
MacOwan, 1051 ! Var. β : Mossel Bay Div. ; Attaquas Kloof, 2000–3000 ft., *Drège* !
CENTRAL REGION : Ceres Div. ; near Ceres, 1500 ft., *Bolus, Herb. Norm. Austr.-
Afr.*, 1356 !

12. **L. ambigua** (Meisn. in Linnæa, xiv. 417, var. *major*) ; a shrub ;
branches short, slender with tubercle-like leaf-scars, blackish ; leaves
alternate, acerose, pungent, 4 lin. long, ⅓ lin. wide ; heads terminal,
4-8-flowered ; bracts 2-3-seriate, elliptic, subacute, 2½–3 lin. long,
1 lin. wide, slightly ribbed outside, glabrous ; calyx pale yellow,

silky-villous outside, longer than the bracts; tube 1 lin. long, sub-globose, constricted above; lobes subequal, elliptic, obtuse, concave, 1⅓ lin. long, 1 lin. wide, densely hairy inside as well as outside; stamens much shorter than the calyx-lobes; scales linear, ⅔ lin. long; ovary elliptic, compressed; style lateral, filiform; stigma penicillate. *Meisn. in DC. Prodr.* xiv. 576; *Drège, Zwei Pfl. Documente,* 82.

VAR. β, **minor** (Meisn. in Linnæa, xiv. 418); leaves very narrowly linear-acerose; heads 2-(rarely 4-) flowered; flowers white, scarcely exceeding the bracts. *Meisn. in DC. Prodr.* xiv. 577; *Drège, Zwei Pfl. Documente,* 82.

COAST REGION: Worcester Div.; Dutoits Kloof, 3000–4000 ft.. *Drège!* Caledon Div.; Genadendal Mountains, 4000 ft., *Galpin,* 4498! Var. β: Worcester Div.; Dutoits Kloof, 4000–5000 ft., *Drège!*

13. **L. ericoides** (Meisn. in DC. Prodr. xiv. 579); an undershrub, up to 1 ft. high, fastigiately branched above; branches slender, hairy, reddish; leaves opposite or alternate, oblong, incurved, 1–3 lin. long, ½ lin. wide, keeled below, concave above, obtuse, terminated by a tuft of short hairs; flowers few in terminal heads or solitary; calyx reddish; tube cylindrical and glabrous below, slightly constricted and hairy above; lobes suborbicular, 1 lin. across, pubescent outside; longer stamens nearly as long as the calyx-lobes; scales clavate, very short; ovary obovoid; style terminal, filiform, slightly hairy above; stigma penicillate. *Drège in Linnæa,* xx. 209. *Gonophylla ericoides, Eckl. & Zeyh. ex Meisn. l.c.*

COAST REGION : Swellendam Div.; near Swellendam, *Mund,* 4! *Zeyher,* 3776! Zuurbraak Mountain, 1000 ft., *Galpin,* 4496! Tradouw Pass, 1200 ft., *Galpin,* 4495! Riversdale Div.; Langeberg Range, near Riversdale, 1080 ft., *Schlechter* 1730!
Galpin, 4495, has the leaves more oval than in the other specimens.

14. **L. globulifera** (Meisn. in Linnæa, xiv. 412); a shrub several feet high; branches slender, glabrous except when quite young; leaves opposite, linear, pungent, 3–4 lin. long, ⅓ lin. wide, glabrous; flowers in terminal involucrate heads about 8 lin. in diam.; bracts 4, ovate-rotundate, acute, 2 lin. long and nearly as broad, glabrous on both surfaces, ciliate on the margin with hairs ¾ lin. long; calyx-tube 2 lin. long, glabrous and inflated below, cylindrical and hairy above; lobes equal, ovate, acute, 1½ lin. long, silky outside; longer stamens a little shorter than the calyx-lobes; scales filiform, ⅓ lin. long, exserted; ovary oblong, compressed; style lateral, filiform; stigma penicillate. *Meisn. in DC. Prodr.* xiv. 576; *Drège, Zwei Pfl. Documente,* 119, *and in Linnæa,* xx. 209. *Gonophylla setosa, Eckl. & Zeyh. ex Meisn. l.c.* *Passerina eriocephala, Thunb. Fl. Cap. ed. Schult.* 375, *not of others.*

VAR. β, **cœrulescens** (Meisn. in DC. Prodr. xiv. 576); branches white-puberulous at the apex; flowers a little larger than in the type; calyx-lobes ovate-oblong, acute. *Gonophylla cœrulescens, Eckl. & Zeyh. ex Meisn. l.c.;* *Drège in Linnæa,* xx. 209.

COAST REGION: Clanwilliam Div.; near Middleburg, in the Groot-Kloof Valley, *Leipoldt,* 897! Tulbagh Div.; Tulbagh, *Pappe!* mountain near Tulbagh Waterfall, 1500 ft., *Bolus,* 5259! 1700 ft., *Bolus,* 5318! *Schlechter,* 7460! Great Winterhoek; *Zeyher,* 1474! Worcester Div.; Worcester, *Zeyher!* in sandy places near the Goudinie, 800–1000 ft., *Drège!* Var. β: Clanwilliam Div.; Brakfontein, *Ecklon & Zeyher,* 80!

15. **L. funicaulis** (Schinz in Bull. Herb. Boiss. iii. 408); an undershrub; branches filiform, villous at first, afterwards glabrous and bright brown; leaves linear-lanceolate, obtuse, 2 lin. long, ciliate and with a terminal tuft of hairs simulating a mucro; flowers in terminal involucrate heads; bracts 2½–3 lin. long, from spathulate to nearly orbicular, long-ciliate (terminal cilia ¾ lin. long); calyx yellow, silky, subtended by basal hairs 1 lin. long; tube 1¼ lin. long, slightly inflated below, funnel-shaped above; lobes ½ lin. long, ovate, obtuse; scales very short, filiform; ovary oblong, glabrous; style glabrous; stigma clavate.

CENTRAL REGION: Ceres Div.; amongst stones at the foot of mountains near Ceres, 1900 ft., *Bolus, Herb. Norm. Austr.-Afr.,* 1357! Mitchells Pass, 2000 ft., *Schlechter,* 9961! Cold Bokkeveld, Klyn Vley, 4500 ft., *Schlechter,* 10066!

16. **L. penicillata** (Meisn. in Linnæa, xiv. 421); a shrub with the appearance of *Passerina filiformis*; branches slender, densely white-silky when young; leaves decussate, 1–2 lin. long, lanceolate, triquetrous, acute, at first tipped and edged with white hairs; flowers terminal, solitary, sessile; calyx scarcely 2 lin. long, white-silky outside; lobes ovate-oblong, acute, slightly toothed; scales short, clavate. *Meisn. in DC. Prodr.* xiv. 579; *Drège, Zwei Pfl. Documente,* 116. *Passerina filiformis, var. depauperata, E. Meyer in Linnæa,* xiv. 422. *P. bruniades and P. brunioides, Eckl. & Zeyh., partly, ex Meisn. in DC. Prodr.* xiv. 579.

COAST REGION: Tulbagh Div.; New Kloof, 3000 ft. *Schlechter,* 7503! Swellendam Div.; between Sparrbosch and Tradouw, 1000–2000 ft., *Drège!* Mossel Bay Div.; on mountains near the Gouritzrivier, *Ecklon ex Meisner.*
CENTRAL REGION: Ceres Div.; Cold Bokkeveld at Wagendrift, 5000 ft., *Schlechter,* 10072!

17. **L. passerinoides** (N. E. Br. in Kew Bulletin, 1901, 132); an undershrub, 4–9 in. high; branches slender, white-tomentose; leaves opposite, adpressed to the stem, lanceolate, concave, strongly keeled, 1 lin. long, ⅓ lin. wide, ciliate and with a terminal tuft of hairs; heads terminal, 2-flowered; bracts 4, ovate, concave, 2 lin. long, 1 lin. wide, densely ciliate along the obtuse top; receptacle with silky hairs nearly 1 lin. long; calyx hairy outside; tube 1½ lin. long, subcylindrical, slightly constricted about the middle; lobes about 1 lin. long, the 2 outer ⅓ lin. wide and acute, the 2 inner wider and obtuse; longer filaments ½ lin. long; scales linear, ¼ lin. long; ovary oblong, glabrous; style thickened and hairy above; stigma capitate.

COAST REGION: Riversdale Div.; Garcias Pass and Muis Kraal, 1200–1500 ft., *Bolus,* 4497! 11372!

18. **L. micrantha** (Schlechter in Engl. Jahrb. xxiv. 451); an erect much-branched shrub, 2–3 ft. high; branchlets filiform, erect, puberulous or villous; leaves opposite, sessile, lanceolate or linear-lanceolate, acute, 1½ lin. long, ¼ lin. wide, glabrous, slightly keeled on the back; flowers solitary in the leaf-axils, very small; calyx white, at the length reddish, glabrous; tube ovoid, ½ lin. long; outer lobes 1 lin. long, ½ lin. wide ovate, acute, inner suborbicular, obtuse, several-nerved; stamens shorter than the calyx-lobes; scales short, fleshy, bilobed; ovary oblong; style lateral, filiform; stigma capitate.

COAST REGION : Bredasdorp Div.; Elim, 400 ft., *Schlechter*, 7702! Riet Fontein Poort, 200 ft., *Bolus*, 8596!

19. **L. axillaris** (Meisn. in Linnæa, xiv. 422); a dwarf much-branched shrub; branches flexuous and interlacing, very slender, puberulous; leaves opposite, 1–2 lin. long, ¼ lin. wide, acute or mucronate, convex on the back, glabrous; flowers axillary, solitary; calyx-tube oblong, 1 lin. long, glabrous; lobes broadly elliptic, obtuse, 1¼ lin. long, 1 lin. wide, glabrous; scales villous, conspicuous; stamens shorter than the calyx-lobes; ovary oblong, glabrous; style excentric, filiform; stigma penicillate. *Meisn. in Hook. Lond. Journ.* ii. 552, *and in DC. Prodr.* xiv. 580. *Radojitskya capensis, Turcz. in Bull. Soc. Imp. Mosc.* 1852, ii. 176, *and in Flora*, 1853, 743. *Passerina brunioides, Eckl. & Zeyh., partly, ex Meisn. in DC. Prodr. l.c.*

SOUTH AFRICA : without locality, *Drège*!
COAST REGION : Swellendam Div.; Zoetendals Vallei, *Krauss*, 1265 bis; on mountains by the Gouritz River, *Ecklon, Zeyher*, 2163!

Imperfectly known species.

20. **L. dubia** (Gandog. in Bull. Soc. Bot. France, lx. 417, 1913); allied to *L. globulifera*, Meisn., from which it differs by the following characters :—branches adpressed-pubescent; leaves muticous, half as long (2½ lin.), usually alternate, not opposite; flowers smaller.

COAST REGION : Cape Div.; Table Mountain, *Debeaux.*

VII. STRUTHIOLA, Linn.

Flowers hermaphrodite. *Calyx-tube* slender, sometimes slightly inflated above; lobes 4, patent. *Petals* 4, 8 or 12, erect, thick and fleshy, equidistant in a single whorl, each one surrounded by hairs. *Stamens* 4, included in the calyx-tube and alternate with its lobes; anthers subsessile, linear, connective shortly produced above the cells; cells contiguous on the inner side. *Hypogynous disc* 0. *Ovary* sessile, 1-celled; style filiform; stigma capitate. *Fruit* small, dry, included in the persistent base of the calyx; pericarp thinly

membranous. *Seed* compressed ; testa crustaceous, shining ; albumen
scanty ; cotyledons rather thick.

Heath-like shrubs or undershrubs ; leaves opposite or more rarely alternate,
small, coriaceous, more or less imbricate ; flowers sessile in the axils of the upper
leaves, solitary, rarely geminate ; bracteoles 2.
DISTRIB. Species about 40, all South African.

  *Petals 4 :
    Leaves ovate-oblong ; calyx-lobes oblong, acute ... (1) **striata.**

    Leaves lanceolate ; calyx-lobes lanceolate, acuminate (2) **tetralepis.**

  **Petals 8 :
   †Calyx glabrous :
    Leaves broad :
     Calyx-tube 6–11 lin. long :
      Leaves patent ; flowers axillary all down
        the branches ... ... (3) **epacridioides.**

      Leaves imbricate ; flowers axillary a long way
        down the branches ... ... (4) **Macowani.**

      Leaves imbricate ; flowers (except in var.
       lanceolata) confined to axils of upper
       leaves :
        Branches glabrous ... ... (5) **ovata.**

        Branches puberulous ... ... (6) **tuberculosa.**

     Calyx-tube 4–5 lin. long :
      Leaves hirsute beneath ... ... (7) **hirsuta.**

      Leaves glabrous beneath, ciliate ... ... (8) **pondoensis.**

      Leaves quite glabrous ... ... (9) **congosta.**

    Leaves narrow :
     Calyx-tube 11 lin. long ... ... (10) **eckloniana.**

     Calyx-tube not more than 7 lin. long :
      Calyx-lobes linear ... ... (11) **lineariloba.**

      Calyx-lobes oblong ... ... (12) **cicatricosa.**

      Calyx-lobes more or less ovate :
       Leaves acute :
        Flowers slightly longer than the leaves (13) **parviflora.**

        Flowers much longer than the leaves :
         Leaves linear-lanceolate, 3 lin. long ... (14) **erecta.**

         Leaves oblong, 6–7 lin. long ... ... (15) **longifolia.**

       Leaves obtuse .. .. (16) **ericoides.**

   ††Calyx hairy :
    Calyx-tube 7–12 lin. long (see also 26, *virgata*, var.
     pubescens) :
     Branches not virgate ... ... (17) **Schlechteri.**

     Branches virgate :
      Leaves oblong, obtuse, glabrous ... ... (18) **leptantha.**

      Leaves oblong, obtuse, pilose ... ... (19) **floribunda.**

      Leaves ovate- or linear-lanceolate, acuminate,
       at first ciliate ... ... (20) **longiflora**

    Calyx-tube 6 lin. long or less :
     Flowers slightly longer than the leaves :
      Calyx pubescent ... ... (21) **flavescens.**

      Calyx densely pilose ... ... (22) **rustiana.**

Flowers much longer than the leaves :
    Calyx-lobes broadly ovate, acute   ...   ... (23) **lucens**.
    Calyx-lobes oblong:
        Leaves linear ... ... (24) **recta**.
        Leaves acerose ... ... (25) **angustifolia**.
    Calyx-lobes ovate, obtuse :
        Leaves oblong ... ... (26) **virgata**.
        Leaves linear-lanceolate ... ... (27) **confusa**.
***Petals 12 :
    Calyx glabrous:
        Calyx-tube 12 lin. long ... ... (28) **leiosiphon**.
        Calyx-tube 8–9 lin. long ... ... (29) **rigida**.
        Calyx-tube 6 lin. long or less :
            Leaves obtuse ... ... (30) **ramosa**.
            Leaves acute :
                Branchlets pubescent ... ... (31) **Galpini**.
                Branchlets villous ... ... (32) **Mundtii**.
    Calyx hairy :
        Calyx-tube 9–10 lin. long:
            Leaves suborbicular ... ... (33) **argentea**.
            Leaves ovate-lanceolate, subacute ... ... (34) **bachmanniana**.
            Leaves lanceolate, acute ... ... (35) **martiana**.
            Leaves oblong, obtuse ... ... (36) **garciana**.
        Calyx-tube 3½ lin. long ... ... (37) **fasciata**.
        Calyx-tube 5–6 lin. long ... ... (38) **tomentosa**.

1. **S. striata** (Lam. Ill. i. 314) ; a subshrub ; branchlets pubescent, becoming glabrous and scarred in age ; leaves opposite, closely imbricate, ovate-oblong, subacute, 5 lin. long, 1½ lin. wide, strongly nerved on the back, ciliate when young, glabrous in age ; bracteoles linear, obtuse, 1½ lin. long ; flowers solitary in the axils of the upper leaves ; calyx-tube pubescent outside, 5 lin. long, cylindrical except close to the apex ; lobes oblong, acute, 1¼ lin. long, ½ lin. wide ; petals 4, half as long as the calyx-lobes, oblong, surrounded by straight hairs of about equal length ; anthers included, slightly callose at the apex ; ovary glabrous. *Lam. Encycl.* vii. 476 (*incl. var. imbricata*); *Wikstr. in Vet. Acad. Handl. Stockh.* 1818, 288 ; *Meisn. in Linnæa,* xiv. 477, *and in DC. Prodr.* xiv. 572 ; *Drège in Linnæa,* xx. 211 ; *Bolus & Wolley-Dod in Trans. S. Afr. Phil. Soc.* xiv. 315. *S. imbricata, Andr. Rep. t.* 113 ; *Pers. Syn.* i. 149 ; *Herb. Amat.* iii. 184 ; *Roem. & Schult. Syst. Veg.* iii. 332. *S. virgata, Wendl. Bot. Beobacht.* 9, *t.* 2, *fig.* 9 ; *Willd. Sp. Pl.* i. 691, *partly* ; *Link, Enum.* i. 111. *S. lateriflora, Hornem. Hort. Hafn.* ii. 955 ; *Roem. & Schult. Syst. Veg.* iii. 330. *S. sulcata, Schmidt in Usteri, Ann.* vi. 120 ; *Wikstr. in Vet. Acad. Handl. Stockh.* 1818, 292. *S. brevifolia, Willd. ex Meisn. in Linnæa,* xiv. 477.

SOUTH AFRICA : without locality, *Sieber,* 184 ! *Pappe,* 49 ! *Thom,* 428 ! 554 ! 995 ! *Ecklon & Zeyher,* 57, *Harvey,* 690 !

COAST REGION : Cape Div. ; Cape Flats, *Burke* ! *Burchell*, 714 ! 8391 ! 8395 !
near Cape Town, 100 ft., *Bolus*, 3857 ! near Raapenburg, *Wolley-Dod*, 413 !
near Millers Point, *Wolley-Dod*, 779 ! Table Mountain, 500–1500 ft., *Mund*, 21 !
Ecklon, 782 ! Caledon Div. ; Hermanuspetrusfontein, 50 ft., *Galpin*, 4512 !
Hawston, 50 ft., *Schlechter*, 9464 ! Swellendam Div. ; thicket sides near Grootvaders Bosch, *Bowie*, 14 ! Uitenhage Div. ; without precise locality, *Bowie*, 9 !

2. **S. tetralepis** (Schlechter in Engl. Jahrb. xxvii. 171) ; plant
about 1 ft. high ; stem branched ; branches virgate, pubescent
when young ; leaves opposite, lanceolate, acuminate, striate on the
back when dry, strongly ciliate on the margins, otherwise glabrous ;
flowers in the axils of the leaves near the apex of the stem ;
bracteoles subulate, 2 lin. long ; calyx pubescent outside ; tube $3\frac{1}{4}$
lin. long, cylindrical, $\frac{1}{5}$ lin. in diam. ; lobes lanceolate, acuminate,
$1\frac{1}{2}$ lin. long, $\frac{1}{4}$ lin. wide ; petals 4, oblong, thick, surrounded by
hairs slightly longer than them ; anthers $\frac{1}{3}$ lin. long, apiculate ;
ovary oblong, glabrous ; style filiform, hairy above.

VAR. β, **glabricaulis** (Schlechter, l.c. 171) ; stem glabrous ; leaves less acute ;
flowers a little smaller.

COAST REGION : Paarl Div. ; French Hoek, 2500 ft., *Schlechter*, 9257 ! Var. β :
Caledon Div. ; Villiersdorp, 900 ft., *Schlechter*, 9902 ! 2100 ft., *Bolus*, 5261 !

3. **S. epacridioides** (C. H. Wright) ; branches long, pilose when
young, finally glabrous ; leaves patent, lanceolate, acuminate, 6 lin.
long, $1\frac{1}{2}$ lin. wide, flat, ciliate on the margin ; flowers in the axils
of leaves for a long distance down the stem ; bracteoles $1\frac{3}{4}$ lin. long,
subulate, slightly recurved, ciliate ; calyx-tube glabrous, 7 lin. long,
almost straight, $\frac{1}{3}$ lin. in diam. ; lobes ovate, 1 lin. long, $\frac{3}{4}$ lin. wide,
obtuse ; petals 8, $\frac{3}{4}$ lin. long, with surrounding hairs nearly as long ;
ovary oblong, glabrous ; style filiform ; stigma penicillate.

SOUTH AFRICA : without locality, *Mund*, 19 !

4. **S. Macowani** (C. H. Wright) ; stem erect, rather stout, not
much branched ; branches slightly pubescent when young ; leaves
opposite, imbricate, elliptic, acute, minutely scaberulous, 4–5 lin.
long, $1\frac{1}{2}$ lin. wide ; flowers in the axils of the leaves a long way
down the branches ; bracteoles oblong, obtuse, ciliate, 2 lin. long ;
calyx-tube 8 lin. long, glabrous, subcylindrical, slightly curved ;
lobes almost rotundate, obtuse, 1 lin. long ; petals 8, oblong, thick,
$\frac{3}{4}$ lin. long, surrounded by hairs of about equal length ; anthers
with obtuse connective above the cells ; ovary oblong, glabrous ;
style filiform, 5 lin. long ; stigma penicillate.

COAST REGION : Humansdorp Div. ; Kruisfontein Mountains, 900 ft., *Galpin*,
4505 ! Albany Div. ; near Grahamstown, *MacOwan*, 14 ! Howisons Poort,
Schönland.

5. **S. ovata** (Thunb. Prodr. 76) ; an undershrub ; branches
alternate, sometimes fastigiate at the summit of the stem, glabrous ;
leaves opposite, ovate, acute, up to 6 lin. long and $2\frac{1}{2}$ lin. wide,

imbricate, erecto-patent, glabrous ; flowers in the axils of the upper
leaves ; bracteoles 2, lanceolate, concave, keeled, acuminate, 2½ lin.
long ; calyx-tube glabrous, 10–11 lin. long, slender, cylindrical,
slightly swollen at the top ; lobes ovate, acuminate, 2½ lin. long,
1 lin. wide, glabrous ; petals 8, oblong, ¾ lin. long, surrounded by
hairs shorter than themselves ; stamens oblong, apiculate, 1 lin.
long ; ovary oblong, glabrous. *Thunb. Fl. Cap. ed. Schult.* 382 ;
Andr. Bot. Rep. t. 119 ; *Lodd. Bot. Cab.* 141 ; *Willd. Sp. Pl.* i. 693 ;
Ait. Hort. Kew. ed. 2, i. 272 ; *Roem. & Schult. Syst. Veg.* iii. 333 ;
Wikstr. in Vet. Acad. Handl. Stockh. 1818, 290 ; *Meisn. in Linnæa,*
xiv. 471 (*incl. var. longiflora, but excl. var. breviflora*) *and in DC.
Prodr.* xiv. 569 (*incl. var. myrsinites*). *S. myrsinites, Lam. Ill.* i. 314 ;
Poir. Encycl. vii. 478. *Belvala myrsinites, O. Kuntze, Rev. Gen.
Pl.* iii. ii. 280.

VAR. β, **lanceolata** (Meisn. in DC. Prodr. xiv. 569) ; leaves lanceolate, 1-nerved
or apparently nerveless, about 4 lin. long and 1 lin. wide. *S. lanceolata, Retz.
Obs.* iii. 26. *S. erecta, var. lanceolata, Meisn. in Linnæa,* xiv. 470.

VAR. γ, **scariosa** (Meisn. in DC. Prodr. xiv. 569) ; leaves elliptic or lanceolate,
3 lin. long, 1–1½ lin. wide, 1–3-nerved, scarious on the margin.

SOUTH AFRICA : without locality, *Thom* !

COAST REGION : Clanwilliam Div. ; between Lange Valley and Olifants River,
Drège ! Tulbagh Div. ; Winter Hoek, 1700 ft., *Bolus,* 5264 ! Worcester Div. ;
Dutoits Kloof, 2000–3000 ft., *Drège* ! Stellenbosch Div. ; Stellenbosch, *Zeyher* !
Swellendam Div. ; Zuurbraak Mountain, 1000 ft., *Galpin,* 4506 ! Langeberg Range
near Swellendam, 1500 ft., *Bolus,* 8093 ! Riversdale Div. ; between Little Vet
River and Garcias Pass, *Burchell,* 6894 ; Uniondale Div. ; on a rocky hill near
Haarlem, *Burchell,* 5027 ! Humansdorp Div. ; north side of Kromme River,
Burchell, 4844 ! Uitenhage Div. ; Van Stadens Mountains, *Toynbee* ! by a rivulet
between Maitland River and Van Stadens River, *Burchell,* 4630 ! and without
precise locality, *Zeyher,* 378 ! Port Elizabeth Div. ; Krakakamma, *Zeyher,* 572 !
Algoa Bay, *Cooper,* 3073 ! between Krakakamma and the upper part of Leadmine
River, *Burchell,* 4590 ! near Port Elizabeth, *Bolus,* 1264 ! Albany Div. ; Grahams-
town, *Misses Daly & Sole,* 478 ! Var. β : Worcester Div. ; Dutoits Kloof, *Drège* !
Cape Div. ; Cape Flats, *Burchell,* 8382 ! Caledon Div. ; Genadendal, 2000–3000 ft.,
Drège ! Houw Hoek, 1500 ft., *Schlechter,* 7362 ! *Zeyher,* 3732 ! Var. γ : Tulbagh
Div. ; Winter Hoek, *Zeyher* ! *Bolus,* 5262 ! Swellendam Div. ; near Swellendam,
Ecklon & Zeyher, 58 ! and without locality, *Ludwig.*

6. **S. tuberculosa** (Lam. Encycl. vii. 479) ; branches puberulous,
leaf-scars close and prominent ; leaves imbricate, closely placed,
elliptic or ovate, obscurely 1–3-nerved, glabrous, 3–4 lin. long,
1½–2 lin. wide ; calyx quite glabrous, 5–6 (rarely 8) lin. long ;
lobes ovate, acute, scarcely 1 lin. long ; petals 8, twice as long as
the surrounding hairs. *Lam. Ill.* i. 314 ; *Roem. & Schult. Syst.* iii.
333 ; *Wikstr. in Vet. Acad. Handl. Stockh.* 1818, 291 ; *Meisn. in
DC. Prodr.* xiv. 569. *S. Vahlii, Wikstr. in Vet. Acad. Handl. Stockh.*
1818, 291. *S. ovata, var. breviflora, Meisn. in Linnæa,* xiv.
471. *S. subulata, Vahl ex Wikstr. in Vet. Acad. Handl. Stockh.*
1818, 291?

COAST REGION : *Ludwig, Ecklon,* 56 and 627 ; *Zeyher,* 3737 γ and δ from
Grootvaders Bosch, Uitenhage and Krakakamma, ex *Meisner.*
I have not seen a specimen of this species.

7. S. hirsuta (Wikstr. in Vet. Acad. Handl. Stockh. 1818, 290) ; an erect shrub, 4–5 ft. high (*Burchell*) ; branches virgate, hirsute ; leaves elliptic, acute, 3 lin. long, 1½ lin. wide, imbricate, glabrous above, hirsute below ; bracteoles 1½ lin. long, oblong ; flowers solitary in the axils of the leaves in the upper half of the branches, fragrant (*Burchell*) ; calyx white, glabrous ; tube 5 lin. long, slender, slightly increasing in diameter upwards ; lobes ovate, acute, 2 lin. long, 1 lin. wide ; petals 8, oblong, seated in a ring of bristles ; anthers oblong ; ovary clavate. *Drège, Zwei Pfl. Documente*, 123 ; *Meisn. in Linnæa*, xiv. 476 (*excl. var. glabrescens*), and in *DC. Prodr.* xiv. 569. *S. grandis, Bartl. ex Meisn. in Linnæa*, xiv. 476.

SOUTH AFRICA: without locality, *Mund & Maire*, 25 ! 157 ! *Burke* ! *Wallich* ! *Reeves* !

COAST REGION : Riversdale Div. ; between Garcias Pass and Krombeks River, *Burchell*, 7164 ; banks of the Vet River, *Muir*, 288 ! George Div. ; at the edge of woods and in woods at George, 1000 ft., *Drège* ! near Lange Vallei, west end, *Burchell*, 5699 ! in the forest near Touw River, *Burchell*, 5713 ! Montague Pass, 1200 ft., *Young in Herb. Bolus*, 5531 ! Knysna Div. ; between Knysna and Plettenberg Bay, *Pappe* ! *Burchell*, 5356 ! near Bitou River, *Burchell*, 5305 ! Uniondale Div. ; banks of rivulet near Edmonton, *Burchell*, 5120 ! Humansdorp Div. ; Kruisfontein Mountain, 900 ft., *Galpin*, 4507 !

8. S. pondoensis (Gilg) ; stem branched ; branches at first pilose, soon glabrescent, finally rough with the projecting scars of fallen leaves ; leaves ovate-lanceolate, acute, slightly contracted at the base, 4 lin. long, 1½ lin. wide, ciliate on the margin ; flowers in the axils of the upper leaves ; bracteoles 2, lanceolate, 1⅓ lin. long, ½ lin. wide, concave, densely ciliate ; calyx glabrous outside ; tube cylindrical, slender, 5 lin. long ; lobes deltoid, acute, 1 lin. long, ¾ lin. wide at the base ; petals 8, half as long as the calyx-lobes, thick, surrounded by hairs slightly shorter than themselves ; anthers oblong, ¾ lin. long, obtuse ; ovary oblong, compressed, glabrous ; style filiform ; stigma penicillate.

EASTERN REGION : Pondoland ; without precise locality, *Bachmann*, 719 !

9. S. congesta (C. H. Wright) ; a much-branched shrub ; branches glabrous, at length rough with the prominent scars of fallen leaves, rather slender ; leaves elliptic, 4 lin. long, just over 1 lin. wide, obtuse or subacute, glabrous, very minutely denticulate ; flowers in the axils of the uppermost leaves ; bracteoles 1¼ lin. long, induplicate, with broad membranous ciliate margins and a terminal tuft of hairs ; calyx-tube glabrous, 4 lin. long, ⅓ lin. in diam. ; lobes ovate, 1¼ lin. long, 1 lin. wide, obtuse ; petals 8, ½ lin. long, thick, with a few surrounding hairs half their length ; anthers crowned by a short truncate dark-coloured connective ; ovary oblong, glabrous ; style filiform, 3 lin. long ; stigma penicillate.

EASTERN REGION : Pondoland ; in a damp valley near Murchison, *Wood*, 3030 !

10. S. eckloniana (Meisn. in DC. Prodr. xiv. 568); branches virgate, glabrous, with the leaf-scars rather distant and only slightly prominent ; leaves lanceolate, acuminate, 6 lin. long, about 1 lin. wide, glabrous; flowers borne in the axils of leaves for some distance down the stem ; bracteoles 3 lin. long, oblong, acuminate, glabrous ; calyx-tube glabrous, 11 lin. long, $\frac{1}{4}$ lin. wide ; lobes lanceolate, 2 lin. long, $\frac{1}{2}$ lin. wide ; petals 8, $\frac{1}{2}$ lin. long, subulate shorter than the surrounding hairs ; anthers oblong, shortly apiculate ; ovary oblong, glabrous ; style filiform, 7 lin. long ; stigma penicillate.

SOUTH AFRICA : without locality, *Thom* !
COAST REGION: Caledon Div. ; Zwart Berg, 1500 ft., *Galpin*, 4511 ! by the Zondereinde River, *Burchell*, 7543 ! Swellendam Div. ; Grotvaders Bosch, *Mund* ! summit of a mountain near Swellendam, *Burchell*, 7425 ! Riversdale Div. ; Garcias Pass, *Burchell*, 6936/2 ! 6982 ! 7052 ! *Galpin*, 4504 ! River Gouritz, *Ecklon & Zeyher*, 59 ! George Div. ; on the Cradock Berg near George, *Burchell*, 5967 !

11. S. lineariloba (Meisn. in Linnæa, xiv. 468) ; an undershrub ; branches tetragonal at the apex, quite glabrous ; leaves opposite, erecto-patent, at length with the internodes 2 lin. long, acerose, acute, 3 lin. long, glabrous ; bracteoles 2, linear, acute, 2 lin. long ; flowers in the upper half of the stem ; calyx glabrous ; tube cylindrical, slightly inflated above, 4 lin. long ; lobes linear, acute, 2 lin. long, $\frac{1}{2}$ lin. wide ; petals 8, oblong, about $\frac{1}{2}$ lin. long, surrounded by hairs of equal length ; anthers oblong, inserted a short distance down the tube ; ovary oblong, glabrous. *Drège, Zwei Pfl. Documente*, 73 ; *Meisn. in DC. Prodr.* xiv. 568 (*incl. var. glabra*). *S. erecta, Curt. Bot. Mag. t.* 222 ? *S. angustifolia, Steud. ex Meisn. in Linnæa*, xiv. 469. *S. juniperina, Retz. Obs.* iii. 26. *S. stricta, Donn, Hort. Cantab. ed.* 6, 40.

COAST REGION: Clanwilliam Div. ; Blue Berg, 3000–4000 ft., *Drège* !

12. S. cicatricosa (C. H. Wright) ; erect, much-branched ; branches slender, at first pilose, leaf-scars prominent ; leaves opposite, subulate, 3 lin. long, $\frac{1}{4}$ lin. wide, acute, tipped by a tuft of hairs ; flowers in the axils of the upper leaves ; bracteoles 2 lin. long, oblong, ciliate and with a terminal tuft of hairs ; calyx-tube glabrous, 7 lin. long, slender ; lobes $1\frac{1}{2}$ lin. long, $\frac{1}{3}$ lin. wide, oblong, acute, with a terminal tuft of hairs ; petals 8, oblong, $\frac{1}{2}$ lin. long, surrounded by hairs of about equal length ; anthers with acute connective ; ovary oblong, glabrous ; style filiform ; stigma small.

SOUTH AFRICA : without locality, *Thom*, 577 !

13. S. parviflora (Bartl. ex Meisn. in Linnæa, xiv. 467) ; an undershrub, 1–2 ft. high (*Burchell*) ; branches tetragonal at the apex, canescent when young ; leaves slightly spreading, acerose, subacute, glabrous, sometimes sparingly ciliate and with a short terminal hair when young, $2\frac{1}{2}$ lin. long, $\frac{1}{2}$ in. wide at the base ; bracteoles 2, linear, long, acute ; flowers in the axils of the uppermost leaves, glabrous; calyx-tube 3½ lin. long, slightly widened near the apex ; lobes ovate-lanceolate, acute, 1 lin. long; petals 8, oblong, about half as long as the lobes, surrounded by hairs of equal length; anthers oblong, inserted a short distance down the calyx-tube ; ovary glabrous. *Meisn. in Hook. Lond. Journ.* ii. 455, *and in DC. Prodr.* xiv. 568 ; *Krauss, Beitr. Fl. Cap- und Natal.* 143. *S. lutescens, Eckl. & Zeyh. ex Meisn. in DC. Prodr.* xiv. 568. *S. glabra, Zeyh. ex Meisn. in Linnæa,* xiv. 468.

SOUTH AFRICA : without locality, *Thom,* 800 ! *Wallich*! *Mund & Maire* ! COAST REGION : Riversdale Div. ; hills near Zoetemelks River, *Burchell,* 6743 ! Knysna Div. ; Plettensberg Bay, *Pappe*! between Knysna and the mouth of the Knysna River, *Burchell,* 5384 ! Humansdorp Div. ; near Humansdorp, 300 ft., *Galpin,* 4515 ! *Bolus,* 1263 ! Uitenhage Div. ; Uitenhage, *Zeyher,* 21 ! *Pappe* ! Zuureberg, 2000–3000 ft., *Drège* ! Albany Div. ; Grahamstown, *MacOwan* ! *Tyson,* 1467 ! Stones Hill near Grahamstown, *Bennie,* 66 ! near Tea Fontein, between Riebeck East and Grahamstown, *Burchell,* 3495 ! Bathurst Div. ; between Blaauw Krantz and Kowie River, *Burchell,* 3870 ! near Port Alfred, between Riet Fontein and Kowie River, *Burchell,* 3989 !

14. **S. erecta** (Linn. Mant. i. 41) ; a glabrous undershrub ; branches tetragonal at the apex; leaves opposite, slightly spreading, linear-lanceolate, acute, 3 lin. long, ½ lin. wide ; flowers in the axils of leaves in the upper part of the branches ; bracteoles linear, acute, 2 lin. long ; calyx glabrous ; tube 4–5 lin. long, slightly swollen above ; lobes ovate-oblong, acuminate, 1½ lin. long, ½ lin. wide ; petals 8, oblong, ⅓ lin. long, surrounded by hairs of equal length ; anthers included. *Wendl. Bot. Beobacht.* 9, *t.* 2, *fig.* 10 ; *Lodd. Bot. Cab. t.* 74 ; *Wikstr. in Vet. Acad. Handl. Stockh.* 1818, 289 ; *Willd. Sp. Pl.* i. 692 ; *Lam. Encycl.* vii. 478 (*excl. var.* β) ; *Drège, Zwei Pfl. Documente,* 105 ; *Meisn. in DC. Prodr.* xiv. 568 (*incl. var. vulgaris*) ; *Bolus & Wolley-Dod in Trans. S. Afr. Phil. Soc.* xiv. 315. *S. glabra, Roem. & Schult. Syst.* iii. 331. *S. pendula, Salisb. Prodr.* 282. *S. subulata, Lam. Ill.* i. 314. *S. tetragona, Retz. Obs.* iii. 25. *Belvala dodecandra, O. Kuntze, Rev. Gen. Pl.* iii. ii. 280.

SOUTH AFRICA : without locality, *Sieber,* 64 ! *Pappe*! *Forster* ! *Villet* ! *Thom,* 588 ! 668 ! *Burke* ! *Harvey*! *Cooper,* 2295 ! COAST REGION : Worcester Div. ; Breede River Valley, near Bains Kloof, 800 ft., *Bolus,* 2918 ! Cape Div. ; Flats and hills around Cape Town, 50–1000 ft., *Zeyher* ! *Ecklon,* 28 ! *Milne,* 157 ! *MacOwan & Bolus, Herb. Norm. Austr.-Afr.,* 249 ! 954 ! *Burchell,* 208 ! 440/1 ! *Wolley-Dod,* 340 ! 614 ! 983 ! *Drège* ! *Bolus,* 2917 ! near Simons Town, *Bolus,* 4951 ! Simons Bay, *MacGillivray,* 558 ! 559 ! *Wright* ! Caledon Div. ; hill side near Caledon, 900 ft., *Galpin,* 4514 ! Palmiet River, *Bolus,* 4191 ! Albany Div. ; Mayors seat, *Miss Daly,* 121 !

15. **S. longifolia** (C. H. Wright) ; branches erect, virgate, quad- rangular, glabrous, leaf-scars prominent ; leaves oblong, 6–7 lin. long, ½ lin. wide, acute, glabrous ; flowers in the axils of the leaves a long way down the branches ; bracteoles 2½ lin. long, oblong, obtuse, keeled, glabrous, margins membranous ; calyx-tube glabrous, 6 lin. long, cylindrical below, inflated above ; lobes ovate,

1½ lin. long, 1 lin. wide, acute and thickened at the apex; petals 8; anthers with produced acute connective; ovary oblong; style 5 lin. long, filiform; stigma penicillate.

COAST REGION : Caledon Div.; Zoetemelks Valley, *Burchell*, 7578 !

16. **S. ericoides** (C. H. Wright); branches at first densely pubescent and with short internodes, finally glabrous and with slightly prominent leaf-scars; leaves strictly 4-ranked, oblong, 2½ lin. long, ⅔ lin. wide, obtuse, ciliate when young; flowers axillary in the upper part of the stem; bracteoles ½ lin. long, lanceolate, obtuse, midrib thick, margins membranous, ciliate; calyx-tube glabrous, 4½ lin. long, slightly widening upwards; lobes ovate, 1 lin. long, ¾ lin. wide, obtuse; petals 8, about half as long as the calyx-lobes, thick, slightly longer than the surrounding hairs; anthers with pointed connective; ovary oblong, glabrous; style filiform, 3 lin. long; stigma penicillate.

COAST REGION : Riversdale Div.; Milkwoodfontein, 600 ft., *Galpin*, 4509 !

17. **S. Schlechteri** (Gilg) ; a much-branched shrub; branches at first pubescent, finally glabrous, and rough with leaf-scars; leaves imbricate, oblanceolate, 3 lin. long, ¾ lin. wide, with a single row of rather distant cilia on the margin and a tuft of hairs at the apex; flowers in the axils of the upper leaves; bracteoles oblong, slightly longer than the leaves, about ¼ lin. wide, induplicate; calyx-tube pubescent, 7 lin. long, slender; lobes 2 lin. long, 1 lin. wide, oblong, acute; petals 8, purplish, about ¾ lin. long, thick, surrounded by whitish hairs rather shorter than themselves; anthers about ½ lin. long, apiculate; ovary oblong, glabrous; style filiform; stigma penicillate.

CENTRAL REGION : Calvinia Div.; Oorlogs Kloof, 2200 ft., *Schlechter*, 10960 !

18. **S. leptantha** (Bolus in Trans. S. Afr. Phil. Soc. xvi. 142) ; stem branched; branches slender, glabrous; leaves opposite, oblong, obtuse or subacute, 4 lin. long, 1 lin. wide, thick, flat above, convex beneath, glabrous; bracteoles like the leaves, but only 1½ lin. long; flowers in the axils of the upper leaves, pale yellow; calyx-tube slender, nearly 1 in. long, pubescent, cylindrical, slightly inflated above; lobes ovate, obtuse, 2 lin. long, 1¼ lin. wide; petals 8, oblong, thick, 1 lin. long, surrounded by hairs half their length; anthers 4, oblong, obtuse, ¾ lin. long; ovary compressed, oblong, glabrous; style filiform; stigma feathery.

COAST REGION : Clanwilliam Div.; Lange Kloof, *Schlechter*, 8049 ! Blue Berg, *Schlechter*, 8448 ! Piquetberg Div.; Pikeniers Kloof, 850 ft., *Schlechter*, 4938 ! Malmesbury Div.; Darling, *Bachmann*, 400 !
CENTRAL REGION : Calvinia Div.; near Nieuwoudtville, *Leipoldt*.
WESTERN REGION : Little Namaqualand; in stony places near Ookiep, 3200 ft., *Bolus in MacOwan & Bolus, Herb. Norm. Austr.-Afr.*, 688 !

19. **S. floribunda** (C. H. Wright); stem branched; branches at first pubescent; leaves opposite, approximate, oblong, obtuse, 3½ lin. long, ¾ lin. wide, at first pilose, finally verrucose on the back; flowers in the axils of the leaves a long way down the branches; bracteoles 3 lin. long, ⅓ lin. wide, oblong, obtuse, long-ciliate; calyx-tube pubescent, 8½ lin. long, slender; lobes oblong, obtuse, 2½ lin. long, nearly 1 lin. wide; petals 8, oblong, ¾ lin. long, about as long as the surrounding hairs; connective obtuse; ovary oblong, glabrous; style filiform, 4½ lin. long; stigma penicillate.

COAST REGION: Clanwilliam Div.; Zekoe Vley, *Schlechter,* 8506!

20. **S. longiflora** (Lam. Ill. i. 314, t. 78); an undershrub; branches slender, tetragonal and puberulous above; leaves imbricate, ovate-lanceolate or linear-lanceolate, shortly acuminate, sheathing at the base, at first ciliate, finally glabrous, striate on the back, 4 lin. long, 1½ lin. wide; flowers in the axils of the upper leaves; bracteoles 2½ lin. long, lanceolate, concave, margins membranous, ciliate; calyx-tube pubescent, about 8 lin. long, slender, slightly widened upwards; lobes ovate-oblong, obtuse, 2 lin. long; petals 8, oblong, 1 lin. long, longer than their surrounding hairs; anthers 1 lin. long, with the connective produced and acute above; ovary oblong, ½ lin. long, glabrous. *Lam. Encycl.* vii. 476; *Wikstr. in Vet. Acad. Handl. Stockh.* 1818, 292; *Meisn. in Linnæa,* xiv. 479, and *in DC. Prodr.* xiv. 570; *Drège, Zwei Pfl. Documente,* 98; *Bolus & Wolley-Dod in Trans. S. Afr. Phil. Soc.* xiv. 315. *S. pubescens, Sims in Bot. Mag. t.* 1212; *Retz. Obs.* iii. 26. *S. glauca, Nois. ex Meisn. in DC. Prodr.* xiv. 570. *S. rubra, Donn, Hort. Cantab. ed.* iv. 31.—*Thymelæa foliis oppositis cruciatis, etc., Burm. Rar. Afr. Pl.* 127, t. 47, fig. 1.

SOUTH AFRICA: without locality, *Villet! Mund! Hooker! Reeves!*
COAST REGION: Worcester Div.; Pienaars Kloof, *Burke!* Paarl Div.; between Paarl and Lady Grey Railway Bridge, under 1000 ft., *Drège!* Cape Div.; Flats and hills around Cape Town, 100–3000 ft., *Bolus,* 4587! *Ecklon,* 784! *Burke! Tyson,* 1440! *Wolley-Dod,* 615! 1036! *Wilms,* 3596! 3599! 3600! *Burchell,* 709! *Mund,* 23! *Galpin,* 4516! Simons Bay, *Wright!* near Simonstown, *Bolus,* 4052! *Wolley-Dod,* 612! 940! Stellenbosch Div.; between Lowrys Pass and Jonkers Hoek, *Burchell,* 8327! between Stellenbosch and Cape Flats, *Burchell,* 8355! Caledon Div.; Genadendal, 3500 ft., *Schlechter,* 10318!
KALAHARI REGION: Orange Free State; without precise locality, *Cooper,* 3082! Basutoland, without precise locality, *Cooper,* 3088! (both these localities are probably erroneous).

21. **S. flavescens** (Gilg); branches reddish, finely pubescent when young, soon glabrescent; leaves imbricate, lanceolate, 3½–6 lin. long, 1 lin. wide, obtuse, densely white-ciliate on the margin when young, otherwise glabrous, coriaceous; flowers in the axils of the upper leaves and only slightly exserted from them; bracteoles 2, ovate, obtuse, 1 lin. long, white-ciliate at the apex, margins hyaline below; calyx pubescent outside; tube 3½ lin. long, slightly widening upwards, ½ lin. in diam. at apex; lobes broadly ovate, acute, ½ lin.

long; petals 8, oblong, rather shorter than the calyx-lobes, surrounded by hairs rather longer than themselves; anthers oblong, acute, ½ lin. long; ovary oblong, glabrous; style filiform, shorter than the calyx-tube; stigma penicillate.

COAST REGION: Cape Div.; Devils Peak, 1900 ft., *Kuntze*!

22. **S. rustiana** (Gilg in Engl. Jahrb. xix. 270); branches at first silky-pilose; leaves adpressed to the branches, lanceolate, concave, obtuse, 3 lin. long, ½ lin. wide, densely white-ciliate on the margins and at the apex; flowers in the axils of the upper leaves: bracteoles 2, linear, obtuse, 2 lin. long, ciliate on the margin and with a terminal tuft of hairs; calyx-tube cylindrical, 6 lin. long, lowest third glabrous, rest densely pilose; lobes 1½ lin. long, ½ lin. wide, obtuse, with a terminal tuft of hairs; petals 8, oblong, about half as long as the calyx-lobes, surrounded by hairs as long as themselves; anthers ½ lin. long, acute; ovary oblong, glabrous; style filiform; stigma penicillate.

COAST REGION: Worcester Div.; on mountains near the Hex River, *Bolus*, 5800! Riversdale Div.; near Riversdale, *Rust*, 560!

The acerose pungent appearance of the leaves, mentioned in the original description, is due to their rolling up when dry and having a terminal tuft of hairs.

23. **S. lucens** (Lam. Encycl. vii. 477); branches virgate, pubescent when young; leaves imbricate, lanceolate or oblong, acuminate, about 6 lin. long, 1 lin. wide, longitudinally multistriate, ciliate when young, glabrous in age; flowers in the axils of the upper leaves; bracteoles 2, oblong, 2 lin. long, ciliate; calyx-tube pubescent outside, about 6 lin. long, slender below, slightly inflated above; lobes broadly ovate, acute, 1 lin. long; anthers oblong, apiculate; petals 8, oblong, ¾ lin. long, surrounded by hairs of equal length; ovary oblong; style filiform, hairy; stigma subcapitate. *Wikstr. in Vet. Acad. Handl. Stockh.* 1818, 287; *Roem. & Schult. Syst. Veg.* iii. 332; *Meisn. in Linnæa*, xiv. 478, *and in DC. Prodr.* xiv. 570; *Bolus & Wolley-Dod in Trans. S. Afr. Phil. Soc.* xiv. 315. *S. virgata, Houtt. Handl.* vii. 364, *t.* 40, *fig.* 2; *Willd. Sp. Pl.* i. 691. *S. ciliata, Lam. Ill.* i. 314; *Lam. Encycl.* vii. 477, *partly. S. pubescens, Steud. ex Meisn. in Linnæa*, xiv. 478. *Belvala lucens, O. Kuntze, Rev. Gen. Pl.* iii. ii. 280.

SOUTH AFRICA: without locality, *Bergius*! *Villet*! *Forster*! *Pappe*! *Harvey*. 688!
COAST REGION: Cape Div.; Table Mountain, *Ecklon*, 65! 785! *Drège*! *MacGillivray*, 557! *Bolus*, 2919! 4670! *Tyson*, 2380! Devils Peak, *Wolley-Dod*, 586! *Schlechter*, 54! above Wynberg Range, *Wolley-Dod*, 613! Simonstown, 1000 ft., *Schlechter*, 315! Cape Flats, near Rondebosch, *Burchell*, 160! Camps Bay, *Burchell*, 297! Muizenberg, *Tyson*! Port Elizabeth Div.; near Port Elizabeth, *Tyson*, 2182!

24. S. recta (C. H. Wright); stem erect, slightly branched at the apex, pilose when young, glabrous and grey when old; leaves opposite, linear, obtuse, 4 lin. long, $\frac{1}{3}$ lin. wide, at first slightly pilose, soon glabrous; flowers in the axils of the upper leaves; calyx-tube pubescent, 6 lin. long, slender curved; lobes oblong, acute, $1\frac{1}{2}$ lin. long, $\frac{1}{3}$ lin. wide; petals 8, clavate, $\frac{1}{2}$ lin. long, shorter than the surrounding hairs; anthers shortly apiculate; ovary oblong, 1 lin. long, glabrous; style filiform, 4 lin. long; stigma penicillate.

COAST REGION: Swellendam Div.; Swellendam, 800–2000 ft., *Mund*, 25!

25. S. angustifolia (Lam. Encycl. vii. 477); branches densely white-woolly; leaves subadpressed, acerose, obtuse, convex on the back, 1–3-nerved, 2–3 lin. long, margins and apex white-ciliate; calyx pubescent, thrice as long as the leaves; tube slender; lobes small, oblong, patent; petals 8, half as long as the calyx-lobes, longer than the scanty hairs. *Lam. Ill.* i. 314; *Roem. & Schult. Syst.* iii. 331; *Meisn. in Linnæa,* xiv. 466, *and in DC. Prodr.* xiv. 567.

COAST REGION: Clanwilliam Div.; Zeekoe Vley, *Zeyher,* 3733. Pikeniers Kloof and Ezels Bank, *Drège.*

26. S. virgata (Linn. Mant. i. 41); branches at first white-pubescent, finally glabrous and with elevated leaf-scars; leaves-oblong, obtuse, 4 lin. long, $\frac{3}{4}$ lin. wide, at first densely white-ciliate on the margin, longitudinally sulcate on the back when dry; bracteoles similar to the leaves but half their size; calyx-tube pubescent, 6 lin. long, slightly inflated above; lobes ovate, obtuse, $1\frac{1}{2}$ lin. long; petals 8, oblong, obtuse, $\frac{1}{2}$ lin. long, surrounded by hairs of about equal length; anthers obtusely apiculate; ovary compressed, oblong, glabrous, $\frac{1}{2}$ lin. long; style filiform; stigma small, penicillate. *Lam. Encycl.* vii. 476; *Lam. Illustr.* i. 314; *Wikstr. in. Vet. Acad. Handl. Stockh.* 1818, 286; *Thunb. Fl. Cap. ed. Schult.* 382; *Roem. & Schult. Syst. Veg.* iii. 330; *Drège, Zwei Pfl. Documente,* 68, 73, 98, 102, 108, 110, 114; *Meisn. in Linnæa,* xiv. 464, *and in DC. Prodr.* xiv. 567. *S. virgata, var. linnæana, Meisn. in Linnæa,* xiv. 464, *and in DC. Prodr.* xiv. 567. *S. virgata, var. genuina, Meisn. in Linnæa,* xiv. 464. *S. ciliata, Andr. Bot. Rep. t.* 149, *not of Lam. S. ciliata, var., Andr. Bot. Rep. t.* 139. *Passerina Zeyheri, Spreng. ex Meisn. in Linnæa,* xiv. 464. *S. glauca, Sieb. ex Presl, Bot. Bemerk.* 108. *S. glauca, Lodd. Cat.* 33? *S. incana, Lodd. Cab. t.* 11. *S. tuberculosa, Vahl ex Wikstr. in Vet. Acad. Handl. Stockh.* 1818, 287. *Belvala virgata, O. Kuntze, Rev. Gen. Pl.* iii. ii. 280.

VAR. β, **pubescens** (Meisn. in Linnæa, xiv. 464); a more slender plant than the type; branches conspicuously white hairy; calyx pilose outside, up to twice as long as the leaves. *Drège, Zwei Pfl. Documente,* 74; *DC. Prodr.* xiv. 567. *S. pubescens, Retz. Obs.* iii. 26. *S. virgata, Sm. Exot. Bot.* i. 80, *t.* 46.

SOUTH AFRICA: without locality, *Wallich*! *Zeyher*! Var. β: *Thom,* 363! 696! *Wallich*! *Zeyher,* 1482!

COAST REGION: Vanrhynsdorp Div. ; between Heerenlogement and Knagas Berg, under 1000 ft., *Drège* ! Clanwilliam Div. ; Zeekoe Vley, *Schlechter*, 8578 ! mountains near Olifants River, *Bolus*, 5799 ! Malmesbury Div. ; Mamre, near Groene Kloof, 300 ft., *Bolus*, 4322 ! Tulbagh Div. ; Saron, 700 ft., *Schlechter*, 7865 ! Winter Hoek, *Zeyher* ! Worcester Div. ; Hex River Mountains, Axellsfarm, *Rehmann*, 2707 ! mountains, Hex River Valley, *Tyson*, 805 ! Worcester, *Zeyher* ! Stellenbosch Div. ; Stellenbosch, *Zeyher* ! Bredasdorp Div. ; near Elim, 300 ft., *Bolus*, 7854 ! *Schlechter*, 7705 ! Swellendam Div. ; on dry hills near Breede River, *Burchell*, 7457 ! Var. *β* : Clanwilliam Div. ; sand flats near Ezels Bank, *Drège* ! Malmesbury Div. ; between Groene Kloof and Dassen Berg, *Drège* ! Paarl Div. ; between Paarl and Lady Grey Bridge, *Drège* ! Cape Div. ; Tyger Berg, *Drège* ! Swellendam Div. ; Buffeljagts River, *Gill* !

CENTRAL REGION : Ceres Div. ; between Karroo Poort and Zoutpansdrift, 2000 ft., *Pearson*, 5016 !

27. **S. confusa** (C. H. Wright) ; erect, much-branched from the base ; branches at first pilose ; leaves linear-lanceolate, 3 lin. long, obtuse, tipped with a bunch of hairs, ciliate, soon glabrescent ; flowers in the axils of the uppermost leaves ; bracteoles $1\frac{3}{4}$ lin. long, oblong, obtuse, ciliate ; calyx-tube sparingly pubescent, 6 lin. long, gradually widened upwards ; lobes ovate, obtuse, $1\frac{1}{2}$ lin. long, $\frac{3}{4}$ lin. wide ; petals 8, half as long as the calyx-lobes, surrounding hairs of equal length ; anthers with blunt shortly produced connective ; ovary oblong, glabrous ; style filiform ; stigma penicillate.

COAST REGION : Tulbagh Div. ; mountains near the waterfall, 1200 ft., *Bolus*, 5263 ! *Pappe* ! Witzen Berg, behind Steendahl, 2000 ft., *Bolus*, 5378 ! Cape Div. ; beyond Raapenberg Vley, *Woolley-Dod*, 340 !

This species resembles *S. erecta*, Linn., but differs in having a pubescent calyx with obtuse lobes.

28. **S. leiosiphon** (Gilg) ; 3 ft. high (*Burchell*), much-branched ; branches at first pilose, finally glabrous and rough with prominent leaf-scars ; leaves lanceolate, concave, 6 lin. long, $1\frac{1}{2}$ lin. wide, pilose when young, afterwards glabrous and tuberculate on the back ; flowers in the axils of the uppermost leaves ; bracteoles 3 lin. long, $\frac{1}{2}$ lin. wide, densely ciliate ; calyx-tube glabrous, 12 lin. long, cylindrical, $\frac{1}{3}$ lin. in diam. ; lobes ovate, acuminate, $2\frac{1}{2}$ lin. long, $1\frac{1}{4}$ lin. wide ; petals 12, nearly 1 lin. long, oblong, surrounded by hairs nearly as long ; ovary 1 lin. long, oblong, glabrous ; style filiform, 9 lin. long ; stigma small.

COAST REGION : Caledon Div. ; tops of the mountains of Bavians Kloof near Genadendal, *Burchell*, 7730 !

29. **S. rigida** (Meisn. in DC. Prodr. xiv. 570) ; branches slender, densely leafy, terete, puberulous, leaf-scars prominent ; leaves opposite, at first imbricate, finally spreading, rigid, lanceolate, rather acute, convex on the back, 3–5 lin. long, $\frac{2}{3}$–1 lin. wide, contracted at the base, densely ciliate on the involute margins, otherwise glabrous ; flowers glabrous, 8–9 lin. long ; calyx-tube dilated in the throat ; lobes ovate-oblong, 1 lin. long ; petals 12, yellow, twice as long as the surrounding hairs.

COAST REGION : Swellendam Div. ; Breede River, between Sebastians Bay, Rhinoster Fontein and Port Beaufort, *Garnot*.

30. **S. ramosa** (C. H. Wright); much-branched; branches at first pubescent, finally glabrous, leaf-scars small; leaves oblong-lanceolate, $3\frac{1}{2}$ lin. long, 1 lin. wide, densely white-ciliate when young, at length glabrous, longitudinally sulcate when dry; flowers in the axils of the upper leaves; bracteoles 1 lin. long, oblong, with a strong midrib and membranous margins; calyx-tube glabrous, 5 lin. long, rather inflated and ribbed above; lobes ovate, obtuse, about 1 lin. long; petals 12, oblong, $\frac{2}{3}$ lin. long, with surrounding hairs of about equal length; ovary oblong, glabrous; style filiform, as long as the calyx-tube; stigma penicillate.

SOUTH AFRICA : without locality, *Mund*!
COAST REGION : Tulbagh Div. ; Witzenberg Range, *Zeyher*!

31. **S. Galpini** (C. H. Wright); branches at first pubescent, finally glabrous, with the leaf-scars only slightly prominent; leaves rather closely placed, lanceolate, $3\frac{1}{2}$ lin. long, 1 lin. wide, acute, at first densely white-ciliate on the margin, finally quite glabrous and shining; flowers densely placed in the axils of the uppermost leaves; bracteoles ovate, acute, about 1 lin. long, $\frac{2}{3}$ lin. wide, densely ciliate, herbaceous; calyx-tube glabrous, 6 lin. long, slightly curved, gradually widening upwards; lobes ovate, subacute, $1\frac{1}{2}$ lin. long, $\frac{2}{3}$ lin. wide, thin for the genus; petals 12, half as long as the calyx-lobes, rather slender, about as long as the surrounding hairs; connective of anthers acute; ovary oblong, glabrous; style filiform, 4 lin. long; stigma penicillate.

COAST REGION : Riversdale Div. ; Milkwoodfontein, 600 ft., *Galpin*, 4508!

32. **S. Mundtii** (Eckl. ex Meisn. in DC. Prodr. xiv. 572); a shrub, 1–2 ft. high; branches sparingly villous; leaves imbricate, ovate-oblong, or lanceolate, very acute, 5 lin. long, about 2 lin. wide, strongly striate on the back, floccose on the margins at least when young; flowers in the axils of the uppermost leaves; calyx glabrous; tube slender, 6 lin. long, subcylindrical; lobes ovate, cuspidate, nearly 1 lin. long; petals 12; anthers oblong, acuminate, $\frac{1}{4}$ lin. long, included; ovary ovoid, $\frac{1}{3}$ lin. long. *S. hirsuta, var. glabrescens, Meisn. in Linnæa*, xiv. 476. *S. striata, Eckl. & Zeyh. partly ex Meisn. in DC. Prodr.* xiv. 572.

SOUTH AFRICA : without locality, *Mund, Ludwig*.
COAST REGION : Tulbagh Div. ; without precise locality, *Pappe*! *Drège*, 7337!
Caledon Div. ; Klein River Mountains, *Ecklon & Zeyher*, 50!

33. **S. argentea** (Lehm. Del. Sem. Hort. Hamburg. 1831, 6, 7); an undershrub 1–2 ft. high; branches at first pubescent, finally with raised scars; leaves suborbicular or shortly ovate, obtuse, about 4 lin. long, densely imbricate and at first adpressed, densely (but deciduously) white-ciliate on the thickened margin, otherwise glabrous, silvery beneath; flowers numerous in the axils of the upper leaves; bracteoles oblong, obtuse, ciliate; calyx-tube pube-

scent, slender, 10 lin. long ; lobes suborbicular, obtuse, 1¼ lin. in
diam. ; petals 12, nearly 1 lin. long, surrounded by hairs ½ lin. long ;
anthers oblong with prolonged acute connective ; ovary ½ lin. long,
oblong, glabrous ; style lateral near the apex, filiform.   *Meisn. in
DC. Prodr.* xiv. 571. *S. dregeana, Meisn. in Linnæa,* xiv. 472 ;
Drège, Zwei Pfl. Documente, 63, 64, 122, *and in Linnæa,* xx. 211.

VAR. β, **oblongata** (Meisn. in DC. Prodr. xiv. 571); leaves ovate or oblong,
with many fine conspicuous nerves on the back. *S. aurea, Eckl. & Zeyh. ex
Meisn. l.c.*
VAR. γ, **laxior** (Meisn. in DC. Prodr. xiv. 571) ; leaves ovate-oblong, up to
5 lin. long, less densely imbricate than in the type and var. β, slightly spreading,
recurved at the apex. *S. formosa, Eckl. & Zeyh. ex Meisn. l.c.*

SOUTH AFRICA : without locality, *Lehmann* ! *Thom,* 55 ! *Harvey* ! Var. β :
Niven.
COAST REGION : Albany Div. ; *Cooper,* 22 ! 1556 ! 3074 ! Var. β : Swellendam
Div. ; between Zuurbraak and Buffeljagts River Drift, *Burchell,* 7265 ! Rivers-
dale Div. ; Stille Bay, *Muir in Herb. Galpin.,* 5315 ! Oudtshoorn Div. ; Cango,
Atherstone ! Albany Div. ; near Tea Fontein, between Riebeck East and
Grahamstown, *Burchell,* 3486 ! near Grahamstown, 2000 ft., *Glass in MacOwan,
Herb. Austr.-Afr.,* 1526 ! *MacOwan,* 13 ! *Schönland,* 3 ! Brak Kloof, *Mrs. White.*
32 ! and without precise locality, *Ecklon & Zeyher,* 628, *Cooper,* 27 ! Var. γ :
Stellenbosch Div. ; Stellenbosch, *Zeyher* ! Knysna Div. ; on sand hills near the
west end of Groene Vallei, *Burchell,* 5655 ! Uitenhage Div. ; between Kraka-
kamma and Van Staadens Berg, *Ecklon, Ecklon & Zeyher,* 47. Albany Div. ;
Zwart Hoogte, *Ecklon & Zeyher* !
CENTRAL REGION : Prince Albert Div. ; Zwartberg Range, near Vrolyk, *Drège* !

34. **S. bachmanniana** (Gilg in Engl. Jahrb. xix. 270) ; stems at
first villous, finally glabrous ; leaves imbricate, ovate-lanceolate,
subacute, rounded at the base, at first silky-villous especially on the
margin, 6 lin. long, 2 lin. wide ; flowers in the axils of the upper
leaves ; bracteoles 2 lin. long, oblong, ciliate ; calyx-tube pubescent
outside, 9 lin. long, slender, cylindrical ; lobes 2 lin. long, ¾ lin.
wide, ovate, subacute ; petals 12, conical, pale yellow, ½ lin. long,
surrounded by hairs slightly shorter than themselves ; anthers 1 lin.
long, apiculate ; ovary oblong, 1 lin. long, glabrous ; style filiform.

SOUTH AFRICA : without locality, *Mund & Maire* !
COAST REGION : Malmesbury Div. ; near Hopefield, *Bachmann,* 2037.

35. **S. martiana** (Meisn. in DC. Prodr. xiv. 570) ; branches
long, slender, pubescent ; leaves alternate, imbricate, lanceolate,
acute, semiamplexicaul, 5–6 lin. long, 1–1¼ lin. wide, sparingly pilose,
ciliate, at length glabrous ; bracteoles small, narrowly linear, obtuse,
ciliate ; calyx 9–10 lin. long ; tube slender, rather pilose, campanulate above ; lobes oblong, acute ; petals 12, twice as long as the
surrounding hairs.

SOUTH AFRICA : without locality, *Niven.*

36. **S. garciana** (C. H. Wright) ; stem erect, simple or sparingly
branched, at first pilose, reddish ; leaves oblong, obtuse, 6 lin. long,
1 lin. wide, at first densely white-ciliate on the margins and with a

terminal tuft of white hairs; flowers in the axils of the upper leaves; bracts 6 lin. long, $1\frac{1}{4}$ lin. wide, lanceolate, acuminate, densely white-ciliate on the margin; calyx-tube pubescent outside, 10 lin. long, $\frac{1}{2}$ lin. diam., cylindrical; lobes lanceolate, acuminate, $1\frac{1}{2}$ lin. long, $\frac{1}{2}$ lin. wide, hairy outside; petals 12, clavate, 1 lin. long, surrounded by hairs of about equal length; anthers oblong, acuminate, 1 lin. long; ovary oblong, $\frac{1}{2}$ lin. long, glabrous; style filiform; stigma penicillate.

COAST REGION: Riversdale Div.; near Garcias Pass, *Burchell,* 7152! Humansdorp Div.; Kruisfontein, near Humansdorp, *Galpin,* 4510!

37. S. fasciata (C. H. Wright); fasciculately branched; branches densely woolly when young, leaf-scars not very prominent; leaves opposite, linear, $3\frac{1}{2}$ lin. long, $\frac{1}{3}$ lin. wide, obtuse; flowers in the axils of the uppermost leaves and just protruded from them; bracts lanceolate, acute, very densely woolly on the margins; bracteoles 1 lin. long, $\frac{1}{6}$ lin. wide, woolly on the margins; calyx-tube $3\frac{1}{2}$ lin. long, pubescent; lobes ovate, acute, 1 lin. long, $\frac{2}{3}$ lin. wide; petals 12, oblong, subacute, slightly longer than the surrounding hairs; anthers apiculate; ovary oblong, glabrous; style filiform; stigma small.

COAST REGION: Swellendam Div.; between Zuurbraak and Buffeljagts River Drift, *Burchell,* 7266!

38. S. tomentosa (Andr. Bot. Rep. t. 334); an undershrub; branches slender, at first tomentose, at length covered with raised scars; leaves closely imbricate, oval-oblong, subobtuse, up to 5 lin. long and 3 lin. wide, coriaceous, the lowermost sometimes smaller and linear-lanceolate, 3–5-nerved on the back, when young canescent, sometimes almost glabrous in age, the uppermost slightly sheathing; flowers in the axils of the upper leaves; calyx hairy; tube 5–6 lin. long, cylindrical; lobes oblong, subobtuse, about $\frac{1}{2}$ lin. long, hairy; petals 12, rather longer than the calyx-lobes, surrounded by slightly shorter hairs; anthers included, oblong, acute, about $\frac{1}{2}$ lin. long; ovary oblong. *Ait. Hort. Kew. ed. 2,* i. 272; *Roem. & Schult. Syst.* iii. 333; *Wikstr. in Vet. Acad. Handl. Stock.* 1818, 292; *Meisn. in DC. Prodr.* xiv. 571. *S. villosa, Wikstr. in Vet. Acad. Handl. Stock.* 1818, 288. *S. chrysantha, Eckl. & Zeyh. ex Meisn. in DC. Prodr.* xiv. 572, *not of Lichtenstein.*

SOUTH AFRICA: without locality, *Niven!*
COAST REGION: Stellenbosch Div.; Hottentots Holland, 800–2000 ft., *Mund,* 22! Caledon Div.; mountains, Hermanuspetrusfontein, 300 ft., *Galpin,* 4513! Lowrys Pass, 1000 ft., *MacOwan in Herb. Norm. Austr.-Afr.,* 1951! mountains near Hemel en Aarde, *Zeyher,* 3740, Bredasdorp Div.; Elim, 2500 ft., *Schlechter,* 7742! Swellendam Div.; without precise locality, *Ecklon.*

Imperfectly known species.

39. S. chrysantha (Lichtenst. ex Roem. & Schultes, Syst. iii. 333); branches pilose when young; leaves imbricate, ovate-oblong,

subobtuse, 3-nerved, densely pilose when young ; calyx pubescent ? ;
tube long, filiform ; petals 12, surrounded by golden hairs.   *Meisn.*
in DC. Prodr. xiv. 570.

COAST REGION : Tulbagh Div. ; Witzenberg Range, *Lichtenstein.*
Meisner compares this with *S. tomentosa,* Andr.

40. S. eckloniana (Gandog. in Bull. Soc. Bot. France, lx. 419,
1913, not of Meisn.) ; an undershrub ; branches tortuous, divaricate
or patent ; leaves twice as long as those of *S. virgata,* but narrower,
more floccose, longer than the flowers ; calyx-tube hirtellous ; lobes
elliptic, glabrous.

SOUTH AFRICA : without locality, *Ecklon & Zeyher.*

41. S. dodecapetala (Bartl. ex Meisn. in Linnæa, xiv. 475) ;
branches stout, 1 ft. or more high, tetragonous and hirsute above,
terete and glabrous below ; leaves opposite, quadrifarious, imbricate,
ovate-oblong, obtuse, slightly recurved at the apex, 6 lin. long or
more, obscurely sulcate, midrib prominent, secondary nerves obscure,
silky-ciliate when young, at length glabrous ; flowers sessile, axillary near the ends of the branches, solitary, twice as long as the
leaves ; bracteoles linear, obtuse, ciliate, 3 lin. long ; calyx-tube
filiform, inflated above, silky outside ; lobes ovate, obtuse ; petals 12,
surrounded by hairs of equal length.   *Meisn. in DC. Prodr.* xiv.
571.

VAR. β, **Kraussii** (Meisn. in DC. Prodr. xiv. 571) ; branches simple ; leaves
3–5-nerved, the upper ovate-oblong, 6–7 lin. long, lower 8–10 lin. long, 3–4 lin.
wide, slightly sheathing ; calyx 12–14 lin. long, sparingly puberulous ; tube
slender ; lobes 1 lin. long.   *S. dodecapetala, Meisn. in Hook. Lond. Journ. Bot.*
ii. 455 ; *Krauss, Beitr. Fl. Cap- und Natal.* 143.

COAST REGION : Swellendam Div. ; Outeniqua, near Tradouw, *without collector's
name.*   Var. β : Cape Div. ; Devils Mountain, *Krauss,* 763.

VIII. GNIDIA, Linn.

Flowers hermaphrodite.   *Calyx-tube* cylindrical, circumscissile
above the ovary ; lobes 4, patent.   *Petals* 4, 8 or 12, smaller than
the calyx-lobes, membranous or fleshy.   *Stamens* 8 in 2 whorls, the
upper shortly exserted, rarely imperfect ; anthers oblong or linear ;
connective narrow.   Hypogynous *disc* none or very shortly annular.
Ovary sessile, 1-celled ; style filiform ; stigma capitate or penicillate.
Fruit dry, small, enclosed in the persistent base of the calyx ;
pericarp membranous.   *Seed* with crustaceous testa ; albumen
scanty, rarely thick and fleshy ; cotyledons rather thick.

Virgate or heath-like shrubs ; leaves usually small or narrow, opposite or
alternate ; flowers white, yellow, red or violet, capitate at the ends of the
branches, rarely spicate or solitary and axillary, bracteate.

DISTRIB. Species about 90 in Tropical and South Africa.

Perfect anthers 4 ; staminodes 4 :
Petals 4 　... 　... 　... 　... 　... 　... 　... (1) **harveyana.**
Petals 8 　... 　... 　... 　... 　... 　... 　... (2) **anomala.**

Perfect anthers 8 ; staminodes 0 :
　*Petals 4 :
　　Petals fleshy, laciniate 　... 　... 　... 　... (8) **pulvinata.**
　　Petals fleshy, entire or emarginate :
　　　Involucral leaves much broader than the cauline 　(4) **microcephala.**
　　　Involucral leaves not very different from the cauline :
　　　　Leaves glabrous, alternate 　... 　... 　... (5) **pinifolia.**
　　　　Leaves glabrous, opposite (see also 9, *gemini-flora*) :
　　　　　Leaves orbicular 　... 　... 　... 　... (6) **orbiculata.**
　　　　　Leaves ovate 　... 　... 　... 　... 　...(7) **oppositifolia.**
　　　　Leaves hairy, alternate 　... 　... 　... 　... (8) **tomentosa.**
　　　　Leaves hairy (at least when young), opposite :
　　　　　Petals 1¼ lin. long 　... 　... 　... 　... (9) **geminiflora.**
　　　　　Petals minute :
　　　　　　Flowers 5 lin. long ... 　... 　... 　... (10) **Burmanni.**
　　　　　　Flowers under 3 lin. long ... 　... 　... (11) **wikstrœmiana.**
　Petals membranous :
　　Calyx hairy :
　　　Leaves glabrous :
　　　　Leaves linear, obtuse ... 　... 　... 　... (12) **penicillata.**
　　　　Leaves narrowly lanceolate, acuminate 　... (13) **quadrifaria.**
　　　　Leaves lanceolate, pungent 　... 　... 　... (14) **styphelioides.**
　　　　Leaves ovate, acuminate 　... 　... 　... (15) **myrtifolia.**
　　　　Leaves ovate-oblong, subobtuse 　... 　... (16) **sonderiana.**
　　　Leaves hairy :
　　　　Leaves obtuse, adpressed-silky :
　　　　　Calyx 5 lin. long 　... 　... 　... 　... (17) **tenella.**
　　　　　Calyx 3–4 lin. long ... 　... 　... 　... (18) **chrysophylla.**
　　　　Leaves acute, pilose 　... 　... 　... 　... (19) **humilis.**
　　Calyx glabrous :
　　　Leaves alternate :
　　　　Branches glabrous 　... 　... 　... 　... (20) **juniperifolia**
　　　　Branches pubescent when young 　... 　... (21) **subulata.**
　　　　Branches adpressed-tomentose 　... 　... (22) **variegata.**
　　　Leaves opposite :
　　　　Leaves more or less ovate :
　　　　　Branches erect, parallel 　... 　... 　... (23) **Flanagani.**
　　　　　Branches divergent ... 　... 　... 　... (24) **coriacea.**
　　　　Leaves linear- to oblong-lanceolate :
　　　　　Calyx-tube narrow above 　... 　... 　... (25) **parviflora.**
　　　　　Calyx-tube wide above :
　　　　　　Calyx-lobes oblong 　... 　... 　... (26) **decurrens.**
　　　　　　Calyx-lobes ovate ... 　... 　... 　... (27) **Galpini.**
　**Petals 8, fleshy :
　　Calyx glabrous ... 　... 　... 　... 　... 　... (28) **pallida.**

Calyx hairy :
 Leaves glabrous :
  Flowers axillary ... ... (29) **thesioides.**
  Flowers on very short axillary branches ... (30) **fastigiata.**
  Flowers racemose, axillary ... ... (31) **racemosa.**
  Flowers distinctly terminal :
   Flowers solitary :
    Leaves linear-subulate ... ... (32) **linoides.**
    Leaves elliptic-oblong ... ... (33) **Cayleyi.**
   Flowers clustered :
    Leaves opposite :
     Leaves oblong-lanceolate, upper wider (34) **obtusissima.**
     Leaves linear-oblong, uniform ... (35) **ericoides.**
    Leaves alternate :
     Flowers less than 6 lin. long :
      Calyx sparsely hairy :
       Petals as long as calyx-lobes ... (36) **parvula.**
       Petals ½ as long as calyx-lobes ... (37) **polystachya.**
      Calyx densely hairy :
       Leaves 3 lin. long. ... ... (38) **scabra.**
       Leaves 4–9 lin. long ... ... (39) **setosa.**
     Flowers 10 lin. long ... ... (40) **Woodii.**
 Leaves hairy :
  Leaves lanceolate ... ... (41) **Baurii.**
  Leaves linear-lanceolate :
   Flowers capitate ... ... (42) **stellatifolia.**
   Flowers solitary ... ... (43) **sparsiflora.**
  Leaves linear-oblong :
   Flowers in dense clusters ... ... (44) **nodiflora.**
   Flowers few ... ... (45) **strigillosa.**
  Leaves ovate or oval :
   Flowers not much longer than the leaves... (46) **sericea.**
   Flowers much longer than the leaves :
    Branches virgate ... ... (47) **denudata.**
    Branches diffuse :
     Calyx densely tomentose ... ... (48) **Leipoldtii.**
     Calyx adpressed-silky ... ... (49) **nitida.**
***Petals 8, membranous :
 Calyx glabrous ... ... (50) **Meyeri.**
 Calyx hairy :
  Branches spreading :
   Leaves linear, pilose ... ... (51) **multiflora.**
   Leaves oval, 12 lin. wide, hairy ... ... (52) **ovalifolia.**
   Leaves oval, 1½ lin. wide, canescent... ... (53) **imbricata.**
   Leaves oblong, 1½ lin. wide, glabrous :
    Calyx pubescent ... ... (54) **caniflora.**
    Calyx silky-villous ... ... (55) **inconspicua.**
  Branches erect :
   Leaves subulate ... ... (56) **Francisci.**
   Leaves linear-lanceolate ... ... (57) **scabrida.**
****Petals 12 ; calyx hairy ... ... (58) **Cephalotes.**

1. G. harveyana (Meisn. in Linnæa, xiv. 437) ; branches sparingly branched above, puberulous at the apex ; leaves opposite or in verticils of 4 each, oval, very obtuse, 4½–5½ lin. long, 2–3 lin. wide, thickly coriaceous, 3–5-nerved, slightly concave, at first adpressed silky-pilose, at length glabrous except sometimes at the base ; flowers geminate at the ends of the branches ; calyx 7–8 lin. long, white-silky outside ; tube cylindrical ; lobes oblong, subobtuse ; petals 4, linear-oblong, subacute, waxy, yellow, glabrous ; upper anther exserted. *Meisn. in DC. Prodr.* xiv. 581.

SOUTH AFRICA : without locality, *Drège.*

2. G. anomala (Meisn. in Linnæa, xiv. 435) ; a much-branched undershrub ; branches at first white-pilose, finally glabrous and with slightly raised leaf-scars ; upper leaves opposite, lower alternate, oval or ovate-oblong, concave, 3½ lin. long, 1½ lin. wide, obtuse, silky on both sides, 3–5-nerved ; bracts similar to the leaves ; flowers few ; calyx silky outside ; tube 6½ lin. long, slender ; lobes oblong-lanceolate, acute, 2½ lin. long, 1 lin. wide, margins involute ; petals 8, geminate, clavate ; upper stamens absent, lower with anthers ¾ lin. long ; ovary oblong, glabrous ; style filiform, 4 lin. long ; stigma small. *Meisn. in DC. Prodr.* xiv. 580; *Bolus and Wolley-Dod in Trans. S. Afr. Phil. Soc.* xiv. 315. *G. sericea, Drège in Linnæa,* xx. 208. *G. argentea, Eckl. ex Meisn. in DC. Prodr.* xiv. 581.

SOUTH AFRICA : without locality, *Wallich! Stanger!*
COAST REGION : Cape Div. ; stream beyond Pauls Berg, *Wolley-Dod,* 2929 ! Simons Bay, *Wright!* lower part of Table Mountain, *Burchell,* 8437 ! Stellenbosch Div. ; near Lowrys Pass, Hottentots Holland, 1300 ft., *Zeyher,* 3764, *Bolus,* 4190 ! Caledon Div. ; Houw Hoek, 1200 ft., *MacOwan, Herb. Norm. Austr.-Afr.,* 248 ! *Scott-Elliot,* 1113 ! *Burchell,* 8146 ! on Donker Hoek Mountain, *Burchell,* 7046 ! Bredasdorp Div. ; Elim, 500 ft., *Schlechter,* 7657 ! Riversdale Div. ; Riversdale, *Rust,* 599 ! George Div. ; between Zwart Vallei and the west end of Lange Vallei, *Burchell,* 5693 ! Knysna Div. ; between Goukamma River and Groene Vallei, *Burchell,* 5619 !

3. G. pulvinata (Bolus in Trans. S. Afr. Soc. xvi. 142) ; a shrub 3–4 ft. high ; branches diffuse, rigid, at first pilose, soon glabrous, verrucose, leaf-scars prominent ; leaves crowded, opposite, oblong-lanceolate, 5 lin. long, ¾. lin. wide, incurved, obtuse, thickly coriaceous, densely white-pilose when young, at length minutely verrucose ; flowers in terminal 3–5-flowered clusters, exinvolucrate ; calyx densely tomentose, dull purple, ribbed ; tube subcylindrical, 8 lin. long ; lobes ovate or lanceolate, acute, 1½ lin. long, 1 lin. wide ; petals 4, divided into many fleshy processes intermixed with hairs and almost closing the mouth of the calyx-tube ; anthers oblong, obtuse, ½ lin. long ; ovary ovoid, with a terminal tuft of hairs ; style nearly as long as the calyx-tube ; stigma subcapitate. *De Wild. Pl. Nov. Herb. Hort. Then.* i. 205, *t.* 46, *figs.* 1–9.

COAST REGION : Bredasdorp Div. ; on the mountains between Caledon and Elim, about 600 ft., *Bolus,* 9238 ! near Koude River, *Schlechter,* 9619 ! Riversdale Div. ; Gysmans Hoek, *Muir* 502 !

4. G. microcephala (Meisn. in DC. Prodr. xiv. 589) ; stems many, about 1 ft. high, probably annual from a woody base, virgate, glabrous ; leaves alternate, subulate-linear, 4–6 lin. long, slightly incurved, channelled above, faintly keeled below ; heads terminal, many-flowered ; bracts ovate-lanceolate, 3 lin. long, outer 1 lin. wide, inner 1½ lin. wide, glabrous except sometimes on the margins ; calyx-tube with a basal glabrous swelling ½ lin. long, deciduous part hairy outside and 2 lin. long ; lobes oblong, hairy outside, obtuse, 1 lin. long, ¼ lin. wide or less ; petals 4, clavate ; ovary ovoid, glabrous except for a tuft of long silky hairs from the base ; style short ; stigma capitate. *Rendle in Trans. Linn. Soc. ser.* 2, iv. 40 ; *Pearson in Dyer, Fl. Trop. Afr.* vi. i. 225. *G. apiculata, Gilg in Engl. Jahrb.* xix. 263. *Gnidiopsis microcephala, Van Tieghem in Bull. Soc. Bot. France,* xl. (1893), 76.

KALAHARI REGION : Orange River Colony ; Besters Vlei near Witzies Hoek, 5300 ft., *Bolus*, 8244 ! Transvaal ; Magaliesberg, *Burke*, 96 ! *Zeyher*, 1492 ! Wonderboompoort, *Rehmann*, 4530 ! near Pretoria, *Burtt-Davy*, 685 ! 2541 ! Rustenberg, *Miss Pegler*, 977 ! near Lydenburg, *Wilms* ! Abbotts Hill, Barberton, 3500 ft., *Galpin*, 1011 ! Witbank, Middelburg distr., *Gilfillan in Herb. Galpin*, 7236 !
EASTERN REGION : Natal ; Biggarsberg, 4000 ft., *Wood*, 846 ! grassy flat near Lambonjwa River, Tugela district, *Wood*, 3447 ! near Pietermaritzburg, *Wilms*, 2247 ! Howick, *Mrs. Hutton*, 423 ! and without precise locality, *Cooper*, 3080 ! *Gerrard*, 1389 ! Swaziland ; Havelock Concession, 4000 ft., *Saltmarshe in Herb. Galpin*. 1011 !
Also in Tropical Africa.

5. G. pinifolia (Linn. Sp. Pl. ed. i. 358, not of Linn. f.) ; branches forked or fastigiate, glabrous, rather slender, leaf-scars rather close, moderately prominent ; leaves alternate, acerose, pungent-acuminate, up to 8 lin. long and ⅓ lin. wide, glabrous, involucral rather wider ; heads many-flowered ; calyx densely hairy outside ; tube 6 lin. long, the lower quarter inflated and triangular in section, upper part subcylindrical ; lobes oblong or obovate, obtuse, 2 lin. long, 1⅓ lin. wide ; petals 4, 1 lin. long, anther-like, fleshy, densely hairy all over ; anthers oblong, obtuse, ⅔ lin. long ; ovary shortly stipitate, glabrous ; style nearly as long as the calyx-tube ; stigma capitate. *Berg. Descr. Pl. Cap.* 122 ; *Thunb. Prodr.* 76, *and Fl. Cap. ed. Schult.* 379 ; *Andr. Rep. t.* 52 ; *Lodd. Bot. Cab. t.* 7 ; *Bot. Reg. t.* 19 ; *Bot. Mag. t.* 2016 ; *Wikstr. in Vet. Acad. Handl. Stockh.* 1818, 310 ; *Meisn. in Linnæa,* xiv. 445, *and in DC. Prodr.* xiv. 589 ; *Drège, Zwei Pfl. Documente,* 83, 105. *G. radiata, Linn. Mant.* 67 ; *Lodd. Bot. Cab. t.* 29 ; *Wendl. Bot. Beobacht.* 15, *t.* 2, *fig.* 12. *G. Schlechteri, Gandog. in Bull. Soc. Bot. France,* lx. 417, 1913. *Canalia daphnoides, Schmidt in Flora,* 1830, 555.—*Rapunculus foliis nervosis linearibus, Burm. Rar. Afr. Pl.* 112, *t.* 41, *fig.* 3.

VAR. β, **ochroleuca** (Bot. Reg. t. 624) ; more slender than the type ; leaves patent, obtuse, bright green ; involucral leaves stellately arranged ; calyx yellow outside except at the purple base ; petals nearly as long as the calyx-lobes. *Meisn. in DC. Prodr.* xiv. 590. *G. ochroleuca, Lodd. Bot. Cab. t.* 1184.

SOUTH AFRICA: without locality, *Thom*, 634! 780! 806! *Cooper*, 3081! Var. β: cultivated specimens.

COAST REGION: Piquetberg Div. ; between Kromme River and Pietersfontein, *Drège*! Paarl Div. ; French Hoek Kloof, *Drège*! Cape Div. ; hills and flats near Cape Town, *Ecklon*, 110! 358! *Zeyher*! *Mund*, 13! *Bolus*, 2921! 3701! *Schlechter*, 1037! *Ecklon & Zeyher*! *MacOwan in MacOwan & Bolus, Herb. Norm. Austr.-Afr.*, 247! *Pappe*, 37! *Rogers*, 2419! *Wolley-Dod*, 610! *Wilms*, 3595a! *MacOwan*, 2471! *Burchell*, 316! 381! 8409! 8428! *Drège*! Simons Bay, *MacGillivray*, 627! *Wright*! Stellenbosch Div. ; Lowrys Pass, *Galpin*, 3156! Caledon Div. ; near Caledon, *Thom*, 1009! *Rogers*, 11024! Hangklip, *Zeyher*, 3758! Houw Hoek Mountains, 1500 ft., *Burchell*, 8136! *Galpin*, 4525! Zwartberg, 2000 ft., *Galpin*, 4526! Bredasdorp Div. ; Rietfontein, *Schlechter*, 10592! Queenstown Div. ; without precise locality, *Cooper*, 3088 bis! Knysna Div. ; Milwood, *Tyson*, 3128! and without precise locality, *Zeyher*, 4752! Var. β : Western District, without locality, *Cooper*, 3081!

KALAHARI REGION : Transvaal ; Magalies Berg, *Burke*! Basutoland ; without precise locality, *Cooper*, 3086! (both localities doubtful).

6. G. orbiculata (C. H. Wright); an erect shrub; branches corymbose, erect, glabrous; leaves opposite, orbicular, 2–3 lin. in diam., shortly cuspidate, 1-nerved, glabrous; flowers few at the apex of the branches; calyx tomentose outside, pale gamboge (*Burchell*); tube 8 lin. long, subcylindrical, ribbed; lobes orbicular, 1½ lin. in diam.; petals 4, anther-like, very thick, ⅔ lin. long, shortly stalked; anthers oblong, ⅔ lin. long, obtuse.

SOUTH AFRICA: without locality, *Thom*, 162!

COAST REGION : Uniondale Div. ; in damp places by the Aapies River, in Long Kloof, *Burchell*, 4945!

7. G. oppositifolia (Linn. Sp. Pl. ed. i. 358), a much-branched shrub up to 12 ft. high; branches virgate, slender, glabrous, reddish when young; leaves decussate, longer than the internodes, ovate or ovate-lanceolate, acute, 5 lin. long, 2 lin. wide, quite glabrous, 1-nerved beneath, the upper often reddish especially at the margin; flowers in clusters of 4–6 at the branches; calyx pubescent outside; tube 8 lin. long, articulated below, slightly widened upwards, ribbed; lobes obovate, obtuse, 1½ lin. long, 1 lin. wide; petals 4, fleshy, 2-lobed in front, 1 lin. long; anthers ⅓ lin. long, much narrower than the petals, all included; ovary ¼ lin. long, compressed, densely hairy at the top; style half as long as the calyx-tube; stigma capitate. *Lam. Encycl.* ii. 766, *Ill. t.* 291, *fig.* 2; *Bot. Reg. t.* 2; *Bot. Mag. t.* 1902; *Andr. Bot. Rep. t.* 225. *Wikstr. in Vet. Acad. Handl.* 1818, 312; *Ait. Hort. Kew. ed.* 2, ii. 413; *Drège, Zwei Pfl. Documente*, 79, 85, 89; *Meisn. in Linnæa*, xiv. 431, *and in DC. Prodr.* xiv. 586; *O. Kuntze, Rev. Gen. Pl.* iii. ii. 281; *Bolus & Wolley-Dod in Trans. S. Afr. Phil. Soc.* xiv. 316. *G. lævigata, Thunb. Prodr.* 76, *and Fl. Cap. ed. Schult.* 379; *Andr. Bot. Rep. t.* 89. *G. latifolia, Hort. ex Meisn. in Linnæa*, xiv. 432. *Passerina lævigata, Linn. Amœn. Acad.* iv. 312, *and Sp. Pl. ed.* ii. 513. *Nectandra lævigata, Berg. Descr. Pl. Cap.* 134. *Gnidiopsis oppositifolia, Van Tiegh. in Bull. Soc. Bot. France*, xl. (1893) 76.—*Thymelæa foliis planis acutis, etc., Burm. Rar. Afr. Pl.* 137, *t.* 49, *fig.* 3.

SOUTH AFRICA : without locality, *Sieber*! *Villet*! *Grey*! *Bowie*! *Harvey*, 681!
COAST REGION : Tulbagh Div. ; Tulbagh, *Pappe*, 21 ! Waterfall, *Tyson*, 1473!
Mitchells Pass, *Pearson*, 3519 ! Worcester Div. ; Dutoits Kloof, *Drège*! Hex
River Vley, *Tyson*, 806 ! Paarl Div. ; Paarl Mountain, *Drège*! Cape Div. ;
Muizenberg, *Burke*! *Bolus*, 4645! Table Mountain, *Drège*! *Burchell*, 559 !
Kuntze! Vley north of Constantia Berg, *Wolley-Dod*, 665 ! Steen Berg, 900 ft.,
Dümmer, 985 ! Simons Bay, *Wright*! Vlagge Berg, *Schlechter*, 189 ! Stellenbosch
Div. ; Stellenbosch, *Ecklon & Zeyher*! Lowrys Pass, *Burchell*, 8253 ! *MacOwan*,
Herb. Norm. Austr.-Afr., 246 ! Swellendam Div. ; mountains near Swellendam,
2000–4000 ft., *Mund*, 16 ! 17 ! *Burchell*, 7400 ! Riversdale Div. ; Garcias Pass,
900 ft., *Galpin*, 4528 ! Paardeberg, *Muir in Herb. Galpin.* 5321 ! Corente River
Farm, *Muir in Herb. Galpin*, 5320 ! George Div. ; on the Cradock Berg, near
George, *Burchell*, 5947 ! 5958 ! Knysna Div. ; on mountains near Millwood,
Tyson, 3127 ! Humansdorp Div. ; Kruisfontein, *Galpin*, 4527 ! near Humansdorp,
Kennedy ! Uitenhage Div. ; Uitenhage, *Zeyher*, 208 ! Albany Div. ; along the
rivulet at Grahamstown, *Zeyher*, 891 ! *Burchell*, 3543 ! *MacOwan* !
EASTERN REGION : Griqualand East; Fort Donald, *Tyson*, 1639 !

8. **G. tomentosa** (Linn. Sp. Pl. ed. i. 358); an erect shrub;
branches terete, dark purple, pubescent when young, leaf-scars
prominent ; leaves alternate, ovate-lanceolate or ovate-oblong,
5 lin. long, 2 lin. wide, subobtuse, pilose when young, afterwards
flat and glabrous above, verrucose and 3–5-nerved beneath ; flowers
in terminal heads ; calyx densely silky outside ; tube 8–10 lin. long,
slightly widened upwards ; lobes ovate, subobtuse, 2 lin. long, 1 lin.
wide ; petals 4, anther-like, slightly bilobed, 1 lin. long, glabrous :
anthers oblong, ½ lin. long ; ovary ovoid, hairy at the top ; style
7 lin. long ; stigma small. *Wikstr. in Vet. Acad. Handl. Stockh.*
1818, 317 ; *Thunb. Fl. Cap. ed. Schult.* 381. *G. glandulosa, Hayne
ex Meisn. in Linnæa*, xiv. 439. *G. pubescens, Berg. Descr. Pl. Cap.*
124 ; *Meisn. in Linnæa*, xiv. 438, *and in DC. Prodr.* xiv. 581 ;
Bolus & Wolley-Dod in Trans. S. Afr. Phil. Soc. xiv. 315. *G. punctata, Lam. Encycl.* ii. 765 ; *Drège, Zwei Pfl. Documente*, 88. *G. scabra,
Thunb. Fl. Cap. ed. Schult.* 380. *Calycosericos typica, Eckl. & Zeyh.
ex Meisn. in DC. Prodr.* xiv. 581.

SOUTH AFRICA : without locality, *Villet* ! *Burke* ! *Mund* ! *Grey* ! *Milne* !
COAST REGION : Cape Div. ; Table Mountain, *Pappe* ! *Drège* ! *Ecklon*, 359 !
Zeyher, 4734 ! *Burchell*, 576 ! Devils Mountain, 1900 ft., *Burchell*, 8457 !
MacOwan, 560 ! *Miss Kensit* ! below Constantia Berg, *Wolley-Dod*, 654 ! slope on
Fish Hoek, *Wolley-Dod*, 429 ! Steen Berg, *Wolley-Dod*, 1144 ! 1148 ! Elsje Peak,
Wolley-Dod, 2993 ! Muizenberg, near Kalk Bay, 1600 ft., *Bolus*, 3908 ! Simons
Bay, *Wright* ! Caledon Div. ; Houw Hoek, 2000 ft., *Schlechter*, 9401 ! mountain
near Palmiet River, *Ecklon & Zeyher* !

9. **G. geminiflora** (E. Meyer ex Meisn. in Linnæa, xiv. 441); a
much-branched undershrub ; branches rather slender, slightly puberulous at the apex ; leaves opposite, lanceolate, 4 lin. long, 1 lin.
wide, concave, the uppermost slightly hairy, the others quite
glabrous ; flowers in pairs near the apex of the branches ; calyx
silky outside ; tube 7 lin. long, ribbed, slightly inflated below ;
lobes ovate, acute, 2¼ lin. long, 1 lin. wide ; petals 4, fleshy, bifid,
1½ lin. long, glabrous ; ovary compressed, slightly puberulous

above; style excentric, filiform, 3 lin. long; stigma capitate. *Drège, Zwei Pfl. Documente,* 95 ; *Meisn. in DC. Prodr.* xiv. 585.

VAR. β, **brevifolia** (Meisn. in Linnæa, xiv. 442); uppermost leaves ovate-lanceolate, 3 lin. long, the lower smaller, ovate, remote ; flowers paler ; calyx 6–8 lin. long ; lobes 1 lin. long. *Meisn. in DC. Prodr.* xiv. 586. *G. geminiflora, Drège, Zwei Pfl. Documente,* 114.

COAST REGION : Vanrhynsdorp Div. ; Gift Berg, 1000–2000 ft., *Phillips,* 7457 ! Clanwilliam Div. ; Clanwilliam, 300 ft., *Schlechter,* 8010 ! on mountains around Kromme River, Cederberg Range, 2900 ft., *Bolus,* 5810 ! Tulbagh Div. ; Witzenberg, behind Steendahl, 2000–2500 ft., *Bolus,* 5474 !

CENTRAL REGION : Calvinia Div. ; Nieuwoudtville, *Leipoldt,* 9385 ! Willems River, *Leipoldt,* 140 ! 872 ! Var. β : Ceres Div. ; between Hex River mountains and the warm Bokkeveld, 3000–4000 ft., *Drège* !

WESTERN REGION : Little Namaqualand ; between Zwartdoorn River and Groen River, under 1000 ft., *Drège* !

10. **G. Burmanni** (Eckl. & Zeyh. ex Meisn. in DC. Prodr. xiv. 583) ; an undershrub about 8 in. high ; branches slightly spreading, rather slender, pilose, leaf-scars small but prominent ; leaves opposite, lanceolate, acute, 3 lin. long, $\frac{2}{3}$ lin. wide, flat, midrib obvious beneath, adpressed silky-pilose on both surfaces ; flowers terminal, geminate ; calyx silky outside ; tube 4 lin. long, slender, inflated below, subcylindrical above ; lobes lanceolate, acute, 1 lin. long ; petals 4, minute, clavate ; anthers oblong, $\frac{1}{2}$ lin. long ; ovary ovoid ; style nearly as long as the calyx-tube ; stigma small. *Gnidiopsis Burmanni, Van Tiegh. in Bull. Soc. Bot. France,* xl. (1893) 76.

COAST REGION : Cape Div. ; at the foot of Table Mountain, *Ecklon & Zeyher,* 11 ! Lion Mountain, *Ecklon & Zeyher,* 13 ! and 85, *ex Meisner,* eastern side of the Lions Rump, *Burchell,* 146 ! kloof between the Lions Head and Table Mountain, *Burchell,* 278 ! on the plain between Cape Town and Table Mountain, *Burchell,* 80 ! Stellenbosch Div. ; Lowrys Pass, 600 ft., *Schlechter,* 7812 ! and without precise locality, *Mund & Maire* !

11. **G. wikstrœmiana** (Meisn. in Linnæa, xiv. 434) ; a dwarf shrub ; branches patent, short, rigid ; leaves opposite, oblong-lanceolate, subacute, 2 lin. long, $\frac{1}{2}$ lin. wide, with adpressed silky hairs on both surfaces ; flowers in small heads near the ends of the branches ; calyx silky outside ; tube 2 lin. long, ovoid below, subcylindrical above ; lobes oblong, obtuse, $\frac{2}{3}$ lin. long, $\frac{1}{4}$ lin. wide ; petals 4, subulate, minute ; anthers oblong ; ovary ovoid, with an apical tuft of hairs ; style cylindrical, $1\frac{1}{4}$ lin. long ; stigma slightly swollen. *Drège, Zwei Pfl. Documente,* 55 ; *Meisn. in DC. Prodr.* xiv. 582. *G. stricta, Wikstr. in Vet. Acad. Handl. Stockh.* 1818, 315. *Passerina stricta, Thunb. Prodr.* 75, *and Fl. Cap. ed. Schult.* 377.

CENTRAL REGION : Graaff Reinet Div. ; Sneeuw Berg Range, 4000–5000 ft., *Drège,* 7369 !

12. **G. penicillata** (Lichtenst. ex Meisn. in Linnæa, xiv. 448) ; a subshrub ; branches at first reddish and pubescent, finally glabrous and nearly smooth ; leaves opposite, linear, obtuse, up to 6 lin. long

and nearly 1 lin. wide, at first densely white-ciliate on the margins, otherwise glabrous; flowers 2–6 at the apex of the branches; calyx silky outside; tube cylindrical, 4 lin. long, rather slender; lobes ovate, acute, 3 lin. long, 2 lin. wide; petals 4, membranous, bifid; ovary compressed laterally; anthers oblong, $\frac{2}{3}$ lin. long; fruit lenticular, brown with dark small spots. *Drège, Zwei Pfl. Documente,* 84, *and in Linnæa,* xx. 209; *Meisn. in DC. Prodr.* xiv. 582; *Bolus & Wolley-Dod in Trans. S. Afr. Phil. Soc.* xiv. 315.

COAST REGION: Paarl Div.; Toll Bar Vley, French Hoek, *Grey*! Cape Div.; Smitswinkel Vley, *Wolley-Dod.* 768! 1241! Simons Bay, *Wright*! Klaver Vley, near Simons Town, 800 ft., *Bolus,* 7014! Cape Point, 800 ft., *Schlechter,* 7310! Stellenbosch Div.; Lowrys Pass, 1000–2000 ft., *Drège*! Hottentots Holland, 1000–3000 ft., *Mund,* 13! *Zeyher,* 3742! Caledon Div.; Vogelgat, 200 ft., *Schlechter,* 9567! Zwart Berg, *Ecklon & Zeyher*!

13. G. quadrifaria (C. H. Wright); a much-branched shrub; branches slender, reddish, at first pubescent, leaf-scars moderately prominent; leaves approximate, subopposite, narrowly lanceolate, acuminate, 5 lin. long, nearly 1 lin. wide, distinctly 4-ranked, glabrous, margins inflexed above, 3-nerved beneath; flowers few at the apex of the branches; calyx yellow, pubescent outside; tube 4 lin. long; lobes ovate-lanceolate, 2 lin. long, 1 lin. wide; petals 4, membranous, about $1\frac{1}{2}$ lin. long, 1 lin. wide; anthers oblong, $\frac{1}{2}$ lin. long, the upper exserted on stout filaments $\frac{1}{2}$ lin. long; ovary oblong, hairy at the apex; style nearly as long as the calyx-tube; stigma capitate.

COAST REGION: Humansdorp Div.; Kruisfontein Mountains, 1000 ft., *Galpin,* 4518!

14. G. styphelioides (Meisn. in Linnæa, xiv. 453); a subshrub; branches obscurely angled, pubescent when young, with moderately prominent leaf-scars; leaves opposite, lanceolate, pungent, 5–8 lin. long, 1–1$\frac{1}{2}$ lin. wide, flat, glabrous, 3–5-nerved, involucral wider; flowers 1–3, terminal, sessile; calyx yellow, puberulous outside, pubescent inside; tube 4$\frac{1}{2}$ lin. long, funnel-shaped; lobes oblong, acute, 2 lin. long, 1 lin. wide; petals 4, membranous, half as long as the lobes; upper anthers exserted, all oblong, 1$\frac{1}{2}$ lin. long, obtuse; ovary ovate, compressed, with a terminal tuft of hairs; style excentric, nearly as long as the calxy-tube; stigma capitate. *Drège, Zwei Pfl. Documente,* 137; *Meisn. in DC. Prodr.* xiv. 587. *Epichroxantha pungens, Eckl. & Zeyh. ex Meisn. l.c., and E. simplex, Eckl. & Zeyh. ex Meisn. l.c. partly. Gnidiopsis styphelioides, Van Tiegh. in Bull. Soc. Bot. France,* xl. (1893) 76.

COAST REGION: Uitenhage Div.; Zuurberg Range, northern slope, 2500–3500 ft., *Drège* ex *Meisner,* between Coega and Zondereinde Rivers, *Ecklon & Zeyher*! Van Stadens River Mountains, *Zeyher,* 267! 3753! *Pappe,* 24! *Ecklon,* 58! Van Stadens Hoogte, *MacOwan,* 2056! Algoa Bay, *Forbes*! and without precise locality, *Cooper,* 1475! Port Elizabeth Div.; at the upper part of the Maitland River, *Burchell,* 4618! between Krakakamma and the upper part of Maitland

River, *Burchell,* 4597 ! Port Elizabeth, *Miss Cherry,* 911 ! *West,* 99 ! Albany Div. ;
without precise locality, *Atherstone,* 90 ! Alexandria Div. ; sandy slopes between
Bushmans River and de Begha, *Bennie,* 743 ! Bathurst Div. ; between Kasuga
River and Port Alfred, *Burchell,* 3979 ! at the mouth of the Great Fish River,
western side, *Burchell,* 3744 !

15. G. myrtifolia (C. H. Wright) ; a densely branched shrub ;
branches short, slender, reddish, at first hirsute, soon glabrescent,
leaf-scars small ; leaves opposite, approximate, ovate or ovate-
oblong, 5 lin. long, 2–2½ lin. wide, acuminate, coriaceous, finely
verrucose on the margins of the upper part, otherwise glabrous,
3–5-nerved beneath ; flowers few at the ends of the branches ;
calyx pubescent outside ; tube 5 lin. long, narrowly funnel-shaped ;
lobes ovate-lanceolate, acute and thick at the apex, 2 lin. long,
1 lin. wide ; petals 4, membranous, 1 lin. long, ¾ lin. wide ; anthers
oblong, obtuse, ⅓ lin. long, the upper exserted on filaments ½ lin.
long ; ovary ovoid, hairy at the top ; style longer than the calyx-
tube, stout ; stigma papillose.

Coast Region : East London Div. ; plains near Cove Rock, East London, 50 ft.,
Galpin, 3177 ! hill near Kwenquea River mouth, 300 ft., *Galpin,* 5803 !

16. G. sonderiana (Meisn. in DC. Prodr. xiv. 587) ; a slender
undershrub ; branches densely patent-pubescent ; branchlets densely
white-villous ; leaves opposite, coriaceous, ovate-oblong, attenuate,
subobtuse, margins long white-villous, nerves obscure ; flowers
subsolitary, terminal ; calyx loosely pilose ; lobes ovate, acute,
half as long as the tube ; petals membranous, half as long as the
calyx-lobes. *Epichroxantha villosa, Eckl. & Zeyh. ex Meisn. l.c.*

Coast Region : Caledon Div. ; amongst rocks, Babylons Tower, *Ecklon,* 82 !

17. G. tenella (Meisn. in Hook. Lond. Journ. Bot. ii. [1843],
554) ; stem herbaceous? erect, simple, slender ; leaves opposite,
subimbricate, subadpressed, oval-oblong, obtuse, subconcave, densely
covered with silky hairs, 2–4 lin. long and 1 lin. wide, increasing
upwards to 4–5 lin. by 3 lin. ; heads 2–4-flowered ; calyx 5 lin.
long, hairy outside ; lobes oval, obtuse, scarcely 1 lin. long ; petals 4,
membranous, entire, rather more than half as long as the calyx-
lobes. *Drège in Linnæa,* xx. 208. *G. albicans, var. tenella, Meisn.*
in DC. Prodr. xiv. 584.

Var. *β,* **elatior** (C. H. Wright) ; a sparingly branched shrub, about 1 ft. high ;
leaves oval or oblong, obtuse, 4 lin. long, 2 lin. wide ; flowers rather larger than
in the type. *G. sericea, var. villosissima, Meisn. in Linnæa,* xiv. 436 ; *Drège,*
Zwei Pfl. Documente, 81, 188. *G. anomala, var. villosissima, Meisn. l.c. G.*
albicans, var. elatior, Meisn. in DC. Prodr. xiv. 584.

South Africa : without locality, Var. *β* : *Bergius,* 300 ! *Harvey,* 682 ! *Zeyher,*
1498 !
Coast Region : Cape Div. ; Constantia Berg, 2000 ft., *Krauss,* 776, Var. *β* :
Worcester Div. ; Dutoits Kloof, 2000–3000 ft., *Drège* ! Cape Div. ; Claremont

and near Cape Town, below 100 ft., *Bolus*, 3756! Cape Flats, *MacOwan*! Stellenbosch Div. ; Palmiet River, *Ecklon*, 10! Hottentots Holland, *Pappe*!
CENTRAL REGION : Var. *β* : Ceres Div. ; Cold Bokkeveld, 3500 ft., *Schlechter*,
8930!

I have not seen *G. albicans*, var. *grandiflora*, Meisn. in DC. Prodr. xiv. 584,
from the Stellenbosch mountains (*Ecklon*, 52 and 91), but it seems to differ from
var. *elatior* only in its larger size, its flowers reaching a length of 8 or 9 lines.

18. G. chrysophylla (Meisn. in DC. Prodr. xiv. 584) ; branches
slender, glabrous ; branchlets silky ; leaves opposite, imbricate,
obtuse, nerveless, densely adpressed golden-silky, the upper scarcely
larger ; heads 2–3-flowered ; calyx 3–4 lin. long, silky ; lobes ovate,
obtuse ; petals 4, rather fleshy, ovate-oblong, entire, half as long as
the calyx-lobes ; ovary puberulous at the apex.

COAST REGION : Bredasdorp Div. ; Kars River, *Ecklon*, 86, *Pappe*, 15!

19. G. humilis (Meisn. in DC. Prodr. xiv. 586) ; an undershrub
up to 1 ft. high ; branches loosely pilose, slender, slightly spreading ;
leaves opposite, sessile, herbaceous, oblong, acute at both ends, 3–6
lin. long, 1½–2 lin. wide, pilose, hairs fugacious on the surface,
more numerous and persistent longer on the margins, obscurely
1-nerved ; flowers geminate and terminal or solitary in the uppermost axils ; calyx pilose outside ; tube 3 lin. long, oblong below,
rather widely funnel-shaped above ; lobes 1 lin. long, ovate ; petals 4,
membranous, ovate, half as large as the lobes ; anthers much
longer than broad ; ovary ovoid, hairy at the top ; style as long as
the calyx-tube ; stigma capitate. *Bolus & Wolley-Dod in Trans.
S. Afr. Phil. Soc.* xiv. 316.

SOUTH AFRICA : without locality, *Bergius*!
COAST REGION : Cape Div. : Table Mountain, *Ecklon*, 89, *Bolus*, 4497! *and in
Herb. Norm. Austr.-Afr.*, 1354! lower plateau, *Wolley-Dod*, 919! Simons Bay,
Wright, 36! Stellenbosch Div. ; mountains of Lowrys Pass, *Burchell*, 8210!
Caledon Div. ; Houw Hoek, *Schlechter*, 7398!

20. G. juniperifolia (Lam. Encycl. ii. 765) ; a shrub turning
blackish when dry, entirely glabrous ; branches slightly angular,
leaf-scars conspicuous ; leaves scattered, linear-subulate, 5–6 lin.
long, plano-convex, smooth ; flowers terminal, solitary or more
frequently geminate, surrounded by a cluster of ordinary leaves ;
calyx glabrous, 3–4 lin. long ; tube dilated above ; lobes acute,
nearly as long as the tube.

SOUTH AFRICA : without locality, *Sonnerat.*
This is not the same as *G. juniperifolia* described by Meisner in DC. Prodr.
xiv. 587, which is *G. subulata*, Lam.

21. G. subulata (Lam. Encycl. ii. 765) ; a shrub about 1 ft. high,
much-branched ; branches 4–5 in. long, divided at the apex, pubescent when young ; leaves linear-subulate, slightly concave on the
upper surface, convex on the back, scattered, 5–6 lin. long,
glabrous ; flowers 2–3, terminal, scarcely as long as the bracts ;

receptacle pilose; calyx glabrous; tube 4 lin. long, narrowly
funnel-shaped, articulated just below the middle; lobes ovate,
acuminate, 1½ lin. long, 1 lin. wide; petals 4, hyaline, lanceolate,
1 lin. long; upper stamens exserted on filaments ½ lin. long; ovary
oblong, compressed, shortly hairy at the apex; style stout, as long
as the calyx-tube; stigma capitate. *Drège, Zwei Pfl. Documente,*
107; *Meisn. in Linnæa,* xiv. 449, *and in DC. Prodr.* xiv. 587;
Bolus & Wolley-Dod in Trans. S. Afr. Phil. Soc. xiv. 316. *G. biflora,
Thunb. Fl. Cap. ed. Schult.* 380; *Wikstr. in Vet. Acad. Handl.
Stockh.* 1818, 314. *G. simplex, Linn. Mant.* 67; *Thunb. Fl. Cap. ed.
Schult.* 380; *Bot. Mag. t.* 812. *G. juniperifolia, Meisn. in Linnæa,*
xiv. 450, *and DC. Prodr.* xiv. 587, *incl. var. uncinata*; *O. Kuntze,
Rev. Gen. Pl.* iii. ii. 280. *G. viridis, Berg. Descr. Pl. Cap.* 125.
G. pinifolia, Linn. f. Suppl. 225. *G. acerosa, Gmelin, Syst.* 633.
G. Sparrmanni, Martyn in Mill. Gard. Dict. ed. ix. *n.* 10. *G. aurea,
Steud. ex Meisn. in Linnæa,* xiv. 451. *Epichroxantha juniperifolia
and E. simplex, Eckl. & Zeyh., partly, ex Meisn. in DC. Prodr.* xiv. 587.
*Gnidiopsis juniperifolia, and G. subulata, Van Tiegh. in Bull. Soc. Bot.
France,* xl. (1893) 76.—*Thymelæa æthiopica Passerinæ foliis, Breyn.
Cent.* 10, *t.* 6.

VAR. β, **pubigera** (C. H. Wright); leaves congested at the apex of the
branches, at first ciliate on the margin and pilose on the upper surface.
G. juniperifolia, var. pubigera, Meisn. in Drège, Zwei Pfl. Documente, 87, and
in DC. Prodr. xiv. 587.

SOUTH AFRICA: without locality, *Bergius!* *Sparrman, Sieber,* 186! *Forster!
Reeves!* *Grey!* *Harvey,* 680!

COAST REGION: Cape Div.; hills and flats around Capetown, *Burchell,* 27!
124! 152! 725! *Bolus,* 3069! 3807! 4646! *Wolley-Dod,* 330! 666! 996.
Zeyher, 3752! *Galpin,* 4521! *Pappe,* 2! 8! *Ecklon,* 363! *Schlechter,* 218! *Mund,*
12! *Kuntze!* Simons Bay, *Wright!* *MacGillivray,* 596! *Milne,* 147! Stellen-
bosch Div.; near Stellenbosch, *Drège!* between Lowrys Pass and Jonkers Hoek,
Burchell, 8328! between Stellenbosch and Somerset West, *Drège.* Caledon
Div.; Houw Hoek, *Zeyher,* 3750! *Schlechter,* 9388! between Villiersdorp and
French Hoek, 1300 ft., *Bolus,* 5270! mountains of Baviaans Kloof near Gena-
dendal, *Burchell,* 7802! on Donker Hoek Mountains, *Burchell,* 7971! 7999!
Riversdale Div.; hills near Zoetemelks River, *Burchell,* 6780; near waterfall at
Garcias Pass, *Burchell,* 7019! Var. β: Paarl Div.; Paarl Mountains, 1000–
2000 ft., *Drège!* Caledon Div.; Appels Kraal, near the Zondereinde River,
Zeyher, 3749!

22. **G. variegata** (Gandog. in Bull. Soc. Bot. France, lx. 417,
1913); an undershrub, 1 ft. or more high; branches adpressed-
tomentose, very densely leafy; leaves imbricate, straight, linear,
mucronate, keeled, 4–5 lin. long; involucre present; flowers densely
capitate, purple and pale yellow outside; calyx 5–6 lin. long,
pubescent; lobes ovate, 1 lin. wide.

COAST REGION: Port Elizabeth Div.; Port Elizabeth, *Laidley in E. S. C. A.
Herb.,* 488.

Allied to *G. subulata,* Lam. (*G. dimidiata,* Gandog.).

23. **G. Flanagani** (C. H. Wright); an erect undershrub, up to
1 ft. high; branches erect, straight, glabrous; leaves opposite,

ovate-lanceolate, up to 6 lin. long and $2\frac{1}{2}$ lin. wide, acute, glabrous, 3-nerved ; flowers in terminal clusters ; calyx glabrous ; tube 5 lin. long, slightly inflated below, widened above ; lobes ovate, acute, $1\frac{1}{2}$ lin. long, 1 lin. wide ; petals 4, membranous, 1 lin. long, $\frac{2}{3}$ lin. wide ; anthers $\frac{1}{3}$ lin. long, upper exserted on short thick filaments ; ovary oblong, compressed ; style as long as the calyx-tube, rigid ; stigma penicillate.

COAST REGION : Komgha Div. ; grassy hills near Keimouth, *Flanagan*, 621 ! EASTERN REGION : Transkei ; grassy slopes, Kentani, 50 ft., *Miss Pegler*, 32 ! Pondoland ; grassy places between Umkwani and Omsakabo, *Tyson*, 2636 !

This is at once distinguished from *G. coriacea*, Meisn., by its branches being erect and parallel, instead of divergent.

24. G. coriacea (Meisn. in Linnæa, xiv. 454) ; a much-branched shrub, glabrous in all its parts ; branches short, divergent, slender, terete, smooth, leaf-scars inconspicuous ; leaves opposite, sessile, ovate-oblong to nearly lanceolate, acute, submucronate, up to 9 lin. long and $2\frac{1}{2}$ lin. wide, obtuse at the base, 1-nerved on the upper side, 3–5-nerved beneath, dark brown when dry ; involucral leaves slightly larger ; flowers 2–4 in terminal heads shorter than the involucre ; calyx glabrous ; tube 4 lin. long, slightly widening upwards, smooth ; lobes ovate-oblong, $1\frac{1}{2}$ lin. long, 1 lin. wide, acute ; petals 4, subhyaline, elliptic, half as long as the lobes ; upper anthers exserted ; ovary ovoid, compressed, with a terminal tuft of hairs ; style excentric, as long as the calyx-tube ; stigma capitate. *Drège, Zwei Pfl. Documente*, 135 ; *Meisn. in DC. Prodr.* xiv. 586. *Epichroxantha ovata, Eckl. & Zeyh. ex Meisn. in DC. Prodr.* xiv. 586. *Gnidiopsis coriacea, Van Tiegh. in Bull. Soc. Bot. France*, xl. (1893) 76.

COAST REGION : Uniondale Div. ; Long Kloof, between Avontuur and the sources of the Keurbooms River, *Burchell*, 5040 ! *Bolus*, 2444 ! on a rocky hill near Haarlem, *Burchell*, 5021 ! Uitenhage Div. ; Zuur Berg Range, between Enon and Drie Fontein, 2000–3000 ft., *Drège* ! Van Stadens Berg, *Zeyher*, 3754 ! and without precise locality, *Zeyher*, 355 ! Albany Div. ; Bothas Hill, *MacOwan*, 191 !

25. G. parviflora (Meisn. in Linnæa, xiv. 453) ; a much-branched undershrub, about 9 in. high, quite glabrous ; leaves opposite, linear-lanceolate, acute, flat, 5 lin. long, 1 lin. wide, midrib not very conspicuous beneath ; flowers solitary or more rarely 2–4 together, shorter than the leaves ; calyx glabrous, 3 lin. long ; tube inflated below, narrow above ; lobes one-third as long as the tube, ovate, acute ; petals 4, membranous, small ; anthers subglobose ; ovary ovoid ; style slightly exserted ; stigma capitate. *Meisn. in DC. Prodr.* xiv. 588, *incl. var. debilis ; Drège, Zwei Pfl. Documente*, 76. *G. lanceolata and G. polygalæfolia, var. lanceolata, Lichtenst. ex Meisn. in DC. Prodr.* xiv. 588. *Epichroxantha debilis, Eckl. & Zeyh. ex Meisn. l.c. Gnidiopsis parviflora, Van Tiegh. in Bull. Soc. Bot. France*, xl. (1893) 76.

COAST REGION : Piquetberg Div. ; Piquet Berg, 1500–3000 ft., *Drège* ex *Meisner*. Caledon Div. ; on sides of mountains near Genadendal and Appels Kraal by the Zondereinde River, *Zeyher*, 1437b !

CENTRAL REGION: Ceres Div. ; between Witzenberg Range and Skurfdeberg, *Zeyher,* 1487 !

26. **G. decurrens** (Meisn. in Linnæa, xiv. 451) ; an undershrub, under 1 ft. high, much-branched from the base, quite glabrous ; branches ascending or slightly spreading, bearing very slender obscurely 4-angled branchlets, leaf-scars rather prominent ; leaves opposite, linear-lanceolate, 4 lin. long, ½-1 lin. wide, acute, flat, 1-nerved ; flowers terminal ; calyx glabrous ; tube 2½ lin. long, narrow at the very base, widely campanulate and 1½ lin. wide above ; lobes oblong, obtuse, 2 lin. long, ⅔ lin. wide ; petals 4, membranous, half as long as the lobes ; anthers oblong, obtuse, ½ lin. long ; style as long as the calyx-tube ; stigma rather small, capitate. *Drège, Zwei Pfl. Documente,* 88, 89 ; *Meisn. in DC. Prodr.* xiv. 588 ; *O. Kuntze, Rev. Gen. Pl.* iii. ii. 280 ; *Bolus & Wolley-Dod in Trans. S. Afr. Phil. Soc.* xiv. 316. *G. biflora, Thunb. Prodr.* 76, *and Fl. Cap. ed. Schult.* 380 ? *Epichroxantha biflora, Eckl. & Zeyh. ex Meisn. in DC. Prodr.* xiv. 588. *Gnidiopsis decurrens, Van Tiegh. in Bull. Soc. Bot. France,* xl. (1893) 76.

SOUTH AFRICA : without locality, *Wallich* ! *Grey* ! *Forbes* !
COAST REGION : Cape Div. ; Table Mountain, *Burchell,* 422 ! *Rogers,* 1072 ! *Drège* ! *Wolley-Dod,* 1726 ! *Galpin,* 4520 ! *Bolus,* 4194 ! *Ecklon & Zeyher* ! *Cooper,* 2296 ! Red Hill, *Wolley-Dod,* 3020 ! Wynberg, 800–900 ft., *Mund,* 11 ! *Wolley-Dod,* 411 ! Devils Mountain, *Pappe,* 30 ! *Drège, Wilms,* 3583 ! *Wolley-Dod,* 417 ! Stellenbosch Div. ; Lowrys Pass, *Schlechter,* 7801 ! *Kuntze* ! Caledon Div. ; Houw Hoek, 900 ft., *Phillips,* 51 ! Albany Div. ; Grahamstown, *Glass,* 664 ! *Misses Daly & Sole,* 482 !

27. **G. Galpini** (C. H. Wright) ; diffusely branched ; branches glabrous, reddish, wiry ; leaves opposite, oblong-lanceolate, acuminate, 6 lin. long, 1½ lin. wide, glabrous, punticulate below, margins involute above ; flowers geminate at the apex of the branches ; calyx glabrous ; tube 3½ lin. long, cylindrical below, rather widely funnel-shaped above ; lobes ovate, obtuse, 1½ lin. long, 1 lin. wide ; petals 4, membranous, elliptic, obtuse, 1 lin. long, ⅔ lin. wide, hyaline ; anthers shortly oblong, obtuse, ½ lin. long, the upper exserted on short stiff filaments ; ovary oblong, compressed, with a terminal tuft of hairs ; style as long as the calyx-tube, wiry ; stigma penicillate.

COAST REGION : Riversdale Div. ; Garcias Pass, 1200 ft., *Galpin,* 4519 !

This resembles *G. styphelioides,* Meisn., but differs in having a glabrous calyx.

28. **G. pallida** (Meisn. in Linnæa, xiv. 442) ; branch nearly 1 ft. long ; branchlets nearly filiform, quite glabrous, internodes 1–2 lin. long, leaf-scars tooth-like ; leaves alternate, erect or slightly spreading, linear-lanceolate, acute, 4–6 lin. long, ½-1 lin. wide, flat or with recurved margins, ashy-green, glabrous, 1-nerved, upper leaves nearly verticillate ; flowers surrounded at the base with white hairs ; calyx 6–7 lin. long, glabrous ; tube slender, almost

filiform at the middle, narrowly funnel-shaped above ; lobes ovate or ovate-oblong, acuminate, 1½ lin. long, sparingly hairy outside ; petals 8, rather fleshy, oblong, obtuse, ⅔ lin. long, inserted in pairs ; anthers half as long as the petals, the upper exserted. *Drège, Zwei Pfl. Documente,* 154 ; *Meisn. in DC. Prodr.* xiv. 588.

EASTERN REGION : Pondoland ; in grassy places near the mouth of the Umtentu River, *Drège,* 4667.

29. **G. thesioides** (Meisn. in Linnæa, xiv. 457) ; branches slender, at first pilose, terete, straight, corymbosely branched above, leaf-scars small, but rather prominent ; leaves alternate, rarely opposite or whorled, linear or nearly lanceolate, rather obtuse, flat, 1-nerved, glabrous, 3–9 lin. long, ⅓–2 lin. wide ; flowers solitary (rarely geminate) in the axils of the upper leaves ; calyx densely hairy outside ; tube about 3 lin. long, inflated below, narrowly funnel-shaped above, ⅔ lin. in diam. at the top ; lobes rounded, ½ lin. long and wide ; petals 8, clavate, half as long as the lobes ; anthers about as large as the petals, obtuse ; ovary compressed, hairy at the apex ; style 1½ lin. long ; stigma subcapitate ; fruit surrounded by the ovoid pubescent base of the calyx. *Meisn. in DC. Prodr.* xiv. 590.

VAR. β, **laxa** (Meisn. in DC. Prodr. xiv. 590) ; stems very slender, internodes 4–8 lin. long ; leaves narrower than in the type, distant. *G. thesioides, Drège, Zwei Pfl. Documente,* 136. *G. cærulea, Eckl. & Zeyh. ex Meisn. l.c.*

VAR. γ, **condensata** (Meisn. in DC. Prodr. xiv. 590) ; stem and branches short, straight, rigid ; leaves subcoriaceous, densely crowded, somewhat adpressed.

SOUTH AFRICA : without locality, *Forster* ! *Bowker* ! Var. γ : *Ludwig ex Meisner.*

COAST REGION : Riversdale Div. ; by the Zoetemelks River, *Burchell,* 6606 ! 6807 ! near Riversdale, 650 ft., *Bolus,* 11368 ! Uitenhage Div. ; near Uitenhage, *Burchell,* 4267 ! Addo, *Zeyher,* 3773 ! Albany Div. ; Dassies Klip, between Grahamstown and Port Elizabeth, *Bolus,* 2679 ! Bathurst Div. ; Trapps Valley, *Miss Daly,* 610 ! Albany Div. ; Grahamstown, *Glass,* 549 ! Sandy Drift, *Miss Daly,* 58 ! at Kurukuru River, *Burchell,* 3528 ! between Zwartwater Poort and the east end of Zwartwater Berg, *Burchell,* 3446 ! and without precise locality. *Zeyher,* 847 ! Fort Beaufort Div. ; Kat River, *Bartelo* ! King Williamstown Div. ; near King Williamstown, 1500 ft., *Tyson,* 2233 ! *Flanagan,* 2176 ! Komgha Div. ; Kei River, near Komgha, 1800 ft., *Flanagan,* 1363 ! British Kaffraria ; without precise locality, *Cooper,* 29 ! 424 ! Var. β : Cape Div. ; Vyges Kraal River, *Wolley-Dod,* 3409 ! near Claremont, *Dümmer,* 1670 ! Swellendam Div. ; near Swellendam, *Zeyher,* 1499 ! Riversdale Div. ; Garcias Pass, *Bolus,* 11370 ! dry hills near Spiegel River, *Burchell,* 7204 ! Uitenhage Div. ; near Bontjes River, *Drège,* 7353.

30. **G. fastigiata** (Rendle in Trans. Linn. Soc. ser. 2, iv. 41) ; a subshrub, up to 10 in. high ; branches ascending, pubescent ; leaves alternate, approximate, suberect, lanceolate, up to 5 lin. long and 1 lin. wide, acute, 1-nerved, glabrous ; flowers solitary or several on short axillary branches ; bracteoles 1½ lin. long, oblong, pubescent ; pale blue (*Sankey*), silky-pubescent ; tube 4 lin. long, spindle-shaped in the lower half, funnel-shaped above ; lobes ovate, obtuse, ¾ lin. long ; petals 8, ½ lin. long, oblong, obtuse, alternating with tufts of

short hairs ; ovary ovoid, compressed, hairy ; style lateral, variable in length ; stigma small. *Pearson in Dyer, Fl. Trop. Afr.* vi. i. 222. *G. Holstii, Gilg in Engl. Jahrb.* xix. 257, *partly ; var. kilimandscharica, Gilg in Engl. Pfl. Ost-Afr. C.* 283.

KALAHARI REGION : Orange Free State ; Harrismith, *Sankey,* 224 ! Transvaal ; summit of Saddleback Mountain, Barberton, 5000 ft., *Galpin,* 1088 !
EASTERN REGION : Natal ; Tabamhlope, *Wylie in Herb. Wood,* 10528 !
Also in Tropical Africa.

31. **G. racemosa** (Thunb. Prodr. 76) ; a glabrous shrub, 2 ft. high ; branches sparse, long, green ; branchlets filiform, virgate, sub-fastigiate, lax, short ; leaves ovate, obovate or obovate-lanceolate, 5–6 lin. long, 2–3 lin. wide, obtuse, glabrous, unequal, erecto-patent, the lower larger, subpetiolate ; flowers racemose, axillary towards the apex of the branches ; calyx hirsute, 5 lin. long ; petals 8, anther like, glabrous, yellow ; anthers yellow, sessile. *Thunb. Fl. Cap. ed. Schult.* 380 ; *Wikstr. in Vet. Acad. Handl. Stockh.* 1818, 315 ; *Drège, Zwei Pfl. Documente,* 102 ; *Meisn. in Linnæa,* xiv. 456, *and in DC. Prodr.* xiv. 590.

COAST REGION : Malmesbury Div. ; Laauwskloof, under 1000 ft., *Drège.*

32. **G. linoides** (Wikstr. in Vet. Acad. Handl. Stockh. 1818, 316) ; plant up to 8 in. high, branched from the base ; branches very slender, glabrous ; leaves opposite, linear-subulate, 3 lin. long, ¼ lin. wide, acute ; flowers terminal, solitary ; calyx-tube pubescent, 2½ lin. long, the lower third ovoid and firm, the upper part long-funnel-shaped and fleshy ; lobes oblong, acute, 1½ lin. long, ⅓ lin. wide, pubescent outside ; petals 4, bilobed, ¼ lin. long ; anthers oblong ; ovary oblong, ⅓ lin. long, hairy at the apex ; style nearly as long as the calyx-tube. *Drège, Zwei Pfl. Documente,* 87 ; *Meisn. in Linnæa,* xiv. 449, *and DC. Prodr.* xiv. 582, *incl. var. major* ; *Drège in Linnæa,* xx. 209 ; *Bolus & Wolley-Dod in Trans. S. Afr. Phil. Soc.* xiv. 316. *Calycoseris linearifolia, Eckl. & Zeyh. ex Drège in Linnæa,* xx. 209. *Gnidiopsis linoides, Van Tiegh. in Bull. Soc. Bot. France,* xl. (1893) 76.

SOUTH AFRICA : without locality, *Zeyher,* 69 ; *Harvey,* 421 !
COAST REGION : Worcester Div. ; on the mountains above Worcester, *Rehmann,* 2430 ! Paarl Div. ; Paarl Mountains, 1000–2000 ft., *Drège* ! French Hoek, 3000 ft., *Schlechter,* 9309 ! Cape Div. ; by a swamp in Orange Kloof, *Wolley-Dod,* 2524 ! Stellenbosch Div. ; Lowrys Pass, 2000 ft., *Schlechter,* 7243 ! Caledon Div. ; Grietjesgat, near the Palmiet River, 700 ft., *Bolus,* 4192 ! on the shore of the River Zondereinde, near Appelskraal, *Zeyher,* 3774 ! Riversdale Div. ; along the river at Garcias Pass, *Burchell,* 7042 !
KALAHARI REGION : Transvaal ; Pretoria, *Burtt-Davy,* 685 ! Rustenberg, *Miss Pegler,* 977 !
Meisner says that the type is herbaceous and the variety *major* suffruticose, but that his specimens of the latter are imperfect.

33. **G. Cayleyi** (C. H. Wright) ; plant 2–8 in. high, woody ; branches straight, slender, pubescent ; leaves opposite, elliptic-oblong, acute, 3½ lin. long, ¾ lin. wide, glabrous ; flowers solitary

at the apex of the branches ; calyx adpressed-silky outside ; tube
$2\frac{1}{2}$ lin. long, ovoid below, funnel-shaped above ; lobes elliptic,
$1\frac{1}{2}$ lin. long, $\frac{3}{4}$ lin. wide, acute ; petals 8, minute, much smaller
than the anthers ; anthers oblong, $\frac{1}{3}$ lin. long, obtuse ; ovary compressed, glabrous ; style excentric, filiform, nearly as long as the
calyx-tube ; stigma penicillate.

SOUTH AFRICA : without locality, *Herb. Caley in Herb. Kew*!

This much resembles *G. parvula*, Wolley-Dod, but differs in the small petals.
A note on the sheet states that it agrees with a specimen in the Berlin Herbarium,
collected on Table Mountain by Bergius.

34. G. obtusissima (Meisn. in Linnæa, xiv. 432) ; branches virgate,
pilose when young, leaf-scars prominent, rather distant ; leaves
opposite, oblong-lanceolate, obtuse, 4 lin. long, $1-1\frac{1}{2}$ lin. wide,
coriaceous, convex or complicate, glabrous, midrib prominent
beneath ; involucral leaves rather longer, ovate : calyx pubescent
outside ; tube 6 lin. long, slightly widened upwards, ribbed ; lobes
ovate, obtuse, 2 lin. long, $1\frac{1}{2}$ lin. wide ; petals 8, half as long as the
lobes, oblong, obtuse, rather fleshy ; anthers oblong, acute, 1 lin.
long. *Meisn. in DC. Prodr.* xiv. 589. *G. dichotoma, Gilg in Engl.
Jahrb.* xix. 264, *and in Engl. & Prantl, Pflanzenfam.* iii. 6 A, 227.
G. obtusifolia, Bartl. ex Meisn. in Linnæa, xiv. 461. *Gnidiopsis
obtusissima, Van Tiegh. in Bull. Soc. Bot. France,* xl. (1893) 76.

COAST REGION : Caledon Div. ; Hangklip, 3000–4000 ft., *Mund & Maire* ! near
Caledon, 3000–4000 ft., *Ecklon & Zeyher, 97 ex Meisner.* Knysna Div. ; near
Bitou River, *Burchell,* 5297 ! Uitenhage Div. ; between Sunday River and
Coega River, *Zeyher ex Meisner.*

WESTERN REGION : Little Namaqualand ; Roodeberg, 3000–4000 ft., *Drège,*
7359, *ex Meisner.*

35. G. ericoides (C. H. Wright) ; a dwarf heath-like shrub ;
branches erect, stout, at first pubescent ; leaves opposite, approximate, linear-oblong, 3 lin. long, $\frac{1}{3}$ lin. wide, obtuse, slightly pubescent
at first, soon quite glabrous ; flowers clustered at the ends of the
branches ; calyx adpressed-silky outside ; tube 6 lin. long, ovoid
below, contracted at the middle, funnel-shaped above, strongly
ribbed ; lobes ovate, acute, 2 lin. long, $1\frac{1}{4}$ lin. wide ; petals 8,
anther-like, oblong, obtuse, 1 lin. long, glabrous ; anthers $\frac{1}{2}$ lin.
long, obtuse ; ovary ovoid, hairy at the apex ; style filiform, 3 lin.
long ; stigma penicillate.

COAST REGION : Riversdale Div. ; Tygerfontein, 600 ft., *Galpin,* 4523 !

36. G. parvula (Wolley-Dod in Journ. Bot. 1901, 401) ; dwarf,
branched from the base ; branches up to 4 in. long, virgate, glabrous ;
leaves alternate or subopposite, longer than the internodes, sessile
or very shortly petioled, erect, narrowly lanceolate, acute, glabrous,
1-nerved, 3–5 lin. long, $\frac{1}{2}-1$ lin. wide ; flowers 6–8 in an apical
cluster ; calyx sparingly hairy outside ; tube 4 lin. long, ovoid
below, cylindrical above, slender ; lobes ovate, obtuse, 1 lin. long ;

petals 8, oblong, as long as the lobes ; anthers small ; ovary compressed, glabrous ; style 2½ lin. long ; stigma capitate. *Bolus &
Wolley-Dod in Trans. S. Afr. Phil. Soc.* xiv. 316.

COAST REGION : Cape Div.; by the Signal Station on Lions Mountain near
Cape Town, *Wolley-Dod*, 2928 !
This species much resembles *Arthrosolen laxus*, E. Meyer.

37. G. polystachya (Berg. Descr. Pl. Cap. 123) ; a shrub, 1–4 ft.
high, of variable habit ; branches sometimes long and simple, at
others short and corymbosely or racemosely arranged, pilose when
young, leaf-scars small, but rather prominent ; leaves alternate,
closely placed, linear-lanceolate, subacute, 4 lin. long, ½ lin. wide,
1-nerved, flat or slightly keeled, quite glabrous ; flowers in clusters
of 6–∞ at the ends of the branches ; calyx pilose ; tube 3½ lin.
long, narrowly ovoid and strongly ribbed in the lower half, narrowly
funnel-shaped above ; lobes ovate, obtuse, 1½ lin. long, ¾ lin. wide ;
petals 8, half as long as the lobes, anther-like, shortly stalked,
emarginate ; anthers ½ lin. long ; ovary oblong, compressed, hairy
at the apex ; style of variable length ; stigma capitate. *Thunb. Fl.
Cap. ed. Schult.* 380 ; *Bot. Mag. t.* 8001 ; *Gard. Chron.* 1907, xli. 294,
fig. 120. *G. carinata, Thunb. Prodr.* 76 ; *Wikstr. in Vet. Acad.
Handl. Stockh.* 1818, 312 ; *Meisn. in DC. Prodr.* xiv. 588 ; *Bolus &
Wolley-Dod in Trans. S. Afr. Phil. Soc.* xiv. 316. *G. pinifolia,
Wendl. Beobacht.* 15, *t.* 2, *fig.* 11, *not Linn. G. simplex, Andr. Bot.
Rep. t.* 70 ; *Drège, Zwei Pfl. Documente,* 30, 123 ; *Herb. Amat.* ii.
128. *G. imberbis, Dry. in Ait. Hort. Kew. ed.* 2, ii. 412 ; *Bot. Mag.
t.* 1463 ; *Lodd. Bot. Cab. t.* 1958. *Thymelina simplex, Hoffmansegg,
Verzeich.* 198. *Daphne squarrosa, Linn. Sp. Pl. ed.* i. 358?—
*Thymelæa capitata, lanuginosa, foliis creberrimis, etc., Burm. Pl. Afr.
Rar.* 134, *t.* 49, *fig.* 1 (*flowers drawn as pedicelled*).

VAR. β, **congesta** (C. H. Wright) ; habit much more dense than in the type ;
flowers in the axils of the upper leaves, less distinctly capitate ; calyx 3 lin.
long.

SOUTH AFRICA : without locality, *Thunberg, Thom*, 49 !
COAST REGION : Cape Div. ; Bosky Dell, near Simonstown, *Bodkin in Herb.
Wolley-Dod*, 870 ! Caledon Div. ; Hawston, *Schlechter*, 9465 ! Hermanuspieters-
fontein, *Bolus*, 9842 ! Robertson Div. ; Kochmans Kloof, 800 ft., *Mund*, 15 !
Riversdale Div. ; Plattebosch, *Muir in Herb. Galpin.*, 5322 ! near Milkwood
Fontein, 600 ft., *Galpin*, 4524 ! Gysmans Hoek, *Muir*, 500 ! Mossel Bay Div. ;
rocky and sandy hills near the landing place at Mossel Bay, *Burchell*, 6291 !
Rogers, 4200 ! between Little Brak River and Hartenbosch, *Burchell*, 6218 !
Driefontein and Mossel Bay, under 500 ft., *Drège* ! George Div. ; George, *Bolus*,
2441 ! Uitenhage Div. ; Zwartkops River, under 100 ft., *Drège* ! valley and hills
of the Zwartkops River from Villa Paul Mare to Uitenhage, 50–500 ft., *Zeyher*,
3771 ! 3772 ! Redhouse, *Mrs. Paterson*, 2344 ! Port Elizabeth Div. ; Port
Elizabeth, *Burchell*, 4876 ! Algoa Bay, *Forbes* ! Bathurst Div. ; at the source of
the Kasuga River, *Burchell*, 3907 ! between Kasuga River and Port Alfred,
Burchell, 3965 ! mouth of the River Kowie, *MacOwan*, 835 ! Albany Div. ;
Grahamstown, *Misses Daly & Sole*, 506 ! Rockcliffe near Sidbury, *Miss Daly*,
787 ! and without precise locality, *Zeyher*, 904 !
KALAHARI REGION : Var. β : Orange River Colony ; Mont aux Sources, 9500–
11000, *Flanagan*, 2023 ! *Evans*, 761 !

38. G. scabra (Thunb. Prodr. 74) ; erect, shrubby, 1 ft. or more
high ; branches subverticillate, at first pubescent, soon glabrous,
leaf-scars small and prominent ; leaves alternate or subopposite,
linear-lanceolate, acute, glabrous, about 3 lin. long and ½ lin. wide ;
flowers few, terminal or in the axils of the uppermost leaves ; calyx
densely hairy outside, white (*Bolus*) ; tube 4 lin. long, slightly
inflated below, subcylindrical above, faintly ribbed ; lobes broadly
ovate, about 1½ lin. long ; petals 8, subulate, 1 lin. long ; anthers
⅕ lin. long, obtuse ; ovary compressed, hairy at the apex ; style
excentric, nearly as long as the tube ; stigma capitate.   *Thunb. Fl.
Cap. ed. Schult.* 380 ; . *Wikstr. in Vet. Acad. Handl. Stockh.* 1818,
313 ; *Meisn. in Linnæa,* xiv. 460, *and in DC. Prodr.* xiv. 585.
G. priestleyæfolia, Eckl. & Zeyh. ex Meisn. in DC. Prodr. xiv. 585.
Gnidiopsis scabra, Van Tiegh. in Bull. Soc. Bot. France, xl. (1893) 76.

SOUTH AFRICA : without locality, *Thunberg, Thom,* 232 ! 276 ! *Zeyher,* 1496 !
COAST REGION : Worcester Div. ; Tafelberg, Hex River, *Pappe,* 10 ! *Bolus,*
5811 ! near Touws River, *Bolus, Herb. Norm. Austr.-Afr.,* 1092 ! Swellendam
Div. ; between Kochmans Kloof and Gouritz River, *Pappe,* 19 !
CENTRAL REGION : Ceres Div. ; between Hottentots Kloof and Karroo Poort,
Pearson, 4809 ! Verkerde Vley, *Rehmann,* 2848 ! Calvinia Div. ; near stream-bed
in plains on Roggeveld above Blaukrantz Pass, 2500 ft., *Pearson,* 4983 ! 4985 !
bed of Doorn River, *Pearson,* 3889 ! Karieboomfontein, 2500 ft., near dam,
Pearson, 4982 !

39. G. setosa (Wikstr. in Vet. Acad. Handl. Stockh. 1818,
315) ; an erect shrub ; branches terete, erecto-patent, glabrous ;
leaves alternate, lanceolate, 4–9 lin. long, ½ lin. wide, acute or
pungent, 1- (or inconspicuously 3-) nerved, glabrous ; flowers in
terminal clusters which elongate into spikes ; calyx silky outside,
densely so at the base ; tube 3½ lin. long, subcylindrical : lobes
oblong, obtuse, ¾ lin. long, ½ lin. wide ; petals 8, subulate, slightly
shorter than the calyx-lobes ; anthers oblong, ⅓ lin. long ; ovary
hairy at the apex ; style 3 lin. long ; stigma subcapitate.   *Meisn.
in Linnæa,* xiv. 447, *and in DC. Prodr.* xiv. 590.   *G. caledonica,
Eckl. & Zeyh., and G. stricta, Eckl. & Zeyh.* (*partly*) *in DC. Prodr.*
xiv. 590.   *Passerina setosa, Thunb. Prodr.* 75, *and Fl. Cap. ed.
Schult.* 376.

COAST REGION : Cape Div. ; Paardeberg, *Thunberg.* Caledon Div. ; mountains
between Zwartberg and Ganze Kraal, *Zeyher,* 3768 ! Riversdale Div. ; without
precise locality, *Rust,* 146 !

40. G. Woodii (C. H. Wright) ; erect ; branches virgate, at first
with a few long hairs, soon glabrous, leaf-scars small ; leaves
alternate, lanceolate or oblong-lanceolate, 10 lin. long, 1–1¼ lin.
wide, acuminate, quite glabrous, 1-nerved ; flowers few at the apex
of the branches ; calyx hirsute outside, yellow (*Wylie*) ; tube 8 lin.
long, slightly inflated below, cylindrical above ; lobes ovate, 2 lin.
long, ¾ lin. wide, acute ; petals 8, lanceolate, thick, 1 lin. long,
⅕ lin. wide ; anthers oblong, obtuse, ½ lin. long ; ovary oblong,
hairy at the apex ; style 6 lin. long ; stigma capitate.

41. G. Baurii (C. H. Wright) ; a diffuse undershrub ; branches
slender, weak, pilose at first, leaf-scars small but prominent ; leaves
opposite, lanceolate, acuminate, 6 lin. long, $1\frac{1}{2}$ lin. wide, adpressed-
silky on the undersurface, glabrous above ; flowers geminate, at the
ends of the branches ; calyx silky outside ; tube 4 lin. long, ovoid
below, narrowly funnel-shaped above ; lobes ovate, about 1 lin. long
and $\frac{2}{3}$ lin. wide, acute ; petals 8, oblong, obtuse, $\frac{1}{4}$ lin. long, rather
thick ; anthers shortly oblong, very small, the upper exserted on
short slender filaments ; ovary oblong, hairy at the top ; style
slender, shorter than the calyx-tube ; stigma small.

42. G. stellatifolia (Gandog. in Bull. Soc. Bot. France, lx. 417,
1913) ; pubescent, subincanescent ; branches virgate, 2 ft. long ;
leaves linear-lanceolate, acute, not or obscurely nerved, the lower
erect, upper verticillate, spreading-stellate, keeled on the back ;
inflorescence capitate, involucrate ; flowers shorter than the leaves ;
calyx-lobes ovate, adpressed-tomentose outside, yellow.

Differs from *G. nodiflora*, Meisn., in the indumentum, the obscurely 3-nerved
patent upper leaves and the calyx being less hairy outside.

43. G. sparsiflora (Bartl. ex Meisn. in Linnæa, xiv. 462) ; a
shrub, 1 ft. or more high ; branches few, suberect, branched at the
apex ; leaves alternate, subimbricate, adpressed, linear-lanceolate,
acute, subhirsute when young, finally glabrous, 3 lin. long,
attenuate at the base ; flowers solitary in the axils of the congested
upper leaves ; calyx 6 lin. long, silky-tomentose ; tube filiform ;
lobes a quarter as long as the tube, broadly ovate, subacute ; petals
8, membranous ?, oblong, glabrous, one-third as long as the calyx-
lobes. *Meisn. in DC. Prodr.* xiv. 591.

44. G. nodiflora (Meisn. in Linnæa, xiv. 458, incl. var. *verticillata*) ;
an undershrub, up to 1 ft. high, branched from the base ; branches
erect, terete, densely pilose above, glabrous below, leaf-scars not
prominent ; leaves alternate, subopposite or verticillate, linear-
oblong, acute, rarely obtuse, 3–6 lin. long, 1 lin. wide, usually flat,
at first densely hairy, glabrescent ; flowers in sessile terminal or
axillary clusters ; calyx densely silky outside ; tube $2\frac{1}{2}$ lin. long,
inflated below ; lobes ovate, acute, $\frac{3}{4}$ lin. long, $\frac{1}{2}$ lin. wide ; petals 8,
linear, half as long as the calyx-lobes ; anthers $\frac{1}{6}$ lin. long,

rounded at both ends; ovary compressed, pubescent; style as long
as the calyx-tube; stigma capitate.   *Meisn. in DC. Prodr.* xiv. 584;
Drège, Zwei Pfl. Documente, 124, 145, *and in Linnæa,* xx. 209;
O. Kuntze, Rev. Gen. Pl. iii. ii. 280.   *G. cæspitosa, Eckl. & Zeyh. ex
Drège in Linnæa,* xx. 209.

SOUTH AFRICA : without locality, *Mund* ! *Krebs,* 289 !
COAST REGION : Vanrhynsdorp Div. ; Olifants River, *Ecklon & Zeyher* ! Riversdale Div. ; Paardeberg, *Muir in Herb. Galpin.,* 5319 ! Riversdale, *Rust,* 148 ! and
without precise locality, *Zeyher,* 3768 ! George Div. ; near George, *Burchell,*
5996 ! Zuur Flats, *Tyson,* 1470 ! Hoogekraal River, on mountains under 1000 ft.,
Drège, 3539 ! Knysna Div. ; Homtini Pass, 800 ft., *Galpin,* 4522 ! near Vlugt,
Bolus, 2442 ! Uitenhage Div. ; Uitenhage, *Zeyher* ! between Addo and Kraka-
kamma ; *Zeyher,* 3770 ! Bathurst Div. ; between Blue Krantz and Kaffir Drift
Military Post, *Burchell,* 3702 ! between Kasuga River and Port Alfred, *Burchell,*
3957 ! Albany Div. ; Grahamstown, *Glass* ! *Misses Daly & Sole,* 277 ! 507 ! 848 !
Rockcliffe near Sidbury, *Miss Daly,* 839 ! Queenstown Div. ; mountains near
Queenstown, 3600–4000 ft., *Galpin,* 1570 ! Stockenstrom Div. ; Willowsdale,
Scully, 81 ! Old Katberg Pass, *Galpin,* 2409 ! East London Div. ; East London,
Rattray, 118 ! King Williamstown Div. ; Perie Forest, *Kuntze* ! Komgha Div. ;
grassy hills near Komgha, *Flanagan,* 326 ! British Kaffraria ; without precise
locality, *Cooper,* 118 ! 403 ! Mossel Bay Div. ; Ruyterbosh, *Mrs. Britton,* 155 !
CENTRAL REGION : Somerset Div. ; Bosch Berg, *MacOwan,* 1880 ! and without
precise locality, *Bowker* !
EASTERN REGION : Transkei ; between Gcua River and Bashee River, *Drège* !
Kentani, *Miss Pegler,* 1156 ! Tembuland ; Little Bush, 4000 ft., *Royffe,* 172 !
Natal ; near Newcastle, *Wilms,* 2249 ! Pinetown, *Wood,* 5489 !

Meisner in DC. Prodr. l.c. states that his variety *verticillata* cannot be
maintained.

45. **G. strigillosa** (Meisn. in Linnæa, xiv. 459); a dwarf shrub,
branched from the base ; branches erect, slender, pubescent ; leaves
opposite, linear-oblong, acute, 5 lin. long, 1 lin. wide, with long
straight scattered hairs when young ; flowers few at the apex of the
branches ; calyx densely long-hairy outside ; tube 4 lin. long ; ovoid
below, cylindrical above ; lobes oblong, acute, the outer 2 lin. long,
the inner rather shorter ; petals 8, short, subulate ; anthers small ;
style 2½ lin. long ; stigma small.   *Drège, Zwei Pfl. Documente,* 47 ;
Meisn. in DC. Prodr. xiv. 583 ; *O. Kuntze, Rev. Gen. Pl.* iii. ii. 281.
G. caffra, Eckl. & Zeyh. ex Meisn. in DC. Prodr. xiv. 584.

COAST REGION : Swellendam Div. ; Swellendam, *Kuntze* ! Cathcart Div. ;
Blesbok Flats, near Windvogel Mountains, *Drège* !

46. **G. sericea** (Linn. Syst. ed. xii. 272); an erect shrub ; branches
densely silky-pubescent, finally with raised leaf-scars ; leaves
densely imbricate, uniform, opposite, oval, 3 lin. long, 1¼ lin. wide ;
densely adpressed silky on both surfaces when young, the upper
soon becoming glabrous, veins obscure ; flowers few at the apex of
the branches, surrounded by an involucre of about 4 leaves similar
to the lower ones ; calyx pubescent outside ; tube 4½ lin. long,
slightly inflated below, subcylindrical above ; lobes ovate, 1 lin.
long, ½ lin. wide, subobtuse ; petals 8, subulate, ⅓ lin. long ; anthers
oblong, small ; ovary ovoid, compressed ; style excentric, as long as

the calyx-tube ; stigma subcapitate. *Wikstr. in Vet. Acad. Handl. Stockh.* 1818, 318 ; *Wendl. Bot. Beobacht.* 16, *t.* 2, *fig.* 13 ; *Thunb. Fl. Cap. ed. Schult.* 381 ; *Lam. Encycl.* ii. 766, *and Ill. t.* 291, *fig.* 3 ; *Meisn. in Linnæa,* xiv. 435, *and in DC. Prodr.* xiv. 583 *(incl. var. vulgaris partly).* *Passerina sericea, Linn. Sp. Pl. ed.* 2, 513. *P. Thunbergii, Wikstr. in Vet. Acad. Handl. Stockh.* 1818, 343. *Nectandra sericea, Berg. Descr. Pl. Cap.* 131. *Thymelina sericea, Hoffmannsegg, Verzeich.* 199, *in obs.—Thymelæa sericea, foliis oblongis, etc., Burm. Rar. Afr. Pl.* 135, *t.* 49, *fig.* 2.

VAR. β, **hirsuta** (Meisn. in DC. Prodr. xiv. 583) ; leaves oval-oblong, obtuse, 4 lin. long, 1–1½ lin. wide, the uppermost wider, all densely silky-hirsute. *G. sericea, var. vulgaris, Meisn. l.c., partly. G. nana and G. phylicifolia, Eckl. & Zeyh. ex Meisn. l.c. G. cyanea, Burch. Trav. S. Afr.* i. 255, 257. *G. azurea, Meisn. in Linnæa,* xiv. 435, *and in DC. Prodr.* xiv. 583. *G. sericea, Bolus & Wolley-Dod in Trans. S. Afr. Phil. Soc.* xiv. 316.

SOUTH AFRICA : without locality, *Forster* ! Var. β : *Pappe* ! *Forster* ! *Mund,* 20 ! *Harvey,* 680 !
COAST REGION : Cape Div. ; Rosebank near Cape Town, *Bolus,* 3699 ! Worcester Div. ; Touws River Railway Station, 3000 ft., *Bolus,* 7452! Swellendam Div. ; near Swellendam, *Zeyher,* 3760! 3761! Albany Div. ; Grahamstown, *MacOwan* ! *Schönland,* 1713 ! Var. β : Cape Div. ; near Cape Town, *Burchell,* 480 ! Simons Bay, *Wright*! Devils Peak, above Newlands, *Wilms,* 3586 ! by waterfall, *Wolley-Dod,* 3418; Caledon Div. ; Hemel in Aarde, 1000 ft., *Schlechter,* 10376 ! Riversdale Div. ; between Garcias Pass and Krombeks River, *Burchell,* 7173 ! Paardeberg, *Muir in Herb. Galpin.,* 5317 ! banks of Vet River, *Muir,* 287 ! Mossel Bay Div. ; between Mossel Bay and Zout River, *Burchell,* 6325 ! between Zout River and Dwyka River, *Burchell,* 6362 ! west bank of Great Brak River, *Burchell,* 6152 ! George Div. ; tops of mountains, *Ecklon & Zeyher,* 20 ! Knysna Div. ; between Keurbooms River and Bitou River, *Burchell,* 5261 ! Plettenberg Bay, *Zeyher,* 1483 ! *Pappe* ! Uniondale Div. ; Lange Kloof, *Bolus,* 1764 ! Uitenhage Div. ; Uitenhage, *Zeyher,* 975 ! Albany Div. ; Grahamstown, *Misses Daly & Sole,* 483 ! Fort Beaufort Div.; near Fort Beaufort, *Cooper,* 3078 ! Stockenstrom Div. ; Katberg, *Shaw,* 116 ! *Baur,* 1066 ! *Galpin,* 1733 ! King Williamstown Div. : Perie, 4000 ft., *Sim,* 1317 !
CENTRAL REGION : Ceres Div. ; Hottentots Kloof, *Pearson,* 4910 ! Sutherland Div. ; Roggeveld Mountains, *Burchell,* 1316 !

Variety *glabrescens,* Meisner in DC. Prodr. xiv. 583, is a doubtful plant.

47. G. denudata (Lindl. in Bot. Reg. t. 757) ; an erect shrub, up to 12 ft. high ; branches virgate, moderately slender, densely pilose ; leaves opposite, ovate-oblong, obtuse, approximate, 6 lin. long, 2–2½ lin. wide, densely pilose at first on both surfaces, finally almost glabrous on the back ; flowers in terminal and lateral clusters ; calyx yellowish, densely pilose outside; tube 9 lin. long, ovoid and discoloured below, gradually widening above to 1 lin. in diam.; lobes oval, acute, glabrous inside, 2 lin. long, 1 lin. wide; petals 8, ½ lin. long; anthers twice as long as the petals; ovary compressed, glabrous except at the apex; style excentric, 5½ lin. long ; stigma subcapitate. *Spreng. Syst. Cur. Post.* 152 ; *Drège, Zwei Pfl. Documente,* 123; *Meisn. in Linnæa,* xiv. 441, *and in DC. Prodr.* xiv. 585. *G. tomentosa, Hook. in Bot. Mag. t.* 2761, *excl. syn. Berg. et Thunb. G. virescens, Eckl. & Zeyh. Herb.* 11, *ex Meisn. in DC. Prodr.* xiv. 585.

SOUTH AFRICA : without locality, *Mund* !
COAST REGION : Swellendam Div. ; Swellendam mountains, *Pappe,* 17 ! George
Div. ; near George, *Tyson,* 3010 ! and *in MacOwan & Bolus, Herb. Norm. Austr.
Afr.,* 978 ! *Rogers,* 4293 ! *Drège* ! in the forest near Tuuw River, *Burchell,* 5714 !
Knysna Div. ; near Knysna, *Bolus,* 2443 ! 3714 ! *Newdigate,* 71 ! Humansdorp
Div. ; forest at Elands River, Zitzikamma, 500 ft., *Galpin,* 4529 !

48. G. Leipoldtii (C. H. Wright) ; much-branched ; branches
diffuse, spreading, slender, pubescent when young, finally glabrous
and with prominent leaf-scars ; leaves opposite, ovate-oblong, acute,
7 lin. long, 3 lin. wide, densely adpressed-silky on both sides,
1–3-nerved below ; flowers 2–6 at the ends of the branchlets ; calyx
densely tomentose ; tube 7 lin. long, slightly ribbed, inflated below,
subcylindrical above ; lobes oval, 1½ lin. long, 1 lin. wide ; petals 8,
⅓ lin. long, anther-like, but not very thick, emarginate ; anthers as
long as the petals, but narrower, obtuse ; ovary ovoid, hairy at
the top ; style filiform, 5 lin. long ; stigma clavate.

CENTRAL REGION : Calvinia Div. ; Nieuwoudtville, Willems River, and Bokke-
veldt Mountains, 2000–3000 ft., *Leipoldt,* 882 ! Somerset East Div. ; on mountain
sides near Somerset East, 4000 ft., *Bolus,* 1764 !

49. G. nitida (Bolus) ; a diffusely branched shrub ; branches
slender but rigid, at first hirsute, soon glabrescent, obscurely tetragonous, leaf-scars small but prominent ; leaves opposite, congested
at the ends of the branches, oval, rounded or subacute, 3 lin. long,
1 lin. wide, at first adpressed-silky, finally glabrous, margins inflexed
above, obscurely 3-nerved ; flowers in pairs at the ends of the
branches ; calyx densely silky outside, yellow (*Bolus*) ; tube 7 lin.
long, inflated below, narrowly funnel-shaped above ; lobes 2 lin.
long, 1¼ lin. wide, subacute ; petals 8, anther-like, 1¼ lin. long,
obtuse ; anthers linear, ⅘ lin. long, obtuse ; ovary compressed,
glabrous ; style filiform, 4 lin. long ; stigma penicillate.

WESTERN REGION : Little Namaqualand ; in stony places near Ookiep, 3200 ft.,
Bolus, in Bolus & MacOwan, Herb. Norm. Austr.-Afr., 689 !

50. G. Meyeri (Meisn. in Linnæa, xiv. 443) ; branches slender,
terete, glabrous, ascending, leaf-scars distant, not very prominent ;
leaves sessile or subsessile, alternate, linear-lanceolate, acute, 4–6 lin.
long, 1 lin. wide, glabrous, nerves inconspicuous ; heads terminal, up
to 8-flowered ; calyx 8–12 lin. long, yellow (*Meisner*), glabrous ;
tube slightly widened upwards ; lobes 2½ lin. long, 1 lin. wide,
subacute ; petals 8, membranous, lanceolate, 1½ lin. long, slightly
hairy ; upper anthers exserted ; ovary oblong, hairy at the apex ;
style nearly as long as the calyx-tube ; stigma capitate. *Drège,
Zwei Pfl. Documente,* 95 ; *Meisn. in DC. Prodr.* xiv. 588.

VAR. β, **pilosa** (C. H. Wright) ; receptacle pilose ; calyx-lobes slightly hairy
outside.

CENTRAL REGION : Calvinia Div. ; Nieuwoudtville, *Leipoldt in Herb. Bolus,*
13387 ! Murraysburg Div. ! near Murraysburg, 4300 ft., *Tyson,* 345 !

WESTERN REGION : Little Namaqualand ; between Uitkomst and Geelbeks
Kraal, 2000–3000 ft., *Drège* ! Riet Kloof near Bowesdorp, 2500 ft., *Schlechter*,
11181 ! near Vanrhynsdorp, 2000 ft., *Leipoldt*, 887 !
EASTERN REGION : Var. β : Pondoland ; without precise locality, *Bachmann*,
908 !

51. G. multiflora (Bartl. ex Meisn. in Linnæa, xiv. 462) ; an
erect subshrub, about a span high, densely branched above ; branchlets silky-pilose ; leaves alternate, subimbricate, erect, linear, subobtuse, slightly concave above, pilose, prominently 3-nerved beneath,
upper slightly wider and more obtuse ; flowers 2–5, terminal ; calyx
white-silky, 4–5 lin. long ; lobes ovate, obtuse, scarcely 1 lin. long ;
petals 8, membranous ?, lanceolate, subacute, yellow ; anthers included. *Meisn. in DC. Prodr.* xiv. 591.

SOUTH AFRICA : without locality or collector's name.

52. G. ovalifolia (Meisn. in Linnæa, xiv. 455) ; a tree up to 40 ft.
high (*Mudd*) ; branchlets densely and patently pilose, cylindrical,
internodes up to 2 in. long ; leaves opposite, oval, acute, up to 2 in.
long and 1 in. wide, membranous, densely adpressed-hairy on both
surfaces ; petiole about 1 lin. long ; flowers in terminal leafless
ebracteate spikes ; calyx densely silky outside ; tube 4 lin. long,
almost cylindrical ; lobes oblong-lanceolate, acute, 1¼ lin. long ;
petals 8, membranous, linear ; anthers linear, ½ lin. long ; ovary
densely pilose at the apex ; style half as long as the calyx-tube ;
stigma subcapitate. *Drège, Zwei Pfl. Documente*, 159 ; *Meisn. in
DC. Prodr.* xiv. 591 ; *Wood, Natal Pl. t.* 248 ; *Sim, For. Fl. Cape
Col.* 300. *Wikstrœmia ovalifolia, Decne in Jacquem. Voy. Bot.* 146.

COAST REGION : British Kaffraria ; Dontsah, *Sim.* 2208 !
KALAHARI REGION : Transvaal ; Mac Mac, *Mudd* !
EASTERN REGION : Transkei ; Kentani, *Miss Pegler*, 166 ! Tembuland ; Engcoba
Mountain, 4000 ft., *Flanagan*, 2772 ! *Henkel* ! Griqualand East ; by streams in
woods on the Zuurberg Range, 3000 ft., *Tyson*, 1775 ! Natal ; Umzimkulu River,
amongst bushes and in plantations near the mouth of the river, *Drège* ! Durban,
Gueinzius ! Inanda, *Wood*, 332 ! 501 ! 8024 ! and without precise locality, *Gerrard*,
328 !

" *Uhloso* " of the Kaffirs (*Henkel*).

53. G. imbricata (Linn. f. Suppl. 225) ; a dwarf much-branched
shrub ; branches slender, at first puberulous ; leaves oval or oblong,
about as long as the internodes, 5 lin. long, 1–1½ lin. wide, obtuse,
canescent, at length glabrous, distinctly 3-nerved ; flowers few in
terminal heads ; calyx densely hairy outside ; tube 6 lin. long, without a differently coloured basal area, subcylindrical, ribbed ; lobes
oval, rounded at the apex, 1½ lin. long, 1 lin. wide ; petals 8, more
than half as long as the lobes ; anthers ⅔ lin. long ; ovary compressed,
½ lin. long ; style excentric, filiform ; stigma subcapitate. *Wikstr.
in Vet. Acad. Handl. Stockh.* 1818, 318 ; *Thunb. Fl. Cap. ed. Schult.*
381 ; *Lodd. Bot. Cab. t.* 890 ; *Drège, Zwei Pfl. Documente*, 109, *and*

in Linnæa, xx. 208 ; *Meisn. in Linnæa*, xiv. 437, *and in DC. Prodr.*
xiv. 584 *(incl. vars. genuina and incana). G. incana, Eckl. & Zeyh. ex
Meisn. in DC. Prodr.* xiv. 585.

COAST REGION : Clanwilliam Div. ; Lange Vallei, under 1000 ft., *Drège*!
Zeyher, 15 ! 1497 ! *Schlechter*, 8040 ! Zandveld, *Leipoldt*, 848 ! Piquetberg
Div. ; Piqueniers Kloof, *Zeyher*, 3763 ! Cape Div. ; Cape Town, *Mrs. Paterson*,
78 !

54. G. caniflora (Meisn. in Linnæa, xiv. 440) ; branches slender,
short, erect, at first tomentose, smooth ; leaves alternate or subopposite, oblong or lanceolate, obtuse, 4–5 lin. long, 1–1½ lin. wide,
subcoriaceous, glabrous, margins involute, nerves not conspicuous ;
clusters 2–4-flowered, terminal or subterminal ; calyx pubescent
outside ; tube 7 lin. long, slightly widened above, ribbed ; lobes
ovate, obtuse, about 1 lin. long, glabrous inside, long-pilose outside ;
petals 8, nearly as long as the calyx-lobes, lanceolate ; anthers
oblong, ½ lin. long ; ovary compressed, glabrous ; style excentric,
filiform, 5 lin. long ; stigma penicillate. *Drège, Zwei Pfl. Documente*, 71 ; *Meisn. in DC. Prodr.* xiv. 585. *G. taxifolia, Eckl. & Zeyh.
ex Meisn. in DC. Prodr.* xiv. 585.

COAST REGION : Vanrhynsdorp Div. ; Giftberg, *Drège*.   Swellendam Div. ;
Gouritz River, *Ecklon & Zeyher*, 23 ! Riversdale Div. ; near Riversdale, 600 ft.,
Bolus, 11369 !

55. G. inconspicua (Meisn. in Linnæa, xiv. 433) ; dwarf, much-
branched ; branches short ; branchlets divaricate, adpressed strigil-
lose when young, leaf-scars prominent ; leaves decussate, approximate, spreading or reflexed, oblong, subacute, 2–5 lin. long, 1–3 lin.
wide, at first adpressed pilose, glabrescent, flat ; flowers geminate ; calyx silky-villous, 4–5 lin. long ; lobes narrow, subobtuse,
lilac ; petals 8, membranous, white. *Meisn. in DC. Prodr.* xiv.
582.

SOUTH AFRICA : without locality, *Drège*, 7349.

56. G. Francisci (Bolus in Trans. S. Afr. Phil. Soc. xvi. 399) ;
an undershrub about 9 in. high, sparingly branched from the base ;
branches tetragonal, simple or with a few branches, slender, at first
pilose, soon becoming glabrous ; leaves opposite, erect, longer than
the internodes, subulate, acuminate, 4½ lin. long, ½ lin. wide,
channelled above, convex beneath, with a deciduous terminal tuft
of hairs ; heads terminal, about 5-flowered ; bracts shorter and
wider than the leaves, silky ; calyx silky outside, cream-coloured
(*Bolus*) ; tube 6 lin. long, ovoid in the lower third, narrowly
funnel-shaped above ; lobes ovate, acute, 1½ lin. long, 1 lin. wide ;
petals 8, lanceolate, 1 lin. long, ⅓ lin. wide, rather fleshy ; anthers
oblong, ½ lin. long ; ovary ovate, compressed, with a conspicuous
terminal tuft of hairs ; style filiform, 3 lin. long ; stigma penicillate.

COAST REGION : Oudtshorn Div. ; in rocky places on the summit of the Zwart
Berg Range, near the Pass, 5600 ft., *Bolus*, 11631 !

The 8 petals are all free ; not united in pairs as suggested by "petala bipartita"
in the original description.

57. **G. scabrida** (Meisn. in Linnæa, xiv. 446) ; a shrub, up to
4 ft. high (*Burchell*) ; branches simple, straight, at first puberulous ;
leaves linear-lanceolate, 6 lin. long, $1\frac{1}{2}$ lin. wide, acute, suberect,
pilose and rather long-ciliate when young, minutely scabrous in age ;
flowers at the ends of the branches ; involucral leaves up to 9 lin.
long ; calyx gamboge-coloured (*Burchell*) ; tube 8 lin. long, basal line
glabrous and ribbed, remainder woolly-pubescent ; lobes linear-
lanceolate, 2 lin. long, $\frac{2}{3}$ lin. wide, woolly-pubescent outside, glabrous
within ; petals 8, lanceolate, acuminate, 1 lin. long ; anthers 8,
oblong, much shorter than the petals ; ovary ovate ; style half
as long as the calyx-tube ; stigma clavate. *Drège, Zwei Pfl.
Documente*, 117 ; *Meisn. in DC. Prodr.* xiv. 581. *G. elongata, Eckl.
& Zeyh. ex Meisn. in DC. Prodr.* xiv. 581.

SOUTH AFRICA : without locality, *Bowie* !
COAST REGION : Worcester Div. ; mountains near Worcester, *Ecklon & Zeyher*,
3 ! Cape Div. ; Devils Mountain, *Ecklon*. Stellenbosch Div. ; Hottentots
Holland, 1500–2500 ft., *Mund* ! Swellendam Div. ; on a mountain peak near
Swellendam, *Burchell*, 7404 ! George Div. ; on the Cradock Berg, near George,
Burchell, 5927 ! *Drège*, 7356.

58. **G. Cephalotes** (Lichtenst. ex Meisn. in DC. Prodr. xiv. 581) ;
branches at first pilose, finally glabrous and with prominent leaf-
scars, stout ; leaves approximate, alternate or subopposite, lanceolate, obtuse, at first pilose, finally minutely scabrous ; bracteoles
oblong, obtuse, scabrous, $2\frac{1}{2}$ lin. long, $\frac{2}{3}$ lin. wide ; flowers in the
axils of the uppermost leaves ; calyx silky outside ; tube 5 lin.
long, slightly inflated above ; lobes elliptic, 4 lin. long, 2 lin. wide ;
petals about 12 in one series, $\frac{1}{3}$ lin. long, lanceolate, slender ;
longer stamens with anthers about half as large as the shorter ;
style filiform, hairy ; stigma clavate. *G. grandiflora, Meisn. l.c.* 582.
*Cryptadenia elongata, and var., Eckl. & Zeyh., and Calycosericos
argentea, Eckl. & Zeyh. ex Drège in Linnæa*, xx. 210.

COAST REGION : Stellenbosch Div. ; Hottentots Holland, 1500–2500 ft., *Zeyher*,
3743 ! *Mund*, 1 ! Lowrys Pass, 4500 ft., *Schlechter*, 7219 ! Caledon Div. ; mountains at Kleinriver Kloof, 1000–3000 ft., *Zeyher*, 3745 ! River Zondereinde,
Herb. Salisbury ! Houw Hoek, 2000 ft., *Schlechter*, 9402 ! Bredasdorp Div. ; near
Elim, 600 ft., *Bolus*, 8595 !

Imperfectly known species.

59. **G. acutifolia** (Wikstr. in Vet. Akad. Handl. Stockh. 1818,
315) ; quite glabrous ; leaves opposite, elliptic, acute ; flowers

2–3 terminal ; calyx glabrous ; lobes lanceolate, subobtuse. *Meisn. in DC. Prodr.* xiv. 592.

SOUTH AFRICA : without locality, *Thunberg.*

Near *G. subulata*, Lam.

60. G. argentea (Thunb. Prodr. 76) ; a glabrous erect cinereofuscous shrub, about 1 ft. high ; branches short, sparse, nodulose ; leaves alternate, obovate, acute, thinly silvery tomentose, obscurely silky, sessile, approximate especially towards the ends of the branches, erecto-patent, 1 in. long ; flowers capitate, silky outside, purple within ; petals 8, fleshy, exserted. *Thunb. Fl. Cap. ed. Schult.* 381 ; *Wikstr. in Vet. Acad. Handl. Stockh.* 1818, 318 ; *Meisn. in DC. Prodr.* xiv. 591.

SOUTH AFRICA : without locality or collector's name.

This may be a *Lasiosiphon.*

61. G. dimidiata (Gandog. in Bull. Soc. Bot. France, lx. 417, 1913) ; glabrous, much-branched ; branches divaricate, sparingly leafy ; leaves linear, obtuse, keeled, patulous or retrorsely falcate, 2–3 lin. long ; involucral leaves scarcely dilated ; flowers capitate ; calyx 3½–4 lin. long, sparingly adpressed-pubescent outside ; lobes orbicular, small, ½ lin. wide.

SOUTH AFRICA : without locality, *Ecklon & Zeyher.*

This may be a form of *G. subulata*, Lam. (*G. juniperifolia*, var. *uncinata*, Meisn.).

62. G. flava (Lindl. ex Steud. Nomencl. ed. 2; i. 697, name only).

63. G. grandiflora (Willd. Enum. Hort. Berol. Suppl. 21, name only).

64. G. nana (Wikstr. in Vet. Acad. Handl. Stockh. 1818, 316) ; an erect undershrub, about 6 in. high ; branches few fastigiate ; leaves alternate, imbricate, linear, scarcely 1 in. long, convex, obtuse, at first pilose, finally glabrous, upper purplish, with white hairs ; flowers many, terminal ; calyx hirsute-tomentose ; lobes lanceolate, acute ; petals many ; stamens 4. *Meisn. in DC. Prodr.* xiv. 591. *Struthiola nana*, *Murr. Syst.* 164 ; *Linn. f. Suppl.* 128 ; *Thunb. Prodr.* 76, *and Fl. Cap. ed. Schult.* 383.

COAST REGION : Tulbagh Div. ; on tops of mountains, Roodezand, *Thunberg.* An anomalous plant, which may prove to be a species of *Struthiola.*

65. G. Rehmannii (Durand & B. D. Jackson in Ind. Kew. Suppl. i. 186).

This is an error for **Gymnosporia Rehmannii**, Szyszyl., Pl. Rehmann. ii. (1884) 34.

IX. LASIOSIPHON, Fresen.

Flowers hermaphrodite, 5-merous. *Calyx-tube* slender, hairy outside, often with longer hairs in the lower part, circumscissile above the base ; lobes patent. *Petals* 5, membranous, smaller than the calyx-lobes, sometimes minute, rarely absent, entire or divided. *Stamens* 10, in two series ; anthers subsessile, dorsifixed near the base, linear or oblong, obtuse. Hypogynous *disc* none or very small. *Ovary* 1-celled ; style slender ; stigma capitate or small. *Fruit* dry, small, enclosed in the persistent base of the calyx. *Seeds* with scanty albumen or none ; cotyledons rather thick.

Small or large shrubs ; leaves opposite or alternate ; flowers in terminal sessile or stalked heads or spikes.

DISTRIB. Species about 35, chiefly in Tropical and South Africa ; five or six in Madagascar and one in India.

Leaves linear-oblong, 2 lin. long, distant... ... ... (1) **microphyllus.**
Leaves much larger, longer than the internodes :
  Calyx-tube 4–8 lin. long :
    Leaves more or less silvery tomentose :
      Tomentum persistently dense :
        Leaves lanceolate or oblanceolate, 6–12 lin. long (2) **splendens.**
        Leaves oblong-lanceolate, 5 lin. long ... ... (3) **canoargentea.**
      Tomentum less dense in age ... ... (4) **Burchellii.**
    Leaves not silvery tomentose :
      Branches spreading :
        Leaves glabrous, oblong, acute ... ... (5) **Wilmsii.**
        Leaves glabrous, oblong, obtuse ... ... (4) **Burchellii,** var.
                                    glabrifolius.
        Leaves hispidulous, oblong, obtuse ... ... (6) **deserticola.**
        Leaves adpressed-silky, spathulate, acute ... (7) **macropetalus.**
        Leaves adpressed-silky, spathulate-oblong, mu-
          cronulate ... ... (8) **dregeanus.**
      Branches erect :
        Petals minute, tooth-like ... ... (9) **similis.**
        Petals 1 lin. long, oblong ... ... (10) **linifolius.**
  Calyx-tube 9–15 lin. long :
    Flower-heads sessile :
      Leaves adpressed-silky on both surfaces :
        Petals tooth-like or obsolete ... ... (12) **anthylloides.**
        Petals obvious :
          Calyx-tube spreading-pilose ... ... (11) **polyanthus.**
          Calyx-tube adpressed-pubescent ... ... (13) **meisnerianus.**
      Leaves glabrous :
        Leaves linear-lanceolate or subspathulate, 1½–2¼
          lin. wide ... ... (14) **pulchellus.**
        Leaves oblanceolate, 2½ lin. wide ... ... (15) **triplinervis.**
        Leaves linear, 1½ lin. wide ... ... (16) **caffer.**
    Flower-heads distinctly stalked :
      Petals lanceolate, acute ... ... (17) **Kraussii.**
      Petals oblong, obtuse ... ... (18) **hoepfnerianus.**

1. **L. microphyllus** (Meisn. in DC. Prodr. xiv. 593); a dwarf undershrub, branched from the base; branches slender, rigid, glabrous; leaves alternate, rather distant, adpressed to the stem, linear-oblong, obtuse, glabrous, about 2 lin. long and ⅛ lin. wide; involucral leaves broadly ovate, acute, 3 lin. long, 1⅓ lin. wide, puberulous; flowers in terminal heads; calyx-tube 7 lin. long, with long silky hairs below, puberulous or glabrous above; lobes elliptic, obtuse, 1½ lin. long, 1 lin. wide; petals small; anthers oblong; ovary shortly stalked, ovoid, glabrous; style nearly as long as the calyx-tube. *Gnidia microphylla, Meisn. in Linnæa,* xiv. 432; *Drège, Zwei Pfl. Documente,* 94. *G. Kuntzei, Gilg in O. Kuntze, Rev. Gen. Pl.* iii. ii. 280.

SOUTH AFRICA : without locality, *Tyson!*
CENTRAL REGION : Somerset Div.; Somerset, *Bowker,* 56! 90! Tarka Div. ; Tarka River, 4500 ft., *MacOwan,* 1801! Richmond Div. ; Bok Fontein, *Burke,* 511! *Zeyher,* 1489! Middelburg Div.; Middelburg Road, *Kuntze!* Colesberg Div. ; Colesberg, *Shaw!*
WESTERN REGION: Little Namaqualand; near the mouth of the Orange River, *Drège.*

2. **L. splendens** (Endl. Gen. Pl. Suppl. iv. ii. 67); a much-branched shrub ; branches at first densely adpressed-silky, leaf-scars small ; leaves alternate, lanceolate or oblanceolate, acute, up to nearly 12 lin. long and 3 lin. wide, densely clothed with adpressed silvery silky hairs ; involucral leaves scarcely wider than the cauline ; flowers in terminal sessile heads; calyx 8 lin. long, subcylindrical slightly inflated above, densely silky outside, hairs longer in the lower part ; lobes oblong, obtuse, 2½ lin. long, ⅔ lin. wide, rather thick, yellow (*Wood*); petals ½ lin. long, ovate, membranous ; anthers oblong, obtuse, ½ lin. long ; ovary oblong, compressed, glabrous ; style as long as the calyx-tube ; stigma capitate. *Meisn. in DC. Prodr.* xiv. 595. *Gnidia splendens, Meisn. in Linnæa,* xiv. 428 ; *Drège, Zwei Pfl. Documente,* 159.

KALAHARI REGION : Transvaal ; near Lydenburg, *Wilms,* 1297! Swaziland; Hlatikulu, *Miss Stewart,* 2522!
EASTERN REGION : Pondoland ; without precise locality, *Bachmann,* 904! Griqualand East; Mount Malowe, 3000 ft., *Tyson,* 2056! and in *Bolus & MacOwan, Herb. Austr.-Afr.,* 1229! Natal; near Durban, *Drège! Wood,* 114! Rooi Kopjes near Durban, *Wood!* near Umkomanzi River, *Wood,* 3159! Pietermaritzburg, *Wilms,* 2248! Vryheid, *Sim,* 2921! 2931! near Colenso, *Rehmann,* 7187! Krantz Kloof, *Wood,* 1155! and without precise locality, *Gerrard,* 207! Zululand ; Ungoya, 1000–2000 ft., *Wood,* 5690! Umhlatuzi, *Sim,* 2936!

3. **L. canoargentea** (C. H. Wright); a much-branched shrub ; branches densely adpressed white-silky when young, leaf-scars small but prominent ; leaves alternate, oblong-lanceolate, 5 lin. long, 1⅓ lin. wide, acute, densely clothed (especially on the lower surface) with adpressed silvery hairs ; involucral leaves twice as broad as the cauline ; flowers many in a terminal cluster ; calyx pubescent outside, with longer hairs below ; tube 6 lin. long ; lobes oblong, obtuse, 1½ lin. long, 1 lin. wide ; petals minute, tooth-like ; anthers

thrice as long as the petals; anthers oblong, obtuse, ½ lin. long; ovary oblong, hairy at the apex; style slender, rigid, about as long as the calyx-tube; stigma capitate.

KALAHARI REGION: Transvaal; Witte Kranz, near Lydenburg, *Wilms*, 1298! on the sides of mountains near Lydenburg, *McLea in Herb. Bolus*, 3020!

4. **L. Burchellii** (Meisn. in DC. Prodr. xiv. 594, var. *villosus*); a much-branched shrub; branches villous; leaves alternate, lanceolate, 6 lin. long, 1 lin. wide, the upper rather wider, densely adpressed-pubescent on both surfaces, somewhat glabrescent; heads sessile, terminal, many-flowered; calyx-tube densely villous outside, 7 lin. long, slightly inflated below; lobes oval, obtuse, 2 lin. long, 1 lin. wide; petals minute, clavate; anthers oblong, obtuse, ½ lin. long; ovary oblong, compressed, glabrous; style nearly as long as the calyx-tube; stigma capitate.

VAR. β, **glabrifolius** (Meisn. in DC. Prodr. xiv. 594); leaves linear, obtuse, glabrous; involucral leaves oblong, densely pilose.
VAR. γ, **angustifolius** (C. H. Wright); resembling var. *glabrifolius*, but leaves linear, acute, densely adpressed-hairy beneath.

COAST REGION: Cathcart Div.; Windwogel Mountain, *Cooper*, 245!
KALAHARI REGION: Orange River Colony; Bethlehem, *Richardson*! Basutoland; without precise locality, *Cooper*, 3085! Bechuanaland; between Matlowing River and Takun, *Burchell*, 2201! 2203! Mashowing River, near Takun, *Burchell*, 2256! Var. β: Transvaal; Vaal River, *Zeyher*, 1490! *Burke*, 85! Var. γ: Transvaal; Barberton Mountains, *Burtt-Davy*, 347!
EASTERN REGION: Griqualand East; sides of mountains, Ingeli, 5500 ft., *Tyson*, 1322! 1990!

5. **L. Wilmsii** (C. H. Wright); a much-branched shrub; branches at first pilose, reddish, leaf-scars small; leaves alternate, very shortly petiolate, oblong, acute, 7 lin. long, 1¾ lin. wide, glabrous, coriaceous, midrib prominent beneath; involucral leaves rather larger than the cauline, pilose; flowers numerous in terminal heads, very sweet-scented in the evening (*Cooper*); calyx densely silky outside; tube 6–7 lin. long, cylindrical; lobes orange (*Cooper*), oblong, obtuse, 2 lin. long, about 1 lin. wide; petals minute; anthers ⅔ lin. long, thrice as long as the petals, obtuse; ovary oblong, glabrous; style about as long as the calyx-tube; stigma capitate.

KALAHARI REGION: Orange River Colony; Witteberg, *Rehmann*, 3943! Basutoland; without precise locality, *Cooper*, 696! Transvaal; by the Vaal River near Kloete, *Wilms*, 1299! by the Crocodile River in Lydenburg District, *Wilms*, 1299b.
Resembling *L. anthylloides*, Meisn., but differing in having quite glabrous leaves.

6. **L. deserticola** (C. H. Wright); a much-branched shrub; branches spreading, woody, pubescent when very young, soon glabrous, grey, leaf-scars small but prominent; leaves usually clustered at the ends of the branches, oblong, obtuse, 3 lin. long,

1 lin. wide, acute, subcoriaceous, hispidulous; involucral leaves rather wider, velvety; flowers 5–7, terminal; calyx densely silky outside, basal part furnished with long hairs; tube 6 lin. long, subcylindrical; lobes oblong, rounded at the apex, 1½ lin. long, 1 lin. wide, yellow; petals very small, membranous, hyaline; anthers oblong, obtuse, ½ lin. long, larger than the petals; ovary small, oblong, glabrous; style much shorter than the calyx-tube, rigid; stigma capitate. *L. pulchellus, var. dasyphyllus, Meisn. in DC. Prodr.* xiv. 595. *Gnidia pulchella, var. dasyphylla, Meisn. in Linnæa,* xiv. 425; *Drège, Zwei Pfl. Documente,* 61, 63. *G. deserticola, Gilg in Engl. Jahrb.* xix. 263.

Coast Region: Clanwilliam Div.; near Wuppenthal, 1900 ft., *Bolus,* 9084! Bidouw Berg, 2000 ft., *Schlechter,* 8695! Ladismith Div.; Groot River, Caroo, *Mund & Maire!* Oudtshoorn Div.; Oudtshoorn, *Tyson! Miss Britton,* 19! East London Div.; Cambridge, *Miss Wormald,* 81!
Central Region: Calvinia Div.; Nieuwoudtville, *Leipoldt in Herb. Bolus,* 9386! Ceres Div.; between Little Doorn River and Great Doorn River, *Burchell,* 1208! Prince Albert Div.; Gamka River, *Burke,* 241! *Zeyher,* 1491! Zwartbulletje, on stony places and by the Gamka River, 2500 ft., *Drège!* Jansenville Div.; near the Sundays River, 1500–2000 ft., *Drège!* Graaf Reinet Div.; Graaf Reinet, 2600 ft., *Bolus,* 2188!

7. **L. macropetalus** (Meisn. in DC. Prodr. xiv. 594); a small diffusely branched shrub; branches at first pubescent, rather slender, leaf-scars small but prominent; leaves alternate, more or less spathulate, adpressed-silky. when young, less densely so when old, acute, 4–8 lin. long, 2–2½ lin. wide; involucral leaves similar to the cauline; flowers in terminal heads; calyx silky outside; tube 6 lin. long, subcylindrical; lobes oblong, obtuse, 1½ lin. long, ⅔ lin. wide; petals 5, membranous, 1 lin. long, ⅔ lin. wide; anthers linear, 1 lin. long; ovary oblong, glabrous; style filiform; stigma capitate. *Sim, For. Fl. Cape Col.* 302; *Wood, Natal Pl. t.* 262. *Gnidia macropetala, Meisn. in Hook. Lond. Journ.* ii. 553; *Krauss, Beitr. Fl. Cap- und Natal.* 142.

Eastern Region: Natal; near Durban, *Krauss,* 237! *Peddie! Wood in MacOwan & Bolus, Herb. Norm.* 1028! Umlazi, *Mudd!* Inanda, *Wood,* 36! Intschanga, *Rehmann,* 7887! Vryheid, *Sim,* 2947! Dumisa Station, Alexandra District, *Rudatis,* 651! and without precise locality, *Mrs. Saunders! Nelson,* 11! *Gerrard,* 27!

8. **L. dregeanus** (Endl. Gen. Suppl. iv. ii. 67); branches adpressed-villous at the apex; leaves alternate, spreading, spathulate-oblong, obtuse, mucronulate, 5–8 lin. long, 3–4 lin. wide, adpressed-pilose on both surfaces; involucral leaves not wider than the cauline, rather silky, ciliate; calyx silky-villous; tube 6–8 lin. long; lobes obtuse; petals minute, ovate. *Meisn. in DC. Prodr.* xiv. 594. *Gnidia dregeana, Meisn. in Linnæa,* xiv. 426; *Drège, Zwei Pfl. Documente,* 160.

Eastern Region: Natal; Durban, *Drège.*
Habit and leaves as in *L. meisnerianus,* Endl., but pubescence less thick and calyx much shorter.

9. L. similis (C. H. Wright); stems about 3 in. high, arising from a woody rootstock, pubescent; leaves alternate, lanceolate, acute, 6 lin. long, 1½ lin. wide, hairy on both surfaces, midrib conspicuous, lateral nerves about 2 on each side; involucral leaves ovate, acute, 2½ lin. wide; flowers in terminal heads; calyx adpressed-silky outside; tube 6 lin. long, slightly inflated below; lobes shortly elliptic, obtuse, 1½ lin. long, 1 lin. wide; petals minute, tooth-like; anthers oblong, obtuse, ⅔ lin. long; ovary oblong, glabrous; style filiform, shorter than the calyx-tube; stigma capitate.

KALAHARI REGION : Transvaal ; Warmbaths, *Miss Leendertz,* 1314 !

Resembling *L. Krausii,* Meisn., but differing in having sessile flower-heads and much smaller petals.

10. L. linifolius (Decne in Jacquem. Voy. Bot. 148) ; a shrub; branches diffuse, rigid, glabrous, leaf-scars prominent; leaves lanceolate, acute, subpungent, 7–11 lin. long, 2 lin. wide; flowers about 10 in terminal clusters; involucral bracts ovate-lanceolate, acute, pungent, 6 lin. long, 2½ lin. wide; calyx densely tomentose outside; tube about 4 lin. long, slender, inflated below, subcylindrical above; lobes patent, ovate, obovate or rotundate, about 1½ lin. long and 1 lin. wide; petals 1 lin. long, oblong, obtuse, entire, glabrous; anthers oblong, the upper exserted and larger than the lower; ovary oblong; style excentric, slightly shorter than the calyx-tube; stigma small. *Walp. Ann.* i. 587 ; *Meisn. in DC. Prodr.* xiv. 595, *incl. vars. pubescens und glabrata* ; *Sim, For. Fl. Cape Col.* 302. *Gnidia capitata, Linn. f. Suppl.* 224 ; *Thunb. Prodr.* 76, *and Fl. Cap. ed. Schult.* 378, *partly* ; *Willd. Sp. Pl.* ii. 426 ; *Wikstr. in Vet. Acad. Handl. Stockh.* 1818, 314 ; *Drège, Zwei Pfl. Documente,* 138 ; *Meisn. in Linnæa,* xiv. 424. *G. transvaaliensis, Gilg in De Wild. Pl. Nov. Herb. Hort. Then.* i. 206, *t.* 46, *figs.* 10–16. *Dais linifolia, Lam. Encycl.* ii. 255, *and Ill.* ii. 492, *t.* 368, *fig.* 3 ; *Pers. Syn.* i. 471. *Passerina involucrata, Spreng. ex Meisn. in DC. Prodr.* xiv. 595.

COAST REGION : Uitenhage Div. ; on plains near Uitenhage, *Bowie* ! Addo, *Zeyher,* 210 ! Zuurberg Range, *Cooper,* 3089 ! Albany Div. ; near Grahamstown, *Burchell,* 3607 ! *MacOwan* ! *Misses Daly & Sole,* 355 ! *Atherstone* ! between Zwartewater Poort and the east end of Zwartewater Berg, *Burchell,* 3448 ! Rockcliffe near Sidbury, *Miss Daly,* 761 ! Brak Kloof, *Mrs. White,* 1064 ! and without precise locality, *Cooper,* 1525 ! King Williamstown Div. ; King Williamstown, 1500 ft., *Sim,* 2948 ! Queenstown Div. ; plains, Queenstown, *Galpin,* 1560 ! *Cooper,* 3084 ! Imvane, *Baur,* 80 ! Zwartkei River, *Baur,* 100 !

CENTRAL REGION : Somerset Div.; Klein Bruintjes Hoogte, *Drège* ! Bosch Berg, *Bolus,* 1749 !

KALAHARI REGION : Orange River Colony ; Bethlehem, *Richardson* ! and without precise locality, *Cooper,* 3087 ! Transvaal; Magalies Berg, *Zeyher,* 1495 ! near Pretoria, *McLea in Herb. Bolus.,* 3112 ! *Miss Leendertz,* 405 ! *Wilms,* 1304 ! around Barberton, 2000–4000 ft., *Galpin,* 495 ! Maquasi Hills, *Nelson,* 235 ! Skinners Court Experiment Station, 4500 ft., *Burtt-Davy,* 5044 ! Lydenberg, *Wilms,* 1305 ! Crocodile River, *Quintas,* 229, King's Farm, between Venters Poort and Middelburg, *Bolus,* 9759 !

EASTERN REGION : Tembuland ; Bazeia, 2000 ft., *Baur,* 537 ! Umtata, *Baur,* 538 ! Griqualand East ; Kokstad, 4300 ft., *Tyson,* 1112 ! 1531 ! and in *MacOwan*

& Bolus, Herb. Austr.-Afr., 1228 ! Natal; near Newcastle, *Wilms*, 2251 ! Weenen County, *Wood*, 3446 !

Meisner's two varieties cannot be maintained. The pubescence on the involucral leaves seems to be more or less deciduous and cannot, therefore, be relied upon as a varietal character.

11. **L. polyanthus** (Gilg in Engl. Jahrb. xix. 265) ; an undershrub, more or less branched from the base ; branches rather stout, pilose at first, leaf-scars prominent ; leaves alternate, oblong, oval or ovate, acute, 15 lin. long, 5 lin. wide, 5-nerved, densely adpressed-silky on both surfaces ; involucral leaves similar to the cauline but more densely silky ; heads terminal, sessile, up to 40-flowered ; calyx spreading-pilose outside; tube about 1 lin. long, slender, sub-cylindrical ; lobes yellow, oblong-lanceolate, or oblong, 3 lin. long, 1½ lin. wide; petals minute, oblong, ⅓ lin. long; anthers oblong, twice as long as the petals ; ovary oblong ; style slightly shorter than the calyx-tube ; stigma capitate.   *Gnidia polyantha, Gilg, l.c.*.

EASTERN REGION: Griqualand East; Mount Currie, 5000 ft., *Tyson*, 1586! Zuurberg, 5000 ft., *Tyson*, 1563 ! 1849 ! Mount Malowe, *Tyson*, 2130! Tembuland ; saddle between Iggakancu and Bazeia Mountains, 2500–3000 ft., *Baur*, 646! Natal; Dumisa Station, 2000 ft., *Rudatis*, 425 ! Durban, *Krauss*, 368 ! Bothas Hill, *Tyson*! *Wood*, 4420! Normandieu, *Sim*, 2933! Noodsberg, *Wood*, 115 ! 4135 ! 5277! Camperdown, *Rehmann*, 7848! Drakensberg Range, Bushmans River, 6000–7000 ft., *Evans*, 52! top of Mahwaga, 6000–7000 ft., *Evans*, 523 ! and without precise locality, *Gerrard*, 1392!

12. **L. anthylloides** (Meisn. in DC. Prodr. xiv. 595, incl. var. *vulgaris*) ; a shrub of variable habit ; branches at first densely villous, leaf-scars small but prominent ; leaves alternate, imbricate, lanceolate or oblong, adpressed-silky on both surfaces, about 6 lin. long and 2 lin. wide ; flowers in dense terminal sessile heads ; calyx-tube 12–15 lin. long, densely woolly outside, gradually widened upwards ; lobes elliptic, obtuse, 3 lin. long, 2 lin. wide ; petals minute and tooth-like, or obsolete ; anthers oblong, obtuse, ⅔ lin. long ; ovary oblong ; style filiform, nearly as long as the calyx-tube ; stigma capitate. *Bot. Mag. t.* 7303 ; *Sim, For. Fl. Cape Col.* 302 ; *Wood, Natal Pl. t.* 270. *Passerina anthylloides, Linn. f. Suppl.* 225; *Thunb. Prodr.* 76, *and Fl. Cap. ed. Schult.* 377 ; *Wikstr. in Vet. Acad. Handl. Stockh.* 1818, 347 ; *Drège, Zwei Pfl. Documente,* 128, 143 ; *Meisn. in Linnæa,* xiv. 392, *and in Hook. Lond. Journ.* ii. 551. *Dais sericea, Lam. Encycl.* ii. 767. *D. anthylloides, Eckl. & Zeyh. ex Meisn. in DC. Prodr.* xiv. 530, 596. *Arthrosolen anthylloides, C. A. Mey. in Bull. Phys.-Mat. Acad. Pétersb.* i. (1843) 359. *Gnidia tomentosa, Eckl. ex Meisn. in DC. Prodr.* xiv. 596. *G. virescens, Wikstr. in Vet. Acad. Handl. Stockh.* 1818, 317.

VAR. β, **macrophylla** (Meisn. in DC. Prodr. xiv. 596); leaves oblong or oval-oblong, obscurely 3-nerved, the upper densely silky on both sides, the lower pilose, up to 15 lin. long and 3 lin. wide. *Passerina anthylloides, var. macrophylla, Meisn. in Linnæa,* xiv. 393; *Drège, Zwei Pfl. Documente,* 152, 154.

VAR. γ, **glabrescens** (Meisn. in DC. Prodr. xiv. 596); leaves oblong, 3–8 lin. long, 3–5 lin. wide, 3-nerved, subacute or obtuse, the upper sparingly silky, the lower sparingly pubescent beneath. *Passerina anthylloides, var. glabrescens, Meisn. in Linnæa,* xiv. 393; *Drège, Zwei Pfl. Documente,* 159.

COAST REGION : Riversdale Div. ; between Little Vet River and Garcins Pass, *Burchell,* 6877 ! George Div. ; mountains near George, *Alexander* ! Knysna Div. ; Knysna, *Tyson,* 3043 ! between Plettenberg Bay and Knysna, *Burchell,* 5355 ! between Goukamina River and Groene Valley, *Burchell,* 5606 ! Humansdorp Div. ; near Kruisfontein, 800 ft., *Galpin,* 4530 ! Uitenhage Div. ; Van Stadens Berg, *Drège* ! and without precise locality, *Zeyher,* 269 ! Port Elizabeth Div. ; Port Elizabeth, *Miss West,* 42 ! between Port Elizabeth and Van Stadens River, 400 ft., *Bolus,* 1941 ! Albany Div. ; near Grahamstown, *Burchell,* 3608 ! *MacOwan,* 311 ! Sidbury, *Burke* ! Kleinemond, *Mrs. White* ! Mayors Seat, *Miss Daly,* 122 ! and without precise locality, *Cooper,* 1521 ! Bathurst Div. ; between Blue Krantz and Kaffir Drift Military Post, *Burchell,* 3717 ! King Williamstown Div. ; King Williamstown, 1500 ft., *Sim,* 2949 ! Komgha Div. ; between Zandplaat and Komgha, 2000–3000 ft., *Drège.* Var. β : Port Elizabeth Div. ; Algoa Bay, *Forbes* ! Albany Div. ; Grahamstown, *MacOwan* ! Var. γ : Uitenhage Div. ; Van Stadens Berg, *Zeyher,* 3765 ! Stockenstrom Div. ; summit of Katberg, *Shaw* ! King Williamstown Div. ; Buffalo River, *Cooper,* 69 !

CENTRAL REGION : Graaf Reinet Div. ; Compass Berg, *Shaw* ! Var. γ : Somerset Div. ; without precise locality, *Bowker* !

KALAHARI REGION : Transvaal ; near Pretoria, 4000 ft., *McLea in Herb. Bolus,* 3084 ! Basutoland ; without precise locality, *Cooper,* 696 !

EASTERN REGION : Natal ; without precise locality, *Gerrard,* 76 ! *Plant,* 36 ! *Cooper,* 1232 ! Zululand ; Umhlafuzi, *Sim,* 2937 ! Var. β : Pondoland ; between St. Johns River and Umsikaba River, under 1000 ft., *Drège!* between Umtentu River and Umzimkulu River, under 500 ft., *Drège,* Natal ; Inanda, 1800 ft., *Wood,* 34 ! Durban, *Grant* ! *Krauss,* 282 ! *Sanderson* ! *Wood,* 4948 ! *Peddie* ! Nottingham, *Buchanan,* 136 ! Bothas Hill, *Wood,* 4420 ! and without precise locality, *Cooper,* 3083 ! *Sutherland* ! *Gerrard,* 206 ! Var. γ : Natal ; Durban, *Drège* !

13. **L. meisnerianus** (Endl. Gen. Suppl. iv. ii. 67) ; a much-branched shrub ; branches rather stout, pubescent at first, leaf-scars not very prominent ; leaves alternate, spathulate-oblong or lanceolate, narrowed at the base, up to 10 lin. long and 2 lin. wide, upper scarcely larger, adpressed-silky on both surfaces ; flowers in terminal heads ; calyx-tube 9–12 lin. long, cylindrical, with long straight hairs below, pubescent above ; lobes lanceolate, obtuse, 2 lin. long, ¾ lin. wide ; petals minute, bifid, membranous ; anthers oblong, obtuse, ⅔ lin. long ; ovary oblong, glabrous ; style nearly as long as the calyx-tube ; stigma capitate. *Meisn. in DC. Prodr.* xiv. 594. *L. Meisneri, Sim, For. Fl. Cape Col.* 302, *t.* 153, *fig.* 4. *Gnidia cuneata, Meisn. in Linnæa,* xiv. 427 ; *O. Kuntze, Rev. Gen. Pl.* iii. ii. 280. *G. argentea, Zeyh. ex Meisn. l.c.* 428.

VAR. β, **spathulatus** (Meisn. in DC. Prodr. xiv. 594); leaves obovate- or spathulate-lanceolate, 3–5 lin. wide, rounded or very shortly acuminate. *Dais argentea, Eckl. & Zeyh. ex Meisn. in DC. Prodr.* xiv. 594. *Gnidia cuneata, var. spathulata, Meisn. in Linnæa,* xiv. 427 ; *Drège, Zwei Pfl. Documente,* 142.

VAR. γ, **angustifolius** (Meisn. in DC. Prodr. xiv. 594); leaves narrowly lanceolate, attenuate at both ends, scarcely spathulate, 2–3 lin. wide, subacute. *Gnidia cuneata, var. angustifolia, Meisn. in Linnæa,* xiv. 427 ; *Drège, Zwei Pfl. Documente,* 138.

COAST REGION : Uitenhage Div. ; without precise locality, *Bowie* ! Albany Div. ; Bothas Berg, 2000 ft., *MacOwan,* 368 ! Hell Poort, *MacOwan* ! between Sidbury

and the Bushman River, *Burchell*, 4198! Fish River, 400–500 ft., *Baur*, 1084!
Var. β: Uitenhage Div.; without precise locality, *Zeyher*, 732! 3766! Peddie
Div.; Fish River, hills near Trumpeters Drift, *Drège*!
CENTRAL REGION: Somerset Div.; at Blyde River, *Burchell*, 2965! west side
of Bruintjes Hoogte, 3300 ft., *Bolus*, 1779! Var. γ: Somerset Div.; near Little
Fish River and Great Fish River, 2000–3000 ft., *Drège*!
EASTERN REGION: Natal; Vryheid, *Sim*, 2932! Krantz Kloof, *Kuntze*! Clair-
mont, *Kuntze*!

Meisner in *DC. Prodr.* xiv. 594 has transposed the references to the two
varieties in Drège's *Documente*, and also the localities connected with them.

14. **L. pulchellus** (Decne in Jacquem. Voy. Bot. 149); branches
virgate, slender, glabrous; leaves alternate, linear-lanceolate or
subspathulate, mucronate or obtuse, 6–12 lin. long, 1½–2¼ lin. wide,
glabrous; involucral leaves 8–12, spathulate, scarcely wider than
the cauline; heads sessile or shortly stalked, many-flowered; calyx
densely adpressed-silky, 1 in. long; lobes narrowly oblong, 2 lin.
long. *Walp. Ann.* i. 587; *Meisn. in DC. Prodr.* xiv. 594, *incl. var.
glabratus, but excl. var. dasyphyllus. Gnidia pulchella, and var. glabrata,
Meisn. in Linnæa*, xiv. 425; *Drège, Zwei Pfl. Documente*, 44, 149.
Dais argentea, var. depressa, Eckl. & Zeyh. ex Meisn. in DC. Prodr.
xiv. 595.

COAST REGION: Stockenstrom Div.; in grassy places, Kat Berg, 3000–4000 ft.,
Drège.
EASTERN REGION: Pondoland; between Umtata River and St. Johns River,
Drège.

15. **L. triplinervis** (Decne in Jacquem. Voy. Bot. 149); a much-
branched shrub; branches glabrous, slender, leaf-scars small but
prominent; leaves alternate, oblanceolate, very shortly cuspidate,
up to 12 lin. long and 2½ lin. wide, quite glabrous, midrib prominent
beneath; involucral leaves ovate, 4 lin. long, 2½ lin. wide, acute,
ciliate; flowers numerous in terminal clusters; calyx-tube silky
outside, hairs longer below, cylindrical, 10–15 lin. long; lobes
oblong, obtuse, 2½ lin. long, 1½ lin. wide; petals 5, membranous,
obovate, 1 lin. long, ½ lin. wide; anthers linear, nearly 1 lin. long;
ovary ovate, compressed; style longer than the calyx-tube; stigma
small. *Meisn. in DC. Prodr.* xiv. 595. *Gnidia triplinervis, Meisn.
in Linnæa*, xiv. 429. *Dais Owenii, Eckl. & Zeyh. ex Meisn. in DC.
Prodr.* xiv. 595.

COAST REGION: King Williamstown Div.; Perie Mountains, 2000 ft., *Tyson*,
1041! *Sim*, 2951! sides of Buffalo Mountain, *Tyson*, 604!
EASTERN REGION: Pondoland; banks of the Umkwani River, near the sea,
Tyson, 2643! near the mouth of the Umtentu River, *Drège*! and without
precise locality, *Bachmann*, 901! Griqualand East; by streams near Clydesdale,
2500 ft., *Tyson*, 1159! 2015! and in *MacOwan & Bolus, Herb. Austr.-Afr.*, 1227!
Natal; Umgeni, *Rehmann*, 7464!

16. **L. caffer** (Meisn. in DC. Prodr. xiv. 593); a small undershrub; branches slender, erect, glabrous, terete; leaves alternate,
erect, linear, acute, 6–9 lin. long, 1½ lin. wide, glabrous; heads
sessile, terminal, few-flowered; involucral leaves ovate, acuminate,
2 lin. wide; calyx densely silky outside; tube 10 lin. long, slender,

cylindrical; lobes oblong, obtuse, 4 lin. long, 1¼ lin. wide; petals
⅓ lin. long, linear; ovary oblong, compressed, hairy at the top;
style as long as the calyx-tube; stigma capitate. *Wood in Trans.
S. Afr. Phil. Soc.* xviii. 218.

KALAHARI REGION : Orange River Colony; near Witzies Hoek, 5400 ft., *Bolus*,
8245! *Flanagan*, 1887! Harrismith, *Sankey*, 169! and without precise locality,
Cooper, 831! Transvaal; near Lydenburg, *Atherstone*! *Wilms*, 1306! near
Barberton, *Bolus*, 9761! High Veldt near Belfast, 6500 ft., *Bolus*, 12269!
Jenkins! Waterval Boven, *Rogers*, 245! Magaliesberg, *Zeyher*, 1488! Carolina
District, 1 mile north of Robinsons, *Burtt-Davy*, 2977! and without precise
locality, *Mrs. Stainbank*, 3641!

EASTERN REGION: Natal; Little Tugela District, near Hoffenthal Mission
Station, and Tabamhlope, *Wood*, 3448! Van Reenans Pass, 5000-6000 ft., *Wood*,
4520! Vryheid, *Sim*, 2923! and without precise locality, *Gerrard*, 1391!

17. **L. Kraussii** (Meisn. in DC. Prodr. xiv. 596); a very variable
plant; stems stout, herbaceous from a woody base, glabrous or
pubescent; leaves lanceolate to .oval, 8–18 lin. long, 1–6 lin. wide,
quite glabrous to densely pilose, acute or obtuse; flowers in a
terminal peduncled head; bracts ovate, acute, silky; calyx-tube
7–9 lin. long, cylindrical, silky outside and with long white hairs at
the base; lobes oval, obtuse, 2 lin. long, 1 lin. wide; petals half as
long as the calyx-lobes, lanceolate, acute, membranous; anthers
oblong, ½ lin. long; ovary ovate, compressed, hairy, especially at the
top; style about as long as the calyx-tube; stigma capitate, *Engl.
Hochgebirgsfl. Trop. Afr.* 310: *Wood & Evans, Natal Pl.* iii.
t. 256; *Pearson in Dyer, Fl. Trop. Afr.* vi. i. 231; *Sim, For. Fl.
Cape Col.* 302. *L. affinis, Kotschy & Peyr. Pl. Tinn.* 39, *t.* 19 B.
L. djuricus, Gilg in Engl. Jahrb. xix. 269. *Gnidia kraussiana,
Meisn. in Hook. Lond. Journ. Bot.* ii. (1843), 552, 553, *incl. vars.
pubescens and glabrata*; *Gilg in Engl. Pfl. Ost-Afr. C.* 283.
G. djurica, Gilg in Engl. Jahrb. xix. 268. *G. dschurica, Gilg in
Engl. & Prantl, Pflanzenfam.* iii. 6 A, 228. *G. usinjensis, Gilg in
Engl. Jahrb.* xix. 269; *Engl. Pfl. Ost-Afr. l.c.*

KALAHARI REGION: Orange River Colony; Bethlehem, *Richardson*! Harrismith,
Sankey, 170! Besters Vlei, near Witzies Hoek, 5400 ft., *Bolus*, 8426! *Flanagan*,
1871! and without precise locality, *Cooper*, 832! Transvaal; near Pretoria,
4500 ft., *Miss Leendertz*, 280! *Rehmann*, 4531! *McLea in Herb. Bolus*, 5809!
Burtt-Davy, 786; Waterval Boven, *Rogers*! Magaliesberg, *Burke*! Piet Retief,
Burtt-Davy, 1914! Lydenburg, *Wilms*, 1801! near Barberton, *Bolus*, 9758!

EASTERN REGION: Tembuland; Mount Bazeia, *Baur*, 254! Pondoland;
without precise locality, *Bachmann*, 897! Griqualand East; Maclear, plains at
base of Tent Kop, 5600 ft., *Galpin*, 6826! Kokstadt, 4300 ft., *Tyson*, 1098! and
in *Bolus & MacOwan, Herb. Norm. Austr.-Afr.*, 458! Natal; near Durban,
Sutherland! Inanda, *Wood*, 177! 188! between Pietermaritzberg and Greytown,
Wilms, 2246! Vryheid, *Sim*, 2920! Dumisa, *Rudatis*, 490! Table Mountain,
Krauss, 455! and without precise locality, *Gerrard*, 585! *Mrs. Hutton*, 349!
Cooper, 3090!

Meisner (in *DC. Prodr.* xiv. 596) has described three varieties, viz., *pubescens,
glabratus* and *angustifolius*, which cannot be maintained owing to the impossibility
of defining them by constant characters, as pointed out by Pearson, *l.c.* Bolus,
8246, has densely hairy and quite glabrous leaves in the same gathering. The
species is polymorphic to a high degree.

Also in Tropical Africa.

18. **L. hoepfnerianus** (Vatke ex Gilg in Engl. Jahrb. xix. 268);
stems many from a woody rootstock, at first densely pilose, soon
glabrous; leaves alternate, lanceolate, acute, 6 lin. long, 1 lin. wide,
at first silky, midrib prominent beneath; involucral leaves ovate,
4–5 lin. long, 2–2½ lin. wide, densely silky; heads terminal,
peduncled, many-flowered; calyx densely silky outside; tube about
9 lin. long; lobes obovate, 1½–2 lin. long, 1 lin. wide, yellow; petals
membranous, oblong, obtuse, ⅔ lin. long; anthers linear, obtuse,
⅔ lin. long; ovary oblong; style shorter than the calyx-tube;
stigma capitate. *Pearson in Dyer, Fl. Trop. Afr. vi. i. 233.
Gnidia hoepfneriana, Gilg in Jahrb. xix. 268; Hiern in Cat. Afr.
Pl. Welw. i. 925, partly.*

KALAHARI REGION: Transvaal; Vereeniging, *Leslie*!
This may be only an extreme form of *L. Kraussii,* Meisn.
Also in Tropical Africa.

Imperfectly known species.

19. **L. macranthus** (Gandog. in Bull. Soc. Bot. France, lx. 418,
1913); branches virgate, pubescent or laxly pilose, green; leaves
straight, not imbricate, oblong-ovate, mucronulate, slightly attenuate
below, nerved, rather lax, about 2 lin. wide, glabrous above, pubescent beneath; flowers capitate; calyx about 15 lin. long; lobes
yellow when dry, elliptic-obovate.

COAST REGION: Riversdale Div.; Lange Berg Range, *Schlechter,* 1904.
Near *L. anthylloides,* Meisn.

20. **L. oblongifolius** (Gandog. in Bull. Soc. Bot. France, lx. 418,
1913); a divaricately branched woolly undershrub; leaves narrowly
oblong, acute, nerved; flowers capitate, blackish when dried; calyx
nearly 10 lin. long, tomentose outside, glabrous within, whitish.

COAST REGION: Port Elizabeth Div.; Port Elizabeth, *Laidley, E. S. C. A.
Herb.* 485.
This is said to resemble a large form of *L. anthylloides,* Meisn.

X. ENGLERODAPHNE, Gilg.

Flowers hermaphrodite, 4-merous. *Calyx* subcylindrical, coloured,
glabrous; tube slightly constricted below the middle, not articulated; lobes patent. *Petals* bipartite, rather fleshy, shorter than
the calyx-lobes. *Stamens* 8, the upper slightly exserted; anthers
with inconspicuous connective. *Ovary* sessile, 1-celled, very hispid;
style excentric.

DISTRIB. A single species extending to Tropical Africa.

1. **E. leiosiphon** (Gilg in Engl. Jahrb. xix. 274); a shrub, up to
10 ft. high (*Galpin*), subscandent (*Sim*), glabrous, internodes up to
1 in. long; leaves opposite, very shortly petioled, ovate-oblong, up

to 1 in. long and $\frac{1}{2}$ in. wide, subacute, rounded at the base, quite glabrous, pinnately nerved; flowers in terminal few-flowered ebracteate heads; calyx glabrous; tube 6 lin. long, ampulliform below, funnel-shaped above; lobes ovate-oblong, acute, 2 lin. long, 1 lin. wide at the base; petals 8, membranous, oblong, acute, irregularly dentate; anthers oblong, $\frac{3}{4}$ lin. long, acute; ovary ovate, compressed, with straight white hairs on the upper half as long as itself; style excentric, half as long as the calyx-tube; stigma capitate. *Engl. Pfl. Ost-Afr. C.* 284; *Pearson in Dyer, Fl. Trop. Afr.* vi. i. 238; *J. M. Wood in Trans. S. Afr. Phil. Soc.* xviii. 219. *Gnidia subcordata, Meisn. in Linnæa,* xiv. 430, *and in DC. Prodr.* xiv. 586; *Drège, Zwei Pfl. Documente,* 148.

COAST REGION : East London Div.; Buffalo River, at second creek, *Galpin,* 1845! East London, *Miss Wormald,* 85! *Rattray,* 82! King Williamstown Div. ; King Williamstown, 1200 ft., *Sim,* 308! Komgha Div. ; near Komgha, 1500 ft., *Flanagan,* 318!
EASTERN REGION : Transkei; Kentani, *Miss Pegler,* 175! Tembuland; at the edge of woods near Morley, 1000–1500 ft., *Drège,* 4670. Natal; Pietermaritzburg, *Sim ex Wood.*

Also in Tropical Africa.

XI. SYNAPTOLEPIS, Oliv.

Flowers hermaphrodite. *Calyx-tube* elongated, cylindrical, at length swollen below, not articulated; lobes 5, patent. *Petals* united into a short erect entire or obscurely 5-lobed ring at the base of the calyx-lobes. *Stamens* 10 in two distinct whorls at the upper end of the calyx-tube, the upper opposite the calyx-lobes and not exceeding the corolla ; anthers oblong with wide connective. Hypogynous *disc* of minute scales or shortly cup-shaped, very thinly membranous. *Ovary* shortly stalked, 1-celled, 1-ovuled ; style long, slender ; stigma cylindrical or capitate. *Fruit* dry, enclosed in the persistent more or less fleshy base of the calyx ; pericarp crustaceous. *Seed* exalbuminous ; testa membranous ; cotyledons fleshy.

Erect or climbing shrubs ; leaves alternate, membranous or subcoriaceous ; flowers solitary or fascicled in the leaf-axils, sessile or shortly stalked.

DISTRIB. 7 species in Tropical and South Africa.

Petals glabrous ; hypogynous disc of small scales ... (1) **Kirkii.**
Petals fimbriate ; hypogynous disc cup-shaped, lobed ... (2) **oliveriana.**

1. **S. Kirkii** (Oliv. in Hook. Ic. Pl. t. 1074); a much-branched shrub; branches divaricate, terete, slender, blackish, lenticillate; leaves opposite, ovate or elliptic-lanceolate, acute or subacute, about 12 lin. long and $5\frac{1}{2}$ lin. wide, mucronulate, rounded at the base, subcoriaceous, glabrous, very shortly petiolate ; flowers few in axillary fascicles; calyx-tube 6–8 lin. long, slender, cylindrical below, slightly widened above; lobes oblong, $2\frac{1}{2}$ lin. long, nearly 1 lin. wide, obtuse; petals united into a glabrous ring $\frac{1}{3}$ lin. high;

filaments short; anthers oval; ovary shortly stalked, glabrous, with a ring of minute scales at its base; style shorter than the calyx-tube; stigma small. *Pearson in Dyer, Fl. Trop. Afr.* vi. i. 247.

EASTERN REGION: Delagoa Bay; 3 miles north-west of Lorenzo Marquez, 80 ft., *Bolus*, 9762!

Also in Tropical Africa.

2. **S. oliveriana** (Gilg in Engl. Jahrb. xix. 276); a spreading bush, about 2–3 ft. high (*Monteiro*); branches divaricate, terete, blackish, lenticillate, glabrous; leaves opposite, ovate or elliptic, up to 8 lin. long and 6 lin. wide, obtuse or subacute, rounded at the base, glabrous, shortly petioled, secondary nerves obscure; flowers white, pendulous, solitary or in pairs, axillary; bracteoles very small; calyx-tube funnel-shaped, 5 lin. long, glabrous; lobes lanceolate, 2 lin. long, ½ lin. wide; petals united into a ring ⅓ lin. high, fimbriate; filaments short; anthers oval; ovary ovoid, sessile, surrounded at the base by a wrinkled glabrous cup-shaped disc. *S. Kirkii, Gilg in Engl. & Prantl, Pflanzenfam.* iii. 6 A, 231, *fig.* 81, F–J.

EASTERN REGION: Delagoa Bay, *Monteiro*, 45! *Schlechter*, 12165!

XII. PEDDIEA, Harv.

Flowers hermaphrodite, 4–5-merous. *Calyx* deciduous; tube sub-cylindrical, naked in the throat; lobes short, patent. *Petals* none. *Stamens* 8 or 10, in two series included in the calyx-tube; filaments very short; anthers oblong, with very small connective. Hypogynous *disc* cup-shaped, entire or lobed. *Ovary* shortly stalked, 2-celled, cells 1-ovuled; style shortly filiform; stigma capitate or more or less saucer-shaped. *Fruit* naked; exocarp fleshy; endocarp hard, 1-(rarely 2-) seeded. *Seed* exalbuminous, rarely with scanty albumen; testa membranous; cotyledons broad, rather fleshy.

Glabrous shrubs; leaves alternate, subcoriaceous or membranous; flowers in terminal umbels.

DISTRIB. 8 species in Tropical Africa and 1 in South Africa.

1. **P. africana** (Harv. in Hook. Journ. Bot. ii. 266, t. 10, 1840); an erect shrub; branches glabrous, at length greyish; leaves alternate, sometimes clustered at the ends of the branches, ovate-oblong, up to 4½ in. long and nearly 2 in. wide, obtuse, membranous; peduncle ½–1½ in. long, slender, ebracteate or with a few basal bracts, 6–12-flowered; pedicels 3–6 lin. long; calyx-tube 6 lin. long, glabrous; lobes ovate, 2 lin. long, 1 lin. wide, revolute, acute; anthers ⅓ lin. long; hypogynous disc ½ lin. long, membranous, fimbriate; ovary ovoid, glabrous or hairy at the apex; ovules pendulous. *Meisn. in DC. Prodr.* xiv. 528; *Drège in Linnæa*, xx. 209; *Wood, Natal Pl. t.* 87, *and in Trans. S. Afr. Phil. Soc.* xviii. 219; *Sim, For. Fl. Cape*

Col. 301, *t.* 153, *fig.* 2. *P. Dregei and P. Harveyi, Meisn. in DC. Prodr.* xiv. 528. *Cyathodiscus umbellatus, Hochst. in Flora,* 1842, 240. *Cestrum umbellatum, E. Meyer in Drège, Zwei Pfl. Documente,* 154, 159. *Psilolena umbellata, Presl, Bot. Bemerk.* 102 ; *Walp. Ann.* i. 589.

COAST REGION: Komgha Div. ; Gwenkala River and near Komgha, 1800–2000 ft., *Flanagan,* 693 !
KALAHARI REGION: Transvaal ; margin of woods, Upper Moodies, Barberton, 4500 ft., *Galpin,* 962 ! Woodbush, *Grenfell,* 10 ! *Rehmann,* 5948 ! Potatobosch, 4750 ft., *Burtt-Davy,* 1195 !
EASTERN REGION : Transkei ; Kentani, 1000 ft., *Miss Pegler,* 491 ! Pondoland ; between Umtentu River and Umzinkulu River, under 500 ft., *Drège,* near Fort Donald, *Sim,* 2394 ! Griqualand East ; Zuurberg Range, 4500 ft., *Tyson,* 2807 ! Natal ; Durban, *Peddie ! Plant,* 14 ! *Cooper,* 1229 ! *Drège ! Krauss,* 427 ! *Wood,* 1013 ! Inanda, *Wood,* 580 ! Pietermaritzburg, *Cordukes,* 31 ! Dumisa, *Rudatis,* 652 ! Kettlefontein, *Cooper,* 1210 ! Howick, *Mrs. Hutton,* 10 ! Ngoma Forest, Vryheid, *Sim,* 2939 ! and without precise locality, *Gerrard,* 327 ! *Sanderson,* 301 ! 532 ! *Cooper,* 3092 ! Zululand ; Umlalazi, Ngoya Forest, *Sim,* 2940 ! Qudeni Forest, 6000 ft., *Wood,* 7902 ! Swaziland ; Forbes Reef, *Burtt-Davy,* 2798 ! and without precise locality, *Miss Stewart* !

The floral differences upon which Meisner founded his two species are probably due to sexual variation, and not to constant specific characters. The bracts near the base of the peduncle are caducous.

ORDER CXIX. PENÆACEÆ.

(By E. L. STEPHENS.)

Flowers regular, hermaphrodite. *Perianth* inferior, persistent, accrescent, 4-lobed ; tube cylindrical ; lobes equal, valvate or reduplicate-valvate in bud. *Stamens* four, inserted in the throat of the perianth and alternating with its lobes ; filaments usually not longer than the anthers ; anthers basifixed or rarely versatile ; cells 2, introrse, adnate to the connective and often shorter than it, margin of the valves membranous. *Ovary* free, 4-celled, cells opposite the lobes of calyx, without a central column ; style terete and terminated by a 4-lobed capitate stigma ; or with 4 wings or angles alternating with the cells, topped by 4 oblong cruciform lobes, and with four minute cushion-like stigmatic surfaces in the angles between the lobes below the apex of the style ; or rarely terete with 4 ridges at the base and 4 upright cruciform lobes at the apex, and 4 minute cushion-like stigmatic surfaces in the angles between the lobes. *Ovules* anatropous, 2 or 4 in each cell, basal, erect ; or 4 in each cell, inserted half-way down on an axile placenta, 2 erect and 2 pendulous. *Capsule* loculicidal. *Seeds* by abortion 1–2 in each cell, erect or more rarely pendulous, according to the position of the ovules, exalbuminous ; testa smooth, glossy, sometimes puncticulate ; tegumen membranous ; hilum furrowed at the base ; funicle very short, white, swollen; raphe filiform, extrorse, coloured; funicle and raphe separable from the seed; embryo conic-ovoid, with the massive hypocotyl terminated towards the hilum by the truncate concave radicle, and towards the chalaza by 2 very minute cotyledons.

Small shrubs or undershrubs of an ericoid habit; leaves decussate, often imbricate, entire, coriaceous, feather-veined, flat or rarely ericoid or partially revolute, sessile or very shortly petiolate; stipules when present minute, awl-like or auriculate or sometimes glandular; setæ or minute scales often present in the axils of the leaves; flowers sessile or on very short pedicels, solitary or in much reduced racemes, lateral or approximated at the ends of the branches; bracts leaf-like, or differing from the leaves in colour and form; bracteoles opposite, in one or more pairs, sometimes very caducous.

DISTRIB.   Genera 5, species 21, endemic in the coast region of South Africa.

Tribe I. PENÆÆ.—*Ovules* 2 or 4 in each cell, basal, erect (in *Brachysiphon rupestris* occasionally 2 erect and 2 pendulous in each cell).

   I. **Penæa.**—*Perianth-tube* subequal to the limb or twice as long; lobes erect, valvate in bud. *Style* 4-winged or 4-angled, rarely cylindrical, topped by 4 cruciform lobes; stigmatic surfaces 4, minute, cushion-like, in the angles between lobes of the style just below its apex.

  II. **Brachysiphon.**—*Perianth-tube* subequal to the limb or nearly twice as long; lobes erect, or reflexed only in the faded flower, reduplicate-valvate in bud. *Style* filiform, terete or cylindrical; stigma small, capitate, obscurely 4-lobed.

 III. **Sarcocolla.**—*Perianth-tube* about 3 times as long as the limb; lobes reflexed on opening, reduplicate-valvate in bud. *Stamens* exserted. *Style* filiform, terete; stigma capitate, obscurely 4-lobed.

Tribe II. ENDONEMEÆ.—*Ovules* 4 in each cell, inserted half-way down on an axile placenta, 2 erect and 2 pendulous.

  IV. **Glischrocolla.**—*Flowers* approximated at the tips of the branches. *Perianth-tube* about 3 times as long as the limb. *Filaments* erect, very short.

   V. **Endonema.**—*Flowers* lateral. *Perianth-tube* about 4 times as long as the limb. *Filaments* folded in bud, becoming erect as the flower opens, subulate above the fold, ligulate below.

I. PENÆA, Linn.

Perianth-tube cylindrical or ovate-cylindric, twice as long as the limb or subequal to it, subequal to the bracts; lobes erect, valvate in bud. *Stamens* shorter than the perianth-lobes; filaments very short, flattened; anthers with thick fleshy 2-lobed connective, bearing a small obliquely-placed cell on the inner side of each lobe near the base; margin of the valves fimbriate. *Ovary and style* as long as the perianth; *ovary* 4-angled, glabrous or minutely scabrous; ovules 2 or 4 in each cell, basal erect; *style* usually with 4 longitudinal membranous wings or 4-angled, easily separable into 4 parts, wings or angles alternating with the cells and topped by 4 flat oblong lobes, arranged in a cruciform manner (stigmata of other authors); *or style* terete, with 4 longitudinal ridges at the base and

4 upright lobes at the apex; *stigmatic surfaces* in each case 4, rather minute, cushion-like, in the angles between the lobes just below the apex of the style.

Low much-branched undershrubs; branches quadrangular in the younger parts, becoming terete lower down, upper more or less thickly beset with leaves, lower often defoliated; flowers either lateral in the axils of leaf-like bracts, or approximated at the tips of the branches in the axils of the upper leaves, which generally pass suddenly into broader and somewhat shorter concave leaf-like coloured bracts; whole inflorescence then greenish-yellow, or tinged to a rosy or brownish-red; bracteoles usually 2, concave, shorter and narrower than the bracts.

DISTRIB. Species 13, endemic.

§ 1. EUPENÆA.—Ovary glabrous.   Style winged.
   Perianth-lobes deltoid, acute or obtuse :
      Lower leaves ovate, acute, upper cordate, acumi-
         nate ... ... (1) **mucronata.**
      Leaves ovate or elliptical, subacute or obtuse,
         3–6 lin. long... ... ... (2) **ovata.**
      Leaves lanceolate or elliptical, obtuse, 6–15 lin.
         long, 2½–5 lin. broad ... ... (3) **myrtoides.**
      Leaves linear, obtuse, 9–14 lin. long, 1½–2 lin.
         broad ... ... (4) **myrtifolia.**
      Leaves linear-lanceolate or lanceolate, very
         acute, 5–10 lin. long ... ... (5) **Cneorum.**
   Perianth-lobes subulate :
      Leaves linear-lanceolate or elliptical, very acute,
         3–12 lin. long ... ... (6) **acutifolia.**
§ 2. STYLAPTERUS.—Ovary scabrous.   Style wingless.
   Style angular, topped by four flat oblong lobes :
      Flower 1½–2 lin. long ... ... (7) **fruticulosa.**
   Flower 3–4 lin. long :
      Leaves flattened, elliptical ... ... (8) **candolleana.**
   Leaves ericoid, channelled beneath :
      Bundle of short brown setæ in axil of leaf at
         each side ... ... (9) **ericoides.**
      Bundle of long woolly whitish-brown hairs in
         axil of leaf at each side ... ... (10) **barbata.**
   Style terete, angled at the base only :
      Leaves acicular.   Style topped by four upright
         ovate lobes ... ... (11) **ericifolia.**
      Leaves flat, ovate or elliptical.   Apex of style
         unknown ... ... (12) **dubia.**

1. **P. mucronata** (Linn. Sp. Pl. ed. ii. 162); a small erect much-branched undershrub, ½–1½ ft. high, branching mainly at the base; branches finely pubescent in the younger parts; lower leaves erect or spreading, ovate, acute, 4–6 lin. long, 3–4 lin. broad, upper often very close-ranked, erect-reflexed or spreading, cordate to sub-cordate or rarely ovate, acuminate, wide at the base, with prominent dorsal nerve, 2–4 lin. long, 1½–3 lin. broad, passing suddenly into bracts at the apex of the flowering shoot; stipules 2, very minute, auriculate, continued up into a subulate point, minutely pubescent; bracts shorter and wider than the leaves, sessile or contracted into

G 2

a short stalk, cordate-auriculate,·cuspidate, deciduous ; bracteoles linear-lanceolate or oblong, acuminate, with a membranous fimbriate margin, deciduous; perianth 3–4 lin. long ; tube ovate-cylindric, nearly twice as long as the limb ; lobes deltoid, acute ; ovary glabrous ; style topped by 4 flat oblong cruciform lobes with a wing decurrent from each. *Thunb. Prodr.* 30, *and Fl. Cap.* 149 ; *Meerb. Pl. Rar. Depict. t.* 51, *fig.* 2 (*by error fig.* 3 *in text* ; *very bad fig.*) ; *Vent. Hort. Malm. t.* 87 (*figs.* 3 *and* 4 *bad*) ; *Lam. Encycl.* vi. 539 ; *Lodd. Bot. Cab. t.* 1770 ; *A. Juss. in Ann. Sci. Nat.* 3me *sér.* vi. 22, *t.* 1, *fig.* 1 ; *A.DC. in DC. Prodr.* xiv. 484 (*excl. vars.*) ; *Gilg in Engl. & Prantl, Pflanzenfam.* iii. 6 *A*, 212, *and fig.* 73, S–V ; *Drège in Linnæa*, xx. 207, 208 (*also var. microphylla*) ; *Endl. Gen. Suppl.* iv. ii. 73 (*also var. Dregei*).

SOUTH AFRICA : without locality, *Burmann*, 50 ! 74 ! *Spielhaus* ! *Bergius* ! *Banks* ! *Sieber*, 60 ! *Roxburgh* ! *Forster* ! *Bowie*, 38 ! *Drège* ! *Zeyher*, 279 ! *and many others* !

COAST REGION : Paarl Div. ; Simons Berg, *Drège* ! Worcester Div. ; Dutoits Kloof, *Drège*, 8160 ! Cape Div. ; Table Mountain, *Thunberg* ! *Burchell*, 623 ! *Ecklon*, 622 ! 623 ! *Drège*, 104 ! *Zeyher*, 3726 ! 3727 ! *Bergius*, 330 ! *Ludwig* ! *Bolus*, 2915 ! *Bachmann*, 305 ! *Diels*, 104 ! Devils Peak, *Krauss* ! *Wilms*, 3601 ! *Wolley-Dod*, 619 ! Camps Bay, *Burchell*, 394 ! 894 ! False Bay, *Reynauld* ! Simons Bay, *Perrottet*, 557 ! *Boivin*, 557 ! *Wright* ! *MacGillivray*, 587 ! *Krauss* ! near Cape Town, *Knoop*, 15 ! *Burchell*, 442 ! *Castlenau*, 464 ! 465 ! Cape Flats, Doornhoogte, *Drège*, 78 ! Wynberg, *Burchell*, 863 ! Constantia, *Ecklon & Zeyher*, 4 ! 5 ! Stellenbosch Div. ; mountains near Stellenbosch, *Drège*, 8159 ! Hottentots Holland mountains, *Diels*, 1289 ! Caledon Div. ; without precise locality, *Verreaux* ! Palmiet River, *Penther*, 1596 ! Zwart River, *Drège*, 3726 ! Houw Hoek Mountains, *Drège*, 53 ! near Caledon, *Penther*, 1599 ! Zwartberg and near the hot springs, *Ecklon & Zeyher*, 6 ! mountains near Genadendal, *Drège* ! *Burchell*, 7612 ! 8618 ! Bredasdorp Div. ; Elim, *Schlechter*, 7635 ! Swellendam Div. ; without precise locality, *Verreaux* ! mountains near Swellendam, *Ecklon & Zeyher*. Riversdale Div. ; near Riversdale, *Rust*, 157 ! 589 ! between Little Vet River and Garcias Pass, *Burchell*, 6860 ! 6866 ! Garcias Pass, *Galpin*, 4533 ! *Phillips*, 304 ! Knysna Div. ; Plettenbergs Bay, *Mund* ! Uitenhage Div. ; Van Stadens River, *Ecklon*, 279 !

I have reduced vars. *microphylla* and *Dregei*, as there is a series connecting them with the type. They are probably forms of sterile localities.

Gueinzius' 185, marked "Port Natal," was probably collected at the Cape.

2. **P. ovata** (Eckl. & Zeyh. ex A. Juss. in Ann. Sc. Nat. 3me sér. vi. 22, name only) ; an erect much-branched undershrub, 1–2 ft. high ; branches glabrous ; leaves ovate or elliptical, subacute or obtuse, with prominent dorsal nerve, 3–6 lin. long, 1½–3 lin. broad, passing suddenly into bracts at the apex of the flowering shoot ; stipules 2, minute, auriculate, continued up into a subulate point : setæ 1–3 in the axil at each side of the leaf : bracts shorter and wider than the leaves, sessile or contracted into a stalk, cordate or ovate, widely auriculate, obtuse or acute, rarely acuminate, deciduous : bracteoles linear or spathulate, acute, with a membranous margin, deciduous ; perianth 3–4 lin. long ; tube ovate-cylindric, twice as long as the limb ; lobes deltoid, acute or obtuse ; ovary glabrous : style topped by 4 flat oblong cruciform lobes, with a wing decurrent from each lobe. *A.DC. in DC. Prodr.* xiv. 484. *P. mucronatæ*,

var. *Dregei affinis, A. Juss. l.c.   P. affinis, Endl. Gen. Suppl.* iv. ii.
73. *P. sp., E. Meyer in Drège, Zwei Pfl. Documente,* 84, 210.

VAR. β, **intermedia** (Endl. Gen. Suppl. iv. ii. 73); leaves longer and narrower
than in the type, elliptical or lanceolate, acute or obtuse, narrowing to the base,
5–7 lin. long, 2–2½ lin. broad. *A.DC. in DC. Prodr.* xiv. 484. *P. ovata, Drège
in Linnæa,* xx. 207. *P. myrtoides, var. multiflora, Krauss in Flora,* 1845, 77,
name and locality only. P. myrtilloides, Meisn. in Hook. Lond. Journ. 1843, 556,
not of Thunb. P. sp., E. Meyer in Drège, Zwei Pfl. Documente, 117, 210. *P.
sp., A. Juss. in Ann. Sc. Nat.* 3me. sér. vi. 22.

VAR. γ, **concinna** (E. L. Stephens); branches virgate, very erect; leaves erect,
very regularly arranged, giving the plant a very trim appearance, slightly shorter
and proportionately broader than in the type; flowers and bracts smaller than in
the type; perianth 2–2½ lin. long.

SOUTH AFRICA: without locality, *Burmann!* var. β: *Thunberg! Verreaux!
Masson!*

COAST REGION: Swellendam Div.; mountains near Swellendam, *Ecklon &
Zeyher,* 7! Mossel Bay Div.; mountain slopes, Robinson Pass, *Taylor in
MacOwan & Bolus, Herb. Norm. Austr.-Afr.,* 1070! Knysna Div.; Outeniqua
Mountains, *Bolus,* 1562! between Knysna and Plettenbergs Bay, *Pappe!*
Uitenhage Div.; Van Stadens Berg, *Drège,* 8157! *Ecklon & Zeyher,* 1! *Zeyher,* 3727
partly! Hallack in Herb. Galpin, 3011! *Zeyher,* 279! near Elands River, *Ecklon &
Zeyher!* Port Elizabeth ·Div.; at the upper part of Maitland River, *Burchell,*
4613! Var. β: Stellenbosch Div.; around Somerset West, *Ecklon & Zeyher,* 2!
Mossel Bay Div.; Attaqua's Kloof, *Drège,* 8156a! Uitenhage Div.; by streams
on Winterhoek Mountains, *Krauss,* 1213! Var. γ: George Div.; on the
Cradock Berg, near George, *Burchell,* 5968! 6017! Humansdorp Div.; Kruisfontein
Mountains, *Galpin,* 4535! Kromme River, in a high-lying valley, *Drège,* 8158!

P. myrtilloides, Thunb. Fl. Cap. 149, seems to be either var. *intermedia* or
P. myrtoides, but the description leaves this doubtful. The specimens named
P. myrtilloides in Thunberg's own herbarium are partly var. *intermedia,* partly
P. myrtoides and partly *P. acutifolia.*

3. **P. myrtoides** (Linn. f. Suppl. 122); an erect much-branched
undershrub, 1–2 ft. high; branches glabrous; leaves erect or
suberect, lanceolate or elliptical, obtuse, with a more or less
prominent dorsal nerve, 6–15 lin. long, 2½–5 lin. broad, passing
suddenly into bracts at the apex of the flowering shoot; stipules 2,
minute, auriculate, continued up into a subulate point; setæ 1–2,
in the axil at each side of the leaf; bracts shorter and wider than
the leaves, sessile or contracted into a stalk, outer lanceolate to
hastate, inner cordate, subacuminate, apex obtuse, deciduous;
bracteoles oblong-ovate or linear, obtuse, with a membranous
slightly fimbriate margin, deciduous; perianth 3–4 lin. long; tube
ovate-cylindric, twice as long as limb; lobes deltoid, obtuse; ovary
glabrous; style topped by 4 flat oblong cruciform lobes with a wing
decurrent from each. *Lam. Encycl.* vi. 539; *E. Meyer in Drège,
Zwei Pfl. Documente,* 117, 210, *name only; A. Juss. in Ann. Sc. Nat.*
3me *sér.* vi. 22, *excl. syn., name only; A. DC. in DC. Prodr.* xiv. 485.
P. myrtilloides, Thunb. Prodr. 30, *and Fl. Cap. ed. Schult.* 149?
P. ovata, var. colorata, Eckl. & Zeyh. ex Drège in Linnæa, xx. 207,
*name only. P. ovata, var. mucronata, Eckl. & Zeyh. ex A.DC. in
DC. Prodr.* xiv. 485.

SOUTH AFRICA: without locality, *Thunberg! Mund & Maire,* 22! *Burmann!
Drège!*

COAST REGION : Swellendam Div. ; by the Zondereinde River, *Zeyher* !
mountains of Swellendam, Tradouw, and Grootvaders Bosch, *Bowie* ! Mossel
Bay Div. ; without precise locality, *Zeyher*, 1949 ! George Div. ; mountains
near George, *Burchell*, 6026 ! 6069 ! *Bowie* ! *Galpin*, 4531 ! *Bolus*, 2439 ! *Drège*,
3548 ! *Ecklon & Zeyher* ! near Touw River, *Burchell*, 5723 ! 5780 ! Knysna Div. ;
mountain-slopes near Knysna, *Newdigate*, 31 ! and in *MacOwan, Herb. Austr.-
Afr.*, 1644 ! Uniondale Div. ; Lange Kloof, *Zeyher*, 18 ! Humansdorp Div. ;
Zitzikamma, *Pappe* ! Uitenhage Div. ; Elands River Mountains, *Zeyher*, 3723 !

The original description in Linn. f. Suppl. was probably made from a specimen
of Thunberg's, which I have not seen. I have here taken Drège's specimens as
types. It is difficult to separate small-leaved specimens of this species from
large-leaved forms of *ovata* ; typical plants of the two are quite distinct, but some
of the intermediate forms cannot be placed definitely in either species.

The varietal name "*mucronata*" quoted by De Candolle is probably due to a
slip of the pen and stands for "*colorata*," there being no specimen named var.
mucronata in the Berlin Herbarium.

4. **P. myrtifolia** (Endl. Gen. Suppl. iv. ii. 73, name only) ; an erect
much-branched undershrub, 1–2 ft. high ; branches glabrous or
rarely puberulous in the youngest parts ; leaves rather close-ranked
near the apex of the shoot, erect or suberect, linear, obtuse, with a
distinct dorsal nerve, 9–14 lin. long, 1½–2 lin. broad, passing
suddenly into bracts at the apex of the flowering shoot ; stipules 2,
very minute, auriculate, continued up into a subulate point ; setæ
several in the axil at each side of the leaf ; bracts shorter and
wider than the leaves, sessile or contracted into a short stalk, outer
lanceolate to hastate, inner ovate-auriculate or rhomboid, obtuse,
deciduous ; bracteoles oblong or oblong-obovate, obtuse or acute, with
a membranous slightly fimbriate margin, deciduous ; perianth 3–4 lin.
long ; tube ovate-cylindric, twice as long as limb ; lobes deltoid,
obtuse or acute ; ovary glabrous ; style topped by 4 flat oblong
cruciform lobes, with a wing decurrent from each. *A.DC. in DC.
Prodr.* xiv. 485. *P. Cneorum, Meisn. in Hook. Lond. Journ.* 1843, 556 ;
Krauss in Flora, 1845, 77. *Penæa an diversa a P. myrtoide ? A. Juss.
in Ann. Sc. Nat.* 3me sér. vi. 22, *name only*.

SOUTH AFRICA : without locality, *Roxburgh* !
COAST REGION : George Div. ; rather moist places on the mountains, *Drège*,
8155b (partly) ! in the forest and at its edge between bushes, *Drège*, 8155a ! in
damp places and by streams between Touw River and Kaymans River, *Burchell*,
5780 ! on the banks of the Notsinakama River, *Krauss* ! between Malgaten River
and Great Brak River, *Burchell*, 6143 ! Uitenhage Div. ; without locality,
Zeyher, 338 !

5. **P. Cneorum** (Meerb. Pl. Rar. Depict. t. 51, fig. 3 in plate,
2 in text) ; an erect much-branched undershrub, 1–2 ft. high ;
branches puberulous in the youngest parts ; leaves erect or
spreading, or rarely reflexed, linear-lanceolate or lanceolate, very
acute, dorsal nerve distinct, 5–10 lin. long, 1½–2½ lin. broad,
passing suddenly into bracts at the apex of the flowering shoot ;
stipules 2, minute, auriculate, continued up into a subulate
point ; seta solitary in the axil at each side of the leaf ; bracts

shorter and wider than the leaves, sessile or contracted into a short stalk, outer lanceolate or hastate, inner cordate or cordate-auriculate, acuminate-acute, deciduous; bracteoles linear or linear-lanceolate, acuminate-acute, with a membranous fimbriate margin, deciduous; perianth 3–4 lin. long; tube ovate-cylindric, twice as long as the limb; lobes deltoid, acute; ovary glabrous; style topped by 4 flat oblong cruciform lobes, with a wing decurrent from each. *Lam. Ill.* i. 317, *and Encycl.* vi. 541; *not of Krauss or Meisner; A.DC. in DC. Prodr.* xiv. 485.

SOUTH AFRICA: without locality, *Mund & Maire! Roxburgh! Gueinzius!*
COAST REGION : Stellenbosch Div. ; around Somerset West, *Ecklon & Zeyher*, 2 ! Swellendam Div. ; banks of the Zondereinde River, near Appels Kraal, Eksteens, and neighbouring mountains, *Zeyher!* Uitenhage Div., *Zeyher*, 62 !

6. **P. acutifolia** (A. Juss. in Ann. Sc. Nat. 3me sér. vi. 22); an erect much-branched undershrub, 1–2 ft. high ; branches glabrous ; leaves erect or spreading, linear-lanceolate or elliptical, acute at the base, very acute at the apex, dorsal nerve usually prominent, 12–13 lin. long, 2½–1 lin. broad, passing suddenly into bracts at the apex of the flowering shoot; stipules 2, very minute, auriculate, with the apex continued up into a subulate point ; setæ 2–3 in the axil at each side of the leaf ; bracts shorter and wider than the leaves, sessile or contracted into a stalk, ovate or rhomboid, acuminate-acute, deciduous ; bracteoles linear-lanceolate or lanceolate, acuminate-acute, with a membranous fimbriate margin, deciduous ; perianth 3–4 lin. long ; tube not or scarcely exceeding the limb ; lobes subulate, deltoid-acuminate ; ovary glabrous ; style topped by 4 flat oblong cruciform lobes, with a wing decurrent from each. *A.DC. in DC. Prodr.* xiv. 485.

SOUTH AFRICA : without locality, *Thunberg! Mund & Maire! Burmann!*
COAST REGION : George Div. ; Cradock Berg and other mountains near George, *Drège*, 8155b (partly) ! *Burchell*, 5913 ! 6018 ! *Alexander! Bowie! Galpin*, 4532 ! *Bolus*, 8691 ! Montague Pass, *Penther*, 1601 !

7. **P. fruticulosa** (Linn. f. Suppl. 121) ; a small, much-branched undershrub, ½–1½ ft. high ; branches erect or decumbent, glabrous or sparsely scabrous ; leaves equally distributed, erect or suberect, sessile or narrowed at the base into a very short stalk ; dorsal nerve obscure except at the extreme base ; lower leaves obovate, sometimes orbicular, obtuse to subacute ; upper obovate or orbicular to elliptical, ovate or rhomboidal, obtuse to subacute or acute, 3–1½ lin. long, 1¼–¾ lin. broad ; stipules 2, minute, awl-like or gland-like ; two groups of several short setæ in the axil at each side of the leaf ; setæ in the axils of the upper leaves, bracts and bracteoles usually persistent, setæ of the bracteoles forming a circlet on the apex of the pedicel ; bracts of the form and size of ordinary foliage leaves ; bracteoles 6 or 4 to each flower, outer pair semicoriaceous, inner scarious, ovate or obovate, cuspidate-acute, margin entire or fimbriate, 1½–2 lin. long, 1–1½ lin. broad, deciduous ; perianth

1½–2½ lin. long, subcampanulate ; tube cylindrical, slightly exceeding the limb; lobes deltoid-acute or acuminate ; ovary scabrous ; style topped by 4 flat oblong cruciform lobes, angled beneath each lobe. *Thunb. Prodr.* 30, *and Fl. Cap. ed. Schult.* 149 ; *Lam. Encycl.* vi. 540 ; *E. Meyer in Drège, Zwei Pfl. Documente,* 112, 210 ; *Krauss in Flora,* 1845, 77. *Stylapterus fruticulosus, A. Juss. in Ann. Sc. Nat.* 3ᵐᵉ *sér.* vi. 23 ; *A.DC. in DC. Prodr.* xiv. 486, *excl. var.*

SOUTH AFRICA: without locality, *Sparrman,* 19! 34! *in Herb. Linnæus*! *Burmann*! *Masson in Herb. Lambert,* 8686! *Lichtenstein*! *Roxburgh*! *Mund & Maire*! *Drège*! *Boivin*! *Zeyher,* 4750! *Wallich*! COAST REGION : Cape Div. ; Simons Bay, *Wright*! *Alexander*! *Boivin*! dunes in the neighbourhood of False Bay and Table Bay, *Schlechter,* 805! *Reynauld*! *Zeyher*! *Drège,* 112! *Bergius,* 327! Sandy Flats near Riet Vlei, *Thunberg*! Vygekraal Farm, *Wolley-Dod,* 978! Cape Flats, *Ecklon,* 624! *Wolley-Dod,* 1341! *Burchell,* 161! 718! *Bowie*! East slopes of Table Mountain near Constantia, *Zeyher,* 242! 1481! *Ecklon & Zeyher,* 3! Caledon Div. ; on stony hills at Grabouw and near Palmiet River, *Bolus,* 4189!

8. **P. candolleana** (E. L. Stephens in Kew Bulletin, 1911, 358); an undershrub, about 1 ft. high, " alpine, procumbent on the rocks " (*Niven*) ; lower parts with a reddish bark, defoliated ; leaves close-ranked on the upper parts, erect or more often spreading, narrowed at the base into a short petiole, elliptical, obtuse to acute, usually with a slight central furrow beneath, 3–4 lin. long, 1–1½ lin. broad, exstipulate ; a few minute brown setæ in the axils at each side of leaves, bracts and bracteoles ; bracts of the form and size of ordinary foliage leaves, persistent ; bracteoles 2, deciduous, rarely seen, leaving a scar on each side of the apex of the pedicel, elliptical or lanceolate, with a coriaceous centre and membranous margin, shorter than the bracts ; perianth narrowly cylindric-ovate, 3½–4 lin. long ; tube twice as long as the limb ; lobes deltoid, acuminate ; style topped by 4 flat oblong cruciform lobes, angled beneath each lobe.

SOUTH AFRICA : without locality or collector's name in *Herb. De Candolle*! COAST REGION : Stellenbosch Div. ; Stellenbosch, *Niven* (British Museum)!

9. **P. ericoides** (Endl. Gen. Suppl. iv. ii. 73); an erect much-branched undershrub, 1–2 ft. high, of heathlike habit; branches glabrous or sparsely scabrous ; leaves close-ranked, equally distributed, erect or spreading, sessile, linear-lanceolate, acute, acerose, with a narrow central channel beneath, dorsal nerve obscure, 2½–4 lin. long, ¼–½ lin. broad, exstipulate ; bundle of short brown setæ in the axil at each side of the leaf, and in the axil of bracts and bracteoles ; bracts of the form and size of ordinary foliage-leaves, persistent ; bracteoles 2 (?), very caducous, seen only in the young buds, leaving a scar on each side of the apex of the pedicel, linear-lanceolate, leathery, not channelled beneath, about ¼–½ the length of the leaves ; perianth cylindric-ovate, 3½–4 lin. long ; tube 8-ribbed, hardly exceeding the limb ; lobes linear-deltoid or subulate, acuminate ; ovary scabrous ; style topped by 4 flat oblong cruciform

lobes, angled beneath each lobe. *Gilg in Engl. & Prantl, Pflanzenfam.*
iii. 6 A, 212. *Stylapterus ericoides, A. Juss. in Ann. Sc. Nat.* 3me *sér.*
vi. 23 ; *A.DC. in DC. Prodr.* xiv. 487.

SOUTH AFRICA : without locality, *Roxburgh* (Herb. Delessert) ! *Niven* (British
Museum) !

10.· P. barbata (Endl. Gen. Suppl. iv. ii. 73) ; an erect much-
branched undershrub, 1–2 ft. high ; branches glabrous or sparsely
scabrous ; leaves close-ranked, equally distributed, erect or suberect,
sessile, lanceolate to subulate, with the margin much thickened
beneath, leaving a broadly lanceolate central channel, with a very
narrow upturned edge above, dorsal nerve obscure, 3–5 lin. long,
½–1 lin. broad, exstipulate with a bundle of long simple pale brown
hairs in the axil at each side of the leaf, falling after the leaves ;
bracts of the form and size of ordinary foliage leaves, and provided
with similar bundles of hairs, persistent ; bracteoles 2, rarely seen,
linear-lanceolate, leathery, not channelled beneath, one-third the
length of the leaves, caducous, leaving a scar on each side of the
apex of the pedicel ; perianth 2–3 lin. long ; tube cylindric-ovate,
8-ribbed, hardly exceeding the limb ; lobes deltoid-acute or acuminate ; ovary scabrous ; style with 4 flat oblong cruciform lobes,
angled beneath each lobe. *Gilg in Engl. & Prantl, Pflanzenfam.* iii.
6 A, 212. *Stylapterus barbatus, A. Juss. in Ann. Sc. Nat.* 3me *sér.* vi.
23 ; *A.DC. in DC. Prodr.* xiv. 486.

SOUTH AFRICA . without locality, *Roxburgh* (Herb. Delessert) !

11. P. ericifolia (Gilg in Engl. & Prantl, Pflanzenfam. iii. 6 A,
212, fig. 73, O) ; an erect much-branched undershrub, 1–1½ (?) ft. high,
of a heathlike habit ; branches glabrous or sparsely scabrous
continued beyond the flowering region ; leaves close-ranked, equally
distributed, erect or spreading, sessile, acicular, flattened-trigonal,
very acute, 5–6 lin. long, ¼–½ lin. broad, with a bundle of very
minute brown setæ or a minute pulvinus on each side of their axils,
exstipulate ; flowers lateral ; bracts of the form and size of ordinary
foliage leaves, with setæ in their axils, persistent ; pedicel 4 lin.
long, with a pair of scars (probably of very caducous bracteoles),
accompanied by minute setæ, at the base of the flower ; perianth
cylindric-ovate, 3–4 lin. long ; tube 8-ribbed, twice as long as the
limb ; lobes deltoid-acuminate ; ovary minutely scabrous ; style
filiform, expanded at the base into 4 short longitudinal ridges
alternating with the cells of the ovary, and at the apex into 4
upright ovate cruciform lobes (" 4-lobed stigma " of other authors),
also alternating with the cells of the ovary ; stigmatic surfaces as
in other species 4, rather minute, cushion-like, in the angles between
the lobes. *Brachysiphon ericæfolius, A. Juss. in Ann. Sc. Nat.* 3^{mc}
sér. vi. 24 ; *A.DC. in DC. Prodr.* xiv. 487.

SOUTH AFRICA : without locality, *Roxburgh* (Herb. Delessert) ! *Niven* (British
Museum) !

12. **P. dubia** (E. L. Stephens in Kew Bulletin, 1910, 237); erect branched undershrub, about 1 ft. high (?); branches minutely scabrous; leaves erect, sessile, ovate or elliptical, acute or mucronate-acute, with dorsal nerve obscure except at the extreme base, $2\frac{1}{2}$–$3\frac{1}{2}$ lin. long, $1\frac{1}{2}$–2 lin. broad, exstipulate, with 1–2 minute brown setæ at each side of the axil; flowers lateral, sessile; bracts of the form and size of ordinary foliage leaves; bracteoles 2 pairs at the base of the flower; outer pair at right angles to the leaf, semicoriaceous, concave, oblong-acute or mucronate; inner pair parallel with the leaf, membranous, ovate, acuminate-acute; tube of persistent perianth (at fruiting stage) 3 lin. long; lobes erect to one-third of their length, then reflexed, deltoid, acute, 1 lin. long; ovary in fruit slightly rough, with ill-defined projections; style filiform, scabrous, minutely tuberculate, with 4 small obscure horizontal ridges at its base, which alternate with the cells and are sometimes bifid by a continuation of the furrow between them; apex of style not seen; seeds 2 in each cell, basal.

SOUTH AFRICA : without locality, *Roxburgh* (Herb. Delessert)!

The affinities of this species are doubtful, the solitary specimen being in the fruiting stage and showing neither the stamens nor the upper half of the style. I have placed it next *P. ericifolia* because of the projections at the base of the style, which may correspond to the more definite ridges at the base of the style in that species. Its general habit is that of a small-leaved *P. ovata*, but in its inflorescence it resembles the three preceding species. In the number and form of its bracts it approaches *P. fruticulosa*, and its perianth is that of *P. ericifolia* except for the reflexed lobes—a point which, however, is probably not important, as the perianth-lobes in any species of *Penæa* may occasionally reflex in the fruiting stage.

Imperfectly known species.

13. **P. macrosiphon** (Gandog. in Bull. Soc. Bot. France, lx. 420, 1913); an undershrub about 6 in. high; stems simple; leaves imbricate, straight not spreading, ovate, cuspidate, rounded (not sheathing) at the base, about 3 lin. wide, midrib prominent beneath; flowers about 5 lin. long, pedicellate.

COAST REGION : Cape Div. ; Retrait near Cape Town, *Bonomi.*

Allied to *P. mucronata*, Linn., from which it is distinguished by its larger size and erect, not spreading, leaves.

II. BRACHYSIPHON, A. Juss.

Perianth-tube cylindrical or ovate-cylindric, nearly twice as long as the limb or subequal to it, longer than or subequal to the bracts; lobes erect or reflexing only in the faded flower, reduplicate-valvate in bud. *Stamens* shorter than the perianth-lobes; filaments very short; anthers as long as the filaments, with cells adnate on a thick lobed connective and usually shorter than it. *Ovary* and *style* at first as long as the perianth-tube, later surpassing it; ovary four-angled, glabrous or minutely scabrous; ovules 2 or 4 in each cell, basal, erect (in *B. rupestris* occasionally

4 in each cell, inserted half-way down on a median placenta, 2 erect
and 2 pendulous) ; style filiform, terete or cylindrical ; stigma small,
obscurely 4-lobed.

Low much-branched undershrubs ; branches quadrangular or subquadrangular
in the younger parts, becoming terete lower down, upper more or less thickly
beset with leaves, lower often defoliated ; flowers in scanty bracteate racemes,
in the axils of the upper leaves, usually gathered in many-flowered terminal
clusters, rarely solitary ; bracts leaf-like, coloured, all fertile or the lower sterile ;
bracteoles 2, as long as or nearly as long as the bracts, but narrower ; inflorescence tinged to a rosy red.

DISTRIB.   Species 5, endemic.

Erect undershrubs, 1–2 ft. high :
  Leaves glabrous ; bracts all fertile :
    Leaves sessile, broadly ovate to obovate :
      Perianth-tube ovate-cylindric ; ovary minutely
        scabrous ...   . (1) **imbricatus.**
      Perianth-tube narrow cylindrical ; ovary glabrous... (2) **acutus.**
    Leave shortly petiolate, narrowly spathulate   ... (3) **Mundii.**
  Leaves denticulate-ciliate ; lower bracts sterile   ... (4) **speciosus.**
Dwarf, partly decumbent undershrub, under 6 in. high (5) **rupestris.**

1. **B. imbricatus** (A. Juss. in Ann. Sc. Nat. 3me sér. vi. 25,
t. 2, fig. 3) ; an erect much-branched undershrub, 1–1½ ft. high ;
branches glabrous, upper branches and ' leaves sometimes with a
hoary bloom, lower defoliated ; leaves sessile, widely ovate to
elliptical or obovate, acute, glabrous, upper leaves erect, imbricate,
lower suberect or spreading, almost or quite as broad as long, 4–5 lin.
long, 3–4 lin. broad, with a minute bundle of dark-brown setæ at each
side of the axil ; stipules when present minute, subulate, incurved ;
flowers in groups of 2 to many at the top of the branches ; pedicels
less than 1 lin. long ; bracts all fertile, obovate, acute to mucronate,
narrower and slightly shorter than the leaves, tinged with red ;
bracteoles oblong-linear, acute or obtuse, denticulate-ciliate or
entire, as long as the bract, inserted on the pedicel at or near the
base of the flower, caducous ; perianth 4 lin. long, pale to deep
pink ; tube tinged with red (*Stephens*), ovate-cylindric, scarcely
longer than the limb ; lobes widely ovate, acute or obtuse ; stamens
slightly more than 1 lin. long, cells much shorter than the
connective ; ovary minutely scabrous.   *A.DC. in DC. Prodr.* xiv.
487.   *B. fucatus, Gilg in Engl. Jahrb.* xviii. 520, *figs.* L–N, *and
in Engl. & Prantl, Pflanzenfam.* iii. 6 A, 212, *fig.* 73, L–N.   *Penæa
imbricata, Grah. in Bot. Mag. t.* 2809.   *P. squamosa, Linn. Sp. Pl. ed.*
ii. 162 ; *Thunb. Fl. Cap. ed. Schult.* 149.   *P. Sarcocolla, Berg. Descr.
Fl. Cap.* 35 ; *Thunb. Fl. Cap. ed. Schult.* 150.   *P. fucata, Linn.
Mant. Alt.* 199 ; *Lam. Ill.* i. 317, *t.* 78, *fig.* 1.   *P. fuscata, Lam.
Encycl.* vi. 540.

SOUTH AFRICA : without precise locality, *Burmann ! Roxburgh ! Mund !
Mund & Maire ! Verraux ! Sieber ! Martin ! Ludwig ! Lalande ! Pappe !*
  COAST REGION : Cape Div. ; among the rocks on the summit of Table Moun-
tain, *Bergius ! Ecklon,* 621 ! *Bachmann,* 311 ! *Krauss ! Wilms,* 3602 ! *Cooper,*

3594! *Zeyher,* 4743! *Diels,* 36! near Cape Town, *Burchell,* 8403! Kloof and
other high places on Table Mountain, *Ecklon & Zeyher,* 12! Devils Peak, *Drège,*
239! *Bolus in Herb. Norm. Austr.-Afr.,* 362! *Bolus,* 3293! *Wolley-Dod,* 1327!
Wilms, 3602a!

2. **B. acutus** (A. Juss. in Ann. Sc. Nat. 3me sér. vi. 25); an
erect much-branched undershrub, about 1 ft. high; branches
minutely scabrous, upper corymbose, lower defoliated; leaves
imbricate towards the apex of the branches, sessile, widely ovate
to elliptical or obovate, acute to acuminate, glabrous, 3–3½ lin.
long, 1½–2½ lin. broad, with 1–2 short setæ sometimes present at
each side of the axil; stipules when present minute, wart-like or
subulate; flowers in groups of 1–6 at the top of the branches;
bracts all fertile, obovate, acuminate, somewhat smaller than the
leaves, tinged with red; bracteoles lanceolate, linear-spathulate or
linear, acuminate, slightly longer than the bracts; perianth 4½–8
lin. long, pink, tube more deeply coloured than the lobes; tube
narrow cylindrical, in flowering longer than the bracts, 3–5 lin.
long, less than 1 lin. in diam.; lobes ovate, acute to mucronate,
1½–3 lin. long, 1–2 lin. broad; stamens ½–¾ lin. long, cells as long
as the connective; filaments ligulate; ovary glabrous. *A.DC. in
DC. Prodr.* xiv. 488. *Penæa acuta, Thunb. Fl. Cap. ed. Schult.* 150.
Sarcocolla acuta, Kunth in Linnæa, v. 678.

SOUTH AFRICA : without precise locality, *Lehmann* ! *Masson* ! *Bowie* !
COAST REGION : Caledon Div.; Zwart Berg, *Alexander* ! *Ecklon & Zeyher* !
Zeyher, 3728! *Pappe* ! hills between Caledon and Elim, *Bolus,* 8485 ! Klein
River Mountains, *Ecklon & Zeyher,* 14 ! Mountains near Onrust River, *Schlechter,*
9489 ! Swellendam Div. ; without precise locality, *Harvey* !

3. **B. Mundii** (Sond. in Linnæa, xxiii. 102); a small branched
undershrub ; branches glabrous, upper with a hoary bloom, lower
defoliated ; leaves thick, slightly channelled dorsally, spathulate,
glabrous, narrowed at the base into a short petiole, erect or recurved,
1½–2 lin. long, blade elliptical or ovate, acute or shortly mucronulate, up to 1 lin. broad ; petiole channelled above, ½ to 1 lin. long,
with several minute setæ at each side of the axil, exstipulate ;
flowers solitary or 2–3 at the top of the branch ; pedicels 1–1½ lin.
long ; bracts all fertile, obovate, a little longer than the leaf,
tinged with red ; bracteoles inserted half-way up the pedicel,
slightly shorter than the bracts ; perianth pink, becoming red
(*Sonder*), 4½–5½ lin. long ; tube broadly cylindrical, three times as
long as the limb ; lobes ovate, acute, 1–1½ lin. long ; stamens less
than 1 lin. long ; anthers broadly cordate, cells shorter than the
connective ; ovary glabrous. *A.DC. in DC. Prodr.* xiv. 488.

COAST REGION : Swellendam Div. ; mountains near Swellendam, *Mund* (Cape
Herbarium) !

4. **B. speciosus** (Sond. in Linnæa, xxiii. 103) ; an erect undershrub, 1–2 ft. high, branching mainly from the base; branches long,
glabrous ; leaves erect, imbricate, sessile, broadly obovate to cuneate
or rotund, rounded or emarginate at the apex, cartilaginous or

denticulate-ciliate at the margin, with a dorsally-projecting mucro tipped by a gland, represented by a pit in the older leaves, $2\frac{1}{2}$–4 lin. long, 3–5 lin. broad, passing gradually into bracts at the top of the flowering shoots, bearing several minute setæ in their axils, which pass in the axils of the bracts into long branched hairs; stipules, when present, subulate, very short or rarely up to 1 lin. long; flowers 5 to many in showy terminal heads $\frac{3}{4}$. to $2\frac{1}{2}$ in. in diam.; pedicels 1–1$\frac{1}{2}$ lin. long; bracts fimbriate, with a gland-tipped mucro like that of the leaves, lower sterile, very widely obovate and leaf-like, median sterile, broadly cuneate, membranous along the margin, 6–9 lin. long, 5–7 lin. broad, upper fertile, membranous, narrowly cuneate to oblong, 7–8 lin. long, 2–4 lin. broad; bracteoles inserted half-way up the pedicel, oblong or linear, fimbriate, tipped with a glandular mucro less prominent than that of the leaves, nearly or quite as long as the bracts, 1–1$\frac{1}{2}$ in. broad; perianth flesh-coloured (*Bolus*), 8–11 lin. long; tube cylindrical, dilated at the throat, nearly twice as long as the limb, tinged with red; lobes ovate, acute or obtuse, 3–4 lin. long, 1$\frac{1}{2}$–2$\frac{1}{2}$ lin. broad, reflexed when fading; anthers sessile or subsessile, cells slightly shorter than the connective; ovary glabrous. *A.DC. in DC. Prodr.* xiv. 488; *Gilg in Engl. & Prantl, Pflanzenfam.* iii. 6 A, 212.

SOUTH AFRICA: without locality, *Ecklon & Zeyher* ! *Bowie* !
COAST REGION: Caledon Div.; stony places on the Klein River Mountains, *Ecklon & Zeyher*, 11! sandy places on the mountains near Hemel en Aarde, *Zeyher*, 3725! hills between Caledon and Elim, *Bolus*, 8484! top of a mountain near Hermanus, *Bolus*, 9825! Bredasdorp Div.; mountains near Elands Kloof, *Schlechter*, 9745!

5. **B. rupestris** (Sond. in Linnæa, xxiii. 101); a dwarf under-shrub; partly decumbent, of an alpine rhododendroid habit, under 6 in. high; branches arising from a short thick rootstock, glabrous, thickly beset with leaves above, defoliated below, 2–4 in. long; leaves very close-ranked, imbricate, erect or spreading, sessile or very shortly petiolate, obovate, obtuse, with the dorsal nerve not evident, 4–6 lin. long, 2–3 lin. broad; petiole $\frac{1}{2}$ lin. long, with several very minute setæ in its axil; stipules when present minute, subulate; flowers in 2 4-flowered terminal racemes, each flower in the axil of a leaf-like bract; racemes sometimes overtopped by a subterminal branch; pedicels about 1 lin. long, with two small scars (probably of caducous bracteoles) half-way down, each scar bearing axillary setæ; bracts rather smaller than the leaves, tinged with red; perianth about 5 lin. long, rosy-red (*Sonder*); tube oblong, 8-ribbed, a third longer than limb; lobes ovate; stamens $\frac{3}{4}$ lin. long, cells shorter than the connective; ovary glabrous; ovules usually 2 in each cell, erect, sometimes 4 in each cell, inserted half-way down on an axile placenta, 2 erect and 2 pendulous, raphe extrorse in all; seeds erect or pendulous, according to their position as ovules. *A.DC. in DC. Prodr.* xiv. 488.

COAST.REGION: Caledon Div.; stony places on Klein River Mountains, *Zeyher* ! clefts of rocks on the seashore, near Hermanus, *Bolus*, 9827!

III. SARCOCOLLA, Kunth.

Perianth-tube cylindrical, about three times as long as the limb ; lobes reduplicate-valvate and erect in bud, becoming reflexed as the flower opens. *Stamens* exserted at the top of the perianth-tube, projecting for their whole length, erect, about two-thirds the length of the limb ; anthers oblong, as long as the filament or slightly shorter, cells subequal to the connective, margins of the valves entire. *Ovary* and *style* at first as long as the perianth-tube, later projecting beyond it ; style slender, terete ; stigma capitate, 4-lobed ; ovules 2 or 4 in each cell, inserted at the base, erect.

Much-branched undershrubs, branching mainly from the base ; branches erect or semi-decumbent, glabrous ; upper branches and leaves often covered with a whitish waxy bloom secreted by the glandular stipules and by the apical gland of the leaf, lower branches defoliated ; leaves flat, imbricated, sessile, usually tipped with a more or less distinct gland, passing into bracts at the apex of the flowering shoot ; stipules small, pyramidal or auriculate, often glandular, becoming black ; minute brown setæ present in the axils of the leaves ; flowers solitary and terminal, or gathered into terminal heads ; bracts imbricate, coloured, flat or more or less convex in the younger stages, becoming markedly convex as the fruit swells, resinous, lower sterile, tipped with a gland less prominent than that of the leaf, with glandular stipules like those of the leaf at each side of their axil ; bracteoles narrowly linear or linear-spathulate, acute, acuminate or cuspidate-acute, ciliate or entire, shorter than the bracts, inserted on the perianth at or near its base ; bracts and bracteoles yellowish or tinged with pink ; perianth with a yellow tube and rosy-pink limb.

DISTRIB. Species 3, endemic.

Flowers in terminal heads :
  Leaves 5–8 lin. long ; lower and median bracts 5–10
    lin. broad   ...   . (1) squamosa.
  Leaves 2–5 lin. long ; lower and median bracts 1½–4
    lin. broad   ...   . (2) minor.
Flowers terminal, solitary ...   . (3) formosa.

1. S. squamosa (Endl. Gen. Suppl. iv. ii. 74) ; an undershrub, 1½–2 ft. high ; leaves ovate to elliptical or obovate, obtusely mucronate, with a glandular tip, varying much in size, sometimes as broad as long, 5–8 lin. long, 3–8 lin. broad ; flowers gathered into terminal heads of about half-a-dozen, sometimes less, very exceptionally 1 or 2 ; lower bracts sterile, leaf-like, larger than the leaves ; median bracts fertile, broadly obovate to cuneate, entire or ciliate, often emarginate and mucronate at the apex, larger than the leaves and longer than the lower bracts, 5–10 lin. broad ; upper bracts fertile, cuneate, membranous, ciliate, often emarginate and mucronate at the apex, longer and narrower than the median ; perianth-tube 9–15 lin. long, 1½–3 lin. in diam. ; lobes ovate, obtuse or acute, 3–6 lin. long, 2½–4 lin. broad. *A.DC. in DC. Prodr.* xiv. 489. *Penæa squamosa, Linn. Mant. Alt.* 331 : *Bot. Reg. t.* 106 ; *Lam.*

Encycl. vi. 540 ; *A. Juss. in Ann. Sc. Nat.* 3*ᵐᵉ sér.* vi. 25.   *P. Sarco-colla, Linn. Sp. ed.* ii. 162 ; *Lam. Encycl.* vi. 538.   *P. tetragona, Berg. Descr. Pl. Cap.* 36.   *Sarcocolla Linnæi, A. Juss. in Ann. Sc. Nat.* 3*ᵐᵉ sér.* vi. 25, *t.* 2, *fig.* 4.

SOUTH AFRICA: without locality, *Bergius*! *Burmann,* 160! *Thunberg*! *Roxburgh*! *Niven*! *Masson*! *Thom,* 362! *Robertson*! *Bowie*!
COAST REGION : Cape Div. ; Table Mountain, *Alexander*! South-west slopes of Devils Peak, *Mund*! *Krauss*! *Wilms,* 3603! 3604! Lions Head, at the foot of the mountain and by Green Point, *Drège*! Muizenberg Mountains, *Zeyher,* 278! 3722a! *MacOwan & Bolus,* Herb. *Norm. Austr.-Afr.,* 387! *Bolus,* 2916! *Wilms,* 3603! *Wolley-Dod,* 572! 1121! *Scott-Elliot,* 1194! Nordhoek, *MacOwan,* 2384! Stellenbosch Div. ; Hottentots Holland Mountains, *Lichtenstein*! *Mund*! Lowrys Pass, *Burchell,* 8188! *Drège*! *Rogers*! Caledon Div. ; mountains near Grietjes Gat, between Lowrys Pass and the Palmiet River, *Ecklon & Zeyher,* 8! 9! Houw Hoek Mountains, *Burchell,* 8031! 8099! 8147! *Schlechter,* 7369! mountains of Baviaans Kloof, near Genadendal, *Burchell,* 7843! *Verreaux*! above Van Rynevelds Valley, *Diels,* 1398! mouth of the Klein River, *Zeyher,* 3722! near Caledon,*Kuntze*!

2. **S. minor** (Zeyh. ex A.DC. in DC. Prodr. xiv. 489) ; an under-shrub, ¾–1½ ft. high ; leaves obovate, obtusely mucronate, with the gland at the tip less distinct than in *S. squamosa,* often not developed, 2–5 lin. long, 1½–4 lin. broad ; inflorescence as in *S. squamosa,* but usually with fewer (occasionally one or two) flowers, and bracts and perianth narrower and slightly shorter ; bracts 1½–4 lin. broad ; perianth-tube 6–12 lin. long, 1½–2½ lin. in diam. ; lobes 3 4½ lin. long, 1½–2½ lin. broad.   *S. fucata, var. minor, Eckl. & Zeyh. ex A.DC. l.c.*

SOUTH AFRICA: without locality, *Burmann*! *Masson*! *Nelson*! *Lalande*!
COAST REGION: Cape Div. ; Camps Bay, *Schlechter,* 161! Glencairn, *Stephens*! mountains above Simons Bay, *Ecklon & Zeyher*! *Wright*! *Milne,* 177! *McGillivray,* 595! *Elliot,* 16! sandy slopes beyond Millers Point, *Wolley-Dod,* 2287!

3. **S. formosa** (A. Juss. in Ann. Sc. Nat. 3ᵐᵉ sér. vi. 25, t. 2, fig. 4, f.) ; an undershrub, 1–1½ ft. high ; leaves obovate or elliptical, obtusely mucronate, with the gland at the tip less distinct than in *S. squamosa,* sometimes absent, 3½–5 lin. long, 2½–3½ lin. broad ; flowers solitary, terminal, subtended by 3–6 pairs of decus-sate bracts ; lower bracts leaf-like, obovate, 3½ lin. long, 3 lin. broad ; median bracts longer, leaf-like, oblong or oblong-obovate, partly ciliate or entire, acute or with an obtuse or acute mucro, 4½–10 lin. long, 2–3½ lin. broad ; upper bracts longer, oblong-obovate, oblong-obcordate, or linear-spathulate, semicoriaceous to membranous, acute to acuminate, or with an acute or semiacute mucro, margin partly ciliate or entire, usually infolded on each side of the recurved apex, giving it an acuminate appearance, 6½–10 lin. long, 1–3 lin. broad ; perianth-tube 10–16 lin. long, 1½–3 lin. in diam. ; lobes acute or acuminate, 3½–6½ lin. long, 1½–4 lin. broad. *A.DC. in DC. Prodr.* xiv. 489.

SOUTH AFRICA : without locality, *Bergius*! *Chamisso*! *Sonnerat*!
COAST REGION : Cape Div. ; at the foot of Lions Head and by Green Point,

Drège! Slopes above Kamps Bay, *Dümmer*, 118 ; Mountains above Simons Bay,
Elliott! *Boivin*, 555 ! *Wright*, 128 ! *Ecklon & Zeyher*, 10! *Alexander*! Mountains
above Klaver Vley and Smitswinkel, *Wolley-Dod*, 1492 !

IV. GLISCHROCOLLA, Endl.

Perianth-tube oblong, 4-ribbed, about three times as long as the
limb ; lobes erect, becoming reflexed in the faded flower, reduplicate-
valvate in bud. *Stamens* shorter than perianth-lobes ; filaments
very short ; anthers cordate-ovate ; cells oblong, longer than the
connective, with a membranous margin. *Ovary* and *style* as long
as the perianth, glabrous ; ovules 4 in each cell, inserted half-way
down on an axile placenta, 2 erect and 2 pendulous, raphe extrorse
in all ; style filiform ; stigma obsoletely 4-lobed (*A. Jussieu*). *Fruit*
unknown.

Probably a low shrub ; branching dichotomous ; branches glabrous, obscurely
4-angled in the younger parts by the decurrent leaf bases, becoming terete lower
down ; leaves ovate, close-ranked, large for the order, passing into bracts at the
apex of the flowering shoot ; flowers approximated at the tips of the branches ;
peduncles dichotomous ; pedicels very short.

DISTRIB.   Species 1, endemic.

1. G. lessertiana (A.DC. in DC. Prodr. xiv. 490) ; leaves erect,
imbricate, hardly more than 1 lin. apart, ovate, obtuse, with a
narrow yellowish margin and prominent dorsal nerve, $\frac{3}{4}$–1 in. long,
$\frac{1}{2}$ in. broad, passing into bracts at the apex of the flowering shoot ;
stipules (when present) 2, very minute, awl-like ; a minute crested
scale at each side in the axil of the leaf ; bracts leaf-like coloured,
sterile bracts rather larger than the leaves, fertile bracts longer but
narrower than the leaves ; bracteoles 2, coloured, coriaceous, linear-
oblong, slightly shorter than bracts, about 1 lin. broad ; perianth
about 1$\frac{1}{2}$ in. long ; tube four times as long as the limb, nerves
alternating with the lobes ; lobes ovate, obtuse, entire or sinuate, 3–4
lin. long ; stamens half as long as the lobes ; filaments shorter than
the anther, thick ; connective tuberculate. *Gilg in Engl. & Prantl*,
Pflanzenfam. iii. 6 A, 212, *fig.* 73, H–K. *Sarcocolla lessertiana*,
A. Juss. in Ann. Sc. Nat. 3me *sér.* vi. 26. *Penæa formosa*, *Thunb.*
Fl. Cap. ed. Schult. 149 (*from description and original specimen in*
Thunberg's herbarium).

COAST REGION : Stellenbosch Div. ; French Hoek Mountains, *Masson* !

V. ENDONEMA, A. Juss.

Perianth-tube long, cylindrical, about four times as long as the
limb ; lobes thickened to form a central ridge on the inner side,
erect, becoming reflexed as the flower fades, valvate in bud. *Stamens*
nearly as long as the perianth-lobes or slightly longer ; filaments

folded in bud, becoming erect as the flower opens, subulate above the fold, ligulate below, almost as long as the anthers or longer ; anthers ovate, inflexed in bud, then erect and introrse ; cells rather shorter than the connective, dehiscing marginally. *Ovary and style* as long as the perianth, glabrous ; ovules 4 in each cell, inserted half-way down on an axile placenta, 2 erect and 2 pendulous, raphe extrorse in all ; style filiform ; stigma obscurely 4-lobed. *Fruit and seeds* of order ; seeds erect or pendulous according to their position when ovules.

Small erect shrubs or undershrubs ; branches quadrangular or subquadrangular in the younger parts, becoming terete lower down, upper more or less thickly beset with leaves, lower often defoliated ; flowers lateral, solitary or apparently solitary in the axils mainly of the upper leaves ; branches continued beyond the flowering region ; bracts (bracteoles) in two or three pairs below each flower.

DISTRIB. Species 2, endemic.

Leaves broad, flat ; bracts oblong ; anthers slightly
    longer than filaments ... ... (1) **Thunbergii.**

Leaves narrow-linear, revolute ; bracts rotund,
    caducous ; anthers shorter than filaments ... (2) **retzioides.**

1. **E. Thunbergii** (A. Juss. in Ann. Sc. Nat. 3me sér. vi. 27) ; a small, loosely branched shrub ; branches long, glabrous, or shortly puberulous and subquadrangular in the younger parts ; leaves erect or spreading, usually imbricate, ovate or elliptical, obtuse, with the central nerve slightly prominent beneath and a few slightly-marked lateral oblique veins, 6–10 lin. long, 3–7 lin. broad , exstipulate ; with a minute crested scale usually present in the axil at each side in both leaves and bracts ; flowers apparently solitary in the axils of the upper leaves, in reality each terminating a much reduced raceme, whose very short peduncle bears three pairs of decussate bracts, which are either sterile or enclose an axillary bud ; this bud usually remains dormant, but (very exceptionally) may grow out into another short peduncle, bearing a terminal flower and furnished with 2, 1 or no pairs of bracts, according to whether it was borne in the axil of the first, second or third pair on the original peduncle ; bracts 8, decussate, coriaceous to scarious, obovate to oblong or ovate, obtuse to acute, golden, shorter than the perianth-tube, each pair longer than the one below it ; perianth golden, projecting beyond the leaves, about 1 in. long ; tube three times as long as the limb ; lobes tinged with red, oblong or ovate, obtuse or acute, 3–4 lin. long ; stamens about two-thirds the length of the perianth-lobes ; anther slightly longer than the filament ; capsule 6 lin. long ; seeds 2 lin. long. *Sond. in Linnæa,* xxiii. 104 ; *A.DC. in DC. Prodr.* xiv. 491. *E. lateriflora, Gilg in Engl. & Prantl, Pflanzenfam.* iii. 6 A, 210. *Penæa lateriflora, Linn. f. Suppl.* 122 ; *Murr. Syst. Veg. ed.* xiv. 154 ; *Thunb. Prodr.* 30, *and Fl. Cap. ed. Schult.* 150, *and Naturf. Magaz. Berlin,* i. *t.* 3, *fig.* 2, *ex A. Juss. l.c.* ; *Lam. Encycl.* vi. 540 ; *Eckl. & Zeyh. ex Sond. in Linnæa,* xxiii. 104. *Geissoloma lateriflorum, Drège in Herb. Berol. ex Sond. l.c.*

98   PENÆACEÆ (Stephens).   [*Endonema.*

SOUTH AFRICA : without precise locality, *Burmann* ! *Niven* ! *Bojer* ! *Roxburgh* ! *Zeyher*, 176 !
COAST REGION : Caledon Div. ; Zoetemelks Valley, *Thunberg* ! at or near Genadendal, *Ecklon & Zeyher*, 17 ! *Bolus*, 5396 ! *Schlechter*, 9837 ! 10305 ! Mountains of Baviaans Kloof, near Genadendal, *Burchell*, 7782 ! Swellendam Div. ; Grootvaders Bosch, *collector unknown, in the Cape Herbarium* ! Uitenhage Div. ; Van Stadens Berg, *Ecklon & Zeyher* !

2. **E. retzioides** (Sond. in Linnæa, xxiii. 103) ; a much-branched glabrous undershrub or small shrub ; branches corymbose, 4-angled above, becoming terete lower down ; lower branches always defoliated ; leaves erect, rigid, much longer than the internodes, narrowly linear, acute, revolute at the margin, minutely puncticulate on the upper surface, with the nerve prominent beneath, 9–15 lin. long, 1 lin. broad, exstipulate, with 1 or 2 short setæ in the axils at each side of the leaves and bracts ; setæ of bracts persistent as a circlet on the apex of the pedicel ; flowers solitary on short pedicels in the axils of the upper leaves, usually only 2 or 4 on each twig ; flowers and bracts 4, golden, decussate on the pedicel at the base of the flower, sterile or very exceptionally with an arrested minute axillary bud, convolute, suborbicular, lower with a short acumen, 1–2 lin. long, upper with usually rounded apex, sometimes emarginate and mucronate, 2½–3 lin. long, caducous or falling very soon after the flower opens ; perianth golden, up to 1¾ lin. long ; tube about four times as long as the limb ; lobes deeply tinged with red, lanceolate, 4–5 lin. long ; stamens as long as the perianth-lobes or slightly longer ; anther shorter than the filament ; capsule 8 lin. long ; seeds 2–2½ lin. long. *A.DC. in DC. Prodr.* xiv. 491 ; *Gilg in Engl. & Prantl, Pflanzenfam.* 210, *fig.* 73, *A–G. Sarcocolla retzioides, Eckl. & Zeyh. ex Sond. l.c.* 104.

SOUTH AFRICA : without locality, *Ecklon & Zeyher* !
COAST REGION : Caledon Div. ; Houw Hoek, among stones on the mountains, *Schlechter*, 7337 ! mountains near Genadendal, *Roser* ! *Burchell*, 7712 ! *Pappe* ! *Tyson* ! on the summits of the Zwartberg Range near Tygerhoek. *Vigne in MacOwan, Herb. Austr.-Afr.*, 1646 ! Swellendam Div. ; banks of the Zondereinde River, near Appels Kraal, and on the neighbouring mountains, *Zeyher*, 3724 ! *Ecklon & Zeyher*, 15 !

ORDER CXIX. A. **GEISSOLOMACEÆ.**

(By E. L. STEPHENS.)

Flowers hermaphrodite, regular. *Perianth* inferior, persistent, gamosepalous, 4-partite to the base ; segments ovate, mucronate, imbricate in bud, with the lateral segments internal, one enfolding the other, and the external segments at right angles to the axis, the margin of the inner enfolding the outer. *Stamens* 8, perfect, inserted at the base of the perianth, 4 opposite the perianth-lobes

and 4 alternating with them, the latter slightly shorter; filaments
free, ligulate, rather shorter than the perianth; anthers many
times shorter than the filament, ovoid, bilobed at the base, erect,
versatile; connective scarcely manifest; cells subintrorse, dehiscing
longitudinally. *Ovary* free, sessile, superior, syncarpous, 4-carpellary,
4-celled; cells alternating with the lobes of the calyx, each narrowly
winged; style terminal, formed by the continuation and union of
the wings of the cells, pyramidal-acuminate, easily separable into
4 parts, and terminated by 4 very small stigmas, which alternate
with the calyx-lobes. *Ovules* 2 in each cell, collateral, pendulous
from the apex, anatropous. *Fruit* a capsule, 4-celled, dehiscing by
4 longitudinal sutures running down the wings of the cells. *Seeds*
by abortion solitary in each cell, ovate, subcompressed, smooth,
hanging from the top of the cell by a short funicle, furrowed at the
upper extremity between the hilum and the micropyle, and slightly
inflated on each side of the furrow; albumen fleshy; embryo central,
straight, almost as long as the albumen; cotyledons long, linear,
fleshy; radicle superior, short, cylindrical, obtuse; plumule inconspicuous.

A low shrub; leaves decussate, sessile, entire, penninerved, ovate, exstipulate,
silky-pilose in bud, soon glabrous; flowers apparently solitary, in reality in short
much reduced bracteate racemes, axillary along the upper twigs; bracts 6,
decussate, persistent; pedicels short, bracteolate or ebracteolate.

DISTRIB. Genus 1, monotypic.

Differs from the *Penæaceæ* by the æstivation of the perianth, by the doubled
number of stamens, by the form of the anthers, and by the albuminous seed and
embryo with short radicle and long cotyledons.

I. GEISSOLOMA, Lindl.

Characters as for the order.

1. **G. marginatum** (A. Juss. in Ann. Sc. Nat. 3ᵐᵉ sér. vi. 27,
t. 4, fig. 6); an erect shrub, 2–3 ft. high (*Stephens*); branches
4-angled and pilose to puberulous above, becoming terete and
glabrous lower down; internodes lengthening lower down; leaves
imbricate on the younger parts, in bud ovate, acute, silky-pilose,
older leaves ovate or ovate-subcordate, acute to obtuse or obtuse-
mucronulate, margin thickened beneath, 6–12 lin. long, 3–8 lin.
broad; bracts ovate, acute or acuminate, penninerved, margin
membranous, entire or fimbriate; first pair lateral, dorsally pilose,
1–1½ lin. long, second pair dorsally pilose or glabrous, 2–3 lin. long,
third pair glabrous, 3–4 lin. long; basal bracteoles (where present)
sometimes coalesced; perianth membranous; segments ovate or
elliptical, subulate to cuspidate, with fine parallel nerves, pink,
5–6 lin. long, half as broad; filaments obtusely angled on the inner
side, about 3 lin. long; capsule 4–5 lin. long, slightly woody; seeds
2 lin. long, yellow. *A.DC. in Prodr.* xiv. 492; *Sond. in Linnæa,* xxiii.
105; *Gilg in Engl. & Prantl, Pflanzenfam.* iii. 6 A, 207. *Penæa
marginata, Linn. Mant. Alt.* 199; *Thunb. Fl. Cap. ed. Schult.* 150.
P. mucronata, Vent. Hort. Malm. t. 87, *fig.* 1, *flower only.*

SOUTH AFRICA : without locality *Burmann*! *Masson*! *Thunberg*! *Scholl*, 615!
1128! *Mund*!
COAST REGION : Swellendam Div. ; in the forest on the mountains at Voormans-
bosch, *Ecklon & Zeyher*, 16! *Zeyher*, 3729! Riversdale Div. ; Garcias Pass,
Burchell, 6948! *Phillips*, 305! *Stephens*! on the Kampsche Berg, near Garcias
Pass, *Galpin*, 4539! on the Langeberg Range at Corente River Farm, *Muir*, 170!

The bracts of the racemes are either sterile or enclose an axillary bud ; this bud
usually remains dormant, but may grow out into another short bracteolate or
ebracteolate peduncle.

ORDER CXIX. B. **LORANTHACEÆ.**

(By T. A. SPRAGUE.)

Flowers regular or zygomorphic, hermaphrodite or unisexual,
3–6-merous. *Calyx* superior, gamosepalous, lobed or truncate,
sometimes obsolete. *Corolla* superior, polypetalous or gamopetalous,
petaloid or sepaloid, valvate in bud. *Stamens* as many as and
opposite the petals or corolla-lobes, and inserted on them ; anthers
usually 2-celled, sometimes divided into numerous small cells, which
may be arranged irregularly (*Viscum*) or in 2 or 4 vertical rows
(*Loranthus*). *Disc* superior, annular, or absent. *Ovary* inferior,
usually without a distinct placenta and ovule; style simple or
absent ; stigma not or hardly lobed. *Fruit* baccate (in all the
African species), crowned by the persistent calyx when the latter
is present ; pericarp sticky. *Seed* solitary, albuminous or exalbuminous, without a distinct testa ; embryo fairly large, terete or
angled, with distinct hypocotyl and 2 (more rarely 3–6) cotyledons.

Chlorophyll-containing shrubs or more rarely herbs, parasitic on other plants ;
very rarely trees ; leaves opposite, ternate or alternate, simple, entire, exstipulate,
sometimes reduced to mere scales or teeth ; inflorescence racemose or cymose ;
flowers often large and brightly coloured (*Loranthus*), or small, greenish and
inconspicuous (*Viscum*).

DISTRIB. Genera 27, chiefly tropical and subtropical ; species about 1000.

I. **Loranthus.**—*Flowers* hermaphrodite. *Calyx* present, though sometimes
reduced to a small rim. *Style* long.

II. **Viscum.**—*Flowers* unisexual. *Calyx* obsolete. *Style* short or none.

I. LORANTHUS, Linn.

Flowers hermaphrodite. *Calyx* more or less lobed, or truncate,
sometimes very short, occasionally provided inside at the base with
a fleshy annular thickening (*intramarginal ring*). *Corolla* polypetalous or, more frequently, gamopetalous, regular or zygomorphic ;
tube often split unilaterally for some distance downwards when
the flower expands. *Filaments* united in their lower part with the
petals ; anthers introrse, not versatile. *Style* filiform, or gradually

thickened upwards in the upper part and then rather suddenly contracted into a narrow neck below the stigma (*skittle-shaped*) ; stigma truncate or more or less capitate. *Fruit* baccate, usually globose, ovoid or ellipsoid, crowned by the persistent calyx. *Seed* albuminous ; embryo straight, terete.

Green leafy shrubs, parasitic on Dicotyledons, seldom on Coniferæ or Mono-cotyledons, often very brittle, even in a living state ; leaves opposite, ternate or alternate, penninerved, or several-nerved from the base ; inflorescence (in the African species) a raceme, spike, umbel or head ; subtending bract of each flower situated at the apex of the pedicel when the latter is present ; flowers often large and brightly coloured.

DISTRIB.   Species about 500, all Old World, mostly tropical and subtropical.

Corolla polypetalous :
  Flowers tetramerous ; petals under ¾ in. long :
    Anthers transversely septate :
      Receptacle less than ¾ lin. long; disc hardly dis-
        tinguishable from the base of the style   ... (1) **Woodii.**
      Receptacle 1 lin. long or more ; disc distinct ... (2) **subcylindricus.**
    Anthers not transversely septate ... ... (3) **garcianus.**
  Flowers pentamerous ; petals 1½–2¼ in. long :
    Style with a double bend shortly above the base ... (4) **undulatus.**
    Style without a double bend ... ... (5) **kalachariensis.**
Corolla gamopetalous :
  Filaments not produced into a tooth in front of the
    anther :
    Anthers transversely septate :
      Corolla tomentose outside with verticillate
        branched hairs : calyx equalling the re-
        ceptacle ... ... (6) **ovalis.**
      Corolla scurfy outside with stellate hairs ; calyx
        much shorter than the receptacle ... ... (7) **glaucus.**
    Anthers not transversely septate :
      Corolla-tube not split unilaterally :
        Corolla-lobes revolute ... ... (8) **elegans.**
      Corolla-lobes erect or reflexed :
        Corolla villous with silky subadpressed hairs (9) **Dregei.**
        Corolla glabrous :
          Upper part of filament much thickened,
            spirally coiled in the expanded flower (10) **Wyliei.**
          Upper part of filament not thickened :
            Corolla under 1½ in. long ; lobes erect,
              cohering ... ... (11) **quinquenervis.**
            Corolla about 2¼ in. long ; lobes reflexed,
              free ... ... (12) **Galpinii.**
      Corolla-tube split unilaterally :
        Umbels terminating leafy short-shoots, which
          are perulate at the base :
          Pedicels hispid ... ... (13) **Zeyheri.**
        Pedicels glabrous :
          Style skittle-shaped above :
            Bract longer than, or at least equalling
              the receptacle and calyx ; disc over-
              topping the calyx ... ... (14) **Moorei.**

Bract shorter than the receptacle and
    calyx ; disc a little shorter than the
    calyx ... ... (15) **natalitius.**

    Style filiform ... ... (16) **minor.**

    Umbels axillary ... ... (17) **Bolusii.**

Filaments produced into a tooth in front of the anther :
  Flowers pentamerous ; style skittle-shaped above :
    Corolla-lobes reflexed :
      Corolla glabrous :
        Apical swelling of bud with concave faces,
          distinctly 5-ribbed ... ... (18) **rubromargina-**
                                      tus.

        Apical swelling of bud with flat or convex
          faces, not distinctly ribbed ... ... (19) **oleæfolius,** var.
                                        Forbesii.

      Corolla more or less hairy ... ... (19) **oleæfolius.**

    Corolla-lobes erect :
      Corolla glabrous ... ... (20) **kraussianus.**

    Corolla puberulous :
      Receptacle and calyx urceolate ... ... (20) **kraussianus,**
                                        var. **puberulus.**

      Receptacle and calyx campanulate ... ... (21) **prunifolius.**

Flowers tetramerous ; style filiform ... ... (22) **Schlechteri.**

1. **L. Woodii** (Schlechter & Krause in Engl. Jahrb. li. 454, partly) ;
glabrous ; branches terete, nodose, 1–2 lin. in diam. 6 in. below the
apex, rough, greyish-brown ; branchlets slightly flattened, longitudinally wrinkled in a dried state ; internodes ¼–¾ in. long ; leaves
opposite or subopposite, petioled, ovate or ovate-oblong, more rarely
obovate-oblong, ¾–1½ in. long, ¼–¾ in. broad, obtuse or rounded at
the apex, obtuse or cuneate at the base, thinly coriaceous, dull,
margin crispate, midrib slightly raised especially on the lower
surface, lateral nerves 3–4 on each side, irregular, rather oblique,
anastomosing well within the margin, slightly raised on both surfaces
or hardly visible on the lower ; petiole 1–2 lin. long, very slightly
winged ; racemes both terminal and axillary, 6–10-flowered ;
rhachis 5–10 lin. long, pale green ; pedicels spreading, very oblique
at the apex, ½–1 lin. long ; bract unilateral, inserted obliquely with
reference to the pedicel, suborbicular, hooded, ⅜ lin. long, very
strongly umbonate on the back ; flowers pink and white (*Wood*),
tetramerous, polypetalous ; receptacle and calyx together narrowly
campanulate, ¾ lin. long ; calyx subtruncate, about ⅛ lin. long ;
corolla quadrangular in bud, the angles corresponding with the
edges of the petals, constricted about the middle, broadest below
the constriction, 2½–3 lin. long ; petals 3 lin. long, midrib raised on
the outer surface, claw erect, oblong, ¾ lin. long, ⅜ lin. broad,
produced above on the inner surface into a small rounded flap on
which the filament is inserted, limb spreading or reflexed shortly
above its base, oblong-linear, narrowed towards the apex ; stamens
pink ; filaments erect, 1 lin. long, broadened into the base ; anthers
linear, over 1 lin. long, dehiscing laterally, divided transversely

into 20 or fewer small cells arranged in 2 vertical rows, connective produced into a subulate point above the uppermost cells; no distinct disc present; style quadrangular, much thickened at the base, about $2\frac{1}{2}$ lin. long; stigma truncate; berry ellipsoid, about 3 lin. long, scarlet. *L. Sandersoni, Harv. ex Benth. et Hook. f. Gen. Pl.* iii. 208, *name only; Wood, Handb. Fl. Natal,* 115, *name only.*

EASTERN REGION: Natal; Krauns Kloof, on a tree at the Tower Rock, very rare, *Sanderson,* 697! Zululand; Ungoya Forest, 1000 ft., on *Burchellia capensis,* R. Br., *Wood,* 3874! Ngoya, *Wylie in Herb. Wood,* 7469!

Sanderson, 697, has a slightly longer petal-limb and anthers than typical *L. Woodii,* but undoubtedly belongs to that species.

The description of *L. Woodii* given by Schlechter and Krause, l.c., agrees on the whole with *Wood,* 3874, but includes characters drawn from *Rudatis,* 904, which is here referred to *L. subcylindricus.*

2. **L. subcylindricus** (Sprague); very closely allied to the preceding species, from which it differs in the following characters: leaves lanceolate or oblanceolate, $1-2\frac{3}{4}$ in. long, $\frac{1}{3}-\frac{3}{4}$ in. broad, margin slightly crispate; petiole 1–3 lin. long, winged; rhachis up to 1 in. long; pedicels 1–2 lin. long, not oblique or only slightly oblique at the apex; bract inserted in the same line with the pedicel, ovate, very concave, $\frac{1}{2}$ lin. long, umbonate on the lower half; receptacle subcylindric, 1 lin. long or more; petals $4\frac{1}{2}-5\frac{1}{2}$ lin. long, claw $1-1\frac{1}{4}$ lin. long, $\frac{1}{2}-\frac{3}{4}$ lin. broad, with a thickened central band on the inner surface; filaments $1\frac{3}{4}$ lin. long; anthers $1\frac{1}{2}-2\frac{1}{4}$ lin. long, cells 18–26; disc quadrangular, surrounding the base of the style, free from or adnate to it; style $3\frac{3}{4}-4$ lin. long; berry oblong-ellipsoid. *L. Woodii, Schlechter & Krause in Engl. Jahrb.* li. 454, *partly.*

EASTERN REGION: Natal; Alexandra District, Umtwalumi, on *Ochna arborea,* Burch., *Rudatis,* 904! Zululand; Nkandhla, 4000–5000 ft., *Wylie in Herb. Wood,* 9013!

3. **L. garcianus** (Engl. in Engl. Jahrb. xl. 539); glabrous; stem and branches very stout, gnarled, $\frac{1}{2}$ in. in diam. at the base, bark smooth, ash-grey, lenticels transversely oblong; branchlets stout, buff-coloured, nodose, $2\frac{1}{2}$ lin. in diam. at the base, wrinkled in a dried state; internodes $\frac{1}{3}-\frac{3}{4}$ in. long; leaves opposite or subopposite, petioled, obovate, $1-1\frac{3}{4}$ in. long, $\frac{3}{4}-1\frac{1}{4}$ in. broad, rounded at the apex, obtuse at the base, fleshy, thickly coriaceous in a dried state, dull, midrib and lateral nerves slightly raised on the upper surface, less distinct on the lower, lateral nerves very oblique, irregular, about 3 on each side; petiole 4–7 lin. long; racemes longer than the leaves, about 30-flowered, peduncle about 1 in. long (*Engler*), rhachis 2–3 in. long; pedicels $1-1\frac{1}{4}$ lin. long, stout, much thickened at the base; bract unilateral, ascending, very concave, truncate, $\frac{5}{8}-\frac{3}{4}$ lin. long, slightly umbonate on the lower half; flowers tetramerous (rarely pentamerous), polypetalous; receptacle and intramarginal ring together campanulate, $1\frac{1}{2}$ lin. long; calyx erect,

truncate, $\frac{1}{6}$ lin. long, adpressed to the intramarginal ring; intramarginal ring very prominent, projecting $\frac{1}{8}$ lin. above the calyx;
corolla quadrangular in bud, the angles corresponding with the
edges of the petals, slightly curved below the oblique apex; petals
oblong-linear, narrowed towards the apex, 8 lin. long, claw erect,
oblong, $1\frac{3}{4}$ lin. long, papillate on its inner edges, with a strongly
thickened central band on the inner surface, limb reflexed at nearly
1 lin. above its base; stamens erect; filaments $2\frac{1}{2}$ lin. long, slightly
narrowed upwards; anthers linear, slightly narrowed towards the
apex, nearly $4\frac{1}{2}$ lin. long, not divided transversely, connective
produced into a small cusp; disc quadrangular, $\frac{1}{4}$ lin. high; style
quadrangular, with concave faces, 6 lin. long; stigma truncate.

KALAHARI REGION : Transvaal ; Komati Poort, *Kirk,* 75 !
EASTERN REGION : Portuguese East Africa, Ressano Garcia, about 1000 ft.,
Schlechter, 11921 !

The flowers are usually tetramerous ; only one pentamerous one was observed.
The petals appear at first sight as if jointed to the intramarginal ring. They are
in reality bent inwards at right angles above the intramarginal ring and inserted
between this and the disc.

4. **L. undulatus** (E. Meyer in Drège, Zwei Pfl. Documente, 92, 93,
200, name only); glabrous; branches stout, greyish-brown, inconspicuously lenticellate ; long-shoots ascending or patulous, slender, rather
nodose, otherwise smooth, brown ; short-shoots 1–5 lin. long, bearing
1 or 2 pairs of leaves, and terminated by an umbel ; leaves opposite,
sessile, coriaceous, faintly 3-nerved, nerves parallel ; leaves of the
long-shoots : upper ones linear-oblong, 1–1$\frac{3}{4}$ in. long, 2$\frac{1}{2}$–5$\frac{1}{2}$ lin.
broad, very obtuse at the apex, subcordate at the base ; middle
ones shorter, oblong or ovate-oblong ; lower ones ovate-oblong or
broadly ovate, rounded at the apex, cordate at the base ; leaves of
the short-shoots broadly ovate, cordate at the base, 4–12 lin. long,
4–8 lin. broad ; umbels 2-flowered ; peduncle 1$\frac{1}{2}$–3 lin. long ;
pedicels stout, 2–3 lin. long ; bract unilateral, broadly oblong or
ovate-oblong, truncate, $\frac{3}{4}$–1$\frac{1}{2}$ lin. long, thickly keeled, ciliate above ;
flowers pentamerous, polypetalous; receptacle obconical, 4–5 lin.
long ; calyx ascending, truncate, $\frac{1}{2}$–$\frac{5}{8}$ lin. long ; corolla orange,
strongly arcuate and clavate above in the bud ; petals 1$\frac{3}{4}$–2 in.
long, claws inserted obliquely on the receptacle, about 4 lin. long,
1$\frac{3}{4}$–2$\frac{1}{2}$ lin. broad, with undulate margins, interlocking by means of
about 4 pairs of oblique fleshy ridges which descend from the adnate
portion of the filament, claw of the odd petal the narrowest, strongly
curved outwards, lateral claws the broadest; limb subspathulate-
linear, obtuse, twisted about the middle, upper part 1$\frac{1}{8}$–1$\frac{1}{4}$ lin.
broad, middle part $\frac{5}{8}$–$\frac{7}{8}$ lin. broad, lower part gradually broadened
into the base ; filaments inserted at the apex of the claws, erect,
curved in their upper part ; anthers oblong-linear, over 4 lin. long,
not divided transversely, connective truncate ; style with a double
bend shortly above the base, minutely pilose above ; stigma depressed-capitate, $\frac{5}{8}$–$\frac{3}{4}$ lin. in diam. *Harv. in Harv. & Sond. Fl. Cap.* ii.

577 ; *Engl. in Engl. Jahrb.* xx. 130 ; *Schinz in Bull. Herb. Boiss.* iv.
App. iii. 54 ; *Sprague in Dyer, Fl. Trop. Afr.* vi. i. 278.   *Plicosepalus
undulatus, Van Tiegh. in Bull. Soc. Bot. France,* xli. 504, 540.

VAR. β, **angustior** (Sprague) ; leaves shortly petioled or sessile, oblong or linear-
oblong, rounded or obtuse at the base ; receptacle about 3 lin. long.

WESTERN REGION : Little Namaqualand ; between Holgat River and the Orange
River, 1000–1500 ft., *Drège*! between Verleptpram and the mouth of the Orange
River, *Drège*! Var. β : Great Namaqualand ; Sandverhaar, 3100 ft., on *Acacia
Giraffae,* Burch., and *A. horrida,* Willd., *Pearson,* 4694!

Also in Tropical Africa. It has not been possible to determine from dried
material whether the odd petal is anticous or posticous. This is a point which
should be investigated in the field.

5. **L. kalachariensis** (Schinz in Bull. Herb. Boiss. iv. App. iii. 53) ;
glabrous ; branchlets ascending, elongated, brown, slightly lenticellate, nearly 1½ lin. in diam. 6 in. below the apex ; leaves opposite
or alternate, oblong, oblong-linear or linear, 1½–3½ in. long, 5–10
lin. broad, rounded at the apex, cuneate, obtuse or rounded at the
base, rigidly coriaceous, glabrous, 3–5-nerved ; petiole ½–2½ lin.
long ; umbels axillary, solitary or fascicled, sessile or peduncled,
2–6-flowered ; peduncle up to 5 lin. long ; pedicels 5–6 lin. long ;
bract saucer-shaped with an ovate or oblong limb, dorsal margin
⅝–1½ lin. long, strongly umbonate or keeled, ventral margin ⅛ lin.
long ; flowers pentamerous, polypetalous ; receptacle obconical,
2–3 lin. long ; calyx ascending, ½–⅝ lin. long, truncate ; corolla red
or deep pink, strongly arcuate and clavate above in the bud ; petals
2½ in. long or less, claws inserted obliquely on the receptacle,
3½–11 lin. long, with straight margins, interlocking by means of
4–8 (usually 5) pairs of very oblique folds which descend from the
adnate part of the filament ; limb reflexed shortly above the base
and coiling into loose spirals, subspathulate-linear ; filaments inserted at the apex of the claws, erect, curved in their upper part,
about 1 in. long, tapering slightly upwards ; anthers linear, 3½–7 lin.
long, not divided transversely ; style without a double bend at the
base ; stigma capitate, ⅝ lin. in diam. *Sprague in Dyer, Fl. Trop.
Afr.* vi. i. 280. *L. Fleckii, Schinz, l.c. L. Pentheri, Schlechter in
Journ. Bot.* 1898, 376. *L. Dinteri, Schinz in Bull. Herb. Boiss.* 2ᵐᵉ
sér. i. 869 (1901). *L. splendens, N. E. Br. in Kew Bulletin,* 1909,
136. *L. curviflorus, Schinz, Pl. Menyharth.* 42 ; *Rendle in Journ.
Bot.* 1905, 52 ; *not of Benth.*

KALAHARI REGION : Transvaal : Waterberg Division ; Pietpotgieters Rust,
Transvaal Colonial Herb., 4557!

Also in Tropical Africa.

6. **L. ovalis** (E. Meyer in Drège, Zwei Pfl. Documente, 92, 200,
name only) ; branches terete, pale greyish-brown, 1–1½ lin. in diam.
6 in. below the apex ; branchlets spreading, the youngest parts densely
clothed with verticillate-branched hairs ; leaves alternate, petioled,
obovate or obovate-oblong, 5–13 lin. long, 2½–7 lin. broad, rounded
at the apex, narrowed into the base, coriaceous, clothed in a young

state with verticillate-branched hairs which fall off leaving a dense velvety covering of stellate hairs, nerves not visible, or the midrib sometimes slightly raised on the lower surface ; petiole 1–3 lin. long ; umbels axillary, solitary, sessile, 2-flowered, or flowers solitary ; pedicels $1\frac{1}{2}$–$1\frac{3}{4}$ lin. long, like the bract, receptacle and calyx, densely tomentose with verticillate-branched hairs ; bract ascending, unilateral, ovate-oblong or narrowly oblong, 1–$1\frac{1}{2}$ lin. long ; flowers tetramerous ; receptacle and calyx together campanulate, $1\frac{1}{2}$–$1\frac{3}{4}$ lin. long ; calyx shallowly lobed, $\frac{3}{4}$–$\frac{7}{8}$ lin. long ; corolla slightly clavate above in the bud, slightly enlarged at the base, about $1\frac{1}{2}$ in. long, villous-tomentose outside with verticillate-branched hairs ; basal swelling ellipsoid, $1\frac{1}{2}$–$1\frac{3}{4}$ lin. long ; tube splitting unilaterally for a short distance downwards, studded inside with conical papillæ, especially on the adnate part of the filaments ; lobes more or less reflexed, linear-spathulate, 4–$4\frac{1}{4}$ lin. long, upper part $\frac{1}{2}$–$\frac{5}{8}$ lin. broad ; stamens erect ; filaments $1\frac{1}{2}$–2 lin. long, linear, thickened towards the apex ; anthers lanceolate-linear, 1–$1\frac{3}{8}$ lin. long, $\frac{1}{4}$–$\frac{1}{3}$ lin. broad, divided transversely into numerous small cells arranged in 4 vertical rows, connective produced beyond the cells, cells 5 in each of the inner rows, 6 in each of the outer ; disc quadrangular ; style narrowed upwards ; stigma ellipsoid, $\frac{1}{8}$ lin. long. *Harv. in Harv. & Sond. Fl. Cap.* ii. 575 ; *Engl. in Engl. Jahrb.* xx. 84 ; *Engl. in Engl. & Prantl, Pflanzenfam. Nachtr.* 1 *zu* ii.–iv. 131 ; *Schinz in Bull. Herb. Boiss.* iv. *App.* iii. 54. *Septulina ovalis, Van Tiegh. in Bull. Soc. Bot. France,* xlii. 263.

WESTERN REGION : Great Namaqualand ; Aris Drift on the Orange River, on *Tamarix* sp., *Schenck,* 258 ! Gais, on the lower Orange River, on *Tamarix* sp., *Schenck,* 341 ! Little Namaqualand ; between the Holgat River and the Orange River, *Drège* ! Kaus Mountains, *Drège* ! sandy ravine below Doornpoort, on *Lycium* sp., *Pearson,* 6010 ! bed of the Kuboos River, on *Tamarix* sp., *Pearson,* 6070 ! sandy plain 5 miles north of Anenous, on *Lycium* sp., *Pearson,* 6185 !

Loranthus ovalis, E. Meyer, is closely related to *L. glaucus,* Thunb., from which it differs in the long verticillate-branched hairs which clothe the fully developed corolla, and in the calyx, which is as long as the receptacle. Very young flower-buds of *L. glaucus* are rather thinly covered with short verticillate-branched hairs which disappear by the time the corolla is fully developed.

A fruiting specimen collected in Little Namaqualand between Bitterfontein and Stinkfontein, on *Zygophyllum* sp. (*Pearson,* 5534 !) may belong to *L. ovalis.* The berry is oblong-ovoid, crowned by the persistent calyx, which considerably overtops the disc. It is difficult to distinguish *L. ovalis* from *L. glaucus* in the absence of flowers.

7. **L. glaucus** (Thunb. Prodr. 58) ; branches terete, pale brown, with a glaucous appearance due to a covering of minute scale-like stellate hairs, rather slender, 1 lin. or less in diam. 6 in. below the apex ; branchlets spreading or ascending, like the leaves, pedicels, receptacle and calyx, densely scurfy with stellate hairs and scales ; leaves alternate, petioled, oblanceolate-oblong or obovate-oblong, more rarely obovate or elliptic, $\frac{1}{2}$–$1\frac{1}{2}$ in. long, $1\frac{1}{2}$–6 lin. broad, obtuse or rounded at the apex, gradually narrowed into the base, coriaceous, midrib slightly raised on the lower surface ; petiole

$\frac{1}{2}$–$1\frac{1}{2}$ lin. long ; umbels axillary, sessile or more rarely, three together, the middle one peduncled, the lateral ones sessile, 2- or 4-flowered, or flower solitary ; peduncle 2 lin. long or less ; pedicels 1–$1\frac{1}{2}$ lin. long ; bract suberect, unilateral, oblong-linear, 1–$1\frac{1}{4}$ lin. long ; flowers tetramerous ; receptacle and calyx together narrowly campanulate, 1–$1\frac{1}{4}$ lin. long ; calyx distinctly 4-lobed, distinctly shorter than the receptacle, $\frac{1}{4}$–$\frac{3}{8}$ lin. long ; corolla more or less clothed in a young state with short verticillate-branched hairs which rub off quickly, disclosing a scurfy covering of stellate hairs, quadrangular in bud with concave faces, slightly clavate and more densely scurfy above, slightly enlarged at the base, 1–$1\frac{1}{2}$ in. long ; basal swelling ellipsoid, 1$\frac{1}{2}$ lin. long ; flower expanding in two stages, the lobes first separating from one another and becoming reflexed, and the tube afterwards splitting unilaterally for a short distance downwards ; tube studded inside with conical papillæ ; lobes oblong-spathulate, 2$\frac{3}{4}$–4 lin. long, upper part hardly $\frac{1}{2}$ lin. broad ; stamens erect ; filaments inserted about $\frac{1}{2}$ lin. above the base of the corolla-lobes, 1–$1\frac{3}{4}$ lin. long ; anthers linear, about 1 lin. long, divided transversely into numerous small cells arranged in 4 vertical rows, connective produced beyond the cells, cells 5 in each of the inner rows, 6–7 in each of the outer ; disc quadrangular, papillate ; style narrowed upwards ; stigma ellipsoid, $\frac{1}{8}$ lin. long ; berry oblong-ellipsoid, 3–3$\frac{1}{2}$ lin. long, scaly, crowned by the persistent calyx, which hardly overtops the disc. *Thunb. Fl. Cap. ed.* 2, ii. 251 ; *Fl. Cap. ed. Schult.* 295 ; *Cham. & Schlecht. in Linnæa,* iii. 208 ; *Eckl. & Zeyh. Enum.* 358 ; *Drège, Zwei Pfl. Documente,* 63, 94, 108, 113 ; *Harv. in Harv. & Sond. Fl. Cap.* ii. 575 ; *Engl. in Engl. Jahrb.* xx. 84 ; *Engl. in Engl. & Prantl, Pflanzenfam. Nachtr.* 1 *zu* ii.–iv. 131 ; *Schinz in Bull. Herb. Boiss.* iv. *App.* iii. 53. *L. canescens, Burch. Trav.* ii. 90 ; *DC. Prodr.* iv. 304. *L. Burchellii, Eckl. & Zeyh. Enum.* 358, *excl. syn. L. longitubulosus, Engl. & Krause in Engl. Jahrb.* li. 455, *fig.* 1. *Septulina glauca, Van Tiegh. in Bull. Soc. Bot. France,* xlii. 263.

COAST REGION: Van Rhynsdorp Div. ; Ebenezer, *Drège*! Heerenlogement, on *Rhus* sp., *Zeyher*! Clanwilliam Div. ; by the Olifants River, *Bachmann,* 358! Malmesbury Div.; near Hopefield, between Matjesfontein and Hazenkraal, *Bachmann,* 1854! between Phezanthoek and Tantjesfontein, *Bachmann,* 1907! Saldanha Bay, *Thunberg, Grey*! *Zeyher,* 2280! *Bolus,* 12824! between Groene Kloof and Saldanha Bay, *Drège*! Worcester Div. ; near the Hex River, *Zeyher,* 2281!

CENTRAL REGION: Ceres Div. ; Verkerde Vley, *Rehmann,* 2847! between Little Doorn River and Great Doorn River, *Burchell,* 1210! Prince Albert Div. ; between Dwyka River and Zwartbulletje, *Drège*! Graaff Reinet Div. ; south side of the Snowy Mountains, *Burke*! Murraysburg Div. ; near Murraysburg, on *Lycium* sp., *Tyson,* 117! Beaufort West Div. ; Gouph and Winterveld, *Zeyher,* 2281! Richmond Div. ; Uitvlugt, *Zeyher,* 753! Britstown Div.; Rhenoster Poort, by the Brak River, *Burchell,* 2219/5!

WESTERN REGION: Great Namaqualand ; Keetmanshoop, *Fleck,* 312a! Inachab, *Dinter,* 915! Little Karas Mountains, on the Us River, on *Phaeoptilon* sp., *Engler,* 6662; Geiab River, near Kanus, *Dinter,* 3071 ; mouth of the Orange River, *Schenck,* 234! Little Namaqualand ; near the mouth of the Orange River, *Drège* ; Van Rhynsdorp Div. ; near Leislap, *Zeyher,* 753?! Karee Berg, *Schlechter,* 8290!

The excellent figure and description of *Loranthus longitubulosus*, Engl. & Krause, leave no room for doubt that it is identical with the common S. African *L. glaucus*, Thunb. The authors do not mention the transverse septation of the anthers, but this may be easily overlooked.

Loranthus Burchellii, Eckl. & Zeyh., is a mere form of *L. glaucus*, Thunb., but *L. glaucus*, var. *Burchellii*, DC., which was supposed by Ecklon and Zeyher to be synonymous, is typical *L. elegans*, Cham. & Schlecht. *L. glaucus*, DC. Prodr. iv. 303, is also a synonym of *L. elegans*.

Thunberg described the flowers of *L. glaucus* as pentamerous: none but tetramerous flowers have been observed by the writer.

8. **L. elegans** (Cham. & Schlecht. in Linnæa, iii. 209); practically glabrous; branches terete, greyish-brown, 1–1½ lin. in diam. 6 in. below the apex; branchlets spreading or ascending; internodes ¼–1 in. long; leaves ternate, opposite or alternate, petioled, linear-lanceolate or narrowly lanceolate, ¾–2½ in. long, 2–8 lin. broad, obtuse at the apex, narrowed into the base, thinly coriaceous, very obliquely penninerved, nerves slightly raised, especially on the lower surface; petiole 1–2 lin. long; inflorescence a 3–6-flowered umbel, or a very short raceme bearing one or two flowers near the middle of the rhachis and an umbellate group of three or four flowers at its apex; umbels or racemes axillary, solitary or geminate, leafless or bearing at the base a pair of leaves or a whorl of three leaves; peduncle or rhachis 1½–6 lin. long; pedicels 1–2½ lin. long, like the bract very minutely puberulous with brownish stellate hairs; bract elliptic-oblong or triangular-ovate from a saucer-shaped base, ⅝–¾ lin. long, ventral margin hardly ⅛ lin. long; flowers pentamerous, apparently expanding in two stages, the corolla-lobes separating for a short distance in the first place, and becoming more or less revolute, while the stamens remain erect; in the second stage the corolla-lobes separate from one another as far down as the insertion of the stamens and coil up spirally, while the stamens break off shortly above their insertion; the corolla-tube eventually splits down unilaterally or irregularly nearly to its base; receptacle and calyx together narrowly campanulate, 1⅜–1½ lin. long; calyx subtruncate or undulate, ¼–⅜ lin. long; corolla linear in bud, inconspicuously pentagonal, 1¼ in. long or more, slightly constricted 2½–3½ lin. above the base, slightly enlarged towards the apex; lobes linear, about 1 in. long; filaments inserted 4¼–5½ lin. above the base of the corolla, narrowed upwards, 6–7 lin. long; anthers linear, subtruncate, 3½–5½ lin. long, not transversely divided; disc ⅙ lin. high; style filiform; stigma capitate, slightly oblique. *Drège in Linnæa*, xix. 663; *Sprague in Kew Bulletin*, 1914, 362. *L. schlechtendalianus, Schultes, Syst. Veg.* vii. 1635; *Krauss in Flora,* 1844, 432; *Meisn. in Hook. Lond. Journ. Bot.* ii. 539. *L. croceus, E. Meyer in Drège, Zwei Pfl. Documente*, 63, 109, 139, 200, *name only*. *L. glaucus, DC. Prodr.* iv. 303, *not of Thunb.*; *var. Burchellii, DC. l.c.* *L. oleæfolius, Eckl. & Zeyh. Enum.* 358; *Harv. in Harv. & Sond. Fl. Cap.* ii. 576; *Benth. & Hook. f. Gen. Pl.* iii. 209; *Engl. in Engl. Jahrb.* xx. 83; *Engl. & Prantl, Pflanzenfam.* iii. 1, 187, *t.* 126, *fig. L–N*; *not of Cham. & Schlecht.* *L. oleifolius*,

Marloth, Fl. S. Afr. i. 167, *t.* 38, *fig. A. L. speciosus, Engl. in Engl.
& Prantl, Pflanzenfam. Nachtr.* 1 *zu* ii.–iv. 131, *not of Dietr.
Lichtensteinia elegans, and L. speciosa, Van Tiegh. in Bull. Soc. Bot.
France,* xlii. 254. *Moquinia rubra, A. Spreng. Tent. Suppl. Syst.
Veg.* 9 (1828) ; *Griesselich in Linnæa,* v. 421.

COAST REGION : Clanwilliam Div. ; Olifants River, on *Acacia* sp., *Zeyher,* 2282 !
Lange Vallei, *Drège.* George Div. ; Outeniqualand, *Niven* 8 ! Uitenhage Div. ;
by the Zwartkops and Bushmans Rivers, *Zeyher,* 107 ! by the Zwartkops River,
Burchell, 4429 ! *Zeyher,* 2283 ! *Krauss,* 1217 ; and without precise locality, *Cooper,*
1507 ! Alexandria Div. ; Olifants Hoek near the Bushmans River, *Pappe* ! Albany
Div. ; near Grahamstown, on *Acacia* sp., *MacOwan,* 568 ! Bushmans River, *Drège* !
Queenstown Div. ; base of Shepstone Berg, 4500 ft., *Galpin,* 1812 ! Cathcart Div. ;
Goshen, Windvogelsberg, on *Acacia horrida,* Willd., *Baur,* 930 ! King Williams-
town Div. ; Berlin, on *Acacia horrida,* Willd., *Durban,* 45 ! Komgha Div. ; near
Komgha, *Flanagan,* 727 ! Div. ? Caledons Kluft, *Mund* !
CENTRAL REGION : Prince Albert Div. ; between Dwyka River and Zwart-
bulletje, *Drège* ! Beaufort West Div. ; by the Gamka River, *Burke* ! *Zeyher,* 752 !
Somerset Div. ; *Bowker* ! · *Atherstone,* 13 ! 106 ! Blyde River, *Burchell,* 2972 !
Little Fish River, *Hutton* ! near Somerset East, 2100 ft., on *Acacia* sp.,
Schlechter, 2702 ! Graaff Reinet Div. ; Graaff Reinet, *Burchell,* 2115 ! *Bolus,* 72 !
near Monkey Ford, *Burchell,* 2887 ! Murraysburg Div. ; near Murraysburg,
Tyson, 347 !

Harvey, *Fl. Cap.* ii. 576, erroneously identified *Loranthus oleæfolius,* Cham. &
Schlecht., with *L. elegans,* Cham. & Schlecht., and described the true
L. oleæfolius as a new species, *L. namaquensis,* Harv.

9. **L. Dregei** (Eckl. & Zeyh. Enum. 358) ; young parts rusty-
tomentose with verticillate-branched hairs, soon becoming
glabrous ; branches terete, greyish brown, 1 1½ lin. in diam. 6 in.
below the apex ; internodes ½–2 in. long ; leaves opposite, oblong or
elliptic, rounded at the apex, cuneate, obtuse or rounded at the
base, 1–2½ in. long, ½–1½ in. broad, coriaceous or thinly coriaceous,
usually glabrous on both surfaces when adult ; petiole 1½–4 lin. long,
rusty-pubescent in a young state, usually glabrous when adult ;
heads axillary, fascicled, 2–6-flowered, mainly produced on the
older parts of the branches after the leaves which subtend them
have fallen ; peduncle 2–8 lin. long, pubescent or tomentose ;
bract unilateral, broadly elliptic or suborbicular, very concave, 1¼–
1¾ lin. long ; flowers pentamerous, expanding in two stages, the
corolla-lobes in the first place separating below while remaining
united above ; in the second stage the upper parts of the corolla-
lobes separate from one another and become reflexed, while the
thickened upper parts of the filaments coil up sharply and break
off, bearing the anthers with them ; receptacle and calyx together
subcylindric, 2–2¼ lin. long ; receptacle densely villous with simple
adpressed hairs ; calyx truncate, 1¼ lin. long, sometimes minutely
toothed, sparingly adpressed-pilose, ciliate ; corolla about 2 in. long,
densely adpressed-villous outside ; upper part oblong-ovoid in the
bud, acute, about 2½ lin. long ; tube orange-red (*Wood*), about
10 lin. long, not split unilaterally, glabrescent near the base,
slightly swollen above the calyx, and broadened upwards above the
swelling, which is about 2½ lin. long ; lobes yellow-green (*Wood*),

about 1¼ in. long, reflexed above the lower third, linear, gradually broadened upwards, upper part linear-lanceolate ; filaments inserted 3–3½ lin. above the base of the corolla-lobes, about 9½ lin. long, breaking above the lower third when the flower expands, the lowermost 3½ lin. straight, linear, the remainder much thickened, involute ; anthers linear, 2-horned at the apex, 1½–2 lin. long, not divided transversely ; disc ⅙ lin. high ; style filiform ; stigma ellipsoid, ⅜ lin. long, slightly bifid ; berry oblong, tapering to an obtuse apex, covered with white silky hairs, and crowned with the persistent calyx, dull pink when ripe, 5 lin. long, 3 lin. in diam. (*Wood*). *Harv. in Harv. & Sond. Fl. Cap.* ii. 575 ; *Engl. in Engl. Jahrb.* xx. 84, 104 ; *Engl. Pfl. Ost-Afr. C.* 166 ; *M. S. Evans in Nature,* li. 236 ; *Schinz in Mém. Herb. Boiss. no.* 10, 31 ; *Wood, Natal Pl.* iv. *t.* 312 ; *Sprague in Dyer, Fl. Trop. Afr.* vi. i. 311. *L. Dregei, forma subcuneifolia, Engl. l.c.* 104, *as to the South African specimens. L. Dregei, forma obtusifolia, Engl. l.c.* 105, *as to the South African specimens. L. roseus, Klotzsch in Peters, Reise Mossamb. Bot.* 177 ; *Schinz in Mém. Herb. Boiss. no.* 10, 32. *L. oblongifolius, E. Meyer in Drège, Zwei Pfl. Documente,* 148, 149 (*angustifolius by error*), 159, 200. *Erianthemum Dregei, Van Tiegh. in Bull. Soc. Bot. France,* xlii. 248.

COAST REGION : Komgha Div. ; near Komgha, *Flanagan in MacOwan, Herb. Austr.-Afr.,* 1527 ! British Kaffraria, *Cooper,* 142 !
KALAHARI REGION : Transvaal ; near Lydenburg, *Wilms,* 200 ! Port Shepstone, *Burtt-Davy,* 2413 ! and without precise locality, *Sanderson* !
EASTERN REGION : Transkei : Kentani, *Miss Pegler,* 418 ! Tembuland ; Morley. *Drège* ! Pondoland ; between Umtata River and St. Johns River, *Drège* ! Natal : near Durban, *Drège, Peddie* ! *Sanderson,* 203 ! *Gerrard,* 641 ! Inanda, *Wood,* 445 ! Alexandria District, *Rudatis,* 266 ! and without precise locality, *Gerrard ! Cooper,* 1223 ! 2458 ! *Sanderson,* 552 ! Delagoa Bay ; on *Trichilia emetica,* Vahl, *Monteiro,* 15 ! Lourenço Marques, *Schlechter,* 11564 !

Also in Tropical Africa, where nine distinct varieties occur in addition to the type. According to M. S. Evans (*l.c.*), the flowers of *L. Dregei* are visited by sunbirds, which insert their beaks into the slits between the corolla-lobes and cause the flower to open with a jerk. As the flower opens, the anthers are broken sharply off, and fly away, scattering their pollen as they go. The point of breakage is evidently at the junction between the slender and thickened parts of the filament, judging from dried specimens, and it is apparently the sudden coiling up of the thickened part which causes the explosive opening of the flower.
A specimen collected by Medley Wood (4467) on *Acacia* sp., near Qumbeni, 3000–4000 ft., Natal, apparently represents an undescribed variety. The leaves differ from the type in being subcordate at the base, and in retaining a covering of stellate and verticillate-branched hairs on the lower surface when fully developed. The bract is rather larger (1¾ lin. long). The receptacle is surrounded by a very dense band of ascending hairs which increase its apparent diameter so that the combined receptacle and calyx appear to be flask-shaped instead of subcylindric.
A second variety may be represented by *Galpin,* 708, from near Barberton, Transvaal, and *Rehmann,* 6470, from Houtbosh. This resembles the above, but has short stout peduncles and longer flowers.

10. **L. Wyliei** (Sprague) ; branches subterete, nodose, ash-coloured, rather slender, about 1 lin. in diam. 6 in. below the apex, glabrous ; branchlets minutely and rather densely pilose with very short un-

branched hairs; internodes $\frac{1}{4}$–$\frac{1}{2}$ in. long; leaves alternate, petioled, oblanceolate or oblong-oblanceolate, 8–13 lin. long, 3–5$\frac{1}{2}$ lin. broad, obtuse or rounded at the apex, narrowed into the base, thinly coriaceous, glabrous, penninerved, nerves slightly raised on both surfaces, lowest pair very oblique, running subparallel to the midrib; petiole $\frac{1}{2}$–1 lin. long, minutely pilose on the upper surface; fascicles axillary, 2–3-flowered, or flowers solitary; pedicels $\frac{1}{2}$–$\frac{5}{8}$ lin. long; bract cupular, subtruncate with a broadly ovate dorsal lobe, minutely pilose outside, coarsely ciliate, slightly umbonate below the lobe, dorsal margin 1$\frac{1}{8}$ lin. long, ventral margin $\frac{3}{4}$ lin. long; flowers pentamerous; receptacle and calyx together narrowly cylindric, 3 lin. long, $\frac{5}{8}$ lin. in diam.; receptacle slightly thicker than the calyx-tube, glabrous; calyx 2 lin. long, 5-toothed, ciliate, otherwise glabrous, teeth triangular, $\frac{1}{8}$ lin. long; corolla about 2 in. long, glabrous, linear in bud, clavate above, acute, apical swelling about 3$\frac{1}{2}$ lin. long; tube with a slight suprabasal swelling which extends above the calyx, very narrow for 1$\frac{1}{2}$ lin. above the swelling, and then broadened gradually to the apex; lobes about 1$\frac{1}{4}$ in. long, reflexed at the middle, lower part linear from a broad base, upper part oblanceolate-linear; filaments inserted 4$\frac{1}{2}$ lin. above the base of the corolla-lobes, lower part erect, filiform, 4 lin. long, upper part much thickened, coiling spirally when the flower expands; anthers linear, nearly 2 lin. long, not divided transversely, connective truncate; disc acutely lobed; style filiform; stigma ovoid, $\frac{5}{8}$ lin. long.

EASTERN REGION : Zululand ; Ngoya, *Wylie in Herb. Wood*, 7468 !

L. Wyliei is most nearly allied to the Tropical African *L. Menyharthii*, Engl. & Schinz, from which it may be distinguished by the smaller glabrous leaves. It bears a considerable resemblance to *L. quinquenervis*, Hochst., from which it differs in the filaments and reflexed corolla-lobes.

11. **L. quinquenervis** (Hochst. in Flora, 1844, 432); glabrous; branches ash-coloured, 1–1$\frac{1}{2}$ lin. in diam. 6 in. below the apex; branchlets smooth, greenish; internodes $\frac{1}{4}$–$\frac{3}{4}$ in. long; leaves alternate, petioled, elliptic, ovate or suborbicular, $\frac{3}{4}$–2$\frac{1}{2}$ in. long, $\frac{1}{2}$–1$\frac{1}{2}$ in. broad, obtuse or rounded at the apex, subcuneate into the base, coriaceous, 5-nerved from shortly above the base, nerves slightly raised, especially on the upper surface; petiole 1–3 lin. long; umbels axillary, solitary or more rarely 3 together, 2–8-flowered; peduncle very short and stout, expanded above into a pyramidal receptacle divided into sockets, hardly 1 lin. long including the receptacular portion; pedicels $\frac{1}{2}$ lin. long; bract cupular, produced into a short truncate or 2-toothed lobe, dorsal margin $\frac{3}{4}$ lin. long, bearing a large flattened elliptic umbo beneath the lobe, ventral margin $\frac{1}{2}$ lin. long; flowers pentamerous, expanding in two stages; in the first, the lower portions of the corolla-lobes separate from one another, while the upper portions remain connate; in the second, the upper part of the limb splits unilaterally, the lobes remaining connate in a single piece; receptacle and calyx together cylindric, 1$\frac{1}{8}$ lin. long; calyx $\frac{5}{8}$ lin. long, truncate; corolla red, with white

bands, about $1\frac{1}{4}$ in. long, linear in bud; tube with an oblong or
ellipsoid swelling about $1\frac{1}{2}$ lin. long immediately above the calyx,
very narrow for $1\frac{1}{2}$ lin. above the swelling, then slightly broadened
to the apex; lobes $7\frac{1}{2}$–$8\frac{1}{2}$ lin. long, erect, linear, slightly broadened
in the uppermost $1\frac{1}{2}$ lin.; filaments inserted $2\frac{1}{2}$–4 lin. above the
base of the corolla-lobes, filiform, $2\frac{1}{2}$ lin. long, inflexed or involute
when the flower expands, with two minute or almost obsolete teeth
$\frac{1}{3}$–$\frac{1}{2}$ lin. below the apex on the ventral surface; anthers linear,
1 lin. long, not divided transversely, connective truncate; disc $\frac{1}{4}$ lin.
high, slightly lobed; style filiform; stigma subglobose, $\frac{1}{4}$ lin. in diam.;
unripe berry oblong-obovoid, 4 lin. long. *L. quinquenervius, Harv.*
in Harv. & Sond. Fl. Cap. ii. 578; *Wood, Natal Pl.* iii. *t.* 295. *L.*
tenuiflorus, Harv. l.c. in syn., not of Hook. f.

COAST REGION: East London Div.; in bush along sea-coast, *Galpin*, 1832!
Komgha Div.; by the Kei River, *Flanagan*, 327! and *in MacOwan, Herb. Austr.-*
Afr., 1648! Kaffraria; Krielis Country, *Bowker*, 556! British Kaffraria,
Cooper, 60!
CENTRAL REGION: Albert Div.; *Cooper*, 1761!
KALAHARI REGION: Transvaal, *Sanderson*!
EASTERN REGION: Transkei; Kentani, *Miss Pegler*, 1517! Natal; near Durban,
Grant! *Sutherland*! *Gerrard*, 639! *Wood*, 1653! and without precise locality,
Gerrard, 71! *Cooper*, 2457! *Sanderson*, 698!

According to Medley Wood, l.c., the corolla of *L. quinquenervis* is inflated
and pinky-white at base, lower part of the central portion bright scarlet, then
pinky-white to where the stamens are inserted, then another band of white,
remainder scarlet.

12. **L. Galpinii** (Schinz MSS.); glabrous; branches rather stout,
very nodose, greyish-brown, rather densely lenticellate, the leafless
parts bearing the inflorescences 2–$2\frac{1}{4}$ lin. in diam.; branchlets
smooth, brown, about $1\frac{1}{2}$ lin. in diam. 6 in. below the apex; internodes $\frac{1}{2}$–$1\frac{1}{4}$ in. long; leaves opposite, petioled, oblong-lanceolate,
straight or slightly curved, 3–$4\frac{1}{4}$ in. long, 5–10 lin. broad, obtuse
at the apex or apiculate, gradually narrowed into the base, rigidly
coriaceous, penninerved, lateral nerves oblique, more or less raised
on both surfaces or indistinct, midrib distinctly raised; petiole
4–5 lin. long; umbels axillary, solitary or geminate on the leafless
parts of the branches, or borne singly on the branchlets, 2-flowered;
peduncle stout, $1\frac{1}{2}$–3 lin. long, receptacular portion bordered by a
thin margin $\frac{1}{3}$ lin. broad; pedicels about 1 lin. long, very stout;
bract subcupular, 2-lipped, dorsal lip erect, $1\frac{1}{4}$–$1\frac{1}{2}$ lin. long, rounded,
ventral lip patulous, $\frac{3}{8}$ lin. long, truncate; flowers pentamerous;
receptacle and calyx together campanulate, much expanded above
the middle, 3 lin. long; receptacle $1\frac{3}{4}$ lin. long, $1\frac{1}{4}$ lin. in diam.;
calyx cupular, truncate, $1\frac{1}{4}$–$1\frac{1}{2}$ lin. long, $2\frac{1}{4}$ lin. in diam.; corolla
yellow (*Galpin*), about 3 lin. long, linear in bud, very slightly clavate
above; tube pentagonal, not inflated at the base, gradually broadened
from below the middle to the apex, not splitting unilaterally; lobes
$1\frac{3}{4}$ in. long, $1\frac{1}{4}$ lin. broad, linear-oblanceolate, acute, reflexed $2\frac{1}{2}$–3
lin. above their base; stamens crimson (*Galpin*); filaments inserted
$1\frac{1}{2}$ lin. above the base of the corolla-lobes, erect, about 10 lin. long,

under $\frac{1}{2}$ lin. broad; anthers linear, slightly broadened upwards 8 lin. long, connective truncate; disc sunk, pentagonal, $\frac{3}{8}$ lin. high; stigma broadly ovoid, $\frac{3}{4}$ lin. long.

KALAHARI REGION: Transvaal; Kaap River valley, Barberton, on *Sclerocarya caffra*, Sond., *Galpin*, 896!

Allied to the Tropical African *L. panganensis*, Engl., which has tetramerous flowers.

13. **L. Zeyheri** (Harv. in Harv. & Sond. Fl. Cap. ii. 576); branches stout, brown, shortly pilose, $1\frac{1}{2}$–2 lin. in diam. 6 in. below the apex, bearing short-shoots in the axils of the fallen leaves; branchlets shortly hispid, about 1 lin. in diam.; internodes $\frac{1}{2}$–$1\frac{1}{4}$ in. long; short-shoots perulate at the base, shortly and densely hispid, bearing 3–5 pairs of leaves and terminated by an umbel; leaves opposite, petioled, obovate or oblanceolate, the lower broader and shorter than the upper, 1–2 in. long, 5–9 lin. broad, obtuse or rounded at the apex, gradually narrowed into the base, shortly pilose in a young state, at length becoming nearly glabrous, very obliquely penninerved, the lower 3–5-nerved from shortly above the base, nerves raised on both surfaces; petiole $\frac{1}{2}$–$2\frac{1}{2}$ lin. long; umbels 3–6-flowered; pedicels 2–3 lin. long, shortly hispid, thickened at the apex and base; bract unilaterally developed from a platter-shaped base, erect, oblong or lanceolate oblong, $1\frac{1}{4}$–$2\frac{1}{2}$ lin. long, $\frac{5}{8}$ lin. broad, keeled, margins inflexed, pilose outside, equalling or overtopping the calyx; flowers pentamerous; receptacle and calyx together campanulate, $1\frac{3}{8}$–$1\frac{5}{8}$ lin. long; calyx truncate, ciliate, margin patulous, $\frac{1}{4}$–$\frac{3}{8}$ lin. long, intramarginal ring $\frac{1}{8}$ lin. high; corolla white and yellow (*Miss Pegler*), 2–$2\frac{1}{2}$ lin. long, linear in bud, very acute, broadest about the middle; tube slightly enlarged at the base, splitting unilaterally downwards to the middle; lobes erect, narrowly oblanceolate-spathulate, about $7\frac{1}{2}$ lin. long, upper part linear-oblanceolate, 5 lin. long, $\frac{5}{8}$ lin. broad, middle part $1\frac{1}{2}$ lin. long, $\frac{3}{8}$ lin. broad, lower part 1 lin. long, $\frac{5}{8}$ lin. broad at the base, narrowed upwards; filaments inserted at the base of the corolla-lobes, inflexed, 3–4 lin. long, flattened, narrowed upwards, with a ventral oblong thickening $\frac{1}{3}$ lin. long at the apex; anthers linear, 3 lin. long; disc pentagonal, hardly lobed, shorter than the calyx; style skittle-shaped above, thickened part 3–$3\frac{1}{4}$ lin. long; stigma ellipsoid, nearly $\frac{1}{2}$ lin. long; berry ellipsoid or obovoid, 4 lin. long. *Engl. in Engl. & Prantl, Pflanzenfam. Nachtr.* 1 *zu* ii.–iv. 131. *Acranthemum Zeyheri, Van Tiegh. in Bull. Soc. Bot. France*, xlii. 255.

KALAHARI REGION: Transvaal; Magaliesberg Range, by the Crocodile River, on *Dodonaea thunbergiana*, Eckl. & Zeyh., and *Acacia* sp., *Zeyher*, 168! 751! *Worsdell*! Hang Klip Berg between Matlalas Hills and Makapans Poort, *Baines*! near Rustenburg, *Miss Pegler*, 987! Houtbosh, *Rehmann*, 6466! Modderfontein on *Acacia caffra*, Willd., and *A. horrida*, Willd., *Conrath*, 330!

Also recorded as collected by Burke, by the Gamka River, Beaufort West Div., but this is probably a mistake due to misplacement of labels.

14. L. Moorei (Sprague); closely allied to the preceding species, from which it differs in the following characters: branches, short-shoots, leaves and inflorescence glabrous ; leaves glaucous ; umbels about 6-flowered ; bract unilaterally developed from a saucer-shaped base, more or less foliaceous, 3–6 lin. long, $\frac{3}{4}$–1$\frac{1}{2}$ lin. broad, or non-foliaceous, 1$\frac{1}{2}$–2 lin. long, flat, not keeled ; receptacle and calyx together broadly campanulate, 1$\frac{3}{8}$ lin. long ; calyx 5-toothed, $\frac{1}{8}$–$\frac{1}{6}$ lin. long ; corolla 2$\frac{1}{2}$ in. long ; tube distinctly swollen at the base ; lobes linear-lanceolate, 7 lin. long, nearly $\frac{3}{4}$ lin. broad ; filaments deflexed, 2$\frac{1}{2}$ lin. long ; anthers nearly 3 lin. long ; disc $\frac{3}{8}$–$\frac{1}{2}$ lin. longer than the calyx ; thickened part of style 4$\frac{1}{2}$ lin. long.

KALAHARI REGION : Transvaal ; near Barberton, *Moore* !

L. Moorei may be distinguished from *L. natalitius,* Meisn., by the glaucous leaves, and the bract, which is longer than or at least as long as the receptacle and calyx.

15. L. natalitius (Meisn. in Hook. Lond. Journ. Bot. ii. 539); glabrous ; branches fairly stout, brown, about 1$\frac{1}{2}$ lin. in diam. 6 in. below the apex, bearing short-shoots in the axils of the fallen leaves ; branchlets $\frac{3}{4}$–1 lin. in diam. ; short-shoots perulate at the base, bearing 2 or more pairs of leaves, and terminated by an umbel : leaves opposite or alternate on the branches and branchlets, opposite on the short-shoots, petioled, oblanceolate or obovate, 1–2 in. long, 4–10 lin. broad, obtuse or rounded at the apex, narrowed into the base, coriaceous, obliquely penninerved, nerves slightly raised on both surfaces ; petiole 1–4 lin. long ; umbels 3–6-flowered ; pedicels 3–5 lin. long ; bract broadly and shortly ovate from a saucer-shaped base, dorsal margin $\frac{5}{8}$–$\frac{3}{4}$ lin. long, subtruncate, sparingly ciliate, broadly umbonate to within $\frac{1}{4}$ lin. below the apex, ventral margin $\frac{1}{4}$ lin. long ; flowers pentamerous or occasionally hexamerous (*Harvey*) ; receptacle and calyx together shortly and widely cam-panulate, 1$\frac{1}{4}$–1$\frac{1}{2}$ lin. long ; calyx patulous, truncate, irregularly split, nearly $\frac{1}{2}$ lin. long including the intramarginal ring, which is under $\frac{1}{8}$ lin. high ; corolla waxy-white tipped with orange-yellow, linear in bud, very acute, slightly enlarged at the base, 2$\frac{1}{2}$–2$\frac{3}{4}$ in. long ; tube splitting unilaterally downwards to about the middle ; lobes erect, lanceolate-linear, 10–11 lin. long, $\frac{3}{4}$–$\frac{7}{8}$ lin. broad ; filaments inserted $\frac{1}{2}$ lin. above the base of the corolla-lobes, deflexed or inflexed, gradually narrowed upwards, 4–4$\frac{1}{2}$ lin. long, scarlet in lower portion, orange-yellow upwards (*Wood*), with a ventral oblong thickening $\frac{1}{2}$ lin. long at the apex ; anthers linear, 3–3$\frac{1}{2}$ lin. long, connective produced $\frac{1}{12}$–$\frac{1}{8}$ lin. above the cells ; disc pentagonal, not lobed, $\frac{3}{8}$ lin. high, a little shorter than the calyx ; style skittle-shaped above, thickened part 5$\frac{1}{2}$ lin. long ; stigma ellipsoid, $\frac{1}{3}$ lin. long, more or less distinctly bilobed or trilobed. *Harv. Thes. Cap.* i. 19, *t.* 30 ; *Harv. in Harv. & Sond. Fl. Cap.* ii. 576 ; *Wood, Natal Pl.* iv. *t.* 374 ; *Engl. in Engl. & Prantl, Pflanzen-fam. Nachtr.* 1 *zu* ii.–iv. 131. *Acranthemum natalitium, Van Tiegh. in Bull. Soc. Bot. France,* xlii. 255.

EASTERN REGION: Natal; Table Mountain, *Krauss*, 208! *Gueinzius*! near
Durban, *Sanderson*, 194! near Itafamasi, *Wood*, 748! Umlaas, *Wood*, 9631; and
without precise locality, *Gerrard*, 223! *Mrs. Saunders*!

16. **L. minor** (Sprague); branches slender, ash-coloured or greyish-
brown, under 1 lin. in diam. 6 in. below the apex, bearing short-
shoots in the axils of the fallen leaves; branchlets pale brown, $\frac{1}{2}$–$\frac{3}{4}$
lin. in diam.; short-shoots perulate at the base, minutely puberu-
lous, bearing 2–3 pairs of leaves, and terminated by an umbel;
leaves opposite, petioled, ovate-lanceolate or ovate, $\frac{3}{4}$–$1\frac{1}{4}$ in. long,
4–6 lin. broad, obtuse or rounded at the apex, obtuse or cuneate at
the base, thinly coriaceous, glabrous, 3-nerved from shortly above
the base, nerves inconspicuous; petiole $\frac{3}{4}$–2 lin. long, slender;
umbels 2–5-flowered; pedicels 3–4 lin. long, slender; bract uni-
laterally developed from a platter-shaped base, $\frac{1}{2}$–$\frac{5}{8}$ lin. long, oblong,
very concave, sometimes almost cymbiform, obtuse, rounded or
truncate at the apex, thickly and obtusely keeled; flowers pentamerous; receptacle and calyx together campanulate, $1\frac{1}{4}$ lin. long;
calyx $\frac{1}{4}$ lin. long including the intramarginal ring, subtruncate;
corolla linear in bud, broadest about the middle, slightly clavate
above, acutely acuminate, up to 2 in. long; filaments $2\frac{1}{2}$–3 lin. long,
with a ventral oblong thickening $\frac{3}{8}$ lin. long at the apex; anthers
linear, $2\frac{1}{2}$–$3\frac{1}{2}$ lin. long, connective rounded; disc much shorter than
the calyx; style filiform; stigma ovoid, nearly $\frac{1}{2}$ lin. long; berry
red, obovoid, 4–5 lin. long. *L. natalitius, var. minor, Harv. in
Harv. & Sond. Fl. Cap.* ii. 576 (*misplaced under L. Zeyheri*); *Wood,
Handb. Fl. Natal,* 115.

EASTERN REGION: Natal; Mooi River, *Gerrard*, 1434! by the Umtwalumi
River, on *Clausena inaequalis*, Benth., *Gerrard & McKen*, 1863! Umzinyati, *Wood*,
1320! Alexandra District, Dumisa, *Rudatis*, 1120! Zululand, *Quleni Forest*,
6000 ft., *Davis in Herb. Wood*, 8608!

Distinguished from the preceding species, to which it has hitherto been referred
as a variety, by the minutely puberulous short-shoots, the smaller, less coriaceous
leaves, the smaller, very slender corolla, the disc much shorter than the calyx,
and the filiform style. The corolla is white, tipped with yellow (*Gerrard*), or
red and pink (*Wood*).

17. **L. Bolusii** (Sprague); glabrous; branches terete, ash-coloured,
densely lenticellate; branchlets subangular, brown, about $\frac{3}{4}$ lin. in
diam. 6 in. below the apex; internodes $\frac{1}{3}$–1 in. long; leaves opposite
or alternate, petioled, oblong-lanceolate or lanceolate, $1\frac{1}{2}$–$2\frac{3}{4}$ in. long,
5–8 lin. broad, minutely apiculate or obtuse at the apex, narrowed
into the base, rigidly coriaceous, trinerved from shortly above the
base, nerves slightly raised on the upper surface, less evident on the
lower; petiole $1\frac{1}{2}$–$2\frac{1}{2}$ lin. long; umbels axillary, solitary, shortly
peduncled, 4–5-flowered; peduncle stout, $\frac{1}{2}$–$\frac{3}{4}$ lin. long; pedicels
$\frac{1}{2}$–$\frac{5}{8}$ lin. long; bract cupular, produced dorsally into a deltoid lobe,
dorsal margin $\frac{3}{4}$ lin. long, equalling the calyx, ventral margin $\frac{1}{2}$ lin.
long; flowers pentamerous; receptacle and calyx together turbinate,
$\frac{7}{8}$ lin. long; calyx truncate, $\frac{1}{3}$ lin. long; corolla 10–11 lin. long;

tube splitting unilaterally or irregularly, with a suprabasal ellipsoid swelling $1\frac{1}{4}$ lin. long commencing $\frac{1}{4}$ lin. above the base, constricted for about $\frac{3}{4}$ lin. above the swelling and then broadened to the apex; lobes erect, linear-spathulate, 5–6 lin. long, $\frac{1}{3}$ lin. broad; filaments inserted $1\frac{1}{2}$ lin. above the base of the corolla-lobes, deflexed, tapering, about $2\frac{3}{4}$ lin. long; anthers linear, $\frac{7}{8}$ lin. long; disc pentagonal, not lobed, $\frac{1}{8}$ lin. high; style filiform; stigma depressed-globose, $\frac{1}{3}$ lin. in diam.

EASTERN REGION: Delagoa Bay, 18 miles from Lourenço Marques, *Bolus*, 9764!

A very distinct species, apparently allied to *L. Lugardi*, N. E. Brown, a native of Ngamiland. The corolla resembles that of *L. rondensis*, Engl., but the latter belongs to a group of species characterized by involute filaments with transverse grooves on the dorsal surface.

18. L. rubromarginatus (Engl. in Engl. Jahrb. xl. 535); glabrous; stem very thick, up to 7 lin. in diam., much swollen at the nodes: branches fairly stout, greyish-brown, nodose, nearly 2 lin. in diam. 6 in. below the apex; short-shoots borne on the stem and branches in the axils of the fallen leaves, extremely contracted, cushion-like, bearing leaves and inflorescences; leaves petioled, elliptic-oblong or elliptic, 1–$1\frac{1}{2}$ in. long, 5–14 lin. broad, obtuse or rounded at the apex, obtuse at the base, coriaceous with cartilaginous margins, obliquely penninerved, nerves about 4 on each side of the midrib, raised on both surfaces; petiole 1–3 lin. long: umbels fascicled, 3–4-flowered; peduncle stout, 1 lin. long or less, with thin partitions between the sockets of the pedicels; pedicels $\frac{1}{2}$ lin. long; bract ovate-cupular, dorsal margin $\frac{7}{8}$–1 lin. long, ventral margin $\frac{3}{8}$–$\frac{1}{2}$ lin. long; flowers pentamerous; receptacle and calyx together campanulate, $1\frac{5}{8}$–$1\frac{7}{8}$ lin. long; calyx subtruncate, $\frac{5}{8}$–$\frac{7}{8}$ lin. long; corolla purplish-crimson (*Galpin*), 2 in. long in bud or less, clavate above, inflated at the base, constricted above the basal swelling, broadened to the middle and again narrowed to the base of the apical swelling, apical swelling oblong, $2\frac{1}{2}$ lin. long, pentagonal with concave faces, 5-ribbed, basal swelling ellipsoid, $2\frac{1}{2}$–3 lin. long; tube splitting unilaterally downwards to below the middle; lobes reflexed below the middle, spathulate, $4\frac{1}{2}$–5 lin. long, $\frac{7}{8}$–1 lin. broad; filaments inserted $\frac{1}{2}$ lin. below the base of the corolla-lobes, deflexed, produced in front of the anther into a deltoid obtuse tooth $\frac{1}{8}$ lin. long, $2\frac{1}{2}$–$3\frac{1}{4}$ lin. long excluding the tooth; anthers narrowly oblong, $1\frac{1}{2}$–$1\frac{3}{4}$ lin. long, truncate; disc $\frac{1}{8}$–$\frac{1}{6}$ lin. high, pentagonal, hardly lobed: style skittle-shaped above, thickened part $3\frac{1}{4}$–4 lin. long, broadest below the middle; stigma ovoid or ellipsoid, $\frac{1}{2}$ lin. long or less. *L. glabriflorus, Conrath in Kew Bulletin, 1908, 226.*

KALAHARI REGION: Transvaal; Magaliesberg Range at Buffelspoort, very common on *Faurea* sp., *Protea* sp., and *Combretum* sp., *Engler*, 2837a! Upper Moodies, near Barberton, on *Chrysophyllum magalismontanum*, Sond., *Galpin*, 1084! near Witpoortje, on *Acacia* sp., *Conrath*, 331!

Allied to *L. oleæfolius*, Cham. & Schlecht., from which it differs in the glabrous corolla with a distinctly ribbed apical swelling.

19. **L. oleæfolius** (Cham. & Schlecht. in Linnæa, iii. 209); branches terete, nodose, $2\frac{1}{2}$–$3\frac{1}{2}$ lin. in diam. 1 ft. below the apex, pale brown, buff-coloured or ashy-grey, glabrous; branchlets minutely and densely puberulous in a young state, soon becoming glabrous, rather slender, 1–$1\frac{1}{2}$ lin. in diam. 6 in. below the apex; internodes $\frac{1}{4}$–2 in. long; leaves opposite, subopposite or alternate, petioled, elliptic-oblong, ovate, lanceolate, obovate or oblanceolate, $\frac{3}{4}$–$3\frac{1}{2}$ in. long, $\frac{1}{4}$–$1\frac{1}{4}$ in. broad, obtuse or rounded at the apex, obtuse or cuneate at the base, rather thickly coriaceous, minutely puberulous in a young state, quickly becoming glabrous, dull, penninerved, lateral nerves rather oblique, 4–5 on each side, anastomosing far from the margin, more or less raised on both surfaces in a dried state; petiole 1–3 lin. long; umbels axillary, solitary or fascicled, 3–4-flowered, minutely puberulous; peduncle stout, $\frac{3}{4}$–2 lin. long; pedicels $\frac{5}{8}$–1 lin. long; bract ovate-cupular, dorsal margin $\frac{7}{8}$–$1\frac{1}{4}$ lin. long, ventral margin $\frac{3}{8}$–$\frac{5}{8}$ lin. long; flowers pentamerous, expanding in two stages: in the first the corolla-lobes separate from one another and become reflexed, the filaments remaining closely pressed together round the style, in the second the corolla-tube splits unilaterally to about the middle, and the filaments become deflexed; receptacle and calyx together campanulate, wide-mouthed, $1\frac{1}{2}$–2 lin. long; calyx subtruncate or repand, $\frac{5}{8}$–$\frac{7}{8}$ lin. long; corolla red, tipped with green (*Pearson*), $1\frac{1}{2}$–$1\frac{3}{4}$ in. long in bud, minutely puberulous outside, basal swelling ellipsoid, $1\frac{3}{4}$–$2\frac{3}{4}$ lin. long, apical swelling oblong-ellipsoid, obtusely pentagonal with flat or slightly convex faces, subtruncate or apiculate from a rounded apex, 2–$2\frac{1}{4}$ lin. long, $1\frac{1}{8}$–$1\frac{1}{4}$ lin. in diam.; tube constricted above the basal swelling, then broadened to the middle and again narrowed to the base of the apical swelling; lobes reflexed below the middle, spathulate, $3\frac{1}{2}$–$4\frac{1}{2}$ lin. long, upper part 2 lin. long, $\frac{7}{8}$ lin. broad; filaments inserted $\frac{1}{2}$–$\frac{3}{4}$ lin. below the base of the corolla-lobes, deflexed or inflexed, $2\frac{1}{4}$–$2\frac{3}{4}$ lin. long, produced in front of the anther into a tooth $\frac{1}{4}$–$\frac{3}{8}$ lin. long; anthers narrowly oblong, truncate or very slightly emarginate, $1\frac{1}{4}$–$1\frac{1}{2}$ lin. long; disc $\frac{1}{6}$–$\frac{1}{4}$ lin. high, not lobed; style skittle-shaped above, thickened part $2\frac{3}{4}$–$3\frac{1}{4}$ lin. long; stigma depressed-globose or obovoid, $\frac{1}{3}$–$\frac{3}{8}$ lin. long, $\frac{3}{8}$–$\frac{1}{2}$ lin. in diam.; berry orange (*Pearson*), ellipsoid, 5 lin. long. *DC. Prodr.* iv. 304, *Schultes, Syst. Veg.* vii. 1634; *E. Meyer in Drège, Zwei Pfl. Documente*, 93, 96; *Sprague in Kew Bulletin*, 1914, 362. *L. speciosus, F. G. Dietr. Lexik. Gaertn. Nachtr.* iv. 473. *L. Lichtensteinii, Herb. Willd. ex Cham. & Schlecht. in Linnæa*, iii. 209, *in synonymy*. *L. Meyeri, Presl, Bot. Bemerk.* 76, *name only*; *Engl. & Gilg in Warb. Kunene-Samb. Exped.* 228; *var. ligustrifolius, Hiern in Cat. Afr. Pl. Welw.* i. 932; *var. inachabensis, Engl. in Engl. Jahrb.* xl. 535. *L. namaquensis, Harv. in Harv. & Sond. Fl. Cap.* ii. 577; *Schinz in Bull. Herb. Boiss.* iv. *App.* iii. 54; *N. E. Brown in Kew Bulletin*, 1909, 135; *De Wild. Pl. Nov. Herb. Hort. Then. t.* 78; *Sprague in Dyer, Fl. Trop. Afr.* vi. i. 361; *var. ligustrifolius, Engl. in Engl. Jahrb.* xx. 120. *L. bumbensis, Hiern in Cat. Afr. Pl. Welw.* i. 933. *Lichtensteinia oleæfolia, Wendl.*

Coll. Pl. ii. 4, *t.* 39.   *Tapinanthus namaquensis, Van Tiegh. in Bull. Soc. Bot. France,* xlii. 267.

VAR. β, **Forbesii** (Sprague) ; corolla glabrous.

VAR. γ, **Leendertziæ** (Sprague) ; corolla densely villous or coarsely pilose with rusty septate hairs ; peduncle, pedicels and bract villous or coarsely pilose ; pedicels sometimes suppressed.

CENTRAL REGION : Prieska Div. ; Prieska, by the Orange River, *Lichtenstein* ; near the Orange River between Gariep Station and Shallow Ford, *Burchell,* 1648 !

WESTERN REGION : Great Namaqualand ; Naiams, *Schinz,* 284 ; Gauas, *Schinz,* 288 ! Kuibes, *Schinz,* 289 ; *Schenck,* 377 ; Bethany, *Schenck,* 394, 395 ; north of Keetmanshoop, *Fenchel,* 154 ; Karas Mountains, *Fenchel,* 153 ; Sandverhaar, on *Parkinsonia africana,* Sond., *Pearson,* 4440 ! Little Bushmanland ; Ramonds Drift, Orange River, *Schlechter* ! Bushmanland ; banks of the Orange River near Abbasis, *Pearson,* 3005 ! sandy plains between Ougrabies and Kweekfontein, on *Parkinsonia africana,* Sond., *Pearson,* 3796 ! Little Namaqualand ; by the Groen River, *Drège* ! Orange River, near Verleptpram, *Drège* ! near Modderfontein, *Whitehead* ! *Bolus,* 9450 ! and without precise locality, *Wyley,* 73 !

KALAHARI REGION : Bechuanaland ; near Kuruman, on *Acacia detinens,* Burch., *Marloth,* 1075 ! Var. γ : Transvaal ; Pietpotgieters Rust, on *Acacia* sp., *Miss Leendertz,* 1142 ! in thickets near Badsloop, *Schlechter,* 4291 ! Crocodile River, *Burke* ! Doorn River, on a quince bush, *Nelson,* 538 !

EASTERN REGION : Var. β : Delagoa Bay, *Forbes* ! Morakwen, *Junod,* 457 !

Harvey, following Ecklon & Zeyher, wrongly identified *Lichtensteinia oleæfolia,* Wendl., with *Loranthus elegans,* Cham. & Schlecht., and described the latter under the name *Loranthus oleæfolius* (Fl. Cap. ii. 576). He described the true *Loranthus oleæfolius,* Cham. & Schlecht., as a new species, *L. namaquensis,* Harv. An historical account of *L. elegans* and *L. oleæfolius* is given in Kew Bulletin, 1914, 359.

L. oleæfolius, Cham. & Schlecht., as defined above, is polymorphic in regard to the shape and size of the leaves, and the indumentum of the corolla. Var. β *Forbesii* appears to differ from the type in nothing but the glabrous corolla ; but Miss Leendertz's specimen, which is the type of var. γ *Leendertziæ,* has densely villous inflorescences and sessile flowers, and has a very distinct facies. It is, however, connected with typical *L. oleæfolius* by the specimens collected by Burke and Nelson, which have much less hairy inflorescences and distinctly pedicelled flowers.

20. **L. kraussianus** (Meisn. in Hook. Lond. Journ. Bot. ii. 539) ; glabrous ; branches rather slender, hardly 1½ lin. in diam. 1 ft. below the apex, greyish-brown, conspicuously lenticellate ; branchlets slender, subangular, smooth, green, turning black on drying ; internodes ¾–1½ in. long ; leaves opposite, subopposite or alternate, conspicuously petioled, ovate or lanceolate, 1¼–3 in. long, ½–1½ in. broad, obtuse or rounded at the apex, obtuse or more or less cuneate at the base, thinly coriaceous, slightly glossy, obliquely penninerved, the second pair of nerves from the base more conspicuous than the rest, running subparallel to the midrib, the upper part of the leaf often appearing trinerved in consequence, nerves raised on both surfaces in a dried state ; petiole 4–6 lin. long ; umbels axillary, solitary on the branchlets, solitary or geminate on the branches, 4–8-flowered ; peduncle 1½–4 lin. long ; pedicels 2–2¾ lin. long, very oblique at the apex ; bract erect, unilaterally developed from a saucer-shaped base, broadly oblong, truncate or rounded, very con-

cave, $\frac{1}{2}$–$\frac{5}{8}$ lin. long, strongly umbonate; flowers pentamerous or hexamerous (*Wood*), expanding in two stages: the lower parts of the corolla-lobes first separating from one another and becoming bowed outwards, the whole corolla afterwards splitting unilaterally downwards to the middle of the tube, the lobes remaining more or less connate by their upper parts, and the filaments becoming deflexed; receptacle and calyx together urceolate, 1–1$\frac{1}{8}$ lin. long; calyx subtruncate, $\frac{1}{8}$–$\frac{1}{4}$ lin. long including the intramarginal ring, which is $\frac{1}{16}$–$\frac{1}{12}$ lin. high; corolla bright red externally, a little lighter within, dull greenish-yellow at the apex (*Wood*), 1$\frac{1}{2}$–2 in. long in bud, basal swelling subglobose, ellipsoid or oblong-ovoid 1$\frac{1}{2}$–2$\frac{3}{4}$ lin. long, apical swelling oblong, 2$\frac{1}{2}$ lin. long, rounded or truncate, pentagonal, strongly 5-ribbed, tube constricted for about 1 lin. above the basal swelling, then broadened upwards to the middle and again narrowed to the base of the apical swelling; lobes erect, spathulate-linear, 3$\frac{1}{2}$–4 lin. long, lower part soft, narrowed upwards, $\frac{1}{2}$–$\frac{5}{8}$ lin. broad at the base, $\frac{3}{8}$ lin. broad at the apex, flat, slightly ribbed outside, upper part with a hard inner layer, 2–2$\frac{3}{8}$ lin. long, $\frac{3}{8}$–$\frac{1}{2}$ lin. broad, narrowly boat-shaped as seen from the side, strongly keeled; filaments inserted at the base of the corolla-lobes, deflexed or inflexed, produced in front of the anther into an obtuse or rounded tooth $\frac{1}{8}$–$\frac{1}{6}$ lin. long, 2–2$\frac{1}{2}$ lin. long excluding the tooth; anthers oblong-linear, 1$\frac{1}{4}$ lin. long; disc $\frac{1}{4}$ lin. high, distinctly lobed; style skittle-shaped above, thickened part 2–2$\frac{1}{4}$ lin. long; stigma depressed-globose, $\frac{1}{4}$ lin. long, $\frac{1}{3}$ lin. in diam.; berry obovoid, 3–3$\frac{1}{2}$ lin. long. *Engl. in Engl. Jahrb.* xx. 109; *M. S. Evans in Nature,* li. 235; *Wood, Natal Pl.* i. *t.* 76. *Tapinanthus kraussianus, Van Tiegh. in Bull. Soc. Bot. France,* xlii. 267.

Var. β, **transvaalensis** (Sprague); branchlets pale brown in a dried state; leaves rigidly coriaceous; pedicels 1$\frac{1}{2}$–1$\frac{3}{4}$ lin. long; receptacle and calyx together $\frac{7}{8}$ lin. long; corolla-lobes 3$\frac{3}{4}$ lin. long, upper part 1$\frac{3}{4}$ lin. long.

Var. γ, **puberulus** (Sprague); like the type, but inflorescence and corolla puberulous; receptacle and calyx together $\frac{7}{8}$ lin. long; corolla-lobes 3–3$\frac{1}{2}$ lin. long, upper part 1$\frac{3}{4}$ lin. long.

Coast Region : Var. γ: Komgha Div.; near Keimouth, *Flanagan,* 25!
Kalahari Region : Var. β : Transvaal; near Barberton, on *Combretum Kraussii,* Hochst., and other trees, *Galpin,* 879!
Eastern Region : Transkei; Kentani, *Miss Pegler,* 292! Natal; near Durban, *Krauss,* 125! *Gueinzius! Gerrard & McKen,* 640! *Wood,* 195! 1724! *Cooper,* 2461! *Conrath,* 755! on *Chætachme aristata,* Planch., *Evans;* Mount Moreland, *Wood,* 1383! Alexandra District, Dumisa, *Rudatis,* 1246! and without precise locality, *Gerrard,* 94! on *Sapindus* sp., *Cooper,* 2459! *Mrs. Saunders!*

21. **L. prunifolius** (E. Meyer in Drège, Zwei Pfl. Documente, 140, 142, name only); vegetative parts glabrous; branches 1$\frac{1}{2}$–2 lin. in diam. 1 ft. below the apex, the younger ones brown, much wrinkled in a dried state, the older greyish-brown, conspicuously lenticellate; branchlets rather slender, smooth, pale green, often turning black on drying; internodes $\frac{1}{2}$–2$\frac{1}{2}$ lin. long; leaves opposite or subopposite,

conspicuously petioled, ovate or ovate-lanceolate, 1–3 in. long, $\frac{2}{3}$–1$\frac{3}{4}$ in. broad, obtuse or rounded at the apex, obtuse or subcuneate at the base, coriaceous, dull, penninerved, nerves irregular, patulous or rather oblique, more or less raised on both surfaces in a dried state or hardly visible; petiole 3–10 lin. long; umbels borne on the branchlets, axillary, solitary, 6–10-flowered; peduncle 1$\frac{1}{2}$–4 lin. long, like the pedicels, bracts, receptacle and calyx minutely and densely pilose with rusty hairs; pedicels 1$\frac{1}{2}$–2$\frac{1}{4}$ lin. long, rather oblique at the apex; bract unilaterally developed from a saucer-shaped or platter-shaped base, ovate or oblong, $\frac{5}{8}$–$\frac{7}{8}$ lin. long, slightly thickened on the back; flowers pentamerous, expanding as in *L. kraussianus*; receptacle and calyx together campanulate, 1$\frac{1}{4}$–1$\frac{1}{2}$ lin. long; calyx shortly 5-toothed or repand, $\frac{1}{3}$–$\frac{1}{2}$ lin. long including the teeth and the intramarginal ring, which is $\frac{1}{12}$–$\frac{1}{8}$ lin. high; corolla orange, 1$\frac{1}{2}$–2 in. long, basal swelling oblong-ovoid or ellipsoid, 2–3$\frac{1}{2}$ lin. long, apical swelling oblong, 2$\frac{1}{4}$ lin. long, pentagonal, 5-ribbed; tube constricted for about 1 lin. above the basal swelling, then broadened upwards to the middle and again narrowed to the base of the apical swelling; lobes erect, spathulate-linear, 4–5$\frac{1}{2}$ lin. long, lower part soft, distinctly ribbed outside, upper part with a hard inner layer, 2–2$\frac{1}{4}$ lin. long, narrowly boat-shaped as seen from the side, keeled; filaments inserted at the base of the corolla-lobes, deflexed or inflexed, produced in front of the anther into a tooth $\frac{1}{8}$–$\frac{1}{6}$ lin. long, 2$\frac{1}{2}$–3 lin. long excluding the tooth; anthers oblong-linear, 1$\frac{3}{8}$–1$\frac{1}{2}$ lin. long; disc $\frac{1}{6}$–$\frac{1}{4}$ lin. high, hardly lobed; style skittle-shaped above, thickened part 2$\frac{1}{2}$–3 lin. long; stigma sub-globose, $\frac{3}{8}$ lin. in diam.; berry obovoid, 3$\frac{1}{2}$–4 lin. long. *Harv. in Harv. & Sond. Fl. Cap.* ii. 578; *Engl. in Engl. Jahrb.* xx. 109. *Tapinanthus prunifolius, Van Tiegh. in Bull. Soc. Bot. France*, xlii. 267.

Coast Region : Bathurst Div.; Glenfilling. *Drège* ! by the Kowie River, *Ecklon & Zeyher* ! Albany Div. ; Grahamstown, on *Ficus capensis*, Thunb., *Galpin*, 2922 ! Fish River, *Schlechter*, 6109 ! and without precise locality, *Hutton* ! Bedford Div. : in the valley of the Mankasana River, *MacOwan*, 411 ! King Williamstown Div. ; between Keiskamma River and the Buffalo River, *Drège* ! Perie Forest, *Kuntze* ! Komgha Div. ; near Komgha, *Flanagan*, 111 ! British Kaffraria, *Cooper*, 329 ! 543 !

Central Region : Somerset Div. ; near Somerset East, *Bowker* !

L. prunifolius is closely allied to *L. kraussianus*, from which it differs in the shape of the receptacle, the more or less distinctly 5-toothed calyx, the longer corolla-lobes, which are less strongly keeled, and the indumentum of the corolla. The leaves are on the whole broader, thicker and less obliquely nerved than those of *L. kraussianus*. *L. kraussianus* var. *puberulus* resembles *L. prunifolius* in the indumentum of the inflorescence and corolla, but agrees with *L. kraussianus* in other respects.

22. **L. Schlechteri** (Engl. in Engl. Jahrb. xl. 530); branches terete, very nodose, densely lenticellate, ash-coloured, 1$\frac{1}{2}$–1$\frac{3}{4}$ lin. in diam. 6 in. below their apex ; branchlets very short, $\frac{1}{4}$–2 in. long, subangular, pale brown, minutely pilose ; internodes 2–9 lin. long ; leaves opposite, ovate or elliptic, $\frac{1}{2}$–1 in. long, 4–7 lin. broad, obtuse or rounded at the apex, rounded or subcuneate at the base,

coriaceous, glabrous, dull, nerves indistinct; petiole about $\frac{1}{2}$ lin. long, minutely rusty-pilose in a young state; umbels axillary, solitary or fascicled, 3–4-flowered; peduncle $1\frac{1}{2}$ lin. long, puberulous above, with cupular sockets in which the pedicels are inserted; pedicels very slender, $2\frac{1}{4}$ lin. long; bract shorter than the receptacle and calyx, elliptic-ovate from a saucer-shaped base, dorsal margin $\frac{1}{2}$ lin. long, minutely ciliate, ventral margin $\frac{1}{8}$ lin. long; flowers tetramerous, only known in bud; receptacle and calyx together campanulate, $\frac{7}{8}$ lin. long; calyx subtruncate, $\frac{1}{3}$–$\frac{3}{8}$ lin. long, minutely ciliate; corolla 8 lin. long in bud, clavate, tetragonal, apical swelling $2\frac{3}{4}$ lin. long, obtuse; filaments becoming involute when the flower expands, inserted 2 lin. above the base of the corolla, linear, $3\frac{1}{2}$ lin. long, grooved on the inner surface, produced in front of the anther into a triangular very acute tooth, $\frac{3}{8}$ lin. long; anthers linear, 2 lin. long, truncate, the inner lobes $\frac{1}{8}$ lin. shorter than the outer; disc tetragonal, hardly lobed, $\frac{1}{8}$ lin. high; style filiform; stigma ellipsoid, $\frac{3}{8}$ lin. long.

EASTERN REGION : Mozambique ; Macocololo, *Schlechter*, 12061 !
Allied to *L. ramulosus*, Sprague, a native of British East Africa.

II. VISCUM, Linn.

Flowers monœcious or diœcious. *Calyx* absent or represented by a mere rim. *Corolla* regular. *Male flower* trimerous or tetramerous : —*Receptacle* more or less hollowed. *Corolla* polypetalous, but usually appearing gamopetalous owing to the absence of demarcation between the petals and the receptacular tube; petals 3–4, more or less triangular. *Anthers* sessile, adhering by their dorsal surface to the lower part of the petals and the upper part of the receptacular tube; cells numerous, dehiscing introrsely by pores. *Female flower* :—*Corolla* superior, polypetalous ; petals 3–4, deciduous or persistent. *Ovary* inferior ; style short or none ; stigma thick, cushion-shaped. *Berry* crowned by the petals or not. *Seed* albuminous ; embryos 1–3, terete.

Green leafy or, at first sight, leafless shrubs, parasitic on other plants. Branches usually much forked, jointed immediately above the nodes ; internodes often compressed, angled or not ; leaves opposite, well developed or represented by small scales ; inflorescences axillary, or axillary and terminal, consisting of solitary or fascicled flowers or cymules ; cymules peduncled or sessile, 3–9-flowered, with the flowers in one plane ; flowers small, inconspicuous, green, yellow or white ; berries red, orange, yellow or white.

Each axillary branch, of whatever order, has a pair of small scale-leaves at its very base. The scale-leaves, which are placed transversely with respect to the subtending leaf, are also found at the base of each axillary inflorescence. Each cymule and, as a rule, each solitary flower is subtended at its base by a peduncled or sessile pair of bracts (*bracteal cup*).

DISTRIB. Species about 60, all Old World, mostly natives of warm regions.

Section 1. Ploionixia.—Leaves not scale-like.
Berry warted:
Bracteal cup sessile; berry sessile:
Petals of the female flower deltoid-ovate or ovate, shorter than the receptacle; receptacle warted; style and stigma $\frac{1}{6}$–$\frac{1}{3}$ lin. long ... ... (1) **obovatum.**
Petals of the female flower lanceolate-oblong, equalling the receptacle; receptacle not distinctly warted; style and stigma $\frac{3}{8}$ lin. long ... (2) **pulchellum.**
Bracteal cup peduncled; berry pedicelled ... ... (3) **subserratum.**
Berry smooth:
Petals persistent in fruit ... ... (4) **nervosum.**
Petals deciduous:
Berries pedicelled:
Flowers diœcious; female flowers solitary in their bracteal cups; leaves obovate ... ... (5) **obscurum.**
Flowers monœcious; bracteal cups 3-flowered; leaves usually not obovate:
Leaves very thick; nerves not visible ... ... (6) **pauciflorum.**
Leaves thinner; nerves more or less visible:
Leaves $\frac{2}{3}$–$1\frac{1}{4}$ in. long ... ... (7) **Eucleæ.**
Leaves $\frac{1}{4}$–$\frac{2}{3}$ in. long:
Leaves broadly ovate or suborbicular, rounded at the base ... ... (8) **rotundifolium.**
Leaves ovate-oblong, elliptic-oblong or oblong-lanceolate, obtuse or acute at the base ... ... (9) **tricostatum.**
Berries sessile ... ... (11) **Crassulæ.**
Insufficiently known species of this section ... ... (10) **Schæferi.**

Section 2. Aspidixia.—Leaves scale-like.
Stem bushy, much-branched:
Internodes conspicuously flattened:
Internodes $\frac{1}{2}$–$2\frac{3}{4}$ lin. long; berry sessile ... ... (12) **combreticola.**
Internodes $\frac{1}{4}$–1 in. long; berry pedicelled in *V. anceps*, not known in *V. Junodii*:
Male flowers 3 together in each bracteal cup ... (13) **anceps.**
Male flowers solitary ... ... (14) **Junodii.**
Internodes not conspicuously flattened:
Berries smooth:
Berries sessile:
Branchlets slender, $\frac{5}{8}$–$\frac{3}{4}$ lin. in diam. in a dried state; scale-leaves very prominent; branches nodose ... ... (15) **capense.**
Branchlets stout, $\frac{7}{8}$–1 lin. in diam. in a dried state; scale-leaves not very prominent; branches not nodose ... ... (16) **robustum.**
Berries pedicelled ... ... (17) **continuum.**
Berries warted:
Lobes of bracteal cup rounded ... ... (18) **verrucosum.**
Lobes of bracteal cup acute ... ... (19) **rigidum.**
Stem minute, consisting of a single internode ... ... (20) **minimum.**

1. **V. obovatum** (Harv. in Harv. & Sond. Fl. Cap. ii. 579);

branches and branchlets terete, 1–1½ lin. in diam. 6 in. below the apex; branchlets up to 4½ in. long; internodes 4–9 lin. long, the youngest minutely papillate; leaves distinctly petioled, broadly obovate, rounded at the apex, subcuneate into the base, ½–1 in. long, 5–9 lin. broad, coriaceous, glabrous, obscurely 3-nerved, dull; petiole ½–1 lin. long; bracteal cups axillary, sessile, usually three together, the central one of which develops first, bearing 2–3-flowered cymules, with the flowers all male or of both sexes, or single female flowers; bracteal cups of the male cymules boat-shaped, 1–1¼ lin. long, nearly ½ lin. high, coarsely glandular-ciliate on the inner margin, those bearing solitary female flowers shorter, distinctly bilabiate; male cymules 3-flowered; flowers quadrangular, very broadly fusiform; petals deltoid-ovate, hardly ½ lin. long, not distinctly demarcated from the receptacle; anthers 2–4, elliptic or ovate-oblong, ½–⅝ lin. long, ⅜ lin. broad; male flowers of the mixed cymules having the external appearance of female flowers, the petals being distinct from the receptacle; female flowers: receptacle campanulate, ⅝ lin. long, warted; petals deltoid-ovate or ovate, ½–⅝ lin. long, or less; style and stigma together ⅙–¼ lin. long; berry yellow, ellipsoid, 2 lin. long, strongly warted.

COAST REGION: Bathurst Div.; near Barville Park, *Burchell*, 4113!
EASTERN REGION: Natal; near Durban, *Gerrard & McKen*, 659! *Wood*, 1631! and without precise locality, *Sanderson*!

Burchell's specimen has somewhat smaller leaves than typical *V. obovatum* (4–7 lin. long, 3–6 lin. broad), the receptacle of the female flower is strongly warted, the petals ovate-oblong, and the style and stigma are a little longer (¼–⅓ lin. long).

The inflorescence of *V. obovatum* should be studied in the field. There are often three bracteal cups in an axil, the central one bearing three male flowers, the lateral ones a single female flower each. Bracteal cups occur, however, in which there are flowers of both sexes; and the male flowers in these cups are hardly distinguishable externally from the female, possessing a similar receptacle and distinct petals. One female flower examined had an anther on one of the petals.

2. V. pulchellum (Sprague); branches subterete, slender, under 1 lin. in diam. 6 in. below the apex; branchlets very slender, ½–2½ in. long, subangular, minutely papillate; internodes of the branchlets 2½–6 lin. long; leaves distinctly petioled, broadly obovate, rounded or obtuse at the apex, subcuneate into the base, 3–6 lin. long, 2–4 lin. broad, coriaceous, glabrous, obscurely 3-nerved, slightly glossy; petiole ½–1 lin. long; female inflorescence: bracteal cups axillary, sessile, solitary or geminate, bearing solitary female flowers, distinctly bilabiate, lips diverging at a right angle or less, ½–⅝ lin. long, ¼–⅓ lin. high in the middle, coarsely glandular-ciliate on the inner margin; receptacle narrowly campanulate, not distinctly warted, ⅝ lin. long, ⅜–$\frac{7}{16}$ lin. in diam.; petals lanceolate-oblong, subacute, ⅝ lin. long; style and stigma ⅜ lin. long; berry ellipsoid, 2 lin. long, finely warted.

EASTERN REGION: Natal; Tugela River, *Gerrard*, 1649!

Gerrard, 1649, in Herb. Trin. Coll. Dublin, includes one female and two male specimens. The latter have not been described, as they are from different localities, and are possibly not conspecific with the female specimen.

3. V. subserratum (Schlechter in Journ. Bot. 1896, 504); glabrous; branches subterete, fairly stout, about 1¾ lin. in diam. 6 in. below the apex; branchlets up to 4 in. long; internodes of the branchlets ½–1 in. long, compressed, tapering downwards, strongly 6-ribbed, the two ribs below the leaves wing-like; leaves distinctly petioled, obovate or obovate-oblong, rounded or very obtuse at the apex, subcuneate into the base, ⅔–1¼ in. long, 3½–9½ lin. broad, minutely serrulate in a dried state, thickly coriaceous, dull, 3-nerved, nerves prominent on the upper surface, less evident on the lower; petiole ½–¾ lin. long; flowers apparently diœcious; male cymules axillary, solitary or ternate, peduncled, 3-flowered; peduncle nearly ½ lin. long; bracteal cup boat-shaped, hardly 1¼ lin. long; male flowers quadrangular, fusiform, 1½ lin. long; petals deltoid, ¾ lin. long, not distinctly demarcated from the receptacle; anthers oblong, ¾ lin. long, ⅓–⅜ lin. broad; female inflorescence: bracteal cups axillary, solitary or geminate, peduncled, 1-flowered, strongly bilabiate, lips diverging at nearly a right angle, ½ lin. high, minutely glandular-ciliate; peduncle ½–⅝ lin. long; flowers pedicelled; receptacle and pedicel together subcylindric, slightly tapering downwards, ⅞ lin. long; receptacle warted; petals lanceolate-oblong, ¾ lin. long; style and stigma together ⅝ lin. long; stigma capitate; berry ovoid, truncate, coarsely and densely warted, 2 lin. long excluding the pedicel; pedicel stout, broadened into the base of the berry, ¾ lin. long. *V. galpinianum, Schinz in Vierteljahrsschr. Nat. Ges. Zürich*, xlix. 179.

KALAHARI REGION: Transvaal; Barberton, on *Cussonia* sp., *Galpin*, 452! Shilovane, *Junod*, 2246!

EASTERN REGION: Natal; Tugela River, *Gerrard & McKen*, 1654! and without precise locality, *Gerrard*, 1650!

V. subserratum is probably diœcious: *Galpin*, 452, and *Junod*, 2246, bear berries only; and *Gerrard*, 1650, is represented in the Kew Herbarium by two specimens, one bearing male inflorescences exclusively, the other female flowers and berries.

4. V. nervosum (Hochst. ex A. Rich. Tent. Fl. Abyss. i. 338); main stem terete; branches angular; branchlets compressed, 6-ribbed; internodes about ¾ lin. broad towards the apex, slightly tapering downwards; leaves shortly petioled, elliptic or ovate-elliptic, obtuse or rounded at the apex, more rarely subacute or acute, especially in a young state, cuneate into the base, ½–1¾ in. long, ⅓–1¼ in. broad, coriaceous, glabrous, distinctly 3–5-nerved, finely reticulate, dull; nerves more or less raised, especially on the upper surface; petiole ½–1 lin. long; cymules axillary, solitary or one on each side of an axillary branch, peduncled, 3-flowered; flowers monœcious, tetramerous, either all of the same sex or male and female together in the same cymule; peduncles ½–1½ lin. long, those of the male or mixed cymules rather shorter than those of the female; bracteal cup boat-shaped, ⅝–1 lin. long, ⅓–½ lin. broad, minutely ciliate; male flower ellipsoid or obovoid in bud, ⅞–1 lin. long, solid base $\frac{1}{12}$–⅓ lin. long; receptacular tube ¼–⅓ lin. long; petals deltoid-ovate or ovate, ½ lin. long, $\frac{7}{16}$–½ lin. broad at the base;

anthers inserted about the base of the petals, obtusely trigonous, elliptic or suborbicular in outline, $\frac{1}{2}$ lin. long, $\frac{3}{8}-\frac{7}{16}$ lin. broad; female flower: receptacle subclavate, $\frac{7}{8}$ lin. long, $\frac{3}{8}$ lin. in diam., rapidly becoming ellipsoid-oblong and then ellipsoid after pollination; petals slightly patulous, oblong-ovate, about $\frac{3}{8}$ lin. long, $\frac{1}{3}$ lin. broad; style and stigma together $\frac{1}{4}$ lin. long; berry ellipsoid or ovoid, 2 lin. long, smooth, crowned by the persistent petals. *Engl. Hochgebirgsfl. Trop. Afr.* 198; *Engl. Jahrb.* xx. 131 ; *Pfl. Ost-Afr.* C. 167 ; *Schweinf. in Bull. Herb. Boiss.* ix. *App.* ii. 152 ; *Van Tiegh. in Bull. Soc. Bot. France,* xliii. 190 ; *Sprague in Dyer, Fl. Trop. Afr.* vi. i. 397. *V. murchisonianum, Schweinf. ex Baker in Journ. Bot.* 1882, 245, *in obs., name only.*

EASTERN REGION : Pondoland ; Port St. John, *Galpin,* 2886 ! Natal ; Tugela River, *Gerrard,* 1651 ! Enyangweni, *Gerrard & McKen* 1865 ! Ungoya Forest, *Wood,* 3864 ! Great Noodsberg, *Wood,* 4145 ! Umlaas, *Weale!* Alexandra District, Dumisa, on *Rapanea* sp., *Rudatis,* 1044 ! Delagoa Bay, *Mrs. Monteiro!*

Also in Tropical Africa, extending northwards to Eritrea.

5. **V. obscurum** (Thunb. Prodr. 31) ; glabrous ; stem and older branches terete, younger branches and branchlets hexagonal ; internodes $\frac{1}{2}-2\frac{1}{2}$ in. long, the uppermost slightly compressed : leaves distinctly petioled, oblanceolate, narrowly or broadly obovate, rounded at the apex, cuneate into the base, $\frac{1}{2}-1\frac{1}{2}$ in. long, $\frac{1}{4}-\frac{3}{4}$ in. broad, coriaceous, dull, 3 nerved, nerves more distinct on the upper surface, sometimes not visible ; petiole $\frac{1}{4}-3$ lin. long ; bracteal cups terminal and axillary, solitary or ternate, subsessile ; flowers diœcious ; male inflorescence : bracteal cup boat-shaped, $1\frac{3}{4}$ lin. long, $\frac{1}{4}-\frac{5}{8}$ lin. high ; flowers obovoid, quadrangular : petals deltoid, $\frac{1}{2}$ lin. long, $\frac{1}{2}-\frac{3}{4}$ lin. broad at the base ; anthers elliptic-oblong, $\frac{1}{4}-\frac{7}{8}$ lin. long, $\frac{3}{8}-\frac{5}{8}$ lin. broad ; female inflorescence : bracteal cup conspicuously bilabiate, $\frac{3}{4}$ lin. long, $\frac{3}{8}$ lin. high, $\frac{1}{2}$ lin. high in the middle, lips diverging from each other at less than a right angle, with erect dorsal margins ; flowers solitary, pedicelled ; pedicel $1\frac{1}{4}$ lin. long ; receptacle $\frac{7}{8}$ lin. long ; petals erect, slightly patulous above, ovate-oblong, obtuse, hardly $\frac{3}{4}$ lin. long ; style and stigma together $\frac{3}{8}$ lin. long ; berry long-pedicelled, ellipsoid, $2\frac{1}{4}$ lin. long, smooth, yellowish-white or white ; pedicel 2–3 lin. long. *Fl. Cap. ed. Schult.* 154 ; *DC. Prodr.* iv. 285 ; *Drège, Zwei Pfl. Documente,* 129 ; *Harv. in Harv. & Sond. Fl. Cap.* ii. 579 ; *Engl. in Engl. Jahrb.* xx. 131 ; *Bolus & Wolley-Dod in Trans. S. Afr. Phil. Soc.* xiv. 316. *V. pauciflorum, Drège, Zwei Pfl. Documente,* 229, *not of Thunb. V. rotundifolium, Eckl. & Zeyh. Enum.* 357, *not of Linn. f. V. brevifolium, Engl. in Engl. Jahrb.* xx. 131. *V. bivalve, Engl. in Engl. & Prantl, Pflanzenfam. Nachtr.* 1 *zu* ii.–iv. 140. *Aspidixia bivalvis, Van Tiegh. in Bull. Soc. Bot. France,* xliii. 192.

COAST REGION : Cape Div. : Tokay Plantation, on *Olea* sp., *Bolus & Wolley-Dod ;* Swellendam Div. ; Grootvaders Bosch and Duyvels Bosch, *Ecklon & Zeyher,* 2273 ! *Mund* ! George Div : Kaymans Gat, *Drège* ! Uitenhage Div. ; Zwartkops River, *Drège* ! *Ecklon & Zeyher,* 2272 ! 2700 ! near the Zwartkops and Sundays

Rivers, on *Rhus* sp. and *Celastrus* sp., *Zeyher*, 244! near Uitenhage, *Burchell*, 4414! *Cooper*, 1576! near Enon, *Baur*, 106! Albany Div. : *Ecklon & Zeyher*, 2272! *Williamson*! Grahamstown, on *Salix* sp., *MacOwan*, 453! Howisons Poort, *Hutton*! Stockenstrom Div. ; Chumi Berg, *Ecklon & Zeyher*, 2273! Kat River, unknown collector! Cathcart Div. ; Goshen, on *Salix* sp., *Baur*, 933! East London Div. ; in scrub on sea-coast at East London, *Galpin*, 1860!
CENTRAL REGION: Somerset Div. ; near Somerset East, by the Little Fish River, on *Rhus* sp., *MacOwan*, 453 ! 543! Bosch Berg, near Somerset East, *Burchell*, 3211 !
EASTERN REGION : Natal, *Gerrard*, 1654 !

6. **V. pauciflorum** (Linn. f. Suppl. 426) ; glabrous ; branching mainly dichasial ; stem and older branches terete, younger branches and branchlets hexagonal ; internodes $\frac{1}{3}$–$1\frac{1}{2}$ in. long, the uppermost slightly compressed ; leaves subsessile or shortly petioled, elliptic, elliptic-oblong or obovate, apiculate, obtuse or rounded at the apex, obtuse at the base, 4–9 lin. long, 3–5 lin. broad, thickly coriaceous, dull, 3-nerved, nerves just visible in the very young leaves, not visible in the older ones ; petiole up to $\frac{1}{2}$ lin. long ; bracteal cups terminal and axillary, solitary or 2–3 together, peduncled, boat-shaped, distinctly bilabiate, 3-flowered ; peduncle $\frac{1}{2}$–$\frac{5}{8}$ lin. long ; bracts ascending, triangular-ovate, acute, $\frac{5}{8}$–$\frac{3}{4}$ lin. long; flowers monœcious, the central one male, the two lateral female ; male flowers examined abnormal, with a solid receptacle, and the anthers fused into a central mass occupying the position of the stigma in a female flower ; female flowers : receptacle and pedicel together nearly $\frac{3}{4}$ lin. long ; petals ovate-oblong, $\frac{1}{2}$ lin. long ; stigma sessile, $\frac{1}{6}$ lin. long ; berry pedicelled, oblong (*Harvey*), smooth, yellowish-white ; pedicel shorter than the berry (*Harvey*). *Thunb. Fl. Cap. ed. Schult.* 154 ; *DC. Prodr.* iv. 285 ; *Eckl. & Zeyh. Enum.* 357 ; *Harv. in Harv. & Sond. Fl. Cap.* ii. 579, excluding var. *Eucleæ*.

COAST REGION : Van Rhynsdorp Div. ; Heerenlogement, on *Rhus Thunbergii*, Hook., *Zeyher*, 750 ! Tulbagh Div. ; near Tulbagh, *Ecklon & Zeyher*, 2274 ! Paarl Div. ; Paarl Mountain, *Drège* !
CENTRAL REGION : Calvinia Div. ; Onder Bokkeveld, *Ecklon & Zeyher* ! 2274 !
WESTERN REGION : Little Namaqualand ; Granite knoll, Brakdam, *Pearson*, 5656 ! Sneeuwkop, *Pearson*, 5851 !

V. pauciflorum is closely allied to *V. rotundifolium*, Linn. f., and *V. tricostatum*, E. Meyer : it differs from both in the denser, usually markedly dichasial branching, and the larger thick nerveless leaves. The material available for examination was rather poor.
Rehmann, 2466, from mountains above Worcester, has been referred to *V. pauciflorum*, but the branching is mainly racemose and the leaves are distinctly nerved.
Burchell, 1022, from near Tulbagh, also differs in the branching, and is said to have orange-coloured berries.

7. **V. Eucleæ** (Eckl. & Zeyh. Enum. 357) ; glabrous ; branching mainly racemose ; stem and older branches subterete, younger branches and branchlets hexagonal ; internodes $\frac{1}{2}$–$1\frac{1}{4}$ in. long, the uppermost slightly compressed ; leaves subsessile or shortly petioled, elliptic or elliptic-oblong, acute at the apex or apiculate, obtuse at the base, $\frac{2}{3}$–$1\frac{1}{4}$ in. long, 4–7 lin. broad, distinctly 3-nerved ; petiole up to $\frac{1}{2}$ lin. long ; bracteal cups terminal and axillary, solitary or

ternate, peduncled, boat-shaped, 3-flowered; peduncle $\frac{1}{2}$–$\frac{3}{4}$ lin. long; flowers pedicelled (*Ecklon & Zeyher*); berry pedicelled, globose, smooth, 2–2$\frac{1}{2}$ lin. in diam.; pedicel 1–1$\frac{1}{2}$ lin. long. *V. pauciflorum*, *var. Eucleæ, Harv. in Harv. & Sond. Fl. Cap.* ii. 580.

COAST REGION: Malmesbury Div.; near Driefontein, in Groene Kloof, *Ecklon & Zeyher*, 2275! Cape Div.; Hout Bay, on *Euclea racemosa*, Murr., *MacOwan, Herb. Austr.-Afr.*, 1647! Camps Bay, near Oude Kraal, *Pappe*!

Differs from *V. pauciflorum*, Linn. f., to which it was reduced by Harvey, in the branching, the acute, distinctly nerved leaves, and the red berries. None of the specimens examined bore flowers; but judging from the arrangement of the berries, each bracteal cup appears to bear a central male and two lateral female flowers.

8. **V. rotundifolium** (Linn. f. Suppl. 426); a small much-branched shrub, glabrous or nearly glabrous in all its vegetative parts; stem terete, 2–2$\frac{1}{4}$ lin. in diam. $\frac{1}{2}$–1 ft. below the apex of the branchlets; older branches 8- or 12-ribbed, younger branches and branchlets hexagonal, the uppermost branchlets slightly compressed; internodes 2–18 lin. long; leaves sessile or subsessile, broadly ovate (or more rarely suborbicular), obtuse, acute or apiculate at the apex, rounded at the base, 2$\frac{1}{2}$–6 lin. long, 1$\frac{1}{2}$–4 lin. broad, thickly coriaceous, obscurely 3-nerved, brown or blackish in a dried state, margin cartilaginous; cymules axillary, solitary or fascicled, 3-flowered, usually composed of a central male and two lateral female flowers, more rarely of three female flowers; peduncle $\frac{1}{2}$–1$\frac{1}{4}$ lin. long, produced $\frac{1}{10}$–$\frac{1}{5}$ lin. or less beyond the bracteal cup; bracteal cup distinctly lobed, 1$\frac{1}{8}$–1$\frac{3}{8}$ lin. long; lobes ascending, ovate or ovate-oblong, apiculate or subacute, $\frac{5}{8}$–$\frac{3}{4}$ lin. long, glandular-ciliolate; flowers monœcious, tetramerous; male flowers: receptacle and pedicel together oboonical, $\frac{3}{4}$ lin. long; pedicel about $\frac{3}{8}$ lin. long; receptacular tube $\frac{1}{4}$–$\frac{1}{3}$ lin. long; petals marked off from the receptacle by a distinct groove on the outer surface, ovate or ovate-deltoid, slightly unequal, $\frac{1}{2}$–$\frac{5}{8}$ lin. long, $\frac{3}{8}$–$\frac{1}{2}$ lin. broad, the two outer separated $\frac{1}{4}$ lin. from each other at the apex; anthers elliptic, $\frac{1}{4}$–$\frac{3}{8}$ lin. long; female flowers: receptacle and pedicel together 1$\frac{1}{4}$–1$\frac{3}{4}$ lin. long; pedicel $\frac{1}{4}$–$\frac{5}{8}$ lin. long; petals ovate or ovate-oblong, acute, $\frac{1}{3}$–$\frac{3}{8}$ lin. long; stigma projecting $\frac{1}{8}$–$\frac{1}{4}$ lin. above the insertion of the petals; berry pedicelled, red, orange or yellow, ellipsoid, 2$\frac{1}{2}$ lin. long; pedicel $\frac{3}{4}$–1 lin. long. *Thunb. Prodr.* 31; *Fl. Cap. ed. Schult.* 154; *DC. Prodr.* iv. 279; *Harv. in Harv. & Sond. Fl. Cap.* ii. 580; *Engl. in Engl. Jahrb.* xx. 131; *Schinz in Bull. Herb. Boiss.* iv. App. iii. 55; *Van Tiegh. in Bull. Soc. Bot. France*, xliii. 190; *Sprague in Dyer, Fl. Trop. Afr.* vi. i. 403, 1034. *V. glaucum, Eckl. & Zeyh. in S. Afr. Journ.* 1830, 375; *Enum.* 357. *V. Zizyphi-mucronati, Dinter, Deutsch-Südw.-Afr.* 56, *partly.*

COAST REGION: Riversdale Div.; Corente River Farm, *Muir in Herb. Galpin*, 5324! Mossel Bay Div.; on dry hills on the eastern side of the Gouritz River, *Burchell*, 6422! Uitenhage Div.; Zwartkops River, on *Salix* sp., *Ecklon & Zeyher*, 2272! 2276! *Zeyher*, 624! 2701! Port Elizabeth Div.; Port Elizabeth, *Pappe*! Albany Div.; *Ecklon & Zeyher*, 2272! *Williamson*! Grahamstown, on *Euclea*

sp., *MacOwan* ! 1067 ! *Bolton* ! Howisons Poort, *Hutton* ! Queenstown Div. ;
Finchams Nek, *Galpin,* 1820 !
    CENTRAL REGION : Prince Albert Div. ; Gamka River, *Burke* ! Somerset Div. ;
between Zuurberg Range and Klein Bruintjes Hoogte, *Drège* ! Graaff Reinet Div. ;
near Graaff Reinet, *Burchell,* 2935 ! *Bolus,* 73 ! Middelburg Div. ; near Middelburg, *Burchell,* 2808 !
    WESTERN REGION : Little Namaqualand ; near the mouth of the Orange River,
Drège, 7651 ! Van Rhynsdorp Div. ; mountain 3 miles north-east of Stinkfontein,
Pearson, 5688 !
    KALAHARI REGION : Transvaal ; Magaliesberg, *Sanderson* ! Lehlaba River,
Nelson, 557 ! Pretoria, *Rehmann,* 4709 ! Lydenburg, *Wilms,* 571 !
    EASTERN REGION : Natal ; Springvale, *Sanderson* ! Tugela River, *Gerrard,*
1652 ! Mooi River, *Wood,* 4473 ! Weenen, *Wood,* 4473a ! and without precise
locality, *Gerrard,* 1432 !
    Also in Tropical Africa.

9. **V. tricostatum** (E. Meyer in Drège, Zwei Pfl. Documente, 94,
229, name only) ; a globose much-branched shrub, glabrous in all its
vegetative parts ; stem terete, about $2\frac{1}{4}$ lin. in diam. 16 in. below
the apex of the branchlets ; older branches 8- or 12-ribbed, younger
branches and branchlets hexagonal, the uppermost internodes
slightly compressed ; internodes 4–16 lin. long ; leaves subsessile,
spreading, ovate-oblong, elliptic-oblong or oblong-lanceolate, very
acute or apiculate at the apex, obtuse or acute at the base, $4–7\frac{1}{2}$ lin.
long, $1\frac{1}{4}–3\frac{1}{2}$ lin. broad, coriaceous, 3-nerved, green or brownish in a
dried state, margin cartilaginous ; cymules terminal and axillary,
solitary or fascicled, 3-flowered, usually composed of a central male
and two lateral female flowers, more rarely of three female flowers ;
peduncle $\frac{1}{2}–\frac{5}{8}$ lin. long, produced $\frac{1}{8}$ lin. or less beyond the
bracteal cup ; bracteal cup distinctly lobed, $\frac{7}{8}–1\frac{1}{4}$ lin. long ; lobes
ascending, ovate or ovate-oblong, acute or subacute, $\frac{1}{2}–\frac{5}{8}$ lin. long,
minutely glandular-ciliolate ; flowers monœcious, tetramerous ; male
flower : receptacle and pedicel together obconical, $\frac{5}{8}$ lin. long ;
pedicel about $\frac{3}{8}$ lin. long ; receptacular tube about $\frac{1}{4}$ lin. long ; petals
marked off from the receptacle by a distinct groove on the outer
surface, subequal, ovate, obtuse, about $\frac{1}{2}$ lin. long, $\frac{1}{3}$ lin. broad ;
anthers elliptic, $\frac{1}{4}–\frac{1}{3}$ lin. long ; female flower : receptacle and pedicel
together about $1\frac{1}{8}$ lin. long ; pedicel about $\frac{1}{2}$ lin. long ; petals
shortly ovate-oblong, obtuse or rounded, $\frac{3}{8}–\frac{7}{16}$ lin. long, $\frac{1}{3}–\frac{3}{8}$ lin.
broad ; stigma projecting $\frac{1}{6}–\frac{1}{2}$ lin. above the insertion of the petals ;
berry pedicelled, orange-red (*Dinter*), ellipsoid, $1\frac{3}{4}–2$ lin. long ;
pedicel 1 lin. long. *Harv. in Harv. & Sond. Fl. Cap.* ii. 580 ; *Engl.
in Engl. Jahrb.* xx. 131 ; *Van Tiegh. in Bull. Soc. Bot. France,* xliii.
190 ; *Sprague in Dyer, Fl. Trop. Afr.* vi. i. 403, 1034. *V. thymi-
folium, Presl, Epim. Bot.* 251. *V. Zizyphi-mucronati, Dinter,
Deutsch-Südw.-Afr.* 56, *partly.*

    CENTRAL REGION : Hopetown Div. ; *Wyley,* 19 !
    WESTERN REGION : Little Namaqualand ; near the mouth of the Orange River,
Drège ! and without precise locality, *Wyley,* 71 !
    KALAHARI REGION ; Orange River Colony ; by the Orange River, on *Salix* sp.,
Burke, 387 ! *Zeyher,* 387 ! 747 !
    Also in Tropical Africa.

V. tricostatum is very closely allied to *V. rotundifolium*, Linn. f., and might be treated as a variety of it. The type specimens of *V. tricostatum* have ovate-oblong leaves, 2–3½ lin. broad. *Burke*, 387, and *Zeyher*, 387 and 747, have oblong-lanceolate leaves, 1–2 lin. broad, but do not seem to differ in other respects.

10. V. Schæferi (Engl. & Krause in Engl. Jahrb. li. 470); glabrous, yellowish-green in a dried state; branches subterete, sulcate, rather stout; young branchlets slightly compressed; internodes 6–10 lin. long; leaves subsessile, soon deciduous, narrowly ovate-lanceolate or (more rarely) obovate lanceolate, acute at the apex, rather gradually narrowed towards the base, 6–8 lin. long, 1¼–1¾ lin. broad, thickly coriaceous, nerves hardly raised; male inflorescence not known; female inflorescence: bracteal cups axillary, solitary or several together, peduncled, usually 3-flowered; bracts narrowly ovate, acute, ½–¾ lin. long; peduncle stout, 1½–2½ lin. long; petals subovate, acute, 1¼–1½ lin. long, connate at the base; style subcylindric, hardly ½ lin. long; stigma slightly thickened; berries ovoid, 2–2¼ lin. long, 1–1½ lin. in diam., yellowish, smooth, or slightly warted in a dried state.

WESTERN REGION: Great Namaqualand; on the Fish River near Seeheim, *Schäfer*, 465; near Seeheim, on *Mærua Schinzii*, Pax, *Engler*, 6601.

Engler and Krause do not state whether the female flowers and berries of *V. Schæferi* are pedicelled or sessile. The leaves apparently resemble those of the narrow-leaved form of *V. tricostatum*, but the relatively long peduncle and large petals distinguish it from that species.

11. V. Crassulæ (Eckl. & Zeyh. Enum. 357); yellowish-green in a dried state, young branchlets minutely papillate, otherwise glabrous; branches terete, rather stout, 2½ lin. in diam. 6 in. below the apex, succulent, much wrinkled in a dried state; branchlets slightly compressed; internodes ¼–1 in. long; leaves subsessile, obovate or suborbicular, rounded at the apex, obtuse at the base, 3–5 lin. long, 2½–3½ lin. broad, very thick and fleshy, nerves not visible; flowers apparently diœcious; male inflorescence: bracteal cups axillary, solitary or ternate, broadly boat-shaped with rounded ends, 1¾ lin. long, ⅝ lin. deep, glandular-ciliate inside the margin, 3-flowered; flower obovoid, quadrangular; petals deltoid, ½ lin. long, not distinctly demarcated from the receptacle; female inflorescence; bracteal cups terminal and axillary, solitary, boat-shaped with obtuse or rounded ends, 1-flowered, at the time of fruiting 2 lin. long, ¾–⅞ lin. high; berry sessile, ovoid, 3 lin. long, smooth, red; style and stigma together ⅝ lin. long, partially sunk in an apical depression of the berry about ¼ lin. deep. *Harv. in Harv. & Sond. Fl. Cap.* ii. 580; *Engl. in Engl. Jahrb.* xx. 131. *V. Euphorbiæ, E. Meyer in Drège, Zwei Pfl. Documente*, 61, 229.

COAST REGION: Albany Div.; Bothas Berg, near the Great Fish River, on arborescent Crassulaceæ, *Ecklon & Zeyher*, 2277! Great Fish River, near Cookhuis Drift, on *Portulacaria afra*, Jacq., *MacOwan*, 2101!

CENTRAL REGION: Jansenville Div.; Sundays River, on *Euphorbia* sp., *Drège*!

According to MacOwan, "This rare *Viscum* grows plentifully between Cookhouse Drift on Fisch River and Patrys Hoogte, but invariably on *Portulacaria afra*, Jacq., whose younger leaves it strikingly resembles. But for the scarlet berries, few but botanists would detect it. The large arborescent *Crassula portulacea*, Lam., is abundant in the same locality, but though I have examined hundreds of trees, the *Viscum* has never occurred upon any. Some sharp-eyed Boers say it grows on the Groot Noors-doorn (*Euphorbia tetragona*), but this may be a remembrance of the large scarlet berry of *Viscum minimum*, Harv., on another species of *Euphorbia*."

12. **V. combreticola** (Engl. in Engl. Jahrb. xl. 542, *V. combreticolum*); a much-branched shrub, 1–1½ ft. high; stem terete, 3½–4 lin. in diam. at the base, brown and slightly glossy in a dried state; branches and branchlets conspicuously flattened, ribbed in a dried state; internodes broadly linear, tape-like, broadest in their upper part, slightly contracted at the apex, slightly tapering into the base, ⅔–2¾ in. long, 1½–3½ lin. broad, those of the main branches much thickened along the middle; leaves scale-like, inconspicuous; flowers diœcious, tetramerous; male inflorescences axillary, composed of 1–5 3-flowered cymules, each of which is borne by a bracteal cup; bracteal cup subtended by a pair of scale-leaves at the base, sessile, boat-shaped, 1¾ lin. long; lobes ascending, rounded or obtuse, 1 lin. long, glandular-ciliolate; pedicel (solid base of flower) hardly ½ lin. long; receptacular tube ⅜ lin. long; petals alternately deltoid and deltoid-ovate, ¾ lin. long, about ¾ lin. broad; anthers trigonous with a convex outer surface, elliptic in outline, ⅝–¾ lin. long, nearly ½ lin. broad; female inflorescences axillary, composed of 1 or 3 flowers, each of which is borne by a bracteal cup; bracteal cup conspicuously 2-lobed, 1 lin. high; lobes ⅝ lin. long, rounded, glandular-ciliolate, exceeding the receptacle at the time of expansion of the flower; receptacle shortly and broadly obovoid, ⅝ lin. long; petals triangular, over ½ lin. long, $\frac{7}{16}$–½ lin. broad, deciduous; style broadly conical, ⅕ lin. long; stigma projecting ⅓ lin. above the insertion of the petals; berry ellipsoid, 2½ lin. long, red, smooth or slightly warted. *Sprague in Dyer, Fl. Trop. Afr.* vi. i. 404. *V. dichotomum, Harv. in Harv. & Sond. Fl. Cap.* ii. 581, *as to the Magaliesberg specimens, excluding synonyms; De Wild. Pl. Nov. Herb. Hort. Then. t.* 89, *excluding synonyms; not of D. Don.*

KALAHARI REGION: Transvaal; Magaliesberg Range, *Sanderson*! *Burke*, 125! *Zeyher*, 748! *Worsdell*! Buffelspoort, very common on *Combretum* spp., *Engler*, 2840a! near Potgieters Rust, *Bolus*, 11009! Badsloop, *Schlechter*, 4287! Rustenburg District, *Miss Nation*, 320!

Also in Tropical Africa.

13. **V. anceps** (E. Meyer in Drège, Zwei Pfl. Documente, 148, 149, 229, name only); branches and branchlets conspicuously flattened, ribbed in a dried state; internodes oblanceolate-oblong or oblanceolate-linear, tape-like, broadest in their upper part, slightly or conspicuously contracted at the apex, tapering into the base, ⅓–1 in.

long, 1–3 lin. broad, those of the main branches thickened along
the middle ; leaves scale-like, inconspicuous ; flowers diœcious,
tetramerous ; male inflorescences axillary, composed of 1–3 3-flowered
cymules, each of which is borne by a bracteal cup ; bracteal cup
subtended by a pair of scale-leaves at the base, sessile, boat-shaped,
$1\frac{1}{8}$ lin. long ; lobes spreading, rounded, with scarious margins ;
flower $1\frac{1}{4}$ lin. long ; solid base $\frac{1}{8}$ lin. long ; receptacular tube $\frac{3}{8}$ lin.
long ; petals alternately deltoid and deltoid-ovate, $\frac{1}{2}$ lin. long ;
anthers trigonous, elliptic-oblong in outline, $\frac{1}{2}$–$\frac{3}{4}$ lin. long, $\frac{3}{8}$ lin. broad
or less ; female flowers axillary, solitary, each borne by a bracteal
cup ; bracteal cup conspicuously bilobed, $\frac{5}{8}$ lin. high, $\frac{5}{8}$ lin. long ; lobes
$\frac{4}{8}$ lin. long, rounded, with scarious margins, falling $\frac{3}{8}$ lin. short of the
apex of the receptacle ; pedicel not clearly differentiated from the
receptacle in the flower ; receptacle and pedicel together subcylindric,
narrowed to the base in the lower third, $\frac{7}{8}$ lin. long, nearly $\frac{1}{2}$ lin. in
diam. ; petals ovate-oblong, $\frac{1}{2}$ lin. long, deciduous ; style broadly
conical, $\frac{1}{4}$ lin. long ; stigma projecting $\frac{1}{3}$ lin. above the insertion of
the petals ; berry shortly pedicelled, ellipsoid, 2 lin. long, more or
less warted ; pedicel $\frac{1}{2}$ lin. long. *Sprague in Dyer, Fl. Trop. Afr.*
vi. i. 407, *in obs. V. dichotomum, Harv. in Harv. & Sond. Fl. Cap.* ii.
581, *partly, not of D. Don.*

COAST REGION : Komgha Div. ; Kei Kop, on *Acacia horrida,* Willd., *Flanagan,*
197 !
EASTERN REGION : Transkei ; Kentani, *Miss Pegler,* 1505 ! Fort Bowker,
Bowker, 554 ! Tembuland ; Morley, *Drège* ! Pondoland ; between Umtata River
and St. Johns River, *Drège* ! Egossa, *Sim,* 2489 ! Natal ; Alexandra District,
Dumisa, on *Combretum* sp., *Rudatis,* 1066 ! Tugela River, *Gerrard,* 1647 !
Zululand ; Entumeni, on *Clausena inæqualis,* Benth., *Wood,* 3973 !

Viscum dichotomum, Harv. (l.c.), included two diœcious African species,
V. anceps, E. Meyer, and *V. combreticola,* Engl. *V. dichotomum,* D. Don, is a
monœcious species, native of India, Malaya, Indo-China and China ; each
bracteal cup bears either a solitary female flower or a 3-flowered cymule, of which
the central flower is female and the two lateral male.

14. **V. Junodii** (Engl. in Engl. & Prantl, Pflanzenfam. Nachtr. 1,
zu ii.–iv. 140, *V. Jussodii*) ; very closely allied to the preceding species,
from which it differs in the following characters : internodes oblan-
ceolate-oblong or obovate-oblong, $2\frac{1}{2}$–6 lin. long, $\frac{3}{4}$–$1\frac{1}{4}$ lin. broad ; male
flowers axillary, solitary, with or without a bracteal cup ; bracteal
cup broadly trough-shaped, $\frac{1}{4}$ lin. high, $\frac{5}{4}$ lin. long, hardly $\frac{1}{2}$ lin.
broad ; flower $1\frac{3}{8}$ lin. long ; solid base $\frac{1}{3}$ lin. long ; receptacular
tube $\frac{1}{3}$ lin. long ; petals ovate, $\frac{5}{8}$ lin. long ; female inflorescence and
berries not known. *Aspidixia Junodi, Van Tiegh. in Bull. Soc. Bot.
France,* xliii. 193.

EASTERN REGION : Delagoa Bay, *Junod,* 452 !

Possibly a mere form of *V. anceps,* E. Meyer. The internodes seem a little
shorter and the male flowers rather larger, with different proportions. Only a
fragment of the type number was available for examination. The sole specific
character given by Van Tieghem was the flattened stem, which it has in common
with *V. anceps.* Van Tieghem erroneously placed *V. anceps* in a group of species
characterised by monœcious flowers (l.c.).

15. V. capense (Linn. f. Suppl. 426); a densely branched shrub; stem terete, about $2\frac{1}{2}$ lin. in diam. near the base; branches and branchlets conspicuously nodose; branches very numerous, spreading, longitudinally wrinkled in a dried state; branchlets obtusely tetragonal, $\frac{5}{8}$–$\frac{3}{4}$ lin. in diam.; internodes 2–8 lin. long, the uppermost slightly compressed; leaves scale-like, connate, spreading, at length becoming more or less deflexed, deltoid, acute or apiculate, $\frac{1}{2}$–$\frac{3}{4}$ lin. long; lateral scale-leaves broadly subulate, $\frac{1}{3}$–$\frac{3}{8}$ lin. long; flowers diœcious, tetramerous; male inflorescence consisting of 1 or 3 flowers, each of which is borne by a bracteal cup; bracteal cup very shortly peduncled, deeply 2-lobed; lobes diverging at rather less than a right angle, broadly boat-shaped, acuminate or apiculate, $\frac{3}{4}$–$\frac{7}{8}$ lin. long, $\frac{7}{16}$–$\frac{1}{2}$ lin. broad, minutely ciliate; peduncle about $\frac{1}{6}$ lin. long; petals broadly ovate, $\frac{3}{4}$ lin. long, $\frac{5}{8}$ lin. broad; female inflorescence consisting of 1–3 flowers, each of which is borne by a bracteal cup; bracteal cup transverse, very deeply 2-lobed, shortly and broadly boat-shaped, $\frac{5}{8}$ lin. long, $\frac{7}{16}$ lin. broad, $\frac{1}{3}$ lin. deep; lobes diverging at less than a right angle, minutely ciliolate, equalling the receptacle at the time of expansion of the flower; receptacle shortly subcylindric, slightly compressed, rounded at the base, $\frac{5}{8}$ lin. long, $\frac{5}{8}$ lin. broad at the apex; petals narrowly triangular, $\frac{1}{2}$ lin. long, $\frac{3}{8}$ lin. broad at the base; style broadly conical, $\frac{1}{6}$–$\frac{1}{4}$ lin. long; stigma projecting $\frac{3}{8}$ lin. above the insertion of the petals; berry sessile, subglobose, about $1\frac{1}{2}$ lin. in diam., smooth, white. *Thunb. Fl. Cap. ed. Schult.* 154; *DC. Prodr.* iv. 283; *Harv. in Harv. & Sond. Fl. Cap.* ii. 581, *partly*; *Engl. in Engl. Jahrb.* xix. 131, *and* xx. 131; *Schinz in Bull. Herb. Boiss.* iv. *App.* iii. 55; *Bolus & Wolley-Dod in Trans. S. Afr. Phil. Soc.* xiv. 316; *Sprague in Dyer, Fl. Trop. Afr.* vi. i. 409; *Marloth, Fl. S. Afr.* 168, *t.* 38, *fig.* C. *V. sp., Drège, Zwei Pfl. Documente,* 112. *Aspidixia capensis, Van Tiegh. in Bull. Soc. Bot. France,* xliii. 193.

COAST REGION: Piquetberg Div.; Elands Berg, *Wallich*! Worcester Div.; without precise locality, *Ecklon & Zeyher*, 2278! Paarl Div.; near Paarl, *Burchell*, 954! Cape Div.; Hout Bay, *Harvey*, 196! Simons Bay, *Wright*, 590! between Paarden Island, Blauw Berg and Tyger Berg, *Drège*, 7653a! Cape Flats, on *Rhus lucida*, Linn., *Pappe*! *Burchell*, 8521! *Burke*! Steen Berg, *Wolley-Dod*, 807! Swellendam Div.; without precise locality, *Ecklon & Zeyher*, 2278! Uniondale Div.; mountain sides near the west bank of Wagenbooms River, *Burchell*, 4925! Uitenhage Div.; near the mouth of the Zwartkops River, *Zeyher*, 658! 749! and without precise locality, *Cooper*, 1508! *Ecklon & Zeyher*, 2278! Queenstown Div.; Finchams Nek, *Galpin*, 1821!

CENTRAL REGION: Ceres Div.; between Little and Great Doorn Rivers, *Burchell*, 1207! Laingsburg Div.; Witteberg Range, near Matjesfontein, *Rehmann*, 2898! Middelburg Div.; near Middelburg, *Burchell*, 2808/2!

WESTERN REGION: Van Rhynsdorp Div.; between Nieuwerust and Bitterfontein, *Pearson*, 5546!

KALAHARI REGION: Griqualand West Div.; near Griquatown, *Marloth*.

Also in Tropical Africa.

Zeyher, 749, is represented in the Cape Government Herbarium by a mixture of *V. capense* and *V. continuum*. According to Zeyher's labels, part of the material was gathered near the mouth of the Zwartkops River and part by the Gamka River. As *V. capense* is known from the former locality (*Zeyher*, 658) and *V. continuum* from the latter (*Burke*), it is more probable that the specimens and labels of Zeyher, 749, have been mixed, than that the two species occur together in both localities.

16. V. robustum (Eckl. & Zeyh. Enum. 358); whole plant sulphur-yellow (*Ecklon & Zeyher*); branches and branchlets terete, fleshy, rather stout ; branches not nodose ; branchlets $\frac{7}{8}$–1 lin. in diam. ; internodes 2–10 lin. long ; leaves scale-like, connate, spreading, at length becoming deflexed, deltoid, acute, $\frac{1}{2}$–$\frac{3}{4}$ lin. long ; lateral scale-leaves broadly subulate; flowers not known ; berry sessile, subglobose, about $1\frac{1}{2}$ lin. in diam., smooth. *Harv. in Harv. & Sond. Fl. Cap.* ii. 581 ; *Engl. in Engl. Jahrb.* xx. 131. *Aspidixia robusta, Van Tiegh. in Bull. Soc. Bot. France,* xliii. 193.

WESTERN REGION : Little Namaqualand ; T'Kausi (? Kousies or Buffels) River, *Ecklon & Zeyher,* 2279 !

Also collected by *Drège,* 7653b ! The locality given in Drège, Zwei Pfl. Documente, 112, for No. 7653, is in the Cape Division, and no doubt refers to No. 7653a (*Viscum capense*). It is improbable that No. 7653b was collected in the Cape Division.

V. robustum is undoubtedly a distinct species, and not a variety of *V. capense,* Thunb., as suggested by Harvey.

17. V. continuum (E. Meyer in Drège, Zwei Pfl. Documente, 66, 229, name only); branches terete, nodose ; branchlets subangular, very slender, $\frac{3}{8}$–$\frac{1}{2}$ lin. in diam. in a dried state, longitudinally wrinkled ; internodes 3–13 lin. long ; leaves scale-like, connate, spreading, not becoming deflexed, rounded, $\frac{3}{4}$–$\frac{1}{2}$ lin. long, margin scarious ; lateral scale-leaves rounded, hardly $\frac{1}{2}$ lin. long, connate posticously, margin scarious ; flowers diœcious, tetramerous ; male inflorescences axillary, consisting of 1–2 flowers without a bracteal cup, or of a 3-flowered cymule borne by a bracteal cup, or of a 3-flowered cymule with a single flower on each side of it ; bracteal cup subsessile, boat-shaped with rounded ends, $1\frac{1}{2}$ lin. long ; peduncle $\frac{1}{4}$ lin. long ; flowers $1\frac{3}{4}$ lin. long ; solid base $\frac{3}{8}$ lin. long ; receptacular tube $\frac{5}{8}$ lin. long ; petals ovate, $\frac{7}{8}$ lin. long ; anthers trigonous, obovate-oblong in outline, $\frac{7}{8}$–1 lin. long, $\frac{1}{2}$ lin. broad or less ; female flowers axillary, solitary, borne by a bracteal cup ; bracteal cup about 1 lin. long, bilobed, $\frac{5}{8}$ lin. long in the middle ; lobes diverging at a right angle, nearly $\frac{1}{2}$ lin. long, with scarious margins, falling $\frac{1}{4}$ lin. short of the apex of the receptacle ; receptacle and pedicel together $\frac{7}{8}$ lin. long, subcylindric, narrower towards the base ; petals $\frac{1}{2}$ lin. long in the bud ; style broadly conical ; stigma projecting $\frac{1}{3}$ lin. above the insertion of the petals ; berry pedicelled, ovoid, $2\frac{1}{4}$ lin. long, smooth, yellow ; pedicel very stout, $1\frac{1}{4}$ lin. long. *Sprague in Dyer, Fl. Trop. Afr.* vi. i. 410, *in obs. V. capense, Harv. in Harv. & Sond. Fl. Cap.* ii. 581, *partly.*

COAST REGION : Albany Div. ; near Grahamstown, on *Acacia* sp., *MacOwan,* 1142 !
CENTRAL REGION : Prince Albert Div. ; between Driekoppie and Blood River. *Drège* ! Gamka River, *Burke* ! *Zeyher,* 749 ! Somerset Div. ; near Bruintjes Hoogte, on *Acacia* sp., *Burchell,* 3106 ! 3107 !

V. continuum, E. Meyer, was united by Harvey (l.c.) with *V. capense*, from which it differs in the pedicelled berries, and the rounded lateral scale-leaves, which are connate posticously. It is more closely allied to *V. verrucosum*, Harv., from which it may be distinguished by the smooth berries.

18. **V. verrucosum** (Harv. in Harv. & Sond. Fl. Cap. ii. 581); a much-branched shrub; stem terete, about $2\frac{3}{4}$ lin. in diam. $1\frac{1}{2}$ ft. below the apex of the branchlets; branches subterete, longitudinally wrinkled in a dried state; branchlets slightly compressed, about $\frac{5}{8}$ lin. broad near the apex; internodes $\frac{1}{4}$–$1\frac{1}{2}$ in. long; leaves scale-like, connate, spreading, rounded, $\frac{3}{8}$–$\frac{1}{2}$ lin. long, margin scarious; lateral scale-leaves subdeltoid, obtuse or acute, $\frac{3}{8}$–$\frac{1}{2}$ lin. long, connate posticously, margin scarious, ciliate; flowers diœcious, tetramerous; male inflorescence consisting of 1–3 sessile flowers, or of a 3-flowered cymule, borne in a bracteal cup, or of a central 3-flowered cymule with 1–2 flowers on each side; bracteal cup broadly trough-shaped, $1\frac{3}{8}$ lin. long, $\frac{5}{8}$ lin. broad, $\frac{5}{8}$ lin. high; lobes slightly ascending, rounded, nearly $\frac{3}{4}$ lin. long, margin scarious; solid base of flower $\frac{1}{3}$–$\frac{3}{8}$ lin. long; receptacular tube hardly $\frac{3}{4}$ lin. long; petals deltoid-ovate, $\frac{5}{8}$–$\frac{3}{4}$ lin. long, $\frac{11}{16}$–1 lin. broad at the base; anthers trigonous, elliptic-oblong in outline, $\frac{3}{8}$–$\frac{1}{2}$ lin. broad; female inflorescence consisting of 1–3 flowers, with or without bracteal cups; bracteal cup (when present) embracing only the base of the receptacle; receptacle and pedicel together $\frac{7}{8}$ lin. long, obovoid; receptacle coarsely and densely warted; pedicel $\frac{1}{3}$–$\frac{3}{8}$ lin. long; style tetragonal, $\frac{1}{3}$–$\frac{3}{8}$ lin. long; stigma projecting $\frac{7}{16}$ lin. above the insertion of the petals; berry conspicuously pedicelled, yellow, subglobose, $2\frac{1}{2}$ lin. in diam., coarsely and densely warted; pedicel about $\frac{3}{4}$ lin. long. *Engl. in Engl. Jahrb.* xx. 131, *and* xxviii. 385; *Sprague in Dyer, Fl. Trop. Afr.* vi. i. 408.

KALAHARI REGION: Transvaal; Magaliesberg, *Sanderson*! 165! Tropic of Capricorn, *Nelson*, 556!
EASTERN REGION: Natal; Weenen, Mooi River Valley, *Sutherland*! Sinkwasi, *Wood*, 732! Sinkwasi, on *Acacia* sp., *Wood*, 3851! 3999! Tugela River, *Gerrard*, 1653!
Also in Tropical Africa.

19. **V. rigidum** (Engl. & Krause in Engl. Jahrb. li. 471); branches and branchlets terete or with slightly compressed terminal internodes, fairly stout, rigid, slightly striate longitudinally, hardly thickened at the nodes; internodes $\frac{1}{4}$–$\frac{1}{2}$ in. long; leaves minute, scale-like; flowers diœcious, tetramerous; male inflorescence not known; female inflorescence 1-flowered, or more rarely few-flowered: bracteal cup sessile, rather thickly coriaceous, bilobed, lobes broadly ovate, acute, about 1 lin. long, concave, rigid; petals connate at the base, subovate, about 1 lin. long, rather thick, slightly exceeding the bracts; unripe berries ovoid, obtuse, 1 lin. long, finely warted.

WESTERN REGION: Great Namaqualand; on the Us River, near Great Karas, on *Sericocoma shepperioides*, Schinz, *Engler*, 6445.

Engler and Krause state that *V. rigidum* is allied to *V. Menyharthii,* Engl. & Schinz, a native of Portuguese East Africa and Rhodesia. They do not mention whether the berry is sessile or pedicelled.

20. **V. minimum** (Harv. in Harv. & Sond. Fl. Cap. ii. 581); a minute plant; stem consisting of a single internode about ¼ lin. long, bearing at its apex a whorl of 3, or 2 opposite scale-leaves, and a single inflorescence or one terminal and 3 or 2 axillary inflorescences; scale-leaves subdeltoid, ⅓ lin. long; lateral scales ovate, acute, much thickened on the back, ¼ lin. long; bracteal cup peduncled, usually boat-shaped, ⅞–1 lin. long, 3-flowered, the terminal one sometimes 3-lobed and 4-flowered; female flower: pedicel ½ lin. long; receptacle ¾ lin. long; petals deltoid, ⅓–¾ lin. long; berry pedicelled, globose, 3 lin. in diam., smooth, red, crowned by the persistent petals; pedicel stout, thickened upwards, ¾–1 lin. long. *Engl. & Krause in Ber. Deutsch. Bot. Ges.* xxvi. A, 524; *Marloth, Fl. S. Afr.* 168, *t.* 38, *fig.* B. *Aspidixia minima, Van Tiegh. in Bull. Soc. Bot. France,* xliii. 192.

COAST REGION: Uitenhage Div.; near Port Elizabeth, on *Euphorbia* sp., *Kemsley*! on *Euphorbia polygona,* Harv., *Drège*; Albany Div.; near Grahamstown, on *Euphorbia* sp., *MacOwan,* 1229! and without precise locality, *Mrs. Barber,* 226!

Harvey (l.c.) described the stem of *V. minimum* as consisting of a single internode, and this has been confirmed in the two plants examined by the writer. Engler and Krause (l.c.), on the other hand, state that the primary axis bears 2–3 closely crowded pairs of scale-leaves. As they do not mention the lateral scales which occur at the base of the axillary peduncles, it seems possible that they may have regarded them as belonging to the primary axis.

ORDER CXX. **SANTALACEÆ.**

(By A. W. HILL.)

Flowers hermaphrodite or subdiœcious, regular. *Perianth* simple, green or corolline, sometimes fleshy, adnate to the base of the ovary or the disc; segments usually 4–5, valvate, glabrous or with a tuft of hairs on the face. *Stamens* as many as the perianth-segments, inserted at or below their base, anthers dehiscing longitudinally. *Disc* epigynous or perigynous. *Ovary* inferior, 1-celled; ovules 2–3, pendulous from the apex of a free-central placenta; style short or cylindrical; stigma terminal, capitate or 2–3-lobed. *Fruit* indehiscent, dry or fleshy. *Seed* globose or ovoid; testa obsolete; albumen copious; embryo central, oblique; cotyledons usually subterete; radicle superior.

Herbs, shrubs or trees, often parasitic; leaves usually alternate, entire, exstipulate; inflorescence axillary or terminal; flowers minute, usually greenish.

DISTRIB. Species about 400 spread through the temperate and tropical regions of both hemispheres.

Tribe I. THESIEÆ.—*Fruit* dry. *Stamens* equal in number to perianth-segments.

　　I. **Thesium.**—*Flowers* hermaphrodite.

　　II. **Thesidium.**—*Flowers* diœcious.

Tribe II. OSYRIDEÆ.—*Fruit* succulent. *Stamens* equal in number to perianth-segments.

　　III. **Osyridocarpus.**—*Flowers* hermaphrodite. *Perianth-tube* above ovary elongate. *Disc* obscure.

　　IV. **Rhoiocarpus.**—*Flowers* hermaphrodite. *Perianth-tube* above ovary short. *Disc* prominent.

　　V. **Osyris.**—*Flowers* diœcious or subdiœcious. *Perianth-tube* above ovary scarcely present. *Disc* prominent.

Tribe III. GRUBBIEÆ.—*Fruit* drupaceous. *Stamens* twice as many as perianth-segments.

　　VI. **Grubbia.**—*Flowers* hermaphrodite in axillary strobili.

I. THESIUM, Linn.

Flowers hermaphrodite. *Perianth* superior, cup-shaped on a tubular or turbinate receptacle enclosing the ovary; external glands 5, usually present alternating with the perianth-segments; segments 5, valvate, more or less hooded, with or without an apical beard of hairs, margins hairy, papillose, lacinulate or glabrous, sometimes incurved, usually with a tuft of hairs on their face behind the anthers and adhering to the apex of the anther-cells, when absent a ring of hairs occurs at the throat of the perianth-tube at the level of the insertion of the filaments. *Stamens* 5, inserted at the base of the segments or in the perianth-tube; filaments short, slender; anthers ovoid or oblong with two parallel cells dehiscing longitudinally. *Epigynous disc* often conspicuous. *Ovary* inferior, 3-merous; ovules 2–4, pendulous from the apex of a slender flexuous placenta; style cylindrical or almost absent; stigma capitate or obscurely 3-lobed. *Fruit* dry, ellipsoid, globose or obovoid, usually 10-ribbed, with more or less conspicuous reticulation between the ribs, crowned with the persistent perianth. *Seed* similar in shape to the fruit; albumen fleshy; embryo central, usually oblique; radicle as long as or longer than the cotyledons.

Herbs or undershrubs, glabrous or pubescent, usually (if not always) semi-parasitic; leaves in the South African species linear, linear-lanceolate, subulate or reduced to scales or spines, rarely suborbicular; inflorescence a loose or compact terminal or axillary raceme, spike or panicle, frequently cymose, sometimes a small or fairly large dense head, the individual flowers being arranged in the axils of a bract and two or more bracteoles.

DISTRIB. Species about 250, two only South American, the others inhabiting the temperate regions or mountains of the tropical zone of the Old World, about 70 extra-African.

The sections of the genus adopted by De Candolle have been found to be somewhat inconsistent with observed facts, and a rearrangement of the species has been made. Though the present arrangement follows that given in the Prodromus on general lines, new names have been assigned to the sections to prevent confusion with those of De Candolle.

Section 1. IMBERBIA.—Margins of perianth-segments entire, glabrous, papillose, fimbriate or provided with lacinulæ, apical beard absent ; anthers attached to the segments or tube by a tuft of perianth-hairs adhering to their apices.

Sub-section 1. SUBGLABRA.— Perianth-segments glabrous or fringed with minute papillæ ; anthers attached to perianth-segments by a tuft of hairs.

*Flowers usually single in the bract axils and arranged
　in simple terminal spikes or racemes :
Plants with rigid spine-like branches ; leaves spinous,
　spinulose, scale-like or if leafy fugacious :
Leaves spinous or spinulose :
Leaves stout, all spinous :
Leaves folded, decurrent　...　...　... (1) **spinosum.**

Leaves solid, terete, not decurrent ...　... (2) **pungens.**

Upper leaves spinulose, lower linear　...　... (3) **spinulosum.**

Leaves not spinous :
Branchlets spinous ; stems sulcate ; leaves linear,
　fugacious　...　...　...　...　... (4) **lineatum.**

Branchlets spinous ; leaves scale-like　...　... (5) **rigidum.**

Plants not spinous, herbs or sub-shrubs ; leaves herba-
　ceous or fleshy :
Flowers with well-marked perianth-tube :
Plants hairy　...　...　...　...　... (6) **hirsutum.**

Plants glabrous :
Plants with weak straggling branches　... (7) **virens.**

Plants with erect stiff branches :
Flowers pedunculate ; bracts slightly ad-
　nate to peduncle ; leaves hard, acutely
　acuminate　...　...　...　... (8) **costatum.**

Flowers sessile or if shortly pedunculate
　bracts wholly adnate ; leaves herba-
　ceous :
Stems tall ; leaves ½–1¼ in. long　... (9) **Nationæ.**

Stems short ; leaves 3–5 lin. long　... (10) **racemosum.**

Flowers without marked perianth-tube and more
　or less provided with a disc :
Leaves succulent ...　...　...　...　... (11) **crassifolium.**

Leaves herbaceous :
Plants minutely scabrous　...　...　... (12) **disciflorum.**

Plants glabrous :
Flowers pedicellate in axils of leafy bracts ;
　bracteoles very small ; inflorescence in-
　definite ; bracts not adnate to pedicels (13) **namaquense.**

Flowers in more or less definite terminal
　inflorescences ; bracts more or less
　adnate to pedicels ; bracteoles about
　equal in length to bracts :
Perianth-segments glabrous ; much-
　branched herbs or sub-shrubs :
Slender sub-prostrate herbs ; leaves
　narrowly linear, acuminate, as-
　cending　...　...　...　... (14) **acutissimum.**

Stout sub-shrubs ; leaves broadly linear,
　subacute, recurved　...　... (15) **squarrosum.**

Perianth-segments with papillose margins; stout erect shrubs:
  Plants densely leafy; bracts and bracteoles longer than flowers ...   ... (16) **foliosum.**

  Plants with few scattered leaves; bracts
  and bracteoles shorter than flowers (17) **fruticosum.**

**Flowers in axillary cymules arranged in many-flowered
cymose heads, elongated spikes or more or less
dense racemes or panicles of cymules :
  Plant spinous ...   . (18) **dissitiflorum.**

  Plant not spinous :
  †Flowers aggregated in compact terminal many-
  flowered racemose heads :
    Leaves and bracts broadly ovate-lanceolate,
    amplexicaul ...   . (19) **euphorbioides.**

    Leaves acicular, terete ...   . (20) **pinifolium.**

  ††Flowers in axillary cymules forming elongated deter-
  minate spikes or racemes ; plants erect :
    Cymules sessile in bract-axils ; bracts much
    longer than cymules   ...   . (21) **glomeruliflorum.**

    Cymules shortly pedunculate, compact; bracts
    shorter than cymules :
      Stems winged ...   . (22) **angulosum.**

      Stems not winged   ...   . (23) **Susannæ.**

  †††Flowers in indeterminate branched racemes ;
  cymules 3–5-flowered ; peduncles elongated,
  divaricately branched ; plants scandent :
    Leaves and bracts ½–1½ in. long :
      Leaves and bracts broadly linear or linear-
      lanceolate ...   . (24) **triflorum.**

      Leaves fleshy, terete ...   . (25) **scandens.**

      Leaves and bracts less than 3 lin. long...   ... (26) **galioides.**

  ††††Flowers loosely paniculate :
    Plants scabrous   ...   . (27) **asperifolium.**

    Plants glabrous :
      Leaves reduced, about 1 lin. long   ...   ... (28) **corymbulige-**
                                                              rum.

      Leaves well-developed, ⅓–1½ in. long :
        All leaves about ⅓ lin. long, numerous and
        recurved ; bracts only slightly adnate
        to peduncles   ...   . (15) **squarrosum.**

        Lower leaves ½–1½ in. long, all leaves ascend-
        ing ; bracts adnate for some distance to
        peduncles :
          Stems sharply ribbed ; leaves and bracts
          keeled ; cymules lax   ...   ... (29) **floribundum.**

          Stems slightly ribbed ; leaves and bracts
          more or less rounded on back ;
          cymules compact   ...   . (30) **pallidum.**

***Flowers arranged in small terminal cymose heads or
clusters :
  Branches fairly densely leafy ; leaves well-developed ;
  bracts glabrous ; fruits not or scarcely reticulate
  between ribs :
    Branches erect ; perianth without external glands (31) **quinqueflorum.**

Branches spreading ; perianth with conspicuous
    external glands :
    Leaves imbricate ; bracts and bracteoles longer
      than flowers ; inflorescences inconspicuous (32) **cupressoides.**
    Leaves lax ; bracts and bracteoles equal to or
      shorter than flowers ; inflorescences con-
      spicuous ... ... (33) **ericæfolium.**
Branches very sparingly leafy or nearly leafless :
    leaves small and bract-like ; bract-margins finely
    serrulate :
    Stigma subsessile ; bracts equal to or longer than
      flowers :
      External glands conspicuous ; bracts acutely
        acuminate ... ... (34) **nigromonta-**
                                                  num.
      External glands not seen ; bracts acute :
        Perianth-segments glabrous ; lower leaves
          1–2 lin. long ... ... (35) **leptocaule.**
        Perianth-segments with papillose margins ;
          lower leaves up to 6 lin. long ... ... (36) **commutatum.**
    Style $\frac{1}{4}$ lin. or more long ; bracts about half the
      length of flowers :
      Leaves few, scale-like, ovate-triangular, $\frac{3}{4}$ lin.
        long ... ... (37) **nudicaule.**
      Leaves well-developed, oblong-lanceolate, 3 lin.
        long, flat, obtuse or subacute ... ... (38) **schumannia-**
                                                  num.
****Flowers arranged in compact corymbose or umbellate
    inflorescences ; leaves few, scattered :
    Plants slender, rush-like ; leaves very small, subulate,
      rarely a few narrowly linear basal leaves ;
      perianth $\frac{1}{2}$–$\frac{2}{3}$ lin. long, margins of segments
      minutely papillose or subglabrous ; corymbs
      few-flowered :
      Leaves almost absent ; branchlets below corymbs
        leafless ... ... (39) **juncifolium.**
      Leaves subulate, fairly numerous on branchlets
        below corymbs ... ... (40) **virgatum.**
    Plants stout, woody ; leaves somewhat fleshy, linear-
      lanceolate, conspicuous, towards base up to 2$\frac{1}{2}$
      in. long ; perianth $\frac{3}{4}$–1$\frac{1}{2}$ lin. long, margins of
      segments more or less conspicuously papillose ;
      corymbs many-flowered :
      Perianth 1$\frac{3}{4}$ lin. long, fringe of papillæ conspicu-
        ous ; anthers $\frac{3}{8}$ lin. long ; style $\frac{1}{4}$ lin. long ... (41) **occidentale.**
      Perianth $\frac{3}{4}$–1$\frac{1}{4}$ lin. long, papillæ short ; anthers
        $\frac{1}{2}$ lin. long ; style $\frac{1}{4}$ lin. long ... ... (42) **strictum.**

Sub-section 2. FIMBRIATA.—Perianth-segments with a marginal fringe of long
papillæ, or with two lateral lacinulæ, but no apical beard ; anthers with attachment hairs ; plants rigid, woody, spinescent.

    Margins of perianth-segments provided with lacinulæ :
      Stems rigid, spinescent, puberulous ... ... (43) **lacinulatum.**
      Stems flexuous, herbaceous, glabrous ... ... (44) **pleuroloma.**
    Margins of perianth-segments fimbriate with fringe of
      long papillæ :
      Branchlets ascending, crowded, covered with imbri-
        cate adpressed leaves ... ... (45) **horridum.**

Branchlets spreading ; leaves reduced to scales :
  Plants covered with minute hairs ...   . (46) **hystricoides.**

  Plants glabrous   ...   . (47) **Hystrix.**

Section 2. BARBATA.—Perianth-segments with a more or less dense beard
dependent from their apices, margins more or less hairy ; anthers attached to
the segments or tube by a tuft of perianth-hairs adhering to their apices.

*Flowers solitary or in small 3–5-flowered clusters at
  the ends of main and axillary branches ; bracts
  forming an involucre ; leaves reduced to scales,
  rarely narrowly linear or acicular leaves also
  present :
Flowers solitary :
  Plants minutely puberulous   ...   . (48) **sertulariastrum.**

Plants glabrous :
  Plants slender, profusely branched, with scale
  and numerous long narrowly-linear leaves   (49) **paniculatum.**

  Plants stout ; upper leaves scale-like, a few
  stout acicular leaves below ...   . (50) **euphrasioides.**

Flowers in small clusters :
  Anthers exserted   ...   . (51) **micromeria.**

  Anthers included in perianth-tube :
  Bracts much shorter than flowers, ovate-lanceolate, with blackish acuminate tips ; leaves
  adpressed, all subulate-acuminate ...   ... (52) **capituliflorum.**

  Bracts nearly as long as flowers, ovate-elliptic,
  with membranous margins ; leaves somewhat spreading, acicular to triangular-
  lanceolate   ...   . (53) **cuspidatum.**

**Flowers solitary or in groups of 2 or 3 at ends of
  branches ; bracts not forming an involucre ; all
  leaves well-developed :
Anthers included in perianth-tube ...   . (54) **rariflorum.**

Anthers exserted :
  Flowers over 2 lin. long ; style exceeding 1 lin.
  long   ...   . (55) **Zeyheri.**

  Flowers not exceeding 1¼ lin. long ; style ¼–½ lin.
  long :
  Branches erect, fastigiate ; leaves scattered ;
  bracts as long as flowers   ...   . (56) **cytisoides.**

  Branches spreading, densely leafy ; bracts longer
  than flowers ...   . (57) **Burchellii.**

***Flowers in small terminal or subterminal heads or
  clusters ; plants more or less prostrate, much
  branched :
Leaves few, more or less scale-like especially at ends
  of branches :
  Lower leaves not scale-like ; anthers included in
  perianth-tube ...   . (58) **repandum.**

  Leaves nearly all reduced to scales ; anthers
  exserted ...   . (59) **glaucescens.**

Leaves numerous, acicular or linear-lanceolate,
  equally distributed over the stem :
  Stems and leaves scabrid-puberulous   ...   ... (60) **hispidulum.**

Stems and leaves glabrous :
  Leaves acicular, acute, ascending :
    Plants lax, spreading ; anthers exserted ... (61) **prostratum.**
    Plants stout, much branched ; anthers included (62) **acuminatum.**
  Leaves linear-lanceolate, flattened above, re-
    curved, scattered or densely imbricate :
    Plants slender, spreading, much branched ;
      anthers exserted ... ... (63) **selagineum.**
    Plants stout ; branches few, ascending ;
      anthers included :
      Leaves scattered ; stigma subsessile ... (64) **capitellatum.**
      Leaves densely imbricate ; style ¾ lin. long (65) **imbricatum.**
****Flowers arranged in simple terminal spikes, racemes
  or loose paniculate racemes :
  †Flowers in definite simple spikes or racemes ; bracts
    not adnate to peduncles :
    Stems rush-like ; leaves scale-like or rarely linear-
      subulate ; bracts with hyaline, scurfy or finely
      fringed margins :
      Flowers in lax spikes ; bracts with scurfy or
        hyaline margins :
        Apical beard of dense woolly hairs ; anthers in-
          cluded in perianth-tube :
          Bracts and leaves reduced to ovate scales with
            broad scurfy margins ... ... (66) **junceum.**
          Bracts with membranous margins, sometimes
            with slight scurfy edge :
            Bracts subulate, acutely acuminate ; leaves
              linear-subulate ... ... (67) **natalense.**
            Bracts broadly ovate, acute ; leaves reduced
              to ovate scales ... ... (68) **scirpioides.**
        Apical beard of stiff comb-like hairs ; anthers
          partly exserted :
          Slender annuals, 4–5 in. high ; inflorescences
            lax ... ... (69) **paronychioides.**
          Straggling much-branched undershrub, 1 ft.
            high ; inflorescences compact, often
            branched ... ... (70) **flexuosum.**
      Flowers in short more or less dense spikes ; bracts
        with finely fringed margins :
        Stems slender, ascending, brown when dry ... (71) **spartioides.**
        Stems stout, flexuous or prostrate, grey when
          dry ... ... (72) **confine.**
    Stems leafy ; leaves well-developed ; bracts with entire
      or rarely scabrous margins, never scurfy :
      Flowers in compact dense spikes ; style ¼ lin.
        long ; bracts equal to or longer than the
        flowers :
        Plants finely pubescent :
          Leaves rounded on back ; perianth externally
            glabrous ... ... (73) **griseum.**
          Leaves ribbed ; perianth externally hairy ... (74) **transvaalense.**
        Plants glabrous :
          Bracts finely scabrid-puberulous on margins (75) **gnidiaceum.**
          Bracts glabrous on margins :
            Stigma sessile or subsessile ; anthers included (76) **phyllostachyum.**
            Style ¼ lin. long ; anthers exserted ... (77) **impeditum.**

Flowers in elongated lax spikes; style $\frac{1}{3}$ lin. or
  more long; bracts shorter than flowers ... (78) **magalismonta-**
                                               num.
††Flowers in racemes ; bracts adnate to peduncles :
  Bracts equal to or longer than flowers :
    Styles $\frac{1}{2}$ lin. long or more ; leaves scattered :
      Bracts with minutely serrulate or scabrous
        margins :
        External glands conspicuous :
          Perianth 1–1$\frac{1}{4}$ lin. long ... ... (79) **Burkei.**

          Perianth 2 or more lin. long ... ... (80) **orientale.**

        External glands not present :
          Bracts equal in length to flowers, rounded
            on back, minutely serrulate ... ... (81) **macrogyne.**

          Bracts longer than flowers, keeled, with
            membranous scabridulous margins ... (82) **lobelioides.**

    Bracts with entire glabrous margins :
      External glands scarcely visible :
        Bracts keeled ; leaves erect... ... ... (83) **gœtzeanum.**

        Bracts rounded on back ; leaves recurved... (84) **resedoides.**

      External glands conspicuous :
        Plants slender ; leaves narrowly linear,
          curved ... ... (85) **Junodii.**

        Plants stout ; leaves straight :
          Perianth over 2 lin. long ; segments flat,
            without hood ; bracts elliptic-lanceo-
            late, obtuse ... ... (86) **coriarium.**

          Perianth 1$\frac{1}{2}$ lin. long ; segments with
            deep hood ; bracts lanceolate, acute (87) **nigrum.**

    Styles very short ; stigma subsessile ; plants
      densely leafy ... ... (88) **gracillarioides.**

  Bracts much shorter than the flowers ; stigma sessile :
    Plants branching from the base, leafy ; perianth
      1$\frac{1}{3}$–1$\frac{3}{4}$ lin. long ... ... (89) **asterias.**

    Plants branching above, sparingly leafy ; perianth
      $\frac{3}{4}$ lin. long ... ... (90) **polygaloides.**

†††Flowers usually in 3-flowered pedunculate cymules in
  bract-axils arranged in racemes :
  Leaves with recurved tips ... ... (84) **resedoides.**

  Leaves straight :
    Bracts longer than cymules, adnate for some length
      to peduncles ... ... (79) **Burkei.**

    Bracts shorter than cymules, slightly adnate to
      peduncles :
      Racemes elongate, distinct ; style reaching to
        top of anthers ... ... (91) **cornigerum.**

      Racemes short ; style scarcely reaching to base
        of anthers ... ... (92) **palliolatum.**

††††Flowers in loose paniculate racemes :
  Plants lax, with spreading leafy branches ; leaves
    linear-lanceolate ; bracts longer than flowers ... (93) **gypsophiloides.**
  Plants stiff, erect, with erect branches ; leaves
    scattered, narrowly linear ; bracts shorter than
    flowers ... ... (94) **utile.**

*****Flowers in compact rounded heads, short compact
　　spikes or dense corymbose heads or clusters ;
　　leafy subshrubs :
　　Flowers in corymbose heads or clusters :
　　　Bracts conspicuous ; stigma sessile :
　　　　Bracts broadly oblanceolate, foxy-red, with con-
　　　　　spicuous translucent margins　...　... (95) **fallax.**
　　　　Bracts lanceolate, greenish-yellow, margins in-
　　　　　conspicuous ...　...　...　...　... (96) **helichrysoides.**
　　　Bracts inconspicuous ; style elongate　...　... (97) **umbelliferum.**
　　Flowers in rounded heads or compact spikes :
　　　Leaves and bracts covered with a fine pubescence :
　　　　Leaves adpressed to stem ; stigma sessile　... (98) **boissieranum.**
　　　　Leaves spreading ; style elongate :
　　　　　Leaves very dense ; bracts green ; flowers in
　　　　　　rounded heads　...　...　...　... (99) **pubescens.**
　　　　　Leaves more scattered ; bracts ruddy ; flowers
　　　　　　in compact spikes　...　...　... (100) **rufescens.**
　　　Leaves and bracts with scabrous margins...　... (101) **scabrum.**
　　　Leaves and bracts glabrous :
　　　　Stems woody ; flowering branches with re-
　　　　　mote leaves ...　...　...　...　... (102) **polycephalum.**
　　　　Stems herbaceous ; leaves numerous :
　　　　　Leaves imbricate, adpressed, stout, obtuse (103) **micro-**
　　　　　　　　　　　　　　　　　　　　　　cephalum.
　　　　　Leaves somewhat spreading, linear, acute to
　　　　　　acuminate ...　...　...　...　... (104) **pycnanthum.**
　　Leaves glabrous ; bracts with broad hyaline and
　　　more or less scarious or rarely finely scabrid-
　　　puberulous margins :
　　　　Stems prostrate　...　...　...　...　... (105) **ecklonianum.**
　　　　Stems erect :
　　　　　Inflorescences dense spikes :
　　　　　　Leaves recurved ; bracts glabrous ; style
　　　　　　　elongate...　...　...　...　... (106) **sonderianum.**
　　　　　　Leaves erect ; bracts finely scabrid-pube-
　　　　　　　rulous on margins ; stigma subsessile... (75) **gnidiaceum.**
　　　　　Inflorescences globular or corymbose heads ;
　　　　　　leaves erect or slightly spreading :
　　　　　　Style elongate, ⅔–1⅔ lin. long :
　　　　　　　Stems densely leafy especially near flower-
　　　　　　　　heads ...　...　...　...　... (107) **capitatum.**
　　　　　　　Stems with more or less scattered leaves,
　　　　　　　　with very few below flower-heads　... (108) **glomeratum.**
　　　　　　Style short, ¼ lin. long ; perianth-tube
　　　　　　　markedly shorter than the segments... (109) **fimbriatum.**
　　　　　Stigma sessile :
　　　　　　Perianth-tube very short ; segments
　　　　　　　elongate, with a small beard at
　　　　　　　apex and papillose margins...　... (110) **translucens.**
　　　　　　Perianth-tube well-marked ; segments
　　　　　　　with dense woolly beard :
　　　　　　　Bracts broadly ovate-lanceolate ; leaves
　　　　　　　　scattered　...　...　...　... (111) **densiflorum.**
　　　　　　　Bracts lanceolate ; leaves crowded,
　　　　　　　　imbricate ...　...　...　... (112) **carinatum.**

Section 3. PENICILLATA.—Perianth-segments with dense apical beard; the pencil of perianth-hairs behind each anther remains free and does not adhere to the apex of the anther.
Only South African species ... ... (113) **penicillatum.**

Section 4. ANNULATA.—Perianth-segments with a more or less dense apical beard, rarely papillose; perianth-hairs behind anthers absent, but replaced by a ring of short downwardly-directed golden hairs inserted in the tube at the level of the attachment of the filaments.

Flowers in elongated spikes or racemes:
Perianth-segments with papillose margins ... ... (114) **micropogon.**
Perianth-segments bearded:
Anthers included in perianth-tube ... ... (115) **urceolatum.**
Anthers exserted:
Plants with numerous stout branches spreading almost at right angles to stem; style ¼ lin. long ... ... (116) **patulum.**
Plants sparingly branched; branches erect; stigma sessile:
Plants slender; bracts small, subulate; perianth ¾ lin. long ... ... (117) **funale.**
Plants stout; bracts leaf-like; perianth above 1 lin. long ... ... (118) **macro-stachyum.**

Flowers in heads or short spicate clusters:
Flowers in rounded heads or dense compact spikes with imbricate conspicuous bracts; perianth-segments with an apical beard of stiff comb-like hairs:
Bracts distinctly toothed on margins:
Upper leaves numerous, lanceolate, recurved at tip ... ... (119) **diversifolium.**
Upper leaves few, subulate, ascending ... ... (120) **aggregatum.**
Bracts quite entire, usually reddish-brown when dry:
Spikes stout, oblong or subcapitate; bracts broadly ovate; plants leafy:
Spikes mostly oblong; bracts broadly ovate, sharply keeled; upper leaves linear-subulate ... ... (121) **spicatum.**
Spikes mostly depressed-capitate; bracts lanceolate, not sharply keeled; upper leaves linear ... ... (122) **bathyschistum.**
Spikes rather slender, often elongated; bracts linear-lanceolate; plants with few leaves ... (123) **subnudum.**
Flowers in more or less lax heads or spicate clusters; bracts inconspicuous; perianth-segments with dense woolly beard:
Style above ¼ lin. long ... ... (124) **elatius.**
Stigma sessile:
Bracts with fringed margins ... ... (125) **Frisea.**
Bracts with entire margins:
Leaves more or less crowded, adpressed to stem, somewhat fleshy ... ... (126) **annulatum.**
Leaves scattered, spreading:
Leaves narrowly linear; flower-heads globose, few-flowered ... ... (127) **brachygyne.**
Leaves broadly linear; flower-heads ovoid, many-flowered ... ... (128) **Patersonæ.**

1. **T. spinosum** (Linn. f. Suppl. 161); a much-branched woody shrub with very spiny leaves and bracts; branches flexuous, often glaucous-grey, glabrous; leaves spreading at right angles, pungent, triangular-subulate, very acute, $2\frac{1}{2}$–4 lin. long, with the margins closely adpressed together, subdecurrent on the stem, glabrous; bracts like the leaves but a little shorter; flowers solitary in the axil of the bract, pedicellate; bracteoles 2, very small, subulate; pedicel short, rather stout, glabrous; perianth $\frac{3}{4}$ lin. long, with distinct external glands; segments triangular, slightly hooded, about $\frac{1}{2}$ lin. long, with a slight trace of flaps on the margin; anthers exserted, $\frac{1}{8}$ lin. long; disc conspicuous; style $\frac{1}{4}$ lin. long; fruits oblong-ellipsoid, $1\frac{1}{3}$ lin. long, rather faintly 10-ribbed, reticulate between the ribs. *Thunb. Diss. Thes.* 6; *Fl. Cap. ed. Schult.* 208; *Drège, Zwei Pfl. Documente,* 107, 108, 113; *Sond. in Flora,* 1857, 353; *A.DC. in DC. Prodr.* xiv. 654. *T. teretifolium, R. Br. Prodr.* 353; *A.DC. l.c.* 672, *name only.*

South Africa: without locality, *Thunberg*! *Zeyher*, 106! *Nelson in Herb. Banks*! Coast Region: Van Rhynsdorp Div.; near Herrelogement, *Drège* c! *Zeyher*, 1504! Ebenezer below 500 ft., *Drège* b! Clanwilliam Div.; Lange Kloof, 400 ft., *Schlechter*, 8048! Malmesbury Div.; near Saldanha Bay, *Bolus*, 12825! *Bachmann*, 1700! *Ecklon & Zeyher*, 38! between Greene Kloof & Saldanha Bay, *Drège* a! Cape Div.; Cape Flats, near Claremont, *Schlechter*, 536!

2. **T. pungens** (A. W. Hill in Kew Bulletin, 1915, 39); a much-branched shrub, about 2 ft. high; branches crowded, glaucous-grey or greenish when dry, with rather short internodes, glabrous; leaves spreading at right angles, pungent, rigidly subulate with very acute spinous tips, 2–$3\frac{1}{2}$ lin. long, subterete, glaucous-grey with yellowish tips; flowers solitary in the axils of the bracts; bracts similar to the leaves but a little shorter; bracteoles very small at the base of the short pedicel; perianth $\frac{3}{4}$–$1\frac{1}{4}$ lin. long, with conspicuous external glands; segments triangular-ovate, $\frac{1}{2}$–$\frac{3}{4}$ lin. long, slightly hooded, flat, with flaps at the side; anthers exserted, $\frac{1}{8}$–$\frac{1}{4}$ lin. long; disc conspicuous; style $\frac{1}{4}$ lin. long; fruits ellipsoid-globose, $1\frac{1}{3}$ lin. long, 10-ribbed, slightly reticulate between the ribs. *T. spinosum, Drège, Zwei Pfl. Documente,* 68, *partly, not of Linn. f.*

Western Region: Little Namaqualand; between Pedros Kloof and Lily Fontein, 3000–4000 ft., *Drège* (*T. spinosum*, e partly)! Kamiesberg Range, at Kharkamo, *Pearson*, 6684!

3. **T. spinulosum** (A.DC. in DC. Prodr. xiv. 647); a small slender plant about 5 in. high, much-branched from the base; branches ascending or spreading, slender, sharply angular, glabrous; leaves subacicular, slightly angular when dry, acute or subacute, $2\frac{1}{2}$–6 lin. long, fleshy, glabrous; flowers arranged in axillary flexuous racemes; bracts subulate, very acute, about $1\frac{1}{4}$ lin. long, glabrous, nearly as long as the flower; bracteoles 2, arising on each side of the bract, about half as long as the flower, subulate and very acute; flower shortly pedicellate, about $1\frac{1}{3}$ lin. long including

the pedicel ; perianth ¾ lin. long, with conspicuous external glands ; segments triangular, slightly hooded, ⅓ lin. long, glabrous ; anthers exserted, ⅛ lin. long ; style ⅛–⅐ lin. long ; fruits ellipsoid, stipitate, 2 lin. long, prominently 10-ribbed, prominently reticulate between the ribs, glabrous. *T. aristatum, Schlechter in Engl. Jahrb.* xxvii. 116. *T. spinosum, Drège, Zwei Pfl. Documente,* 68, *partly. T. sp., Drège, l.c.* 73.

COAST REGION : Clanwilliam Div. ; Blue Berg, 3000–4000 ft., *Drège,* 8175a ! Tulbagh Div. ; New Kloof, *Schlechter,* 7506 ! Caledon Div. ; Houw Hoek, *Schlechter,* 9387 ! between Fairfield and Elim, *Bolus,* 8599 !

WESTERN REGION : Little Namaqualand ; between Pedros Kloof and Lily Fontein, 3000–4000 ft., *Drège* e, partly !

4. **T. lineatum** (Linn. f. Suppl. 162) ; a woody bush about 1½–3 ft. high ; branches dense, spreading, prominently grooved, glabrous ; leaves mostly few and inconspicuous or sometimes fairly numerous and conspicuous, linear, flattened and fleshy, but mostly linear or occasionally linear-oblanceolate, subterete, subacute, 1–1½ lin. long, fleshy, glabrous ; bracts small and scale-like, triangular-ovate or sublanceolate, glabrous ; pedicel up to about 1 lin. long, with 2 very small bracteoles at the apex ; flowers solitary, white ; perianth 1¼ lin. long ; segments about ¾ lin. long, ovate-lanceolate, subacute, hooded, with slightly inflexed margins ; anthers exserted, scarcely ⅓ lin. long ; filaments ¼ lin. long ; style ¾ lin. long ; fruits white when mature, oblong-ellipsoid, about 4 lin. long, 2 lin. in diam., with a slightly impressed reticulation, not ribbed. *Thunb. Diss. Thes.* 6 ; *Fl. Cap. ed. Schult.* 210 ; *Drège, Zwei Pfl. Documente,* 90 ; *Sonder in Flora,* 1857, 354 ; *A.DC. in DC. Prodr.* xiv. 654, *excl. syn. T. rigidum, Sond.* ; *Baker & Hill in Dyer, Fl. Trop. Afr.* vi. i. 425. *T. ephedroides, A. W. Hill in Kew Bulletin,* 1910, 183. *T. sparteum, R. Br. Prodr.* 353 ; *A.DC. in DC. Prodr.* xiv. 672, *name only.*

SOUTH AFRICA : without locality, *Sparrmann* ! *Thunberg, Burke* !
COAST REGION : Swellendam Div. ; Tradouw, *Mund & Maire* !
CENTRAL REGION : Calvinia Div. ; between Lospers Plaats and Springbok Kuil River, *Zeyher,* 1502 ! Prince Albert Div. ; Gamka River, *Burke,* 145 ! Somerset Div. ; near Somerset East, *Bowker,* 101 ! 136 ! Graaff Reinet Div. ; near Graaff Reinet, 2600 ft., *Bolus,* 80 ! Fraserburg Div. ; between Karee River and Klein Quaggasfontein, near Fraserburg, *Burchell,* 1412 ! Hopetown Div. ; near Hopetown, *Muskett,* 125 !

WESTERN REGION : Great Namaqualand ; Great Karasberg, near Krai Kluft, 5000–6000 ft., *Pearson,* 7807 ! 8287 ! Naruda Süd, 4000 ft., *Pearson,* 7808 ! 8134 ! north of Ramans Drift, 2400 ft., *Pearson,* 4007 ! Spitz Koppjes, *Dinter,* 64 ! Omburo, *Dinter,* 1407 ! Little Namaqualand ; Rattel Poort, *Pearson,* 2977 ! Khamiesberg Range, Twee River Settlement, *Pearson,* 6614 ! near Kasteel Poort, 3000 ft., *Bolus,* 9447 ! Silver Fontein, near Ookiep, 2000–3000 ft., *Drège* a ! Aus, *Marloth,* 5012 ! Schorsteen Berg, *Marloth,* 4108 ! Haazenkrals River, *Drège* b ! Van Rhynsdorp Div. ; Zout River, *Schlechter,* 8140 !

KALAHARI REGION : Griqualand West ; Ashestos Mountains, Kloof village, *Burchell,* 2062 !

Also in German South-West Africa.

The specimen of *T. sparteum, R. Br.,* is preserved in the British Museum and proves to belong to *T. lineatum, Linn. f.*

5. **T. rigidum** (Sond. in Flora, 1857, 354) ; a small much-branched rigid shrub with rigid spine-tipped branchlets ; branchlets divaricate, subterete, scarcely grooved, glabrous and sometimes slightly glaucous ; bark splitting transversely and peeling off; leaves small and scale-like, ovate or ovate-triangular, acute, about $\frac{1}{2}$ lin. long or less, brownish, with scarious hyaline margins, glabrous; bracts and bracteoles similar to the leaves but smaller ; flowers solitary or subsolitary, very shortly pedunculate; perianth $\frac{3}{4}$ lin. long, with a distinct disc within the base ; segments triangular, subacute, $\frac{1}{2}$ lin. long, slightly hooded, glabrous ; anthers exserted, $\frac{1}{6}$ lin. long ; style about $\frac{1}{4}$ lin. long, reaching to the top of the anthers ; fruits ovoid, 10-ribbed, slightly rugose.   *T. lineatum, A.DC. in DC. Prodr.* xiv. 654, *so far as concerns Ecklon & Zeyher,* 39.

COAST REGION : Uitenhage Div. ; Steinbock Flats, north of Winterhoeks Berg, near Karreebosch, *Ecklon & Zeyher,* 39 !

6. **T. hirsutum** (A. W. Hill in Kew Bulletin, 1915, 31) ; root-stock rather slender, branched ; stems fairly numerous, subsimple, erect, slightly angular, rather densely puberulous ; leaves linear, with a very acute cartilaginous apex, flattened, 3–6 lin. long, 1-nerved, fleshy, scabrid-puberulous on the margin ; flowers arranged in bracteate racemes, solitary in the axils of the bracts ; peduncles $\frac{1}{2}$–1$\frac{1}{4}$ lin. long ; bracts adnate to the peduncle up to the base of the flower, usually much longer than (but sometimes equalling) the latter, leaf-like, often puberulous on the midrib as well as the margins ; bracteoles arising from the base of the bract, mostly slightly longer than the flower, a little narrower than (but other-wise similar to) the bracts ; flowers about 2 lin. long; perianth about 1$\frac{1}{2}$ lin. long, with conspicuous external glands ; segments 1 lin. long, hooded, glabrous ; anthers $\frac{1}{3}$ lin. long, slightly exserted from the perianth-tube ; style $\frac{3}{4}$–1 lin. long, reaching to above the tops of the anthers ; fruits broadly ellipsoid, 3 lin. long including the persistent perianth, about 10-ribbed, prominently reticulate between the ribs.

COAST REGION : Queenstown Div. ; Queenstown plains, 3600 ft., *Galpin,* 1585 !
CENTRAL REGION : Somerset Div. ; near Somerset East, 2800 ft., *MacOwan,* 1618 ! Graaff Reinet Div. ; Cave Mountain, 4400 ft., near Graaff Reinet, *Bolus,* 525 !
KALAHARI REGION : Orange River Colony ; Leeuw Spruit and Vredefort, *Barrett-Hamilton* ! Transvaal ; Heidelberg, at Grootvlei Farm, *Gilfillan,* 244 !

7. **T. virens** (E. Meyer in Drège, Zwei Pfl. Documente, 147, 226) ; stems very slender and probably trailing, finely grooved, glabrous ; branches sparingly leafy, very slender ; leaves linear, $\frac{1}{2}$–1 in. long, very acute and cartilaginous at the apex, nearly flat, glabrous ; flowers subsolitary, subtended by a bract and two bracteoles ; bract adnate to the peduncle right up to the base of flower, equalling or up to twice the length of the flower, linear,

subterete, acute, up to ½ in. long, glabrous ; bracteoles as long as or longer than the flower, similar to the bract ; flower nearly 2 lin. long, slender ; perianth 1¼ lin. long, glabrous; segments linear-lanceolate, ½ lin. long, hooded, margin incurved ; anthers exserted, ⅛ lin. long ; style 1 lin. long, reaching to the top of the anthers ; fruit oblong-ellipsoid, about 2 lin. long including the persistent perianth, prominently 10-ribbed, prominently reticulate between the ribs. *A.DC. in DC. Prodr.* xiv. 653.

EASTERN REGION : Tembuland ; Morley, 1000–2000 ft., *Drège!* Natal, *Gerrard*, 1279!

8. T. costatum (A. W. Hill in Kew Bulletin, 1915, 25); root stock slender, erect ; stems few and fairly numerous, branched in the upper part, ribbed or angular, glabrous; leaves laxly arranged, linear, acutely acuminate, 3–6 lin. long, flat, glabrous, midrib prominent on both surfaces ; cymules 1–3-flowered ; bracts shortly adnate to the peduncle, 3–5 lin. long, otherwise similar to the leaves ; peduncle shorter or a little longer than the bract, compressed, rather slender ; bracteoles shorter or longer than the flowers, linear, acute, glabrous ; perianth 1¼–1½ lin. long ; segments triangular-lanceolate, ¾ lin. long, hooded, the margins with a broad flap-like fringe near the base, the hood beaked and slightly papillose ; anthers ⅓ lin. long ; style ¾ lin. long, overtopping the anthers ; fruits campanulate-globose, about 3 lin. long including the persistent perianth, prominently 10-ribbed, coarsely reticulate between the ribs.

VAR. β, juniperinum (A. W. Hill) ; leaves densely crowded on the shoots, more finely pointed than in the type.

KALAHARI REGION : Orange River Colony ; Bethlehem, *Richardson* ! Basutoland ; Leribe, *Dieterlen*, 647 ; Transvaal ; near Pretoria, *Wilms*, 1308a ! Swaziland ; near Bremersdorp, 2600 ft., *Bolus*, 12273 ! near Mbabane, 4700 ft., *Bolus*, 12277!
EASTERN REGION : Griqualand East, *Tyson* ! Natal ; near Camperdown, 3330 ft., *Schlechter*, 3284 ! Weenen, 3500 ft., *Wood*, 3582 ! between Pietermaritzburg and Greytown, *Wilms*, 2252 ! near Pietermaritzburg, *Wilms*, 2254 ! near Emberton, *Schlechter*, 3239 ! Inanda, *Wood*, 1141 ! and without precise locality, *Gerrard*, 1281 ! Var. β : Natal and Zululand, *Gerrard*, 1280 !

9. T. Nationæ (A. W. Hill in Kew Bulletin, 1915, 34); stem slender, subsimple, evidently arising from a rhizome, about 1 ft. high, slightly flexuous, ribbed, glabrous ; leaves linear, with a short acute cartilaginous apex, ½–1¼ in. long, about ½ lin. broad, flat, fleshy, glabrous, midrib conspicuous on the upper surface ; flowers few, very shortly pedunculate ; bracts shortly adnate to the peduncle, much longer than the flowers, flat and very similar to the leaves ; bracteoles as long as or little longer than the flowers, green and like the bracts, but a little narrower ; perianth 1¼ lin. long ; segments triangular-elliptic, ⅝ lin. long, hooded, with papillose margins ; anthers half exserted from the perianth-tube,

⅓ lin. long; style ¾–1 lin. long, reaching to above the base of the anthers.

KALAHARI REGION: Transvaal; Rustenburg, 4500 ft., *Miss Nation*, 266!

10. **T. racemosum** (Bernh. in Flora, 1845, 79); rootstock knotted; stems several, simple or subsimple, suberect, conspicuously angular, glabrous; leaves linear or linear-lanceolate, subacute, 3–5 lin. long, about ¾ lin. broad, flat, 1-nerved, with narrowly cartilaginous margins, glabrous; inflorescence racemose, bracteate; bracts slightly adnate to the peduncle, linear-lanceolate, very acute, slightly keeled, longer than the flowers; peduncle up to 1 lin. long; bracteoles 2, about 1 lin. long, usually shorter than the flowers; perianth 1¼ lin. long, with conspicuous external glands; segments ovate or lanceolate, subacute, 1¼–1½ lin. long, hooded, with incurved slightly papillose margins; anthers about ¼ lin. long, on short filaments; style ½–¾ lin. long, thick, tapering, reaching to the middle of the anthers; fruits not seen. *A.DC. in DC. Prodr.* xiv. 659, *not of Sond. in Flora*, 1857, 357, *nor* 403.

KALAHARI REGION: Swaziland; Emlembo Mountain, Havelock concession, 4000 ft., *Saltmarshe*, 986!
EASTERN REGION: Natal; Table Mountain, *Krauss*, 386! Vryheid, *Sim*, 2924! and without precise locality, *Gerrard*, 1275!

Sonder has caused some confusion, in the first place by placing under this species *Zeyher*, 1500, from the Magalisberg Range. This plant is a bearded species (*Zeyher*, 1500, in Herb. Stockh., and partly in Herb. Kew), and is now referred to *T. Burkei*, A. W. Hill. In the second place he suggests that *T. angulosum*, A.DC., is, according to the description, identical with *T. racemosum*.

11. **T. crassifolium** (Sond. in Flora, 1857, 355); stem erect, much-branched, angular, glabrous; branches spreading or ascending, fairly stout; leaves recurved, linear, subacute, 3–4 lin. long, very convex on the lower surface, V-shaped on the upper surface, glabrous; inflorescence leafy and subspicate, slightly flexuous; bracts and bracteoles subequal or the former longer, both much longer than the flowers, with infolded margins, broad at the base, boat-shaped; flowers subsessile, obconic, almost completely hidden by the bracts; perianth ½ lin. long; segments ⅓ lin. long, triangular, flat, glabrous; anthers and style ¼ lin. long; fruit oblong-ellipsoid, capped by the persistent perianth, shortly contracted at the base, 1⅓ lin. long, prominently 10-ribbed, obscurely reticulate between the ribs. *A.DC. in DC. Prodr.* xiv. 660. *T. spinosum, Jacq.*, not of *Linn. nor Drège, and T. fragile, Link, not of Linn., ex A.DC. l.c.* names only. *T. sedifolium, A.DC. l.c.*, name only.

COAST REGION: Cape Div.; top of Table Mountain, *Bergius*! Caledon Div.; mountains near Grietjes Gat, *Ecklon & Zeyher*, 34! Houw Hoek, 3000 ft., *Schlechter*, 7346! *Bolus*, 9193! near Genadendal, Baviaans Kloof Mountains, *Burchell*, 7753!

T. crassifolium, R. Br. Prodr. 353, quoted in DC. Prodr. xiv. 672, proves from an examination of the specimen preserved in the British Museum to be *T. Frisea*, Linn., var. *Thunbergii*, A.DC.

12. T. disciflorum (A. W. Hill in Kew Bulletin, 1915, 27); stems weak and probably procumbent, much-branched from the base; branches slender, subterete, minutely scabrous, reddish-brown when dry; leaves numerous, linear, acutely mucronate, 2–3 lin. long, convex below, flat above, at length recurved, minutely scabrous, fairly fleshy; flowers solitary; bracts and bracteoles longer than the flowers, the former leaf-like, adnate to the peduncle; peduncle very short; bracteoles 2, about half the length of the bracts, acutely mucronate; perianth-segments triangular, $\frac{1}{3}$ lin. long, hooded, glabrous; anthers exserted from the perianth-tube, $\frac{1}{8}$ lin. long; style $\frac{1}{7}$ lin. long, reaching to the base of the anthers; fruits ellipsoid, capped by the persistent perianth, $1\frac{2}{3}$ lin. long, 1 lin. in diam., ribbed and reticulate, glabrous.

CENTRAL REGION: Graaff Reinet; in grassy places on Tandjes Berg, 4300 ft., *Bolus*, 1967!

13. T. namaquense (Schlechter in Engl. Jahrb. xxvii. 120); stems branched, rather slender, terete; branches slender, flattened and sulcate, glabrous; leaves linear, acute, 3–4 lin. long, rather fleshy, glabrous; flowers solitary, pedicellate; bracts acicular, much longer than the flowers, resembling the leaves; pedicel about $\frac{3}{4}$ lin. long, with 2 small bracteoles at the base, glabrous; bracteoles linear-lanceolate, acute, $\frac{3}{4}$ lin. long; perianth $\frac{3}{4}$ lin. long, glabrous; segments $\frac{1}{3}$ lin. long, triangular, acute, slightly hooded, glabrous; anthers $\frac{1}{8}$ lin. long; filaments $\frac{1}{4}$ lin. long; disc conspicuous, lobed; style $\frac{1}{4}$ lin. long, its apex below the anthers; fruits with prominent ribs.

WESTERN REGION: Van Rhynsdorp Div.; Karree Bergen, 2000 ft., *Schlechter*, 8206!

14. T. acutissimum (A.DC. Esp. Nouv. Thes. 4); stems numerous and slender from a slender rootstock, finely sulcate, very minutely tuberculate; branches slender; leaves acicular, subterete or in the barren shoots flat and with a distinct midrib, acutely mucronate, up to $\frac{1}{2}$ in. long, rather fleshy, glabrous, recurved; inflorescence 1–3-flowered; bract shortly adnate to the peduncle, flat on the upper surface, concave below, apex cartilaginous and acute; bracteoles longer than the flowers, similar to the bracts; perianth $\frac{3}{4}$ lin. long; segments triangular, $\frac{1}{2}$ lin. long, hooded, glabrous; anthers $\frac{1}{8}$ lin. long; disc conspicuous; style $\frac{1}{4}$ lin. long, reaching to the top of the anthers; fruits ellipsoid, capped by the persistent perianth, $1\frac{3}{4}$ lin. long, prominently 10-ribbed, conspicuously reticulate between the ribs. *Sond. in Flora*, 1857, 405; *A.DC. in DC. Prodr.* xiv. 659. *T. Krebsii, A.DC. Esp. Nouv. Thes.* 1, *and in DC. Prodr.* xiv. 656; *Sond. in Flora*, 1857, 403. *T. apiculatum, Sond. in Flora*, 1857, 357; *var. corniculatum, Sond. in Flora*, 1857, 357. *T. corniculatum, A.DC. in D.C. Prodr.* xiv. 662, *partly. T. strictum, Thunb. ex Sond. in Flora*, 1857, 357, *name only.*

VAR. β, **corniculatum** (A. W. Hill) ; flowers crowded and somewhat glomerulate at the ends of the shoots. *T. corniculatum, E. Meyer in Drège, Zwei Pfl. Documente,* 145, 226 ; *A.DC. in DC. Prodr.* xiv. 662, *partly, not var. corniculatum, Sond. in Flora,* 1857, 357.

SOUTH AFRICA : without locality, *Krebs,* 291 !
COAST REGION : Uitenhage Div. ; Zwartkops River, *Zeyher,* 3816 ! 3816a ! 3816b ! 22 ! Port Elizabeth Div. ; near Port Elizabeth, *Burchell,* 4337 ! Redhouse and Van Staadens, *Mrs. Paterson,* 729 ! Albany Div. ; between Kaffir Drift and Port Alfred, *Burchell,* 3781 ! Grahamstown, *Misses Daly & Sole,* 90 ! and without precise locality, *Bowker* ! King Williamstown Div. ; King Williams Town, 1500 ft., *Sim,* 1475 !
CENTRAL REGION : Somerset Div. ; near Somerset East, *MacOwan,* 2218 !
EASTERN REGION : Transkei ; Tsomo, 2000 ft., *Baur,* 476 ! Var. β : Transkei ; Gcua River, *Drège* !

15. **T. squarrosum** (Linn. f. Suppl. 162) ; stem erect or suberect, branched from near the base or in the upper half only, subterete, glabrous ; branches numerous, crowded, angular, purplish when dry ; leaves linear, with subacute cartilaginous tips, up to about 4 lin. long, rather fleshy, glabrous, at length recurved ; peduncles very short and adnate to the bract ; bracts leaf-like, reaching to the top of the flowers ; bracteoles as long as or slightly longer than the flowers ; perianth ¾ lin. long ; segments ovate-lanceolate, ½ lin. long, hooded, glabrous ; anthers exserted, ¼ lin. long ; disc conspicuous ; style very stout and short ; fruit ellipsoid-globose, 2½ lin. long, 1½ lin. in diam., capped by the persistent perianth, stipitate, prominently 10-ribbed, reticulate between the ribs. *Thunb. Prodr.* 40 ; *Fl. Cap. ed. Schult.* 211 ; *A.DC. in DC. Prodr.* xiv. 659 ; *Sond. in Flora,* 1857, 361, 405. *T. multiflorum, A.DC. Esp. Nouv. Thes.* 4. *T. corniculatum, Krauss ex A.DC. in DC. l.c.,* name only, not of *E. Meyer. T. hispidulum, Zeyh. ex A.DC. in DC. l.c.,* name only, not of *Lam. T. paniculatum, Thunb. in Fl. Cap. ed. Schult.* 210, as to spec. in *Herb. Stockh.* ?

SOUTH AFRICA : without locality, sheets α and β in *Herb. Thunberg* !
COAST REGION : Swellendam Div. ; on plains, *Bowie* ! Knysna Div. ; hills near Knysna, 150 ft., *Schlechter,* 5929 ! Uitenhage Div. ; Uitenhage, *Zeyher,* 59 ! Van Stadens Berg, *Zeyher,* 3804 ! Goda Hopsudden, *Sparmann* ! Algoa Bay, *Forbes* ! Van Staadens, near Port Elizabeth, *Mrs. Paterson,* 760 ! Albany Div. ; near Grahamstown, 2500 ft., *Galpin,* 81 ! *Bolus, Herb. Norm. Austr.-Afr.,* 1363 !
CENTRAL REGION : Somerset Div. ; Somerset, *Bowker* ! Bosch Berg, 400 ft., *MacOwan,* 1997 ! 3700 ft., *Bolus,* 281 ! 281b !

16. **T. foliosum** (A.DC. in DC. Prodr. xiv. 656) ; stems erect, much-branched, ribbed, glabrous ; branches ascending or suberect ; leaves linear, obtuse or subobtuse, ⅓–1 in. long, flat on the upper surface, keeled below, glabrous, midrib fairly prominent ; flowers crowded towards the end of the shoots, the latter sometimes subcorymbose ; cymes 1–3-flowered ; bracts boat-shaped, with slightly membranous margins, broader at the apex and thickened, as long as or longer than the peduncles, glabrous ; bracteoles as long as or

longer than the flowers, similar to the bracts ; perianth 1¼ lin. long ;
segments triangular-ovate, ¾ lin. long, fleshy, deeply hooded, with
papillose margins ; anthers exserted, ⅐ lin. long ; style about ¼ lin.
long ; fruits shortly stalked, ellipsoid-globose, capped by the
persistent perianth, about 2½ lin. long including the stalk, 1¼ lin.
in diam., 10–11-ribbed, reticulate between the ribs. *T. capitatum,*
var. interruptum, Krauss in Flora, 1845, 80. *T. Turczaninowii,*
Sond. in Flora, 1857, 354.

COAST REGION : Uniondale Div. ; mountains of Long Kloof, *Drège,* 8164*a* !
Humansdorp Div. ; Clarkson, 1000 ft., *Galpin,* 4544 ! Uitenhage Div. ; Van
Stadens Berg, *Ecklon & Zeyher* ! Winterhoek Mountains, *Krauss,* 1806 ! Port
Elizabeth Div. ; Van Staadens, near Port Elizabeth, *Mrs. Paterson,* 882 ! upper
part of Maitland (Leadmine) River, *Burchell,* 4619 ! Albany Div. ; Salat Kraal,
near Grahamstown, *Zeyher,* 304 ! Bathurst Div. ; vicinity of Bathurst, *Ecklon &*
Zeyher, 27 !

17. **T. fruticosum** (A. W. Hill in Kew Bulletin, 1915, 28) ; a
shrub, 4–6 ft. high ; stem terete, with transversely splitting bark, up
to ¾ in. thick at the base, perfectly simple till 1–2 ft. from the top and
then dividing into 4–6 whip-like 2–3-chotomously divided branchlets,
glabrous ; branchlets ascending or " weeping," slightly angular or
compressed, glabrous ; leaves in young plants present along the
stem from the very root, subsequently deciduous, linear-oblong,
obtuse, recurved, 3–4 lin. long or up to ¾ in. long, fleshy, convex
below, flat or concave above, glabrous ; cymules 1–3-flowered,
loosely arranged in spike-like racemes ; bracts shorter than the
flowers, more or less free from the very short peduncle, up to 2½ lin.
long, rather thick, somewhat boat-shaped, obtuse ; bracteoles 2, half
as long as the flowers or less, linear-lanceolate, acute ; perianth wide
and cup-shaped with broad spaces between the segments, 1 lin.
long ; with glandular disc ; segments broadly triangular, ½ lin. long,
slightly hooded, apex and margins papillose ; anthers rounded,
included in the perianth-tube, ¼ lin. long ; style ⅛ lin. long ; fruits
ellipsoid-globose, contracted at the base, 3 lin. long including the
persistent perianth, prominently 5-ribbed at the base, with less
prominent ribs between, reticulate.

COAST REGION : Uitenhage Div. ; Zuurberg, *Mrs. Paterson,* 35 ! Albany Div. ;
Howisons Poort Hills, 2200 ft., *Galpin,* 2900 ! hills near Grahamstown, *MacOwan,*
2094 ! *Bolus,* 1558 ! *Atherstone,* 58 ! *Cooper,* 56 ! Featherstone Kloof, 2000 ft.,
Schönland, 567 ! Queenstown Div. ; Queenstown, *Cooper,* 3045 !

According to a note by MacOwan the stems in this species are 4–6 ft. high,
flexible, about ¾ in. thick at the base perfectly simple till 1–2 ft. from the top,
then dividing into 4–6 whip-like 2–3-chotomously divided branchlets. Leaves
subsequently deciduous. Remarkable by its weeping flagellate aspect.

18. **T. dissitiflorum** (Schlechter in Engl. Jahrb. xxvii. 118) ; a
small decumbent shrub, much-branched ; branches and branchlets
rigid, terete, glabrous, the latter with spiny tips ; leaves scattered,
subulate-linear, often curled in to the stem when dry, 1½–2 lin. long,

concave on the upper side, convex below, glabrous; flowers in axillary clusters usually about 2 lin. apart, 1–3 together; bracts shorter than the flower-clusters, lanceolate, acute to acuminate, becoming black, caducous; bracteoles 3–6, forming a small involucre, ovate, acute, concave, $\frac{1}{2}$–$\frac{3}{4}$ lin. long; perianth 1 lin. long; segments $\frac{3}{4}$ lin. long, with conspicuous external glands between each, margin slightly involute, glabrous; anthers $\frac{1}{4}$ lin. long; filaments $\frac{1}{4}$ lin. long; style very short; fruit subglobose, contracted at the base, 1$\frac{1}{2}$ lin. long, prominently 10-nerved, finely reticulate between the nerves.

CENTRAL REGION : Ceres Div.; Cold Bokkeveld, *Schlechter*, 8859 !
WESTERN REGION : Calvinia Div. ; Papelfontein, *Schlechter*, 10897 !

19. **T. euphorbioides** (Linn. Mant. Alt. 214); an erect somewhat woody shrub, up to 6 ft. high; branches straight, angular, glaucous; leaves broadly ovate or suborbicular, cordate at the base, mostly adpressed to or encircling the branches, often overlapping, acutely mucronate, $\frac{1}{2}$–1$\frac{1}{4}$ in. long, $\frac{1}{2}$–1 in. broad, rigidly coriaceous, several nerved, glaucous, glabrous; cymes racemose, with large leafy coloured bracts; bracts suborbicular, mostly a little smaller than the leaves, as long as and enclosing the cymes; bracteoles obovate or elliptic, rather longer than the flowers; perianth 1–1$\frac{1}{4}$ lin. long, with distinct external glands; segments lanceolate, obtuse, about 1$\frac{1}{4}$ lin. long, hooded, with incurved margins; anthers pendent, exserted, $\frac{1}{4}$ lin. long; disc rather inconspicuous; style about $\frac{1}{2}$ lin. long; fruit shortly stalked, subglobose, beaked by the persistent style, 2$\frac{1}{2}$ lin. long, with 5 prominent ribs and 5 faint ones between, scarcely reticulate. *Berg. Descr. Pl. Cap.* 74 ; *Thunb. Fl. Cap. ed. Schult.* 211 ; *Sond. in Flora*, 1857, 353. *T. amplexicaule, Linn. Mant. Alt.* 213.

SOUTH AFRICA : without locality, *Thunberg* !
COAST REGION : Tulbagh Div. ; between New Kloof and Elands Kloof, *Drège a* ! Cape Div. ; Cape Flats, *Ecklon & Zeyher*, 40 ! Stellenbosch Div.; Hottentots Holland, 1500 ft., *Ludwig* ! *Bolus*, 4204 ! *Diels*, 1295 ! between Stellenbosch and Cape Flats, *Burchell*, 8351 ! Caledon Div. ; Baviaans Kloof, *Burchell*, 7646 ! Zwart Berg, 200–300 ft., *Zeyher*, 3787 ! *MacOwan*, 2752 ! and *Herb. Norm. Austr.-Afr.*, 764 ! *Pappe* ! *Schlechter*, 5371 ! near Genadendal, *Drège b* ! Donker Hoek Mountain, *Burchell*, 7978 ! Bredasdorp Div. ; Koude River, 900 ft., *Schlechter*, 9737 ! Riversdale Div. ; Garcias Pass, *Phillips*, 324 ! *Galpin*, 4547 ! near the Gouritz River, *Ecklon & Zeyher*, 40 ! Uitenhage Div. ; Zwartkops River, *Zeyher*, 3786 !

20. **T. pinifolium** (A.DC. Esp. Nouv. Thes. 2); stem 5 ft. high, stout and woody, conspicuously ribbed, glabrous; branches rather crowded in the upper part of the stem, ascending or suberect, rather sharply ribbed, glabrous; leaves acicular, terete, obtuse or subobtuse, slightly decurrent on the stem, those on the main stem 1–1$\frac{1}{2}$ in. long, arcuate, those on the flowering branches shorter, glabrous; flowers subglomerate at the ends of the branches; cymes few-flowered; bracts linear, subobtuse, about $\frac{1}{4}$ in. long, boat-

shaped, sharply keeled, with a prominent midrib on the upper
surface; peduncle adnate to the lower part of the bract but
shorter; bracteoles as long as the flowers, similar to the bracts;
perianth 1 lin. long; segments ovate, $\frac{3}{4}$ lin. long, hooded, glabrous
except on the incurved papillose edges; anthers exserted, $\frac{1}{8}$ lin.
long; style $\frac{1}{2}$ lin. long, exceeding the anthers or reaching about
to their middle; fruit ellipsoid, narrowed to both ends, capped
by the persistent perianth, $1\frac{1}{2}$ lin. long, prominently 10-ribbed,
5 of the ribs more prominent than the others. *A.DC. in DC.
Prodr.* xiv. 655. *T. corymbiflorum, Sond. in Flora,* 1857, 354,
404.

COAST REGION: Swellendam Div.; Voormans Bosch, *Zeyher,* 3791! Riversdale
Div.; mountains of Garcias Pass, 1200 ft., *Galpin,* 4545!

21. **T. glomeruliflorum** (Sond. in Flora, 1857, 355); an undershrub up to $1\frac{1}{2}$ ft. high; stem angular or almost winged, glabrous;
branches ascending; leaves linear, broadening towards base, acute,
$\frac{1}{3}$–$\frac{3}{4}$ in. long, flat on the upper surface, rounded on the lower,
glabrous; flowers 3–5 in axillary sessile or very shortly pedunculate
clusters, the latter about half as long as the leaves; bracts and
bracteoles imbricate around and much longer than the flowers,
lanceolate, acute, up to 2 lin. long, prominently keeled; perianth
$\frac{3}{4}$ lin. long, with external glands between the segments; segments
erect, lanceolate, obtuse, $\frac{1}{2}$ lin. long, glabrous; anthers exserted
from the perianth-tube, $\frac{1}{8}$ lin. long, dark-coloured; style very
short. *A.DC. in DC. Prodr.* xiv. 671.

COAST REGION: Swellendam Div.; Swellendam, *Mund!* Riversdale Div.;
between Little Vet River and Garcias Pass, *Burchell,* 6864, partly! mountains,
Garcias Pass, 1000–1500 ft., *Galpin,* 4543! *Bolus,* 11422! Klein Berg, at Platte-
Kloof, *Muir,* 473 (5446 *in Herb. Galpin*)!

22. **T. angulosum** (A.DC. Esp. Nouv. Thes. 2); a shrub, 5 ft.
high; stems erect, broadly winged, wings with cartilaginous margins,
glabrous; branchlets less broadly winged to angular, sometimes
rather slender, glabrous; leaves linear, decurrent on the stem, with
a prominent midrib continuous with the cartilaginous edge of the
wing, acute or subacute, up to about $\frac{1}{2}$ in. long and about $1\frac{1}{2}$
lin. broad, flat, glabrous; cymules arranged in rather elongated
lax racemes; bracts adnate to the peduncle for more or less than
half their length or sometimes almost free, linear-lanceolate, acutely
acuminate; bracteoles shorter than the flowers; perianth $1\frac{3}{4}$–$2\frac{1}{4}$ lin.
long; segments elliptic, 1–$1\frac{1}{4}$ lin. long, hooded, with papillose incurved margins; anthers exserted from the perianth-tube, $\frac{1}{2}$ lin.
long; style $1\frac{1}{4}$–$1\frac{1}{2}$ lin. long, reaching to the top of the anthers;
fruit with a stout yellow stalk about 1 lin. long, subglobose, about
2 lin. long, including the gaping persistent perianth, prominently
10-ribbed, not or only very slightly reticulate between the ribs.
A.DC. in DC. Prodr. xiv. 653; *Sond. in Flora,* 1857, 403.
T. Galpinii, Schlechter in Journ. Bot. 1897, 222.

COAST REGION : Queenstown Div. ; mountains near Queenstown, 4000–4500 ft., *Galpin*, 1654 ! King Williamstown Div. ; Perie Mountains, 3500 ft., *Galpin*, 3273 ! *Sim*, 1533.
EASTERN REGION : Transkei ; Kentani Distr., 1200 ft., *Miss Pegler*, 176 ! Tembuland ; Bazeia Mountain, 3500 ft., *Baur*, 752 ! Natal ; near Durban, *Gueinzius*, 365 ! Inanda, *Wood*, 262 ! hill-side near Bothas, 2200 ft., *Wood*, 5009 ! Emberton, 1970 ft., *Schlechter*, 3238 ! and without precise locality, *Gerrard*, 1277 !

23. T. Susannæ (A. W. Hill in Kew Bulletin, 1915, 40) ; stems erect, woody, with reddish angles, glabrous ; branches ascending or suberect, leafy ; leaves linear-acicular, more or less trigonous, acute, $\frac{1}{2}$–$\frac{3}{4}$ in. long, about $\frac{1}{3}$ lin. thick, glabrous ; flowers in shortly pedunculate or subsessile 3–5-flowered cymules arranged in terminal leafy racemes ; peduncles 1–2 lin. long ; bracts adnate to the apex of the peduncle, much overtopping the flowers, narrowly linear-lanceolate, acute or subacute, sometimes with sub-translucent reddish margins, convex below, concave above, glabrous, entire ; bracteoles about half as long as the free portion of the bracts, otherwise very similar to the latter ; perianth 1–1$\frac{1}{4}$ lin. long, with distinct external glands and an internal lobed disc ; segments broadly triangular-ovate, subacute, $\frac{1}{2}$ lin. long, hooded, fleshy, glabrous, margins scarcely papillose ; anthers included, reaching to the base of the perianth-segments, $\frac{1}{3}$ lin. long ; stigma subsessile ; fruits ellipsoid-globose, 2$\frac{1}{2}$ lin. long, distinctly 10-ribbed, rather fleshy, scarcely reticulate between the ribs.

SOUTH AFRICA : without locality, *Krebs*, 150 ! 175 ! without indication of the collector in *Herb. Kew* !
COAST REGION : Riversdale Div. ; Gysmans Hoek, *Muir*, 359 ! *and in Herb. Galpin*, 5327 !

24. T. triflorum (Thunb. ex Linn. f. Suppl. 162) ; stems much-branched, terete below, upper angular and sulcate, glabrous ; leaves linear or linear-lanceolate, subacute, $\frac{1}{2}$–2 in. long, up to 1$\frac{1}{2}$ lin. broad, flat, 1–3-nerved, or subacicular, somewhat fleshy, glabrous ; inflorescence 1- to several-flowered, cymose ; bracts leafy, often sickle-shaped and somewhat reflexed ; peduncle mostly shorter than the bracts ; bracteoles 2, minute ; flowers subsessile ; perianth 1$\frac{2}{3}$ lin. long ; segments triangular, about 1 lin. long, flat, hooded, glabrous ; anthers exserted, $\frac{1}{4}$ lin. long ; filaments $\frac{2}{3}$ lin. long ; disc prominent, lobed; style $\frac{2}{3}$ lin. long ; fruit ovoid-globose, capped by the persistent perianth, 2 lin. long, 1$\frac{1}{2}$ lin. in diam., coarsely reticulate. *Thunb. Prodr.* 46 ; *Fl. Cap. ed. Schult.* 211 ; *A.DC. in DC. Prodr.* xiv. 661, *partly* ; *Sond. in Flora*, 1857, 354, *partly. T. planifolium, A.DC. Esp. Nouv. Thes.* 5 ; *Sond. in Flora*, 1857, 405.

COAST REGION : Humansdorp Div. ; Gamtoos River, 330 ft., *Schlechter*, 6055 ! Uitenhage Div. ; near the Zwartkops River, *Zeyher*, 50 ! 3794 ! *Ecklon & Zeyher*, 33 ! Redhouse, *Mrs. Paterson*, 400 ! Albany Div. ; Fish River Heights, *Hutton* ! Fort Beaufort Div. ; between the Koonap and Kat Rivers, *Ecklon & Zeyher*, 32 ! British Kaffraria, *Cooper*, 66 !

CENTRAL REGION: Albert Div.; without precise locality, *Cooper*, 1762!
Murraysburg Div.; near Murraysburg, 4000 ft., *Tyson*, 82! Richmond Div.;
Uitvlugt, near Stylkloof, 4000–5000 ft., *Drège*, 8180! Graaff Reinet Div.;
Sneeuwberg Range, 3700 ft., and Karroo near Graaff Reinet, *Bolus*, 2011!
Camdeboo, *Dunn*! Philipstown Div.; near Rietfontein, at Waschbanks River,
Burchell, 2719, 2736!
KALAHARI REGION: Transvaal; Shilovane, *Junod*, 523!
EASTERN REGION: Natal; Mooi River, *Wood*, 4432! Portuguese East Africa:
Inhambane, *Scott*!

25. T. scandens (E. Meyer in Drège, Zwei Pfl. Documente, 133,
226); stem scandent, covered with shiny light-brown bark, glabrous;
branches weak, elongated, probably hanging, slightly flexuous;
leaves large, few, reflexed, fleshy, terete, up to $1\frac{1}{2}$ in. long, about
1 lin. thick, glabrous and closely wrinkled when dry; inflorescence
cymose, about 3-flowered; bracts adnate to the peduncle for 1–3
lin., the lower leaf-like, recurved, up to $\frac{3}{4}$ in. long; peduncle up to
$1\frac{1}{2}$ in. long, subterete, glabrous; bracteoles minute; flowers very
shortly pedicellate, about 1 lin. long; perianth cupular, $\frac{3}{4}$ lin. long;
segments ovate, obtuse, $\frac{2}{3}$ lin. long, slightly hooded, fleshy, flat;
anthers $\frac{1}{8}$ lin. long; filaments $\frac{1}{8}$ lin. long; style about $\frac{1}{2}$ lin. long,
stout, reaching to the top of the anthers; fruit ellipsoid-globose,
about 2 lin. long, wrinkled and somewhat fleshy, red (*Tyson*).
A.DC. in DC. Prodr. xiv. 661; *Sond. in Flora*, 1857, 354.

COAST REGION: Uitenhage Div.; near the Zwartkops River, *Zeyher*, 694!
3795! Zwartkops, *Tyson* in Herb. *Marloth*! Enon, below 100 ft., *Drège*!
Redhouse, *Mrs. Paterson*! Fort Beaufort Div.; between the Koonap and Kat
Rivers, *Ecklon & Zeyher*, 32!

26. T. galioides (A.DC. Esp. Nouv. Thes. 5); stems slender,
grooved, glabrous, much-branched; leaves not seen; inflorescence
cymose, dichotomously branched; bracts adnate to the peduncle for
about $\frac{3}{4}$ lin., linear, subacute, at length recurved, up to $1\frac{1}{2}$ lin.
long, fleshy, glabrous; peduncle $\frac{1}{4}$–$\frac{1}{2}$ in. long; bracteoles 2, about
half as long as the flowers or less, subulate-lanceolate, subacute:
perianth nearly 1 lin. long, campanulate; segments triangular, flat,
$\frac{1}{2}$ lin. long, slightly hooded, glabrous; disc lobed and well marked;
anthers exserted, $\frac{1}{8}$ lin. long; filaments $\frac{1}{4}$ lin. long; style $\frac{1}{4}$ lin.
long. *A.DC. in DC. Prodr.* xiv. 661. *T. triflorum, Thunb. ex*
Sond. in Flora, 1857, 354, *partly*; *A.DC. in DC. l.c., partly*. *Sond.*
in Flora, 1857, 405.

COAST REGION: Stellenbosch Div.; Grietjes Gat, 2000–4000 ft., *Ecklon &*
Zeyher, 23 ! Caledon Div.; near Caledon, *Ecklon & Zeyher*, 30 ! Riversdale Div.;
Corenti River Farm, *Muir*, 5326! near Riversdale, 500–650 ft., *Schlechter*, 1896!
Bolus, 11374! Mossel Bay Div.; hills east of the Gouritz River, *Burchell*, 6412!
Uitenhage Div.; Addo, *Zeyher*, 3797! *Ecklon & Zeyher*, 23! 33! Uitenhage,
Zeyher, 51! Port Elizabeth Div.; Port Elizabeth, *Drège*, 267!
CENTRAL REGION: Graaff Reinet Div.; near Graaff Reinet, 2500 ft.,
Bolus, 281!

27. T. asperifolium (A. W. Hill in Kew Bulletin, 1915, 23);
stems branched, slender, up to about 8 in. high, angular, finely
scabrous; branches ascending, slender, leafy, scabrous; leaves

linear or linear-lanceolate, subacute, up to $\frac{3}{4}$ in. long, with a distinct keeled midrib on the lower surface, minutely scabrous; flowers in lax cymes; bracts more or less lanceolate, flat, acute, as long as or longer than the flowers, finely scabrous on the back and margins; bracteoles shorter than the flowers, very similar to the bracts; perianth $\frac{3}{4}$–$\frac{4}{5}$ lin. long; segments ovate, subacute, $\frac{3}{8}$ lin. long, slightly hooded; disc prominent, lobed; anthers exserted, $\frac{1}{5}$ lin. long; style stout, $\frac{1}{8}$ lin. long; fruits globose, 2 lin. long, prominently ribbed, very sparingly and inconspicuously reticulate between the ribs.

COAST REGION: George Div.; on a hill near George, *Schlechter*, 2358! Queenstown Div.; Table Mountain, *Drège*, 8170b!

28. **T. corymbuligerum** (Sond. in Flora, 1857, 362); a small shrub, up to 1 ft. high; stems several, erect or suberect, woody, glabrous; branches bearing small corymbs of several flowers; leaves small and inconspicuous, linear, acute, about 1 lin. long, fleshy, concave on the upper surface, glabrous; flowers small, in small terminal branched cymes, bracts a little shorter than the flowers, ovate or oblong, acute; bracteoles very small; perianth saucer-shaped, 1 lin. long; segments triangular, $\frac{1}{2}$ lin. long, hooded, flat, glabrous or nearly so on the margin; disc conspicuous; anthers exserted, nearly $\frac{1}{4}$ lin. long; style $\frac{1}{4}$ lin. long; fruits turbinate at the base, scarcely 2 lin. long, rather fleshy and scarcely ribbed. *A.DC. in DC. Prodr.* xiv. 660. *T. polyanthum, Schlechter in Journ. Bot.* 1898, 27.

COAST REGION: Cape Div.; Beacon Hill, *Wolley-Dod*, 2821! roadside towards Chapmans Bay, *Wolley-Dod*, 1551! Caledon; stony heights of the Hartebeest River, *Zeyher*, 3812! Zwart Berg, *Ecklon & Zeyher*, 85! Riversdale Div.; near Garcias Pass, 1800 ft., *Bolus*, 11376!

29. **T. floribundum** (A. W. Hill in Kew Bulletin, 1915, 27); stems erect, branched from about the middle, prominently ribbed, glabrous; branches ascending, slender, flexuous; leaves linear, acute, $\frac{1}{3}$–1 in. long, up to $\frac{1}{2}$ lin. broad, somewhat fleshy, glabrous; flowers paniculate in the upper part of the branches; bracts adnate to the peduncle for nearly a third of their length or less, leaf-like, keeled; cymules 3–5-flowered; bracteoles longer than the flowers, linear-lanceolate, acute; perianth $\frac{3}{4}$ lin. long; segments triangular, $\frac{1}{3}$ lin. long, fleshy, hooded, with more or less papillose margins; disc conspicuous; anthers exserted, $\frac{1}{8}$ lin. long; style stout, very short; fruit oblong-ellipsoid, capped by the persistent perianth, contracted at the base, 2 lin. long, $1\frac{1}{4}$ lin. in diam., very prominently 10-ribbed, reticulate between the ribs, often slightly glaucous.

COAST REGION: British Kaffraria, *Cooper*, 138!
KALAHARI REGION: Transvaal; near Wonderfontein Railway Station, 6000 ft., *Bolus*, 12278! Heidelberg, *Schlechter*!
EASTERN REGION: Tembuland; Bazeia, *Baur*, 336, partly! Griqualand East; Mount Currie, 5200 ft., *Tyson*, 1838! Pondoland; Port St. John, summit of

West Gate, 1200 ft., *Galpin*, 3467! Natal; Port Shepstone, *Rogers*! Dumisa,
1970 ft., *Rudatis*, 472! Malvern, near Durban, 500–600 ft., *Wood*, 4971! near
Newcastle, 3000–4000 ft., *Wood*, 7186! Inanda, 1800 ft., *Wood*, 154! 249! and
without precise locality, *Gerrard*, 352! *Mrs. K. Saunders*!

30. **T. pallidum** (A.DC. Esp. Nouv. Thes. 2); stems few, from
an erect and rather slender rootstock, rather closely sulcate, glabrous,
branched in the upper part; branches erect, angular; leaves linear,
obtuse, $\frac{3}{4}$–$1\frac{1}{2}$ in. long, $\frac{2}{3}$ lin. broad, rather coarsely verrucose, somewhat fleshy, glabrous; inflorescence several-flowered, terminating
the branches; bracts adnate to the peduncle for half their length,
leaf-like, those near the flowers acute and longer than the flowers;
peduncle equalling the bracts, glabrous; bracteoles 2, shorter than
the flowers, lanceolate, subacute; perianth $\frac{3}{4}$–1 lin. long; segments
$\frac{1}{2}$–$\frac{2}{3}$ lin. long, broadly ovate, hooded, margins slightly fringed,
incurved, membranous; anthers exserted from the perianth-tube,
slightly pendulous, about $\frac{1}{4}$ lin. long; style $\frac{1}{4}$–$\frac{1}{3}$ lin. long, stout;
stigma capitate, reaching to the middle of the anthers; fruit
ellipsoid, narrowed to the base, capped by the persistent erect
perianth-segments, $2\frac{1}{2}$ lin. long, $1\frac{1}{4}$ lin. in diam., very prominently
10-ribbed, coarsely reticulate between the ribs, slightly shining.
A.DC. in DC. Prodr. xiv. 654; *Sond. in Flora*, 1857, 404.

COAST REGION: Albany Div.; near Grahamstown, *MacOwan*, 181! Queenstown
Div.; mountain near Zwartkei Bridge, 4000 ft., *Galpin*, 2275!
CENTRAL REGION: Wodehouse Div.; Stormberg Range, *Drège*, 8170a!
EASTERN REGION: Griqualand East; Mount Currie, *Tyson, in MacOwan &
Bolus, Herb. Norm. Afr.-Austr.*, 1230!

31. **T. quinqueflorum** (Sond. in Flora, 1857, 354); a small erect
shrub, 1–2 ft. high, much-branched; branches erect, leafy, finely
pustulate, glabrous; leaves erect-patent, linear, with subacute,
blackish tips, 2–$4\frac{1}{2}$ lin. long, concave above, keeled below, finely
pustulate; inflorescence terminal or subterminal, of cymose 4–5-
flowered umbellules; bracts and bracteoles subspathulate, keeled,
longer than the flowers, margins slightly membranous; perianth 1–$1\frac{1}{2}$
lin. long; segments $\frac{1}{2}$–$\frac{3}{4}$ lin. long, ovate-triangular, obtuse, deeply
hooded, with incurved wavy margins; anthers exserted and situated
under the hood, about $\frac{1}{4}$ lin. long; disc conspicuous; style $\frac{1}{3}$–$\frac{1}{2}$ lin.
long; fruit subglobose, capped by the persistent perianth, $2\frac{1}{2}$ lin.
long, $1\frac{3}{4}$ lin. in diam., 5-winged towards the base, smooth between
the wings. *T. erectiramosum, A.DC. Esp. Nouv. Thes. 2, partly;
A.DC. in DC. Prodr.* xiv. 655, *partly; Sond. in Flora*, 1857, 404.
T. affine, Schlechter in Engl. Jahrb. xxvii. 115.

COAST REGION: Caledon Div.; mountains near Grietjes Gat, *Ecklon & Zeyher,*
21! Houw Hoek, 1200 ft., *Schlechter*, 9394!

Sonder rightly pointed out in his supplementary paper that De Candolle has
confused two different plants under his species *T. erectiramosum.* Only *Ecklon &
Zeyher,* 21, can be referred to *T. quinqueflorum,* which is a quite distinct species,
the other specimens belong to *T. virgatum,* Lam.

32. T. cupressoides (A. W. Hill in Kew Bulletin, 1915, 26); stems bushy, about 8 in. high, much-branched, woody, terete; branches ascending, rather densely leafy, glabrous; leaves linear, acute, 2–3½ lin. long, flat on the upper surface, slightly keeled below, somewhat fleshy, glabrous; flowers clustered at the ends of short axillary branches; bracts and bracteoles lanceolate, longer than the flowers; perianth ¾ lin. long, with a distinct disc and small prominent external glands; segments ovate, obtuse, ½ lin. long, hooded; anthers exserted, ¹⁄₁₀ lin. long; stigma subsessile; fruits not seen.

EASTERN REGION: Natal; Niginya, 5500 ft., *Wylie in Herb. Wood*, 10618!

33. T. ericæfolium (A.DC. Esp. Nouv. Thes. 3); a heath-like much-branched low shrub; stem purplish, subterete, glabrous; branches ascending, rather slender, slightly grooved; leaves linear, keeled, obtuse or subacute, 1½–2 lin. long, glabrous; flowers crowded towards the apex of the branchlets, sessile; bracts rather longer than the flowers, ovate-lanceolate, acute, keeled, concave above; bracteoles similar to the bracts but a little shorter; perianth with conspicuous external glands, ½ lin. long; segments ovate-triangular, about ¼ lin. long; anthers very small; style ⅛ lin. long, reaching to the top of the anthers; fruits small, ovoid, scarcely 1 lin. long, prominently 10-ribbed especially at the base, scarcely reticulate between the ribs. *Sond. in Flora*, 1857, 404; *A.DC. in DC. Prodr.* xiv. 658. *T. ramellosum, Sond. in Flora*, 1857, 362. *T. ericæfolium, var. confertum, A.DC. l.c.* 658. *T. ericoides, R. Br. Prodr.* 353, *name only*; *DC. l.c.* 672.

COAST REGION: Tulbagh Div.; near Tulbagh Waterfall, *Ecklon & Zeyher*, 26! Cape Div.; Orange Kloof, *Dreyius*! Caledon Div.; near Genadendal, Baviaans Kloof, *Burchell*, 7760! Houw Hoek, *Scott-Elliot*, 1132! *Bolus*, 9919! Steenbrass River, 1000 ft., *Schlechter*, 5384! Vogelgat, 1500 ft., *Schlechter*, 10421! Riversdale Div.; Garcias Pass, 1000 ft., *Galpin*, 4542! Uitenhage Div.; Van Stadens Berg, *Drège*, 7174! *Zeyher*, 3805! Uitenhage, *Ecklon & Zeyher*, 273!

The specimen collected by R. Brown and labelled *T. ericoides* is preserved in the British Museum and proves to be *T. ericæfolium*, A.DC.

34. T. nigromontanum (Sond. in Flora, 1857, 361); a small heath-like subshrub branched from the base; branches erect or suberect, slender, subterete; leaves of two kinds, those towards the base of the plant linear and subterete, subobtuse, up to 4 lin. long, fleshy, glabrous; the upper closely adpressed to the shoot, subulate-lanceolate, acute, about 1 lin. long; cymules axillary, often sub-corymbose, with the flowers crowded at the apex, leafy in the lower part; flowers sessile or subsessile in the middle of the bract and two bracteoles; bract acutely acuminate from an ovate base, about the same length as the flower, with minutely hirsute margins; bracteoles with blackened tips, ovate-lanceolate, acute, a little shorter than the flower; perianth saucer-shaped, ¾ lin. long, with conspicuous external glands; segments triangular, thickened and deeply hooded at the apex, margins minutely hairy ; anthers about $\frac{1}{10}$ lin. long, exserted ; filaments $\frac{1}{9}$ lin. long ; disc flat and conspicuous ; style extremely short ; fruit ellipsoid-campanulate, $1\frac{1}{4}$ lin. long, prominently 10-ribbed at the base, finely reticulate between the ribs. *A.DC. in DC. Prodr.* xiv. 657. *T. leptocaule, Sond. in Flora,* 1857, *362, partly. T. leptocaule, Sond., var. glabriusculum, A.DC. in DC. Prodr.* xiv. 658, *partly ; Sond. l.c.* 404, *partly. T. brevifolium, A.DC. Esp. Nouv. Thes.* 3, *partly ; var. glabriusculum, A.DC. l.c.* 3.

CoAST REGION : Cape Div. ; Cape flats, Doornhoogte, *Ecklon & Zeyher,* 17 ! Stellenbosch Div. ; French Hoek, 1500 ft., *Schlechter,* 9336 ! Caledon Div. ; between Zwart Berg and the River Zondereinde, *Zeyher,* 3813 ! Bredasdorp Div. ; Elim, 700 ft.. *Schlechter,* 7628 ! George Div. ; Montagu Pass, 3800 ft., *Schlechter,* 5828 ! near George, 650 ft., *Schlechter,* 2393 ! Knysna Div. ; hills near Knysna, 160 ft., *Schlechter,* 5924 ! Humansdorp Div. ; near Gamtoos River, *Schlechter,* 6040 ! Uitenhage Div. ; between Zwartkops River and Sundays River, *Ecklon & Zeyher,* 28, partly !

Considerable confusion exists as to the specimens to be referred to *T. nigromontanum* and *T. leptocaule* respectively, and the specimens appear to have been mixed under the same number. In *T. nigromontanum* the external glands are conspicuous, while they are apparently absent from *T. leptocaule* ; the acutely acuminate bracts of *T. nigromontanum* and the black-drying flowers also serve as a ready means of distinguishing this species from the more robust *T. leptocaule.* *Rhinostegia brevifolia,* Turcz., which has also been associated with *T. leptocaule* (Sond. in Flora, 1857, 404), appears to be an imperfect specimen and probably should be referred to *T. virgatum.*

35. **T. leptocaule** (Sond. in Flora, 1857, 362) ; rootstock slender, erect ; stem simple for some distance or with several branches from the apex of the rootstock, erect or ascending, subterete or slightly angled, glabrous ; branchlets rather short and twiggy ; leaves of two kinds, those towards the base spreading, acicular, obtuse, 1–2 lin. long, glabrous, the upper very small and closely adpressed to the stem, subulate-lanceolate, glabrous ; flowers crowded at the ends of the branchlets ; bracts often blackish when dry, shortly adnate to the very short peduncle, ovate-lanceolate, fairly acute, $\frac{3}{4}$ lin. long, keeled, with slightly jagged membranous margins ; bracteoles a little shorter than the flower, keeled ; perianth $\frac{3}{4}$ lin. long ; segments $\frac{1}{3}$–$\frac{1}{2}$ lin. long, hooded, glabrous ; anthers $\frac{1}{10}$ lin. long ; disc conspicuous ; style $\frac{1}{8}$ lin. long ; fruits ovoid, $1\frac{1}{2}$ lin. long, prominently 9–10-ribbed at the base, reticulate between the ribs. *A. DC. in DC. Prodr.* xiv. 657, *partly. T. brevifolium, A.DC. Esp. Nouv. Thes.* 3, *partly ; Sond. in Flora,* 1857, 404, *not T. (Rhinostegia) brevifolium, Sond. l.c.* 361, 404.

CoAST REGION : Uitenhage Div. ; between Zwartkops River and Sundays River, *Ecklon & Zeyher,* 28 partly ! near Uitenhage, *Ecklon & Zeyher,* 11 ! Port Elizabeth Div. ; Port Elizabeth, *Herb. E. S. C. A.* 469 ! Div. ? ; Humewood. *Miss Daly,* 1062 ! *Mrs. Paterson,* 614 ! EASTERN REGION : Griqualand East ; Mount Fletcher, *Sim,* 2641 !

36. **T. commutatum** (Sond. in Flora, 1857, 362) ; stem much-branched, subterete, slightly longitudinally wrinkled when dry ;

branches more or less angular, subcorymbosely or paniculately
arranged, glabrous ; leaves of two kinds, the lowermost very few
or absent, acicular, acute, subterete, up to 6 lin. long, black and
arcuate when dry, grooved on the upper surface, upper leaves short,
subulate-lanceolate, decurrent on the branchlets, 1–2 lin. long, subacute, glabrous, with black recurved tips ; flowers few, crowded and
sessile at the ends of the branchlets ; bracts like the upper leaves,
equalling the flowers, narrowly ovate-lanceolate, acute, edges more
or less fimbriate ; bracteoles linear-oblong, shorter than the flowers ;
perianth ¾ lin. long ; segments triangular, hooded, about ½ lin. long,
very shortly papillose ; anthers ⅛ lin. long, exserted ; style ⅛–¼ lin.
long ; fruits broadly ovoid, about 2 lin. long, contracted and prominently 10-ribbed at the base, faintly reticulate between the ribs.
A.DC. in DC. Prodr. xiv. 659.

COAST REGION : Cape Div. ; Simons Berg, *Wolley-Dod*, 3015 ! and without
precise locality, *Harvey*, 715 ! 716 ! Caledon Div. ; Baviaans Kloof, near Genadendal,
Burchell, 7663 ! Zwart Berg, 2620 ft., *Schlechter*, 5550 ! Houw Hoek, 1800 ft.,
Schlechter, 9389 ! Vogel Gat, 2000 ft., *Schlechter*, 9515 ! Bredasdorp Div. ;
between Elim and Fairfield, 1100 ft., *Bolus*, 8600 ! Uitenhage Div. ; without
locality, *Zeyher*, 272 !
CENTRAL REGION : Ceres Div. ; Cold Bokkeveld, Gydouw Berg, 5800 ft.,
Schlechter, 10235 !

37. **T. nudicaule** (A. W. Hill in Kew Bulletin, 1915, 35) ; a
divaricately branched leafless or nearly leafless shrub ; branches
subterete, about 1 lin. thick, glabrous ; upper leaves (or bracts ?)
bract-like, ovate-triangular, acutely acuminate, about ¾ lin. long
and nearly as much broad, scale-like, with blackened tips, fimbriate-
ciliate on the margins, otherwise glabrous ; bracts similar to the
scale-leaves ; bracteoles more or less lanceolate, acute, about half the
length of the flowers, margins very narrowly membranous and
minutely serrulate ; perianth 1 lin. long, with conspicuous external
fleshy glands and an internal disc ; segments triangular-ovate,
hooded, ¾ lin. long, margins finely ciliolate and reflexed ; anthers
exserted from the perianth-tube, ⅙ lin. long ; style ¼ lin. long,
reaching to the middle of the anthers ; fruits stipitate, ellipsoid-
globose, nearly 3 lin. long including the 1 lin.-long stipe, strongly
10-ribbed, coarsely reticulate between the ribs.

COAST REGION : Clanwilliam Div. ; Olifants River, 500 ft., *Schlechter*, 8479 !
Malmesbury Div. ; near Hopefield, *Bachmann*, 15 !

38. **T. schumannianum** (Schlechter in Engl. Jahrb. xxiv. 452) ;
a small shrub, sparingly branched ; stem and branches ascending,
green, slightly angular, glabrous ; leaves oblong-lanceolate, obtuse
or subacute, 1½–3 lin. long, about ¾ lin. broad, rigid, thick, flat on
the upper surface, slightly keeled or convex below, green, with
conspicuous cartilaginous subtranslucent margins ; flowers in small
terminal clusters, becoming blackish when dry ; bracts ovate-lanceolate, about half as long as the flowers, slightly keeled, thick and

fleshy, with a suspicion of minute teeth on the slightly membranous margin; bracteoles more or less linear, very much shorter than the bracts; perianth 1–1¼ lin. long; segments triangular, flat, slightly hooded, ¾ lin. long, with slightly papillose margins; anthers exserted, ¼ lin. long; style ⅜ lin. long, reaching to the base of the anthers; fruits ovoid, about 2½ lin. long, prominently 10-ribbed, laxly reticulate between the ribs, green when dry.

COAST REGION: Cape Div.; Orange Kloof, *Wolley-Dod*, 2623! Vlagge Berg, *Wolley-Dod*, 438! Stellenbosch Div.; Lowrys Pass, 2500 ft., *Schlechter*, 7266!

39. **T. juncifolium** (A.DC. Esp. Nouv. Thes. 1); subshrub, 1–2 ft. high; stem much-branched from the base; branches wiry, more or less dichotomously divided, slender, minutely pustulate, terete, with rather long internodes, glabrous; leaves very small and subulate, closely adpressed to the stem, acute, about ¾ lin. long, mostly rather black when dry, glabrous; inflorescences very slender, subdichotomously branched; cymules 3–6-flowered; peduncle free from the bract; bracts similar to the leaves, the one immediately subtending the flower forming with the two bracteoles a cup-like involucre; bracteoles very small, much shorter than the flower, subulate-lanceolate, acute; flower 1½ lin. long, subturbinate; perianth ¾ lin. long; segments ½ lin. long, triangular, with ciliolate margins, thickened at the apex but not hooded; anthers about ⅙ lin. long; style stout, ¼ lin. long. *A.DC. in DC. Prodr.* xiv. 656; *Sond. in Flora*, 1857, 402. *T. virgatum, Sond. l.c.* 361, *as regards Ecklon & Zeyher* 55 (? *Lam.*). *T. brevifolium, Sond. l.c.* 361, *partly*. *Rhinostegia brevifolia, Turcz. in Bull. Soc. Nat. Mosc.* xvi. 1843, 57, *partly*.

COAST REGION: Clanwilliam Div.; river-bed below Kradouw Krantz, *Pearson*, 5372! Tulbagh Div.; near Tulbagh Waterfall, *Ecklon & Zeyher*, 55! Worcester Div.; Dutoits Kloof, *Drège*, 8167a, partly in Herb. Stockholm and Kew, not in Herb. Boiss. Near De Doorns, Hex River Valley, *Bolus*, 13186!

Some doubt and confusion in the synonymy of *T. juncifolium* are occasioned by the fact that under the number Drège, 8167a, there are two different plants in different herbaria, though they are not always intermixed. In the Boissier Herbarium, 8167a is the type of *T. dregeanum*, A.DC., a form of *T. virgatum*, Lam., but at Kew and Stockholm two different plants are mounted on the same sheet, one of which is *T. juncifolium*, and the other is the same plant as the type of *T. dregeanum*, represented at Berlin by Drège, 8167b. *Rhinostegia brevifolia* was founded on Drège, 8167, and it is therefore doubtful whether it should be wholly referred to *T. juncifolium* or to *T. virgatum*.

40. **T. virgatum** (Lam. Ill. ii. 123); stem much-branched, terete; branches numerous, erect, slender, angular, slightly pustulate, glabrous; leaves acicular, terete, acute, up to 2 in. long, fleshy, glabrous; inflorescences numerous, few-flowered, paniculate-corymbose; peduncle free from the bracts; bracts linear-subulate, acute or subacute, 1–2 lin. long, concave on the lower surface; bracteoles shorter than the flowers, oblong-lanceolate, obtuse, about ⅓ lin. long, with slightly membranous margins, glabrous; perianth ½–¾ lin. long; segments triangular, flat, ⅓–¾ lin. long, deeply hooded,

glabrous except on the slightly papillose margin; anthers exserted, ⅛ lin. long; style very short; fruit ellipsoid-globose, rather attenuated at the base, 2 lin. long, 7-winged at the base, somewhat reticulate between the wings. *Sond. in Flora,* 1857, 361, *excl. Ecklon & Zeyher,* 55. *T. strictum, Thunb. herb.* sheet δ (*right-hand specimen*), sheet γ (*left-hand specimen*). *T. paniculatum, Thunb. Prodr.* 45, *Fl. Cap. ed. Schult.* 210 *as to Prodr. l.c. only, not of Linn.; Willd. Sp. Pl.* i. ii. 1215; *Lam. Ill.* ii. 122; *Sond. in Flora,* 1857, 361; *A.DC. in DC. Prodr.* xiv. 656; *var. compressum, Sond. l.c.* 361; *var. apertum, A.DC. l.c.* 656. *T. dregeanum, A.DC. in DC. Prodr. l.c.* 657. *T. brevifolium, Sond. l.c.* 361, *partly. Rhinostegia brevifolia, Turcz. in Bull. Soc. Nat. Mosc.* xvi. 1843, 57, *partly. T. erectiramosum, A.DC. l.c.* 655, *as to Ecklon & Zeyher,* 36 *in Herb. Berol. and Stockh.*

SOUTH AFRICA: without locality, *Thunberg*! *Drège,* 8165!
COAST REGION: Clanwilliam Div.; Kers Kop, *Schlechter,* 8799! Malmesbury Div.; Moorreesburg, *Bachmann,* 1698! Tulbagh Div.; Tulbagh Waterfall, *Ecklon & Zeyher,* 20! Mitchells Pass, *Schlechter,* 8967! Worcester Div.; Dutoits Kloof, *Drège,* 8167a partly! 8167b! Breede River Valley, *Bolus,* 2935! Paarl Div.; French Hoek, *Schlechter,* 9361! Cape Div.; hills and Flats around Cape Town, *Bergius!* *Forster!* *Burchell,* 82! 457! 483! 534! *Ecklon,* 409! 796! 797! *Zeyher,* 3799! 3809! *Ecklon & Zeyher,* 53! *Harvey,* 713! 719! 420! *Milne,* 306! *Priess!* *Wright!* *Bolus,* 2932! 4569! 7045! and in *Herb. Norm. Austr.-Afr.,* 1358! *Wolley-Dod,* 1710! 2456! *Schlechter,* 37! 252! 349! *Diels,* 67! 1202! Stellenbosch Div.; Lowrys Pass, *Schlechter,* 5355! *Marloth,* 285! *Diels,* 1232! Caledon Div.; near Grietjes Gat, *Ecklon & Zeyher,* 36! Riversdale Div.; Corente River Farm, *Muir in Herb. Galpin,* 5325! George Div.; Montagu Pass, *Schlechter,* 5890! Uitenhage Div., various localities, *Burchell,* 4764! *Zeyher,* 103! 2802! *Ecklon & Zeyher,* 53! *Schlechter,* 6076!
CENTRAL REGION: Ceres Div.; Bokkeveld, *Thunberg!* near Ceres, *Bolus,* 7453! and in *Herb. Norm. Austr.-Afr.,* 1359!

Owing to the type specimen of Linnæus' *T. paniculatum* not having been examined by Thunberg or Sonder, the plants which have passed under this name in every publication since the "Mantissa" are not the true *paniculatum* of Linnæus, which is identical with *T. tenue,* Bernh. Lamarck's name *T. virgatum* has been restored for the species so long known as *T. paniculatum.* It is unfortunate that Sonder in taking up *T. virgatum* included specimens referred by De Candolle to *T. juncifolium.* *T. virgatum* embraces a wide range of forms which merge into one another, and it has not been found possible to keep up *T. dregeanum,* A.DC., as a distinct species.

41. T. occidentale (A. W. Hill in Kew Bulletin, 1915, 35); branches elongated, slightly angular, very sparsely leafy, glabrous; upper leaves adpressed to the stem, linear, acute or subacute, 2–4 lin. long, about ¼ lin. thick, fleshy, flat or slightly concave on the upper surface, glabrous; flowers in lax terminal corymbs about ¾ in. in diam.; bracts much shorter than the flowers, linear-lanceolate, subacute, about 1½ lin. long, slightly keeled, glabrous; bracteoles lanceolate or ovate-lanceolate, less than half the length of the bracts; perianth 1¾ lin. long; segments broadly triangular-ovate, obtuse or subacute, 1 lin. long, with no distinct beard, but the margins thickly fringed all round with stout papillæ; anthers exserted from the perianth-tube, $\frac{3}{8}$ lin. long; style $\frac{1}{2}$ lin. long, reaching nearly to the base of the anthers; fruits not seen.

WESTERN REGION: Little Namaqualand; Modderfontein, *Whitehead* !

42. T. strictum (Berg. Descr. Pl. Cap. 73); a shrub, up to about 4 ft. high; stem erect, branched in the upper part, subterete, glabrous; branches subterete; lower leaves longer than the others, acicular, with a very narrow groove on the upper side, up to $2\frac{1}{2}$ in. long, obtuse or subacute, somewhat fleshy, glabrous, upper leaves smaller and a little recurved, more or less flat on the upper surface; flowers crowded in small dense corymbs at the ends of the branches; bracts shortly adnate to the peduncle, similar to the upper leaves; peduncle $\frac{1}{2}$–1 lin. long, slightly angular; bracteoles linear-lanceolate, shorter than the flowers; perianth $\frac{3}{4}$–$1\frac{1}{4}$ lin. long; segments $\frac{1}{2}$ lin. long, triangular-ovate, apex and margin shortly papillose; anthers $\frac{1}{4}$ lin. long; filaments about the same length; style about $\frac{1}{4}$ lin. long; fruit ovoid-globose, $1\frac{1}{2}$ lin. long, $1\frac{1}{4}$ lin. in diam., 10-ribbed, with well-marked reticulation between the ribs. *Linn. Mant.* 214; *Thunb. Fl. Cap. ed. Schult.* 210; *A.DC. in DC. Prodr.* xiv. 654; *Sond. in Flora*, 1857, 361. *T. euphorbioides, Jacq. ex A.DC. l.c.* 655, *name only, not of Linn. T. robustum, Bernh. in Flora*, 1845, 80, *accord. to Sond. in Flora*, 1857, 361.

COAST REGION: various localities in the following Divisions—Van Rhynsdorp Div.; *Schlechter*, 8344! Clanwilliam Div.; *Ecklon & Zeyher*, 16! *Stephens*, 7219! 7220! 7237! *Diels*, 773! *Pearson*, 5420! 5421! Piquetberg Div.; Pikiniers Pass, *Pearson*, 5152! Malmesbury Div.; *Bachmann*, 1699! Tulbagh Div.; *Ecklon & Zeyher*, 15! *Pappe*! Worcester Div.; *Drège* 8165a! *Rehmann*, 2495! 2496! Paarl Div.; *Drège*, 8166b! *Bolus*, 4064! Cape Div.; *Sparrman*! *Bergius*! *Forster*! *Burchell*, 348! 500! 610! *Ecklon*, 59! 411! *Zeyher*, 3792! 3793! *Ecklon & Zeyher*, 15! *Wahlberg*! *Anderson*! *Pappe*! *Cooper*, 3515! *Wright*! *Wilms*, 3606! *MacGillivray*, 603! *Hooker*, 627! *MacOwan, Herb. Norm. Austr.-Afr.*, 765! *Wolley-Dod*, 438! 2663! *Diels*, 1412! Caledon Div.; Hang Klip, *Mund & Maire*! Houw Hoek, *Schlechter*, 7584! Swellendam Div.; Voormans Bosch, *Zeyher*, 3792! Riversdale Div.; *Burchell*, 6973! *Galpin*, 4541! *Muir*! Humansdorp Div.; Kruisfontein Mountain, *Galpin*, 4548! Albany Div.; Grahamstown, *Zeyher*, 3792!

WESTERN REGION: Little Namaqualand; various localities, *Bolus*, 9442! *Pearson*, 5718! 5951! 6273!

43. T. lacinulatum (A.W. Hill in Ann. Bolus Herb. ined.); a bush, about 1 ft. high, everywhere covered with a very short puberulous indumentum; branches spiny at the tips, rigid; leaves adpressed to the stem, subulate-lanceolate, acute, with brownish-tips, $\frac{3}{4}$–1 lin. long; flowers axillary, solitary, subsessile; bracts small and scaly, with subulate tips, very much shorter than the flowers; bracteoles subulate, cartilaginous, about $\frac{1}{3}$ lin. long; perianth about $\frac{4}{5}$ lin. long; segments $\frac{1}{2}$ lin. long, ovate, subacute, with broad inflexed membranous flaps on the margins, glabrous; anthers about $\frac{1}{4}$ lin. long, exserted from the perianth-tube; style $\frac{1}{8}$ lin. long, conical; fruits ovoid-globose, $1\frac{1}{4}$ lin. long, bright green, 10-ribbed, slightly reticulate between the ribs.

WESTERN REGION : Great Namaqualand ; Great Karasberg, summit of a low hill, *Pearson,* 7805 !

44. T. pleuroloma (A. W. Hill in Kew Bulletin, 1915, 38) ; a prostrate herb or subshrub ; stems glabrous, prostrate, flexuous, ribbed, 10–14 in. long, bearing numerous elongated flexuous flowering branchlets ; leaves scattered, acicular, somewhat fleshy, $1\frac{1}{2}$–$2\frac{1}{2}$ lin. long, abruptly acute ; flowers pedicellate, solitary in the axils of bracts disposed in lax racemose inflorescences ; bracts subulate, shorter than the flowers, $\frac{3}{4}$–1 lin. long, acutely acuminate, glabrous ; bracteoles inconspicuous ; perianth $\frac{3}{4}$ lin. long ; external glands conspicuous ; segments $\frac{3}{8}$ lin. long, ovate with hooded apex, margins provided with two infolded membranous flaps or lacinulæ protecting the anthers ; anthers $\frac{1}{4}$ lin. long, exserted ; style $\frac{1}{4}$ lin. long, reaching to the base of the anthers ; fruit not seen.

CENTRAL REGION : Murraysburg Div. ; near Murraysburg, *Tyson,* 129 *in Herb. Bolus* ! Carnarvon Div. ; Kareeberg Range, *Burchell,* 1566 !

A weak straggling plant externally resembling *T. junceum,* Bernh., but distinct in possessing membranous flaps to the perianth-segments like those of *T. lacinulatum,* A. W. Hill, as well as in the absence of a beard.

45. T. horridum (Pilger in Engl. Jahrb. xliv. 118) ; a low spreading bush ; branches numerous, short, stiff, pungent-pointed, minutely puberulous ; leaves more or less imbricate and adpressed to the stem, linear-lanceolate, acutely acuminate, about 2 lin. long, convex below, concave above, finely scabrid-puberulous on the margin and surface ; inflorescences crowded, 3–4-flowered ; bracts ovate-lanceolate, acute, keeled ; flowers shortly pedicellate ; perianth $\frac{3}{4}$ lin. long ; segments ovate-triangular, acute, fringed at the apex and on the margin ; anthers inserted in the perianth tube and partly exserted, $\frac{1}{4}$ lin. long ; style $\frac{1}{4}$ lin. long ; fruits globose, $\frac{3}{4}$ lin. long, with distinct ribs and reticulation, supported on a pedicel about $\frac{1}{2}$ lin. long.

WESTERN REGION : Calvinia Div. ; west of Hantam Mountains, *Diels,* 718 !

46. T. hystricoides (A. W. Hill in Kew Bulletin, 1915, 31) ; a rigid much-branched shrub, with rigid tapered spine-tipped branchlets ; branchlets straight, contracted at the base, longitudinally wrinkled, minutely puberulous ; leaves very small, triangular, rigidly coriaceous, acute, very slightly puberulous outside ; flowers solitary, distinctly pedicellate within the bracts, with a finely pubescent ribbed receptacle ; bracts ovate, subacute, glabrous ; bracteoles minute ; perianth nearly 1 lin. long ; segments ovate, obtuse, $\frac{3}{4}$ lin. long, with the margin and apex fringed with fairly long hairs ; anthers exserted from the perianth-tube, $\frac{1}{4}$ lin. long ; style thick, $\frac{1}{4}$ lin. long, reaching to about the middle of the anthers, with a somewhat capitate stigma ; fruits ovoid-globose, stipitate, about 2 lin. long, strongly 10-ribbed, finely puberulous, scarcely reticulate between the ribs.

47. T. Hystrix (A. W. Hill in Kew Bulletin, 1915, 31) ; a rigid much-branched shrublet, with rigid tapered spine-tipped branchlets ; branchlets sometimes flexuous, contracted at the base, coarsely and closely longitudinally wrinkled or closely grooved when dry, glabrous ; leaves very small, more or less triangular, thick and coriaceous, slightly mucronate, glabrous ; flowers solitary, short and sessile within the bracts ; receptacle glabrous ; bracts very small, broadly ovate, slightly mucronate, glabrous ; bracteoles minute ; perianth $\frac{3}{4}$ lin. long ; segments triangular-ovate, hooded, $\frac{1}{3}$ lin. long, the margin and apex fringed with fairly long hairs ; anthers exserted from the perianth-tube, $\frac{1}{4}$ lin. long ; filaments nearly as long as the anthers ; style short, reaching to the base of the anthers, $\frac{1}{4}$ lin. long ; fruits ovoid-globose, nearly 2 lin. long, strongly 10-ribbed and glaucous between the ribs.

CENTRAL REGION : Graaff Reinet Div. ; near Graaff Reinet, 2500 ft., *Bolus*, 523 ! Middelburg Div. ; Conway Farm, 3600 ft., *Gilfillan in Herb. Galpin*, 5503 ! KALAHARI REGION : Griqualand West ; Kimberley, Alexandersfontein, 4100 ft., *Galpin*, 7000 !

48. T. sertulariastrum (A. W. Hill in Kew Bulletin, 1915, 41) ; stems much-branched, subterete ; branches rather crowded and ascending, very minutely puberulous ; leaves scattered, a few towards the base of the plant acicular, terete, about $\frac{1}{2}$ in. long, subobtuse, glabrous, the remainder subulate-lanceolate, with blackened very acute tips, about 1 lin. long, glabrous ; flowers up to 3 at the apex of each shoot, sessile within the bracts ; bracts and bracteoles forming an involucre, subequal, linear- or ovate-lanceolate, very acutely acuminate, shorter than the flowers, with black tips and fringed submembranous margins, glabrous ; perianth globose at the base, 1 lin. long ; segments linear-lanceolate, acute, nearly $\frac{2}{3}$ lin. long, with a long hooded apex and dense apical beard ; anthers included in the perianth-tube, $\frac{1}{4}$ lin. long ; stigma subsessile ; fruits ellipsoid, $\frac{3}{4}$ lin. long, prominently 10-ribbed, prominently reticulate between the ribs.

COAST REGION : Caledon Div. ; Paapies Vlei, *Schlechter*, 10448 ! Bredasdorp Div. ; Riet Fontein, *Bolus*, 8597 !

49. T. paniculatum (Linn. Mant. 51, not of Thunb. nor Sond.) ; stems slender, flexuous, terete, glabrous ; branches very slender, spreading almost at right angles, lax, glabrous ; leaves usually few, rarely rather numerous, linear, grooved on the upper surface, acute, $\frac{1}{4}-\frac{3}{4}$ in. long, glabrous ; flowers arranged in lax dichotomously branched cymes, pedicellate ; bracts free from the peduncle, similar to the leaves but shorter ; peduncle slender ; bracteoles several at the apex of the peduncle, subulate-lanceolate, with blackish acute tips, mostly shorter than the flowers ; perianth $\frac{3}{4}$ lin. long, with con-

spicuous external ovoid glands between the segments; segments reflexed, $\frac{1}{2}$ lin. long, flat, with a woolly beard; anthers included in the perianth-tube, $\frac{1}{8}$ lin. long; style more or less sessile; fruits ellipsoid, 2 lin. long, prominently 10-nerved, finely reticulate between the nerves. *Willd. Sp. Pl.* i. ii. 1215, *as to Linn. specimen, not of Thunb.*; *Lam. Ill.* ii. 122, *not of Thunb. T. tenue, Bernh. in Flora*, 1845, 80; *Sond. in Flora*, 1857, 364; *A.DC. in DC. Prodr.* xiv. 666. *T. debile, Spreng. Syst. Veg.* i. 830; *Sond. in Flora*, 1857, 364.

SOUTH AFRICA: without locality; *Tulbagh?* in the Linnean Herbarium! *Chamisso*, 11! without locality or collector, *T. debile*, Spreng., in Herb. Stockholm. COAST REGION: Paarl Div.; French Hoek, 500 ft., *Schlechter*, 9226! Cape Div.; mountains near Cape Town, *Bergius*! *Zeyher*, 4764! *Wolley-Dod*, 494! 3426! *Bolus*, 3946! 3946B! and *Herb. Norm. Austr.-Afr.*, 1361! Cape Flats, *Krauss*, 1807! Stellenbosch Div.; around Somerset West, *Ecklon & Zeyher*, 24! CENTRAL REGION: Somerset Div.; Bruintjes Hoogte, *Burchell*, 3040!

The specimen in the Linnean Herbarium, which forms the type of the Mantissa, and is labelled "*T. paniculatum*" by Linnæus, was probably collected by Tulbagh; it is identical with the plant named *T. tenue* by Bernhardi. The specimen seen by Sonder at Stockholm is an entirely different plant and agrees in part with the plants labelled *T. strictum* by Thunberg in the Upsala Herbarium. *T. debile*, Spreng., appears to be only a straggling form of *T. paniculatum*, Linn., as suggested by Sonder.

50. **T. euphrasioides** (A.DC. Esp. Nouv. Thes. 8); a low slender subshrub, up to 9 in. high, much-branched; main-stem slender, terete, glabrous; branches dichotomously forked, spreading or ascending; leaves few and scattered, the lower about 4 lin. long, linear, terete, the upper subulate-lanceolate, acute, at first adpressed to the stem, at length spreading or recurved, about 1 lin. long, glabrous; flowers solitary and sessile at the apex of the branchlets or of the branches of the cymes; bract and bracteoles forming an involucre, acutely acuminate from an ovate or lanceolate base, about $\frac{2}{3}$ lin. long, glabrous; flower about 2 lin. long; perianth cylindric, $1\frac{1}{4}$ lin. long, with no external glands; segments linear-lanceolate, subacute, $\frac{3}{4}$ lin. long, with a dense apical beard; anthers included in the perianth-tube, $\frac{1}{4}$ lin. long; style $\frac{1}{8}$ lin. long, reaching to the middle of the anthers; fruits broadly ellipsoid, $2\frac{1}{2}$ lin. long including the long persistent perianth, prominently 10-nerved, distinctly reticulate between the nerves, reddish-brown when dry. *A.DC. in DC. Prodr.* xiv. 665. *T. hottentottum, Sond. in Flora*, 1857, 363, 407.

COAST REGION: Tulbagh Div.; Tulbagh Waterfall, 900 ft., *Schlechter*, 9015! Cape Div.; mountains near Cape Town, *Thunberg*, partly! *Masson*! *Zeyher*, 203! *Ecklon & Zeyher*, 37! *Harvey*, 717! Devils Peak, *Bergius*! Stellenbosch Div.; Hottentots Holland, near Lowrys Pass, 1000 ft., *Bolus*, 5563!

The left-hand specimen on the sheet in Thunberg's herbarium marked "*T. strictum, δ*," belongs to *T. euphrasioides*, A.DC.

51. **T. micromeria** (A.DC. Esp. Nouv. Thes. 8); stem ascending, slender, branched in the upper part; branches short, more or less

grooved, glabrous ; leaves small, scattered, lanceolate, acute, recurved, about 1 lin. long, glabrous ; cymules about 3-flowered ; flowers sessile or subsessile; bracts and bracteoles ovate, acute, with black tips, shorter than the flowers, more or less keeled, with membranous minutely ciliate edges ; perianth ⅔ lin. long ; segments oblong-lanceolate, obtuse, ¾ lin. long, fleshy, slightly hooded, apex and margins densely ciliate; anthers exserted from the perianth-tube, ¼ lin. long ; style ¼ lin. long, reaching to the base of the anthers ; fruit ellipsoid, contracted at the base, 2 lin. long including the persistent perianth, prominently 10-ribbed from the base, slightly reticulate between the ribs. *A.DC. in DC. Prodr.* xiv. 665 ; *Sond. in Flora*, 1857, 407. *T. parvifolium, A.DC. Esp. Nouv. Thes.* 3 ; *Sond. in Flora*, 1857, 404 ; *A.DC. in DC. Prodr.* xiv. 657.

Coast Region : Paarl Div. ; between Paarl and French Hoek, *Drège*, 8168 ! Swellendam or George Div. ; collected by *Mund* and distributed by *Ecklon & Zeyher*, 19 !

52. T. capituliflorum (Sond. in Flora, 1857, 363) ; a low spreading heath-like shrub, about 6 in. high ; branches numerous from the apex of a rather slender erect rhizome, somewhat angular, glabrous ; leaves closely adpressed to the stem, with spreading tips, lanceolate, subulate-acuminate, ¾–1 lin. long, convex below, with finely puberulous margins; flowers sessile in small dense clusters at the apex of the slender branches ; bracts ovate-lanceolate, with blackish acuminate tips, shorter than the flowers, keeled, with more or less membranous fringed margins ; bracteoles similar to the bracts but a little shorter ; perianth ¾–1 lin. long ; segments ovate-triangular, acute, ⅓–⅔ lin. long, flat, densely bearded ; anthers ¼ lin. long, exserted from the perianth-tube ; style about ⅛ lin. long ; fruits subglobose, 2 lin. long, including the beak-like closed persistent perianth, prominently 10-ribbed, conspicuously reticulate between the ribs. *A.DC. in DC. Prodr.* xiv. 665.

South Africa : without locality, *Ecklon & Zeyher*, 17 partly ! *Zeyher*, 194 ! Coast Region : Tulbagh Div. ; by Tulbagh Waterfall, *Zeyher* ! *Ecklon & Zeyher*, 18 ! New Kloof, *Schlechter*, 7507 ! Cape Div. ; Slang Kop, *Wolley-Dod*, 3188 ! near Cape Town, *Bergius* ! *Harvey*, 107 ! Kenilworth Race Course, *Bolus*, 7046 ! Camps Bay, *Burchell*, 380 ! Cape Flats, *Burchell*, 8560 ! Caledon Div. ; near Genadendal, *Drège*, 8169 ! Hawston, *Schlechter*, 9472 ! Grabouw, near Palmiet River, *Bolus*, 4205 ! Great Houw Hoek, *Zeyher*, 3811 ! Central Region : Ceres Div. ; near Ceres, *Bolus*, 9651 !

53. T. cuspidatum (A. W. Hill in Kew Bulletin, 1915, 26) ; a low much-branched woody subshrub ; stem much-branched from the apex of a slender suberect rhizome, subterete, glabrous ; branches slightly ribbed ; leaves few and scattered, the lower acicular, with a subacute cartilaginous apex, about ½ in. long, subterete, glabrous ; the upper suberect and more or less adpressed to the stem, triangular-lanceolate, acute, about 2 lin. long ; flowers glomerate at the ends of the branchlets, subcorymbose ; bracts and bracteoles crowded,

the former obovate-elliptic, cuspidate-acuminate, nearly as long as
the flowers, with membranous subscarious margins; bracteoles
similar to the bracts but smaller and a little narrower; perianth
$1-1\frac{1}{4}$ lin. long; segments lanceolate-triangular, acute, hooded, with
a woolly apical beard; anthers $\frac{1}{4}$ lin. long, included in the perianth-
tube; style $\frac{1}{3}-\frac{1}{2}$ lin. long, stout, reaching to or above the base of
the anthers; fruits ellipsoid-globose, long-beaked with the persistent
perianth, 2 lin. long, prominently 10-ribbed especially at the base,
conspicuously reticulate between the ribs. *T. capituliflorum, Sond.,*
var., Sond. in Flora, 1857, 363, *as to Zeyher,* 47; *A.DC. in DC.*
Prodr. xiv. 665.

CoAST REGION: Cape Div.; south-west part of Devils Peak, *Wilms*, 3611!
Caledon Div.; Zwart Berg, *Zeyher,* 47! Bredasdorp Div.; Elim, 500 ft.,
Schlechter, 7664 in *Herb. Bolus!* 7666 in *Herb. Berlin!* Riet Fontein Poort, near
Elim, 150 ft., *Bolus,* 8597!

54. **T. rariflorum** (Sond. in Flora, 1857, 364); stems ascending
or straggling, slender, slightly angular, glabrous; leaves few,
acicular, subacute, $\frac{1}{2}-1$ in. long, about $\frac{1}{2}$ lin. thick, fleshy, glabrous;
flowers solitary or subsolitary at the ends of short lateral cymules;
bracts adnate to the peduncle, as long as or sometimes considerably
longer than the flowers, linear-lanceolate, acute, $1\frac{1}{2}-3$ lin. long, with
narrowly membranous margins, glabrous; bracteoles a little shorter
than the flowers but otherwise very similar to the bracts; perianth
about 1 lin. long; segments $\frac{1}{2}-\frac{2}{3}$ lin. long, ovate, subacute, flat, with
a dense apical beard; anthers inserted at the base of or low down
in the tube, $\frac{1}{6}$ lin long; style $\frac{1}{6}$ lin. long, reaching to the top of
the anthers, or almost wanting; fruits ellipsoid-globose, shortly
contracted at the base, prominently 10-ribbed, slightly reticulate
between the ribs, capped by the persistent closed perianth. *T.*
Maximiliani, Schlechter in Engl. Jahrb. xxvii. 119.

CoAST REGION: Tulbagh Div.; Tulbagh Waterfall, 1500 ft., *Ecklon & Zeyher,*
56! *Schlechter,* 9063! Caledon Div.; Vogelgat, 300 ft., *Schlechter,* 9531!

55. **T. Zeyheri** (A.DC. Esp. Nouv. Thes. 8); a subshrub, more
or less prostrate, much-branched; main stem stout and knotty;
branches short, crowded, closely grooved, with sharp scabrous
ridges; leaves linear, flat on the upper surfaces, with slightly
recurved subacute tips, 3–5 lin. long, keeled, with scabrous keel
and margins; flowers single and terminal or 2–5 in a fairly lax
terminal cyme; bracts shorter than the flowers, linear, acute, 1–2
lin. long, keeled, with scabrous margins; bracteoles sometimes
slightly longer than the bracts, otherwise similar to them; perianth
$2-2\frac{1}{4}$ lin. long; segments linear-lanceolate, subacute, $1\frac{1}{2}$ lin. long,
densely hairy within the apex; anthers partly exserted from the
perianth-tube, $\frac{1}{3}$ lin. long; style $1\frac{1}{8}$ lin. long, reaching nearly
to the top of the anthers; fruits oblong, 3 lin. long, prominently
10-ribbed, conspicuously reticulate between the ribs, green when
dry. *A.DC. in DC. Prodr.* xiv. 666; *Sond. in Flora,* 1857, 407.

T. transgariepinum, Sond. in Flora, 1857, 356.   *T. longirostre, Schlechter in Journ. Bot.* 1897, 345, *et ex A. W. Hill in Kew Bull.* 1910, 186.   *T. Schlechteri, A. W. Hill in Dyer, Fl. Trop. Afr.* vi.–i. 415.

COAST REGION : Queenstown Div. ; doleritic kopje on the bank of Komani River, near Queenstown, 3600 ft., *Galpin,* 2157 !
CENTRAL REGION: Aliwal North Div. ; Elands Hoek, near Aliwal North, 4600 ft., *Bolus,* 285 ! foot of the Witberg Range at Nieuwjaars Spruit, between the Orange and Caledon Rivers, 4000–5000 ft., *Zeyher* !
EASTERN REGION ? locality uncertain, probably near Harrismith, *Wood,* 4818 !
Also in Tropical Africa.

56. T. cytisoides (A. W. Hill in Kew Bulletin, 1915, 27);
stems few from a branched rhizome, erect, strongly sulcate and ribbed, glabrous ; branches erect, somewhat sparingly leafy ; leaves narrowly linear-acicular, acute, 4–6 lin. long, about $\frac{1}{4}$ lin. thick, verrucose when dry with a prominent thick midrib on the upper surface ; flowers solitary at the ends of the shoots, sessile ; bracts partly adnate to the peduncle, linear, about as long as the flowers ; bracteoles shorter than the flowers, linear-subulate, fleshy, glabrous ; perianth $1\frac{1}{6}$ lin. long ; segments ovate-lanceolate, $\frac{3}{4}$ lin. long, hooded, with a small apical beard of a few long hairs ; anthers exserted from the perianth-tube, $\frac{1}{3}$ lin. long ; style $\frac{1}{4}$ lin. long ; fruits ellipsoid, $2\frac{1}{2}$ lin. long, reddish-brown when dry and strongly ribbed, but not reticulate between the ribs.

KALAHARI REGION : Transvaal ; Waterval Onder, *Jenkins* !

57. T. Burchellii (A. W. Hill in Kew Bulletin, 1915, 24) ;
shrubby ; main branches (or stem ?) slightly angular, light-brown, glabrous ; lateral branchlets rather densely leafy ; leaves linear or linear-acicular, with a short cartilaginous acute apex, $3\frac{1}{2}$–$4\frac{1}{2}$ lin. long, subterete or somewhat angular, glabrous ; flowers solitary or subsolitary at the ends of the branchlets ; bracts longer than the flowers, linear, very similar to the leaves ; bracteoles as long as or a little longer than the flowers ; perianth 1 lin. long, with no external glands ; segments ovate-elliptic, subacute, $\frac{3}{4}$ lin. long, hooded, bearded ; anthers exserted from the perianth-tube, $\frac{1}{4}$ lin. long ; style $\frac{1}{2}$ lin. long, reaching to the top of the anthers; fruits ellipsoid, acute at the base, 3 lin. long, strongly 10-ribbed, reticulate between the ribs.

KALAHARI REGION : Bechuanaland ; near the source of Kuruman River, at Little Klibbolikhonni, *Burchell,* 2504 !

58. T. repandum (A. W. Hill in Kew Bulletin, 1915, 39); a
spreading branched shrublet, about 6 in. high ; root erect, slender, with very long light straw-coloured horizontally spreading lateral rootlets ; stems spreading and ascending, 3 or 4 from the apex of the root, prominently ridged, glabrous ; leaves sparse, recurved, acicular, acute, 3–5 lin. long, about $\frac{1}{4}$ lin. thick, glabrous ; flowers 3–4 in small clusters at the ends of the shoots, sessile ; bracts inconspicuous, scarcely half the length of the flowers, ovate or

lanceolate, acutely acuminate, with more or less membranous·
margins, glabrous; bracteoles about ¾ the length of the bracts but
otherwise very similar to them; perianth ¾–1 lin. long; segments
ovate, subacute, flat, not hooded, ⅓ lin. long, margin and apex with
a beard of long hairs; anthers included in the perianth-tube,
slightly over ¼ lin. long; stigma subsessile or supported on a very
short style; fruits oblong-globose, contracted and stipitate at the
base, 1½ lin. long, fairly prominently ribbed, reticulate between the
ribs, reddish-brown when dry.

COAST REGION : Malmesbury Div. ; neighbourhood of Hopefield, between Lilie-
fontein and Rondekuil, *Bachmann*, 2195 !

59. **T. glaucescens** (A. W. Hill in Kew Bulletin, 1915, 29);
shrubby; branches 1–2 ft. long, diffuse, rosy-glaucous, glabrous;
branchlets divaricate, nearly leafless, subterete; leaves very small,
and few, linear, obtuse, ¾–1½ lin. long, rigid, glabrous; flowers few
in clusters at the ends of the branchlets; bracts inconspicuous,
much shorter than the flowers, lanceolate, subacute, concave on the
upper surface, glabrous; bracteoles very small; perianth cylindric,
¾ lin. long; segments lanceolate, subobtuse, ⅔ lin. long, with an
apical beard of long hairs; anthers exserted from the perianth-
tube, ⅓ lin. long; style ¼ lin. long, reaching above the anthers;
fruits ovoid, 2 lin. long, strongly 10-ribbed, glaucous and finely
reticulate between the ribs.

COAST REGION: Swellendam Div. ; dry plains by the Zondereinde River,
Burchell, 7513 !

60. **T. hispidulum** (Lam. ex Sond. in Flora, 1857, 363); stems
much-branched, sulcate, woody, scabrid-puberulous; branchlets
more or less dichotomously forked, scabrid-puberulous; leaves
recurved, acicular, acutely mucronate, about 3½ lin. long, glabrous
or puberulous like the stems; flowers few and crowded at the ends
of the shoots, subsessile; bracts and bracteoles crowded, with
puberulous margins; perianth 1–1⅛ lin. long; segments ⅓–⅔ lin.
long, lanceolate, subacute, hooded, and with a dense apical beard;
anthers included in the perianth-tube, ¼ lin. long; style ⅓ lin. long,
reaching to the base of the anthers; fruits subglobose, 2 lin. long
including the persistent perianth, prominently 10-ribbed, strikingly
reticulate between the ribs. *A.DC. in DC. Prodr.* xiv. 671. *T.
hispidulum, var. glabratum, Sond. in Flora,* 1857, 363, *not T. paniculatum, Thunb. Herb. and Prodr.* 45 (*fide Sond. l.c.*), *nor T. selagineum,
A.DC. fide Sond. in Flora,* 1857, 404. *T. hispidum, Schlechter in Engl.
Jahrb.* xxiv. 452.

VAR. β, **subglabrum** (A. W. Hill in Kew Bulletin, 1915, 30); plants very
minutely hairy; perianth and anthers larger than in the type; glands well
marked. *T. conostylum, Schlechter in Engl. Jahrb.* xxvii. 117.

COAST REGION : Clanwilliam Div. ; Foot of Krakadouw Pass, 2790 ft., *Diels*,
934 ! Piquetberg Div. ; east side of Pikeniers Pass, *Pearson*, 5218 ! foot of
Piquet Berg, 980 ft., *Schlechter*, 5226 ! Tulbagh Div. ; New Kloof, 450 ft.,
Schlechter, 7486 ! near Saron, 2300 ft., *Schlechter*, 10667 ! Tulbagh, *Pappe* ! East

from Wolseley, 740 ft., *Diels*, 1027 ! Tulbagh Kloof, *Ecklon* ! Tulbagh Waterfall, *Ecklon* ! Var. β : Clanwilliam Div. : Blaw Berg, 1200 ft., *Schlechter*, 8451 ! hills near Clanwilliam, 800 ft., *Leipoldt*, 500 !

61. T. prostratum (A. W. Hill in Kew Bulletin, 1915, 38) ; stems numerous, prostrate from the apex of a slender erect rootstock, glabrous, terete ; branchlets ascending, very slender, sparingly leafy ; leaves acicular, subacute, terete, 4–6½ lin. long, fleshy, glabrous ; flowers glomerate at the ends of the branchlets ; bracts equalling or longer than the flowers, linear or linear-lanceolate, subacute, glabrous ; perianth about ½ lin. long ; segments triangular, acute, nearly ½ lin. long, apex and margins above bearded with papillæ ; anthers exserted from the perianth-tube, ⅐ lin. long ; style nearly ¼ lin. long, reaching to the middle or top of the anthers ; fruits ovoid-globose, 1¾ lin. long, conspicuously 10-ribbed, rather delicately reticulate between the ribs.

CENTRAL REGION : Ceres Div. ; Skurfdeberg Range, near Gydouw, 5000 ft., *Schlechter*, 10008 !

62. T. acuminatum (A. W. Hill in Kew Bulletin, 1915, 22) ; rootstock slender, erect, subterete, about 1¾ lin. thick ; stems numerous from the apex of the rootstock, spreading or ascending, subterete, glabrous ; leaves acicular, acute, ⅓–1 in. long, subterete, glabrous ; flowers in rather dense terminal clusters ; bracts triangular-lanceolate, acute, fleshy, 1–1½ lin. long, glabrous ; bracteoles about half the length of the flowers ; perianth-segments ¾ lin. long, linear-lanceolate, with long acicular points, with a densely adpressed beard in the lower part ; anthers included in and at the base of the perianth-tube, ¼ lin. long ; style ⅛ lin. long ; fruits ovoid-globose, about 2 lin. long including the long persistent perianth, rather faintly ribbed and reticulate.

SOUTH AFRICA : without locality, *Hooker*, 608 ! *Reynoud, ex herb. Kunth in Herb. Berlin* !
COAST REGION : Cape Div.; near Noahs Ark Battery, Simonstown, *Wolley-Dod*, 2806 ! 3016 ! Simons Bay, *Wright*, 536 ! hills west of Simonstown, *Wolley-Dod*, 1879 ! Muizen Berg, 1000 ft., *Bolus*, 8040 ! Constantia Berg, 2000 ft., *Schlechter*, 543 ! Steenberg Flats, *Wolley-Dod*, 2741 !

63. T. selagineum (A.DC. Esp. Nouv. Thes. 3) ; branches slender and straggling, slightly angular or subterete, very minutely puberulous ; leaves all recurved, acicular, flattened on the upper surface, subterete, acute, 2 lin. long or less, glabrous ; flowers sessile in small clusters at the apices of the shoots ; bracts and bracteoles about equal in length to the flowers, linear-lanceolate, acute to acuminate ; perianth with conspicuous external glands between the segments, ⅔ lin. long ; segments triangular-ovate, obtuse, ⅜ lin. long, shortly bearded, with papillose margins ; anthers exserted from the perianth-tube, ⅛ lin. long ; style stout, ⅛ lin. long, reaching to the middle of the anthers ; fruit ovoid-ellipsoid, contracted at the base, 1½ lin. long, slightly 10-ribbed, distinctly reticulate between the

ribs. *A.DC. in DC. Prodr.* xiv. 658 ; *Sond. in Flora*, 1857, 404.
T. thunbergianum, A.DC. in DC. Prodr. xiv. 666. *T. paniculatum, Thunb. Herb.*; *Fl. Cap. ed. Schult.* 210, *partly* ; *? not of Prodr.* 45 *nor of Linn.*

COAST REGION : Piquetberg Div. ; Piquetberg, 1725 ft., *Drège*, 8172 ! *Schlechter*, 5237 ! Worcester Div. ; near De Doorns, Hex River Valley, 1700 ft., *Bolus*, 13185 !
CENTRAL REGION : Ceres Div. ; Bockberg in Upsala Herbarium !

64. **T. capitellatum** (A.DC. Esp. Nouv. Thes. 7); shrubby, about 1 ft. high, much-branched ; branches ascending, somewhat angular, glabrous ; leaves linear or linear-lanceolate, acute or subacute, slightly curved, 2–3 lin. long, concave on the upper surface, rigid, fleshy, glabrous ; flowers few in small dense terminal heads ; bracts lanceolate, acute, keeled, fleshy, as long as the flowers, with slightly scabrous membranous greenish margins, becoming quite black with age ; bracteoles a little smaller than the bracts ; perianth tubular, fleshy, 1 lin. long, with conspicuous external glands ; segments ovate, subacute, fleshy, $\frac{1}{2}$ lin. long, with a dense apical beard ; anthers inserted in or at the base of the tube, $\frac{1}{6}$ lin. long ; stigma sessile or subsessile, or about $\frac{1}{10}$ lin. long and reaching to the base of the anthers ; fruits ovoid-ellipsoid, 2–2$\frac{1}{2}$ lin. long, prominently 10-ribbed, finely reticulate between the ribs. *A.DC. in DC. Prodr.* xiv. 664. *T. brachycephalum, Sond. in Flora*, 1857, 360, 406. *T. foveolatum, Schlechter in Engl. Jahrb.* xxvii. 119.

COAST REGION : Paarl Div. ; French Hoek ! 3600 ft., *Schlechter*, 9348 ! Stellenbosch Div. ; *Flor. Ful.* 87 in Stockholm Herbarium ! Caledon Div. ; Zwart Berg, *Ecklon & Zeyher*, 4 ! Bredasdorp Div. ; Elim, 500 ft., *Schlechter*, 7664 !

65. **T. imbricatum** (Thunb. Fl. Cap. ed. Schult. 208) ; a much-branched conifer-like shrub with stout stem and branches closely beset with the persistent bases of the leaves ; branchlets densely leafy, slightly scabrous ; leaves opposite, linear, triangular in section, flat above, sharply keeled below, with a light-coloured cartilaginous acute apex, thick and rigid, 1$\frac{1}{2}$–4 lin. long, about $\frac{1}{4}$ lin. long, with finely scabridulous margins ; flowers in few-flowered terminal heads ; bracts similar to the leaves ; bracteoles narrower, shorter than the flowers ; perianth about 2 lin. long ; segments triangular, acute, 1–1$\frac{3}{4}$ lin. long, bearded inside ; anthers $\frac{1}{3}$–$\frac{1}{2}$ lin. long ; inserted at top of perianth-tube ; style $\frac{3}{4}$ lin. long, reaching to the middle of the anthers ; fruits oblong-ovoid, 3 lin. long, including the thick closed perianth, obtusely 10-ribbed, not or scarcely reticulate between the ribs. *Sonder in Flora*, 1857, 355 ; *A.DC. in DC. Prodr.* xiv. 672 (*among excluded species*). *T. abietinum, Schlechter in Journ. Bot.* 1897, 282.

SOUTH AFRICA : without locality, *Thunberg* !
COAST REGION : Queenstown Div. ; Kliplaat, *Zeyher*, 38 ! summit of the Andriesberg Range, 6600–6800 ft., *Galpin*, 2172 !

CENTRAL REGION : Graaff Reinet Div. ; Oude Berg near Graaff Reinet, 4400 ft.,
Bolus, 637 ! Phillipstown Div. ; near Riet Fontein, Waschbanks River, *Burchell*,
2724 !
EASTERN REGION : Tembuland ; Gat Berg, Intwanazana, 4000 ft., *Baur*, 518 !
Natal ; Niginya, 5500 ft., *Wylie in Herb. Wood*, 10532 !
KALAHARI REGION : Basutoland, *Cooper*, 699 !

66. T. junceum (Bernh. in Flora, 1845, 80) ; stems branched from
the base or in the upper part, whip-like, sulcate, glabrous ; branches
ascending, stiff and often elongated ; lower leaves very few, rarely
present, terete, about ¾ in. long, glabrous ; flowers arranged in
rather slender loose spikes ; bracts ochreate, ovate-lanceolate, very
acute, ¾–1 lin. long, keeled, with broad scurfy jagged membranous
margins, glabrous ; bracteoles 2, shorter than the flower, narrower
than the bracts but otherwise similar ; perianth about 1 lin. long ;
segments triangular, acute, ½–⅔ lin. long, their inner surface white
and fleshy, with a dense apical woolly beard ; anthers included in
the perianth-tube, ¼ lin. long ; stigma sessile or supported on an
extremely short style ; fruits ellipsoid-globose, 2 lin. long, 1¼ lin. in
diam., prominently 10-ribbed, especially in the lower part, faintly
reticulate between the ribs. *A.DC. in DC. Prodr.* xiv. 669, *partly*,
excl. syn. ; *Sond. in Flora*, 1857, 406.

VAR. β, mammosum (A. W. Hill in Kew Bulletin, 1915, 33) ; differing from the
typical form in having a large fleshy protuberance from the inside of each perianth-
segment over the anthers.
VAR. γ, plantagineum (A. W. Hill in Kew Bulletin, 1915, 33) ; characterised by
the dense spikes resembling those of *Plantago maritima.*

COAST REGION : Clanwilliam Div. ; near Brakfontein, *Ecklon & Zeyher*, 7
partly ! Humansdorp Div. ; Zitzikama, *Krauss*, 1804 ! Kruisfontein, 500 ft.,
Galpin, 4551 ! Uitenhage Div. ; Van Stadens Berg, *Ecklon & Zeyher*, 46 ! Albany
Div. ; near Grahamstown, *MacOwan*, 690 ! Bathurst Div. ; near Kaffir Drift,
Burchell, 3868 ! Var. β : Port Elizabeth Div. ; Port Elizabeth, *Mrs. Paterson*,
806 !
EASTERN REGION : Transkei ; Var. γ : Kentani, 1000 ft., *Miss Pegler*, 878,
partly ! near the mouth of the Qolora River, *Miss Pegler*, 1302 !

67. T. natalense (Sond. in Flora, 1857, 358) ; stems few and
erect from a short knotty rhizome, branched in the upper third or
half, nearly leafless, longitudinally sulcate, ¾–1 lin. thick, glabrous ;
leaves very few and closely adpressed to the stem, linear, 1–2 lin.
long, glabrous ; flowers borne on the primary branches, spicate ;
spikes slender, suberect, slightly flexuous towards the apex, loose-
flowered, glabrous ; bracts shorter than the flowers, subulate from
an ovate-triangular base, 1–1½ lin. long, keeled, with scurfy
membranous margins ; bracteoles similar to the bracts but a little
shorter ; perianth about 1½ lin. long ; segments elliptic, subacute,
flat, ¾ lin. long, with a dense woolly apical beard and hairy
margins ; anthers included in the perianth-tube, or partly exserted ;
stigma sessile or supported on an extremely short style ; fruits
rather long-stipitate, ellipsoid, quite 3 lin. long, with numerous
conspicuous ribs, prominently reticulate between the ribs. *T.*

macrostachyum, A.DC. Esp. Nouv. Thes. 6, *partly, and in DC. Prodr.*
xiv. 669, *partly.*

KALAHARI REGION : Transvaal ; Shilovane, *Junod,* 749 ! near Lydenburg,
4800 ft., *Schlechter,* 3953 !
EASTERN REGION : Natal ; near Durban, *Gueinzius,* 407 ! Inanda, *Wood,* 168 !
173 ! 242 ! Dalton, 3300 ft., *Rudatis,* 14 ! Clairmont, 50 ft., *Wood,* 4920 !
Schlechter, 3064 ! and without precise locality, *Gerrard,* 204 !

68. T. scirpioides (A. W. Hill in Kew Bulletin, 1915, 40) ;
stems woody, rather numerous and erect from the apex of a many-
headed woody rhizome, branched from near the base or in the
upper part, conspicuously sulcate, glabrous, leafless ; flowers borne
on loose erect spikes, the latter forming a loose corymb ; bracts
scaly, much shorter than the flowers, broadly ovate, acute or acutely
acuminate, ¾–1 lin. long, keeled, with brown narrowly membranous
margins, glabrous ; bracteoles similar to the bracts but a little
shorter ; perianth 1¼ lin. long ; segments triangular, acute, flat,
¾ lin. long, with a dense apical woolly beard and hairy margins ;
anther included in the perianth-tube, ¼ lin. long ; stigma sessile or
almost so ; fruits slightly contracted at the base, ellipsoid-globose,
3 lin. long, 1⅓ lin. in. diam., rather faintly 10-ribbed, scarcely
reticulate between the ribs.

KALAHARI REGION : Orange River Colony ; Harrismith, 5500 ft., *Sankey,* 249 !
Besters Vlei, near Witszies Hoek, 5400 ft., *Bolus,* 8248 ! and without precise
locality, *Cooper,* 834 !
EASTERN REGION : Griqualand East ; near Kokstad, 1300 ft., *Tyson,* 1505 !
Natal ; by the Mooi River, 4000 ft., *Wood,* 4066 !

This species is nearly allied to *T. natalense,* Sond., but is distinguished by its
stout stiff erect branches and broader leaves and bracts with their brown narrowly
membranous margins. As in *T. natalense,* the beard of the perianth-segments is
composed of flexuous woolly hairs.

69. T. paronychioides (Sond. in Flora, 1857, 359) ; a small
slender plant about 5 in. high ; branches numerous from the apex
of a very slender erect rootstock, slender, prominently verrucose
when dry, otherwise glabrous ; leaves few, linear, obtuse or subacute, 2½–3 lin. long, flat on the upper surface, glabrous ; flowers
loosely spicate ; bracts ochreate, scale-like, broadly ovate, acutely
acuminate, sharply keeled, with very broad hyaline scurfy margins ;
bracteoles 2, similar to the bracts but a little shorter ; perianth
⅗–⅘ lin. long ; segments lanceolate, subacute, about ½ lin. long,
deeply hooded, bearded with stout stiff hairs ; anthers included in
the perianth-tube or slightly exserted, about ¼ lin. long ; stigma
capitate, sessile or subsessile ; fruits ellipsoid-globose, 2 lin. long,
strongly 10-ribbed, faintly reticulate between the ribs. *T. junceum,*
A.DC. in DC. Prodr. xiv. 669, *partly, not of Bernh.*

COAST REGION : Swellendam and George Div. ; collected by *Mund in Ecklon &*
Zeyher's distribution of Thesium, 13 ! George Div. ; hill near George, *Schlechter,*
2428 !

70. T. flexuosum (A.DC. Esp. Nouv. Thes. 6) ; a tall rather straggling plant about 1 ft. high, often with whip-like branches; branches longitudinally sulcate, glabrous ; leaves very few, acicular, subterete, acute, about 5 lin. long, fleshy, glabrous ; flowers in fairly close and rather stout spikes ; bracts spreading, ovate-lanceolate, acutely acuminate, about 1 lin. long, sharply keeled, with broad hyaline scurfy margins ; bracteoles similar to the bracts but a little narrower ; perianth about 1 lin. long ; segments lanceolate, acute, $\frac{1}{2}$–$\frac{2}{3}$ lin. long, with a long hood, bearded with straight hairs ; anthers partly exserted and partly included in the tube, $\frac{1}{4}$ lin. long ; style $\frac{1}{4}$ lin. long, reaching to the middle of the anthers ; fruits ellipsoid-globose, about 2$\frac{1}{2}$ lin. long, beaked by the very long cone-like persistent perianth, prominently 10-ribbed, clearly reticulate between the ribs. *A.DC. in DC. Prodr.* xiv. 669. *T. spicatum, Drège in Linnæa,* xx. 211, *not of Linn. T. junceum, Sond. in Flora,* 1857, 359, *partly, not of Bernh.*

COAST REGION : Constantia Berg, 2700 ft., *Schlechter,* 519! Oudtshoorn Div. ; Oudtshoorn, 1100 ft., *Bolus,* 12275! Humansdorp Div. ; near Gamtoos River, 164 ft., *Schlechter,* 6051! Uitenhage Div. ; near Zwartkops River, *Zeyher,* 184! 3796 partly! *Ecklon & Zeyher,* 9! Redhouse, *Mrs. Paterson,* 52! Bathurst Div. ; Glenfilling, *Drège,* partly! near Round Hill, 1000 ft., *Bolus,* 10652! CENTRAL REGION : Graaff Reinet Div. : mountains near Graaff Reinet, *Bolus,* 813! 1943*! Camdeboo, *Dunn!*

71. T. spartioides (A. W. Hill in Kew Bulletin, 1915, 41); a small wiry plant about 5 in. high ; rootstock about 2$\frac{1}{2}$ lin. thick, with forked branches ; stems several, ascending, very slender, terete, simple or sparingly branched, glabrous ; leaves closely adpressed to the stem, subulate-lanceolate, acutely acuminate, 1–1$\frac{1}{4}$ lin. long, convex on the back, glabrous ; flowers rather crowded towards the ends of axillary branches ; bracts scale-like, ovate-lanceolate, acuminate, about half as long as the flowers, with minutely fringed edges ; bracteoles similar to the bracts but a little narrower ; flowers sessile within the bracts ; perianth $\frac{3}{4}$ lin. long ; segments about $\frac{1}{2}$ lin. long, triangular, acute or acuminate, bearded ; anthers $\frac{1}{4}$ lin. long, exserted from the perianth-tube ; style $\frac{1}{2}$ lin. long, reaching to the middle of the anthers ; fruits small, subglobose, about 1$\frac{1}{2}$ lin. long including the persistent perianth, conspicuously 10-ribbed, finely reticulate between the ribs.

KALAHARI REGION : Transvaal; on the hills near Brug Spruit, 4600 ft., *Schlechter,* 3754!

72. T. confine (Sond. in Flora, 1857, 363) ; stems numerous, branching, arising from a woody rootstock : rootstock slender, subterete, about 1$\frac{1}{2}$ lin. thick ; branches very slender, whip-like, grey, subterete, glabrous ; leaves scale-like, adpressed to or parallel with the stem, subulate-lanceolate, 1–2$\frac{1}{2}$ lin. long, glabrous ; flowers few in short spikes ; bracts shorter than the flowers, lanceolate, long and acutely acuminate, about 1 lin. long, fleshy in the middle, with

membranous jagged margins ; bracteoles similar to the bracts but
a little shorter ; perianth 1¼ lin. long ; segments ¾ lin. long, lanceolate, acute, flat, with a long apical beard ; anthers partly exserted
from the perianth-tube, ⅜ lin. long ; style ½ lin. long, reaching
nearly to the top of the anthers ; fruits ellipsoid, 2 lin. long, prominently 10-ribbed, faintly reticulate between the ribs, more or less
glaucous-green. *A.DC. in DC. Prodr.* xiv. 665.

KALAHARI REGION : Orange River Colony ; near Nieuwjaars Spruit, between
the Orange and Caledon Rivers, *Zeyher* ! Bloemfontein, under shade of trees and
shrubs, *Grey College Herb.* 76 *in Herb. Bolus* ! Leeuw Spruit and Vredefort,
Barrett-Hamilton in Herb. British Museum !

73. T. griseum (Sond. in Flora, 1857, 357) ; whole plant very finely
scabrid-puberulous; stems numerous, short and ascending from a
short rhizome, simple or sparingly branched ; leaves linear, acute or
subacute, 2–5 lin. long, convex below ; flowers arranged in fairly
dense terminal spikes ; bracts lanceolate, subacute, about as long as
the flowers, sometimes almost verrucose ; bracteoles a little shorter
than the bracts, but otherwise very similar ; perianth 1 lin. long ;
segments triangular-lanceolate, subacute, with a few long apical
hairs on the inside ; anthers exserted from the perianth-tube, ¼ lin.
long ; style stout, ¼ lin. long, reaching to the base of the anthers ;
fruits not seen. *A.DC. in DC. Prodr.* xiv. 670.

CENTRAL REGION : Graaff Reinet Div. ; Camdeboo, *Dunn in Herb. Bolus* !
Nieuwjaars Spruit, between the Orange River and Caledon River at the foot of
the Witberg Range, *Zeyher* !

74. T. transvaalense (Schlechter in Journ. Bot. 1897, 432) ;
rhizome often several-headed, stout ; stems slender, ascending,
densely leafy, finely and rather densely and shortly pubescent with
spreading or slightly recurved hairs ; leaves mostly more or less
parallel with the shoot, linear-subulate, very acute, 1½–2 lin. long,
convex on the back, rather minutely but conspicuously ciliate and
puberulous ; flowers in short spikes, overtopped by the same year's
leafy shoot, with the ovary conspicuously pubescent ; bracts longer
than or equal to the flowers, similar to the leaves, often incurved ;
bracteoles shorter than the flowers or about as long ; perianth 1¼
lin. long, slightly pubescent outside ; segments narrowly lanceolate,
acute, ¾ lin. long, with a dense apical beard ; anthers half exserted
from the perianth-tube, ¼ lin. long ; style a little over ¼ lin. long,
reaching to the base of the anthers ; fruits ovoid-globose, 2 lin.
long, pubescent, strongly 10-ribbed, reticulate between the ribs.

KALAHARI REGION : Transvaal ; Jeppes Town Ridges, Johannesburg, 6000 ft.,
Gilfillan in Herb. Galpin, 6068 ! Elandsfontein, 5500 ft., *Gilfillan in Herb. Galpin,*
1419 ! Magaliesberg Range, *Burke* ! Hoekemoer Spruit near Klerksdorp, *Nelson,*
249 ! near Modderfontein, 4500 ft., *Miss Nation,* 39 ! *Jenkins* ! Elsburg, 5900 ft.,
Schlechter, 3542 ! Aapies Poort, *Rehmann,* 4009 ! Klippan, Boschveld, *Rehmann,*
5212 ! near Pretoria, *Burtt-Davy,* 781 ! *Miss Leendertz,* 255 ! near Irene, *Burtt-
Davy,* 2306 !

75. T. gnidiaceum (A.DC. Esp. Nouv. Thes. 7); stems and branches ascending, very densely leafy ; leaves erect, subulate, with very acute rigid cartilaginous whitish apex, flat on the upper surface, slightly rounded below, 1–3 lin. long, usually slightly scabrous ; flowers in rather dense terminal spikes ; bracts shorter than the open flowers, lanceolate or linear-lanceolate, with an acuminate subulate very acute apex, finely scabrid-puberulous on the margin, very slightly keeled ; bracteoles similar to the bracts but narrower ; perianth $1\frac{1}{4}$–$1\frac{3}{4}$ lin. long, with distinct external glands ; segments linear or linear-lanceolate, subacute, $\frac{3}{4}$–$1\frac{1}{4}$ lin. long, flat, with a dense apical beard ; anthers almost wholly included in the perianth-tube, about $\frac{1}{4}$ lin. long ; style very short or almost absent ; fruits ovoid-globose, $2\frac{1}{4}$ lin. long, prominently 10-ribbed, very prominently or almost honeycombed-reticulate. *A.DC. in DC. Prodr.* xiv. 662. *T. imbricatum, E. Meyer in Drège, Zwei Pfl. Documente,* 132, *not of Thunb. T. Dregei, Sond. in Flora,* 1857, 356, 406.

VAR. β, **Zeyheri** (Sond. in Flora, 1857, 356); leaves acute, glabrous, with recurved tips ; bracts rather broader than in the type with almost glabrous margins. *A.DC. in DC. Prodr.* xiv. 662.

COAST REGION : Uitenhage Div. ; Addo, 1000–2000 ft., *Drège* a! *Ecklon,* 8! Albany Div. ; near Grahamstown, *Burchell,* 3541 ! *Bolus,* 1912! *MacOwan,* 45! *Guthrie,* 3319 ! *Schönland,* 57 ! Between Coldstream and Grahamstown, 2000 ft., *Bolus,* 3594! *Krebs,* 386 ! Fort Beaufort Div. ; *Cooper,* 419 ! Queenstown Div. ; Queenstown, *Cooper,* 3095! British Kaffraria ; Keiskamma Hoek, *Cooper,* 236 ! Var. β : Caledon Div. ; Zwart Berg and Zondereinde River, *Zeyher,* 3807 !

76. T. phyllostachyum (Sond. in Flora, 1857, 355); a small shrublet, many-stemmed, erect ; branches leafy, slender, glabrous ; leaves linear, recurved in the upper part, acutely mucronate, $1\frac{1}{2}$–2 lin. long, flat on the upper surface, rounded below, glabrous ; flowers in lax spikes, with a leafy shoot above which turns black when dry ; bracts linear, acute, shorter or longer than the flowers, concave on the upper side, slightly keeled below ; bracteoles about two-thirds as long as to nearly as long as the bracts, but otherwise very similar ; perianth 1–$1\frac{1}{2}$ lin. long, with conspicuous large external glands ; segments linear-lanceolate, obtuse or subacute, $\frac{1}{2}$–1 lin. long, densely bearded in the upper portion ; anthers about $\frac{1}{6}$ lin. long, included in the perianth-tube ; style $\frac{1}{8}$ lin. long or almost absent ; fruits ovoid, $2\frac{1}{4}$ lin. long, conspicuously 10-ribbed, slightly reticulate between the ribs. *A.DC. in DC. Prodr.* xiv. 670.

COAST REGION : Riversdale Div. ; Garcias Pass, *Galpin,* 4540 ! near Riversdale, *Schlechter,* 1886 ! Mossel Bay Div. ; between Duyker River and Gouritz River, *Burchell,* 6394 ! George Div. ; Zwarteberg Range, *Bolus,* 2455 ! Knysna Div. ; Outeniqua Mountains, *Bolus,* 2455 ! Uitenhage Div. ; between Port Elizabeth and the Zwartkops River, *Zeyher,* 3800 !

77. T. impeditum (A. W. Hill in Kew Bulletin, 1915, 32); stems tufted from the apex of a rhizome, slender, simple or sparingly branched, slightly angular, glabrous ; leaves linear, with acute recurved tips, 2–7 lin. long, flat, rather thick, glabrous ; flowers in lax spikes, solitary in the axil of each bract ; bracts slightly adnate

to or free from the peduncle, linear, with acute recurved tips, as
long as or slightly exceeding the flower, flat on the upper surface,
glabrous ; bracteoles half as long as the flowers or slightly more,
otherwise similar to the bracts ; perianth 1–1½ lin. long ; segments
narrowly lanceolate, subacute, ⅔–1 lin. long, densely bearded ;
anthers about ¼ lin. long, exserted from the perianth-tube ; style
¼ lin. long, reaching to a little below or above the base of the
anthers ; fruits ellipsoid-globose, 2⅓ lin. long including the persistent
perianth, prominently 10-ribbed, purplish, distinctly reticulate
between the ribs. *T. sonderianum, Schlechter in Journ. Bot.* 1898,
376, *as to Bolus,* 526.

VAR. β, rasum (A. W. Hill in Kew Bulletin, 1915, 32) ; differs from the type
in the beard of the perianth-segments being very small and of few short hairs.

COAST REGION : Queenstown Div. ; on a kopje near Queenstown, 3600 ft.,
Galpin, 2157, partly ! *Galpin,* 1545 ! *Kolbe*! Hangklip, 5600 ft., *Galpin,* 5856 !
Mund & Maire !
CENTRAL REGION : Graaff Reinet Div. ; Sneeuwberg Range, near Graaff Reinet,
4600 ft., *Bolus,* 526, partly !
KALAHARI REGION : Orange River Colony ; Besters Vlei, near Witszies Hoek,
5400 ft., *Bolus,* 8249 ! Var. β : Transvaal ; near Pretoria, *Rehmann,* 4544 !
Zuikerbosch Rand, *Schlechter,* 3507 !
EASTERN REGION : Natal ; near Durban, 3000 – 4000 ft., *Sutherland*! near
Estcourt, 4920 ft., *Schlechter,* 3357 !

Galpin, 2157, was made the type of *T. longirostre,* Schlechter (= *T. Zeyheri,*
A.DC.) ; the specimen under this number at Kew was sent to Kew by
Mr. Galpin, and though from the same kopje as 2157 in Herb. Schlechter, is not
part of the original gathering and is quite a distinct plant. *Bolus,* 526, in Herb.
Stockholm is identical with this species, though this number is included by
Schlechter under *T. sonderianum.*

78. **T. magalismontanum** (Sond. in Flora, 1857, 358) ; stems
few and slender from a slender rhizome, erect or suberect, often
glaucous, faintly sulcate, glabrous ; branches very sparingly leafy ;
flexuous and bearing scattered flowers over nearly their full length ;
leaves linear, subacute, ½–¾ in. long, often adpressed to the stem,
concave on the side next the stem, mostly glaucous, glabrous ;
flowers solitary in the bract-axils of lax spikes ; bracts ovate or
ovate-lanceolate, acutely acuminate, shorter than the open flowers,
slightly warted outside, glaucous-grey ; bracteoles nearly three-
quarters as long as the bracts and very similar to them but
narrower ; perianth with prominent external glands, infundibuliform, ¾–1½ lin. long ; segments elliptic, obtuse or subacute, hooded,
½ lin. long, bearded, with hairy margins ; anthers partly included
in the perianth-tube, ½ lin. long ; style ⅓–¾ lin. long, reaching to
above the base or about the middle of the anthers ; fruits broadly
ellipsoid, 2¾ lin. long, prominently 10-ribbed, slightly glaucous and
conspicuously reticulate between the ribs. *T. megalismontanum,*
Sond. ex A.DC. in DC. Prodr. xiv. 670.

KALAHARI REGION : Transvaal ; Magaliesberg Range, *Zeyher,* 1501 ! *Burke* !
Zeyher, 1500, partly ! Pretoria, Aapies Poort, *Rehmann,* 4010 ! Derde Poort,
Miss Leendertz, 375 ! Pretoria Kopjes, *Miss Leendertz,* 293 ! Rustenburg,
Miss Nation, 207 ! near Warm Baths, *Bolus,* 12274 !

79. T. Burkei (A. W. Hill in Kew Bulletin, 1915, 24); stems slender, grass-like, sulcate or ribbed, glabrous; leaves linear, acute, about 4 lin. long, slender, glabrous; flowers arranged in lax racemes of 1-flowered cymules; bracts adnate to the apex of the peduncle, linear, acute, about as long as the flowers, with very minutely scabrous margins; bracteoles a little shorter than the bracts; perianth with distinct external glands, 1–1¼ lin. long; segments linear-lanceolate, subacute, ⅔ lin. long, bearded at the apex, margins incurved; anthers exserted from the perianth-tube, ¼ lin. long; style about ½ lin. long, reaching to the top of the anthers; fruits ovoid-globose, 2 lin. long, strongly 10-ribbed, reticulate between the ribs. *T. racemosum, Sond. in Flora,* 1857, 357, *partly, not of Bernh.*

KALAHARI REGION : Griqualand West ; Kimberley, *Marloth* ! Bechuanaland ; near the sources of the Kuruman River, *Burchell,* 2493/1 ! Barolong Territory, *Holub* ! Magaliesberg Range, *Burke* ! *Zeyher,* 1500, partly ! near Pretoria, *Burtt-Davy,* 2535 ! *Rehmann,* 4544 ! Boschveld, between Elands River and Klippan, *Rehmann,* 5013 ! by the Komati River, *Bolus,* 9765 ! near Komati Poort, 1100 ft , *Schlechter,* 11803 ! and without precise locality, *Sanderson* ! Delagoa Bay, Tembé, *Junod,* 325 ! Natal ; near Maritzburg, *Schlechter,* 3288 ! coast land, 1150 ft., *Sutherland* !

This species shows an external resemblance to *T. racemosum,* Benth., under which Sonder has placed Zeyher, 1500. The presence of an apical beard, however, removes *T. Burkei* from that affinity.

80. T. orientale (A. W. Hill in Kew Bulletin, 1915, 36); stems few from the apex of a fairly stout woody rhizome, suberect, angular and sulcate, glabrous; branches suberect, sparingly leafy; leaves linear, acute or subacute, 3–7 lin. long, ½–¾ lin. broad, thick and fleshy, with very finely scabrous cartilaginous margins, mostly with a distinct midrib or more or less keeled on the lower surface, glabrous; flowers few in leafy racemes, solitary in the bracts; bracts adnate to the peduncle, lanceolate or ovate-lanceolate, acutely acuminate, longer than the flowers, with distinctly carti-laginous scabrous margins; bracteoles similar to the bracts but about half as long; perianth urceolate, with distinct external glands, about 2 lin. long; segments linear-lanceolate, obtuse, 1–1½ lin. long, hooded with a dense woolly apical beard; anthers included, reaching to the top of the perianth-tube, ⅜–½ lin. long; style ¾ lin. long, reaching to nearly the top of the anthers; fruits ovoid-ellipsoid, 3 lin. long, conspicuously 10-ribbed, only very slightly reticulate between the ribs.

COAST REGION : Stockenstrom Div. ; Katberg, *Hutton* !
KALAHARI REGION : Basutoland, *Cooper,* 3094 !
EASTERN REGION : Tembuland ; Tabase, near Bazeia, 2500 ft., *Baur,* 336, partly ! Griqualand East ; near Kokstad, 4300 ft., *Tyson,* 3157 !

81. T. macrogyne (A. W. Hill in Kew Bulletin, 1915, 34); stems very short, up to 5 in. high, simple or forked, probably arising from a rhizome, slender, sulcate or angular, glabrous ; leaves linear, with a cartilaginous acute apex, 4–5 lin. long, about ½ or ⅓

lin. broad, slightly keeled below, fairly thick and fleshy, glabrous ;
flowers few, solitary, axillary, very shortly pedunculate ; bracts
adnate to the apex of the peduncle, linear, with acutely subulate
apex, equal to or slightly longer than the flowers, deeply concave
on the upper surface, very minutely serrulate on the margin ;
bracteoles as long as the flower, narrower than the bracts ; perianth
white, 2 lin. long, without external glands ; segments oblong-
lanceolate, obtuse, 1 lin. long, hooded, flat, densely woolly-bearded ;
anthers partly included in the tube and partly exserted, ⅜ lin.
long ; style 1⅛ lin. long, longer than the anthers.

KALAHARI REGION : Orange Free State ; Bethlehem, low - lying veld,
Richardson !

82. **T. lobelioides** (A.DC. Esp. Nouv. Thes. 8) ; stem usually
much-branched in the upper part, sulcate, glabrous ; branches
ascending or suberect, purplish or grey, dull and glabrous ; leaves
linear or linear-lanceolate, fleshy, acute, ½–¾ in. long, flat, 1-nerved
on the upper surface, glabrous ; flowers solitary ; bracts shortly
adnate to the peduncle, much longer than the flowers, keeled ;
bracteoles as long as or slightly longer than the flowers, with
slightly scabrous edges ; perianth 2¼ lin. long ; segments 1¼ lin.
long, oblong-lanceolate, subacute, densely bearded inside ; anthers
slightly exserted, ¼ lin. long ; style ½–1 lin. long, sometimes reaching
to well above the anthers ; young fruits ribbed. *A.DC. in DC.
Prodr.* xiv. 666. *T. recurvifolium, Sond. in Flora,* 1857, 356, 407.

CENTRAL REGION : Stockenstrom Div. ; by the Kat River, near Philipton, *Ecklon
& Zeyher,* 25 !
EASTERN REGION : Natal ; Hoffenthal, 4000 ft., *Wood,* 3574 !

83. **T. gœtzeanum** (Engl. in Engl. Jahrb. xxx. 306) ; a subshrub ;
caudex small, thick ; stems erect, 10–12 in. long, numerous, branched,
glaucous, longitudinally grooved ; leaves linear-lanceolate, apex cartilaginous, 7 lin. long, strongly keeled ; bracts and bracteoles adnate
to the pedicels ; bracteoles 2–2½ lin. long ; perianth white, 1–2 lin.
long ; segments triangular, elongate, obtuse, ¾–1 lin. long, margin
inflexed, with a dense beard of long hairs at the apex and upper
part of the margins ; anthers ¼ lin. long, equal in length to the
filaments ; style ⅜ lin. long ; fruit ovoid, with thick prominent
ribs, 1¾–2 lin. long, 1¼ lin. broad. *Baker and Hill in Dyer, Fl.
Trop. Afr.* vi. i. 418. *T. Schweinfurthii, var. laxum, Engl. Pfl. Ost-
Afr. C.* 168.

KALAHARI REGION : Bechuanaland ; Chooi Desert, *Burchell,* 2340 ! Transvaal ;
Heidelberg, 5350 ft., *Schlechter,* 4792 ! Houtbosch, 4500–4900 ft., *Bolus,* 11158 !
11159 !
Also in Tropical Africa.

84. **T. resedoides** (A. W. Hill in Kew Bulletin, 1910, 187) ; a
herb or subshrub, perennial, forming low bushes ; rootstock woody ;

stems erect, 6–8 in. high, bearing spreading branches 2–4 in. long, angled and grooved; leaves distant, spreading, 3–4 lin. long, subulate, acute, with colourless tips; inflorescences in simple or compound spikes or racemes; flowers either single and sessile in the axils of bracts or borne in short 3-flowered axillary cymes; bracts lanceolate, acute, equal in length to or half as long again as flower, sessile on axis or adnate to the peduncle; bracteoles springing from the bract-axil, shorter than the flower; perianth white, about $1\frac{1}{4}$ lin. long; segments elliptic-ovate, obtuse, $\frac{3}{4}$ lin. long, apex hooded and with upper part of margin densely bearded; filaments $\frac{1}{5}$ lin. long; anthers about $\frac{1}{3}$ lin. long; style $\frac{3}{8}$ lin. long; fruit ovoid, $1\frac{1}{2}$ lin. long, 1 lin. broad, immature, ribs obscure. *Baker and Hill in Dyer, Fl. Trop. Afr.* vi. i. 419. *T. Welwitschii, Gilg in Baum, Kunene-Samb. Exped.* 230, *not of Hiern in Cat. Afr. Pl. Welw.* i. 938.

KALAHARI REGION: Transvaal; Warmbaths, *Miss Leendertz*, 1335! 1353! Also in Tropical Africa.

85. T. Junodii (A. W. Hill in Kew Bulletin, 1915, 33); stems branched from the base; branches ascending, slender, sharply angular or almost winged, sulcate between the angles, glabrous; leaves linear, very acute, $\frac{1}{4}$–$\frac{1}{2}$ in. long, keeled below, flat or nearly so and prominently 1-nerved on the upper surface, glabrous; flowers solitary in the bract axil, subsessile; bract shortly adnate to the very short peduncle, longer than the flower, very similar to the leaves; bracteoles about as long as the flowers, narrowly linear, acute; perianth $1\frac{1}{3}$ lin. long; segments with conspicuous external glands, lanceolate, subacute, $\frac{3}{4}$ lin. long, hooded, bearded at the apex; anther exserted from the perianth-tube, $\frac{1}{4}$ lin. long; style $\frac{3}{4}$–$\frac{4}{5}$ lin. long, reaching to the middle or top of the anthers; fruits elongate-ellipsoid, about 2 lin. long including the persistent perianth, with fairly conspicuous ribs and reticulation.

SOUTH AFRICA: without locality, *Wahlberg*! in Stockholm Herbarium.
KALAHARI REGION: Transvaal; Shilovane, *Junod*, 1301!

86. T. coriarium (A. W. Hill in Kew Bulletin, 1915, 24); stems solitary or subsolitary from a small knotty rhizome, sparingly branched, erect, about 4 in. high, strongly ribbed, somewhat glaucous, glabrous; leaves few, linear, obtuse or subacute, 4–6 lin. long, $\frac{3}{4}$–1 lin. broad, slightly keeled below, slightly glaucous, glabrous, thick and fleshy; flowers white, few in loose racemes, solitary in the bracts; bracts adnate to the short (up to 1 lin. long) peduncle, elliptic-lanceolate, obtuse, more or less boat-shaped, with smooth margins, longer than the flowers, entire; bracteoles about two-thirds as long as the bracts; perianth $2\frac{1}{8}$ lin. long, with very large and conspicuous external glands; segments ovate-lanceolate, obtuse, $1\frac{1}{3}$ lin. long, with a dense woolly apical beard; anthers partly exserted, nearly ½ lin. long; style ½ lin. long, reaching to the base of the anthers; fruit not seen.

KALAHARI REGION : Orange River Colony ; Harrismith, *Sankey*, 223 !

87. T. nigrum (A. W. Hill in Kew Bulletin, 1915, 35) ; stems several from a broad-topped woody rhizome, strongly ribbed and sulcate, glabrous ; leaves broadly linear, acute, straight, ½–¾ in. long, flat, rather thick and fleshy, becoming blackish when dry, glabrous ; flowers solitary in the bract axils, subsessile, arranged in lax spikes ; bracts as long as or longer than the open flower, linear-lanceolate, acute, entire, glabrous ; bracteoles about two-thirds as long as the bracts ; perianth with large conspicuous external glands, about 1½ lin. long ; segments linear-lanceolate, subacute, 1 lin. long, infolded and deeply hooded, densely fringed and bearded to the base ; anthers included towards the top of the perianth-tube, about ¼ lin. long ; style ½ lin. long, reaching to the top or the middle of the anthers ; fruits ovoid-ellipsoid, 3½ lin. long, strongly 10-ribbed, very slightly reticulate between the ribs.

KALAHARI REGION : Orange River Colony ; *Cooper*, 826 ! 1061 !
EASTERN REGION : Natal ; between Pietermaritzburg and Greytown, *Wilms*, 2253 ! Giants Castle, 9000 ft., *Guthrie*, 4954 ! Griqualand East ; near Kokstad, 5000 ft., *Tyson*, 1863 !

88. T. gracilarioides (A. W. Hill in Kew Bulletin, 1915, 29) ; stems branched from the base, up to nearly 1 ft. high ; branches rather densely leafy, ascending, slender, angular, glabrous ; leaves linear-acicular, acute, ¼–½ in. long, scarcely ⅓ lin. thick, keeled on the back, glabrous, with a fairly prominent midrib on the upper surface ; flowers arranged in leafy racemes of cymules at the ends of the shoots ; bracts adnate to the peduncles, very similar to the leaves, with narrow subtranslucent margins, about twice the length of the flowers ; bracteoles as long as or a little longer than the flowers ; perianth urceolate, with prominent external glands, about 1 lin. long ; segments linear-lanceolate, subacute, hooded, ⅔–¾ lin. long, with an apical beard of a few hairs and incurved margins ; anthers at the base of the perianth-segments or almost in the tube, about ⅓ lin. long ; stigma sessile or subsessile ; fruits ovoid, 2 lin. long, finely 10-ribbed, clearly reticulate between the ribs.

KALAHARI REGION : Transvaal ; grassy mountain sides of the Saddleback Range, near Barberton, 4000–5000 ft., *Galpin*, 543 ! Swaziland ; Havelock Concession, 3700 ft., *Saltmarshe in Herb. Galpin*, 1048 !

89. T. asterias (A. W. Hill in Kew Bulletin, 1915, 23) ; rootstock stout, many- or few-headed ; stems erect, branched in the upper part, compressed and angled, purplish when dry, glabrous ; leaves acicular, grooved above, acute, ¼–1 in. long, about ⅓ lin. thick, glabrous ; flowers arranged in lax racemes or racemes of cymules ; bracts partly adnate to the peduncle, linear, acute, up to

2½ lin. long, shorter or longer than the flowers ; bracteoles shorter than the flowers ; perianth 1¾ lin. long, with large rounded external glands ; segments lanceolate, about 1¼ lin. long, bearded at the apex, with finely hairy incurved margins ; anthers ¼ lin. long, exserted ; stigma sessile ; fruits ovoid, 3½ lin. long including the long persistent perianth, strongly 10-ribbed, finely and rather faintly reticulate between the ribs.

KALAHARI REGION : Transvaal ; Shilovane, *Junod,* 749a ! Champs du Sanatorium, *Junod,* 837 ! Aapies Poort, near Pretoria, *Rehmann,* 4013 ! Houtbosch, *Rehmann,* 5958 ! 5959 ! and without precise locality, *Sanderson,* 916 ! Swaziland ; Havelock Concession, 4000 ft., *Saltmarshe in Herb. Galpin,* 1008 !

EASTERN REGION : Natal ; near Murchison, *Wood,* 3003 ! and without precise locality, *Gerrard,* 333 !

90. **T. polygaloides** (A. W. Hill in Kew Bulletin, 1915, 38) ; root apparently annual, short, with longer spreading straw-coloured lateral ones ; stem slender, branched in the upper half or upper two-thirds, very sparingly leafy, with several longitudinal fairly conspicuous ribs, glabrous ; branches erect, slender ; leaves slender, linear, acute, ⅓–½ in. long, blackish when dry, glabrous ; flowers arranged in short racemose 1-flowered cymules ; cymes shortly stalked, with 1 or 2 small leaves and about 3 bracts around the flower ; bracts linear, about as long as the flower ; bracteoles acute, half the length of the flower ; perianth ¾ lin. long ; segments elliptic-lanceolate, subacute, ½–¾ lin. long, with an apical beard and hairy margins ; anthers ⅛ lin. long, exserted, at the base of the segments or almost in the tube ; stigma subsessile ; fruits ovoid, 2 lin. long, reddish-glaucous, rather prominently 10-ribbed and reticulate.

EASTERN REGION : Natal ; Clairmont, *Wood,* 1095 ! in a marsh near Clairmont, *Schlechter,* 2976 !

91. **T. cornigerum** (A. W. Hill in Kew Bulletin, 1915, 25) ; stems elongated and very slender, finely sulcate, glabrous ; branches sometimes flowering nearly their whole length ; leaves linear, very acute, ¾–1¼ in. long, 1-nerved on the upper surface, glabrous ; cymules racemosely arranged in the upper part of or nearly the whole length of the branches, 3- to several-flowered ; bracts shorter than the flowers, horn-like in arrangement, linear, acute, shortly adnate to the peduncle ; bracteoles much smaller than the bracts but otherwise similar ; perianth ¾ lin. long, with prominent ovoid external glands ; segments ⅔ lin. long, ovate-oblong, subacute, hooded, bearded, with papillose margins ; anthers exserted from the perianth-tube, ¼ lin. long ; style about ¼ lin. long, reaching to the top of the anthers ; fruits not seen.

EASTERN REGION : Natal ; Mooi River, 4000–5000 ft., *Wood,* 4487 ! 5344 ! near Stanger, 150 ft., *Wood,* 10193 ! and without precise locality, *Gerrard,* 1278 !

92. **T. palliolatum** (A. W. Hill in Kew Bulletin, 1910, 187) ; a

herb, 8 in. high, sparingly branched above; stems deeply grooved, lax; leaves linear, strap-shaped or concave with distinct midrib, 4–6 lin. long, acute; flowers shortly pedicellate, arranged singly or in 3-flowered cymes in the axils of leafy bracts, forming lax terminal inflorescences; bracts linear, concave, acute, 4 lin. long, adnate to the peduncle for about half its length; bracteoles 2, $1\frac{1}{4}$ lin. long; perianth $1\frac{1}{4}$ lin. long, campanulate; segments $\frac{3}{4}$ lin. long, with elongated hooded apex about $\frac{1}{4}$ lin. long, and an apical beard of thick hairs; filaments $\frac{1}{5}$ lin. long; anthers $\frac{1}{8}$ lin. long; style $\frac{3}{8}$ lin. long; fruit globose, $\frac{3}{4}$ lin. in diam., with main ribs and delicate reticulations. *Baker and Hill in Dyer, Fl. Trop. Afr.* vi. i. 417.

KALAHARI REGION: Transvaal; near Potgieters Rust, on stony and grassy hills, 3700 ft., *Bolus*, 11008!

Also in Tropical Africa.

93. **T. gypsophiloides** (A. W. Hill in Kew Bulletin, 1915, 30); stems fairly slender, sometimes rather copiously branched, finely ribbed, glabrous; branches slender, spreading; leaves linear-lanceolate or broadly linear, very acute, $\frac{1}{3}$–1 in. long, up to $1\frac{1}{4}$ lin. broad or sometimes very narrow, flat, with a prominent midrib, and minutely serrulate margins, glabrous; flowers few at the end of the branchlets; bracts leaf-like, adnate to the peduncle for nearly half its length, keeled; bracteoles as long as or shorter than the flowers, acute; perianth urceolate, swollen in the lower part, $\frac{2}{3}$ lin. long; segments ovate, subacute, $\frac{1}{2}$ lin. long, with a dense apical beard; anthers partly exserted from the perianth-tube, subpendant in pocket-like recesses, nearly $\frac{1}{4}$ lin. long; style $\frac{1}{6}$ lin. long or less, sometimes reaching to the top of the anthers; fruits oblong-ellipsoid, $2\frac{3}{4}$ lin. long, 10-ribbed, strongly reticulate between the ribs.

EASTERN REGION: Natal; Umtwalumi, *Wood*, 573! 3105! near Verulam, *Wood*, 756! and without precise locality, *Gerrard*, 407!

KALAHARI REGION: Transvaal; Queens River Valley, near Barberton, *Galpin*, 758!

94. **T. utile** (A. W. Hill in Kew Bulletin, 1915, 43); stems few from the apex of an erect slender woody rhizome, strongly ribbed and sulcate, glabrous; branches ascending; leaves linear-acicular, acute or subacute, about $\frac{3}{4}$ in. long, $\frac{1}{3}$–$\frac{1}{2}$ lin. broad, with a fairly distinct keeled midrib below, glabrous; flowers solitary or three together in the bract axils at the apex of a short peduncle; bracts shorter than the flowers, linear, subacute, glabrous, entire; bracteoles about half the length of the bracts but very similar to them; perianth about $1\frac{1}{2}$ lin. long; segments elliptic-lanceolate or lanceolate, obtuse, $\frac{3}{4}$–1 lin. long, hooded and bearded at the apex, with papillose margins; anthers exserted from the perianth-tube, $\frac{1}{4}$ lin. long; style $\frac{1}{3}$–$\frac{3}{4}$ lin. long, reaching nearly to the top of the anthers; fruits ellipsoid-ovoid, 3 lin. long, strongly 10–11-ribbed, slightly reticulate between the ribs.

COAST REGION: Cape Div. ; eastern slope of Table Mountain (possibly introduced), *Schlechter*, 485 !
KALAHARI REGION : Transvaal ; near Pretoria, *Rehmann*, 4012 ! 4543 ! 4718 !
Miss Leendertz, 293 ! near Heidelburg, *Schlechter*, 3532 ! Jeppes Town Ridge, Johannesburg, *Mrs. de Jongh in Herb. Galpin*, 1471 ! *Gilfillan in Herb. Galpin*, 6069 ! near Modderfontein, *Miss Nation*, 69 ! 70 ! Middleburg district, at Bronkhorst River, *Wilms*, 1309 !

Miss Olive Nation states that this " stiff upright herb is used by the Kafirs to make brooms."

95. **T. fallax** (Schlechter in Engl. Jahrb. xxvii. 118) ; branches erect, woody, with transversely splitting corky bark ; young branchlets light-green, stout, finely sulcate, glabrous ; leaves flat, thick and coriaceous, linear or linear-lanceolate, obtuse, $\frac{1}{3}$–1 in. long, 1–2 lin. broad, with rather broad cartilaginous subtranslucent margins, glabrous ; flowers in fairly dense solitary capitate clusters at the ends of the shoots ; bracts coloured, oblanceolate, subacute, 3–4 lin. long, with a broad fleshy keel in the upper part, margins subtranslucent and greenish-yellow, glabrous ; bracteoles similar to the bracts but a little shorter and narrower ; perianth $1\frac{1}{2}$ lin. long, fleshy, with conspicuous external glands ; segments linear, subacute, fleshy, $1\frac{1}{4}$ lin. long, with an apical beard of hairs and a fringe on the margins ;.anthers exserted, at the base of the perianth-segments, $\frac{1}{6}$ lin. long ; stigma sessile.

COAST REGION ; Bredasdorp Div.; mountains between Elim and Fairfield, *Bolus*, 8598 ! near Napier, *Schlechter*, 9658 !

96. **T. helichrysoides** (A. W. Hill in Kew Bulletin, 1915, 30) ; stems elongated, up to 2 ft. high, sparingly branched in the upper part, green, with narrow purple angles, glabrous ; branches suberect ; leaves linear, subacute, $\frac{1}{2}$–$1\frac{1}{2}$ in. long, nearly 1 lin. broad, thick and fleshy, flat or slightly concave on the upper surface, with very narrowly cartilaginous margins, glabrous, becoming wrinkled when dry ; flowers in rather dense terminal corymbs ; bracts yellowish-green when dry, lanceolate or oblong-lanceolate, subacute, shortly adnate to the peduncle, reaching to the top of the flower, glabrous and rather fleshy ; bracteoles a little shorter than the bracts but otherwise very similar ; perianth with conspicuous external glands, $1\frac{1}{4}$ lin. long ; segments lanceolate, obtuse, $\frac{4}{5}$–1 lin. long, deeply hooded, with a dense apical beard and infolded hairy margins ; anthers included in the perianth-tube just at the base of the perianth-segments, $\frac{1}{5}$ lin. long ; stigma subsessile ; fruits with a stalk about $\frac{1}{2}$ lin. long, ellipsoid, 3 lin. long, with 5 very prominent and 5 much less prominent nerves, finely reticulate between the nerves, yellowish-green when dry.

COAST REGION : Riversdale Div.; between Garcias Pass and Muis Kraal, 1850 ft., *Bolus*, 11375 !

97. **T. umbelliferum** (A. W. Hill in Kew Bulletin, 1915, 42) ; stems tall, erect, woody, with more or less rounded angles, glabrous ;

branches few, ascending; leaves large and fleshy, subacicular, obtuse, 1–2 in. long, about 1 lin. thick, glabrous; flowers in fairly dense terminal corymbs about ½ in. in diam.; bracts slightly adnate to the very short peduncles, reddish, reaching to about the top of the flowers, with subtranslucent margins, glabrous; bracteoles nearly as long as the bracts but a little narrower; perianth with conspicuous external glands, 1¼ lin. long; segments lanceolate, obtuse, ¾ lin. long, bearded at the apex, with hairy margins, slightly hooded; anthers exserted from the perianth-tube, ¼ lin. long; style ⅜ lin. long, reaching to the middle of the anthers; fruits ovoid-ellipsoid, nearly 2 lin. long, fairly prominently 5-ribbed, with inconspicuous intermediate ribs and reticulation.

COAST REGION: Prince Albert Div.; tops of the mountains of Zwartberg Pass, 4400 ft., *Bolus*, 11633! 12276! *Marloth*, 2489b!

98. **T. boissierianum** (A.DC. in DC. Prodr. xiv. 663); a small shrub, 9 in. high; stems ascending, densely leafy, rather slender, terete, scabrid-puberulous; leaves numerous, imbricate, acicular, flattened on the upper surface, acutely mucronate, 2½–3 lin. long, ¼ lin. thick, fleshy, glabrous; flowers very small, in dense terminal heads 2½–4 lin. in diam.; bracts as long as or longer than the flowers, linear or linear-lanceolate, acutely acuminate, 1¼ lin. long, with slightly hyaline margins; bracteoles similar to the bracts but a little shorter; perianth ⅞ lin. long, with external glands; segments ½–⅔ lin. long, linear-lanceolate, subacute, bearded at the apex; anthers ⅛ lin. long, included in the perianth-tube; stigma sessile; fruits oblong-ellipsoid, 2 lin. long, distinctly but rather finely 10-ribbed, laxly reticulate between the ribs.

SOUTH AFRICA: without locality, *Verreaux in Herb. Boissier*! without collector's name in *Kew Herbarium*!

99. **T. pubescens** (A.DC. Esp. Nouv. Thes. 7); a much-branched subshrub, up to 1 ft. high; stem woody, subterete, shortly pubescent with somewhat reflexed hairs; branches spreading from the main stem at a wide angle, densely leafy, sulcate, pubescent; leaves linear, acute, keeled, 3–4 lin. long, more or less triangular in section, shortly ciliate-pubescent on the margins and keel or sometimes all over both surfaces; flowers in small dense terminal leafy clusters; bracts and bracteoles similar to the leaves but broader; perianth 2 lin. long; segments lanceolate, subacute, 1 lin. long, densely bearded inside; anthers exserted from the perianth-tube, ½ lin. long; style ¾–1 lin. long, reaching to the middle of the anthers; fruits oblong-ellipsoid, contracted at the base, about 3 lin. long, not reticulate. *A.DC. in DC. Prodr.* xiv. 664. *T. hirtulum, Sond. in Flora*, 1857, 359, 406.

COAST REGION: Clanwilliam Div.; Vogel Fontein, 1200 ft., *Schlechter*, 8515! Warm Baths, *Stephens*, 7729! Malmesbury Div.; Modder River, near Groene Kloof, 300 ft., *Bolus*, 4328! Hopefield, 160 ft., *Schlechter*, 5301! *Bachmann*, 1702! Cape Div.; Chapmans Bay, Kommetje sand hills, *Wolley-Dod*, 1634! Millers Point, *Wolley-Dod*, 2845! Div.? Groctepostveld, *Ecklon & Zeyher*, 54.

100. T. rufescens (A. W. Hill in Kew Bulletin, 1915, 40); stems reddish-brown, very slender, sparsely and divaricately branched, sulcate, shortly pubescent with subreflexed hairs; branches rather densely leafy; leaves spreading, linear, acute, 3–4 lin. long, shortly pubescent or nearly glabrous; flowers in rather dense terminal oblong spikes; bracts reddish, linear-lanceolate, acutely acuminate, about as long as or a little longer than the flowers, keeled, with somewhat membranous margins, pubescent on the margin and outside; bracteoles similar to the bracts but a little shorter; perianth minutely pubescent outside, $\frac{4}{5}$ lin. long; segments triangular, subacute, $\frac{3}{8}$ lin. long, with slightly incurved margins, densely bearded inside; anthers half exserted from the perianth-tube, $\frac{1}{4}$ lin. long; style $\frac{3}{5}$ lin. long, reaching nearly to the top of the anthers; fruits ovoid, 2 lin. long, conspicuously ribbed, distinctly reticulate between the ribs.

COAST REGION: Riversdale Div.; in fields near Riversdale, 300 ft., *Schlechter*, 1851!

101. T. scabrum (Linn. Sp. Pl. ed. ii. 302); stems erect, about 1 ft. high, densely leafy, with very short internodes, glabrous; leaves linear, acute, up to 1 in. long, but mostly about $\frac{1}{2}$ in. long, more or less triquetrous, sharply pectinate-serrulate on the angles; flowers in dense terminal subglobose heads about $\frac{1}{2}$ in. in diam.; bracts broadly ovate or ovate-lanceolate, acutely acuminate, with membranous scabrous margins and a broad keel; bracteoles similar, but narrower; perianth about 1 lin. long; segments $\frac{3}{4}$ lin. long, bearded, ovate-lanceolate, acute; anthers included in the perianth-tube, $\frac{1}{4}$ lin. long; style $\frac{1}{4}$–$\frac{1}{2}$ lin. long; fruits oblong-ovoid, $2\frac{1}{4}$ lin. long, rather conspicuously 10-nerved, finely reticulate between the nerves. *Berg. Descr. Pl. Cap.* 72; *Thunb. Fl. Cap. ed. Schult.* 209; *Thunb. Prodr.* 45; *Sond. in Flora*, 1857, 360; *A.DC. in DC. Prodr.* xiv. 663. *T. scabrum, var. denudatum, Sond. in Flora*, 1857, 361; *A.DC. in DC. Prodr.* xiv. 664; *var. gracile, A.DC. l.c.* (?) *T. ciliatum, R. Br. Prodr.* 353; *A.DC. in DC. Prodr.* xiv. 672, *name only*.

COAST REGION: Tulbagh Div.; near Tulbagh Waterfall, *Ecklon*, 14! Saron, 1500 ft., *Schlechter*, 10665! Worcester Div.; mountains near De Liefde, *Drège* a! Paarl Div.; French Hoek, 500 ft., *Schlechter*, 9219! Cape Div.; Table Mountain, *Thunberg*! between Wynberg and Constantia, *Burchell*, 795! Mowbray, *Wilms*, 3605! Devils Peak, 1000 ft., *Bolus*, 2936! *Bergius*! *Büttner*! *Krebs*! Cape Flats near Kenilworth, *Schlechter*, 216! Stellenbosch Div.; between Stellenbosch and Bottelary Hill, *Burchell*, 8338! Caledon Div.; Donker Hoek Mountain, *Burchell*, 7996! Bredasdorp Div.; near Cape Agulhas, *Ecklon*!

CENTRAL REGION: Sutherland Div.: Middle Roggeveld, *Thunberg*!

Sheet *a* of *T. capitatum* in Thunberg's herbarium belongs to this species.

102. T. polycephalum (Schlechter in Engl. Jahrb. xxvii. 120): a low much-branched shrub; older branches with corky transversely splitting bark, younger ones slender, rather elongated, glabrous; leaves few and distant, subacicular, with an obtuse cartilaginous apex, concave above, rounded below, 2–3 lin. long, about $\frac{1}{4}$ lin.

thick, glabrous; flowers in small capitate clusters at the ends of
the branches ; bracts leaf-like, equalling the flowers in length, with
slightly hispid margins ; bracteoles similar to the bracts but a little
shorter ; perianth about 1 lin. long, with prominent external
glands ; segments ovate, subacute, ½ lin. long, with a woolly beard ;
anthers partly exserted from the perianth-tube, ¼ lin. long ; style
reaching to the middle of the anthers, ¼ lin. long ; fruits not seen.

WESTERN REGION : Little Namaqualand ; near Naries, *Bolus*, 9446 ! Van
Rhynsdorp Div. ; Kareeberg Range, *Schlechter*, 8256 !

103. T. microcephalum (A. W. Hill in Kew Bulletin, 1915,
34) ; branches subterete, glabrous; branchlets twiggy; leaves
fairly dense, narrowly lanceolate, subacute, 1¼–2 lin. long, about
½ lin. broad, thick and fleshy, rounded on the lower surface, flat or
concave above, glabrous ; flowers few in a fairly dense cluster at
the apex of each shoot ; bracts shorter than the flowers, reddish,
similar to the leaves, fleshy, finely fimbriate on the lower part of
the margins, otherwise glabrous ; bracteoles nearly as long as the
bracts and slightly keeled ; perianth about 1¼ lin. long ; segments
lanceolate, obtuse, ⅔ lin. long, reddish outside, hooded, with an
apical beard and hairy margins; anthers exserted from pockets
inside the perianth-tube, ¼ lin. long ; style ¼ lin. long, reaching to
just above the base of the anthers ; stigma capitate ; fruits oblong-
ovoid, 2 lin. long, strongly 10-ribbed, coarsely reticulate between
the ribs.

COAST REGION : Worcester Div. ; Matroos Berg, 6560 ft., *Marloth*, 2252 *in
Herb. Bolus* (not in Herb. Marloth) !

104. T. pycnanthum (Schlechter in Engl. Jahrb. xxvii. 120) ; a
small shrublet, up to about 1 ft. high ; branches several from the
base, erect or ascending, subterete, slender, glabrous ; leaves acicular, very acute, 3–5 lin. long, about ¼ lin. thick, with a very
prominent midrib on the upper surface, rounded below, glabrous,
with very narrow subtranslucent margins ; flowers in small terminal
heads ; bracts linear-lanceolate, very acute, a little shorter than the
open flower, keeled, with thinner reddish margins ; bracteoles about
two-thirds as long as the bracts but otherwise similar ; perianth
1¾ lin. long, with distinct external glands ; segments linear-lanceolate, hooded, 1–1¼ lin. long, with an apical beard ; anthers included
in the perianth-tube, ¼ lin. long ; style ⅛–¼ lin. long, reaching to
the base of the anthers ; fruits ellipsoid-globose, 3 lin. long, prominently 10-ribbed, shining, distinctly reticulate between the ribs.

COAST REGION : Worcester Div. ; Goudini, 900 ft., *Schlechter* 9946 ! Hex
River Valley, near De Doorns, *Bolus*, 13187 ! Paarl Div. ; French Hoek, 3500 ft.,
Schlechter, 9353 !

105. T. ecklonianum (Sond. in Flora, 1857, 356) ; stems prostrate or decumbent, elongated, terete, glabrous ; leaves large,
linear or linear-lanceolate, often somewhat falcate, acute or sub-

acute, flat on the upper surface, somewhat convex below, $\frac{1}{2}$–1$\frac{1}{4}$ in.
long, glabrous ; flowers few in a subcapitate cluster at the ends of
the branches ; bracts oblong-lanceolate or lanceolate, acutely
acuminate, 1$\frac{1}{2}$–2$\frac{1}{2}$ lin. long, slightly keeled, with scabrous slightly
membranous edges, reddish ; bracteoles nearly as long as the bracts,
but much narrower ; perianth 1–1$\frac{1}{2}$ lin. long ; segments about $\frac{1}{3}$ lin.
long, lanceolate-triangular, acute, with a woolly apical beard ;
anthers included, at the top of the perianth-tube, $\frac{1}{4}$ lin. long ; style
nearly $\frac{1}{2}$ lin. long, reaching to the middle of the anthers ; fruits
with a short stout stalk, ellipsoid-globose, prominently 10-ribbed or
almost angled, distinctly reticulate between the ribs, 2$\frac{1}{2}$ lin. long
including the persistent perianth. *A.DC. in DC. Prodr.* xiv.
670.

Coast Region : Cape Div. ; Cape Flats, *Ecklon !* *Zeyher*, 4845 ! Rondebosch,
100 ft., *Bolus*, 3920 ! and without precise locality, *Harvey*, 725 !

106. **T. sonderianum** (Schlechter in Journ. Bot. 1898, 376, partly) ;
a rather stout more or less dichotomously branched shrub ; branches
rather densely leafy, ribbed, glabrous ; leaves often all towards one
side of the branch, linear, acutely mucronate, rigid, recurved, 2–3
lin. long, about $\frac{1}{2}$ lin. broad, with a fairly prominent midrib on both
surfaces or rather sharply keeled below, slightly scabridulous on the
margins and keel ; flowers in rather dense thick more or less oblong
spikes ; bracts linear or linear-lanceolate, acute, shorter than the
flowers, with membranous rather jagged or scabrous margins ;
bracteoles as long as or nearly as long as the bracts and rather
thinner and narrower, otherwise similar ; perianth 2$\frac{1}{4}$–2$\frac{3}{4}$ lin. long ;
segments linear-lanceolate, subacute, 1$\frac{1}{2}$ lin. long, hooded, bearded
inside ; anthers $\frac{1}{2}$ lin. long, on short filaments, partly exserted from
tube ; style $\frac{3}{4}$–1 lin. long, reaching to the middle of the anthers ;
fruit ovoid-ellipsoid, 5 lin. long including the long cylindric
persistent perianth, prominently 5-ribbed with much less conspicuous intermediate ribs, reticulate.

Coast Region : Port Elizabeth Div. ; Redhouse, *Mrs. Paterson*, 689 ! Albany
Div. ; near Grahamstown, *MacOwan*, 804 !

Bolus, 526, in herb. Stockholm, on which number, as well as *MacOwan*, 804,
Schlechter based his species, is found to belong to *T. impeditum*, A. W. Hill.

107. **T. capitatum** (Linn. Sp. Pl. ed. ii. 302) ; a small shrub, often
about 9–12 in. high ; stems erect or ascending, sparingly branched :
branches densely leafy, glabrous ; leaves imbricate, parallel with
the stem or spreading, linear, with an acute or subacute cartilaginous
apex, sharply keeled, more or less triangular in section, 4–7 lin.
long, rigid, minutely subtranslucent and scabrous on the margins :
flowers in dense terminal heads ; bracts gradually differentiated
from the upper leaves, obovate to oblong-lanceolate, acutely acuminate, keeled, broad and fleshy in the middle, with rather wide
jagged or toothed margins, glabrous ; bracteoles narrower than the
bracts but otherwise very similar ; perianth 2$\frac{1}{2}$–3$\frac{1}{2}$ lin. long, with

hairy pocket-like depressions behind the anthers ; segments linear-lanceolate, subacute, $1\frac{1}{2}$–2 lin. long, fleshy, densely bearded ; anthers $\frac{1}{3}$–$\frac{1}{2}$ lin. long, included in the perianth-tube ; style $\frac{4}{5}$–$1\frac{2}{3}$ lin. long, reaching to the middle or the top of the anthers ; fruits ellipsoid-globose, $3\frac{1}{2}$ lin. long, including the persistent perianth, rather slenderly 10-ribbed, smooth and not reticulate between the ribs. *Thunb. Diss. Thes.* 9 ; *Fl. Cap. ed. Schult.* 209 ; *Bernh. in Flora,* 1845, 80 ; *Sond. in Flora,* 1857, 360 ; *A.DC. in DC. Prodr.* xiv. 663.

COAST REGION : Malmesbury Div. ; Hopefield, *Buchanan,* 1701! Cape Div. ; various localities, *Thunberg! Forster! Bergius! Ecklon,* 406! *Ecklon & Zeyher! Chamisso,* 10 ! *Wilms,* 3589! *Wolley-Dod,* 3019! 1505! *Wright! Harvey,* 715 ! *Zeyher,* 58! 5! *Drège* b! *Wallich! Hooker,* 624! *Bolus,* 3073 ! 4762 ! *Burchell,* 593! Caledon Div. ; near Grietjes Gat, *Zeyher,* 3788! Humansdorp Div. ; Kruis-fontein, 800 ft., *Galpin,* 4552 !

108. **T. glomeratum** (A. W. Hill in Kew Bulletin, 1915, 29) ; stems ascending, slender, sparingly branched, very slightly scabrid-puberulous ; branches slender, sulcate, glabrous or nearly so ; leaves linear, acute, flat or concave on the upper surface, sometimes keeled below, 3–4 lin. long, glabrous or slightly scabrous on the margins ; flowers in very short racemes or in small terminal heads ; bracts ovate-lanceolate, acute, about as long as the flowers, keeled, with membranous very minutely ciliate margins ; bracteoles about two-thirds the length of the bracts, narrower but otherwise similar ; perianth $1\frac{1}{2}$–$1\frac{3}{4}$ lin. long ; segments $\frac{1}{2}$ lin. long, linear-lanceolate, acute, with a dense apical beard ; anthers included in the perianth-tube, $\frac{3}{4}$ lin. long ; style $\frac{3}{4}$ lin. long, reaching almost to the top of the anthers ; fruits oblong, scarcely 3 lin. long, greenish when dry, subconspicuously 5-ribbed, especially near the base, with nearly invisible intermediate ribs.

COAST REGION : George Div. ; without precise locality, *Bolus,* 2458 ! Uniondale Div. ; Long Kloof, 300 ft., *Schlechter,* 8399 !

109. **T. fimbriatum** (A. W. Hill in Kew Bulletin, 1915, 27) ; a small shrub, about 1 ft. high, branched from the base ; branches ascending, rather rough with the persistent leaf-bases, glabrous ; leaves linear, subacute, closely adpressed to the stem in the lower part, slightly recurved above, 3–4 lin. long, flat on the upper surface, slightly keeled below, glabrous ; flowers arranged in dense terminal capitate clusters 4–5 lin. in diam. ; bracts broadly ovate, acutely acuminate, longer than the flowers, with prominent membranous fimbriate margins ; bracteoles similar to the bracts but much narrower and shorter ; perianth $1\frac{2}{3}$ lin. long, with conspicuous external glands ; segments linear-lanceolate, $1\frac{1}{4}$ lin. long, with a conspicuous hood $\frac{1}{2}$ lin. long, bearded with a few short hairs ; anthers included in the perianth-tube, $\frac{1}{4}$ lin. long ; style $\frac{1}{4}$ lin. long, reaching to the middle of the anthers.

COAST REGION: Tulbagh Div. ; eastern base of the Roodezand Mountains, 490 ft., *Diels,* 1125 !

110. T. translucens (A. W. Hill in Kew Bulletin, 1915, 42) ; root erect, slender, greyish-white ; stems up to 18 in. high, branched from near the base ; branches erect or ascending, angular, glabrous ; leaves straight or incurved, ascending, linear-acicular, acute, 4–6 lin. long, about ⅓ lin. thick, flat and with a prominent midrib on the upper surface, keeled below, glabrous ; flowers arranged in small dense bracteate terminal heads 3–6 lin. in diam. ; bracts longer than the flowers, gradually differentiated from the upper leaves, lanceolate or linear-lanceolate, acutely acuminate, reddish, keeled, with membranous jagged margins about the middle, glabrous ; bracteoles somewhat shorter and narrower than the bracts ; perianth 2–2½ lin. long ; tube very short ; segments linear-lanceolate, acute, with horny terete translucent apices, with a very short beard at the base of the hood and finely papillose margins ; anthers included in the perianth-tube, ¼ lin. long ; stigma subsessile ; fruits not seen.

COAST REGION : Caledon Div. ; Houw Hoek, 2500 ft., *Schlechter*, 7580 ! near Caledon, *Bolus*, without number in Herb. Bolus ! Riversdale Div. ; summit of Kampsche Berg, *Burchell*, 7106 !

111. T. densiflorum (A.DC. Esp. Nouv. Thes. 7) ; a low erect shrublet, about 1 ft. high, branched from near the base or below the middle ; branches erect, subterete, very finely spotted ; leaves rather sparse, ½–1 in. long, linear-acicular, subacute, flat on the upper surface, rounded below, rigid, glabrous ; flowers in numerous small capitate subcorymbose clusters ; bracts ovate or ovate-lanceolate, acute or acuminate, keeled, shorter than the flowers ; bracteoles narrower and shorter than the bracts, with the latter often blackened ; perianth 1½ lin. long ; segments ovate-elliptic, flat, hooded, 1 lin. long, with a dense tuft of apical hairs, margins and inner surfaces more or less hairy ; anthers included in the tube, ¼ lin. long ; stigma sessile ; fruits ellipsoid, contracted at the base, 2½ lin. long, strongly 10-ribbed, distinctly reticulate between the ribs. *A.DC. in DC. Prodr.* xiv. 664 ; *Sond. in Flora,* 1857, 406. *T. densiflorum, var. Linkii, A.DC. in DC. Prodr.* xiv. 664.

COAST REGION : Worcester Div. ; Bains Kloof, 3500 ft., *Schlechter*, 9093 ! Cape Div. ; Table Mountain, *Bergius* ! mountain sides, near Simons Town, 1200 ft., *Bolus* ! Caledon Div. ; Zwart Berg, *Ecklon & Zeyher*, 4 ! near Genadendal, *Bolus*, 7423, partly !

Sonder considered *T. densiflorum,* A.DC., to be a young condition of *T. spicatum,* Linn. This, however, is not the case, as in *T. densiflorum* there is no ring of hairs at the throat of the perianth.

112. T. carinatum (A.DC. Esp. Nouv. Thes. 7) ; a small shrub, about 18 in. high ; stem erect, sparingly branched ; branches densely leafy, glabrous ; leaves mostly parallel with the stem, at length sometimes spreading, linear, acutely mucronate, flat above, keeled below, more or less triangular in section, 5–8 lin. long, finely scabrous on the margins and keel ; flowers stalked, in dense terminal

bracteate heads 4–5 lin. in diam. ; bracts ovate-lanceolate acutely
acuminate, sharply keeled, as long as the flowers, with broad membranous jagged margins in the middle third of their length, glabrous ;
bracteoles a little shorter and narrower than the bracts but other-
wise similar ; perianth $1\frac{1}{4}$–$1\frac{2}{3}$ lin. long ; segments lanceolate, subobtuse, $\frac{3}{4}$–$\frac{4}{5}$ lin. long, horned at the apex, with a dense apical
beard ; anthers included, at the base of the perianth-tube, $\frac{1}{4}$ lin.
long ; stigma sessile or nearly so ; fruits ellipsoid-globose, 3–$3\frac{1}{2}$ lin.
long, prominently 10-ribbed, shining and nearly smooth between the
ribs. *A.DC. in DC. Prodr.* xiv. 663. *T. capitatum,* β, *Thunb.*
Herb. ex A.DC. l.c., not of Linn. T. capitatum, Willd. Herb.
no. 5080, *ex A.DC. l.c., not of Linn. T. assimile, Sond. in Flora,*
1857, 360, 406.

VAR. β, **pallidum** (A. W. Hill) ; leaves paler when dry ; flowers in laxer heads
than in the type, with narrower less membranous bracts. *T. assimile, var. pallidum,*
Sond. in Flora, 1857, 360, 406.

COAST REGION : Clanwilliam Div. ; Packhuis Berg, 3000 ft., *Schlechter,* 8630 !
Koude Berg, 2500 ft., *Schlechter,* 8719 ! Northwards from "Stasi," 1830 ft.,
Diels, 785 ! Piquetberg Div. ; Piquetberg Mountain, 2200 ft., *Bolus,* 13646 ! Cape
Div. ; Simons Bay, *Wright,* 532 ! and without precise locality, *Harvey,* 590 !
Stellenbosch Div. ; Grietjes Gat, 2000–4000 ft., *Zeyher,* 3789 ! *Ecklon & Zeyher,*
3 ! Lowrys Pass, 1000 ft., *Schlechter,* 4837 ! Caledon Div. ; near Vogelgat,
2000 ft., *Schlechter,* 10419 ! near Houw Hoek, 900–2000 ft., *Bolus* ! George Div. ;
Cradock Berg, near George, *Burchell,* 5946 ! Uitenhage Div. ; Winterhoek
Mountain, 2000 ft., *Marloth* ! Var. β : Tulbagh Div. ; near Tulbagh Waterfall,
1000–2000 ft., *Ecklon & Zeyher,* 3 ! Worcester Div. ; Dutoits Kloof, *Drège,* o !
Drakensteen Mountains, near Bains Kloof, 1800 ft., *Bolus,* 4065 ! Breede River
valley, *Bolus,* 2937 ! Cape Div. : near Simonstown, *Wolley-Dod,* 3018 ! Lion
Mountain, *Burchell,* 296 ! and without precise locality, *Bergius* ! *Hooker,* 610 !
Caledon Div. : ; Houw Hoek, *Bolus* !

CENTRAL REGION : Prince Albert Div. ; Zwartberg Pass, 5300 ft., *Bolus,* 11632 !

113. **T. penicillatum** (A. W. Hill in Kew Bulletin, 1915,
37) ; stems erect, longitudinally sulcate, stout, woody, glabrous,
branched in the upper part ; branches erect, subcorymbose ; leaves
linear, acutely mucronate, 1–$1\frac{1}{4}$ in. long, about 1 lin. broad, keeled
below, glabrous, becoming rather coarsely wrinkled when dry ;
flowers in fairly dense terminal corymbs up to 1 in. in diam. ;
bracts linear-oblong, subacute, keeled, as long as the flowers, up to
1 lin. broad, purplish, with subtranslucent margins ; bracteoles
about three-quarters the length of the bracts ; perianth $1\frac{1}{4}$ lin.
long ; segments linear-lanceolate, subacute, about $1\frac{1}{4}$ lin. long, flat,
hooded, with the apex and margin clothed with long hairs ; anthers
$\frac{1}{8}$ lin. long, included in the perianth-tube and inserted in front of
a large tuft of orange-coloured hairs, which are not united to form
a definite ring and are free from the anthers ; style $\frac{1}{4}$ lin. long or
almost absent ; fruit with a short stout stalk, ellipsoid, 3 lin.
long, rather prominently 5-nerved, with inconspicuous intermediate
nerves, conspicuously transversely reticulate between the ribs.

COAST REGION : George Div. ; Cradock Berg, 2500 ft., *Galpin,* 4546 ! Humansdorp Div. ; Storms River, 250 ft., *Schlechter,* 5986 !

114. T. micropogon (A.DC. Esp. Nouv. Thes. 6) ; root perpendicular, wavy, about 2 lin. thick at the apex ; stems several or few from the apex of the root, simple or subsimple, about 6 in. long, moderately leafy, about $\frac{1}{2}$ lin. thick, glabrous; leaves slender, acicular, with slightly recurved subacute tips, 3–5 lin. long, about $\frac{1}{4}$ lin. thick or less, glabrous, black when dry ; flowers arranged in fairly dense slender spikes $\frac{1}{2}$–1 in. long; bracts shorter than the flowers, lanceolate, acutely acuminate, slightly keeled on the back, glabrous, fleshy ; bracteoles a little shorter than the bracts but otherwise very similar ; perianth fleshy, urceolate, about $\frac{3}{4}$ lin. long, with prominent external glands and a ring of inconspicuous throat hairs ; segments lanceolate, $\frac{2}{5}$ lin. long, hooded, fleshy, margins papillose, but not bearded ; disc fairly conspicuous ; anthers exserted from the perianth-tube, $\frac{1}{3}$–$\frac{1}{2}$ lin. long, without attachment hairs ; stigma sessile or supported on a very stout short 3-cornered style ; fruits not seen. *A.DC. in DC. Prodr.* xiv. 669. *T. patentiflorum, Sond. in Flora,* 1857, 357, 406.

COAST REGION : Caledon Div. ; Zwart Berg, near the hot springs, 1000–2000 ft., *Ecklon & Zeyher,* 12!

115. T. urceolatum (A. W. Hill in Kew Bulletin, 1915, 43) ; stems woody, glabrous ; branches spreading, angular, glaucous ; leaves stout, semicircular in section, flat or slightly concave above, linear, subacutely mucronate, $\frac{1}{2}$–$\frac{3}{4}$ in. long, about $\frac{1}{4}$ lin. broad, fleshy, somewhat glaucous, glabrous ; flowers rather large and at first arranged in crowded spikes, at length lax, usually 3 together ; bracts shorter than the flowers, linear-lanceolate or oblanceolate, acutely mucronate, thick and fleshy, boat-shaped ; bracteoles about two-thirds the length of the bracts but otherwise similar ; perianth 1$\frac{1}{2}$ lin. long ; segments $\frac{2}{3}$–$\frac{3}{4}$ lin. long, ovate, subacute, with a dense apical tuft of hairs ; anthers included in the perianth-tube, about $\frac{1}{4}$ lin. long, with a ring of hairs in the tube at the level of their insertion ; style stout, $\frac{1}{8}$–$\frac{1}{4}$ lin. long, reaching to the base of the anthers ; fruits ellipsoid, 2$\frac{1}{4}$ lin. long including the persistent perianth, prominently 10-ribbed, glaucous grey when mature, transversely wrinkled between the nerves.

CENTRAL REGION : Calvinia Div. ; Nieuwoudtville, *Leipoldt in Herb. Bolus,* 9377 !
WESTERN REGION : Little Namaqualand ; on hills near Brakdam, 2000 ft., *Schlechter,* 11138 !

116. T. patulum (A. W. Hill in Kew Bulletin, 1915, 37) ; stems divaricately branched, erect, glabrous ; branches slender, elongated, straight, spreading from the stem at an angle of about 45° ; leaves present near the forks of the main branches, acicular, acute, 3$\frac{1}{2}$–5 lin. long, very slender, sometimes with a few short teeth on the back or on the margins, glabrous ; upper leaves each subtending and slightly adnate to a branch, recurved, linear or linear-lanceolate, subacute, up to 2 lin. long, glabrous ; flowers loosely arranged in elongated flexuous spikes, solitary or 3 together; bracts much shorter than the flower, lanceolate, subacute, fleshy, about 1 lin. long, glabrous; bracteoles about three-quarters the length of the bracts but otherwise similar; perianth 1–1¾ lin. long, with conspicuous external glands; segments lanceolate, subacute, 1–1¼ lin. long, the margins and the apex with long stout hairs, with a well-marked ring of hairs at the throat of the tube; anthers exserted from the perianth-tube, ⅔ lin. long; filaments as long as the anthers; style stout, ¼ lin. long; fruits top-shaped at the base, subglobose, 3¼ lin. long, fairly prominently 10-ribbed, faintly reticulate between the ribs. *T. funale, var. caledonicum, Sond. in Flora,* 1857, 359; *A.DC. in DC. Prodr.* xiv. 668.

COAST REGION: Malmesbury Div.; near Moorees Berg, 500 ft., *Bolus,* 9981! Zwartland and region of Berg River streams, *Ecklon & Zeyher,* 51! Paarl Div.; near Paarl, 300 ft., *Schlechter,* 9207! Cape Div.; Devils Peak, *Bergius!* and without precise locality, *Harvey,* 711, partly!

117. **T. funale** (Linn. Sp. Pl. ed. ii. 302); a very slender woody undershrub branched from the base; branches terete, closely sulcate, glabrous; leaves very few or almost entirely absent, or sometimes especially on the young annual plants fairly numerous, acicular, acute or subacute, slender, ⅓–¾ in. long, glabrous; flowers small arranged in rather lax elongated spikes; bracts narrowly lanceolate, with acute blackish tips, equal in length to the peduncle, glabrous; bracteoles 2, rather shorter than but very similar to the bracts, peduncle stout, perianth more or less urceolate, ¾–1 lin. long, with external glands, and a ring of hairs at the throat; segments ovate-elliptic to linear, subacute, ⅔–¾ lin. long, with a dense apical beard; anthers exserted, ¼ lin. long; stigma sessile; fruits ellipsoid-globose, 1¼ lin. long, strongly 10-ribbed, prominently reticulate between the ribs. *Thunb. Diss. Thes. 7; Fl. Cap. ed. Schult.* 209; *Sond. in Flora,* 1857, 358, *partly; A.DC. in DC. Prodr.* xiv. 668. *T. adpressifolium, Sond. in Flora,* 1857, 358.

COAST REGION: Tulbagh Div.; New Kloof, *Schlechter,* 9034! Worcester Div.: near Bains Kloof, 800 ft., *Bolus,* 2928! Cape Div.; flats and hills around Cape Town, *Thunberg; Burchell,* 964! *Bergius! Ludwig,* 41! *Anderson! Drège! Ecklon & Zeyher,* 49! *Zeyher,* 3810! *Wright,* 533! *Bolus,* 2934! 3871! 3919! 7126! and *Herb. Norm. Austr.-Afr.,* 1362! *Wolley-Dod,* 2142! 2255! 2742! *Schlechter,* 243! 358! *Wilms,* 3608! 3609! Bredasdorp Div.; Elim, 150 ft., *Schlechter,* 9662! Koude River, 700 ft., *Schlechter,* 10451! Swellendam Div.; hills near the Zondereinde River, *Zeyher,* 3803! near Swellendam, *Bolus,* 8092! Riversdale Div.; Garcias Pass, *Burchell,* 7018! George Div.; near George, *Burchell,* 6003! *Schlechter,* 2469! Zwartberg Range, *Bolus,* 2456!

118. **T. macrostachyum** (A.DC. Esp. Nouv. Thes. 6, partly); branches elongated, erect, fairly stout, subterete, glabrous; leaves lanceolate or linear-lanceolate, acute or subacute, 3–4 lin. long, flat and fleshy, glabrous; flowers solitary or 2–3 together in the axils of the bracts, very shortly pedunculate; bracts leaf-like, shortly adnate to the peduncle, lanceolate, acutely acuminate, up to 2 lin.

long, rather rigid, glabrous ; bracteoles very small ; perianth 1⅓ lin. long, urceolate, with conspicuous external glands and a ring of hairs at the throat; segments linear, obtuse, ¾ lin. long, with an apical beard of stiff straight hairs; anthers exserted from the perianth-tube, ¼ lin. long; stigma sessile ; fruits not seen. *A.DC. in DC. Prodr.* xiv. 669, *excl. Gueinzius,* 407. *T. sparteum, Sond. in Flora,* 1857, 358, 406.

SOUTH AFRICA : without precise locality, *Harvey,* 711 ! *Penther* !
COAST REGION : Clanwilliam Div. ; near Brakfontein, *Ecklon & Zeyher,* 7 !
Worcester Div. ; Breede River valley, near Bains Kloof, 800 ft., *Bolus,* 2928 !
Cape Div. ; near Rondebosch, Cape Dunes, below 100 ft., *Bolus,* 3919 !

119. **T. diversifolium** (Sond. in Flora, 1857, 359); stem finely ribbed, woody, glabrous ; branches short, ascending ; leaves de-current on the stem, rather densely arranged, at length recurved, the lower linear, subterete, acute, 6–7 lin. long, about ⅓ lin. thick, the upper linear-lanceolate, subacute, 1½–2 lin. long, flat on the upper surface, slightly keeled below, sometimes very minutely glandular-puberulous on the margins and midrib, otherwise glabrous, thick and fleshy ; flowers arranged in dense ovoid spikes ⅓–½ in. long ; bracts purplish, linear-lanceolate, acutely acuminate, all except the lowest shorter than the flowers, with narrowly membranous minutely serrulate margins, glabrous ; bracteoles similar to the bracts but a little shorter and narrower ; perianth 2 lin. long, with conspicuous external glands and a ring of throat hairs ; segments linear-lanceolate, acute, 1½ lin. long, hooded, with a beard of stout straight apical hairs and hairy margins ; anthers exserted, at the base of the perianth-segments, ¼ lin. long ; stigma sessile ; fruits globose (*Sonder*), ribbed, equalling in length the persistent perianth. *T. spicatum, A.DC. in DC. Prodr.* xiv. 668, *partly.*

COAST REGION : Tulbagh Div. ; mountain plains near Tulbagh Waterfall, 1000–2000 ft., *Ecklon & Zeyher,* 5, partly !

120. **T. aggregatum** (A. W. Hill in Kew Bulletin, 1915, 22); stems and branches ascending, subterete, glabrous ; leaves small, flat, adpressed to the stem, lanceolate or linear-lanceolate, acute, 1½–2 lin. long, rigid, blackish when dry, glabrous ; flowers crowded in terminal heads, the latter subcapitate to oblong-linear ; bracts ovate, acutely acuminate, with a broad fleshy midrib and green subtranslucent jagged-denticulate margins, glabrous ; bracteoles as long as, but much narrower than the bracts, otherwise similar ; perianth 1½–2 lin. long, with a ring of hairs within the throat ; segments about 1⅜ lin. long, linear, subacute, with a stiff stout comb-like apical beard ; anthers exserted from the perianth-tube, ¼ lin. long ; stigma sessile ; fruits ellipsoid, 3 lin. long including the persistent perianth, prominently 10-ribbed, rather delicately reticulate between the ribs.

SOUTH AFRICA : without locality, *Osbeck* ! *Wallich* !
COAST REGION : Vanrhynsdorp Div. ; Windhoek, 1000 ft., *Schlechter*, 8348 !
Clanwilliam Div. ; Lammskraal, 1150 ft., *Diels*, 779 ! Malmesbury Div. ; near
Hopefield, *Bachmann*, 1694 ! 1695 ! near Darling, 100 ft., *Schlechter*, 5337 ! Cape
Div., 80–1000 ft. ; near Cape Town, 100 ft., *Bolus, Herb. Norm. Austr.-Afr.*,
1360 ! *Ecklon*, 793 ! Wynberg, *Bolus*, 2931 ! *Schlechter*, 215 ! 7545 ! Muizenberg,
Bolus, 2933, partly ! Kenilworth, *Bolus*, 7049 ! Herzog House Retreat, *Wolley-
Dod*, 2364 ! near Vygeskraal, *Wolley-Dod*, 2371 ! Liesbeck River, *Bergius* !
Zeyher, 793 ! 4879 !

121. **T. spicatum** (Linn. Mant. Alt. 214); stems simple or
branched, grooved or angular, glabrous; branches erect or suberect; lower leaves erect, acicular, acute, up to 2½ in. long, acute, about
⅗ lin. thick, terete except for a narrow groove on the upper surface,
glabrous; upper leaves adpressed to the stem, with spreading tips,
linear-lanceolate, acute, 3–4 lin. long, concave on the upper surface,
rigid, glabrous ; flowers arranged in dense thick terminal cylindrical
spikes ¾–2 in. long; bracts as long as or shorter than the flowers,
obovate-oblanceolate, acutely acuminate, 2–3 lin. long, 1–1½ lin.
broad, with a black thick fleshy keel and rather membranous
reddish margins, glabrous ; bracteoles linear or linear-lanceolate,
as long as, but much narrower than the bracts ; perianth 1½–2 lin.
long, with a ring of hairs at the throat ; segments linear-lanceolate,
subacute, nearly 2 lin. long, with a stiff adpressed beard of hairs ;
anthers exserted from the perianth-tube, ⅓ lin. long ; stigma sessile.
Sond. in Flora, 1857, 360 ; *A.DC. in DC. Prodr.* xiv. 668, *excl.
Ecklon, Zeyher & Drège and syns. T. diversifolium and T. subnudum,
Sond.*

SOUTH AFRICA : without locality, *Otto* ! *Chamisso*, 14c ! *specimen in the Linnean
Herbarium* !
COAST REGION : Cape Div. ; Devils Peak, *Bergius* ! Table Mountain, *Burchell*,
550 ! 619 ! 1500 ft., *Bolus*, 4206 β ! Liesbeck River, *Bergius* ! Simons Bay,
Wright ! *Ringgold & Rogers* ! Caledon Div. ; near Genadendal, 3400–4600 ft.,
Bolus, 7423 ! without precise locality, *Thom* !

122. **T. bathyschistum** (Schlechter in Engl. Jahrb. xxvii. 116);
stems erect, about 9 in. high, branched in the upper part, glabrous ;
branches erect ; lower leaves acicular, subterete, obtuse, ¾–1¼ lin.
long, ½ lin. thick, fleshy, glabrous ; flowers arranged in short dense
capitate or subcapitate spikes ; heads composed of 3-flowered cymes ;
bracts lanceolate or a few of the lower ovate-lanceolate, with subacute often blackish tips, 2–3 lin. long, rather sharply keeled, with
slightly membranous entire margins, glabrous ; bracteoles linear or
linear-lanceolate, acute, usually nearly as long as the bracts ;
perianth 1¾ lin. long, with a ring of hairs in the throat ; segments
linear or linear-lanceolate, subacute, hooded, 1½ lin. long, with a
beard of stout stiff hairs within the hood, and hairy margins, and a
ring of short hairs at the throat of the tube ; anthers exserted,
inserted at the base of the perianth-segments, ¼ lin. long ; stigma
sessile ; fruits subglobose, capped by the long persistent perianth,
2½ lin. long, distinctly 10-ribbed, slightly reticulate between the ribs.

COAST REGION : Caledon ; Onrust River, 3000 ft., *Schlechter,* 9490 !
T. spicatum, β, in herb. Thunb. very probably should be referred to this
species.

123. **T. subnudum** (Sond. in Flora, 1857, 360) ; stems and
branches elongated, very slender, finely sulcate, glabrous, almost
leafless in the upper parts ; with a few lower leaves long and
acicular, $\frac{3}{4}$–2 in. long, $\frac{1}{3}$–$\frac{1}{2}$ lin. thick, subterete, glabrous ; flowers
in rather slender and often elongated terminal spikes ; bracts linear
or linear-lanceolate, acute, about as long as the flowers, keeled ;
bracteoles similar to the bracts but shorter ; perianth $\frac{3}{4}$–1$\frac{1}{4}$ lin.
long, with a ring of hairs in the throat ; segments linear-lanceolate,
acute, hooded, $\frac{3}{4}$–1 lin. long, with an apical beard of few stout hairs
and hairy margins ; anthers exserted, at the base of the perianth-
segments, $\frac{1}{4}$ lin. long ; stigma sessile ; fruits ovoid-globose, 2 lin.
long, subconspicuously 10-ribbed, very slightly reticulate between
the ribs. *T. spicatum, E. Meyer in Drège, Zwei Pfl. Documente,* 82,
not elsewhere ; A.DC. in DC. Prodr. xiv. 668, *partly, not of Linn. ;
Sond. in Flora,* 1857, 406 *in obs. T. funale, Sond. in Flora,* 1857,
358, *partly, not of Linn.*

VAR. β, **foliosum** (A. W. Hill in Kew Bulletin, 1915, 42) ; stems with scattered
leaves on upper portions.

COAST REGION : Clanwilliam Div. ; Olifants River Valley, *Stephens,* 7304 !
Tulbagh Div. ; Tulbagh Waterfall, *Ecklon & Zeyher,* 56, partly ! New Kloof,
500 ft., *Schlechter,* 9034 ! Worcester Div. ; Dutoits Kloof, *Drège* a ! Paarl Div. :
Paarl Mountain, 600 ft., *Bolus,* 2929 ! Riversdale Div. ; Garcias Pass, 1000 ft.,
Galpin, 4549 ! Oudtshorn Div., *Miss Britten,* 84 ! Uniondale Div. ; Zwartberg
Range near Avontuur, *Bolus,* 2457 ! Uitenhage Div. ; Vanstadens Berg, *Ecklon &
Zeyher,* 6 ! Var. β : Bredasdorp Div. ; Elim, 500 ft., *Schlechter,* 7694 ! Humans-
dorp Div. ; Kruisfontein, 500 ft., *Galpin,* 4550 ! Port Elizabeth Div. ; Port
Elizabeth, *Bolus* !

124. **T. elatius** (Sond. in Flora, 1857, 355) ; a fleshy shrub
several feet high ; branches flexuous, terete, glabrous ; leaves linear,
boat-shaped, keeled, acute or subacute, 3–7 lin. long, about 1 lin.
broad, fleshy, glabrous, at length spreading or recurved ; flowers
white, arranged in rather short or subcapitate spikes ; bracts shorter
than the flowers, lanceolate or linear-lanceolate, subacute, 2–3 lin.
long, at first adjacent to the flower, at length spreading or recurved,
thick and fleshy, sharply keeled, glabrous ; bracteoles 2, soon falling
off, linear, acute, a little shorter than the bracts ; flowers large and
solitary within the bracts, with a short stout pedicel ; perianth
1$\frac{3}{4}$–2 lin. long, with conspicuous external glands ; segments
lanceolate, acute, 1–1$\frac{1}{4}$ lin. long, with a dense woolly apical beard
of long hairs and a dense ring of short stiff brown hairs in the
throat of the tube ; anthers in the tube mainly exserted or partially
included, about $\frac{3}{5}$ lin. long ; style nearly $\frac{1}{2}$ lin. long, the stigma below
the base of the anthers ; fruits contracted at the base, subglobose,
nearly 3 lin. long including the persistent perianth, not very con-

spicuously 10-ribbed, verrucose-reticulate between the ribs. *A.DC. in DC. Prodr.* xiv. 670.

COAST REGION: Van Rhynsdorp Div. ; near the Olifants River at Driefontein, *Zeyher,* 1503 ! Ebenezer, 100 ft., *Diels,* 515 ! Karee Berg, 1500 ft., *Schlechter,* 8216 ! Calvinia Div. ; Onder Bokkeveld, at Papelfontein, 2200 ft., *Schlechter,* 10891 !

125. **T. Frisea** (Linn. Mant. Alt. 213) ; stems fairly slender, ascending, often much-branched ; branches sulcate, glabrous ; leaves linear-acicular, acute, $\frac{1}{4}$–$1\frac{1}{4}$ in. long, about $\frac{1}{3}$ lin. thick, glabrous ; flowers arranged in fairly dense spikes of cymules ; bracts free from the peduncle, linear, acute, up to 2 lin. long, rather sharply keeled below, concave above ; bracteoles about half the length of the flowers, with very narrowly membranous finely toothed margins ; perianth 1–1$\frac{1}{4}$ lin. long, with conspicuous external glands ; segments elliptic, flat, subacute, $\frac{3}{4}$ lin. long, more or less hooded, with a dense woolly apical beard and a ring of throat hairs ; anthers exserted, $\frac{1}{3}$ lin. long ; style $\frac{1}{8}$–$\frac{1}{4}$ lin. long, reaching to the level of the ring of hairs ; fruits ellipsoid-globose, 1$\frac{1}{2}$ lin. long, distinctly 10-ribbed, strongly reticulate between the ribs. *Sond. in Flora,* 1857, 359 ; *A.DC. in DC. Prodr.* xiv. 667. *T. debile, R. Br. Prodr.* 353, *name only* ; *Spreng. Syst.* i. 830 ; *A.DC. l.c.* *T. debile, var. Ecklonis, A.DC. l.c.* *T. amblystachyum, A.DC. Esp. Nouv. Thes.* 6, *and in DC. l.c.* 668. *T. monticolum, Sond. in Flora,* 1857, 359, *and* 406.

VAR. β, **Thunbergii** (A.DC. in DC. Prodr. xiv. 667) ; stems mostly procumbent and subsimple, up to 10 in. long ; leaves much broader and more fleshy than in the type. *T. Frisea, Thunb. Diss. Thes.* 10 ; *Fl. Cap. ed. Schult.* 210, *not of Linn.* *T. crassifolium, R. Br. Prodr.* 353 ; *A.DC. in DC. Prodr.* xiv. 672, *name only.* *T. debile, var. humile, A.DC. in DC. l.c.* 667.

SOUTH AFRICA : without locality, *Banks in herb. Mus. Brit.* !
COAST REGION : Clanwilliam Div. ; Olifants River Mountains, 840 ft., *Schlechter,* 5090 ! Piquetberg Div. ; near Porterville, 600 ft., *Schlechter,* 10738 ! near Piquetberg Road Station, 300 ft.. *Diels,* 169 ! Malmesbury Div. ; Coeraten Berg, near Hopefield, *Bachmann,* 1697 ! near Mamre, 300 ft., *Bolus,* 4329 ! Tulbagh Div. ; near Tulbagh Waterfall, *Ecklon & Zeyher,* 10 ! near Saron, 1000 ft., *Schlechter,* 10623 ! Cape Div. ; region around Cape Town, *Koenig in the Linnean Herbarium* ! *Osbeck* ! *Forster* ! *Bergius* ! *Ecklon,* 795 ! *Harvey* ! 714 ! 723 ! *Bolus,* 2930 ! 3927 ! 4329B ! 7047 ! *Sehlechter,* 240 ! 652 ! *Wilms,* 3607 ! *Wolley-Dod,* 2658 ! 2920 ! *Drège e in Herb. Kew and Stockholm* ! Riversdale Div. ; near Riversdale, 550 ft., *Bolus,* 11377 ! Knysna Div. : Vlugt, *Bolus,* 2459 ! Var. β : Piquetberg Div. ; Verlooren Valley, *Thunberg* ! Malmesbury Div. ; near Hopefield, *Bachmann,* 1696 ! Berg River, *Ecklon & Zeyher,* 45 ! *Zeyher,* 323 ! Cape Div. ; Raapenberg Vley, *Wolley-Dod,* 3359 ! near Sea Point, *Wolley-Dod,* 1735 ! Cape Flats, *Schmieterloh,* 202 ! and without precise locality, *Sparmann* ! *Zeyher,* 155 !

126. **T. annulatum** (A. W. Hill in Kew Bulletin, 1915, 23) ; a small plant probably only up to 4 in. high, branched from near the base ; branches few, ascending, subterete, glabrous, clothed in the lower part with a few persistent blackened leaves ; leaves fairly densely arranged, subparallel to the stem, linear or linear-lanceolate,

acute, 2–3 lin. long, concave or flat on the upper surface, rounded on the lower, thick and fleshy, glabrous; flowers arranged in small dense subglobose heads 3–4 lin. in diam.; bracts linear-oblong, acute, nearly as long as the flowers, fleshy, slightly keeled; bracteoles nearly as long as the bracts but narrower; perianth 1¼ lin. long, with conspicuous external glands and a ring of throat hairs; segments ovate, obtuse, ¾ lin. long, flat, with a dense apical woolly beard; anthers ⅕ lin. long; filaments attached in the tube, but the anthers slightly exserted; stigma sessile; fruits not seen.

COAST REGION: Worcester Div.; Matroosberg, 6500 ft., *Marloth*, 2252 (in Herb. Marloth, not of Herb. Bolus)!

127. **T. brachygyne** (Schlechter in Engl. Jahrb. xxvii. 117); stems very slender, branched, ⅓ lin. thick, slightly angular, glabrous; branches ascending, sparingly leafy : leaves scattered, slender, subterete, acute, 3–6 lin. long, about ⅙ lin. thick, distinctly warted when dry, otherwise glabrous; flowers arranged in small globose terminal clusters about 2 lin. in diam., often with a younger shoot from just below one side of the older clusters; bracts shorter than the flowers, narrowly lanceolate, acutely acuminate, with slightly membranous entire margins, glabrous; bracteoles reaching to the base of the perianth, subulate, acute; perianth ¾ lin. long, with a ring of hairs in the throat; segments oblong-lanceolate, subacute, ⅔ lin. long, rather densely bearded with woolly hairs; anthers exserted from the perianth-tube, about ⅛ lin. long; stigma sessile; fruits ovoid-globose, stipitate, 1½ lin. long, including the stipe and persistent perianth, distinctly 10-ribbed and reticulate.

COAST REGION: Paarl Div.; mountains near French Hoek, 2800–3800 ft., *Schlechter*, 9247, 9297!

128. **T. Patersonæ** (A. W. Hill in Kew Bulletin, 1915, 36); stem about 9 in. high, sparingly branched from the base or near the base, slightly sulcate, glabrous; branchlets ascending; leaves linear, acute, subterete or sometimes somewhat flattened on the upper surface, 2–8 lin. long, thick, glabrous; flowers in 3–5-flowered cymules arranged in dense terminal ovoid spikes about ½ in. long; bracts linear-lanceolate, acute, shorter than the flowers, greenish when dry, glabrous, with narrow membranous entire margins; bracteoles much shorter than the bracts; perianth urceolate, 1¼ lin. long, with conspicuous external glands and a ring of throat hairs; segments lanceolate, subacute, ¾ lin. long, with a dense apical beard of woolly hairs and hairy margins; anthers ¼ lin. long, exserted; stigma sessile; fruits not seen.

COAST REGION: Port Elizabeth Div.; Walmer, *Mrs. Paterson*, 682! 792!

II. THESIDIUM, Sond.

Flowers diœcious.   *Male flowers*: Perianth hypocrateriform; tube short, slender, continuous with the solid receptacle; segments spreading, with a bundle of hairs arising from their base and attached to the back of the anthers. *Stamens* 4; filaments short, inserted at the base of the perianth-segments; anthers small, cells parallel, dehiscing longitudinally. *Style* rudimentary or more often absent. *Female flowers*: receptacle ovoid, adnate to the ovary. *Perianth-tube* very shortly campanulate or scarcely evident; segments 4, rarely 5. *Disc* usually obscure. *Staminodes* rarely present. *Ovary* inferior; style short; stigma obscurely 2–3-lobed; ovules 2–3, pendulous from the apex of a central straight or folded filiform placenta. *Fruit* a nut as in *Thesium* with a fleshy basal ring, small, globose or ovoid, crowned by the persistent perianth, 5-ribbed and conspicuously reticulate, sometimes pitted; endocarp crustaceous. *Embryo* terete in the centre of fleshy albumen, often oblique.

Low shrublets or herbs, much-branched, semiparasitic; leaves alternate, often very small or squamiform, rigid; flowers very small, subsessile, solitary or in 2–3-flowered cymules in the axils of bracts, arranged in slender terminal spikes; bracts and bracteoles in the male plants usually small, in the female often conspicuous, frequently hispid or scabridulous.

DISTRIB. Species 7 or 8, all South African.

Male and female plants similar. Bracts and bracteoles
  scale-like, fleshy:
  Plants very slender; branches delicate, flexuous, pros-
    trate or erect; flowers distant; bracts very small;
    fruit subsessile   ...   . (1) **exocarpæoides.**
  Plants fairly stout; branches flexuous, prostrate;
    flowers distant; bracts keeled, broadly ovate-
    lanceolate, navicular; fruit stipitate   ...   ... (2) **Thunbergii.**
  Plants stout; branches erect, intricate, crowded; flowers
    crowded; bracts with prominent apical keel-wing;
    fruit subsessile   ...   . (3) **fragile.**
Male and female plants dissimilar. Male plants with
  subulate or ovate-lanceolate, herbaceous bracts.
  Female plants with conspicuous leafy bracts:
Male plants:
  Stems and bracts conspicuously hairy; bracts
    crowded, imbricate, curved, longer than the usually
    solitary flowers   ...   . (4) **hirtum.**
  Stems and bracts scabridulous or subglabrous; bracts
    usually spreading:
    Bracts shorter than or subequal to flowers, curved;
      plants slender, subglabrous, annual   ...   ... (5) **minus.**
    Bracts twice-as long as flowers, spreading; woody
      subshrubs:
      Bracts and stem-ribs verruculose or subsca-
      bridulous; bracts keeled; leaves 2–4 lin. long (6) **fruticulosum.**

Bracts and stem-angles conspicuously scabridulous
or scabrid-fimbriate ; bracts winged ; leaves
¾-1 in. long   ...   . (7) **longifolium.**
Female plants :
Stems and bracts densely hairy ; bracts crowded,
imbricate ; plants 4-5 in. high   ...   . (4) **hirtum.**
Stems and bracts verruculose, scabridulous or subglabrous :
Bracts and stems subglabrous ; slender annuals   ... (5) **minus.**
Bracts and stems verruculose or scabridulous ; woody
subshrubs :
Bracts lanceolate, 2-3 lin. long, horny ; flowers
in 2-flowered cymules ; leaves 2-4 lin. long... (6) **fruticulosum.**
Bracts linear-lanceolate, 6-8 lin. long, herbaceous ;
flowers solitary ; leaves ¾-1 in. long ...   ... (7) **longifolium.**

1. **T. exocarpæoides** (Sond. in Flora, 1857, 365) ; rootstock
stout, woody ; stems erect, woody below, stouter in the female
plants, glabrous, green, terete, ribbed, 10-15 in. high ; branches
and branchlets elongate, slender, flexuous especially in the male
plants, conspicuously ribbed, becoming subangled towards the
apex, ribs horny, translucent ; leaves reduced to fleshy subrotund
scales ; flowers solitary, regularly and somewhat sparsely arranged
in elongated flexuous spikes ; bracts and bracteoles broadly subrotund, fleshy, keeled, about ¼ lin. long, much shorter than the
flowers, margins membranous and fimbriate ; male flowers : perianth
about ⅜ lin. long ; tube saucer-like ; segments ovate, spreading,
¼ lin. long ; anthers exserted ; female flowers : perianth ⅜ lin. long ;
segments lanceolate, subacute, margins infolded, subspreading ;
style ⅛ lin. long ; fruit subsessile, globose, ½-¾ lin. long, with conspicuous reticulations. *Thesium microcarpum, A.DC. Esp. Nouv.
Thes.* 5 ; *Sond. in Flora*, 1857, 405. *Thesidium microcarpum, A.DC.
in DC. Prodr.* xiv. 674.

Coast Region : Swellendam Div. ; Hessaquas Kloof, *Zeyher*, 3814 ! Riversdale Div. ; hills near the Gouritz River, 200 ft., *Schlechter*, 5717 ! Albany Div. ;
Bothas Berg, near Grahamstown, 2200 ft., *MacOwan*, 1188 !

2. **T. Thunbergii** (Sond. in Flora, 1857, 364) ; a subshrub, more
or less prostrate, entirely glabrous ; stems 10-15 in. long, ribbed,
angled above, with numerous lateral flexuous branches ; branches
conspicuously ribbed or angled, subtriquetrous, angles horny, translucent ; leaves reduced to fleshy subulate or linear-subulate scales
1-1½ lin. long ; flowers remote on the axillary branchlets ; bracts
subrotund, navicular, subacute or obtuse, fleshy, ½ lin. long ;
bracteoles broadly ovate-lanceolate, acute, navicular, margins membranous, subimbricate, sharply keeled, all shorter than the flowers ;
male flowers solitary in the bract-axils ; perianth ½ lin. long ; segments ¼ lin. long, spreading ; female flowers solitary ; perianth
¼ lin. long ; segments erect, flat, ovate ; disc more or less conspicuous ; style ⅛ lin. long ; fruit ovoid-globose, about ¾ lin. in

diam., pedicellate; pedicel $\frac{3}{4}$ lin. long. *Thesium fragile, Linn. fl. Suppl.* 162; *Murr. Syst. Veg. ed.* xiv. 250; *Willd. Sp. Pl.* i. ii. 1215; *Thunb. Fl. Cap. ed. Schult.* 208, *as to Thunb. Herb. fol. a, Thesium podocarpum, A.DC. Esp. Nouv. Thes.* 5; *Sond. in Flora*, 1857, 405. *Thesidium podocarpum, A.DC. in DC. Prodr.* xiv. 674.

South Africa: without locality, *Osbeck*, 159!

Coast Region: Malmesbury Div.; sand hills at Saldanha Bay or St. Helena Bay, *Thunberg*! Cape Div.; Cape Sand Dunes, *Ecklon & Zeyher*, 29! Knysna Div.; sand hills of Plettenbergs Bay, *Burchell*, 5322! Homtini Pass, 800 ft., *Galpin*, 4553! Uitenhage Div.; strand near Cape Recief, *Ecklon & Zeyher*, 31! *Zeyher*, 642! Port Elizabeth, *Sim*, 2669!

3. **T. fragile** (Sond. in Flora, 1857, 364); a shrub or subshrub, stout, woody, erect, 12–18 in. high, much-branched, entirely glabrous; branches erect or subspreading, dense, intricate, more or less densely covered with flowers, ribbed or subangled; leaves scale-like, fleshy, broadly subrotund, with a thickened keel-wing at the apex; inflorescences more or less crowded; bracts and bracteoles broadly triangular, ovate, navicular, curved, abruptly subacute or acute, with a prominent fleshy keel at the apex, about $\frac{1}{2}$ lin. long; bracteoles with membranous slightly fimbriate margins, $\frac{1}{2}$–$\frac{5}{8}$ lin. long; male flowers: perianth $\frac{3}{4}$ lin. long; tube well marked; segments ovate, obtuse or subacute, hooded, $\frac{3}{8}$ lin. long; anthers $\frac{1}{12}$ lin. long; filaments $\frac{1}{8}$ lin. long; style $\frac{1}{4}$ lin. long, sometimes present, but no rudiment of ovary; female flowers: perianth $\frac{5}{8}$–$\frac{3}{4}$ lin. long; segments $\frac{3}{8}$ lin. long; disc well marked; style stout, $\frac{1}{8}$ lin. long; fruit sessile or subsessile, ovoid, 1–1$\frac{1}{4}$ lin. long, $\frac{3}{4}$–1 lin. broad with conspicuous ribs. *A.DC. in DC. Prodr.* xiv. 674; *Marloth, Fl. S. Afr.* 161, *t.* 37, *fig.* B. *Thesium fragile, Thunb. Herb. fol.* β.

South Africa: without locality, *Drège*! *Harvey*, 539!

Coast Region: Malmesbury Div.; sand hills at Saldanha Bay or St. Helena Bay, *Thunberg*! Cape Div.; Cape dunes, *Ecklon & Zeyher*, 29! Muizenberg, sandy hills near the sea, *Bolus*, 4926! beyond Retreat, near the railway, *Wolley-Dod*, 3653! False Bay, *Marloth*! Kalk Bay, *Bolus*, 4759! Caledon Div.; without precise locality, *Thom*, 963! Riversdale Div.; Garcias Pass, 1200 ft., *Galpin*, 4555!

T. exocarpæoides, *T. Thunbergii* and *T. fragile* are all closely allied, especially the two latter, and it may be that *T. Thunbergii* represents only a varietal form of *T. fragile*. They can, however, be easily separated on the characters of the bracts and the lax or dense inflorescences.

4. **T. hirtum** (Sond. in Flora, 1857, 365); herbaceous above, woody at the base, 4–9 in. high, with stout woody tap root; stems woody below, erect, with numerous erect branches arising from near the base, ribbed, angled, hirsute; male plants: leaves few, linear, acute, 3–6 lin. long, keeled, margins and keel hirsute-scabridulous; branches almost entirely floriferous; flowers arranged in more or less dense spikes; bracts elliptic- or ovate-lanceolate, acute, incurved, sharply keeled, prominently decurrent, 1$\frac{1}{4}$–1$\frac{1}{2}$ lin. long; bracteoles similar to the bracts, hirsute-scabridulous on keel and margins, both

bracteoles and bracts longer than the flowers; flowers solitary or in
3-flowered cymules, sessile in bract-axils; perianth orange when dry,
$\frac{5}{8}$ lin. long; segments triangular-ovate, $\frac{3}{8}$ lin. long; anthers about
$\frac{1}{10}$ lin. long; style rudiment present; female plants : leaves
numerous, linear-lanceolate, acute, somewhat fleshy, sharply keeled,
about 4 lin. long, margins infolded, densely hirsute on keel and
margins; inflorescences 2–4 in. long, dense; bracts and bracteoles
leaf-like, fleshy, keeled, incurved, 4–6 lin. long, keel and margins
horny, densely hirsute-scabridulous, bracteoles with infolded margins,
$1\frac{1}{2}$–3 lin. long; flowers solitary, rarely in 3-flowered cymules;
perianth $\frac{3}{8}$ lin. long; segments $\frac{1}{4}$ lin. long, margins hooded, infolded;
style $\frac{1}{8}$–$\frac{1}{4}$ lin. stout; stigma subtrilobed; fruit globose, about 1 lin.
in diam., pitted between the prominent reticulations. *T. globosum,*
A.DC. in DC. Prodr. xiv. 673. *T. strigulosum, A.DC. in DC. Prodr.*
xiv. 673. *T. hirtulum, Sond. in Flora,* 1857, 405 (*cit. in err.*).
Thesium globosum, A.DC. Esp. Nouv. Thes. 4. *Thesium strigulosum,*
A.DC. Esp. Nouv. Thes. 4 ; *Sond. in Flora,* 1857, 405.

SOUTH AFRICA : without locality, *Thom* !
COAST REGION : Cape Div. ; Simons Bay, *Wright* ! near Simons Town, 1200 ft.,
Bolus, 4689 ! Table Mountain, near Constantia, *Ecklon & Zeyher,* 35 ! Caledon Div. ;
Zwart Berg, *Zeyher,* 3815 ! Bredasdorp Div. ; near Elim, 400–700 ft., *Bolus,* 8602 !
Schlechter, 7642 ! Koude River, 1000 ft., *Schlechter,* 9627 ! 9628 !

5. **T. minus** (A. W. Hill in Kew Bulletin, 1915, 98) ; annual;
tap-root stout; stems numerous, especially in male plants, arising
from the apex of the rootstock, erect or spreading, sparingly
branched above, 3–6 in. long, ribbed, floriferous almost throughout,
subglabrous, leaves at the base of plants narrowly linear-lanceolate
or acicular, with a distinct rib, 3–6 lin. long, conspicuous and
gradually replaced by linear bracts in female and abruptly by
subulate bracts in male plants ; male plants : inflorescences simple
or branched ; bracts and bracteoles ovate-lanceolate or subulate,
navicular, curved, acute, slightly keeled, about equal to or shorter
than the flowers, $\frac{1}{2}$ lin. long, margins translucent, scarcely scabrous,
verruculose ; flowers sessile in axillary glomerules of 3-flowered
cymules ; perianth $\frac{3}{8}$ lin. long; segments spreading $\frac{1}{4}$ lin. long;
female plants : inflorescences simple or branched, rather lax ; stems
usually floriferous in the upper parts ; bracts leaf-like, $1\frac{1}{2}$–$2\frac{1}{2}$ lin.
long, linear or acicular, abruptly acute, erect or spreading, keeled,
margins and keel membranous, subscabridulous ; bracteoles $\frac{5}{8}$ lin.
long, about equal in length to the flowers or shorter, folded, sharply
keeled ; flowers usually solitary, lateral flowers sometimes developing
later, shortly pedunculate ; perianth about $\frac{1}{4}$ lin. long; segments
$\frac{1}{6}$ lin. long, margins more or less undulate ; style $\frac{1}{8}$ lin. long; fruit
globose, $\frac{1}{2}$ lin. in diam., prominently reticulate ; pedicel $\frac{1}{2}$ lin. long.

COAST REGION : Caledon Div. ; Houw Hoek, 1200 ft., *Schlechter,* 9431 ! 9432 !
near Vogelgat, 1200 ft., *Schlechter,* 10415 ♀ ! (?) Bredasdorp Div. ; hills at Riet
Fontein Poort, near Elim, *Bolus,* 8601 ! Riversdale Div. ; near Riversdale, *Rust,*
280 ! Garcias Pass, 1200 ft., *Galpin,* 4554 !

6. **T. fruticulosum** (A. W. Hill in Kew Bulletin, 1915, 99) ; a subshrub, about 1 ft. high ; stems erect or spreading, much-branched in male plants, stout, woody, sharply ribbed or subangled, ribs verruculose or subscabridulous ; leaves near the base, linear-lanceolate, with horny translucent keel and membranous margins, $\frac{1}{4}$–$\frac{3}{4}$ in. long ; male plants : branches numerous, stiff, spreading, with prominent verruculose or subscabrous ribs, entirely and more or less densely floriferous ; bracts leaf-like, lanceolate, acute, about 1$\frac{1}{2}$ lin. long, sharply keeled, flat above, margins and keel translucent, verruculose or subscabridulous ; bracteoles 1$\frac{1}{4}$ lin. long, longer than flowers ; flowers subsessile in axillary 3-flowered cymules ; perianth $\frac{5}{8}$ lin. long ; segments triangular-ovate, $\frac{3}{8}$ lin. long ; anthers about $\frac{1}{12}$ lin. long ; female plants : branches stout, erect, crowded, with lateral axillary inflorescences ; stem-angles prominent, horny, verruculose ; bracts 2–3 lin. long, lanceolate, keel-wing sharp, decurrent, horny and scabrous, margins infolded, scabrous ; bracteoles similar to the bracts, longer than the flowers ; flowers in 2-flowered cymules in the bract-axils, sessile ; perianth about $\frac{3}{8}$ lin. long ; segments erect, $\frac{1}{4}$ lin. long ; style stout, $\frac{1}{8}$ lin. long ; fruit globose, $\frac{3}{4}$ lin. in diam., subsessile, conspicuously reticulate.

SOUTH AFRICA : without locality, *Harvey*, 709 ♂ & ♀ !
COAST REGION : Cape Div. ; Table Mountain, in Groene Kloof, *Galpin*, 4556 (♀) ! Slang Kop, 700 ft., *Wolley-Dod*, 3187 (♂) ! slopes near Buffels Bay, *Wolley-Dod*, 2869 ! Durban Hills, *Guthrie*, 2407 ♂ ! Caledon Div. ; near Vogelgat, 1000 ft., *Schlechter*, 10414 ♂ ! (?)

7. **T. longifolium** (A. W. Hill in Kew Bulletin, 1915, 99) ; a subshrub, about 15 in. high ; stems erect, woody at the base, conspicuously ribbed and angled, ribs and angles scabrid-verruculose ; leaves at the base narrowly linear-lanceolate, acute, keeled, $\frac{3}{4}$–1 in. or more long, margins and keel translucent, scabrid-verruculose or fimbriate ; flowers distributed uniformly and fairly densely along the whole length of the branches ; male plants : bracts spreading, ovate-lanceolate, acute, 1$\frac{1}{2}$ lin. long, margins and keel-wing translucent, scabridulous or scabrid-fimbriate, keel-wing conspicuously decurrent ; bracteoles $\frac{3}{4}$ lin. long, margins infolded ; flowers in axillary glomerules of about 5-flowered cymules ; perianth $\frac{1}{2}$ lin. long ; segments $\frac{3}{8}$ lin. long ; style rudiment often present ; female plants : bracts leaf-like, conspicuous, lanceolate, acute, ascending or slightly recurved, 6–8 lin. long ; bracteoles 4–6 lin. long, with margins infolded, keel-wing in both sharp with margins translucent scabridulous or scabrid-fimbriate ; flowers infolded in the large bracteoles, solitary ; perianth $\frac{1}{2}$ lin. long ; segments $\frac{1}{4}$ lin. long ; disc prominent ; fruit globose, about 1 lin. in diam., stipitate, with conspicuous reticulations.

COAST REGION : Cape Div. ; eastern side of Table Mountain, 1200 ft., *Bolus*, 4607 ♀ ! 4608 ♂ !

The female plant is the largest in any known species and may be easily recognised by its conspicuous leaves and leafy bracts. The male plants resemble those of *T. fruticulosum*, but may be distinguished by their stouter habit with the stem-angles and keel-wings of the bracts markedly scabridulous or scabrid-fimbriate.

Imperfectly known species.

8. **T. leptostachyum** (Sond. in Flora, 1857, 405) ; a subshrub ; branches angled, 8 in. long ; leaves narrowly linear, 2–5 lin. long, ⅓ lin. broad, spreading, acute, with revolute margin ; bracts scarcely shorter than the flowers, ovate, obtuse, concave, subciliate ; bracteoles smaller than the bracts ; flowers (male only known) solitary, sparsely distributed in slender axillary branching spikes 2 in. long ; perianth ⅓ lin. long, subglobose, glabrous ; segments rounded ; style rudiment absent. *A.DC. in DC. Prodr.* xiv. 674. *Thesium leptostachyum, A.DC. Esp. Nouv. Thes.* 5 ; *Sond. in Flora,* 1857, 405.

CoAST REGION : Knysna Div. ; Karratęra River, *Drège,* 8173, in the Vienna Herbarium.

The description of the male plant suggests a close affinity to *T. minus,* and only a drawing of the type specimen at Stockholm has been seen. The solitary flowers, verruculose stem and bracts and bracteoles with ciliated margins suggest that the species should be retained.

III. OSYRIDICARPOS, A.DC.

Flowers hermaphrodite. *Perianth* adnate to the ovary and pro-duced above into a cylindrical tube ; segments 5, valvate, with a tuft of hairs on the face attached to the back of the anther. *Stamens* 5, inserted below the segments ; filaments short, slender ; anthers ovoid, with 2 parallel cells dehiscing longitudinally ; disc indistinct. *Ovary* inferior ; ovules 2–3, pendulous from the tip of a flexuous filiform central placenta ; stigma obscurely lobed. *Drupe* globose, crowned with the persistent perianth.

Undershrubs, with long slender sarmentose sulcate branchlets ; leaves alternate, shortly petioled, oblong or lanceolate, triplinerved ; flowers in terminal racemes produced down into the axils of the leaves, small, green ; bracteoles minute.

DISTRIB. Species 5, four in Tropical Africa.

1. **O. natalensis** (A.DC. in DC. Prodr. xiv. 635) ; a slender much-branched half-climbing shrub, 5–8 ft. high ; branches scattered, somewhat spreading ; stems and branches conspicuously ribbed, glabrous or minutely puberulous especially when young, ribs horny, colourless ; leaves alternate, scattered, spreading, ovate or elliptic-lanceolate, acute, narrowing at the base into a more or less distinct petiole, mid-rib and two lateral parallel veins usually prominent, subcoriaceous, subglabrous or minutely puberulous especially on veins and margins, ½–1½ in. long, 2–8 lin. broad : petiole up to 3 lin. long, usually minutely puberulous ; flowers creamy-white (*Galpin*), solitary or in 3-flowered cymules in the axils of leafy bracts arranged in lax racemes ; bracts ovate-lanceolate, acute, leaf-like : peduncles minutely puberulous ; bracteoles linear, acute or

subulate, minutely puberulous, 1–2 lin. long; perianth $2\frac{3}{4}$–$3\frac{1}{4}$ lin. long, with or without 5 subdependent glandular bodies or callosities at the base and with 5 external ribs continued down the ovary, ribs and perianth externally subglabrous or minutely puberulous; perianth-tube 2–$2\frac{1}{4}$ lin. long; segments ovate-lanceolate, acute, spreading, $\frac{3}{4}$–1 lin. long; stamens partly exserted; filaments inserted just below throat of perianth-tube, $\frac{1}{2}$–$\frac{3}{4}$ lin. long; anthers $\frac{1}{3}$ lin. long, attached at the apex by a tuft of long perianth-hairs; style 2–$2\frac{3}{4}$ lin. long; fruit ovoid or subglobose, $2\frac{1}{2}$–3 lin. long; $2\frac{1}{2}$ lin. in diam. (4–5 lin. in diam., *Harvey*), smooth, with depressed ribs and reticulations. *Harv. Thes. Cap.* ii. 63, *t.* 199; *Marloth, Fl. S. Afr.* i. 161. *Thesium macrocarpum, E. Meyer in Drège, Zwei Pfl. Documente*, 143, 226.

COAST REGION: Albany Div.; near Grahamstown, *Schlechter*, 2667! *Read*! *Bolus*, 1929! Blue Krantz, *Burchell*, 3646! between Blue Krantz and Kowi Poort, *Burchell*, 3662! King Williamstown Div.; Yellow-wood River, 1000–2000 ft., *Drège*! near King Williamstown, *Tyson*, 1008! Komgha Div.; near the Kei River, *Flanagan in MacOwan, Herb. Austr.-Afr.*, 1528! British Kaffraria; without precise locality, *Cooper*, 30!
CENTRAL REGION: Somerset Div.; without precise locality, *Bowker*!
KALAHARI REGION: Transvaal; Houtbosch, *Rehmann*, 5961! near Barberton, *Galpin*, 593! Waterval Onder, *Rogers*!
EASTERN REGION: Transkei; Kentani District, *Miss Pegler*, 3! Natal; Inanda, *Wood*, 612! Van Reenens Pass, *Rehmann*, 7266! and without precise locality, *Gerrard*, 159; *Cooper*, 1188!

The specimens from the Transvaal, and especially that from Waterval Onder, are much more puberulous than those from other regions and the external glandular bodies appear to be absent from the base of the perianth-tube.

IV. RHOIACARPOS, A.DC.

Flowers hermaphrodite. *Perianth-tube* slender, obconic; segments 5, ovate, acute, persistent, with a tuft of hairs attached to the anthers. *Stamens* with slender filaments; anthers 2-celled, ovoid. *Disc* subconcave, with short prominent obtuse lobes. *Ovary* inferior, fleshy; ovules 5, hanging from the apex of the straight cylindric placenta; style cylindric-conical; stigmas 5, minute, alternating with the perianth-segments. *Drupe* ovoid, crowned with the persistent perianth-segments; seed bony.

A shrub; branches stiff, erect, sometimes flexuous and subscandent, quadrangular, with well-marked angles, younger branches almost winged; leaves opposite, flat, margins slightly revolute, sessile, with a prominent midrib, varnished above; flowers in short axillary or terminal racemes or panicles composed of 3-flowered axillary cymules; bracts persistent, small, leaf-like, partly adnate to the peduncles; fruit fleshy, red, edible, one-seeded. Harv. Gen. S. Afr. Pl. ed. ii. 333. *Hamiltonia*, Harv. Gen. S. Afr. Pl. ed. i. 298. *Colpoon*, Berg.; Benth. & Hook. f. Gen. Pl. iii. 225, partly.

DISTRIB. One endemic species.

Bentham and Hooker reduced De Candolle's genus *Rhoiacarpos* to *Colpoon*, Berg., but this latter genus has here been merged with *Osyris*, Linn., since the character of the opposite leaves on which it was mainly based has not been found to hold good. There is also no floral difference between *Colpoon*, Berg., and *Osyris*, Linn.

Rhoiacarpos is distinguished especially in the perianth with its 5 segments and in the style bearing 5 stigmatic surfaces; the segments being persistent on the mature fruit. The sessile definitely opposite leaves also differ markedly from those of *Osyris* in being varnished above and in having their margins slightly revolute.

1. **R. capensis** (A.DC. in DC. Prodr. xiv. 635); leaves opposite, ovate or elliptic, cordate at the base, sessile or subsessile, apex acute or subacute, 1–2½ in. long, ¾–1 in. broad, midrib prominent with lateral veins almost at right angles to it, margins entire or remotely and minutely denticulate, usually revolute, coriaceous, varnished above; inflorescences small, compact, few-flowered, 6–9 lin. long; bracts narrowly elliptic-lanceolate, acute, partly adnate to the short peduncles, keeled, 1–1¼ lin. long, cartilaginous; peduncles ½–¾ lin. long; flowers in 3-flowered cymules, subsessile; perianth 1½ lin. long; segments about 1 lin. long, ovate, acute, stout, coriaceous; disc conspicuous, with fleshy obtuse lobes; stamens exserted; filaments slender, ⅓ lin. long; anthers ¼ lin. long, attached to the perianth-segments by a tuft of hairs; style stout, ½ lin. long; fruit mature (?), globose, about 4 lin. long, large, red, edible (*Burchell*). *Santalum capense, Spreng. ex A.DC. in DC. Prodr.* xiv. 635. *Hamiltonia capensis, Harv. Gen. S. Afr. Pl. ed.* i. 298; *Sond. in Flora,* 1857, 365.

COAST REGION: Mossel Bay Div.: between Mossel Bay and Zout River, *Burchell,* 6336! Uitenhage Div.; Zwartkop River, *Zeyher,* 6! near Uitenhage, *Burchell,* 4260! *Pappe!* near Bontjes River, *Drège,* 2376b! Redhouse, *Rogers,* 3619! Bathurst Div.; near Barville Park, *Burchell,* 4066! 4111! Albany Div.; near Grahamstown, *Burchell,* 3577! Curries Kloof, *MacOwan!* King Williamstown Div.; Tamacha, *Sim,* 1972! East London Div.; at East London, *Galpin,* 1858! British Kaffraria; without precise locality, *Cooper,* 52!

V. OSYRIS, Linn.

Flowers subdiœcious. *Perianth-tube* in the male flowers very short and solid, in the female entirely adnate to the ovary; segments 3–4, valvate, deltoid, with a tuft of hairs on the face attached to the back of the anthers. *Stamens* 3–4, inserted at the base of the segments; filaments rather thick; anther-cells subparallel, dehiscing longitudinally. *Disc* flat, angled between the stamens. *Ovary* inferior; ovules 2–4, pendulous from a short central placenta; style short or long; stigma 3–4-fid. *Fruit* globose, succulent, crowned with the persistent perianth-segments; albumen fleshy; embryo straight or rather curved.

Glabrous shrubs; leaves alternate, subopposite or opposite, narrow or broad; flowers small, in short axillary panicles; bracts solitary, minute.

DISTRIB. Species 9, spread through Southern Europe, India and the whole of Africa.

1. **O. abyssinica** (Hochst in Flora, 1841, i. Intell. 22, name only); a much-branched shrub, 6–8 ft. high, glabrous in all its parts; leaves alternate, subopposite or opposite, shortly petioled, oblong, 1–3 in.

long, mucronate, narrowed to the base, rigidly coriaceous, glaucous, veins (except the midrib) immersed and almost invisible ; male flowers in shortly peduncled axillary umbellate cymes; bracts minute, lanceolate ; pedicels larger than the flowers ; buds globose, ½ lin. in diam. ; perianth-lobes 3 or 4, ovate-triangular, with a conspicuous disc ; female flowers usually solitary ; berry oblong, scarlet, the size of a small pea. *A. Rich. Tent. Fl. Abyss.* ii. 236 ; *A.DC. in DC. Prodr.* xiv. 633 ; *Engl. Hochgebirgsfl. Trop. Afr.* 199 ; *Schinz in Bull. Herb. Boiss.* iv. *App.* iii. 55 ; *Schweinf. in Bull. Herb. Boiss.* iv. *App.* ii. 152 ; *Hiern in Cat. Afr. Pl. Welw.* i. 938 ; *Baker & Hill in Dyer, Fl. Trop. Afr.* vi. i. 433. *O. compressa, A.DC. in DC. Prodr.* xiv. 634 ; *Engl. Jahrb.* xxx. 305, var. β *oblongifolia, A.DC. l.c. Fusanus alternifolia, R. Br. in Salt, Abyss. App.* lxiii., *name only. Colpoon compressum, Berg. Descr. Pl. Cap.* 38, *t.* 1, *fig.* 1. *Fusanus compressus, Lam. Ill.* iii. 435, *t.* 842, *Sond. in Flora,* 1857, 365.

VAR. β, **speciosa** (A. W. Hill) ; differs from the type in robust habit and almost leafless, much exserted inflorescences, with larger bracts and flowers.

COAST REGION : Cape Div. ; Fish Hoek, *Wolley-Dod,* 667 ! Camps Bay, *Zeyher,* 311 ! *Burchell,* 368 ! 842 ! Simons Bay, *Grey* ! *Wright* ! Table Mountain, *Thunberg* ! *MacOwan, Herb. Norm. Austr.-Afr.,* 574 ! *Drège,* a ! and without precis elocality, *Hooker* ! *Armstrong* ! *Wallich* ! *Pappe* ! Stellenbosch Div. ; near Lowrys Pass, *Burchell,* 8284 ! Riversdale Div. ; Garcias Pass, *Phillips,* 399 ! near Zoetemelks River, *Burchell,* 6629 ; Mossel Bay Div. ; near landing place at Mossel Bay, *Burchell,* 6245 ! Knysna Div.; Plettenbergs Bay, *Burchell,* 5327 ! Humansdorp Div. ; north side of Kromme River, *Burchell,* 4853 ! Albany Div. ; Grahamstown, *MacOwan,* 43 ! Broekhuisens Poort 1500–2000 ft., *Galpin,* 22 ! Queenstown Div. ; mountain gullies near Queenstown, 4000 ft., *Galpin,* 1577 ! British Kaffraria and Eastern Districts, *Cooper,* 50 ! 225 ! Var. β : Caledon Div. ; near Houw Hoek, 1100 ft., *Bolus* ! near Hermanus Pieters Fontein, 100 ft., *Bolus* !

CENTRAL REGION : Somerset Div. ; mountain above Commadagga, *Burchell,* 3327 ! Graaf Reinet Div. ; Compass Berg, *Shaw* ! Voor Sneeuw Berg, *Burchell,* 2858 ! Cave Mountain, near Graaff Reinet, 3500 ft., *Bolus,* 207 ! Oude Berg, near Graaff Reinet, 3800 ft., *Bolus,* 207 ! Aliwal North Div. ; Elands Hoek, near Aliwal North, *Bolus,* 194 ! Colesberg Div. ; Colesberg, *Arnott* ! Philipstown Div. ; near Petrusville, *Burchell,* 2696 ! Bavers Pan, *Burchell,* 2715 !

KALAHARI REGION : Orange River Colony ; Doorn Kop, *Burke* ! and without precise locality, *Cooper,* 3096 ! Transvaal ; Pretoria, Aapies Poort, *Rehmann,* 4051 ! Meintjes Kop, 4575–4775 ft., *Burtt-Davy,* 5033 ! Wonderboom Poort, 4600 ft., *Miss Leendertz,* 435 ! 268 ! Moordrift, *Miss Leendertz,* 2168 !

EASTERN REGION : Natal ; Murchison, 1000 ft., *Wood,* 3004 ! Alexandra Dist. : Dumisa, 2300 ft., *Rudatis,* 1128 !

Also in Tropical Africa.

Cooper on his label to No. 50 British Kaffraria adds "Bark Bosch" used for tanning leather, and Bolus makes a similar remark on his sheets 207, and gives the name "Wilde Granaat"—fruits edible.

VI. GRUBBIA, Berg.

Flowers hermaphrodite, sessile, arranged in 3- or 2-flowered axillary cymules or crowded in axillary strobili. *Perianth-tube* short, adnate to the ovary ; segments 4, greenish, ovate or obovate, valvate, densely pilose on the back. *Stamens* 8, 4 longer inserted at the base of the perianth-lobes and 4 shorter alternate with them ; filaments stout, incurved ; anthers small, dehiscing laterally. *Disc* hairy, very slightly prominent. *Ovary* inferior ; style short, fili- form ; stigma emarginate or slightly bifid ; ovules 2, pendulous from a placenta which may be free or more or less adnate to the wall. *Fruits* connate, crowned with disc and style, only one perfect ; exocarp somewhat fleshy ; endocarp often bony. *Seed* ovoid ; embryo linear, embedded in the middle of the albumen, sub- terete ; radicle much longer than the cotyledons.

Heath-like shrubs ; leaves opposite, linear or lanceolate, entire, persistent, with revolute margins ; flowers small, ternate or in strobili in the axils of each pair of opposite leaves.

DISTRIB. Species 4, endemic.

This genus was considered by De Candolle to belong to a distinct Natural Order, *Grubbiaceæ*, which he placed between *Elæagnaceæ* and *Santalaceæ*. Sonder in Harvey and Sonder, Fl. Cap. ii. 325, described the genus under *Hamamelidaceæ*, and Bentham & Hooker, Gen. Pl. iii. 231, placed it in their fourth tribe, *Grubbieæ*, of *Santalaceæ*.

Flowers arranged in 3–2-flowered axillary cymules ; leaves
  linear or linear-lanceolate :
  Bracts bilobed equal to or slightly shorter than the
    flowers, sharply keeled below :
  Branches tomentose or strigulose ; leaves strigulose
    above, subsessile, subcordate or subauriculate at
    the base   ...   . (1) **rosmarinifolia.**
  Branches minutely downy ; leaves glabrous or sub-
    glabrous above, narrowing into definite petioles ... (2) **pinifolia.**
  Bracts usually entire, much shorter than the flowers,
    rounded on the back ...   . (3) **hirsuta.**
Flowers numerous in axillary strobili ; leaves lanceolate ... (4) **stricta.**

1. **G. rosmarinifolia** (Berg. in Vet. Acad. Handl. Stockh. 1767, 36, t. 2 ; Descr. Pl. Cap. 90, t. 2) ; a shrub, 1–5 ft. high ; stems woody, erect, much-branched ; branches erect or subspreading, tomentose or hirsute ; branchlets terete or somewhat angled, hirsute or glabres- cent ; leaves narrowly linear-lanceolate, 4–6 lin. long, rarely longer, ½–1¼ lin. broad, subcordate or subauriculate at the base, margins revolute, hairy and scabrous above, densely tomentose below, with long strigulose hairs along the midrib ; petiole not conspicuous, not more than ¼ lin. long ; bracts hemispherical, membranous, chestnut- brown, sharply keeled below, bilobed to the middle, lobes rounded or subacute, rather shorter than or about equal in length to the cymules and enclosing them, smooth ; flowers sessile, connate ; perianth-segments broadly obovate, acute, curved at the apex with small thickened flanges at the base, densely and conspicuously hairy on the back with long white wavy hairs, projecting ¼ lin. or more beyond the segments ; stamens with filaments about ¼ lin. long ; style filiform, about ¼ lin. long, enveloped in a dense tuft of disc-hairs ; fruits crowned externally by a dense ring of hairs at the apex of the perianth-tube, glabrous below. *Thunb. Fl. Cap. ed. Schult.* 373 ; *Drège, Zwei Pfl. Documente,* 88 ; *Sond. in Harv. & Sond.*

Fl. Cap. ii. 326 ; *A.DC. in DC. Prodr.* xiv. 618 ; *Marloth, Fl. S. Afr.*
i. 162, *t.* 37 E, *figs.* 81 A. *G. sp., Drège, l.c.* 117. *Ophira stricta,*
Linn. Mant. Alt. 229.

COAST REGION : Cape Div. ; Table Mountain, *Thunberg* ! *Cooper,* 2491 ! *Ecklon,*
371 ! *Drège,* a ! *Wilms,* 949 or 3949 ! *Sieber* ! *Harvey* ! *Pappe* ! *Wolley-Dod,* 545 !
2123 ! 3054 ! *Bolus,* 3943 ! *Zeyher,* 2654 ! Bredasdorp Div. ; Elim, *Schlechter,*
7637 ! Swellendam Div. ; without precise locality, *Bowie* ! Riversdale Div. ;
Garcias Pass, *Galpin,* 4558 ! George Div. ; Cradock Berg, *Burchell,* 5961 ! *Galpin,*
4559 ! near George, *Drège,* 8161 !

2. G. pinifolia (Sond. in Harv. & Sond. Fl. Cap. ii. 326) ; a
shrub ; branches woody, erect ; branchlets short, ascending, covered
with a fine downy pubescence ; leaves linear or narrowly elliptic-
linear, margins strongly inrolled, subacute, narrowing at the base
into a definite petiole, 6–8 lin. long, $\frac{1}{3}$–$\frac{1}{2}$ lin. broad, glabrous above
or nearly so, grey when dry, finely pubescent below ; petioles
$\frac{1}{2}$–1 lin. long, finely pubescent, broadening at the base ; flowers in
2–3-flowered cymules ; bracts about equal in length to the flowers,
hemispherical, bilobed for a third of their length, keeled below,
truncate, membranous, chestnut-brown ; perianth-segments $\frac{1}{2}$ lin.
long, fleshy, subacute, concave, with thickened margins and
prominent fleshy protuberances or flanges at the base, densely
hairy on the back with conspicuous white wavy hairs ; style about
$\frac{1}{4}$ lin. long, surrounded by disc-hairs ; fruit crowned externally with
a dense ring of hairs, glabrous below.

COAST REGION : Cape Div. ; Table Mountain, 2800 ft., *MacOwan in Herb. Norm.*
Austr.-Afr., 918 ! Caledon Div. ; mountains near Grietjes Gat, 2000–4000 ft.,
Ecklon and Zeyher !

3. G. hirsuta (E. Meyer in Drège, Zwei Pfl. Documente, 73) ; a
shrub ; branches woody, erect, branches and branchlets ascending,
villous ; leaves linear-lanceolate, with strongly revolute margins,
slightly subauricled at the base 4–5 lin. long, $\frac{1}{3}$–$\frac{1}{2}$ lin. broad, villous
or strigulose-villous above, villous below ; petioles very short, densely
villous ; flowers sessile in 3-flowered cymules ; bracts less than half
the length of the flowers, broadly ovate, rounded, entire or slightly
notched, chestnut-brown, membranous, concave, not keeled ; perianth-
segments obovate or broadly obovate, subacute, $\frac{1}{2}$ lin. long, with
thickened flanges at the base and straight silky, somewhat inconspicuous, white hairs on the back projecting slightly beyond the
apex of the segments ; stamens with filaments about $\frac{1}{3}$ lin. long ;
style $\frac{1}{4}$ lin. long ; disc with a tuft of hairs ; fruits minutely
pubescent over the whole surface. *A.DC. in DC. Prodr.* xiv. 618 ;
Sond. in Harv. & Sond. Fl. Cap. ii. 327.

COAST REGION : Clanwilliam Div. ; Wupperthal, 1500–2000 ft., *Drège* ! Cold
Bokkeveld, near Tweefontein, 5000 ft., *Schlechter,* 10125 !
CENTRAL REGION : Ceres Div.; near Ceres, 1700 ft., *Bolus,* 7454 !

4. G. stricta (A.DC. in DC. Prodr. xiv. 618) ; a shrub, 2–5 ft.
high ; stems erect, woody, quadrangular and striate with sharp

almost winged angles ; branches and branchlets covered with short adpressed hairs ; leaves 1–2½ in. long, 1–7 lin. broad, linear, narrowly elliptic-lanceolate or ovate-lanceolate, acute or subacute, narrowing at base into the petiole, varnished, glabrous but tuberculate above, silky-pubescent beneath, margins more or less revolute ; petioles 1½–3 lin. long, densely covered with adpressed hairs; flowers in ovoid axillary strobili of 15–20 flowers, ripening to a syncarpium ; bracts at the base foliaceous, linear-lanceolate, acuminate, about 1 lin. long, pubescent ; perianth-segments ovate, acute, ⅚ lin. long, with short scarcely projecting hairs on the back ; stamens with incurved filaments ⅓ lin. long ; style about ¼ lin. long, surrounded by a dense tuft of disc-hairs ; syncarpium globose-ovoid, bright red when ripe (*Marloth*), 3–4 lin. long, nuts covered with the large adnate crustaceous disc, one-seeded. *Sond. in Harv. & Sond. Fl. Cap.* ii. 327 ; *Marloth, Fl. S. Afr.* i. 162, *t.* 37 D, *fig.* 81 B. *Taxus tomentosa, Thunb. Fl. Cap. ed. Schult.* 547. *Ophira stricta, Lam. Encycl.* iv. 565, *and Ill. t.* 293, *not of Burm.* ; *Drège, Zwei Pfl. Documente,* 77, 116. *Strobilocarpus diversifolius, Klotzsch in Linnæa,* xiii. 381. *G. latifolia, Schnizl. Ic. Fam. Nat. fasc.* 13, 108, *figs.* 1 *and* 21.

SOUTH AFRICA : without locality, *Forbes*! *Thom*, 174!
COAST REGION : Tulbagh Div. ; between New Kloof and Elands Kloof, 1000–2000 ft., *Drège*, a! Cape Div. ; ridge beyond Smitswinkel Vley, *Wolley-Dod*, 2676! Klaasjegers Berg, *Wolley-Dod*, 2404! Simons Bay, *Wright*! near Simonstown, *Schlechter*, 1093! Stellenbosch Div. ; Lowrys Pass, 1600 ft., *Bolus*, 7319! *Zeyher*! Caledon Div. ; Baviaans Kloof, near Genadendal, *Burchell*, 7717 ! Genadendal, *Drège*, b! *Bolus*, 7424! Swellendam Div. ; on mountains, *Bowie*! near Swellendam, *Bolus*! ridges near the lower part of the Zondereinde River, *Zeyher*, 2656 ! Riversdale Div. ; Paardeburg, *Muir in Herb. Galpin*, 5328! Garcias Pass, *Burchell*, 6946 ! 7157 ! *Galpin*, 4557 ! *Phillips*, 317 ; George Div. ; Cradock Berg, *Burchell*, 5954 !

ORDER CXX. A. **BALANOPHORACEÆ.**

(By C. H. WRIGHT.)

Flowers small, unisexual in the South African genera. *Male flowers* : *Perianth* regular or 2-lipped ; lobes 3–4, valvate. *Stamens* 2–3 ; filaments short ; anthers either 2-celled and bursting by longitudinal slits, or several-celled and bursting by terminal pores ; pollen-grains cubical or globose. *Female flowers* : *Perianth* subglobose or tubular and 3- to many-lobed, or absent. *Disc* sometimes large and cushion-like. *Ovary* inferior, 1- (or at length 3-) celled ; style long or absent; stigma discoid or 3-lobed ; ovules 1–3, pendulous, naked or with a single coat. *Fruits* indehiscent, separate (*Mystropetalon*), or united into syncarpia (*Sarcophyte*). *Seed* with fleshy albumen ; embryo central or apical.

Herbs parasitic on the roots of trees or shrubs, usually brightly coloured; rootstock tuberous ; leaves reduced and scale-like ; inflorescence monœcious and simple (*Mystropetalon*), or diœcious and much-branched (*Sarcophyte*), bracteate and bracteolate or not.

DISTRIB. Genera about 15, species about 50, widely distributed in tropical and subtropical regions.

I. **Sarcophyte.**—*Inflorescences* much-branched, diœcious. *Anthers* 3, many-celled, dehiscing by terminal pores. *Fruits* united into syncarpia.

II. **Mystropetalon.**—*Inflorescences* unbranched, monœcious, male above, female below. *Anthers* 2, 2-celled, dehiscing by longitudinal slits. *Fruits* separate.

1. SARCOPHYTE, Sparrm.

Flowers diœcious. *Male flowers*: *Perianth-tube* short, solid; segments 3, valvate. *Stamens* 3, inserted at the base of the perianth-lobes; filaments short, cylindrical; anthers basifixed, globose, with many cells in the upper part bursting by apical pores. Rudiment of *ovary* none. *Female flowers* usually united into globose heads. *Ovary* ovoid, 1-celled, or at length 3-celled through the protrusion of the placentas; stigma sessile, discoid; ovules 1–3, pendulous. *Fruit* a fleshy syncarpium; endocarp hardened, trigonous, 1-celled, 1-seeded. *Seed* subglobose; albumen fleshy and oily; embryo central, globose.

Herbs parasitic on roots; rootstock tuberous, irregularly lobed; stem erect; leaves reduced to scales, flowers in much-branched panicles.

DISTRIB. Species 2, one in tropical Africa, the other in South Africa.

1. **S. sanguinea** (Sparrm. in Vet. Acad. Handl. Stockh. 1776, 300, t. 7); male plant about 10 in. high; rootstock thick, irregularly lobed, verrucose; stem short, erect; leaves reduced to oblong obtuse or subacute scales up to 9 lin. long and 6 lin. wide; inflorescence much-branched; flowers usually in pairs on short pedicels which are connate below; perianth-segments navicular, almost patent, very thick and fleshy, subacute, 2 lin. long, 1¼ lin. wide; filaments 1½ to nearly 2 lin. long, cylindrical; anthers terminal, scarcely wider than the filaments; female plant very similar to the male, but rather shorter; flowers numerous in subglobose shortly stalked heads about 3 lin. in diam.; ovary 1–3-celled; ovule solitary, pendulous; stigma discoid, sessile; fruit a syncarpium; seed about ½ lin. long. *Griff. in Trans. Linn. Soc.* xix. 339; *Hook. f. in Trans. Linn. Soc.* xxii. 29; *Drège, Zwei Pfl. Documente*, 132, 137, 138; *Wedd. in Ann. Sc. Nat.* 3me sér. xiv. 173, t. 10, figs. 34–38; *Harv. in Harv. & Sond. Fl. Cap.* ii. 574; *Eichler in DC. Prodr.* xvii. 127; *Engl. in Engl. & Prantl, Pflanzenfam.* iii. i. 253, fig. 160; *Marloth, Fl. S. Afr.* i. 170, t. 42, fig. A; *Hutchinson in Kew Bulletin*, 1914, 251; not of *Hemsl. in Dyer, Fl. Trop. Afr.* vi. i. 436. *Icthyosma Wehdemanni, Schlechtend. in Linnæa*, ii. 671, t. 8, and iii. 194.

I. Wehdamanni, Steud. Nomencl. ed. 2, i. 801.   *I. Weinmanni, Harv.*
Gen. S. Afr. Pl. ed. i. 300.   *I. Wiedemanni, Griff. in Trans. Linn.*
Soc. xix. 339.

COAST REGION : Uitenhage Div. ; Addo, *Drège.* Albany Div. ; near Grahams-
town, on roots of *Acacia horrida,* 1800 ft., *MacOwan,* 1204 ! *Schönland* !
CENTRAL REGION : Somerset Div. ; near Little Fish River and Great Fish
River, 2000-3000 ft., *Drège*; at the foot of Bosch Berg, on roots of *Acacia
horrida,* Willd., 3000 ft., *MacOwan, Herb. Austr.-Afr.,* 1800 ! between Zuurberg
Range and Klein Bruintjes Hoogte, 2000-2500 ft., *Drège* !
EASTERN REGION : Transkei ; Kentani, 1200 ft., *Miss Pegler,* 846 !

Miss Pegler has noted on her specimen, "horrible odour ; deep red."

The tropical African plant is separated as *S. Piriei,* Hutchinson in Kew Bulletin,
1914, 252.

II. MYSTROPETALON, Harv.

Flowers monœcious.   *Male flower* : *Perianth* 2-lipped, anterior lip
of one segment, posterior of two.   *Stamens* 2, inserted on the
posterior perianth-segments ; anthers 2-celled, dehiscing longitudinally ; pollen-grains cubical, pentagonal or rarely hexagonal, sometimes fluted at the angles.   *Female flower,* opening before the male :
Perianth small, urceolate or tubular, 3-lobed.   *Staminodes* 2 and
very small, or none.   *Disc* thick, cushion-like.   *Ovary* inferior,
ovoid or elliptical, 1- or 3-celled ; ovules 3, pendulous ; style
columnar ; stigma discoid or 3-lobed.   *Fruit* globose ; pericarp
crustaceous.   *Seed* solitary ; embryo small, apical ; albumen copious.

Fleshy parasitic herbs springing from an irregular nodular rhizome ; leaves scale-
like, fleshy ; flowers in dense spikes, proterogynous, male above, female below,
bracteate and 2-bracteolate.

DISTRIB.   Species 3 in extratropical South Africa.

Bract of male flower oblong ... ... (1) **Thomii.**

Bract of male flower spathulate :
Perianth of female flower tubular, 3-fid ... ... (2) **Polemanni.**

Perianth of female flower subglobose or campanulate,
multifid ... ... (3) **Sollyi.**

1. **M. Thomii** (Harv. Gen. S. Afr. Pl. ed. i. 419) ; a fleshy plant,
about 5 in. high ; leaves scale-like, oblong, obtuse, 1 in. long, 1 lin.
wide ; spike oblong, 2-4 in. long, $\frac{3}{4}$-$1\frac{1}{4}$ in. in diam. ; bract oblong,
obtuse, densely hairy outside, 3 lin. long ; bracteoles two-thirds as
long as the bract, oblong, hairy on the midrib outside ; perianth
red above, yellow below, nearly twice as long as the bract ; anterior
segment spathulate, obtuse, posterior segments slightly longer than
the anterior, slightly concave ; anthers oblong ; pollen-grains cubical
with fluted angles ; female flower : bract oblong, obtuse, densely
hairy outside in the upper part and on the midrib ; bracteoles
shorter than the bract in flower, but nearly twice as long as it in
fruit, induplicate, acute, ciliate on the midrib ; perianth subglobose

or ellipsoid, shortly 3-lobed, glabrous ; ovary ovoid ; style columnar, recurved ; stigma discoid. *Harv. in Ann. Nat. Hist. ser.* 1, ii. 386, *t.* 19, *and in Harv. & Sond. Fl. Cap.* ii. 573 ; *Griff. in Trans. Linn. Soc.* xix. 336 ; *Hook. f. in Trans. Linn. Soc.* xxii. 29 ; *Eichler in DC. Prodr.* xvii. 125 ; *Marloth, Fl. S. Afr.* i. 170, *tt.* 40, 42, *fig.* B ; *Harvey-Gibson in Trans. Linn. Soc. ser.* 2, viii. 143, *tt.* 15–16. *Balanophora capensis, Eckl. & Zeyh. ex Eichler in DC. Prodr.* xvii. 125.

COAST REGION : Caledon Div. ; about Caledon Baths, *Thom, Zeyher! Polemann.*

2. **M. Polemanni** (Harv. Gen. S. Afr. Pl. ed. i. 418) ; a fleshy plant about 5 in. high, reddish in all its parts ; leaves narrowly oblanceolate, obtuse or subacute, up to 1½ in. long and 2 lin. wide ; spike oblong, 2¼ in. long ; male flower : bract spathulate, unguiculate, about 3 lin. long, villous above ; bracteoles about half as long as the bract, lanceolate, acuminate, hairy on the back in the lower half ; anterior perianth-segment about a quarter longer than the bract, spathulate, unguiculate ; posterior segments much longer than the anterior, deeply concave ; anthers elliptic ; pollen-grains cubical with fluted angles ; female flower : bract lanceolate ; bracteoles as long as the bract, oblong, acute ; perianth-tube oblong ; lobes 3, ovate, acute, slightly shorter than the tube ; ovary oblong, seated on the cushion-like disc ; style more slender than in *M. Thomii* ; stigma only slightly enlarged. *Harv. in Ann. Nat. Hist. ser.* 1, ii. 387, *t.* 20, *and in Harv. & Sond. Fl. Cap.* ii. 573 ; *Hook. f. Trans. Linn. Soc.* xxii. 29 ; *Eichler in DC. Prodr.* xvii. 125 ; *Harvey-Gibson in Trans. Linn. Soc. ser.* 2, viii. 143. *Scybalium ? Harv. Gen. S. Afr. Pl. ed.* i. 315.

COAST REGION : Malmesbury Div. ; near Wellington, *Bolus,* 4399 ! Caledon Div. ; on mountain sides near the hot springs at Caledon, 900–1000 ft., on roots of *Protea mellifera,* Thunb., *Bolus,* 7465 ! *and in MacOwan and Bolus,* Herb. Norm. *Austr.-Afr.,* 1364 ! Houw Hoek, on *Aspalathus* sp., *Mrs. Denys, Ecklon ! MacOwan,* 2303 !

3. **M. Sollyi** (Harvey-Gibson in Trans. Linn. Soc. ser. 2, viii. 153, tt. 15–16) ; resembling *M. Thomii,* but differing in the following characters :—male flowers : bract spathulate ; bracteoles one-quarter as long as the bract ; perianth-segments slightly concave ; anterior segment rather shorter than the posterior ; pollen-grains cubical or pentagonal, without fluted angles ; female flower : bract lanceolate ; bracteoles shorter than the bract ; perianth subglobose or campanulate, multifid ; ovary ovoid ; stigma trilobed.

COAST REGION : Caledon Div. ; Caledon Pass, on roots of *Protea* sp., *Mrs. Solly.*

ORDER CXXI. **EUPHORBIACEÆ.**

(By N. E. BROWN, J. HUTCHINSON and D. PRAIN.)

Flowers monœcious, usually regular. *Perianth* occasionally absent from one or both sexes, usually small, often dissimilar in the two sexes, simple, valvate· or imbricate, calycine, rarely petaloid, or double, both outer and inner calycine and imbricate, or the inner petaloid, imbricate, rarely subvalvate, longer or shorter than the outer. *Male* : stamens definite or indefinite (1–1,000) ; filaments free or connate ; anthers 2- (rarely 3–4-) celled ; cells usually parallel, adnate to the connective throughout or free except at the base or apex and erect, divaricate or suspended, rarely superposed ; dehiscence usually longitudinal, rarely porous ; rudimentary ovary present or absent. *Female* : ovary sessile, rarely shortly stipitate, usually 3-, frequently 2- or 4-, very rarely 1- or more than 4-celled ; styles usually as many as and continuous with the carpels, free or more or less connate, erect or spreading, entire or 3-fid or laciniate ; inner face of styles or style-arms usually stigmatic throughout ; ovules in each cell solitary or 2 collateral, pendulous from the inner angle ; funicle often thickened ; disc annular, entire or lobed, or of free contiguous or discrete scales, or none. *Fruit* usually capsular, of 2-valved cocci separating from a persistent axis, or dehiscent and drupaceous, 1–3-celled, or of a single or 2–3 connate nuts. *Seeds* attached laterally near or above the middle of the cell, with or without a caruncle or an arillus ; albumen usually copious, fleshy ; embryo straight, radicle superior ; cotyledons broad, flat, rarely thick, fleshy.

Herbs, shrubs or trees, often with milky juice. Leaves alternate or opposite, simple or rarely compound, sometimes rudimentary, stipulate or exstipulate. Flowers usually small or very small ; inflorescence rather variable.

DISTRIB. Species about 4000, mostly in the tropics of both hemispheres.

JUNODIA, Pax in Engl. Jahrb. xxviii. 22, is *Anisocycla triplinervia*, Diels (*Menispermaceæ*).

Tribe 1. EUPHORBIEÆ.—*Apparent flower* composed of a number of stamens (really male flowers, each consisting of a single stamen jointed to a pedicel and soon falling away from it, with or without a minute rudimentary calyx) mingled with bracteoles, with or without one sessile or stalked ovary (really a sessile or pedicellate female flower, with or without a small or rudimentary or very rarely comparatively large calyx) in their midst, enclosed in a cup-shaped, obconic or 4-angled involucre ; the whole resembling a small male or hermaphrodite flower.   *Ovary* 2–3-celled, with 1 pendulous ovule in each cell.

I. **Synadenium.**—*Involucre* with one continuous rim-like gland, which is quite entire or occasionally with 1–5 cut-like notches dividing it into segments, but not forming equally spaced glands. A bushy shrub, with terete spineless succulent branches and large alternate leaves.

II. **Elæophorbia.**—*Involucre* with 5 separate equally spaced contiguous glands. *Fruit* indehiscent, thickly fleshy, containing one hard 3-celled "stone." A succulent tree, with angular branches armed with spines in pairs.

III. **Euphorbia.**—*Involucre* with 2–8 (usually 4–5) separate and usually equally spaced contiguous or distant glands. *Fruit* separating into 3 lobes or cells, which open down the inner face to liberate the seeds, not fleshy or but slightly so in a few succulent species. Herbs, shrubs or trees, often succulent.

Tribe 2. BUXEÆ.—*Perianth* calycine. *Segments* imbricate or rarely (in extra-African species) absent. *Stamens* in the African genera 4 or 6. *Ovules* 2 in each cell, rarely solitary ; raphe dorsal. *Styles* undivided or slightly bifid. *Cotyledons* various.   Shrubs or trees with opposite entire leaves.

IV. **Buxus.**—*Stamens* 4, opposite the perianth-segments.

V. **Notobuxus.**—*Stamens* 6, two solitary ones opposite the outer perianth-segments, four in pairs opposite the inner segments.

Tribe 3. PHYLLANTHEÆ.—*Sepals* valvate or imbricate in bud, 1–2-seriate. *Petals* when present small and scale-like. *Stamens* 1–2-seriate, the outer opposite the sepals, rarely indefinite in the middle of the flower. *Ovules* 2 in each cell, collateral and usually contiguous ; caruncle usually conspicuous ; raphe ventral. *Cotyledons* much broader than the radicle. *Inflorescence* various, axillary, rarely terminal.   Habit various.

Leaves opposite ; stipules intrapetiolar and sheathing ; stamens very numerous, arranged on an elongated receptacle.

VI. **Androstachys.**—A tall erect tree with diœcious flowers and numerous stamens.

** *Leaves alternate or rarely whorled ; stipules neither intrapetiolar nor sheathing ; stamens rarely numerous ; receptacle not elongated.*

†Sepals of the male flowers valvate in bud.

VII. **Bridelia.**—*Fruits* drupaceous. *Ovary* 2-celled.   Tertiary *nerves* of the leaves usually parallel.

VIII. **Cleistanthus.**—*Fruits* capsular. *Ovary* 3-celled.   Tertiary *nerves* of the leaves not parallel.

††Sepals of the male flowers imbricate in bud.

‡Petals present in the male flowers.

§Flowers diœcious.   Petals larger than the sepals in the male flowers. Disc in the male flowers of separate glands or absent.

IX. **Lachnostylis.**—*Flowers* in axillary fascicles. *Sepals* in the male flowers 5. *Disc-glands* thick, villous.   Rudimentary *ovary* well developed.

X. **Heywoodia.**—*Flowers* in axillary glomerules.   *Sepals* in the male flowers 3.   *Disc* and rudimentary *ovary* absent from the male flowers.

§§Flowers monœcious.   Petals subequal to the sepals.   Disc in the male flowers cupular.

XI. **Andrachne.**—A slender shrub with small leaves.   *Male flowers* fasciculate, female solitary.   *Seeds* wrinkled, estrophiolate, with fleshy albumen.

‡‡Petals absent from the male flowers.
§Flowers pedicellate, in axillary fascicles or solitary.

XII. **Phyllanthus.**—*Disc* in the male flowers outside the stamens, or the disc-glands between the filaments.   Rudimentary *ovary* absent.

XIII. **Fluggea.**—*Disc* as in *Phyllanthus.*   Rudimentary *ovary* well developed, tripartite.

XIV. **Drypetes.**—*Disc* central, entire or undulately lobed, with the stamens inserted around it.   *Fruits* indehiscent.

§§Flowers disposed in cymes, racemes or spikes.
‖Fruits not compressed or winged ; disc usually present in the male flowers.

XV. **Antidesma.**—*Flowers* disposed in slender catkin-like spikes.   *Fruits* small, drupaceous.   *Leaves* alternate.

XVI. **Pseudolachnostylis.**—*Flowers* disposed in cymes.   *Fruits* large, tardily septicidal.   *Leaves* alternate.

XVII. **Toxicodendron.**—*Male flowers* in dense pedunculate or subsessile cymules, female sessile, 1-3 in each leaf-axil.   *Fruits* early dehiscent.   *Leaves* in whorls of 4.

‖‖Fruits compressed, broadly winged ; disc absent from both sexes.

XVIII. **Hymenocardia.**—*Male flowers* in catkin-like spikes, female shortly racemose.   *Anthers* with a conspicuous yellow gland on the back.

Tribe 4.—CROTONEÆ.—*Sepals* usually small, closed or valvate, less often imbricate or open in bud.   *Petals* when present always free, often larger than the sepals ; usually petals 0.   *Stamens* 1-2-seriate ; the outer alternate with the sepals or more usually central and few or indefinite ; sometimes very many.   *Ovules* solitary in each cell.   *Cotyledons* much broader than the radicle.   *Inflorescence* various.

*Anthers reversed and filaments inflexed in bud, becoming erect in the open flower ; petals usually present in the male flower ; racemes or spikes terminal, androgynous or 1-sexual.

XIX. **Croton.**—*Sepals* usually equal, valvate or occasionally slightly imbricate.   *Petals* usually well developed, at times minute or obsolete.

**Anthers erect and filaments straight or twice flexed in bud.
†Petals present in the male flower.
‡Flowers panicled or fasciculate 2-3-chotomously cymose, androgynous with a central female flower.

XX. **Jatropha.**—*Sepals* imbricate, often connate below.   *Petals* usually imbricate, connate below or free.   *Stamens* 10 or fewer.

‡‡Flowers in axillary racemes or fascicles, rarely axillary solitary, androgynous or 1-sexual.

XXI. **Cluytia.**—*Sepals* imbricate. *Petals* free, equal. *Stamens* in a single whorl of 5 below the apex of a central column. *Capsule* septicidally 3-valved. *Flowers* in axillary fascicles or the female solitary, almost always diœcious.

XXII. **Caperonia.**—*Sepals* valvate. *Petals* free, usually 2 smaller than the others. *Stamens* 10 or fewer in 2 whorls on a column. *Capsule* of 3 2-valved cocci. *Flowers* in axillary androgynous racemes with few basal females.

††Petals 0.

‡Sepals not open in bud.

§Styles free or, if united, the column slender and continuous with columella.

‖Flowers in terminal racemes, spikes or heads ; filaments straight or twice flexed in bud, simple.

XXIII. **Cephalocroton.**—*Sepals* valvate, closed in bud. *Stamens* 6–8, 2-seriate ; filaments twice flexed in bud. *Styles* connate below in a short column ; free and multifid above.

‖‖Flowers in axillary, rarely terminal racemes or fascicles or cymules or spikes ; filaments straight in bud, simple.

¶Anther-cells distinct, attached to the filament by their base only.

⊙Anther-cells sessile ; sepals valvate, closed in bud.

XXIV. **Erythrococca.**—*Buds* perulate. *Stamens* 2–60. *Styles* plumosely laciniate or undivided.

XXV. **Micrococca.**—*Buds* naked. *Stamens* 3–30. *Styles* plumosely laciniate. *Capsule* 3-coccous.

XXVI. **Mercurialis.**—*Buds* naked. *Stamens* 8–20. *Styles* undivided. *Capsule* 2-coccous. *Leaves* usually opposite. *Racemes* axillary.

XXVII. **Leidesia.**—*Buds* naked. *Stamens* 3–7. *Styles* undivided. *Capsule* 2-coccous. *Leaves* usually alternate. *Racemes* terminal.

XXVIII. **Seidelia.**—*Buds* naked. *Stamens* 3, rarely 2. *Styles* undivided. *Capsule* 2-coccous. *Leaves* usually alternate. *Flowers* in terminal or axillary fascicles or cymules.

⊙⊙Anther-cells stipitate ; male sepals valvate, closed in bud, female sepals imbricate.

XXIX. **Acalypha.**—*Inflorescence* various. *Stamens* usually 8.

¶¶Anther-cells adnate laterally to a connective throughout their length.

XXX. **Alchornea.**—*Sepals* valvate, closed in bud. *Stamens* usually 8. *Styles* free, very long, entire.

‖‖‖Flowers in terminal panicles ; filaments straight in bud, usually simple ; anther-cells adnate to connective, usually more than 2.

XXXI. **Macaranga.**—*Sepals* in male flower valvate, closed in bud. *Stamens* usually 2–3, occasionally many. *Styles* short, stout.

‖‖‖‖Flowers in terminal panicles ; filaments repeatedly branched ; anther-cells distinct, subglobose, sessile.

XXXII. **Ricinus.**—*Sepals* in male flower valvate, closed in bud ; female calyx spathaceous. *Stamens* very many, up to 1000 ; filaments columnar below, much branched upwards. *Styles* usually 2-fid, more or less plumose.

‖‖‖‖‖Flowers in axillary fascicles or cymules sometimes passing into terminal pseudo-panicles ; male calyx-lobes imbricate ; styles 2-fid.

XXXIII. **Adenocline.**—*Stamens* 2-seriate, peripheral. *Styles* slightly connate below. *Flowers* in cymules. *Herbs.*

XXXIV. **Gelonium.**—*Stamens* central. *Styles* free. *Flowers* glomerulate. *Trees or shrubs.*

§§Styles connate in a column continuous with the body of the carpels ; male calyx-lobes valvate, closed in bud.

XXXV. **Plukenetia.**—*Ovary* 4-carpellary.

XXXVI. **Dalechampia.**—*Ovary* 3-carpellary. *Flowers* in dense involucrate heads.

XXXVII. **Ctenomeria.**—*Ovary* 3-carpellary. *Flowers* in racemes. *Stamens* 30 or more.

XXXVIII. **Tragia.**—*Ovary* 3-carpellary. *Flowers* in racemes or spikes. *Stamens* normally 3.

‡‡Sepals of male flower open in bud.

XXXIX. **Maprounea**—*Calyx-lobes* of male slightly imbricate. *Stamens* monadelphous. *Seeds* caruncled. *Spikes* very dense, ovoid.

XL. **Spirostachys.**—*Calyx-lobes* of male slightly imbricate. *Stamens* monadelphous. *Seeds* not caruncled. *Spikes* very dense, subcylindric.

XLI. **Sapium.**—*Calyx-lobes* of male not overlapping. *Stamens* free. *Seeds* not caruncled. *Spikes* rather lax.

I. SYNADENIUM, Boiss.

Apparent flower consisting of an entire shallowly cup- or saucer-like involucre, with a very spreading or more rarely erect rim-like gland outside of and completely surrounding an inner series of 5 inflexed-erect membranous subquadrate fringe-toothed lobes ; gland usually entire, occasionally having a cut-like notch on one side or divided by 2–5 cut-like notches into unequal or equal segments, but not forming equally-spaced separate glands. *Stamens* (really male flowers without a perianth, as in *Euphorbia*) arranged in 5 groups contained in 5 compartments with membranous walls opposite the lobes of the involucre. *Ovary* (really a female flower with the perianth reduced to a rudimentary rim or of 3 minute or rarely well developed conspicuous lobes, as in *Euphorbia*) pedicellate, 3-celled, often absent ; when present central and its pedicel surrounded by a membranous tube formed by the inner wall of the compartments containing the stamens, lobed and fringed at the top, puberulous ; styles 3, united at the basal part ; stigmas bifid, rarely entire. *Ovule* solitary in each cell, attached to the inner angle at or above the middle of the cell.

Shrubs or small trees, with the young branches fleshy, full of milky juice ; leaves alternate, entire, exstipulate, more or less fleshy, coriaceous when dried ; inflorescence axillary, cymose, cymose-paniculate or umbel-like, with a pair of free persistent or deciduous bracts at the base of each involucre and not or scarcely exceeding its rim-like gland.

DISTRIB. Species 13, all but the following in Tropical Africa.

1. S. arborescens (Boiss. in DC. Prodr. xv. ii. 187); a shrub 3–5 ft. high; branches terete, fleshy and green when young, finally woody, marked with leaf-scars, glabrous; leaves alternate, 2–4 in. long, ¾–1½ in. broad, cuneately obovate, acute or shortly cuspidate at the apex, cuneately tapering from above the middle into a short petiole, wing-keeled on the midrib beneath, glabrous on both sides; umbels axillary and terminal, ¾–1⅓ in. in diam., with 3–5 simple or once-forked rays; peduncles 5–9 lin. long, glabrous; rays 2–5 lin. long, puberulous; bracts under the involucre 1½ lin. long, 1¾–2 lin. broad, cuneately subquadrate, truncate and with a few minute teeth at the apex, puberulous on both sides; involucre 2–3 lin. in diam., broadly funnel- or bowl-shaped, puberulous on the basal part, with a spreading or ascending entire rim-like gland and 5 sub-quadrate fringed puberulous lobes, greenish-yellow; ovary only seen in a very immature state and included in the involucre, densely pubescent; styles 1½ lin. long, very shortly united at the base, deeply bifid at the apex, with spreading tips; capsule not seen. *Hook. f. in Bot. Mag. i.* 7184; *Wood, Natal Pl. iii. t.* 296. *Euphorbia arborescens, E. Meyer in Drège, Zwei Pfl. Documente,* 184, *not of Roxb. E. cupularis, Boiss. Cent. Euphorb.* 23. *E. synadenia, Baill. Adans.* iii. 142.

EASTERN REGION: Natal; in thickets near Umlazi River, below 500 ft., *Drège*, 4634! Inanda at 1800 ft., and in Durban Botanic Garden, *Wood*, 1623! 1651a! woods near Durban, *Wood*, 6377! 8492! Lower Umzimkulu River, 500 ft., *Wood*, and *cultivated specimens*!

Described from a living plant cultivated at Kew. A very poisonous plant. Mr. Wood states that when gathering specimens for the Herbarium, "after taking the precaution of covering his face, keeping at arm's-length from the plant and carefully washing hands and face as soon as the specimens were disposed of, he has felt the effects on the eyelids, nostrils and lips for several hours afterwards." Under cultivation at Kew, however, it does not seem so virulent, as I and others have frequently handled the plant without feeling the slightest effects from so doing.

II. ELÆOPHORBIA, Stapf.

Floral structure exactly as in *Euphorbia*, from which it only differs by its fruit, as follows:—*Fruit* indehiscent, with a thick flesh enclosing a hard bony 3-celled endocarp or "stone," marked with a slender groove down each of the 3 very obtuse angles and with a pore on each face near the apex between the grooves, also, when separated from the flesh, there is an opening at the base by which the central vascular bundle enters the "stone." *Seed* solitary in each cell, sometimes abortive in one or two of the cells; testa thin, crustaceous; albumen copious, somewhat fleshy; cotyledons flat, thick and fleshy.

Trees with succulent angular branches, becoming round and woody with age; leaves alternate, fleshy, entire, with a pair of spines at their base; peduncles axillary, simple or once or perhaps twice forked.

DISTRIB. Species 2, one of them a native of West Tropical Africa.

When not in fruit this genus can scarcely be distinguished from *Euphorbia*, the ovary, however, has thicker and more fleshy walls than are found in *Euphorbia*. Although in *Euphorbia ingens*, E. Meyer, and one or two others, the outer layer of the fruit is more or less fleshy, yet the endocarp always separates into its 3 component cells and is never consolidated into a "stone" as in *Elæophorbia*.

1. E. acuta (N. E. Br.); habit unknown, probably a tree, succulent, only a strip from an angle of a branch, which seems not to be toothed, with flowers and fruit seen; spine-shields separate, about 2 lin. long and $1\frac{1}{2}$ lin. broad, apparently rather soft, bearing a pair of diverging spines $\frac{1}{2}$–$\frac{3}{4}$ lin. long, light brown; peduncles about $1\frac{1}{2}$ lin. long, arising from the axils of the leaf-scars, glabrous, bearing a cyme of 3 (or more?) involucres at the apex; bracts 1–2 lin. long, very broadly rounded or acute at the apex, deeply concave, thin; involucre $\frac{1}{4}$ in. or probably rather more in diam., $1\frac{1}{3}$ lin. deep, rather shallowly and broadly cup-shaped, glabrous outside, with 5 glands and 5 very broad transversely oblong finely toothed lobes; glands contiguous, spreading, $1\frac{1}{2}$–$1\frac{3}{4}$ lin. in their greater diam., somewhat half-circular, with the ends deflexed and the inner margin straight and the outer broadly rounded, in the dried specimen both margins are turned upwards; ovary and styles not seen; fruit $\frac{7}{8}$–1 in. long, ellipsoid, tapering into an acute conical beak at the apex and into a stout stalk at the base, glabrous, with a fleshy outer layer enclosing a bony 3-celled stone or endocarp, the latter marked with a slight furrow down each of the 3 obtuse angles and with a small hole or deep pit on each side near the apex.

SOUTH AFRICA : probably from the Transvaal, *Burtt-Davy*!

Although the material is so scanty, it is sufficient to show the alliance of the plant with *E. drupifera*, Stapf, from the west coast of Africa, from which it distinctly differs in its smaller and beaked fruit; in *E. drupifera* there is scarcely any beak. It was sent to Kew without any information as to locality.

III. EUPHORBIA, Linn.

Apparent flower consisting of a number of stamens (really male flowers, each consisting of a single stamen jointed to a pedicel and soon falling away from it, without or rarely with a minute calyx just above the articulation) mingled with membranous or woolly scales or bracteoles, with or without a stalked or sessile ovary (really a pedicellate or sessile female flower, with or without a minute 3-lobed or very rarely cup-like or tubular calyx at the base of the ovary, but without a membranous tubular involucel surrounding the pedicel) in their midst, contained in a calyx-like cup-shaped involucre, the whole resembling a small hermaphrodite or male flower. *Involucre* a cup with an outer series of 2–8 (usually 4–5) glands, distinct and equally spaced or very rarely united, entire, petal-like, 2-horned or divided into teeth or processes on the outer margin, alternating with an inner series of 4–8 (usually 5) membranous erect or inflexed entire or toothed lobes. *Anthers* 2-celled; cells usually subglobose and more or less diverging, longitudinally dehiscent. *Ovary* wholly or partly included or exserted, 3- (rarely 2-) celled, with a single ovule in each cell, pendulous from the apex of the inner angle; styles 3, rarely 2, free or more or less united, but rarely to the apex, entire or bifid at the tips. *Fruit* a 3- (rarely 2-) celled capsule; cells separating at maturity from the central persistent axis and opening along their inner face into two valves, liberating the seed; inner part of the valves hard or cartilaginous. *Seed* with a thin crustaceous testa, smooth or variously sculptured, usually carunculate at the hilum; embryo straight, with flat cotyledons, enclosed in a thick albumen.

Herbs, shrublets, shrubs or trees, very variable in habit, leafy or leafless, often succulent or cactus-like, with copious milky juice; leaves alternate or the upper or all opposite, entire, toothed or rarely lobed; stipules present or absent, in some of the succulent species often transformed into prickles or spines above a pair of larger spines; involucres solitary and terminal or axillary or in the forks of the stems, or in axillary or terminal clusters or cymes or umbels, which are simple or compound, paniculate, racemosely arranged along the branches in pairs or rarely whorled, very rarely in axillary racemes or arising immediately behind the base of large solitary spines.

DISTRIB. Species about 1000, dispersed throughout the warmer and temperate regions.

This vast genus is remarkable, apart from its curious floral structure, for its great range of variability of habit, and in South Africa there are more distinct types of vegetative variation than in any other region of equal area. Yet however diverse in appearance the various types may be, the floral structure remains remarkably uniform, such variation as exists being chiefly confined to the glands of the involucre, which vary in number and also in their appendages. These differences in the involucral glands, however, grade into one another completely and sometimes vary in the same species and on the same individual, so that some specimens will have petal-like appendages or tooth-like processes to the glands, whilst other specimens of undoubtedly the same species will be entirely without such appendages or processes. Therefore I do not think that these characters can be legitimately used to divide this exceedingly natural genus into either smaller genera or sections, as has been done by some authors.

In Commelin, *Hort. Med. Amstelodam.* i. t. 17, is a very remarkable figure of a species of South African Euphorbia allied to and possibly the same as *E. Gorgonis,* but indeterminable, which under cultivation, doubtless owing to the more humid atmosphere, has developed slender herbaceous, terete, leafy tips to its native-grown, thick, fleshy, tuberculate branches, and one central branch that is entirely slender and terete like those of herbaceous species. Thus demonstrating, in all probability, the effect that dry climatic conditions have through long ages produced by changing a herbaceous perennial into a succulent plant. See the *Gardeners' Chronicle,* 1914, lvi. 230, fig. 91.

The spines of the South African species are of three types :—1, Where the apex of the branch becomes transformed into a sharp spine; this only occurs in *E. lignosa* and *E. spinea,* although one or two others have tapering branches, but they are not acutely spine-tipped. 2, Where the peduncle, either after bearing a flower or being abortive from its origin, becomes transformed into a hard sharp spine; all the species with solitary spines have them formed in this manner. 3, Spines placed in pairs under the leaf or leaf-scar; these only occur on succulent species with angular stems. In books these have been called "stipular spines," but as I have stated in the *Flora of Tropical Africa*, vi. § i. 471, they are always developed under and sometimes at a considerable distance below the leaf or leaf-scar and cannot be stipules in the ordinary sense of the term, I do not understand what their real relationship to the leaf is. Besides these pairs of spines, true stipules are, however, sometimes developed on the succulent species ; sometimes they take the form of small hard persistent auricles, at others are represented by minute points or small spines (well seen in *E. Schinzii*) and are always seated one on each side of the leaf-scar or base of the leaf, but in most South African species they are badly developed or entirely absent.

The formation of a key to the South African species I have found to be extremely difficult, as the distinctive characters that can be utilised are often exceedingly few, not always present at all periods, and so need supplementing by others, and mostly cannot be stated in few words, as there are many cases where it is perfectly obvious to the eye that two or more plants placed side by side are quite distinct species, yet the characters available for a key are so few, that it becomes exceedingly difficult to express what the eye instantly perceives in words that will enable one to discriminate the species when seen separately. This particularly applies to those succulent species I have described as leafless, a term that is not strictly correct, and must be understood to apply to the general appearance of the plant, for during the growing season leaves are present, but they are often so rudimentary and inconspicuous as not to be noticeable, and are usually deciduous. In the following key only characters that are apparently absolute have been made use of. In a few cases where a plant varies or dried specimens do not always give an adequate idea of the plant and it might appear to belong to either of two groups in the key, it has been inserted under both headings to facilitate identification.

A. **Plant herbaceous or sometimes woody below, never succulent nor spiny.**

I. *Stems and branches evident, erect or prostrate. Leaves always present, conspicuous and well developed, except in* 8, E. Pfeilii, 12, E. multifida, *and* 13, graveolens, *where they are small and sometimes deciduous.*

Flowering leaves or bracts with a conspicuous white area
at their base ... ... (15) **phylloclada.**
None of the leaves with a conspicuous white area at their
base :
*Annual or perennial herbs, erect or prostrate ; rootstock
not tuberous :
†Leaves all opposite (except in 14, *E. glaucella*, and
8, *E. Pfeilii*, where the few below the lowest
branch are usually alternate) ; involucres not in
umbels :
‡Leaf-blade from as broad as long to 4 times as
long as broad, variable but not linear nor
linear-lanceolate :
Involucres in axillary pedunculate cymes or
subglobose heads :
Stems erect or ascending, with conspicuous
spreading yellow hairs on their upper
part ; involucres in dense subglobose
heads ... ... (6) **hirta.**
Stems glabrous or finely adpressed-puberulous ;
involucres in small leafy cymes, more
rarely in dense heads :
Stems erect or ascending, often puberulous ;
leaves herbaceous, puberulous beneath
or on both sides ... ... (5) **hypericifolia.**

Stems prostrate or decumbent, always
  glabrous ; leaves coriaceous or some-
  what fleshy, glabrous on both sides ... (4) **livida.**

Involucres solitary and terminal or in the
  forks or at the nodes of the branches, 1 to
  each pair of leaves, often crowded along
  very short axillary branchlets with reduced
  leaves, rarely cymose :

Stems glabrous all round :

Stems and branches prostrate or rarely
  erect : leaves 1–5 lin. long ; involucres
  minute, 1 to each pair of leaves of the
  stems or of very short axillary leafy
  branchlets :

  Annual ... ... (2) **inæquilatera.**

  Perennial ... ... (2) **inæquilatera,**
                              var. *β.*

Stems and branches erect or ascending ;
  involucres solitary in the forks of the
  branches or terminal or sometimes
  cymose :

  Annual, 2–15 in. high ; leaves 3–9 lin.
  (or when better developed up to
  2½ in.) long ... ... (14) **glaucella.**

  Perennial, 9–18 in. high ; leaves 1½–2½
  lin. long ... ... (8) **Pfeilii.**

Stems puberulous on the upper side, at least
  along a middle line, glabrous beneath,
  always prostrate ; leaves 1½–3½ lin. long (1) **prostrata.**

Stems adpressed-pubescent or subtomentose
  all round, erect, 4–8 in. high ; leaves
  3–7 lin. long ... ... (3) **Schlechteri.**

‡‡Leaf-blade 5–12 times as long as broad, linear,
  linear-lanceolate or oblong-linear, glabrous on
  both sides ; erect plants ; involucres solitary
  in the forks of the branches or terminal :

Glands of the involucre with small petal-like
  white appendages ... ... (7) **neopoly-**
                                     cnemoides.

Glands of the involucre without petal - like
  appendages ... ... (14) **glaucella.**

††Leaves on the stem below the umbellate inflorescence
  all alternate and often different in form and
  colour from the opposite flowering leaves or
  bracts ; umbel always terminal, but sometimes
  with 1 or more ray-like branches arising below
  it from the axils of the upper leaves :

Stem-leaves distinctly but finely toothed :

Stem-leaves sessile, at least half as broad at the
  base as elsewhere, lanceolate or oblanceolate,
  pilose ... ... (18) **pubescens.**

Stem-leaves tapering into a petiole or a very
  narrow base, cuneately obovate, glabrous ... (16) **Helioscopia.**

Stem-leaves entire or minutely scabrous on the
  margin or at the apex, but not distinctly
  toothed :

Leaves (including the petiole when present)
  1¼–4 in. long, glabrous :

Leaves sessile or subsessile, linear or linear-
lanceolate, very acute or somewhat pungently pointed ; glands of the involucre
with a short point or horn at each end ... (24) **striata.**

Leaves cuneately tapering into a more or less
evident petiole, obtuse or acute, but not
pungently pointed :
Leaves ⅓-⅔ in. broad, linear, linear-lanceolate or cuneately oblanceolate ; involucre
1-1½ lin. in diam., with entire glands   (33) **kraussiana.**

Leaves ⅔-2 in. broad, elliptic, elliptic-ovate
or oblong-lanceolate ; involucre 2-2¾
lin. in diam. ; glands 2-3-toothed, or
lobed or subentire   ...   . (34) **transvaalensis.**

Leaves (including the petiole when present)
usually less than 1 in. (or in 32, *E. epicyparissias* ; 24, *E. striata* ; and 25, *E.
erythrina*, var. β, up to 1¼ in.) long :
Annual ; leaves tapering into a conspicuous
petiole, obovate, suborbicular or ovate ... (17) **Peplus.**

Perennials ; leaves sessile or very shortly
petiolate, but not tapering into a conspicuous petiole :
‡Leaves ovate, obovate, elliptic, oblong,
cuneately oblong, lanceolate or linear-
lanceolate :
§Leaves obtuse or subobtuse, apiculate :
Leaves more or less reflexed or very
spreading, not imbricate :
Leaves 2½-4 lin. long   ...   ... (19) **dumosa.**

Leaves 4-12 lin. long   ...   ... (32) **epicyparissias**
and vars.

Leaves ascending or ascending-spreading :
Leaves numerous or crowded, 8-40
to an inch of stem, usually more
or less imbricate :
Leaves ovate or broadest below the
middle, puberulous ...   ... (23) **sclerophylla,**
var. β.

Leaves linear-lanceolate to elliptic,
but not broadest below the
middle, glabrous, sometimes
minutely ciliate :
Leaves 2½-6 times as long as
broad, linear-lanceolate to
linear-oblong, with a
straight apiculus at the
apex   ...   . (25) **erythrina.**

Leaves not more (usually less)
than twice as long as broad,
obovate to elliptic-oblong,
with a recurved apiculus at
the apex :
Leaves minutely scabrous at
the margins, not ciliate ;
umbels ½-¾ in. in diam.,
with few involucres   ... (26) **foliosa.**

Leaves entire, very minutely
   ciliate; umbels ¾–1 in.
   in diam., with many
   crowded involucres ... (27) **artifolia.**

Leaves usually 3–8 to an inch of
   stem, lax or not distinctly
   imbricating :
Leaves ovate, 2–3 times as long
   as broad ... ... (20) **ovata.**

Leaves linear-lanceolate to narrowly cuneate-obovate, 3–9
   times as long as broad ... (25) **erythrina,** var. β.

§§Leaves acute and often pungently mucronate, glabrous except in 23, *E.
   sclerophylla,* var. β :
Leaves all regularly much deflexed,
   cordate at the base ... ... (31) **natalensis.**

Leaves very spreading or slightly
   deflexed :
Leaves ovate, slightly cordate at the
   base ; umbel 1½–5 in. in diam.... (21) **albanica.**

Leaves lanceolate, rounded or
   cuneately rounded at the base ;
   umbels ⅔–1 in. in diam. ... (29) **muraltioides.**

Leaves erect or ascending :
Leaves lax or imbricate, ½–2 in. long ;
   umbels often more than 2 in. in
   diam. ... ... (24) **striata** and vars.

Leaves more or less imbricate, ½–¾
   in. long ; umbels ¾–2 in. in
   diam. :
Leaves cordate at the base ; stems
   5–9 in. high ... ... (22) **ruscifolia.**

Leaves rounded at the base ; stems
   6–18 in. high ... ... (23) **sclerophylla.**

‡‡Leaves linear, often narrowly so, or linear
   from a dilated base :
Leaves all regularly deflexed, but often
   with their tips upcurved, sessile :
Leaves acute or roundedly obtuse at the
   apex, ½–3 lin. broad ; involucre
   1½–2½ lin. in diam. :
Stems 2–5 ft. high, with branches
   scattered along it or clustered
   at intervals ; leaves not dilated
   at the base ... ... (32) **epicyparissias**
                                        and vars.

Stems ½–2 ft. high, simple or once
   (rarely twice) umbellately
   branched at the top; leaves
   often, but not always dilated
   and cordate at the base ... (31) **natalensis.**

Leaves truncate and mucronate or
   minutely 3-toothed at the apex ;
   involucre 1¼–1½ lin. in diam. ... (30) **ericoides.**

Leaves erect, ascending or spreading :
   Leaves flat or with incurved (never
   revolute) margins, usually lax or
   distant, ½–2 in. long ; stems simple
   below the inflorescence ...   ... (24) **striata** and vars.
   Leaves flat or with revolute margins,
   very numerous and often crowded ;
   stems usually branching :
      Plant 1½–5 ft. high ; leaves 6–15
         lin. long ; umbel ½–3 in. in
         diam., with 3–8 rays ½–3½ in.
         long   ...   . (32) **epicyparissias.**
      Plant ⅓–1½ ft. high ; leaves 2–6
         (rarely 8–10) lin. long ; umbel
         ⅓–1¼ in. in diam., with usually
         4 rays 1–6 lin. long ...   ... (28) **genistoides.** ‧
**Perennial erect herbs 1–6 in. high, with a tuberous
   rootstock ; leaves alternate on the unbranched
   part of the stems, opposite at the flowering part and
   forkings of the branches, all similar or those on the
   stem not very unlike the flowering leaves or
   bracts, but sometimes reduced or very small :
   Bracts and leaves linear, linear-lanceolate, lanceolate
      or elliptic, glabrous (or in 11, *E. Gueinzii*, var. *β*,
      albovillosa, villose) on both sides ; involucres
      solitary or in lax terminal cymes, not in umbels :
      Glands of the involucre with 3–7 simple or forked
         processes or teeth along the outer margin :
         Involucre 3–4 lin. in diam., with the processes
            of the glands corrugated on their upper
            surface   ...   . (9) **pseudotuberosa.**
         Involucre 4–5 lin. in diam., with the processes
            of the glands not corrugated on their upper
            surface   ...   ..:   ...   . (10) **trichadenia.**
      Glands of the involucre entire or minutely crenu-
         late ; involucre 1–3 lin. in diam.   ...   ... (11) **Gueinzii.**
   Bracts (and probably the leaves) ovate, long-pointed,
      puberulous on both sides ; involucres (including
      the 1–1½ line-long processes of the glands) 5–6
      lin. in diam., in lax cymes   ...   . (12) **multifida.**
   Bracts oblong or oblong-obovate, rounded and
      toothed at the apex, glabrous on the back ;
      involucres 2½–3 lin. in diam., in a terminal
      umbel   ...   . (13) **graveolens.**

   II. *Stemless herbs with a tuberous rootstock.   Leaves petiolate, well developed and
together with the peduncles all radical and dying to the ground in winter.*

Blade of the leaf 4–12 lin. broad, subtruncate or
   shortly and rather abruptly narrowing into the
   petiole ; peduncles ¼–2 in. long :
   Blade of the leaf elliptic, broadest above the base ;
      styles ¾ lin. long, shortly united at the base   ... (80) **pistiæfolia.**
   Blade of the leaf oblong or lanceolate-oblong, often
      broadest at the base ; styles 1–1¼ lin. long, united
      for half their length...   ...   . (81) **tuberosa.**
Blade of the leaf 1–6 lin. broad, linear to elliptic, cuneate
   or very tapering at the base :

Leaves not wavy at the margins ; peduncles 1–5 in.
long ; umbels with 3–15 involucres   ...   ...   (82) **elliptica.**

Leaves wavy at the margins ; peduncles ½–1¼ in.
long, bearing 1 involucre or an umbel of 2–5
involucres   ...   .   (83) **crispa.**

**B. Woody shrubs 5–7 ft. high; branches sometimes tapering
to a point, but not distinctly spine-pointed, leafless or leaves
rudimentary, soon deciduous, alternate.**

Branches with numerous very prominent flowering
tubercles scattered along them ; bark dark brown ;
involucres puberulous ...   .   (35) **frutescens.**

Branches with slightly prominent leaf-scars, but no
prominent flowering tubercles ; bark light ochreous
brown ; involucres glabrous ...   .   (36) **guerichiana.**

**C. Shrublets 6–12 in. high with sharp spine-pointed tips to the
branches, more or less succulent at the young parts, becoming
subwoody. Leaves very small or rudimentary, soon deciduous.**

Leaves and branches alternate ; involucre 3¼–5 lin. in
diam. ...   .   (37) **lignosa.**

Leaves and branches opposite ; involucre 1 lin. in diam.   (38) **spinea.**

**D. Plant distinctly succulent, at least as to the branches, the
trunk or lower part of the arborescent or shrubby species becoming woody. Leaves often rudimentary or small, soon deciduous,
the plant often appearing leafless.**

I. *Plants quite spineless, but sometimes with dried or hard persistent peduncles or
cymes, and in a few species the branches taper to the apex, but never form a sharp
spine.*

*Shrublets, shrubs, bushily branched plants or (in 70, *E.*
Tirucalli) a tree, sometimes appearing more or less
woody when dried ; branches with or without scattered
(rarely closely placed) tubercles or prominent leaf-
scars, stout or terete and slender :
†Leaves, leaf-scars or the scattered tubercles formed by
the leaf-bases and the branches all alternate (see
also 129, *E. clava*) :
Branches with distant conical tubercles (persistent
bases of the petioles) 1–9 lin. prominent scattered
along them :
Branches, bracts and involucres minutely puberulous ; glands of the involucre erect, contiguous, forming a cup with a very shortly and
densely toothed margin   ...   .   (74) **peltigera.**

Branches, bracts and involucres glabrous ; glands
of the involucre not forming a cup, entire
or minutely crenulate on the outer margin :
Leaves sessile upon the rather stout conical
stem-tubercles ; involucres 2½–3½ lin. in
diam. ; styles united into a column 1½–1¾
lin. long, with short bifid tips   ...   ...   (75) **hamata.**

Leaves with rather slender petioles 1–5 lin.
  long ; involucres 1¼–1¾ lin. in diam. ;
  styles united into a column ½–¾ lin. long,
  with entire acute tips ...   . (76) **gariepina.**

Branches cylindric, without distinct conical tubercles,
  but often with prominent leaf-scars scattered
  along them (69, *E. patula*, Mill., may belong
  here. See also specimens with reduced tubercles
  of 75, *E. hamata* and 76, *E. gariepina*) :
Leaves well developed and present at the time
  of flowering or always, mostly 2–6 (or
  occasionally only 1–1½) in. long, linear-lanceolate, narrowly oblong-lanceolate, lanceolate
  or oblanceolate ; shrubs 2–5 ft. high ;
  peduncles withering and persisting, 3–6 in.
  long or sometimes shorter :
Young branches and peduncles puberulous ;
  leaves lanceolate or oblong-lanceolate, ob-
  tuse ; involucres 1¾–2 lin. in diam.   ... (128) **oxystegia.**

Young branches and peduncles glabrous ; involucres 2½–3 lin. in diam. :
Leaves 3–15 lin. broad, cuneately oblanceolate, obtuse or subacute, apiculate ; bracts
  under the involucre rather abruptly and
  shortly acute or obtuse and apiculate ... (126) **bubalina.**

Leaves 3–5 lin. broad, linear-lanceolate, acute ;
  bracts under the involucre gradually
  acute or acuminate   ...   . (127) **tugelensis.**

Leaves often absent from the branches, when
  present rudimentary or ⅛–1 in. long ; shrubs
  or bushes 2–6 ft. high :
Involucres in umbels or cymes, with the
  peduncles or rays 2–12 lin. long :
Dried stems usually 1½–3 lin. (rarely more)
  thick ; rays of the umbel simple, with
  one glabrous involucre ; capsule glabrous (68) **mauritanica.**

Dried stems usually 4–6 lin. thick ; rays of
  the umbels or cymes divided, bearing 3
  or ultimately more puberulous involucres ;
  capsule puberulous   ...   . (71) **dregeana.**

Involucres sessile or in subsessile cymes
  clustered at the ends of the branches :
A tree 10–20 ft. high ; involucre 1½ lin. in
  diam., minutely puberulous or thinly
  tomentose at the upper part outside   ... (70) **Tirucalli.**

Bushes 3–4 ft. high ; dried flowering branches
  2–4 lin. thick ; involucre 2–2½ lin. in
  diam. and together with the capsule
  puberulous or tomentose :
Capsule or mature ovary with its base just
  exserted from the involucre, erect ;
  male involucre purple with a whitish
  tomentum   ...   . (72) **gummifera.**

Capsule and ovary exserted on a recurved
  pedicel ¼–½ in. long and together with
  the male involucre densely tawny-
  tomentose   ...   . (73) **gregaria.**

Plant 3–6 in. high, bushily much branched ;
branches alternate and opposite on the
same plant ; involucre 1–1¼ lin. in
diam., glabrous outside (and see 65, *E.
stapelioides*) ... ... (64) **gentilis.**

††Leaves, leaf-scars and branches (except by abortion)
all opposite ; leaves rudimentary or very small,
½–1½ lin. long :
Glands at the base of the rudimentary leaves or at
the sides of the leaf-scars large and very conspicuous, subglobose, persistent :
Ultimate flowering branches (dried) 1 lin. thick ;
involucres sessile or in small subsessile
terminal cymes ... ... (66) **karroensis.**

Ultimate flowering branches (dried) ⅓–⅔ (rarely
1 lin.) thick ; involucres in terminal pedunculate cymes ... ... (48) **Burmanni.**

Glands at the base of the leaves or at the sides of the
leaf-scars none or very inconspicuous (in a few
species they are present at the base of young
bracts and leaves, but are small and soon disappear) :
Branches terete, puberulous ; involucres subsessile or very shortly pedunculate, in
opposite pairs along or at the ends of the
branches, 1–1¼ lin. in diam., with entire
glands ... ... (67) **spicata.**

Branches more or less angular or terete, rough
(under a lens), especially along the angles,
with more or less evident asperities (not
hairs) upon the surface, apart from wrinkles
caused by shrinkage in drying :
Asperities formed of small laterally compressed
tubercles, which sometimes seem to form
very narrow, wavy, crenulate wings ;
branches forking into terminal cymes :
Primary branches (dried) 1¼–2 lin. thick,
often alternate, distinctly angular and
the asperities very evident ; ultimate
divisions of the cymes 1½–3 lin. long ... (46) **muricata.**

Primary branches (dried) ⅔–1 lin. thick,
opposite, scarcely or obscurely angular
and the asperities often scanty ; ultimate divisions of the cymes usually 3–9
lin. long, sometimes less ... ... (52) **arceuthobioides.**

Asperities formed of minute acute points or
blunt papillæ :
Rudimentary leaves (not the bracts) with
a minute spreading point or angle on
each side at the base ; cymes short,
racemosely arranged ; involucres dark
purple ... ... (60) **caterviflora.**

Rudimentary leaves without points or angles
at their base ; involucres not dark
purple :
Branches slender, terete, divided at their
ends into lax cymes ¾–1¼ in. in diam. (57) **tenax.**

Branches distinctly angular when dried,
with solitary involucres or small cymes
up to 4 lin. in diam. racemosely
scattered along them ...   . (59) **aspericaulis.**

Branches smooth (often wrinkled when dried, but
without tubercles or other asperities as seen
under a lens), glabrous ; leaves and bracts
rudimentary :
Involucres puberulous all over outside, solitary
and terminal or in small terminal and
lateral cymes ¼–½ in. in diam. :
Rudimentary leaves lanceolate or elliptic,
acute or obtuse, apiculate   ...   ... (43) **cibdela.**

Rudimentary leaves spathulate, subtruncate
or broadly rounded at the apex ...   ... (62) **macella.**

Involucres glabrous outside or in 45, *E. Rudolfii,*
minutely puberulous on the basal part,
glabrous above :
‡Cymes or solitary involucres terminal or
apparently so :
Cymes crowded into dense corymbose
masses :
Branches and branchlets of the cymes
(dried) 1–1¼ lin. thick ; bracts very
broadly subcordate-ovate, dark pur-
ple or dark brown ...   . (50) **angrana.**

Branches and branchlets of the cymes
(dried) ½–⅔ lin. thick ; bracts spathu-
late, green ...   . (49) **corymbosa.**

Cymes not densely crowded into masses :
Bracts or leaves sessile, deltoid, deltoid-
ovate or deltoid-hastate, acute or
subacute, shorter than the involucre :
Plant apparently 3–9 in. high :
Leaves and sometimes also the bracts
with an angle or acute spreading point on each side at the
base :
Main stems or branches with
compact or crowded ascending
branchlets 1 in. or less long ... (63) **hastisquama.**

Main stems or branches with
widely spreading lax branch-
lets more than 1 in. long   ... (40) **brachiata.**

Leaves and bracts without angles or
spreading points at the base :
Involucres in somewhat lax cymes
½–1 in. in diam., formed by
the primary branches forking
4–5 times ...   . (47) **perpera.**

Involucres in small sessile cymes
or clusters ¼–⅓ in. in diam. at
the ends of widely spreading
branches or of their second or
rarely third forkings...   ... (41) **chersina.**

Plant more than 9 in. high :
Primary lateral branches 1½–2 lin.

thick, usually with 1-2 pairs of
lateral branchlets below the
forked or simple ends ...   ... (44) **amarifontana.**

Primary lateral branches 1-1½ lin.
thick, repeatedly forking into a
lax cyme 3-8 in. across   ... (45) **Rudolfii.**

Bracts or leaves with a distinct petiole-
like part, spathulate, rounded to
subtruncate at the entire apiculate
or denticulate apex, about as long
as or shorter than the involucre :
Plant 6-8 in. high ; branches diverging,
with internodes 3-12 lin. long ;
ovary exserted 1-2 lin. beyond
the involucre   ...   . (51) **Æquoris.**

Plant 9-12 in. or more high ; internodes mostly ⅔-2¾ in. long :
Branches and branchlets erect, subparallel ; male involucres 1-1¼
lin. in diam.   ...   . (54) **ephedroides.**

Branches and branchlets diverging
from each other at an angle of
50°-70° ; male involucre 1¼-
1½ lin. in diam. :
Styles ¼-⅓ lin. long, free to the
base   ...   . (54) **spartaria.**

Styles ½ lin. long, united into a
short column at the base   ... (56) **rectirama**
(browsed specimens).

‡‡Cymes or (in a few species) solitary involucres
arranged racemosely in pairs along the
branchlets and also often a terminal one,
or sometimes in a subspicate manner near
or at their tips :
Plant 3-6 in. high ; branches thick, very
fleshy, brittle when dried ; cymes or
involucres few, subsessile at the tips
of very short terminal branchlets (and
see 65, *E. stapelioides*)...   ...   ... (64) **gentilis.**

Plant usually more than 1 ft. high, not
very brittle when dried ; cymes in
several or numerous pairs scattered
along the usually elongated branches :
Bracts under the involucres sessile, without a petiole-like part, brown to
blackish-brown :
Cymes subsessile or on peduncles less
than ½ in. long ; branches 1-2½
lin. thick :
Branches diverging from each other
at an angle of 90°-150° ; bracts
not thick and fleshy, deltoid-
ovate, acute, concave ...   ... (39) **decussata.**

Branches diverging from each other
at an angle of 60°-90° ; bracts
thick and fleshy, suborbicular or
oblong, obtuse, apiculate   ... (61) **Mundii.**

Cymes on peduncles $\frac{1}{2}$–1 in. long ;
  branches $\frac{2}{3}$–1 lin. thick, suberect,
  diverging from each other at the
  base at an angle of 45°–50° ... (56) **arrecta.**

Bracts under the involucres spathulate,
  with a distinct petiole-like part or
  obovate and distinctly narrowed to
  the base, green or brown :
  Branches 1$\frac{1}{2}$–2 lin. thick, diverging
    from each other at an angle of
    90°–110° and spreading ; cymes
    sessile or subsessile ... ... (42) **indecora.**

Branches $\frac{2}{3}$–1 lin. thick, diverging from
  each other at the base at an angle
  of 40°–80°, then erect or ascending ; involucres solitary and pedunculate or 2 to several in pedunculate
  cymes :
  Bracts persisting during the flower-
    ing period or maturing of the
    fruit ; western region ... ... (53) **spartaria.**

Bracts quickly falling off or before
  the maturing of the fruit ;
  Kalahari, central and eastern
  regions :
  Plant 2–3 ft. high, with stems
    2–3 lin. thick at the base ... (55) **rectirama.**

Plant 1–2 ft. high, with stems
  1–2 lin. thick at the base :
  Involucre solitary or occasionally
    2–3 on peduncles usually
    1–6 lin. (or more) long ... (58) **rhombifolia.**

Involucres 3 together, sessile in
  a cluster on peduncles 1–2
  lin. long ... ... (58) **rhombifolia,**
                                    var. β.

Involucres three to many in
  1–5-times forked cymes :
  Cymes (including the
    peduncles) $\frac{1}{2}$–1$\frac{1}{4}$ in. long,
    not very lax ... ... (58) **rhombifolia,**
                                      var. γ.
  Cymes (including the
    peduncles) 1–2$\frac{1}{4}$ in. long,
    very lax ... ... (58) **rhombifolia,**
                                      var. δ.

**Stem, rootstock or main body of the plant usually very
  much thicker than the branches, and globose, pear-
  shaped or cylindric, often partly or quite buried in
  the ground, or in a few species unbranched and
  cylindric, or when branched scarcely thicker than
  the branches, or branching at the base in a clump-
  like manner ; branches cylindric, not jointed, and
  together with the stem always covered with closely
  placed tubercles separated (when alive) by impressed
  lines :
†Peduncles persisting several seasons, withering or in a
  few species remaining green :

‡Peduncles usually more than 1 (sometimes 3–7) in.
long, but in 111, *E. albertensis,* 122, *E. tuberculatoides,* 123, *E. tuberculata,* 124, *E. Bolusii,*
125, *E. Macowani* and 130, *É. pubiglans* some
of them are less than 1 in. long :
§Leaves ½–6 in. long, deciduous or always present
(not seen in 124, *E. Bolusii,* but suspected to
be elongated) :
Glands of the involucre entire ; branching plants
½–5 ft. high ; peduncles mostly 2–7 in. long,
sometimes (in dry seasons ?) less :
Tubercles on the stem and branches not very
    conspicuous or prominent and not
    crowded ; glands of the involucre glabrous
    on the upper surface :
    Young branches and peduncles puberulous ;
      involucre 1¾–2 lin. in diam. ...   ...   (128) **oxystegia.**

    Young branches and peduncles glabrous :
      Involucre 2½–3 lin. in diam. ; styles with
       the united part ⅓–½ lin. long :
      Leaves 3–5 lin. broad, linear-lanceolate,
       acute ; bracts under the involucre
       gradually acute or acuminate   ...   (127) **tugelensis.**

      Leaves 3–15 lin. broad, cuneately
       oblanceolate, obtuse or subacute ;
       bracts under the involucre rather
       abruptly and shortly acute or ob-
       tuse and apiculate...   ...   ...   (126) **bubalina.**

    Involucre about 4 lin. in diam. ; styles
      with the united part 1–2¼ lin.
      long   ...   .   (129) **clava.**

    Tubercles on the glabrous stem and branches
      densely crowded, conspicuous, hemi-
      spherical ; glands of the involucre puberu-
      lous on the border of the upper surface...   (130) **pubiglans.**

Glands of the involucre 2-lobed or with 2–7
entire or forked processes or teeth along
their outer margin ; plants ½–2 ft. high
(183, *E. Haworthii,* Sweet, probably
belongs here).
Flowering stems and sometimes the branches
when dried ¾–1½ in. thick :
Peduncles 3–4½ in. long   ...   .   (131) **restituta.**
Peduncles ½–1¼ in. long   ...   .   (123) **tuberculata.**

Flowering branches when dried ¼–½ in. thick :
Involucre 6–7 lin. in diam., villous-
    pubescent outside and on the pro-
    cesses of its glands   ...   .   (125) **Macowani.**

Involucre 3–5 lin. in diam., glabrous or
    nearly so outside and also on the
    processes of its glands :
    Bracts whorled close under the involucre :
      Glands of the involucre distant, 1 lin.
       broad, with 2–4 abruptly reflexed
       entire processes ...   .   (113) **brakdamensis.**
      Glands of the involucre subcontiguous,
       2–2½ lin. broad, with 4–6 spread-
       ing branched processes   ...   ...   (124) **Bolusii.**

Bracts scattered singly along the peduncle ;
  glands of the involucre not contiguous, 1½–2 lin. broad, with 3
  spreading bifid processes ...   ... (112) **filiflora.**

§§Leaves ¾–6 lin. long, deciduous (see also 124,
  E. Bolusii above, of which the leaves are
  unknown) :
Flowering branches 3–15 in. long, arising from
  a thick rootstock ; involucre 5–7 lin. in
  diam. :
Flowering branches when dried 6–9 lin. thick ;
  tubercles very prominent ; leaves 3–5
  lin. long ...   . (122) **tuber-**
                                              culatoides.

Flowering branches when dried 2–4 lin. thick,
  tubercles not very prominent ; leaves
  1–3 lin. long   ...   . (120) **marlothiana.**

Flowering branches ¾–1 in. long, 1½ lin. thick
  when dried, numerous, arising from a thick
  cylindric or subglobose rootstock or main
  stem ; involucre 2–2¼ lin. in diam.   ... (111) **albertensis.**

‡‡Peduncles mostly 1–6 lin. long, but in 109, *E. rudis*,
  110, *E. inelegans*, 115, *E. Braunsii*, and 121,
  E. Muirii, sometimes 7–11 lin. long, and in 120,
  E. marlothiana, and 122, *E. tuberculatoides*,
  varying from 4–24 lin. long (see also species
  which sometimes have short peduncles mentioned
  under the paragraph ‡) ; main body of the plant
  or rootstock rarely rising more than 3–4 in.
  above the ground and sometimes buried in it,
  always much thicker than the branches arising
  from it, obconic, subglobose or cylindric :
Tubercles on the branches or many of them tipped
  with a conical acute hard whitish point ;
  involucre 3½ – 4½ lin. in diam., usually
  pubescent outside   ...   . (114) **namaquensis.**

Tubercles on the branches tipped with a small
  truncate or obtusely conical whitish leaf-scar :
Cup of the involucre pubescent or puberulous
  outside ; main body of the plant or rootstock buried in the ground ; branches erect
  or ascending, simple or branching at or near
  the top, 3–15 lin. long and 3–5 lin. thick :
Undivided part of the involucral glands as
  long as broad, without a short lip or
  turned up margin at its base   ...   ... (120) **marlothiana.**

Undivided part of the involucral glands twice
  as broad as long, with a lip or turned up
  margin at its base   ...   . (121) **Muirii.**

Cup of the involucre glabrous on the outside,
  but sometimes the lobes or glands are
  pubescent or woolly outside :
Tubercles on the branches slightly or very
  distinctly recurved ; body of the plant or
  rootstock buried in the ground, with only
  the branches rising shortly above it ;
  glands of the involucre with 2–4 short
  teeth :

Branches usually clavate, ¾–2 in. long ;
tubercles slenderly conical, 1½–2½ lin.
prominent ; leaves 1½–2 lin. long   ... (103) **hypogæa.**

Branches cylindric, 1–6 in. long ; tubercles
shortly conical, 1 lin. prominent ;
leaves 3–4 lin. long   ...   . (109) **rudis.**

Tubercles on the branches not at all recurved :
Body of the plant or rootstock buried in
the ground ; branches cylindric, erect
or ascending, 6–15 in. long ...   ... (122) **tuber-**
culatoides.

Body of the plant or the top of it rising
above the ground ; branches either
less than 6 in. long or radiately spreading, not erect :
Branches in about 3 series, very short,
cylindric or subglobose, the longest
about 4 lin. long (probably longer
under cultivation), 3–4 lin. apart ;
glands of the involucre entire, with
a cup-like cavity ...   . (102) **brevirama.**

Branches cylindric, clavate or tapering
upwards, usually in more than 3
series or regularly arranged all over
the top of the main body of the
plant, radiating or erect, the
longest varying in different species
from ¾–15 in. long ; glands of the
involucre with teeth or processes
along the outer margin (182, *E.*
procumbens, Mill., may belong here) :
Involucre with a dense mass of woolly
white hairs exserted from it and
its glands either with conspicuous
white appendages or woolly on the
top ...   . (117) **inermis** and
var. β.

Involucre either without exserted
woolly white hairs or if they are
present are not very dense and
the glands are then without conspicuous white appendages :
Outer branches usually 4–15 in. long,
radiating :
Leaves 1½–2½ lin. long ; involucres
4–6 lin. in diam., having
green glands with white pro-
cesses ...   . (118) **Caput-Medusæ.**

Leaves 3–6 lin. long ; involucres
2½–3½ lin. in diam. having
green glands with greenish-
white teeth ...   . (119) **Bergeri.**

Outer or longest branches ¾–3 in.
long, erect or radiating :
Branches when dried mostly ½
in. or more thick at the upper
part, more or less clavate,

crowded a n d completely
covering the top of the main
stem ... ... (115) **Braunsii.**

Branches when dried less than $\frac{1}{2}$
in. thick, cylindric or slightly
narrowing upwards, not
always covering the top of
the main stem to its centre:
Styles united into a column
$\frac{3}{4}$–$1\frac{1}{2}$ lin. long ; glands of
the involucre green or
yellowish-green :
Tubercles on the branches
scarcely prominent ;
peduncles about $1\frac{1}{2}$ lin.
long, remaining green and
fleshy as they persist ... (101) **Huttonæ.**
Tubercles on the branches
distinctly although
slightly prominent ;
peduncles 2–6 lin. long,
withering as they persist :
Main stem subcylindric :
involucre 2 lin. in
diam., with dull olive-
green glands ; ovary
thinly pubescent ... (106) **arida.**

Main stem globose ; in-
volucre $2\frac{1}{2}$–3 lin. in
diam., with bright dark
green glands ; ovary
glabrous ... ... (107) **decepta.**
Styles only shortly united at
their base or for $\frac{1}{4}$–$\frac{1}{2}$ lin. :
Glands of the involucre
brown or chocolate-
coloured :
Branches covering the
whole top of the main
stem and when they
fall away their bases
form tubercles on the
naked part of the
stem ... ... (116) **baliola.**

Branches arranged in many
series around a central
branchless tuberculate
area at the top of the
main stem, not form-
ing tubercles on the
stem when they fall
away ... ... (105) **fusca.**
Glands of the involucre green
or not brown or chocolate-
coloured :
Main stem bearing branches
all over the top, no
central branchless area ;
leaves 8–18 lin. long ... (104) **namibensis.**

Main stem with a flattened
or depressed tubercu-
late central area desti-
tute of branches :
　Peduncles 4–9 lin. long ;
　　involucre 2½–3½ lin.
　　in diam. ... ... (110) **inelegans.**
　Peduncles 1½–5 lin. long ;
　　involucre 2–2¼ lin.
　　in diam. ... ... (108) **crassipes.**
††Peduncles deciduous, never persisting after 1 season :
　Leaves 1½–6 in. long ; stem simple, globose to
　　cylindric, 1½–3 in. thick ... ... (84) **bupleurifolia.**
　Leaves mostly ⅔–1½ in. long :
　　Stem cylindric, unbranched (always ?), 6–12 in.
　　high ; glands of the sessile involucre entire ... (85) **clandestina.**
　　Stem subglobose, with 2–3 series of branches 1½–3
　　in. long ; glands of the shortly pedunculate
　　involucre with 2–3 short teeth on the outer
　　margin ... ... (86) **Davyi.**
　Leaves less or not more than ½ in. long, deciduous :
　　Main stem or body of the plant with a large (or
　　small in 88, *E. basutica*) flat or depressed
　　tuberculate central area destitute of branches
　　at the top (181, *E. parvimamma*, Boiss., and
　　182, *E. procumbens*, Mill., may belong here) :
　　Leaves usually 2–6 lin. long, linear or linear-
　　lanceolate :
　　　Glands of the involucre with processes branch-
　　　ing at their tips ; ovary pubescent ... (87) **ramiglans.**
　　Glands subentire or with small teeth or
　　crenations :
　　　Peduncles 0–1 lin. long ; ovary glabrous,
　　　with styles 1–1¼ lin. long ... ... (93) **gatbergensis.**
　　　Peduncles 1–9 lin. long :
　　　　Glands whitish-green, represented as long
　　　　as broad ; styles with slender tips ... (94) **pugniformis.**
　　　　Glands yellow or greenish-yellow, some-
　　　　times changing to red ; styles with
　　　　very broad tips :
　　　　　Outer series of branches mostly less
　　　　　than 1½ in. long ; ovary puborulous (07) **Flanagani.**
　　　　　Outer series of branches mostly 2 in.
　　　　　or more long :
　　　　　　Branches 4–6 lin. thick, full grown
　　　　　　always radiately spreading ;
　　　　　　ovary pubescent ... ... (96) **passa.**
　　　　　　Branches 2½–3 lin. thick, full grown
　　　　　　at first erect, finally radiately
　　　　　　spreading ; ovary glabrous or
　　　　　　thinly hairy :
　　　　　　　Peduncles 4–9 lin. long ... ... (98) **Franksiæ.**
　　　　　　　Peduncles 1–3½ lin. long ... ... (99) **Woodii.**
　Leaves ½–1½ lin. long, lanceolate to elliptic :
　　Branches 2–8 in. long and ½–1 in. thick ;
　　involucre 1¾–2¼ lin. in diam., with brown
　　glands ... ... (92) **esculenta.**

Branches ⅓–1½ in. long and ¼–½ in. thick:
 Involucre 3½–4 lin. in diam., with greenish
  glands: ovary glabrous   ...   ... (88) **basutica.**

Involucre 2½–2¾ lin. in diam., with dark
 brownish-crimson to bright red glands;
  ovary pubescent   ...   . (95) **Gorgonis.**

Main stem covered with branches, nearly or
 quite to the centre of its top, without or with
 a very small branchless tuberculate area;
 glands of the involucre greenish or yellow:
 Branches laxly scattered over the top of the
  plant; peduncles 1½–2½ lin. long; ovary
  thinly pubescent   ...   . (100) **discreta.**

Branches crowded; peduncles none or not more
 than ½–1 lin. long:
 Branches with age repeatedly branching at
  their tips forming a convex cushion-like
  mass; ovary glabrous ...   . (90) **clavarioides.**

Branches simple or with here and there a
 lateral branch:
 Branches 4–8 lin. thick:
  Branches subcylindric, truncate at the
   apex, closely contiguous and crowded
   into a compact flat-topped mass;
   ovary thinly hairy ...   . (91) **truncata.**

  Branches clavate, rounded at the apex,
   somewhat crowded, but not closely
   contiguous, at least at the tips;
   ovary glabrous   ...   . (88) **basutica.**

 Branches 3–4 (when dried 2–3) lin. thick,
  cylindric, crowded, ovary glabrous ... (89) **Ernesti.**

***Plant dwarf, consisting of a single unbranched or
  occasionally branching globose, obconic or cylindric 8–12-angled stem, not tuberculate or only
  along the angles; cymes puberulous:
 Stem cylindric-oblong, 6–8 in. high, 3–3½ in. thick;
  cymes erect, 1¼–2 in. long and broad, 3–4-times
  forked, on peduncles 3–10 lin. long   ...   ... (155) **valida.**

 Stem obconic or pear-shaped, as thick as long;
  cymes ascending, ½–1 in. long, once or twice
  forked, on peduncles 1–2 lin. long   ...   ... (158) **pyriformis.**

 Stem globose, as thick as long:
  Scars along the angles crowded, ½–¾ lin. apart;
   cymes apparently deciduous; ovary glabrous (159) **obesa.**

  Scars along the angles not crowded, 1½–3 lin.
   apart; cymes persistent or deciduous on
   different plants of the same species:
   Cymes divided close to the base into 2–3 once
    or twice-forked branches and spreading
    over and close to the top of the plant as if
    pressed upon it, at least when in flower;
    ovary puberulous   ...   . (156) **meloformis.**

   Cymes with peduncles ½–1 in. long, erect or
    ascending and standing out from the plant (157) **infausta.**

****Plant very dwarf, consisting of a small clump of
   branches constricted at their origin or renewal of

growth into globose, clavate or cylindric joints, more or less tuberculate or with prominent leaf-scars ; glands of the involucre with 3–5 subulate or finger-like processes :

Branches or joints globose or clavate ; peduncles usually ½–3 in. long, sometimes shorter ; styles 1½ lin. long ...   . (79) **globosa.**

Branches or joints mostly cylindric or slightly tapering upwards, occasionally oblong or sub-globose :

Peduncles ½–3 in. long, simple or forked into 2 or 3 rays ; styles 2–3 lin. long ...   . (78) **ornithopus.**

Peduncles 1–2 lin. long, simple ...   . (77) **tridentata.**

II. *Plants armed with hard sharp spines.*

*Spines not in pairs, solitary or (in 150, *E. horrida* and 151, *E. polygona*) with 1–4 smaller spines clustered at the base of the main spines, simple or stellately branched, formed of modified peduncles :

†Stems or branches tessellately marked into rhomboid areas or covered with conspicuous or slightly evident tubercles arranged in spiral series not forming distinct vertical angles ; spines simple :

Spines and peduncles similar, ½–2 in. long, ½–1 lin. thick at the base, straight ; glands of the involucre entire :

Plant much branched (bushy ?) 1–3 ft. high ; leaves 1–3 in. long ; spines usually grey or brown ...   . (135) **loricata.**

Plant with crowded branches forming a densely spiny hemispherical cushion 4½–6 in. high ; leaves ¾–1¼ in. long ; spines white ...   ... (134) **Eustacei.**

Spines ¼–2¼ in. long, 1–2½ lin. thick at the base, much longer and stouter than the peduncles, curved ; glands of the involucre with 2–8 teeth or processes, ½–¾ lin. long on the outer margin :

Stem unbranched, obconic or cylindric with stout spines arising from stout tubercles (in dried specimens the spines sometimes falsely appear to come from the axils of the tubercles); peduncles 6–12 lin. long, arising behind the spines ...   . (132) **fasciculata.**

Stem or main axis thick, conical, ½–2 ft. high, densely covered nearly to the top with short tuberculate branches intermingled with or bearing spines ; flowering peduncles 3–4 lin. long, arising from the axils of the tubercles at the tips of the branches ...   . (133) **multiceps.**

††Stem with 5–20 distinct acute or tessellately tubercu-late vertical or slightly spiral angles ; spines scattered along the angles :

‡Spines or some of them forked or stellately branch-ing into smaller spines at the apex :

Stem 10–16-angled, uniformly green; peduncle with 1 involucre :

Stalk of spine-clusters 1½–4 lin. long ; spines ½ lin. thick, brown ...   . (152) **stellæspina.**

Stalk of spine clusters ½–2½ lin. long; spines ¾–1 lin. thick, grey ... ... (153) **astrispina.**

Stem about 5–7-angled, transversely banded with pale green and darker green; peduncles usually bearing an umbel of 2–6 involucres, but occasionally only one ... ... (154) **Pillansii.**

‡‡Spines all entire:

Stems 2–4½ in. thick, with 10–20 angles; peduncles 1½–3 lin. long:

Stem-angles 1–1¼ in. prominent; main spines 4–10 lin. long, stout, with 2–4 smaller spines at their base, all crowded ... ... (150) **horrida.**

Stem-angles ½–⅔ in. prominent; main spines 2–4 lin. long, solitary or with 1–2 smaller spines or remains of peduncles at their base (151) **polygona.**

Stems or flowering branches ⅔–2 in. thick:

Stems divided along the angles into very short and very obtuse tubercles by impressed lines (tessellately tuberculate); spines usually reddish or purplish when young, becoming pale brown or grey; peduncles ¾–2 lin. long:

Plant 4–8 in. high; branches ⅔–1 in. thick, 7–10-angled; involucre dark purple ... (141) **submammillaris.**

Plant 1–3 ft. high; branches 1–1¾ in. thick, 7–17-angled; spines ½–¾ lin. thick:

Tubercles on the stem-angles without or with no very evident raised line across them; involucre dull purple; styles united into a column 1 lin. long ... (142) **mammillaris.**

Tubercles on the stem-angles with a very evident raised line across them; involucre green; styles united into a column ½ lin. long ... ... (143) **fimbriata.**

Stems not divided along the angles into tubercles by impressed lines, but the angles sometimes toothed:

Spines frequently or mostly curved, often 1 lin. thick at the base, some smaller:

Stems or branches 1¼–1¾ in. thick, 9–12-angled; spines ½–1¼ in. long, brown, becoming grey ... ... (148) **ferox.**

Stems or branches ⅞–1¼ in. thick, 6–7-angled; spines ½–2½ in. long, dark red or blackish-purple, becoming grey ... (147) **enopla.**

Spines all straight, ¼–1 in. long, ¼–¾ lin. thick at the base:

Plant apparently with a single stem 9–10 in. high, 10–12-angled; angles apparently not very prominent, obtusely rounded ... ... (149) **cucumerina.**

Plant 2–8 in. high, branching at the base into many stems, often forming dense spiny cushion-like masses:

Spines black; stems with 6–9 broadly rounded angles; peduncles 2 lin. (or more ?) long ... ... (136) **atrispina.**

Spines reddish or brown, sometimes becoming grey with age ; stems ¾-1½
in. thick :
Stems transversely banded with dull
green and whitish-green on the
growing part, 8–10-angled ; angles
subacute ...   . (139) **alternicolor.**

Stems uniformly green (without bands)
on the growing part ; involucre
sessile or very shortly pedunculate :
Stems 7-angled ; angles subacute ;
involucral glands dark purple... (137) **pulvinata.**

Stems usually 8–10- (occasionally 7-)
angled ; angles obtuse ; in-
volucral glands greenish-yellow (138) **aggregata.**

Plant 1–9 ft. high, mostly branched at the
base or upper part :
Stems 9–11-angled :
Angles distinctly toothed ; flowering
peduncles 1½–3 lin. long ; involucral glands dark purple or
blackish-purple ; ovary puberulous (144) **cereiformis.**

Angles not toothed ; flowering peduncles 1 lin. (or more ?) long ; ovary
glabrous ...   . (140) **captiosa.**

Stems 4–8- (or in 146, *E. heptagona,*
sometimes up to 10-) angled ; involucral glands purple or purple-
brown :
Plant 4–9 ft. high ; spines light brown ;
flowering peduncles 2–6 lin. long ;
involucre 2 lin. in diam. ...   ... (145) **pentagona.**

Plant up to 2 ft. high ; spines purple-
brown, becoming grey ; flowering
peduncles 3–9 lin. long ; involucre
1½–1½ lin. in diam.   ...   ... (146) **heptagona.**

**Spines in pairs along the angles of the 3–8-angled, or
occasionally flattened 2-angled branches ; glands of
the involucre entire or in 167, *E. Schinzii,* with a
slight notch on the margin :
Spines ¾ 2½ in. long ; branches 2–6 in. in diam., 3-
angled, very deeply constricted into short sagittate-
reniform segments ...   . (173) **grandicornis.**
Spines 1–6 lin. long :
†Trees 10–30 ft. high :
Branches deeply constricted into short conical
or deltoid-ovate segments ; angles with broad
continuous horny margins   ...   . (174) **Cooperi.**

Branches constricted into elongated slightly conical
lanceolate or parallel-sided segments ; spine-
shields quite separate or in 178, *E. Evansii*
and 177, *E. triangularis,* frequently united
into a very narrow continuous horny margin :
Flowering branches either more than 2 in. in
diam., or if less, then with the edges of the
angles between the spine-shields 2½–3 lin.
thick :

R 2

Flowering-eyes 2–4 lin. above the spine-
shields and quite separate from them ;
angles of young branches not more than
1¼ lin. thick at the edges between the
spine-shields ... ... (176) **similis.**

Flowering-eyes nearly or quite touching the
spine-shields :
Branches erect or ascending, forming an
obconic crown ; angles 2½–3 lin. thick
at the edges ; spine-shields of a thin
disintegrating substance and often
spineless, suborbicular or transverse,
rust-coloured ... ... (175) **ingens.**

Branches spreading and upcurved, forming
a rounded crown ; angles 1½–2 lin.
thick at the edges ; spine-shields
horny, longer than broad, free or
united into a horny margin ... ... (177) **triangularis.**

Flowering branches ¼–2 in. in diam., 3–5-angled
or flat and 2-angled, usually less than 1¼
lin. thick at the edges between the spine-
shields :
Flowering branches 1–2 in. in diam. ; spine-
shields sometimes united into a horny
margin ... ... (178) **Evansii.**

Flowering branches ½–1 in. in diam. ; spine-
shields free, never united into a horny
margin ; flowering-eyes touching the
spine-shields :
Flowering branches often 3- (sometimes 2-)
angled, conspicuously toothed ... (179) **grandidens.**

Flowering branches usually 4–5- (sometimes 3-) angled, often rather slightly
toothed ... ... (180) **tetragona.**

††Bushes or shrubs 2–8 ft. high, with the spine-shields
united into a continuous horny margin or
sometimes free :
Stems simple or occasionally sparingly branched,
4–8 ft. high, 5–7-angled, constricted into
ellipsoidal segments ... ... (172) **virosa.**

Stems more or less branched, 2–8 ft. high :
Branches 3–4-angled ; flowering-eyes seated
nearly midway between two pairs of spines (169) **frankiana.**

Branches 4–7-angled ; flowering-eyes seated at
about one-third of the distance up between
two pairs of spines :
Stems more or less glaucous or bluish-green ;
spines 3–6 lin. long, ⅔–¾ lin. thick ; involucre unknown ... ... (170) **cœrulescens.**

Stems green, not glaucous ; spines 1–3 lin.
long, ¼–½ lin. thick, sometimes absent ;
involucre narrowly obconic, yellow ... (171) **Ledienii.**

†††Dwarf plants, less than 1 ft. high and usually only
attaining 3–6 in. above the ground, often with
a thick tuberous rootstock :
Spines subconnate at the base, then diverging ;
tubercles on the angles crowded, spirally
arranged ... ... (161) **mammillosa.**

Spines free, diverging from the base ; tubercles
on the angles sometimes near together or
spirally arranged, but not crowded :
Dried branches (including the teeth) 9-12 lin.
in diam. at the broadest part, 3-4-angled :
Branches not or rarely constricted at intervals ;
spine-shields very small, not extending
to the flowering eyes, with 1 pair of
spines, pale brown ... ... (164) **clavigera.**

Branches constricted at intervals ; spine-
shields broadly extending to the flowering
eyes, with 1 small and 1 large pair of
spines, grey ... ... (165) **enormis.**

Dried branches (including the teeth) 3-8 lin. in
diam. at the broadest part :
Spine-shields united into continuous horny
grey margins ; stems 5-angled, decum-
bent ; involucre not described ... ... (168) **griseola.**

Spine-shields quite separate ; glands of the
involucre yellow or greenish-yellow :
Branches always simple, radiately spreading
from the top of the tuber, tapering to
a more or less stalk-like base :
Branches with only 2 spine-bearing angles,
flattish or slightly concave above,
convex beneath ... ... (162) **stellata.**

Branches with 3-4 spine-bearing angles :
Branches 3 (or 4 ?)-angled, with conical
teeth 2-3 lin. prominent ... (160) **squarrosa.**

Branches 4 angled, with the teeth
slightly or not more than 1 lin.
prominent ... ... (163) **micracantha.**

Branches often branching, erect, 3-5-angled :
Spines dark brown or grey ; capsule
sessile ; styles 1-1¼ lin. long, shortly
united at the base ... ... (167) **Schinzii.**

Spines at first light brown, becoming
grey ; capsule exserted on a pedicel
1½-3 lin. beyond the involucre ;
styles ⅔ lin. long, united to half-way
up ... ... (166) **Knuthii.**

1. **E. prostrata** (Ait. Hort. Kew. ed. i. ii. 139) ; annual ; stems
radiately spreading on the ground, 2-8 in. long, with alternate
branches, usually slightly flattened from above, puberulous on the
upper side, at least along a middle line, with minute curved hairs,
glabrous on the underside ; leaves opposite, shortly petiolate, 1½-3½
lin. long, ½-2 lin. broad, oblong to elliptic or slightly oblong-
obovate, obtuse or rounded at the apex, distinctly or obscurely
toothed, sometimes ciliate, glabrous on both sides or with a few
scattered hairs beneath, chiefly near the apex ; stipules on the
upper side of the stem usually free and on the under side united
into a deltoid-ovate body, toothed at the apex ; inflorescence of
short axillary leafy raceme-like branches, with 1 axillary involucre
to each pair of leaves or sometimes reduced to a cluster of 2-3

involucres on a short peduncle with 2–3 pairs of minute spathulate leaves ; peduncles $\frac{1}{3}$–1 lin. long, glabrous ; involucre $\frac{1}{4}$–$\frac{1}{3}$ lin. long, campanulate, glabrous or very thinly pubescent, with 4 glands and 5 lobes ; glands minute, with the appendage just exceeding their margin or obsolete ; capsule $\frac{1}{2}$–$\frac{2}{3}$ lin. in diam., pubescent with spreading hairs along the somewhat acute angles, glabrous on the sides ; styles minute, bifid ; seeds $\frac{1}{2}$ lin. long, 4-angled, transversely wrinkled, pale reddish. *Boiss. Ic. Euphorb.* 12, *t.* 17, *and in DC. Prodr.* xv. ii. 47 ; *N. E. Br. in Dyer, Fl. Trop. Afr.* vi. i. 510.

COAST REGION : Uitenhage Div. ; Redhouse, *Mrs. Paterson*, 983 ! Albany Div. ; Railway near Grahamstown, *Daly* ! East London Div. ; East London, *Rattray*, 881 !

KALAHARI REGION : Orange River Colony ; Viljoens Drift, *Rogers*, 4807 !

EASTERN REGION : Natal ; Inanda, *Wood*, 80 ! near Pietermaritzburg, *Wilms*, 2257 ! Phœnix, *Schlechter*, 2938 ! Berea, near Durban, *Wood*, 541 !

A native of Tropical America, introduced into other Tropical and Subtropical regions.

2. **E. inæquilatera** (Sond. in Linnæa, xxiii. 105) ; annual ; stems several, prostrate or occasionally apparently erect, 1–6 in. long, spreading from the crown of the root, pinnately branched, often angular, glabrous all round ; leaves opposite, shortly petiolate, 1–5 lin. long, $\frac{1}{3}$–2$\frac{1}{2}$ lin. broad, oblong or obliquely elliptic, obtusely rounded at the apex, obliquely half-cordate at the base, entire or toothed, glabrous on both sides, herbaceous or subcoriaceous, some-times marked with a red spot or blotch ; stipules all free or those on the underside of the stem more or less united, subulate or cut into filiform segments, often as long as or longer than the petioles ; involucres solitary and axillary on very short axillary branchlets, minute, very shortly pedunculate, $\frac{1}{3}$–$\frac{1}{2}$ lin. long and as much in diam., subcampanulate or globose-campanulate, glabrous, with 4 glands and 5 minute ciliate lobes ; glands minute, half surrounded by a narrow entire or 2–3-toothed appendage, sometimes almost obsolete ; capsule $\frac{3}{4}$–1 lin. in diam., glabrous, exserted on a pedicel slightly longer than the involucre and recurved ; styles very minute, bifid ; seeds $\frac{2}{3}$–$\frac{3}{4}$ lin. long, 4-angled, with slight transverse rugosities, pale reddish or whitish. *Walp. Ann.* iii. 358. *E. parvifolia, E. Meyer, and E. setigera, E. Meyer in Drège, Zwei Pfl. Documente,* 184 (*name only*), *and ex Boiss. in DC. Prodr.* xv. ii. 34, 35. *E. sanguinea, Hochst. & Steud. ex Boiss. in DC. Prodr.* xv. ii. 35 (*including vars.* setigera *and* natalensis, *Boiss.*) ; *N. E. Br. in Dyer, Fl. Trop. Afr.* vi. i. 508. *Anisophyllum inæquilaterum, Klotzsch & Garcke in Abhandl. Akad. Berlin,* 1860, 22. *A. Mundii, Klotzsch & Garcke, l.c.* 25. *A. setigerum, Klotzsch & Garcke, l.c.* 29.

VAR. β, **perennis** (N. E. Br.) ; perennial, with a more or less woody rootstock ; leaves oblong to elliptic, sometimes nearly as broad as long, entire or finely and very sharply toothed, usually more coriaceous than in the type, otherwise similar.

COAST REGION : Bedford Div. ; Goba River, *MacOwan*, 1469 ! Fort Beaufort Div. ; Adelaide, *Rogers*, 4494 ! Queenstown Div. ; near Queenstown, *Galpin*, 1950 ! King Williamstown Div. ; near King Williamstown, *Schlechter*, 6122 ! East London

Div. ; near East London, *Rattray*, 882 ! Komgha Div. ; banks of the Kei River, *Drège*, 4618 !
CENTRAL REGION : Calvinia Div. ; Bitterfontein, *Zeyher*, 1541 ! Prince Albert Div. ; by the Gamka River, *Mund & Maire*, 15 ! Jansenville Div. ; Żwartruggens, *Drège*, 8191 ! Graaff Reinet Div. ; Ryneveldt Pass, *Bolus*, 412 ! Murraysburg Div. ; near Murraysburg, *Tyson*, 338 ! Albert Div., *Cooper*, 786 !
WESTERN REGION : Great Namaqualand ; various localities, *Pearson*, 3161 ! 3786 ! 4285 ! 4327 ! 4689 ! 4769 ! 4776 ! 8181 ! Little Namaqualand ; between Holgat River and the Orange River, *Drège*, 2953 !
KALAHARI REGION : Griqualand West ; Kimberley, *Marloth*, 745 ! Orange River Colony ; Thaba Unchu, *Burke* ! Sand River, *Burke* ! Bloemfontein, *Burtt-Davy*, 11849 ! 11850 ! Transvaal ; various localities, *Rehmann*, 6673 ! *Wilms*, 1338 ! *Miss Leendertz*, 18 ! 36 ! *Burtt-Davy*, 51, 1164 ! 1194 ! 1226 ! 1459 ! Var. *β* : Orange River Colony ; Sand River, *Burke*, 507 ! Bechuanaland ; near the sources of the Kuruman River, *Burchell*, 2476 ; between the sources of the Kuruman River and Kosifontein, *Burchell*, 2535 ! Transvaal ; Fourteen Streams, *Burtt-Davy*, 1544 ! Schweizer Reneke, *Burtt-Davy*, 1695 ! Waterval, *Miss Leendertz*, 823 ! hills near Wilge River, *Schlechter*, 3744 ! near Pretoria, *Kirk*, 49 ! and without precise locality, *Zeyher*, 1542 !
EASTERN REGION : Natal ; near Ladysmith, *Gerrard*, 611 ! Clairmont, *Wood*, 1432 ! near Weenen, *Wood*, 4436 ! near Durban, *Miss Owen* ! and without precise locality, *Gerrard*, 60 ! Var. *β* : near Tugela, *Wood*, 3552 ! near Durban, *Miss Owen* !

When working out the Tropical African species of *Euphorbia* I had not investigated those of South Africa, but now that I have done so, I find that all the names above referred to this species certainly belong to one and the same plant. It therefore becomes necessary in accordance with the rule of priority to adopt the name *E. inæquilatera*, Sond., for this species instead of that of *E. sanguinea*, Hochst. and Steud., adopted in the Flora of Tropical Africa, because it is the first name for the plant that was published with a description. *E. inæquilatera* was published in 1850, and although the names *E. parvifolia* and *E. setigera* were published by Drège in 1843, they are mere names in a catalogue, unaccompanied by any description, and the name *E. sanguinea* was not published until 1862, when Boissier gave a description of it, as he then also did of *E. parvifolia* and *E. setigera* for the first time.

Also in Tropical Africa and Arabia.

3. E. Schlechteri (Pax in Engl. Jahrb. xxviii. 26) ; a perennial leafy herb, 6–8 in high ; stems apparently several from the same root, erect, with erect branches, pubescent ; leaves opposite, 3–7 lin. long, 1¼–3 lin. broad, obliquely ovate or ovate-oblong, obtuse, very unequal and half-cordate at the base, minutely toothed or entire, glabrous above, thinly pubescent beneath, apparently purplish along the margins ; involucres solitary, axillary and in the forks of the branchlets, much shorter than the leaves, on peduncles ½ lin. long, somewhat pear-shaped, ¾ lin. in diam., glabrous, with 4 appendaged glands and 5 minute ciliate lobes ; glands ¼–⅓ lin. in their greater diam., transversely elliptic or oblong, with a narrow petaloid entire white appendage on the outer margin ; capsule 1¼ lin. in diam., obtusely 3-angled, thinly pubescent with rather long and somewhat adpressed hairs, exserted and recurved on a pedicel nearly twice as long as the involucre ; styles ¼ lin. long, free to the base and very deeply bifid, with slender segments ; seeds ¾ lin. long, ovoid-oblong, subacute at one end, 4-angled, slightly rugulose, reddish.

248   EUPHORBIACEÆ (Brown).   [*Euphorbia.*

EASTERN REGION : Portuguese East Africa ; Ressano Garcia, near Komati Poort, 1000 ft., *Schlechter*, 11915 !

4. **E. livida** (E. Meyer in Drège, Zwei Pfl. Documente, 184, ex
Boiss. in DC. Prodr. xv. ii. 14); perennial ; stems many from a
woody rootstock, decumbent or prostrate, branched, glabrous;
leaves opposite, shortly petiolate, coriaceous or perhaps somewhat
fleshy, $\frac{1}{3}$–$\frac{3}{4}$ in. long, $2\frac{1}{2}$–$5\frac{1}{2}$ lin. broad, ovate, elliptic or suborbicular,
obtuse or rounded at the apex, obliquely subcordate at the base,
entire, glabrous on both sides ; stipules 2 to each node, ovate,
toothed ; cymes small, axillary, pedunculate ; involucres on peduncles
$\frac{3}{4}$–$1\frac{1}{2}$ lin. long, cup-shaped, $\frac{3}{4}$ lin. in diam., glabrous, with 3–4 glands
and 4–5 minute fringed lobes; glands about $\frac{1}{4}$ lin. in their greater
diam., transversely elliptic, entire, with the appendage merely forming an outer rim to the outer margin ; capsule exserted and recurved,
$1\frac{1}{3}$–$1\frac{1}{2}$ lin. in diam., glabrous ; styles $\frac{1}{3}$ lin. long, free to the base,
erect with bifid recurved tips ; seeds about $\frac{3}{4}$ lin. long, ovoid or
ellipsoid, obscurely 4-angled, slightly rugulose, whitish.

COAST REGION : East London Div. ; by the shore on ,Cove Rocks at East
London, *Galpin*, 7351 ! Komgha Div. ; near the mouth of the Kei River.
Flanagan, 180 !
EASTERN REGION : Transkei ; prostrate on the beach in patches, *Miss Pegler*,
1290 ! Krielis Country, *Bowker*! Pondoland ; between Umsikaba River and
Umtentu River, *Drège*, 4622 ! and without precise locality, *Bachmann*, 756 !
Natal : The Bluff, 260 ft., near Durban, *Wood*, 7933 ! and without precise
locality, *Gerrard*, 1171 ! sea shore at Winkle Spruit, *Miss Franks in Herb. Wood*,
11898 !

5. **E. hypericifolia** (Linn. Sp. Pl. ed. i. 454, and Amœn. Acad. iii.
113); annual, 3–18 in. high, branching at the base or sometimes
simple ; stems simple or alternately branching, puberulous or
glabrous ; leaves opposite, shortly petiolate, $\frac{1}{4}$–$1\frac{1}{2}$ in. long, $\frac{1}{6}$–$\frac{3}{4}$ in.
broad, oblong, oblong-lanceolate, elliptic or ovate, subacute to
rounded at the apex, oblique at the base, minutely toothed or
rarely entire, puberulous beneath or on both sides ; stipules divided
into 2 or more slender fimbriate segments ; cymes axillary, $\frac{1}{6}$–$\frac{1}{2}$ in.
in diam., laxly few- to densely many-flowered, on a peduncle $\frac{1}{10}$–$\frac{3}{4}$
in. long, puberulous or glabrous, often with a pair of leaves at its
apex ; bracts $\frac{1}{2}$–$\frac{2}{3}$ lin. long, lanceolate, acuminate, ciliate or entirely
glabrous ; involucres about $\frac{1}{2}$ lin. long, on peduncles $\frac{1}{2}$–1 lin. long,
cup-shaped, puberulous or glabrous, with 4 appendaged glands and
5 lobes ; glands minute, orbicular or transversely elliptic ; appen-
dages spreading, transversely elliptic, entire, $\frac{1}{4}$–$\frac{1}{3}$ lin. broad, white :
capsule 1 lin. in diam., pubescent or glabrous ; styles deeply bifid ;
seeds ellipsoid, 4-angled, slightly transversely rugose, whitish-grey
to reddish-brown, with a glaucous hue at the apex or all over.
Hook. Exot. Fl. i. *t.* 36 ; *Boiss. in DC. Prodr.* xv. ii. 23 ; *N. E. Br.
in Dyer, Fl. Trop. Afr.* vi. i. 498. *E. pilulifera, Linn. Sp. Pl. ed.*
i. 454, *in Amœn. Acad.* iii. 115 *and Herbarium, excluding reference to
Burmann, not of other authors. E. indica, Lam. Encycl.* ii. 423 ;

Boiss. in DC. Prodr. xv. ii. 22, *incl. var. angustifolia, Boiss. Anisophyllum indicum, Schweinf. Beitr. Fl. Aethióp.* 34. *Phyllanthus obliquus, E. Meyer in Drège, Zwei Pfl. Documente,* 211, *name only, not of Müll. Arg.—Tithymalus americanus flosculis albis, Commelin, Præl. Bot.* 60, *t.* 10.

KALAHARI REGION: Transvaal; Shilovane, *Junod,* 644! Kaap Muiden, near Barberton, *Thorncroft,* 758!

EASTERN REGION: Pondoland; between St. Johns and Umtsikaba Rivers, *Drège,* 4625! Natal; Umhlanga, *Wood,* 1213! near Phœnix, *Wood,* 1802! *Schlechter,* 2899! and without precise locality, *Gerrard,* 56! between Delagoa Bay and Komati River, *Bolus,* 9768!

Also in Tropical and North Africa, the warmer parts of Asia and Malaya, Tropical and North America. Probably introduced into South Africa.

6. **E. hirta** (Linn. Sp. Pl. ed. i. 454, and Amœn. Acad. iii. 114); annual, $1\frac{1}{2}$–16 in. high; stems erect or decumbent at the base, simple or dichotomously branching, rather coarsely pilose with yellow spreading hairs, usually densely at the upper part, thinly below, with an under pubescence of minute curved subadpressed hairs; leaves opposite, $\frac{1}{3}$–2 in. long, $\frac{1}{6}$–$\frac{3}{4}$ in. broad, obliquely lanceolate or ovate or rhomboid-oblong, acute or subobtuse, on one side of the midrib cuneate at the base, on the other rounded, finely serrate, thinly adpressed-pubescent on both sides, more minutely so or sometimes glabrous above; petiole $\frac{1}{2}$–$1\frac{1}{2}$ lin. long; stipules minute, subulate; cymes axillary, $\frac{1}{4}$–$\frac{1}{2}$ in. in diam., globose or dichotomously divided into 2–3 globose heads, with peduncles 1–6 lin. long, puberulous with minute curved hairs; involucres densely crowded, male or bisexual, minute, about $\frac{1}{3}$ lin. long, campanulate, obconic or cup-like, with 4 glands and 5 deltoid acute fringed lobes; glands linear viewed sideways, orbicular and $\frac{1}{8}$–$\frac{1}{12}$ lin. in diam. at the truncate apex as seen from above, with a very minute appendage; capsule exserted and curved to one side, 3-angled, $\frac{1}{2}$ lin. in diam., thinly puberulous with minute adpressed curved hairs; styles free, $\frac{1}{8}$ lin. long, deeply divided into 2 slender, truncate arms; seeds about $\frac{1}{3}$ lin. long, oblong, 4-angled, with slight transverse rugosities, light reddish. *Jacq. Collect.* v. 160, *t.* 11, *fig.* 1; *N. E. Br. in Dyer, Fl. Trop. Afr.* vi. i. 496. *E. capitata, Lam. Encycl.* ii. 422. *E. pilulifera, Jacq. Ic.* iii. 5, *t.* 478, *and Collect.* ii. 361; *Boiss. in DC. Prodr.* xv. ii. 21.

KALAHARI REGION: Transvaal; Bremersdorp, *Burtt-Davy,* 3007! Barberton, *Thorncroft,* 4993! Warmbath, *Walker*!

EASTERN REGION: Natal; Malvern, near Durban, *Wood,* 647! near Durban, *Wood,* 120! a weed in Durban Botanic Garden, *Wood,* 3130! Phœnix, *Schlechter,* 2890! and without precise locality, *Gerrard,* 239! Lourenço Marques, *Quintas,* 178.

Widely distributed throughout the Tropics, probably introduced into South Africa.

7. **E. neopolycnemoides** (Pax & Hoffm. in Engl. Jahrb. xlv. 240); annual, or perhaps sometimes perennial, 2–10 in. high, divided at the base into 2 (or occasionally more) main branches, which are

1-6-times forked, erect, rather slender, glabrous, laxly leafy; leaves all opposite, spreading, very shortly petiolate, $\frac{1}{2}$-$1\frac{1}{8}$ in. long, $\frac{1}{2}$-$1\frac{1}{2}$ lin. broad, linear, acute, obliquely subcordate at the base, with the margins narrowly revolute, glabrous on both sides; involucres solitary in the forks or terminal on the ultimate branchlets, shortly pedunculate, 1 lin. in diam., cup-shaped, glabrous, with 4 appendaged glands and 4-5 minute toothed lobes; glands $\frac{1}{4}$-$\frac{1}{3}$ lin. in their greater diam., transverse, narrowly elliptic, excavated at the top, half encircled by a conspicuous petal-like white or red appendage; capsule about 1 lin. in diam., obtusely 3-angled, exserted on a recurved pedicel, glabrous; styles $\frac{1}{3}$ lin. long, free to the base, erect, collected together, minutely bifid at the apex; seeds about $\frac{3}{4}$ lin. long, ellipsoid, 4-angled, with 3-4 transverse ridges on each face, reddish. *E. arabica, var. latiappendiculata, Pax in Engl. Jahrb.* xliii. 85.

KALAHARI REGION : Transvaal; Queens River Valley and Kaap Valley, near Barberton, *Galpin,* 757! Warmbath, *Burtt-Davy,* 5563! *Bolus,* 12280! Komati Poort, *Schlechter,* 11736! near the Magalaqueen (Nyl) River, between Nylstroom and Naboomfontein, *Schlechter,* 4278!
EASTERN REGION : Delagoa Bay, *Junod,* 140!

This belongs to a small group of African and North American species which closely resemble one another, but seem clearly distinct when carefully examined. From *E. arabica,* Hochst. & Steud., it is at once distinguished by its much larger gland-appendages and very much longer styles. *E. Eylesii,* Rendle, and *E. leshumensis,* N. E. Br., are also similar; from the former it differs by its annual habit, mode of branching and smaller gland-appendages, &c., and from the latter, besides other differences, by always dividing into 2 main branches at the node formed by the seed-leaves, whilst in *E. leshumensis* the seed-leaf node and 2-3 nodes above it never appear to produce branches.

8. **E. Pfeilii** (Pax in Engl. Jahrb. xxiii. 534); a perennial herb, 9-18 in. high, with the stems repeatedly forked from the base, slender, 1-1$\frac{1}{2}$ lin. thick at the base, rigidly herbaceous at the upper part, sometimes naked below, glabrous ; leaves very small, opposite, rather thick and fleshy, including the short petiole 1$\frac{1}{2}$-2$\frac{1}{2}$ lin. long, $\frac{1}{2}$-$\frac{2}{3}$ lin. broad, linear-oblong, entire and apiculate or minutely toothed at the apex, glabrous on both sides, probably glaucous ; involucre solitary, at first terminal, but by the outgrowth of branches immediately beneath it, is left in the forks of the branches, where it ripens seed and then falls away, sessile, $\frac{3}{4}$ lin. in diam., cup-shaped, glabrous or with some minute crisped hairs outside, with 4 glands and 5 rectangular minutely toothed lobes; glands minute, transversely oblong, without an appendage ; capsule about $\frac{1}{2}$ in. in diam., thinly covered with minute curved hairs, exserted on a pedicel $\frac{3}{4}$-1$\frac{1}{2}$ lin. long, at first curved to one side, ultimately erect; styles $\frac{1}{4}$-$\frac{1}{3}$ lin. long, free to the base, erect, bifid at the apex ; seeds not seen. *E. anomala, Pax in Bull. Herb. Boiss.* 2me *sér.,* viii. 636, *not of Boiss. nor of Salzmann.*

WESTERN REGION: Great Namaqualand ; Stolzenfels Reitfontein, *Graf Pfeil,* 91! in a river-bed at Buchholzbrunn, 3250 ft., *Pearson,* 3658! Inachab, *Dinter,* 15!

9. E. pseudotuberosa (Pax in Bull. Herb. Boiss. 2me sér. viii. 637);
a perennial herb; rootstock a large fleshy tuber, producing one to
several elongated fleshy neck-like subterranean stems, from which
the leaf-stems arise; leaf-stems few or many, annual, ½–2 in. high,
simple or forked, glabrous or thinly and minutely puberulous;
leaves opposite under the involucres or at the forkings of the stem,
alternate elsewhere, sessile, ½–1½ in. long, ¾–2 lin. broad, linear or
lanceolate, acute, with infolded margins or concave-channelled,
glabrous on both sides, more or less glaucous, thinly coriaceous;
stipules none; involucre ¼–½ in. in diam., solitary and terminal or in
the forks of the stems or 2–3 in a lax cyme, cup-shaped, glabrous
or minutely puberulous outside, with 4–5 glands and 5 transversely
rectangular fringed puberulous lobes; glands 1½–2 lin. across,
palmately divided into 3–5 rather stout recurved-spreading segments ½–1 lin. long, corrugated on their upper surface and along
the outer portion of the 2-lipped undivided part; capsule 3–3½ lin.
in diam., tricoccous, glabrous, much exserted on a puberulous
pedicel exceeding the involucre by 1½–2 lin.; styles 1½–2 lin. long,
united for one-third to one-half their length, with recurved-
spreading arms, slightly thickened and minutely 2-lobed at the
apex; seeds not seen.

KALAHARI REGION : Transvaal ; near Rustenburg, 4000 ft., *Miss Pegler*, 934 !
near Pretoria, *Fehr*, 43 ! *Miss Leendertz*, 239 ! Hartebeest Poort, Pretoria District,
4500 ft., *Burtt-Davy*, 9819 ! Groonkloof Valley, *Mogg* ! Six-miles Spruit, near
Pretoria, *Schlechter*, 4794 ! Smitskraal, *Burtt-Davy*, 9942 ! 11284 ! 12889 ! and
without precise locality, *Zeyher*, 1530 partly !

Very similar in habit and appearance to *E. trichadenia*, Pax, and easily
mistaken for that species, but may be readily distinguished by its corrugated
involucre-glands and much more exserted ovary and capsule.

10. E. trichadenia (Pax in Engl. Jahrb. xix. 125); a perennial
herb; rootstock a tuber with an elongated neck, producing annual
herbaceous stems 1–4 in. high, branching from the base, puberulous
or glabrous; leaves opposite at the flowering nodes and forkings of
the stem, alternate elsewhere, sessile, thinly coriaceous or perhaps
slightly fleshy, ¾–2½ in. long, ½–2½ lin. broad, linear, acute, usually
slightly curved, often longitudinally folded, glabrous on both sides,
sometimes ciliate on the narrow cartilaginous margins; lowest
leaves and sometimes those under the involucres (bracts) much
smaller, lanceolate, linear-lanceolate or scale-like; stipules none;
involucres solitary in the forkings of the stems or sometimes 3–5
in small terminal cymes, shortly pedunculate, 4–5 lin. in diam.,
cup-shaped, glabrous or minutely puberulous outside, with 5 glands
and 5 transversely rectangular or subquadrate fringed lobes;
glands 1⅓–1¾ lin. long, 1½–2 lin. broad, palmate or somewhat fan-
shaped, deeply divided into 3–7 (or more?) linear or filiform
segments ¾–1½ lin. long, once or twice forked at the apex, flat
or channelled but not corrugated on their upper surface, with the
undivided basal part concave or 2-lipped from the inner margin

being inflexed; capsule 5–5½ lin. in diam., glabrous, exserted on an erect pedicel equalling or exceeding the involucre; styles united into a column 1–1¾ lin. long, with revolute arms ¾–1¼ lin. long, minutely 2-lobed at the tips; seeds 1½–2 lin. in diam., globose, acutely pointed at one end, thinly and minutely subrugulose with what appear to be irregular agglutinated masses of minute hairs. *N. E. Br. in Dyer, Fl. Trop. Afr.* vi. i. 523. *E. benguelensis, Pax in Bull. Herb. Boiss.* vi.̇ 741. *E. subfalcata, Hiern, Cat. Afr. Pl. Welw.* i. 948. *E. Gossweileri, Pax in Engl. Jahrb.* xliii. 88.

KALAHARI REGION : Transvaal; tops of mountains at Rietfontein in Zoutpansberg district, *Miss Leendertz,* 872 ! Warmbath, *Burtt-Davy,* 5337 ! Pilgrims Rest, *Greenstock* ! between Komati River Drift and Crocodile River, *Bolus,* 9766 ! and without precise locality, *Burke*! *Zeyher,* 1539 partly !

Also in Tropical Africa. Growths springing up after the vegetation has been burnt are very short and totally unlike those normally developed.

11. **E. Gueinzii** (Boiss. in DC. Prodr. xv. ii. 71); a tuberous-rooted perennial, diœcious; stems often several, 2–6 in. high, simple or branched, herbaceous, thinly to thickly covered with rather long spreading hairs or occasionally glabrous ; leaves lax or numerous, alternate below, opposite on the flowering part or at the forkings of the stem, subsessile or very shortly petiolate, ¼–1¼ in. long, 1–5½ lin. broad, linear, lanceolate or oblong-lanceolate, or elliptic or ovate-lanceolate, acute, cuneate or rounded at the base, usually glabrous on both sides, rarely with a few long hairs beneath or on both sides ; bracts like the leaves ; involucres in terminal leafy cymes or solitary in the forks of the branches, on peduncles ½–4 lin. long, glabrous or pubescent outside, unisexual, 1–3 lin. in diam., cup-shaped, with 5 glands and 5 subquadrate or oblong-ovate toothed or ciliate lobes; glands ⅓–1¼ lin. in their greater diam., transverse, cuneately oblong or somewhat half-circular, with the inner margin rounded and the outer straight and entire or minutely crenulate; capsule about 2¾ lin. in diam., tricoccous, hairy or glabrous, exserted on a recurved pedicel 1½–3 lin. beyond the involucre ; styles very shortly united at ʻthe base, ½–¾ lin. long, stout, with spreading deeply bifid tips ; seeds 1½ lin. long, stoutly oblong, slightly 4-angled, with a groove along one angle, obscurely sculptured, pale greyish-white.

VAR. β, **albovillosa** (N. E. Br.); stems, both sides of the leaves and the involucres all densely hairy or villous ; otherwise as in the type. *E. albovillosa, Pax in Engl. Jahrb.* xxxiv. 373.

KALAHARI REGION : Basutoland, *Cooper,* 945 ! Orange River Colony ; Harrismith, *Sankey,* 20 ! Transvaal ; roadsides at Carolina, *Nicholson,* 4594 ! Ermelo, *Tennant,* 6936 ! near Robinson, *Burtt-Davy,* 2981 ! Embabaan, *Burtt-Davy,* 2854 !
EASTERN REGION : Griqualand East ; near Clydesdale, *Tyson,* 2692 ! Natal ; near Durban, *Sutherland* ! *Sanderson,* 214 ! Ladysmith, *Gerrard,* 612 ! Klip River, *Sutherland* ! Clairmont Flats, *Wood,* 697 ! 1715 ! 3402 ! *Schlechter,* 3034 ! near Phœnix, *Schlechter,* 3025 partly (mixed with *Phyllanthus maderaspatensis*) ! and without precise locality, *Sanderson, Gerrard,* 523 ! Var. β : Natal ; Inchanga, *Schlechter,* 3245 ! Dumisa, *Rudatis,* 718 !

12. E. multifida (N. E. Br.); evidently a perennial herb, about 4–6 in. high, with a tuberous or woody rootstock producing annual stems, only one seen, about $1\frac{1}{4}$ lin. thick, branching into a 3-rayed umbel at the top, with 1 ray below it, glabrous; rays $1\frac{1}{3}$–$2\frac{1}{4}$ in. long, $\frac{1}{2}$–$\frac{3}{4}$ lin. thick, once (or perhaps twice) forked, with the secondary rays $\frac{3}{4}$–$1\frac{1}{2}$ in. long, puberulous; leaves not seen, all fallen, but probably like the bracts and evidently alternate; bracts opposite, sessile, 3–6 lin. long, 2–3 lin. broad, broadly ovate, acuminute, puberulous on both sides; involucre $\frac{1}{2}$ in. in diam., broadly and rather shallowly cup-shaped, puberulous outside, with 5 pectinate glands and 5 transversely rectangular fringed puberulous lobes; glands not contiguous, spreading, $1\frac{1}{2}$ lin. in their greater diam., unequally two-lipped, with 5–6 subulate simple or forked processes 1–$1\frac{1}{2}$ lin. long on their crinkled outer margin, puberulous all over the outer surface, but glabrous on the inner or upper surface of the gland itself; ovary and capsule not seen, the involucres examined being male.

EASTERN REGION: probably Natal, described from a single specimen in the Natal Herbarium, without locality or collector's name, but bearing the number 10483!

Similar in habit and closely allied to *E. trichadenia,* Pax, but evidently distinct.

13. E. graveolens (N. E. Br.); branches 3–6 in. long, probably arising from a tuberous rootstock, curved at the base in all the examples seen, 1–$1\frac{1}{4}$ lin. thick, glabrous; leaves 4–6 on a branch and alternate, with a whorl of 4–5 at the base of the umbel, ascending, 2–6 lin. long, 1–$1\frac{1}{2}$ lin. broad, lanceolate to elliptic-lanceolate, acute, narrowed at the sessile base, glabrous, deciduous, when the branches appear leafless, those of the whorl all fallen from the specimens seen; umbel 4–5-rayed, often with 1–2 rays below, in the axils of the upper leaves; rays $\frac{1}{3}$–$\frac{3}{4}$ in. long, bearing 1 involucre and a pair of bracts 1–2 lin. below it, glabrous; bracts 2 lin. long, 1–$1\frac{1}{4}$ lin. broad, oblong or oblong-obovate, toothed at the rounded apex, very concave from the margins being incurved, glabrous on the back, pubescent on the inner surface; involucre $2\frac{1}{2}$–3 lin. in diam., broadly cup-shaped, glabrous outside, pubescent within, purplish, with 4 glands and 5 transversely rectangular deeply toothed lobes; glands deflexed, 1–$1\frac{1}{4}$ lin. in their greater diam., either transversely elliptic or suborbicular, with or without a slight notch on the upper margin and with 4–6 teeth $\frac{1}{4}$–$\frac{1}{3}$ lin. long on the lower margin or divided into two broad denticulate lobes, apparently green or yellowish-green; ovary glabrous, exserted from the involucre on a curved pedicel; styles united into a column about $\frac{1}{4}$ lin. long, with short radiating very broadly cuneate arms or stigmas, channelled down their face, contiguous and forming a disc $\frac{1}{2}$–$\frac{2}{3}$ lin. in diam.; capsule and seeds not seen.

WESTERN REGION: Little Namaqualand; between Stinkfontein and Garies, *Pearson,* 5579! Van Rhynsdorp Div.; Karroo at Bakhuis (Bak Oven), *Pillans,* 5486!

The dried specimens of this plant, when placed in boiling water for examination, give forth an extremely disagreeable odour.

14. E. glaucella (Pax in Bull. Herb. Boiss. vi. 737); annual, 2–15 in. high, in very small plants unbranched, when more fully developed with about 3–6 ascending main branches, 1–6-times forked, glabrous; leaves all opposite or some of the lower alternate, with a whorl of 3 at the apex of the main stem, petiolate, glabrous on both sides; petiole $\frac{1}{2}$–4 lin. long; blade $\frac{1}{2}$–2$\frac{1}{2}$ in. long, 1–4 lin. broad, linear, linear-lanceolate, oblong, elliptic or suborbicular, obtuse and apiculate or minutely 3-toothed or acute at the apex, otherwise entire, rounded or cuneate at the base; stipules very minute or absent; involucres solitary in the forks of the branches or terminal, shortly pedunculate, $\frac{3}{4}$–1 lin. in diam., somewhat sub-globosely cup-shaped, glabrous, with 3–5 (usually 4) glands and 4–5 short oblong or subquadrate bifid or subentire fringed lobes; glands $\frac{1}{3}$–$\frac{2}{3}$ lin. in their greater diam., transversely oblong or elliptic, entire, with a narrow firm (not petal-like) rim along the outer margin, green; capsule 1$\frac{1}{4}$–1$\frac{1}{2}$ lin. in diam., often rather longer than broad, exserted on a slender pedicel and thinly sprinkled with minute stout adpressed hairs or glabrous; styles about $\frac{1}{4}$ lin. long, free to the base, bifid at the apex; seeds 1–1$\frac{1}{4}$ lin. long, oblong, subtruncate at each end, dorsally flattened, labyrinthically tuber-culate-rugose or scrobiculate-tuberculate, dark brown (perhaps when immature) or bluish-white, with a pale yellowish or whitish caruncle. *N. E. Br. in Dyer, Fl. Trop. Afr.* vi. i. 514. *E. kwebensis, N. E. Br. in Kew Bulletin,* 1909, 137.

WESTERN REGION: Bushmanland; on broken ground west of Pella, 2500 ft., *Pearson,* 3551! Great Namaqualand; various localities, 3250–4300 ft., *Pearson,* 3661! 3728! 3740! 3756! 4041! 4596! 4638! 4677! 4756!

Also in Tropical Africa.

15. E. phylloclada (Boiss. in DC. Prodr. xv. ii. 66); annual or perennial, with radiately spreading or prostrate branches 1–6 in. long, in the larger specimens repeatedly branched, glabrous; leaves and bracts on the branches all opposite, sessile, often crowded and imbricate on the flowering parts, forming head-like masses, $\frac{1}{5}$–$\frac{2}{3}$ in. long and as much in breadth, broadly cordate-ovate or orbicular-ovate, obtuse or acute, mucronate, coriaceous or perhaps somewhat fleshy, with a narrow subcartilaginous white or reddish margin, green, with a large white or whitish area at the base or occupying half the leaf on one side of the midrib; leaves on the young main stem below the primary branches alternate, $\frac{1}{3}$–1$\frac{1}{4}$ in. long, spathu-lately obovate or orbicular-obovate, obtuse, apiculate, cuneately tapering into a petiole half to two-thirds as long as the blade; stipules none; involucres solitary at the usually very crowded flowering nodes, subsessile, 1 lin. in diam., and rather longer than broad, campanulate, thin or somewhat membranous, apparently whitish or purplish, with 4 glands and 5 quadrate or rectangular

3-toothed ciliate lobes ; glands ⅓–⅔ lin. in their greater diam., transversely linear-oblong, with a conspicuous petaloid slightly crenulate whitish or purplish appendage ; capsule far exserted, recurved, 1½ lin. long, 1¼ lin. in diam., oblong, glabrous ; styles ¼–⅓ lin. long, free, deeply bifid ; seeds 1 lin. long, 4-angled, with a small caruncle, very minutely tuberculate-scabrous, grey. *N. E. Br. in Dyer, Fl. Trop. Afr.* vi. i. 494. *E. peploides, E. Meyer in Drège, Zwei Pfl. Documente,* 184, *ex Boiss. in DC. Prodr.* xv. ii. 66, *not of Gouan. E. hereroensis, Pax in Engl. Jahrb.* x. 35.

WESTERN REGION : Great Namaqualand ; stony slopes 12 miles west of Zandverhaar, *Pearson,* 4272 ! Little Namaqualand ; between Verleptpram and the mouth of the Orange River, *Drège* !

Also in Tropical Africa.

This is readily distinguished from all other species by the habit and the remarkable variegation of its leaves.

16. **E. Helioscopia** (Linn. Sp. Pl. ed. i. 459, and Amœn. Acad. iii. 124) ; an annual herb, ½–1 ft. high, glabrous in all parts ; stem usually with a pair of branches arising from its base, sometimes simple, usually dividing into a 5-rayed umbel 2–6 in. in diam. at the top ; primary rays dividing into 3 secondary rays, which are 1–3-times forked ; leaves alternate on the stem, whorled or opposite on and at the base of the umbel, sessile or tapering into a petiole, ½–1½ in. long, ¼ 1 in. broad, cuneately obovate, rounded at the apex, denticulate ; involucre solitary in the forks of the rays and terminal, subsessile, 1–1¼ lin. in diam. and 1 lin. deep, cup shaped, yellowish-green, with 4 glands and 5 oblong or subquadrate minutely toothed lobes ; glands ½ lin. in their greater diam., transversely oblong, entire, flat, yellowish-green ; ovary obtusely trigonous, conspicuously 3-grooved between the angles, glabrous, exserted on a curved pedicel 1⅓–1¾ lin. long ; styles about ⅓ lin. long, free almost to the base, not diverging, shortly 2-lobed at the apex ; capsule 1¾ lin. in diam. and 1¼ lin. long, 3-lobed as seen from above, smooth ; seeds 1–1⅙ lin. long, ellipsoid or subglobose, with a raised network all over their surface, dark brown, with a small yellowish caruncle. *Bernhardi in Flora,* 1845, 86, *and in Krause, Beitr. Fl. Cap und Natal.* 150 ; *Boiss. in DC. Prodr.* xv. ii. 136.

SOUTH AFRICA : without locality, *Thunberg* ! *Harvey* ! 604 !

COAST REGION : Cape Div. ; Nordhoek Forest, *Miln,* 196 ! Simons Bay, *Wright,* 449 ! Mowbray, near Cape Town, *Wilms,* 3629 ! Wynberg, *Wallich* ! *Masson* ! near Cape Town, *Tyson,* 2275 ! Claremont, *Schlechter,* 736 !

CENTRAL REGION : Graaff Reinet Div. ; around Graaff Reinet, *Bolus,* 456 ! *Thornton,* 198 ! Colesberg Div. ; around Colesberg, *Shaw* !

A weed of cultivation introduced from Europe.

17. **E. Peplus** (Linn. Sp. Pl. ed. i. 456, and Amœn. Acad. iii. 117) ; annual, 3–15 in. high, erect, branching near or at the base ; branches glabrous ; leaves alternate on the stem and branches, but opposite on the branches of the umbel, thinly herbaceous, ¼–¾ in. long, ⅙–½ in. broad, all more or less petiolate, and ovate,

obovate or suborbicular, obtuse, tapering into the petiole or those on
the umbel (or bracts) sessile and broadly deltoid-ovate and somewhat
acute, all entire, glabrous; umbel usually 3-rayed, with or without
one or more ray-like branches below it, or the whole plant repeatedly
forking; rays ½–6 in. long, once or several times forked, glabrous;
involucre solitary in the forks or axillary or terminal, excluding
the horns about ⅔ lin. in diam. and nearly as deep, cup-shaped,
glabrous, green, with 4 glands and 5 oblong very minutely toothed
lobes; glands about ¼ lin. in their greater diam., transversely
oblong with a subulate horn at each end about ¼ lin. long; capsule
about 1 lin. in diam., trigonous, with 2 very narrow wings along
each angle, glabrous, exserted on a pedicel 1¼–1½ lin. long; styles
minute; seeds ¾ lin. long, oblong, slightly 6-angled, with a series
of 3–4 deep pits on 4 of the faces and one deep pit on each of the
other two faces, grey, tipped with a yellowish caruncle. *Boiss. in
DC. Prodr.* xv. ii. 141; *Bolus & Wolley-Dod in Trans. S. Afr. Phil.
Soc.* xiv. 318.

South Africa: without locality, *Thunberg! Mund! Mrs. Barber,* 291!
Coast Region: Cape Div.; waste ground near Cape Town, *Tyson,* 2274!
Simons Bay, *Wright,* 451! Devils Peak, *Wilms,* 3628! shore of Gordon Bay,
Diels, 1277! Newlands Avenue, *Wolley-Dod,* 2231! railway near Diep River,
Wolley-Dod, 1222! near Claremont, *Schlechter,* 788! Komgha Div.; borders of
woods on Prospect Farm, near Komgha, *Flanagan,* 245!
Central Region: Graaff Reinet Div.; about Graaff Reinet, *Thornton,* 199!
Bolus, 84!
Kalahari Region: Transvaal; Pretoria, *Miss Leendertz,* 398!
Eastern Region: Transkei; Krielis Country, *Bowker,* 291! Natal; Durban
Flat, *Wood,* 3925!

A weed of cultivation, introduced from Europe and now widely spread over
South Africa.

18. **E. pubescens** (Vahl, Symb. ii. 55); a perennial herb, 1–2 ft.
high; stem 2–3 lin. thick, simple or with ray-like branches below
the umbel, glabrous and when in full flower often naked below;
leaves numerous, alternate, with a whorl under the umbel, sessile,
¾–1¾ in. long, 3–5 lin. broad, somewhat oblanceolate, acute, subcordate at the base, finely serrulate, pilose all over on both sides or
only on the marginal part above; umbel 4–6-rayed, with or without
rays in the axils of the leaves below it, bright yellowish-green
(*Wolley-Dod*); rays when fully developed 4–12 in. long, 2–4-times
branched; bracts variable, some like the leaves, others much
smaller and ovate or rhomboid-ovate or elliptic-lanceolate, acute,
¼–¾ in. long, 2–5 lin. broad, finely serrulate, varying from glabrous
on both sides to pilose like the leaves, even on the same specimen;
involucre sessile, unisexual or bisexual, 1–1¼ lin. in diam., cup-
shaped, glabrous, with 4 glands and 5 oblong ciliate lobes; glands
⅓–⅔ lin. in their greater diam., transverse, elliptic-oblong, entire,
yellow or yellowish-green; capsule 1½ lin. in diam., subglobose,
slightly 3-grooved, minutely tuberculate, glabrous, exserted about
½ lin. beyond the involucre on a recurved or finally erect pedicel;

styles ⅔ lin. long, very shortly united at the base, entire, usually
collected together (not spreading) and directed downwards; seeds
1 lin. long, ellipsoid, smooth or minutely tuberculate. *Jacq. Eclog.*
i. 98, *t.* 66; *Boiss. in DC. Prodr.* xv. ii. 134. *E. platyphyllos, var.
literata, Bolus & Wolley-Dod in Trans. S. Afr. Phil. Soc.* xiv. 319.

COAST REGION: Cape Div.; in a marshy field near Maitland Bridge,
Wolley-Dod, 3202! by the railway near Salt River, *Wolley-Dod,* 3022!

A native of the south of Europe, the Orient and North Africa, whence it has
been introduced. This was originally, from imperfect material, thought to be
E. platyphyllos, Linn., var. *literata,* Koch, but from a statement upon a label
subsequently received that it is a perennial, and from the stoutness of the stem
and from its habitat, it would appear to be rather *E. pubescens,* Vahl, which only
seems to differ from *E. platyphyllos* by being a perennial and growing in humid
places. The seeds of *E. platyphyllos* are described as smooth, and those of
E. pubescens as minutely tuberculate, but I find specimens of *E. pubescens* with
seeds as smooth as those of *E. platyphyllos* and its varieties, and, in one instance,
smooth and tuberculate seeds in the same capsule.

19. **E. dumosa** (E. Meyer in Drège, Zwei Pfl. Documente, 184, ex
Boiss. in DC. Prodr. xv. ii. 168); apparently a dwarf shrublet, 1 ft.
or more high, with a woody main stem about 1 lin. thick, repeatedly
dividing in an umbellate manner; branches 3–4 in each umbel-like
group, slender, ¾ lin. or less thick, umbellately branching, leafy
above, naked and rough from prominent leaf-scars below, very
minutely puberulous on the younger parts, brown or ochreous-
brown; leaves alternate, with 3–5 in a whorl at the base of each
umbel of branches or flowers, numerous, sessile, very spreading or
deflexed, 2½–4 lin. long, 1–2½ lin. broad, obovate-oblong or somewhat cuneately oblong, obtuse, apiculate, minutely serrulate at the
apical part, glabrous on both sides; flowering-rays 1, 2 or 3 in an
umbel, ½–1 in. long, slender, glabrous, each bearing a pair of bracts
and 1 involucre; bracts 1½–2 lin. long, 2–3 lin. broad, somewhat
half-circular, obtuse, apiculate, entire or minutely denticulate,
glabrous on both sides; involucre sessile, 1½–1¾ lin. in diam.,
outside very minutely cup-shaped, glabrous, with 4 glands and 5
subquadrate minutely ciliate lobes; glands ⅔–¾ lin. in their greater
diam., transversely oblong or crescentic-oblong, with 2 horns about
¼ lin. long; capsule about 1½ lin. in diam. (immature), glabrous,
exserted on a recurved pedicel ½ lin. longer than the involucre;
styles free to the base, erect or ascending, ½ lin. long, bifid at the
apex; seeds not seen.

EASTERN REGION: Pondoland; near the Umsikaba River, *Drège,* 4619! and
without precise locality, *Bachmann,* 754!

The peculiar umbellate manner in which the stem and branches repeatedly divide,
and the small spreading or deflexed leaves and horned glands readily distinguish
this species from its allies.

20. **E. ovata** (E. Meyer in Drège, Zwei Pfl. Documente, 184, ex
Boiss. in DC. Prodr. xv. ii. 167); stems several and probably
annually produced from a perennial rootstock, 3–6 in. high, varying
from pilose with rather long spreading hairs to nearly or quite

glabrous; leaves alternate, lax or crowded and somewhat imbricate, ascending or somewhat spreading, very shortly petiolate, 4–8 lin. long, 2–5 lin. broad, ovate or lanceolate-ovate, obtuse, slightly cordate or rounded at the base, flat, from pilose with rather long hairs to glabrous on both sides; umbels 3–5-rayed, 1–2 in. in diam.; rays $\frac{3}{4}$–$1\frac{3}{4}$ in. long, once-forked, glabrous; bracts 3–5 lin. long, 5–8 lin. broad, somewhat reniform-ovate, obtuse, not apiculate, glabrous; involucres on peduncles $\frac{1}{3}$–$\frac{1}{2}$ lin. long, cup-shaped, glabrous, with 4 glands and 5 subquadrate notched lobes; glands $\frac{2}{3}$–1 lin. in their greater diam., transverse, crescent-shaped, with a short horn at each end; ovary exserted on a recurved pedicel, glabrous; styles free to the base, radiately spreading, $\frac{1}{2}$ lin. long, bifid to half-way down, with revolute tips; capsule and seeds not seen. *Tithymalus ovatus, Klotzsch & Garcke in Abhandl. Akad. Berlin*, 1860, 97.

COAST REGION : Stockenstrom Div. ; Kat Berg (or according to Drège's original label, on a mountain between the Kat and Klipplaat Rivers), 4000–5000 ft., *Drège*, 3561 !

This is closely allied to *E. sclerophylla*, Boiss., but is dwarfer, with more herbaceous and less rigid stems and leaves ; the stems and leaves are usually clothed with long conspicuous hairs, quite different from the minute pubescence on *E. sclerophylla*, var. *puberula*, and the leaves are broader and without the pungent point characteristic of that species. No collector besides Drège seems to have found it.

21. **E. albanica** (N. E. Br.) ; a perennial herb, with many erect annual stems, 10–12 in. high, arising from a woody rootstock; stems rather slender $\frac{1}{2}$–1 lin. thick, very minutely puberulous, with a 3–4-rayed umbel at its apex and several axillary rays below, forming a subcorymbose panicle ; leaves alternate, with a whorl of 3–4 at the base of the umbel, numerous but not crowded, very spreading or slightly deflexed, those at the upper part of the stem $\frac{2}{3}$–$\frac{3}{4}$ in. long and $3\frac{1}{2}$–4 lin. broad, the others gradually decreasing in size downwards, all elongate-ovate, acute, slightly cordate at the very shortly petiolate or subsessile base, glabrous on both sides, with the midrib not very conspicuous and the veins obsolete ; rays of the umbel 2–4 in. long and those below the umbel as long or shorter, all once-forked, slender, glabrous ; bracts 3–5 lin. long, 4–7 lin. broad, sessile, broadly triangular-cordate, abruptly very acute, glabrous ; involucres sessile, $1\frac{3}{4}$ lin. in diam., cup-shaped, glabrous, with 4 glands and 5 oblong or ovate obtuse minutely ciliate lobes ; glands $\frac{3}{4}$–1 lin. in their greater diam., including the short horns at the ends of the somewhat crescent-shaped body, probably yellow or yellowish-green ; ovary glabrous, exserted on a recurved pedicel $\frac{3}{4}$ lin. beyond the involucre ; styles united at the very base only, $\frac{1}{2}$ lin. long, very spreading, deeply bifid, with diverging recurved tips ; fruit and seeds not seen.

COAST REGION : Albany Div. ; Brookhuisens Poort, near Grahamstown, *MacOwan*, 657 !

22. E. ruscifolia (N. E. Br.); stems many from a perennial rootstock, erect, simple, 5–9 in. high, thinly and minutely puberulous or subglabrous, rather densely leafy throughout; leaves alternate, with a whorl of 3 at the base of the umbel, ascending or suberect, more or less imbricating, very shortly petiolate, $\frac{1}{4}$–$\frac{3}{4}$ in. long, $\frac{2}{3}$–4$\frac{1}{2}$ lin. broad, varying on different stems of the same plant from linear-lanceolate to ovate, acute and shortly aristate-mucronate at the apex, distinctly cordate at the base in the broader and rounded in the narrower leaves, slightly revolute at the margins, glabrous on both sides; umbel $\frac{3}{4}$–1$\frac{1}{2}$ in. in diam., 3-rayed; rays $\frac{1}{2}$–1 in. long, shortly once- or twice-branched, glabrous; bracts 3–5 lin. long, 3$\frac{1}{2}$–7 lin. broad, broadly deltoid-cordate, acute, apiculate, glabrous; involucre shortly pedunculate, 1$\frac{1}{2}$ lin. in diam., cup-shaped, glabrous, with 4 glands and 5 subquadrate minutely toothed lobes; glands about $\frac{2}{3}$ lin. in their greater diam., somewhat crescent-shaped or transversely oblong, with a short horn at each end; ovary glabrous, exserted $\frac{1}{2}$–$\frac{2}{3}$ lin. from the involucre on a recurved pedicel; styles free, radiately spreading, $\frac{1}{2}$ lin. long, deeply bifid at the apex; capsule and seeds not seen. *E. sclerophylla, var. ruscifolia, Boiss. in DC. Prodr.* xv. ii. 169. *E. aculeata, E. Meyer in Drège, Zwei Pfl. Documente,* 184, *ex Boiss. l.c., not of Forskal.*

EASTERN REGION: Transkei; between Kei River and Geua (Gekau) River, *Drège,* 4621! Krielis Country, *Bowker*!

Nearly allied to *E. sclerophylla,* Boiss., but rather different in appearance, and the cordate-based leaves and larger cordate bracts readily distinguish it from that species. The name *E. aculeata* is a mere catalogue designation, published without a description, and is long antedated by *E. aculeata,* Forsk., besides being quite inapplicable to this plant.

23. E. sclerophylla (Boiss. Cent. Euphorb. 37, and in DC. Prodr. xv. ii. 169); stems usually many from a perennial rootstock, erect, $\frac{1}{2}$–1$\frac{1}{2}$ ft. high, $\frac{1}{2}$–1 lin. thick at the base, woody, simple or branching, leafy throughout or naked with prominent leaf-scars below, puberulous or glabrous; leaves alternate, scattered or crowded and somewhat imbricate, ascending, very shortly petiolate or subsessile, coriaceous, subrigid, 4–8 lin. long, 1$\frac{1}{2}$–2$\frac{1}{2}$ (rarely 3) lin. long, lanceolate, ovate-lanceolate or ovate, acute, with a rather pungent point, rounded at the base, glabrous on both sides; umbel terminal, 4–5-rayed, $\frac{3}{4}$–2 in. in diam.; rays $\frac{1}{2}$–2 in. long, once or twice forked or occasionally with a series of several pairs of barren bracts, glabrous; bracts 1$\frac{1}{2}$–4 lin. long, 2–5 lin. broad, very broadly ovate or transversely elliptic-ovate, acute, with a subpungent apiculus, broadly rounded or cuneately rounded at the base, coriaceous or subrigid, glabrous; involucres sessile, 1$\frac{1}{4}$–1$\frac{1}{2}$ lin. in diam., cup-shaped, glabrous, with 4 glands and 5 subquadrate or oblong slightly denticulate lobes; glands $\frac{1}{3}$–$\frac{3}{4}$ lin. in their greater diam., somewhat crescent-shaped or narrowly transversely oblong, with a short horn at each end; capsule about 1$\frac{3}{4}$ lin. in diam., tricoccous, glabrous, exserted about $\frac{1}{2}$ lin. beyond the involucre on a

curved pedicel ; styles free, $\frac{1}{2}$ lin. long, bifid at the apex, spreading ;
seeds not seen. *E. myrtifolia, E. Meyer in Drège, Zwei Pfl. Documente,* 184, *ex Boiss. in DC. Prodr.* xv. ii. 169, *and E. sclerophylla,
var. myrtifolia, Boiss. l.c.* 169. *Tithymalus multicaulis, Klotzsch &
Garcke in Abhandl. Akad. Berlin,* 1860, 98.

VAR. β, **puberula** (N. E. Br.); leaves puberulous on both sides, otherwise as in
the type.

COAST REGION : Albany Div. ; near Grahamstown, *Burchell,* 3545 ! *Krebs,* 296 !
MacOwan, 17 ! *Williamson!* *Schlechter,* 2616 ! *Schönland,* 40 ! *Ecklon & Zeyher,*
Euphorb. 11, *Burtt-Davy,* 11592 ! *Cooper,* 20 ! *Misses Daly & Cherry,* 1027 ! near
Assegai Bosch, *Drège,* 3563 ! Var. β : Bathurst Div. ; at Riet Fontein, between
Kasuga River and Port Alfred, *Burchell,* 3961 !

24. **E. striata** (Thunb. Prodr. 86, and Fl. Cap. ed. Schult. 406) ;
stems several, annually produced from a perennial rootstock, erect,
8–22 in. high, herbaceous, simple or branching into a panicle at the
inflorescence, striate, glabrous; leaves alternate, usually laxly
scattered, sometimes few, more rarely closely placed, erect or
ascending, rarely spreading or deflexed, sessile or subsessile,
coriaceous, $\frac{1}{2}$–2 in. long, $\frac{1}{2}$–3 lin. broad, linear, linear-lanceolate or
ovate-lanceolate, very acute and somewhat pungent at the apex,
slightly narrowed or rounded at the base, flat or with incurved
(never revolute) margins, those of the whorl at the base of the
umbel often broadly ovate or rhomboid-ovate, acute or long-pointed,
glabrous on both sides ; umbel 3–5-rayed, 1–4 in. in diam., with or
without axillary rays below it or sometimes the umbel is wanting
and all the rays are axillary and alternate, forming a panicle; rays
$\frac{1}{3}$–3$\frac{1}{2}$ in. long, or when barren or proliferous often much longer,
glabrous ; bracts 2–4 lin. long, 3–7 lin. broad, half-circular, rhom-
boid-reniform or rhomboid-ovate, obtuse or acute, apiculate, glabrous ;
involucre subsessile or pedunculate, 1$\frac{1}{2}$–2$\frac{1}{4}$ lin. in diam., cup-shaped,
glabrous outside, with 4 glands and 5 oblong or oblong-ovate ciliate
lobes ; glands $\frac{3}{4}$–1$\frac{1}{4}$ lin. in their greater diam., more or less crescent-
shaped or transversely oblong, with a point or short horn at each
end, rarely entire ; capsule 1$\frac{3}{4}$–2$\frac{1}{4}$ lin. in diam., tricoccous, glabrous,
exserted $\frac{1}{3}$–1$\frac{1}{4}$ lin. beyond the involucre on a curved pedicel ; styles
$\frac{1}{4}$–$\frac{1}{2}$ lin. long, free to the base, radiately spreading, bifid at the
apex ; seeds 1$\frac{1}{4}$ lin. long, oblong or ellipsoid-oblong, with a small
caruncle at one end and a depressed ring at the other, smooth,
slate-grey or blackish. *E. pungens, E. Meyer in Drège, Zwei Pfl.
Documente,* 184, *ex Boiss. in DC. Prodr.* xv. ii. 170, *not of Lam.
Tithymalus capensis, Klotzsch & Garcke in Abhandl. Akad. Berlin,*
1860, 98.

VAR. β, **cuspidata** (Boiss. in DC. Prodr. xv. 170) ; leaves laxly scattered, $\frac{3}{4}$–1$\frac{1}{2}$
in. long, 1$\frac{1}{2}$–3 lin. broad, lanceolate, linear-lanceolate or ovate-lanceolate,
pungently acute, rounded or narrowed at the base ; umbels $\frac{1}{2}$–1$\frac{1}{2}$ in. in diam.,
sometimes head-like with exceedingly short rays, at others with rays $\frac{1}{2}$–1$\frac{1}{4}$ in.
long ; glands of the involucre with their horns so recurved or revolute that they
appear to be merely concave or notched or subentire along their outer margin
when viewed from above. *E. cuspidata, Bernh. in Flora,* 1845, 86, *and in Krauss,
Beitr. Fl. Cap- and Natal.* 150.

VAR. γ, **brachyphylla** (Boiss. in DC. Prodr. xv. ii. 170) ; leaves more numerous and more closely placed than in the type or var. β, 3-7 lin. long, ½-1¼ lin. broad, linear-lanceolate, pungently acute, narrowed at the base ; umbels 1-4 in. in diam., with rays ¾-2¾ in. long ; bracts 1½-2¾ lin. long, 1¾-3½ lin. broad, rhomboid-ovate, pungently acute ; glands of the involucre crescent-shaped, very obtuse or with very short points at the ends.

SOUTH AFRICA : without locality, *Thunberg*! *Drège*, 4623 !
COAST REGION: Cape Div. ; at the foot of Lion Mountain, *Schlechter*, 79 ! Bathurst Div. ; between Port Alfred and Kaffir Drift, *Burchell*, 3841 ! Uitenhage Div. ; near Uitenhage, *Burchell*, 4255 ! *Schlechter*, 2525 ! Coega, *Rogers*, 2115 ! Albany Div. ; various localities, *Burchell*, 3633 ! *Burtt-Davy*, 11562 ! *MacOwan*, 327 ! 328 ! *Burke*! *Miss Daly*, 607 ! 838 ! *Miss Sole*, 389 ! Queenstown Div. ; various localities, *Drège*, 3562 ! *Galpin*, 1574 ! 2613 ! King Williamstown Div. ; near Peelton, *Cooper*, 109 ! East London Div. ; East London, *Wood*, 3353 ! Cambridge, *Miss Wormald*, 77 ! Komgha Div. ; near Komgha, *Flanagan*, 236 ! Eastern Frontier, *MacOwan*, 328 ! *Hutton*! Var γ : Albany Div. ; near Grahamstown, *MacOwan*, 19 ! Trapps Valley, *Miss Daly*, 669 ! Bedford Div. ; near Bedford, *Miss Nicol*, 13 ! Rietfontein, between Kasuga River and Port Alfred, *Burchell*, 3960 ! King Williamstown Div. ; Frankfort, *Sim*, 1450 ! British Kaffraria, *Cooper*, 123 !
CENTRAL REGION : Somerset Div. ; Somerset East, *Bowker*!
KALAHARI REGION : Orange River Colony ; Harrismith, *Sankey*, 233 ! Bethlehem, *Richardson* ! Basutoland ; Leribe, *Dieterlen*, 351 ! Transvaal ; various localities, *Burke*! *Zeyher*, 1538 ! *Ecklon & Zeyher*, Euphorb. 10 ! *Rehmann*, 4551 ! *Wilms*, 1333 ! 1333a ! *Burtt-Davy*, 1995 ! 5046 ! 5493 ! 9057 ! *Miss Nation*, 12 ! *Miss Leendertz*, 362 ! 892 ! 977 ! 1702 ! 2305 ! *Rademacher*, 7300 ! *Rogers*, 396 ! 1202 ! *Miss Haagner* ! *Tennant*, 6921 ! 6942 !
EASTERN REGION : Transkei ; Krielis Country, *Bowker*, 245 ! Kentani district, *Miss Pegler*, 1393 ! 1798 ! Tembuland ; Bazcia, *Baur*, 230 ! Griqualand East ; near Kokstad, *Tyson*, 1094 ! and in *MacOwan & Bolus*, Herb. Norm. Austr.-Afr., 452 ! Natal ; various localities, *Sanderson* ! *Wood*, 518 ! 4780 ! 7459 ! *Miss Franks in Herb. Wood*, 12196 ! *Wilms*, 2256 ! *Schlechter*, 3164 ! 3373 ! *Rogers*, 1133 ! Var. β : Natal ; various localities, *Krauss*, 441 ! *Sanderson* ! *Gerrard*, 764 ! 1169 ! *Wood*, 195 ! 518 ! 6522 ! 8682 ! *Rudatis*, 441 ! *Schlechter*, 3048 ! 3164 ! Var. γ : Transkei ; Krielis Country, *Bowker*!

The varieties *ouspidata* and *brachyphylla* are distinct from the type in appearance and may possibly prove to be distinct species, but there seem to be some intermediate forms, which require to be studied in the living state. The variety *brachyphylla* has the appearance of being a hybrid between *E. striata* and *E. sclerophylla* and only occurs in the general region where both species grow.

25. **E. erythrina** (Link, Enum. Pl. Hort. Berol. ii. 12) ; a perennial shrublet, ½-2 ft. high ; stems few or several, erect, simple or branched, leafy throughout or becoming naked at the base, glabrous ; leaves very numerous, rather crowded, alternate, ascending, subcoriaceous, often imbricate, 2-8 lin. long, ½-1½ lin. broad, usually linear-lanceolate or narrowly cuneate-oblanceolate, sometimes linear-oblong, obtuse or acute, with a straight apiculus at the apex, tapering below into a very short petiole not revolute at the margins, glabrous ; umbel terminal, ½-2½ in. in diam., 3-5-rayed, with a whorl of 3-5 oblong or ovate or broadly rhomboid leaves 2-5 lin. long and 1-5 lin. broad at its base ; rays ¼-2 in. long, usually once or twice forked, sometimes simple and occasionally the secondary rays bear a succession of 3-4 pairs of barren (always ?) bracts along them, glabrous ; bracts 2-3½ lin. long, 2½-6 lin. broad, transversely rhomboid, very obtuse or rounded and apiculate at the

apex, glabrous, green or often purple at the margins, glabrous; involucre 1¼–1½ lin. in. diam., glabrous, with 4–5 glands and 5 small ovate or oblong entire or slightly toothed lobes; glands ⅔–1 lin. in their greater diam., transverse, more or less crescent-shaped, acute or with a short horn at each end; capsule 1¾–2 lin. in diam. and about as long, narrowing at the apex, glabrous, exserted on a recurved pedicel about 1½ lin. long; styles ⅓–⅔ lin. long, free or united for one-third to one-half their length, erect or collected together or perhaps spreading when young, subentire, minutely notched or bifid at the apex; seeds ¾–1 lin. in diam., subglobose or ellipsoid, smooth, grey or dark brown. *Spreng. Syst. Veg.* iii. 798; *Boiss. in DC. Prodr.* xv. ii. 169, *incl. var. Burchellii, Boiss. E. striata, Eckl. & Zeyh. ex Boiss. in DC. Prodr.* xv. ii. 169. *Tithymalus erythrinus, Klotzsch & Garcke in Abhandl. Akad. Berlin*, 1860, 91.

VAR. β, **Meyeri** (N. E. Br.); leaves laxly scattered, ascending or spreading, ½–1¼ in. long, 1–2½ lin. broad; rays of the umbel 1–3¼ in. long, 1–3-times forked; bracts 2½–5 lin. long, 4–8 lin. broad, otherwise as in the type. *E. Meyeri, Boiss. Cent. Euphorb.* 35, *and in DC. Prodr.* xv. ii. 146. *E. dilatata, E. Meyer in Drège, Zwei Pfl. Documente*, 184, *name only, not of Hochstetter nor of Torrey & Gray. Tithymalus apiculatus, Klotzsch & Garcke in Abhandl. Akad. Berlin*, 1860, 94.

SOUTH AFRICA: without locality, *Forster*! *Mund & Maire*, 65! 248! Var. β, *Mund & Maire*!

COAST REGION: Cape Div.; various localities near Cape Town, *Bergius*! *Burchell*, 458! 8530! *Ecklon*, 303! *Wallich*! *Wright*, 448! *Harvey*, 445! *Wolley-Dod*, 1293! 1793! 3044! 3161! *Diels*, 784! 1201! *Schlechter*, 987! 1315! *Dümmer*, 10! 37! Caledon Div.; Genadendal, *Roser*! Swellendam Div.; on mountains, *Pappe*! Riversdale Div.; near Zoetemelks River, *Burchell*, 6715! Garcias Pass, *Galpin*, 4564! Uniondale Div.; between Avontuur and Klip River, *Drège*! Albany Div.; near Teafontein between Riebeek East and Grahamstown, *Burchell*, 3496! Var. β: Malmesbury Div.; Malmesbury, *Schlechter*, 5348! Paarl Div.; Paarl Mountain, *Drège*, 2197! Cape Div.; mountains near Cape Town, *Ecklon & Zeyher*, Euphorb. 14!

26. E. foliosa (N. E. Br.); a dwarf shrublet, 6–8 in. high, woody at the older parts, much-branched at the base; branches rather crowded, erect, simple, glabrous or very minutely puberulous, densely leafy or the basal part naked and rough with prominent leaf-scars; leaves alternate, with whorls of 4–5 under the umbels, crowded, ascending and more or less imbricate, subsessile, 1–3½ lin. long, ¾–2 lin. broad, somewhat obovate, oblong, elliptic or elliptic-oblong, obtuse, with a recurved apiculus at the apex, minutely scabrous (not ciliate) at the margin, rather thick in texture, glabrous on both sides; umbels terminal, 3-rayed, ½–¾ in. in diam.; rays 2–7 lin. long, simple or once-forked; bracts 1½–2¼ lin. long, 2¼–3½ lin. broad, somewhat rhomboidal-half-circular, very obtuse, with a recurved apiculus, entire or obscurely scabrous at the margins; involucres sessile, 1¼–2 lin. in diam., cup-shaped, glabrous outside, puberulous within, with 4 glands and 5 erect oblong deeply bifid lobes; glands ½–¾ lin. in their greater diam., transverse, with the inner margin broadly rounded, and the outer margin nearly straight and minutely denticulate or concave and produced into a slight

tooth on each side of the notch ; capsule 1¾–2 lin. in diam., very
obtusely trigonous, glabrous, exserted on a curved pedicel ⅔ lin.
beyond the involucre ; styles free, ½–⅔ lin. long, moderately stout,
bifid at the apex, spreading or radiating ; seeds about 1 lin. long,
ellipsoid or subglobose, with a depressed ring at the end opposite the
hilum, smooth ; caruncle large. *Tithymalus foliosus, Klotzsch &
Garcke in Abhandl. Akad. Berlin,* 1860, 67.

COAST REGION: Cape Div. ; Cape Flats, *Ecklon & Zeyher,* Euphorb. 12 ! on
dunes not far from the shore, *Zeyher*!

Very near *E. erythrina,* but dwarfer and differs by its smaller densely crowded
leaves, which are broader in proportion to their length and more obovate, the
umbels are also smaller and more head-like.

27. **E. artifolia** (N. E. Br.) ; plant apparently about 8–10 in. (or
less) high, with probably several main stems arising from a
perennial rootstock, only one such stem has been seen, which
consists of a simple naked stem 3 in. long and 1¼ lin. thick,
marked with numerous leaf-scars, bearing at its apex an umbel
of 7 simple or divided erect branches 3–5 in. long and ½–⅔ lin.
thick, densely leafy, glabrous, purplish ; leaves alternate, crowded,
spreading, more or less imbricate, scarcely petiolate, rather thick
in substance, 1½–2 lin. long, ¾–1½ lin. broad, elliptic or elliptic-
oblong, obtusely rounded at the apex, with a recurved apiculus,
entire, with thickened margins, glabrous, very minutely ciliate,
having a glaucous appearance when dried ; umbels ¾–1 in. in diam.,
compact or dense, 3–5-rayed ; rays (including the flowers) 4–6 lin
long, 3–5-flowered ; bracts at the base of the umbel like the leaves,
those under the involucres, suborbicular, 1½–2 lin. long and broad,
with a recurved apiculus ; involucre 1–1½ lin. in diam. ; cup-shaped,
with 4 glands, glabrous ; glands spreading, ½–⅔ lin. in their greater
diam., transverse, crescent-shaped, with the short horns deflexed
and slightly incurved ; capsule about 2 lin. in diam., exserted on
a pedicel shortly exceeding the involucre and recurved, glabrous,
apparently glaucous ; styles free to the base, radiately spreading,
⅓ lin. long, bifid for about half their length, with diverging tips ;
seeds subglobose, nearly 1 lin. in diam., smooth, dark grey, with a
proportionately very large obtusely conical dull yellow aril about
⅔ lin. long at one end.

COAST REGION : Riversdale Div. ; near Milkwoodfontein, 600 ft., *Galpin,* 4562 !

28. **E. genistoides** (Berg. Descr. Pl. Cap. 146) ; a dwarf shrublet,
⅓–1½ ft. high, sometimes branching at the base into numerous
simple stems, sometimes with the stems branching in the upper
part or irregularly, woody at the lower part; branches rather
slender, puberulous to glabrous on the younger parts, leafy ; leaves
alternate, usually rather crowded, shortly petiolate or subsessile,
spreading or ascending, 2–9 lin. long, ¼–1¼ lin. broad, linear, acute
or obtuse, mucronate, rounded at the base, with revolute margins,
glabrous ; umbels terminal, ½–1 in. in diam., usually 4-rayed, often

with some rays in the axils of the leaves below the umbel; rays
1-6 lin. long, glabrous or puberulous, simple with a pair of bracts
and 1 involucre, or once or twice forked, with 3-7 pairs of bracts
and 3-7 involucres; bracts 1½-3 (rarely 4) lin. long, ¾-3 (rarely
up to 4½) lin. broad, usually rhomboid-ovate, but varying from
ovate-lanceolate to transversely rhomboid, acute or acuminate or
obtuse and apiculate, glabrous or rarely puberulous; involucre
1¼-2 lin. in diam., obconic-cup-shaped, distinctly ribbed, glabrous
or puberulous, with 4 glands and 5 oblong or ovate minutely
toothed or ciliate lobes; glands ½-1 lin. in their greater diam.,
transverse, crescent-shaped or somewhat 2-lobed, with acute or
obtuse points, which are apparently sometimes reflexed or recurved;
ovary obtusely trigonous, exserted on a recurved pedicel about
1 lin. long, glabrous or puberulous; styles ½ lin. long, united at
the base, with spreading bifid and often thickened tips. *Linn.*
Mant. Alt. 564; *Lam. Encycl.* ii. 430; *Willd. Sp. Pl.* ii. 908;
Thunb. Prodr. 86, *and Fl. Cap. ed. Schult.* 405; *Spreng. Syst. Veg.*
iii. 790 (*excl. syn. E. spartioides, Jacq.*); *Bernhardi in Flora,* 1845,
86, *and in Krauss, Beitr. Fl. Cap- und Natal.* 150; *Boiss. in DC.*
Prodr. xv. ii. 167, *incl. vars. major and leiocarpa.* *E. taxifolia and*
E. linifolia, Burm. Prodr. Fl. Cap. 14. *Galarhœus genistoides, Haw.*
Syn. Pl. Succ. 144. *Tithymalus genistoides, Klotzsch & Garcke in*
Abhandl. Akad. Berlin, 1860, 97. *T. revolutus, Klotzsch & Garcke,*
l.c. 99.

VAR. β, puberula (N. E. Br.); leaves more or less puberulous, otherwise as in
the type.

VAR. γ, corifolia (N. E. Br.); leaves laxly scattered along the stems, small,
1½-2½ lin. long, ⅓-⅔ lin. broad, oblong-linear, mucronate-acute, with revolute
margins, varying on the type specimen from thinly and minutely puberulous to
glabrous; otherwise as in the type. *E. corifolia, Lam. Encycl.* ii. 431; *Pers.*
Syn. ii. 16; *Spreng. Syst. Veg.* iii. 798; *Boiss. in DC. Prodr.* xv. ii. 168.

SOUTH AFRICA: without locality, *Thunberg! Thom! Var. β, Thunberg! Mund!*
Harvey, 444! Var. γ, *Sonnerat in Herb. Lamarck!*

COAST REGION: Clanwilliam Div.; Zeekoe Vley, *Schlechter,* 8496! mountains
near the Olifants River, *Schlechter,* 5091! Malmesbury Div.; between Mamre and
Dassen Berg, *Drège!* Zwartland, *Ecklon & Zeyher,* Euphorb. 2! near Malmesbury,
Bolus, 4358! Tulbagh Div.; Mitchells Pass, *Schlechter,* 8937! *Bolus!* Paarl Div.;
Drakensteen Mountains, *Drège,* 8193! Cape Div.; various places near Cape
Town, *Burchell,* 836! *Prior! Ecklon & Zeyher! Pappe! Bolus,* 2942! *Wolley-Dod,*
1458! 2773! and with var. β, 3385! *Dümmer,* 482! *Diels,* 1156! *Wilms,* 3626!
Wright! and mixed with var. β, *Schlechter,* 1381! Stellenbosch Div.; near
Lowrys Pass, *Schlechter,* 1118! Stellenbosch, *Marloth,* 4893! Swellendam Div.;
Zuurbraak, *Galpin,* 4563! Var. β: Malmesbury Div.; Hopefield, *Bachmann,* 85!
Tulbagh Div.; New Kloof, *Drège!* Cape Div.; Lion Mountain, *Drège,* 8192!
Schlechter, 1381! *Wolley-Dod,* 3104! near Cape Town, *Prior!* Simons Bay,
Wright, 447! Var. γ: Malmesbury Div.; near Hopefield, *Bachmann,* 1044!
(*Bachmann,* 1985, from near Hopefield seems to be a glabrous form of this
variety.)

CENTRAL REGION: Ceres Div.; between Ceres and Leeuwfontein, *Pearson,* 3248!
WESTERN REGION: Van Rhynsdorp Div.; Gift Berg, *Phillips,* 7388! 7389!

29. **E. muraltioides** (N. E. Br.); stems several from a perennial
woody rootstock, 10-15 in. high, ⅔-1¼ lin. thick, subumbellately
branching at the upper part, naked and simple or sometimes with

a few branches on the lower part; branches more slender, rather densely leafy throughout, very minutely puberulous; leaves alternate, very spreading or slightly deflexed, subpetiolate, somewhat rigid, 3–6 lin. long, $\frac{3}{4}$–1$\frac{1}{2}$ lin. broad, lanceolate, acute, somewhat pungently mucronate, veinless, except for the midrib, glabrous on both sides; umbel 3–5-rayed; rays $\frac{1}{3}$–$\frac{3}{4}$ in. long, simple or once-branched, sometimes naked below the flowering-bracts, with the whorl of bracts at the base like the flowering-bracts, sometimes bearing several alternate leaves like those on the stem or with two pairs of opposite barren flowering bracts besides those under the flowers and the whorl of bracts at the base of the umbel lanceolate and nearly like the stem-leaves; flowering-bracts 1$\frac{3}{4}$–2$\frac{1}{4}$ lin. long, 2$\frac{1}{2}$–3 lin. broad, very broadly deltoid-ovate, acute, or subreniform-ovate and very abruptly and shortly acute, glabrous on both sides; involucre about 1$\frac{1}{2}$ lin. in diam., cup-shaped, glabrous, with 4 glands and 5 oblong minutely toothed lobes; glands $\frac{1}{2}$–$\frac{2}{3}$ lin. in their greater diam., transverse, somewhat crescent-shaped with a horn at each end; capsule 1$\frac{1}{3}$–1$\frac{1}{2}$ lin. in diam., glabrous, very obtusely three-lobed, glabrous, exserted $\frac{3}{4}$ lin. beyond the involucre on a curved pedicel; styles free, widely spreading from their base, $\frac{1}{3}$ lin. long, bifid at the apex; seeds 1 lin. long, ellipsoid, smooth, dark brown.

COAST REGION: Albany Div.; Brookhuisens Valley, near Grahamstown, *MacOwan*, 320! 642! near Grahamstown, *Glass*, 665!

30. **E. ericoides** (Lam. Encycl. ii. 430); plant 6–18 in. high, probably branching at the base; stems or branches erect, somewhat woody, $\frac{1}{2}$–1$\frac{1}{2}$ lin. thick, often branching in a whorled or subumbellate manner, leafy throughout or naked at the base, glabrous; leaves crowded, alternate, reflexed, with their tips upcurved, sessile, 1$\frac{1}{2}$–6 lin. long, $\frac{1}{4}$–$\frac{1}{2}$ lin. broad, linear, with revolute margins, truncate and mucronate or minutely 3-toothed at the apex, glabrous; umbel terminal, sometimes head-like and $\frac{1}{2}$–$\frac{2}{3}$ in. in diam., with exceedingly short rays, at others 1–2 in. in diam., with rays $\frac{1}{2}$–1$\frac{1}{4}$ in. long; rays once or twice forked, glabrous; bracts 2–3 lin. long, 2$\frac{1}{2}$–5 lin. broad, transversely rhomboid, obtuse and slightly notched at the apex, with a minute apiculus in the notch, glabrous; involucre 1$\frac{1}{4}$–1$\frac{1}{2}$ lin. in diam., cup-shaped, glabrous, with 4 glands and 5 oblong lobes truncate and minutely toothed at the apex; glands $\frac{1}{2}$–$\frac{3}{4}$ lin. in their greater diam., transverse, more or less crescent-shaped, with a short horn at each end and the horns sometimes straight, sometimes incurved; capsule about 2 lin. in diam., glabrous, exserted on a recurved or finally erect pedicel 1$\frac{1}{4}$–2$\frac{1}{4}$ lin. long; styles $\frac{1}{2}$–$\frac{2}{3}$ lin. long, free, subentire or minutely notched at the apex; seeds about 1 lin. long, elliptic-oblong, black or dark grey, with a small yellow caruncle. *Spreng. Syst. Veg.* iii. 797; *Boiss. in DC. Prodr.* xv. ii. 168, *and Ic. Euphorb.* 23, *t.* 111. *Tithymalus confertus, Klotzsch & Garcke in Abhandl. Akad. Berlin,* 1860, 94.

SOUTH AFRICA : without locality, *Sonnerat* ! *Mund & Maire*, 27 !
COAST REGION : Swellendam Div. ; between Swellendam and the Buffeljagts
River, *Pappe* ! George Div. ; near George, *Prior* ! Knysna Div. ; Plettenbergs Bay
or Outeniqua Mountains, *Drège* ! Uitenhage Div. ; Van Stadens River, *Burchell*,
4651 ! Van Stadens Berg, *Ecklon & Zeyher*, Euphorb. 5 ! Port Elizabeth Div. ;
Port Elizabeth, *Drège*, 81 ! Baakens River Valley, *Mrs. Paterson*, 843 ! Bathurst
Div. ; near Kaffir Drift, *Burchell*, 3772 ! Eastern Frontier, *Hutton* !

31. E. natalensis (Bernhardi in Flora, 1845, 86, and in Krauss,
Beitr. Fl. Cap- und Natal. 150) ; perennial ; stems several from a
woody rootstock, $\frac{1}{2}$–$1\frac{1}{2}$ or rarely up to $2\frac{1}{4}$ ft. high, simple or with
a pair or whorl of 3–4 branches near the top, woody below, densely
leafy above, naked at the base, glabrous or puberulous ; leaves
sessile, deflexed and more or less imbricating downwards, straight
or with the tips upcurved, 3–10 lin. long, $\frac{1}{2}$–3 lin. broad, linear,
with or without a dilated cordate base, or elongated ovate, cordate
at the base, acute or subobtuse, mucronate, revolute at the
margins, the whorl under the umbel often larger, flatter and more
oblong, glabrous on both sides or rarely minutely puberulous
beneath ; umbel terminal, head-like with very short rays and
$\frac{1}{2}$–1 in. in diam. or up to 2 in. in diam. with once or twice-forked
glabrous or puberulous rays up to $1\frac{1}{2}$ in. long ; bracts 2–$4\frac{1}{2}$ lin.
long, $1\frac{3}{4}$–$5\frac{1}{2}$ lin. broad, varying from broadly ovate to ovate-
subrhomboid, usually broader than long, obtuse or slightly notched
at the apex, apiculate, glabrous ; involucre $1\frac{1}{2}$–$2\frac{1}{2}$ lin. in diam.,
cup-shaped, glabrous, with 4 glands and 5 oblong acutely bifid,
toothed or subentire lobes ; glands $\frac{1}{2}$–$1\frac{1}{2}$ lin. in their greater diam.,
varying from crescent-shaped and two-horned with the two horns
often incurved towards each other, to transversely oblong and
denticulate on the outer margin ; capsule $1\frac{3}{4}$–$2\frac{1}{4}$ lin. in diam.,
slightly narrowing upwards, glabrous, exserted on a recurved
pedicel $1\frac{1}{2}$–3 lin. long ; styles $\frac{2}{3}$–$\frac{3}{4}$ lin. long, usually upcurved from
the decurved ovary, united at the base or for one-third of their
length, bifid at the apex, with spreading tips ; seeds 1–$1\frac{1}{8}$ lin. long,
ellipsoid, smooth, black, with a small yellowish-white caruncle.
Boiss. in DC. Prodr. xv. ii. 170 ; *Wood, Natal Pl.* iv. *t.* 302.

KALAHARI REGION : Transvaal ; Ermelo, *Burtt-Davy*, 5402 !
EASTERN REGION : Tembuland ; Bazeia mountains, *Baur*, 511 ! Pondoland,
Bachmann, 751 ! 753 ! Griqualand East ; near Kokstad, *Tyson*, 1541 ! *Haygarth
in Herb. Wood*, 4195 ! Natal ; various localities, *Sutherland* ! *Gerrard*, 524 ! 1172 !
Sanderson, 65 ! 212 ! *Krauss*, 434 ! *Wood*, 6 ! 238 ! 382 ! 1407 ! 1429 ! 8666 !
9599 ! *and in Herb. Norm. Austr.-Afr.*, 1027 ! *Schlechter*, 3086 ! 5770 ! *Kuntze* !
Miss Doidge ! *Kolbe* !

This is very similar to *E. ericoides*, Lam., but may be easily distinguished from
that species by its acute leaves, larger involucre and more united bifid styles. It
is variable in appearance, and some specimens resemble *E. epicyparissias*,
E. Meyer, from which it is distinguished by its numerous shorter and usually
simple or once or twice umbellately branched stems. *E. epicyparissias* is 2–5 ft.
high and much branched.

32. E. epicyparissias (E. Meyer in Drège, Zwei Pfl. Documente,
184, ex Boiss. in DC. Prodr. xv. ii. 168) ; a shrub, 2–5 ft. high ;

stem simple at the base, branching above; branches often clustered in groups, densely leafy throughout or naked and rough with prominent leaf-scars below, with dark reddish bark, minutely puberulous or glabrous on the young parts; leaves alternate, rather crowded, sessile, spreading, deflexed or upcurved, $\frac{1}{3}$–1 (very rarely 1$\frac{1}{4}$) in. long, $\frac{1}{4}$–1$\frac{1}{2}$ (very rarely 2–3) lin. broad, linear or oblong-linear, obtuse or acute, mucronate, entire or rarely with the margins at the apex minutely denticulate, flat or with revolute margins, glabrous on both sides, those of the whorl at the base of the umbel often larger, oblong to ovate and up to 4$\frac{1}{2}$ lin. broad; umbels terminal $\frac{3}{4}$–3 in. in diam., 3–8-rayed, and often with rays arising from the axils of the leaves below the umbel; rays $\frac{2}{5}$–3$\frac{1}{4}$ in. long, 1–4-times forked, glabrous, when short the umbel is usually congested and head-like; bracts 1$\frac{1}{2}$–6 lin. long, 2–8 lin. broad, broadly rhomboid-ovate or somewhat half-circular, very obtuse, apiculate, glabrous; involucres 1$\frac{1}{2}$–2$\frac{1}{2}$ lin. in diam., cup-shaped, glabrous outside, with 4–5 glands and 5 oblong subentire emarginate or shortly bifid ciliate lobes; glands $\frac{2}{3}$–1$\frac{1}{4}$ lin. in their greater diam., transverse, usually more or less crescent-shaped, with 2 straight or incurved horns $\frac{1}{4}$–$\frac{1}{2}$ lin. long, but sometimes with the horns very reduced or obsolete; capsule 1$\frac{3}{4}$–2 lin. in diam., tricoccous, glabrous, exserted on a pedicel $\frac{2}{3}$–2 lin. beyond the involucre; styles free or very shortly united at the base, $\frac{3}{4}$–1$\frac{1}{4}$ lin. long, bifid at the apex, usually upcurved from the decurved ovary and immature fruit; seeds rather more than 1 lin. long, oblong, with a large depression at the caruncular end and a small ring-like marking at the opposite end, smooth, blackish-grey when quite ripe. *Bernhardi in Flora*, 1845, 87, *and in Krauss, Beitr. Fl. Cap- und Natal.* 150. *E. involucrata, E. Meyer in Drège, Zwei Pfl. Documente,* 184, *ex Boiss. in DC. Prodr.* xvii. 168 *incl. var. megastegia. E. Bachmanni, Pax in Engl. Jahrb.* xxiii. 535. *Tithymalus involucratus, Klotzsch & Garcke in Abhandl. Akad. Berlin,* 1860, 91.

VAR. β, **puberula** (N. E. Br.) ; leaves minutely toothed or entire at the margins about the apex ; involucre (at least on the upper part) more or less puberulous ; ovary puberulous or pubescent.

VAR. γ, **Wahlbergii** (N. E. Br.) ; branches more slender than in the type or var. β, $\frac{1}{2}$–$\frac{3}{4}$ lin. thick, rather laxly leafy ; leaves entire or minutely denticulate ; umbels small and lax, $\frac{1}{2}$–1 in. in diam., 3–5-rayed, often proliferous ; bracts 2–2$\frac{1}{4}$ lin. long, 1$\frac{1}{2}$–3 lin. broad, broadly ovate or transverse and subrhomboid, acute or very obtuse, apiculate ; involucres 1$\frac{1}{4}$–1$\frac{1}{2}$ lin. in diam. ; glands $\frac{3}{4}$ lin. in their greater diam. ; capsule 1$\frac{1}{4}$ lin. in diam., glabrous ; styles $\frac{1}{2}$–$\frac{3}{4}$ lin. long ; otherwise as in the type. *E. Wahlbergii, Boiss. in DC. Prodr.* xv. ii. 169, *and Ic. Euphorb.* 23, *t.* 112. *E. epicyparissias as to letter c, E. Meyer in Drège, Zwei Pfl. Documente,* 184. *Tithymalus epicyparissias, Klotzsch & Garcke in Abhandl. Akad. Berlin,* 1860, 88.

SOUTH AFRICA : without locality, *Thunberg*! *Krebs*! *Zeyher*, 1540!

COAST REGION : Oudtshoorn Div. ; near Oudtshoorn, *Miss Taylor*! George Div. ; near George, *Burchell*, 5993! *Drège*, 8194! *Schlechter*, 5856! Knysna Div. ; near Goukamma River, *Burchell*, 5590! Uniondale Div. ; Lange Kloof, *Thunberg*! between Haarlem and Avontuur, *Burchell*, 5036! Uniondale, *Krauss*, 1727! Alexandria Div. ; between Hoffmans Kloof and Driefontein, *Drège*! Albany Div. ;

Slay Kraal, *Burke*! near Grahamstown, *Zeyher*, 1016! *MacOwan*, 84! *Atherstone*, 70! *Bolton*! *Rogers*, 55! 230! Coldstream, *Misses Daly & Sole*, 258! 258a! Stockenstrom Div.; by the Kat River near Philipton, *Ecklon & Zeyher*, Euphorb. 8! between Kat River and Klipplaat River, *Drège*, 3560! Kat Berg, *Drège*! *Hutton*! near Stockenstrom, *Scott-Elliot*, 265! Queenstown Div.; various localities, *Baur*, 34! *Cooper*, 242! *Galpin*, 2186! 2683! Komgha Div.; near Keimouth, *Flanagan*, 380! Var. β: George Div.; near George, *Prior* (*Alexander*)!

CENTRAL REGION: Somerset Div.; Bosch Berg, *MacOwan*, 1763! Graaff Reinet Div.; descent of the Voor Sneeuwberg, *Burchell*, 2846! Murraysberg Div.; near Murraysburg, *Tyson*, 238! Richmond Div.; between Richmond and Brak Valley River, *Drège*, 857! Molteno Div.; Broughton, near Molteno, *Flanagan*, 1636!

KALAHARI REGION: Orange River Colony; Vet River, *Burke*! Basutoland; Leribe, *Dieterlen*, 675a! Transvaal; various localities, *Wilms*, 1334! 1335! *Mudd*! *Miss Leendertz*, 1067!

EASTERN REGION: Transkei; Krielis Country, *Bowker*! Kaffrarian Mountains, *Mrs. Barber*! Pondoland, *Bachmann*, 755! *Drège*, 4620a! Natal; various localities, *Wood*, 907! *Wilms*, 2255! *Rudatis*, 678! 813! Var. β: Transkei; Kentani district, by streams, *Miss Pegler*, 460! Var. γ: Pondoland; between Umtata River and St. Johns River, *Drège*! Natal; various localities, *Cooper*, 3152! *M'Ken*, 650! *Rehmann*, 7992! *Wood*, 148; 5284! *Schlechter*, 6720!

I am quite unable to find any distinction between *E. epicyparissias* and *E. involucrata* as maintained by Boissier, either in structure or appearance, which latter is remarkably uniform, except in var. γ, which seems only to differ in its weaker and more lax habit. The form named *E. Bachmanni* by Pax is not distinct from *E. involucrata*, var. *megastegia*, Boiss., and merely consists of very luxuriant specimens of the plant, with larger bracts than usual.

33. **E. kraussiana** (Bernhardi in Flora, 1845, 87); somewhat shrubby, 1½–3 ft. high; stems 1–1⅔ lin. thick, somewhat woody and naked below, umbellately branching above, 1–1½ lin. thick; branchlets 3–4 in a whorl, simple or branching like the stem, ascending, leafy throughout or naked below, glabrous; leaves alternate, with a whorl of 4–5 at the base of the umbel, lax or crowded, spreading, 1¼–4 in. long, ⅛–⅔ in. broad, linear, linear-lanceolate, cuneate-oblanceolate or elongated obovate, obtuse or acute, apiculate, cuneately tapering into a short petiole, glabrous on both sides; umbels terminating some or all of the branchlets, 3–5-rayed; rays ⅔–4 in. long, once to 3-times forked, glabrous; bracts ⅕–½ in. long and as much in breadth, deltoid-ovate, acute or obtuse, apiculate, subtruncate at the base, glabrous; involucre pedicellate, 1–1½ lin. in diam., cup-shaped, glabrous, with 5 glands and 5 erect oblong slightly notched or entire lobes; glands contiguous, ⅓–½ lin. in their greater diam., transversely oblong, entire, in dried specimens usually with the inner margin slightly inflexed, but not or scarcely so when alive, not sublunate as originally described; ovary globose-trigonous, glabrous, exserted on a recurved pedicel 2–2½ lin. long; styles 1–1⅓ lin. long, united into a column ⅓–½ lin. long below, with spreading deeply bifid tips; capsule about 2 lin. in. diam.; seeds 1¼ lin. long, ellipsoid, smooth, very dark brown or blackish (not seen ripe), with a small yellowish caruncle. *Tithymalus truncatus, Klotzsch & Garcke in Abhandl. Akad. Berlin*, 1860, 75.

VAR. **erubescens** (N. E. Br.); stem 1–2 ft. high, 1⅓–2 lin. thick, herbaceous, not woody below; bracts ⅕–¾ in. long and broad; otherwise as in the type.

E. erubescens, E. Meyer in Drège, Zwei Pfl. Documente, 184, *ex Boiss. in DC. Prodr.* xv. ii. 116. *Tithymalus Meyeri, Klotzsch & Garcke in Abhandl. Akad. Berlin,* 1860, 75.

SOUTH AFRICA : without locality, *Thunberg*! *Krebs*, 292 ! Var. *β* : *Krebs*, 293 ! COAST REGION : George Div. ; woods near George, *Burchell*, 5845 ! *Alexander* (*Prior*)! Knysna Div. ; near Knysna Ford, *Burchell*, 5533 ! Little Umtini River and Karatra, *Schlechter*, 5891 ! Uitenhage Div. ; various localities, *Mrs. Paterson*, 936 ! *Marloth*, 4888 ! *Ecklon & Zeyher*, Euphorb. 13 ! *Zeyher*, 3858 ! Port Elizabeth Div. ; near Port Elizabeth, *Drège*, 19 ! Bathurst Div. ; woods near Port Alfred, *Schlechter*, 2762 ! Trapps Valley, *Miss Daly*, 608 ! Albany Div. ; near Grahamstown, *Atherstone*, 5 ! *Williamson* ! near Sidbury, *Miss Daly*, 782 ! Bedford Div. ; near Bedford, *Miss Nicol*, 48 ! Stockenstrom Div. ; summit of Kat Berg, *Shaw* ! Stutterheim Div. ; forest at Fort Cunninghame, *Galpin*, 2476 ! King Williamstown Div. ; Perie forest, *Galpin*, 3281 ! *Schönland*, 865 ! *Kuntze* ! East London Div. ; near East London, *Rattray*, 668 ! British Kaffraria, *Cooper*, 176 ! Var. *β* : Uitenhage Div. ; Zuurberg Range, *Drège*, 2347 ! Bathurst Div. ; near Port Alfred, *Schönland*, 787 ! 807 ! *Potts*, 207 ! Albany Div. ; Blue Krantz, *Burchell*, 3639 ! near Grahamstown, *Drège* ! *MacOwan*, 291 !

CENTRAL REGION : Var. *β* : Somerset Div. ; banks of the Great Fish River, *Burchell*, 3254 ! near Somerset, *Bowker* !

KALAHARI REGION : Var. *β* : Transvaal ; Houtbosch, *Rehmann*, 5913 ! Rimers Creek, near Barberton, *Thorncroft*, 768 !

EASTERN REGION : Tembuland ; Bazeia, *Baur*, 28 ! Natal ; margins of woods near Pietermaritzburg, *Krauss*, 256 ! near Richmond, *Sanderson*, 844 ! Fairfield, *Rudatis*, 208 ! Var. *β* : Natal ; between the Umzimkulu and Umkomanzi Rivers, *Drège*, 4617 ! Tugela, *Gerrard*, 1173 ! near Byrne, *Wood*, 346 ! near Enon and near Richmond, *Wood*, 1851 ! Verulam, *Wood*, 767 ! Weenen County, *Sutherland* ! and without precise locality, *Gerrard*, 1627 ! *Sanderson* ! *Cooper*, 1161 !

Partly described from fresh material preserved in fluid (*Drège*, 19). I am doubtful if *E. erubescens* should be even distinguished as a variety, the specimens constituting it may be only young growths, which later assume the typical form, as they seem to grade into one another. This, however, can only be decided by those who can examine living plants.

34. E. transvaalensis (Schlechter in Journ. Bot. 1896, 394) ;

a shrub, 2–5 ft. high, usually with forked branches ; young branches herbaceous, becoming woody with age, terete, glabrous, hollow, terminating in a 3–4-rayed umbel, the peduncle of which at first appears to be a continuation of the branch, but the latter gradually thickens and persists, becoming 2–4½ in. long, and 1½–2½ lin. thick, whilst the peduncle of the umbel scarcely thickens and falls off at the end of the season ; leaves 2–3 and alternate on the basal part of each new branch, with a whorl-like cluster apparently at its middle, really at its apex or base of the peduncle and another whorl of 3–4 under the umbel ; petiole ¼–1¼ in. long, glabrous ; blade 1¼–4¼ in. long, ⅔–2 in. broad, oblong-lanceolate, elliptic or elliptic-ovate, obtuse, rounded or subacute at the apex, cuneately narrowed into the petiole, glabrous on both sides or thinly pubescent beneath ; stipules none ; umbel solitary, terminal, with a peduncle 2–8 in. long and 3–4 simple or once or twice forked rays, sometimes accompanied by a similar ray at its base, glabrous, deciduous ; bracts ½–1¼ in. long, ovate, rhomboid-ovate, subreniform or orbicular, obtuse, acute or slightly notched at the apex, apiculate, abruptly rounded or cuneately contracted into a short petiole, glabrous on both sides or thinly pubescent beneath ; involucre 2–2¾ lin. in diam., cup-shaped, glabrous outside, with 5 pubescent stripes opposite the glands within, with 4–5 glands and 5 transversely oblong or subquadrate ciliate lobes ; glands ⅔–1¼ lin. in their greater diam., transversely oblong, obtusely or acutely 2–3-toothed or lobed or subentire, rugulose or pitted-rugulose on the upper surface ; capsule about ⅓ in. in diam., glabrous, exserted on an erect pedicel 2½–5 lin. long ; styles ¾–1 lin. long, united at the basal part, then ascending-spreading, entire or minutely bifid ; seeds 2¼ lin. long, ellipsoid or ellipsoid-oblong, stoutly apiculate at one end, with scattered flat-topped tubercles or irregular raised markings of a different tint from the ground colour. *N. E. Br. in Dyer, Fl. Trop. Afr.* vi. i. 530. *E. Galpini, Pax, and E. ciliolata, Pax in Bull. Herb. Boiss.* vi. 742, 743 ; *and E. ciliolata in Baum, Kunene-Samb. Exped.* 284. *E. Gœtzei, Pax in Engl. Jahrb.* xxviii. 420.

KALAHARI REGION : Transvaal ; Kaap River Valley, near Barberton, 2000 ft., *Galpin*, 1198 ! *Thorncroft*, 594 ! Potgieters Rust, *Miss Leendertz*, 1432 ! near Lydenburg, *Wilms*, 1336 !

Also in Tropical Africa. *Galpin* 1198 is the type of *E. transvaalensis*, Schlechter, published in 1896, and of *E. Galpini*, Pax, published in 1898.

35. E. frutescens (N. E. Br.) ; a much-branched woody bush, 5 ft. high, apparently quite leafless and spineless, unisexual ; branches alternate, ascending or ascending-spreading, the ultimate 3–10 in. long, 1–1½ lin. thick, rigid, woody, glabrous, dark brown, with numerous alternate flowering tubercles scattered along them ; leaves none or reduced to the merest rudiments, not seen ; bracts at the base of the involucres ⅓–½ lin. long, broadly obovate, minutely ciliate ; involucres clustered on the flowering tubercles, sessile, 2 lin. in diam. and 1 lin. deep, obconic-cup-shaped, minutely puberulous all over on the outside including the backs of the glands and puberulous within, with 5 glands and 5 subquadrate or transversely oblong ciliate lobes, apparently white ; glands subcontiguous, spreading, ¾–1 lin. in their greater diam., transversely oblong or elliptic-oblong, entire, smooth and glabrous on the upper side ; female plant not seen.

WESTERN REGION : Little Namaqualand (Little Bushmanland) ; lower mountain slopes at Aus, 3000 ft., *Pearson*, 4714 !

This, by its perfectly leafless nodose-tuberculate woody branches, is one of the most distinct of the South African species of *Euphorbia* and resembles no other from that region.

36. E. guerichiana (Pax in Engl. Jahrb. xix. 143) ; a shrub about 6–7 ft. high, apparently woody, with slender straight erect or ascending branches, the younger ¾–1½ lin. thick, with slightly prominent scars at the nodes, usually nearly or quite leafless at the time of flowering ; leaves alternate, very small, shortly petiolate, soon deciduous, 1½–3 lin. long, ⅔–1¼ lin. broad, oblong, obovate or lanceolate, obtuse or acute, rather thick in texture, minutely

puberulous on both sides; flowering branchlets developed in the axils of the fallen leaves, 1–2½ lin. long, minutely puberulous, bearing a few small leaves and bracts and a terminal involucre; bracts obovate, sessile, thinner than the leaves; involucre 2 lin. in diam., cup-shaped, glabrous, hermaphrodite or male with a rudimentary ovary, with 5 glands and 5 subquadrate slightly notched denticulate or subciliate lobes; glands ¾–1 lin. in their greater diam., transversely elliptic-oblong, entire; perfect ovary partly (or shortly?) exserted, erect, glabrous, with a very small 3-lobed calyx at its base; lobes very short, emarginate; styles nearly 1 lin. long, united at the base for a quarter of their length, then recurved-spreading with slightly thickened entire tips; rudimentary ovary with minute erect styles scarcely ¼ lin. long; capsule ¼ in. in diam., very obtusely 3-angled, smooth, exserted on a curved pedicel 1½–1¾ lin. long; seeds 1¾ lin. long, compressed-ellipsoid, smooth, glabrous.

WESTERN REGION: Great Namaqualand; rocky places south of Korekas, *Gürich*, 73! Little Namaqualand; hills at I'us, *Schlechter*, 11433!

37. E. lignosa (Marloth in Trans. Roy. Soc. S. Afr. i. 316 and 317, fig. 2); a dwarf densely much-branched bush, forming a hemispheric cushion up to 3 ft. in diam., and ½–1 ft. high; main stem very stout, somewhat globose at the crown from which the branches arise, descending as a thick carrot-shaped root; main branches very numerous, repeatedly divided into short alternate spine-like or sharp-pointed branchlets, soft and succulent when young, becoming rigid, glabrous; leaves alternate, only present on very young branches, 1½–5 lin. long, lanceolate or linear-lanceolate or sometimes obovate, acute, tapering near the base into a short petiole, longitudinally folded, recurved, very minutely puberulous; flowering branches with 1 terminal sessile involucre or with 1 sessile and 1 or 2 lateral involucres on peduncle-like branches ⅓–½ in. long, forming a 2–3-flowered cyme, very minutely puberulous; bracts about 2–2½ lin. long and 1½–2 lin. broad, orbicular-obovate and apiculate, or broadly cuneate, subtruncate and minutely toothed, very minutely puberulous on both sides; involucre 3½–5 lin. in diam., shallowly cup-shaped, very minutely puberulous, with 5 glands and 5 transverse rectangular lobes fringed with numerous subulate teeth along the truncate top; glands obliquely cup-like or funnel-shaped, with 4–9 entire or bifid channelled finger-like processes ¼–½ lin. long along the outer margin, glabrous in the cavity, very minutely puberulous outside; capsule sessile, erect, with its base embraced by the remains of the involucre, 2½ lin. in diam., globose, very minutely puberulous; styles 1–1¼ lin. long, united nearly to the top, with 3 spreading bifid arms ⅛–¼ lin. long; seeds 1¼ lin. long, ovoid, acute at one end, somewhat 3-angled, smooth, glabrous.

WESTERN REGION: Great Namaqualand; stony ground near Tschaukaib, 2900 ft., *Marloth*, 4637! Kleinfontein, *Marloth*, 5070! Akam River, *Pearson*, 4159! plains and mountain sides at Schakalskuppe, 4900–5600 ft., *Pearson*, 4160!

Great Karas Berg region at Aus, *Pearson*, 8030! and on hill slopes south of
Krai Kluft, *Pearson*, 8110! *Range*, 819!

38. E. spinea (N. E. Br.); leafless bush, $\frac{1}{2}$–$\frac{3}{4}$ ft. high, with rigid
spine-tipped branches 1–2 lin. thick, opposite and diverging from
each other at an angle of 100°–165°, the primary and secondary being
3–6 in. long and the ultimate $\frac{1}{2}$–2 in. long, all ending in an acute
spine, glabrous, probably glaucous, with a somewhat succulent bark,
which at the tips or on old branchlets dries and separates into
rings by the formation of transverse cracks all round the branch;
leaves very rudimentary, deciduous, opposite, $\frac{1}{3}$ lin. long, sessile or
nearly so, ovate or deltoid-ovate, acute, glabrous, blackish-purple;
cymes lateral on the ultimate branchlets, small and dense, sessile or
subsessile, 2–3 lin. long and $2\frac{1}{2}$–4 lin. in diam., 3- to several-
flowered; bracts scale-like, $\frac{1}{3}$ lin. long and about one-third as long as
the involucre, linear-oblong, obtuse, glabrous, purple; involucre
1 lin. in diam. and as much in depth, but appearing longer than
broad, somewhat urceolate, being slightly constricted under the
glands, glabrous outside, pubescent within, purple, with 5 glands
and 5 minute oblong ciliate lobes; glands not quite contiguous,
ascending-spreading, $\frac{1}{3}$–$\frac{1}{2}$ lin. in their greater diam., transversely
oblong or elliptic-oblong, two-lipped, from the inner margin being
turned up, concave in front of it, entire; ovary and capsule not
seen.

CENTRAL REGION: Calvinia Div.; plentiful on sandy plains about 12 miles
north-east of Klipplaat, 1700 ft., *Pearson*, 3296!

WESTERN REGION: Great Namaqualand; common among rocks near Dabaigabis,
Pearson, 4380! on a saline flat north of Ganus, 3300 ft., *Pearson*, 4585!

Both Namaqualand specimens are stated to be only 6–9 inches high, whilst the
Calvinia specimen is stated on the label to be 2–3 ft. high. The latter, however,
is so exactly like the Namaqualand specimens that I suspect a mistake has been
made in the labelling of it, and that the label probably belongs to the allied
E. decussata, E. Meyer, which does grow in Calvinia and is 2–3 ft. high; a specimen
of it was also received at Kew from Prof. Pearson without a label.

39. E. decussata (E. Meyer in Drège, Zwei Pfl. Documente,
184, ex Boiss. in DC. Prodr. xv. ii. 74, excl. syn.); a divaricately
much-branched leafless bush, 2–3 ft. high, diœcious; main stems
3–4 lin. thick; branches opposite, diverging from each other at an
angle of 90°–150°, rigid, 1–$2\frac{1}{2}$ lin. thick, articulated at the base,
straight or slightly curved, apparently more or less succulent when
young, or with a succulent bark, tapering to a more or less acute
but scarcely spine-like apex, glabrous; leaves rudimentary, opposite,
scale-like, sessile, persisting for a short period, $\frac{1}{2}$–$\frac{2}{3}$ lin. long, broadly
deltoid-ovate, acute, finally recurved, concave, rigid, glabrous, at first
green, becoming dark brown or blackish; cy es very short, com-
pact, $\frac{1}{4}$–$\frac{1}{3}$ in. in diam., opposite, one to four pairs to a branch or
rarely solitary; bracts like the leaves; involucres sessile, 1–$1\frac{1}{4}$ lin.
in diam., cup-shaped, glabrous outside, with 5 glands and 5 oblong
subentire or toothed and minutely ciliate lobes; glands $\frac{1}{3}$–$\frac{2}{3}$ lin. in

their greater diam., transversely elliptic-oblong, entire; capsule 1½ lin. in diam., obtusely 3-lobed, with a small disc-like calyx at its base, glabrous, minutely white-dotted, very shortly exserted from the involucre on a recurved pedicel, erect when immature; styles shortly united at the base, ⅓–½ lin. long, spreading, with bifid revolute tips; seeds nearly 1 lin. long, conical, acute, truncate at the base, slightly 4-angled, tuberculate-rugose, grey. *E. Tirucalli, Thunb. Prodr.* 86, *and Fl. Cap. ed. Schult.* 405, *partly, as to sheet* 2, *left-hand specimen in Thunberg's Herbarium.*

SOUTH AFRICA : without locality, *Thunberg* !
CENTRAL REGION : Calvinia Div. ; between Lospers Plaats and Sprinkbok Kuil River, *Zeyher*, 1533 ! mountain pass south of Klipplaat, *Pearson*, 3853 ! ravine at Loeriesfontein, *Pearson*, 4858 ! Ceres Div. ; between Gansfontein and Pappekuil, *Pearson*, 3685 !
WESTERN REGION : Great Namaqualand ; opposite Sendlings Drift, *Pearson*, 6105 ! Great Karas Berg region, *Pearson*, 8117 ! 8258 ! 8347 ! 8348 ! Little Namaqualand ; hills by the Koussie (Buffels) River, *Drège*, "392 ?" (not 3926 as quoted by Boissier) ! *Krapohl, Marloth*, 4895 ! plain between Aggenys and Pella, *Pearson*, 3580 ! slopes north of Middelkraal, *Pearson*, 5618 ! Van Rhynsdorp Div. ; near Bakhuis, *Pearson*, 5468 !

40. **E. brachiata** (E. Meyer in Drège, Zwei Pfl. Documente, 184, ex Boiss. in DC. Prodr. xv. ii. 74 partly) ; a dwarf much-branched shrublet, leafless and spineless ; branches and branchlets opposite or forking, jointed and more or less constricted at their origin, diverging from each other at an angle of 100°–180°, fleshy when young, glabrous, rusty-brown when dried ; leaves rudimentary, deciduous, opposite, sessile, abruptly recurved, 1 lin. long, deltoid, acute, angular on each side at the base, concave-channelled down the face, rigid when dried, glabrous ; cymes terminal, ¼–⅓ in. in diam., consisting of two diverging branchlets about 1 lin. long, each bearing 3 sessile involucres ; bracts much shorter than the involucres, about ½ lin. long and ¾ lin. broad when flattened, broadly deltoid, apiculate, concave, glabrous ; involucres sessile, 1 lin. in diam., cup-shaped, glabrous, with 4–5 glands and 5 oblong fringed lobes ; glands about ⅔ lin. in their greater diam., transversely oblong, entire ; ovary and capsule not seen.

COAST REGION : Van Rhynsdorp Div. ; hills near Ebenezer, *Drège*, 2948 !

Under the name of *E. brachiata*, Drège distributed two distinct but allied species, both of which seem to be included in Boissier's description. I have taken as the type, *E. brachiata*, E. Meyer, letter "a," which in E. Meyer's Herbarium at Lübeck bears the number 2948, as quoted by Boissier and is named "*E. brachiata*" in E. Meyer's handwriting, none of the specimens seen of this bear any ovary or capsule upon them. Under letter "b" an entirely different species was distributed, described below as *E. perpera*, N. E. Br., which was collected over 200 miles further north, near the Orange River and bears female flowers and fruit, and it is from this species that I believe Boissier to have described the styles and fruit under *E. brachiata*, no specimen of it, however, exists in Meyer's Herbarium at Lübeck. The plant figured by Burmann, *Pl. Afr. Rar. t.* 5, also quoted by Boissier under *E. brachiata*, belongs to a totally different species (*E. arceuthobioides*, Boiss.), which grows in a different region.

41. **E. chersina** (N. E. Br.); a dwarf much-branched bush, apparently less than 1 ft. high, leafless and spineless, with a succulent bark; primary branches or stems of the specimens seen $3\frac{1}{2}$–6 in. long, $2\frac{3}{4}$–$3\frac{1}{2}$ lin. thick, dichotomously branched so as to form a flattish-topped cyme 5–6 in. across, terete, glabrous, perhaps glaucous, but not papillate nor asperate; branches opposite, diverging from each other at an angle of 90°–130°, stout, with internodes $\frac{1}{2}$–2 in. long; leaves rudimentary, opposite, persisting for a short time, sessile, scale-like, $\frac{3}{4}$ lin. long, deltoid-ovate, acute, concave, glabrous, spreading or recurved, blackish; cymes terminal on the ultimate forkings of the branches and stem, sessile, very small (about $\frac{1}{4}$–$\frac{1}{3}$ in. in diam.) and dense, nearly all fallen from the specimens seen; bracts $\frac{1}{2}$ lin. long, scale-like, broadly deltoid-ovate, subacute, adpressed to the involucres or finally spreading, keeled, glabrous, reddish-brown or blackish-brown; involucres sessile, 1 lin. in diam., $\frac{3}{4}$ lin. deep, cup-shaped, glabrous outside, with 5 glands and 5 broadly cuneate ciliate lobes, apparently purple; glands about $\frac{1}{2}$ lin. in their greater diam., transverse, narrowly oblong, entire, with the inner margins turned up, forming a slight lip; ovary oblong, obtusely 3-angled, glabrous, with a small obtusely 3-lobed calyx at its base, finally just exserted from the involucre on a short pedicel, erect; styles $\frac{1}{2}$ lin. long, very shortly united at the base, ascending-spreading, minutely 2-lobed at the apex; capsule and seeds not seen.

WESTERN REGION: Great Namaqualand; Angra Pequena, *Pearson & Galpin*, 7584! *Marloth*, 4638.

42. **E. indecora** (N. E. Br.); a bush, 2–3 ft. high, succulent at the young parts and with a succulent bark, leafless and spineless, unisexual; branches opposite, diverging from each other at an angle of 90°–100°, terete, slightly curved, 3–12 in. long, simple or branching like the stems, tapering to the apex, but not spine-pointed, $1\frac{1}{2}$–2 lin. thick when dried, with internodes 1–2 in. long, not papillate nor asperate, glabrous, perhaps glaucous; leaves rudimentary, soon deciduous, opposite, recurved-spreading, $\frac{2}{3}$ lin. long, spathulate or obovate-spathulate, with a very short broad stalk and rounded or transverse very obtuse blade, concave, glabrous, green; cymes lateral, apparently few, scattered along the young branches, sessile, 3- (or perhaps more) flowered; bracts like the leaves; involucres not seen quite mature, $\frac{2}{3}$ lin. in diam. and $\frac{3}{4}$ lin. deep, but probably slightly larger when fully developed, subcylindric or very slightly obconic, glabrous outside, puberulous within, with 5 glands and 5 subquadrate very minutely ciliate lobes; glands $\frac{1}{4}$ lin. in their greater diam., transversely oblong, entire; ovary glabrous, shortly stalked, perhaps ultimately exserted from the involucre; styles $\frac{1}{2}$–$\frac{2}{3}$ lin. long, shortly united at the base, acutely bifid at the apex.

WESTERN REGION: Little Namaqualand; on sandy plains sloping towards the Orange River, between Dabainoris and Houms Drift, *Pearson*, 3387!

This much resembles *E. decussata*, E. Meyer, and, as in that species, the bark of the dead branchlets cracks transversely and breaks up into rings, but besides its smaller and more cylindric involucres is readily distinguished by its green spathulate rudimentary leaves and bracts, which are more quickly deciduous than they are in *E. decussata*.

43. **E. cibdela** (N. E. Br.); a bush, 2–3 ft. high, leafless and spineless, with the young branches succulent and the older with a succulent bark, glabrous, green, apparently not glaucous, unisexual; branches opposite, diverging from each other at an angle of 50°–60°, straight, 1–2 lin. thick, with internodes ¾–1¼ in. long, all attaining to the same general level or the lower overtopping the main and upper branches, not asperate nor papillate; leaves rudimentary and soon deciduous, opposite, recurved-spreading, 1–1⅓ lin. long, ⅓ lin. broad and nearly as thick, narrowly lanceolate or elliptic, acute or obtuse, apiculate, narrowed into a short petiole at the base, glabrous, minutely ciliate on the margins of the petiole; male involucres solitary and terminal and also in small terminal and lateral cymes, with a pair of leaf-like bracts at their base, sessile or subsessile, 1½ lin. in diam. and 1 lin. deep, cup-shaped, puberulous outside and pubescent with much longer hairs within, with 5 glands and 5 rectangular ciliate lobes; glands rather distant, ¼–½ lin. in their greater diam., transversely oblong or elliptic-oblong, entire, with the inner margin turned up and a slight depression in front of it; female plant not seen.

WESTERN REGION : Great Namaqualand ; hills at Schakalskuppe, 4900–5600 ft., *Pearson*, 4428 !

44. **E. amarifontana** (N. E. Br.); a succulent leafless and spineless bush, 1 ft. or more high, divaricately branched; main stems 2 lin. thick (dried); main branches opposite, 1½–2 lin. thick, diverging from each other at an angle of 60°–100°, with internodes ¾–2 in. long, tapering upwards and usually with one pair of lateral branchlets below the forked or simple ends, sometimes simple, without any branchlets, straight, each pair gradually shorter, so that the whole forms a flat-topped corymbose cyme at the end of each main stem 8–12 in. in diam., glabrous, not papillate nor asperate; leaves rudimentary, opposite, persisting for some time, ½ lin. long, sessile, ovate or deltoid-ovate, acute, recurved, glabrous, dark brown or blackish; cymes very small and compact, sessile, terminating the ultimate branchlets; bracts scale-like, ⅓ lin. long, broadly ovate, subacute, adpressed to the involucre, glabrous, dark brown; involucre 1 lin. long, ¾ lin. in diam., or perhaps larger, somewhat ovoid-campanulate, glabrous outside, with 5 glands and 5 broadly cuneate or subspathulate-obovate ciliate lobes; glands ⅓–⅔ lin. in their greater diam., transversely oblong, entire, somewhat 2-lipped from the inner margin being turned up or inwards; ovary, capsule and seeds not seen.

WESTERN REGION : Van Rhynsdorp Div. ; Bitterfontein, *Zeyher*, 1534 ! between Bitterfontein and Stinkfontein, *Pearson*, 5532 !

Of this species I have been able to examine only one involucre, which is past the flowering stage and has its glands ascending, but probably when alive and in full flower they would be spreading and so the diameter of the involucre would be larger than above stated. Its mode of branching at the inflorescence and cuneate involucre-lobes well distinguish it from most of its allies.

45. E. Rudolfii (N. E. Br.); a leafless and spineless unisexual bush, apparently succulent or with a succulent bark, probably 1 ft. or more high, with the main branches repeatedly forking at the top into large lax cymes 3–8 in. across; in the specimens seen the pieces are 9–12 in. high, with the main stems $1\frac{1}{2}$–$2\frac{1}{2}$ lin. thick, and the slender branches $\frac{1}{2}$–$\frac{3}{4}$ lin. thick, terete, glabrous, not papillate nor asperate, with internodes mostly 1–$2\frac{1}{2}$ in. long, the ultimate shorter; leaves and bracts rudimentary, opposite, scale-like, $\frac{1}{2}$ lin. long, sessile, deltoid, acute, spreading, blackish or dark brown; male involucres solitary or 3 together in a sessile cyme at the tips of the ultimate branchlets, campanulate, $1\frac{1}{4}$ lin. or less in diam. and 1 lin. deep, very minutely puberulous outside on the lower part, distinctly puberulous within, with 5 glands and 5 rather linear-oblong ciliate lobes; female plant not seen.

Western Region: Van Rhynsdorp Div. ; on hills near Bitterfontein, 1300 ft., *Schlechter*, 11047 ! between Bitterfontein and Stinkfontein, *Pearson*, 5533 !

46. E. muricata (Thunb. Prodr. 86 and Fl. Cap. ed. Schult. 405); a much-branched leafless and spineless succulent shrub, $1\frac{1}{2}$–2 ft. high; only upper portions of three main branches or stems bearing a few lateral branches seen; branches more or less constricted or jointed at their origin, opposite or sometimes alternate (except at the inflorescence) from only one branch at each node being developed, diverging from the stem from which they arise at an angle of 20°–35°, terete or (when dried) sometimes very distinctly 6-angled, with concave sides between the angles, scabrous, especially on the older parts, with very small crowded laterally compressed tubercles or crenations, glabrous; leaves rudimentary and scale-like, opposite, sessile, about $\frac{1}{2}$ lin. long, ovate or deltoid-ovate, acute or obtuse, recurved and concave-channelled at the apical part, dark brown, soon deciduous; cymes terminal, $\frac{1}{2}$–$1\frac{1}{2}$ in. in diam., consisting of 2 opposite diverging branchlets $\frac{1}{2}$–1 in. long, each forking once or twice into shorter branchlets bearing 3 involucres, but flowers are wanting on the specimens and are not described by Thunberg; bracts scale-like, sessile, $\frac{1}{4}$–$\frac{1}{2}$ lin. long, very broadly ovate, acute to very obtuse, dark brown.

Coast Region: Van Rhynsdorp Div. ; on hills at Atties, *Pearson*, 5459 ! Clanwilliam Div. ; between the Olifants River and Bockland, *Thunberg* !

Boissier has erroneously considered this to be identical with two other species which he associated with it under *E. brachiata*, E. Meyer, and states that the scabrous epidermis is only due to shrinkage in drying. This, however, is not the case, for the thin laterally compressed and somewhat minute tubercles are evidently structural, and neither they nor the angles and grooves present on the stem and branches of *E. muricata* are to be found upon any dried specimens of

E. brachiata, E. Meyer, or any other South African species living or dried that I have seen. Besides the nature of its scabrous stems, this species is readily distinguished from the true *E. brachiata,* E. Meyer, by the often alternate and very much less divergent branches and different inflorescence, probably the involucres will also be found to differ. Thunberg's type consists of two small branches and Pearson's specimen of one branch only; the latter is identical with the type except that the stem shows only a slight trace of the six angles which are conspicuous on that of Thunberg.

47. E. perpera (N. E. Br.); a much-branched succulent shrub or shrublet, leafless and spineless; branches and branchlets all opposite or forking, more or less constricted or jointed at their origin, diverging from each other at an angle of not more than 60°–75°, fleshy when young, glabrous; leaves rudimentary, deciduous, scale-like, opposite, sessile, spreading or with abruptly recurved tips, ½ lin. long and more in breadth, very broadly deltoid, acute, scarcely channelled, glabrous; cymes terminal, ¼–1 in. in diam., formed of 2 diverging branchlets ¼–½ in. long, each once or twice forking into shorter branchlets, the ultimate bearing 3 involucres; bracts shorter than the involucres, scale-like, ½ lin. long, deltoid-ovate, acute or subobtuse, glabrous; involucres sessile, ¾ lin. in diam., cup-shaped, glabrous, with 5 glands and 5 minute oblong or spathulate-oblong slightly ciliate lobes; glands ¼ lin. in their greater diam., transversely oblong or elliptic-oblong; ovary ellipsoid, trigonous, narrowing at the apex, glabrous; styles ⅓–⅔ lin. long, united at the basal half, ascending-spreading above, with bifid tips; capsule about 1½ lin. long and 1⅓ lin. in diam., trigonous, exserted just beyond the involucre; seed about 1 lin. long, pyramidal, subtruncate at the base, obscurely 4-angled, densely covered with very minute whitish tubercles on a dark brown ground. *E. brachiata,* letter *b,* E. *Meyer in Drège, Zwei Pfl. Documente,* 184, *name only; and Boiss. in DC. Prodr.* xv. ii. 74, *partly (excl. Drège,* 2948, *and E. muricata, Thunb.*). *Arthrothamnus brachiatus, Klotzsch & Garcke in Abhandl. Akad. Berlin,* 1860, 62.

WESTERN REGION : Little Namaqualand; along the Orange River, between Verleptpram and its mouth, *Drège*!

This plant bears a superficial resemblance to *E. muricata,* Thunberg, but the peculiar tubercles, grooves and angles on the stems of that species are quite absent from this and it grows in a different region, about 200 miles farther south. Boissier has associated this and 3 other perfectly distinct species under the name of *E. brachiata,* E. Meyer, and I believe his description of the flowers and fruit under that name are taken from this species. No specimen of *E. perpera* exists in E. Meyer's Herbarium, his type of *E. brachiata* being quite distinct from it.

48. E. Burmanni (E. Meyer in Drège, Zwei Pfl. Documente, 184, ex Boiss. in DC. Prodr. xv. ii. 75); a bushy plant, 1–2 ft. high (*Phillips*), with the younger parts succulent, spineless and almost leafless, diœcious; main stems or branches about 2 lin. and the flowering branches about ⅔–1 lin. thick when dried, glabrous, green, not glaucous, drying greyish-green; branches opposite, diverging at an angle of 55°–75°, ascending and forming at the inflorescence an irregular corymb of cymes, conspicuously marked at the nodes with large persistent dark brownish-red stipular glands; leaves opposite, spreading, deciduous, 1–1½ lin. long, spathulate, puberulous on the upper side of the proportionately broad petiole, otherwise glabrous ; blade ⅓–½ lin. in diam., rhomboid or suborbicular-rhomboid, or those on the flowering part transversely oblong, acute to obtuse, flat or concave; inflorescence terminal, consisting of 3 branches or of 2–3 pairs of branches once or twice forked, forming a cyme or a raceme of cymes 1–2 in. long and 1–1⅓ in. in diam., with the ultimate branches each bearing 3 crowded involucres; bracts of the same form and size as the leaves; involucres sessile, unisexual, 1¼–1¾ lin. in diam., cup-shaped or very broadly and shallowly obconic, minutely puberulous, with 5 glands and 5 subquadrate toothed lobes; glands ½–¾ lin. in their greater diam., transversely oblong, entire, dull yellowish-green; capsule 1½–1¾ lin. in diam., obtusely trigonous, minutely velvety, exserted on a pedicel equalling the involucre in length, erect; styles free to the base, ⅓ lin. long, spreading, minutely notched at the apex and with a line-like channel down the face; seeds nearly 1 lin. long, oblong, slightly pointed at one end, 4-angled, irregularly tuberculate on the faces, glabrous. *E. viminalis, Burm. Prodr. Pl. Cap.* 14, *ex Boiss. in DC. Prodr.* xv. ii. 75, *not of Linn. E. Tirucalli, Thunb. Prodr.* 86 *and Fl. Cap. ed. Schult.* 405, *as to one specimen on sheet* 3 *of his Herbarium. E. biglandulosa, Willd. Enum. Pl. Hort. Berol. Suppl.* 27, *not of Desfontaines nor of Boiss. Arthrothamnus Ecklonii, Klotzsch & Garcke, l.c.* 63, *as to description and Ecklon, Euphorb.* 25, *but not* 23 *& * 24. *A. Bergii, Klotzsch & Garcke, l.c.* 63.

SOUTH AFRICA : without locality, *Thunberg* !
COAST REGION : Van Rhynsdorp Div. ; Gift Berg, *Phillips,* 7637 ! Clanwilliam Div. ; near Clanwilliam, *Leipoldt,* 227 ! Doorn River, *Schlechter,* 8058 ! Piquetberg Div. ; mountain sides, Het Kruis. *Misses Stephens & Glover,* 8752 ! Pikeniers Pass, *Pearson,* 5211 ! Malmesbury Div. ; near Darling, *Bachmann,* 402 ! 1045 ! *Marloth,* 4014 ! Moorreesburg, *Bachmann,* 1046 ! Prezant Hoek. near Hopefield, *Bachmann,* 2181 ! near Groene Kloof, *Bolus,* 4360 ! Tulbagh Div. ; New Kloof. *MacOwan,* 3225 ! *and in Herb. Austr.-Afr.,* 2002 ! *Diels,* 183 ! near Tulbagh Waterfall, *Schlechter,* 1418 ! Cape Div. ; Blue Berg. *Drège* ! Robertson Div. ; Montagu, 400 ft., *Marloth,* 3270 !
CENTRAL REGION : Ceres Div. ; Ceres Karoo, *Marloth,* 5102 !
WESTERN REGION : Little Namaqualand : Kamaggas. *Schultze,* 178 ! and without precise locality, *Alston in Herb. Marloth,* 4894 !
Partly described from a living cultivated plant.

Of the type sheets of *Arthrothamnus Ecklonii,* Klotzsch & Garcke, in the Berlin Herbarium, *Euphorbia* 25, Ecklon & Zeyher, is the only specimen of the three quoted by Klotzsch & Garcke that agrees with their description, the others have not the characteristic stipulary glands. This specimen bears the locality number 82, indicating that it was collected near Uitenhage, but this locality is almost certainly an error, as no other collector has found this species so far to the eastward, and many species of this group are so similar that it may easily have been mistaken for an eastern species. Several errors of this kind have been made in the distribution of Ecklon & Zeyher's plants.

The type of *Arthrothamnus Tirucalli,* Klotzsch & Garcke (*Euphorbia* 22 Ecklon & Zeyher), consists of two branches of *Euphorbia mauritanica* and two of

E. Burmanni ; the locality number 85, upon the label, indicates that they were collected on the eastern slopes of Table Mountain, but I believe this only to refer to the branches of *E. mauritanica*, as *E. Burmanni* appears not to occur there.

49. E. corymbosa (N. E. Br.) ; a branching leafless and spineless succulent shrub, 1 or more ft. high, diœcious ; main stems $1\frac{3}{4}$–2 lin. thick (dried) ; branches opposite or occasionally alternate, terete, glabrous, the terminal ending in rather dense corymbose cymes ; leaves rudimentary and soon deciduous, probably like the bracts in size and form, not seen ; corymbs all terminal, $1\frac{1}{4}$–$2\frac{1}{4}$ in. in diam., densely many-flowered, composed of numerous small cymes ; bracts 1 lin. long, spreading, spathulate, with the small transverse concave subtruncate or triangular blade much shorter than the broad linear petiole, glabrous, green ; ultimate cymes of the male plant $2\frac{1}{2}$–4 lin. in diam., consisting of very short branchlets bearing 3–7 sessile involucres 1–$1\frac{1}{4}$ lin. in diam., obconic, glabrous outside and within, with 5 glands and 5 minute subquadrate toothed lobes ; glands contiguous, spreading, $\frac{1}{2}$–$\frac{2}{3}$ lin. in their greater diam., transversely oblong or subreniform ; female plant not seen.

COAST REGION : Riversdale Div. ; near Albertina, *Muir* !

Well distinguished from its allies by the small cymes being crowded into rather large flat-topped corymbs.

50. E. Angræ (N. E. Br., *angrana* by error on p. 232 ante) ; a much-branched leafless and spineless bush, perhaps dwarf ; main branches $\frac{1}{4}$ in. thick, the ultimate about 1 lin. thick, crowded, repeatedly forked, subcylindric or slightly compressed, constricted at the nodes, with internodes or joints $\frac{1}{2}$–1 in. long, smooth, glabrous ; leaves rudimentary, opposite, scale-like, $\frac{3}{4}$–$\frac{3}{4}$ lin. long, 1–$1\frac{1}{4}$ lin. broad, sessile, very broadly subcordate-ovate or subreniform, subacute or apiculate, rather thick and fleshy, puberulous on the upper side, glabrous beneath, reddish-brown ; cymes rather dense, terminal, formed of the very numerous and very short articulations into which the branches divide, each ultimate joint with 1–3 involucres at its apex ; bracts like the leaves ; involucres sessile, $1\frac{1}{2}$–2 lin. in diam., cup-shaped or somewhat obconic, glabrous outside, pubescent within, with 5 (or perhaps sometimes more) glands and 2 or more small fleshy 3-toothed lobes, sometimes replaced by reduced glands, or absent ; glands irregular in size, contiguous, spreading, $\frac{1}{2}$–1 lin. in their greater diam., transverse, subreniform or the smaller subelliptic, minutely pitted ; capsule about 2 lin. in diam., rather deeply 3-lobed, glabrous, exserted on a pedicel slightly longer than the involucre ; styles free to the base, at first erect, afterwards spreading, $\frac{1}{2}$ lin. long, shortly bifid, with slender diverging lobes at the apex ; seeds not seen.

WESTERN REGION : Great Namaqualand ; Angra Pequena, *Galpin & Pearson* 7549 !

51. E. æquoris (N. E. Br.) ; a dwarf succulent leafless and spineless shrublet, 8 or 9 in. high and 4–6 in. in diam., much-branched, with thick fleshy or subtuberous roots; branches opposite or repeatedly forked, $\frac{1}{2}$–2 lin. thick (dried), terete, with internodes 2–6 lin. long, glabrous; leaves rudimentary, soon deciduous, opposite, $\frac{1}{2}$–$\frac{3}{4}$ lin. long, spathulate, with a short broad or linear petiole abruptly dilated into a transverse elliptic or elliptic-oblong blade, subtruncate or very broadly rounded and subapiculate at the apex, longitudinally folded or channelled, glabrous or with a few hairs along the midrib on the upper side; flowers solitary at the ends of the ultimate branchlets, but with a flowering branchlet growing out from the axil of one or both bracts; bracts about as long as the involucre, like the leaves; involucre sessile, $\frac{3}{4}$–1 lin. in diam. and $\frac{2}{3}$–$\frac{3}{4}$ lin. deep, cup-shaped, glabrous outside, pubescent within, but the pubescence not or scarcely visible when wetted, with 5 glands and 5 very small ovate ciliate lobes; glands scarcely contiguous, shortly stalked, $\frac{1}{4}$–$\frac{1}{3}$ lin. in their greater diam., transversely oblong or elliptic-oblong, entire; ovary and capsule obtusely 3-angled, glabrous, with a small acutely 3-lobed calyx at the truncate base, exserted and curved to one side on a slender pedicel 1–2 lin. beyond the involucre; styles united into a slender column $\frac{1}{4}$ lin. long, with spreading arms $\frac{1}{4}$ lin. long, minutely or shortly bifid at apex; capsule about 2 lin. in diam., smooth; seeds 1$\frac{1}{4}$ lin. long, ovoid, truncate at the base, shortly acute at the apex, with 3 slight ridge-like angles, rugulose, of a somewhat steel-grey colour.

CENTRAL REGION : Middelburg Div. ; plains at Rosmead Junction, 4000 ft., *Sim in Herb. Galpin*, 5626! Schoombie, *Trollip*! Colesberg or Hanover Div. ; on the great plain between Colesberg and Hanover, 4500 ft. *Bolus*, 2201!

52. **E. arceuthobioides** (Boiss. Cent. Euphorb. 20, and in DC. Prodr. xv. ii. 76) ; a much-branched succulent bush, $\frac{3}{4}$–1 ft. high, leafless and spineless, unisexual ; main stems 1$\frac{1}{2}$–2 lin. thick at the base ; main branches in 3–4 pairs, opposite, $\frac{1}{2}$–1 lin. thick when dried, variably diverging, dichotomously and trichotomously forking into numerous branchlets, often variably curved, terete, usually slightly rough from minute papilla-like and mostly laterally compressed tubercles ; leaves rudimentary, opposite, sessile or subsessile, often recurved, $\frac{2}{3}$–1 lin. long, broadly ovate, oblong or oblong-obovate, very obtuse or subapiculate, usually with a small tooth or angle on each side near the base, slightly concave and puberulous on the upper surface, convex and glabrous on the back, dark brown, persistent ; cymes once or twice forked, their ultimate branchlets 1$\frac{1}{2}$–9 lin. long, 1-flowered ; bracts shorter than the involucre, $\frac{3}{4}$–1 lin. long, somewhat obovate or subspathulate or like the leaves ; involucre unisexual, 1$\frac{1}{4}$–1$\frac{3}{4}$ lin. in diam. cup-shaped, very minutely puberulous at the very base, otherwise glabrous outside, with 5 glands and 5 oblong fringe-toothed lobes ; glands $\frac{1}{3}$–$\frac{3}{4}$ lin. in their greater diam., transversely oblong, entire ; capsule 2 lin. in diam., glabrous, with a circular disc-like calyx at its base and just exserted beyond the involucre ; styles very shortly united at the base, with radiating arms $\frac{2}{5}$ lin. long, bifid to below the middle ; seeds about

1 lin. long, ovoid, subobtuse at the apex, truncate at the base, rugose. *E. Tirucalli, Thunb. Prodr.* 86, *and Fl. Cap. ed. Schult.* 405, *partly, as to the right-hand specimen on sheet* 3 *of his Herbarium, not of Linn. E. scopiformis (serpiformis), Boiss. in DC. Prodr.* xv. ii. *errata and* 75, *partly. Arthrothamnus Ecklonii, Klotzsch & Garcke in Abhandl. Akad. Berlin,* 1860, 63, *as to Ecklon & Zeyher,* 23, *and* 24 *partly, not as to* 25, *nor description. A. scopiformis, Klotzsch & Garcke, l.c.* 63.—*Tithymalus tuberosus, aphyllus, &c. Burm. Rar. Afr. Pl.* 11, *t.* 5.

COAST REGION : Clanwilliam Div. ; near Brakfontein, *Ecklon & Zeyher,* Euphorb. 24 partly ! Tulbagh Div. ; Piquetberg Road (Gouda), *Schlechter,* 4850 ! Cape Div. ; near Cape Town, *Thunberg* ! *Harvey,* 25 ! Riet Valley, *Bergius* ! Camps Bay, *Prior* ! Green Point, *Prior* ! Lion Mountain, *Froembling,* 124 ! above the road beyond Sea Point, *Wolley-Dod,* 1777 ! Cape Flats, *Mund & Maire* ! Stellenbosch Div. ; near Somerset West, *Ecklon & Zeyher,* Euphorb. 23 ! Lowrys Pass, *Schlechter,* 7215 !

According to reports received from Dr. Marloth and Major Wolley-Dod, this species would appear to be now very scarce in the vicinity of Cape Town, and the female plant is decidedly rare, as only two of the specimens seen are of that sex.

The specimen collected by Bergius in Riet Valley is the type of *Arthrothamnus scopiformis,* Klotzsch & Garcke, and is identical with the type of *Euphorbia arceuthobioides,* Boiss. Both of these names were published in 1860, and in the same year the name *E. scopiformis,* Boiss., was also published. But as this name seems to have been actually published at a later date than that of *E. arceuthobioides,* and as Boissier included under it (according to the specimens he quotes) two distinct species, one of them being *E. arrecta,* N. E. Br., it seems advisable to adopt the name *E. arceuthobioides* in preference to that of *E. scopiformis.* Boissier, under *E. arceuthobioides,* quotes the collectors of the type as *Ecklon & Zeyher,* 76, he has here, however, mistaken the locality number for the distribution number, which should be *Ecklon & Zeyher,* 24 partly, the number 76 which appears on the label is merely the locality number, and indicates that the plant was collected near Brakfontein, by the Olifants River, see *Linnæa,* xix. 583 and 589. The other part of *Ecklon & Zeyher,* 24, belongs to *E. rhombifolia,* var. *cymosa,* N. E. Br., and comes from another region.

Some specimens collected on Hex River Mountains in Worcester Div. (*Drège,* 8204, *Tyson,* 649) are nearly related to *E. arceuthobioides,* but appear to be distinct, the material at my disposal is, however, insufficient to decide this point. The Hex River plant requires to be compared in the living state with *E. arceuthobioides* to form a correct opinion.

53. E. spartaria (N. E. Br. in Dyer, Fl. Trop. Afr. vi. i. 558) ; a

succulent, leafless and spineless bush or shrublet, ¾–1 ft. or more high, trichotomously branched, diœcious ; branches opposite, in 2–3 pairs, diverging from each other at an angle of 40°–70°, terete, ⅔–1¼ lin. thick, with internodes ¾–2 in. long, slightly curved-ascending, glabrous, not asperate or papillate ; leaves rudimentary, opposite, only seen as bracts under the involucres and shorter than or about equalling them, ¾–1 lin. long, obovate-spathulate, with the broad petiole rather gradually dilated into the obovate or orbicular-obovate blade, subacute, very obtusely rounded or minutely 3-toothed at the apex, concave-channelled and minutely puberulous down the face, usually persisting during the flowering period or whilst the fruit is maturing ; cymes small, of 3 involucres, either on a pair of branchlets ¼–½ in. long at the apex of the branches or on 2–3 pairs of short branchlets forming a raceme, with the central involucre sessile and the lateral on short branchlets; involucres $1\frac{1}{4}$–$1\frac{1}{2}$ lin. in diam. in the male and 1–$1\frac{1}{4}$ lin. in diam. in the female plant, obconic-cup-shaped, glabrous outside, pubescent within, with 5 glands and 5 subquadrate toothed lobes; glands contiguous or nearly so in the male, distinctly separated in the female, $\frac{1}{4}$–$\frac{3}{4}$ lin. in their greater diam., transversely oblong, with a slight depression in front of the inner margin, entire; capsule $2\frac{1}{4}$ lin. in diam., with 3 rounded lobes, glabrous, with its base exserted just above the involucre and provided with a small disc-like 3-angled calyx; styles free to the base and spreading on the top of the ovary, $\frac{1}{3}$ lin. long, bifid to nearly half-way down with diverging tips; seeds 1–$1\frac{1}{4}$ lin. long, ovoid, truncate at the base, acute at the apex, obscurely 4-angled, rugose, dark grey.

WESTERN REGION : Little Namaqualand ; hills near Steinkopf, *Schlechter*, 11381! near Kasteel Poort, *Bolus*, 9443! Klip Kalk, *Pearson*, 6521! Van Rhynsdorp Div. ; Zout River, *Schlechter*, 8146!

Also in Tropical Africa.

E. spartaria is closely allied to *E. ephedroides*, but the much more divergent branches give it a distinct appearance, and the involucres are somewhat larger, whilst it also seems to be a more slender plant. The female plant of *E. ephedroides* is unknown.

54. **E. ephedroides** (E. Meyer in Drège, Zwei Pfl. Documente, 184, ex Boiss. in DC. Prodr. xv. ii. 75) ; a leafless and spineless succulent shrub or shrublet, trichotomously or dichotomously branched, diœcious, only the male plant seen ; branches opposite, erect and subparallel, somewhat bunched together, with internodes $\frac{3}{4}$–$2\frac{3}{4}$ in. long, terete or slightly compressed, the stouter of those seen about 1 lin. thick, arising from stems or main branches 2 lin. thick, not asperate or tuberculate ; leaves rudimentary, opposite, soon deciduous, only present in the specimens seen as bracts under and as long as or longer than the involucre, 1–$1\frac{1}{4}$ lin. long, spathulate, with a rather narrow linear petiole, very abruptly dilated into a transversely elliptic blade, very obtusely rounded or subtruncate and dorsally subapiculate at the apex, channelled down the face, glabrous, green ; involucres sessile, solitary at the apex of the ultimate branchlets or sometimes 3 in small cyme, with the central one in the fork between the 2–3 lin.-long branchlets bearing the lateral involucres, 1–$1\frac{1}{8}$ lin. in diam. and $\frac{2}{3}$ lin. deep, cup-shaped, glabrous outside, densely pubescent within, with 5 glands and 5 subquadrate toothed lobes ; glands subcontiguous or slightly separated, $\frac{1}{3}$–$\frac{1}{2}$ lin. in their greater diam., transversely oblong, with the inner margin nearly straight and turned up, forming a slight lip and the outer margin very broadly rounded, entire ; female involucres and fruit not seen.

WESTERN REGION : Little Namaqualand ; between Koper Berg and Kookfontein (Goodemans Kraal ex Boissier), *Drège*! and without precise locality, *Alston in Herb. Marloth*, 4876!

This species is well distinguished from its allies, with the exception of *E. spartaria*, N. E. Br., by the bracts being as long as the involucre, as well as by its erect subparallel branches.

Burchell, 1724 (not 1424 as printed), is erroneously quoted by Boissier under this species, but not only are the habit and appearance totally different, but the male involucres are larger and the bracts shorter than the involucres in Burchell's plant, for which see 55, *E. rectirama*, N. E. Br.

55. **E. rectirama** (N. E. Br.); a succulent, leafless and spineless bush, up to 3 ft. high (*Burchell*), with many branching stems 2–3 lin. thick at the base, diœcious; branches opposite, diverging from each other at an angle of 45°–65°, erect or ascending, terete, glabrous, with internodes $\frac{3}{4}$–3 in. long; leaves opposite, rudimentary, soon deciduous, $\frac{3}{4}$–1 lin. long, spathulate, with the linear or oblong-linear petiole abruptly dilated into a suborbicular or very broadly ovate or transversely elliptic blade, rounded or subacute and apiculate at the apex, channelled and more or less pubescent at the bottom of the channel down the face, otherwise glabrous; cymes opposite and usually racemosely arranged along the branches, $\frac{1}{2}$–1$\frac{1}{2}$ in. long, once or twice forked, with 2–4 involucres or sometimes the upper are undivided with only 1 involucre; bracts like the leaves; involucres solitary, sessile, in the male plant 1$\frac{1}{4}$–1$\frac{1}{2}$ lin. in diam. and 1 lin. deep, in the female $\frac{3}{4}$–1 lin. in diam. and $\frac{3}{4}$ lin. deep, cup-shaped, glabrous outside, pubescent within, with 5 glands and 5 subquadrate toothed lobes; glands of the male $\frac{1}{3}$–$\frac{2}{5}$ lin. and of the female $\frac{1}{4}$–$\frac{1}{3}$ lin. in their greater diam., transversely oblong or elliptic, entire; ovary finally exserted on a pedicel just above the involucre, glabrous; styles $\frac{1}{2}$ lin. long, shortly united at the base, with spreading bifid arms; capsule 1$\frac{1}{2}$–2 lin. in diam., 3-lobed, smooth; seeds 1 lin. long, ovoid, pointed at one end, obscurely 4-angled, tuberculate-rugose, grey.

KALAHARI REGION: Griqualand West; between Spuigslang Fontein and the Vaal River, *Burchell*, 1724! Klipfontein, *Burchell*, 2633! Warrenton, *Miss Adams*, 158! Orange River Colony; Amandelboom, *Burtt-Davy*, 10096! Bloemfontein district, *Burtt-Davy*, 11850!

56. **E. arrecta** (N. E. Br.); a leafless and spineless succulent bush, probably 2 ft. or more high, compactly much-branched above, unisexual; main stems or branches 1$\frac{1}{4}$–1$\frac{1}{2}$ lin. thick (dried); lateral branches more slender, opposite, diverging from each other at the base at an angle of 45°–50°, then erect, elongated and by forking into cymes or with opposite pairs of cymes racemosely scattered along them forming a panicle 1 ft. long and 3–4 in. broad, glabrous; leaves rudimentary, scale-like, opposite, $\frac{2}{3}$–$\frac{3}{4}$ lin. long, deltoid-ovate, often angular on each side at the base, and there contracted into an exceedingly short petiole or subsessile, obtuse, apiculate; stipular glands none; cymes usually twice forked, each fork or branchlet bearing 1 involucre; bracts under the involucre sessile, broadly ovate or oblong, obtuse, apiculate; involucre sessile, 1–1$\frac{1}{4}$ lin. in diam., cup-shaped, glabrous outside, with 5 glands and 5 oblong or subquadrate minutely toothed

and ciliate lobes ; glands $\frac{1}{2}$–$\frac{2}{3}$ lin. in their greater diam., transversely oblong, entire, very slightly concave ; ovary and capsule not seen. *E. scopiformis* (*serpiformis by error*), *Boiss. in DC. Prodr.* xv. *errata and 75 partly, as to Zeyher* 1535 *only, excl. all synonyms.*

COAST REGION : Clanwilliam Div.; Berg Valley, *Zeyher*, 1535! Possibly *Schlechter*, 9940, from Brand Vley in Worcester Div. may belong here.

The name *E. scopiformis* as applied by Boissier, according to the only three specimens quoted, includes two perfectly distinct species, being founded upon a specimen collected by *Bergius* in Riet Valley, which is the type of *Arthrothamnus scopiformis*, Klotzsch & Garcke, and another collected by *Ecklon & Zeyher* (Euphorb. 24) near Brakfontein in Clanwilliam Div., both of which belong to *E. arceuthobioides*, whilst the distinct Berg Valley plant (*Zeyher* 1535) appears to have been added afterwards, as its characteristics do not appear in the description. As the name *scopiformis* was not originally applied to this plant (*Zeyher* 1535) and belongs to the plant described by Boissier as *E. arceuthobioides* before or at about the same time, I deem it best to abolish the name *E. scopiformis* altogether and apply a fresh name as above.

57. E. tenax (Burchell, Trav. S. Afr. i. 219) ; a leafless, spineless, succulent bush, about 2 ft. high, diœcious, only the male plant seen, dichotomously and trichotomously much-branched ; branches mostly opposite, diverging from each other at an angle of 55°–80°, 1–1$\frac{1}{4}$ lin. thick, terete, not angular, rough from minute hardened papillæ, all rising to about the same general level and forming a somewhat flat-topped corymb of small cymes, green, scarcely milky (*Burchell*) ; leaves rudimentary, scale-like, opposite, recurved from about the middle, about $\frac{1}{2}$ lin. long and broad, subquadrate or oblong, very obtuse or subtruncate, very minutely apiculate, minutely puberulous on the upper, glabrous on the lower side, chocolate-coloured, soon deciduous ; cymes small, in pairs at the ends of the branches, forming there a lax cyme $\frac{3}{4}$–1$\frac{1}{4}$ in. in diam., with or without 1–3 pairs of other cymes racemosely spaced along the branch below them ; bracts scale-like, firm, $\frac{1}{2}$–$\frac{2}{3}$ lin. long, oblong, very obtuse, apiculate, chocolate-coloured ; involucre 1$\frac{1}{3}$–1$\frac{1}{2}$ lin. in diam., cup-shaped, minutely papillate or subglabrous outside, with 5 glands and 5 transversely rectangular toothed lobes ; glands $\frac{1}{2}$–$\frac{3}{4}$ lin. in their greater diam., transversely elliptic or elliptic-oblong, entire, apparently yellowish or greenish ; ovary and capsule not seen.

CENTRAL REGION : Ceres Div. ; in stony places on Hangklip, near the Ongeluks River, *Burchell*, 1219 !

This species has been overlooked by all authors and omitted from the Index Kewensis.

58. E. rhombifolia (Boiss. Cent. Euphorb. 19) ; plant unisexual ; stems probably growing in bush-like clumps, 1–2 ft. high, succulent, spineless, leafless, 1$\frac{1}{2}$–2 lin. thick at the base, simple or with 1–5 distant pairs of slender branches 5–18 in. long, bearing 6–12 distant pairs of peduncles (or, in the varieties, cymes) in a racemose manner, glabrous ; leaves rudimentary and soon deciduous, opposite, $\frac{1}{2}$–1 lin. long, occasionally oblong, but usually spathulate, with a short broad petiole and an ovate, cordate-ovate or transversely elliptic blade, obtuse or acute, mostly recurved, usually slightly puberulous on the upper side at the base, otherwise glabrous; peduncles opposite, the pairs ½–3 in. apart, usually 1–6 lin. long, but occasionally longer, each bearing only 1 involucre (rarely 2–3), glabrous; bracts under the involucres like the leaves in size and form, soon deciduous; involucre sessile, unisexual, 1¼–2 lin. in diam., cup-shaped, glabrous outside, with 5 glands and 5 sub-quadrate fringe-toothed lobes; glands ½–1 lin. in their greater diam., transversely elliptic or elliptic-oblong, entire, usually larger in the male than in the female involucre; ovary erect, with a conspicuous disc-like calyx at its base, glabrous, exserted on a pedicel not exceeding the involucre; styles ½–⅔ lin. long, very shortly united at the base, with spreading bifid arms. *Boiss. in DC. Prodr.* xv. ii. 75, *and Ic. Euphorb.* 16, *t.* 46. *E. racemosa, E. Meyer in Drège, Zwei Pfl. Documente,* 184, *ex Boiss. in DC. Prodr.* xv. ii. 75.

Var. β, **triceps** (N. E. Br.); involucres 3 together, sessile on peduncles 1–2 lin. long; otherwise as in the type.

Var. γ, **cymosa** (N. E. Br.); involucres in several- or many-flowered cymes, which (including the 1–4 lin.-long peduncles) are ½–1¼ in. long, with internodes or joints 1–4 lin. long; occasionally the involucre is puberulous outside; otherwise as in the type. *Arthrothamnus cymosus, Klotzsch & Garcke in Abhandl. Akad. Berlin,* 1860, 63.

Var. δ, **laxa** (N. E. Br.); involucres four to several in very lax cymes, which (including the peduncles) are 1–2½ in. long, with internodes or joints 5–14 lin. long.

SOUTH AFRICA : without locality, *Drège,* 8217 !

COAST REGION : Uitenhage Div. ; Uitenhage, *Hutton* ! Fort Beaufort Div. ; Kunap River, *Baur,* 1046 ! Queenstown Div. ; mountains near Imbarne River, mingled with var. β, *Cooper,* 318 ! Finchams Nek, *Galpin,* 1598 ! Queenstown, *Rogers,* 4050 ! Var. γ : Humansdorp Div. ; Kabeljaauw, near Humansdorp, *Burtt-Davy,* 12044 ! Uitenhage Div. ; Coegakammas Kloof, *Zeyher,* 863 ! Ecklon & Zeyher, Euphorb. 24 partly ! Redhouse, *Mrs. Paterson,* 722 ! Port Elizabeth Div. ; near Port Elizabeth, *Kemsley,* 289! *Drège,* 5 ! Albany Div.; Bushmans River Poort, *Galpin,* 2975 ! King Williamstown Div. ; near King Williamstown, *Flanagan,* 1754 ! Var. δ : Bathurst Div. ; near Port Alfred, *Galpin,* 2959 ! Komgha Div. ; among rocks along the Chichaba River, *Flanagan,* 838 ! Div. ? near Biesjesfontein, *MacOwan,* 1612!

CENTRAL REGION : Graaff Reinet Div. ; mountains near Graaff Reinet, *Bolus,* 191 ! Aberdeen Div. : Hamerkuil, in the Camdeboo, *Drège,* 690 ! Beaufort West Div. ; Nieuwveld, *Drège,* 8118, partly! Var. β : Graaff Reinet Div. ; Sneeuwberg Range, *Wyley* ! Var. γ : Prince Albert Div. ; Gamka River, *Burke* ! Sand River Mountains, *Marloth,* 4394 ; Somerset Div. ; Bosch Berg, *MacOwan* ! Aberdeen Div. ; in the Camdeboo, *Dunn in Herb. Bolus,* 6259 !

EASTERN REGION : Natal ; Mooi River Thorns, beyond Greytown, *Wood,* 4336 ! and without precise locality, *Gerrard,* 1170 ! Var. δ : Natal ; Mooi River Valley, *Sutherland* !

Specimens (*Gilfillan in Herb. Galpin,* 6070), collected on a ridge at Zwart Krans Caves, Krugersdorp, in the Transvaal, have the smaller habit and general appearance of *E. rhombifolia,* but the leaves are persistent and more fleshy, like those of *E. rectirama.* Possibly it may prove to be distinct, but requires to be compared with living plants of both species.

This is either a very variable plant or more than one species is here included under this name, but from the material seen I have been unable to obtain

characters that are of specific value. The four forms above characterised are
distinct in appearance, but grade into each other in such a manner that any specific
difference they may possess vanishes in the dried material, and if specifically
different they must be characterised from living plants. The different forms
occur in the same geographical area.

I am quite unable to distinguish *E. rhombifolia* and *E. racemosa.* The former
was founded upon *Drège*, 8217, and a young immature growth figured for it, but
the distributed specimens of *Drège*, 8217, that I have seen, consist of weak
branches corresponding to those of Boissier's figure and flowering branches
corresponding to the type of *E. racemosa* in *E.* Meyer's Herbarium, and *Gerrard's*
1170 collected in Natal are partly exactly as in the figure of *E. rhombifolia*, partly
as in *E. racemosa.* As no description of *E. racemosa* was published until 1862,
when Boissier first described it in *De Candolle's Prodromus*, I take the earlier
published name of *E. rhombifolia* for this plant, although certainly not so
applicable. Boissier has described the peduncles of *E. racemosa* as bearing a cyme
of 5–7 involucres, but this is quite inaccurate as to the type of *E. racemosa* in
E. Meyer's Herbarium, in which the peduncles bear only 1–3 involucres, and in
none of the specimens that I have seen, which are conspecific with that type, are
more than 3 involucres borne upon any peduncle, usually there is only one.
In the original description (not in *DC. Prodr.*) Boissier quotes "*Ecklon & Zeyher*
nos. 23, 83" as belonging to this species, but this specimen is *E. arceuthobioides,*
Boiss., and the quotation really refers to one specimen only, of which the number
is 23, the added number 83 is merely the locality number and not the number of
a separate specimen (see *Linnæa*, xix. 583 and xx. 258). *Arthrothamnus cymosus,*
Klotzsch & Garcke (*Ecklon & Zeyher*, 24 partly), is wrongly referred to *E. decussata,*
E. Meyer, by Boissier in *DC. Prodr.* xv. ii. 75. *E. rhombifolia* is readily distinguished
from its nearest allies by its quickly deciduous bracts.

59. **E. aspericaulis** (Pax in Engl. Jahrb. xxviii. 26); a small
leafless and spineless succulent bush, apparently 1–1½ ft. high,
diœcious; branches alternate or perhaps sometimes opposite,
diverging from the stem at an angle of 25°–40°, 1–1½ lin. thick,
6-angled, with slight furrows between the angles, rough from
minute hard papillæ, glabrous; leaves opposite, rudimentary, about
1 lin. long, oblong, obtuse, with the apical half deflexed, dark brown
or purple-brown, glabrous; cymes opposite, or from one axil, racemosely scattered along the branches, up to about 4 lin. in diam., at
first with 1, finally producing 3–5 involucres; bracts about ⅔ lin.
long and nearly or quite as broad, oblong, very obtuse and apiculate
or subtruncate at the apex, rather flexible; involucres unisexual,
about 1½ lin. in diam., cup-shaped, glabrous outside, with 4–5 glands
and very small oblong or subquadrate slightly toothed lobes; glands
½–¾ lin. in their greater diam., transverse, oblong or elliptic-oblong,
entire, apparently yellowish or greenish; ovary and capsule not
seen.

CENTRAL REGION: Calvinia Div. ; Hantam Mountains, *Meyer*!

Allied to *E. muricata*, Thunb., by its rough branches, but the habit is different,
the branches more slender and the papillæ upon them much more minute and
even, not compressed into irregular crenulations as in that species.

60. **E. caterviflora** (N. E. Br.); a leafless, spineless, succulent
shrublet, apparently about 9–12 in. high, mostly trichotomously
branched, diœcious, only a male specimen seen; main stems or

branches 1½–2 lin. thick (dried), lateral branches ⅔–1 lin. thick, opposite, diverging from each other at an angle of about 70°–80° and rising in a corymbose manner to about the same level, distinctly 6-angled, with slight furrows between the angles, rough from minute hard papillæ; leaves rudimentary, opposite, abruptly re-flexed from their base, 1–1¼ lin. long and as much in breadth, triangular-subhastate in outline from a minute spreading point or angle at their base,.fleshy, purplish-black (dried), glabrous; cymes small, about ⅓–½ in. long and ¼–⅓ in. in diam., in opposite pairs racemosely arranged along the branches; bracts about ⅔ lin. long, very broadly obovate or suborbicular, sometimes with a minute tooth on one or both sides at or below the middle, subapiculate, fleshy, glabrous, dark purple; involucre about 1½ lin. in diam., cup-shaped, glabrous, dark purple, with 5 glands and 5 subquadrate minutely toothed lobes; glands ⅓–½ lin. in diam., transversely elliptic or suborbicular, entire, dark purple; ovary and capsule not seen.

CENTRAL REGION: Murraysburg Div.; on stony slopes of mountains near Murraysburg, 4000 ft., *Tyson,* 167! Beaufort West Div.; Nieuwveld, *Drège,* 8218 partly!

This species is nearly allied to *E. aspericaulis,* Pax, but the leaves and bracts are totally different in shape and much more fleshy, and the involucres and their glands are dark purple. Possibly a specimen collected at Melrose, Eastpoort, in Bedford Div. by *Burtt-Davy,* 12281, may belong here, but is too imperfect for determination.

61. E. Mundii (N. E. Br.); a succulent bush, 1 ft. or more high, leafless and spineless, unisexual; main stems 1½–3 lin. thick; branches opposite, diverging from each other at an angle of 60°–90°, articulated at their origin, mostly simple, not spine-pointed, glabrous and neither papillate nor asperate; leaves rudi-mentary, scale-like, opposite, sessile, recurved, ½–1 lin. long and as much in breadth, suborbicular, broadly ovate or oblong, some-times angular on each side near the base, obtuse, minutely apiculate, fleshy, rigid, persisting for a time, dark brown or reddish; cymes rather numerous, opposite, 4–6 lin. long and about as broad, racemosely arranged along the branches, at first with 3, but often ultimately producing several involucres, at least in the male plant, and their internodes or joints as thick as long and somewhat bead-like, glabrous; bracts shorter than the involucres, like the leaves in form and colour, glabrous; involucre 1¼–1⅓ lin. in diam., somewhat campanulate and as long as broad in the male, obconic-cup-shaped and broader than long in the female, glabrous, with 5 glands and 5 subquadrangular or transversely rectangular minutely toothed lobes; glands not contiguous, ⅓–½ lin. in their greater diam., transversely oblong or elliptic-oblong, entire; capsule not longer than broad, 1½–1¾ lin. in diam., obtusely 3-lobed, glabrous, exserted (as is also the ovary) on a pedicel about ¼ lin. longer than the involucre; styles very shortly united at the base,

with ascending-spreading deeply bifid straight tips ; seeds 1 lin.
long, somewhat ovoid, pointed at one end, slightly 4-angled, rugose,
dark grey. *E. decussata, Marl. ?* in *Wissensch. Ergebn. Deutsch.
Tiefsee-Exped.* ii. iii. 225, *fig.* 87, *not of E. Meyer. Arthrothamnus
densiflorus, Klotzsch & Garcke in Abhandl. Akad. Berlin,* 1860, 62.

SOUTH AFRICA : without locality, *Thunberg* !
COAST REGION : Robertson Div. ; Montagu, *Marloth,* 2805 ! Oudtshoorn Div. ;
Karoo near the Olifants River, *Mund & Maire* !
CENTRAL REGION : Laingsburg Div. ; near Laingsburg, *Marloth,* 3904 ! Matjes-
fontein, *Marloth,* 4884 ! Beaufort West Div. ; Nieuwveld Mountains, near
Beaufort West, *Marloth,* 4878 ! Prince Albert Div. ; stony hills near Prince
Albert, *Bolus,* 11634 ! Somerset Div. ; Sheldon, *Hutton* !

This species is quoted by Boissier in *DC. Prodr.* xv. ii. 75, as a synonym of
E. decussata, E. Meyer, but although it has a slight resemblance to that plant, it
is not so stout, its branches are less divergent, apparently less rigid and
have not such a tendency to end in somewhat spine-like points, whilst the bracts,
instead of gradually tapering from base to apex, are suborbicular or oblong,
obtuse and apiculate, not so thin nor concave, but thick, fleshy and somewhat
concave or flat on the upper surface ; also the plant is a native of a different
region.

As the name *densiflora* has been used for another species it cannot be retained
for this one. The type specimens from which my description is made are *Marloth,*
2805, 3904 and 4878, which are identical with the type of *Arthrothamnus densiflorus,* collected by Mund, after whom it is named.

62. **E. macella** (N. E. Br.) ; a succulent, leafless and spineless shrub
" 4 ft. high " (*Burchell*), unisexual ; branches opposite, diverging
from one another at an angle of 40°–60°, terete, smooth, wrinkled
when dried, glabrous ; leaves opposite, soon deciduous, only seen as
bracts under the involucres, $\frac{3}{4}$ lin. long, spathulate, with a rather
broad petiole and a short abruptly dilated terminal part, broadly
rounded or subtruncate at the apex, minutely puberulous near the
base, glabrous elsewhere, when young, with small round glands at
their base ; cymes racemosely arranged along the branchlets, 3–4
lin. in diam., simple, with 3 involucres or once forked and bearing
7 involucres, the lateral on pedicel-like cyme-branches $\frac{1}{2}$–$\frac{3}{4}$ lin. long,
with a pair of bracts close under each ; involucre $1\frac{1}{4}$–$1\frac{1}{3}$ lin. in
diam., cup-shaped, minutely puberulous all over outside, with 5
glands and 5 rectangular toothed lobes ; glands not contiguous,
$\frac{2}{5}$–$\frac{1}{2}$ lin. in their greater diam., transverse, narrowly oblong, entire,
convex, " green " (*Burchell*) ; female plant not seen.

COAST REGION : Mossel Bay Div. ; near Little Brak River, *Burchell,* 6197/^{2} !

63. **E. hastisquama** (N. E. Br.) ; apparently about 3–6 in. high,
with numerous crowded trichotomously divided branches, glabrous ;
branches opposite, $\frac{1}{2}$–1 lin. thick (dried), terete ; leaves rudimentary,
scale-like, opposite, sessile, $\frac{1}{2}$ lin. long, deltoid-hastate, with the
apex and basal lobes acute, recurved, blackish-brown, glabrous ;
involucre solitary at the apex of the terminal branchlets, sessile,
about 1 lin. in diam., cup-shaped, glabrous outside, pubescent within,
with 5 glands in the male (and perhaps 4 in the female) and 5 very

small oblong or subquadrate fringed lobes; glands in the male unequal (always?), $\frac{1}{3}$-$\frac{1}{2}$ lin. in their greater diam., transversely elliptic-oblong or subreniform, convex; female or hermaphrodite involucre not seen.

COAST REGION : Uitenhage Div. ; fields by the Zwartkops River, 40 ft., *Zeyher*, 1099 ! 3854 ! *Ecklon & Zeyher*, Euphorb. 25 !

All the specimens seen are very imperfect and damaged by insects.

64. **E. gentilis** (N. E. Br.); dwarf, succulent, much-branched, leafless and spineless, 3–6 in. high, unisexual; branches numerous, rather crowded, erect or ascending, alternate and opposite on the same plant, 1$\frac{1}{2}$–3 lin. thick on the dried specimens, probably much stouter when alive, glabrous and perhaps glaucous, very brittle when dried; leaves none or reduced to mere ovate rudiments less than $\frac{1}{4}$ lin. long, opposite or alternate; bracts under the involucre $\frac{1}{2}$ lin. long, narrowly oblong-obovate or spathulate, rounded at the apex, slightly toothed, glabrous on the back, puberulous on the upper surface; involucres clustered at the apex of the branchlets, usually in opposite pairs and shortly subspicate, sessile, 1–1$\frac{1}{4}$ lin. in diam. and $\frac{2}{3}$ lin. deep, cup-shaped, glabrous outside, with 5 glands and 5 subquadrate ciliate lobes; glands contiguous, spreading, $\frac{1}{2}$–$\frac{2}{3}$ lin. in their greater diam., transversely elliptic-oblong, entire; female plant not seen.

CENTRAL REGION : Calvinia Div. ; Bitterfontein, *Zeyher*, 1531 ! Grauwater, 1700 ft., *Pearson*, 3271 !

WESTERN REGION : Van Rhynsdorp Div. ; hills near Zout River, 500 ft., *Schlechter*, 8136 !

This species is very brittle when dried and in that character is very much like *E. spicata*, E. Meyer, and *E. karroensis*, N. E. Br., which it also resembles in appearance. But is readily distinguished from *E. spicata* by its glabrous stems, and from *E. karroensis* by the very different and often alternate rudimentary leaves and absence of conspicuous stipulary glands.

65. **E. stapelioides** (Boiss. Cent. Euphorb. 26, and in DC. Prodr. xv. ii. 91); very dwarf, leafless and spineless, succulent, 3–4 in. high ; main stems branching close to the ground ; branches opposite, ascending or suberect, terete, jointed, with internodes $\frac{1}{4}$–$\frac{1}{2}$ in. long, in the dried specimen 2 lin. thick and very brittle, breaking up into joints at the nodes ; leaves opposite, rudimentary, sessile, $\frac{1}{2}$–$\frac{2}{3}$ lin. long, broadly deltoid, acute, blackish ; flowers not seen, but Boissier states that they are produced near the apex and describes them as follows :—bracts 2 under each involucre and about equalling it, ovate, mucronulate ; involucre sessile, 1$\frac{1}{2}$ lin. long and in diam., hemispherical, with transversely oblong entire glands and short subentire hairy lobes.

WESTERN REGION : Little Namaqualand ; near the mouth of the Orange River, *Drège*, 8199 !

No specimen of this species exists in E. Meyer's Herbarium at Lübeck, and the type in Herb. Bunge, now in the Paris Herbarium, is reduced to two very small fragments without flowers, but with two leaves present, most of the specimen having been broken off and lost, for it is evidently a very brittle plant when dried, easily disarticulating at the nodes. Therefore, until rediscovered, it is impossible to give a better description. The position Boissier assigns to it between *E. hamata,* Sweet, and *E. clavarioides,* Boiss., is very misleading, as it certainly has no similarity to either of those species. Its real affinity is with *E. karroensis,* N. E. Br., *E. spicata,* E. Meyer, and *E. gentilis,* N. E. Br., being evidently very similar to the latter in size and appearance, but the leaves are quite different. In my opinion there is nothing whatever about the plant that suggests the slightest resemblance to any known *Stapelia.*

66. E. karroensis (N. E. Br.) ; a dwarf compactly much-branched bush, succulent, spineless, diœcious, only male specimens seen ; branches opposite, erect or suberect and more or less bunched together, articulated at their origin and often at the nodes ; main stems or branches 3 lin. and the flowering branches ¾–1¼ lin. thick when dried, drying dark brown, glabrous, with large stipular glands at the nodes ; leaves opposite, rudimentary, deciduous, ⅔–¾ lin. long, spathulate, recurved at the tips, puberulous on the upper side of the proportionately broad petiole, glabrous elsewhere ; blade suborbicular, obtuse, about ½ lin. in diam. ; cymes terminal, about ¼ in. in diam., consisting of a central subsessile involucre and 2 lateral branches ⅔–1 lin. long, each bearing 3 involucres ; bracts like the leaves in size and form ; involucres 1 lin. in diam., cup-shaped, minutely puberulous, with 5 glands and 5 subquadrate fringe-toothed lobes ; glands ½–⅔ lin. in their greater diam., transversely oblong, entire ; rudimentary ovary pedicellate, glabrous. "*E. Burmanni, E. Mey.?*" *in Drège, Zwei Pfl. Documente,* 184, *name only. E. Burmanni, var. karroensis, Boiss. in DC. Prodr.* xv. ii. 75. *Arthrothamnus Burmanni, Klotzsch & Garcke in Abhandl. Akad. Berlin,* 1860, 62.

WESTERN REGION : Van Rhynsdorp Div. ; near Hol River, *Drège,* 2947 !

This plant was distributed by Drège as "*E. Burmanni,* E. Mey.?" and was considered by Boissier to be a variety of that species, but it is evidently a stouter and more succulent plant, with shorter and more erect branches, a much less compound inflorescence, and it dries dark brown instead of greyish-green. I have not seen female involucres, but as the rudimentary ovary in the male involucres is glabrous, probably the perfect ovary and capsule are also glabrous.

67. E. spicata (E. Meyer in Drège, Zwei Pfl. Documente, 184, ex Boiss. in DC. Prodr. xv. ii. 97) ; dwarf, succulent, much-branched from the base, 2–6 in. high, leafless and spineless ; branches opposite or sometimes solitary at a node, in the fragmentary specimens seen 1–4 in. long, 1–2 lin. thick, terete, subscabrous or harshly puberulous with very short stiff spreading hairs or points ; leaves rudimentary, soon deciduous, opposite or occasionally alternate, about ½ lin. long, ovate, acute, harshly puberulous ; cymes (including the peduncles) 1½–2 lin. long, in a few (or possibly several) opposite crowded pairs arranged in a subspicate manner at the tips of the branches, each with 3 involucres ; bracts under the involucres ½–⅔ lin. long, spathulate, with a broad petiole and transversely oblong

blade or linear-oblong with the blade scarcely distinct from the
petiole, truncate at the apex, minutely papillate on the back and
with a few short hairs on the upper surface ; involucres sessile,
unisexual, 1–1¼ lin. in diam., cup-shaped, thinly pubescent outside,
at least on the upper part, with 4–5 glands and 5 oblong or subquadrate toothed lobes ; glands ½–⅔ lin. in their greater diam.,
transversely oblong, entire ; ovary sessile, included, pubescent ;
styles exserted from the involucre, ⅔ lin. long, united for half their
length, with spreading bifid arms ; capsule and seeds not seen.

WESTERN REGION: Little Namaqualand ; Silverfontein, near Ookiep, *Drège*,
2946 ! and without precise locality, *Marloth*, 4892 !

This is well distinguished from all its allies by its subscabrous branches. Boissier
describes the branches as papillose-hairy, but that does not seem to me a correct
description, the hairs are minute and resemble short stiff conical acute points,
unlike any that I have seen on any other species. Boissier quotes *Zeyher*, 1531,
as being *E. spicata*, but that number as represented by the specimens I have seen
is a glabrous plant and belongs to *E. gentilis*, N. E. Br.

68. **E. mauritanica** (Linn. Sp. Pl. ed. i. 452, and Amœn. Acad.
iii. 111) ; a succulent, spineless shrub, 3–4 ft. high, branching
throughout, leafy only on the young growths ; branches alternate,
erect, terete, smooth, marked or sometimes somewhat tubercled
with alternate leaf-scars, the younger or flowering branches ⅛–¼ in.
thick, becoming thicker with age, glabrous, green, not glaucous ;
leaves sessile, soon deciduous, ¼–1 in. long, 1–3 lin. broad, linear-
lanceolate or lanceolate, acute, glabrous, green ; umbels terminal,
simple 3–5-rayed, when young with about 3 leaf-like very deciduous
bracts at the base ; rays or peduncles 2–6 lin. long, each bearing
1 involucre, and when immature a pair of very deciduous ovate
acute deeply concave bracts 1½–2 lin. long and 1½ lin. broad,
glabrous ; involucre in dried specimens about ¼ in. in diam. and
⅛ in. deep, cup-shaped, glabrous, with 5 glands and 5 subquadrate
bifid or emarginate ciliate lobes ; glands 1¼–1⅓ lin. in diam.,
suborbicular, entire, "yellowish-green" (*Miller*) ; ovary exserted on
a curved pedicel 2–3 lin. long, glabrous ; styles ⅔–1½ lin. long,
united at the basal third, bifid at the apex ; capsule 2¾–3½ lin. in
diam., 3-lobed as seen from above, with subobtuse angles ; seeds
1½–1¾ lin. long, oblong, grey, speckled with black. *Mill. Gard.
Dict. ed.* viii. *no.* 16 ; *Ait. Hort. Kew. ed.* 1, ii. 137 ; *Willd. Sp. Pl.*
ii. 889, *and Enum. Pl. Hort. Berol.* 502 ; *Thunb. Prodr.* 86, *and Fl.
Cap. ed. Schult.* 405 ; *Poir. Encycl. Suppl.* ii. 610 ; *Spreng. Syst. Veg.*
iii. 788 ; *Boiss. in DC. Prodr.* xv. ii. 94 ; *Berger, Sukk. Euphorb.* 26 *and*
20, *fig.* 3 ; *Marloth in Wissensch. Ergebn. Deutsch. Tiefsee-Exped.*
ii. iii. 234, 295, *figs.* 94 *and* 119. *E. mauritiana, Bernh. in Flora,* 1845,
87, *and in Krauss, Beitr. Fl. Cap- und Natal.* 150. *E. Tirucalli,*
Thunb. Prodr. 86, *and Fl. Cap. ed. Schult.* 405 *partly, as to sheet* 4
of his Herbarium, not of Linn. *E. phymatoclada, Boiss. Cent. Euphorb.*
24, *and in DC. Prodr.* xv. ii. 95. *E. Hydnoræ, E. Meyer, and*
E. melanosticta, E. Meyer in Drège, Zwei Pfl. Documente, 184, *ex*

U 2

Boiss. in DC. Prodr. xv. ii. 95. *Tithymalus virgatus, Haw. Syn. Pl. Succ.* 139. *T. Zeyheri, Klotzsch & Garcke in Abhandl. Akad. Berlin,* 71. *T. brachypus, Klotzsch & Garcke, l.c.* 74.—*T. aphyllus Mauritaniæ, Dill. Hort. Eltham.* 384, *t.* 289.

VAR. β, **namaquensis** (N. E. Br.) ; a bush, 2–3 ft. high, with the flowering branchlets much more slender than in the type and in dried specimens ¼–1 lin. thick ; leaves 1½–3 lin. long, ⅓–⅔ lin. broad ; involucre 2½–3 lin. in diam.; styles ⅗–¾ lin. long ; otherwise as in the type.

SOUTH AFRICA : without locality, *Thunberg!*
COAST REGION : Clanwilliam Div. ; Clanwilliam, *Leipoldt*, 237! Olifants River Mountains and Valley, *Pearson*, 7332! *Diels*, 1147! Malmesbury Div. ; near Theefontein, *Bachmann*, 1041! Hopefield, *Bachmann*, 2154! Worcester Div. ; near Worcester, *Marloth*, 4877 ! Cape Div. ; various localities, *Bergius*! *Zeyher*! *Ecklon & Zeyher*, Euphorb. 22 partly! and 26 ! *Pappe*! *Krauss*, 1733 ! *Wolley-Dod*, 1629 ! Stellenbosch Div. ; near Gordons Bay, *Treleaven in MacOwan, Herb. Austr.-Afr.*, 1953! Caledon Div. ; Zoetemelks River, *Gill* ! Swellendam Div. ; Barrydale, *Galpin*, 4566! Riversdale Div. ; Tygerfontein, *Galpin*, 4560! Oudtshoorn Div. ; Oudtshoorn, *Miss Britten*! Mossel Bay Div. ; near Little Brak River, *Burchell*, 6197/3! *Rogers*, 4222! Uitenhage Div. ; various localities, *Burchell*, 4227 ! *Drège*, 124 ! *Mrs. Patterson*, 88 partly! 246 ! 713 ! 714! 723 ; Port Elizabeth Div.; near Port Elizabeth, *Sim*, 2668! Albany Div. ; various localities, *Bowker*! *Williamson*! *Galpin*, 174 ! *Mrs. Whyte*, 1060! Fort Beaufort Div. ; Adelaide, *Hutton*, 617 ! Queenstown Div. ; Andriesberg Range, near Bailey, 5000 ft., *Galpin*, 2236 ! Eastern Frontier, *MacOwan*, 363 !
CENTRAL REGION : Calvinia Div.; Karoo flats west of Calvinia, *Diels*, 655 ! between Blaukrantz Pass and Karieboemfontein, *Pearson*, 3481 ! Laingsburg Div. ; Matjesfontein, *MacOwan*, 3314 ! 3326 ! *and in Herb. Austr.-Afr.*, 1954 ! Somerset Div. ; near Somerset East, *MacOwan*, 363 ! Cradock Div. ; near Cradock, *Burtt-Davy*, 9846 ! Graaff Reinet Div. ; near Graaff Reinet, *Bolus*, 805 ! Beaufort West Div. ; Nieuweveld Mountains, near Beaufort West, *Drège* ! Carnarvon Div. ; north exit of Karreeberg Poort, *Burchell*, 1578 ! Carnarvon, *Schönland* ! Colesberg Div. ; Naauwpoort, *Rogers*, 1029 !
WESTERN REGION : Great Namaqualand ; between Ramans Drift and Warmbad, *Pearson*, 4204 ! 4435 ! Little Namaqualand ; Kaus Mountains, *Drège*, 2945 ! Garies, *Alston* (fasciated growth)! (also probably, Great Karasberg Range, *Pearson*, 8073). Van Rhynsdorp Div. ; Ebenezer, *Drège*, 2943 ! 8215 ! Karreeberg Range, *Schlechter*, 8311 ! Gift Berg, *Phillips*, 7391 ! Bokkeveld, *Diels*, 570 ! Var. β : Great Namaqualand ; Great Karasberg Range, *Pearson*, 8279 ! 8345 ! 8346 ! south of Warmbad, *Pearson*, 4432 ! Little Namaqualand ; various localities, *Pearson*, 3055 ! 3628 ! 3845 ! *Alston in Herb. Marloth*, 5103 ! *Marloth*, 4879 !

This plant has a very wide range and varies somewhat in appearance, but I cannot find any distinctive characters in the dried specimens to separate more than the one variety. The leaf-scars are usually narrow and somewhat crescent-like, but sometimes (by a corky growth ?) they enlarge into dark brown or blackish tubercles, producing the form known as *E. melanosticta*, E. Meyer. This form and the typical one, however, occur on the same plant. *E. phymatoclada*, Boiss., and *E. Hydnoræ*, E. Meyer, are both founded upon the same specimen, *Drège*, 2943.

69. **E. patula** (Mill. Gard. Dict. ed. viii. no. 11) ; succulent and spineless ; stem 6–7 in. high, tapering upwards, producing at the top a few tapering branches, spreading on every side, not (" scaly ") tuberculate and bearing at their tips several small narrow deciduous leaves ; flowers unknown.

SOUTH AFRICA : formerly cultivated at Chelsea.

Haworth refers this plant to his *Dactylanthes patula*, but Miller's description does not seem to accord with that plant, which is a synonym of *E. ornithopus,*

Jacq. Can Miller's plant be a small weak form of *E. mauritanica*, Linn., with spreading branches ? He describes the branches as not "scaly," by which I suppose he means they are not tuberculate, since those of *E. Caput-Medusæ*, Linn. are described as scaly.

70. **E. Tirucalli** (Linn. Sp. Pl. ed. i. 452) ; a spineless, succulent tree, 15–20 ft. high, diœcious ; branches and branchlets alternate or opposite or in clusters of 2–7 at the ends of the branches they arise from and distinctly jointed to them, diverging or subparallel in brush-like masses, more or less deciduous ; ultimate branchlets $2\frac{1}{2}$–$3\frac{1}{2}$ (when dried $1\frac{1}{4}$–2) lin. thick, cylindric, very obtuse or subtruncately rounded at the apex, glabrous, rather light green, with very fine whitish striations, and marked with very small leaf-scars ; leaves alternate, $\frac{1}{4}$–$\frac{1}{2}$ in. long, $\frac{1}{2}$–$\frac{3}{4}$ lin. broad, linear or linear-lanceolate, acute, glabrous ; involucres in sessile clusters at the apex of the branchlets, cup-shaped, $1\frac{1}{2}$ lin. in diam., minutely puberulous or thinly sub-tomentose on the upper part outside, with 5 glands and 5 trans-versely subrectangular or subquadrate toothed lobes ; glands sub-contiguous or separate, $\frac{2}{3}$–$\frac{3}{4}$ lin. in their greater diam., transversely oblong or elliptic-oblong, entire, flat, with a slightly raised margin when fresh ; ovary subglobose, minutely tomentose, without a calyx at its base, finally exserted on a stout pedicel ; styles united into a very stout column $\frac{1}{4}$ lin. long, with very spreading arms $\frac{1}{2}$ lin. long, divided to half-way down or nearly to the base into 2 widely spread-ing-recurved tips ; capsule $\frac{1}{3}$ in. in diam., slightly and very obtusely 3-lobed as seen from above, minutely puberulous, exserted on a pedicel $\frac{1}{3}$ in. long and curved to one side ; seeds 2 lin. long, ellipsoid, smooth, glabrous, dark brown, with whitish margins around the small white caruncle and along the brown suture extending from it. *Boiss. in DC. Prodr.* xv. ii. 96 ; *T. Thoms. in Speke, Journ. Nile, Append.* 646, *and Trans. Linn. Soc.* xxix. 144 ; *Volkens in Notizbl. Königl. Bot. Gart. Berlin,* ii. 263 ; *Talbot, Forest Fl. Bombay,* ii. 434 *and* 435, *fig.* 487 ; *Berger, Sukk. Euphorb.* 20, *fig.* 3, ii. ; *Sim, For. Fl. Port. E. Afr.* 104, *t.* 84, *fig.* 2. *E. media, N. E. Br. in Dyer, Fl. Afr.* vi. i. 556 ; *Zimmermann in Der Pflanzer,* 1912, 636, *t.* 8–9.—*Ossifragra lactea, Rumph. Herb. Amboin.* vii. 62, *t.* 29. *Tiru-calli, Rheede, Hort. Malabar.* ii. 85, *t.* 44.

KALAHARI REGION : Transvaal ; Moorddrift, *Miss Leendertz,* 2245 ! near Potgieters Rust, *Madge,* 8443 ! *Marloth,* 5146 ! Waterberg district, *Burtt-Davy,* 1700 ! near Mafutane, *Bolus,* 12279 ! Komati Poort, *Rogers* !

EASTERN REGION : Transkei ; in cultivation at Columba Mission, *Miss Pegler,* 1000 ! Tembuland ; Bashee River, *Bowker* ! near Mganduli, ex *Miss Pegler.* Natal ; Groenberg, *Wood,* 1339 !

In the Flora of Tropical Africa I considered this plant to be distinct from *E. Tirucalli,* Linn. At that time I had not seen any flowering specimen of the undoubted typical *E. Tirucalli,* either from India or other country where it is known to be cultivated, and could not reconcile the dense crowded heads of involucres of the male plant (the female not having been seen) from Tropical Africa, with the very different appearance of the female involucres figured in Rheede's Hort. Malabar., upon which Linnæus founded the species. Since then, however, female specimens of the Natal plant were received, the male being identical with the male of the Tropical African plant. Subsequently I have seen one Indian specimen and some from Mauritius of undoubted *E. Tirucalli,* Linn., all female, and these prove to be in every way identical with the female specimens of the Natal plant; so that there can no longer be any doubt that the plant is a native of the eastern side of Africa, from German East Africa southwards to Tembuland, and that it was probably introduced from Portuguese East Africa by the Portuguese into India, where it is used for hedges and, I am informed, does not often flower. The female flowers are sometimes few and lax, as represented by Rheede, and sometimes numerous, forming a dense head, as in the male plant. In Natal the plant is being extensively worked for rubber.

71. **E. dregeana** (E. Meyer in Drège, Zwei Pfl. Documente, 184, ex Boiss. in DC. Prodr. xv. ii. 95); a succulent leafless spineless bush, 3-6 ft. high and as much or more in diam., very much-branched at the base; main stems 2-3 in. (*Pearson*), $\frac{3}{4}$-1$\frac{1}{4}$ in. (*Marloth*) thick at the base, erect, branching; flowering branches alternate, cylindric, $\frac{1}{3}$-$\frac{1}{2}$ in. thick, with prominent leaf-scars, very minutely puberulous on the young parts, becoming glabrous, or sometimes apparently quite glabrous, whitish-green, sometimes covered with a dry whitish exudation when dried; leaves alternate, rather rudimentary, only present on very young branchlets, soon deciduous, sessile, or subsessile, 1$\frac{1}{2}$-4 lin. long, 1$\frac{1}{4}$-2$\frac{1}{4}$ lin. broad, ovate or ovate-deltoid, acute or obtuse, channelled down the face, recurved, somewhat fleshy, minutely puberulous; umbels or cymes terminal, solitary or 2 to several more or less clustered at the ends of short alternate erect branchlets, 1$\frac{1}{2}$-3 in. in diam., 3-rayed; rays $\frac{3}{4}$-3 in. long, usually 1-3-times forked, but occasionally simple, minutely puberulous; bracts sessile, 2-3$\frac{1}{2}$ lin. long, 3$\frac{1}{2}$-6 lin. broad, nearly half-circular or rounded-subrhomboid when flattened out, obtuse or apiculate, concave, minutely puberulous on both sides; involucre sessile, 3-4 lin. in diam. and about 1$\frac{3}{4}$ lin. deep, sub-globose-cup-shaped, puberulous outside, glabrous within, with 5 glands overtopped by 5 erect subquadrate slightly fringed or very shortly and densely ciliate lobes; glands 1$\frac{1}{4}$-1$\frac{1}{2}$ lin. long and 1$\frac{1}{2}$-1$\frac{3}{4}$ lin. broad when flattened out, somewhat ovate-triangular or deltoid, with the part adnate to the involucre acute or subacute and the outer part subtruncate or crenulate or slightly lobed and wavy, incurved in dried specimens, but probably more or less deflexed or spreading and concave when alive, puberulous on the back, glabrous on the inner or upper surface; capsule sessile, 3 lin. long, 4-4$\frac{1}{2}$ lin. in diam., somewhat 6-angled, with 3 of the angles subacute, minutely puberulous; styles 1-1$\frac{1}{4}$ lin. long, united for $\frac{1}{2}$-$\frac{3}{4}$ of their length, with erect or recurved-spreading bifid (or entire?) tips; seeds 2 lin. long, oblong, somewhat 4-angled, slightly pointed at one end, smooth, pale brown in the examples seen, but perhaps unripe. *Berger, Sukk. Euphorb.* 26; *Marloth in Trans. Roy. Soc. S. Afr.* iii. 124. *E. elastica, Marloth in Trans. Roy. Soc. S. Afr.* ii. 37 (*not of Jumelle nor of Altiramo & Rose*).

CENTRAL REGION: Calvinia Div.; Roggeveld, *Marloth*, 3905!
WESTERN REGION: Great Namaqualand; Obib, *Range*, 582! Tsaukaib, *Range*, 418! *Schultzse*, 392! Little Namaqualand; at Silverfontein and frequent between the Koussie (Buffels) River and the Orange River, *Drège*, 2942! Nixons Cutting,

near Klipfontein, 3000 ft., *Bolus*, 9444 ! rocky slopes at Eenriet, *Pearson*, 4066 ! Ratel Poort Mountain, *Pearson*, 2971 ! Namies. *Alston* ! Bushmanland, *Rogers* ! near Anenous, covering large areas of country, *Marloth*, 4684 ! between Anenous and Chubiesis, *Pearson*, 5974 !

I have dissected and compared the type specimen of *E. elastica* with that of *E. dregeana* and can find no difference between them. This species is used for rubber.

72. E. gummifera (Boiss. Cent. Euphorb. 26, and in DC. Prodr. xv. ii. 97) ; a leafless spineless bush, forming large clumps 3–4 ft. high, succulent at the younger parts, woody below, diœcious, with a very disagreeable odour ; flowering branches erect or ascending, $2\frac{1}{2}$–5 lin. thick, slightly angular from raised lines decurrent from the slightly prominent leaf-scars, glabrous, or, in the female plant, minutely tomentose for about $\frac{1}{4}$ in. immediately under the inflorescence at the apex, more or less covered with a dry gummy or resinous exudation (not velvety-pruinose as described by *Boissier*) ; leaves minute, rudimentary, scale-like, fleshy, recurved, dark red, only seen on the tips of growing branches, soon deciduous ; male plant with 1 or more dense sessile clusters of involucres at the tips of the branches, about $\frac{1}{4}$–$\frac{1}{3}$ in. in diam. ; female plant with a few sessile involucres at the tips ; involucres sessile, about $\frac{1}{4}$ in. in diam., less when dried, cup-shaped, apparently reddish or purplish, minutely white-tomentose, with 4–5 glands and 5 subquadrate or rounded entire lobes ; glands about $\frac{2}{3}$ lin. in their greater diam., transverse, reniform or half-orbicular, entire, apparently dark red or purple ; capsule about $\frac{1}{2}$ in. in diam., obtusely 3-angled, very minutely and not densely puberulous, probably somewhat fleshy when alive, exserted on a pedicel not longer than the involucre, erect ; styles $\frac{3}{4}$ lin. long, stout, channelled, bifid at the apex, spreading ; seeds 3 lin. long, and $3\frac{1}{2}$ lin. broad, somewhat subquadrate, compressed dorsally, with a ridge down the back, smooth, at first brown, becoming white when perfectly ripe. *Boiss. in DC. Prodr.* xv. ii. 97 ; *Marloth in Trans. Roy. Soc. S. Afr.* i. 316. *E. sessiliflora, E. Meyer in Drège, Zwei Pfl. Documente*, 184, *name only, not of Roxb.*

WESTERN REGION: Great Namaqualand ; Rotkuppe, near Bethany, *Range*, 55 ! a predominant plant on sandy plains at Gorup (Garub), *Pearson*, 4174 ! 4175 ! Namib Desert, near Tschaukaib, *Marloth*, 4636 ! Little Namaqualand ; near the Orange River, *Drège*, 2944 !

73. E. gregaria (Marloth in Trans. Roy. Soc. S. Afr. ii. 36, t. 1, fig. 7) ; a leafless spineless bush, forming dense clumps 3–6 ft. (or sometimes more) high and 3–20 ft. in diam., succulent at the younger parts, diœcious ; main stems $1\frac{1}{4}$–2 in. thick (*Marloth*), alternately branching ; dried flowering branches 2–4 lin. thick, erect or ascending, terete, not angular, with inconspicuous and not prominent leaf-scars, glabrous, or, in the female plant, minutely tomentose under the inflorescence for $\frac{1}{4}$–$\frac{1}{3}$ in. at the apex only, everywhere covered with a dry gummy (or waxy ?) exudation,

greyish, white or green when dried ; leaves represented by minute
rudimentary scales, alternate, only seen at the tips of the branches,
soon deciduous; male plant with a subglobose cluster of several- or
many-flowered subsessile cymes $\frac{1}{3}$–$\frac{1}{2}$ in. in diam., at or near the
apex of the branches; female plant with only 2–3 (or more?)
subsessile involucres at the tips ; involucres 2–2$\frac{1}{2}$ lin. in diam.,
cup-shaped, minutely tawny-tomentose, with 4–5 glands and 4–5
short transverse entire lobes, and with a pair of dull reddish scale-
like bracts at their base; glands $\frac{2}{3}$–1 lin. in their greater diam.,
half-orbicular, reniform or transversely elliptic, entire ; capsule 8–11
lin. in diam., globose or orbicular-ovoid, rounded or subacute at
the apex, obscurely 3–5-angled, 3–5-celled, exserted and pendulous
from the involucre on a stout abruptly recurved pedicel $\frac{1}{4}$–$\frac{1}{2}$ in.
long and covered with a minute dense tawny tomentum ; styles 3–5,
about $\frac{3}{4}$ lin. long, very stout, recurved, channelled, bifid at the
apex ; seeds 3–3$\frac{1}{2}$ lin. long and 2$\frac{3}{4}$–3 lin. thick, usually obtusely
3-angled, sometimes scarcely angular, smooth, white, with or
without a round blackish-brown spot on each side or sometimes
brownish, with a brown line along the middle angle. " *Aggenys
Euphorbia,*" *Pearson in Ann. S. Afr. Mus.* ix. 9, 13, 15.

WESTERN REGION : Great Namaqualand ; Kuibis, *Pearson, Marloth,* 4683 !
near Sandverhaar, between Gobas and Keetmanshoop, Holoog and Sabiesis, ex
Pearson, Great Karasberg Range, *Pearson,* 8075 ! 8084 ! 8564 ! Bushmanland :
Abbasis, *Pearson* ; Wortel, 2700 ft., *Pearson,* 3339 ! Pella, *Pearson,* 5037 ! Aggenys,
Pearson, 3338 ! 3538 !

74. **E. peltigera** (E. Meyer in Drège, Zwei Pfl. Documente, 184,
ex Boiss. in DC. Prodr. xv. ii. 91) ; plant about 1–1$\frac{1}{4}$ ft. high,
branching in a bushy manner with a thick underground rootstock,
very similar in habit and appearance to *E. hamata,* Sweet ; branches
$\frac{1}{4}$–$\frac{1}{2}$ in. thick, with conical spreading leaf-bearing alternate tubercles
1$\frac{1}{2}$–9 lin. prominent, bounded by slight furrows, minutely puberulous on the young parts, green ; leaves alternate, sessile on the
tubercles, $\frac{1}{4}$–$\frac{3}{4}$ in. long, $\frac{1}{4}$–$\frac{1}{2}$ in. broad, elliptic or ovate, acute or
obtuse, slightly folded lengthwise, minutely puberulous on both
sides, deciduous ; bracts 3 in a whorl, forming a cup $\frac{2}{3}$–$\frac{3}{4}$ in. in
diam. surrounding a single involucre at the apex of the branches
and branchlets, 4–6 lin. long and 6–10 lin. broad when flattened
out, transverse, somewhat rhomboid-elliptic or suborbicular, broadly
rounded or subtruncate, apiculate at the apex, minutely puberulous
on both sides, green ; involucre sessile, much shorter than the
bracts, unisexual, cup-shaped, minutely puberulous, 2$\frac{1}{2}$–3 lin. in
diam., with 5 glands and 5 erect oblong fringed lobes; glands
broadly cuneate, 1$\frac{1}{2}$ lin. broad, suberect, rather closely contiguous,
forming a cup, covered with very short rather crowded simple or
divided teeth on its truncate margin, green ; female involucres and
fruit not seen.

WESTERN REGION : Little Namaqualand ; near the Orange River, below 1000 ft.,
Drège, 2951 ! Steinkopf, *Kling,* 156 !

Partly described from fresh material in fluid (*Kling*, 156), contributed by
Dr. S. Schönland. This species very closely resembles *E. hamata*, Sweet, but is
easily distinguished by its stems, bracts and involucres being minutely puberulous
and by the very different glands of the involucre. Drège's specimens are mere
scraps, but must have been taken from a very luxuriant plant, as the tubercles
are ¾ in. long, on Kling's specimen they are much shorter.

75. E. hamata (Sweet, Hort. Suburb. Lond. 107, and Hort. Brit. ed.
i. 356); succulent, bushily branched, sometimes forming large dense
masses up to 1 ft. high, unisexual, with an elongated or oblong tuber-
like rootstock 1–2 in. (or more) in diam.; branches 3–6 (or when dried
1½–4) lin. thick, fleshy, somewhat 3-angled, with somewhat distant
conical tubercles scattered along them, glabrous, green, becoming
grey on the old parts; tubercles 1–8 lin. prominent, horizontally
spreading or slightly recurved, bounded by a slight furrow and
obtusely 3-ribbed down the decurrent part, bearing leaves when
young and after their fall at first somewhat acute, but with age the
apex of the tubercle withers and becomes obtusely rounded; leaves
⅓–⅔ in. long, ¼–½ in. broad, sessile on the tubercles, ovate, elliptic
or lanceolate, obtuse or acute, rounded or broadly cuneate at the
base, flat or slightly folded lengthwise, very slightly fleshy, glabrous
on both sides, green, soon deciduous; bracts usually 3 in a whorl,
forming a cup around a solitary involucre at the apex of the
branches and branchlets, 3–5 lin. long, 2–5 lin. broad, usually suborbicular and obtuse and apiculate at the apex, or occasionally broadly
ovate and acute, glabrous, apparently yellow or yellowish; involucre
sessile, much shorter than the bracts, 2½ 3½ lin. in diam., unisexual,
cup-shaped, glabrous, with 5 glands and 5 fringed lobes; glands
distant, probably spreading or deflexed, usually smaller than the
lobes, ¾–1 lin. long and broad, suborbicular, with the outer margin
minutely crenulate (not toothed) or entire, and the inner slightly
turned up, apparently yellow in some specimens and red in others,
possibly changing with age; capsule sessile, ¼ in. in diam., subglobose, glabrous; styles united into a column 1½–1¾ lin. long,
with short spreading bifid tips; seeds 1½ lin. long, ovoid, obtuse
and with 3 slight radiating keels at the apex, minutely rugulose,
grey. *Boiss. in DC. Prodr.* xv. ii. 91; *Berger, Sukk. Euphorb.* 124.
E. antiquorum, E. Meyer in Drège, Zwei Pfl. Documente, 184, *name
only, not of Linn. E. cervicornis, Boiss. Cent. Euphorb.* 27, *and in
DC. Prodr.* xv. ii. 90; *Berger, Sukk. Euphorb.* 123 *and* 124, *fig.* 33.
Dactylanthes hamata, Haw. Syn. Pl. Succ. 133. *Medusea hamata,
Klotzsch & Garcke in Abhandl. Akad. Berlin*, 1860, 61.—*Euphorbium
caule rotundo, folius hamatis, &c., Burm. Rar. Afr. Pl.* 14, *t.* 6, *fig.* 3.

SOUTH AFRICA : without locality, *Thunberg* ! *Zeyher*, 1529 !
COAST REGION : Van Rhynsdorp Div. ; Heerenlogement, *Zeyher*, 1530 ! Clanwilliam Div. ; near Clanwilliam, *Leipoldt in Herb. MacOwan*, 3183 ! *Marloth*,
4883 ! *Schlechter*, 8429 ! Piquetberg Div. ; Piquet Berg, *Drège*, 2950 ! *Bolus*,
8298 !

76. E. gariepina (Boiss. Cent. Euphorb. 28, and in DC. Prodr. xv.
ii. 91) ; a compact succulent bush, often about 6–9 in., but sometimes

1–2 ft. high; branches of the main shoots alternate, 1–3 lin. thick (dried), cylindric, with small conical tubercles or processes (persistent bases of the petioles, not spines as originally described) $\frac{1}{2}$–3 lin. long scattered along them, glabrous, smooth, ascending or ascending-spreading, all attaining to about the same height in a corymbose manner; ultimate branchlets in the male plant bearing 1 or more 2–3-rayed umbels, with or without 1 or more single rays below the umbel, in the female plant terminating in a single sessile involucre; leaves only present on the young branches, alternate, petiolate, somewhat fleshy, glabrous; petiole 1–5 lin. long, rather slender, its basal part persisting after the fall of the remainder; blade $2\frac{1}{2}$–7 lin. long, $\frac{3}{4}$–2 lin. broad, varying from linear-lanceolate to elliptic, acute or obtuse and sometimes recurved at the apex, more or less tapering at the base, longitudinally folded; rays of the umbel $\frac{1}{8}$–$1\frac{1}{2}$ (usually $\frac{1}{2}$–1) in. long, glabrous, bearing a pair of bracts and 1 involucre; bracts 1–4 lin. long and as much in breadth, obovate-oblong to orbicular, rounded and apiculate or rarely denticulate at the apex, glabrous, those under the involucre usually larger than those at the base of the umbel; involucre unisexual, $1\frac{1}{2}$–2 lin. long and $1\frac{1}{3}$–$1\frac{3}{4}$ lin. in diam. in the male, rather shorter in the female, campanulate or slightly contracted at the mouth, glabrous outside, puberulous within, with 5 glands and 5 ovate or oblong acute or bifid or toothed puberulous lobes; glands spreading, not quite contiguous, $\frac{1}{3}$–$\frac{3}{4}$ lin. in their greater diam., two-lipped, transversely elliptic or with the inner lip somewhat triangular and subacute, entire, smooth on the upper surface, in dried specimens the lips are sometimes much incurved and form a concave upper surface to the gland; capsule erect, with its base just exserted from the involucre, $2\frac{1}{2}$–3 lin. in diam., globose, more or less distinctly 9-ribbed, glabrous; styles united into a column $\frac{1}{2}$–$\frac{3}{4}$ lin. long, with rather slender recurved entire acute arms $\frac{1}{2}$–$\frac{2}{3}$ lin. long; seeds $1\frac{1}{4}$ lin. long, ovoid, shortly pointed at one end, smooth.

WESTERN REGION: Great Namaqualand: various localities on the Great Karasberg Range, 5000–5500 ft., *Pearson*, 7810! 7811! 8118! 8228! 8259! river bed 12 miles south of Warmbad, 2400 ft., *Pearson*, 4283! 4640! Little Namaqualand: common on the lower mountain slopes at Aggenys, 3200 ft., *Pearson*, 3540! by the Orange River near Verleptpram, *Drège*, 8214!

77. **E. tridentata** (Lam. Encycl. ii. 416); plant dwarf, succulent, spineless, branching from the base; branches ascending or somewhat spreading, 1–6 in. long, $\frac{1}{3}$–$\frac{1}{2}$ in. thick, cylindric or slightly tapering upwards, tessellately tuberculate with hexagonal flattish tubercles $\frac{1}{4}$–$\frac{1}{3}$ in. in diam., having a slightly prominent whitish leaf-scar, glabrous, dull green; leaves sessile, soon deciduous, 2–3 lin. long, $1\frac{1}{2}$–2 lin. broad, elliptic or elliptic-oblong, acute, dark green, with a reddish minutely toothed margin; peduncles 3–4 at the ends of the branches, about 2 lin. long, bearing a pair of ovate or elliptic bracts and 1 involucre, glabrous; involucre about $\frac{1}{2}$–$\frac{2}{3}$ in. in diam., cup-shaped, glabrous, with 5 glands and 5 transversely oblong

toothed and ciliate inflexed purplish lobes; glands subcontiguous, about 2½ lin. in diam. across the tips, very concave at the basal part, divided into 3–4 spreading finger-like corrugated white processes 1–1½ lin. long; ovary pedicellate, scarcely exserted, with styles ¼ in. long, united for two-thirds of their length, with entire spreading tips. *DC. Pl. Grass. t.* 144; *Poir. Encycl. Suppl.* ii. 607. *E. anacantha, Ait. Hort. Kew. ed.* 1, ii. 136; *Willd. Sp. Pl.* ii. 888; *Lodd. Bot. Cab. t.* 220; *Bot. Mag. t.* 2520; *Spreng. Syst. Veg.* iii. 787; *Boiss. in DC. Prodr.* xv. ii. 86; *Berger, Sukk. Euphorb.* 107. *Dactylanthes anacantha, Haw. Syn. Pl. Succ.* 132. *Medusea tridentata, Klotzsch & Garcke in Abhandl. Akad. Berlin,* 1860, 61.—*Euphorbium anacanthum squamosum, &c., Isnard in Act. Paris,* 1720, 387, no. 12, and 392, t. 11. *Euphorbium erectum aphyllum, &c., Burm. Rar. Afr. Pl.* 16, t. 7, *fig.* 2. *Euphorbium africanum caule squamoso, &c., Bradl. Hist. Succ. Pl. Dec.* 5, 12, t. 45.

SOUTH AFRICA: without locality, *Herb. Lamarck!*

Described from Lamarck's type and the figures above quoted.

78. **E. ornithopus** (Jacq. Fragm. 76, t. 120, fig. 2); plant (excluding the peduncles) 2–3 in. high, succulent, spineless, irregularly branching close to the ground, dimorphic; branches procumbent or straggling, often one over another, jointed, with 3–5 laxly spiral series of acute conical tubercles, mostly 1 2 lin. prominent, glabrous, dull green or purplish; stem-joints in one form (which although bearing bisexual involucres, only some of them appear to prove fertile) mostly cylindric and 1–4 in. long, 3–5 lin. thick excluding the tubercles, or some of them ovoid or subglobose and less than 1 lin. long; in another form (in which nearly all the involucres appear fertile) they are subglobose, oblong or shortly cylindric, ½–1¼ in. long; leaves rudimentary, deciduous, 1–2½ lin. long, ½–¾ lin. broad, lanceolate, acute, glabrous; peduncles of the long-jointed form solitary, terminal, 1½–3 in. long, ¾–1 lin. thick, bearing 3–4 small alternate bracts below the middle, a pair or whorl of 3 larger elliptic bracts 2–2½ lin. long and 1½ lin. broad at its apex and 1 involucre, or forking into a 2–3-rayed cyme or umbel with rays ¾–1½ in. long, each with 1 involucre, glabrous; peduncles of the short-jointed form 1–3 at the apex of the branches ½–1½ in. long, 1–2 lin. thick, simple or forking into 2–3 rays, otherwise as in the long-jointed form; involucre (including the glands) 5–6 lin. in diam., obconic-cup-shaped, glabrous, green, with 4 glands and 5 inflexed-connivent subquadrate ciliate lobes; glands ¼ in. long and broad, ascending-spreading, deeply divided into 3–4 subulate finger-like lobes with white-margined pits along the inner side, and an oblong white lobe inflexed over the cavity in the basal part of the gland; ovary exserted and curved to one side; styles 2–3 lin. long, united to about the middle, ascending-spreading above, with dilated and somewhat 2-lobed tips; capsule erect, ⅓ in. in diam., with 3 slight rounded lobes glabrous. *Willd. Enum. Pl. Hort. Berol.*

501 ; *Poir. Encycl. Suppl.* ii. 610 ; *Spreng. Syst. Veg.* iii. 787 ; *Boiss. in DC. Prodr.* xv. ii. 87 ; *Berger in Monatsschr. für Kakt.* xv. 62 *with fig., and Sukk. Euphorb.* 106, *fig.* 28. *E. patula, Sweet, Hort. Suburb. Lond.* 107, *and Hort. Brit. ed.* i. 356, *not of Miller. Dactylanthes patula, Haw. Syn. Pl. Succ.* 132. *Medusea patula, Klotzsch & Garcke in Abhandl. Akad. Berlin,* 1860, 61.

SOUTH AFRICA : without locality, *cultivated specimens* !

Described from living plants cultivated at Kew, received from South Africa without indication of locality. This species is closely allied to *E. globosa,* Sims, but is decidedly different in its elongated cylindric stem-joints, which even where they are subglobose are different in appearance, and the involucres seem always to have only 4 glands, whilst in *E. globosa* there are constantly 5. There are certainly two forms of this plant, which, whilst not strictly unisexual, seem to have a tendency to be so. The short-jointed form when out of flower looks specifically distinct from the long-jointed form, but the flowers are identical, and by its shorter and stouter peduncles and by usually perfecting fruit, I am inclined to believe it to represent the female form of the plant, although the long-jointed form also develops fruit. It has been in cultivation for over 100 years, yet no wild specimens seem to have been collected.

79. **E. globosa** (Sims in Bot. Mag. t. 2624) ; plant (excl. the flowers) 1–3 in. high, succulent, spineless, consisting of a cluster of branches superposed or connected together like beads on a string : branches or joints usually depressed-globose, thicker than long and $\frac{1}{2}$–1 in. in diam., but sometimes ellipsoid, obovoid or clavate and $\frac{1}{2}$–1 in. long, $\frac{1}{3}$–1$\frac{1}{4}$ in. thick, marked in a somewhat tessellate manner by impressed lines into irregularly 6-angled flattish or slightly prominent tubercles, with a slightly raised leaf-scar at their centre, glabrous, dull green or purplish where exposed to the sun, not glaucous, becoming pale grey or brownish with age ; leaves rudimentary, deciduous, $\frac{3}{4}$–1$\frac{1}{2}$ lin. long, lanceolate, acute, erect or spreading ; peduncles terminal, some not more than 1–2 lin. long, bearing 1 involucre and usually perfecting fruit, others (and the more numerous) $\frac{1}{2}$–3 in. long, bearing about 2 minute leaves near the base and a whorl of 3–4 larger ovate or elliptic-lanceolate acute leaves 2–2$\frac{1}{4}$ lin. long and 1 lin. broad under the solitary involucre at the apex, occasionally they fork 1–3 times into a lax cyme and bear three to several involucres, male or perfecting fruit, glabrous : involucre (including the glands) $\frac{1}{2}$–$\frac{2}{3}$ in. in diam., obconic, glabrous, green, with 5 glands and 5 inflexed-connivent quadrate entire minutely ciliate lobes ; glands ascending-spreading, 2–3 lin. long and nearly quite as broad, deeply divided into 3–4 subulate green segments with minute white-margined pits scattered along their upper side and a white and pitted margin to the cavity in the united basal part, which has a small white lobe folded over it ; capsule very obtusely and slightly 3-lobed, exserted and curved to one side, glabrous ; styles 1$\frac{1}{2}$ lin. long, united for about half their length, slightly spreading above and slightly thickened at the entire tips. *A.DC. Sept. Not. Pl. Rar. Gener.* 24, *t.* 5 ; *Boiss. in DC. Prodr.* xv. ii. 87 ; *Berger, Sukk. Euphorb.* 104, *fig.* 27 ; *Marloth in*

Wissensch. Ergebn. Deutsch. Tiefsee-Exped. ii. iii. 249, *fig.* 103, 2. *Dactylanthes globosa, Haw. in Phil. Mag.* 1823, 382. *Medusea globosa, Klotzsch & Garcke in Abhandl. Akad. Berlin,* 1860, 61.

COAST REGION: Uitenhage Div.; near Redhouse, *Mrs. Paterson,* 670! near Zwartkops, *Marloth*! Albany Div.; near Grahamstown, *Brett in Herb. Dümmer,* 627! *Rogers*! Bedford Div.; Melrose, Eastpoort, *Burtt-Davy,* 12290! and *cultivated specimens*!

Described from living plants cultivated at Kew.

80. **E. pistiæfolia** (Boiss. in DC. Prodr. xv. ii. 93); tuber elongated, fusiform, 1–1¼ in. in diam., bearing a crown of leaves and flowers at the apex; leaves all in a radical rosette, apparently pressed upon the ground, ¾–2 in. long, ⅓–1 in. broad, elliptic, obtuse or notched at the apex, rather abruptly narrowing into the petiole, glabrous on both sides, reticulately veined beneath (*Burchell*); peduncles often numerous, ¼–1 in. long, usually bearing a small cyme or umbel of 3–5 involucres, pubescent or glabrous; bracts under the involucres 1¼–2½ lin. long, 1⅓–2¼ lin. broad, obovate, very obtuse to subacute, puberulous, those at the base of the cyme larger and up to 4 lin. long; involucres 1¾–2½ lin. in diam., the male larger than the female, cup-shaped, puberulous outside, with 5 glands and 5 subquadrate toothed lobes; glands of the male ¾–1¼ and of the female ½–¾ lin. in their greater diam., transverse, oblong or elliptic, entire, with a slight depression near the inner margin; capsule 2¼ lin. in diam., subglobose-tricoccous, velvety pubescent, subsessile, but exserted from the involucre; styles ¼ lin. long, shortly united at the base, spreading, deeply bifid at the tips, with diverging lobes; seeds 1–1⅓ lin. long, ellipsoid, abruptly acute at one end, obscurely 4-angled, minutely tuberculate, dark grey. *Tithymalus Eckloni, Klotzsch & Garcke in Abhandl. Akad. Berlin,* 1860, 68.

COAST REGION: Tulbagh Div.; near Wolseley, *Diels,* 1002! Caledon Div.; hills near Caledon, *Bolus*! Bredasdorp Div.; near Elim, *Guthrie,* 3816! *Bolus,* 7851! Swellendam Div.; near Swellendam, *Ecklon & Zeyher,* Euphorb. 16! *Drège,* 8195! Riversdale Div.; near Zoetemelks River, *Burchell,* 6678! 6704! near Riversdale, *Bolus,* 11378! *Muir,* 208! Zoetemelksfontein, *Muir in Herb. Galpin,* 5330! Tygerfontein, *Galpin,* 4561!

81. **E. tuberosa** (Linn. Sp. Pl. ed. i. 456, and Amœn. Acad. iii. 117); a diœcious perennial herb, 1½–3 in. high, with an elongated perpendicular tuber ⅓–1 in. thick; leaves all radical, ascending-spreading, subcoriaceous, quite glabrous or puberulous on the petiole and midrib beneath; petiole ½–1½ in. long; blade ¾–2 in. long, 4½–12 lin. broad, oblong or somewhat lanceolate-oblong, usually slightly narrowing from the subtruncate, subcordate or more or less cuneate base to the obtuse, rounded or emarginate apex, sometimes crisped at the margins; peduncles 1–7 to a plant, ¾–2 in. long, naked below, with a pair or whorl of 3–4 sessile bracts at the base of the umbel, glabrous or puberulous; bracts 2–3 lin.

long, elliptic-obovate, elliptic-oblong or suborbicular, rounded at the apex, glabrous or pubescent on the underside; umbel with 2–4 simple 1-flowered rays $\frac{1}{6}$–$\frac{2}{3}$ in. long, glabrous or puberulous; involucre unisexual, sessile, 2–2$\frac{1}{2}$ lin. in diam. and 1$\frac{1}{4}$–1$\frac{3}{4}$ lin. deep when dried, cup-shaped, glabrous or puberulous, with 5 glands and 5 subquadrate toothed lobes; glands about 1 lin. in their greater diam., transversely elliptic-oblong, entire; ovary puberulous, at first more or less included, in fruit exserted on a pedicel about as long as the involucre; styles 1–1$\frac{1}{4}$ lin. long, united for half their length, with spreading 2-lobed tips, puberulous or glabrous; capsule 3–3$\frac{1}{2}$ lin. in diam., depressed globose-trigonous; seeds 1$\frac{1}{2}$ lin. long, ovoid, subacute at one end, rugulose all over, faintly greenish-grey when ripe, reddish when immature. *Lam. Encycl.* ii. 428; *Willd. Sp. Pl.* ii. 905; *Thunb. Prodr.* 86, *and Fl. Cap. ed. Schult.* 405; *Haw. Miscell. Nat.* 185; *Spreng. Syst. Veg.* iii. 797; *Bernh. in Flora*, 1845, 87, *and in Krauss, Beitr. Fl. Cap- und Natal.* 150; *Boiss. in DC. Prodr.* xv. ii. 93; *Marloth in Wissensch. Ergebn. Deutsch. Tiefsee-Exped.* ii. iii. 317, *fig.* 131, 1. *Tithymalus tuberosus, Haw. Syn. Pl. Succ.* 137 *and Rev. Pl. Succ.* 62; *Klotzsch & Garcke in Abhandl. Akad. Berlin*, 1860, 68.—*T. tuberosus, acaulos, &c. Burm. Rar. Afr. Pl.* 9, *t.* 4. *T. humilis, foliis Lapathi, Buxb. Cent.* ii. 27, *t.* 23.

COAST REGION : Van Rhynsdorp Div. ; Kanagas Berg, *Zeyher*, 3855 ! Malmesbury Div. ; near Hopefield, *Bachmann*, 653 ! 1047 ! 2030 ! Tulbagh Div. ; Saron, *Schlechter*, 7876 ! Cape Div. ; near Tyger Berg, *Burchell*, 969 ! various places around Cape Town, *Sparrman* ! *Thunberg* ! *Burchell*, 8416 ! 8579 ! *Krauss*, 1726, *Pappe* ! *Harvey*, 27 ! *Drège*, 175 ! *Burke* ! *Prior* ! *Wolley-Dod*. 2994 ! *Schlechter*, 821 ! *Marloth*, 4882 ! *Dümmer*, 79 ! Simons Bay, *Wright*, 450 ! Honey Klip, *Mund & Maire* ! Stellenbosch Div. ; Lowrys Pass, *Schlechter* !

82. **E. elliptica** (Thunb. Prodr. 86, and Fl. Cap. ed. Schult. 405) ; a stemless tuberous-rooted herb, unisexual; tuber ellipsoid, fusiform or somewhat tapering, $\frac{3}{4}$–2 in. thick, dark brown ; leaves all radical, deciduous; petiole $\frac{1}{2}$–4 in. long ; blade 1–4 in. long, 1–6 lin. broad, linear to elliptic-lanceolate, obtuse or occasionally acute and somewhat folded together at the apex, tapering at the base or from below the middle into the petiole, entire, glabrous on both sides, of a very dark and somewhat bluish-green above, paler beneath ; peduncles in the axils of the leaves, 1–5 in. long, minutely puberulous, with an umbel of 3–5 rays $\frac{1}{4}$–1 in. long at their apex and a whorl of 3–5 bracts at the base of the umbel ; rays with 1–3 involucres; bracts sessile, 2–5 lin. long, 1–3 lin. broad, obovate, ovate or elliptic, acute or obtuse and apiculate, glabrous above, puberulous beneath ; involucres 2–3 lin. in diam., cup-shaped, puberulous outside, green, with 5 dark purple-brown or blackish-brown glands and 5 small oblong fringed lobes ; glands not contiguous, spreading, $\frac{3}{4}$–2 lin. in their greater diam., transverse, suboblong, subelliptic or very broadly and shortly wedge-shaped, entire, usually nearly straight on the outer margin ; capsule erect,

sessile, ¼ in. in diam., trigonous-subglobose, pubescent; styles ¾ lin. long, united into a column for half their length, then spreading and deeply bifid, with widely diverging lobes at the apex; seeds 1½ lin. long, ellipsoid or subglobose, obscurely pointed at one end, reticulately rugose, olive-brown. *Bernh. in Flora*, 1845, 87, *and in Krauss, Beitr. Fl. Cap- und Natal.* 150; *Boiss. in DC. Prodr.* xv. ii. 93. *E. silenifolia, Sweet, Hort. Brit. ed.* i. 356. *Tithymalus silenifolius, Haw. Rev. Pl. Succ.* 61. *T. Bergii and T. longepetiolatus, Klotzsch & Garcke in Abhandl. Akad. Berlin,* 1860, 68. *T. attenuatus and T. ellipticus, Klotzsch & Garcke, l.c.* 69.

SOUTH AFRICA: without locality, *Thunberg*! COAST REGION: Cape Div.; various localities, *Burchell*, 94! 8447! *Bergius*, 168! *Ecklon & Zeyher*, 18! *Krauss*, 1728! *Harvey*, 440! *Pappe*! *Prior*! *Wilms*, 3625! *Scott-Elliot*, 1058! *Pillans*! *Phillips*, 713! *Dümmer*, 362! *Rogers*, 2377! 2578! *Marloth*, 4414! 4887! Caledon Div.; at the foot of the Zwart Berg, *Bolus*, 7474! Caledon, *Marloth*, 4890! Knysna Div.; near Knysna River Ford, *Burchell*, 5534! Humansdorp Div.; Kruisfontein, *Galpin*, 4565! Uitenhage Div.; Van Stadens Berg, *Ecklon & Zeyher*, 19! Winterhoek Mountains, *Krauss*, 1729! Port Elizabeth Div.; sand dunes near Port Elizabeth, *Bolus*, 2680! *Mrs. Paterson*, 1143! *Drège*! Albany Div.; Rockcliffe, near Sidbury, *Miss Daly*, 825! near Grahamstown, *Misses Daly & Sole*, 197! Bathurst Div.; between Kasuga River and Port Alfred, *Burchell*, 3956!

83. **E. crispa** (Sweet, Hort. Brit. ed. i. 356); a stemless tuberous-rooted herb, unisexual; tuber ellipsoid or ovoid, sometimes divided at the top (from injury?) into cylindric subterranean branches, dark brown; leaves all radical, deciduous; petiole ½–2 in. long; blade ½–2 in. long, 2½–5 lin. broad, lanceolate to elliptic, obtuse and usually slightly folded together at the apex, more or less cuneate at the base, wavy on the margin, entire, glabrous on both sides, or pubescent beneath; peduncles in the axils of the leaves, ½–1½ in. long, glabrous or pubescent, with an umbel of 2–5 rays in the male and a solitary involucre or a 2–3-rayed umbel in the female, and 2–5 bracts at the base of the umbel or involucre; rays with 1 involucre; bracts sessile, 1½–2¼ lin. long, 1¼–2¼ lin. broad, obovate or elliptic-oblong to suborbicular, obtuse or broadly rounded and more or less apiculate at the apex, glabrous or pubescent; involucre sessile between the bracts, 2½–3 lin. in diam., cup-shaped, glabrous or pubescent, with 5 glands and 5 oblong toothed and ciliate lobes; glands not contiguous, spreading, ½–1 lin. in their greater diam., transverse, suboblong, borne upon short flat stalks, apparently greenish-yellow; capsule erect, ¼ in. in diam., trigonous-subglobose, pubescent, with a puberulous crenate ring at its base and seated on a short puberulous pedicel much shorter than the involucre; styles about ¾ lin. long, united into a column for nearly half their length, with spreading deeply bifid arms; seeds 1¾–2 lin. long, oblong or subglobose-oblong, abruptly acute at one end, truncate at the other, varying from nearly smooth with a ridge or ring of tubercles near the truncate end to irregularly tuberculate all over, dark brown, opaque. *E. elliptica, var. undulata, Boiss. in DC. Prodr.* xv. ii. 93. *Tithymalus crispus, Haw. Rev. Pl. Succ.* 61.

COAST REGION: Clanwilliam Div.; foothills of the Cold Bokkeveld Mountains opposite Warm Baths, *Stephens*, 7217! lower slopes of Olifants River Mountains, *Stephens*, 7225! Malmesbury Div.; near Hopefield, *Bachmann*! Zwartland, *Bachmann*, 1048!

CENTRAL REGION: Calvinia Div.; Hantam, *Meyer*! Ceres Div.; Karroo Poort, *Alston in Herb. Marloth*, 5279!

WESTERN REGION: Van Rhynsdorp Div.; Gift Berg, *Phillips*, 7392! 7634!

This is so closely related to *E. elliptica* that I feel doubtful if it ought to be maintained as distinct from that species. But as Haworth, who described from living plants, considered it to be different, and Dr. Marloth, who must be well acquainted with *E. elliptica* in the living state, has recently sent it to Kew as a supposed new species, I retain it. The chief differences appear to be the wavy or crisped margins of its proportionately shorter and broader leaves, fewer involucres to a peduncle and apparently different colour of the glands of the involucre. The seeds seem also to be larger, more oblong and differently sculptured, but I have only seen seeds from one specimen of *E. elliptica*. It requires further investigation.

84. **E. bupleurifolia** (Jacq. Hort. Schoeubr. i. 55, t. 106, and Fragm. 68, t. 101, fig. 2); plant unisexual; stems subglobose, obovoid or cylindric, rising 1–9 in. above the ground, $1\frac{1}{2}$–$2\frac{1}{4}$ in. thick, tuberculate, glabrous, brown or brownish-green; tubercles crowded, $1\frac{1}{2}$–2 lin. prominent, transverse, subrhomboid-trigonous, with a leaf-scar at the apex; leaves in a tuft at the apex of the stem, $1\frac{1}{2}$–6 in. long, $\frac{1}{6}$–$\frac{2}{3}$ in. broad, spathulate-lanceolate, acute, tapering into a rather long petiole, entirely glabrous or the petiole more or less puberulous, deciduous; peduncles several to a plant, solitary in the axils of the leaves, $\frac{1}{2}$–$2\frac{1}{4}$ in. long, bearing a pair of bracts and 1 involucre at the apex, puberulous or velvety; bracts 4–$5\frac{1}{4}$ lin. long, $4\frac{1}{2}$–10 lin. broad, suborbicular to reniform, very obtuse, apiculate, entirely glabrous or puberulous at the base, forming a cup around the involucre, green; involucre sessile, unisexual, 3–5 lin. in diam., cup-shaped, glabrous or puberulous, with 5 glands and 3 rectangular or subquadrate lobes cut to their middle into several segments; glands 1–2 lin. in their greater diam., transverse, oblong or somewhat half circular, 2-lipped from the inner margin being incurved and forming a slight pocket-like cavity, entire or minutely crenulate; capsule $4\frac{1}{2}$ lin. in diam., obtusely trigonous, glabrous, erect, exserted on a puberulous pedicel as long as the involucre; styles united in a column $1\frac{1}{2}$–2 lin. long, with spreading bifid tips $\frac{2}{3}$–1 lin. long; seeds 2 lin. long, ovoid, pointed, smooth, pale greyish-brown. *Willd. Sp. Pl.* ii. 888; *Ait. Hort. Kew. ed.* 2, iii. 158; *Poir. Encycl. Suppl.* ii. 609; *Spreng. Syst. Veg.* iii. 787; *Bernh. in Flora,* 1845, 87, *and in Krauss, Beitr. Fl. Cap- und Natal.* 150; *E. Meyer in Drège, Zwei Pfl. Documente,* 184 *as to letter "a" not "b"*; *Boiss. in DC. Prodr.* xv. ii. 92; *Hook. in Bot. Mag.* t. 3476; *Goebel, Pflanzenbiol. Schilderung.* i. t. 1, fig. 1; *Berger, Sukk. Euphorb.* 125. *E. proteifolia, Boiss. in DC. Prodr.* xv. ii. 92. *Tithymalus bupleurifolius, Harv. Syn. Pl. Succ.* 138; *Klotzsch & Garcke in Abhandl. Akad. Berlin,* 1860, 81.

COAST REGION: Albany Div.; near Grahamstown, *Bolton*! *Misses Daly & Sole,* 308! Queenstown Div.; near Queenstown, 3600–3800 ft., *Galpin,* 1562!

mountain summit near Bongolo Nek, 5000 ft., *Galpin*, 7951 ! East London Div. ;
near East London, *Wood*, 3251 ! Komgha Div. ; near Komgha, 2000 ft., *Flanagan*,
1315 ! Eastern Frontier, *MacOwan*, 649 ! British Kaffraria, *Cooper*, 151 !
EASTERN REGION : Transkei ; Krielis Country, *Bowker* ! Kentani district, near
Black Rock Cove, *Miss Pegler*, 649 ! Tembuland ; mountains around Bazeia, *Baur*,
250 ! Natal ; Camperdown, *Gerrard*, 1174 ! near Durban, *Drège* ; near Pietermaritzburg, *Krauss*, 106 b ; Inanda, *Wood*, 228 ! and without precise locality,
Gerrard, 655 !

This well marked species forms a connecting link between the tuberous-rooted
group to which *E. tuberosa* and *E. elliptica* belong and the group with succulent
tuberculate stems, through such species as *E. clandestina* and *E. Davyi*. For it is
evident that the succulent stem of *E. bupleurifolia* is merely an above ground
development of the subterranean tubers of the *E. tuberosa* group, its large
deciduous leaves and the character of the inflorescence being quite of the same
nature.

85. **E. clandestina** (Jacq. Hort. Schoenbr. iv. 43, t. 484) ; stem
solitary, unbranched (always ?), $\frac{1}{2}$-1 ft. (or perhaps more) high,
1-1$\frac{1}{2}$ in. thick, cylindric or cylindric-clavate, fleshy, spineless,
covered with slightly recurved or ascending stout tubercles 2-4 lin.
prominent, in several spiral series, and bearing when in flower a
crown of leaves at the apical part with solitary sessile involucres
surrounded by and included in a cup formed of bracts in their axils,
minutely puberulous on the upper part, becoming glabrous below ;
leaves terminating the tubercles, $\frac{3}{4}$-1$\frac{1}{2}$ in. long, 1-2 lin. broad,
linear or linear-oblanceolate, acute and slightly recurved at the
apex, tapering to the base, more or less longitudinally folded, somewhat fleshy, minutely puberulous beneath, at least at the basal
part ; bracts 7-9, imbricate ; inner series 2 lin. long, 1-1$\frac{1}{2}$ lin.
broad, ovate-oblong, obtuse, apiculate, glabrous, ciliate, green ;
outer series gradually smaller and more or less puberulous on the
back, purple ; involucre sessile, about half as long as the inner
bracts, about 1$\frac{1}{3}$ lin. in diam., cup-shaped, minutely puberulous,
with 5-7 glands and 5 subquadrate minutely toothed lobes ; glands
$\frac{1}{3}$-$\frac{2}{3}$ lin. in their greater diam., transverse, narrowly oblong, entire,
puberulous on the under side, glabrous on the upper, yellow ; capsule
sessile, twice as long as the involucre, 2$\frac{1}{2}$ lin. in diam., subglobose,
slightly narrowing at the top, with 3 slight furrows, minutely
puberulous ; styles free, $\frac{1}{4}$ lin. long, spreading, shortly bifid at the
apex ; seeds not seen.　*Poir. Encycl. Suppl.* ii. 610 ; *Spreng. Syst.
Veg.* iii. 787 ; *Boiss. in DC. Prodr.* xv. ii. 92 ; *Berger, Sukk. Euphorb.*
125.

COAST REGION : Swellendam Div. ; Hessaquas Kloof, *Zeyher*, 3847 ! Riversdale
Div. ; Plattebosch, *Muir*, 282 !

86. **E. Davyi** (N. E. Br.) ; very dwarf, succulent, spineless ;
body or main stem of the plant subglobose to elongated-obconic or
obovoid, with the greater part buried in the ground and only rising
1-1$\frac{1}{2}$ in. above it, about 2$\frac{1}{4}$ in. in diam. in the only specimen seen,
subtruncate or broadly rounded at the top, covered with large
rhomboid tubercles and bearing a lax crown of branches in 2-3

series around the top, but none at the central part, glabrous, light
green at the top, brown at the sides ; tubercles on the top of the
plant laterally compressed, 5–9 lin. long, 3–6 lin. broad and 3–4½
lin. prominent, becoming with age as they pass to the sides of the
stem compressed from above and twice as broad as long ; branches
arising between the tubercles, erect, curved at the base, finally
deciduous, 1½–3 in. long, ½–¾ in. thick (including the tubercles),
cylindric, leafy at the tips when growing, covered with conical
tubercles 1½–2 lin. prominent and very much smaller than those on
the main stem ; leaves terminating the tubercles of the branches
(none seen on the main stem), ⅔–1⅛ in. long, ¾–1¾ lin. broad, linear
or linear-lanceolate, acute, with a minute dorsal point at the apex,
shortly narrowed to the sessile base, entire, longitudinally folded,
fleshy, glabrous ; peduncles solitary in the axils of the tubercles
among the leaves at the apex of the stem, 1–1½ lin. long, becoming
2–5 lin. long in fruit, bearing 1 involucre and 4–6 deciduous oblong
acute scale-like bracts ⅔–1¼ lin. long, minutely ciliate on their
margins ; involucres ¼ in. in diam., cup-shaped, glabrous, with
5 glands and 5 transverse subrectangular toothed or fringed lobes ;
glands not contiguous, apparently green, 1 lin. in their greater
diam., transversely elliptic-oblong, with the inner margin turned up
(perhaps flat when alive) and with 2–3 subulate teeth ⅓–½ lin. long
or one of them shorter on the outer margin ; capsule sessile, not
seen mature, apparently about 3–3½ lin. in diam., obtusely somewhat 3-lobed, glabrous ; styles stout, united into a column about
1¼ lin. long, with thick spreading channelled stigmas ½ lin. long.

KALAHARI REGION : Transvaal ; near Pretoria, *Kirk*, 48 ! Warm Bath, in
Waterberg district, *Burtt-Davy*, 5562 ! Springbok Flats, *Burtt-Davy*, 2196 !

87. **E. ramiglans** (N. E. Br.) ; plant probably similar in habit
to *E. Gorgonis*, Berger, but only a few branches seen, evidently
taken from a larger body ; branches 1–1¼ in. long, ½–¾ in. thick
(dried), covered with crowded rhomboid conical tubercles 1½–2 lin.
prominent, tipped with a whitish leaf-scar, glabrous ; leaves in a
small tuft at the apex of the branches, 3½–4½ lin. long, linear,
obtuse or subacute, channelled down the face, glabrous ; peduncles
solitary in the axils of the tubercles at the tips of the branches,
1½–2 lin. long, bearing 1 involucre and 2 or 3 small and very
deciduous bracts ; involucre 3½–4 lin. in diam., shallowly cup-
shaped, glabrous outside, with 5 glands and 5 erect subquadrate
2-toothed lobes and filled with white-woolly stamens ; glands
spreading, subcontiguous, ¾–1¼ lin. in their greater diam., transversely oblong, with 4–6 filiform processes ⅔–¾ lin. long on the
outer margin, very much-branched at their tips ; ovary sessile,
sprinkled with rather long hairs ; styles united to the top into a
slender column 1¾ lin. long, with spreading cuneate and slightly
2-lobed stigmas ⅓ lin. long ; capsule and seeds not seen.

WESTERN REGION : Little Namaqualand ; without precise locality, *Bolus*,
9448 !

88. **E. basutica** (Marloth in Trans. Roy. Soc. S. Afr. i. 408, t. 27, fig. 6, excl. syn.); dwarf, succulent, spineless; main body of the plant subglobose or obconic, nearly buried in the ground, and excepting a small central tuberculate area covered over the top with very numerous (50–60) branches arranged in 4–5 series, forming a clump about 4 in. in diam. (not sparingly branched as originally described); branches $\frac{3}{4}$–$1\frac{1}{2}$ in. long, $4\frac{1}{2}$–6 lin. thick, simple or occasionally branching, more or less clavate, covered with transversely rhomboid or subhexagonal tubercles $1\frac{1}{2}$–2 lin. in diam. and $\frac{1}{2}$–$\frac{3}{4}$ lin. prominent, glabrous, green at the upper part, brown below; leaves rudimentary, soon deciduous, 1 lin. long, $\frac{1}{2}$ lin. broad, lanceolate or narrowly elliptic, obtuse, very concave above as if longitudinally folded, very convex beneath, thick and fleshy; involucres solitary in the axils of tubercles near the apex of the branches, subsessile or on peduncles $\frac{1}{2}$–1 lin. long, rather shallowly basin- or somewhat funnel-shaped, $3\frac{1}{2}$–4 lin. in diam., glabrous, with 5 glands and 5 transversely rectangular deeply toothed lobes; glands spreading, $1\frac{1}{3}$–2 lin. in their greater diam., subcontiguous or slightly separated, transversely oblong, with about 6 short entire or denticulate teeth along the outer margin, and the entire inner margin more or less turned up, forming a slight cavity in front of it, somewhat greenish; ovary sessile, included in the involucre, glabrous; styles united into a stout column $\frac{2}{3}$–$\frac{3}{4}$ lin. long, scarcely exceeding the lobes of the involucre, their very stout tips exserted, recurved-spreading, entire, broadly cuneate-obovate, channelled down the face, $\frac{2}{3}$ lin. long and as much in breadth.

KALAHARI REGION: Basutoland; Maseru (not Leribe as originally stated), *Mrs. A. Dieterlen*, 415!

Described partly from a branch in fluid accompanied by an excellent life-sized photograph from the type plant, communicated by Dr. E. P. Phillips, and partly from flowers in fluid from the same plant, kindly lent by Dr. Marloth and from which he made his original description. Dr. Marloth has, however, unfortunately identified this plant with *E. Caput-Medusæ* of *DC. Plantes Grasses*, t. 150, from which it is entirely different, that plant being *E. Bergeri*, N. E. Br., with different leaves and flowers and much longer persistent peduncles. Also at the time when the plant figured by De Candolle was introduced, Basutoland was practically an unknown country, and the plant could not have come from there.

89. **E. Ernesti** (N. E. Br.); plant not rising more than 1–$1\frac{1}{2}$ in. above the ground, succulent, spineless, leafless; main stem obconic or subcylindric, 1–$2\frac{1}{2}$ in. thick, partly buried in the ground, covered on the top nearly or quite to the centre with numerous branches in several series, glabrous, dull green or tinted with purple; branches radiately spreading or the inner ascending, quite unbranched, with the inner series $\frac{1}{8}$–$\frac{1}{4}$ in. long, subglobose, the others increasingly longer up to $2\frac{1}{2}$ in. long, $\frac{1}{4}$–$\frac{1}{3}$ in. thick, cylindric, tessellately tuberculate, glabrous, green; tubercles $1\frac{1}{2}$–2 lin. in diam., rhomboid or hexagonal, not very prominent, marked with a whitish leaf-scar; leaves rudimentary, $\frac{1}{4}$–$\frac{2}{3}$ lin. long, sessile, deltoid-ovate, oblong or suborbicular, under cultivation becoming 1–$1\frac{1}{3}$ lin. long and spathu-

late, with an ovate blade, acute, fleshy, glabrous, soon withering or deciduous; involucres sessile or subsessile on the central area of the top of the main stem and near the apex of the branches, $2\frac{1}{2}$–3 lin. in diam., cup-shaped, glabrous, with 5 glands and 5 sub-quadrate or transversely oblong finely toothed lobes and surrounded at the base by 3–5 obovate or oblong-obovate subtruncate finely toothed bracts; glands rich dazzling golden-yellow (*Galpin*), $\frac{3}{4}$–1 lin. in their greater diam., transversely oblong, concave, with 2–8 short teeth on the outer margin; capsule sessile, 3–$3\frac{1}{2}$ lin. in diam., very obtusely 3-lobed, glabrous; styles $\frac{1}{2}$–$\frac{3}{4}$ lin. long, stout, united into a column for half their length, with stout cuneate-obovate recurved-spreading tips; seeds 2 lin. long, ovoid, acute, sometimes very slightly tuberculate, at others thickly and roughly tuberculate, dark olive-brown in the examples seen, but not quite ripe.

COAST REGION : Queenstown Div. ; near Queenstown, on stony plains around the town and on a dry stony plateau on Hospital Hill, 3500–3600 ft., *Galpin*, 8066 !

Partly described from living plants sent to Kew by Mr. E. E. Galpin, who states that the flowers are very beautiful and conspicuous.

E. Ernesti differs from *E. clavarioides*, Boiss., by its main stem rising distinctly above the ground and not being covered to the centre with branches, also by the branches being unbranched and all developed entirely above the ground, whilst in *E. clavarioides* the branches are partly buried in the ground and with age become repeatedly branched at their tips. The flowers also of *E. Ernesti* are rather larger and more brilliantly coloured than those of *E. clavarioides*.

90. **E. clavarioides** (Boiss. Cent. Euph. 25, and in DC. Prodr. xv. ii. 91); plant forming a cushion-like mass 2–3 in. high and 4–12 in. in diam. of densely crowded branches, succulent, spineless, leafless; rootstock or main stem thick and fleshy, obconic, buried in the ground and covered to the centre of its flattened top with crowded branches, which are also partly or wholly buried in the ground, as with age (after attaining a length of 1–2 in.) they branch at the tips and the branchlets again divide in a similar manner, becoming in turn buried with the development of the plant; younger branchlets $\frac{1}{3}$–$\frac{2}{3}$ in. thick, at first subglobose, becoming cylindric or clavate and finally greatly thickened, very obtusely rounded at the apex, tessellately tuberculate, dull green or purplish-tinted; tubercles $1\frac{1}{2}$–2 lin. in diam., about $\frac{1}{2}$ lin. prominent, rhomboid or hexagonal, very obtusely and broadly rounded-conical, with a whitish and not at all impressed leaf-scar at their apex; leaves rudimentary, soon deciduous, fleshy, $\frac{1}{2}$–1 lin. long, $\frac{1}{3}$–$\frac{1}{2}$ lin. broad, sessile, ovate or lanceolate, subacute, channelled down the face, glabrous; involucres sessile at the tips of the branches, surrounded by 3–5 suborbicular or broadly obovate ciliate thin scale-like bracts, male and bisexual occurring on the same plant, cup-shaped, $2\frac{1}{2}$–3 lin. in diam. and about 1 lin. deep, glabrous, with 5 glands and 5 subquadrate or transversely rect-angular ciliate lobes; glands distant, greenish-yellow (*Galpin*), $\frac{1}{2}$–1 lin. in their greater diam., transversely oblong, subentire or

minutely toothed or with subulate processes up to $\frac{1}{4}$ lin. long, on their outer margin; capsule sessile, about $\frac{1}{3}$ in. in diam., slightly and very obtusely 3-lobed or trigonous, glabrous; styles free nearly to the base, radiating, $\frac{1}{2}$ lin. long, very stout, broadly wedge-shaped or with suborbicular stigmas, channelled down the face; seeds $1\frac{3}{4}$ lin. in diam., globose-ovoid, abruptly acute, truncate at the base, minutely tuberculate on the dorsal side and with a broad smooth stripe on the ventral side, glabrous, dark-brown.

COAST REGION: Queenstown Div.; on a dry ridge near the summit of a mountain on the east side of Bongolo Nek, 12 miles from Queenstown, 5000 ft., *Galpin*, 7950!

CENTRAL REGION: Graaff Reinet Div.; on the Sneeuwberg Range, 4000–5000 ft., *Drège*, 8200!

The above description is chiefly made from living plants sent to Kew by Mr. E. E. Galpin of Queenstown, who states that his "7950 grows only in one spot, so far as I know, on the top of a high mountain." Drège's specimens, on which the species was founded, are merely branches or branchlets, but these in some cases are distinctly branched at the tips as in the Queenstown plant, and in size, shape, tuberculation and flowers seem quite to accord with that plant.

91. **E. truncata** (N. E. Br.); plant very dwarf, not rising more than $1\frac{1}{2}$–3 in. above the ground, succulent, leafless and spineless, diœcious, consisting of a thick fleshy obconic body tapering into the root-system and buried nearly to the top in the ground, producing on the top a crowded mass of branches to the very centre, all rising to the same level, forming a flat-topped or slightly convex (with age) mass 3–8 in. (or perhaps more) in diam.; branches on the outer part 1–$2\frac{3}{4}$ in. and those at the centre about $\frac{1}{3}$ in. long, $\frac{1}{3}$–$\frac{2}{3}$ in. thick, simple or the outer branched at their tips, subcylindric or more or less compressed by mutual pressure, tessellately tuberculate, truncate at the apex when alive or when dried often obtusely rounded, glabrous, green or more or less brownish where exposed to the sun at the tips only, brown elsewhere; tubercles at first about as long as broad and $1\frac{1}{3}$–$1\frac{1}{2}$ lin. in diam. becoming broader than long, rhomboid and arranged in crowded spirals, not very prominent; leaves rudimentary, soon deciduous, 1–$1\frac{1}{2}$ lin. long, $\frac{2}{3}$ lin. broad, oblong or lanceolate, acute, channelled down the face, fleshy, glabrous; involucres sessile in the axils of the tubercles at the apex of the branches, about 1 lin. long and $2\frac{1}{2}$ lin. in diam., cup-shaped, glabrous, with 5 glands and 5 transversely rectangular toothed lobes; glands not contiguous, 1 lin. in their greater diam., transversely oblong, with a depression across their centre and 3–5 teeth $\frac{1}{8}$–$\frac{1}{4}$ lin. long on their outer margin, apparently yellow; ovary subsessile, trigonous, without a calyx, thinly sprinkled with rather long hairs; styles $\frac{3}{4}$ lin. long, very shortly united at the base, very stout, much dilated or suborbicular and channelled at the spreading tips.

KALAHARI REGION: Transvaal; near Lydenburg, *Wilms*, 1339! Standerton, *Burtt-Davy*, 1953! Irene, *Miss Leendertz*, 670! Johannesburg, *Miss Leendertz*, 1873!

EASTERN REGION: Natal, Estcourt, *Kolbe*!

Partly described from a living plant (*Burtt-Davy,* 1953) received at Kew in October, 1904. This species is closely allied to *E. clavarioides,* Boiss., but seems to be well distinguished by the branches being usually quite simple, rather stouter when dried, with less prominent tubercles, and (when alive) truncate at the top, but becoming rounded when dried, all reaching to nearly the same level, forming a flat or slightly convex top to the plant. The ovary also is thinly hairy ; whilst in *E. clavarioides* the branches form a more or less convex cushion-like mass and are constantly branched at their tips, obtusely rounded at the apex when alive, and the ovary is glabrous. According to a note, this species is widely distributed in the Transvaal and Orange River Colony, growing in clayey soil on the borders of Vleys. Goats eat it greedily. It has a strong odour like that of mice.

92. **E. esculenta** (Marloth in Trans. Roy. Soc. S. Afr. i. 319) ; very dwarf, succulent, spineless and leafless ; main stem buried in the ground nearly to the top, club-shaped or obconic, 4–8 in. thick, with the central part of the flat or slightly depressed top covered with conical acute tubercles and the outer part bearing a rosette (in old plants attaining to 18 in. in diam.) of very numerous crowded branches in several series, glabrous, green on the young parts, becoming pale brown with age ; branches radiately spreading, 2–8 in. long, ½–1 in. thick, cylindric or the outermost tapering from the base to the obtuse apex, quite unbranched, tessellately tuberculate ; tubercles densely crowded in many spirals, flattish, scarcely or but very slightly prominent, 1½–3 lin. in diam., usually a little longer than broad, obovate-rhomboid or subhexagonal, with a minute central leaf-scar ; leaves rudimentary, minute, soon deciduous ; peduncles clustered at the ends of the branches, ½–2 lin. long, stout, with 2–3 alternate and 2 opposite bracts and 1 involucre, glabrous, not persisting more than one season ; bracts oblong or oblong-spathulate, glabrous, ciliate ; involucres 1¾–2¼ lin. in diam., broadly and shallowly cup-shaped, glabrous except on the back of the lobes, with 5–6 glands and 5–6 broad transversely oblong ciliate lobes and densely filled with white-woolly bracteoles ; glands distant, reflexed and closely pressed to the involucre, ½–1 lin. in their greater diam., transverse and somewhat reniform or irregular in outline, more or less deeply fissured and sometimes divided into 2 or more parts, rather thick and fleshy, convex, slightly corrugated, brown ; ovary sessile, glabrous ; styles united into a column about 1 lin. long with recurved-spreading arms ½–⅔ lin. long, with very large stigmas deeply channelled down the face ; capsule and seeds not seen. *Marloth in Wissensch. Ergebn. Deutsch. Tiefsee-Exped.* ii. iii. *fig.* 102, 2.

CENTRAL REGION : Jansenville Div. ; Klipplaat, *Marloth,* 4162 ! common also in Graaff Reinet and Aberdeen Div., *Marloth* ! Willowmore Div. ; near Willow-more, *Brauns* !

Described from living plants sent to Kew by Dr. S. Schönland and branches from the type plant and a photograph from Dr. Marloth. According to Dr. Marloth, this plant affords " a very nutritious food for stock in times of drought and formerly was occasionally roasted in the ashes for human use. The involucres are sweet-scented, like violets.".

93. **E. gatbergensis** (N. E. Br.) ; very dwarf, succulent, spineless ; body of the plant 1¼–2½ in. (or perhaps more) in diam., subglobose

(or obconic?), flattened at the top, with a crown of radiating branches surrounding a large central area densely covered with conical tubercles, glabrous; branches in about 3 series, radiating, the outer. $\frac{2}{3}$–$1\frac{1}{4}$ in. long and 3–4 lin. thick (dried), the inner gradually shorter, cylindric, very obtuse, covered with small slightly prominent tubercles about $1\frac{1}{2}$ lin. in diam.; leaves soon deciduous, 1–3 lin. long, linear-lanceolate, acute, spreading or erect, only present at the tips of growing branches, glabrous; involucres sessile or on peduncles $\frac{1}{2}$–1 lin. long, surrounded by 3–4 subquadrate toothed membranous scale-like bracts, produced at the circumference of the central area and on the very young branches, about 3 lin. in diam., cup-shaped, glabrous, with 5 glands and 5 subquadrate many-toothed lobes; glands 1–1$\frac{1}{3}$ lin. in their greater diam., transversely oblong or elliptic-oblong, sometimes nearly or quite entire, at others with few or several short teeth along the outer margin, probably concave, as in dried specimens the outer and inner margins are often infolded, apparently greenish or greenish-yellow; ovary sessile, glabrous; styles united into a column $\frac{2}{3}$–$\frac{3}{4}$ lin. long, with spreading broadly cuneate arms $\frac{1}{2}$–$\frac{2}{3}$ lin. long, slightly notched at the apex; capsule and seeds not seen.

EASTERN REGION : Tembuland ; near Gat Berg, 3000–3500 ft., *Baur*, 251 !

94. **E. pugniformis** (Boiss. in DC. Prodr. xv. ii. 92); dwarf, succulent, spineless; main body of the plant subglobose, partly buried in the ground, truncate at the top, with a large slightly depressed branchless central area covered with tubercles and producing flowers and branches around its circumference; branches in about 2–3 series, radiately spreading or slightly ascending, $\frac{1}{2}$–$1\frac{1}{4}$ in. long, 3–4 lin. thick at the base, cylindric, but slightly tapering towards the tips, covered with numerous rhomboid or subhexagonal tubercles about 1–1$\frac{1}{2}$ lin. in diam., marked with a dusky leaf-scar, glabrous; leaves 2–3 lin. long, linear-lanceolate, acute, deciduous; involucres produced around the margin of the central area of the main body of the plant, on peduncles 1–2 lin. long, apparently shallowly basin-shaped, with 5 entire elliptic whitish-green glands; capsule sessile, somewhat bluish, glabrous; styles rather long, apparently slender; seeds small, oblong, glabrous, brown. *E. procumbens, Sweet, Hort. Suburb. Lond.* 107, *and Hort. Brit. ed.* i. 356, *not of Miller, nor of N. E. Br. nor Berger. E. Caput-Medusæ, var.* δ, *Linn. Sp. Pl. ed.* i. 452.—*Euphorbium humile procumbens, &c. Burm. Rar. Afr. Pl.* 20, *t.* 10, *fig.* 1.

SOUTH AFRICA : without locality, *Burmann.*

This species is only known from Burmann's figure, no specimens that have been seen agree with it in the possession of entire ovate or roundish glands to the involucre, which are represented in the figure as being as long as or longer than broad. It clearly belongs to the group where the branches do not cover the top of the main stem and seems near to *E. Gorgonis*, but the longer leaves and shape and colour of the glands of the involucre clearly distinguish it from that species.

95. E. Gorgonis (Berger in Engl. Jahrb. xlv. 230) ; very dwarf, succulent, spineless ; main body of the plant globose or obconic, 2–4 in. in diam., with a crown of short radiating branches in 3–5 series around a branchless flat or depressed central area or disc 1–2 in. in diam. ; disc covered with acute conical tubercles 1–2½ lin. prominent and as much in diam., glabrous, dull green or more or less tinged with purplish ; branches ½–1 in. long or under cultivation up to 2 in. long, 3–5 lin. thick, cylindric or the younger globose, covered with small 5–6-angled conical acute tubercles 1¼–2 lin. in diam. and ½–1 lin. prominent, glabrous, green or tinged with purplish, not glaucous ; leaves rudimentary, only present on the young growth and soon deciduous, ½–1 lin. (or under cultivation up to 1½ lin.) long, ½–⅔ lin. broad, lanceolate or elliptic, acute, glabrous ; peduncles 2–5 lin. long, solitary in the axils of the tubercles of the disc and branches, erect, stout, bearing 1 involucre and 3–5 minute, scale-like ciliate bracts, glabrous ; involucre 2½–2¾ lin. in diam., cup-shaped, glabrous, with 5 glands of a rich dark purple-brown and 5 short broadly rounded or transversely oblong dull purplish white-ciliate lobes ; glands vertically deflexed, or deflexed-spreading, ¾–1¼ lin. in diam., suborbicular, with the side margins reflexed and so often appearing very broadly ovate, with a notch on the upper margin and the lower margin entire or notched or as if the apical part were notched and pinched together and produced into 2 acute or short subulate teeth or with 3–7 small irregular teeth scattered along it, minutely pitted, varying even on the same plant (see note below), usually rich dark crimson or brownish crimson, but under cultivation sometimes bright red on the same plant at different seasons, with the subulate points yellow ; capsule sessile or sub-sessile, 2 lin. in diam., subglobose, slightly 3-grooved, thinly sprinkled with hairs when young, sometimes nearly or quite glabrous when ripe ; styles united into a column about ½ lin. long, with spreading broadly cuneate-obcordate or 2-lobed arms ½–⅔ lin. long, channelled down their face, green ; seeds 1¼ lin. long, ovoid, acute at one end, minutely tuberculate, except along a narrow space down the ventral side, blackish-grey.

COAST REGION : Uitenhage Div. ; hills between the Sundays River and Zwart-kops River, *Zeyher* ! The Fountain, *Palmer* ! Port Elizabeth Div. ; North End Hill, near Port Elizabeth, *Mrs. Paterson*, 2144 ! Albany Div. ; near Grahamstown, *MacOwan*, 3269, and in *Herb. Austr.-Afr.*, 1957 ! *Schönland*, 48 !

Described from a living plant collected at The Fountain by Mr. C. N. Palmer and sent to me by Mr. Burtt-Davy, who stated that it is "abundant, and the juice is used locally for the making of bird-lime ; stock of all kinds is said to be very fond of the plant." This species admirably illustrates the effect of our insular damp climate upon South African plants. When received at the end of June, 1911, I planted it in the open air at Kew, where it flowered in August. Although the summer was one of the driest on record, with no rain for 40 days, the short normal native-grown branches immediately began to elongate, and, as well as the new ones which formed, produced leaves 2–3 times as large as those that were native-grown, giving the plant an entirely different appearance as compared with that which it had when originally received. The English-grown branches are about 2½ times as long as those found on South African specimens.

The glands of the involucre of MacOwan's specimens differ from the others in having 3–7 teeth along their outer margin, but this may only be a seasonal variation, as I have found that the glands on my cultivated plant are not only variable in this toothing, but one year were of a brilliant red instead of the brownish-crimson of the previous and following times of flowering.

In Commelin, *Hort. Med. Amstelod.* i. 33, fig. 17, under the name, Planta lactaria africana, a species of *Euphorbia* is represented, which may possibly be *E. Gorgonis.* This figure is a very remarkable and instructive one. It represents an imported plant, whose branches, under the influence of the moister European climate, have become much elongated and taper upwards into quite slender leafy portions, resembling those of the ordinary herbaceous species, whilst from the central part of the main body of the plant arises a branch, evidently developed in Europe, which is quite slender, terete, without a trace of tubercles and bearing leaves at the upper part. This branch is exactly like, and (if cut from the plant) would be mistaken for a young shoot of one of the herbaceous species, and seems to demonstrate that these dwarf many-branched thick fleshy species of South Africa, during a very long period of a very slow modification from a moist to a very dry climate, have gradually been evolved from perennial herbaceous species, whose stems or branches have become more and more dwarfed and succulent, and whose rootstock has become very much thickened into a succulent storehouse of water and food as the climate has become dryer, see the *Gardeners' Chronicle,* 1914, lvi. 230, fig. 91.

Upon this plate of Commelin's, Isnard in *Mém. Acad. Roy. des Sciences, Paris,* 1720, 386, founds his *Euphorbium anacanthum, angusto Polygoni folio,* but the two varieties he has placed under it belong to *E. Caput-Medusæ,* Linn., and are certainly quite distinct from Commelin's plant.

96. **E. passa** (N. E. Br.); dwarf, succulent, spineless; main body of the plant globose, obconic or subcylindric, 2–4 in. thick, producing at the top numerous radiately spreading branches around the circumference of a truncate or slightly depressed tuberculate central area; branches 1–8 (or under cultivation sometimes becoming 6–14) in. long, 4–6 lin. thick, cylindric, tuberculate, bright green; tubercles rhomboid, 4–6-angled, 1½–3 lin. long, 1½–2 lin. broad and ½–¾ lin. prominent; leaves 1½–4 lin. long, narrowly linear, acute, concave above, convex beneath, green; involucres on peduncles 1–6 lin. long, solitary in the axils of the tubercles of the central area of the main stem, and at the tips of some of the branches, often very numerous, 3½–4 lin. in diam, cup-shaped, glabrous, with 5 glands and 5 broadly rounded or transversely oblong fringed lobes; glands horizontally spreading, 1 lin. in their greater diam., transversely oblong or elliptic-oblong, entire to slightly crenulate on the outer margin, yellow, under cultivation often changing to orange or to bright red; ovary acutely 3-angled, pubescent with rather long hairs, subsessile or very shortly pedicellate, included in the involucre; styles united in a column ⅔ lin. long, with very broad cuneate or obcordate recurved-spreading stigmas, notched at the tips, glabrous. *E. pugniformis, Baker in Saund. Ref. Bot.* iii. *t.* 161, *not of Boiss. E. procumbens, N. E. Br. in Bot. Mag. t.* 8082, *and of Berger, Sukk. Euphorb.* 118, *excl. all syns. except the above, not of Miller.*

EASTERN REGION: Natal; Scottsburg, *Pole Evans!* Umzumbi, 50–150 ft., *Wood!* and without precise locality, *Cooper!*

Described from a living plant and specimens preserved in fluid. The history of this plant is somewhat interesting. It was introduced into cultivation by Mr. T. Cooper in 1862, and from a plant cultivated by himself the figure in *Refugium Botanicum*, t. 161, was made, and not (as Mr. Cooper himself informed me) from a plant in the collection of Mr. Wilson Saunders as there stated. Some time after, Mr. Cooper sold this plant to Mr. Justus Corderoy, and 36 years later Mr. Corderoy's plant was figured in the *Botanical Magazine* at t. 8082, so that both figures were actually made from the same individual at a long interval. The plant subsequently passed into the possession of Kew. At the time I wrote the account in the *Botanical Magazine* I was not aware of all this, and took the identification as given in *Refugium Botanicum* to be correct, without investigation, but used for the *Botanical Magazine* the older name *E. procumbens*, Mill., quoted as a synonym by Mr. Baker. As Mr. Cooper did not recollect where he collected the plant, it was supposed that he might have got it somewhere in Cape Colony, where he was about 1860 and, therefore, it might be Miller's plant. But Dr. J. Medley Wood has recently sent to Kew a drawing, photographs and branches in fluid of plants collected near Scottsburg and at Umzumbi, in Natal, which are in every way absolutely identical with the plant introduced by Mr. Cooper, who probably got the plant from near the same locality, as he was in Durban, Natal, in 1862. In Miller's time, however, Natal was an unexplored land, and the plant he described could not have come from that country, and cannot be the same species, for these plants are mostly very local in their range. The same remark also applies to the plant figured by Burmann upon which the name *E. pugniformis* was established by Boissier, which also differs from the Natal plant in having whitish-green flowers and very different styles. Dr. Wood states that a living plant found near Scottsburg or the Umkomaas River and taken to Pretoria and there planted on the rockery near the Botanical Laboratory in Aug., 1913, had by April, 1914, completely changed its appearance. When first planted at Pretoria it was quite normal, and a drawing of it was then made, which shows the plant to have had between 30 and 40 branches, varying from 1–3 in. long, arranged in about 3 series. When Dr. Wood visited Pretoria eight months later, the plant then bore 140 branches in many series, of which the inner were 4–6 and the outer 9–14½ in. long. This luxuriant growth being doubtless due to change of soil, climate and elevation.

97. **E. Flanagani** (N. E. Br.); very dwarf, succulent, spineless; body of the plant apparently subcylindric or cylindric-obconic and only rising 1 or 2 in. above the ground, 1½–2 in. thick, and with a crown of 3 or 4 series of branches around the flattened tuberculate top; branches erect or ascending, ½–1½ in. long, when dried 2½–3 lin. thick, with somewhat tooth-like tubercles, glabrous; leaves 3–5 lin. long, ⅓–½ lin. broad, linear, acute, channelled down the face, erect, glabrous; peduncles very numerous, nearly covering the top of the plant inside the branches, about 2 lin. long, bearing 1 involucre and a pair of linear-oblong bracts 1¼–1½ lin. long, ciliate at the apex; involucre ¼ in. in diam., shallowly and broadly cup-shaped, glabrous, with 4–5 glands and 5 subquadrate or transverse denticulate lobes; glands distant, spreading, ¾–1¼ lin. in their greater diam., transversely oblong, subentire or minutely and irregularly toothed on the outer margin, apparently yellow; ovary sessile, puberulous; styles 1 lin. long, united into a column for half their length, with broadly cuneate 2-lobed spreading arms, glabrous; capsule and seeds not seen.

COAST REGION: Komgha Div.; near Keimouth, *Flanagan*, 1800!

This is readily distinguished from *E. Woodii*, N. E. Br., by its much shorter branches and hairy ovary.

98. E. Franksiæ (N. E. Br.) ; dwarf, succulent, spineless ; main body of the plant buried in the ground, cylindric or cylindric-obconic, 1⅓–1¾ in. thick, covered with small broadly conical closely placed tubercles about ¾–1 lin. prominent on the upper part, passing into the root below ; branches in about 3 series around the top (not at the centre) of the plant, erect or ascending, ¾–2 in. long, 2½–3 lin. thick when alive, ¾–1½ lin. thick when dried, cylindric, covered with rhomboid-oblong tubercles about 2 lin. long, 1¼–1½ lin. broad and ½ lin. prominent, or under cultivation, some-times without distinct tubercles, glabrous, green, not glaucous ; leaves 1–3 lin. long, linear or the smaller oblong-ovate, acute, more or less channelled down the face, glabrous ; peduncles arising from the central part of the top of the main body of the plant inside the crown of the branches, 4–9 lin. long, erect, bearing 3–4 bracts and a solitary involucre, glabrous, deciduous ; bracts 1½–1¾ lin. long, 1–1½ lin. broad, oblong or subquadrate, very obtuse, ciliate, concave, submembranous, deciduous ; involucre 3½–4½ lin. in diam., broadly cup-shaped, glabrous outside and within, with 5 glands and 5 broadly rounded ciliate lobes ; glands not contiguous, horizontally spreading, flat, 1¼–1½ lin. in their greater diam., transversely elliptic-oblong, or very shortly and often somewhat crenately toothed or subentire on the outer margin, dull greenish-yellow ; ovary sessile, with a few thinly scattered straight hairs or almost glabrous ; styles united into a column ½–¾ lin. long, with stout cuneate entire or crenate arms ½ lin. long and ½–¾ lin. broad ; ripe capsule and seeds not seen.

EASTERN REGION : Natal ; Camperdown, 2000 ft., *Miss Franks in Herb. Wood,* 11727 !

This differs from *E. Woodii,* N. E.. Br., in its longer peduncles, and flat horizontally spreading involucral glands, which are different in colour, have much shorter teeth on their outer margin, and are not notched on the inner margin nor longitudinally grooved.

99. E. Woodii (N. E. Br.) ; succulent, spineless ; body of the plant buried in the ground, 3–6 in. long, 1⅓–3 in. thick, cylindric or ovoid-cylindric, truncate at the top, with the central area covered with conical acute tubercles and the margin bearing a crown of 3–4 series of branches, glabrous, dull green ; branches at first erect, finally radiately spreading, 1–3½ in. long, 3–3½ (or when dried 1–2) lin. thick, cylindric or slightly tapering to the apex, scarcely tuber-culate, but marked by depressed lines into rhomboid or elongated areas 1½–4 lin. long and 1–1½ lin. broad, each bearing a leaf or leaf-scar close to the upper end, glabrous, green or dull purple where exposed to the sun ; leaves 1–3 lin. long, linear or linear-lanceolate, acute, fleshy, spreading, glabrous, persisting for some time ; peduncles usually numerous, solitary from the axils of the outer tubercles of the central area of the stem or from the youngest branches, 1–3½ lin. long, bearing about 5 bracts and 1 involucre, glabrous ; bracts 1–1½ lin. long, oblong, ovate, obovate, or suborbicular, concave, submembranous, obtuse or acute, glabrous, ciliate ; involucre 3–4 lin. in diam., broadly and rather shallowly cup-shaped, glabrous outside, with rather woolly stamens inside, with 5 glands and 5 transverse subrectangular toothed or ciliate lobes ; glands separate or contiguous, deflexed-spreading, convex, often with a longitudinal groove from the upper margin, $\frac{3}{4}$–1$\frac{1}{2}$ lin. in their greater diam., suborbicular or transversely elliptic or elliptic-oblong, with 5–10 minute but distinct often recurved teeth on their outer margin and the inner margin entire or slightly notched, dull yellow; capsule sessile, 3–3$\frac{1}{2}$ lin. in diam. and rather shorter than broad, obtusely 3-lobed, glabrous ; styles united into a column, $\frac{1}{2}$–1 lin. long, with broad cuneate spreading free or connate arms $\frac{1}{3}$–$\frac{2}{3}$ lin. long, $\frac{1}{2}$–$\frac{3}{4}$ lin. broad, entire, channelled down the middle ; seeds 1$\frac{3}{4}$ lin. long, elliptic-ovoid, acute at one end, truncate at the other, with a slight keel on each side and one down the back, slightly subreticulate-rugulose on the dorsal and nearly smooth on the ventral side.

EASTERN REGION : Natal ; Clairmont Flat and Durban Flat, 10–50 ft., *Wood*, 4090! 11803! Umgeni, *Wood*, 12612!

100. **E. discreta** (N. E. Br.) ; very dwarf, succulent, spineless ; body of the plant globose, in the dried specimens seen about 1 in. in diam., but probably attaining to a larger size, laxly covered to the centre with numerous ascending and spreading branches and between their bases marked out into large hexagonal areas or flattish tubercles, glabrous ; branches 1–2$\frac{1}{2}$ in. long and 1$\frac{1}{2}$–5 lin. thick when dried, cylindric, covered with rhomboid tubercles 1$\frac{1}{4}$–3 lin. long, 1$\frac{1}{4}$–2 lin. broad and $\frac{1}{2}$–$\frac{2}{3}$ lin. prominent, glabrous ; leaves rudimentary, but present along the whole length of the branches, $\frac{2}{3}$–1$\frac{1}{4}$ lin. long, oblong or oblong-lanceolate, acute, spreading, glabrous ; peduncles solitary in the axils of the tubercles at the tips of the branches, 1$\frac{1}{2}$–2$\frac{1}{4}$ lin. long, bearing 1 involucre and 6–7 membranous oblong ciliate concave bracts $\frac{3}{4}$–1 lin. long and $\frac{1}{2}$ lin. broad ; involucre about 3$\frac{1}{2}$ lin. in diam., broadly cup-shaped, glabrous, with 5 glands and 5 subquadrate ciliate lobes ; glands subcontiguous, reflexed, 1$\frac{1}{4}$–1$\frac{3}{4}$ lin. in their greater diam., transversely elliptic-oblong, with 7–8 very small teeth or crenulations on the outer margins ; ovary sessile, very thinly pubescent with a few spreading and apparently deciduous hairs ; styles free to the base, $\frac{1}{2}$ lin. long, stout, ascending-spreading, with dilated slightly notched tips; capsule and seeds not seen.

EASTERN REGION : Pondoland ; on the banks of the Umzimkulu River near the sea shore, *Bachmann*, 757 !

This has some affinity to *E. Woodii*, N. E. Br., but the body of the plant is much smaller, the leaves and styles are much shorter, and the glands of the involucre appear to be more deflexed.

101. **E. Huttonæ** (N. E. Br.) ; succulent, spineless, leafless, probably similar in habit to *E. Caput-Medusæ* or *E. esculenta*, but only

4 branches have been seen, ¾–2 in. long, 4–5 lin. thick, cylindric, tessellately tuberculate, glabrous, simple or slightly branching near the apex; tubercles 2–2½ lin. long, 1¼–2½ lin. broad and scarcely or but slightly prominent, rhomboid or subhexagonal, marked with a white leaf-scar; leaves minute, rudimentary, soon deciduous; peduncles few on each branch, solitary in the axils of the tubercles at the tips of the branches, about 1½ lin. long and 1 lin. thick, bearing about 5 bracts and 1 involucre; bracts subquadrate or sub-orbicular, about ¾ lin. in diam., entire, glabrous, deciduous; involucre about ¼ in. in diam., broadly and rather shallowly cup-shaped, glabrous, with 5 glands and 5 rather short transversely oblong toothed lobes; glands contiguous or nearly so, apparently deflexed, 1–1¼ lin. in their greater diam., transverse, rounded on the inner margin, truncate and toothed or subentire on the outer margin, but not produced into processes, apparently greenish or yellowish-green; ovary sessile, glabrous; styles united to the apex into a column nearly 1 lin. long, with spreading cuneate subtruncate stigmas; capsule and seeds not seen.

COAST REGION : Albany Div. ; Carlisle Bridge, *Mrs. Hutton* !

102. **E. brevirama** (N. E. Br.); very dwarf, succulent, spineless, but with persistent remains of the peduncles on its branches; body of the plant in the specimen seen 2 in. in diam., obconic, flat at the top and there marked out by depressed lines into octagonal flattened areas with a very short point at their centre on the outer part, gradually passing into much smaller hexagonal and very shortly and obliquely conical tubercles at the central part, producing around the outer part of the top about 3 series of very short branches, glabrous, fleshy, dull green ; branches arising between the octagonal areas, about 2–4 lin. apart, the outer or longest about ⅓ in. long, the others shorter, cylindric and 2½–3 lin. thick, but dilated and stouter at the base, obtuse, covered with small convex hexagonal tubercles about 1 lin. in diam., finally withering and deciduous or bending down on the main body ; leaves very rudimentary and soon deciduous, ¼ lin. long, ovate, acute ; peduncles 2–4 to a branch in a season, arising from the axils of the tubercles at the ends of the branches, spreading, 2–5 lin. long, withering and persisting, bearing 3–4 minute ciliate deciduous bracts near or at the middle, distant from the solitary involucre, glabrous, green ; involucre 1½–1¾ lin. in diam. and about as deep, campanulate, glabrous outside, dull reddish, with 5 glands and 5 rounded minutely puberulous lobes ; glands distant, about ½ lin. in diam., erect, circular as seen from above, entire, fleshy, with a rather deep central cavity, green ; ovary sessile, glabrous ; styles united into a column ¾–1¼ lin. long, with revolute arms ⅓–1 lin. long, according to maturity.

COAST REGION : Jansenville Div. ; near Klipplaat, *Schönland,* 1716 !

Described from a living plant sent to Kew by Dr. Schönland in Oct., 1912.

103. E. hypogæa (Marloth in Trans. Roy. Soc. S. Afr. ii. 37, t. 1, figs. 2–3); plant succulent, very dwarf, with the large fleshy oblong or turnip-shaped (?) main body and primary branches buried in the ground and only the branchlets appearing above the surface; branchlets $\frac{3}{4}$–2 in. long and (including the tubercles) 5–10 lin. thick, clavate, cylindric-clavate or sometimes cylindric nearly to the base, not angular, covered with spreading or recurved cylindric-conical tubercles $1\frac{1}{2}$–$2\frac{1}{2}$ lin. long and $\frac{3}{4}$–1 lin. thick at their base, glabrous; leaves rudimentary, soon deciduous, $1\frac{1}{2}$–2 lin. long, $\frac{1}{3}$ lin. broad, ascending-recurved, linear, obtuse, longitudinally folded, with the margins touching, glabrous; peduncles 1–3 to a branch, solitary in the axils of the tubercles at their tips, 4–5 lin. long, bearing 1 involucre and 3–4 prominent scars of fallen bracts, glabrous, persisting; involucre about $\frac{1}{4}$ in. in diam., campanulate, glabrous, with the cup part greenish-red or purple and with 5 dark green glands and 5 transversely oblong fringe-toothed whitish lobes (*Marloth*); glands distant, variable, sometimes about 1 lin. long and as much in breadth and somewhat circular in general outline, with a rather deeply concave notch forming 2 short teeth or horns on the outer margin, and the upper surface slightly convex or nearly flat, at others with the entire or glandular part transversely oblong or somewhat crescent-shaped and flattish or slightly concave, with 2–3 filiform processes $\frac{1}{3}$–$\frac{1}{2}$ lin. long on the outer margin; ovary and capsule quite sessile, the former quite included in the involucre; capsule $1\frac{3}{4}$ lin. in diam., obtusely 3-angled, glabrous; styles united into a column 1–$1\frac{1}{2}$ lin. long, with spreading arms $\frac{2}{3}$–$\frac{3}{4}$ lin. long; thickened at the apex; seeds not seen.

CENTRAL REGION: Beaufort West Div.; in clayey soil on the Nieuwveld Mountains, near Beaufort West Div., *Marloth*, 4692! Victoria West Div.; without precise locality, *Armstrong*!

Described from Dr. Marloth's type specimen, dried and in fluid. The glands of the involucre are very variable on different specimens.

104. E. namibensis (Marloth in Trans. Roy. Soc. S. Afr. i. 318 and 317, fig. 3); dwarf, succulent, glabrous; main stem subglobose or ovoid, 3–8 in. high, 3–6 in. in diam., partly buried in the ground, bearing numerous erect or spreading branches about $\frac{1}{4}$–$\frac{1}{2}$ in. apart all over the upper half, naked below, whitish (*Marloth*), whitish-brown when dried; branches $\frac{3}{4}$–2 in. long, 5–6 lin. thick at the base, cylindric or slightly narrowing to the apex, spirally tuberculate; ultimately deciduous; tubercles rhomboid, $1\frac{1}{4}$–3 lin. long, 1–$1\frac{3}{4}$ lin. broad and $\frac{1}{2}$–$\frac{3}{4}$ lin. prominent; leaves $\frac{2}{3}$–$1\frac{1}{2}$ in. long, linear, acute, longitudinally folded and in that condition $\frac{1}{3}$–$\frac{2}{3}$ lin. broad, glabrous, somewhat fleshy, only present on the growing branches, soon deciduous; peduncles solitary in the axils of the tubercles, more or less clustered at the tips of the branches, 1–$2\frac{1}{2}$ (in fruit 3–5) lin. long and bearing 1 involucre, withering and persisting; involucre $2\frac{1}{2}$–$3\frac{1}{4}$ lin. in diam., cup-shaped, glabrous, with 5–6 glands and 5–6 transversely oblong fringed lobes; glands transverse, 2-lipped, with 2–5 simple or forked processes $\frac{2}{3}$–$\frac{3}{4}$ lin. long on the outer margin, apparently green ; ovary not exserted from the involucre ; capsule on a pedicel about as long as the involucre or shortly exserted from it, $\frac{1}{4}$ in. in diam., obtusely 3-angled, glabrous ; styles $\frac{1}{2}$–$\frac{3}{4}$ lin. long, united to the middle, with recurved-spreading rather stout bifid tips ; seeds 2 lin. long, ovoid, acute with a few slight scattered tubercles, brown.

WESTERN REGION : Great Namaqualand ; Namib desert near Tschaukaib, about 31 miles from Angra Pequena, *Marloth*, 4635 ! *Schultze* ! Gorup (Garub), *Pearson*, 4459 ! Schakalskuppe, *Pearson*, 4161 !

105. **E. fusca** (Marloth in Trans. Roy. Soc. S. Afr. ii. 38) ; very dwarf, succulent, leafless and spineless, resembling *E. Caput-Medusæ* ; main stem globose, up to 6 in. in diam., tuberculate, thickly covered (except at the apex) with numerous radiating and ascending branches, glabrous ; branches $\frac{3}{4}$–2 in. long, $1\frac{1}{2}$–5 lin. thick, cylindric, tessellately tuberculate ; tubercles not very prominent, rhomboid-hexagonal, with a small whitish leaf-scar ; leaves rudimentary, soon deciduous ; peduncles clustered at the tips of the branches, $2\frac{1}{2}$–4 lin. long, bearing a few bracts and 1 involucre, glabrous, withering and persisting ; upper pair of bracts 1 lin. long, spathulate-obovate, ciliate ; involucre 3–$3\frac{1}{2}$ lin. in diam., very shallowly cup-shaped, scarcely 1 lin. deep, glabrous, with 5–6 glands and 5–6 transversely oblong fringed and ciliate lobes ; glands not contiguous, 1–$1\frac{1}{4}$ lin. in their greater diam., transversely oblong or elliptic-oblong, with 5–7 subulate processes $\frac{1}{2}$–$\frac{3}{4}$ lin. long on their outer margin, brown ; stamens shortly hairy below the articulation ; ovary and capsule sessile, covered with rather long spreading hairs ; styles $\frac{2}{3}$ lin. long, united for half their length into a stout 3-grooved column, with stout spreading minutely 2-lobed tips ; ripe capsule and seeds not seen.

CENTRAL REGION : Britstown Div. ; near Britstown, and Steynsburg Div. ; near Steynsburg, *Marloth*, 4682 !

KALAHARI REGION : Griqualand West ; near Kimberley, *Marloth*, 4682 !

Described from excellent photographs and a dried flowering branch of the type and flowers preserved in fluid, lent to Kew by Dr. Marloth, who states in his original description that the peduncles are "non-persistent." This, however, is an error, for they certainly persist for two or more seasons.

106. **E. arida** (N. E. Br.) ; dwarf, succulent, spineless, but with persistent withered peduncles ; main body of the plant $1\frac{3}{4}$–2 in. thick, cylindric, rising at least 2 in. above the ground and covered for that distance with numerous erect or spreading branches 1–$1\frac{1}{2}$ in. long, 4–5 lin. thick, dilated to 6 lin. at the very base and slightly tapering upwards, cylindric, tuberculate, finally deciduous, dull dark green, at the upper part of the main stem separated from one another by flattish or slightly convex 5–6-angled areas about 4 lin. in diam., with a slightly raised point at their centre, at the lower part of the stem becoming transverse and much broader

than long, ultimately forming narrow spaces between the transverse scars of the fallen branches; tubercles on the branches $2\frac{1}{2}$–$3\frac{1}{2}$ lin. long, $1\frac{1}{4}$–$1\frac{1}{2}$ lin. broad, $\frac{1}{2}$ lin. prominent, elongated rhomboid, slightly produced and tipped with a white leaf-scar at the upper part; leaves 1–$1\frac{1}{2}$ lin. long, linear or linear-lanceolate, subobtuse, fleshy, deeply channelled down the face, glabrous, dull green; peduncles arising in the axils of the tubercles at the tips of the branches, 3–4 lin. long, bearing 4–5 linear or linear-spathulate obtuse ciliate deciduous bracts round the base of the solitary involucre, dull green; involucre 2 lin. in diam., campanulate, glabrous outside, dull green, dusted with reddish, with 5 glands and 5 subquadrate toothed ciliate lobes; glands not contiguous, $\frac{2}{3}$–$\frac{3}{4}$ lin. in their greater diam., transversely elliptic-oblong, convex above when fresh, becoming concave when dried, with 2–4 short teeth or processes on their outer margin, dull olive-green; ovary sessile, very thinly sprinkled with hairs; styles united to the top into a column $1\frac{1}{3}$–$1\frac{1}{2}$ lin. long, with knob-like radiating stigmas about $\frac{1}{5}$ lin. long.

CENTRAL REGION: Britstown Div.; near De Aar, *Schönland*!

Described from a living plant sent by Dr. S. Schönland to Kew in 1911, where it flowered in June, 1913. It is evidently nearly allied to *E. fusca*, Marloth, but the main stem is more cylindric, the glands are not brown, but dull olive-green, and the styles are twice as long, and united nearly to their apex into a much more slender column.

A specimen collected at Vesta, near De Aar (*Hopper*, 4409), is probably the same species, but I have not seen flowers of it.

107. E. decepta (N. E. Br.); very dwarf, succulent, leafless and spineless, but the branches are beset with the rigid remains of the peduncles, not spine-tipped; body of plant globose, 3–4 in. in diam., partly buried in the ground, marked by depressed lines into irregular hexagonal flattened areas about $\frac{1}{2}$ in. in diam., with a small compressed-conical (as if pinched between the thumb and finger) tubercle at the centre of each; from between these areas arise all over the top and sides of the body numerous ascending and spreading cylindric branches $\frac{1}{2}$–$1\frac{1}{2}$ in. long and mostly about $\frac{1}{3}$ in. thick and $\frac{1}{4}$–$\frac{1}{2}$ in. apart, covered with spirally arranged slightly prominent rhomboid tubercles $1\frac{1}{2}$–$2\frac{1}{4}$ lin. in diam.; body and branches glabrous, dull olive-grey-green, with a somewhat brownish tint; leaves very rudimentary and soon deciduous, $\frac{1}{4}$–$\frac{3}{4}$ lin. long, ovate or deltoid-ovate, obtuse, fleshy; peduncles solitary in the axils of the tubercles of the branches, 2–6 lin. long, bearing 1 involucre and about 4 very small scale-like deciduous bracts, glabrous, withering, becoming rigid, but not spine-pointed, persisting; involucre $2\frac{1}{2}$–3 lin. in diam. and $1\frac{1}{2}$ lin. deep, cup-shaped, glabrous and green tinged with purple outside, with a tuft of hairs opposite the glands within, with 5 glands and 5 transversely oblong ciliate lobes pubescent on the back; glands distant, $\frac{3}{4}$–$1\frac{1}{4}$ lin. in their greater diam., transversely oblong or elliptic-oblong, convex, with 3–6 short or minute teeth on the outer margin, bright dark green; ovary sessile, included in the involucre, glabrous; styles united into a stout column 1–1¼ lin. long, longitudinally grooved, with stout recurved-spreading stigmas ½ lin. long, exserted. *E. Caput-Medusæ, E. Meyer in Drège, Zwei Pfl. Documente,* 184, name only, not of *Linn.*

CENTRAL REGION: Beaufort West Div. ; Willowmore side, *Brauns,* 1712 !

A specimen collected by *Burke* on the south side of the Snowy Mountains, in Richmond Div. and another collected on the Camdeboo Mountains in Albert Div. by *Drège* (871), may belong here.

Described from a living plant collected by Dr. Brauns and sent to Kew by Dr. Schönland in September, 1911. The stem and branches are so much like dry ground in colour, that at a short distance they would not be easily detected. In appearance this species is so extremely like *E. namibensis,* Marloth, that when out of flower it might easily be mistaken for that species. The odour of the milky juice is very disagreeable.

108. **E. crassipes** (Marloth in Trans. Roy. Soc. S. Afr. i. 318 and 317, fig. 4) ; main body of the plant 4–6 in. long and thick, often half buried in the ground, globose-cylindric, flattened at the apex and bearing numerous fleshy branches 1½–2½ in. long and 5–7 lin. thick, forming a rosette 6–8 in. in diam., the lower branches gradually shrivel and expose the bare stem ; leaves not described ; peduncles 1½–5 lin. long, bearing 1 involucre and several bracts, moderately stout, glabrous, persistent and hardening but not spine-tipped (*Marloth*) ; bracts about 1 lin. long, spathulate-lanceolate, incurved and somewhat concave, ciliate, deciduous ; involucre 2–2¼ lin. in diam., cup-shaped, glabrous, with 5 glands and 5 transversely subrectangular fringed lobes ; glands not contiguous, horizontally spreading, transversely elliptic, concave, with 3–5 subulate processes ¼ lin. long or less on their outer margin ; ovary sessile, included in the involucre, glabrous ; styles exserted, 1 lin. long, united for half their length, with spreading channelled but not bifid tips.

CENTRAL REGION: Beaufort West Div. ; in stony ground of the Karroo, near Beaufort West, *Marloth,* 4397, partly !

The above description as to the plant itself is from Dr. Marloth's original description, but all relating to the peduncles and flowers is from fresh material from the type preserved in alcohol. But it is evident from the material in Dr. Marloth's Herbarium, that some error has been made with regard to the localities given for this plant at the original place of publication. The type sheet in Dr. Marloth's Herbarium contains, firstly, a very much enlarged drawing without name or label of a flowering branchlet of *E. crassipes,* of which a reduced facsimile is reproduced in the figure quoted above. This drawing quite agrees with the flowers in fluid kindly lent by Dr. Marloth. Secondly, a dried specimen and a photograph of it, labelled "*Euphorbia crassipes,* Marloth. Prince Albert. No. 4397." This specimen differs totally from the description in the main body being cylindric and not flat-topped, but covered to the centre with erect or ascending branches, those at the sides being spreading. The peduncles are nearly twice as long and far more slender than those of *E. crassipes,* but probably the measurements "15–20 mm." given for the peduncles of *E. crassipes* by Dr. Marloth are taken from this specimen, which I describe below as *E. albertensis,* N. E. Br., and have assumed that the Beaufort West locality is the only one that rightly belongs to *E. crassipes.*

109. E. rudis (N. E. Br.); plant forming a hemispherical mass, spineless, but with the hardened remains of peduncles on the branches; rootstock or main stem more or less buried in the ground, thick and fleshy, from which arises a mass of numerous branches, apparently also partly buried in the ground with age, glabrous; dried branches 1–6 in. long, $\frac{1}{3}$–$\frac{2}{3}$ in. thick, stouter when alive, usually branching at or near the apex, cylindric, covered with small crowded conical rhomboid tubercles about 2 lin. long, 1–1$\frac{1}{2}$ lin. broad and 1 lin. prominent, spreading or with a slight tendency to recurve at the hardened whitish leaf-scarred apex perhaps glaucous when alive, as dried branches are covered with a thin whitish waxy layer; leaves in a small tuft at the very apex of the growing tips, soon deciduous, recurved-spreading, $\frac{1}{4}$–$\frac{1}{3}$ in. long, linear-lanceolate, subacute, longitudinally folded, with more or less wavy margins, glabrous; peduncles solitary in the axils of the tubercles at the tips of the branches, $\frac{1}{3}$–$\frac{3}{4}$ in. long, simple or once or twice forked, glabrous; bracts alternate, all scattered on the peduncle some distance below the involucre, none immediately under it, 1–1$\frac{1}{2}$ lin. long, oblong-linear, concave, deflexed-spreading; involucre 1$\frac{1}{2}$–2$\frac{1}{2}$ lin. in diam., cup-shaped, glabrous, with 4–6 glands and 4–6 subquadrate or transversely oblong fringe-toothed lobes; glands distant, $\frac{1}{2}$–1 lin. in their greater diam., shortly stalked or subsessile, transverse, rather deeply concave or somewhat 2-lipped, with usually 2 (sometimes 3–4) subulate horns arising from the back of the outer margin; capsule sessile, erect, 3–3$\frac{1}{2}$ lin. in diam., shallowly and very obtusely 3-lobed, with a few hairs on its upper part, otherwise glabrous, glaucous; styles united into a column $\frac{3}{4}$ lin. long, with ascending-spreading arms $\frac{1}{4}$–$\frac{1}{3}$ lin. long, entire and not thickened at the apex; seeds 1$\frac{3}{4}$ lin. long, oblong or slightly conical-oblong, truncate at one end, abruptly acute at the other, with an acute ridge down the back and one on each margin with a short ridge between at the base on the dorsal side, and 2 subcontiguous ridges composed of 2 pairs of short ridges on the ventral side, whitish or pale greyish-white.

WESTERN REGION : Great Namaqualand ; near Gabis, 2800 ft., *Pearson*, 4310 ! between Dabaigabis and Gabis, 2500 ft., *Pearson*, 4369 ! on stony slopes at Grundoorn, 3300 ft., *Pearson*, 4360 ! Great Karas Berg, at Kuibis, *Pearson*, 8014 ! and common on sandy plains north-east of Naruda Sud, *Pearson*, 8141 !

110. E. inelegans (N. E. Br.); dwarf, succulent, spineless and leafless; body of the plant globose, up to 3$\frac{1}{2}$ in. or perhaps more in diam., bearing about 5 series of closely placed ascending branches forming a crown around a depressed central area 1$\frac{1}{4}$–1$\frac{1}{2}$ in. in diam., which is covered with acute conical tubercles but destitute of branches, glabrous, olive-brown; branches all erect and attaining to about the same level, none spreading, the outermost series about 3 in. long, the others gradually shorter, 5 lin. thick, cylindric, densely tuberculate, somewhat dull green or greyish-green, tinged with purple where exposed to the sun; tubercles rhomboid, 1$\frac{1}{2}$–4 lin. long, 1–2 lin. broad and near their apex $\frac{1}{3}$–$\frac{2}{3}$ lin. prominent, tipped with a white leaf-scar; leaves very rudimentary, $\frac{3}{4}$ lin. long, broadly ovate, acute, soon deciduous; peduncles arising in the axils of the tubercles at the tips of the branches, erect or ascending, 4–9 lin. long, $\frac{2}{3}$ lin. thick, bearing 3–5 bracts and 1 involucre, glabrous, withering and persisting; bracts all alternate, the upper-most a little below the base of the involucre, scale-like, thin, soon deciduous, about 1 lin. long, oblong or spathulate-obovate, concave, ciliate, green or reddish-brown; involucres $3\frac{1}{2}$ lin. in diam., broadly obconic-cup-shaped, with 5 glands and 5 transversely rectangular fringe-toothed pubescent lobes, otherwise glabrous, green dotted with red; glands distant, spreading, $1\frac{1}{4}$–$1\frac{3}{4}$ lin. across their tips, shortly stalked, transversely oblong or very broadly wedge-shaped, flat or slightly convex, fleshy, with 2–4 subulate teeth, $\frac{1}{4}$–$\frac{2}{3}$ lin. long, along their outer margin, olive-green above, reddish beneath; stamens woolly below the articulation; capsule sessile, erect, 3–$3\frac{1}{2}$ lin. in diam., obtusely trigonous, thinly covered with rather long spreading woolly hairs; styles united into a column $\frac{1}{3}$–$\frac{1}{2}$ lin. long, with arms $\frac{2}{3}$–$\frac{3}{4}$ lin. long, at first erect, finally spreading, with revolute dilated or obovate entire or slightly notched channelled tips; seeds $1\frac{3}{4}$ lin. long, oblong-ovoid, acute at one end, truncate at the ·other, slightly 4-angled, with the angles forming a slight shoulder below the point, sides very slightly rugose.

KALAHARI REGION: Griqualand West; near Kimberley, *Moran*! and in *Herb. Schönland*, 1718!

Described from a living plant sent to Kew by Dr. Schönland in October, 1912. The flowers, owing to their white spreading filaments, the white woolly hairs on the involucre and the dull olive-green of the glands, have, as a whole, a peculiar greyish appearance when viewed at a short distance.

111. **E. albertensis** (N. E. Br.); main body of the plant cylindric, in the dried specimen 4 in. long and about $1\frac{1}{2}$ in. thick, fleshy, scarcely tuberculate, bearing on the upper part, to the very centre, numerous fleshy branches with persistent remains of the peduncles upon them, glabrous; branches not crowded, but with distinct spaces between them, those at the apex erect or ascending, the others more or less spreading, about $\frac{3}{4}$ in. long and $1\frac{1}{2}$ lin. thick when dried, and according to a photograph 3–4 lin. thick when alive, cylindric, with small slightly prominent rhomboid tubercles; leaves rudimentary, $\frac{3}{4}$–$1\frac{1}{2}$ lin. long, linear-lanceolate, fleshy, chan-nelled or longitudinally folded, glabrous, soon deciduous; peduncles arising singly from the axils of the tubercles along the upper part of the branches, $\frac{1}{2}$–1 in. long, slender and not more than $\frac{1}{2}$ lin. thick at the base, straight, bearing at the apex 3–4 closely placed or somewhat whorled bracts and 1 involucre, glabrous; bracts $\frac{3}{4}$–1 lin. long, linear-spathulate, acute, incurved or concave, cilio-late; involucre on a distinct pedicel 1–$1\frac{1}{2}$ lin. long above the bracts and often making an angle with the peduncle, about 2–$2\frac{1}{4}$ lin. in diam., somewhat obconic-cup-shaped, glabrous, with 5 spreading

glands, which from a photograph appear to be transversely elliptic with 3–4 small teeth on their outer margin.

CENTRAL REGION : Prince Albert Div. ; near Prince Albert, *Marloth,* 4397 partly !

This plant is quoted by Marloth under *E. crassipes,* but is certainly distinct from the plant he describes by that name, and except perhaps in the length given of the pedicels in no way is represented by the description of that species ; the absence of the flat top to the stem at once excluding it from the group to which *E. crassipes* belongs.

112. E. filiflora (Marloth in Trans. Roy. Soc. S. Afr. iii. 123, t. 8, fig. 3) ; plant 8–12 in. high, succulent, spineless, but bearing the dried remains of long peduncles ; main stem slightly clavate or subcylindric, 3–4 in. thick, covered with conical slightly recurved tubercles and bearing along the upper part numerous short thick branches, glabrous ; branches ascending, 2–3¼ in. long, 4–9 lin. in diam., cylindric, covered with rhomboid straight or slightly recurved tubercles ; leaves in a small cluster at the apex of the branches, soon deciduous, ¾–1¼ in. long, linear ; peduncles solitary in the axils of the tubercles at the tips of the branches but withering and persisting on the older parts, erect, 2–3½ in. long, ½ lin. thick, glabrous, bearing 1 involucre and the scars of 3–5 fallen bracts ; involucre ¼–½ in. long and 3–5 lin. in diam., elongated-obconic, glabrous outside and within, with 5 glands and 5 erect oblong or rectangular lobes deeply toothed at their apex ; glands not contiguous, 2 lin. long and as much in breadth across the tips of the segments, divided to two-thirds of the way down into 3 diverging linear segments, bifid and recurving at their tips, greenish-yellow ; ovary (only seen in an immature condition) subsessile, glabrous ; styles united into a column 2½ (or perhaps more) lin. long, with shortly bifid arms 1 lin. long, not dilated nor thickened at the tips.

WESTERN REGION : Little Namaqualand ; near Concordia, *Krapohl in Herb. Marloth,* 5119 !

I have also seen a specimen belonging to the Albany Museum from a plant cultivated at Grahamstown, stated to have been originally sent from Matjesfontein in Laingsburg Div., by Dr. Purcell. I suspect, however, that some error has been made as to the locality.

113. E. brakdamensis (N. E. Br.) ; succulent, spineless, but with erect or ascending persistent remains of hardened peduncles, only some dried branches seen, 2½–5 in. long and 4–5 lin. thick including the tubercles, usually simple or occasionally sparingly branched, glabrous and apparently pale green or glaucous, covered with elongated rhomboid conical tubercles 2½–3 lin. long, 1 lin. broad and 1–1½ lin. prominent, slightly recurved ; leaves erect, ½–1 in. long, about ⅓ lin. broad, linear, obtuse or subacute, entire, with incurved margins or channelled down the face, fleshy, glabrous and perhaps somewhat glaucous ; peduncles erect or ascending, slightly curved, 1–2½ in. long, glabrous, bearing 4–6 bracts and 1 involucre at the apex; upper bracts forming a whorl close under the involucre, narrowly elliptic to suborbicular, 2–2½ lin. long, 1–2½ lin. broad, concave, rather thin, glabrous, ciliate, green, deciduous; involucre about 4½ lin. in diam. and 3 lin. deep including the lobes, campanulate, glabrous outside and with a few hairs opposite each gland within, with 5 glands and 5 transversely rectangular denticulate lobes ciliate on their margins; glands distant, 1½–2 lin. long to the tips of the processes, 1 lin. broad, broadly obovate or orbicular-obovate minutely pitted, with 2–4 abruptly reflexed linear entire processes ½ lin. long on the outer margin; ovary sessile, included, sprinkled with spreading hairs; styles 2½ lin. long, united to half-way up, with the thickened diverging tips, glabrous; capsule and seeds not seen.

WESTERN REGION : Little Namaqualand ; hills at Brackdam, 1600 ft., *Schlechter*, 11123 !

In appearance the branches of this species very closely resemble those of *E. filiflora*, Marloth, but the involucre is only about half as deep and the glands are different, the habit may also be distinct. I have seen numerous branches, but no main stem.

Specimens collected between Mamre and Saldanha Bay in Malmesbury Div., *Drège*, 8202a, represent a very closely allied species, but are flowerless.

114. E. namaquensis (N. E. Br.); plant 3–8 in. or perhaps sometimes up to 2 ft. (*Pearson*) high above ground, succulent, spineless; main stem or body of the plant 1½–2½ in. thick, obconic or cylindric, tapering into a long tough root that descends very deeply into the ground, fleshy, covered to the apex with very numerous rather crowded erect branches, glabrous, green; branches 1–5 in. long, 2–3½ lin. thick when dried, probably stouter when alive, simple or with short branchlets, sometimes cylindric and obtuse, sometimes elongated and tapering to a rather slender and somewhat spine-like point, tuberculate, glabrous; tubercles laxly spiral, rather scattered, ½–2 lin. prominent, cylindric-conical, spreading or recurved, often tipped with a hard whitish conical acute point; leaves only present on the growing tips, 4–8 lin. long, linear and channelled down the face or linear-terete from the margins being infolded, ½–⅔ lin. thick, acute, glabrous, sometimes (always?) glaucous or white in the channel or upper surface, soon deciduous; peduncles solitary in the axils of the tubercles at the tips or racemosely arranged along the branches, 2–6 lin. long, sometimes simple and bearing only 1 involucre, sometimes divided or branching into 2–3 secondary peduncles 3–5 lin. long, often resembling slender branchlets and more or less tuberculate, glabrous or the secondary peduncles thinly pubescent along the upper side; involucre 3½–4½ lin. in diam., broadly and rather shallowly cup-shaped, pubescent or glabrous outside, with 5 glands and 5 transversely quadrangular toothed or ciliate lobes; glands distant, spreading, ¾–1 lin. in their greater diam., shortly but distinctly stalked, 2-lipped, with a shallow transverse depression or cavity in front of the inner margin and the outer lip convex or in dried specimens sometimes infolded, divided on the outer margin into 2–4 simple or occasionally forked teeth or subulate processes $\frac{1}{3}$–1 lin. long, usually revolute; capsule sessile or subsessile, $4\frac{1}{2}$–5 lin. in diam., obtusely 3-lobed, but the lobing nearly evanescent at the basal part, tomentose or densely pubescent; styles united into a column $\frac{1}{2}$–1 lin. long, with spreading entire rather slender lobes $\frac{1}{3}$–$\frac{1}{2}$ lin. long, channelled down the face; seeds 2–2$\frac{1}{3}$ lin. long, ovoid to very broadly ovoid, abruptly pointed at one end, faintly tuberculate or almost smooth.

WESTERN REGION : Great Namaqualand ; plains north of and between Ganus and Grundoorn, 3300 ft., *Pearson*, 4367 ! 4502 ! between Noachebeb and Grundoorn, *Pearson*, 7809 ! Great Karasberg region, 4000 ft., *Pearson*, 8256 ! Little Namaqualand ; between Aggenys and Pella, *Pearson*, 2992 ! south of Tweefontein, 2700 ft., *Pearson*, 3049 ! Dabeep, *Pearson*, 6228 ! 6229 !

115. E. Braunsii (N. E. Br.) ;

very dwarf, succulent, leafless and spineless, but with hardened persistent remains of the peduncles (not spines) on the branches; plant, in the only two specimens seen, consisting of numerous branches densely crowded upon the upper part of an obconic central stem about 1$\frac{1}{2}$ in. thick and completely covering its apex, all attaining to the same level and forming an obconic mass (excluding the root) 2–3$\frac{1}{2}$ in. high (half of which is buried in the ground) and 3$\frac{1}{2}$–5 in. in diam., but perhaps attaining to larger dimensions ; branches all erect or suberect, crowded, simple or sparingly branched, 1–3 in. long, $\frac{2}{3}$–1$\frac{1}{2}$ in. thick, cylindric, thickening upwards, obtusely rounded at the apex, covered with flattish scarcely prominent 4–6-angled tubercles 2$\frac{1}{2}$–3 lin. in diam., each with a slightly prominent white leaf-scar, glabrous, dull grey-green ; peduncles solitary in the axils of the tubercles around the apex of the branches, at first about 1–1$\frac{1}{2}$ lin. long, but apparently developing to 3–8 lin. long, 1 lin. thick, glabrous, bearing 1 involucre and a few scattered bracts, withering and persisting, becoming hard and white, but not spine-tipped ; bracts $\frac{3}{4}$–1$\frac{3}{4}$ lin. long, oblong or linear-oblong, obtuse, ciliate with long fine white hairs, green, dusted with brown ; involucre about 2$\frac{1}{2}$–3 lin. in diam., cup-shaped, glabrous outside, green, dusted with brown, with 4–5 glands and 5 transversely oblong lobes ciliate with long white and somewhat cottony hairs ; glands distant, $\frac{1}{3}$–$\frac{3}{4}$ lin. in their greater diam., transversely elliptic or reniform, depressed or concave at the centre and with 1–4 filiform processes $\frac{1}{3}$–$\frac{1}{2}$ lin. long on their outer margin, dull green, with reddish-tinted processes ; ovary immature in the specimens seen, sessile, glabrous ; styles 1$\frac{1}{2}$ lin. long, united into a column for two-thirds of their length, entire and scarcely thickened at the tips.

CENTRAL REGION: Aberdeen Div. ; without precise locality, *Brauns* ! *Pillans* !

Described from living plants collected by Dr. H. Brauns and sent to Kew by Dr. S. Schönland.

116. E. baliola (N. E. Br.); succulent, spineless; main body of
the plant globose or subcylindric, in the specimen seen 4 in. high
and 3 in. in diam., covered with transversely diamond-shaped
tubercles about ½ in. in their greater diam. and ¼ in. prominent,
arranged in numerous crowded spirals, and formed by the persistent remains of the deciduous branches, at first grey, becoming
brown; branches covering the whole top of the plant, not absent
from the centre, erect or ascending, ¾–2 in. long, 2–3 lin. thick in
the dried specimen, probably twice as thick when alive, cylindric,
scarcely or not at all tuberculate, but marked out by impressed
lines into elongated areas 1–3 lin. long and ¾–1 lin. broad when
dried, scarcely or not at all prominent, marked at their apex with
a conspicuous white round leaf-scar, glabrous; leaves not seen;
peduncles 2–4 clustered at the apex of the branches, 1–3 lin. long,
bearing about 4 bracts and 1 involucre, glabrous, persisting and
withering; bracts 1 lin. long, ⅔–¾ lin. broad, obovate, concave,
thin, ciliate with rather long hairs; involucre 2–2½ lin. in diam.
and 1–1¼ lin. deep, shallowly cup-shaped, glabrous outside and
within, but apparently woolly within from being filled with very
woolly white stamens, with 5 glands and 5 transversely oblong
lobes ciliate with white woolly hairs; glands distant, apparently
deflexed-spreading, ¾–1 lin. in their greater diam., transversely
oblong, with a slight depression in front of the inner margin and
3–4 subulate teeth ¼–⅓ lin. long along the outer margin, dark
velvety brown; ovary included in the involucre, obconic, trigonous,
densely covered with white woolly hairs; styles ½ lin. long, very
shortly united at the base, stout, with entire spreading tips.

WESTERN REGION: Great Namaqualand; Great Karas Berg Range, on slopes
between Krai Kluft and Naruda Sud, 5200 ft., growing among stones, *Pearson*,
8005!

In this species the tubercles on the main body of the plant are formed from the
persistent bases of the branches, from which the remainder has withered and
fallen away; this character, so far as I am aware, separates this species from all
others at present known. From *E. fusca*, Marloth, it differs by the branches
covering the whole top of the plant to the centre and the filaments of the stamens
below the joint are twice as long and the hairs on them are much longer and more
woolly.

117. E. inermis (Mill. Gard. Dict. ed. viii. no. 13); succulent, spineless, leafless; body or main stem short and thick, producing a crown
of three or more series of crowded branches around the central
flattened or depressed tuberculate obconic area at the top, and not
rising much above ground level, tessellately tuberculate; branches
1½–10 in. long or under cultivation much longer, 5–6 lin. thick,
ascending or ascending-spreading, cylindric, tessellately tuberculate,
glabrous, dull green; tubercles rhomboid, 2½–5 lin. long, 1¾–2¾ lin.
broad and ¾ lin. prominent, shortly and obtusely conical, with a
small white leaf-scar; leaves minute, rudimentary, soon deciduous,
¼ lin. long and broad, ovate, acute; peduncles solitary in the axils
of the tubercles at the tips of the branches, 1½–2 lin. long, stout,
bearing about 4 bracts and 1 involucre, glabrous, sometimes per-

sistent; bracts ⅔–1 lin. long, scale-like, ovate or oblong, entire or ciliate; involucre 4–4½ lin. in diam. (2½–3 lin. when dried), cup-shaped, white, with 5 (rarely 4) glands and 5 transversely rect-angular ciliate lobes, glabrous on the cup outside and within, but pubescent on the back of the lobes and filled with woolly-white bracteoles; glands glabrous, variable on the same plant, 1¼–1½ lin. long, 1–1½ lin. broad across the tips, sometimes ovate-oblong and very shortly bifid, with the lobes denticulate at the apex, some-times divided to below the middle, with two diverging lobes cut like the horns of a reindeer, in both cases with the united glandular part dark green, revolute at the sides and the lobes white, some-times divided nearly or quite to the base into 2 much-branched lobes and entirely white, without any glandular dark green part; ovary sessile, woolly-white at the apex, thinly sprinkled with ascending or spreading hairs or rarely glabrous; styles united to the apex into a slender column 1½–2 lin. long; stigmas about ½ lin. long, more or less deeply bifid, with cuneate lobes, radiating and contiguous forming a sort of disc; capsule sessile, trigonous-subglobose, 2½ lin. in diam., glabrous or with a few hairs; seeds 1½ lin. long, ellipsoid, slightly 4-angled, subtruncate at the base, apiculate at the apex, very minutely tuberculate on the dorsal sides smooth on the ventral, dark brown, and (in the only example seen) with some whitish bodies (exudations?) at the angles, perhaps not constant. *E. viperina, Berger in Monatsschr. Kakt.* xii. 39, *and Sukk. Euphorb.* 114; *Hemsley in Bot. Mag. t.* 7971.

VAR. β, **laniglans** (N. E. Br.); glands of the involucre woolly, otherwise as in the type and perhaps only an abnormal form.

COAST REGION: Uitenhage Div.; Coega, *Rogers*! 115! Port Elizabeth Div.; Zwartkops, *Zeyher*, 1098! *Marloth*, 4872! 4897! Redhouse, near Port Elizabeth, *Mrs. Paterson*, 579! near Port Elizabeth, *Drège*! and *cultivated specimens*!
CENTRAL REGION: var. β, Jansenville Div.; near Klipplaat, *Marloth*, 5270!

The figure in the Botanical Magazine merely represents a rooted branch of this plant and gives no idea of its real habit, which is somewhat like that of *E. Caput-Medusæ.*

Described partly from a living plant sent to Kew by Mrs. T. V. Paterson, partly from dried material.

118. **E. Caput-Medusæ** (Linn. Sp. Pl. ed. i. 452, and Amœn. Acad. iii. 110); plant dwarf, succulent, subleafless and spineless, consisting of a globose main body up to 6 or 8 in. in diam., covered (except at the centre) with very numerous crowded branches, forming a hemispheric cushion, in adult plants 1½–2 ft. in diam., uniformly green; branches erect and 2–4 in. long at the central part, ascending-spreading or spreading and curving upwards at the circumference and 6–15 in. long, ⅔–1 in. thick, cylindric or some-what cylindric-clavate, very obtuse, covered with tubercles 2½–4 lin. long, 1½–2½ lin. broad and 1½–2 lin. prominent, obliquely conical, acute; leaves very small, soon deciduous, 1½–2¼ lin. long, ⅓–½ lin. broad, linear, acute or obtuse, thick and fleshy, fl at or slightly concave-channelled above, very convex or obtusely keeled

beneath ; flowers solitary in the axils of the tubercles, clustered at the apex of the branches; peduncles $\frac{1}{2}$–5 lin. long, stout, bearing 5–7 bracts and 1 involucre, glabrous; bracts about $\frac{1}{8}$ in. long, broadly ovate or elliptic, obtuse, ciliate ; involucre $\frac{1}{2}$ in. or rather more (or when dried 4–5 lin.) in diam., broadly and shallowly cup-shaped, with 5 glands and 5 transversely oblong or sub-quadrate connivent lobes, subtruncate and closely toothed at the top, glabrous on the back, pubescent on the inner face, red or purple, at least at their tips ; glands $1\frac{3}{4}$–2 lin. long and $1\frac{1}{4}$–3 lin. broad, palmately divided to half-way down into 3–6 linear segments entire or toothed at the tips, glabrous, with the undivided part green, minutely pitted, and the segments white ; ovary sessile, obtusely 3-angled, glabrous ; styles $1\frac{1}{4}$–$1\frac{1}{2}$ lin. long, united to the middle or nearly to the top, with erect shortly bifid tips, slightly exserted beyond the lobes of the involucre ; capsule and seed not seen. *Linn. Syst. ed.* 12, ii. 330; *Mill. Gard. Dict. ed.* viii. *no.* 7 ; *Ait. Hort. Kew. ed.* 1, ii. 135 *incl. all vars. but excluding the reference to Breyne, Ic. Rar. Pl.* 29, *t.* 19 ; *Lodd. Bot. Cab. t.* 1315 ; *Spreng. Syst. Veg.* iii. 787 ; *Boiss. in DC. Prodr.* xv. ii. 86 ; *K. Schumann in Monatsschr. Kakt.* viii. 54 *and* 53 *with fig.* ; *Berger in Mon-atsschr. Kakt.* ix. 91 *with fig., and Sukk. Euphorb.* 110, *fig.* 29 ; *Marloth in Wissensch. Ergebn. Deutsch. Tiefsee-Exped.* ii. *part* 3, *t.* 9 ; *N. E. Br. in Kew Bulletin,* 1912, 246, *with fig.   E. Fructus-Pini, Mill. Dict. ed.* viii. *no.* 10. *E. Medusæ, Thunb. Prodr.* ii. 86, *and Fl. Cap. ed. Schultes,* 404. *E. tessellata, Sweet, Hort. Suburb. Lond.* 107. *E. Fructus-Pini, var. geminata, Sweet, Hort. Brit. ed.* 1. 356. *E. Commelini, DC. Cat. Pl. Hort. Monspel.* 110 ; *Spreng. Syst. Veg.* iii. 787.   *Medusea Fructus-Pini, M. major and M. tessellata, Haw. Syn. Pl. Succ.* 134, 135, *and Klotzsch & Garcke in Abhandl. Akad. Berlin,* 1860, 61.—*Tithymalus aizoides Africanus simplici squamato caule, Commelin, Prælud. Bot.* 57, *fig.* 7. *Euphorbium procumbens ramis plurimis, &c. and E. procumbens ramis geminatis, &c., Burm. Rar. Afr. Pl.* 17, 18, *tt.* 8, 9. *Euphorbium anacanthum angusto Polygoni folio, as to vars.* 1 *and* 2 *only, Isnard in Mém. L'Acad. Roy. des Sciences, Paris,* 1720, 386, 387.

COAST REGION : Cape Div. ; mountains near Cape Town, *Thunberg* ! *Bergius* ! Kasteels Berg, *Zeyher,* 5082 ! shore near Green Point, *Pappe* ! mountain slopes near Sea Point, *Wilms,* 3624 ! *MacOwan, Herb. Austr.-Afr.,* 1530 ! *Dümmer,* 133 ! Lions Head, *Pillans* ! *Diels,* 119 !

When establishing this species, Linnæus included under it and the varieties he makes no less than five distinct species ; three are included in the references he gives under the species, apart from the varieties. His description is quite inadequate for identification and there is no specimen of it or the varieties in his Herbarium, but from his quotation of the *Hortus Cliffortianus* and *Hortus Upsaliensis,* he must have had in view a plant in cultivation. Therefore, as the cultivated plant would be likely to be one occurring near Cape Town, I have taken the only species of this group growing near that locality as being the true *E. Caput-Medusæ,* Linn. It is the plant that is certainly meant by three or four of the references given by Linnæus and also as understood by some later authors. I describe from a living plant about 18 inches in diameter collected on the slopes of Lions Head, near Cape Town and kindly sent to Kew especially for the purpose by Mr. Pillans of Cape Town. It flowers in July and August.

When grown from seed, a large globose body partly buried 'in the ground is developed, from which the numerous branches arise, as in all other species of this group. But when one of the branches is rooted, the globose main body is not formed, but it becomes a cylindric or clavate stem rising some inches or even up to 2 feet above the ground and produces branches at the top. It is upon branches rooted in this manner that *E. Commelinii*, DC., and *E. Caput-Medusæ*, var. *major*, Ait., were founded and supposed to represent distinct species, the habit being different. From the description given by Miller, Aiton, and Haworth of *E. Fructus-Pini* (*E. Caput-Medusæ*, var. *geminata*, Ait.), I can find no character to separate it from typical *E. Caput-Medusæ*, and I have not seen any specimen of it. Gardeners of the period seem to have distinguished it as the "Little Medusa's Head," whilst the typical *E. Caput-Medusæ* was known as "Medusa's Head"; so that it is probable that it was of a somewhat smaller growth than the type. Although Miller's description seems to have been made from a rooted branch, the plant when originally introduced was probably either only a small specimen of *E. Caput-Medusæ* or of a sex different from the plants then in cultivation. A young plant in cultivation at Kew in January, 1915, answered exactly to Miller's description of *E. Fructus-Pini*. In the 1st edition of Aiton's *Hortus Kewensis*, var. *minor* seems founded upon the plant chiefly intended by Isnard under the reference quoted by Aiton, which was certainly *E. Caput-Medusæ*. Whilst in the 2nd edition, under var. *minor*, Aiton only refers to the plant figured in De Candolle, *Plant. Grass.* t. 150, which is another species, see under *E. Bergeri*.

119. E. Bergeri (N. E. Br.); dwarf, succulent, spineless; main body of the plant subglobose or obconic, thick, fleshy, bearing a large number of radiating branches on the upper part, glabrous; branches 3–9 in. long, ⅓–¾ in. thick, usually simple, cylindric, tessellately tuberculate, often curved, glabrous, green; tubercles rhomboid, mostly 3–4 lin. long and about 2½ lin. broad, ½–1 lin. prominent; leaves 3–6 lin. long, ½–1 lin. broad, linear-spathulate, acute, channelled down the face, glabrous, deciduous; peduncles solitary in the axils of the tubercles at the ends of the branches, 2–5 lin. long, stout, bearing 1 involucre and about 4 bracts, sometimes persisting and thickening after the fall of the flower and becoming clavate or branch-like; involucre 2½–3½ lin. in diam., cup-shaped, with 5 glands and 5 transversely rectangular or subquadrate finely toothed lobes; glands spreading, 1–1½ lin. in their greater diam., transversely oblong, green, with 3–7 short or subulate greenish-white teeth on the outer margin; ovary included in the involucre, glabrous; styles about 1¼ lin. long, united for about half their length, with slightly spreading thickened cuneate stigmas, notched at the apex; capsule and seeds not seen. *E. Caput-Medusæ, Lam. Encycl.* ii. 416; *DC. Pl. Grass. t. 150, not of Linn. E. fructuspina, Sweet, Hort. Suburb. Lond.* 107. *E. Fructus-Pini, Sweet, Hort. Brit. ed.* i. 356, *excl. var. β, not of Miller. E. parvimamma, Berger in Monatsschr. Kakt.* ix. 92, *with fig., and Sukk. Euphorb.* 113, *fig.* 30, *not of Boiss.*

SOUTH AFRICA : without locality, *cultivated specimens !*

The above description is made partly from dried material, partly from a living specimen, both taken from the type plant of *E. parvimamma*, Berger, kindly sent to me by Mr. Alwin Berger. But it is certainly not the *E. parvimamma* of Boissier. *E. Bergeri* is only known from cultivated plants. It is very similar in appearance to *E. Caput-Medusæ*, Linn., but the branches are rather more slender and the involucre-glands quite different. It may possibly be a hybrid.

120. E. marlothiana (N. E. Br.); rootstock or main body of the plant subterranean, clavate or elongated-obconic, often subglobosely thickened at the apex, from which arises several erect or ascending succulent branches 3–15 in. long, these are at first about 3–5 lin. thick, covered with rhomboid or oblong flattish tubercles only ½–¾ lin. prominent, with age they get buried in the sand, become clavate like the main body and up to 1 in. (or more?) thick, and in turn produce slender branches at their apex in like manner; leaves 1–3 lin. long, oblong or linear, spreading, channelled, acute, with recurved tips, fleshy, glabrous, soon deciduous; peduncles 1–4 at a time to a branch, at the tips, withering and often persisting, but not becoming hard or woody, ½–2 in. long, erect, glabrous, bearing 1 involucre; bracts deciduous, not seen; involucre sessile between the bracts, 6–7 lin. in diam., broadly and rather shallowly bowl-shaped below the 5 glands, with 5 large erect subquadrate toothed purplish lobes 1¼ lin. long and broad, thinly puberulous outside; glands not contiguous, spreading, divided to slightly below the middle into 2–6 linear white (becoming reddish with age) processes dilated and slightly branched at their tips, with the undivided part broadly wedge-shaped, without a lip or turned-up margin at the base, 1–1¼ lin. long and about the same in breadth across the top, with revolute sides, green; ovary and capsule sessile, the former included in the involucre, subglobose, pubescent; styles united into a column about 1½ lin. long, with short arms dilated and 2 lobed at the tips; entire capsule and seeds not seen.

COAST REGION: Cape Div.; on sandhills, near Neu Eisleben, in the heart of the Cape Flats between Wynberg and Somerset Strand, about 10 miles from the nearest railway station, flowering in October and November, *Marloth*, 5733!

This species is allied to *E. Muirii*, N. E. Br., the two evidently having the same general habit. But *E. Marlothii* distinctly differs from *E. Muirii* in having fewer flowers to a branch, peduncles withering in a different manner, larger involucres and the glands have their undivided part as long as broad, without a turned-up lobule at the base, whilst in *E. Muirii* the undivided part is twice as broad as long, with a turned-up lobule at the base, and the gland is much shorter than in *E. Marlothii* and apparently different in colour.

121. E. Muirii (N. E. Br.); rootstock or main body of the plant buried in the ground, not seen; stems or branches 6–8 in. (or perhaps more) high, 3½–4½ lin. thick when dried, evidently much stouter when alive, erect, simple or with a whorl of 3–7 branches at or near the top, succulent, cylindric, covered with rhomboid tubercles 2½–6 lin. long, 1½–2 lin. broad and ½–1 lin. prominent, tipped with a white leaf-scar, spineless, but often with (sometimes without) some persistent peduncles, glabrous; leaves erect, 2½–5½ lin. long, ½–¾ lin. broad, linear, acute, fleshy, apparently channelled down the face and keeled on the back, glabrous; peduncles 4–11 lin. long, solitary in the axils of the tubercles, usually forming an umbel-like cluster at the tips of the branches, glabrous, bearing 4–6 bracts and 1 involucre; bracts 1½–2 lin. long, oblong to elliptic-obovate, concave, entire, glabrous outside, thinly pubescent within, ciliate; involucre 4–4½ lin. in diam., probably larger when alive, broadly cup-shaped, with 5 glands and 5 large transversely elliptic toothed lobes, thinly pubescent outside and in front of the glands and on the lobes within; glands 1⅓–2 lin. in diam., broadly cuneate, palmatifidly divided to about half-way down into 4–6 abruptly reflexed linear segments, dilated and entire or slightly toothed at their apices; undivided part entirely glandular, most minutely rugulose, not pitted, apparently yellowish-green, with the inner margin turned up, forming a small lip; segments apparently white; ovary sessile, included, glabrous; styles 1½ lin. long, united into a column nearly to the apex, with slightly spreading thickened subentire tips; capsule sessile, 3 lin. long and 4½ lin. in diam., very obtusely and shallowly 3-lobed as seen from above, glabrous; seeds 1¾–2 lin. long, ovoid, usually slightly keeled on the back, minutely scabrid-tuberculate all over, blackish-brown.

COAST REGION : Robertson Div.; Sand Berg, *Pearson*, 2261 ! Riversdale Div. ; Albertina, *Muir* ! and *Muir*, 174, from Stil Bay, probably also belongs here, but the specimen seen is without flowers or fruit.

Dr. Muir informed me the above-ground stems or branches of this species are solitary.

122. **E. tuberculatoides** (N. E. Br.); stems up to 18 in. long, ½–⅔ in. thick when dried, unbranched, cylindric, succulent, spineless, but with persistent hardened peduncles, and covered with rhomboid shortly conical tubercles 3–4 lin. long, 2–3 lin. broad and 1–1½ lin. prominent, glabrous; leaves erect, 4–5 lin. long, ⅓–½ lin. broad, linear, acute, channelled down the face, glabrous, deciduous; peduncles solitary in the axils of the tubercles, clustered at the apex of the stems, ¼–1¼ in. long, erect or ascending, glabrous, bearing 1 involucre ·and 4 deciduous bracts at the apex; bracts about 1½ lin. long and 1 lin. broad, obovate or orbicular-obovate, concave, glabrous, falling off as the involucre opens; involucre (dried) about 5 lin. in diam., obconic-cup-shaped, glabrous, with 5 glands and 5 |orbicular-subquadrate subentire or very minutely toothed lobes; glands shortly but distinctly stalked, 1½–1¾ lin. long, 1¾–2½ lin. broad, broadly cuneate, recurved at the sides and palmately divided to the middle into 3–5 linear segments, entire or bifid at their revolute tips, with the gland part apparently dark red? and segments white; ovary sessile, included, trigonous, glabrous; styles united into a column 2½–3 lin. long, with 3 thickened entire or bifid erect arms ¼–½ lin. long; capsule not seen.

COAST REGION : Malmesbury Div.; region of Hopefield, road to Theefontein, *Bachmann*, 1042 ! Grootfontein, *Bachmann*, 1043 ! near Groene Kloof (Mamre), *Bolus*, 4359 ! and without precise locality, *Grey* !

Closely allied to *E. tuberculata*, Jacq., but distinguished by its more slender habit, shorter leaves, smaller involucre and glabrous ovary. One specimen has a few weak hairs on the ovary, but they are very different in character from the stiff hairs which thickly cover the ovary of *E. tuberculata*. Also the tubercles on the stem appear to be smaller, the involucre is paler in colour when dried, its lobes smaller, more subquadrate, much more finely toothed and apparently purple.

123. E. tuberculata (Jacq. Hort. Schoenbr. ii. 43, t. 208);
succulent, 1½-2 ft. high, with numerous branches from a thick
rootstock or main stem wholly or partly buried in the ground,
spineless, but the branches beset with the hardened persistent
remains of the peduncles, not true spines, only dried branches seen,
3-18 in. long, ½-¾ in. thick, evidently stouter when alive, unbranched,
cylindric or slightly thickened upwards, erect, covered with rhomboid conical tubercles 3-4 lin. long, 1-3 lin. broad and 1-3½ lin.
prominent, glabrous; leaves ½-2 in. long, ¾-1⅓ lin. broad, linear,
channelled, down the face from incurved margins, acute, soon
deciduous, glabrous, somewhat fleshy; peduncles solitary in the
axils of the tubercles at the tips of the branches, persistent and
hardening, at first ½-¾ in. long, apparently elongating as the fruit
ripens up to 1¼ in. or perhaps more, bearing about 4 bracts and
1 involucre at the apex, glabrous; bracts 1½-2 lin. long and 1-2
lin. broad, elliptic or elliptic-obovate, obtuse, concave, glabrous,
ciliate; involucre 5-7 lin. in diam., broadly bowl-shaped, glabrous
outside and within, with 5 glands and 5 subquadrate toothed or
nearly entire and minutely ciliate lobes; glands closely sessile on
the cup of the involucre, 1½-2½ lin. long, 1¾-3 lin. broad at the
tips, broadly cuneately palmatifid, divided to half-way down into
3-6 linear spreading processes, recurved at the tips, slightly dilated
or once or twice shortly bifid at the apex, glabrous, green and not
pitted on the undivided or glandular part and the processes white;
ovary sessile, quite included in the involucre, thickly pubescent
with erect stiffish or somewhat bristle-like hairs; styles exserted,
2¾-3 lin. long, united nearly to the apex, with very short slightly
spreading thickened or bifid tips; capsule not seen. *Willd. Sp. Pl.*
ii. 887; *Poir. Encycl. Suppl.* ii. 609; *Spreng. Syst. Veg.* iii. 787;
Boiss. in DC. Prodr. xv. ii. 86; *Berger, Sukk. Euphorb.* 109;
Marloth in Trans. Roy. Soc. S. Afr. iii. 123, t. 8, fig. 2. *Dactylanthes tuberculata, Haw. Syn. Pl. Succ.* 133. *Medusea tuberculata,
Klotzsch & Garcke in Abhandl. Akad. Berlin,* 1859, 61.

COAST REGION : Clanwilliam Div. ; near Clanwilliam, *Leipoldt in MacOwan,
Herb. Austr.-Afr.*, 2003 ! *Marloth,* 4880 ! Lange Kloof, *Schlechter,* 8389 ! Malmes-
bury Div. ; between Mamre and Saldanha Bay, *Drège,* 8202 ! Piquetberg Div. ;
Kopje near Het Kruis, *Misses Stephens & Glover,* 8751 ! Uitenhage Div. ; Cannon
Hill, near Uitenhage, *MacOwan,* 3286 partly !

WESTERN REGION : Little Namaqualand ; sand flats between Driefontein and
Heeren Logement, *Pearson,* 6718 !

The tubercles and leaves as represented by Jacquin are larger than on wild
specimens, from which the measurements above given are taken.

With regard to the specimens distributed by MacOwan as coming from Cannon
Hill, that locality is undoubtedly an error, and I have no doubt the specimens
were really collected near Clanwilliam, an opinion that is confirmed by the
presence of pieces of *E. Macowani,* N. E. Br., mixed with it on all the sheets of
this distribution that I have seen, as the latter species has only been collected at
Clanwilliam.

124. E. Bolusii (N. E. Br.); succulent, spineless, but with
persistent ascending remains of the peduncles; habit probably

something like that of *E. tuberculata*, Jacq., but only 3 dried branches have been seen, which resemble those of that species, but are apparently much shorter, 2–2½ in. long, 5–6½ lin. thick, covered with elongated subrhomboid tubercles 3–4 lin. long, 1½ lin. broad and 1½–2 lin. prominent, compressed-conic, slightly recurved ; leaves not seen, soon deciduous ; peduncles solitary in the axils of the tubercles at the tips of the branches, ½–1 in. long, erect or ascending, glabrous, bearing a whorl of 4 bracts and 1 involucre at its apex ; bracts elliptic, obtuse, concave, glabrous, not ciliate, probably not more than 2 lin. long, but only seen in young bud, deciduous ; involucre 4½ lin. in diam., broadly cup-shaped or obconic, glabrous or nearly so outside and within, with 5 glands and 5 subquadrate lobes, toothed at their truncate apices ; glands subcontiguous, broadly cuneate-palmatifid, 1½–1¾ lin. long and about 2½ lin. across the tips, divided to more than half-way down into 4–6 apparently white segments ¾ lin. long, which are bifid to the middle and their flat lobes diverging, recurved and toothed something like a reindeer's horn, quite glabrous, and the gland occupying the whole of the undivided part, not pitted ; ovary sessile, included in the involucre, glabrous ; styles united to the apex into a slender column 2½ lin. long, minutely 3-lobed at the apex, exserted, glabrous ; capsule and seeds not seen.

KALAHARI REGION : Transvaal ; near Middelburg, *Bolus*, 9767 !

125. E. Macowani (N. E. Br.) ; succulent, spineless, but the branches beset with ascending or suberect hardened remains of peduncles, not spines ; only dried branches seen, 3½–8 in. long, 5–8 lin. thick, probably much stouter when alive, cylindric, slightly thickened upwards, covered with elongated subrhomboid tubercles ¼–½ in. long, 1½–2 lin. broad, and about 1 lin. prominent, glabrous ; leaves suberect, 1–1½ in. long, 1–2 lin. broad near the obtuse or acute apex, tapering gradually to the base, linear-spathulate, entire, fleshy, glabrous ; peduncles solitary in the axils of the tubercles at the tips of the branches, suberect, ½–1¾ in. long, glabrous, bearing 5 bracts and 1 involucre at the apex ; bracts 2–3 lin. long, 1–2½ lin. broad, obovate, obtuse or rounded at the apex, about 3 of them pubescent on the back and ciliate, glabrous within, the others glabrous on both sides and scarcely ciliate ; involucre about 7 lin. in diam., broadly cup-shaped, villous-pubescent outside, glabrous within, with 5 glands and 5 rather large subquadrate ciliate lobes, fringed-toothed at the top ; glands 2½ lin. long and 3½–4 lin. broad across the tips, broadly cuneate, palmately divided to two-thirds of the way down into 5–7 primary linear forked segments, with very shortly bifid or trifid tips, villous-pubescent on both sides, but with a transverse excavated glabrous gland at the base of the united part, apparently purplish, tips of the segments white ; ovary subsessile, included in the involucre, glabrous ; styles 2 lin. long, exserted, united almost to the apex into a slender glabrous column articulated close to the base, minutely trifid at the apex ; capsule subsessile, about 5 lin. in diam., globose-trigonous, glabrous ; seeds 2 lin. long, subterete, slightly conical, truncate at the base, very abruptly and shortly conic-acute at the apex, very rough with minute irregular tubercles, grey.

COAST REGION : Clanwilliam Div. ; near Clanwilliam, *Schlechter*, 8419 ! Mac-Owan, 3286 partly (wrongly labelled as from Cannon Hill in Uitenhage Div.) !

126. E. bubalina (Boiss. Cent. Euphorb. 26, and in DC. Prodr. xv. ii. 90) ; a succulent shrub, 2–5 ft. high (*Burchell*), spineless ; branches 5–9 lin. thick, terete, subtuberculate, glabrous, green ; leaves alternate, scattered along the young branches, very spreading, 1½–6 in. long, ¼–1½ in. broad, cuneate-oblanceolate or oblong-lanceolate, obtuse or subacute, apiculate, cuneately tapering to a narrow sessile base, not distinctly petiolate, slightly folded length-wise, glabrous on both sides ; peduncles solitary in the axils of the leaves, spreading or ascending, 1–6 in. long, ½–1 lin. thick, once forked or bearing an umbel of 3 simple or once- or twice-forked rays ½–2 in. long at the apex, with a pair or a whorl of 3 bracts at the base of the umbel and 1 or 2 very reduced alternate leaves below, withering and persisting for 2 or 3 years, glabrous ; bracts ½–¾ in. long, ½–⅔ in. broad, deltoid or deltoid-ovate, obtuse or acute, apiculate, subtruncate at the base, glabrous on both sides, green, usually edged with red ; involucre sessile, 2½–3 lin. in diam., cup-shaped, glabrous, with 5 glands and 5 subquadrate subentire or slightly toothed puberulous lobes ; glands 1¼–1¾ lin. in their greater diam., transverse, oblong or very broadly cuneate-oblong, entire, green, like the bracts ; capsule closely sessile, about ⅓ in. in diam., sub-globose-trigonous, slightly narrowing upwards, glabrous ; styles ¾–1 lin. long, erect, united into a column for half their length, bifid at the apex, green ; seeds 2–2¼ lin. long, ovoid, acute, slightly 2-keeled on the apical part, very minutely papillate-tuberculate or sub-reticulate to nearly smooth. *Berger, Sukk. Euphorb.* 121 *and* 122, *fig.* 32. *E. oxystegia, Baker in Saund. Ref. Bot.* iii. *t.* 209, *not of Boiss. E. clava, E. Meyer in Drège, Zwei Pfl. Documente,* 184, *name only, not of Jacq. E. laxiflora, O. Kuntze, Rev. Gen. Pl.* iii. 286.

COAST REGION : Bathurst Div. ; near Port Alfred, *Burchell*, 3994 ! *Rogers*, 156 ! Trapps Valley, *Miss Daly*, 718 ! King Williamstown Div. ; near the Buffalo River, *Drège*, 4615 ! East London Div. ; river bank, East London, *Galpin*, 3110 ! British Kaffraria, *Cooper*, 132 ! 3150 ! and *cultivated specimens* !

Described from a living plant introduced by T. Cooper, and cultivated at Kew. In the Refugium Botanicum the glands of the involucre are inaccurately figured and described as yellow.

127. E. tugelensis (N. E. Br.) ; plant 2–3 ft. high (*Gerrard*) ; only small pieces of young branches 1½–3 in. long seen, evidently fleshy, 2–3 lin. thick when dried, glabrous ; leaves alternate, ascending, sessile, 2½–6¼ in. long, 3–5 lin. broad, broadly linear or linear-lanceolate, acute, narrowed at the base, glabrous on both sides ; peduncles solitary in the axils of the leaves, 2–6 in. long,

336   EUPHORBIACEÆ (Brown).   [*Euphorbia.*

glabrous, bearing a whorl or cluster of 3 bracts at the apex and
1 involucre or ultimately a small 2–3-rayed umbel, with a pair of
bracts under each of the lateral involucres when they are developed,
naked below ; bracts sessile, ½–1 in. long, ⅓–⅔ in. broad, deltoid-
ovate, acute or acuminate, often obtusely subangular at the base,
glabrous on both sides ; rays 3–4 lin. long ; involucres sessile, 2¾–3
lin. in diam., cup-shaped, glabrous, with 5 glands and 5 transversely
oblong entire very minutely ciliate and puberulous lobes ; glands
not contiguous, 1–2 lin. in their greater diam., spreading, transversely oblong, and, from the inner margin being turned up or
inwards, somewhat 2-lipped, entire ; ovary sessile, glabrous ; styles
united below into a column ⅓ lin. long, with ascending-spreading
arms ½ lin. long, slightly notched at the apex ; capsule and seeds
not seen.

EASTERN REGION : Natal ; near the Tugela River, *Gerrard*, 1626 !

Possibly a small specimen, in the Natal Herbarium, collected by Mrs. K.
Saunders in Alexander County, may belong to this species. Better material is
required to make a more perfect description.

128. E. oxystegia (Boiss. Cent. Euphorb. 27, and in DC. Prodr.
xv. ii. 90) ; stem or branches (only a few scraps ½–3 in. long seen)
succulent, cylindric, possibly branches from a shrub, ¼–½ in. thick,
slightly tuberculate, marked with scars of fallen leaves and bearing
the persistent remains of the peduncles, spineless, puberulous or
minutely velvety ; leaves in a lax rosette at the apex of the stems
or scattered, spreading, 1–4 in. long, 3–7 lin. broad, lanceolate or
oblong-lanceolate, obtuse, sometimes apiculate, tapering at the base
into a puberulous petiole, otherwise glabrous on both sides, apparently slightly fleshy, deciduous ; peduncles solitary in the axils of the
leaves, erect, 2½–6 lin. long, rather slender and scarcely ½ lin. thick,
velvety-puberulous, bearing a pair or a whorl of 3–5 bracts at the
base of the terminal simple 2–3-rayed umbel, and 1–2 bracts and
sometimes a single ray below it, persisting for 2 or more years ;
bracts 2½–5½ lin. long, 2–3½ lin. broad, those on the peduncle
lanceolate, those at the base of the umbel and under the involucres
all free, not connate as originally described, rhomboid and as broad
as long, acute, recurved at the apex, remarkably angular at or
below the middle on each side, puberulous on both surfaces, apparently sometimes red or purplish ; rays ¾–1⅓ in. long, each with
1 involucre, puberulous ; involucre sessile, 1¾–2 lin. in diam., cup-
shaped, puberulous, with 5 glands and 5 oblong or subquadrate
ciliate or toothed lobes ; glands ⅔–1 lin. in their greater diam.,
transversely oblong or very broadly cuneate, entire, puberulous
on the back, glabrous above ; ovary not seen. *Berger, Sukk.
Euphorb.* 123. *E. bupleurifolia, E. Meyer* b (*not* a), *in Drège, Zwei
Pfl. Documente*, 184, *name only, not of Jacq.*

WESTERN REGION : Little Namaqualand ; between Goodmans Kraal and Kaus
(near Koper Berg), *Drège*, 4616 ! Kamaggas, *Whitehead* ! near Spektakel, *Bolus*,
9445 !

129. E. Clava (Jacq. Ic. i. 9, t. 85 and Collect. i. 104); stem erect, simple and clavate or columnar when young, afterwards branching, 1–4 ft. high, 1¼–2¼ in. thick, with a long tapering basal part; branches erect, subparallel, about 1 in. thick, cylindric, covered with hexagonal broadly conical tubercles ½–⅔ in. in diam., and 2–3 lin. prominent, the older part of the main stem tessellately marked by impressed lines into elongated rhomboid scarcely prominent areas produced at their apex into a slight tubercle, glabrous, green, not glaucous; leaves only present on the upper part of the stem and branches, alternate, sessile, spreading or slightly deflexed, deciduous, 1½–5½ in. long, 1½–3 lin. broad, linear or linear-lanceolate, acute, more or less tapering to the base, slightly folded lengthwise, glabrous, green; peduncles solitary in the axils of the leaves or tubercles, usually 3–7 in. long, sometimes shorter, about ½–¾ lin. thick, minutely puberulous, with 3–6 alternate bracts scattered along them and a whorl of 3 forming a cup around the solitary involucre at their apex, withering and persisting several seasons, but not forming spines; lower bracts 3–5 lin. long, 1½–3 lin. broad, lanceolate to elliptic, acute, those under the involucre 3–5 lin. long, 1½–4 lin. broad, transversely rhomboid-ovate or elliptic-subrhomboid, abruptly apiculate or acute, all sessile, glabrous or minutely puberulous at their very base, green; involucre sessile within the bracts, about ⅓ in. in diam., cup-shaped glabrous or rarely minutely puberulous outside, minutely puberulous in front of the glands within, green, with 5 dark green glands and 5 suborbicular or transversely elliptic minutely toothed or ciliate lobes; glands contiguous, 1½–1¾ lin. in their greater diam., transverse, narrowly oblong or subreniform, entire; capsule sessile, about ⅓ in. in diam., glabrous; styles 2–3¼ lin. long, united into a column 1–2¼ lin. long with spreading bifid arms ¾–1½ lin. long; seeds 2 lin. long, ellipsoid, shortly pointed at one end, rugulose all over, dark brown. *Ait. Hort. Kew. ed.* 1, ii. 136; *Willd. Sp. Pl.* ii. 888; *Spreng. Syst. Veg.* iii. 787; *Boiss. in DC. Prodr.* xv. ii. 89; *Berger, Sukk. Euphorb.* 120 (*excl. syn. Treisia Clava, Haw.*). *E. canaliculata, Lam. Encycl.* ii. 417. *E. coronata, Thunb. Prodr.* ii. 86 *and Fl. Cap. ed. Schult.* 404; *Boiss. in DC. Prodr.* xv. ii. 89; *Berger, Sukk. Euphorb.* 121. *E. clavata, Salisb. Prodr.* 389. *E. radiata, E. Meyer ex Boiss. in DC. Prodr.* xv. ii. 90 (*not of E. Meyer's Herbarium nor of Thunb.*). *Treisia tuberculata, Haw. Suppl. Pl. Succ.* 65.—*Euphorbium acaulon, erectum, &c., Burm. Rar. Afr. Pl.* 12, *t.* 6, *fig.* 1. *Tithymalus aizoides africanus simplici squamato caule chamænerii folio, Commelin, Prælud. Bot.* 24 *and* 58, *t.* 8.

SOUTH AFRICA: without locality, *Thunberg*! *Drège*! *Herb. Lamarck*! and cultivated specimens!

COAST REGION: Uitenhage Div.; near the Zwartkops River, *Zeyher*, 1101! 3851! Cannon Hill, near Uitenhage, *MacOwan*, 3179! Port Elizabeth Div.; near Port Elizabeth, *Drège*, 40! Aloes, *Burtt-Davy*, 14299! Karoo, *Gill*! Albany Div.; Grahamstown (cultivated?), *Schönland*, 1600!

Described from a living plant cultivated at Kew. Specimens of this plant have been distributed by Drège and described by Boissier under the name of

"*E. radiata*, E. Meyer." But it is not the plant to which E. Meyer gave that name. The type of *E. radiata*, E. Meyer, in E. Meyer's Herbarium at Lübeck, with the name written in his own handwriting, and a note added by him to the effect that it differs from *E. tuberculata* by the glands of the involucre being toothed, is *Drège*, 2941, collected in Little Namaqualand, Aug. 12, 1830, and is quite distinct from the plant distributed by Drège as "*E. radiata*, E. Meyer," of which no specimen exists in E. Meyer's Herbarium, so that Drège must have had the two species mixed, only giving one of them to E. Meyer for naming purposes. I have seen in two of the sets of Drège's plants, however, a section of the stem of the true *E. radiata*, E. Meyer, distributed with the specimens of *E. Clava*, Jacq., under the name "*E. radiata*, E. Meyer," but other sets have no such sections. As E. Meyer did not publish descriptions with his names, and Boissier describes a different plant from that which E. Meyer intended to bear the name of *E. radiata*, and as Thunberg had long before described another plant as *E. radiata*, I have described the true *E. radiata*, E. Meyer, below as *E. restituta*. *E. Clava* and *E. restituta* grow in totally different regions 500 miles apart. I have dissected a flower of the type specimen of *E. coronata*, Thunb., in Thunberg's Herbarium, and find it to be in every way identical with *E. Clava*, Jacq.

E. Haworthii, Sweet, Hort. Brit. ed. i. 356 (not of 357), founded upon *Treisia Clava*, Haw. Syn. Pl. Succ. 131, is quoted by Boissier as a synonym of *E. Clava*, Jacq., but as Haworth describes the glands as being "pectinate-serrate" it cannot be *E. Clava*, Jacq.

130. E. pubiglans (N. E. Br.); stem succulent, spineless, simple or sparingly branched, in the specimens seen 3½–12 in. high, 1½–2 in. thick at the base, where it is abruptly rounded into the root and gradually tapers upwards to ¾–1 in. thick at the very obtuse apex, covered with densely crowded tubercles, glabrous, pale greyish-green on the young growth, becoming grey; tubercles crowded, arranged in 13 spirals, 2–3½ lin. in their greater diam. and 1½–3 in their lesser, rhomboid or 6-angled at the base, very obtuse, subhemispherical, 1¼–2½ lin. prominent; leaves ¾–1½ in. long, linear, with margins inrolled or folded together and in that state about ¾ lin. broad, glabrous; peduncles ⅔–2½ in. long, puberulous, with 3–5 small lanceolate acute bracts scattered along them and a whorl of 5 large bracts at the apex, forming a perfectly circular very flattened or plate-like cup ¾–1 in. in diam. around and closely embracing the involucre, glabrous except at the base around the involucre, apparently purplish; involucre about ¼ in. in diam., minutely puberulous outside, with 5 glands and 5 rather large broadly rounded minutely toothed and ciliate lobes rising in a blunt cone ¾–⅔ lin. above them; glands contiguous, 1¼–1½ lin. in their greater diam., transverse, oblong, entire, distinctly puberulous on the upper surface along the outer and inner borders; ovary subsessile, subglobose, slightly 6-angled, puberulous; styles ⅔–¾ lin. long, united into a column for ⅓–½ of their length, straight, at first slightly diverging, but scarcely exserted beyond the lobes of the involucre, after fertilisation becoming parallel and closed together, quite entire at the apex, puberulous on the basal part; capsule and seeds not seen.

COAST REGION: Port Elizabeth Div.; near Port Elizabeth, *Drège*!

Described from living plants with shrivelled flowers attached and fresh flowers preserved in fluid; sent to Kew in Sept. 1912 by Mr. I. L. Drège, of Port

Elizabeth. It is allied to *E. Clava*, Jacq., but the manner in which the base of the upwardly tapering stem is abruptly rounded into the root, its very much smaller, very crowded tubercles, shorter peduncles, pubescent glands and straight and much shorter styles abundantly distinguish it. In the Cape Herbarium, mingled with *E. Clava*, Jacq., are 2 pieces of this species under *Zeyher*, 3851, collected near the Zwartkops River in Uitenhage Div.

131. E. restituta (N. E. Br.); stem erect, 1 ft. or more high, 1¼–1¾ in. thick when dried, branching, succulent, cylindric, covered with stout conical recurved tubercles 3–5 lin. prominent, with a leaf-scar at their apex, glabrous; leaves 1–2¼ in. long, linear, acute, longitudinally folded, glabrous, soon deciduous; peduncles 3–4½ in. long, variably curved, bearing several bract-scars scattered along them and a terminal 5-rayed umbel about 1 in. in diam., glabrous, withering and persisting for several seasons; umbel-rays ½ in. long, bearing 1 involucre and scars of a pair of bracts 2 lin. below it, glabrous; involucre about ⅓ in. in diam., cup-shaped, glabrous, with 4 glands and 5 transversely rectangular toothed puberulous lobes; glands about 1½ lin. in diam., subelliptic or suborbicular, slightly concave, with 5–6 processes ½–⅔ lin. long along the outer margin, thickened or slightly lobed at their tips, apparently yellow; ovary exserted on a pedicel curved to one side, glabrous; styles about ⅔ lin. long, united for half their length, with stout recurved-spreading dilated and deeply channelled tips; capsule and seeds not seen. *E. radiata, E. Meyer in Drège, Zwei Pfl. Documente, 184, name only, not of Thunb. nor of Boiss.*

WESTERN REGION: Little Namaqualand; between Zwartdoorn River and Groen River, *Drège*, 2941; hills near Stinkfontein (near Garies), *Schlechter*, 11098! between Stinkfontein and Garies, *Pillans*, 5579!

This is the plant which E. Meyer intended to bear the name *E. radiata*, as confirmed by the type specimens named by himself in his Herbarium at Lubeck, and is entirely different from the plant which Drège afterwards distributed as *E. radiata*, E. Meyer; see under *E. Clava*, Jacq. As the name *E. radiata* had already been used by Thunberg for a totally different plant, and Boissier has mistakenly described another plant for the present species, it appears to be advisable in the face of so much confusion to give it a new name.

132. E. fasciculata (Thunb. Prodr. 86 and Fl. Cap. ed. Schult. 404); stem apparently solitary, unbranched, succulent, spiny, erect, ¼–1 ft. high, 1½–3 in. (or more?) thick, cylindric, covered with large hexagonal broadly and shortly ovoid-conic tubercles ¼–½ in. prominent, each bearing a stout spine (in dried specimens the spines often falsely appear to arise in the axils of the tubercles) and immediately behind the base of the spine a triangular depression, at the apex of which the tubercle terminates in a short stout conical slightly deflexed point, glabrous; spines (modified peduncles) solitary, rigid, woody, incurved-erect or spreading, 1–2 in. long, 1½–2½ lin. thick, straight or variably curved, glabrous, apparently pale brown or purple when young, pale grey or whitish with age; leaves only seen on the peduncles and possibly not developed on the stem, sessile, ¼–1¼ in. long, ½–1½ lin. broad, linear or linear-lanceo-

z 2

late, obtuse, channelled down the face, fleshy, glabrous, soon deciduous; peduncles arising from the depressions immediately behind the spines (not from the axils of them) apparently for several seasons in succession, always shorter than the spines $\frac{1}{2}$–1 in. long, bearing 1–3 involucres and a few leaves or bracts; involucres unisexual, $\frac{1}{4}$–$\frac{1}{3}$ in. in diam., cup-shaped or somewhat obconic, glabrous or thinly puberulous on the upper part, with 5 glands and 5 subquadrate or transversely subrectangular deeply fringed puberulous lobes; glands (including their processes) $1\frac{1}{2}$–$1\frac{3}{4}$ lin. in their greater diam., transversely oblong, with their inner margin inflexed, forming a small lip and their outer with 3–8 entire or bifid linear processes $\frac{1}{2}$–$\frac{3}{4}$ lin. long, apparently longer on the male involucre than on the female; capsule $\frac{1}{4}$ in. in diam., with rounded angles, exserted on a pedicel about as long as the involucre, glabrous; styles $\frac{2}{3}$ lin. long, stout, united up to the large transversely oblong spreading stigmas; seeds 2 lin. long, oblong, with 3 (4?) slight angles, scabrid-tuberculate, whitish. *Boiss. in DC. Prodr.* xv. ii. 177. *E. Schœnlandii, Pax in Jahrb. Ges. Vaterl. Kult.* lxxxii. ii. 24, *and Fedde, Repert.* i. 59.

COAST REGION : Clanwilliam Div. ; Karroo between the Olifants River and Bockland Berg, *Thunberg* ! Clanwilliam (Woodsfield), specimen in Albany Museum, cultivated at Grahamstown, *collector not stated* !
WESTERN REGION : Van Rhynsdorp Div.; Attys, 400 ft., *Schlechter*, 8089 ! near Van Rhynsdorp, *Marloth*, 2696 (photograph only) !

Thunberg has mistaken the stout spines for branches and described it as unarmed, with the branches aggregated at the apex. In the original description in his Prodromus he does not mention the flowers, but later, in Schultes' edition of his Flora Capensis, the plant is described as having the peduncles collected near the apex of the plant, and bearing a simple umbel of flowers. This description is also erroneous, and has been repeated by Boissier, who, however, states that the species is wanting in Thunberg's Herbarium. Thunberg's type is at present in his Herbarium, and consists of a small plant sliced down the middle into two portions ; the specimen is not in flower, but 3 detached umbels belonging to another species are fixed to the same sheet ; these are evidently the peduncles and umbels described by Schultes. These umbels, however, belong to some species allied to *E. mauritanica*, Linn., and have no connection with *E. fasciculata*, Thunb. The other specimens and the photograph quoted above are most certainly identical with *E. fasciculata*, Thunb., and have enabled me to give a more complete description of it ; the Woodsfield specimen is the type of *E. Schœnlandii*, Pax. The manner in which the peduncles arise from a slight cavity behind the spines is very remarkable and quite unlike the mode of flowering in any other species I have seen. The spines are more erect and incurved on some plants than on others.

133. **E. multiceps** (Berger in Monatsschr. Kakt. xv. 182, with fig.); succulent, spiny, resembling a cone, $\frac{1}{4}$–2 ft. high and 3–10 in. in diam. at the base, formed of a thick fleshy axis, covered with very numerous densely crowded horizontally spreading branches, those at the basal part in large specimens 2–$3\frac{1}{2}$ in. long and $\frac{3}{4}$–$1\frac{1}{4}$ in. thick, the others gradually decreasing in size, somewhat clavate-cylindric, covered with crowded rhomboid or hexagonal tubercles 1–2 lin. prominent, glabrous, dull green ; leaves $\frac{1}{4}$–$\frac{1}{2}$ in. long and $\frac{1}{2}$–1 lin. broad under cultivation, cuneate-

linear or subspathulate-linear, obtuse, apiculate, glabrous, green, soon deciduous; spines (modified peduncles) solitary, some arising from the axis of the cone, others from the branches, ascending, $\frac{1}{3}$–$2\frac{1}{2}$ lin. long and 1–$2\frac{1}{2}$ lin. thick, straight or curved, angular, with prominent bract-scars, glabrous; flowering peduncles 3–4 lin. long, solitary in the axils of the tubercles at the tips of the branches, bearing 1 involucre and several very deciduous bracts; involucre $2\frac{1}{2}$–3 lin. in diam., with 5 glands and 5 subquadrate ciliate lobes; glands distant, $\frac{3}{4}$–$1\frac{1}{4}$ lin. in their greater diam., transversely oblong, concave, minutely punctate, with 2–4 linear or oblong recurved or sub-revolute lobes or processes $\frac{1}{4}$–$\frac{1}{3}$ lin. long on the outer margin, truncate or minutely notched at their tips; capsule sessile, $2\frac{3}{4}$–3 lin. in diam., slightly 3-grooved, glabrous; styles united into a column $\frac{1}{2}$ lin. long, with flattened spreading arms $\frac{1}{2}$ lin. long, revolute and notched at the dilated tips; seeds $1\frac{1}{2}$ lin. long, somewhat conical, truncate at the base, acute at the apex, con-stricted at the middle, 4-angled, very indistinctly rugulose, dark brown. *Berger, Sukk. Euphorb.* 109; *Marloth in Wissensch. Ergebn. Deutsch. Tiefsee-Exped.* ii. iii. 248, *fig.* 102, 1; *N. E. Br. in Kew Bulletin,* 1912, 246, *with fig.*

CENTRAL REGION: Laingsburg Div.; common near Matjesfontein, *Marloth,* 5105! *Bolus,* 13414!

WESTERN REGION: Little Namaqualand; on flat ground at Ratel Kraal, near Ookiep, locally common, *Good in Herb. Pearson,* 5047!

Partly described from a living plant sent to Kew without indication of locality, by Prof. Pearson in 1911.

134. **E. Eustacei** (N. E. Br. in Kew Bulletin, 1913, 122 with fig.); plant forming hemispherical cushions $4\frac{1}{2}$–6 in. high and 9–12 in. in diam., composed of numerous crowded succulent branches armed with long white spines, diœcious; branches $2\frac{1}{2}$–$4\frac{1}{2}$ in. long, $\frac{3}{4}$–$\frac{3}{4}$ in. thick, cylindric, obtuse, tessellately marked by depressed lines into 6-angled or rhomboid areas (not prominent tubercles) about 2 lin. long and 2–3 lin. broad, glabrous, light green; leaves petiolate, $\frac{2}{3}$–$1\frac{1}{2}$ in. long, 2–$4\frac{1}{2}$ lin. broad, oblanceolate, obtuse or subacute, mucronate, tapering from about the middle into a petiole 2–3 lin. long, glabrous to the eye, but under a lens very minutely puberulous on both sides, deciduous; spines (modified peduncles) solitary, $\frac{3}{4}$–2 in. long, ascending or spreading, rigid, very white; peduncles solitary in the axils of the rhomboid areas, when flowering $\frac{3}{4}$–$1\frac{1}{4}$ in. long, bearing 1 involucre and 2–3 bracts near the middle and a whorl of 3–4 under the involucre, minutely and thinly puberulous, light green, becoming transformed into spines after the fall of the flower or fruit; bracts sessile, spreading, $1\frac{1}{2}$–$2\frac{1}{2}$ lin. long, 1–$1\frac{3}{4}$ lin. broad, oblong or obovate-oblong, very obtuse or submarginate, minutely apiculate, those at the middle minutely puberulous on the back, those under the involucre larger than the others and glabrous; involucre sessile within the bracts, 2–3 lin. in diam., with the male usually larger than the female, cup-shaped,

glabrous or very minutely puberulous, with 5 glands and 5 sub-quadrate toothed lobes; glands spreading, not contiguous, $\frac{3}{4}$–$1\frac{1}{4}$ lin. in their greater diam., transverse, broadly cuneate-subrect-angular, entire; ovary sessile and included in the involucre, capsule just exserted from it, $\frac{1}{4}$ in. in diam., depressed subglobose, 3-grooved, very minutely velvety-puberulous; styles united into a column $\frac{3}{4}$ lin. long, with spreading arms $\frac{2}{3}$ lin. long, bifid, with diverging lobes at the tips; seeds about $1\frac{1}{2}$ lin. long, ovoid, faintly and minutely rugulose, greyish. *N. E. Br. in Gard. Chron.* 1913, liv. 355, *fig.* 129.   *E. Hystrix, Marloth in Wissensch. Ergebn. Deutsch. Tiefsee-Exped.* ii. iii. 288 *and* 287, *fig.* 114, *not of Jacq.*

CENTRAL REGION: Laingsburg Div.; near Matjesfontein, *Pillans*!

Described from living plants sent to Kew by Mr. Eustace Pillans, which flowered in February, 1914.

135. **E. loricata** (Lam. Encycl. ii. 416); a branching spiny shrub, 1–3 ft. high, with well developed leaves; stems and branches $\frac{1}{3}$–$\frac{1}{2}$ in. thick, fleshy, cylindric, slightly tuberculate in spiral series, glabrous; leaves clustered at the tips of the branches, 1–3 in. long, $1\frac{1}{2}$–3 lin. broad, linear or linear-lanceolate, obtuse and minutely apiculate at the apex, tapering to the narrow sessile base, glabrous on both sides; spines (modified peduncles) numerous, rigid, ascending, solitary, $\frac{1}{2}$–2 in. long, grey or brown; peduncles $\frac{1}{2}$–2 in. long, when young bearing three broadly ovate or orbicular-obovate bracts $2\frac{1}{2}$–3 lin. long and $2\frac{1}{2}$ lin. broad surrounding one involucre at the apex; involucre sessile within the bracts, $\frac{1}{4}$ in. in diam., cup-shaped, glabrous, with 5 glands and 5 subquadrate fringed lobes; glands about 1–$1\frac{1}{4}$ lin. in their greater diam., transversely oblong, entire, green; capsule erect on a pedicel about as long as the involucre, $3\frac{1}{2}$ lin. in diam., globose, with 3 very slight furrows, not at all lobed, glabrous; styles united into a column about $\frac{1}{2}$ lin. long, with spreading bifid arms $\frac{3}{4}$ lin. long; seeds $1\frac{1}{2}$ lin. long, ellipsoid-oblong, rugulose, greyish-brown. *Poir. Encycl. Suppl.* ii. 607; *Pers. Syn.* ii. 11.   *E. Hystrix, Jacq. Hort. Schoenbr.* ii. 43, *t.* 207; *Willd. Sp. Pl.* ii. 885 (*excl. references to Petiver & Plukenet*); *Ait. Hort. Kew. ed.* 2, iii. 157; *Spreng. Syst. Veg.* iii. 786; *Boiss. in DC. Prodr.* xv. ii. 90; *Berger, Sukk. Euphorb.* 120.   *E. armata, Thunb. Prodr.* 86 *and Fl. Cap. ed. Schult.* 402.   *Treisia Hystrix, Haw. Syn. Pl. Succ.* 131.   *Alhagi Maurorum, var., Nees in Ecklon & Zeyh. Enum. S. Afr. Pl.* 252, *not of Medic.*

SOUTH AFRICA: without locality, *Thunberg*!

COAST REGION: Van Rhynsdorp Div.; Gift Berg, *Phillips*, 7636! Clanwilliam Div.; Lange Valley, *Drège*, 2955! Olifants River Valley, *Schlechter*, 7989! *Diels*, 356! between Olifants River and Kanakas Berg, *Zeyher*, 1532! near Wupperthal, *MacOwan*, 3223! and *Herb. Austr.-Afr.*, 1955! *Bolus*, 9088! near Clanwilliam, *Leipoldt*, 670! 681!

CENTRAL REGION: Prince Albert Div.; Sand River Mountains, *Marloth*, 4395!

136. **E. atrispina** (N. E. Br.); habit unknown, unisexual; only a living branch 2 in. long and $\frac{3}{4}$ in. thick seen, succulent, spiny,

cylindric, with 6–9 broadly rounded ribs or angles, slightly prominent on the upper part scarcely so below, and slightly crenate,
glabrous, dark dull green, with a more or less evident whitish
scurfy coating, probably produced by a thin waxy exudation;
leaves rudimentary, ¼–⅓ lin. long, deltoid or deltoid-ovate, acute,
dark brown, deciduous; spines (modified peduncles) solitary, 3–10
(usually 5–6) lin. long, very spreading, bearing 2 or more very
minute scale-like bracts, glabrous to the eye, but under a lens
thinly and most minutely powdery-puberulous, black; peduncles
about 2 lin. long, otherwise like the spines, bearing 3 bracts and
1 involucre at their apex; bracts spreading, ½–⅔ lin. long, ovate,
acute, dark purple or purple-brown; female involucre about 1¼ lin.
in diam. and not quite 1 lin. deep, male probably larger or deeper,
cup-shaped, glabrous outside, rather thinly pubescent within, with
5 glands and 5 subquadrate finely toothed rather large lobes;
glands not contiguous, erect in the only dried flower seen, ½ lin. in
their greater diam., transversely elliptic, not pitted; ovary sessile,
glabrous; styles united into a column 1 lin. long, with stout
spreading arms ⅓ lin. long, very broadly cuneate, bilobed and ½ lin.
broad across their stout diverging tips; fruit and seeds not seen.

CENTRAL REGION : Prince Albert Div. ; near Prince Albert, *Pearson* !

Described from a small living plant sent in 1912 to Kew by Prof.
H. H. W. Pearson. It is closely allied to *E. heptagona*, Linn., and similar to that
species in the only flower seen, but the appearance of the plant is so different that
I do not think it can possibly be a variety of it. The angles, instead of being
acute and distinctly triangular, are, in *E. atrispina*, merely broadly rounded
crenations in transverse sections, the colour of the stem is of a very dark dull
green, with a thin whitish scurfy or powdery coating, and the spines are black. A
dead plant of what I believe to be this species was subsequently sent to me by
Mr. Pillans from Prince Albert Div. This plant seems to indicate that *E. atrispina*
is a dwarf species only attaining a height of 3–4 inches.

137. **E. pulvinata** (Marloth in Trans. Roy. Soc. S. Afr. i. 315
and 317, fig. 1); plant forming a dense spiny cushion-like mass
about 6 in. high, composed of crowded succulent leafless spiny
branches, sometimes several plants are aggregated into masses 2–4 ft.
in diam., diœcious; branches 1–5 in. long, 1–1⅓ in. thick, at first
globose, becoming cylindric, constantly 7-angled in the specimens
seen, very obtusely rounded at the tips with the apex slightly
depressed, green, glabrous; angles subacute, slightly crenate, with
broad triangular furrows about 1½–2 lin. deep between them, which
flatten with age; leaves rudimentary, 1 (or under cultivation 2) lin.
long, linear-lanceolate, acute, deciduous, leaving small white scars;
spines (modified peduncles) ¼–½ in. long, solitary and irregularly
scattered along the angles, usually ¼–½ in. apart, sometimes more
closely placed, with a few minute scale-like bracts scattered along
them when young, glabrous, dull red, becoming brown or grey with
age; involucres clustered at the apex of the branches, sessile or
very shortly pedunculate, in the male plant about 2 lin. and in the
female about 1½ lin. in diam., shortly and broadly cup-shaped,

glabrous, dark red-purple, with 5 glands and 5 transversely oblong toothed or subentire lobes ; glands distinct, $\frac{1}{2}$–$\frac{2}{3}$ lin. in their greater diam., transversely elliptic-oblong, entire, dark purple ; ovary sessile, but exceeding the involucre, glabrous ; styles $\frac{3}{4}$–1 lin. long, stout, united for nearly or quite half their length, then spreading, 2-lobed at the apex ; capsule about $2\frac{1}{4}$ lin. in diam., obtusely 3-angled ; seeds $1\frac{1}{3}$ lin. long, ovoid or somewhat pear-shaped, smooth, glabrous, pale brown.

COAST REGION : Queenstown Div. ; common on the sides and summits of mountains near Queenstown, 4000–5000 ft., *Galpin,* 2527 ! *Phillips, Marloth,* 4372. Cathcart Div. ; near Cathcart, ex *Galpin.*

CENTRAL REGION : Somerset Div. ; near Bushmans River Station, *Rogers* ! Cradock Div. ; near Cradock, *Burtt-Davy,* 7988 ! *Miss Murray* ! Aliwal North Div. ; near Aliwal North, *Burtt-Davy,* 5874 !

Described from living plants sent to Kew by Mr. E. E. Galpin, who originally discovered it. One plant, collected at the top of a mountain, where it had been subjected to grass-fires, had the ends of the branches modified into truncately conical tips, with the edges of the angles singed and discoloured so as to form large ovate or somewhat rhomboid pale brownish-white areas ; just as if the angles had been cut off with a knife, but leaving the spines, giving the plant quite a different appearance from those uninjured by the fire.

138. E. aggregata (Berger, Sukk. Euphorb. 92) ; a tufted spiny succulent, 2–3 in. high ; stems erect, $\frac{3}{4}$–1$\frac{1}{4}$ in. thick, usually 8–9- (sometimes 7-) angled, crowded, succulent, grass-green on the growing part, brown below ; angles obtuse, subentire, about 1 lin. prominent, with broadly triangular grooves between them at the tips, becoming less prominent and the grooves flattened below ; leaves rudimentary, $\frac{2}{3}$–1 lin. long, shortly linear or linear-lanceolate, acute, fleshy, soon deciduous ; spines (modified peduncles) solitary, $\frac{1}{4}$–$\frac{1}{3}$ in. long, minutely puberulous and reddish or purplish when young, absent from some parts of the stems ; involucres subsessile at the apex of the stems, campanulate, with spreading greenish-yellow glands. *E. enneagona, Berger in Monatsschr. Kakt.* xii. 109.

COAST REGION : without locality, *cultivated specimen* !

Described from a living portion of the type, sent to Kew by Mr. Berger, but I have not seen flowers. It is closely allied to *E. pulvinata,* Marloth, but the stems are not so stout, the angles more numerous, smaller, more obtuse and less prominent, the green colour is brighter and the spines are shorter and puberulous.

139. E. alternicolor (N. E. Br.) ; dwarf, succulent, spiny and leafless, forming a tufted cushion-like mass 3–4 in. high ; stems erect, about 1–1$\frac{1}{4}$ in. apart at their tips, 2$\frac{1}{2}$–3$\frac{1}{2}$ in. long, 1–1$\frac{1}{6}$ in. thick, 8–10- (often 9-) angled, glabrous, alternately marked with dull green and whitish-green transverse bands about 1 lin. broad ; angles subacute, 1–1$\frac{1}{2}$ lin. prominent, with slightly toothed crests, separated by broadly triangular grooves having an impressed line down their middle ; leaves rudimentary, $\frac{3}{4}$–1$\frac{1}{4}$ lin. long, linear, acute, glabrous, withering and persisting for a time, then deciduous ; spines (modified peduncles) solitary, very spreading, $\frac{1}{3}$–$\frac{1}{2}$ in. long,

rather slender, about $\frac{2}{3}$ lin. thick at the base, reddish or brown, minutely puberulous, bearing 2-3 minute scale-like bracts; flowers and fruit not seen.

SOUTH AFRICA: without locality, *Pillans*!

Allied to *E. aggregata*, Berger, and *E. pulvinata*, Marloth, but readily distinguished from all by the tessellately variegated stems and minutely puberulous spines.

140. **E. captiosa** (N. E. Br.); habit probably similar to that of *E. heptagona*, Linn.; plant about 1 ft. high, sparingly branching from the base; stems erect, succulent, spiny, $1\frac{1}{2}$ in. or more thick, 10-angled in the only portions of a dried stem seen, glabrous, apparently green and not glaucous; angles apparently obtuse and not very prominent, with an impressed line down the centre of each of the grooves between them; leaves rudimentary, $\frac{3}{4}$-1 lin. long, linear-lanceolate, thick and fleshy, minutely puberulous and ciliate, soon deciduous; spines (modified peduncles) solitary, spreading, closely placed, $1-1\frac{1}{2}$ lin. apart, $\frac{3}{4}$-1 in. long, $\frac{2}{3}-\frac{3}{4}$ lin. thick, glabrous, apparently purplish or reddish, becoming grey, bearing 5 or 6 minute scattered scales or bracts; peduncles solitary, clustered at the apex of the stems, about 1 lin. long on the specimen seen, but probably elongating with age and becoming transformed into spines at the fall of the flower or fruit, bearing 4–5 minute scale-like bracts scattered along them, and a whorl of 4 larger bracts immediately under the solitary involucre at the apex; larger bracts $\frac{3}{4}-1\frac{1}{4}$ lin. long, $\frac{1}{2}$-1 lin. broad, oblong, very obtuse or slightly notched at the apex, apiculate, slightly keeled down the back, adpressed to and the longer about equalling the involucre, minutely ciliate, thinly puberulous on the inner face, glabrous on the back; involucre on dried specimens $1\frac{3}{4}$ lin. in diam., obconic, glabrous outside on the cup and the 5 glands, puberulous within and down the centre of the backs of the 5 subquadrate or transversely rectangular lobes; glands not contiguous, erect in the dried specimen, $\frac{3}{4}$ lin. in their greater diam., transversely elliptic, entire, deeply wrinkled or corrugated on their upper or inner surface, especially at the outer margin; ovary sessile, glabrous; styles united into a column $\frac{1}{2}$ lin. long, with stout spreading arms $\frac{1}{2}$ lin. long, bifid at their tips; capsule and seeds not seen.

CENTRAL REGION: Aberdeen Div.; near Aberdeen, *Schönland*, 1661!

Described from a specimen preserved in the Albany Museum, which was cultivated and flowered in the Museum grounds in September, 1904.

141. **E. submammillaris** (Berger, Sukk. Euphorb. 95); plant succulent, spiny, much-branched, in the specimens seen forming a small bushy mass 4–8 in. high, perhaps ultimately making a dense clump; branches (excluding the spines) $\frac{3}{4}$-1 in. thick, erect, 7–10-angled, with grooves about 2 lin. deep between the angles, glabrous, deep green, without markings and not glaucous;

angles subacute, slightly and obtusely toothed, with transverse impressed lines 1½–2 lin. apart, between the teeth, scarcely distinguishable in dried specimens; leaves rudimentary, soon deciduous, about 1½ lin. long, and ½ lin. broad, linear-lanceolate, very acute, spreading, glabrous, green, reddish at the margins and tips; spines (abortive peduncles) solitary between the teeth along the angles, 5–10 lin. long, slender, pale brown, bearing 3–4 minute leaves or bracts scattered along the upper part; flowering peduncles 1½–2 lin. long, glabrous, purple, with 5–6 oblong acute or obtuse apiculate glabrous minutely ciliate purple bracts 1–1½ lin. long; involucre solitary, 2½ lin. in diam., obconic-campanulate, glabrous below, minutely puberulous at the upper part, dark purple, with 5 glands and 5 transversely oblong toothed puberulous lobes; glands contiguous, spreading, 1–1⅓ lin. in their greater diam., transversely oblong, entire, minutely pitted-rugulose on the upper surface. *E. cereiformis, var. submammillaris, Berger in Monatsschr. Kakt.* xii. 106, *with fig.*, 125.

South Africa: locality and collector unknown, *cultivated specimens*!

Described partly from Mr. Berger's type specimen and partly from living plants cultivated at Kew.

142. **E. mammillaris** (Linn. Sp. Pl. ed. i. 451, and Amœn. Acad. iii. 108); adult plants 1–3 ft. high, succulent, leafless, more or less spiny, erect, branching in a clustered or more or less whorled manner, glabrous, dull green, becoming greyish-brown with age, diœcious; branches 1–1¾ in. thick, cylindric, with 8–17 tessellately tuberculate vertical angles; tubercles 6-angled, 2 lin. long, 2–4½ lin. broad and 1½ lin. prominent, those at the tips of the branches hemispheric, broadening with age and becoming very obtusely convex-conical, with a central whitish leaf-scar, but without a very evident transverse keel-line across the centre; leaves rudimentary, scale-like, soon deciduous, ½–⅔ lin. long, oblong or broadly ovate, acute or subobtuse, glabrous, but often minutely denticulate-ciliate; spines (modified peduncles) solitary in the axils of the tubercles, 4–7 lin. long, usually more or less clustered in whorl-like groups at intervals along the stem and branches, grey; flowering peduncles solitary in the axils of the tubercles, clustered at the apex of the branches, ¾–1 lin. long, bearing 4–9 closely placed small bracts and 1 involucre, glabrous; bracts about 1 lin. long, oblong or obovate-oblong, with recurved acute tips, usually minutely toothed at the apical part, minutely ciliate, otherwise glabrous, dull purple; male involucres about 2 lin. long and 2½–2¾ lin. in diam., and the female about 1–1⅓ lin. long and 1½–1¾ lin. in diam. in fresh specimens, smaller when dried, cup-shaped, glabrous or very minutely puberulous outside under the glands, with 5 or in the female sometimes 4 glands and 5 transversely rectangular toothed or fringed lobes; glands not contiguous, those of the male involucre when fresh 1–1¼ (in dried flowers ¾–1) lin. in their greater diam., of the female involucre ⅔–¾ lin. in their greater diam., in both

transversely elliptic-oblong, entire, with a slight depression in
front of the slightly raised inner margin, pitted-rugose, dark
purple in the female and of a lighter colour (always?) in the male;
capsule (immature) about 2 lin. in diam., subglobose, without a calyx,
exserted on a pedicel about as long as the involucre, glabrous or
perhaps minutely puberulous on the top when young; styles united
into a column 1 lin. long, with spreading or recurved-spreading stout
bifid arms ½–⅔ lin. long; seeds not seen. *Lam. Encycl.* ii. 414; *Willd.
Sp. Pl.* ii. 883; *Boiss. in DC. Prodr.* xv. ii. 88; *Goebel, Pflanzenbiol.
Schilderung.* i. t. 1, *fig.* 2.   *E. erosa, Berger, Sukk. Euphorb.* 90, *not of
Willdenow.   E. cereiformis, K. Schum. in Monatsschr. Kakt.* viii. 55, *not
of Linn.—Tithymalus aizoides africanus, validissimis spinis, &c., Commel.
Prælud. Bot.* 59, *fig.* 9.   *Euphorbium polygonum, aculeis longioribus,
&c., Isnard in Mém. Acad. Roy. des Sciences, Paris,* 1720, 386.

COAST REGION : Uitenhage Div. ; among shrubs on Cannon Hill, near Uitenhage,
MacOwan, 3143, and in *Herb. Norm. Austr.-Afr.*, 1956 ! hills near the Zwartkops
River, *Zeyher*, 3848 ! Redhouse, *Mrs. Paterson*, 721 ! 1009 ! 2099 ! Port Elizabeth
Div. ; hills near Port Elizabeth, *Marloth*, 4669 ! *Drège*, 40ᵃ ! The Creek, *Mrs.
Paterson*, 2095 !

CENTRAL REGION : Somerset Div. ; Bushmans River Station, *Rogers* !

Described from living plants and dried flowers sent to Kew by Mrs. Paterson
and Mr. I. L. Drège. This species seems to have been founded by Linnæus
entirely upon the figure in Commelin's *Præludia Botanica.* No specimen of it
exists in his Herbarium and probably he never saw the plant. Later authors
appear to have mistaken another closely allied species (*E. fimbriata,* Scop.) for it.
But it is quite clear that Commelin's figure represents the Uitenhage and Port
Elizabeth plant, and not that figured by Berger (*Sukk. Euphorb.* p. 91) as
E. mammillaris, which is a native of a more westerly part of Cape Colony and is
commonly cultivated under that name. But the true *E. mammillaris* is a stouter
and softer plant, with more prominent and more hemispherical tubercles. The
synonymy above quoted belongs, I believe, to this plant, but the following may
partly belong to *E. fimbriata,* Scop., but it is now quite impossible to determine,
viz.—*E. mammillaris, Mill. Gard. Dict. ed.* viii. *no.* 8 ; *Thunb. Prodr.* 86, *and
Fl. Cap. ed. Schult.* 403 ; *Ait. Hort. Kew. ed.* 1, ii. 134 ; *Haw. Syn. Pl. Succ.* 128.

Of *E. mammillaris* I have seen both male and female involucres in a dried state,
but only the male preserved in fluid. When dried flowers are boiled for dissection,
the male is decidedly much larger than the female and its glands appear to be of a
much paler and possibly greenish colour. I have found both unisexual and
bisexual involucres on the same plant, but do not know if this is always the case.

A young plant sent to Kew by Mrs. Paterson indicates that the seedling first
forms a globose stem about 1 in. in diam., from which arise the numerous
cylindric erect stems that ultimately form the clump-like growth of the plant.

143. E. fimbriata (Scop. Delic. Insub. iii. 8, t. 4); plant 1–3 ft.
high, succulent, leafless, more or less spiny, erect, branching in a
clustered or more or less whorled manner, diœcious; branches
ascending or erect, but sometimes decumbent and rooting at the
base, ⅞–1⅓ in. thick, 7–12-angled, glabrous, green, becoming light
brown with age; angles not spirally arranged, 1–1½ lin. prominent,
tessellately divided by impressed lines into 6-angled transversely
oblong tubercles 1½–2 lin. long, 2¾–4 lin. broad, very broadly and
obtusely subconical, with a central whitish leaf-scar and a slight
but distinct transverse raised line across their middle; leaves

rudimentary, soon deciduous, 1–1¼ lin. long, ⅓–¾ lin. broad, ovate or elliptic-lanceolate, acute; spines (modified peduncles) solitary in the axils of the tubercles, 3–8 lin. long, horizontally spreading, more or less clustered in whorl-like groups at irregular distances along the stems and branches, green when young, changing to red and finally grey, bearing about 4 minute deciduous bracts; flowers clustered at the apex of the branches, solitary in the axils of the tubercles; peduncles in the male plant 1–1½ lin. long, bearing 1 involucre and 3–4 oblong obtuse entire bracts ¾–1 lin. long; in the female plant 0 and the involucre sessile, surrounded by the bracts; involucre of the male plant 2½–3 lin. in diam., of the female 1¼–1½ lin. in diam., cup-shaped, glabrous, green, with 5 glands and 5 transversely rectangular denticulate lobes; glands ¾–1¼ lin. in their greater diam. in the male and about ½–¾ lin. in the female plant, transversely oblong or with their inner margin nearly straight and the outer forming a semicircle, entire or minutely crenulate, rugose, green; ovary sessile, subglobose, scarcely angular, without a calyx at its base, glabrous; styles 1 lin. long, united for half their length into a stout column, with stout spreading deeply bifid tips, green; fruit not seen. *E. scopoliana,* Steud. Nom. ed. 2, 615; *Boiss. in DC. Prodr.* xv. ii. 87. *E. enneagona, Haw. Misc. Nat.* 184, *and Syn. Pl. Succ.* 128; *Spreng. Syst. Veg.* iii. 786. *E. mammillaris, Berger in Monatsschr. Kakt.* xii. 109, *with figs.,* and *Sukk. Euphorb.* 90, *fig.* 22, *incl. var. spinosior, not of Linn.*

SOUTH AFRICA : without locality, *cultivated plants* !
COAST REGION : Worcester Div. ; near Worcester, *Pillans* !

Described from living plants cultivated at Kew. This species seems to be the one generally cultivated under the name of *E. mammillaris,* and I think that, besides the synonymy above given, it is probable that some of the references to *E. mammillaris* mentioned in the note under that species may belong here. I retain Scopoli's name for this plant, as it was published in 1787, whilst *E. fimbriata* of Roth was not published until 1801 although maintained by preference by Boissier as a species of his own. *E. enneagona,* Haw., was published in 1803. *E. mammillaris,* Thunb. Prodr. 86 and Fl. Cap. ed. Schult. 403, which is stated to have been collected at Hantam, may belong to this species, but the description is quite insufficient to determine, and the colour of the flowers is not mentioned. There is no specimen of it in Thunberg's Herbarium. I believe that I am right in referring *E. enneagona* of Haworth to this species, his description fairly agrees and the yellowish-green involucre, as described by him, distinguishes it at once from *E. cereiformis,* to which species Berger has referred it. The pendulous branches mentioned by Haworth are occasionally seen in this and allied species, and I think are due to a want of sufficient water at the time of their formation.

144. **E. cereiformis** (Linn. Sp. Pl. ed. i. 451, excl. all syn.); stems 2–3 ft. high, 1–2 in. thick, erect, branching, succulent, spiny, 9–11-angled, deep green, not glaucous; angles acute, with triangular grooves ¼ in. deep between them, toothed; teeth small, 1½–2 lin. apart, when young tipped with rudimentary recurved fleshy lanceolate acute leaves, usually 1–1½ (but sometimes up to 4) lin. long, minutely ciliate and usually with a few hairs on the back; spines (modified peduncles) solitary, 2½–5 lin. long, very spreading, needle-

like, reddish-brown, becoming grey with age, usually with a few minute bracteoles, absent from some parts of the stem; peduncles (flower-bearing spines) clustered at the apex of the stems, usually 1⅓–3 lin. long, sometimes up to ¾ in. long, bearing 1 involucre and several small scale-like ovate dull-purple bracts; involucre unisexual, 2 lin. in diam., cup-shaped, most minutely puberulous on the upper part, dull purple, with 5 glands and 5 transverse toothed lobes woolly on their inner surface; glands ¾–1 lin. in their greater diam., transversely oblong, minutely pitted, blackish-purple; ovary at first subsessile, finally exserted on a pedicel about as long as the involucre, globose, puberulous, dull purple; styles 1½ lin. long, rather stout, united for half their length, with bifid spreading tips. *Linn. Amœn. Acad.* iii. 108, *excl. all syn.*; *Mill. Gard. Dict. ed.* viii. *no.* 9; *Ait. Hort. Kew. ed.* 1, ii. 134, *and ed.* 2, iii. 156; *Thunb. Prodr.* 86, *and Fl. Cap. ed. Schult.* 403; *Willd. Sp. Pl.* ii. 883; *Haw. Syn. Pl. Succ.* 129; *Spreng. Syst. Veg.* iii. 786; *Boiss. in DC. Prodr.* xv. ii. 88, *partly, and Ic. Euphorb. t.* 48; *Berger, Sukk. Euphorb.* 96, *fig.* 24 (*excl. syn. E. enneagona*). *E. erosa and E. odontophylla, Willd. Enum. Pl. Hort. Berol. Suppl.* 27, 28; *Link, Enum. Pl. Hort. Berol.* ii. 9; *Spreng. Syst. Veg.* iii. 786. *E. polygonata, Lodd. Bot. Cab. t.* 1334. *Treisia erosa, Haw. Suppl. Pl. Succ.* 66. *E. echinata, Salm-Dyck, Hort. Dyck.* 342. *E. cereiformis, var. echinata, Salm-Dyck, ex Boiss. in DC. Prodr.* xv. ii. 88.

SOUTH AFRICA?: without locality, *cultivated specimens*!

Described from living plants. This species was introduced into cultivation about or before 1730 and is said to have come from South Africa, and according to Thunberg it grows on the Karoo, but its native locality still appears to be unknown, no modern collector having found it. The plant figured in Commelin, *Rar. Plant. Hort. Amstelod.* i. 21, *t.* 11 as *Euphorbium Cerei effigie*, etc. and wrongly quoted by Linnæus under *E. officinarum*, may possibly be intended to represent *E. cereiformis*; if this proves to be the case, then this species will not be a South African plant, as Commelin states that it comes from near Salee on the coast of Morocco. When founding *E. cereiformis*, Linnæus quotes as synonyms references to four authors, of which that of Isnard belongs to *E. officinarum*, Linn.; that of Boerhaave doubtfully belongs to *E. cereiformis*, and according to his synonymy certainly includes *E. stellæspina*, Haw.; that of Burmann belongs to *Crassula pyramidalis*, Linn. f., and those of Morison and of Plukenet belong to *E. stellæspina*, Haw. *E. cereiformis, Lam. Encycl.* ii. 414, is described as having nearly sessile flowers, placed among the spines at the summit of the stems, this may therefore possibly be the female plant of *E. fimbriata*, Scop.

145. **E. pentagona** (Haw. in Phil. Mag. 1828, 187); a succulent spiny shrub, 4–9 ft. high, "forming a dense rounded bush" (*Wood*), unisexual, with the stems and branches bearing clusters or whorls of branches at intervals 4–18 in. apart, all erect and somewhat closely packed, ⅔–1¼ in. thick or perhaps thicker when old, usually 5–6- (occasionally 4- or 7-) angled, glabrous, green, becoming grey; angles acute, 1–2 lin. prominent, very slightly toothed or nearly even, with broad triangular grooves between them, each marked with an impressed line down the centre; leaves rudimentary, spreading, 1–2 or under cultivation up to 3½ lin. long, ½ lin. broad,

linear or linear-lanceolate, acute, slightly channelled down the face, glabrous, withering and persisting for a time (at least under culti- vation), then deciduous; spines (modified peduncles) solitary, regularly scattered along the angles or sometimes few or nearly or quite absent, 3–7 (rarely up to 10) lin. long, bearing 2–3 minute alternate scale-leaves, glabrous, light brown; peduncles solitary, clustered at the apex of the branches, 2–6 lin. long, with 2–3 minute alternate bracts on the lower part, and a whorl of 3 at its apex, and developing 1–3 involucres, very minutely puberulous or perhaps sometimes glabrous, light brown, ultimately persistent and trans- formed into spines; bracts under the involucre spreading, 1–1¾ lin. long, ⅔–1 lin. broad, oblong or oblong-obovate, obtuse or rounded at the apex, minutely puberulous on the upper side, glabrous on the back, dull purple; involucre (male) 2 lin. in diam., cup-shaped, entirely dark dull purple, with 5 glands and 5 sub- quadrate or transversely oblong minutely ciliate lobes ; glands not contiguous ⅔–1 lin. in their greater diam., transversely oblong or elliptic, entire ; stamens densely white-pubescent ; ovary and capsule not seen. *Boiss. in DC. Prodr.* xv. ii. 89 ; *Berger, Sukk. Euphorb.* 93. *E. heptagona, Berger, Sukk. Euphorb.* 93 *and* 94, *fig.* 23, *not of Linn. E. tetragona, Sim, For. Fl. Cape Col.* 316, *partly, and t.* 141, *fig.* iii. 2 *and* 3, *not of Haw.*

SOUTH AFRICA: without locality, *Bowie,* and *cultivated specimens*! COAST REGION: Albany Div. ; hill sides near Alicedale, *Marloth,* 4373 ! 4380! King Williamstown·Div. ; Quarry Hill, near King Williamstown, *Galpin,* 8101 ! East London Div. ; dry rocky places near East London, *Rattray,* 382 ! banks of the Nahoon River, about 3 miles from its mouth, *Wood*! Komgha Div. ; rocky places near Keimouth, *Flanagan,* 2344 !

Described from living plants cultivated at Kew and from living specimens sent to Kew by Messrs. Galpin and Rattray. In the Kew Herbarium is preserved a drawing of a branch (probably a rooted cutting) of this plant introduced by Bowie in 1823. This drawing represents a very poor shrivelled and spineless branch, and although doubtless the species, is evidently not the actual specimen, described by Haworth, since he describes its spines. *Galpin's* 8101 is stouter than the other living specimens seen, but is quite the same as the Alicedale and Keimouth plants.

146. E. heptagona (Linn. Sp. Pl. ed. i. 450, and Amœn. Acad. iii. 109); erect, succulent, branching, spiny and leafless, up to 2 ft. high, unisexual; stems and branches ¾–1¾ in. thick, varying in stoutness in different individuals, 5–10- (but frequently 6–8-) angled, glabrous, green or slightly glaucous in a wild state ; angles 1–2½ lin. prominent, stout, subacute, nearly even or slightly crenulate at the edges, separated by broad triangular grooves, each marked with an impressed straight or more rarely slightly zigzag line down the centre ; leaves rudimentary, ½–1 lin. long, deltoid or deltoid- lanceolate, acute or acuminate, dark purple-brown, sessile, their base neither seated upon nor forming a small tubercle, but even with the general level of the crests of the angles ; spines (modified peduncles) solitary, regularly scattered along the angles, very spreading,

straight, ¼–1 in. long, ½–⅔ lin. thick at the base, glabrous, purple-
brown, becoming grey with age, bearing some minute sessile ovate
acute scales or bracts; peduncles solitary, ¼–¾ in. long, glabrous
(or perhaps sometimes puberulous), purple-brown, bearing a whorl
of 3 or more rarely 2 ovate or elliptic bracts ½–¾ lin. long
close under the solitary involucre at the apex; involucre 1⅓–1½
in. in diam., campanulate, glabrous outside, pubescent within,
purple-brown, with 5 glands and 5 rather large subquadrate or
transversely oblong toothed lobes; glands not contiguous, erect
or ascending in dried specimens, ¾ lin. in their greater diam., transversely oblong or elliptic; ovary sessile, glabrous; styles united
into a column ¾–1 lin. long, with spreading broadly cuneate bifid
arms ⅓ lin. long, and ¼ lin. broad across their tips; capsule and
seeds not seen. *Mill. Gard. Dict. ed.* viii. *no.* 6; *Lam. Encycl.* ii.
414; *Ait. Hort. Kew. ed.* 1, ii. 134; *Willd. Sp. Pl.* ii. 883; *Thunb.
Prodr.* 86, *and Fl. Cap. ed. Schult.* 403; *Pers. Syn. Pl.* ii. 10;
Haw. Syn. Pl. Succ. 128; *Spreng. Syst. Veg.* iii. 786; *Boiss. in
DC. Prodr.* xv. ii. 88. *E. enopla, Berger, Sukk. Euphorb.* 93, *not of
Boiss. E. Morinii, Berger, Sukk. Euphorb.* 98. *Anthacantha
desmetiana, Lem. in Ill. Hort.* 1858, *Miscell.* 64.—*Euphorbium
capense, spinis longis simplicibus, Bradley, Hist. Succ. Pl. Dec.* 2. 4,
t. 13. *Euphorbium heptagonum, spinis longissimis, in apice frugiferis,
Boerhaave, Ind. alter Pl. Lugd. Bat.* i. 258, *with fig.*

VAR. β, dentata (N. E. Br.); angles of the stem or branches distinctly toothed
between the spines and the centre of each of the grooves between the angles
marked with a zigzag (not straight) impressed line, otherwise as in the type.
E. enopla, var. dentata, Berger, Sukk. Euphorb. 95.

SOUTH AFRICA: without locality; the type and var. β, *cultivated specimens!*
CENTRAL REGION: Prince Albert Div.; near Prince Albert, *Krège in Herb.
Bolus*, 12909! Graaff Reinet Div.; near Graaff Reinet, *Bolus*, 811! 603
(monstrous form)! *Marloth*, 4598!

Described from living plants cultivated at Kew. This species has been in
cultivation since before 1717, at which period it was introduced by the Dutch
into Leyden Botanic Garden and figured by Boerhaave. Subsequently the same
specimen was figured by Bradley, for he states that there was only one plant at
Leyden. From Bradley's figure, however, two branches represented in Boerhaave's
figure are missing, which between the period of the two drawings being made
were probably removed for propagation. The Graaff Reinet and Prince Albert
specimens are more spiny than the cultivated plant, but seem otherwise to be the
same species.

I have not seen the type of *E. Morinii*, Berger, but a plant sent to Kew as
being that species is certainly not distinct from *E. heptagona*, nor do I find any
character in Mr. Berger's description to separate it from the latter species.

147. E. enopla (Boiss. Cent. Euphorb. 27, and in DC. Prodr. xv.
ii. 89); plant about 1 ft. high, succulent, very spiny, leafless,
bushily branching at the base and above, diœcious; branches 5–9
in. long, ⅞–1¼ (or when dried about ½–⅔) in. in diam., 6–7-angled,
glaucous-green or greyish-green, except at the very tips, where they
are light green tinged with purplish; angles straight, scarcely
crenulate, obtusely rounded, about ¼ in. prominent, separated

by an acute groove; spines (modified peduncles) solitary, stout, rigid, $\frac{1}{3}$–2$\frac{1}{2}$ in. long, $\frac{1}{2}$–1 lin. thick at the base, 2–3 lin. apart, in vertical rows along the angles, somewhat ascending-spreading, usually more or less curved, glabrous, at first dark red, then blackish-purple, finally grey; flowering peduncles in the male plant 4–12 lin. long, in the female 2$\frac{1}{2}$–7 lin. (or perhaps more) long, $\frac{3}{4}$–1 lin. thick, bearing 1 involucre and 4–6 very small bracts, glabrous, dark red; bracts scale-like, mostly clustered around the base of the involucre and much shorter than it, much broader than long, broadly rounded; involucre about 1$\frac{1}{2}$–2$\frac{1}{2}$ lin. in diam., 1–1$\frac{1}{4}$ lin. deep, usually larger in the male than in the female plant, cup-shaped, glabrous, dark red, with 5 glands and 5 subquadrate or oblong minutely toothed lobes; glands distant, erect in the male, inflexed or pressed against the ovary in the female, $\frac{3}{4}$–1$\frac{1}{4}$ lin. in their greater diam., transversely oblong or broadly rounded, entire; immature capsule sessile, but with the greater part exserted from the involucre, globose, faintly grooved, glabrous; styles 1 lin. long, united into a stout column for two-thirds of their length, with stout slightly spreading bifid tips; ripe fruit and seeds not seen.

CENTRAL REGION: Jansenville Div.; Karoo, near Waterford and near Aberdeen Road, *Drège*, 4! Willowmore Div.; stony places on Witte Poort Mountains, 2000–3000 ft., *Drège*, 8207!

Described from living branches bearing flowers, which exactly agree with the type, and flowers in formalin, sent to Kew by Mr. I. L. Drège of Port Elizabeth, grandson of the original discoverer. Boissier, in *DC. Prodr.* xv. ii. 89, has erroneously described *E. enopla* as a shrub 2–3 ft. high, but the specimens and labels give no evidence of this, and in the original description the height is not mentioned. A photograph of the wild plant sent to Kew by Mr. Drège represents it as less than 1 ft. high. The *E. enopla* of Berger is quite a different plant, see *E. heptagona*, Linn.

148. **E. ferox** (Marloth in Trans. Roy. Soc. S. Afr. iii. 122, t. 8, fig. 1); a succulent very spiny leafless plant; stems forming clumps $\frac{3}{4}$–2 ft. in diam., simple or sparingly branching at the base or at the ground-level, individually 3–10 in. long, of which 1–6 in. in the specimens seen is buried in the ground, 1$\frac{1}{4}$–1$\frac{3}{4}$ in. thick, 9–12-angled, light dull green, not glaucous; angles 1$\frac{1}{2}$–2$\frac{1}{2}$ lin. prominent, obtuse, nearly even; spines (modified peduncles) solitary, numerous, arranged along the angles, mostly 1$\frac{1}{2}$–3 lin. apart, $\frac{1}{2}$–1$\frac{1}{4}$ in. long, stout and $\frac{3}{4}$–1 lin. thick at the base, very rigid, woody, straight or variably curved and more or less horizontally spreading except the few at the apex, brown, glaucous, becoming grey, persisting throughout on the part above ground; leaves rudimentary, 1 lin. long, linear-oblong, subacute, concave-channelled down the face, glabrous, deciduous; peduncles about 5–6 to a stem, apical, erect, 2–3 lin. long, bearing 1 involucre and about 6–9 brown scale-like bracts, those at the base of the involucre $\frac{2}{3}$–1 lin. long, oblong, obtuse or lacerate, glabrous, sparsely ciliate, the others smaller; involucre sessile within the bracts, 1–1$\frac{1}{4}$ lin. in diam., cup-shaped, glabrous, apparently purplish and very minutely white-dotted, with 5 glands

and 5 subquadrate ciliate lobes; glands $\frac{1}{2}$ lin. in their greater diam., transversely oblong or subreniform, entire, "green" (*Marloth*); ovary included; capsule with its base just exserted from the involucre, $\frac{1}{4}$ in. in diam., globose, not angular or grooved, smooth, glabrous; styles about $\frac{3}{4}$–$1\frac{1}{2}$ lin. long, united for half their length, with spreading minutely bifid arms.

CENTRAL REGION : Jansenville Div. ; near Klipplaat, *Marloth*, 5147 ! near Waterford, *Drège*!

Described from a living plant sent to Kew by Mr. I. L. Drège, of Port Elizabeth, in Aug. 1912. Called "Voetangel" by the Dutch.

149. **E. cucumerina** (Willd. Sp. Pl. ii. 886); stem apparently unbranched, 9–10 in. high, $1\frac{1}{2}$ in. or more thick, cylindric, apparently 10–12-angled, succulent, spiny ; angles apparently very slightly prominent and obtusely rounded, leafless ; spines (modified peduncles) solitary, $\frac{1}{4}$ in. (or more?) long; peduncles solitary, few at the apex of the stem, as long as the spines, bearing 1 rather small involucre. *Poir. Encycl. Suppl.* ii. 608 ; *Pers. Syn. Pl.* ii. 11 ; *Spreng. Syst. Veg.* iii. 786 ; *Boiss. in DC. Prodr.* xv. ii. 178.—*Euphorbe concombre*, *Le Vaillant, Second Voy. Afrique,* ii. 160, *t.* 6.

WESTERN REGION : Little Namaqualand, between Groene River and Koper Berg, *Le Vaillant.*

This plant is only known from Le Vaillant's figure, in which the peduncles are represented as sometimes having two small spine-like branches, which may be intended to represent either setaceous or subulate bracts or spine-like branches such as *E. stellæspina* has.

150. **E. horrida** (Boiss. Cent. Euphorb. 27 and in DC. Prodr. xv. ii. 89); stem or stems probably erect, cylindric, $3\frac{1}{2}$–$4\frac{1}{2}$ in. in diam. when dried, very deeply many- (in the specimen seen 14-) angled, with the central solid part $1\frac{1}{2}$–2 in. thick and the angles 1–$1\frac{1}{4}$ in. prominent, wing-like, 3–5 lin. thick at the base and $2\frac{1}{2}$–3 lin. thick at the margin in dried sections, with very numerous densely crowded spines (modified peduncles) along their edges, glabrous ; main spines stout, 4–10 lin. long and $\frac{3}{4}$–1 lin. thick at the base, very rigid, straight, solitary, with shorter spines 2–4 lin. long about their base, at first puberulous, finally glabrous ; leaves not seen, probably rudimentary ; peduncles arising beside the bases of the spines, $2\frac{1}{2}$–3 lin. long, bearing several bracts and 1 involucre, puberulous ; bracts scale-like, $\frac{1}{2}$–1 lin. long, oblong, obtuse, more or less keeled on the back, puberulous ; involucre 2 lin. in diam., cup-shaped, puberulous, with 5 glands and 5 rather large transversely rectangular or subquadrate finely toothed lobes ; glands $\frac{3}{4}$–1 lin. in their greater diam., transversely elliptic, entire ; ovary and capsule not seen.

CENTRAL REGION : Willowmore Div. ; Witte Poort Mountains, *Drège*, 8212 !

Probably this plant is dioecious, as the involucres examined appeared to be male, with a rudimentary ovary, they are, however, immature. Drège's specimens are very imperfect, consisting of sections and fragments of the angles only.

151. E. polygona (Haw. Misc. Nat. 184, and Syn. Pl. Succ. 129); diœcious; stems simple or slightly branching at the base, perhaps several from the same root, succulent, leafless, spiny or nearly spineless, erect, 1–2 ft. high, 3–4 in. thick, when very young 7-angled, with age developing 10–20 angles, glabrous, green and slightly glaucous on the young growth in wild plants, not glaucous when cultivated under glass, becoming grey with age; angles vertical or slightly spiral, acute, slightly crenulate, often wavy, separated by acute furrows about $\frac{2}{3}$ in. deep; leaves rudimentary, $\frac{1}{2}$–$1\frac{1}{2}$ lin. long, oblong-lanceolate or deltoid-lanceolate, acuminate, rigid and hard, soon deciduous, dark reddish-brown or blackish; spines (modified peduncles) solitary or 2–3 from a flowering-eye, scattered along the angles, 2–4 lin. long, bearing a few minute scattered bracts, dark purple or blackish-brown, becoming grey; flowers arising at and near the apex of the stems, often one on each side and at the base of a previously formed spine; peduncles 1–2 lin. long, bearing 1 involucre and several bracts, dull purple; upper bracts $1\frac{1}{2}$–$2\frac{1}{4}$ lin. long, 1–2 lin. broad, obovate, obtusely rounded at the apex, glabrous above, minutely puberulous beneath, minutely ciliate; lower bracts much smaller and oblong; involucre unisexual, $2\frac{1}{2}$–$3\frac{1}{2}$ lin. in diam. and $1\frac{1}{2}$ lin. deep, cup-shaped, nearly or quite glabrous outside, dark purple, with 5 glands and 5 rounded minutely toothed lobes; glands not quite contiguous; slightly sloping outwards, 1–$1\frac{3}{4}$ lin. in their greater diam., transversely elliptic or elliptic-oblong, dark purple; capsule $2\frac{1}{4}$–3 lin. in diam., globose or very slightly 3-lobed as seen from above, velvety-pubescent, erect, exserted on a pedicel not exceeding the involucre; styles united into a column about $\frac{1}{2}$ lin. long, with spreading arms of the same length, minutely bifid at the tips; seeds $1\frac{1}{2}$ lin. long, ovoid, acute at one end, obscurely 4-angled, smooth, brown. *Spreng. Syst. Veg.* iii. 786; *Boiss. in DC. Prodr.* xv. ii. 88; *Berger, Sukk. Euphorb.* 99–100, *fig.* 25.

COAST REGION: Uitenhage Div.; Red Hill, *Mrs. Paterson*, 1173! Port Elizabeth Div.; Commadagga, *Miss Sangster*! near Port Elizabeth, *Drège*, 7! *Mrs. Paterson*, 1143! and *cultivated plants*!

Described from living plants cultivated at Kew and others sent to Kew by Mr. I. L. Drège and Mrs. Paterson. Boissier and Berger both state that the plant grows to 4–5 ft. high, but no example I have seen has been more than 20 inches high, and Mr. Drège writes that out of thousands he has seen none have been more than 2 ft. high.

152. E. stellæspina (Haw. in Phil. Mag. 1827, 275); stems erect, branching at the base, ultimately forming dense clumps, succulent, leafless, spiny, 8–18 in. high, $1\frac{1}{4}$–3 in. in diam., 10–16-angled, with the grooves between the angles 2–3 lin. deep, green, not transversely banded nor glaucous, becoming brown with age; angles rather obtuse or subacute, tuberculate-toothed; teeth short, conical, often somewhat deflexed, $2\frac{1}{2}$–3 lin. apart; leaves rudimentary and soon deciduous, $1\frac{1}{2}$–5 lin. long, linear or linear-lanceolate, acute, fleshy, puberulous ; spines (modified peduncles) solitary between the teeth along the angles, 2–5 lin. long, stout, thickened and branching at the apex into a whorl of 3–5 sharp spines $1\frac{1}{2}$–4 lin. long and $\frac{1}{2}$ lin. thick, rigid, puberulous when young, becoming glabrous, brown ; involucre sessile at the apex of the peduncle and surrounded by the whorl of very young spines arising at its base, $1\frac{3}{4}$–2 lin. in diam., cup-shaped, puberulous, with 5 glands and 5 subquadrate fringed lobes ; glands distant, $\frac{2}{3}$–$\frac{3}{4}$ lin. in their greater diam., transverse, reniform or obtusely cordate ; ovary not seen. *Boiss. in DC. Prodr.* xv. ii. 89 ; *Berger, Sukk. Euphorb.* 99.—*Tithymalus africanus spinosus Cerei effigei, Moris. Hist.* iii. 345 ; *Pluk. Almagest.* ii. 370, i. *t.* 231, *fig.* 1.

CENTRAL REGION : Beaufort West Div. ; near Beaufort West (photograph), *Marloth!* Jansenville Div. ; near the Sundays River, *Drège*, 8213, ex *Boissier.* Carnarvon Div. ; Boter Leegte, *Alston*, 2600 !

WESTERN REGION : Little Namaqualand ; plains south of Nieuwefontein, 2300 ft., *Pearson*, 3362 !

Partly described from living plants cultivated at Kew. In the Kew Herbarium is a drawing of this species, labelled "Received from Mr. Bowie in 1822," which was probably made from the plant upon which Haworth based his description. The plant figured by Plukenet and described by Morison is referred to *L. cereiformis* by Linnæus, but it certainly belongs to this species, and not to *E. cereiformis* as understood by Linnæus himself and all later authors.

153. E. astrispina (N. E. Br.) ; stem of the only specimen seen (perhaps a young plant) succulent, spiny, leafless, 6 in. high, $2\frac{1}{2}$ in. in diam., unbranched, cylindric, 10-angled, glabrous, dull green tinted with brown above the middle, dull grey-brown or earth-colour below ; angles separated by acute channels about $\frac{1}{4}$ in. deep and divided by impressed lines into transversely oblong 6-angled areas which rise into short conical tubercles ; leaves rudimentary, soon deciduous, rigidly fleshy, $1\frac{1}{2}$ lin. long, oblong-lanceolate, acute, recurved, entire, minutely ciliate ; spines (modified peduncles) solitary between the tubercles along the angles, stout, 0–3 lin. long, branching at the apex into 4–6 radiating spines $2\frac{1}{2}$–7 lin. long and $\frac{3}{4}$–1 lin. thick, grey ; flowering peduncles all at the apex of the stem, stout, mostly 1–2 lin. long, bearing 5–8 scale-like oblong obtuse minutely puberulous bracts and 1 involucre surrounded by the very immature spines or sometimes without them ; involucre 2 lin. in diam., 1 lin. deep, cup-shaped, minutely puberulous outside, with 5 glands and 5 transversely oblong toothed lobes ; glands contiguous or nearly so, $\frac{3}{4}$–1 lin. in diam., transverse, oblong or subreniform, entire, minutely pitted ; capsule sessile, $2\frac{1}{2}$ lin. in diam. in the example seen, but perhaps immature, slightly conical, obscurely 3-angled, glabrous ; styles 1 lin. long, united to their middle, with stout spreading bifid tips ; seeds not seen.

CENTRAL REGION : Beaufort West Div. ; Willowmore Side, *Brauns*, 1711 !

This plant is nearly allied to *E. stellæspina*, Haw., but has much shorter spines. It so closely resembles some of the cylindric species of *Echinocactus*, that when I first saw the specimen I mistook it for a member of that genus.

154. E. Pillansii (N. E. Br. in Kew Bulletin, 1913, 122, with fig.);
succulent, spiny, leafless, branching at the base, dioecious; stems
or branches 2–6 in. high and 1¼–2 in. thick, very obtusely 7-angled or
the branches when very young 5-angled, with the faces between the
angles nearly flat, marked with a central impressed line, so that
the angles are scarcely prominent except at the apex, glabrous,
transversely banded with very pale green and darker green, and
the bands alternating on each side of the impressed line, horizontal
(not oblique); angles with small teeth ½–⅔ lin. prominent and 2 lin.
apart at the apical part, becoming slight crenations or almost
vanishing with age;. leaves very rudimentary, soon deciduous, ½ lin.
long, deltoid, acute; spines (modified peduncles) scattered along the
angles, solitary, 4–7 lin. long, ¾–1 lin. thick, rigid, simple or with
2–6 diverging (not horizontally spreading) spines 2–4 lin. long at
their apex, glabrous, grey; flowering peduncles erect at the tips of
the branches, 4–6 lin. long, ½–1¼ lin. thick, simple and bearing
only 1 involucre or subumbellately branching and bearing a cluster
of 2–6 involucres on branches 2½–3 lin. long, glabrous, green;
bracts on the peduncle ½–⅔ lin. long, deltoid or deltoid-oblong,
subacute, very deciduous, those under the involucre 1 lin. long and
broad, subquadrate, apiculate, glabrous; involucre (only males seen)
2½ lin. in diam., cup-shaped or slightly obconic, glabrous, pale green,
with 5 dark green glands and 5 subquadrate toothed pale green
lobes; glands spreading, contiguous or nearly so, 1 lin. in their
greater diam., transversely elliptic or with the inner margin nearly
straight, entire; stamens nearly glabrous, pale greenish-yellow;
ovary and capsule not seen.

COAST REGION : Ladismith Div. ; near Doorn Kloof River, between Muis Kraal
and Ladismith, *Pillans*! and in *Herb. Bolus*, 12543 !

Described from a living plant sent to Kew by Mr. N. S. Pillans, which flowered
in Dec. 1912. Allied to *E. stellæspina*, Haw., but at once distinguished from that
species by the fewer angles, stouter spines and transverse pale greenish bars on
its stems.

155. E. valida (N. E. Br.); stem solitary, about 6–7 in. high,
3–3½ in. thick, in the specimens seen unbranched, succulent, leafless
and spineless, but bearing on the upper part the hard woody remains
of numerous branched cymes, cylindric-oblong in old plants, probably subglobose when young, not or scarcely depressed at the
rounded apex, 8-angled above, becoming cylindric at the base,
glabrous, dull green or purplish-green, or transversely marked with
pale green lines or narrow bands, becoming entirely brown on the
basal part, unisexual ; leaves rudimentary, about 1 lin. long, deltoid
or deltoid-ovate, acute, soon deciduous ; cymes 1¼–2 in. long and in
diam., arising from the angles at the apex, gradually developing
3 branches, which become 2 to 4 times forked, on peduncles 3–10
lin. long, 1–1½ lin. thick, erect or standing out from the stem,
minutely puberulous, green, becoming woody, brown and persisting
for several years ; bracts few and scattered on the peduncle, with a

pair under each involucre, about 1 lin. long, narrowly oblong, obtuse, apiculate, puberulous; involucre sessile within the bracts, $1\frac{3}{4}$ lin. in diam., cup-shaped, very minutely puberulous outside, rather dull green, with 5 glands and 5 reddish-tipped subquadrate ciliate lobes; glands subcontiguous, erect, $\frac{2}{3}$–$\frac{3}{4}$ lin. in their greater diam., reniform, slightly pitted-rugulose, dingy green or olive-green; ovary and fruit not seen.

CENTRAL REGION: Jansenville Div.; near Waterford, *Drège*!

Described from living plants sent by Mr. I. L. Drège of Port Elizabeth to Kew in Aug. 1912, where it flowered in June, 1913. A photograph accompanied the specimens, showing it growing as a solitary plant upon the plain.

156. **E. meloformis** (Ait. Hort. Kew. ed. 1, ii. 135); plant solitary, unbranched, succulent, subglobose, depressed at the apex, 2–6 in. in diam., diœcious, unbranched, usually 8-angled, leafless, spineless, but in the male plant with persistent hardened remains of the flower-cymes, glabrous, usually marked, with oblique transverse light green and purple-brown or darker green bands, rarely entirely green; angles vertical or spiral, obtuse or subacute, obscurely crenate, with leaf- or cyme-scars $1\frac{1}{2}$–3 lin. apart; leaves rudimentary, soon deciduous, $\frac{1}{2}$–$1\frac{1}{2}$ lin. long, linear, channelled, acute, tipped with a short subulate point, minutely ciliate; cymes arising at the centre of the apex of the plant, spreading over and pressed down near the surface, those of the male $\frac{1}{3}$–$2\frac{1}{2}$ in. long, divided at 1–3 lin. above the base into 3 spreading once- or twice-forked rays, those of the female sessile or subsessile, $\frac{1}{4}$ $\frac{1}{2}$ in. long, divided close to the base into 2–3 simple or once-forked rays, puberulous in both sexes, often persistent on the male, deciduous from the female plant; bracts about 1 lin. long, oblong, obtuse, apiculate, minutely ciliate; male involucre about 2 lin. and the female 1–$1\frac{1}{2}$ (or in mature fruit 2) lin. in diam., cup-shaped, puberulous outside, green, with 5 glands and 5 transversely oblong or subquadrate ciliate lobes; glands $\frac{1}{4}$–$\frac{2}{3}$ lin. in their greater diam., transversely oblong or elliptic or with the inner margin somewhat excavated, entire, light green and minutely punctate in both sexes; capsule sessile, about $\frac{1}{4}$ in. in diam., very obtusely trigonous, minutely puberulous; styles $\frac{1}{2}$–$\frac{2}{3}$ lin. long, very shortly united at the base, with broadly cuneate spreading 2-lobed tips, $\frac{1}{2}$–$\frac{2}{3}$ lin. broad; seeds $1\frac{1}{2}$ lin. long, conical-ovoid, acute, smooth, greyish-brown. *Willd. Sp. Pl.* ii. 886; *Wendl. Collect.* i. 42, *t.* 12; *Poir. Encycl. Suppl.* ii. 608; *Boiss. in DC. Prodr.* xv. ii. 87, *partly*; *Goebel, Pflanzenbiol. Schilderung.* i. *t.* 1, *fig.* 3; *N. E. Br. in Kew Bulletin*, 1912, 301, *with fig.* *E. pomiformis*, *Thunb. Prodr.* ii. 86, *and Fl. Cap. ed. Schult.* 403. *E. meloniformis*, *Link, Enum. Pl. Hort. Berol.* ii. 9; *Spreng. Syst. Veg.* iii. 788.—*Euphorbe à cote de melon*, *Le Vaillant, Second Voy. Afr.* iii. 23, *t.* 11 *bis.*?

COAST REGION: Uitenhage Div.; near the Zwartkops River, *Thunberg*; Zwartkops Hills, *Mrs. Paterson*, 970! near Redhouse, *Mrs. Paterson*! Port Elizabeth Div.; near Port Elizabeth, *Drège*! also cultivated plants! Albany Div.; West Hill, near Grahamstown, *Beanie*, 555! *Rogers*!

358   EUPHORBIACEÆ (Brown).   [*Euphorbia.*

Described from living plants cultivated at Kew. Judging from specimens in cultivation the female plant appears to be scarcer than the male.

There can be no doubt that this is the plant originally described by Aiton as *E. meloformis,* in spite of the fact that soon after its introduction another species (*E. infausta,* N. E. Br.) was figured under the name *E. meloformis* by Desfontaines and De Candolle and has subsequently been mistaken for it. For not only does Aiton correctly describe the peduncles as "at first trichotomous, thereafter dichotomous, rarely simple," which character at once distinguishes it from *E. infausta,* N. E. Br., but in the British Museum is preserved an excellent drawing of the plant, made by Masson himself, who introduced it. In Wendland's figure of the female plant the cyme-branches are not represented so depressed upon the top of the plant as they are in nature.

157. **E. infausta** (N. E. Br.) ; stem succulent, leafless and spineless, subglobose or obovoid, up to 3½ in. in diam., depressed at the apex, producing subglobose branches on its sides, usually 8- (but sometimes up to 12-) angled, diœcious, the male plant usually (but not always) with persistent hardened remains of the cymes, glabrous, bright deep green, sometimes somewhat shining, marked with oblique transverse darker green (or purplish) stripes, which are sometimes not very conspicuous ; angles vertical or spiral, somewhat acute, with nearly flat and slightly concave (not convex) sides, nearly even or faintly crenulate at the margin, with the leaf- or cyme-scars about 2½ lin. apart ; leaves rudimentary, soon deciduous, scarcely 1½ lin. long, linear or linear-lanceolate, acute : cymes of the male plant erect or ascending (not depressed on the plant), at first with the peduncles simple, ⅓–1 in. long, and bearing only 1 involucre, finally developing at its apex 1–3 erect or ascending rays ¼–1 in. long, each with 1 terminal involucre or with 2–3 involucres scattered along them, green, velvety-puberulous, often persistent and hardening ; cymes of the female plant similar, but much shorter, always deciduous ; involucre 1¼–2 lin. in diam., cup-shaped, puberulous outside, light green, with 5 glands and 5 transversely rectangular or subquadrate ciliate lobes ; glands ⅖–1¼ lin. in their greater diam., transversely oblong, slightly convex, light green, minutely punctate ; ovary sessile, glabrous ; styles very shortly united at the base, with the free part ½ lin. long, rather stout, two-lobed at the apex ; capsule and seeds not seen. *E. meloformis, Desf. in Ann. Mus. Hist. Nat. Paris,* i. 200, *t.* 16, *fig.* 2, *and in Konig and Sims, Ann. Bot.* i. 122, *t.* 2 ; *DC. Pl. Grass. t.* 139 ; *Andr. Bot. Rep.* x. *t.* 617 ; *Haw. Syn. Pl. Succ.* 129 ; *Lodd. Bot. Cab. t.* 436 ; *Boiss. in DC. Prodr.* xv. ii. 87, *partly ; Berger, Sukk. Euphorb.* 101 *and* 103, *fig.* 26, *not of Aiton.*

SOUTH AFRICA : without locality, cultivated specimens, *Herb. Haworth* ! *Pillans in Herb. Bolus,* 10684 !

For more than 100 years this plant has been mistaken for *E. meloformis,* having been introduced soon after that species and figured as it. But it is readily distinguished from *E. meloformis* by its more freely branching habit, less evidently banded stems, more distinctly crenate angles, especially on young plants, and particularly by the outstanding erect or ascending peduncles, which only branch at their apex and do not divide near their base into a trichotomous or dichotomous cyme pressed down on the top of the stem as in *E. meloformis.* I have failed to discover the locality in which it grows. The involucre and glands of the female plant are smaller than those of the male. I have at the present time *E. infausta* and *E. meloformis* in cultivation, growing side by side, and there can be no question as to their distinctness when thus seen.

158. E. pyriformis (N. E. Br.); succulent, spineless, diœcious; body of the plant or branches pear-shaped, 1½–2 in. long or high, 1⅙–1¾ in. in diam. at the top, tapering downwards, depressed at the top, 8-angled, glabrous, somewhat greyish-green, very indistinctly marked with oblique transverse darker green bands; angles spiral or perhaps sometimes vertical, somewhat obtuse, from their sides being rounded, faintly crenulate, about ¼ in. prominent, with acute grooves between them; leaves very rudimentary, deltoid, acute, soon deciduous; cymes (only male plant seen) arising from the depressed centre, ½–1 in. long, not depressed upon the plant, dividing 1–2 lin. above the base, usually into two (or sometimes only one) simple or once-forked branches, puberulous, green, deciduous, each bearing one involucre and 1 or 2 pairs of oblong obtuse apiculate bracts about 1 lin. long; involucre 1¾ lin. in diam., cup-shaped, puberulous, green, with 5 glands and 5 subquadrate toothed lobes; glands subcontiguous, ¾–1 lin. in their greater diam., transverse, oblong, entire light green, minutely punctate; ovary and capsule not seen.

SOUTH AFRICA; without locality, *cultivated specimen* !

Described from a living plant, long cultivated by Mr. Justus Corderoy of Blewbury, Didcot, and now at Kew. The plant is an old one, divided close to the ground into 5 radiating and 1 erect pear-shaped branches, but the small body from which these originate may have been a seedling plant that had become injured at its apex and so branched instead of forming a simple stem.

It is closely allied to *E. meloformis,* and its male flowers are not distinguishable from those of that species, female flowers have not been seen. But in its much smaller size, not due to age, for the plant is an old one, pear-shaped branches, much attenuated downwards, smaller and more obtuse stem-angles, and deciduous (not persistent) cymes, it is clearly distinct.

159. E. obesa (Hook. f. in Bot. Mag. t. 7888); plant diœcious, only female specimens seen, very similar to *E. meloformis,* subglobose or ellipsoid, 8-angled, glabrous, grey-green, marked with numerous transverse dull purple bands formed of fine lines; angles stout, subacute, finely crenate, with sunken eye-like scars of fallen peduncles between the crenations about 1 lin. apart and their sides marked with faint grooves very obliquely crossing the transverse purple bands; leaves not seen, evidently rudimentary and soon deciduous; peduncles few, at the apex of the plant, about 1 lin. long, bearing 1 involucre and a few imbricating ovate-oblong ciliate bracts, puberulous, deciduous; involucre 2 lin. in diam., cup-shaped, minutely puberulous outside, with 5 glands and 5 subquadrate subentire lobes; glands distinct, about ½ lin. in their greater diam., subquadrate or transversely rectangular, slightly concave (not channelled); ovary subsessile, glabrous; styles stout, united for half their length, with spreading revolute channelled tips. *Berger, Sukk. Euphorb.* 102.

CENTRAL REGION : Graaff Reinet Div. ; near Graaff Reinet, *MacOwan*, 3183 !

160. **E. squarrosa** (Haw. in Phil. Mag. 1827, 276) ; a very dwarf succulent tuberous-rooted perennial, spiny, leafless ; tuber (or main stem) ovoid or oblong, 1¼ in. or more thick, producing many branches at its apex ; branches radiately procumbent 1½–3 in. long, 7–9 lin. in diam. including the teeth, 3–(4 ?)-angled, more or less twisted, "convex on the under side " (*Haworth*), glabrous, dark green, without paler markings ; angles very deeply cut into rather slender widely spreading cylindric-conical tubercles 2–3 lin. long and 1½–2 lin. thick at the base, each tipped with a pair of spines ½–1½ lin. long, not very divergent, light brown ; leaves rudimentary, minute, scale-like, roundish-cordate, soon deciduous ; cymes solitary in the axils of the tubercles, subsessile, 1–3-flowered ; bracts minute, scale-like, shorter than the involucre, oblong, obtuse, dark reddish ; involucres about 2 lin. in diam., glabrous, green, suffused with purple under the 5 glands and apparently with purplish lobes ; glands contiguous, ¾–1 lin. in their greater diam., transversely oblong, green. *Boiss. in DC. Prodr.* xv. ii. 81.

SOUTH AFRICA : without locality, *Bowie*, cultivated specimen !

Described partly from a specimen from a plant cultivated at Kew in 1875, which was believed to be one of the original plants introduced by Bowie, and partly from two very fine coloured drawings preserved at Kew, made in 1824 and 1827 from the plants from which Haworth prepared his description and which were introduced by Bowie in 1823. No other collector appears to have found it. In habit it is very similar to *E. stellata*, Willd., and *E. micracantha*, Boiss., but the long tubercles on the angles of the branches readily distinguish it from both.

161. **E. mamillosa** (Lem. Illustr. Hort. ii. Miscell. 69, in note) ; succulent, leafless, spiny, about 5–6 in. high, very much branched ; branches tuberculate ; tubercles crowded, spirally arranged, elongated, conical, dilated at the base, somewhat compressed laterally, cylindric at the apex, bearing a very small roundish-deltoid scale-like leaf and a pair of spines, which are subconnate at their base, then diverging ; flowers axillary, only seen in bud, not described. *Boiss. in DC. Prodr.* xv. ii. 80. *Anthacantha mamillosa, Lem. l.c.* 69.

COUNTRY UNKNOWN ; probably from South or Tropical Africa.
No specimen seen. Can it be *E. squarrosa*, Haw. ?

162. **E. stellata** (Willd. Sp. Pl. ii. 886) ; dwarf, succulent, spiny, leafless, with a tuberous rootstock ; tuber elongated-oblong, 3–6 in. long, 1–2½ in. thick, producing a tuft of radiating branches at its apex ; branches procumbent, 2–6 in. long, 4–7 lin. broad, 2-angled, slightly concave or flattish on the upper side, convex or very obtusely keeled on the lower side, often tapering into a stalk at the base, toothed at the angles, glabrous, green or purplish-brown, with a feather-like whitish-green variegation along the upper side ; spines

in pairs 1½–3 lin. apart, 1–2 lin. long, diverging, on suborbicular spine-shields, brown or greyish ; cymes solitary in the axils of the teeth at the ends of the branches, on peduncles 1–2½ lin. long, on some plants bearing only 1 bisexual involucre, on others ultimately bearing 3 involucres, of which the central is male and soon deciduous, the two lateral hermaphrodite and persistent, glabrous ; bracts very small, scale-like ; involucre 1¾–2 lin. in diam., somewhat hemispheric-cup-shaped in the male, broadly obconic in the hermaphrodite, glabrous, with 5 glands and 5 transversely oblong toothed lobes ; glands ⅔–1 lin. in their greater diam., narrowly transversely oblong, entire, dull yellow ; capsule as seen from above acutely triangular, 2½ lin. in diam. and about twice as broad as long, glabrous, exserted on a recurved pedicel 2–2½ lin. beyond the involucre ; styles about ⅓ lin. long, seated on a small nipple, free to the base, spreading, bifid at the apex ; seeds 1 lin. long, ellipsoid, smooth. *Poir. Encycl. Suppl.* ii. 608 ; *Pers. Syn. Pl.* ii. 11 ; *Spreng. Syst. Veg.* iii. 786 ; *Boiss. in DC. Prodr.* xv. ii. 81. *E. procumbens, Meerburg, Pl. Rar. t.* 55, *not of Mill. E. uncinata, DC. Pl. Grass. t.* 151 ; *Ait. Hort. Kew. ed.* 2, iii. 155 ; *Pers. Syn. Pl.* ii. 10 ; *Willd. Enum. Pl. Hort. Berol.* 500 ; *Poir. Encycl. Suppl.* ii. 608 ; *Spreng. Syst. Veg.* iii. 786 ; *Boiss. in DC. Prodr.* xv. ii. 81 ; *Berger, Sukk. Euphorb.* 38 ; *Marloth in Wissensch. Ergebn. Deutsch. Tiefsee-Exped.* ii. iii. 249, *fig.* 103, 1. *E. radiata, Thunb. Prodr.* 86, *and Fl. Cap. ed. Schult.* 403. *E. Scolopendria, Donn, Hort. Cantab. ed.* iii. 88. *E. Scolopendrea, Haw. Syn. Pl. Succ.* 126.

COAST REGION : Uitenhage Div. ; near the Zwartkops River, *Thunberg* ! *Marloth,* 4889 ! hills between the Sunday and Zwartkops Rivers, *Zeyher,* 1100 ! near Uitenhage, *Burke*! Redhouse, *Mrs. Paterson,* 949 ! Port Elizabeth Div. ; near Port Elizabeth, *Drège,* 17 ! Peddie Div. ; Line Drift, *Sim,* 6284 ! and *cultivated specimens* !

It is possible that this species may occasionally produce branches having 3–4 angles, as the Line Drift specimen, *Sim,* 6284, consists of a plant with about 10 branches attached to the tuber, which are all 2-angled, and undoubtedly this is *E. stellata,* but on the same sheet are a few detached branches with 3–4 angles, and in that respect they agree better with *E. micracantha,* but the spines are stout, as in *E. stellata,* and not nearly so slender as in *E. micracantha.*

163. **E. micracantha** (Boiss. Cent. Euphorb. 25, and in DC. Prodr. xv. ii. 80) ; a very dwarf succulent tuberous-rooted perennial, leafless and spiny ; tuber large, fleshy, cylindric-oblong, 5–6 in. long, 1¼–2¾ in. thick, producing many branches at its apex ; branches 1½–5½ in. long, ¼–½ in. thick when dried, radiately spreading, 4-angled, with slightly concave sides, glabrous, green without markings ; angles slightly toothed ; spines in pairs 2–4 lin. apart, diverging, 1½–3 lin. long, rather slender, on short rounded or slightly transverse spine-shields, grey ; leaves rudimentary, scale-like, about ½ lin. long, ovate, obtuse or subacute, concave, glabrous, soon deciduous ; peduncles solitary in the axils of the spine-shields, at the apical part of the branches, ½–1 lin. long, ultimately 3-flowered, with the lateral involucres on very short peduncles,

glabrous ; bracts scale-like, $\frac{1}{2}$–$\frac{3}{4}$ lin. long, ovate or oblong, obtuse, glabrous ; male involucre about 1, female about 2 lin. in diam., cup-shaped, glabrous outside, apparently purple on some specimens and greenish-yellow on others, with 5 glands and 5 subquadrate minutely toothed lobes ; glands $\frac{1}{2}$–$\frac{2}{3}$ lin. in the male and about 1 lin. in the female involucre in their greater diam., transverse, narrowly oblong, entire, rugulose ; capsule about 1$\frac{1}{2}$ lin. long and 2$\frac{1}{2}$ lin. in diam., deeply 3-lobed as seen from above, with acute angles, glabrous, exserted on a recurved pedicel 2–3 lin. beyond the involucre ; styles shortly united at the base, spreading, rather slender, shortly bifid at the apex ; seeds subglobose, about $\frac{3}{4}$ lin. in diam., smooth. *E. tetragona, Baker in Saunders, Refug. Bot. i. t. 39, not of Haw. E. Gilberti, Berger, Sukk. Euphorb. 39, fig. 9.*

SOUTH AFRICA : without locality, cultivated specimen, *Cooper* !
COAST REGION : Bathurst Div. ; between Blue Krantz and the sources of Kasuga River, *Burchell*, 3901 ! Albany Div. ; Fish River Rand, *Hutton*, 494 ! Fort Beaufort Div. ; between Fish River and Fort Beaufort, *Drège*, 8206c ! Div. ? ; Sheldon, *Hutton*, 488a !
CENTRAL REGION : Somerset Div. ; between the Zuurberg Range and Klein Bruintjes Hoogte, *Drège*, 8206a !

This is distinguished from *E. stellata*, Willd., by its 4-angled branches and more slender spines.

164. **E. clavigera** (N. E. Br.) ; dwarf, succulent, spiny and leafless, unisexual, with the habit of *E. stellata*, Willd., but only branches have been seen ; branches clavate, 3–6 in. long and $\frac{3}{4}$–1 in. in diam. (including the teeth) at the apical part, thence gradually tapering to a stalk-like base 2–3 lin. thick, apparently more or less decumbent or radiately spreading, 3-angled, glabrous, green, not glaucous ; angles apparently much compressed and wing-like, deeply toothed ; teeth 2–3 lin. prominent and 5–9 lin. apart, deltoid, bearing a pair of diverging pale brown spines 3–5 lin. long at their summits, on small spine-shields, which sometimes extend a short distance down the teeth above or below the spines ; leaves very rudimentary and scale-like, $\frac{1}{4}$–$\frac{1}{2}$ lin. long, deltoid ; cymes solitary in the axils of the stem-teeth, subsessile, with 3 involucres ; bracts 1$\frac{1}{2}$ lin. long and as much in breadth, broadly ovate, obtuse, with membranous very minutely toothed margins, glabrous ; male involucres sessile, $\frac{1}{4}$ in. in diam., cup-shaped, glabrous outside and within, with 5 glands and 5 transversely rectangular ciliate lobes ; glands contiguous, 1–1$\frac{1}{2}$ lin. in their greater diam., rather narrowly subreniform, two-lipped, from the inner margin being turned up or inwards, entire, probably yellow or yellowish-green ; female plant not seen.

KALAHARI REGION : Swaziland ; common on sandy open ridges near Bremmers- dorp, 1800–2200 ft., *Burtt-Davy*, 3010 !

165. **E. enormis** (N. E. Br.) ; dwarf, succulent, spiny and leafless ; rootstock obconic or somewhat carrot-shaped, 3–4 in. thick, fleshy,

subterranean, producing numerous branches at its apex; branches erect, $1\frac{1}{4}$–$5\frac{1}{2}$ in. long and up to 1 in. in diam. at the broadest parts, 3–4-angled, the shorter more or less clavate, the others with 1–4 constrictions, so that they falsely appear to be somewhat jointed, glabrous; angles compressed, irregularly toothed, with the larger teeth collected upon the more dilated parts of the branches, 2–5 lin. apart and 1–3 lin. prominent, deltoid; spine-shields extending above the spines to the flowering-eyes and below them into acute points, but not forming a continuous horny margin to the angles, bearing 2 pairs of spines, a pair close to the flowering-eye $\frac{1}{4}$–1 lin. long or sometimes quite rudimentary, and a pair at the apex of the tooth 2–4 lin. long, diverging, grey; leaves quite rudimentary, $\frac{1}{8}$–$\frac{1}{3}$ lin. long, $\frac{1}{2}$–$\frac{2}{3}$ lin. broad, transverse, truncate, often represented by a raised line; flowers and fruit not seen.

KALAHARI REGION: Transvaal; Pietersburg, *Marloth*, 5144!

Although the specimen seen is without flowers, it is so distinct from all the other South African species that there is no difficulty in distinguishing it. In habit and general appearance it somewhat resembles *E. clavigera*, but the spines are more rigid in texture, grey instead of pale brown, and the spine-shield extends in a rather broad band to the flowering-eye and there bears a pair of small spines, which is not the case in *E. clavigera*. The same characters, as well as its much stouter spines and branches, likewise separate it from *E. Knuthii*.

166. **E. Knuthii** (Pax in Engl. Jahrb. xxxiv. 83); very dwarf, succulent, leafless, spiny, 3–6 in. high; rootstock a tuber, with a short or elongated neck, producing many branches at the ground level; branches simple or branched, 2–6 in. long, $\frac{1}{4}$–$\frac{1}{2}$ in. in diam. when dried, 3–4-angled, glabrous; angles rather deeply sinuate-toothed, with the teeth $\frac{1}{4}$–$\frac{1}{2}$ in. apart and 1–2 lin. prominent, deltoid or the upper margins nearly truncate and the lower sloping; spine-shields 1–3 lin. long, narrow, variably decurrent on the lower margin of the tooth, but not forming a continuous margin to the angles and bearing a pair of diverging spines 2–4 lin. long, with or sometimes without 2 small prickles directed inwards at their base, at first light brown, finally grey; leaves rudimentary, $1\frac{1}{2}$–2 lin. long, sessile, lanceolate, very acute, recurved-spreading, glabrous, soon deciduous; flowering-eyes usually at or near the base of the stem-teeth each producing but one peduncle $1\frac{1}{2}$ lin. long, bearing 1 involucre and a pair of bracts at its apex, glabrous; bracts $\frac{3}{4}$ lin. long, oblong, obtuse or slightly toothed at the apex, scale-like, green, glabrous; involucre $1\frac{3}{4}$–2 lin. in diam., cup-shaped, glabrous, apparently green, with 5 glands and 5 transversely rectangular toothed lobes; glands nearly or quite contiguous, about $\frac{3}{4}$ lin. in their greater diam., transverse, oblong or narrowly oblong, entire, apparently yellowish or green; capsule 2–$2\frac{1}{3}$ lin. in diam., rather deeply tricoccous, glabrous, exserted on a recurved pedicel usually $1\frac{1}{2}$–3 lin. (in one case, perhaps abnormally, 4 lin.) beyond the involucre; styles united into a stout column $\frac{1}{3}$ lin. long, with spreading bifid arms $\frac{1}{3}$ lin. long; seeds about $1\frac{1}{4}$ lin. long, ellipsoid or subglobose, smooth, brown.

EASTERN REGION: Portuguese East Africa; Ressano Garcia, 1000 ft., *Schlechter,* 11949!
Closely allied to *E. squarrosa,* Haw., and *E. Schinzii,* Pax.

167. **E. Schinzii** (Pax in Bull. Herb. Boiss. vi. 739); dwarf, succulent, spiny, leafless, 4–6 in. high, compactly much branched; branches usually 4- (or occasionally 5-) angled, about 4–5 lin. square, slightly channelled down the sides, glabrous; angles with opposite teeth or lobes ¼–⅔ in. apart, with their upper margin nearly truncate and that below the spines sloping; spine-shields narrow, extending below the spines ⅓–¾ of the way to the tooth below, but not forming a continuous horny margin, dark brown; spines in 2 pairs to each tooth, one pair at the base of the tooth, minute, rarely more than ½ lin. long, sometimes wanting, another pair at the apex of the tooth 2–6 lin. long, diverging, dark brown or grey; involucres 3 together, sessile in the axils of the teeth, 1½ lin. in diam., cup-shaped, glabrous, bright yellow, with 5 glands and 5 broadly obovate fringed lobes; glands ½–1 lin. in their greater diam., transverse, narrowly oblong, with the inner margin turned up into a slight ridge, entire or with a slight notch at the middle on both margins; ovary sessile, included in the involucre, glabrous; styles 1–1¼ lin. long, shortly united at the base, minutely bifid or subentire at the apex; capsule sessile, partly exserted, 1½ lin. in diam., 3-lobed as seen from above; seeds immature, about 1 lin. long, apparently 4-angled, with slightly rugose faces. *Pax in Engl. Jahrb.* xxxiv. 82; *N. E. Br. in Dyer, Fl. Trop. Afr.* vi. i. 567.

KALAHARI REGION: Transvaal; hills near Pretoria, *Burtt-Davy,* 538! 9818! 9836! 10427! *Rehmann,* 4347, *Miss Leendertz,* 168! *Engler,* 2794! *Miss Doidge,* 5983! *Galpin,* 6974! Potgeiters Rust, *Miss Leendertz,* 1155! Berea Ridge, near Barberton, *Galpin,* 1297! Magaliesberg Range, near Rustenberg, *Miss Pegler,* 933! Pietersburg and near Chlunis Poort, *Marloth,* 5145!

Also in Tropical Africa.

168. **E. griseola** (Pax in Engl. Jahrb. xxxiv. 375); a succulent bush, spiny and leafless; branches less than 5 lin. in diam., elongated, decumbent, 5-angled; angles sinuate-toothed, with continuous horny grey margins; teeth about 4–5 lin. apart; spines 3–4 lin. long, in pairs, with a pair of minute prickles at their base, grey, with black tips; involucre not described; capsule about 1 lin. long and twice as much in diam., deeply 3-lobed, as seen from above, with keeled lobes. *See N. E. Br. in Dyer, Fl. Trop. Afr.* vi. i. 578.

KALAHARI REGION: Bechuanaland; Lobatsi, *Marloth,* 3413.

I have not seen this plant. Dr. Marloth informs me that he did not retain a specimen of it.

169. **E. franckiana** (Berger, Sukk. Euphorb. 78 and 79, fig. 19); a succulent leafless spiny bush, 2–3 ft. (or more?) high; branches constricted into segments 1–3 in. long, and 1–1¼ in. in diam.,

3-4-angled, glabrous, light green on the young parts, becoming light greyish-green ; angles acute, with flat faces between them on the 3-angled branches, and compressed, with deeply channelled faces between them on young 4-angled branches, ultimately growing into flat faces, more or less sinuate-toothed, sometimes wavy, with continuous or interrupted horny grey margins ; spines in pairs $\frac{1}{4}$–$\frac{1}{2}$ in. apart, diverging, 2–4 lin. long, at first brown, becoming grey, with dark brown or blackish tips ; leaves very rudimentary, scale-like, $\frac{1}{4}$–$\frac{1}{2}$ lin. long and $\frac{1}{2}$ lin. broad, soon deciduous ; flowering-eyes seated nearly midway between the spine-pairs.

COUNTRY UNKNOWN, but possibly from South Africa. Described from a living branch from the type, kindly sent to Kew by Mr. Alwin Berger,

170. **E. cœrulescens** (Haw. in Phil. Mag. 1827, 276) ; a succulent spiny leafless bush, 2–3 (or perhaps more) ft. high, branching throughout ; branches in clusters or somewhat whorled, spreading, $1\frac{1}{4}$–2 in. thick, slightly constricted into rounded oblong or elongated segments $1\frac{1}{2}$–3 in. long, 4–5-angled, with slightly concave sides, dark green, more or less glaucous, at least on the younger parts ; angles sinuate-toothed, with continuous or occasionally interrupted horny and at first pale brown finally grey margins ; spines $\frac{1}{4}$–$\frac{1}{2}$ in. long, in pairs, rather stout, diverging, dark brown ; flowers not seen. *E. virosa, Boiss. in DC. Prodr.* xv. ii. 83 (*excl. reference to Paterson*), *and E. virosa and var. cœrulescens, Berger, Sukk. Euphorb.* 80–82, *fig.* 20, *not of Willd.*

SOUTH AFRICA : without locality, cultivated specimen, *Bowie* !
CENTRAL REGION : Somerset Div., without locality, with *Viscum Crassulæ*, Eckl. & Zeyh., growing upon it, photograph only, *Drège* !

I have not seen flowers of this species nor any dried specimen that I can without doubt refer to it, except a portion of the type plant (still in cultivation at Kew) dried by myself. But it is possible that specimens collected by the late Dr. Bolus in Uitenhage Div. and stated to be common there, which were distributed under no. 1872, may belong to this species. It has the same stout spines, but appears less branched, with longer intervals between the constrictions of the stems, which are 6-angled. It is a plant that requires investigation from living material. *E. cœrulescens* was introduced into Kew Gardens by Bowie in 1823, and probably most of the specimens of it cultivated elsewhere were derived from cuttings of the original plants.

171. **E. Ledienii** (Berger, Sukk. Euphorb. 80) ; a succulent spiny leafless bush, 4–6 ft. high, erect ; branches up to $2\frac{1}{4}$ (when dried $\frac{1}{2}$–$1\frac{1}{2}$) in. in diam., 4–7-angled, slightly constricted at varying intervals, glabrous, green ; angles compressed, slightly or conspicuously sinuate-toothed, separated by concave faces or grooves $\frac{1}{2}$–$\frac{3}{4}$ in. broad ; spine-shields separate or connected into a horny brown border, even on the same branch, 1–$1\frac{1}{2}$ lin. broad ; spines in pairs 3–9 lin. apart, diverging, 1–3 lin. long or sometimes rudimentary or absent, without prickles at their base, dark brown ; leaves rudimentary, scale-like, $\frac{1}{2}$ lin. long, broadly deltoid-ovate, soon deciduous ; flowering-eyes 1–2 lin. above the spine-pairs and touching or

enclosed in the spine-shields; cymes usually 3 together at each
flowering-eye, sessile or on peduncles up to $1\frac{1}{2}$ lin. long, each with
3 involucres, the central male, the lateral hermaphrodite; bracts
$\frac{1}{2}$–$\frac{3}{4}$ lin. long, scale-like, ovate or ovate-oblong, obtuse or subacute,
glabrous; involucre 2–$2\frac{1}{2}$ (when dried $1\frac{1}{2}$–2) lin. in diam., cup-
shaped or campanulate, glabrous, with 5 glands and 5 subquadrate
toothed lobes, all bright yellow; glands contiguous or subcontiguous,
$\frac{3}{4}$–$1\frac{1}{4}$ lin. in their greater diam., transversely oblong, entire, often
with the ends deflexed; capsule 3–$3\frac{1}{2}$ lin. in diam., exserted $1\frac{1}{2}$–2
lin. beyond the involucre, 3-angled, with a distinct keel down each
angle and a disc-like calyx at its base; styles united into a column
$\frac{3}{4}$–1 lin. long, with bifid spreading or recurved arms $\frac{1}{4}$–$\frac{3}{4}$ lin. long;
seeds immature in the specimens seen.

VAR. β, **Dregei** (N. E. Br.); involucre narrowly funnel-shaped or obconic
$1\frac{1}{2}$–2 (when dried $\frac{3}{4}$–1) lin. in diam.; otherwise as in the type, but the fruit is
unknown. *E. Ledienii, N. E. Br. in Bot. Mag. t.* 8275. *É. canariensis, Thunb.
Prodr.* 86, *and Fl. Cap. ed. Schult.* 403, *not of Linn.*

COAST REGION: Uitenhage Div.; Karoo-like hills between the Sundays and
Zwartkops Rivers, *Zeyher,* 1097! Zwartkops, *Marloth,* 4891! Redhouse, *Mrs.
Paterson,* 88a! 88b! 720! 880! Port Elizabeth Div.; near Port Elizabeth,
Drège, 42! near Bethelsdorp, *Mrs. Paterson,* 2019! Div.? Norvals Poort, *Rogers,*
2035! also *cultivated plants*! Var. β: Humansdorp Div.; near Zeekoe River,
Thunberg! Port Elizabeth Div.; near Port Elizabeth, *Drège*!

Described partly from living plants and flowers in fluid. The form figured in
the Botanical Magazine, and of which a specimen with flowers in fluid has also
been sent to Kew by Mr. I. L. Drège of Port Elizabeth. seems distinctly to differ
from the type in its narrower and proportionately more elongated funnel-shaped
involucres. It may be specifically distinct, but living plants with fruit and seeds
require to be compared before this point can be decided. The specimen of
E. canariensis in Thunberg's Herbarium appears to be the same as the variety
Dregei, but is without flowers, so that good flowering specimens from the locality
where Thunberg collected it are required to confirm the identification.

172. **E. virosa** (Willd. Sp. Pl. ii. 882); a succulent spiny leafless
bush formed of a clump of erect stems, 5–7 (or according to *Paterson*
up to 15) ft. high and 2–3 in. in diam., simple or sparingly branched
at the upper part, 5–7-angled, constricted at intervals of $1\frac{1}{2}$–3 in.,
so that the angles appear to be broadly scolloped, green, with a
bluish tint, probably glaucous; angles not spirally twisted, separated
by concave channels about $\frac{3}{4}$ in. deep, slightly sinuate-toothed, with a
continuous horny margin $1\frac{1}{2}$–2 lin. broad; spines in pairs $\frac{1}{4}$–$\frac{1}{2}$ in. apart,
2–6 lin. long, stout, widely diverging, straight or slightly curved,
brownish-grey with darker tips; leaves rudimentary, transverse,
about $\frac{1}{2}$ lin. long and 2 lin. broad, truncate, soon deciduous;
flowering-eyes seated 2–3 lin. above the spine-pairs and nearer the
pair of spines above than below them on the specimen seen; flowers
not seen, but according to Paterson's figure, each flowering-eye
produces but one 3-flowered cyme, on a peduncle 2–3 lin. long and
2 lin. thick, with involucres about 4 lin. in diam., having contiguous
transversely oblong glands. *Pers. Syn.* ii. 10; *Poir. Encycl. Suppl.*
ii. 607; *Spreng. Syst. Veg.* iii. 786. *Euphorbia sp., Paterson, Narra-*

tive of four Journeys into the country of the Hottentots, 62, *tt.* 8–9.

WESTERN REGION: Little Namaqualand; near the Orange River, without precise locality, *Paterson*, and at Viols Drift, *Rogers*, 3383!

E. virosa is at present most imperfectly known, as the only specimen I have seen that I think must certainly belong to it, is a fragment about 2 in. long from the top of a stem, without flowers, collected at Viols Drift, which lies to the west of the locality where Paterson found it.

The only other specimens seen, which may or may not belong to *E. virosa*, are:—(1) One collected by Dr. Marloth (4687) at Tsarras, in Great Namaqualand. This consists only of the marginal portion of one of the flowering scollops of an angle, accompanied by a photograph of the plant, which seems to quite agree with Paterson's figure in appearance. In this specimen 2–3 cymes are produced at each flowering-eye, on stout peduncles $\frac{1}{2}$–1 lin. long; involucre sessile, about 3 lin. in diam., obconic-cup-shaped, glabrous outside, with 5 glands and 5 broadly rounded or transversely oblong minutely toothed lobes; glands contiguous, $1\frac{1}{4}$–$1\frac{1}{2}$ lin. in their greater diam., transversely oblong, entire; ovary and capsule not seen, only the central male involucre being developed. This is the plant mentioned and figured as *E. Dinteri*, *Marloth in Wissensch. Ergebn. Deutsch. Tiefsee-Exped.* ii. iii. 52, 291, 313, *Karte* 8, *not of Berger*. (2) Two other specimens, collected by Prof. H. H. W. Pearson (8022, 8085) on the Great Karasberg Range, in Little Namaqualand, and figured by him in *Annals of the Bolus Herbarium*, i. 42, as *E. virosa*, differ from the Orange River plant in the following particulars. The stems are much more slender, being only $1\frac{1}{2}$ in. in diam. at the thickest part and less than 1 in. at the apex, corresponding to the apex of the Orange River plant, which measures 2 in. in diam. at the same point. The leaves are stated by Prof. Pearson in a letter to be "about $\frac{1}{4}$–$\frac{1}{3}$ in. long, oblong or slightly elliptic with a broad base and almost acute apex"; there are no leaves on the specimens, but this description does not agree with those on the Orange River specimen. Each flowering eye produces only one 3-flowered cyme on a peduncle $\frac{1}{2}$–1 lin. long; involucre sessile, about $1\frac{3}{4}$ lin. in diam., campanulate, glabrous outside, apparently yellow, with 5 glands and 5 broadly rounded or transversely oblong minutely toothed lobes; glands erect in the dried specimen, $\frac{3}{4}$–1 lin. in their greater diam., transversely oblong, entire; capsule about 2 lin. long and $3\frac{1}{2}$ lin. in diam., subacutely triangular, glabrous, erect, exserted $\frac{3}{4}$–1 lin. beyond the involucre; styles about $\frac{1}{2}$ lin. long, united into a column almost to the apex, with minute spreading emarginate stigmas $\frac{1}{5}$–$\frac{1}{4}$ lin. long; seeds about 1 lin. long, oblong, apparently 4-angled, but immature, smooth, areolate-reticulate.

From the above it will be noted that both Marloth's and Pearson's plants differ in certain particulars from *E. virosa*, and I think it probable that they belong to two other distinct, but closely allied species. This, however, can only be decided by a careful comparison of ample material of good flowering and fruiting specimens of all three plants.

173. E. grandicornis (Goebel, Pflanzenbiol. Schilderung. i. 42, 59 and 63, figs. 15 (as *E. grandidens*), and 26, 29 and 30); a stout succulent leafless bush, 2–6 ft. high, much branched from the base, armed with very long spines; branches erect or ascending, very deeply constricted into subsagittate-ovate or sagittate-reniform segments 2–5 in. long and 2–6 in. in diam., 3-angled, with the solid central part $\frac{3}{4}$–1 lin. thick, glabrous, green, not glaucous; angles wing-like, 1–$2\frac{1}{2}$ in. broad and $\frac{1}{6}$–$\frac{1}{4}$ in. thick, wavy, with continuous horny greyish-white margins; leaves rudimentary, minute, scale-like; spines very stout, $\frac{1}{2}$–$2\frac{1}{2}$ in. long and 1–2 lin. thick at the base, in pairs $\frac{1}{2}$–$1\frac{1}{4}$ in. apart, widely diverging, greyish or pale brown;

flowering-eyes seated midway between the spine-pairs; involucres 3 together, all sessile, or with peduncles not more than $\frac{1}{3}$ lin. long (but possibly ultimately elongating), $2\frac{1}{2}$ lin. in diam., cup-shaped, glabrous, with 5 glands and 5 subquadrate denticulate lobes; glands contiguous, 1–1$\frac{1}{3}$ lin. in their greater diam., transverse, subreniform-oblong, entire, rugulose on the upper surface, yellowish; ovary and capsule not seen, all the involucres being male. *Neubert, Deutsches Gart.-Mag.* 1893, 291, *with fig.*; *N. E. Br. in Hook. Ic. Pl.* xxvi. *t.* 2531, 2532; *Pax in Engl. Jahrb.* xxxiv. 74; *Berger, Sukk. Euphorb.* 52.

EASTERN REGION: Zululand, *Stone*! *Marriott*! and *cultivated specimens*!

Described from living plants cultivated at Kew since 1876. The involucres produced have all been males, but at their base are rudimentary lateral involucres containing very young female flowers, which have not developed at Kew. Possibly *E. breviarticulata*, Pax, a native of German East Africa is not distinct from this.

174. **E. Cooperi** (N. E. Br. ex Berger, Sukk. Euphorb. 83 and 84, fig. 21); a succulent leafless spiny tree, 10–15 ft. high; trunk becoming naked and cylindric below, 6–8 in. thick; branches ascending, curved at their basal part, 5–6-angled, deeply constricted into conic-ovate or somewhat heart-shaped segments 2–6 in. long, and 1$\frac{1}{2}$–3 in. in diam., with the small central solid part not more than $\frac{3}{4}$–1 in. thick in the younger branches, glabrous; angles wing-like, with triangular channels $\frac{3}{4}$–1$\frac{1}{2}$ in. deep between them, their margins with a continuous horny nearly even grey border; leaves rudimentary, scale-like, about $\frac{1}{2}$ lin. long and 1 lin. broad, trans-verse, apiculate; spines 1$\frac{1}{2}$–4 lin. long, in pairs $\frac{1}{4}$–$\frac{3}{4}$ in. apart, widely diverging, grey, with blackish tips; flowering-eyes 1$\frac{1}{2}$–4 lin. above the spine-pairs; cymes 1–3 from the same eye, sessile, each with 3 involucres, glabrous; bracts about 1$\frac{1}{2}$ lin. long and 2 lin. broad, rounded, concave, usually minutely denticulate; involucres all sessile and the middle one male, lateral fertile, 2$\frac{1}{2}$–3 lin. in diam., cup-shaped, glabrous, with 5 glands and 5 erect short trans-versely rectangular fringed lobes; glands contiguous, 1$\frac{1}{3}$–1$\frac{3}{4}$ lin. in their greater diam., narrowly transverse oblong, very minutely rugulose on the upper surface; capsule about $\frac{1}{4}$ in. long and 4$\frac{1}{2}$–6 lin. in diam., exserted on a stout pedicel, curved to one side, deeply 3-lobed seen from above, with laterally compressed lobes, glabrous, dark purple on the apex and along the angles, having a somewhat fleshy calyx at its base, with 3 deltoid-ovate acute lobes about 1 lin. long; cell-walls about $\frac{1}{4}$ lin. thick, woody; styles 1 lin. long, united for two-thirds of their length, with spreading arms, bifid at the apex; seeds 1$\frac{1}{2}$ lin. in diam., globose, with a raised line in a very slight furrow on one side, and a small pit at one end, light grey.

KALAHARI REGION : Transvaal; Buffelspoort Farm, near Sterkstroom, in Rustenburg District, *Burtt-Davy*, 5993! Komati Poort, *Rogers*, 2504! Potgeiters Rust District, *Marloth*, 5143!

EASTERN REGION : Natal ; Umgeni Valley, *Cooper* !

Described from a living plant cultivated at Kew, originally brought by Mr. T. Cooper from Natal in 1862, and from living material supplied by Mr. J. Burtt-Davy. This appears in the Kew *Hand-List of Tender Dicotyledons*, 1900, 295, as "*E. Cooperi,* Hort.," my description then drawn up not having been published.

175. **E. ingens** (E. Meyer in Drège, Zwei Pfl. Documente, 184, ex Boiss. in DC. Prodr. xv. ii. 87) ; a tree, 20–30 ft. high, succulent, leafless, spiny, branching in a broadly obconical manner ; branches erect or ascending, straight, subparallel, all attaining to about the same general level ; flowering branches 4–7-angled, constricted into segments 3–6 in. long, $1\frac{1}{4}$–3 in. (or perhaps more) in diam., with the solid central part $\frac{3}{4}$–1 in. in diam. ; angles wing-like, $2\frac{1}{4}$–3 lin. thick at the obscurely crenate or sinuate margin, $\frac{3}{4}$–$1\frac{1}{4}$ in. broad, deep green ; leaves rudimentary and scale-like, 1–$1\frac{1}{2}$ lin. long, obovate or broadly ovate, acute, with a hard rigid dark brown auricle (stipule) on each side at the base, glabrous, soon deciduous ; spine-shields $\frac{2}{3}$–$\frac{3}{4}$ in. apart, 2–$2\frac{1}{2}$ lin. in diam., suborbicular or transversely elliptic or reniform, usually poorly developed and formed of a thin rust-coloured disintegrating substance, spineless or with a pair of reduced spines $\frac{1}{4}$–1 lin. long ; flowering-eyes nearly or quite contiguous to the spine-shields, each with 3 cymes on stout peduncles 1–$1\frac{1}{2}$ lin. long, bearing 3 involucres, all at first sessile, the lateral ultimately on very short branches ; bracts $1\frac{1}{2}$–$2\frac{1}{2}$ lin. long, 2–3 lin. broad, very broadly rounded, obtuse or subacute, concave, glabrous ; involucres 4–5 lin. in diam., cup-shaped or somewhat obconic, glabrous outside, pale green, with 5 glands and 5 transversely oblong or subquadrate fringed lobes ; glands contiguous, $1\frac{1}{2}$–$2\frac{1}{2}$ lin. in their greater diam., somewhat half circular in outline when seen from above, and from beneath somewhat triangular with rounded auricles at the base, thick and fleshy, with a sharp ridge along their inner margin, thence sloping to the acute edge of the outer margin, smooth, but in dried flowers more or less wrinkled, "light green" (*Marloth*) ; ovary at first subsessile, with a conspicuous 3-lobed calyx at its base, becoming exserted in young fruit on a stout pedicel as long as the involucre, glabrous ; calyx-lobes in fruit very broadly cuneate or transversely rectangular at the basal part and produced into 2–3 linear-filiform segments 1–2 lin. long ; styles united into a column $\frac{3}{4}$ lin. long, with radiating arms $\frac{3}{4}$–1 lin. long, subentire or minutely 2-lobed at the apex ; capsule erect, 4–5 lin. in diam., with the outer substance evidently somewhat fleshy, glabrous ; seeds $1\frac{3}{4}$–2 lin. long, ellipsoid, with a slight groove down the ventral side, and a slight keel down the dorsal, very faintly and minutely tuberculate as seen under a lens, brown. *E. Cooperi, Berger, Sukk. Euphorb.* 84, *fig.* 21 *only, not as to description. E. grandidens, Adlam in Gard. Chron.* 1886, xxvi. 720, *fig.* 139, *not of Haw.*

KALAHARI REGION : Transvaal ; near Barberton, *Pole Evans,* 2919 ! 2931 ! Pruizen Farm, Potgieters Rust, *Burtt-Davy,* 2200 ! 5658 !

EASTERN REGION : Natal ; in woods near Durban, *Drège*, 4614 ! Inchanga, *Marloth*, 5111 ! and probably a flowerless specimen from steep rocky hillsides near Camperdown, *Burtt-Davy*, 10434 !

Of this species, Drège only collected a few transverse sections and strips from the angles of the branches, which have been badly eaten by insects. The flowers of his specimen are very young, with neither stamens nor ovary exserted from the involucre, but in the form, size and glands of the involucre, and in the very distinct calyx under the ovary and in the styles, it exactly agrees with the Inchanga and Transvaal plants, of which latter I have seen good fruiting specimens, but none in young flower. There are no leaves upon Drège's specimen, but upon the Camperdown and Transvaal specimens they are as described above. Upon a plant brought from Inchanga in Natal by Dr. R. Marloth and cultivated by him at Cape Town, the well-developed leaves are $\frac{3}{4}$-1 in. long and 5-6 lin. broad, sessile, oblong-obovate, obtusely rounded and mucronate or slightly toothed at the apex.

176. E. similis (Berger, Sukk. Euphorb. 69 and 70, with fig.) ; a tree, 20–30 or more ft. high, succulent, spiny, leafless or with well-developed foliage leaves ; branches erect, subparallel, probably forming an obconic crown, fleshy, usually 4- (sometimes 5-) angled, slightly constricted into parallel-sided segments 6–18 in. long, 2–5 in. square, becoming thicker with age, deep green, not tinted with blue nor glaucous on the younger parts ; angles wing-like, rather thin and not more than $1\frac{1}{4}$ lin. thick at the edge on the younger branches, except at the spine-shields, straight or wavy, nearly even or slightly sinuate-toothed at the margins, when young separated by broad triangular channels $1\frac{1}{4}$–$1\frac{1}{2}$ in. deep, with age growing out into flat faces, their sides marked by a slightly prominent longitudinal rib nearly midway between the centre and margin, from which other slightly prominent ribs obliquely ascend to the spine-shields ; leaves sometimes rudimentary and scale-like, $1\frac{1}{2}$ lin. long, deltoid, subulate-acuminate and recurved, sometimes developed into a linear-cuneate or cuneate-lanceolate sessile foliage-leaf $\frac{3}{4}$–$3\frac{1}{2}$ in. long, $1\frac{1}{2}$–8 lin. broad, with a short subulate point at the apex, deciduous, with small hard auricle-like persistent or deciduous blackish-brown stipules at the base ; spine-shields $\frac{3}{4}$–$1\frac{1}{4}$ in. apart, $1\frac{1}{4}$–$1\frac{1}{2}$ lin. long and $1\frac{1}{2}$–$1\frac{3}{4}$ lin. broad, suborbicular, bearing a pair of diverging and distinctly deflexed spines $1\frac{1}{2}$–2 lin. long, blackish ; flowering-eyes 2–4 lin. above the spine-shields and quite separate from them, with 1–2 small blackish-brown scales about or covering them ; flowers and fruit not seen. *N. E. Br. in Dyer, Fl. Trop. Afr.* vi. i. 591. *E. natalensis, Hort. ex Berger, Sukk. Euphorb.* 71, *not of Bernh.*

SOUTH AFRICA ? Described from a living plant long cultivated at Kew !

The native country of this plant is unknown, but as the name "*E. natalensis*" has been applied to it in gardens, it may possibly have been introduced from Natal by Mr. T. Cooper about 1862. It was in cultivation at Kew in 1873.

177. E. triangularis (Desf. Cat. Hort. Paris, ed. 3, 339, name only) ; a tree, 15–20 ft. high, having a naked cylindric trunk with a trace of 4-angles, and a rounded crown of curved ascending-spreading branches at the top ; branches succulent, in whorl-like

groups, up to 3–4 ft. long and 1½–4 in. in diam., 3–5-angled,
deeply constricted into segments 3–12 in. long, with parallel sides
or gradually tapering upwards from a broader base, with the solid
central part about ¾ in. thick and the wing-like angles about
⅔–1¾ in. broad at the broadest part, 1½–2 lin. thick at the margin,
sinuate-toothed or toothless, green, not glaucous; leaves very small,
soon deciduous, sessile, 3–3½ lin. long, 2½–3 lin. broad, cordate-
ovate to cordate-orbicular, obtuse or subacute, glabrous; spine-
shields sometimes separate, sometimes united into a continuous or
interrupted horny brown or greyish margin to the angles even on
the same branch, narrow; spines 1½–4 lin. long, in pairs 4–9 lin.
apart, widely diverging, brown, becoming grey; flowering-eyes
2–3 lin. above the spine-pairs and touching or surrounded by the
horny spine-shields, each producing 2–3 cymes on peduncles 1–1½
lin. long, bearing 3 involucres, glabrous; bracts scale-like, ½–1 lin.
long, broadly ovate or suborbicular, obtusely rounded at the apex,
the larger slightly keeled, glabrous; involucres 2–2½ lin. in diam.,
cup-shaped, glabrous, with 5 glands and 5 subquadrate or transversely rectangular toothed lobes; glands ¾–1 lin. in their greater
diam., rather rigid, transversely oblong or elliptic-oblong, convex
from their ends being recurved, rather deeply and somewhat reticulately pitted or labyrinthally wrinkled, yellow; capsule 3–4 lin.
in diam., with 3 obtusely rounded lobes as seen from above,
glabrous, exserted on a curved pedicel 1¼–2 lin. beyond the
involucre; styles united into a column ⅓–1 lin. long, with rather
slender spreading arms ½–¾ lin. long, thickened and channelled at
the revolute tips; seeds 1⅓–1½ lin. long, oblong, equally obtuse
at each end, somewhat keeled down the inner face, smooth, dark
brown. *Berger, Sukk. Euphorb.* 57. *E. grandidens, Sim, For. Fl.
Cap. Col.* 317, *t.* 141, *fig.* 1, *not of Haw.*

COAST REGION: Uitenhage Div.; Redhouse, *Mrs. Paterson,* 88, partly! East
London Div.; near East London, *Rattray,* 383! *Marloth,* 5100! *Galpin,* 3108!
Komgha Div.; near Komgha, *Flanagan,* 1704! also *cultivated specimens!*

EASTERN REGION: Transkei; Kentani district, *Miss Pegler,* 1203! 1419 partly!
Natal; Amanzimtote, *Miss Franks in Herb. Wood,* 11866!

Partly described from a very old living plant, long cultivated at Kew, and probably obtained from the Jardin des Plantes early in the 19th century. The wild
specimens above quoted undoubtedly belong to this species.

178. **E. Evansii** (Pax in Engl. Jahrb. xliii. 86); a tree with a
bushy crown of succulent spiny leafless branches at the top;
branches flat and 2-angled or 3–4-angled, in the specimens seen
6–18 in. long, 1–1¾ in. in diam., glabrous, light green; angles
wing-like, broader than the solid central part, slightly or distinctly
sinuate-toothed; teeth ⅓–⅔ in. apart, very broadly triangular and
about equally sloping upwards and downwards; spine-shields
narrow, about 1 lin. broad at the middle and either confluent into
a narrow horny border to the angles or (if free) about equally
produced above and below the spines, light brown, bearing a
pair of very widely diverging dark brown slender spines 2–4

lin. long, with or without a pair of minute points near their base, one on each side of the leaf-scar; leaves very small, sessile, very spreading or recurved, 1½–3 lin. long and as much in breadth, broadly rounded or slightly notched at the apex, fleshy, glabrous; flowering-eyes 1½–2½ lin. above the spine-pairs, enclosed in or free from the horny margins formed by the spine-shields; flowers not seen, according to Pax the involucre is solitary, very shortly pedunculate, about 2 lin. in diam., with transverse "ovate" (elliptic-oblong?) rugulose glands.

KALAHARI REGION : Transvaal ; Low Veld near Barberton, *Evans* (ex *Pax*), and without precise locality, *Burtt-Davy,* 5657 !
EASTERN REGION : Natal ; Marian Hill, *Landauer* !

I have not seen the type and describe from the specimens indicated as seen, which, from the original description, I believe to belong to this species. It is closely allied to *E. triangularis,* Desf.

179. **E. grandidens** (Haw. in Phil. Mag. 1825, 33); a tree, growing to 30 ft. high, with a trunk up to 3 ft. in diam. (? girth), with the branches and branchlets more or less clustered in whorls ; main branches cylindric, with a crown of secondary branches at their ends, naked below ; secondary branches and branchlets succulent, spiny, leafless, ½–¾ in. in diam., acutely 3–4-angled, rather deeply sinuate-toothed, glabrous, green, deciduous ; spine-shields at the apex of the teeth ⅓–1¼ in. apart, small, ovate, more or less acutely pointed at each end, bearing a pair of divergent spines ¼–3 lin. long or sometimes nearly obsolete, and often a pair of minute prickles above them, grey ; leaves rudimentary, minute, scale-like, deltoid, soon deciduous ; flowering-eyes in the axils of the spine-shields and not embraced by them, producing 1 subsessile or very shortly pedunculate 3-flowered cyme ; bracts minute, shorter than the involucre, scale-like, very obtuse, glabrous; involucres 2–2½ lin. in diam., cup-shaped, glabrous, pale green, with 5 glands and 5 subquadrate fringed lobes ; glands contiguous, slightly deflexed when mature, ¾–1¼ lin. in their greater diam., transversely oblong, entire, rugose, yellowish-green ; capsule about ⅓ in. in diam., of 3 laterally compressed lobes as seen from above, glabrous, exserted on a curved pedicel 1–2½ lin. beyond the involucre ; styles free nearly to the base, spreading, ¾ lin. long, slender, very shortly 2-lobed at the apex ; seeds subglobose, about 1¼ lin. in diam., smooth, brown. *Boiss. in DC. Prodr.* xv. ii. 82 ; *Goebel, Pflanzenbiol. Schilderung.* i. 64, *fig.* 31 (*not* 41, *fig.* 15) ; *Berger, Sukk. Euphorb.,* 47–48, *fig.* 12. *E. arborescens, Salm-Dyck, and E. magnidens, Haw. ex Salm-Dyck, Hort. Dyck.* 104–105, *names only.*

SOUTH AFRICA : without locality, cultivated specimens, *Bowie* !
COAST REGION : East London Div. ; Buffalo River Valley, 10–12 miles above the mouth of the River, *Wood* ! First Creek, Queens Park, *Rattray* (photograph only) !
EASTERN REGION : Transkei ; Kentani, near Kobonqubo, *Miss Pegler,* 1419 partly ! in woods near Columba, *Miss Pegler,* 1518 ! Natal ; Botanic Garden, Durban, *Wood,* 9129 !

Described partly from a descendant of the type plant, which, introduced by Bowie in 1822, is still flourishing at Kew, where a drawing of the original plant from which Haworth described is also preserved. The plant figured by Sim, in the *Forest Flora of Cape Colony*, t. 141, fig. 1, as *E. grandidens* is *E. triangularis*, Desf.

180. **E. tetragona** (Haw. in Phil. Mag. 1827, 276); a tree, up to 40 ft. high, sometimes with a single trunk up to 6 in. thick, sometimes also with 3–5 trunk-like branches ascending from near the base, slightly 6–8-angled, each with a short broad crown of spreading or ascending-spreading succulent spiny leafless branches and branchlets, usually clustered in whorl-like groups, naked below from the branches being deciduous; main branches at first about 1, ultimately 2 in. or more thick, usually 4–6-angled; flowering branchlets $\frac{2}{3}$–1 in. thick, usually 4–5- (sometimes 3-) angled, with concave or nearly flat sides, with or without a few constrictions, rather light green; angles slightly toothed, with the teeth $\frac{1}{4}$–$\frac{3}{4}$ in. apart, armed with spines on young and the lower branches of old trees, but sometimes on old trees the spine-shields are nearly or quite spineless; leaves rudimentary, scale-like; spine-shields $1\frac{1}{2}$–4 lin. long, separate, not forming a continuous horny margin to the angles, lanceolate, cuneate or obovate, bearing a pair of widely diverging spines 1–6 lin. long, without prickles at their base, light brown, finally grey; flowering-eyes touching or partly or wholly included in the spine-shields; cymes solitary, with a peduncle about 1 lin. long, bearing 3 involucres and some small scale-like bracts; involucres $1\frac{1}{2}$–2 lin. in diam., obconic, yellow, with 5 transverse narrowly oblong entire yellow glands; ovary subglobose, exserted on a pedicel not exceeding the involucre, light green; styles free to the base, apparently about $\frac{1}{3}$ lin. long, ascending-spreading, subentire at the apex, dull ochreous. *Boiss. in DC. Prodr.* xv. ii. 84; *Berger, Sukk. Euphorb.* 58; *Marloth in Wissensch. Ergebn. Deutsch. Tiefsee-Exped.* ii. iii. 57, *fig.* 6.

COAST REGION : Albany Div. ; Zuurberg Range, near Alicedale, *Marloth*, 4381 ! Queenstown Div. ; valley of the Zwart Kei River near its junction with the White Kei River, abundant, *Galpin*, 8100 ! East London Div. ; dry banks near East London, *Rattray*, 381 ! *Marloth*, 4381 ! also *cultivated specimens* !

According to a note received from Mr. Galpin, this is the only arborescent species in Queenstown Division, and he states that it also occurs in Cathcart Div., King Williamstown Div., and Tembuland, but I have not seen specimens from any of these regions.

E. tetragona of Sim's *Forest Flora of Cape Colony*, 316, t. 141, fig. iii., is a mixture of at least 2 and probably 3 distinct species, of which only that represented on t. 141, fig. iii. 4, appears to belong to the true *E. tetragona*.

As I have not seen flowers of this species, my description of them is from a drawing made in 1880 from the type plant described by Haworth, all the specimens quoted are barren.

Described partly from the type plant (or a portion of it) which still flourishes at Kew, partly from notes and photographs kindly supplied by Mr. E. E. Galpin.

Imperfectly known species.

181. E. parvimamma (Boiss. in DC. Prodr. xv. ii. 86 and 92
under *E. pugniformis*); stem very short, fleshy, thickened at the
obconic apex, producing numerous short stellately radiating
branches scarcely 2 in. long; tubercles minute, depressed, roundish-
ovate, crowded, 2–3 lin. long (breadth and prominence not stated);
leaves scarcely 1½ lin. long, ovate-spathulate, mucronate. Flowers
unknown.

SOUTH AFRICA : without locality, originally described from a cultivated plant of
which nothing more is known.

So far as I have been able to ascertain, no specimen of this species exists in the
Boissier or any other Herbarium. But between 1866 and 1870, in the collections
of Mr. T. Cooper and Mr. W. Wilson Saunders, I saw plants cultivated under the
name of *E. parvimamma*, which I now recognise to have been *E. inermis*, Mill.,
and to a certain extent Boissier's description agrees with that species, but as he
states that the branches are scarcely 2 in. long, whilst in *E. inermis* (especially
under cultivation) they are usually much longer than that, I think it cannot have
been *E. inermis* that he was describing and, therefore, the identification of
E. parvimamma must remain doubtful. However, it cannot possibly have been
the plant Berger has described as *E. parvimamma*, since the latter has long
branches, with larger tubercles and leaves twice as long as those of Boissier's
plant.

182. E. procumbens (Mill. Gard. Dict. ed. viii. no. 12); dwarf,
succulent, spineless; main body of the plant not more than 3 in.
high; branches spreading on the ground, seldom more than 6 in.
long, with square tubercles, leafless; flowers not described. *Medusea
procumbens, Haw. Syn. Pl. Succ.* 134, *excl. reference to Burmann*;
Klotzsch & Gürcke in Abhandl. Akad. Berlin, 1860, 61.

SOUTH AFRICA : originally described from cultivated plants.

Nothing is known of this plant beyond the imperfect descriptions of Miller and
Haworth, given above. The plant that has been supposed to be this species is
E. passa, N. E. Br., which see.

183. E. Haworthii (Sweet, Hort. Brit. ed. i. 356, not of 357);
stem or branches succulent, spineless, tuberculate; leaves linear-
lanceolate; peduncles pubescent, persistent; bracts cuneate-obovate,
subentire; glands of the involucre (described as petals by *Haworth*)
pectinate-serrate. *Treisia Clava, Haw. Syn. Pl. Succ.* 131, *excl.
all syn.*

SOUTH AFRICA. Described by Haworth from a cultivated plant.

As Haworth describes the glands of the involucre as being toothed, the plant he
had could not have been *E. Clava* of Jacquin, which has entire glands.

Excluded species.

E. spartioides (Jacq. Hort. Schoenbr. iv. 44, t. 486); by some
error this is stated by Jacquin to be a native of South Africa, but
the plant represented by him is merely a form of **E. Cupani**, Guss.,
a native of Sicily.

E. pendula (Link, Enum. Pl. Hort. Berol. ii. 10) ; stems long and pendulous, forking at distant intervals, about 2 lin. thick, terete, succulent, glabrous ; leaves opposite, rudimentary, minute, deltoid, acute, closely adpressed to the branches. *Boiss. in DC. Prodr.* xv. ii. 76 ; *Berger, Sukk. Euphorb.*, 19 *and* 20, *fig.* 1.

This plant has been supposed to be a native of South Africa, but I have not seen any specimens from that region which at all resemble it. I believe it to be **Sarcostemma brunonianum**, Wight and Arnott, a native of India. It is a very old garden plant, whose flowers are unknown. During my 42 years' knowledge of the plant at Kew, I have never seen it in flower, so that it evidently flowers very rarely under cultivation. In January of this year (1915), however, a friend forwarded to me a sketch of a flower, which developed upon a plant of "*Euphorbia pendula,*" cultivated by another lover of succulent plants. This drawing undoubtedly represents the flower of an Asclepiad and apparently of the genus *Sarcostemma,* but is too imperfect to confirm the opinion above expressed of its specific identity.

E. viminalis (Linn. Sp. Pl. ed. i. 452, and Amœn. Acad. iii. 110 ; Mill. Gard. Dict. ed. viii. no. 15) is **Sarcostemma viminale**, R. Br.

IV. BUXUS, Linn.

Flowers monœcious. *Disc* 0. *Male flower : Perianth-segments* 4, imbricate, in 2 series. *Stamens* 4, opposite the perianth-segments ; filaments absent (in the S. African species) or present, free, fleshy ; anthers introrse ; cells parallel, dehiscing longitudinally. Rudimentary *ovary* absent or more usually present. *Female 'flower · Perianth-segments* 4–6, strongly imbricate, the outer smaller. *Ovary* 3-celled ; styles short, thick, usually distant from one another ; ovules 2 in each cell, pendulous from the apex of the cells ; raphe dorsal ; micropyle facing the axis. *Capsule* ovoid, loculicidal ; valves 2-horned with the persistent styles ; pericarp woody ; endocarp cartilaginous. *Seeds* oblong or ellipsoid, with a small strophiole ; testa crustaceous, shining, usually black ; albumen rather fleshy ; cotyledons oblong.

Much-branched trees or shrubs, usually glabrous ; leaves evergreen, opposite, shortly petiolate, entire, penninerved ; racemes or cymes axillary, sessile or shortly pedunculate ; bracts resembling the sepals ; flowers sessile or shortly pedicellate, the terminal one female, the remainder male.

DISTRIB. About 21 species, 6 from temperate and montane regions of the northern hemisphere, 3 from Tropical Africa, 1 each in Madagascar and South Africa, the remainder West Indian.

1. B. Macowani (Oliv. in Hook. Ic. Pl. t. 1518) ; a shrub or tree, attaining 30 ft. high, with a trunk up to 1 ft. in diam. ; branchlets angular, minutely puberulous, soon becoming quite glabrous ; leaves subsessile, obovate or oblanceolate, rounded or obtuse at the apex, cuneate at the base, ½–1 lin. long, 3½–6 lin. broad, rigidly coriaceous, shining on both surfaces, without distinct nerves; flowers monœcious, axillary, with about 2 lateral shortly pedicellate or subsessile males, and a solitary central shortly pedicellate female; male flowers: bracts much shorter than the perianth, broadly ovate, coriaceous, glabrous; perianth-segments 4, broadly ovate, minutely ciliolate; anthers large, sessile; rudimentary ovary absent; female flowers: pedicels covered with triangular imbricate glabrous bracts; perianth-segments 4, ovate; ovary glabrous; styles ascending, 1½ lin. long, incurved in the young fruit; fruits oblong-ellipsoid, about 3½ lin. long; styles at length diverging. *Sim, For. Fl. Cape Col.* 321, *t.* cxlv. *fig.* i.; *Hutchinson in Kew Bulletin,* 1912, 55. *Buxella Macowani, Van Tiegh. in Ann. Sci. Nat.* 8me sér. v. 326.

COAST REGION : King Williamstown Div. ; Perie Forest, *Tyson*! East London Div. ; in primitive woods near Kwelegha, *Hutchins in MacOwan, Herb. Norm. Austr.-Afr.,* 916 ! Chalumna, *MacOwan,* 2900 ! East London, *Sim,* 2150 ! *Flanagan,* 1727 ! *Ricketts in Natal Gov. Herb.,* 3863 ! East London Park, *Wood in Herb. Galpin,* 3127 ! 3128 ! British Kaffraria; without precise locality, *Hutchins*!

V. NOTOBUXUS, Oliv.

Flowers monœcious. *Petals* absent. *Disc* 0. *Male flower: Sepals* 4, in 2 series, obovate, boat-shaped. *Stamens* 6, in 2 series, the outer of 2, each opposite an outer sepal, the inner series of 4, each 2 opposite an inner sepal ; anthers sessile, ovoid, dehiscing longitudinally. Rudimentary *ovary* absent. *Female flower: Sepals* 4. *Ovary* 3-celled ; styles 3, divergent, stigmatose inside ; ovules 2 in each cell, pendulous. *Fruit* capsular, dehiscing loculicidally ; valves 2-horned. *Seeds* oblong, keeled, black and shining.

Small trees or shrubs ; leaves opposite, entire, chartaceous, shortly petiolate, penninerved ; flowers subfasciculate or in short cymes, the female solitary, terminal, sessile or subsessile, with a few lateral, shortly pedicellate or subsessile male.

DISTRIB. Species 2, the following and one from East Tropical Africa.

1. **N. natalensis** (Oliv. in Hook. Ic. Pl. t. 1400); a shrub, glabrous ; branches longitudinally sulcate, light green ; internodes 1–2 in. long ; leaves obovate-elliptic or oblong, cuneate at the base, obtuse or emarginate at the apex, 2½–4½ in. long, 1–2 in. broad, entire, thinly coriaceous, smooth and slightly shining on both surfaces ; lateral nerves 3–5 on each side, looped a considerable distance from the margin, distinct on both surfaces ; petiole 1–2 lin. long ; cymes axillary, few-flowered ; male flowers : sepals obovate, obtuse ; anthers 1⅓ lin. long ; female flowers : sepals ovate-lanceolate,

coriaceous, obtuse, about 1½ lin. long, rather densely pubescent within the margin ; ovary ovoid, glabrous ; styles spreading, with a stigmatic groove inside; capsule about 5 lin. long; seeds very black and shining, 3½ lin. long. *Sim, For. Fl. Cape Col.* 320, *t.* cxlv. *fig.* ii. ; *Hutchinson in Kew Bulletin,* 1912, 55.

EASTERN REGION: Transkei ; Manubi forest, Kentani district, *Miss Pegler,* 1258! and without precise locality, *Worsdell*! Pondoland ; West Gate, Port St. John, 750 ft., *Galpin,* 3471! Egossa, *Sim,* 2424! 2427! Natal ; Inanda, *Wood,* 374! 1357! Durban, *Schlechter,* 2797! Tongaat, *Cooper,* 3465! *Gerrard,* 20! The Bluff, near Durban, *Wood,* 5790! 11946! and without precise locality, *Mrs. Saunders*!

VI. ANDROSTACHYS, Prain.

Flowers diœcious. *Petals* and *disc* absent. *Male flowers : Calyx* composed of 2–5 bract-like free spirally arranged sepals. *Stamens* very many, spirally arranged on a prolonged axis ; lowest filaments very short and recurved, the remaining anthers sessile ; anthers elongated, more or less applied to the axis ; cells distinct, adnate to the connective, dehiscing longitudinally. Rudimentary *ovary* absent. *Female flowers : Calyx* 5-partite : segments ovate, acuminate, imbricate. *Ovary* 3–4-celled, densely pilose ; styles connate into a pilose column ; stigmas 3, spreading ; ovules 2 in each cell. *Capsule* breaking up into three or four 2-valved cocci ; endocarp crustaceous. *Seeds* compressed, ovate ; albumen fleshy ; cotyledons flat, much broader than the radicle.

A tall erect tree ; branchlets more or less silky-hairy, angular and articulated ; leaves opposite, decussate, long-petiolate, coriaceous, more or less silky beneath ; stipules large, coriaceous, connate, interpetiolar and intrapetiolar, forming a sheath enclosing the flowers and young leaves ; flowers axillary on silky-hairy peduncles, male 3-nate, female solitary in each leaf-axil ; male peduncles usually more or less connate, the lateral rather shorter than the central.

DISTRIB. A solitary species extending into the tropical part of Portuguese East Africa.

1. **A. Johnsonii** (Prain in Kew Bulletin, 1908, 439) ; a tall hard-wooded tree, providing valuable timber, branchlets angular and articulated ; leaves opposite, decussate, ovate, obtuse, rounded or subcordate at the base, 1¼–2 in. long, 1–1½ in. broad, entire, rigidly coriaceous, glabrous above, more or less densely woolly-hairy below ; stipular sheath ¾ in. long, silky-pubescent outside ; petioles ¼–⅓ in. long, silky-hairy ; flowers yellow ; peduncles ¼–⅓ in. long ; male flowers : sepals petaloid, lanceolate, with retuse or 2-lobed tips, long-pilose outside ; staminal axis ½–1 in. long ; female flowers : calyx-segments ¼ in. long, silky ; capsule depressed, ½ in. long ; seeds ⅓ in. long, ¼ in. broad ; testa brown and shining. *Prain in Kew Bulletin,* 1909, 201, *with figs., and* 1912, 307–8 ; *Hutchinson in Dyer,*

Fl. Trop. Afr. vi. i. 741, 1049.   *Weihea* (?) *subpeltata, Sim, For. Fl. Port. E. Afr.* 66, *t.* lxi. A.

EASTERN REGION : Portuguese East Africa ; Inhambane, *O'Neill*! Lebombo Mountains, *Sim*! Swaziland : Ubombo, *Warner*, 7009!

Occurs also in Tropical South East Africa.

VII. BRIDELIA, Willd.

Flowers monœcious or rarely diœcious. *Petals* present. *Male flower : Sepals* 5, valvate. *Petals* 5, usually small and scale-like, clawed or spathulate, the limb often toothed. *Disc* entire or sinuately lobed. *Stamens* 5 ; filaments connate in their lower part into a column in the centre of the flower, the upper parts of the filaments free and spreading ; anther-cells parallel, dehiscing longitudinally. Rudimentary *ovary* inserted at the apex of the staminal column, entire or divided. *Female flower : Sepals* often narrower than those of the male. *Disc* double, the outer annular ; the inner often cupular and embracing the ovary. *Ovary* 2- (rarely 3-) celled ; styles distinct or shortly connate at the base, bilobed or subentire ; ovules 2 in each cell. *Fruit* a small berry or drupe ; exocarp fleshy or pulpy ; endocarp crustaceous or hardened into 2 (or by abortion 1) pyrenes. *Seeds* often solitary in each pyrene ; albumen usually fleshy, deeply excavated on the inner face ; cotyledons broad and thin.

Shrubs or trees with alternate petiolate entire leaves ; tertiary nerves mostly parallel ; flowers small, in axillary glomerules, rarely in spicate clusters, the male numerous and subsessile, the female fewer or solitary, sometimes distinctly pedicellate ; bracts small and scaly ; berries or drupes ovoid or globose, small, smooth.

DISTRIB. About 41 species, mostly from Tropical Asia, ranging through the Malay Archipelago to New Caledonia and Australia ; a few in Madagascar, 16 in Tropical Africa, 6 of which occur in South Africa.

Fruits 2-celled ; leaves densely and softly pubescent on
both surfaces or below only :
    Leaves permanently hairy above ; sepals of the
      male flowers densely pubescent outside   ... (1) **mollis.**
    Leaves glabrous above from the beginning ; sepals
      of the male flowers glabrous   ...   . (2) **angolensis.**
Fruits 2-celled ; leaves glabrous :
    Lateral nerves of the leaves not extended to the
      margin to form a marginal nerve, more or less
      reticulate ...   . (3) **cathartica.**
    Lateral nerves of the leaves extended to the margin
      and forming a marginal nerve ...   . (4) **Schlechteri.**
Fruits 1-celled ; leaves glabrous or hairy :
    Leaves rusty-pubescent below ; veins usually promi-
      nent on the lower surface   ...   . (5) **ferruginea.**
    Leaves glabrous or minutely puberulous below ; veins
      scarcely prominent   ...   . (6) **micrantha.**

1. B. mollis (Hutchinson in Kew Bulletin, 1912, 100); a small tree, about 15 ft. high; branchlets brown-tomentose when young, at length becoming glabrous; leaves broadly elliptic or obovate-rotundate, rounded or truncate and sometimes slightly cordate at the base, $1\frac{1}{2}$–$4\frac{1}{2}$ in. long, $1\frac{1}{4}$–$3\frac{1}{2}$ in. broad, coriaceous, shortly and rather densely pubescent below; lateral nerves 9–12 on each side, branched towards the margin, but finally reaching it and forming a marginal nerve; tertiary nerves close, slightly prominent below; veinlets not or scarcely prominent below; petiole 2–$2\frac{1}{2}$ lin. long, densely pubescent; stipules lanceolate or subulate-lanceolate, 2–3 lin. long, 1 lin. broad, densely pubescent; male flowers subsessile or shortly pedicellate; sepals ovate-triangular, $1\frac{1}{2}$ lin. long, 1 lin. broad, pubescent outside; petals broadly obovate, slightly toothed in the upper half, 1 lin. long, $\frac{3}{4}$ lin. broad, glabrous; disc broad and flat, glabrous; staminal column $\frac{3}{4}$ lin. long, the free part of the filaments $\frac{1}{3}$ lin. long, very slender towards the apex; anthers $\frac{2}{3}$ lin. long, glabrous; rudimentary ovary $\frac{1}{2}$ lin. long, much swollen at the base, glabrous; female flowers very shortly pedicellate; sepals ovate-lanceolate, $1\frac{1}{2}$ lin. long, rusty-pubescent; petals oblong-lanceolate, $\frac{2}{3}$ the length of the sepals; disc cupular, lobed, glabrous; ovary glabrous; fruits 2-celled, transversely ellipsoid or subglobose, about 4 lin. in diam. *Hutchinson in Dyer, Fl. Trop. Afr.* vi. i. 612. *B. stipularis, Müll. Arg. in DC. Prodr.* xv. ii. 499, *partly (as to Kirk's Zambesi specimen) not of Blume.*

KALAHARI REGION: Transvaal; Warm Bath, Waterberg district, *Burtt-Davy*, 2404! 5603! Makapans Berg, at Streydpoort, *Rehmann*, 5393! Macalisberg Range, *Burke*! near Rustenburg, *Miss Pegler in Herb. Bolus*, 1063!

Occurs also in the Zambesi basin, Tropical Africa.

2. B. angolensis (Welw. ex Müll. Arg. in Journ. Bot. 1864, 327); a small tree, 15–20 ft. high; trunk 4–8 in. in diam. at the base, bare to a height of 6–7 ft., then with crowded branches and foliage; branchlets pubescent; leaves broadly elliptic, oblong-elliptic or ovate-elliptic, obtuse, somewhat truncate or rounded at the base, $2\frac{1}{2}$–4 in. long, $1\frac{1}{2}$–$2\frac{1}{2}$ in. broad, rigidly coriaceous, glabrous and dull above, with the nerves and veins densely pubescent below; lateral nerves 14–16 on each side, branching towards the margin and finally reaching it, and forming a marginal nerve, prominent below, impressed above; tertiary nerves close, prominent below; veinlets forming a deep network on the lower surface; petiole very thick, wrinkled, 3–4 lin. long, pubescent; stipules ovate–lanceolate; bracts ovate-deltoid, strongly keeled, brown-villous on the outside; male flowers subsessile; sepals ovate, obtuse, glabrous; petals conspicuous, obovate, 3–5-toothed; disc broad and flat, slightly rugose; rudimentary ovary subentire; young female flowers not seen; disc lobed, glabrous; fruit 2-celled, globose, 4 lin. in diam.; seeds flattened on the inner side; testa shining. *Müll. Arg. in DC. Prodr.* xv. ii. 496; *Hiern in Cat. Afr. Pl. Welw.* i. 953; *Hutchinson*

380   EUPHORBIACEÆ (Hutchinson).   [*Bridelia.*]

in Dyer, Fl. Trop. Afr. vi. i. 615. *B. angolensis, vars. typica and Welwitschii, Gehrm. in Engl. Jahrb.* xli. *Beibl.* 95, 31.

EASTERN REGION : Inhambane, *Schlechter* !
Occurs also in Angola, Tropical Africa.

3. **B. cathartica** (Bertol. f. Illustr. Mozambiq. 16, n. 13, t. 6); a shrub, about 6 ft. high; branchlets slender, glabrous, with very short internodes; leaves elliptic, oblong-elliptic or narrowly obovate, rounded or obtuse at the base, rounded or very obtusely pointed at the apex, 1–3½ in. long, ½–2 in. broad, papery, glabrous on both surfaces, more or less glaucous beneath ; lateral nerves looped before reaching the margin, 5–10 on each side ; tertiary nerves reticulate ; petiole 1–2 lin. long, wrinkled, glabrous or pubescent; stipules lanceolate or linear-lanceolate, acute, 1–1½ lin. long, sparingly pubescent ; bracts pubescent; male flowers : sepals ovate, subacute, glabrous ; petals suborbicular or obovate, entire or slightly toothed ; disc thin ; rudimentary ovary deeply bifid ; female flowers : sepals triangular, subacute, glabrous ; petals large, shortly clawed, entire ; disc enclosing the ovary, toothed, glabrous outside ; ovary ovoid, glabrous ; styles very short, bilobed ; fruits 2-celled, ellipsoid, 4½ lin. long, 4 lin. broad, black when dry. *Müll. Arg. in DC. Prodr.* xv. ii. 502 ; *Pax in Engl. Pfl. Ost-Afr. C.* 237 ; *Hutchinson in Dyer, Fl. Trop. Afr.* vi. i. 617.

KALAHARI REGION : Transvaal ; Komati Poort, *Kirk*, 100 ! Barberton, *Pole Evans*, 2945 !
EASTERN REGION : Portugese East Africa ; Ressano Garcia, *Schlechter*, 11890 !

Extends through the Zambesi basin to the Rovuma River in German East Africa.

4. **B. Schlechteri** (Hutchinson in Kew Bulletin, 1914, 249) ; branches rather slender and marked with prominent lenticels, young branchlets elongated, slightly pubescent at the nodes ; leaves oblong-oblanceolate, subacute, slightly narrowed to a rounded or obtuse base, 1–2¼ in. long, ½–1¼ in. broad, papery, glabrous on both surfaces, glaucous beneath ; lateral nerves continued to the margin, arcuate, very slender, distinct on both surfaces ; tertiary nerves parallel, slender ; petiole 1–1¼ lin. long, black, wrinkled, slightly pubescent; stipules subcaducous, subulate-lanceolate, acuminate, 2 lin. long, pubescent; male flowers shortly pedicellate ; sepals ovate, obtuse, about 1 lin. long, ½ lin. broad, glabrous ; petals suborbicular, about half as long as the sepals, rather fleshy, obscurely toothed at the apex ; disc orbicular, flat and fleshy ; staminal column slender, ⅔ lin. long; anthers small ; rudimentary ovary more or less subulate ; female flowers subsessile, pubescent around the base ; sepals as in the male but slightly larger ; petals as in the male ; outer disc fleshy, inner disc thinner and deeply lobed, glabrous ; ovary ellipsoid, smooth ; styles 2, free to the base, deeply bilobed ; young fruits ellipsoid, apparently 2-celled.

EASTERN REGION : Portuguese East Africa ; Inyamasan, *Schlechter,* 12065 !

This species has also been collected by Rogers (4551) at Beira, Port. E. Africa, which is beyond the area of the Flora Capensis.

5. B. ferruginea (Benth. in Hook. Niger Fl. 511); a shrub or small tree, 10–15 ft. high ; branchlets ferruginous-pubescent ; leaves elliptic and oblong-elliptic, very shortly acuminate, rounded at the base, 2–4 in. long, 1–2½ in. broad, coriaceous, glabrous or almost so above, pubescent on the nerves and veins below ; lateral nerves extending to the margin but often branching, slightly oblique, 7–9 on each side ; tertiary nerves wavy, somewhat lax ; petiole 1½–3 lin. long, tomentose ; stipules lanceolate, acute, about 3 lin. long, pubescent ; bracts ovate-lanceolate, pubescent ; male flowers : sepals ovate-lanceolate, subacute and very slightly hooded at the apex, pubescent outside with long adpressed hairs ; petals small, obovate, toothed ; disc thick, wrinkled ; rudimentary ovary subentire ; female flowers : sepals as in the male ; petals lanceolate, acute ; disc bottle-shaped, enclosing the ovary, lobed, long-pubescent outside ; ovary-ovoid ; styles 2, bipartite ; fruit 1-celled, ovoid-oblong, 3–4 lin. long, 2½ lin. in diam.   *Hiern in Cat. Afr. Pl. Welw.* i. 954, *excl. var.* ; *Gehrm. in Engl. Jahrb.* xli. *Beibl.* 95, 39 ; *Hutchinson in Dyer, Fl. Trop. Afr.* vi. i. 619.   *B. micrantha, var. ferruginea, Müll. Arg. in DC. Prodr.* xv. ii. 498 ; *De Wild. & Durand in Bull. Herb. Boiss.,* 2me *sér.* i. 46 ; *De Wild. Miss. É. Laurent,* 128, *and Études Fl. Bas- et Moyen-Congo,* ii. 276.   *B. speciosa, var. kourousensis, Beille in Bull. Soc. Bot. Fr.* lv. *Mém.* viii. 68.   *Gentilia Chevalieri, Beille, l.c.* 71.

EASTERN REGION : Delagoa Bay, *Schlechter* !

Widely spread in Tropical Africa.

6. B. micrantha (Baill. Adansonia, iii. 164); a tree, 20–40 ft. high ; branchlets pubescent when young, soon becoming quite glabrous, with conspicuous scattered lenticels ; leaves elliptic, oblong-elliptic or obovate, shortly acuminate and subacute or obtuse at the apex, slightly cuneate or rounded at the base, 2–7 in. long, 1–3 in. broad, slightly coriaceous, often shining and glabrous above, glabrous or minutely puberulous below ; lateral nerves ascending, slightly oblique, 8–16 on each side, continued to the margin and forming a marginal nerve, prominent below ; tertiary nerves more or less inconspicuous ; petiole stout, 2–5 lin. long, wrinkled, densely pubescent or puberulous ; stipules caducous ; bracts pubescent ; male flowers shortly stalked ; sepals triangular, subsessile, ½ lin. long, pubescent towards the base ; petals small, obovate-cuneate, trilobed or dentate ; disc thick and fleshy ; rudimentary ovary truncate or trifid, very short ; female flowers sessile : sepals as in the male ; petals ovate, entire ; disc almost enclosing the ovary, ciliate, densely pilose outside ; ovary ovoid, glabrous ; styles 2, 2-lobed, glabrous ; fruits ovoid or

ellipsoid, 2–3 lin. in diam., 1-celled. *Müll. Arg. in DC. Prodr.* xv.
ii. 498 (*excl. var. ferruginea with syn.*); *De Wild. et Durand, Reliq.
Dewevr.*, 200; *De Wild. Étud. Fl. Bas- et Moyen-Congo*, i. 275:
ii. 276; *Sim, For. Fl. Cape Col.* 317; *Hutchinson in Dyer, Fl.
Trop. Afr.* vi. i. 620. *B. gambecola, Baill. Adansonia*, i. 79;
Gehrm. in Engl. Jahrb. xli. *Beibl.* 95, 40. *B. stenocarpa, Müll.-Arg.
in Flora*, 1864, 515; *Pax in Engl. Jahrb.* viii. 61; *Pax in Bolet.
Soc. Brot.* x. 157; *Gehrm. l.c. B. ferruginea, var. gambecola,
Hiern in Cat. Afr. Pl. Welw.* i. 954. *Candelabria micrantha, Hochst.
in Flora*, 1843, 79.

KALAHARI REGION: Transvaal; Rehbokdraai, near Barberton, *Burtt-Davy*,
1626! Barberton, *Burtt-Davy*, 8003! 8035! Shilovane, *Junod*, 682! near
Haenertsburg, Zoutpansberg, *Legat*, 4031! Tzaneen Estate, *Burtt-Davy*, 2563!
Charter, 4213! Potatobosch, *Eastwood*, 2432!

EASTERN REGION: Natal; in woods near Durban, *Krauss*, 133! *Plant*, 43!
Gueinzius! *Gerrard*, 527! Mount Edgecumbe, *Wood*, 1146! Inanda, *Wood*, 1334!
1702! Nolote River, *Gerrard*, 25! and without precise locality, *Gerrard*, 372!

Widely spread in Tropical Africa.

VIII. CLEISTANTHUS, Hook. f.

Flowers monœcious or diœcious. *Petals* usually present. *Male
flower*: *Calyx-segments* 5, valvate. *Petals* 5 or rarely absent, scale-
like, small. *Disc* entire or sinuately lobed. *Stamens* 5, as in
Bridelia; anther-cells parallel, dehiscing longitudinally. Rudimentary *ovary* in the middle of the staminal column, often divided.
Female flower: *Calyx* less deeply divided than that of the male.
Disc double, the inner part cupular, surrounding the ovary. *Ovary*
3-celled; styles distinct, bifid; ovules 2 in each cell. *Capsule*
globose or depressed-globose, often 3-lobed, splitting into three
2-valved cocci. *Seeds* 2 in each cell (or 1 by abortion); albumen
scanty; cotyledons thick and fleshy, rarely thin, often more or
less plicate.

Trees or shrubs; leaves alternate, with reticulate tertiary nerves; flowers small,
in axillary fascicles or very short racemes.

DISTRIB. About 50 species, mostly Indian and Malayan, a few in New Caledonia,
Australia, Madagascar and Tropical Africa.

1. C. Schlechteri (Hutchinson); branches slightly sulcate, glabrous,
greyish; young branchlets shortly pubescent with brown hairs; leaves
oblong or oblong-elliptic, rounded at both ends or almost subcordate
at the base, ½–1 in. long, 3–5 lin. broad, entire, chartaceous, glabrous
on both surfaces; lateral nerves 4–6 on each side, distinct on both
surfaces, arcuate, prominent below; petiole ½–1¼ lin. long, shortly
pubescent; stipules deciduous; flowers monœcious; racemes very
short and few-flowered, about half male and half female; pedicels

fairly slender, up to ⅔ in. long, puberulous; buds clavate-ellipsoid, obtuse, about 2 lin. long; sepals linear or oblong-linear, obtuse, 2 lin. long, minutely puberulous; petals small, about ⅕ the length of the sepals, subspathulate; disc thick, undulate, glabrous; anthers 1⅕ lin. long; rudimentary ovary small, tripartite; female flowers similar to the male; ovary subglobose, black, glabrous; styles 3, somewhat slender, bifid, glabrous; ripe fruits not seen. *Securinega Schlechteri, Pax in Engl. Jahrb.* xxviii. 18.

Eastern Region : Delagoa Bay; Lorenzo Marques, *Schlechter,* 11524 !

This is undoubtedly a species of *Cleistanthus,* and very closely allied to *C. Holtzii,* Pax, but distinguished by its much smaller leaves.

IX. LACHNOSTYLIS, Turcz.

Flowers diœcious. *Petals* present. *Male flower: Sepals* 5, imbricate, subequal. *Petals* 5, slightly larger than the sepals. *Disc-glands* thick, villous. *Stamens* 5; filaments connate for half their length, spreading in the upper part around a villous tripartite rudimentary ovary; anther-cells parallel, dehiscing longitudinally. *Female flower: Sepals* and *petals* as in the male. *Disc* annular, thick, villous. *Ovary* ovoid-globose, 3-celled, tomentose; *styles* short, recurved-spreading, shortly bifid; ovules 2 in each ovary-cell. *Cupsule* breaking up into 2-valved cocci, pericarp thick and crustaceous. *Seeds* subglobose, smooth; albumen scanty; cotyledons broad, much contorted-plicate.

A much-branched shrub; leaves rather small, alternate, coriaceous, entire; flowers fasciculate in the leaf-axils, the male shortly pedicellate, the female fewer than the male or subsolitary, long pedicellate.

Distrib. Species 1, endemic.

1. **L. capensis** (Turcz. in Bull Soc. Nat. Mosc. xix. ii. 503); a shrub, about 7 ft. high, much branched; branches twiggy, tomentose or pubescent when young, at length glabrous and covered with numerous conspicuous lenticels; leaves oblanceolate, somewhat rounded and mucronate at the apex, gradually narrowed to the base, ½–2 in. long, ⅓–¾ in. broad, rather rigidly coriaceous, with entire cartilaginous margins, rather closely and conspicuously reticulate on both surfaces especially above, puberulous on both surfaces when young, at length glabrous; lateral nerves 5–7 on each side, spreading, looped and branched near the margin, conspicuous on both surfaces; petiole about 1½ lin. long, often slightly pubescent; stipules lanceolate, membranous, scarcely 1 lin. long, sparingly pubescent or glabrous; flowers diœcious, in axillary fascicles, shortly pedicellate; pedicels woolly-tomentose; male flower: sepals 5, ovate-elliptic, obtuse, woolly-pubescent, concave on the inside, 1½ lin. long, 1 lin. broad; petals obovate, narrowed to the base, slightly larger than the

384

EUPHORBIACEÆ (Hutchinson).   [*Lachnostylis.*

sepals, striately nerved, nearly glabrous; disc-glands 5, large, woolly-pubescent; stamens 5; filaments connate into a column for half their length, glabrous; column 1 lin. long, the free part 1¼ lin. long; anthers ⅓ lin. long; rudimentary ovary tripartite, segments bilobed, villous; female flower: pedicels up to ¾ in. long, tomentulose; sepals as in the male but nearly glabrous; petals narrower than in the male; disc thick, annular, villous; ovary ovoid-globose, 1½ lin. in diam., tomentose; style spreading, ¾ lin. long, recurved at the tip; capsule trilobed, about 4 lin. in diam., rather densely pubescent; seeds (immature?) wrinkled. *Sond. in Linnæa*, xxiii. 132; *Baill. Étud. Gèn. Euph.* 663; *Sim, For. Fl. Cape Col.* 313, *t.* cxlii. *fig.* i.   *Cluytia hirta, Linn. f. Suppl.* 432; *Vahl, Symb.* ii. 101; *Thunb. Prodr.* 53; *Fl. Cap. ed. Schult.* 272. *Clutia acuminata, Thunb. Prodr.* 53; *Fl. Cap. ed. Schult.* 272, *excl. syn. Lachnostylis minor, Sond. in Linnæa*, xxiii. 132. *Lachnostylis hirta, Müll. Arg. in DC. Prodr.* xv. ii. 224, *incl. vars. genuina, acuminata and minor, Müll. Arg.*

SOUTH AFRICA: without precise locality, *Hallach*, 11!
COAST REGION: Swellendam Div.; dry hills near Breede River, *Burchell*, 7478! George Div.; Kaimans Gat, *Alexander (Prior)*! Mossel Bay Div.; Mossel Bay, *Thunberg*! Knysna Div.; Kaatjes Kraal, *Burchell*, 5213! Plettenberg Bay, *Pappe*! Knysna Forest, *McNaughton in Herb. Galpin*, 2593! *Bowie*! Humansdorp Div.; Diep Valley near Humansdorp, *Bolus*, 2396! between Twee Fontein and Essenbosch, *Burchell*, 4823! Uitenhage Div.; between Galgebosch and Melk River, *Burchell*, 4758! Vanstadens River, *MacOwan*, 1942! *Mrs. Paterson*, 1996! Zwartkops River, *Alexander (Prior)*! Krakakamma and Uitenhage, *Zeyher*; and without precise locality, *Zeyher*, 783! Port Elizabeth Div.; near Port Elizabeth, *Burchell*, 4364! Elands River, *Zeyher*, 3618! Zuurbraak, *Schlechter*, 2123! and without precise locality, *Ecklon & Zeyher*, 105, 3! 34, 2, 1!

X. HEYWOODIA, Sim.

Flowers dioecious. *Petals* present. *Male flower: Sepals* 3, unequal, imbricate. *Petals* 5, about twice the size of the sepals, strongly imbricate, membranous, slightly clawed, entire. *Disc* 0. *Stamens* 5, inserted in the middle of the flower; filaments very short, connate at the base; anthers 2-celled, cells parallel, dehiscing longitudinally. Rudimentary *ovary* absent. Young *female flowers* not known. *Capsules* in axillary fascicles of 3–4, pedicellate, breaking up into 2-valved cocci; exocarp wrinkled and yellow when dry, rather thin, connected with the central axis by 6–7 strands of fibres between the septa; endocarp somewhat thicker, firmly crustaceous. *Seeds* about 4 lin. long, smooth.

A large glabrous tree with alternate petiolate entire shining leaves and lax venation, male flowers small, in dense axillary glomerules; female in axillary fascicles of 3–4, pedicellate.

DISTRIB. Monotypic, endemic.

1. **H. lucens** (Sim, For. Fl. Cape Col. 326, t. 140); a tree, 20–50 ft. high; branches covered with ashy bark; young branchlets rather slender, obtusely angled or subterete, glabrous; leaves elliptic or ovate-elliptic, cuneate at the base, gradually and obtusely pointed at the apex, 2½–4 in. long, 1–2 in. broad, entire, rigidly coriaceous, glabrous and shining on both surfaces; lateral nerves 4–5 on each side, prominent on both surfaces, looped and branched some distance from the margin, diverging from the midrib at an angle of 45°; veins very lax, distinct on both surfaces; petiole ¼–½ in. long, glabrous; male glomerules nearly ¼ in. in diam.; bracts suborbicular, membranous, up to ¾ lin. broad, glabrous; sepals 3, unequal, orbicular, ½–1 lin. in diam., membranous, convex outside, brown and a little thicker in the upper part, glabrous; petals 5, orbicular, slightly clawed, very convex on the outside, about twice the size of the sepals, membranous, glabrous; stamens 8; filaments very short, connate at the base; anthers 2-celled, cells distinct, parallel, dehiscing longitudinally, 1 lin. long; female flowers not known; capsule seen only after dehiscence, evidently ovoid-ellipsoid, about ½ in. long; exocarp yellow and wrinkled when dry, crustaceous, glabrous, with 6–7 strands of fibres connecting it between the septa with the central axis; endocarp yellow, firmly crustaceous, nearly ½ lin. thick; seeds lanceolate-ovoid, about 4 lin. long, light brown, smooth and shining, with a narrow groove down the back when dry.

EASTERN REGION: Transkei; Dwessa Forest, *Sim,* 2594! Pondoland; Port St. John, on the river bank, *Galpin,* 3486!

Native name "Nebelele" and known as Cape or Black Ebony.

XI. ANDRACHNE, Linn.

Flowers monœcious. *Petals* present. *Male flowers: Sepals* 5 or 6, membranous. *Petals* subequal to the sepals. *Disc* cupular, crenulate. *Stamens* 5 or 6, alternating with the petals; filaments free or connate towards the base; anthers erect; cells parallel, distinct, dehiscing longitudinally. Rudimentary *ovary* tripartite. *Female flowers: Sepals* and *petals* often more coriaceous than those of the male. *Disc* as in the male. *Ovary* 3-celled; styles very short, bifid or bipartite; ovules 2 in each cell. *Capsule* dry, trilobed, the lobes keeled, breaking into 2-valved cocci. *Seeds* wrinkled, estrophiolate; albumen fleshy; embryo curved, with a long radicle; cotyledons flat and broad.

Diffuse herbs, undershrubs or slender shrubs; leaves alternate, petiolate, membranous, usually small: flowers pedicellate, axillary, the males often fasciculate, the female solitary.

DISTRIB. About 20 species, in the warmer parts of the Northern Hemisphere.

1. **A. ovalis** (Müll. Arg. in DC. Prodr. xv. ii. 233); a shrub, about 5 ft. high; branches terete, glabrous; leaves ovate, ovate-elliptic or obovate, obtuse at the apex, cuneate at the base, $\frac{3}{4}$–2 in. long, $\frac{1}{3}$–1$\frac{1}{4}$ in. broad, entire, rigidly membranous, bright green, glabrous on both surfaces; lateral nerves 4–5 on each side, looped and branched well within the margin, slightly prominent on both surfaces; petiole 2–5 lin. long, terete, glabrous; stipules ovate-triangular, acute, minutely ciliolate, $\frac{1}{2}$ lin. long; flowers monœcious; male pedicels 1–2$\frac{1}{2}$ lin. long, glabrous; sepals 5, oblong-obovate, obtuse, 1 lin. long, $\frac{3}{4}$ lin. broad, membranous, glabrous, margins jagged or ciliolate; petals obovate, rounded at the apex, narrowed to the base, membranous, glabrous, a little smaller than the sepals; disc cupular, membranous, crenulate on the margin, glabrous; stamens connate in their lower part; rudimentary ovary tripartite, the segments swollen at the tips, very sparingly pubescent; female pedicels 4–7 lin. long, glabrous, thickened towards the apex; sepals, petals and disc more coriaceous than those of the male; ovary sparingly setulose; styles extremely short, bifid; capsule trilobed, lobes slightly keeled, about 5 lin. in diam., reticulate; seeds wrinkled. *Phyllanthus ovalis, Sond. in Linnæa,* xxiii. 135; *Drège, Zwei Pfl. Documente,* 142. *P. dregeanus, Scheele in Linnæa,* xxv. 585. *P. capensis, Spreng. ex Baill. Adansonia,* iii. 163. *Cluytia ovalis, Scheele in Linnæa,* xxv. 583; *Baill. Adansonia,* iii. 153. *Andrachne capensis, Baill. Adansonia,* iii. 163. *A. dregeana, Baill. l.c.* 164. *Cluytia Galpini, Pax in Bull. Herb. Boiss.* vi. 736, *partly.*

SOUTH AFRICA : without precise locality, *Zeyher,* 246; *Harvey*! *Ecklon & Zeyher,* 31!

COAST REGION: George Div.; George, *Schlechter,* 2362! *Prior*! Knysna Div.; Kaatjes Kraal, *Burchell,* 5216! 5226! Uitenhage Div.; near Uitenhage, *Burchell,* 4272! *Alexander (Prior)*! *Zeyher*! Springfields, *Paterson,* 1034a! 2197! Georgetown, *Alexander (Prior)*! Albany Div.; Blue Krantz, *Burchell,* 3644! Zwartkops River, *Zeyher,* 3819! King Williamstown Div.; Buffalo River, *Drège*! and without precise locality, *Sim,* 2229! Komgha Div.; near Keimouth, *Schlechter,* 6188!

KALAHARI REGION : Transvaal; Houtbosch, *Rehmann,* 5922! Umvoti Creek. near Barberton, *Galpin,* 961!

EASTERN REGION : Tembuland ; between Morley and Umtata River, *Drège,* 8220! Natal; near Mount West, *Schlechter,* 6826! and without precise locality, *Gerrard,* 1161! 1162! *Cooper,* 1108!

XII. PHYLLANTHUS, Linn.

Flowers monœcious or sometimes diœcious. *Petals* absent. *Disc* always present in the African species. *Male flowers*: *Sepals* 4–6, free or slightly joined at the base, imbricate, when 4 or 6 often in 2 series. *Disc* of separate glands (annular in 1. *P. discoideus*):

glands usually fleshy, smooth or more or less warted. *Stamens* 2–6, in the middle of the flower; filaments free or connate, or some free and the others connate; anthers 2-celled, oblong or rounded, cells parallel and dehiscing longitudinally, or diverging from the apex, the line of dehiscence then appearing transverse and often continuous between the cells; connective often slightly produced. Rudimentary *ovary* absent. *Female flowers:* Sepals as in the male but often larger. *Disc* hypogynous, usually saucer-shaped or cupular, entire, variously toothed or lobed, or rarely of separate glands. *Ovary* usually 3-celled, sessile or rarely slightly stipitate; styles 3, rarely absent, free or partially connate, bifid or bilobed (rarely entire), the arms slender and sometimes swollen at the apex; ovules 2 in each cell. *Capsule* dry or more rarely fleshy, dehiscent or sub-indehiscent, breaking up into 2-valved cocci. *Seeds* trigonous, convex on the back and often longitudinally sulcate or pitted, without a strophiole; testa membranous or crustaceous; albumen fleshy; embryo straight or slightly incurved; cotyledons flat and straight or rarely flexuous.

Herbs, shrubs or trees of various habit; leaves alternate in all the African species, entire, often distichous, the flowering branchlets frequently simulating pinnate leaves; flowers small, axillary, the males mostly numerous and fasciculate, usually pedicellate, the females few and mostly solitary.

DISTRIB. About 450 species, spread throughout the tropical regions of both hemispheres.

Phyllanthus pervilleanus, Müll. Arg. in Linnæa, xxxii. 13; *Kirganelia pervilleana,* Baill. Adansonia, ii. 50, is attributed in the Index Kewensis to South Africa, but seems to be confined to the Mascarene Islands.

Kirganelia elegans, Juss. (= *Phyllanthus Casticum,* Soyer-Willemet) is quoted by Baillon (Adansonia, iii. 165) from the Cape, and Sonnerat's specimen is marked so in the Jussieu Herbarium in Paris. Baillon suggests that it is cultivated at the Cape; but it seems more probable that the label is wrong, and that the specimen is really from Mauritius.

*Stamens 4–5; filaments free to the base or only one or
    two connate and the remainder free:
  Large woody much-branched shrubs or trees:
    Flowers diœcious:
      Disc of the male flowers annular, entire   ... (1) **discoideus.**

      Disc of the male flowers composed of separate
        glands ... ... (2) **flacourtioides.**

    Flowers monœcious:
      Branches conspicuously verrucose; flowering
        branchlets not produced in the axils of
        flowerless branches ... ... (3) **verrucosus.**

      Branches smooth or very slightly verrucose;
        flowering branchlets produced in the axil of
        a leafy flowerless shoot ... ... (4) **reticulatus.**

Annual herbs or small rhizomatous undershrubs :
Branchlets and often the leaves on both sides
finely scabrid-puberulous :
Plants 2–4 in. high with crowded branchlets ... (5) **parvulus.**

Plants slender, 6–12 in. high, laxly branched ... (6) **humilis.**

Branchlets and leaves quite smooth :
Pedicels capillary, about ¾ in. long ...   ... (7) **nummu-**
                                                       larifolius.
Pedicels not or scarcely capillary, about 1 lin.
long or less :
Flowering branchlets conspicuously flexuous (8) **pentandrus.**

Flowering branchlets not flexuous ...   ... (9) **Burchelli.**

**Stamens 3 ; filaments quite free ; leaves ovate or
ovate-orbicular   ...   . (10) **glaucophyllus.**

***Stamens 3 or 2 ; filaments connate to the apex :
Stamens 3 ; plants never densely pubescent or
subtomentose :
Female disc composed of distinct separate glands :
Leaves oblanceolate or linear-lanceolate, appre-
ciably broad and mostly with strongly
marked nerves   ...   . (11) **madera-**
                                                  spatensis.

Leaves linear, usually long and very narrow,
without visible nerves   ...   . (12) **incurvus.**

Female disc annular, entire, toothed or lobed :
Flowers diœcious ; a shrub with fairly large
ovate leaves and rather large flowers   ... (13) **myrtaceus.**

Flowers monœcious ; small herbs with small
more or less elliptic or suborbicular leaves
and small flowers :
Sepals 5 ...   . (14) **meyerianus.**

Sepals 6 :
Branchlets finely asperulate ; leaves mostly
about ½ in. long   ...   . (15) **asperulatus.**

Branchlets smooth ; leaves usually smaller
than in the above :
Disc of the female flowers pectinately
10-toothed   ...   . (16) **Niruri.**

Disc of the female flowers entire or
slightly undulate :
Stigmas subentire and almost sessile
on the ovary ; leaves very small
and fairly dense   ...   . (17) **delagoensis.**

Stigmas on fairly long spreading styles,
sometimes bifid ; leaves fair-sized
and mostly very laxly arranged ... (18) **heterophyllus.**

Stamens 2 ; plants densely pubescent or sub-
tomentose ...   . (19) **cinereoviridis.**

1. **P. discoideus** (*Müll. Arg. in Linnæa*, xxxii. 51) ; a tree, 30–
50 ft. high or sometimes a shrub ; branches finely sulcate, subterete ;
lateral flowering branchlets up to 2½ in. long, spreading, puberulous

or rarely glabrous ; leaves ovate-elliptic to obovate-oblanceolate,
rounded or very shortly acuminate at the apex, varying from
rounded to cuneate at the base, 1–4 in. long, $\frac{3}{4}$–1$\frac{1}{2}$ in. broad,
thinly chartaceous or membranous, glabrous on both surfaces
except the sometimes puberulous midrib below ; lateral nerves
6–12 on each side, spreading, slightly raised on both surfaces ;
veins slender and rather close ; petiole 1$\frac{1}{2}$–2 iin. long, tomentulose
or glabrous ; stipules soon falling off, oblong-linear, 1$\frac{1}{2}$–2$\frac{1}{2}$ lin. long,
membranous, glabrescent ; flowers diœcious ; males numerous, in
fascicles in the axils of the leaves, females similarly arranged but
only about 2 in each fascicle ; male pedicel very slender, 1–1$\frac{1}{2}$ lin.
long, glabrous or sparingly pubescent, female stouter, 2–3 lin. long,
glabrous or rather densely puberulous ; bracts large, ovate-orbicular,
membranous, soon falling off ; male flowers : sepals 4, obovate-
oblong, $\frac{1}{2}$ lin. long, glabrous ; disc fleshy, entire, glabrous ; stamens
4 ; filaments free to the base ; anthers dehiscing longitudinally ;
female flowers : sepals 4, larger than those of the male ; disc large,
fleshy, flat, entire, very minutely puberulous ; ovary ellipsoid,
glabrous ; styles connate at the base, the free parts spreading and
flattened, bilobed ; capsule 3–4-lobed, 4 lin. in diam., glabrous ;
seeds triquetrous, rounded on the back, smooth and blackish.
Müll. Arg. in DC. Prodr. xv. ii. 416 ; *Ficalho, Pl. Uteis*, 249 ; *Pax
in Bolet. Soc. Brot.* x. 157 ; *Hiern in Cat. Afr. Pl. Welw.* i. 960 ;
S. Moore in Journ. Linn. Soc. xl. 192 ; *Hutchinson in Dyer, Fl. Trop.
Afr.* vi. i. 707. *P. anomalus, Müll. Arg. l.c.* 418 (*as to Kirk's
specimen from Lake Nyasa*) ; *Pax in Engl. Pfl. Ost-Afr. C.* 230.
P. amapondensis, Sim, For. Fl. Cape Col. 325, *t.* cxli. *fig.* ii. *Cicca
discoidea, Baill. Adansonia,* i. 85. *Securinega bailloniana, Müll.
Arg. l.c.* 451. *Diasperus discoideus, O. Kuntze, Rev. Gen. Pl.* ii.
599. *Fluggea obovata, Baill. Adansonia,* ii. 41 (*not Xylophylla
obovata, Willd.*). *F. major, Baill. Étude Gén. Euphorb.* 593, *and
Adansonia,* ii. 42, *and* iii. 164. *F. nitida, Pax in Engl. Jahrb.* xix.
76, *and in Engl. Pfl. Ost-Afr. C.* 236. *F. bailloniana, Pax, l.c.
in obs.*

SOUTH AFRICA : without precise locality, *Lalande* !
EASTERN REGION : Delagoa Bay, *Junod*, 59 ! Lorenzo Marques, *Schlechter*,
11598 ! 11634 ! Natal ; Umcomaas, *Gerrard*, 1176 ! Pondoland ; Egossa Forest,
Sim, 2608 !

Very common and widely spread in Tropical Africa.

2. **P. flacourtioides** (Hutchinson in Kew Bulletin, 1915, 48) ;
shrubby ; branches terete, glabrous ; young branchlets quite glabrous,
blackish when dry ; leaves elliptic or oblong-elliptic, more or less
rounded at both ends, $\frac{1}{3}$–2 in. long, $\frac{1}{4}$–1 in. broad, entire, thinly
chartaceous, glabrous ; lateral nerves about 6 on each side of the
midrib, arcuate, slender, distinct on both surfaces ; veins laxly
reticulate below ; petiole about 2 lin. long, glabrous ; stipules linear-
lanceolate, acute, somewhat membranous, about 2 lin. long, glabrous,
with a distinct midrib and rather hyaline margins ; flowers probably

dioecious; males fasciculate at the ends of the shoots; pedicels 1½ lin. long, slightly thickened towards the apex, glabrous; sepals 4, obovate, rounded at the apex, ¾ lin. long, ⅔ lin. broad, glabrous; disc-glands small, thin, rounded, smooth; stamens 4; filaments free, about half the length of the anthers, the latter ellipsoid, ½ lin. long, dehiscing at the side; female flowers apparently subsolitary towards the base of the young shoots; pedicels 4½ lin. long in the fruiting stage, glabrous; sepals 4, broadly ovate, obtuse, 1¼ lin. long, 1 lin. broad, submembranous, glabrous; disc annular, small; young fruits trilobed, glabrous; styles connate in their lower half, the free part abruptly recurved and bilobed to near the base.

EASTERN REGION : Delagoa Bay ; Lorenzo Marques, 150 ft., *Schlechter*, 11598 ! 11634 !

3. P. verrucosus (Thunb. Prodr. 24); a much-branched shrub, 4–8 ft. high; branches straight, covered with a grey bark, verrucose with conspicuous lenticels; lateral twigs increasing in length towards the base of each main shoot; flowering branchlets very much abbreviated, each with a cluster of 3–5 leaves; leaves obovate or obovate-elliptic, rounded and often slightly retuse at the apex, obtuse or subcuneate at the base, 2–6 lin. long, 1¼–4 lin. broad, thinly chartaceous, glabrous on both surfaces, often somewhat glaucous below, distinctly and rather closely reticulate below; petiole ¾–1 lin. long, glabrous; stipules small, laciniate, membranous, puberulous; flowers monoecious, one female and several males in each cluster; male pedicels capillary, up to 3 lin. long, glabrous; sepals usually 6, rarely 5 or 4, obovate, about ½ lin. long, membranous, glabrous; disc-glands contiguous, fleshy, spread over the top of the torus; stamens usually 6; filaments free, inserted between the disc-glands; anthers rounded, about ⅓ lin. broad; female pedicels stouter than those of the male, about ½ in. long, often becoming recurved, glabrous or slightly hairy towards the apex; sepals as in the male but larger; disc thick and fleshy, slightly toothed, glabrous; ovary smooth; stigmas large and fleshy, reflexed, bilobed; capsule prominently reticulate; seeds smooth. *Thunb. Fl. Cap. ed. Schult.* 500; *Drège, Zwei Pfl. Documente*, 136. *Müll. Arg. in Linnæa*, xxxii. 5, and *in DC. Prodr.* xv. ii. 333. *Pleiostemon verrucosum, Sond. in Linnæa*, xxiii. 136. *Securinega verrucosus, Sim, For. Fl. Cape Col.* 325, *t.* xvi. *fig.* v.

SOUTH AFRICA : without precise locality, *Harvey* ! *Ecklon & Zeyher*, 29 ; *Krauss.*

COAST REGION : Humansdorp Div. ; Gamtoos River, *Thunberg* ! Uitenhage Div. ; Elands River, *Zeyher* ! near Addo, *Zeyher*, 857 ! 3820 ! *Pappe* ! between Galgebosch and Melk River, *Burchell*, 4771 ! near Uitenhage, *Burchell*, 4451 ! *Alexander* (*Prior*) ! *Penther* ! *Schlechter*, 2533 ! Zuurberg Range, *Drège* a ! *Drège*, 2375 ! Bathurst Div. ; Trapps Valley, *Daly*, 620 ! Glenfilling, *Drège* ! Albany Div. ; Blue Krantz, *Burchell*, 3630 ! near Grahamstown, *MacOwan*, 510 ! *Schönland*, 597 ! Howisons Poort, *Hutton* ! Karega River, *Ecklon & Zeyher*, 9 ! Fish River Hill, *Burke* ! Fort Beaufort Div. ; Konap, *Baur*, 1080 ! East London Div. ; near mouth of Kefani River, *Galpin*, 5806 ! Komgha Div. ; near Komgha, *Flanagan*, 404 ! Gwenkale, River, *Flanagan* ! British Kaffraria ; without precise locality, *Cooper*, 312 ! 1904 !

4. **P. reticulatus** (Poir. Encycl. v. 298) ; an erect much-branched shrub ; stems pubescent, at length nearly glabrous ; flowering branchlets sometimes produced in fascicles, but more often solitary, up to 2½ in. long, densely crisped-pubescent ; leaves oblong or elliptic, mostly rounded at both ends, ½–1¼ in. long, 3½–7 lin. broad, somewhat membranous, glabrous or crisped-pubescent ; lateral nerves 7–9 on each side ; petiole pubescent ; stipules lanceolate, acute ; flowers monœcious, axillary, one female and two or three males in each fascicle ; pedicel slender, up to 2 lin. long, slightly pubescent or glabrous ; male flowers : sepals 5, ovate-elliptic, 1-nerved, about ¾ lin. long ; disc-glands 5, obovate, flattened, smooth ; stamens 5 ; two or three of the filaments partially connate, the others free and shorter ; anthers dehiscing at the side, cells cohering at their tips, diverging at the base ; female flowers : sepals as in the male ; disc of separate glands similar to those of the male ; ovary depressed-globose, glabrous ; styles very short, erect, and crowded ; fruit fleshy or coriaceous, 8–16-seeded, about 3 lin. in diam., seeds irregularly trigonous, punctulate. *Pax in Engl. Pfl. Ost-Afr. C.* 236 ; *Durand & De Wild. in Bull. Soc. Bot. Belg.* xxxvii. 103 ; *De Wild & Durand, Reliq. Dewevr.* 205 ; *De Wild. Miss. É. Laurent,* 127, *Études Fl. Bas- et Moyen-Congo,* i. 275 ; ii. 268 ; *Gibbs in Journ. Linn. Soc.* xxxvii. 469 ; *Hutchinson in Dyer, Fl. Trop. Afr.* vi. i. 700. *Phyllanthus multiflorus, Willd. Sp. Pl.* iv. 581. *P. pricurianus, Müll. Arg. in Linnæa,* xxxii. 12. *P. alaternoides, Reichb. ex Baill. Adansonia,* i. 83. *P. reticulatus, var. genuinus, Mull. Arg. in DC. Prodr.* xv. ii. 344 ; *Hiern in Cat. Afr. Pl. Welw.* i. 958. *Anisonema reticulatum, A. Juss. Euph. t.* 4, *fig.* 11. *Kirganelia reticulata, Baill. Étude Gén. Euphorb.* 613. *K. prieuriana, Baill. Adansonia,* i. 82. *Diasperus reticulatus, O. Kuntze, Rev. Gen. Pl.* ii. 600.

VAR. β, **glaber** (Müll. Arg. in Linnæa, xxxii. 12) ; entirely glabrous, otherwise similar to the typical plant. *Müll. Arg. in DC. Prodr.* xv. ii. 345 ; *Hiern, l.c.* ; *Hutchinson, l.c.* 701. *Phyllanthus polyspermus, Schumach. & Thonn. Beskr. Guin. Pl.* 416. *Kirganelia prieuriana, var. glabra, Baill. Adansonia,* i. 83. *K. zanzibariensis, Baill. l.c.* ii. 48.

KALAHARI REGION : Transvaal ; near Barberton, *Pole Evans,* 2942 ! *Burtt-Davy,* 8040 ! between Thabina and Sutherlands Middle Veld, *Burtt-Davy,* 5443 ! Shiluvane, *Junod,* 654 !

EASTERN REGION : Delagoa Bay ; *Sanderson,* 574 ! near Lorenzo Marques, *Bolus,* 9769 ! 9770 ! Var. β : Lorenzo Marques, *Bolus,* 9771 ! *Junod,* 54 ! and without precise locality, *Forbes,* 48 !

Widely spread in the Tropics of the Old World.

5. **P. parvulus** (Sond. in Linnæa, xxiii. 132) ; a small plant (annual?), 2–4 in. high, branched from the base, with a long tapering taproot ; branches slender, minutely puberulous ; leaves ovate-elliptic or elliptic-lanceolate, subacute, rounded at the base, 2–2½ lin. long, 1¼–1½ lin. broad, dull and very slightly scaberulous on both surfaces, with one or two indistinct pairs of lateral nerves ; petiole ⅓ lin. long ; stipules subulate, very acute, ½ lin. long,

glabrous; flowers monœcious, solitary, the males in the upper part, the females in the lower part of each branchlet; male flowers: pedicel $\frac{1}{2}$ lin. long, glabrous; sepals 5, broadly ovate, obtuse, with a distinct midrib and membranous margin; disc-glands flat and thin; more or less transversely oblong, smooth; stamens 5; filaments free; anthers dehiscing at the side, with the line of dehiscence continuous between the cells; female flowers: pedicel $\frac{3}{4}$ lin. long, rather stout, glabrous; sepals 5, ovate, obtuse, $\frac{3}{5}$ lin. long, $\frac{1}{2}$ lin. broad, with a broad midrib and membranous margin, glabrous; disc saucer-shaped, thin, quite entire, glabrous; ovary lobed, smooth; styles short, spreading over the top of the ovary, shortly bilobed, not swollen at the tips; capsule trilobed, depressed, $1\frac{3}{4}$ lin. in diam., smooth and shining; seeds finely and densely pitted on the back. *P. tenellus, var. scabrifolius, Müll. Arg. in Linnæa,* xxxii. 7 ; *var. parvulus, Müll. Arg. in DC. Prodr.* xv. ii. 339.

KALAHARI REGION : Transvaal ; Aapies River, *Burke* ! *Zeyher,* 1508 !

6. **P. humilis** (Pax in Engl. Jahrb. x. 34); woody, much-branched from the base; branches slender, finely puberulous; leaves ovate or ovate-oblong, acute or subacute, rounded or almost truncate at the base, $1\frac{1}{3}$–4 lin. long, $\frac{3}{4}$–$2\frac{1}{2}$ lin. broad, thinly coriaceous, finely puberulous on both surfaces, without visible nervation; petiole $\frac{1}{4}$ lin. long, puberulous; stipules subulate-lanceolate, very acute, nearly glabrous; flowers monœcious; males few, solitary, axillary; pedicel $\frac{1}{3}$ lin. long, glabrous; sepals 5, obovate, scarcely $\frac{1}{3}$ lin. long, glabrous, membranous; disc-glands flat, transversely oblong, smooth; stamens 5; filaments free; female flowers more numerous than the males, axillary, solitary; pedicel scarcely $\frac{1}{2}$ lin. long, glabrous; sepals 5, as in the male but larger; disc small, slightly undulately lobed, glabrous; ovary smooth; styles short, spreading over the top of the ovary, bilobed; capsules depressed-globose, $1\frac{1}{2}$ lin. in diam., smooth; seeds triquetrous, rounded on the back, $\frac{2}{3}$ lin. long, with about 10 lines of minute dots on the back and also on the flattened sides.

KALAHARI REGION : Griqualand West ; near Griquatown, *Orpen* ! Albania, *Mrs. Barber,* 24 ! Kuruman, *Marloth,* 1087 ! Transvaal ; Roodepoort, near Warm Bath, *Bolus,* 12283 ! Boshveld, *Rehmann,* 5337 ! Pretoria Hills, 4500 ft., *Miss Leendertz,* 35 ! *Burtt-Davy,* 3970 ! and without precise locality, *Sanderson* ! Orange River State ; rocky hills near the Modder River, *Mrs. Barber,* 7 ! Bloemfontein, *Mrs. Potts,* 473 !

7. **P. nummulariæfolius** (Poir. Encycl. v. 302, partly); a small slender branched undershrub, up to $1\frac{1}{2}$ ft. high ; stems and branches glabrous; leaves obovate or suborbicular, sometimes subacutely mucronate, rounded or subacute at the base, 3–9 lin. long, 2–5 lin. broad, membranous, glabrous on both surfaces, sometimes glaucous below ; lateral nerves 5–7 on each side, distinct on the lower surface ; veins hardly visible ; petiole $\frac{1}{2}$–$\frac{3}{4}$ lin. long, glabrous ; stipules lanceolate, subacute, $1\frac{1}{4}$ lin. long, entire, glabrous ; flowers monœcious, in axillary fascicles on the young branchlets, females

often solitary or subsolitary; pedicels very slender, up to ¾ in.
long, glabrous; male flowers: sepals 5, broadly obovate, ½ lin. long,
¾ lin. broad; disc of 5 obovate truncate fleshy glands alternating
with the sepals; stamens 5, free; anthers rounded, dehiscing at
the side; female flowers: sepals 4–5 (usually 5), ovate-elliptic,
obtuse at both ends, ½ lin. long, ⅓ lin. broad, entire, membranous,
with a conspicuous greenish-yellow midrib; disc saucer-shaped,
entire, glabrous; ovary globose, glabrous; styles 3, free, slender,
bipartite nearly to the base, spreading or reflexed; capsule depressed-
globose, 1 lin. in diam., smooth; seeds convex on the back, testa
brown, minutely and closely pitted.   *Willd. Sp. Pl.* iv. 584; *Müll.
Arg. in Linnæa,* xxxii. 8, *and in DC. Prodr.* xv. ii. 337; *Pax in
Engl. Pfl. Ost-Afr. C.* 236; *Hutchinson in Dyer, Fl. Trop. Afr.* vi. i.
710.   *Menarda nummularifolia, Baill. Étude Gén. Euphorb.* 609.
Anisonema multiflorum, Thomson in Speke, Nile Journ. App. 647, *not
of Wight.   Diasperus nummulariæfolius, O. Kuntze, Rev. Gen. Pl.* ii.
600.

KALAHARI REGION : Transvaal ; Houtbosh, *Rehmann,* 5920 ! near Lydenburg,
Wilms, 1312 ! *Schlechter,* 3952 ! Crocodile River, *Wilms,* 1313 ! Tzaneen Estate,
Zoutpansberg, *Pole Evans,* 4018 ! Mac Mac Creek, *Mudd* !
EASTERN REGION : Natal ; near Durban, *Wood,* 270 ! near Pietermaritzburg,
Wilms, 2271 !

Occurs also in Eastern Tropical Africa and in Madagascar.

8. **P. pentandrus** (Schumach. & Thonn. Beskr. Guin. Pl. 419) ; a
much-branched herb, up to 18 in. high ; stem woody, subterete,
glabrous ; flowering branchlets very slender, more or less flexuous,
glabrous or very minutely and sparingly asperulate ; leaves oblong-
lanceolate or linear-lanceolate, subacute, obtuse at the base, up to
¾ in. long and ⅓ in. broad, thin, glabrous on both surfaces ; lateral
nerves about 5 on each side, slightly prominent below ; petiole
about ½ lin. long, glabrous ; stipules lanceolate, tapered to a fine
point, ½ lin. long, glabrous ; flowers monœcious, males 2–3 together
in the axils of the lower leaves, females solitary in the upper leaf-
axils ; pedicel slender, short, glabrous ; male flowers : sepals 5,
ovate or elliptic, subacute, small, broadly 1-nerved, margin
membranous, glabrous ; disc-glands 5, flat, smooth and thin ;
stamens 5 ; filaments free ; anthers dehiscing at the side ; female
flowers: sepals as in the male ; disc flat, entire, glabrous ; ovary
lobed, minutely beaded ; styles spreading horizontally, very short,
bifid ; capsule depressed-globose, about 1 lin. in diam., scarcely
lobed ; seeds marked with 5–6 longitudinal lines of dots on the
back.   *Müll. Arg. in DC. Prodr.* xv. ii. 336 (*incl. vars.*); *Hiern in
Cat. Afr. Pl. Welw.* i. 957 ; *Pax in Engl. Pfl. Ost-Afr. C.* 236 ; *De
Wild. Études Fl. Bas- et Moyen-Congo,* i. 275 ; *Hutchinson in
Dyer, Fl. Trop. Afr.* vi. i. 710.   *P. piluliferus, Fenzl in Flora,*
1844, 312, *name only* ; *Benth. in Hook. Niger Flora,* 510.   *P. linoides,
Hochst. ex Baill. Adansonia,* i. 84, *name only.   P. linifolius, Vahl ex
Baill. l.c., name only.   P. deflexus, Klotzsch in Peters, Reise Mossamb.*

Bot. 104.   *P. dilatatus, Klotzsch, l.c.* 106.   *P. scoparius, Welw.*
Apont. 591, *no.* 110.   *P. tenellus, var., Pax in Baum, Kunene-Samb.*
Exped. 282, *not of Roxb.*   *P. Niruri, N. E. Br. in Kew Bulletin,* 1909,
139, *not of Linn.*   *Menarda linifolia, Baill. l.c., name only.*
Diasperus pentandrus, O. Kuntze, Rev. Gen. Pl. ii. 600.

KALAHARI REGION : Transvaal ; Komati, *Bolus,* 9772 ! Klippan, *Rehmann,*
5338 ! Houtbosh, *Rehmann,* 5916 !
EASTERN REGION : Delagoa Bay, *Schlechter,* 11970 ! 12156 ! *Scott* ! Lorenzo
Marques, *Mrs. Howard,* 58 ! Chinyandjana, *Junod,* 493 ! and without precise
locality, *Junod,* 185 !

Widely spread in Tropical Africa.

9. P. Burchelli (Müll. Arg. in Linnæa, xxxii. 7) ; an annual, up
to 6½ in. high, branched from the base ; stems grooved, glabrous ;
flowering branchlets 1½–2 in. long, slender, smooth ; leaves oblong-
lanceolate, very obtuse at both ends, 2–5 lin. long, 1–2 lin. broad,
thin, often more or less glaucous-green on the upper surface when
dry, glabrous, with indistinct lateral nerves ; petiole about ¼ lin. long ;
stipules subulate, acute, ⅓ lin. long, glabrous ; flowers monœcious,
axillary, the males in the lower, the females in the upper part of
the branchlets ; male flowers : pedicel ¾ lin. long, glabrous ; sepals 5,
suborbicular, ⅓ lin. broad, membranous, glabrous ; disc-glands flat,
rounded, very thin, smooth ; stamens 5 ; filaments free ; anthers
dehiscing at the side ; female flowers : pedicels ¾ lin. long, glabrous ;
sepals 5, broadly ovate, subacute, about ½ lin. long, membranous,
with a broad thicker midrib ; disc saucer-shaped, entire, thin,
glabrous ; ovary lobed, smooth ; styles 3, spreading horizontally
from the base, bipartite, with linear segments not swollen at the
tips. *Müll. Arg. in DC. Prodr.* xv. ii. 340.   *P. garipensis, E. Meyer*
in Drège, Zwei Pfl. Documente, 93.   *P. tenellus, vars. natalensis,*
garipensis and exiguus, Müll. Arg. in Linnæa, xxxii. 7, *and in DC.*
Prodr. xv. ii. 338–9.

COAST REGION : Komgha Div. ; Keimouth, *Flanagan,* 52 ! near Kefani River
mouth, *Galpin,* 5805 !
WESTERN REGION : Great Namaqualand ; Tiras, *Schinz,* 890 ! Little Namaqua-
land ; Orange River. near Verleptpram, *Drège* ! and without precise locality, *Wyley* !
KALAHARI REGION : Griqualand West ; Asbestos Mountains, at the Kloof
village, *Burchell,* 2041 ! Asbestos Hills, *Marloth,* 2073 ! Bechuanaland ; Chue Vley,
Burchell, 2383 ! Transvaal ; Potgeiters Rust, *Miss Leendertz,* 1272 ! Boshveld,
Klippan, *Rehmann,* 5337 ! near Pretoria, *Burtt-Davy,* 3970 ! Barberton, *Galpin,*
566 !
EASTERN REGION : Transkei : Kentani, *Miss Pegler,* 1165 ! Pondoland ; Port
St. John, *Galpin,* 3436 ! Natal ; Coast land, *Sutherland* ! Durban Botanic
Gardens, *Rogers,* 873 ! near Durban, *Schlechter,* 2793 ! *Wood,* 270 ! *Gerrard,*
59 ! 649 ! *Sanderson,* 137 ! by the Umlaas River, *Krauss,* 336 ! Berea, *Wood,*
813 ! near Isipingo, *Schlechter,* 2988 !

10. P. glaucophyllus (Sond. in Linnæa, xxiii. 133) ; stems
numerous, slender, up to 1 ft. long, arising from a woody rhizome,
flattened, glabrous ; leaves ovate or ovate-elliptic, mucronate,
rounded or slightly cordate at the base, 3–9 lin. long, 2–5 lin.

broad, entire, rigidly chartaceous, glabrous and glaucous on both
surfaces; lateral nerves 5–7 on each side, looped and branched
some distance from the margin, distinct below; petiole very short,
½ lin. long, glabrous; stipules ovate or ovate-lanceolate, auriculate
at the base, a little longer than the petiole, membranous, brown,
glabrous; flowers monœcious, the males solitary towards the ends
of the shoots, the females solitary in the lower parts of the shoots;
male pedicels very slender, nearly 2 lin. long, glabrous; sepals 6,
unequal, oblong, obtuse, entire, ¾ lin. long, ¼ lin. broad; disc-
glands 6, large, flat, orbicular, peltately attached; stamens 3;
filaments quite free; anthers dehiscing at the side; female pedicels
about 2 lin. long in fruit; sepals larger than in the male; disc
flat and annular, slightly lobed, glabrous; ovary 6-lobed, glabrous;
styles free, bilobed, lobes very slender; capsule depressed-globose,
3- or obscurely 6-lobed, 1½–1¾ lin. in diam., glabrous. *Müll. Arg.
in DC. Prodr.* xv. ii. 393. *P. glaucophyllus, var. major, Müll. Arg.
in Flora,* 1864, 514, *and in DC. l.c.*; *Pax in Engl. Pfl. Ost-Afr.
C.* 236; *Hutchinson in Dyer, Fl. Trop. Afr.* vi. i. 713.

VAR. β, **suborbicularis** (Hutchinson); leaves suborbicular.

COAST REGION: Var. β : British Kaffraria; Kaffrarian Mountains, *Mrs. Barber,*
39!

KALAHARI REGION: Orange River Colony, *Cooper,* 902; Transvaal; Magaliesberg, *Zeyher,* 1509! *Burke!* near Pretoria, *Scott Elliot,* 1407! *Rehmann,* 4719!
Schlechter, 4789!

EASTERN REGION: Pondoland; *Bachmann,* 811! Natal; Drakensberg, at
Ingagaorly, *Rehmann,* 7020! Inanda, *Wood,* 466! between Pinetown and
Umbilo, *Rehmann,* 8070! Nototé River, *Gerrard & M'Ken,* 16! near Camperdown,
Schlechter, 3283! and without precise locality, *Sanderson,* 447! *Cooper,* 3143!

11. **P. maderaspatensis** (Linn. Sp. Pl. ed. i. 982); a woody
undershrub, very variable in habit; flowering branchlets sharply
angular, almost winged or variously compressed, glabrous or
minutely asperulate; leaves linear-lanceolate or oblanceolate,
variable in size, up to 1½ in. long and ½ in. broad, rather rigidly
membranous or thinly chartaceous, glabrous or slightly asperulate;
lateral nerves about 7 on each side of the midrib, usually prominent
below; petiole short, glabrous; stipules lanceolate, acuminate,
broad and auriculate at the base, nearly 1½ lin. long, ¾ lin. broad
at the base, membranous, glabrous; flowers monœcious, the male
and female often together in the leaf-axils, or more often the
females solitary, much larger than the males; pedicels about 1 lin.
long; male flowers: sepals 6, oblong-lanceolate, ⅔ lin. long,
membranous, glabrous; disc-glands 6, thin and smooth, flat;
stamens 3; filaments connate; anthers dehiscing longitudinally;
female flowers: sepals 6, 2-seriate, 3 outer ovate-elliptic or rounded,
slightly apiculate, 1 lin. long, ¾ lin. broad, coriaceous, glabrous,
3 inner slightly narrower, with narrow membranous margins; disc
of 6 separate flat thin glands; ovary lobed; styles 3, distinct,
suberect, thick, bifid; capsule depressed-globose, 1½ lin. in diam.,
3-lobed, smooth; seeds trigonous, convex on the back, ¾ lin. long

and broad, and marked with 9–10 fine lines of dots on the back. *Willd. Sp. Pl.* iv. 575; *Roxb. Fl. Ind. ed. Carey & J. Roxb.* iii. 654; *Grah. Cat. Bomb. Pl.* 180; *Wight, Ic. t.* 1895, *fig.* 3; *Dalz. & Gibs. Bomb. Pl.* 233; *Benth. Fl. Hongk.* 311, *and Fl. Austral.* vi. 103; *Müll. Arg. in Linnæa,* xxxii. 19, *and in DC. Prodr.* xv. ii. 362; *Schweinf. Beitr. Fl. Aethiop.* 37; *Oliv. in Trans. Linn. Soc. ser.* 2, *Bot.* ii. 349; *Hook. f. Fl. Brit. Ind.* v. 292; *Watt, Dict. Econ. Prod. India,* vi. i. 221; *Penzig in Atti Congr. Bot. Genova,* 1892, 360; *Pax in Engl. Hochgebirgsfl. Trop. Afr.* 282, *and in Engl. Pfl. Ost-Afr. C.* 236; *Trim. Handb. Fl. Ceylon,* iv. 20; *Schweinf. in Bull. Herb. Boiss.* vii. *App.* ii. 302; *Hiern in Cat. Afr. Pl. Welw.* i. 959; *Prain, Bengal Pl.* 935; *Cooke, Fl. Bombay,* ii. 586; *S. Moore in Journ. Linn. Soc.* xl. 192; *Hutchinson in Dyer, Fl. Trop. Afr.* vi. 722. *P. andrachnoides, Willd. Sp. Pl.* iv. 575. *P. cuneatus, Willd. Enum. Pl. Hort. Berol. Suppl.* 65. *P. Thonningii, Schum. & Thonn. Beskr. Guin. Pl.* 418. *P. javanicus, Poir. ex Spreng. Syst. Veg.* iii. 21. *P. gracilis, Roxb. Fl. Ind.* iii. 654. *P. obcordatus, Willd. ex Roxb. l.c.* 656. *P. longifolius, Sond. in Linnæa,* xxiii. 135, *not of Lam. P. incurvus, Sond. l.c., not of Thunb. P. heterophyllus, E. Meyer ex Sond. l.c., in syn., name only. P. Gueinzii, Müll. Arg. in Linnæa,* xxxii. 18, *and in DC. Prodr.* xv. ii. 363. *P. venosus, Hochst. ex A. Rich. Tent. Fl. Abyss.* ii. 254; *Schweinf. Beitr. Fl. Aethiop.* 37. *P. arabicus, Hochst. ex Baill. Adansonia,* i. 86. *P. vaccinioides, Klotzsch in Peters, Reise Mossamb. Bot.* 105, *not of Sond. Diasperus maderaspatensis, O. Kuntze, Rev. Gen. Pl.* ii. 600.

COAST REGION : Bathurst Div.; Glenfilling, *Drège*! Albany Div. ; near Grahamstown, *Schlechter,* 2680! between Grahamstown and Coldstream, *Bolus,* 10653! East London Div. ; East London, *Rattray,* 687! Komgha Div. ; near Komgha, *Flanagan,* 659! Kei Hill, *Schönland,* 1358! British Kaffraria ; Krielis Country, *Bowker,* 2641! and without precise locality, *Cooper,* 292! 3139!
CENTRAL REGION : Somerset East Div.; Little Fish River, *Scott Elliot,* 536!
WESTERN REGION : Great Namaqualand ; Schaf riverbed at Seeheim, *Pearson,* 3739! plateau west of Gobas Station, *Pearson,* 3741! Akam River, *Pearson,* 4766!
KALAHARI REGION : Griqualand West ; various localities, *Burchell,* 1739! 1786! 2021! Orange River Colony ; Draai Fontein, *Rehmann,* 3647! Bechuanaland ; Mafeking, *Duparquet,* 23! Transvaal ; near Pietersburg, *Schlechter,* 4364! Fourteen Streams, *Burtt-Davy,* 1586! Springbok Flats, *Burtt-Davy,* 1195! Potchefstroom, *Burtt-Davy,* 1817! 9697! Christiana, *Burtt-Davy,* 12847! near Crocodile River, *Schlechter,* 3915! Houtbosh, *Rehmann,* 5917! near Lydenburg, *Wilms,* 1314! Komati Poort, *Schlechter,* 11785! *Kirk,* 71!
EASTERN REGION : Transkei Div. ; Kentani, *Miss Pegler,* 1139! 1145 partly! Tembuland ; near Emgwali River, *Bolus,* 10280! Pondoland ; Port St. John, *Galpin,* 3429! Griqualand East ; near Clydesdale, *Tyson,* 2862! Natal ; near Durban, *Wahlberg*! *Gerrard,* 1177! *Gueinzius,* 162! Inanda, *Wood,* 1142! near Glencoe, *Wood,* 5177! near Phoenix, *Schlechter,* 3025! Clairmont, *Schlechter,* 3052! near Newcastle, *Wilms,* 2272! and without precise locality, *Sutherland*! *Gerrard,* 3741 1011!
Widely distributed throughout the warmer regions of the Eastern Hemisphere.

12. **P. incurvus** (Thunb. Prodr. 24); a woody undershrub with whip-like branches arising from a rhizome ; branches somewhat flattened, obtusely angular, often greenish when dry, glabrous ; leaves linear or linear-lanceolate, acute or subacute, $\frac{1}{6}$–$1\frac{1}{4}$ in. long,

½–3 lin. broad, coriaceous, 1-nerved, mostly without visible side nerves, glabrous; stipules obliquely ovate-lanceolate, acute, subcordate at the base, up to 1 lin. long, glabrous, reddish; flowers usually monœcious, one female and one or two males in each leafaxil, rarely diœcious; male pedicels ¼–⅓ lin. long, reddish, glabrous; sepals 6, obovate, reddish, glabrous, with narrowly membranous margins; disc-glands 6, rounded, nearly smooth; stamens 3; filaments connate to the apex; anthers free from each other, dehiscing longitudinally; female flowers: pedicels 1 lin. long or slightly more in fruit, at length recurved, reddish; sepals as in the male but larger and more coriaceous, with prominent hyaline margins especially in the fruiting stage; disc composed of 6 separate rounded smooth glands; ovary smooth; styles somewhat spreading, bilobed nearly to the base, free, not swollen at the tips; fruits depressed, 6-lobed, 1½–1¾ lin. in diam., smooth; seeds triquetrous, rounded on the back, marked with numerous longitudinal lines of dots. *Thunb. Fl. Cap. ed. Schult.* 499; *Müll. Arg. in DC. Prodr.* xv. ii. 362. *P. longifolius, Lam. Ill. t.* 756, *fig.* 3; *Müll. Arg. l.c.* 361. *P. genistoides, Sond. in Linnæa,* xxiii. 134. *P. multicaulis, Müll. Arg. in Linnæa,* xxxii. 18, *incl. var. parvifolius, Müll. Arg., and in DC. l.c.* 360. *P. multicaulis, vars. genuinus and parvifolius, Müll. Arg. in DC. l.c.* 360, 361. *Diasperus glaucophyllus, var. stenophyllus, O. Kuntze, Rev. Gen. Pl.* iii. 285.

13. **P. myrtaceus** (Sond. in Linnæa, xxiii. 134); a shrub; branches flattened and slightly winged, glabrous; flowering branchlets flexuous, compressed; leaves ovate or rarely ovate-

elliptic, obtuse, more or less rounded at the base, $\frac{3}{4}$–$1\frac{1}{4}$ in. long,
$\frac{1}{3}$–$\frac{1}{4}$ in. broad, submembranous, glabrous on both surfaces, glaucous
below; lateral nerves 6–7 on each side, slender, looped; veins not
distinct; petiole $\frac{1}{2}$–$\frac{3}{4}$ lin. long, glabrous; stipules lanceolate,
acuminate, about 1 lin. long, denticulate, glabrous; flowers
diœcious; males axillary, solitary or in pairs; pedicel filiform,
up to 2 lin. long, glabrous; sepals 6, in two series, subequal, outer
ovate, inner slightly obovate, obtuse, 1 lin. long, $\frac{1}{2}$–$\frac{3}{4}$ lin. broad,
entire, membranous, glabrous; disc-glands 6, nearly flat and
orbicular, slightly warted above; stamens 3; filaments connate
in the lower two-thirds into a slender column; anthers subglobose,
cells diverging, dehiscing obliquely; female flowers solitary; pedicel
3–4 lin. long, glabrous; sepals 6, in two series, outer ovate-elliptic,
subobtuse, 2 lin. long, $1\frac{1}{4}$ lin. broad, inner obovate, rounded at the
apex, $2\frac{1}{2}$ lin. long, $1\frac{3}{4}$ lin. broad, more distinctly veined than the
outer; disc thick, undulately lobed, glabrous; ovary smooth; styles
spreading, slender, bilobed, lobes terete, not swollen at the tips;
complete capsule not seen; seeds smooth (*Sonder*). *Müll. Arg. in
DC. Prodr.* xv. ii. 397; *S. Moore in Journ. Linn. Soc.* xl. 192;
Sim, For. Fl. Cape Col. 325; *Hutchinson in Dyer, Fl. Trop. Afr.*
vi. i. 726. *P. revolutus, E. Meyer in Drège, Zwei Pfl. Documente,*
153, 211; *Sond. l.c.* 135, *name only. P. Bachmanni, Pax in Engl.
Jahrb.* xxiii. 520. *Diasperus myrtaceus, O. Kuntze, Rev. Gen. Pl.*
ii. 600.

EASTERN REGION : Kentani Div.; Cebe Forests, *Miss Pegler,* 1119 ! Pondoland;
Umsikaba River, *Drège*! and without precise locality, *Bachmann,* 167 ! Natal ;
near Durban, *Gueinzius,* 520 ; coast-land, *Sutherland*! Inanda, *Wood,* 1586 !
Highlands of Natal and Emyati, *Gerrard,* 1162 ! Nototi River, *Gerrard,* 33 !
Amanzimtoti, *Miss Franks in Herb. Wood,* 11910 ! Oakford, Umhloti River,
Rehmann, 8533 ! and without precise locality, *Gerrard,* 379 ! 1924 ! *Drège,* 4629 !

Occurs also in South-Eastern Tropical Africa.

14. **P. meyerianus** (Müll. Arg. in Linnæa, xxxii. 42); stems
suberect, evidently from a rhizome, woody, glabrous; lateral
flowering branchlets numerous, solitary or rarely 2–3 together,
slender, spreading, densely leafy, glabrous; leaves ovate-elliptic or
oblong-lanceolate, rounded at both ends, $1\frac{1}{2}$–5 lin. long, 1–2 lin.
broad, glabrous, with 3–4 pairs of fairly distinct lateral nerves ;
petiole $\frac{1}{4}$ lin. long, glabrous; stipules subulate or subulate-filiform,
acute, about $\frac{1}{2}$ lin. long, glabrous; flowers monœcious, usually one
or two females to each branchlet, the remainder males; male
flowers : pedicels $\frac{3}{4}$ lin. long, glabrous; sepals 5, broadly obovate,
rounded at the apex, $\frac{3}{4}$ lin. long, $\frac{2}{3}$ lin. broad, membranous, glabrous ;
disc-glands 5, rounded, flat, strongly warted ; stamens 3 ; filaments
connate to the apex ; anthers dehiscing at the side ; female flowers :
sepals 5, oblong or oblong-elliptic, rounded at the apex, about $\frac{3}{4}$ lin.
long and $\frac{2}{3}$ lin. broad, membranous, glabrous; disc slightly lobed
(nearly annular), flat, reddish, glabrous; ovary smooth; styles 3,
spreading from the base, slender, bilobed, not swollen at the tips ;

capsule depressed-globose, slightly 6-lobed ; seeds almost smooth.
P. Niruri, Drège, Zwei Pfl. Documente, 151, *not of Linn.   P. Woodii,
Hutchinson in Kew Bulletin,* 1914, 336.

EASTERN REGION : Pondoland ; Egossa, *Sim,* 2515 ! Mengana Cutting, between
Umtata and Port St. John's, *Bolus,* 10279 ! *Flanagan,* 2498 ! between St. Johns
River and Umsikaba River, *Drège* a ! between Umtentu River and Umzimkulu
River, *Drège* b ! Natal ; near the Tugela River, *Gerrard,* 1925 ! Clairmont, *Wood,*
1765 ! near Pinetown, *Wood,* 5303 ! Friedenau Farm, near Dumisa Station,
Rudatis, 317 ! near Maritzburg, *Schlechter,* 3307 !

In describing this plant as *P. Woodii,* I have unfortunately added to the
synonymy, on account of a mistaken determination in the Berlin herbarium.   A
specimen loaned by Prof. Engler was named "P. Meyerianus, Müll. Arg.!!" and
for some time I took this to be an authentic representative of that species ; but
the type of *P. meyerianus* is the "P. Niruri" of Drège's Catalogue, which was
included by me as a synonym under *P. Woodii.*   The Berlin specimen named
P. meyerianus is exactly *P. myrtaceus,* Sond.

15. P. asperulatus (Hutchinson in Kew Bulletin, 1919, ined.) ; an
annual, 9–12 in. high, branched in the upper third ; stem nude in
the lower part, straw-coloured, minutely and sparingly scabrid-
puberulous ; branchlets leafy, compressed-angular, finely asperulate
on the angles ; leaves oblong-elliptic, rounded at both ends, 4–6
lin. long, 2–4 lin. broad, membranous and probably sensitive,
glabrous, glaucous-green when dry ; lateral nerves 5–6 on each side
of the midrib, scarcely visible above, fairly prominent below ; petiole
slender, about ½ lin. long ; stipules subulate-lanceolate, acute, membranous, ⅓ lin. long ; flowers monœcious, the males in the lower
part, the females in the upper part of the branchlets ; male pedicels
extremely short ; sepals 6, obovate, rounded at the apex, ½ lin. long,
1-nerved, membranous, hyaline, glabrous ; disc-glands 6, flat, rounded,
smooth ; stamens 3 ; filaments very short, connate into a column ;
anthers transverse, and so apparently dehiscing transversely ; female
flowers very shortly pedicellate ; sepals 6, subacute, membranous,
greenish, with hyaline margins, glabrous ; ovary smooth ; styles
short, spreading from the base, with swollen bifid apices ; capsules
soon dehiscing, not seen entire, with very thin crustaceous walls ;
seeds triquetrous, rounded on the back, closely and longitudinally
sulcate.

KALAHARI REGION : Transvaal ; Komati Poort, 1000 ft., Dec., *Schlechter,*
11866 !

16. P. Niruri (Linn. Sp. Pl. ed. i. 981) ; an annual up to 1 ft.
high ; stem closely sulcate, smooth ; flowering branchlets up to
4 in. long, compressed or slightly winged, smooth ; leaves oblong or
oblong-elliptic, rounded at both ends, 3–6 lin. long, 1½–3 lin. broad,
membranous, glabrous on both surfaces ; lateral nerves about 5 on
each side, sometimes nearly invisible ; petiole very short ; stipules
lanceolate-subulate, membranous, glabrous ; flowers monœcious,
solitary males in the lower, females in the upper parts of the
branchlets ; female pedicel about ¾ lin. long when in fruit, glabrous ;

400   EUPHORBIACEÆ (Hutchinson).   [*Phyllanthus.*

male flowers : sepals 6, small, 1-nerved, with membranous margins ;
disc-glands 6, small ; stamens 3 ; filaments connate ; female flowers :
sepals 6, larger than in the male ; disc thin and flat, about 10-
toothed ; ovary subglobose, smooth ; styles very short, suberect,
bifid ; capsule depressed-globose, more or less trilobed, about 1¼ lin.
in diam. ; seeds with about 6 longitudinal lines on the back.
Willd. Sp. Pl. iv. 583 ; *Müll. Arg. in Linnæa,* xxxii. 43, *and in DC.
Prodr.* xv. ii. 406 ; *Roxb. Fl. Ind. ed. Carey & J. Roxb.* iii. 659 ;
Hook. f. Fl. Brit. Ind. v. 298 ; *Pax in Engl. Pfl. Ost-Afr. C.* 236, *in
Bolet. Soc. Brot.* x. 157, *and in Bull. Soc. Bot. Belg.* xxxvii. 103 ;
Schweinf. Beitr. Fl. Aethiop. 37 ; *Trim. Handb. Fl. Ceylon,* iv. 23 ;
Hiern in Cat. Afr. Pl. Welw. i. 960 ; *De Wild. & Durand, Contr.
Fl. Congo,* i. 48, *and Reliq. Dewevr.* 204 ; *De Wild. Miss. É. Laurent,*
127, *and Études Fl. Bas- et Moyen-Congo,* ii. 266 ; *Rendle in Journ.
Linn. Soc.* xxxvii. 210 ; *Gibbs, l.c.* 469 ; *Beille in Bull. Soc.' Bot.
France,* lv. *Mém.* viii. 57, *partly* ; *Hutchinson in Dyer, Fl. Trop. Afr.*
vi. i. 731. *Urinaria indica, Burm. Thes. Zeyl. t.* 93, *fig.* 2. *Nym-
phanthus Niruri, Lour. Fl. Coch.* 545. *Diasperus Niruri, O. Kuntze,
Rev. Gen. Pl.* ii. 600.—*Kirganeli, Rheede, Hort. Malab.* x. *t.* 15.

KALAHARI REGION : Orange River Colony ; Drackensberg Range, near Harrismith, *Cooper,* 1019 !

A common weed in most tropical countries.

17. **P. delagoensis** (Hutchinson in Kew Bulletin, 1919, ined.) ;
stems probably arising from a woody rhizome, branched from the
base, sulcate, glabrous ; branchlets fairly leafy ; leaves broadly elliptic
or oblong-elliptic, obtuse at both ends, 1½–2 lin. long, ½–1¼ lin.
broad, subcoriaceous, glabrous ; stipules subulate, membranous ;
flowers monœcious, the males in the lower part, the females in the
upper part of the branchlets ; male pedicels short and slender ;
sepals 6, unequal, lanceolate, subacute, up to ½ lin. long, glabrous ;
disc-glands very small and rounded, smooth ; stamens 3 ; filaments
connate into a very short column ; anthers large and dehiscing
transversely and having the appearance of being 4-celled ; female
flowers with a pedicel ½–¾ lin. long ; sepals 6, as in the male but
somewhat larger and more equal ; disc saucer-shaped, slightly
undulate, fleshy, glabrous ; stigmas subentire and almost sessile
on the ovary ; fruits and seeds not seen.

EASTERN REGION : Delagoa Bay ; Lorenzo Marques, *Schlechter,* 11663 !

18. **P. heterophyllus** (E. Meyer in Drège, Zwei Pfl. Documente,
211) ; a small spreading much-branched undershrub ; branches
rather sharply angular, glabrous ; leaves orbicular or very broadly
elliptic, subobtusely mucronate, 1¼–4 lin. long, 1¼–2½ lin. broad,
rigidly coriaceous, glabrous and dull on both surfaces ; lateral
nerves not visible ; petiole about ⅓ lin. long ; stipules subulate-
lanceolate, very acute, ½–¾ lin. long, glabrous ; flowers monœcious,
solitary or geminate ; male pedicel ¾ lin. long, glabrous ; sepals 6,

lanceolate, subobtuse, glabrous; disc-glands 6, small, rounded, smooth; stamens 3; filaments connate to the apex; anthers dehiscing obliquely; female flowers shortly pedicellate; sepals 6, oblong-lanceolate, subacute, about 1 lin. long, glabrous, becoming reflexed in fruit; disc fleshy, undulate, glabrous; ovary smooth; styles slightly spreading, a little swollen and sometimes bifid at the apex; capsule 3-lobed, about 1¼ lin. in diam.; seeds almost smooth on the back. *Müll. Arg. in Linnæa*, xxxii. 43, *and in DC. Prodr.* xv. ii. 405. *P. incurvus, Sond. in Linnæa*, xxiii. 135, *excl. syn. P. heterophyllus, E. Meyer; Baill. in Adansonia*, iii. 165, *not of Thunb. P. capensis, Spreng. ex Sond. l.c. (in syn.), name only. P. andrachniformis, Pax in Bull. Herb. Boiss.* 2ᵐᵉ sér. viii. 634.

SOUTH AFRICA: without precise locality, *Thunberg*! *Verreaux*!
COAST REGION: Mossel Bay Div.; Mossel Bay, *Gulline in Herb. Guthrie*, 4373! Uitenhage Div.; near the Zwartkops River, *Ecklon & Zeyher*, 599! *Zeyher*, 20! 3821! near Uitenhage, *Schlechter*, 2503! Port Elizabeth Div.; Port Elizabeth, no collector's name, *in Herb. Stockholm*, 30! Bathurst Div.; near Barville Park, *Burchell*, 4101! Albany Div.; Round Hill, *Bolus*, 10654! Bothas Berg, *MacOwan*, 1430! Div.? *Drège*, 8222a!
EASTERN REGION: Natal; near Durban, *Schlechter*, 2986! Dumisa, *Rudatis*, 778! Umgeni, *Rehmann*, 8811! near Clairmont, *Schlechter*, 3052!

19. **P. cinereoviridis** (Pax in Engl. Jahrb. xliii. 76); a shrub; branches whitish, terete, smooth, slightly sulcate; ultimate flowering branchlets rather densely pubescent or almost tomentose, much abbreviated, about 1 lin. long; leaves greyish-green, orbicular or ovate-orbicular, rounded at both ends or sometimes slightly emarginate at the apex, 2½–5 lin. long, 2–4½ lin. broad, rather rigidly chartaceous, minutely pitted and shortly pubescent on both surfaces, trinerved from the base, the middle nerve with 2–3 lateral ones, all slightly raised on both surfaces; petiole 1–2 lin. long, sparingly crisped-pubescent; stipules ovate, thick and fleshy, tomentose; flowers dioecious; male inflorescence axillary, pedunculate, with about 6 sessile flowers in a cluster at the apex of each peduncle; bracts ovate, tomentose; peduncle about 1 lin. long, tomentose; sepals 4, in two series, small, suborbicular, glabrous inside, tomentose outside; disc small, saucer-shaped, entire or slightly undulately lobed, glabrous; stamens 2; filaments connate, very short; anthers large and fleshy, dehiscing at the side and apparently transversely owing to the position of the anthers; female flowers not known.

WESTERN REGION: Great Namaqualand; Gubub, *Dinter*, 893!

Imperfectly known species.

20. **P. minus**, Linn. ex Sond. in Linnæa, xxiii. 135, name only.

EASTERN REGION: Natal; Port Natal, *Gueinzius*.

XIII. FLUGGEA, Willd.

Flowers diœcious. *Petals* absent. *Male flowers*: *Sepals* 5, imbricate. *Stamens* 5, alternating with the disc-glands; filaments free; anthers introrse, erect, cells distinct, dehiscing longitudinally. Rudimentary *ovary* large, deeply 2–3-partite. *Female flowers*: *Calyx* as in the male. *Disc* flat, annular, entire or nearly so. *Ovary* 3-celled; styles connate at the base or nearly absent, recurved or spreading, bifid; ovules 2 in each cell. *Fruit* slightly fleshy; pericarp thin. *Seeds* subtriquetrous, rounded on the back; testa thick and crustaceous; albumen scanty; embryo incurved with broad and flat cotyledons.

Shrubs or trees; leaves alternate, petiolate, entire ; flowers small, in axillary fascicles, male often numerous, female few or solitary.

DISTRIB. About 10, in the tropics generally.

1. **F. microcarpa** (Blume, Bijdr. 580); a shrub or small tree, about 12 ft. high; branches and branchlets usually angular, glabrous; leaves elliptic or obovate, obtusely pointed or the smaller emarginate, cuneate, subacute or obtuse at the base, up to 2½ in. long and 1¼ in. broad, rather rigidly membranous, glabrous on both surfaces, often glaucous below; lateral nerves 5–9 on each side, slightly prominent on the lower surface ; veins fine and close; petiole 1½–3 lin. long, glabrous; stipules lanceolate, margin slightly jagged; flowers diœcious; male flowers numerous, in axillary fascicles; pedicel slender, 1½–2 lin. long, glabrous; sepals 5, unequal, the larger obovate, 1 lin. long and ⅔ lin. broad, the smaller oblong, ½ lin. long, ¼ lin. broad, margins membranous and jagged or ciliate, otherwise glabrous; disc-glands usually large and fleshy, somewhat angular, with 3 or 4 depressions or pits on the outside, glabrous; stamens 5: filaments longer than the rudimentary ovary; anthers ½ lin. long; rudimentary ovary tripartite; segments free for about two-thirds of their length, terete, reflexed and often thickened in the upper part, a small second lobe often developed at the bend, glabrous; female flowers numerous; calyx as in the male; disc annular, glabrous; ovary ovoid, glabrous; styles connate at the base, or rarely nearly absent, bilobed, lobes spreading, flattened or subterete; fruits white, depressed-globose, about 2 lin. in diam.; styles persistent; seeds shining, with several longitudinal lines of pits on the back and sides. *Hook. f. Fl. Brit. Ind.* v. 328; *Hiern in Cat. Afr. Pl. Welw.* i. 961; *S. Moore in Journ. Linn. Soc.* xl. 193; *Hutchinson in Dyer, Fl. Trop. Afr.* vi. i. 736. *F. virosa, Baill. Étude Gén. Euphorb.* 595. *F. abyssinica, Baill. l.c. F. angulata, Baill. Adansonia,* i. 80. *F. obovata, Baill. l.c.* ii. 41; *Pax in Engl. Pfl. Ost-Afr. C.* 235,

and in Engl. Jahrb. xxviii. 418 ; xxx. 338 ; *Schweinf. in Bull. Herb. Boiss.* vii. *App.* ii. 299 ; *De Wild. & Durand, Contrib. Fl. Congo,* i. 48, *and Reliq. Dewevr.* 205 ; *De Wild. Études Fl. Katanga,* 79 ; *Th. & Hél. Durand, Syll. Fl. Congol.* 483. *F. obovata, var. luxurians, Beille in Bull. Soc. Bot. France,* lv. *Mém.* viii. 55. *F. senensis, Klotzsch in Peters, Reise Mossamb. Bot.* 106. *Phyllanthus virosus, Roxb. ex Willd. Sp. Pl.* iv. 578. *P. lucidus, Hort. ex Willd. Enum. Pl. Hort. Bot. Berol.* 329. *P. angulatus, Schum. & Thonn. Beskr. Guin. Pl.* 415. *P. dioicus, Schum. & Thonn. l.c.* 416. *P. polygamus, Hochst. ex A. Rich. Tent. Fl. Abyss.* ii. 256. *Xylophylla obovata, Willd. Enum. Pl. Hort. Bot. Berol.* 329. *Securinega abyssinica, A. Rich. l.c.* ; *Schweinf. Beitr. Fl. Aethiop.* 38. *S. obovata, Müll. Arg. in DC. Prodr.* xv. ii. 449 ; *Benth. Fl. Austral.* vi. 115. *Acidoton obovatus, O. Kuntze, Rev. Gen. Pl.* ii. 592.

SOUTH AFRICA : without precise locality, *Lalande in Herb. Paris* !
KALAHARI REGION : Transvaal ; Barberton, *Galpin,* 686 ! Komati Poort *Schlechter,* 11796 ! 11797 ! Potgieters Rust, *Miss Leendertz,* 1244 ! Swaziland, *Burtt-Davy,* 1062 ! Bremmerdorp, ridge above M'Kathalaage Spruit, *Burtt-Davy,* 3017 !
EASTERN REGION : Natal ; near Durban, *Gerrard,* 2153 ! Zululand ; Ginginhlovu, *Haygarth in Herb. Wood,* 12578 !

Occurs also in India, and widely spread in Tropical Africa.

XIV. DRYPETES, Vahl.

Flowers diœcious. *Petals* absent. *Male flowers* : Buds mostly globose. *Sepals* 4 or 5, broad, imbricate, usually coriaceous. *Stamens* 3– ∞, inserted around the base of a central flat concave or rarely cupular disc ; filaments free ; anthers erect, often large, cells parallel and dehiscing longitudinally. Rudimentary *ovary* not present or rarely represented by a small conical production in the middle of the disc. *Female flowers* : *Calyx* as in the male. *Disc* hypogynous, annular or cupular. *Ovary* 1–4-celled ; styles short or absent ; stigmas thick, flattened, bifid or undivided and more or less reniform ; ovules 2 in each cell. *Fruits* globose, ellipsoid or ovoid, indehiscent ; pericarp thick, woody, 1–4-celled. *Seeds* solitary by abortion ; albumen fleshy : embryo straight ; cotyledons flat and broad.

Trees or shrubs ; leaves alternate, coriaceous or chartaceous, entire or toothed ; stipules caducous, rarely persistent ; male and female flowers fasciculate in the axils of the leaves, or produced on the older branches or the stem, usually pedicellate.

DISTRIB. Species over 70, chiefly in the Old-World Tropics, a few in the West Indies and South America. The reasons for uniting *Drypetes,* Vahl (1807) with *Cyclostemon,* Blume (1825) have been fully discussed by the writer in Dyer's Flora of Trop. Afr. vi. i. 674.

2 D 2

Flowers produced in dense clusters on the older branches
    which are devoid of leaves or on the trunk ; stamens
    20–25 ...   . (1) **natalensis.**
Flowers produced in the axils of the leaves on the young
    shoots ; stamens 4–15 :
    Male and female flowers with pedicels ⅓–¾ in. long ;
        male sepals 5 ; stamens 15 ...   . (2) **arguta.**
    Male flowers sessile or with a pedicel 1–1½ lin. long ;
        female flowers in the fruiting stage with a
        pedicel up to 4 lin. long ; male sepals and
        stamens 4   ...   . (3) **Gerrardii.**

1. **D. natalensis** (*Hutchinson*); a shrub or small tree; young
branchlets thinly pubescent or nearly glabrous; leaves elliptic or
oblong-elliptic, acute, rounded and very slightly unequal-sided at
the base, 3–6 in. long, 1–2½ in. broad, very coarsely and sharply
dentate or sometimes nearly entire or only slightly denticulate,
thinly and rather rigidly coriaceous, glabrous and glossy on both
surfaces ; lateral nerves 6–8 on each side, looped and branched far
from the margin, prominent on both surfaces; veins lax ; petiole
3–4 lin. long, slightly pubescent and wrinkled : flowers of both
sexes fasciculate, springing from the older branches or the trunk :
male flowers crowded ; pedicels about 3 lin. long, pubescent ; sepals
broadly imbricate, rounded, about ¼ in. long, almost membranous,
puberulous outside; stamens 20–25, inserted around a fleshy
glabrous lobed disc ; anthers 1¾ lin. long; female flowers : pedicel
¼ in. long, puberulous ; sepals as in the male ; disc fleshy, much
crenulated, glabrous ; ovary tomentose, 3-celled ; styles 3, spreading,
fleshy, spathulate. *Cyclostemon natalense, Harv. Thes. Cap.* ii. 64,
t. 200; *Müll. Arg. in DC. Prodr.* xv. ii. 483; *Sim, For. Fl. Cape
Col.* 315.

EASTERN REGION : Natal ; near Durban, *Sanderson*, 105 ! *Gerrard*, 13 ! 62 !
724 ! 862 ! *Wood*, 116 ! 1426 ! 1756 !

Harvey states on the authority of Sanderson that the flowers are yellow and
very fetid and offensive. M'Ken states on his label that "the fruits occur on the
thick stems down to the roots."

2. **D. arguta** (*Hutchinson*); branchlets slender, grooved or
angular, pubescent, at length becoming nearly glabrous : leaves
oblong-lanceolate or rarely ovate-lanceolate, long and acutely more
or less gradually acuminate, rounded, subcordate and slightly
unequal-sided at the base, 2–3½ in. long, ¾–1½ in. broad, sharply
dentate, chartaceous, glabrous on both surfaces, slightly shining :
lateral nerves 7–11 on each side, looped and much branched some
distance from the margin, like the tertiary nerves prominent on
both surfaces; petiole 1–1½ lin. long, shortly pubescent; stipules
lanceolate or linear-lanceolate, acute, about ¼ in. long, scarious,
slightly ciliate, otherwise glabrous ; flowers axillary on the young
shoots, solitary or the males sometimes in pairs ; male flowers :
pedicel ⅓–¾ in. long, shortly pubescent ; sepals 5, broadly ovate,

obtuse, up to 1¾ lin. long and broad, unequal, coriaceous, puberulous on both surfaces, shortly and rather densely ciliate; stamens 15, inserted around the disc; filaments 1½ lin. long, glabrous; anthers ⅔ lin. long; disc undulately lobed, rather thin, lobes projecting between the filaments, pubescent; female flowers: pedicel about ⅓ in. long, shortly pubescent; sepals similar to those of the male but larger; disc fleshy, densely rusty-tomentose; ovary 2-celled, densely rusty-hirsute; styles 2, spreading, puberulous, split down the inner face and with two broad spreading fleshy lobes; fruits 2-celled, ellipsoid, about ¾ in. long, brown, tomentulose; seeds 6 lin. long, longitudinally striate. *Cyclostemon argutus, Müll. Arg. in DC. Prodr.* xv. ii. 485.

EASTERN REGION: Pondoland; St. Johns, *Flanagan,* 2518! Natal; near Durban, *Gueinzius,* 88! Westville, *Wood!* Inanda, *Wood,* 631! 632! 1125! Tongaat, *Gerrard,* 19! *Cooper,* 3479! Tugela, *Gerrard,* 1630! Delagoa Bay; Lorenzo Marques, *Schlechter,* 11638!

3. **D. Gerrardii** (Hutchinson); branches shortly pubescent, at length nearly glabrous; flowering branchlets subterete, yellowish-tomentose; leaves lanceolate or elliptic-lanceolate, gradually acuminate to an obtuse apex, slightly rounded or subcuneate and unequal sided at the base, 1½–3½ in. long, ¾–1½ in. broad, thinly coriaceous or chartaceous, remotely crenate or subentire, rather densely pubescent on both sides of the midrib especially in the lower half, otherwise glabrous, shining on the upper surface, dull or slightly shining below; lateral nerves 6–8 on each side, spreading and freely anastomosing, distinct on both surfaces; veins prominent on both sides, especially below, rather close; petiole 2–3 lin. long, yellowish-tomentose; stipules deciduous; flowers dioecious, male axillary, several in each fascicle, female axillary and solitary, pedicels of each about 1 lin. long, yellowish-pubescent or tomentose; male flower: sepals 4, obovate spathulate, 1 lin. long, shortly pubescent outside, densely ciliate towards the rounded apex; stamens 4; filaments nearly 1 lin. long, glabrous; anthers rounded, ½ lin. long, glabrous; disc fleshy, cupular, rather long, irregularly lobed, glabrous, with a small protuberance in the middle; young female flowers not seen; disc annular; fruits scarcely lobed, obovoid-globose, about 4 lin. long, yellowish-tomentose. *Cyclostemon argutus, Sim, For. Fl. Cape Col.* 314, *t.* cxliii. *fig.* ii.; *not of Müll. Arg.*

EASTERN REGION: Pondoland; Umhlagela, *Bachmann,* 804! Intsubane, *Bachmann,* 796! Egossa, *Sim,* 2420! Natal; Tugela, *Gerrard,* 1629!

XV. ANTIDESMA, Linn.

Flowers dioecious. *Petals* absent. *Male flowers: Calyx* deeply 3–5-lobed or partite; lobes or segments imbricate. *Disc* subentire or consisting of distinct glands alternating with the stamens. *Stamens* 2–5, opposite the calyx-lobes; anthers inflexed in bud, erect in the open flower; cells distinct at the apex of a thickened and often curved connective. Rudimentary *ovary* small. *Female flowers: Calyx* as in the male. *Disc* often shortly embracing the base of the ovary. *Ovary* 1- (very rarely 2-) celled; styles 3, very short, usually 2-lobed; ovules 2 in the cell. *Drupes* small, often oblique. *Seed* solitary by abortion (very rarely 2), without a strophiole; albumen fleshy; cotyledons flat, broad.

Trees or shrubs; leaves alternate; flowers small, racemose or spicate; racemes or spikes often very slender, solitary or more rarely subpaniculate towards the ends of short branchlets; bracts small, usually ciliate.

DISTRIB. About 80 species, distributed throughout the warmer regions of the Old World.

1. **A. venosum** (Tul. in Ann. Sci. Nat. 3me sér. xv. 232); a shrub or tree about 30 ft. high; leaves usually more or less elliptic or oblanceolate, somewhat rounded or subacute at the base and apex, sometimes slightly emarginate, up to 6 in. long and 2½ in. broad, thinly coriaceous or almost membranous, glabrous or pubescent above, varying from thinly pubescent to tomentose below; lateral nerves usually 7 on each side, more or less distinctly looped, slightly impressed above, prominent below; petiole up to 3 lin. long, pubescent; stipules lanceolate, entire, acute, mostly more or less tomentose; male spikes up to 6 in. long, tomentose or pubescent; bracts very small; calyx 3–5-partite, more or less pubescent or tomentose; disc subglabrous; stamens 3–5; rudimentary ovary pilose or subglabrous; female racemes 2–5 in. long; bracts as in the male; fruits pedicellate, ellipsoid, slightly flattened, glabrous, about 4 lin. long and 2½ lin. broad. *Harv. Thes. Cap.* ii. 45; *Müll. Arg. in DC. Prodr.* xv. ii. 260; *Pax in Engl. Pfl. Ost-Afr. C.* 237, *and in Baum, Kunene-Samb. Exped.* 282; *Durand & Schinz, Études Fl. Congc* 243; *Hiern in Cat. Afr. Pl. Welw.* i. 965; *De Wild. & Durand, Contr. Fl. Congo,* i. 49, *and Reliq. Dewevr.* 206; *De Wild. Études Fl. Katanga,* 79, *and Miss. É. Laurent,* 128; *Th. & Hél. Durand, Syll. Fl. Congol.* 484; *Sim, For. Fl. Cape Col.* 313, *t.* cxiii. *fig.* iv.; *Hutchinson in Dyer, Fl. Trop. Afr.* vi. i. 646. *A. venosum, var. thouarsianum, Tul. l.c.* 234. *A. bifrons, Tul. l.c.* 229. *A. rufescens, Tul. l.c.* 231. *A. boivinianum, Baill. Adansonia,* ii. 45. *A. natalensis, Harv. Thes. Cap.* ii. 45, *t.* 169, *by error. A. membranaceum, var. molle, Müll. Arg. l.c.* 261; *Hiern, l.c.; Rendle in Journ. Linn. Soc.* xxxvii. 211; *S. Moore in Journ. Linn. Soc.* xl. 194. *A. nervosum, De Wild. Études Fl. Bas- et Moyen-Congo,* ii. 270. *A. venosum, forma glabrescens, De Wild. Études Fl. Katanga,* 79. *A. fusco-cinerea, Beille in Bull. Soc. Bot. France,* lv. *Mém.* viii. 64. *A. Sassandræ, Beille, l.c.* lvi. *Mem.* viii. 123.

KALAHARI.REGION: Transvaal; Avoca, near Barberton, *Galpin,* 1249! between Louws Creek and Adamanda Mine, Barberton district, *Burtt-Davy,* 2812!

Tzaneen Estate, Zoutpansberg, *Burtt-Davy,* 5265 ! Komati Poort, *Schlechter,* 11840 ! between Spitz Kop and Komati River, *Wilms,* 206 !
EASTERN REGION : Natal ; Berea, near Durban, *Wood,* 208 ! 769 ! 9130 ! *Drège ! Gueinzius !* M'Ken, 296 ! *Plant,* 25 ! near Umlaas River, *Krauss,* 138 ! Inanda, *Wood,* 392 ! 1068 ! Dumisa, *Rudatis,* 525 ! 601 ! and without precise locality, *Gerrard,* 279 ! *Cooper,* 1230 !

Very common in Tropical Africa.

XVI. PSEUDOLACHNOSTYLIS, Pax.

Flowers dioecious.   *Petals* absent.   *Male flower : Sepals* 5, imbricate.   *Disc* annular, glabrous.   *Stamens* 5 ; filaments connate into a column for about half their length ; anther-cells parallel, slightly unequal, introrse, dehiscing longitudinally.   Rudimentary *ovary* in the middle of the staminal column often divided.   *Female flower : Calyx* as in male.   *Disc* annular, closely embracing the base of the ovary, often irregularly toothed.   *Ovary.* 3-celled ; styles 3, diverging, very slightly connate at the base, bilobed ; ovules 2 in each cell.   *Capsule* globose or depressed-globose, entire, tardily septicidal ; endocarp thick and bony.   *Seeds* solitary in each cell, ellipsoid, strophiolate ; albumen fleshy ; cotyledons flat and broad.

Trees or shrubs ; leaves alternate, petiolate, simple ; male flowers in axillary few-flowered pedunculate cymes surrounded by membranous bracts ; female flowers axillary, solitary or subsolitary.

DISTRIB.   Species 4, occurring in Southern Tropical Africa, the following extending northwards to Lake Victoria.

1. **P. maprouneæfolia** (Pax in Engl. Jahrb. xxviii. 20) ; a shrub or small tree ; branchlets terete, tomentose when young, becoming glabrous ; leaves ovate or ovate-elliptic, rounded or very slightly and obtusely pointed at the apex, rounded or slightly cuneate at the base, 1½–3 in. long, 1–1½ in. broad, slightly coriaceous, glabrous on both surfaces except on the lower half of the midrib which is sometimes pubescent below ; lateral nerves 6–7 on each side, looped, slightly prominent on the lower surface ; petiole 3–5 lin. long, slightly wrinkled, brownish-pubescent ; stipules deciduous ; male flowers in axillary pedunculate cymes ; peduncle 2–3 lin. long, more or less densely pubescent ; flowers clustered, subsessile ; bracts ovate or oblong, pubescent ; sepals 5, ovate, obtuse, 1½ lin. long, 1¼ lin. broad, pubescent outside, glabrous within ; disc annular, slightly wrinkled ; stamens 5 ; filaments connate in the lower part ; anthers 1 lin. long, glabrous ; rudimentary ovary very small ; female flowers solitary ; peduncle up to 3 lin. long, pubescent ; sepals oblong-lanceolate, obtuse, 2½ lin. long, 1¼ lin. broad, distinctly 5–6-nerved on the outside, pubescent ; disc cupular, toothed, glabrous ; ovary densely brown-villous ; styles 3, connate at the base, lobed, glabrous, lobes rather thick ; fruit globose, ¾ in. in

diam. ; pericarp about ¼ lin. thick, coarsely wrinkled when dry, yellow-drab ; endocarp bony and very hard, yellow, nearly ½ lin. thick ; seeds ellipsoid, shining, with a yellowish strophiole at the base, longitudinally striate, 3½ lin. long, 2¼ lin. broad. *Hutchinson in Dyer, Fl. Trop. Afr.* vi. i. 672, *and in Hook. Ic. Pl. t.* 3011.

EASTERN REGION : Transvaal ; Warm Baths, *Burtt-Davy,* 2151 ! between Thabina and Izaneen, on slope of a Kopje near Groot Letaba, *Burtt-Davy,* 5286 ! near Izaneen Estate, *Burtt-Davy,* 5560 !

Occurs also in Tropical Africa.

XVII. TOXICODENDRUM, Thunb.

Flowers diœcious. *Petals* and *disc* absent. *Male flowers : Sepals* 5–6, short, irregularly imbricate. *Stamens* indefinite ; filaments very short, densely inserted on a broad receptacle ; anthers large, erect, dehiscing longitudinally. Rudimentary *ovary* absent. *Female flowers : Sepals* 6, in two series, the outer smaller. *Ovary* hairy, 3–4-celled ; styles 3–5, connate in the lower half, the free parts thick and wrinkled (densely papillose). *Capsule* depressed-globose, hard, dehiscing into 2-valved cocci. *Seeds* with a shining crustaceous testa ; albumen fleshy ; embryo straight, with broad flat cotyledons.

A small tree or shrub ; leaves in whorls of 4, lanceolate, entire, rigidly coriaceous, penninerved ; flowers axillary, males in dense pedunculate or subsessile cymules, females sessile, 1–3 between each leaf-axil.

DISTRIB. Monotypic, endemic.

Toxicodendron acutifolium, Benth. in Journ. Linn. Soc. xvii. 214, is *Xymalos monospora,* Baill. (cf. C. H. Wright in Dyer, Fl. Cap. v. i. 493).

1. T. **capense** (Thunb. in Vet. Akad. Handl. Stockh. 1796, 188, t. 7) ; a shrub or small tree with corky branches ; branchlets grooved or angular when dry, finely puberulous, soon becoming glabrous ; leaves in whorls of 4, lanceolate, rounded and sometimes emarginate at the apex, gradually narrowed to the base, 2½–4 in. long, ½–1⅓ in. broad, very rigidly coriaceous, entire, at first slightly puberulous on both surfaces, soon becoming glabrous ; lateral nerves 8–10 on each side, rather faint ; petiole ¼–½ in. long, rather stout and finely puberulous ; male inflorescence axillary, densely cymulose, shortly pedunculate or subsessile, about one-sixth the length of the leaves ; peduncle up to ½ lin. long, rather densely and shortly pubescent ; pedicels about 1 lin. long, shortly pubescent ; sepals 5, rounded, much shorter than the stamens, somewhat tomentulose outside ; stamens numerous, inserted on a broad receptacle ; filaments very short ; anthers large, erect ; female flowers sessile, 1–3 between each leaf-axil ; sepals 6 in two series, the outer smaller, all suborbicular and tomentulose ; ovary ovoid,

tomentulose ; styles 3 to 5, connate in the lower half, free portions thick and densely papillose ; capsule usually 4-celled, depressed-globose, 8-lobed, about ½ in. long, scarcely 1 in. in diam. ; exocarp crustaceous ; endocarp bony, about ⅓ lin. thick ; seeds black and shining, 5 lin. long. *Pax in Engl. & Prantl, Pflanzenfam.* iii. 5, 32, *t.* 20, *figs. D.-F.* ; *Sim, For. Fl. Cape Col.* 326. *Jatropha globosa, Gærtn. Fruct.* ii. 122, *t.* 109, *fig.* 3. *Hyænanche globosa, Lamb. Descr. Cinch.* 1797, 52, *t.* 10 ; *A. Juss. Tent. Euphorb.* 41 ; *Harv. Gen. S. Afr. Pl. ed.* i. 303 ; *ed.* ii. 341 ; *Baill. Étud. Gén. Euphorb.* 567, *t.* 23, *figs.* 29–39 ; *Müll. Arg. in DC. Prodr.* xv. ii. 480.— *Croton foliis crassis, venosis, etc., Burm. Rar. Pl. Afr.* 122, *t.* 45.

COAST REGION : Vanrhynsdorp Div. ; Gift Berg, *Thunberg, Drège* ♂ and ♀ ! *Zeyher,* 1527 ! *Fryer in MacOwan, Herb. Austr.-Afr.,* 1529 ! Wind Hoek, *Schlechter,* 8072 ! Clanwilliam Div. ; on the mountains, *Scott Elliot* ! near Clanwilliam, *Prior* !

Harvey states that the pounded capsules (nuts) are used to poison the carcases of lambs whose bodies are used as bait in destroying hyænas, hence the generic name *Hyænanche* applied by Lambert.

The Gift (poison) Berg is so called on account of the prevalence of this shrub upon it. According to Dr. E. P. Phillips, "It is stated that animals drinking water which has collected beneath these bushes and into which some leaves have fallen die from the effects."

XVIII. HYMENOCARDIA, Wall.

Flowers diœcious, apetalous. *Disc* 0. *Male flowers : Calyx-segments* or lobes 4–5, rarely 6, imbricate. *Stamens* usually 5, opposite the calyx-segments ; filaments rarely exceeding the calyx, connate at the base into a very short column or free ; anthers large, at first reflexed, usually adorned with a conspicuous yellow or golden-coloured gland on the back, cells distinct, parallel, dehiscing longitudinally. Rudimentary *ovary* as long as the filaments, bifid or entire. *Female flowers :* Sepals distinct, narrow, caducous. *Ovary* 2-celled, compressed in the opposite plane to that of the septum ; styles 2, free from the base, long, subterete or flattened in the same plane as the ovary, undivided, papillose or glabrous ; ovules 2 in each cell. *Capsule* flat, of 2 compressed very broadly winged cocci, separating from a central axis ; pericarp crustaceous, reticulate ; endocarp membranous. *Seeds* usually solitary in each coccus, compressed ; testa thin ; albumen not copious ; cotyledons thin, broad and flat ; radicle long.

Trees or tall shrubs ; leaves alternate, petiolate, membranous or coriaceous, entire, penninerved ; male inflorescence a catkin-like spike, female shortly racemose.

DISTRIB. Species 8, one in India, the remainder Tropical African.

1. **H. ulmoides** (Oliv. in Hook. Ic. Pl. t. 1131); a shrub or tree
with the habit of *Ulmus*; branchlets slender, glabrous or very
sparingly hairy when young; leaves ovate, ovate-elliptic or lanceolate, obtusely acuminate, rounded or subacute at the base, about
1¼ in. long, 6–8 lin. broad, rather thinly coriaceous, glabrous on
both surfaces; lateral nerves 3–6 on each side, inconspicuous;
petiole slender, 2–4 lin. long, sparingly pubescent; stipules soon
falling off; male spikes racemosely arranged on short lateral
branchlets crowded towards the ends of the branches, about ½ in.
long; axis very slender, puberulous; bracts broadly spathulate,
ciliate; calyx deeply 5-lobed; lobes rounded, densely ciliate;
anthers with a small gold-coloured gland on the back; rudimentary
ovary equalling the filaments, entire, glabrous; female racemes
axillary; bracts oblong, pubescent; calyx-lobes 5, linear, subequal,
membranous, about 1½ lin. long; ovary oblong, compressed, emarginate, glabrous, with a few scattered yellow glands; styles distinct,
linear, flattened, glabrous, about 2 lin. long; fruits broadly obovate
or suborbicular, apex emarginate, base rounded or acute, wings
joined at the base and decurrent on the stipe, ½–¾ in. in diam.,
glabrous, membranous; pedicels 2–3 lin. long, glabrous; stipe 1–2
lin. long. *Pax in Engl. Pfl. Ost-Afr. C.* 236, *and in De Wild. &
Durand, Contr. Fl. Congo,* i. 49; *Hiern in Cat. Afr. Pl. Welw.* i.
966; *De Wild. & Durand, Reliq. Dewevr.* 207; *De Wild. Miss. É.
Laurent,* 127, *and Études Fl. Bas- et Moyen-Congo,* ii. 269; *Hutchinson in Dyer, Fl. Trop. Afr.* vi. i. 648 *H. Poggei, Pax in Engl.
Jahrb.* xv. 528; *Durand & Schinz, Études Fl. Congo,* 243. *H.
ulmoides, var. capensis, Pax in Engl. Jahrb.* xxviii. 22.

EASTERN REGION : Delagoa Bay; Matola, *Schlechter,* 11725!

Occurs also in Tropical Africa, from Angola through French and Belgian Congos
to the east coast.

Pax distinguished *H. ulmoides,* var. *capensis,* by the smaller and more obtuse
leaves, but Schlechter's specimen, the type of this supposed variety, is identical
with the male type sheet described by Oliver.

XIX. CROTON, Linn.

Flowers monœcious, rarely diœcious; petals in male flowers usually
present, in female flowers usually rudimentary or 0; disc glandular
in male flowers, glandular or annular in female flowers. *Male :
Calyx* deeply divided into 5, rarely 4 or 6 valvate or narrowly imbricate segments, or sometimes the sepals free. *Petals* as long as
calyx-lobes or sepals, or shorter. *Disc-glands* opposite calyx-segments,
adnate to their base or free and alternate with outer filaments.
Stamens 5–∞, inserted on a usually pilose receptacle; filaments free, in
bud inflexed, at length erect; anthers in bud with their apices pointing
downwards; anther-cells 2, parallel; dehiscence longitudinal. Rudi-

mentary *ovary* 0. *Female : Calyx* in flower with segments usually smaller than in the male, in fruit somewhat accrescent. *Petals* when present rarely as large as in males ; often much reduced or obsolete and replaced by tufts of hairs, occasionally altogether absent. *Disc-glands* as in males or at times connate in a ring. *Ovary* usually 3-celled, occasionally 2- or 4-celled, ovules solitary in each cell ; styles usually recurved, 1–3 times 2-fid or -partite, rarely multifid. *Capsule* normally 3-dymous, breaking up into 2-valved cocci ; rarely subindehiscent. *Seeds* smooth, strophiolate ; strophiole small ; testa crustaceous or woody ; albumen copious, fleshy ; cotyledons broad, flat.

Trees, shrubs or rarely herbs, of varied habit, usually beset with stellate hairs or orbicular scales ; leaves usually alternate, rarely opposite or verticillate, entire or toothed, rarely lobed, penninerved throughout, or palmately 3–∞ -nerved from the base, usually with 2 prominent glands at apex of petiole ; flowers in spikes or racemes ; in androgynous inflorescences usually with many males above and a few basal females, occasionally with both sexes mixed. Bracts small.

DISTRIB. Species about 600, throughout the tropics of both hemispheres, with a few species extending or confined to subtropical regions.

Young shoots and leaves beneath stellate-pubescent but
not lepidote ; leaves long-petioled, 5-nerved at base,
the midrib thereafter up to 4-jugately penninerved ;
female petals minute or obsolete :
Leaves pubescent above, basal glands obsolete ; female
calyx not accrescent : a shrub   ...   . (1) **rivularis.**

Leaves glabrous, or nearly so, above, basal glands large ;
female calyx not accrescent ; trees :
Basal glands at point of junction of petiole and
blade ; capsule ⅜ in. across   ...   . (2) **sylvaticus.**
Basal glands ₁₀—₁₂ in. below point of junction of
petiole and blade ; capsule ¾–1 in. across   ... (3) **Gubouga.**
Young shoots and leaves beneath lepidote ; leaves 6–11
jugately penninerved throughout ; female calyx not
accrescent :
Petals in female flower wanting ; leaves very shortly
petioled, stellate-pubescent above ...   . (4) **Menyharti.**
Petals in female flowers well developed ; leaves dis-
tinctly petioled, fragrant :
Leaves persistently stellate-puberulous above ; basal
glands at apex of petiole distinct, but hidden
by base of leaf   ...   . (5) **subgratissimus.**
Leaves quite glabrous above :
Lateral nerves of leaves indistinct or impressed
above :
Leaves ovate-lanceolate, acute, not over 2 in.
long ; lateral nerves faintly impressed above,
not visible beneath ; basal glands if present
on upper side of petiole below base of leaf   (6) **gratissimus.**
Leaves elliptic, obtuse or subacute, rarely
under 2½ in. long lateral nerves somewhat
impressed above and raised beneath ; basal
glands at apex of petiole distinct, often
stipitate, hidden by base of leaf   ...   ... (7) **zambesicus.**
Lateral nerves of leaves distinctly raised above,
not visible beneath ; basal glands obsolete ... (8) **pseudopulchellus.**

1. **C. rivularis** (E. Meyer in Drège, Zwei Pfl. Documente, 176, name [*rivulare*]; Müll. Arg. in Linnæa, xxxiv. 112); shrub, 2–6 ft. high; branches cylindric, stellate-pubescent; flowering twigs densely tawny stellate-tomentose ; leaves alternate, distinctly petioled, membranous, palmately 3–5-nerved at the base, midrib 3–4-jugately penninerved, ovate or ovate-lanceolate, obtuse or acute, base rounded, margin shortly rather irregularly serrate, 1½–3 in. long, ¾–1¾ in. wide, dark green, sparingly persistently stellate-pubescent above, softly stellate-pubescent or -tomentose beneath ; petiole slender, densely stellate-pubescent, ½–1 in. long, usually without glands at point of junction with blade ; stipules minute, caducous ; racemes androgynous, up to 4 in. long, at ends of twigs, with 12–20 scattered solitary or paired male flowers above and 3–4 scattered solitary female flowers below ; rhachis densely stellate-tomentose ; bracts small, ovate, stellate-pubescent ; male pedicels 1 lin. long, female pedicels ultimately 2–3 lin. long ; male sepals 5, ovate, obtuse, densely stellate-tomentose, ½ lin. long ; petals orbicular, densely white-villous at margins, thinly pubescent externally ; disc-glands small : stamens 10–15, filaments slender, glabrous ; receptacle villous : female sepals at first 1½ lin. long, ovate-lanceolate, at length accrescent and in fruit 5 lin. long, 2 lin. wide ; petals rudimentary or obsolete ; disc 5-lobed, glabrous ; ovary densely tomentose with stellate hairs ; styles 2-partite to the base, the segments linear, sometimes again 2-partite, glabrous ; capsule subglobose, ½ in. across, distinctly 3-sulcate and readily breaking up into 3 2-valved cocci ; densely stellate-tomentose ; seeds oblong, subtriquetrous, angled on inner face, rounded on the back, smooth. *DC. Prodr.* xv. ii. 602 : *Sim, For. Fl. Cape Col.* 311, *t.* 141, *fig.* 4. *C. dubius, Spreng. ex Eckl. & Zeyh. in Linnæa,* xx. 213, *name* ; *Baill. Adansonia,* iii. 154, *name. Oxydectes rivularis, O. Kuntze, Rev. Gen. Pl.* ii. 612 *and* iii. ii. 289.

COAST REGION: Uitenhage Div. ; Bushmans River, *Zeyher,* 854! near Addo, 1000–2000 ft., *Drège,* b! *Zeyher,* 330! *Ecklon & Zeyher,* 28! 349! Tzamoes, 400 ft., *Schlechter,* 2593! Enon and on the Zuureberg Range, near Bontje's River, 1000–2000 ft., *Drège,* a (2363)! near Uitenhage, *Prior*! Port Elizabeth Div. ; Van Stadens River, *Verreaux*! *Lelande*! Springfields, *Mrs. Paterson,* 2191! Sandflats, *Rogers,* 187! Bathurst Div. ; Port Alfred, 2000 ft., *Burchell,* 3797! *Potts,* 194! *Galpin,* 3044! *Schönland,* 793! *Burtt-Davy* 7890! between Port Alfred and Kaffirs Drift, *Burchell,* 3849! Glenfilling, *Drège,* bb! Albany Div. : Blue Krantz, *Burchell,* 3643! near Grahamstown, *Bowie*! *Ecklon & Zeyher*! *Williamson*! *MacOwan,* 203! *Bolus,* 2682! Howison's Poort, *Mrs. Hutton*! and without precise locality, *Miss Bowker,* 500! Bedford Div. ; Kagaberg, *Scott Elliot,* 664! King Williamstown Div. ; banks of the Buffalo River, *Drège,* c! 4633! *Zeyher*! King Williamstown, 1500 ft., *Hutton,* 95! *Tyson,* 968! Peddie Div. ; Fredericksburg, *Gill*! Komgha Div. ; near Komgha, 2000 ft., *Flanagan,* 715! near the Kei River, 2000 ft., *Schlechter,* 6248! British Caffraria ; without precise locality, *Cooper,* 58! *Pappe*!

EASTERN REGION: Transkei ; Kentani, at Blackpool, 1000 ft., *Miss Pegler,* 174! 189! Tembuland ; Perie Forest, *Kuntze.* Pondoland ; Mount Ayliff, *Schlechter*! Natal ?; without precise locality, *Cooper,* 3141!

2. **C. sylvaticus** (Hochst. apud Krauss in Flora, 1845, 82) ; tree 60 ft. high, trunk 3 ft. in. diam.; branches sulcate, glabrous ;

flowering twigs rusty stellate-puberulous; leaves alternate, long-petioled, thinly chartaceous, palmately 5-nerved at the base, midrib 4-jugately penninerved, ovate, rarely ovate-lanceolate, somewhat caudate-acuminate, base rounded or slightly cordate, margin more or less distinctly serrulate, 1½–5 in. long, 1–4 in. wide, dark green, glabrous above, sparingly stellate-pubescent on the nerves beneath; petiole slender, 1–4 in. long, sparingly stellate-pubescent, with 2 very prominent glands at point of junction with blade; stipules linear, stellate-pubescent, caducous; racemes androgynous, up to 6 in. long, at ends of twigs, with many male flowers and a few females accompanying throughout the raceme; rhachis sulcate, rather closely stellate-pubescent; bracts subulate-lanceolate, several-flowered; male pedicels 1–1¼ lin. long, female pedicels under 1 lin. long; male sepals 5, narrow-lanceolate, obtuse, 1 lin. long, membranous, thinly stellate-pubescent externally; petals 5, narrow oblong, as long as sepals, glabrous externally, the tip and margin within white-villous; disc-glands small; stamens 15–20; filaments slender, glabrous; receptacle villous; female sepals like the males; petals rudimentary; disc deeply 5-lobed, glabrous; ovary tomentose with stellate hairs; styles 2-partite to the base, the segments linear, glabrous; capsule subglobose, ⅓ in. across, shortly stellate-pubescent; seeds subglobose, smooth. *Sond. in Linnæa,* xxiii. 120; *Baill. Adansonia,* iii. 154; *Müll. Arg. in DC. Prodr.* xv. ii. 602; *Sim, For. Fl. Cape Col.* 310, *t.* 138; *S. Moore in Journ. Linn. Soc.* xl. 195; *Hutchinson in Dyer, Fl. Trop. Afr.* vi. i. 771. *C. Stuhlmanni, Pax in Engl. Jahrb.* xix. 80. *Oxydectes sylvatica, O. Kuntze, Rev. Gen. Pl.* ii. 613. *Claoxylum* (?) *sphærocarpum, O. Kuntze, l.c.* iii. ii. 284.

COAST REGION : Komgha Div. ; Coast Kloofs from the Kei River westwards to East London, *Flanagan,* 2528 ! near the mouth of the Kei River, *Flanagan,* 2368 ! *Schlechter,* 6203 !
EASTERN REGION : Pondoland ; St. Johns, *Sim,* 2435 ! Natal ; woods by the Umlaas River, *Krauss,* 142 ! near Durban, *Gueinzius* 82 ! *Gerrard,* 148 ! *Wood,* 10929 ! *Mrs. Saunders* ! Inanda, *Wood,* 758 ! Berea, *Wood,* 7887 ! 11548 ! Winkel Spruit, *Steyner in Herb. Wood,* 11901 ! Clairmont, *Kuntze* ! Dumisa, 1700 ft., *Rudatis,* 522 ! Zululand ; "very common," *Gerrard & McKen,* 1 !

Also in Eastern Tropical Africa.
The bark of this tree is used in Gazaland as a fish-poison (*Moore*).

3. **C. Gubouga** (S. Moore in Journ. Linn. Soc. Bot. xl. 196); tree 15–50 ft. high ; branches cylindric, distinctly lenticellate, glabrous ; flowering twigs sparingly grey stellate-pubescent, soon glabrous ; leaves alternate, long-petioled, thinly chartaceous, palmately 5-nerved at the base, midrib 3–5-jugately penninerved, ovate or ovate-lanceolate, usually caudate-acuminate, base truncate or slightly cordate, margin irregularly crenate-serrate, 2–4½ in. long, ¾–2½ in. wide, medium green, at first sparingly stellate-pubescent, but soon quite glabrous and finely rather closely verruculose above, sparsely grey stellate-pubescent with soft long-rayed hairs beneath : petiole slender, ¾–1¾ in. long, at first sparingly softly stellate-

pubescent, but soon glabrous, with 2 very prominent glands $\frac{1}{12}-\frac{1}{20}$ in.
below point of junction with blade ; stipules linear-lanceolate,
sparingly stellate-pubescent, caducous ; racemes androgynous, usually
$2\frac{1}{2}$–3 in. sometimes up to 5 in. long, at ends of twigs, with 12–20
male flowers in the upper third and 8–12 females in the lower third ;
rhachis angular, rather closely to sparingly grey stellate-pubescent ;
bracts linear-lanceolate, stellate-pubescent ; male pedicels 2 lin.
long, female pedicels $1\frac{1}{4}$–$1\frac{1}{2}$ lin. long ; male sepals 5, oblong-ellipsoid, obtuse, $1\frac{1}{3}$ lin. long, membranous, glabrous on both sides,
rather densely ciliate in the upper half ; petals 5, narrow elliptic,
2 lin. long, glabrous externally, densely ciliate ; disc-glands fleshy,
glabrous ; stamens 15–16, filaments slender, glabrous ; receptacle
villous ; female sepals 5, ovate-oblong, obtuse, $2\frac{1}{2}$ lin. long, shortly
ciliate in the upper half and with a few stellate hairs in the centre
externally ; petals obsolete ; disc crenulate, fleshy, glabrous ; ovary
densely white-tomentose with stellate hairs ; styles deeply 2-partite,
the segments linear and again 2-partite nearly to the base, glabrous,
fleshy, slightly wrinkled ; capsule broadly ellipsoid, not lobed, $1\frac{1}{4}$ in.
long, $\frac{3}{4}$–1 in. across, when ripe shortly sparingly stellate-puberulous
or nearly glabrous ; seeds oblong-ellipsoid, $\frac{3}{4}$ in. long, $\frac{1}{3}$ in. across,
testa dull, greyish-brown, smooth. *Hutchinson in Dyer, Fl. Trop.
Afr.* vi. i. 766.

KALAHARI REGION : Transvaal ; Lydenburg Dist., Selabi, on the banks of the
Olifants River, 1600 ft., *Pole-Evans*, H. 17024 !
EASTERN REGION : Portuguese East Africa ; Lourenço Marques, in the low
veld north of Delagoa Bay, *Maberly.*

Also in Eastern Tropical Africa.

The bark and the seeds of this tree are said to be effective as remedies in cases
of malarial bilious fever (*Maberly*). In the open *C. Gubouga* is usually a shrub
or small tree 15–30 ft. high, in sheltered spots a large tree 30–50 ft. high (*Pole-
Evans*). Its nearest ally in our area is *C. sylvaticus*, Hochst., which has, however,
much smaller fruits. The capsule appears to be indehiscent and is usually 2-seeded,
sometimes 1-seeded. In Gazaland the bark is used as a fish-poison (*Moore*).

4. **C. Menyharti** (Pax in Bull. Herb. Boiss. vi. 733) ; shrub, 3–6
ft. high ; twigs slender, at first densely stellate-tomentose, at length
glabrous ; leaves alternate, very shortly petioled, thinly papery,
penninerved, oblong-ovate or oblong-elliptic, obtuse and slightly
emarginate, base shallow-cordate and slightly unequal, margin entire,
$\frac{1}{2}$–2 in. long, $\frac{1}{4}$–$\frac{3}{4}$ in. wide, rather pale green and thinly permanently
stellate-pubescent above, densely clothed beneath with silvery scales
and also thinly stellate-pubescent ; lateral nerves 6–9 on each
side, very slightly impressed above and distinctly raised beneath ;
petiole about 1 lin. long, densely stellate-pubescent and sparingly
scaly, without glands at point of attachment with leaf ; stipules
minute, subulate ; racemes androgynous, $\frac{3}{4}$–$2\frac{1}{2}$ in. long, 5–15-
flowered at the ends of lateral shoots, rhachis densely stellate-
tomentose and scaly ; male pedicels $\frac{1}{6}$–$\frac{1}{4}$ in. long ; sepals ovate, 1 lin.
long, coriaceous, sparingly stellate-pubescent and densely scaly

outside, glabrous within; petals narrow oblanceolate, obtuse or subacute, ciliate and shortly pubescent within, glabrous outside; disc-
glands minute; stamens 15; filaments slightly hairy; receptacle
pubescent; female pedicels half as long as male or shorter; female
calyx lobed only two-thirds its depth, scaly but hardly tomentose
outside, slightly pubescent within; petals 0; disc annular, thick,
glabrous; ovary densely clothed with large orbicular scales, but
without stellate hairs; styles 3–4, deeply 2-lobed, lobes linear, entire
or again 2-fid, incurved, glabrous; capsule very faintly 3-lobed, $\frac{1}{4}$ in.
long, densely scaly; seeds ellipsoid, smooth. *C. Menyhartii,
Hutchinson in Dyer, Fl. Trop. Afr.* vi. i. 753. *C. pulchellus, Müll.
Arg. in DC. Prodr.* xv. ii. 572, *partly and as regards syn. Klotzsch
only; not of Baill. C. kwebensis, N. E. Br. in Kew Bulletin,* 1909,
140. *Argyrodendron bicolor, Klotzsch in Peters, Reise Mossamb. Bot.*
102 : *not Croton bicolor, Roxb.*

EASTERN REGION : Natal ; on the Tugela River, *Gerrard,* 1460 !
Also in Tropical Africa.

5. **C. subgratissimus** (Prain in Kew Bulletin, 1913, 79); large
shrub or small tree, 16–25 ft. high; twigs slender, angled, covered
with silvery scales when young; leaves fragrant, alternate, distinctly
to long-petioled, coriaceous, penninerved, ovate-lanceolate, acuminate,
emarginate, base minutely narrow-cordate, margin entire, $1\frac{1}{4}$–$3\frac{1}{2}$ in.
long, $\frac{1}{2}$–$1\frac{1}{2}$ in. wide, pale green, dull and uniformly persistently
stellate-puberulous above, densely clothed with silvery scales with a
few rusty scales intermixed beneath; lateral nerves 12–14 a side,
neither impressed above nor raised beneath; petiole $\frac{1}{2}$–$1\frac{1}{2}$ in. long,
densely silvery and rusty lepidote, with 2 sessile glands underneath
at point of attachment with blade; stipules subulate, usually short,
occasionally very long, scaly; racemes androgynous, 1–4 in. long, at
ends of shoots; rhachis densely silvery and rusty scaly; bracts
subulate-lanceolate, scaly; male pedicels 1–2 lin. long, densely scaly;
sepals ovate, obtuse, $1\frac{1}{4}$ lin. long, coriaceous, densely scaly outside,
closely puberulous within; petals ovate, rather shorter than sepals,
scaly externally, margins villous, glabrous within; disc-glands thick,
glabrous; stamens 15–20, filaments thinly pubescent below; receptacle pilose; female sepals as in male; petals oblong-lanceolate,
obtuse, scaly outside, hairy within; disc minute; ovary densely
clothed with fringed scales; styles spreading, 6–8-partite, glabrous;
capsule very slightly 3-lobed, $\frac{1}{4}$ in. long, densely scaly; seeds smooth.
Prain in Dyer, Fl. Trop. Afr. vi. i. 1050. *C. gratissimus, Pax in
Engl. Jahrb.* x. 35, *not of Burch.*

KALAHARI REGION : Bechuanaland ; Lobatsi, *Marloth,* 3331! Transvaal ;
Wonderboom Poort, *Rehmann,* 4552! *Miss Leendertz,* 270! *Rogers,* 233 ! *Burtt-
Davy,* 1849! Magaliesberg Range, *Zeyher,* 2767! Pretoria, *Fehr,* 54!

Also in Hereroland and Tropical North-west Bechuanaland.

Very nearly allied to *C. gratissimus,* Burch., which it most resembles in general
facies, and to *C. zambesicus,* Müll. Arg., which it most resembles as regards flowers,
but equally distinct from both in having leaves which are persistently stellate-
puberulous above.

6. C. gratissimus (Burch. Trav. S. Afr. ii. 263); shrub or small tree, 4–20 ft. high; twigs slender, angular, covered with silvery and rusty scales when young; leaves fragrant, alternate, distinctly to long-petioled, coriaceous, penninerved, ovate-lanceolate, acuminate or acute, emarginate, base minutely narrow-cordate, margin entire, $1\frac{1}{4}$–2 in. long, $\frac{1}{3}$–$\frac{2}{3}$ in. wide, pale green, quite glabrous, polished above, densely clothed with silvery and rusty scales intermixed beneath; lateral nerves 10–12 on each side, neither impressed above nor raised beneath; petiole $\frac{1}{4}$–1 in. long, densely clothed with silvery and rusty scales with basal glands usually very minute or absent and when distinct then always sessile and attached to upper portion of petiole slightly below its attachment to the blade; stipules subulate, very short; racemes androgynous, up to $2\frac{1}{2}$ in. long, at the ends of shoots; rhachis densely silvery and rusty scaly; bracts subulate-lanceolate, scaly; male pedicels 1–$1\frac{1}{2}$ lin. long, densely scaly; sepals ovate, obtuse, $1\frac{1}{2}$ lin. long, coriaceous, outside densely scaly, closely puberulous within; petals ovate, as large as sepals, outside scaly, margin villous, within glabrous; disc-glands thick, glabrous; stamens 10–15; filaments thinly pubescent below; receptacle pilose; female sepals as in the male; petals oblong-lanceolate, obtuse, scaly outside, hairy within; disc minute; ovary densely clothed with fringed scales; styles 3, erect, very stout, each 2-lobed, glabrous; capsule very slightly 3-lobed, $\frac{1}{3}$ in. long, densely scaly; seeds smooth. *Sond. in Linnæa,* xxiii. 119; *Baill. Adansonia,* iii. 154; *Müll. Arg. in DC. Prodr.* xv. ii. 516; *Sim, For. Fl. Cape Col.* 311; *Prain in Dyer, Fl. Trop. Afr.* vi. i. 1051. *C. microbotrys, Pax in Engl. Jahrb.* x. 35. *Oxydectes gratissima, O. Kuntze, Rev. Gen. Pl.* ii. 611.

KALAHARI REGION : Griqualand West; Klipfontein, *Burchell,* 2154! 2631! Bechuanaland; near the sources of the Kuruman River, *Burchell,* 2493/2! near Kuruman, 4200 ft., *Marloth,* 1078! Takun, *Lemue.* Transvaal; Magaliesberg Range, *Burke! Zeyher,* 1513! *Engler,* 2764! Rhenoster Poort, *Nelson,* 93! Warmbath, *Miss Leendertz,* 1367! *Burtt-Davy,* 2185! 2623! 5565! Wonderboom Poort, *Galpin,* 6989! Walmaranstad, *Hull,* 6081! near Nazareth, 4500 ft., *Schlechter,* 4478!

EASTERN REGION : Natal; near Greytown, 4000 ft., *Wood,* 10035! Drakenberg Range, *Wahlberg.*

Also in Hereroland and in Tropical North-eastern Bechuanaland.

7. C. zambesicus (Müll. Arg. in Flora, 1864, 483); large shrub or small tree, 16–25 ft. high; twigs angular, covered with silvery and rusty scales with or without an accompanying sparse stellate-pubescence; leaves fragrant, alternate, distinctly to long-petioled, firmly membranous, penninerved, elliptic-lanceolate, obtuse or slightly narrowed to the apex, emarginate, base narrow shallowly cordate, margin entire, $2\frac{1}{2}$–$4\frac{1}{2}$ in. long, $\frac{3}{4}$–$1\frac{3}{4}$ in. wide, green, glabrous, dull above, densely clothed with silvery and a few rusty scales beneath; lateral nerves 12–14 a side, slightly impressed above and raised beneath; petiole $\frac{1}{2}$–$1\frac{1}{2}$ in. long, densely silvery lepidote and sometimes also sparingly stellate-pubescent with 2 usually distinctly stipitate glands underneath at point of attach-

ment with blade ; stipules subulate, usually short, occasionally very
long, scaly ; racemes androgynous, 1–4 in. long, at ends of shoots ;
rhachis densely silvery and rusty scaly ; bracts subulate-lanceolate,
scaly ; male pedicels 1–2 lin. long, densely scaly ; sepals ovate,
obtuse, 1¼ lin. long, coriaceous, densely scaly outside, closely
puberulous within ; petals ovate, as large as sepals, scaly externally, margins villous, sparingly pilose within ; disc-glands thick,
glabrous ; stamens 15–20, filaments thinly pubescent below ; receptacle pilose ; female sepals as in male ; petals oblong-lanceolate,
obtuse, scaly outside, hairy within ; disc minute ; ovary densely
clad with fringed scales ; styles spreading, 4–6-partite ; segments
1–3-fid, glabrous above, puberulous on back below ; capsule distinctly 3-lobed, ⅔ in. long, densely scaly. *Müll. Arg. in DC. Prodr.*
xv. ii. 515 ; *Pax in Engl. Pfl. Ost-Afr. C.* 237 ; *De Wild. Études
Fl. Bas- et Moyen-Congo,* ii. 278 ; *N. E. Br. in Kew Bulletin,* 1909,
139 ; *Hutchinson in Dyer, Fl. Trop. Afr.* vi. i. 758. *C. welwit-
schianus, Müll. Arg. in Journ. Bot.* 1864, 338, *and in DC. Prodr.*
xv. ii. 515 ; *Hiern in Cat. Afr. Pl. Welw.* i. 970 ; *De Wild. & Durand
in Bull. Herb. Boiss. 2me sér.* i. 47 ; *Th. & Hél. Durand, Syll. Fl.
Congol.* 481. *C. Antunesii, Pax in Engl. Jahrb.* xxiii. 523, *and in
Baum, Kunene-Samb. Exped.* 282. *C. amabilis, N. E. Br. in Kew
Bulletin,* 1909, 140 ; *not of Müll. Arg. Oxydectes welwitschiana, and
O. zambesica, O. Kuntze, Rev. Gen. Pl.* ii. 613.

KALAHARI REGION : Transvaal ; Potgieters Rust, *Miss Leendertz,* 1428 !

EASTERN REGION : Natal ; Tugela, *Gerrard,* 1460 ! Delagoa Bay ; Ressano
Garcia, 1000 ft., *Schlechter,* 11938 !

Also in Tropical Africa.

This species occurs in Great Namaqualand, north of the Tropic and has been
collected as far south as Rehoboth ; it may therefore yet be met with in our
Western Region. It is nearly allied to *C. gratissimus,* Burch., and may be only a
luxuriant form of that plant, but can usually be readily distinguished by its larger
leaves, dull on the upper surface, with distinctly stipitate basal glands at the apex
of the petiole ; also to *C. subgratissimus,* Prain, but is very readily distinguished
therefrom by the glabrous upper surface of the leaves.

8. **C. pseudopulchellus** (Pax in Engl. Jahrb. xxxiv. 371) ; shrub,
6–12 ft. high ; twigs slender, covered with rust-coloured scales when
young ; leaves fragrant, subverticillate, distinctly petioled, firmly
papery, penninerved, lanceolate or elliptic-lanceolate, obtuse and
slightly emarginate, base rounded or wide-cuneate, margin entire,
½–2 in. long, ¼–¾ in. wide, dark green, quite glabrous above, densely
clothed beneath with silvery scales and rusty scales intermixed ;
lateral nerves 9–12 on each side, slightly raised above, not visible
beneath ; petiole up to 1 in. long, scaly like undersurface of leaf
which is without glands at its attachment to the petiole ; stipules
minute, subulate ; racemes androgynous, much abbreviated, forming
small dense corymbs, ⅓ in. across, at the tips of the shoots ; male
pedicels about ⅛ in. long when flower opens ; sepals ovate, 1 lin. long,
coriaceous, densely scaly externally, the margins puberulous within ;

petals oblong-lanceolate, obtuse, with villous margins, otherwise
glabrous on both surfaces; disc-glands minute; stamens 15–18;
filaments explanate and sparingly hairy below; receptacle pubescent;
female sepals as in male; petals rather narrower and shorter than
in male; disc thin, glabrous; ovary densely clothed with large,
orbicular scales; styles 2-partite almost to the base, the arms subterete, glabrous; capsule 3-lobed, $\frac{1}{3}$ in. long, densely scaly; seeds
ellipsoid, brown.   *Hutchinson in Dyer, Fl. Trop. Afr.* vi. i. 757.
C. pulchellus, Müll. Arg. in DC. Prodr. xv. ii. 572, *partly, but excluding
the Madagascar plant and syn. Klotzsch*; *not of Baill.*

EASTERN REGION : Delagoa Bay ; Morakwen, *Junod*, 361 a ! 516 !

Also in Tropical East Africa.

XX. JATROPHA, Linn.

Flowers monœcious, rarely diœcious; petals usually present; disc
of 5 free glands or sometimes in male, often in female annular.
Male: Sepals 5, imbricate, often slightly connate below. *Petals* 5,
usually imbricate, connate below in a tube or free throughout,
rarely obsolete. *Stamens* in our species 8, 2-seriate; filaments
connate below in a column, the 5 outer opposite the petals usually
shorter than the 3 inner; anthers erect, their cells parallel, contiguous; dehiscence longitudinal.   Rudimentary *ovary* 0.   *Female:
Sepals* as in male.   *Petals* as in male.   *Ovary* 2–3- (rarely 4–5-)
celled; ovules solitary in each cell; style connate below, spreading
above, often shortly 2-partite.   *Capsule* ovoid or subglobose, breaking
up into 2–5, usually 3, 2-valved cocci; endocarp hard, thin and
crustaceous or thickened and woody.   *Seeds* carunculate, ovoid or
oblong; caruncle often deeply lobed : testa crustaceous; albumen
fleshy; cotyledons broad, flat.

Herbs, often with thick perennial rootstocks, or shrubs; leaves alternate,
usually scattered, occasionally clustered, petioled or sessile, entire or digitately
rarely pinnately lobed, hairs when present simple, sometimes glandular; stipules
setaceously lobed or partite, rarely rigid and spiny; flowers disposed in terminal
dichotomous cymes, the female solitary, terminal, the male lateral.

DISTRIB. Species about 160, generally distributed throughout the tropics and
extending beyond the tropics into North America and South Africa.

Shrubs with woody subaerial stems; leaves distinctly
    petioled :
  Stems 5 ft. high or higher ; leaves ovate-cordate, long-
    petioled ; male petals united in their lower half ... (1) **Curcas.**
  Stems 3 ft. high or shorter ; male petals free :
    Leaves wide ovate or oblong, nearly as broad as long,
      often 3-lobed, less often 5-lobed or entire ;
      petioles long ; stipules dissected   ...   ... (2) **variifolia.**

Leaves lanceolate or oblong-lanceolate, about thrice
as long as broad, base usually hastately 2-lobu-
late ; petioles short ; stipules small, gland-like (3) **capensis.**

Herbs with soft subaerial stems springing from a persistent rootstock :
Subaerial herbaceous stems tufted at the apices of
their vertical hypogæal stems, themselves rising
from a deeply buried dauciform swollen rootstock ;
leaves shortly petioled, runcinate-pinnatifid ;
stipules entire or 2-fid, eglandular ; sepals with
hyaline denticulate edges connate half-way in a
campanulate calyx ... ... (4) **erythropoda.**

Subaerial herbaceous stems springing directly from
thickened more or less horizontal woody rootstocks ;
sepals very shortly connate at the base only :
Margins of leaves quite eglandular ; leaves sessile ;
stipules entire or 2-fid, eglandular :
Margins of bracts and sepals entire and eglandular ;
leaves sparingly hirsute, very variable in
outline, mainly 3-lobed, but often with entire
and with 5–7-lobed intermixed ... ... (5) **natalensis.**

Margins of bracts and sepals closely glandular-
denticulate ; leaves glabrous, ovate-lanceolate to linear-lanceolate much longer than
broad, entire or at times 1–2-dentate near
base on each side ... ... (6) **lagarinthoides.**

Margins of leaves, bracts and sepals closely glandular-
denticulate or ciliate or setulose :
Marginal glands of leaves, bracts and sepals sessile ;
leaves sessile, always entire :
Leaves quite glabrous ; stipules setaceous, entire
or 2-fid, eglandular, deciduous ... ... (7) **latifolia.**

Leaves more or less hirsute ; stipules dissected,
persistent, their lobules gland-tipped ... (8) **hirsuta.**

Marginal glands of leaves, bracts and sepals stipitate ; leaves usually distinctly lobulate, rarely
entire or subentire :
Cymes hardly overtopping the leaves ; leaves
sessile or very shortly petioled, their apices
and the tips of the lobes and lobules acute ;
petals silky outside :
Stems and leaves softly densely hirsute ; leaves
distinctly pinnately lobed, their marginal
glands very long-stalked ; male petals
slightly united below ... ... (9) **Woodii.**

Stems and leaves rather harshly sparsely
pilose ; leaves spuriously subpalmately
lobed, rarely entire or subentire, their
marginal glands short-stalked ; male
petals free ... ... (10) **Zeyheri.**

Cymes considerably overtopping the leaves ;
leaves distinctly petioled, the tip of lobes
and lobules obtuse, denticulate ; petals
glabrous outside ... ... (11) **Schlechteri.**

1. J. Curcas (Linn. Sp. Pl. ed. i. 1006); shrub of considerable size; twigs stout, glabrous; leaves long-petioled, firmly papery, ovate-rotund, obtuse or subacute, base wide-cordate, margin entire to undulate or 5-lobulate, $3\frac{1}{2}$–6 in. long, 3–$5\frac{1}{2}$ in. wide, 5–9-nerved from the base, glabrous and distinctly reticulately veined on both surfaces; petiole shallowly channelled above, glabrous, $3\frac{1}{2}$–7 in. long; stipules small; cymes much shorter than the leaves; peduncle $1\frac{1}{2}$–2 in. long, glabrous; bracts lanceolate or linear, entire, $2\frac{1}{2}$ lin. long or less, sparsely pubescent; male sepals ovate-elliptic, subacute, glabrous, entire, under 2 lin. long; petals oblong-obovate, connate in their lower half, densely hairy within, 3–$3\frac{1}{2}$ lin. long; disc-glands free, columnar, cylindric, glabrous; stamens 8; outer filaments almost free, inner connate; female sepals 2 lin. long.; petals quite free, oblong, obtuse, 3 lin. long, entire, hairy within near the middle; disc deeply 5-lobed, glabrous; ovary glabrous; styles short, connate at the base; capsule ellipsoid, faintly lobed, about 1 in. long; seeds oblong, $\frac{3}{4}$ in. long, $\frac{1}{2}$ in. wide. *Jacq. Hort. Vindob.* iii. 36, *t.* 63; *Müll. Arg. in DC. Prodr.* xv. ii. 1080 *and in Mart. Fl. Bras.* xi. ii. 487, *t.* 68; *Baker, Fl. Maurit.* 322; *Pax in Engl. & Prantl, Pflanzenfam.* iii. 5, 75, *fig.* 45; *in Engl. Pfl. Ost-Afr. C.* 240; *and in Engl. Pflanzenr. Euphorb. Jatroph.* 77; *Hiern in Cat. Afr. Pl. Welw.* i. 968; *Hutchinson in Dyer, Fl. Trop. Afr.* vi. i. 791. *J. acerifolia, Salisb. Prodr.* 389. *Curcas purgans, Medik. Ind. Pl. Hort. Manhern.* i. 90; *Baill. Étud. Gén. Euphorb.* 314, *t.* 19, *figs.* 10, 11. *C. indica, A. Rich. in Ramon de la Sagra, Fl. Cub. Fanerog.* iii. 208. *C. Adansonii, Endl. ex Heynh. Nomencl.* 176. *Castiglionia lobata, Ruiz & Pav. Fl. Peruv. Prodr.* 139, *t.* 37.

Cultivated by the natives in the Transvaal and in Natal.

A native of tropical America, but now widely spread as a cultivated and naturalised species, throughout Africa, the Mascarene Islands and South-Eastern Asia.

2. J. variifolia (Pax in Engl. Pflanzenr. Euphorb. Jatroph. 54); shrub, up to 3 ft. high; twigs stout, cylindric, glabrous; leaves distinctly to long-petioled, thinly leathery, wide ovate, rarely oblong, acute, base rounded, truncate or shallow cordate, margin entire or 3- (less often 5-) lobed, lobes acute, 3–5 in. long, 2–6 in. wide, bright green, glabrous on both surfaces, always distinctly 5-nerved at the base; petiole $\frac{1}{4}$–2 in. long, glabrous; stipules split up into numerous setaceous gland-tipped lobules, caducous; cymes rather lax, many-flowered, just overtopping the full-grown leaves; peduncles $1\frac{1}{2}$–$2\frac{1}{2}$ in. long, slender, glabrous; primary branches $\frac{1}{2}$–$\frac{3}{4}$ in. long, very slender, glabrous; bracts split up into numerous setaceous gland-tipped lobules, 1 lin. long; male sepals ovate-oblong, obtuse, glabrous, entire, connate below, under 1 lin. long; petals oblong, obtuse, $\frac{1}{3}$ in. long, quite free, glabrous; disc-glands free; stamens 8, filaments of both series united below in a very short glabrous column, shortly free above; female sepals oblong-lanceolate, acute, $\frac{1}{8}$ in. long: petals linear-spathulate, $\frac{1}{4}$ in. long, quite free, glabrous: ovary glabrous:

styles 3, 2-fid above ; capsule 3-dymous, $\frac{2}{3}$ in. across, cocci subglobose ; seeds oblong, $\frac{1}{4}$ in. long, $\frac{1}{8}$ in. wide. *J. triloba, E. Meyer
in Drège, Zwei Pfl. Documente,* 194 ; *not of Cerv. J. heterophylla,
Pax in Engl. Jahrb.* xxviii. 25 ; *not of Heyne. J. capensis, Sim, For.
Fl. Cape Col.* 310, *partly and as to syn. Drège only ; not of Linn. f.*

CoAST REGION : East London Div. ; Fort Jackson, 800 ft., *Rattray,* 1252 !
Komgha Div. ; near the mouth of the Kei River, *Flanagan,* 2336 !
KALAHARI REGION : Transvaal ; Komati Poort, 1000 ft., *Schlechter,* 11798 !
EASTERN REGION : Transkei ; Bashee River, *Drège,* 4631 ! *Bowker,* 466 !
516 ! Delagoa Bay ; near Lourenço Marques, *Bolus,* 9773 !

3. J. capensis (Sond. in Linnæa, xxiii. 118) ; shrub, up to 3 ft.
high ; twigs stout, obscurely 4-gonous, glabrous ; leaves shortly to
distinctly petioled, thinly leathery, lanceolate or oblong-lanceolate,
acute, base wide cuneate, truncate or subcordate usually more or
less distinctly hastately lobulate with lobules oblong obtuse rarely
subacute and rarely as long as the width of the lamina, margin
elsewhere entire, $1\frac{1}{2}$–3 in. long, $\frac{1}{2}$–1 in. wide, basal lobules never
over $\frac{1}{2}$ in. long, usually much shorter, often on lowest leaves of a
shoot obsolete on one or both sides, dark green, glabrous on both
surfaces, when lobulate distinctly 3-nerved at the base ; petiole
$\frac{1}{4}$–$\frac{3}{4}$ in. long, glabrous ; stipules very small, gland-like ; cymes rather
dense, few-flowered, just overtopping the leaves ; peduncles $\frac{3}{4}$–$1\frac{1}{4}$ in.
long, slender, glabrous ; bracts lanceolate, entire, $1\frac{1}{2}$ lin. long,
glabrous ; male sepals ovate-lanceolate, acute, glabrous, entire,
connate below, 1 lin. long ; petals spathulate-lanceolate, obtuse, $\frac{1}{4}$ in.
long, quite free, glabrous ; disc-glands free ; stamens 8 ; filaments
of both series united below in a glabrous column, free above ; female
sepals $1\frac{1}{2}$ lin. long ; petals as in male, but $\frac{1}{3}$ in. long ; disc-glands
free ; ovary glabrous ; styles 3, connate below, 2-fid above ; capsule
ellipsoid, faintly lobed, about $\frac{1}{2}$ in. long ; seeds oblong, $\frac{1}{3}$ in. long,
$\frac{1}{5}$ in. wide. *Baill. Adansonia,* iii. 149 ; *Müll. Arg. in DC. Prodr.*
xv. ii. 1084 ; *Sim, For. Fl. Cape Col.* 310 (*excl. syn. Drège*), *t.* 16,
fig. 9 ; *Pax in Engl. Pflanzenr. Euphorb. Jatroph.* 54. *Croton capensis,
Linn. f. Suppl.* 422 ; *Thunb. Prodr.* 117, *and Fl. Cap. ed. Schult.*
546 ; *Willd. Sp. Pl.* iv. 554 ; *Geisel. Crot. Monogr.* 75 ; *Spreng. Syst.*
iii. 875 ; *E. Meyer in Drège, Zwei Pfl. Documente,* 176 ; *Krauss in
Flora,* 1845, 82 ; *Eckl. & Zeyh. in Linnæa,* xx. 213 ; *Harv. Gen.
S. Afr. Pl. ed. Hook.* 336.

CoAST REGION : Uitenhage Div. ; near Uitenhage, *Prior* ! *Fraser* ! Zwartkops
River, *Thunberg* ! *Drège,* 8219 a ! *Zeyher,* 735 ! between Uitenhage and the
Coega River, *Burchell,* 4417 ! Enon, 1000–2000 ft., *Drège,* 8219 b ! Little Winter-
hoek, *Drège,* 8219 c ! Addo, 1000–2000 ft., *Ecklon & Zeyher,* 27 ! Winterhoek,
Krauss, 1722 ! Port Elizabeth Div. ; Hankey, *Mrs. Paterson,* 6 ! Albany Div. ;
Grahamstown, *Bowker* ! *Williamson* ! *MacOwan,* 459 ! Fish River Heights, *Mrs.
Bowker* ! *Hutton* ! *Scott Elliot,* 882 ! Queens Road, *Schönland,* 936 ! Bothas Hill,
2000 ft., *Schlechter,* 6098 ! Alicedale, *Rogers,* 4626 ! Fort Beaufort Div. ; Kunap
River, *Baur,* 1037 ! Queenstown Div. ; without locality, *Hutton* ! Peddie Div. ;
along the Keiskamma River, at Pinedrift, *Sim,* 2581 !
CENTRAL REGION : Jansenville Div. ; near Jansenville, 1500 ft., *Bolus,* 1660 !
and Karroo, without precise locality, *Gill* !

4. J. erythropoda (Pax & K. Hoffm. in Engl. Pflanzenr. Euphorb. Jatroph. 66); herb up to 5 in. high ; rootstock deeply buried, large, red, astringent, 4 in. long, 1¼ in. thick, giving off one or more apical underground perennial vertical stems 4–6 in. long, whence at the surface spring several slender branching herbaceous leafy stems, glabrous below, softly sparsely white-pubescent above ; leaves all shortly petioled, rather firm, irregularly runcinate-pinnatisect, lobes linear or linear-lanceolate again runcinately incised, margin hyaline and spinulose-denticulate, 2–2½ in. long, 1½–2 in. wide, lobes ½–1 in. long, ½–1 in. wide, lobules ¼–⅓ in. long, glabrous on both surfaces ; petiole ¼–⅓ in. long, softly shortly white-pubescent ; stipules setaceous or subhyaline, persistent, entire or 2-fid ; cymes rather open, few-flowered ; peduncles up to ½ in. long, softly white-pubescent ; primary branches ¼ in. long, white-pubescent ; bracts small, lanceolate, hyaline-denticulate, ⅛ in. long ; male sepals connate to the middle in a campanulate tube, above free, triangular, hyaline-denticulate, ⅛ in. long ; petals glabrous, linear-spathulate or narrow oblong, ¼ in. long ; disc-glands free ; stamens 8, the 5 of the outer series almost free, the 3 inner connate below ; female calyx like male, but rather larger ; petals as in male, but over ⅓ in. long ; hypogynous glands free ; ovary glabrous ; styles 3 ; capsule somewhat depressed-globose, ½ in. across, ⅓ in. long, hardly sulcate, glabrous, verruculose. *Hutchinson in Dyer, Fl. Trop. Afr.* vi. i. 783.

KALAHARI REGION : Bechuanaland ; Chooi Desert, *Burchell*, 2351 ! 2353 ! near Serowe, *Schönland*, 1655 !

Also in German South-West Africa.

5. J. natalensis (Müll. Arg. in Flora, 1864, 485) ; herb up to 2 ft. high ; rootstock stout, perennial ; stems leafy, simple or branched, sparingly pilose with white spreading hairs below, more densely towards the top ; leaves sessile, firm, very variable, occasionally all entire, sometimes all lobed, more often a few entire, the majority lobed ; entire leaves lanceolate, ovate-lanceolate or oblong, lobed leaves usually 3-lobed, less often 5–7-lobed, apices of leaf and lobes acute, slightly hyaline-mucronulate, base cuneate, margin hyaline, quite entire or hyaline-denticulate, 4–5½ in. long, 1½–3 in. wide, rather sparingly beset with white spreading hairs on the nerves on both surfaces ; stipules subulate, ₁/₁₀ in. long, usually entire, not glandular, very caducous ; cymes rather lax, many-flowered, often hardly overtopping the leaves ; peduncles softly and sparingly hirsute, 1½–4 in. long ; primary branches hirsute, ¾–1¼ in. long ; bracts lanceolate, entire, or the lowest with 2–3 hyaline non-glandular teeth on each side, lowest ⅓ in., uppermost ¼ in. long ; male sepals lanceolate, sparingly hairy outside or glabrous, margins quite entire ; petals yellow, spathulate-lanceolate, ¼ in. long, free ; disc-glands free ; stamens 8 ; filaments of both series united in a column ; female sepals like the male but rather larger ; petals as in male but larger ; hypogynous glands free ; ovary glabrous ; styles

3, 2-lobed at the apex; capsule 3-dymous, ovate, $\frac{1}{2}$ in. across; cocci oblong, glabrous; seeds wide ovate-oblong, $\frac{1}{3}$ in. long, $\frac{1}{4}$ in. wide. *DC. Prodr.* xv. ii. 1083; *Wood, Natal Pl. t.* 242; *Pax in Engl. Pflanzenr. Euphorb. Jatroph.* 65.

EASTERN REGION : Natal ; Tugela River, near Colenso, 3000–4000 ft., *Gerrard,* 1633! *Wood,* 3391! *Rehmann,* 7168! *Dimock-Brown,* 7746! Weenen County, Mooi River Valley, 2000–3000 ft., *Sutherland*! Griffins Hill, Eastcourt, 3000–4000 ft., *Rehmann,* 7313! *Wood,* 3391! Newcastle, Arnolds farm, *Rehmann,* 7048! near Ladysmith, 4000 ft., *Wood,* 4243! 7950! *Rehmann,* 7137! *Rogers,* 682!

6. J. lagarinthoides (Sond. in Linnæa, xxiii. 118); herb up to 8 in. high; rootstock stout, perennial ; stems leafy, simple or branched, usually several from the same base, hispid-pilose upwards ; leaves sessile, papery, lanceolate to linear-lanceolate, acute or obtuse, base narrowed or rounded, margin entire or occasionally 1-dentate (very rarely 2-dentate) near the base on each side, 1–2$\frac{1}{2}$ in. long, $\frac{1}{6}$–$\frac{1}{3}$ in. wide, glabrous on both surfaces; stipules setaceous, entire or 2-fid, not gland-tipped, subpersistent ; cymes rather dense, few-flowered, hardly overtopping the leaves ; peduncles patently hirsute, $\frac{1}{4}$–$\frac{1}{3}$ in. long ; primary branches patently hirsute, $\frac{1}{6}$–$\frac{1}{3}$ in. long ; bracts linear-lanceolate, margins closely lacerate-ciliate, glabrous, up to $\frac{1}{5}$ in. long ; male sepals lanceolate, closely glandular-denticulate, glabrous ; petals spathulate-oblong, obtuse, $\frac{1}{4}$ in. long, free; disc-glands free ; stamens 8, filaments of both series united below in a column, the outer 5 free in their upper half, the inner 3 free in their upper fourth ; female sepals like the male sepals ; petals as in male but larger ; ovary glabrous, tubercular-scabrous throughout; styles 3, united below, 2-lobed above ; capsule very slightly sulcate, oblong-ovate, hardly 3-dymous, $\frac{1}{2}$ in. across ; cocci narrow oblong, tubercular-scabrous, glabrous ; seeds narrow-oblong, $\frac{1}{3}$ in. long, $\frac{1}{5}$ in. wide. *Baill. Adansonia,* iii. 149 ; *Müll. Arg. in DC. Prodr.* xv. ii. 1088 ; *Pax in Engl. Pflanzenr. Euphorb. Jatroph.* 64, *fig.* 25.

VAR. β, **cluytioides** (Prain) ; leaves larger, firmer, ovate-lanceolate to lanceolate, 1–3 in. long, $\frac{2}{3}$–$\frac{3}{4}$ in. wide, in all other respects as in the type. *J. cluytioides, Pax & K. Hoffm. l.c.* 65. *J. latifolia, var. stenophylla, Pax, l.c.* 133.

KALAHARI REGION : Transvaal ; Magaliesberg Range, *Burke,* 65 mainly ! *Zeyher,* 1574! near Pretoria, *Fehr* ! *Schlechter,* 3703! Derde Poort, *Miss Leendertz,* 377 ! Irene, *Miss Leendertz,* 677! Pinedene, near Irene, *Burtt-Davy,* 2313 ! Var. β : Transvaal ; Magaliesberg Range, *Burke,* 65 partly ! near Middelburg, *Schlechter,* 3790 ! Witbank, *Gilfillan in Herb. Galpin,* 7237 ! between Porter and Trigardsfontein, *Rehmann,* 6596 ! 6661 !

A very distinct species of which the plant described as *J. cluytioides* would appear to be the more usual form. The two forms appear, however, to pass into each other and to be conditions of one species rather than distinct varieties.

7. J. latifolia (Pax in Engl. Jahrb. xxiii. 531); herb up to 1 ft. high ; rootstock stout, perennial ; everywhere glabrous ; stem leafy, simple or sparingly branched ; leaves sessile, subcoriaceous, oblong or ovate, acute, base rounded, margin closely and minutely glandular-denticulate, 2$\frac{1}{2}$–4$\frac{1}{2}$ in. long, 1–1$\frac{1}{2}$ in. wide ; stipules setaceous, entire or 2-fid, caducous ; cymes rather dense, many-flowered, hardly over-topping the leaves; peduncles 1 in. long ; primary branches $\frac{1}{4}$ in. long ; bracts lanceolate with margin closely and minutely glandular-denticulate, up to $\frac{1}{4}$ in. long ; male sepals lanceolate, closely and minutely glandular-denticulate, $\frac{1}{6}$ in. long, $\frac{1}{6}$ in. wide; petals spathulate-oblong, obtuse, $\frac{1}{3}$ in. long, free ; disc-glands free ; stamens 8, filaments of both series united below in a column, free above; female sepals lanceolate, similar to the male sepals ; petals as in the male but larger ; ovary glabrous ; styles 3, 2-fid above; capsule 3-dymous, $\frac{1}{2}$ in. across ; cocci oblong ; seeds narrow oblong, $\frac{1}{4}$ in. long, $\frac{1}{6}$ in. wide. *Pax in Engl. Pflanzenr. Euphorb. Jatroph.* 61.

VAR. β, **angustata** (Prain); leaves narrow lanceolate, acute, base rounded or truncate, margin closely and minutely glandular-denticulate, $2\frac{1}{2}$–$4\frac{1}{2}$ in. long, $\frac{1}{3}$–$\frac{1}{2}$ in. wide.

VAR. γ, **swazica** (Prain); leaves obovate, obtuse or subacute, base somewhat cuneate, margin entire or sparingly and minutely glandular-denticulate, 3–5 in. long, $1\frac{1}{2}$–2 in. wide.

KALAHARI REGION : Transvaal ; Lydenburg, *Wilms*, 1311 ! Waterval Onder, *Middelberg* ! Heidelberg, *Schlechter* ! near Barberton, 2000–3000 ft., *Bolus*, 9775 partly ! Var. β : near Barberton, 2000–3000 ft., growing with the type, *Bolus*, 9775 partly ! *Galpin*, 526 ! Elands Hoek, *Rogers*, 386 ! Rivulets, *Rogers*, 431 ! Var. γ : Swaziland ; near Bremersdorp, 2600 ft., *Bolus*, 12296 ! ridge between Bremersdorp and Macnabs, *Burtt-Davy*, 2942 !
EASTERN REGION : Natal ; near Durban, *Sutherland* ! *Wilms*, 2274 !

Very nearly allied to *J. hirsuta* but readily distinguished by the glabrous leaves and the eglandular stipules.

The plant named by Dr. Pax *J. latifolia*, var. *stenophylla* (Engl. Pflanzenr. Euphorb. Jatroph. 133), is *J. lagarinthoides*, var. *cluytioides*.

8. **J. hirsuta** (Hochst. apud Krauss in Flora, 1845, 82) ; herb up to 10 in. high ; rootstock stout, perennial ; stems leafy, simple or sparingly branched, glabrous or nearly so below, sparingly pilose above ; leaves sessile, subcoriaceous, subobovate or elliptic, acute or subobtuse, base cuneate, very rarely rounded, margin closely and minutely glandular-denticulate, 2–$2\frac{1}{3}$ in. long, $1\frac{1}{4}$–$1\frac{1}{2}$ in. wide, villous on both surfaces, but especially on the nerves beneath, with soft spreading white hairs ; stipules divided into 4–6 filiform gland-tipped segments ; cymes rather dense, many-flowered, hardly over-topping the leaves ; peduncles 1 in. long ; primary branches $\frac{1}{4}$ in. long ; bracts lanceolate, margins closely and minutely glandular-denticulate, up to $\frac{1}{4}$ in. long ; male sepals lanceolate, closely and minutely glandular-denticulate, villous externally ; petals spathu-late-oblong, obtuse, $\frac{1}{3}$ in. long, free ; disc-glands free ; stamens 8, filaments of both series united below in a long column, free above ; female sepals lanceolate, similar to male sepals ; petals as in male but larger ; ovary hirsute ; styles 3, 2-lobed above ; capsule 3-dymous, $\frac{1}{2}$ in. across ; cocci wide-oblong, hirsute ; seeds oblong, $\frac{1}{4}$ in. long, $\frac{1}{5}$ in. wide. *Sond. in Linnæa*, xxiii. 118 ; *Baill. Adansonia*, iii. 149 ; *Müll. Arg. in DC. Prodr.* xv. ii. 1088 ; *Wood, Natal Pl. t.* 71 ; *Pax in Engl. Pflanzenr. Euphorb. Jatroph.* 62, *fig.* 24.

VAR. β, **glabrescens** (Prain); stems as in the type; leaves as in the type but less pubescent when young, nearly glabrous when fully developed; ovary sparingly pubescent; capsule quite glabrous. *J. glabrescens, Pax & K. Hoffm. l.c.*
VAR. γ, **oblongifolia** (Prain); stems up to 1 ft. high, densely hirsute throughout; leaves oblong, 3–5 in. long, 1½–2 in. wide, with pubescence as in the type; ovary hirsute; cocci hirsute.

KALAHARI REGION : var. γ: Transvaal ; Lydenburg Distr., Doorn Hoek, 4300 ft., *Burtt-Davy,* 7275 ! Witklip, 5800 ft., *Burtt-Davy,* 7257 ! Swaziland ; *Miss Stewart,* ·8972 !

EASTERN REGION: Natal ; near Umlaas River, *Krauss,* 364 ! near Durban, *Gueinzius,* 12 ! *Sanderson,* 13 ! *Schlechter,* 3023 partly ! near Maritzburg, *Wilms,* 2164 ! Camperdown, *Rehmann,* 7813 ! Inanda, *Rehmann,* 8346 partly ! Var. β : Natal ; Clairmont, *Wood,* 4656 ! 5795 ! 6518 ! Inanda, *Wood,* 211 ! *Rehmann,* 8346 mainly ! near Durban, *Schlechter,* 3023 partly ! *Gerrard,* 381 !

The difference between *J. hirsuta,* Hochst., and *J. glabrescens,* Pax & K. Hoffm., is confined to the degree of pubescence. The variety here distinguished as var. *oblongifolia* is as regards shape of leaves indistinguishable from *J. latifolia,* Pax, but the leaves and stems are even more hirsute than in *J. hirsuta,* Hochst. The original gathering of *J. hirsuta* (Krauss, 364) includes both the type and var. *glabrescens.*

·9. **J. Woodii** (O. Kuntze, Rev. Gen. Pl. iii. ii. 287) ; herb up to 1½ ft. high ; rootstock stout, perennial ; stems leafy, sparingly branched, rather densely hirsute with spreading hairs ; leaves sessile, rather soft, usually pinnately 5–7-lobed, rarely almost entire, base rather narrow-cuneate, lobes entire, acute, margin glandular-setulose, setæ rather long, 2½–4 in. long, 3–5 in. across, lobes ¾–2 in. long, ⅓–½ in. wide, densely hirsute, especially on the nerves on both sur faces, nerves underneath prominent ; stipules dissected, lobules glanduliform, ₁₀ in. long ; cymes rather lax, usually many-flowered, hardly overtopping the leaves ; peduncles 1–1½ in. long, very densely hirsute with white soft spreading hairs ; primary branches ½–¾ in. long, densely and softly hirsute ; bracts linear-lanceolate, ¼ in. long or less, margins glandular-ciliate ; male sepals narrow-lanceolate, softly hirsute with white hairs externally, glabrous within, margins glandular-setulose, ⅓ in. long ; petals oblong, obtuse, silky outside, ⅔ in. long, slightly united at the base ; disc-glands free ; stamens 8, 2 seriate, united in a column for over three-fourths of their length ; female sepals like the male ; petals as in the male but free at the base, nearly ½ in. long ; hypogynous glands free ; ovary densely hirsute ; styles 3, 2-lobed at the apex ; capsule oblong, slightly 3-dymous, ⅔ in. long, ½ in. wide, pubescent ; seeds narrow oblong, ⅔ in. long, ⅕ in. wide. *J. Woodii, var. Kuntzei, Pax in Engl. Pflanzenr. Euphorb. Jatroph.* 66, *excl. Rehmann,* 5333.

VAR. β, **vestita** (Pax in Engl. Jahrb. xliii. 84) ; herb up to 6 in. high ; stems and leaves more densely hirsute with soft white hairs ; leaves smaller, under 2 in. long. *Pax in Engl. Pflanzenr. Euphorb. Jatroph.* 66, *fig.* 26. *J. lanata, Harr. Mss. in herb. T.C.D.*

EASTERN REGION: Natal ; Ladysmith, 3000–4000 ft., *Kuntze* ! *Wood,* 4242 ! Biggarsberg, *Rehmann,* 7104 ! Var. β : Natal : Ladysmith, *Engler,* 2725 ! and without precise locality, *Gerrard,* 14 ! 656 !

The variety here recognised is probably not a valid one ; it may be only a dwarf condition or a young state, or both, of the typical plant. Pax states that the leaves of his variety, besides being smaller than in the type, are sometimes entire or subentire or little lobed. This is true, but the circumstance does not afford a distinctive character, for the same thing is true of the leaves in some of the Biggarsberg specimens of the type collected by Rehmann (7004 only, not 5333, which, as Pax elsewhere correctly states, is *J. Zeyheri*). The species is very closely allied to *J. Zeyheri* and appears to be the representative of *J. Zeyheri* to the south of the Drakensberg ; it is, however, readily distinguished by the more completely monadelphous stamens and by the distinctly adherent male petals. The character based on the venation of the leaves, pinnate in *J. Woodii*, palmate in *J. Zeyheri*, relied on by Kuntze and accepted by Pax, although roughly useful, does not always hold good in distinguishing the two species.

10. **J. Zeyheri** (Sond. in Linnæa, xxiii. 117) ; herb up to 1½ ft. high ; rootstock stout, perennial ; stems leafy, sparingly branched, sparingly hirsute with spreading hairs ; leaves sessile or very shortly petioled, firm, rather variable, spuriously palmately 3–5-lobed, base wide-cuneate, lobes more or less runcinately lobulate, acute, margin with short glandular ciliæ, 2½–4 in. long, 3–5 in. across, lobes 1½–3 in. long, ¼–¾ in. wide, more or less pilose on the nerves, especially beneath, nerves underneath prominent ; petiole not exceeding ⅛ in. in length ; stipules dissected, lobules glanduliferous, $\frac{1}{10}$ in. long ; cymes rather lax, usually many-flowered, hardly over-topping the leaves ; peduncles 1–1½ in. long, softly pubescent ; primary branches ½–¾ in. long, softly pubescent ; bracts linear-lanceolate, ¼ in. long or less, margin glandular-ciliate ; male sepals narrow lanceolate, margin glandular-ciliate, ⅕ in. long ; petals yellow, silky outside, spathulate-lanceolate, ¼ in. long ; disc-glands free ; stamens 8, 2-seriate, shortly monadelphous ; female sepals like the male but ¼ in. long ; petals like the male but ⅓ in. long ; hypogynous glands free ; ovary densely hirsute ; styles 3, 2-lobed at the apex ; capsule oblong, slightly 3-dymous, ½ in. long, ⅓ in. wide, pubescent ; seeds narrow-oblong, ⅓ in. long, ⅙ in. wide. *Baill. Adansonia,* iii. 149 ; *Müll. Arg. in DC. Prodr.* xv. ii. 1088 ; *Pax in Engl. Pflanzenr. Euphorb. Jatroph.* 68.

Var. β, **platyphylla** (Pax l.c. 68) ; lobes of leaves rather shorter and broader, little if at all lacinulate, otherwise as in the type. *J. Woodii, Pax l.c.* 66, *as to Rehmann,* 5333 *only ; not of O. Kuntze. J. brachyadenia, Pax & K. Hoffm. l.c.* 66, *partly.*

Var. γ, **subsimplex** (Prain) ; leaves at least some, occasionally almost all entire, ovate or ovate-lanceolate or lanceolate, acute, the others more or less 3-lobed, otherwise as in the type. *J. brachyadenia, Pax & K. Hoffm. l.c.* 66, *partly.*

KALAHARI REGION : Orange River Colony ; Witteberg Range, *Mrs. Bowker,* 657 ! Bechuanaland ; between Mafeking and Ramoutsa, *Lugard* ! Doornbult, *Burtt-Davy,* 10975 ! Transvaal ; Mooi River, *Burke,* 183 ! *Zeyher,* 1515 ! Magaliesberg Range, *Burke* ! *Zeyher* ! Moord Drift, *Miss Leendertz,* 2208 ! Potgieters Rust, *Miss Leendertz,* 1217 ! *Rogers,* 2383 ! 2503 ! Waterval, *Miss Leendertz,* 822 ! Groot Letaba, *Swierstra,* 2184 ! Var. β : Transvaal ; Klippan, *Rehmann,* 5333 ! near Nylstroom, *Burtt-Davy,* 2117 ! near Eerstelling Goldfields, *Λelson,* 372 ! Harte-beeste Fontein, *Nelson,* 119 ! Shilouvane, *Junod,* 747 partly ! Var. γ : Trans-vaal ; Shilouvane, *Junod,* 747 partly ! Potgieters Rust, *Rogers,* 2501 ! Pangkop Siding, *Burtt-Davy,* 7046 ! Warmbath, *Bolus,* 12297 ! *Burtt-Davy,* 2635 ! 7062 ! *Rogers,* 1529 ! near Nylstroom, *Burtt-Davy,* 2116 !

It is probable that further field study may show that neither of the proposed varieties is valid. The original specimens of *Junod*, 747, on which *J. brachyadenia*, Pax & K. Hoffm., is based, belong in some cases to var. *platyphylla*, Pax, and *Nelson*, 119, shows the same transition. At the same time, the distinction between var. *platyphylla*, the type of which Pax has referred to *J. Woodii* as well as to *J. Zeyheri*, is quite arbitrary, and the truth appears to be that, exactly as in *J. Woodii* and in *J. natalensis*, there is in *J. Zeyheri* every transition from a lanceolate or ovate-lanceolate or ovate entire leaf to a lyrate- or runcinate-pinnatifid one.

11. **J. Schlechteri** (Pax in Engl. Jahrb. xxviii. 24); herb up to 1½ ft. high; rootstock stout, perennial; stems leafy, sparingly branched, sparingly hirsute with spreading hairs; leaves distinctly petioled, firm, spuriously palmately to distinctly pinnately 3–7-lobed, base wide-cuneate, lobes more or less runcinately lobulate, obtuse, margin denticulate, the teeth triangular and each with a short glandular apical seta, 2½–4 in. long, 2–3 in. across, lobes ¾–1¼ in. long, ⅓–¾ in. wide, more or less pilose on the nerves especially beneath, nerves underneath prominent; petiole ¾–1 in. long; stipules dissected, lobules glanduliferous, ½ in. long; cymes rather lax, usually many-flowered, considerably overtopping the leaves; peduncles 1–1½ in. long, softly pubescent; primary branches ½–¾ in. long, softly pubescent; bracts linear-lanceolate, ¼ in. long or less, margin glandular-ciliate; male sepals narrow lanceolate, margins glandular-ciliate, ⅕ in. long; petals spathulate-lanceolate, glabrous, ⅕ in. long; disc-glands free; stamens 8, 2-seriate, shortly monadelphous; female sepals like the male but ⅙ in. long; petals like the male but ¼ in. long; hypogynous glands free; ovary densely hirsute; styles 3, 2-lobed at the apex; capsule oblong, slightly 3-dymous, ½ in. long, ⅓ in. wide, pubescent; seeds narrow-oblong, ⅓ in. long, ⅙ in. wide. *Pax in Engl. Pflanzenr. Euphorb. Jatroph.* 67.

KALAHARI REGION: Transvaal; Komati Poort, 1000 ft., *Schlechter*, 11799!

Closely resembling *J. Zeyheri*, Sond., *J. Schlechteri* is easily distinguished by its uniformly petioled leaves with obtuse denticulate lobes and lobules, by its laxer cymes with smaller flowers and by its glabrous petals.

XXI. CLUYTIA, Linn.

Flowers diœcious, very rarely monœcious; petals present in both sexes; disc of free lobulate glands at base of the sepals in both sexes or very rarely in male flowers only. *Male: Sepals* 5, imbricate. *Petals* usually almost as long as sepals, clawed, with one or more glands at the base within. *Stamens* 5; filaments connate around and supporting a dilated rudimentary ovary; anthers short, dorsifixed; dehiscence longitudinal. *Female: Sepals* 5, imbricate. *Petals* as in male. *Ovary* 3-celled; ovules solitary in each cell;

styles free or shortly connate at the base, 2-fid or 2-lobed. *Capsule* small, subglobose, breaking up septicidally into 3 entire or 2-fid valves ; septa thin, free or adnate to the columella. *Seeds* ovoid, carunculate ; testa crustaceous, black and shining ; albumen fleshy ; cotyledons broad, flat.

Shrubs or undershrubs. Leaves alternate, entire, usually small. Flowers small ; male pedicelled in axillary fascicles; female with longer pedicels and often solitary.

DISTRIB. Species about 60, 20 of them in Tropical Africa.

*Petals of male flowers each 1-2-glandular ; glands usually
    adnate to claw, occasionally free from the petal and
    arising from the fundus of the calyx :
  †Leaves sessile or subsessile, rarely (*pterogona, impedita*)
    pellucid-punctate :
    Leaves with margin involute, glabrous ; ovary
      glabrous :
      Twigs puberulous ; leaves 3-4 times as long as
        broad ...   . (1) **ericoides.**
      Twigs glabrous ; leaves less than twice as long as
        broad ...   . (2) **nana.**
    Leaves with margin either flat or revolute :
      Leaves pubescent ; ovary tomentose :
      Petals of female flower glandular at base, like
        the male :
      Leaves small, hardly longer than broad   ... (3) **tomentosa.**
      Leaves medium-sized, twice as long as broad (4) **marginata.**
      Petals of female flower eglandular :
        Leaf-margin quite flat ...   . (5) **sericea.**
        Leaf-margin more or less revolute :
        Leaves 1-3 in. long, slightly revolute   ... (6) **Katharinæ.**
        Leaves ½-1 in. long, usually much revolute (7) **pubescens.**
  Leaves glabrous ; ovary glabrous :
    Stems and twigs winged ; wings membranous,
      erose-denticulate ; leaves pellucid-punctate (8) **pterogona.**
    Stems and twigs cylindric, or if fluted the
      ridges or wings coriaceous and quite entire :
    ‡Leaves opaque :
      Leaf-margin quite smooth or very faintly
        scabrous :
        Leaf-margin usually very markedly revo-
          lute :
          Leaves ⅓-⅔ in. long   ...   . (9) **polifolia.**
          Leaves ⅙-¼ in. long   ...   . (10) **brevifolia.**
        Leaf-margin usually quite or nearly flat :
        Stems simple or subsimple, several from
          a woody base...   ...   . (11) **virgata.**
        Stems copiously branching   ...   ... (12) **laxa.**
      Leaf-margin distinctly finely denticulate
        scabrous :
        Leaf-margin more or less revolute :
        Leaves rather large, ⅔-1½ in. long :
          Leaves obovate-oblong, ½-⅔ in. wide... (13) **africana.**

Leaves lanceolate to obovate-lanceolate,
$\frac{1}{3}-\frac{1}{5}$ in. wide ... ... (14) **Alaternoides.**
Leaves very small, obovate-oblong, $\frac{1}{5}$ in.
long, $\frac{1}{12}$ in. wide ... ... (15) **imbricata.**
Leaf-margin flat :
Internodes much shorter than the close-
set leaves ... ... (16) **rubricaulis.**
Internodes about as long as the leaves ... (17) **ovalis.**
‡‡Leaves pellucid-punctate, their margins quite
flat ... ... (18) **impedita.**
††Leaves distinctly petioled ; ovary glabrous :
Leaf-margin more or less revolute ; stems prostrate ;
leaves pellucid-punctate ... ... (19) **alpina.**
Leaf-margin flat ; stems erect :
Leaves pellucid-punctate :
Capsule smooth ; twigs and leaves glabrous ... (20) **glabrescens.**
Capsule warted-punctate :
Twigs and leaves at first somewhat pubescent,
at length glabrous :
Petiole $\frac{1}{4}$ in. long or shorter ; twigs and
leaves not verrucose ... ... (21) **Galpini.**
Petiole $\frac{1}{2}$ in. long or longer ; twigs and leaves
verrucose ... ... (22) **pulchella.**
Twigs and leaves beneath persistently velvety-
pubescent ... ... (23) **mollis.**
Leaves not pellucid-punctate, more or less pube-
scent ... ... (24) **affinis.**
*Petals of male flowers each 3-more-glandular ; glands
very rarely adnate to claw, usually free from the petal
and scattered over fundus of calyx :
†Leaves pellucid-punctate :
Leaf-margin flat ; stems considerably branched ;
ovary glabrous ... ... (25) **natalensis.**
Leaf-margin slightly revolute :
Stems considerably branched :
Ovary glabrous :
Leaves obtuse ... ... (26) **platyphylla.**
Leaves acute or shortly acuminate ... ... (27) **dregeana,**
Ovary pubescent ... ... (28) **hirsuta.**
Stems simple or slightly branched, several from a
woody base :
Fruiting pedicel several times longer than the
capsule ; venation of leaves not raised
beneath ; ovary usually more or less hirsute (29) **disceptata.**
Fruiting pedicel not or only slightly longer than
the capsule ; ovary always quite glabrous :
Venation of leaves not raised beneath :
Leaves cuneate or rounded at the base ... (30) **monticola.**
Leaves all subcordate at the base ... ... (31) **cordata.**
Venation of leaves distinctly raised beneath ;
upper leaves cuneate, lower subcordate at
the base ... ... (32) **heterophylla.**
††Leaves not pellucid-punctate ; ovary glabrous :
Leaves dull, their margin flat :
Leaves distinctly petioled, 4–5 times as long as
broad ; twigs rather sharply angled ... ... (33) **daphnoides.**

Leaves sessile or nearly so, 1½–2 times as long as
　broad ; twigs subcylindric :
　Stems prostrate ; twigs finely pubescent ; leaves
　　glabrous except for a few hairs along midrib
　　above ...　 ...　 ...　 ...　 ...　 ... (34) **vaccinioides**.

Stems erect :
　Twigs and leaves finely pubescent 　...　 ... (35) **Thunbergii**.

　Twigs and leaves quite glabrous 　...　 ... (36) **crassifolia**.

Leaves polished, their margin revolute ; all parts
　glabrous ...　 ...　 ...　 ...　 ...　 ... (37) **polygonoides**.

1. **C. ericoides** (Thunb. Prodr. 53 [*Clutia*]) ; an undershrub,
1–2 ft. high ; branches stoutish, again branching, twigs spreading,
puberulous ; leaves sessile, coriaceous, opaque, linear-lanceolate,
acute, base cuneate, margin involute, ⅓–½ in. long, ₁⁄₁₀ in. wide, erect,
subimbricate, shining, quite glabrous, convex below and concave
above ; flowers dioecious, male usually in pairs, female solitary from
perulate axillary swellings, scales numerous, dark-brown ; pedicels
shorter than the calyx, puberulous ; male sepals oblong or ovate,
obtuse, with a large 1–3-lobate basal gland ; petals cuneate-obovate,
clawed, with a solitary small gland ; rudimentary ovary very short,
rounded, glabrous ; female sepals narrower and larger than the
male, with usually an entire basal gland ; petals as in male, but
eglandular ; ovary glabrous ; styles free, 2-partite ; capsule ¼ in.
across, subglobose, smooth ; seed black, shining. *Willd. Sp. Pl.* iv.
880 ; *Pers. Syn.* ii. 636 ; *Poir. Encyc. Suppl.* ii. 303 ; *Thunb. Fl.
Cap. ed. Schult.* 270 ; *Spreng. Syst.* iii. 48 ; *E. Meyer in Drège, Zwei
Pfl. Documente,* 174, *partly* ; *Sond. in Linnæa,* xxiii. 121, *partly* ;
Dietr. Syn. v. 455 ; *Baill. Adansonia,* iii. 151, *partly* ; *Müll. Arg. in
DC. Prodr.* xv. ii. 1055, *partly and excl. cit. Bot. Reg.* ; *Pax in Engl.
Pflanzenr. Euphorb. Cluyt.* 81, *partly and as to fig.* 19 *H, J only, and
excl. cit. Bot. Reg.* ; *Prain in Kew Bulletin,* 1913, 384. *C. ericoides,
var. minor, Krauss in Flora,* 1845, 82.

Var. β, **pachyphylla** (Prain, l.c. 385) ; undershrub, 1–2 ft. high ; branches
stout, simple or sparingly branched ; when branched, branchlets ascending, twigs
puberulous ; leaves narrowly ovate-lanceolate, acute or acuminate, base wide
cuneate, ½–⅔ in. long, ⅙–¼ in. wide, more or less convex below, usually concave,
but at times flat, except for the involute margin above. *C. ericoides, E. Meyer,
l.c. partly* ; *Sond. l.c. partly* ; *Baill. l.c. partly* ; *Müll. Arg. l.c. partly* ; *Pax l.c.
as to fig.* 26 *A–E only* ; *hardly of Thunb. C. ambigua, Pax & K. Hoffm. l.c.* 82—
the state with leaves flat above. C. pachyphylla, Spreng. Mss. in sched. Zeyh.

Var. γ, **tenuis** (Sond. l.c. 122) ; an undershrub, 1–2 ft. high ; stems slender ;
branches rather numerous, very slender, fastigiate, twigs puberulous ; leaves
linear, ⅓–⅔ in. long, ₁⁄₁₆ in. wide, convex below, usually concave but occasionally
flat except for the involute margin above. *Baill. l.c.* 151 ; *Prain, l.c.* 385. *C.
tenuifolia, Willd. l.c.* ; *Pers. l.c.* ; *Poir. l.c.* 302 ; *Spreng. Syst.* iii. 49 ; *Dietr.
Syn.* v. 455 ; *Baill. l.c.* 152 ; *Müll. Arg. l.c.* ; *Pax l.c. C. ericoides, Ait. Hort.
Kew. ed.* 2, v. 423 ; *Edw. Bot. Reg. t.* 779, *excl. syn. Thunb. and Willd.* ; *Sond.
l.c. partly, as to ‘ Knoblauch’ loc. only* ; *Müll. Arg. l.c. as to Bot. Reg. cit.* ; *Pax
l.c. as to Bot. Reg. cit.* ; *not of Thunb. C. gracilis, Baill. l.c.* 151.

SOUTH AFRICA: Var. γ, without locality, *Miss Cole*! and cultivated specimens!
COAST REGION: Piquetberg Div.; Mount Congo, *Mund & Maire*! Malmesbury
Div.; near Hopefield, *Bachmann*, 1933! 1934! 1935! 1936! Worcester Div.;
Hex River Mountains, 2000–3000 ft., *Drège*, 262! Cape Div.; various locali-
ties, *Thunberg*! *Hesse*! *Oldenburg*! *Lichtenstein*! *Drège*, 8232 a partly! and d!
Krebs! *Dümmer*, 71 a partly! *Ecklon*! *Wolley-Dod*, 2661! *Burchell*, 965! *Ecklon*!
Ecklon & Zeyher! Stellenbosch Div.; Lowrys Pass, 400 ft., *Schlechter*, 7814
partly! Caledon Div.; near Genadendal, *Burchell*, 7677 mainly! *Bolus*, 7425!
Klein River Mountains, 1000–3000 ft., *Ecklon & Zeyher*, 58 partly! Grabouw,
Palmiet River, *Guthrie*, 3819! Swellendam Div.; near Swellendam, *Bowie*!
Riversdale Div.; Zandhoogde, *Muir*, 299 (*Galpin*, 5334)! George Div.; Cra-
docks Pass, *Prior*! Outeniqua Mountains, *Krauss*, 1716! Knysna Div.; between
Groene Valley and Zwart Valley, *Burchell*, 5675! Uitenhage Div.; Van Stadens-
berg, *Ecklon & Zeyher*, 57! *Scott Elliot*, 309 partly! Winterhoek, *Krauss*, 1717!
Port Elizabeth Div.; Cockscomb, *Mrs. Paterson*, 2047! Albany Div.; near
Grahamstown, *Hutton*! Coldspring, *Glass*, 365! Var. β: Cape Div.; various
localities, *Bergius*! *Miss Cole*! *Roxburgh*! *Burchell*, 966! 8426! 8482! *Harvey*,
5840! *Rehmann*, 975! 2198! *Drège*! *Ecklon*, 115! 200! *Prior*! *Andersson*!
Dümmer, 71 a partly! 1232! *Zeyher*, 3827 partly! *Schlechter*, 770! *Wolley-Dod*,
1109! 2480! 2518! *Robertson*! Stellenbosch Div.; Lowrys Pass, 400 ft.,
Burchell, 8279! *Schlechter*, 7814 partly! 7815! Hottentots Holland, *Ecklon*,
944! *Ecklon & Zeyher*, 50! *Zeyher*, 3827 partly! Uitenhage Div.; between
Maitland and Van Stadens River, *Burchell*, 4628! Van Stadensberg, *Drège*, 8232 e!
Ecklon & Zeyher, 57 partly! *Scott Elliot*, 309 partly! Var. γ: Cape Div.; near
Cape Town, *Roxburgh*! *Masson*! Caledon Div.; Baviaans Kloof near Gena-
dendal, *Burchell*, 7677 partly! Knoflooks Kraal, *Ecklon & Zeyher*, 56! Klein
River Mountains, 1000–3000 ft., *Ecklon & Zeyher*, 58 mainly! Swellendam Div.:
Sparrbosch, *Drège*, 8231! Puspas Valley and Voormansbosch, *Ecklon & Zeyher*,
65! by the River Zonder Einde, *Zeyher*, 3826! near Swellendam, *Bowie*! *Bolus*,
Herb. Norm. Austr.-Afr. 1365! Riversdale Div.; Garcias Pass, 1000–1400 ft.,
Bolus! *Schlechter*, 2198! *Galpin*, 4570! Kleinberg, *Muir*, 501!
CENTRAL REGION: Prince Albert Div.; Zwartberg, near Vrolyk, 3000–4000 ft.,
Drège, 8232 f (fide Sonder).

2. **C. nana** (Prain in Kew Bulletin, 1913, 386); a dwarf undershrub, 3–4 in. high; branches stout, again intricately branching,
twigs quite glabrous; leaves sessile, coriaceous, ovate, acute, base
rounded, margin involute, very closely imbricate, ⅛ in. long, ₁₂ in.
wide, convex subcarinate below, deep concave above, quite glabrous;
flowers diœcious, male solitary; female not seen; pedicels very
short, glabrous; male sepals ovate, cucullate, obtuse, with a large
4-lobate basal gland; petals ovate, obtuse, clawed, with a solitary
small gland at the junction of the lamina and claw; rudimentary
ovary very short, rounded, glabrous.

KALAHARI REGION: Orange River Colony; Mont aux Sources, 10,000 ft.,
Mann in Herb. Marloth, 2870!

3. **C. tomentosa** (Linn. Mant. Alt. 299 [*Clutia*]); a shrub, 2–4 ft.
high; twigs fastigiate, cylindric, densely grey-tomentose; leaves
coriaceous, sessile or nearly so, elliptic or oblong or orbicular, usually
obtuse, base rounded, margin flat, ascending and often imbricate,
⅙–⅓ in. long, ⅛–¼ in. wide, densely grey-pubescent on both surfaces;
flowers diœcious, white, male usually, female almost always solitary
in the leaf-axils; pedicels short, pubescent; male sepals densely

pubescent, narrow-obovate, $\frac{1}{6}$ in. long, with a 3–5-lobate basal
scale ; petals shortly clawed, rounded ovate, as long as the calyx,
hairy externally, with an undivided basal gland ; rudimentary
ovary dilated at the tip, glabrous ; female sepals ovate-oblong, with
an undivided basal scale ; petals rather longer than the calyx,
linear-oblong, tomentose externally, with a 2-lobed basal scale ;
ovary tomentose ; styles short, 2-fid ; capsule $\frac{1}{3}$ in. long, nearly as
broad, densely woolly pubescent ; seeds black, shining. *Lam.
Encycl.* ii. 54 ; *Willd. Sp. Pl.* iv. 881, *excl. syn. Thunb.* ; *Pers. Syn.* ii.
636 ; *Spreng. Syst.* iii. 49 ; *Sond. in Linnæa,* xxiii. 131 ; *Dietr.
Syn.* v. 455 ; *Baill. Étud. Gén. Euphorb.* 331, *t.* 16, *figs.* 20, 21, *and
in Adansonia,* iii. 152 ; *Müll. Arg. in DC. Prodr.* xv. ii. 1053, *incl.
var. elliptica ; Pax in Engl. Pflanzenr. Euphorb. Cluyt.* 76, *incl. var.
elliptica wholly and var. marginata partly ; Prain in Kew Bulletin,*
1913, 386. *C. marginata, E. Meyer in Drège, Zwei Pfl. Documente,*
174, *b only ; Sond. in Linnæa,* xxiii. 130, *partly and as to Drège's
Swellendam plant only. Penæa tomentosa, Thunb. Prodr.* 30, *and
Fl. Cap. ed. Schult.* 150. *Geissoloma tomentosum, Juss. in Ann. Sc.
Nat.* 3me sér. vi. 27.

4. **C. marginata** (E. Meyer in Drège, Zwei Pfl. Documente, 174 ;
a only : name) ; a shrub 3–4 ft. high ; twigs spreading, angular,
densely grey-tomentose ; leaves firmly papery, shortly petioled,
obovate, acute, base gradually cuneate, margin flat, spreading,
not imbricate, $\frac{1}{2}$–1 in. long, $\frac{1}{4}$–$\frac{1}{3}$ in. wide, densely grey-pubescent
on both surfaces ; petiole $\frac{1}{12}$–$\frac{1}{10}$ in. long ; flowers dioecious, white,
male usually 1–3, female solitary in the leaf-axils ; pedicels short,
hoary pubescent ; male sepals densely pubescent, ovate-oblong,
$\frac{1}{5}$ in. long, with a 7–9-lobate basal scale ; petals distinctly
clawed, rhomboid, sparingly hairy externally, with a minute
gland ; rudimentary ovary cylindric, sparingly hairy throughout ;
female sepals ovate-oblong, with a simple basal scale ; petals
as long as calyx, oblong, pubescent externally, with a 2-lobed
basal scale ; ovary tomentose ; styles short, 2-fid ; capsule $\frac{1}{3}$ in.
long, nearly as broad, rather sparingly shortly tomentose ; seeds
black, shining. *Sond. in Linnæa,* xxiii. 130, *mainly, but excl. Drège's
Swellendam plant ; Baill. Adansonia,* iii. 152 ; *Prain in Kew
Bulletin,* 1913, 388. *C. tomentosa, var. marginata, Müll. Arg. in DC.
Prodr.* xv. ii. 1053, *wholly ; Pax in Engl. Pflanzenr. Euphorb. Cluyt.,*
76, *partly.   C. incanescens, Hort. ex Prain in Kew Bulletin,* 1913, 388.

COAST REGION: Ladismith Div. ; Kannaland, between Cogmans Kloof and the Gouritz River, *Ecklon & Zeyher*, 67 ! George Div. ; Montagu, *Marloth*, 2831 ! Albany Div. ; without precise locality, *Bowie*! CENTRAL REGION : Graaf Reinet Div. ; near Graaf Reinet, *Bowie* ! Beaufort West Div. ; Nieuwevelds Range near Beaufort West, 3000–5000 ft., *Drège*, a !

The specimens collected by Marloth in 1903 agree well with those of Ecklon and Zeyher from Kannaland and those of Drège from Beaufort West. Specimens cultivated at Berlin from seed of Marloth 2831 agree exactly with *C. incanescens*, Hort. Kew.

5. **C. sericea** (Müll. Arg. in DC. Prodr. xv. ii. 1053) ; a shrublet, $\frac{1}{2}$–1 ft. high, much branched ; twigs fastigiate, silvery-silky ; leaves sessile, firmly membranous, lanceolate, acuminate, narrowed to the base, margin flat, close set and subimbricately spreading, $\frac{1}{2}$–$\frac{2}{3}$ in. long, $\frac{1}{8}$–$\frac{1}{5}$ in. wide, densely shining silvery silky on both faces ; flowers diœcious, whitish ; male as well as female usually solitary in the leaf-axils ; pedicels short, silky ; male sepals densely silky, narrow-ovate, $\frac{1}{6}$ in. long, with a 3-lobed basal scale ; petals clawed, rounded-obovate, shorter than the calyx, glabrous, with a 2-lobed basal scale ; rudimentary ovary silky at the base ; female sepals oblong-ovate, acute, $\frac{1}{6}$ in. long, with a simple basal scale ; petals without a basal scale, silky externally ; ovary densely tomentose ; styles short, 2-fid ; seeds nearly black, shining. *Pax in Engl. Pflanzenr. Euphorb. Cluyt.* 75, *fig.* 24 *A* ; *Prain in Kew Bulletin*, 1913, 389.

COAST REGION : Malmesbury Div. ; Groene Berg, *Mund & Maire*, 28 ! Malmesbury, *Kässner* !

6. **C. Katharinæ** (Pax in Engl. Pflanzenr. Euphorb. Cluyt. 58) ; a shrub, 3–5 ft. high ; twigs softly silky with spreading hairs ; leaves sessile or nearly so, membranous, lanceolate, somewhat blunt, gradually narrowed to the base, margin slightly revolute, 1–3 in. long, $\frac{1}{4}$–$\frac{1}{3}$ in. wide, when young closely white silky with longish adpressed hairs, at length somewhat glabrescent ; flowers diœcious, white, male usually in pairs, female usually solitary ; pedicels in both sexes very short ; male sepals oblong, silky, with a 3–4-partite basal scale ; petals rhomboid, acute, glabrous, long-clawed, about as long as the sepals, with a minute simple basal scale ; rudimentary ovary narrow, glabrous ; female sepals wide-ovate, acute, silky, with a 2–5-lobed basal scale ; petals oblong, acute, shorter than the sepals, sparingly hairy or glabrous, without a basal scale ; ovary densely silky ; styles free, glabrous, shortly 2-fid ; capsule $\frac{1}{5}$ in. across, subglobose, densely silky-velvety ; seeds brownish-black, shining. *Prain in Kew Bulletin*, 1913, 389. *C. sericea, Harv. Mss. in T.C.D.* ; *not of Müll. Arg.*

COAST REGION : Queenstown Div. ; Elands Berg, *Cooper*, 264 ! 265 ! EASTERN REGION: Pondoland ; Fakus Territory, *Sutherland* ! Griqualand East ; Mount Currie, 5200 ft., *Tyson*, 1321 ! Natal ; Drakensberg Range, 4000–6000 ft., *Sutherland* ! Polela and summit of Mahwaga, 6000–7000 ft., *Evans*, 522 !

7. **C. pubescens** (Thunb. Prodr. 53 [*Clutia*]) ; a shrublet, usually

$\frac{1}{2}$–1 ft. high ; twigs pubescent ; leaves sessile or nearly so, firmly membranous, lanceolate or linear or sometimes low down on the stem ovate or ovate-lanceolate, somewhat blunt, narrowed to the base, margin usually distinctly revolute, $\frac{1}{2}$–$\frac{3}{4}$ (rarely 1–1$\frac{1}{4}$) in. long, $\frac{1}{12}$–$\frac{1}{8}$ (rarely $\frac{1}{6}$–$\frac{1}{4}$) in. wide, pubescent on both surfaces especially when young, occasionally at length glabrescent; flowers diœcious, white or reddish, male from 1–4, female solitary; pedicels in male equalling, in female rather shorter than the calyx ; male sepals densely pubescent, narrow-obovate, with a 3-partite basal scale ; petals narrow-obovate, glabrous, about as long as the sepals, 1-glandular at the base ; rudimentary ovary very short, glabrous ; female sepals lanceolate, obtuse, with a usually 3-partite but often 4-partite, occasionally 5-partite basal scale ; petals narrow-obovate, almost as long as the sepals, slightly hairy or glabrous, without a basal scale ; ovary densely pubescent ; styles free, glabrous, 2-fid ; capsule $\frac{1}{6}$ in. across, subglobose, densely pilose : seeds brownish-black, shining. *Thunb. Fl. Cap. ed. Schult.* 270 ; *Krauss in Flora*, 1845, 82 ; *Sond. in Linnæa*, xxiii. 124, *incl. var.* β *glabrata ; Baill. in Adansonia*, iii. 152, *incl. var.* β *glabrata* ; *Müll. Arg. DC. Prodr.* xv. ii. 1053 ; *Pax in Engl. Pflanzenr. Euphorb. Cluyt.* 80 ; *Prain in Kew Bulletin*, 1913, 389. *C. acuminata, E. Meyer in Drège Zwei Pfl. Documente* 174, *partly and as to letter* b *only* ; *not of Linn. f. nor of Thunb. C. humilis, Bernh. ex Krauss l.c.* 81. *C. eckloniana, Müll. Arg. l.c.* 1054. *C. Rustii, Knauf, Geogr. Verbr. Cluytia*, 49, 54. *C. glabrata, Pax l.c. C. intertexta, Pax l.c. C. fallacina, Pax l.c.*

Coast Region : Van Rhynsdorp Div. : Gift Berg, *Phillips*. 7393 ! Piquetberg Div. ; Nieuweland, *Zeyher*, 3323 partly ! Piqueniers Kloof, 2000 ft., *Schlechter*, 7958 ! 7959 ! Tulbagh Div. ; near Tulbagh Waterfall, 1000–2000 ft., *Ecklon & Zeyher*, 56 ! Paarl Div. ; Great Britain Rock, *Wilms*, 3618 ! Paarl, *Krauss*; *Wilms*, 3018 ! Cape Div. ; mountains near Cape Town. *Thunberg* ! *Oldenburg* ! *Roxburgh* ! *Mund & Maire* ! *Bergius* ! *Drège*, 209 ! *Elliott* ! *Marloth*, 23 ! *Dümmer*, 97 c ! *Ecklon & Zeyher*, 60 ! *Harvey* ! *Prior* ! *Pappe* ! *Krauss*, 1711 ! *Frömbling*, 1358 ! *Wolley-Dod*, 2514 ! *Frau Polemann* ! Stellenbosch Div. : between Stellenbosch and Cape Flats, *Burchell*, 8363 ! Caledon Div. ; between Bot River and the Zwart Berg, *Ecklon & Zeyher*, 68 ! Swellendam Div. : near Swellendam ! *Bowie* ! on the Tradouw Berg, *Bowie* ! Riversdale Div. ; Riversdale, *Rust*, 170,! George Div. ; near George, *Bowie* ! Fort Beaufort Div. : Adelaide, *Marloth*, 4928 b !

Central Region : Prince Albert Div. ; Great Zwart Berg Range, *Drège*, 8230 partly ! Cradock Div. ; near Mortimer, 2600 ft.. *Miss Kensit* !

8. C. pterogona (Müll. Arg. in DC. Prodr. xv. ii. 1048, incl. both vars.) ;

a shrub, 1–3 ft. high, much branched ; twigs spreading, finely warted, glabrous ; stems, branches and twigs membranously winged ; wings narrow, finely denticulate and lacerate ; leaves subsessile, firmly membranous, warted, punctate, linear-lanceolate or elliptic-linear, apex subobtuse or shortly triangular, mucronulate and with the margin denticulate, sometimes flat, at others revolute, narrowed in the lower fourth to the base, margin except at the apex from slightly to markedly revolute, hyaline but not denticulate, $\frac{1}{4}$–1$\frac{1}{4}$ in. long, $\frac{1}{16}$–$\frac{1}{8}$ in. wide, glabrous on both surfaces ; flowers diœcious, white, male clustered, female paired or solitary ; pedicels in both sexes shorter than the calyx, glabrous; male sepals obovate, verrucose-punctate, margins entire, glabrous, with a 2–3-lobate basal gland; petals cuneate-obovate, entire, 2-glandular at the base; rudimentary ovary glabrous, cylindric with a much dilated discoid tip; female sepals ovate-lanceolate, margin denticulate at the apex, glabrous, with a 2–3-lobate basal gland; petals narrow-obovate, glabrous, margin denticulate at the apex, eglandular; ovary glabrous, warted-punctate; styles free, 2-fid; capsule glabrous, warted, ⅙ in. across; seeds black, shining. *Pax in Engl. Pflanzenr. Euphorb. Cluyt.* 78, *fig.* 26 *F, incl.* both vars; *Prain in Kew Bulletin,* 1913, 389. *C. alaternoides, Willd. Sp. Pl.* iv. 879, *partly*; *not of Linn. C. alaternoides,* γ [*angustifolia*], *E. Meyer in Drège, Zwei Pfl. Documente,* 174, *as to b only. C. alaternoides,* γ *lanceolata,* ββ *revoluta, Sond. in Linnæa,* xxiii. 128, *in part only and excl. all locs. C. polygonoides, var. heterophylla, Krauss in Flora,* 1845, 82, *and var. angustifolia, Krauss l.c., as to Cape loc. only*; *not C. polygonoides, Linn. C. polifolia, Sond. l.c.* 124 ; *Baill. Adansonia,* iii. 151; *not of Jacq.*

Coast Region : Paarl Div. ; between Paarl Mountain and Paardeberg, under 1000 ft., *Drège* (C. alaternoides, γ angustifolia, b)! Cape Div. ; many localities, *Bergius*! *Masson*! *Sieber*! *Krebs*! *Lichtenstein*! *Spielhaus*! *Mund & Maire*! *Ecklon & Zeyher,* 62 partly! *Krauss*! *Zeyher,* 3823 ! *Harvey*! *Bolus,* 3725 ! *Schlechter,* 1305! *Wilms,* 3616 mainly ! 3617 ! *Wolley-Dod,* 1729! 1956 ! 2639 ! 2744 ! *Dümmer,* 1189 ! 1977 !

9. C. polifolia (Jacq. Hort. Schoenbr. ii. 67, t. 250); a shrub, up to 4 ft. high ; stems slender, woody, much branched ; twigs rather spreading, glabrous, leaves sessile, thinly coriaceous, opaque, lanceolate or linear-lanceolate, obtuse but mucronulate, base cuneate, margin not scabrous, always revolute, ⅓–⅔ in. long, $\frac{1}{12}$–$\frac{1}{10}$ in. wide, quite glabrous on both surfaces, rather pale greyish-green especially beneath ; internodes faintly angular, under ⅛ in. long; flowers diœcious, white, subsessile, male in 2–3-flowered glomerules, female solitary ; pedicels slender, glabrous ; male sepals oblong-obovate, obtuse, with 3–4 small basal glands; petals obovate, cuneate and 2-glandular at the base ; rudimentary ovary turbinate, glabrous ; female sepals ovate-lanceolate, larger than the male, with a large 2–4-lobate basal scale ; petals cuneate-obovate, eglandular ; ovary glabrous ; styles free, 2-fid ; capsule subglobose, ⅙ in. across ; seeds black, shining. *Willd. Sp. Pl.* iv. 2, 880 ; *Pers. Syn.* ii. 636 ; *Poir. Encyc. Suppl.* ii. 302 ; *Spreng. Syst.* iii. 49 ; *Dietr. Syn.* v. 455 ; *Prain in Kew Bulletin,* 1913, 390. *C. polifolia, a genuina, Müll. Arg. in DC. Prodr.* xv. ii. 1049 ; *Pax in Engl. Pflanzenr. Euphorb. Cluyt.* 77. *C. acuminata, E. Meyer in Drège, Zwei Pfl. Documente,* 174, *letter a only*; *E. Meyer ex Sond. in Linnæa,* xxiii. 125 ; *not of Linn. f. nor of Thunb. C. teretifolia, Sond. l.c.* 124 ; *Baill. Adansonia,* iii. 152. *C. polifolia,* β *teretifolia, Müll. Arg. l.c.*; *Pax l.c. C. polifolia,* γ *cinerascens, Müll. Arg. l.c.*; *Pax l.c. C. meyeriana, Müll. Arg. l.c.* 1055 ; *Pax l.c.* 79. *C. brevifolia, Sond. l.c., in small part and as to Drège* 8230 *only. C. polifolia,* δ *brevifolia, Pax l.c.* 77, *in small part and as to Diels* 595.

SOUTH AFRICA: cultivated specimens !
COAST REGION: Van Rhynsdorp Div. ; Gift Berg, 1000-2000 ft., *Phillips,*
7390 ! Western descent from the Bokkeveld, *Diels,* 595 ! Clanwillian Div. ; Cedarberg Range, on Sneeuw Kop, 4500 ft., *Bodkin* ! *Pearson & Pillans,* 5825 ! at
Pakhuis Pass, *Bolus,* 9089 ! Oliphant River, *Penther,* 889 ! Stormvlei, near
Wupperthal, *Leipold,* 495 ! Piquetsberg Div. ! Piqueniers Kloof, 950-1000 ft.,
Schlechter, 4936 ! 7938 ! 7939 ! Malmesbury Div. ; Paardeberg, *Ecklon & Zeyher,*
61 partly ! near Mooreesberg, 500 ft., *Bolus,* 9980 ! near Hopefield, *Bachmann,*
112 ! Worcester Div. ; Dutoits Kloof, 1000-2000 ft., *Drège* ! Swellendam Div. ;
hills near Riet Kuil, *Ecklon & Zeyher,* 64 ! *Zeyher,* 3835 ! near Swellendam,
Mund ! Riversdale Div. ; Muis Kraal, near Garcias Pass, 1500 ft., *Galpin,* 4568 !
George Div. ; near George, *Mund* ! Knysna Div. ; Plettensbergs Bay, *Bowie* !
Krebs ! Uniondale Div. ; without precise locality, *Newdegate,* 2463 ! Uitenhage
Div. ; Van Stadens Berg, *Drège,* 8234 a ! Port Elizabeth Div. ; Van Stadens
River, *Mrs. Paterson,* 884 !
CENTRAL REGION : Prince Albert Div. ; Great Zwart Berg Range, *Drège,* 8230
partly ! at Zwart Berg Pass, 3600 ft., *Bolus,* 12285 !

10. **C. brevifolia** (Sond. in Linnæa, xxiii. 125, excl. Drège 8230);
a shrub, up to 3 ft. high ; stems slender, woody, much branched ;
twigs subfastigiate, glabrous ; leaves very shortly petioled, thinly
coriaceous, opaque, linear, obtuse but mucronulate, base cuneate,
margin not scabrous, much revolute, $\frac{1}{6}-\frac{1}{4}$ in. long, $\frac{1}{20}-\frac{1}{16}$ in. wide,
quite glabrous on both surfaces, dark green ; internodes faintly
angular, under $\frac{1}{8}$ in. long ; flowers diœcious, white, subsessile, male
in 2-3-flowered glomerules, female solitary ; pedicels slender,
glabrous ; male sepals oblong-obovate, obtuse, with a 3-4-lobed basal
scale ; petals obovate, cuneate and 2-glandular at the base ; rudimentary ovary turbinate, glabrous ; female sepals ovate-lanceolate,
larger than the male, with a large 2-4-lobate basal scale ; petals
cuneate-obovate, eglandular ; ovary glabrous ; styles free, 2-fid ;
capsule subglobose, $\frac{1}{4}$ in. across ; seeds black, shining. *Baill. Adansonia,* iii. 153, *excl. syn. E. Meyer; Prain in Kew Bulletin,* 1913,
391. *C. polifolia,* δ *brevifolia, Müll. Arg. in DC. Prodr.* xv. ii. 1049 :
Pax in Engl. Pflanzenr. Euphorb. Cluyt. 77, *mainly.*

COAST REGION : George Div. ; near George, *Bowie* ! Humansdorp Div. ; Tweefontein, *Burchell,* 4818 ! Zitzikamma, *Ecklon & Zeyher,* 61 mainly ! *Pappe* !
Kruisfontein Mountain, 1000 ft., *Galpin,* 4569 ! near Humansdorp, *Rogers,* 2907
partly ! Uitenhage Div. ; Van Stadens Berg, *Drège,* 8237 ! *Zeyher,* 374 ! 3834 !
Ecklon & Zeyher, 63 ! near the Zwartkops River, *Ecklon,* 601 ! *Zeyher* ! Port
Elizabeth Div. ; Redhouse, *Mrs. Paterson,* 661 ! near Port Elizabeth, *Prager,*
106 b ! *Mrs. Holland* !

A very distinct plant, considered by Müller to be a variety of *C. polifolia,* with a
limited and compact distribution. The localities Grahamstown and Stellenbosch
cited by Sonder are due to a misapprehension as to the provenance of his
specimens ; the plants from the Van Rhynsdorp and Clanwilliam divisions
referred here by Pax do not belong.

11. **C. virgata** (Pax & K. Hoffm. in Engl. Pflanzenr. Euphorb.
Cluyt. 71) ; a small erect shrub, 1-1½ ft. high ; stems slender,
several from a stout woody rootstock, virgate, simple or sparingly
branched near the top, glabrous; leaves sessile, coriaceous, opaque,
oblong-obovate to obovate-lanceolate, obtuse but mucronulate, base
rounded, margin slightly scabrous, usually flat, $\frac{3}{4}$-1 in. long, $\frac{1}{5}-\frac{1}{4}$ in.

wide, quite glabrous on both surfaces, rather dark green above, pale beneath; internodes faintly angular, $\frac{1}{5}$–$\frac{1}{3}$ in. long; flowers diœcious, greenish-white, pedicelled, male solitary or in pairs, female solitary; male sepals obovate, obtuse, with a 3-lobate basal scale; petals wide-obovate, cuneately unguiculate, 1-glandular at the base; rudimentary ovary turbinate, glabrous; female sepals elliptic-lanceolate, larger than the male, with a 3-lobate basal scale; petals oblong-obovate, eglandular; ovary glabrous; styles free, 2-fid; capsule subglobose, $\frac{1}{6}$ in. across; seeds black, shining. *Prain in Kew Bulletin,* 1913, 391.

KALAHARI REGION : Transvaal ; Amsterdam, *Buchanan,* 2970 ! 3185 ! Barberton, *Thornoroft,* 3935 ! Umlomati Valley, near Barberton, 4000 ft., *Galpin,* 1368 ! Swaziland ; Dalriach, near Mbabane, 4900 ft., *Bolus,* 12286 !
EASTERN REGION : Pondoland ; without precise locality, *Bachmann,* 782 ! 809 ! Natal ; Umgoti, *Gerrard* ! Ingoma, *Gerrard,* 1163 ! Klip River, 3500–4500 ft., *Sutherland* !

12. C. laxa (Eckl. ex Sond. in Linnæa, xxiii. 128); a small erect shrub up to 2 ft. high; stems slender, woody, much branched; twigs rather spreading, glabrous; leaves sessile, coriaceous, opaque, lanceolate, obtuse but mucronulate, base rounded, margin slightly scabrous, usually flat, $\frac{1}{4}$–$\frac{1}{2}$ in. long, $\frac{1}{8}$–$\frac{1}{4}$ in. wide, quite glabrous on both surfaces, rather pale green especially beneath; internodes faintly angular, under $\frac{1}{8}$ in. long; flowers diœcious, white, subsessile, male in few-flowered glomerules, female all solitary, rather densely aggregated towards end of twigs; male sepals obovate, obtuse, with a 3-lobate basal scale; petals wide-obovate, cuneately unguiculate, 2-glandular at the base; rudimentary ovary turbinate, glabrous; female sepals elliptic-lanceolate, larger than the male, with a 3-lobate basal scale; petals oblong-obovate, eglandular; ovary glabrous; styles free, 2-fid; capsule subglobose, $\frac{1}{6}$ in. across; seeds black, shining. *Prain in Kew Bulletin,* 1913, 392. *C. Alaternoides, Sims, Bot. Mag. t.* 1321 ; *Ait. Hort. Kew. ed.* 2, v. 422, *partly* ; *not of Linn. C. alaternoides, β intermedia, Sond. l.c., mainly* ; *Baill. Adansonia,* iii. 150. *C. alaternoides, γ lanceolata, aa planifolia, Sond. l.c., excl. syn. Willd. C. alaternoides, ζ lanceolata, Baill. l.c.* ; *Müll. Arg. in DC. Prodr.* xv. ii. 1048. *C. alaternoides, γ angustifolia,* 1 *lanceolata, Pax in Engl. Pflanzenr. Euphorb. Cluyt.* 70, *fig.* 22, *B* ; *not γ angustifolia, E. Meyer.*

SOUTH AFRICA: cultivated specimens !
COAST REGION : Riversdale Div. ; Garcias Pass, *Phillips,* 370 ! Oudtshoorn Div. ; near Oudtshoorn, *Miss Britten,* 89 ; Knysna Div. ; near Knysna, *Newdegate* ! Uitenhage, *Ecklon & Zeyher,* 42 ! Elands Kloof, *Ecklon* ! *Ecklon & Zeyher,* 59 ! Port Elizabeth, *Bolus,* 2243 ! *Mrs. Paterson,* 1109 ! Walmer, *Mrs. Paterson,* 832 ! Albany Div. ; at Soutars Post, *Burchell,* 3504 ! Grahamstown, *Williamson* ! *MacOwan,* 27 ! *Rogers,* 66 ! 3995 ! Stones Hill, *Schönland,* 72 ! Curries Kloof, *Schönland,* 576 ! Kabousie, *MacOwan,* 325 ! Harveys Post, *Galpin.* 78 ! Queenstown Div. ; Hangklip Mountain, 5500–6600 ft., *Galpin,* 1621 ! 1622 ! Stormberg Range, *Wyley* ! Stutterheim Div. ; Fort Cunynghame, *Sim,* 2180 ! Komgha Div. ; near the mouth of the Kei River, 200 ft., *Flanagan,* 1149 ! British Kaffraria ; without precise locality, *Cooper,* 78 ! 79 !

KALAHARI REGION : Transvaal; Lydenburg, *Wilms*, 1318 ! near Barberton,
3000 ft., *Galpin*, 934 !
EASTERN REGION : Transkei ; Kentani, 1000 ft., *Miss Pegler*, 1250 ! Natal ;
Dumisa, 2500 ft., *Rudatis*, 679 ! between Pietermaritzburg and Greytown, *Wilms*,
2270 !

13. C. africana (Poir. Encyc. Suppl. ii. 302, syn. Willd. excl.
[*Clutia*]) ; an undershrub, up to 2 ft. high ; stems thick, woody, usually
considerably branched ; twigs ascending, glabrous ; leaves sessile,
thickly coriaceous, opaque, oblong-obovate, obtuse but mucronulate,
base cuneate or less often rounded, occasionally minutely cordate,
margin scabrous and slightly revolute, $1\frac{1}{2}$–$1\frac{3}{4}$ in. long, $\frac{1}{2}$–$\frac{2}{3}$ in. wide,
quite glabrous on both surfaces, usually drying dark brown ; inter-
nodes $\frac{1}{3}$–$\frac{2}{5}$ in. long : flowers diœcious, yellow, in both sexes solitary,
pedicelled ; pedicels glabrous, $\frac{1}{6}$ in. long, male slender, female in
fruit rigid and stoutish ; male sepals obovate, obtuse, with a 3-lobate
basal scale ; petals wide-obovate, cuneately unguiculate, 2-glandular
at the base ; rudimentary ovary turbinate, glabrous ; female sepals
elliptic-lanceolate, larger than the male, with a 3-lobate basal scale ;
petals oblong-obovate, eglandular ; ovary glabrous ; styles free, 2-fid ;
capsule subglobose, $\frac{1}{6}$ in. across, glabrous ; seeds black, shining.
Prain in Kew Bulletin, 1913, 393.   *C. alaternoides, Linn. Sp. Pl.
ed. i.* 1042 *partly, and ed.* ii. 1475 *partly and as to syn. Comm. only* ;
Lam. Encyc. ii. 54 *partly and as to syn. Comm. only* ; *Thunb. Prodr.* 53,
and Fl. Cap. ed. Schult. 270, *mainly* ; *E. Meyer in Drège, Zwei Pfl.
Documente*, 174, *a, d* ; *Krauss in Flora*, 1845, 82.   *C. daphnoides,
Willd. Hort. Berol.* 52, *excl. t.* 52, *and Sp. Pl.* iv. ii. 880, *excl. syn.
Thunb.* ; *not of Lam.   C. alaternoides, var. major, Krauss l.c.* ; *Müll.
Arg. in DC. Prodr.* xv. ii. 1047 ; *Pax in Engl. Pflanzenr. Euphorb.
Cluyt.* 68, *mainly.   C. alaternoides, a latifolia, Sond. in Linnæa*,
xxiii. 127, *partly* ; *Müll. Arg. l.c.   C. floribunda, Baill. Étud. Gén.
Euphorb. Atl.* 30, *t.* xvi. *figs.* 1–5, *fide Pax.   C. heterophylla, Baill.
Adansonia*, iii. 150, *mainly, but excl. syn. Bernh.* ; *not of Thunb.
C. alaternoides, γ genuina, b. oblongata, Müll. Arg. l.c.*, 1048.—*Alater-
noides africana telephii legitimi imperati folio, Comm. Hort. Amstel.*
ii. 3, *t.* 2.

COAST REGION : Clanwilliam Div. ; Cedarberg Range, near the Honey Valley
and the Koudeberg, 2800–4000 ft., *Drège*, 906 ! *Diels*, 906 ! Piquetberg Div. ;
near Piquetberg, *Drège*, 8228 a ! Oliphants River, near Warm Baths, *Stephens*,
7223 ! *Phillips*, 7254 ! Paarl Div. ; Paarl Mountains and by the Berg River near
Paarl, *Drège* ! Cape Div. ; numerous localities, *Sparrmann* ! *Thunberg* ! *Bergius* !
Mund & Maire ! *Lichtenstein* ! *Drège*, a ! *Burchell*, 260 ! *Ecklon* ! *Ecklon & Zeyher* !
Prior ! *Pappe* ! *Hooker*, 616 ! *Harvey*, 24 ! 112 ! *Wright*, 452 ! *Dubuc* ! *Bolus*,
4586 ! *Miss Cole* ! *Rehmann*, 1394 ! 2028 ! *Wolley-Dod*, 608 ! 2743 partly ! 2799 !
Wilms, 3612 ! *Dümmer*, 27 ! 97 ! 1449 ! 1451 ! Stellenbosch Div. ; Hottentots
Holland, *Mund & Maire* !

14. C. Alaternoides (Linn. Sp. Pl. ed. i. 1042, excl. syn. Burm. ♂
and syn. Comm. [*Clutia*]) ; undershrub up to 2 ft. high ; stems rather
slender, woody, considerably branched ; twigs ascending, glabrous ;
leaves sessile, coriaceous, opaque, lanceolate or narrowly obovate-
lanceolate, obtuse but mucronulate, base gradually narrowed, margin

finally scabrous and usually distinctly revolute, $\frac{2}{3}$–$1\frac{1}{3}$ in. long, $\frac{1}{6}$–$\frac{1}{2}$ in. wide, quite glabrous on both surfaces, dark green above, paler beneath ; internodes faintly angular, $\frac{1}{6}$–$\frac{1}{3}$ in. long ; flowers diœcious, white, pedicelled, male in few-flowered glomerules, female solitary ; pedicels glabrous, male slender, $\frac{1}{6}$ in. long, female in fruit rigid and stoutish, $\frac{1}{3}$ in. long ; male sepals obovate, obtuse, with a 3-lobate basal scale ; petals wide-obovate, cuneately unguiculate, 2-glandular at the base ; rudimentary ovary turbinate, glabrous ; female sepals elliptic-lanceolate, larger than the male, with a 3-lobate basal scale ; petals oblong-obovate, eglandular ; ovary glabrous ; styles free, 2-fid ; capsule subglobose, $\frac{1}{3}$ in. across ; seeds black, shining. *Linn. Sp. Pl. ed.* ii. 1475, *excl. syn. Burm.* ♂ *and syn. Comm.* ; *Burm. f. Prodr. Fl. Cap.* 31 ; *Lam. Encyc.* ii. 54, *excl. syn. Comm. Ait. Hort. Kew. ed.* 1, iii. 419, *and ed.* 2, v. 422, *partly* ; *Willd. Hort. Berol.* 50, *t.* 50, *and Sp. Pl.* iv. ii. 879, *partly* ; *Pers. Syn.* ii. 636 ; *Spreng. Syst.* iii. 49 ; *E. Meyer in Drège, Zwei Pfl. Documente,* 174, *as to aa partly and as to c* ; *Dietr. Syn.* v. 455 ; *Baill. Adansonia,* iii. 150, *as to syn. Willd.* ; *Prain in Kew Bulletin,* 1913, 395. *C. polygalæfolia, Salisb. Prodr.* 390. *C. alaternoides,* β *intermedia, Sond. in Linnæa,* xxiii. 128, *as to syn. Burm., but excl. syn. E. Meyer. C. alaternoides,* γ *lanceolata, Sond. l.c., as to syn. Willd. but excl. syn. Bot. Mag. and syn. Eckl. and all specimens. C. alaternoides,* ε *angustifolia,* α *longifolia, Müll. Arg. in DC. Prodr.* xv. ii. 1048 ; *not var. angustifolia, E. Meyer. C. alaternoides,* γ *angustifolia,* 1 *lanceolata, Pax in Engl. Pflanzenr. Euphorb. Cluyt.* 70, *in part only* ; *not var. angustifolia, E. Meyer, nor var. lanceolata, Sond. C. angustifolia, Burch. Mss.* ; *not of Knauf.*

VAR. β, brevifolia (E. Meyer ex Sond. in Linnæa, xxiii. 128) ; a shrub, 6–10 ft. high ; stems and twigs as in the type ; leaves as in the type, but never exceeding $\frac{1}{2}$ in. in length, usually shorter ; internodes usually distinctly angular, sometimes strongly winged. *C. alaternoides, Thunb. Prodr.* 53, *and Fl. Cap. ed. Schult.* 270, *partly hardly* ; *of Linn.* : ; *Prain l.c.* 396. *C. alaternoides,* β *[brevifolia], E. Meyer in Drège, Zwei Pfl. Documente,* 174, *as to a only* (*name*). *C. alaternoides,* γ *genuina, c brevifolia, Müll. Arg. in DC. Prodr.* xv. ii. 1048. *C. alaternoides,* γ *genuina, e imbricata, Müll. Arg. l.c.* ; *Pax in Engl. Pflanzenr. Euphorb. Cluyt.* 70. *C. alaternoides,* β *genuina,* 3 *elliptica, Pax l.c. partly* ; *not of Müll. Arg. C. angulata, Burch.,* and *C. myrtifolia, Burch. ex Prain in Kew Bulletin,* 1913, 396.

VAR. γ, angustifolia (E. Meyer ex Sond. in Linnæa, xxiii. 128, in part only) ; shrub ; stems and twigs as in type ; leaves not exceeding $\frac{1}{2}$ in. in length, very distinctly revolute ; internodes distinctly angular or almost winged, but wings not toothed. *Prain, l.c.* 396. *C. alaternoides,* γ *[angustifolia], E. Meyer in Drège, Zwei Pfl. Documente,* 174, *as to a only. C. alaternoides,* γ *lanceolata,* ββ *revoluta, Sond. l.c. in part only. C. alaternoides,* ε *angustifolia, c leptophylla, Müll. Arg. in DC. Prodr.* xv. ii. 1048 ; *Pax in Engl. Pflanzenr. Euphorb. Cluyt.* 70.

SOUTH AFRICA : cultivated specimens !
COAST REGION : Clanwilliam Div. ; Cedar Berg, near Hennig Vlei, 2800 ft., *Diels,* 894 ! Piquetberg Div. ; Piqueniers Kloof, 1100 ft., *Schlechter,* 4966 partly ! Worcester Div. ; Dutoits Kloof, 2000–3000 ft., *Drège,* c ! 1836 ! Paarl Div. ; Wellington, *Miss Doidge* ! Cape Div. ; various localities, *Burmann* ! *Oldenburg,* 296 partly ! *Elliott* ! *Harvey,* 24 ! *Masson* ! *Spielhaus* ! *Rehmann,* 1271 mainly ! *Mund & Maire* ! *Reeves* ! *Lalande* ! *Prior* ! *Drège,* aa ! 374 ! *Pappe* ! *Wilms,* 3613 !

3616 partly ! *Burchell*, 867 ! *Ecklon* ! *Wolley-Dod*, 2743 ! Caledon Div. ; Caledon, 1000–2000 ft., *Mund* ! *Zeyher* ! *Spielhaus* ! Swellendam Div. ; near Swellendam, *Bowie* ! Riversdale Div. ; Platte Kloof, *Muir*, 468 ! George Div. ; near George, *Bowie* ! *Rogers*, 4295 ! Knysna Div. ; *Knysna, Newdegate* ! near the Goukamma River, *Burchell*, 5582 ! Little Homtini River, 250 ft., *Schlechter*, 5892 ! Uitenhage Div. ; Van Stadens Berg, *Zeyher*, 832 ! Bathurst Div. ; between Blauw Krantz and Kowie River, *Burchell*, 3886 ! between Bushmans River and the Karuga River, *Zeyher* ! Albany Div. ; near Grahamstown, *MacOwan*, 27 ! Var. β : Paarl Div. ; Paarl Mountain, 1000–2000 ft., *Drège*, β a ! Cape Div. ; east side of Table Mt., *Roxburgh* ! *Ecklon* ! *Lalande* ! Stellenbosch Div. ; Stellenbosch, *Krebs* ! Hottentots Holland Mountains, *Zeyher* ! Caledon Div. ; Zwartberg, 1000–2000 ft., *Ecklon* ! Swellendam Div. ; near Swellendam, *Bowie* ! *Mund & Maire*, 235 ! George Div. ; George, *Bowie* ! *Burchell*, 6038 ! *Prior* ! *Rogers*, 4295 ! Knysna Div. ; at Knysna, *Burchell*, 5543 ! *Bowie* ! Uitenhage Div. ; Van Stadensberg, *Ecklon* ! Albany Div. ; near Grahamstown, *Zeyher*, 1019 ! *Scott Elliot*, 1029 ! Var. γ : Mossel Bay Div. ; west bank of Great Brak River, *Burchell*, 6154 ! Dreifontein, *Drège*, a !

EASTERN REGION : Transkei ; Kentani, Lobinguba, 1000 ft., *Miss Pegler*, 1250 !

15. **C. imbricata** (E. Meyer in Drège, Zwei Pfl. Documente, 174, letter a only [name]) ; a small erect shrub, 2 ft. high ; stems slender, woody, much branched ; twigs intricately spreading, glabrous ; leaves sessile, coriaceous, opaque, obovate-oblong, obtuse but slightly mucronulate, gradually narrowed to the base, margin slightly scabrous, somewhat revolute, $\frac{1}{5}$ in. long, $\frac{1}{12}$ in. wide, quite glabrous, glaucous on both surfaces ; internodes terete, very short, and leaves densely imbricate ; flowers dioecious, white, pedicelled, male in few-flowered glomerules, female solitary ; pedicels glabrous ; male sepals obovate, obtuse, with a 3-lobate basal scale ; petals wide-obovate, cuneately unguiculate, 2-glandular at the base ; rudimentary ovary turbinate, glabrous ; female sepals elliptic-lanceolate, larger than the male, with a 3-lobate basal scale ; petals oblong-obovate, eglandular ; ovary glabrous ; styles free, 2-fid ; capsule subglobose, $\frac{1}{8}$ in. across ; seeds black, shining. *Sond. in Linnæa*, xxiii. 125 ; *Pax in Engl. Pflanzenr. Euphorb. Cluyt.* 83 ; *Prain in Kew Bulletin*, 1913, 398.

WESTERN REGION : Little Namaqualand ; Khamiesberg Range, between Pedros Kloof and Lelie Fontein, *Drège*, a ! 3030 ! near summit of Beacon Hill, *Pearson*. 6710 partly ! near stream in Groene Kloof, *Pearson*, 6617 !

16. **C. rubricaulis** (Eckl. ex Sond. in Linnæa, xxiii. 128) ; a small shrub up to 2 ft. high ; stems firm, woody, considerably branched ; twigs ascending, glabrous ; leaves sessile, densely imbricate ; coriaceous, opaque, oblong-ovate, obtuse but mucronulate, base rounded or wide-cuneate, margin scabrous, flat, $\frac{1}{2}$ in. long, $\frac{1}{4}$ in. wide, quite glabrous, glaucous on both surfaces ; internodes terete, very short ; flowers dioecious, white, pedicelled, male in few-flowered glomerules, female solitary ; pedicels glabrous, male slender, $\frac{1}{6}$ in. long, female in fruit rigid, stoutish, $\frac{1}{8}$–$\frac{1}{4}$ in. long ; male sepals obovate, obtuse, with a 3-lobate basal scale ; petals wide-obovate, cuneately unguiculate, 2-glandular at the base ; rudimentary ovary turbinate, glabrous ; female sepals elliptic-lanceolate, larger than the male, with a 3-lobate basal scale ; petals oblong-obovate, eglandular : ovary

glabrous ; styles free, 2-fid ; capsule subglobose, $\frac{1}{6}$ in. across ; seeds black, shining. *Prain in Kew Bulletin*, 1913, 399. *C. alaternoides* β [*brevifolia*], *E. Meyer in Drège, Zwei Pfl. Documente*, 174, *letters c, d, e, f, g* ; *not* β *brevifolia, E. Meyer ex Sond. C. polygonoides, Krauss in Flora*, 1845, 82 ; *not of Linn., nor of Thunb. and hardly of Willd. or of Sond. C. polygonoides,* β *foliis utrinque glaucis, Sond. l.c. C. glauca, Pax in Ann. Naturhist. Hofmus. Wien*, xv. 50. *C. alaternoides,* δ *microphylla, Müll. Arg. in DC. Prodr.* xv. ii. 1048, *partly and as to syn. Eckl. only ; Pax in Engl. Pflanzenr. Euphorb. Cluyt.* 70, *partly and as to 2, glauca only.*

VAR. β, **microphylla** (Prain in Kew Bulletin, 1913, 400) ; prostrate, fastigiately intricately branching ; leaves as in the type but much smaller, $\frac{1}{8}-\frac{1}{4}$ in. long $\frac{1}{16}-\frac{1}{8}$ in. wide. *C. alaternoides, var. microphylla, Müll. Arg. in DC. Prodr.* xv. ii. 1048, *partly ; Pax in Engl. Pflanzenr. l.c. as to forma typica only, fig. 22. C. alaternoides,* β [*brevifolia*], *E. Meyer in Drège, l.c., letter b only ; not* β *brevifolia, E. Meyer ex Sond. C. polygonoides, Sond. l.c. ; Baill. Adansonia,* iii. 153, *excl. syn. Willd. and Burm. ; not of Linn. and hardly of Willd. or of Krauss. C. gnidioides, Willd., C. microphylla, Burch., and C. polygonoides, var. curvata, E. Meyer ex Prain, l.c.*

VAR. γ, **grandifolia** (Prain in Kew Bulletin, 1913, 400) ; erect, sparingly branched ; twigs ascending ; leaves obovate, narrowed to the cuneate base, $\frac{3}{4}-1\frac{1}{4}$ in. long, $\frac{1}{4}-\frac{3}{8}$ in. wide. *C. alaternoides, Thunb. Prodr.* 53, *and Fl. Cap. ed. Schult.* 470, *partly ; E. Meyer in Drège, Zwei Pfl. Documente*, 174, *letter b only ; not of Linn. C. polygonoides, Willd. Hort. Berol.* 51, *t.* 51, *and Sp. Pl.* iv. ii. 879 ; *Pers. Syn.* ii. 636 ; *Ait. Hort. Kew. ed.* 2, v. 422 ; *Spreng. Syst.* iii. 49 ; *Dietr. Syn.* v. 455 ; *Baill. Adansonia,* iii. 153, *as to syn. Willd. only ; not of Linn. C. polygonoides, var. grandifolia, Krauss in Flora*, 1845, 82. *C. alaternoides,* γ *genuina, Müll. Arg. l.c., as to b oblongata and d elliptica only ; Pax l.c.* 68, *as to 1 grandifolia only.*

VAR. δ, **tenuifolia** (Prain l.c. 400) ; erect or prostrate, fastigiately intricately branching ; leaves lanceolate or linear, acute, base cuneate, $\frac{1}{4}-\frac{3}{8}$ in. long, $\frac{1}{12}-\frac{1}{8}$ in. wide. *C. imbricata, E. Meyer in Drège, Zwei Pfl. Documente*, 174 *partly and as to letter b* (*male*) *only ; not of E. Meyer ex Sond. l.c.* 125. *C. tenuifolia, Sond. l.c.* 123 ; *not of Willd. C. alaternoides,* ε *angustifolia, Müll. Arg in DC. Prodr.* xv. ii. 1048, *as to b brachyphylla only ; Pax l.c.* 70, *as to b, brachyphylla in part. C. thymi-folia, Willd. Mss. in Herb. Berol.*

SOUTH AFRICA: Var. γ, cultivated specimens!

COAST REGION : Clanwilliam Div. ; Wupperthal, *Wurmberg* (*Drège* β, letter e)! Piquetberg Div. ; Oliphants River Mountains behind Warm Baths, *Phillips*, 7253 ! Tulbagh Div. ; New Kloof, *Burchell*, 1016! Winterhoek Mountain, 3500 ft., *Bolus*, 5356 ! Saron, 1000 ft., *Schlechter*! Worcester Div. ; mountains near Worcester, *Rehmann*, 2538 partly ! Paarl Div. ; Drakenstein Range, 3000–4000 ft., *Drège* β, letter c ! Cape Div. ; various localities, *Ecklon*, 416 ! *Krebs*! *Prior*! *Bergius*! *Sparrmann*! *Harvey*! *Drège* β, letter d ! Stellenbosch Div. ; Stellen-bosch, *Prior* ! Hottentots Holland Mountain, 1000 ft., *Diels*, 1310 ! Caledon Div. ; Baviaans Kloof near Genadendal, *Drège* β, letter f ! *Ecklon* ! Caledon, *Prior* ! Houw Hoek, *Bolus*, 9937 partly ! near Greitjesgat, *Ecklon* ! Riversdale Div. ; Albertina, *Muir* ! Mossel Bay Div. ; Attaquas Kloof, *Gill* ! George Div. ; near George, 1000 ft., *Burchell*, 6007 ! *Bowie* ! *Drège* β, letter g ! *Penther*, 1597 ! *Schlechter*, 2240 ! Plettenbergs Bay, *Bowie* ! Humansdorp Div. ; near Humansdorp, *Rogers*, 2907 partly ! 2934 ! 2994 ! Uitenhage Div. ; without precise locality, *Zeyher*, 977 ! Port Elizabeth Div. ; near Port Elizabeth, *Ecklon & Zeyher* ! *Ecklon,* 977 ! *Holub* ! *Mrs. Paterson*, 1109 ! 2135 ! *Drège*, 414 ! Var. β : Malmesbury Div. ; Hope-field, *Bachmann*, 115 ! 943 ! Worcester Div. ; Dutoits Kloof, 2000–4000 ft., *Drège* β, letter b ! near Worcester, *Rehmann*, 2538 partly ! Stellenbosch Div. ; Hottentots Holland Mountains, 1000 ft., *Zeyher*, 3831 ! *Prior* ! Caledon Div. ; various

localities, *Burchell*, 7667 ! 8151 ! 8625 ! *Scott Elliot*, 1115 ! *Ecklon*, 52 partly !
Lichtenstein ! *Miss Cole* ! Swellendam Div. ; without precise locality, *Bowie* !
Var. γ : Clanwilliam Div. ; Cedarberg Range, 3000 ft., *Diels*, 870 ! Cape Div. ;
various localities, *Tulbagh*, 127 *in Herb. Linnæus* ! *Banks & Solander* ! *Bergius* !
Oldenburg, 296 partly ! *Robertson* ! *Lalande* ! *Sieber*, 148 ! *Forbes* ! *Prior* ! *Dubuc* !
Reeves ! *Miss Cole* ! *Rehmann*, 974 ! 1271 partly ! 1393 ! 2029 ! *Wolley-Dod*, 1209 !
Schlechter, 716 ! 977 ! *Dümmer*, 1638 ! *Rogers*, 11222 ! *Burchell*, 260 partly !
Boivin, 733 ! *Krebs*, 103 ! *Lichtenstein* ! *Ecklon*, 603 ! *Zeyher*, 3822 ! *Drège*, b !
138 ! *Diels*, 110 ! *Spielhaus* ! *Fuller* ! *Wilms*, 3613 ! 3614 ! 3615 ! Caledon Div. ;
Caledon, *Ecklon*, 449 ! Mossel Bay Div. ; Attaquas Kloof, *Gill* ! Humansdorp
Div. ; Kruisfontein Mountains, 800 ft., *Galpin*, 4592 ! Port Elizabeth Div. ;
Algoa Bay, *Forbes* ! Var. δ : Van Rhynsdorp Div. ; Giftberg, 1000–2000 ft.,
Phillips, 7387 ! 7395 ! Clanwilliam Div. ; Kakadouw Pass. 3900 ft., *Diels*, 928 !
Malmesbury Div. ; near Hopefield, *Bachmann*, 944 ! between Hopefield and
Langebaan, *Bachmann*, 2079 ! 2080 ! *Bolus* ! Tulbagh Div. ; near Tulbagh Water-
fall, *Ecklon & Zeyher* ! Cape Div. ; without precise locality, *Tulbagh*, 113 *in Herb.
Linnæus* ! *Lichtenstein* ! *Forbes*, 88 ! Stellenbosch Div. ; Lowrys Pass, 500 ft.,
Schlechter, 1191 ! Caledon Div. ; Klein River Mountains, 1000–3000 ft., *Ecklon &
Zeyher*, 64 ! near Caledon, *Bolus*, 8501 ! Swellendam Div. ; without precise
locality, *Mund & Maire* ! Riversdale Div. ; without precise locality, *Rust*, 550 !
Mossel Bay Div. ; near Little Brak River, *Rogers*, 4213 !

CENTRAL REGION : Var. δ : Ceres Div. ; slopes at Hottentots Kloof, *Pearson*,
4897 ! Prince Albert Div. ; Zwart Berg Pass, 5000 ft.. *Bolus*, 12288 !

WESTERN REGION : Var. δ : Little Namaqualand ; Khamiesberg Range, between
Pedros Kloof and Lelie Fontein, *Drège* (*C. imbricata*, b only) ! near the summit
of Beacon Hill, *Pearson*, 6710 partly !

17. **C. ovalis** (Sond. in Linnæa, xxiii. 129) ; a small shrub, 1½ ft.
high ; stems firm, woody, sparingly branched ; twigs ascending,
glabrous ; leaves sessile, oval or ovate-oblong, obtuse or retuse
but mucronulate, base rounded, margin scabrous, flat, ½ in.
long, ⅛ in. wide, quite glabrous ; internodes more or less angled, as
long as the leaves ; flowers diœcious, white, pedicelled, female solitary ; pedicels glabrous, stoutish, rigid, ⅙ in. long ; sepals elliptic-
lanceolate, with a 3-lobate basal scale ; petals oblong-obovate,
eglandular ; ovary glabrous ; styles free, 2-fid ; capsule subglobose,
⅙ in. across ; seeds black, shining. Male flowers not seen. *Baill.
Adansonia*, iii. 153 ; *Müll. Arg. in DC. Prodr.* xv. ii. 1047 ; *Pax in
Engl. Pflanzenr. Euphorb. Cluyt.* 71 ; *Prain in Kew Bulletin*, 1913,
402. *C. alaternoides, β genuina*, 3 *elliptica, Pax l.c.* 70, *partly* ; *not
of Müll. Arg.*

COAST REGION : Piquetberg Div. ; slopes near Piqueniers Kloof, 1200 ft.,
Schlechter, 4966 ! Tulbagh Div. ; mountains near Tulbagh Waterfall, *Ecklon &
Zeyher*, 52 !

18. **C. impedita** (Prain in Kew Bulletin, 1913, 402) ; an undershrub ; stems rigid, erect, rather copiously virgately branched in the
upper half, 1½–2 ft. high, terete, even when young quite glabrous in
every part ; leaves shortly petioled, firmly papery, close set, more or
less imbricate, all obovate, truncate or retuse, base gradually cuneate,
margin flat, ⅓–½ in. long, ¼–⅓ in. wide near the apex, pale green,
glabrous, pellucid-punctate and warted, midrib faintly seen, nerves
obscure ; petiole ¹⁄₁₂ in. long or less ; flowers diœcious, male alone
seen, solitary or in pairs, pink ; pedicels very short, under ¹⁄₁₂ in.

long; male sepals suborbicular, rather fleshy, with a 2–4-lobed basal scale; petals obovate, gradually narrowed to the base, and there with a very minute gland; rudimentary ovary ovoid, glabrous.

SOUTH AFRICA : without locality, *Prior*!
COAST REGION : Queenstown Div.; Andriesberg, near Bailey, 6400 ft., *Galpin*, 2026! Cathcart Div.; Bontebok Flats, *Sim*, 2543!

19. **C. alpina** (Prain in Kew Bulletin, 1913, 403); undershrub; stems several, prostrate from a woody rootstock, 4–12 in. long, giving off several ascending or prostrate branches, 2–6 in. long, again branching; twigs angled or slightly winged, sparingly and softly tawny hirsute; leaves petioled, membranous, pellucid-punctate, ovate, obtuse, base rounded or truncate, margin revolute, $\frac{1}{5}$–$\frac{1}{2}$ in. long, $\frac{1}{5}$–$\frac{1}{3}$ in. wide, adpressed hirsute on the midrib above, otherwise glabrous on both surfaces; nerves not visible; petiole $\frac{1}{8}$–$\frac{1}{4}$ in. long, pubescent; flowers diœcious, green, male 2-nate in the leaf-axils; pedicels very short, surrounded at the base by small ovate hyaline scales with margins ciliate towards the base; male sepals ovate, obtuse, within eglandular; petals spathulate, eglandular, but each with a minute gland attached within its insertion; rudimentary ovary cylindric, glabrous.

CENTRAL REGION : Barkly East Div.; Witteberg Range, on Ben Macdhui, 9300 ft., *Galpin*, 6827!

20. **C. glabrescens** (Knauf in Engl. Jahrb. xxx. 340); a shrub up to 10 ft. high; twigs glabrous, not warted; leaves distinctly petioled, thinly papery, pellucid-punctate but not warted, elliptic-lanceolate, acute, base cuneate, margin flat, 1$\frac{1}{2}$–2$\frac{1}{2}$ in. long, $\frac{3}{4}$–1 in. wide, dark green, glabrous on both surfaces, dull; petiole $\frac{1}{4}$–$\frac{1}{3}$ in. long; flowers diœcious, white, male in few-flowered fascicles, female 1–3; pedicels up to $\frac{1}{4}$ in. long, female elongated in fruit and at length $\frac{1}{2}$ in. long; male sepals obovate, obtuse, punctate but not warted, glabrous, with a 3-lobate basal scale; petals obovate-spathulate, with a simple basal scale; rudimentary ovary hardly dilated at the tip, glabrous; female sepals rather larger and firmer than the male, with a 2-lobate basal scale; petals without a basal scale; ovary glabrous; styles free, shortly 2-fid; capsule $\frac{1}{5}$ in. across, subglobose, glabrous, not warted; seeds black, shining. *Prain in Kew Bulletin*, 1913, 403. *C. abyssinica, Pax in Engl. Pflanzenr. Euphorb. Cluyt.* 56, *partly*; *Hutchinson in Dyer*, *Fl. Trop.* vi. i. 807, *partly*; *not of Jaub. & Spach.*

EASTERN REGION : Zululand; Nkandhla, 4000–6000 ft., *Wylie in Herb. Wood*, 8874!

Also in Eastern Tropical Africa.

21. **C. Galpini** (Pax in Engl. Pflanzenr. Euphorb. Cluyt. 54, under *C. pulchella*); shrub, 3–5 ft. high; twigs smooth, when young puberulous; leaves distinctly petioled, rather firmly membranous,

pellucid-punctate but not warted, ovate, acute, 1 in. long, ½–⅔ in. wide, when young puberulous beneath, soon glabrous, pale green; petiole ⅛–¼ in. long; flowers diœcious, white, male in few-flowered fascicles, female usually solitary, sometimes in threes; pedicels short, female elongated in fruit, at length up to ⅙–⅓ in. long; male sepals oblong-ovate, not warted, with a 3-lobate basal scale; petals deltoid-ovate, narrowed to a rather wide claw, with a simple basal scale; rudimentary ovary rather dilated at the tip, glabrous; female sepals rather firmer than the male with a 3-lobate basal scale; petals as in male with a smaller or obsolete basal scale; ovary glabrous; styles free, shortly 2-fid; capsule ⅕ in. across, subglobose, warted-punctate; seeds black, shining. *Pax in Bull. Herb. Boiss.* vi. 736, *as to name only, but excl. descr. of ♂ fl. and excl. Galpin* 961; *Prain in Kew Bulletin,* 1913, 403. *C. pulchella,* γ *ovalis, Müll. Arg. in DC. Prodr.* xv. ii. 1046. *C. pulchella, forma genuina* (*in small part*) *and forma ovalis, Pax in Engl. Pflanzenr. Euphorb. Cluyt.* 54, 55.

KALAHARI REGION : Transvaal; Pretoria and neighbourhood, *Rehmann,* 4231! 4287! *Miss Leendertz,* 532! *Bolus,* 10839! *Wilms,* 1320 partly! *Kirk,* 50! *Burtt-Davy,* 7477! Wonderboom Poort, *Rehmann,* 4549! Heidelberg, *Miss Leendertz,* 1031! Boschveld, *Rehmann,* 4871! Elandsfontein, near Johannesburg, 5500 ft., *Gilfillan in Herb. Galpin,* 1426! Rustenberg, 4500 ft., *Miss Nation,* 52! 202! Barberton, 3000 ft., *Thorncroft,* 1943! and without precise locality, *Wahlberg!*

22. **C. pulchella** (Linn. Sp. Pl. ed. i. 1042 [*Clutia*]); a shrub up to 8 ft. high; twigs warted, when young slightly adpressed-hairy; leaves distinctly petioled, membranous, warted and punctate, ovate or ovate-oblong, rarely suborbicular, subacute or obtuse, base cuneate, margin flat, 1½–2½ in. long, ½–1¼ in. wide, glabrous; petiole ⅓–1 in. long; flowers diœcious, white, male in few-flowered fascicles from perulate axillary swellings, female usually paired; pedicels up to ¼ in. long, female elongated in fruit and at length ½ in. long or longer; male sepals oblong-ovate, warted-punctate, with a 3-lobate basal scale; petals deltoid-ovate, narrowed to a wide claw, each with a simple basal scale; rudimentary ovary somewhat widened upwards, glabrous; female sepals rather firmer than male, with a 2–3-lobate basal scale; petals as in male, with a small or obsolete basal scale; ovary glabrous; styles free, shortly 2-fid; capsule ¼ in. across, subglobose, warted-punctate; seeds black, shining. *Linn. Sp. Pl. ed.* ii. 1475; *Burm. f. Prodr. Cap.* 27 *bis* [31]; *Lam. Encycl.* ii. 54; *Willd. Sp. Pl.* iv. ii. 881; *Pers. Syn.* ii. 636; *Thunb. Prodr.* 53, *and Fl. Cap. ed. Schult.* 271; *Spreng. Syst.* iii. 49; *A. Juss. Euph. Gen. Tent. t.* 6, *fig.* 21; *Ait. Hort. Kew. ed.* 1, iii. 420, *and ed.* 2, v. 423; *Bot. Mag. t.* 1945; *E. Meyer in Drège, Zwei Pfl. Documente,* 174, *a only; Krauss in Flora,* 1845, 81; *Sond. in Linnæa,* xxiii. 129; *Dietr. Synops.* v. 455; *Baill. Étud. Gén. Euphorb. t.* 16, *fig.* 6–19, *and in Adansonia,* iii. 153; *O. Kuntze, Rev. Gen. Pl.* iii. ii. 284; *Prain in Kew Bulletin,* 1913, 404. *C. cotinifolia, Salisb. Prodr.* 390. *C. pulchella, a genuina, Müll. Arg. in DC. Prodr.* xv. ii. 1045.

C. pulchella, var. obtusata, Müll. Arg. l.c. 1046, *in part*; *not of Sond.*
C. pulchella, forma genuina (excl. syn. C. Galpini and all the Transvaal
localities), forma macrophylla (excl. syn. Müll. Arg.), and forma
obtusata (in part only), Pax in Engl. Pflanzenr. Euphorb. Cluyt. 54.

VAR. β, obtusata (Sond. l.c.) ; twigs not warted, when young slightly adpressed-
hairy ; leaves papery, punctate but not warted, rounded or ovate-obtuse, rarely
ovate-subacute, ¼–3 in. long, ¼–1½ in. wide, glabrous or sparingly adpressed-hairy.
Müll. Arg. in DC. Prodr. xv. ii. 1046 *mainly* ; *Prain, l.c.* 405. *C. pulchella,*
E. Meyer in Drège l.c., b only. C. pulchella, forma microphylla, Pax in Ann.
Naturhist. Hofmus. Wien, xv. 49. *C. pulchella, forma genuina (Rehmann* 5912
only), forma microphylla, forma macrophylla (as to Natal plant only), and forma
obtusata (mainly), Pax in Engl. Pflanzenr. l.c.

VAR. γ, Franksiæ (Prain l.c. 405) ; twigs not warted, persistently softly pilose
with spreading hairs ; leaves thinly membranous, punctate but not warted, ovate,
subacute, ¾–1 in. long, ½–¾ in. wide, glabrous except on the nerves above, softly
persistently pilose with spreading hairs beneath.

SOUTH AFRICA : Many cultivated specimens !
COAST REGION : Cape Div. ; various localities, *Sparrmann* ! *Oldenburg,* 294 !
Thunberg ! *Sonnerat* ! *Brown & Solander* ! *Roxburgh* ! *Mund & Maire* ! *Wallich,*
313 ! *Krebs* ! *Thom* ! *Bowie* ! *Ecklon,* 195 ! 415 ! 606 ! *Drège* ! *Zeyher* ! *Burchell,*
295 ! *MacGillivray,* 612 ! *Prior* ! *Harvey,* 232 ! 498 ! *MacOwan,* 165 & *Herb.*
Norm. Austr.-Afr. 767 ! *von Düben* ! *Krauss,* 1709 ! 1710 ! 1714 ! *Rehmann,* 973 !
1389 ! 1390 ! 1391 ! 1392 ! *Wolley-Dod,* 865 ! 1105 in herb. Brit. Mus. ! *Wilms,*
3619 ! *Diels,* 48 a ! *Marloth,* 45 ! *Dümmer,* 26 ! 59 ! George Div. ; near George,
900 ft., *Bowie* ! *Schlechter,* 2411 ! Silver River, *Penther,* 911 ! Montagu Pass,
Rehmann, 169 ! 170 ! Knysna Div. ; Knysna forests, *Laidley,* 434 ! Humansdorp
Div. ; Storms River Forest, Zitzikamma, 500 ft., *Galpin,* 4574 ! Albany Div.,
Broekhuizens Poort, 1500–2000 ft., *Galpin,* 24 ! Howisons Poort, *Ecklon,* 905 !
Bedford Div. ; Bedford, *Weale* ! Var. β : Humansdorp Div. ; banks of the
Gamtoos River, *Prior* ! Springfields, *Mrs. Paterson,* 2187 ! 2188 ! Albany Div. ;
near Grahamstown, *Bolton* ! *Miss Bowker* ! *Zeyher,* 28 ! 539 ! 905 ! *Binnie,* 612 !
Prior ! *MacOwan,* 165 partly ! *Mrs. White* ! *Misses Daly & Sole,* 299 ! Alexandria
Div. ; between Hoffmanskloof and Dreifontein, *Drège,* 8224 a ! Bathurst Div. ;
Bathurst, *Rogers,* 3502 ! Stockenstrom Div. ; Stockenstrom River, *Pappe,* 2373 !
Queenstown Div. ; Stormberg Range, 5000–6000 ft., *Drège,* 8224 b ! Table
Mountain, 6000–7000 ft., *Drège,* 8224 c ! *Zeyher,* 3824 ! *Ecklon & Zeyher,* 42 !
near Queenstown, 3800 ft., *Fraser* ! *Galpin,* 1572 ! Sterkstroom, *Rogers,* 4051
(forma *microphylla,* Pax) ! 4052 ! Kingwilliamstown Div. ; Kingwilliamstown,
Krook, 916 (forma *microphylla,* Pax) ! Komgha Div. ; sandflats near the Kei
River, *Flanagan,* 222 ! British Kaffraria ; without locality, *Cooper,* 80 ! 81 !
CENTRAL REGION : Var. β : Somerset Div. ; Bruintjes Hoogte, *Burchell,* 2993 !
Bosch Berg, *MacOwan,* 165 partly ! Somerset East, *Bowker* ! *Scott Elliot,* 508 !
Tarka Div. ; near Tarka, *Shaw* !
KALAHARI REGION : Var. β : Orange River Colony ; Witteberg Range, Kadzie
Berg, *Rehmann,* 3988 ! Basutoland ; Leribé, *Dieterlin,* 258 ! Transvaal ; Hout-
bosch, *Rehmann,* 5910 ! 5912 !
EASTERN REGION : Pondoland ; Umzimkulu River, *Bachmann,* 797 ! Natal ;
Friedenau, *Rudatis,* 943 partly ! Ifapalal, *Rudatis,* 945 ! Durban, *Gerrard,* 1162 !
Sutherland ! Morburg, 300 ft., *Rogers* ! Kearsney, 1000 ft., *Engler,* 2617 ! Var. β :
Transkei ; Kentani, 1000 ft., *Miss Pegler,* 228 ! 745 ! Colossa, near Indutwa,
Krook, 895 ! Tembuland ; Bazeia, *Baur,* 171 ! Pondoland ; between Umtata River
and St. Johns River, *Drège* b ! 4635 ! Port St. John, *Galpin,* 3437 ! various
localities, *Bachmann,* 758 ! 759 ! 761 ! 798 ! Griqualand East ; Nalogha, *Krook,*
941 ! Natal ; Friedenau, *Rudatis,* 846 ! 943 partly ! 984 ! 995 ! Durban, *Guein-*
zius ! *Wood,* 6508 ! Inanda, *Wood,* 142 ! 593 ! Attercliffe, *Sanderson,* 411 !
Gerrard, 58 ! *Gerrard & McKen,* 42 ! Umbilo Waterfall, *Rehmann,* 8121 ! Colenso,
Krook, 877 ! Tugela, *Gerrard,* 543 (forma *microphylla,* Pax) ! Umzimkulu, *Krook,*
945 (forma *microphylla,* Pax) ! Var. γ : Natal ; Amanzimtoti, *Miss Franks in*
Herb. Wood, 11912 ! 12606 ! Zululand ; Ongoa, *Gerrard,* 2151 !

23. C. mollis (Pax in Engl. Jahrb. xix. 112) ; a shrub ; branches and twigs terete, not warted, persistently tawny-velvety with soft hairs ; leaves distinctly petioled, membranous, warted and punctate, ovate-oblong, ovate or ovate-subcordate, subacute or acute, base cuneate, truncate or subcordate, margin flat, 2–5½ in. long, 1½–3½ in. wide, sparingly pubescent with spreading hairs on the nerves above, densely velvety on the nerves and sparingly pubescent elsewhere beneath ; petiole pubescent, ½–1¼ in. long ; flowers diœcious, yellow, male short-pedicelled, in few-flowered fascicles from perulate axillary swellings ; female usually paired ; pedicels in flower ½ in. long, pubescent with spreading hairs, elongated in fruit and at length 1 in. long or longer ; male sepals narrow-ovate, acute, warted-punctate and pubescent externally, with a 3-lobate basal scale ; petals spathulate-lanceolate, much longer than the sepals, narrowed to a wide claw, eglandular, but each with a large stalked sub-2-lobed gland attached within the insertion of the petals ; rudimentary ovary very short, cylindric, glabrous ; female sepals firmer than the male, with a 2–3-lobate basal scale ; petals much firmer than the male, hardly longer than the sepals, with a small or obsolete basal scale ; ovary glabrous ; styles free, shortly 2-fid ; capsule ⅓ in. across, subglobose, strongly warted-punctate ; seeds black, shining. *Engl. Pflanzenr. Euphorb. Cluyt.* 55 ; *Hutchinson in Dyer, Fl. Trop. Afr.* vi. i. 808 : *Prain in Kew Bulletin,* 1913, 406. *C. leuconeura, Pax in Engl. Jahrb. l.c. C. abyssinica, var. usambarica, Pax in Engl. Pflanzenr. l.c.* 57 *and var. ovalifolia, Pax l.c.*

EASTERN REGION : Natal ; Alfred County, near Murchison, 2000 ft.. *Wood,* 1990!

Also in Tropical East Africa.

24. C. affinis (Sond. in Linnæa, xxiii. 126) : a shrub, erect, 8–10 ft. high, much branched : twigs spreading, usually terete or slightly angled, grey- or tawny-pubescent ; leaves distinctly petioled, firmly membranous, opaque, lanceolate-spathulate or oblong-obovate or linear-elliptic, obtuse or subacute, narrowed from beyond the middle to the base, margin flat, those of main-branches 2–3 in. long, ½–1 in. wide, of small twigs about 1 in. long, ¼–⅓ in. wide, pubescent or puberulous on both surfaces, with rather distinct main-nerves, usually rather paler beneath ; petiole ⅕–½ in. long, pubescent ; flowers diœcious, yellowish, male 4–6 together in axillary clusters, female 2–3 ; pedicels 1/12 in. long in the male, in the female up to ⅛ in. long in fruit, rather rigid, pubescent ; male sepals narrow-oblong, obtuse, pubescent, with a 3-lobate basal scale ; petals rounded-obovate, with a much thickened claw, 2-glandular at its base ; rudimentary ovary cylindric, glabrous ; female sepals rather narrower than the male, with a 3-lobate basal scale ; petals narrower than the male, eglandular ; ovary glabrous ; styles free, 3-fid : capsule ovoid, glabrous, puncticulate and slightly warted ; seeds black, shining. *Müll. Arg. in DC. Prodr.* xv. ii. 1050, *incl. var.* β; *Pax in Engl. Pflanzenr. Euphorb. Cluyt.* 57, *incl. var.* β; *Prain in Kew Bulletin*, 1913, 406. *C. hirsuta, Eckl. & Zeyh. ex Sond. l.c.*; *not of E. Meyer. C. pubescens, Eckl. & Zeyh., partly, ex Sond. l.c.*; *not of Thunb., nor of Willd. C. phyllanthifolia, Baill. Adansonia*, iii. 153. *C. retusa, Thunb. ex Prain in Kew Bulletin*, 1913, 406; *not of Linn.*

SOUTH AFRICA: without locality, *Thunberg*! *Krebs*!

COAST REGION: Stellenbosch Div.; Hottentots Holland Mountains, *Ecklon*! Swellendam Div.; near Swellendam, *Roxburgh*! *Niven*! *Mund & Maire*, 108! 311! *Ecklon & Zeyher*, 43! George Div.; near George, *Burchell*, 6042! *Bowie*! *Prior*! Zwart River, *Penther*, 876! Oakford, *Rehmann*, 559! Montagu Pass, 1200 ft., *Young in Herb. Bolus*, 5533! *Rehmann*, 168! Uniondale Div.; Long Kloof, *Ecklon & Zeyher*! Humansdorp Div.; Humansdorp, *Rogers*, 2922! 2965! Uitenhage Div.; Van Stadens Berg, *Zeyher*! *Ecklon & Zeyher*! Zwartkops River, *Zeyher*, 3828! Port Elizabeth Div.; Port Elizabeth, *Laidley*, 470! Albany Div.; near Grahamstown, 2000 ft., *Bolton*! *MacOwan*, 1! *Galpin*, 31! *Hutton*! *Cooper*, 19! *Drège*, 8226 c! *Ecklon*, 884! *Atherstone*, 478! *Williamson*! *Schönland*, 585! *Misses Daly & Sole*, 262! Bluekrantz, *Burchell*, 3640! Assegai Bush, *Baur*! Bedford Div.; Bedford, *Weale*, 14! Stockenstrom Div.; Kat Berg, 3000–4000 ft., *Drège*, 8226 a! Queenstown Div.; near Queenstown, *Cooper*, 3141 bis! Stutterheim Div.; Fort Cunynghame, *Sim*, 2177! King Williamstown Div.; Perie Mountains, 2500 ft., *Galpin*, 3266! *Godfrey*, 107! King Williamstown, *Scott Elliot*, 11003! East London Div.; East London, *Rattray*, 60! Komgha Div.; Prospect Farm, *Flanagan*, 281 partly! British Kaffraria; without precise locality, *Cooper*, 76! 77!

KALAHARI REGION: Transvaal; Houtbosch, 5000 ft., *Rehmann*, 5909! *Bolus*, 10980! Mac Mac, *Mudd*! Burghers Pass, 5000 ft., *Burtt-Davy*, 1561! Lydenburg, *Wilms*, 1315! Spitzkop, *Wilms*, 1130!

EASTERN REGION: Transkei: Kentani, 1000 ft., *Miss Pegler*, 8! Natal; Friedenau, *Rudatis*, 1248! Zululand; Ingoma, *Gerrard*, 1159! Qudeni, *Wood*, 7849!

25. C. natalensis (Bernh. ex Krauss in Flora, 1845, 81); a shrub 2–4 ft. high, much branched; twigs fastigiate, rather slender, sparingly tawny-pubescent, soon glabrous; leaves shortly to distinctly petioled, firmly papery, pellucid-punctate, but not warted, linear- to oblong-lanceolate, acute, base cuneate, margin flat, $\frac{3}{4}$–2 in. long, $\frac{1}{6}$–$\frac{1}{3}$, rarely $\frac{1}{2}$ in. wide, when young pubescent on both sides, soon glabrescent and at times quite glabrous, pale green; petiole $\frac{1}{4}$–$\frac{1}{2}$ in. long; flowers diœcious, yellowish, male in fascicles of 2–6, female solitary or in pairs; pedicels longer than the calyx, pilose, male capillary, female firm, in fruit $\frac{1}{6}$–$\frac{1}{5}$ in. long; male sepals rounded-ovate, pubescent outside, with a 3–5-lobed basal scale; petals about as long as sepals, wide rounded-obovate, eglandular but with 30–50 glands in the fundus of the calyx; rudimentary ovary slender, narrow-cylindric, glabrous; female sepals ovate, subacute, pubescent outside, rather larger than the male sepals, with a usually 2-lobed (less often 1- or 3-lobed) basal scale; petals shorter than the calyx, obovate-oblong, without a basal scale; ovary glabrous; styles free, shortly 2-fid; capsule $\frac{3}{5}$ in. across, warted-punctate; seeds black, shining. *Sond. in Linnæa*, xxii. 127, *incl. var. glabrata*; *Baill. Adansonia*, iii. 150, *incl. var. glabrata*;

Müll. Arg. in DC. Prodr. xv. ii. 1052, *incl. var. glabrata* ; *O. Kuntze*, *Rev. Gen. Pl.* iii. ii. 284 ; *Pax in Engl. Pflanzenr. Euphorb. Cluyt.* 64, *incl. var. glabrata*; *Prain in Kew Bulletin*, 1913, 407. *Cluytia* n. 8225 *and* n. 8226 *b, not a, E. Meyer, in Drège, Zwei Pfl. Documente*, 174.

COAST REGION : Alexandria Div. ; Oliphants Hoek Forest, *Ecklon & Zeyher* ! Albany Div. ; Grahamstown, 2000 ft., *MacOwan* ! Queenstown Div. ; Klippaat River, near Shiloh, *Drège*, 8225 ! Klaas Smits River, *Baur*, 57 ! Zwartkei River, *Cooper*, 262 ! 263 ! *Galpin*, 2684 !
CENTRAL REGION : Aliwal North Div. ; by the Orange River near Aliwal North, 4300 ft., *Drège*, 8226 b !
KALAHARI REGION : Orange River Colony ; Caledon River, *Burke* ! *Zeyher*, 1512 ! *Wahlberg* ! Harrismith, *Sankey*, 234 ! Bethlehem, *Richardson* ! *Flanagan*, 2110 ! Besters Vlei near Witzies Hoek, *Bolus*, 8247 ! Ladybrand, *Rogers*, 820 ! Basutoland ; Leribe, *Dieterlin*, 321 ! Transvaal ; *Sanderson* ! *Rehmann*, 3946 ! Lydenburg Dist. ; near Lydenburg, *Wilms*, 1316 mainly ! 1317 ! 1318 ! Waterfall River, *Wilms*, 1317 a ! Standarton, *Rehmann*, 6790 ! at Maquabie near Amersfoot, *Burtt-Davy*, 4111 ! Crocodile Valley, *Burtt-Davy*, 7639 ! Carolina, *Rademacher*, 7477 ! Johannesburg ; near Roodepoort, *Rand*, 985 !
EASTERN REGION : Tembuland ; Cala, 4000 ft., *Miss Pegler*, 1462 ! Griqualand East ; Banks of the Umzimkulu River near Handcocks Drift, 2500 ft., *Tyson*, 2790 ! *and in MacOwan & Bolus, Herb. Norm. Austr.-Afr.* 766 ! Tsitsa River, 3400 ft., *Krook*, 881 ! *Schlechter*, 6371 ! 6379 ! Natal ; Table Mountain, *Krauss* ! Howick, *Junod*, 301 ! Van Reenens Pass, 5500–6000 ft., *Kuntze* ! *Miss Franks in Herb. Wood*, 12198 ! near Currys Post, 4000 ft., *Wood*, 3569 ! Dumisa, *Rudatis*, 677 ! banks of the Tugela River, 3000–4000 ft., *Wood* ! near Acton Homes, 3000–4000 ft., *Wood*, 3584 ! near Mooi River, 4000–5000 ft., *Wood*, 6183 ! near New-castle, 3000–4000 ft., *Wood*, 8615 ! Klip River, *Gerrard*, 788 ! and without precise locality, *Sanderson*, 71 ! *Fraser* ! *Gerrard*, 1159 ! Zululand ; Entumeni, 2000 ft., *Wood*, 3730 !

26. C. platyphylla (Pax & K. Hoffm. in Engl. Pflanzenr. Euphorb. Cluyt. 74); a small shrub ; stems 1 ft. high or less, fastigiately branched ; twigs angular, tawny pubescent at first as are the young leaves, at length almost glabrous ; leaves shortly petioled, close-set, ovate or suborbicular, obtuse, base rounded or subcordate, margin slightly recurved, $\frac{3}{4}$–2 in. long, $\frac{2}{3}$–1 in. wide, firmly papery to subcoriaceous, when young faintly pellucid-punctate, not warted ; midrib and main-nerves beneath very distinct ; flowers diœcious, greenish-white, male 1–3, female not seen ; pedicels up to $\frac{1}{4}$ in. long, capillary, pubescent ; male sepals rounded, pubescent outside, obtuse, with a 4–5-lobate basal scale ; petals rhomboid-orbicular, shortly clawed, glabrous, eglandular, but each with 6–8 free glands attached to fundus of calyx within their point of origin ; rudimentary ovary glabrous, cylindric, expanded at the tip. *Prain in Kew Bulletin*, 1913, 407.

EASTERN REGION : Natal ; Dumisa, Fairfield, 2500 ft., *Rudatis*, 81 !

27. C. dregeana (Scheele in Linnæa, xxv. 583); a shrub, 3–8 ft. high, much branched ; branches spreading ; twigs distinctly angular, at first finely pubescent, at length glabrous or nearly so ; leaves sessile or nearly so, from ovate to lanceolate, shortly acuminate or

acute, less often obtuse, base rounded or wide-cuneate, margin
slightly recurved, $\frac{2}{3}$–$\frac{3}{4}$ (rarely 1–1$\frac{1}{2}$) in. long, $\frac{1}{4}$–$\frac{1}{2}$ (rarely $\frac{2}{3}$) in. wide,
subcoriaceous or coriaceous; when young faintly pellucid-punctate
and with pellucid veins, when mature opaque or only with the veins
faintly pellucid, not warted; midrib rather distinct, nerves obscure;
at first sparingly pubescent but soon nearly to quite glabrous;
flowers dioecious, greenish-white, male 1–3, female usually solitary;
pedicels up to $\frac{1}{4}$ in. long, at first pubescent, male capillary, female
rigid, stouter, angled; male sepals ovate-oblong, obtuse, punctate
but not warted, with a 3–5-lobate basal scale; petals cuneate-
obovate, shortly clawed, base 2–3-glandular; rudimentary ovary
cylindric, glabrous; female sepals like the male but rather larger;
petals as in the male but with base 1-glandular; ovary glabrous;
styles distinctly connate at the base, 2-fid above; capsule $\frac{1}{6}$ in. wide,
subglobose, distinctly warted-punctate. *Baill. Adansonia*, iii. 153;
Prain in Kew Bulletin, 1913, 408. *C. heterophylla, Sond. in Linnæa*,
xxiii. 128, *mainly, but excl. var. hirsuta and syn. Bernh.*; *not of
Thunb. C. sonderiana, Müll. Arg. in DC. Prodr.* xv. ii. 1051, *incl.
both vars*; *Pax in Engl. Pflanzenr. Euphorb. Cluyt.* 72, *incl. all three
vars., but excl. Krook*, 915, *and excl. syn. C. heterophylla, Pax in Ann.
Naturhist. Hofmus. Wien*, xv. 49; *not of Thunb., nor of Sond.
C. similis, Pax in Engl. Pflanzenr. l.c.* 66, *partly, and as to Bachmann,*
750 *only*; *not of Müll. Arg.*

COAST REGION : George Div. ; plains of George, *Bowie* ! Uitenhage Div. ; near
Uitenhage, *Krebs*, 298 ! *Drège*, 8229 ! Ylands River, *Ecklon & Zeyher* ! Van
Stadensberg River, *Ecklon & Zeyher*, 34 partly ! Addo, *Ecklon*, 604 ! *Ecklon &
Zeyher*, 46 ! *Zeyher*, 3829 ! Zuur Berg, *Prager*, 104 ! Alexandria Div. ; Salem,
Deacon, 7377 ! *Webster*, 7531 ! Albany Div. ; Grahamstown, *Rogers*, 1547 !
Zwartehoogde, *Ecklon & Zeyher* ! Blauwkrantz near Grahamstown, *Galpin*, 257
mainly ! British Kaffraria ; Bolaaoo, 1500 ft., *Sim*, 1456 !
EASTERN REGION : Transkei ; Kentani, 1000 ft., *Miss Pegler* ! Natal ; Marien
Hill, at the Trappist Colony, *Landauer* ! and without precise locality, *Wahlberg* !
Gerrard, 727 !

28. C. hirsuta (E. Meyer in Drège, Zwei Pfl. Documente, 174 ;
et ex Sond. in Linnæa, xxiii. 129); a shrub, 2–4 ft. high, often
considerably branched; branches ascending, twigs distinctly angular,
sparingly to rather densely persistently pubescent; leaves distinctly
petioled, from ovate to lanceolate, acute or acuminate, hardly ever
obtuse, base narrow- to wide-cuneate, margin slightly recurved,
$\frac{2}{3}$–1 in. long, $\frac{1}{4}$–$\frac{1}{2}$ in. wide, membranous or papery, very distinctly
and persistently pellucid-punctate and with pellucid veins, not
warted; midrib and nerves rather distinct especially beneath,
sparingly to densely persistently pubescent on both surfaces, but
especially beneath; petiole $\frac{1}{6}$ in. long; flowers dioecious, yellowish-
white; male 1–3, female usually solitary; pedicels in both sexes
short, $\frac{1}{8}$ in. long, male capillary, female rigid, stouter and angled, all
persistently pubescent; male sepals ovate-oblong, obtuse, pubescent,
punctate but not warted, with a 3–5-lobate basal scale; petals
cuneate-obovate, glabrous, eglandular, but each with 4–5 free glands

attached to fundus of calyx within their point of origin ; rudimentary
ovary cylindric, pubescent ; female sepals like the male, but with a
2–3-lobed basal scale ; petals obovate, base with or without a gland ;
ovary densely pilose ; styles free, shortly 2-fid ; capsule $\frac{1}{5}$ in. across,
subglobose, pubescent ; seeds black, shining. *Müll. Arg. in DC.
Prodr.* xv. ii. 1046 ; *Pax in Engl. Pflanzenr. Euphorb. Cluyt.* 73 ;
Prain in Kew Bulletin, 1913, 408. *C. affinis, Baill. Adansonia*, iii.
150, *partly and as to syn. E. Meyer only ; not of Sond. C. hetero-
phylla, Pax in Ann. Naturhist. Hofmus. Wien*, xv. 49 ; *not of Thunb.
C. Schlechteri, Pax in Engl. Jahrb.* xxxiv. 373. *C. hybrida, Pax &
K. Hoffm. in Engl. Pflanzenr. l.c.* 60. *C. heterophylla, var. hirsuta,
Sond. in Linnæa,* xxiii. 129. *C. sonderiana, var. pubescens, Pax in
Engl. Pflanzenr. l.c.* 73, *as to Krook,* 915, *only ; not of Müll. Arg.*

VAR. β, **robusta** (Prain, l.c. 409); stem 4–5 ft. high, much stouter, usually
little branched ; leaves firmly papery to subcoriaceous, those of the stem $1\frac{3}{4}$–$1\frac{1}{4}$ in.
long, $\frac{1}{2}$–$\frac{3}{4}$ in. wide, those of the branches $\frac{1}{2}$–$\frac{3}{4}$ in. long, $\frac{1}{4}$–$\frac{1}{3}$ in. wide. *C. dregeana,
Müll. Arg. l.c.* 1051 ; *Pax in Engl. Pflanzenr. l.c.* 74 ; *not of Scheele. C. hirsuta,
O. Kuntze, Rev. Gen. Pl.* iii. ii. 284 ; *hardly of E. Meyer. C. Krookii, Pax in Ann.
Naturhist. Hofmus. Wien,* xv. 49, *and in Engl. Pflanzenr. l.c.* 74, *excl. syn. C.
Schlechteri.*

COAST REGION : Uitenhage Div. ; Zuurberg Range near Bontjes River, 2000 ft.,
Drège, 2310 ! Albany Div. ; Grahamstown, *Bolton* ! *Williamson* ! Bothas Hill,
MacOwan, 497 partly ! Blauwkrantz Bridge, *Galpin,* 257 partly ! Howisons Poort,
Hutton ! King Williamstown Div. ; near the Kei River, *Krook,* 915 ! East London
Div. ; East London, *Rattray,* 689 ! Var. β : Uitenhage Div. ; Addo, *Ecklon* !
Stockenstroom Div. ; Kat Berg, *Hutton* ! British Caffraria ; near Tarka River,
Cooper, 367 ! 368 ! Yellow Woods, 1500 ft., *Sim,* 1454 !
CENTRAL REGION : Somerset Div. ; Bruintjes Hoogte, *Burchell,* 3076/2 ! Var. β :
Somerset Div. ; Bosch Berg, *MacOwan,* 497 partly !
KALAHARI REGION : Orange River Colony ; Harrismith, *Sankey,* 235 ! *Krook,*
933 ; Besters Vley near Witzies Hoek, 5600 ft., *Bolus,* 8250 mainly ! Transvaal ;
Heidelberg, *Rand,* 1216 ! *Miss Leendertz,* 1043 ! *Rogers,* 120 ! near Lake Chrissie,
Scott Elliot, 1580 ! Ermelo, *Miss Leendertz,* 3072 ! *Burtt-Davy,* 5416 !
EASTERN REGION : Transkei ; hills near the Kei River. 1700 ft., *Schlechter,* 6237 !
Tembuland ; Umtata, *Holy Cross Convent,* 236 ! Pondoland ; Dorkis, *Bachmann,*
799 ! Griqualand East ; near Clydesdale, *Tyson,* 2134 (specimens in some cases
monœcious) ! Mafube, *Jacottet,* 337 ! Natal ; Inchanga, *Engler,* 2712 ! 2712 a !
Marloth, 4079 ! Greytown, *Wilms,* 2269 ! Dumisa, 2000–2400 ft., *Rudatis,* 129 !
677 ! Howick, 1000 ft., *Junod,* 242 ! Pinetown, *Rehmann,* 7975 ! Maritzburg,
Rehmann, 7073 ! 7614 ! Krantz Kloof, 1300 ft., *Schlechter,* 3181 ! and without precise locality, *Wahlberg* ! *Mrs. Sanders* ! Var. β : Pondoland ; Insizwa Range,
Krook, 899 ! and without precise locality, *Bachmann,* 1115 ! Natal ; Inanda, 1800 ft.,
Wood, 186 ! 371 ! 5761 ! 11101 ! Ellesmere, 2000 ft., *Rudatis,* 648 ! Charlestown,
6000 ft., *Kuntze* ! Van Reenens Pass, 6000 ft., *Kuntze* ! Coldstream, *Rehmann,*
6880 ! Mountain Prospect, *Rehmann,* 6999 ! Majuba Hill, *Rogers,* 83 !

29. **C. disceptata** (Prain in Kew Bulletin, 1913, 410) ; an undershrub ; stems usually simple, occasionally branched, several from a
woody rootstock, erect, 8 in. to 2 ft. high, somewhat angular above,
terete below, when young thinly adpressed-hairy ; leaves shortly
petioled, when young membranous, soon firmly papery, pellucid-
punctate and sparingly warted, lower often orbicular, usually ovate-
oblong, upper (or sometimes all) ovate-lanceolate, acute, base wide-
to narrow-cuneate, margin slightly recurved, $\frac{3}{4}$–$1\frac{1}{4}$ in. long, $\frac{1}{3}$–1 in.

wide, pale to medium green, when young sparingly hirsute on both sides, soon glabrous or nearly so, distinctly reticulately veined beneath ; petiole $\frac{1}{12}$ in. long below, almost obsolete above, pubescent ; flowers diœcious, greenish-white, male 2–4, female solitary or sometimes in pairs ; pedicels up to $\frac{1}{3}$ in. long in flower, pubescent, male capillary, female rigid, but very slender and often in fruit nearly $\frac{1}{2}$ in. long ; male sepals ovate-oblong, obtuse, punctate but not warted, with a 3–5-lobate basal scale ; petals rounded-ovate, narrowed to a rather wide claw, eglandular, but each with about five free glands attached within their point of origin ; rudimentary ovary subcylindric, glabrous ; female sepals oblong, punctate, with a 2-lobed basal scale ; petals as in the male, eglandular or with a solitary basal gland ; ovary hirsute ; styles distinctly connate at the base, 2-bifid above ; capsule $\frac{1}{8}$ in. wide, less than half as long as the slender pedicel, sparingly setose to quite glabrous, not at all warted. *C. pulchella, Wood in Wood & Evans, Natal Pl. i. 68, t. 84 ; not of Linn. C. heterophylla, Pax in Engl. Pflanzenr. Euphorb. Cluyt. 66 partly and as to syn. Wood ; not of Thunb.*

EASTERN REGION : Griqualand East ; near Kokstad, 4300 ft., *Tyson,* 1114 ! *and in MacOwan & Bolus, Herb. Austr.-Afr.* 1234 ! Natal ; near Durban, *Sanderson,* 661 ! *Gerrard,* 278 ! *Wood,* 38 ! 4944 ! Inanda, *Wood,* 120 ! *Rehmann,* 8407 ! Claremont, *Schlechter,* 2942 ! Marburg, 300 ft., *Rogers,* 536 partly !

30. **O. monticola** (S. Moore in Journ. Linn. Soc. Bot. xl. 197) ; an undershrub ; stems simple, several from a woody rootstock, erect, 6 in. to 2 ft. high, somewhat angular, all parts at all stages quite glabrous ; leaves sessile, when young membranous, soon firmly papery, pellucid-punctate and warted, ovate-oblong to ovate-lanceolate, usually obtuse below and acute above, but at times all obtuse or all acute, base wide- to narrow-cuneate, margin slightly recurved, $\frac{1}{2}$–1$\frac{1}{2}$ in. long, $\frac{1}{3}$–1 in. wide, pale green, quite glabrous, reticulations beneath visible but not raised ; flowers diœcious, greenish-white to yellow ; male 3–9, female solitary ; pedicels of male capillary, up to $\frac{1}{3}$ in. long, glabrous, of female rigid, stout, angled, even in fruit under $\frac{1}{8}$ in. long ; male sepals ovate-oblong, obtuse, warted-punctate, with a 3–5-lobate basal scale ; petals rhomboid, narrowed to a rather wide claw, with about 5 glands adnate to the base of the claw ; rudimentary ovary cylindric, glabrous ; female sepals oblong, warted-punctate, with a 2-lobed basal scale ; petals as in male but eglandular or with a single basal gland ; ovary glabrous ; styles nearly free, 2-fid ; capsule $\frac{1}{4}$ in. wide, longer than the short stout pedicel, glabrous, strongly warted-punctate ; seeds shining, black. *Hutchinson in Dyer, Fl. Trop. Afr.* vi. i. 803 ; *Prain in Kew Bulletin,* 1913, 410. *C. heterophylla, Pax in Engl. Pflanzenr. Euphorb. Cluyt.* 66, partly and as to syn. Schinz only. *Middelbergia transvaalensis, Schinz ex Pax, l.c.*

KALAHARI REGION : Orange Free State ; Harrismith, *Sankey,* 236 ! Besters Vlei, near Witzies Hoek, 5600 ft., *Bolus,* 8250 partly ! Transvaal ; Zoutpansberg Range, *Mudd* ! Eerstelling, *Miss Leendertz,* 885 ! Waterberg Dist. ; Farm Portugal,

Maguire, 6185 ! near Nylstrom, *Burtt-Davy*, 2038 ! 2049 ! Houtbosch, *Rehmann*, 6303 ! near Lydenberg, *Wilms*, 1319 ! 1320 partly ! Waterval Onder, *Frau Middelberg*, 1 ! Waterval Boven, 4800 ft., *Rogers*, 204 ! 629 ! Shilovane, 3500 ft., *Junod*, 1263 ! Saddle Back Range, 3500–4000 ft., *Galpin*, 415 ! Farm Fairview, *Burtt-Davy*, 4077 ! Ermelo, *Tennant*, 6917 !
EASTERN REGION : Natal ; Van Reenen, 5500 ft., *Miss Franks in Herb. Wood*, 12195 ! Vryheid, *Burtt-Davy*, 11420 ! between Pietermaritzburg and Greytown, *Wilms*, 2268 ! Laingsnek, *Rehmann*, 6947 ! Zululand ; Entumeni, 2000 ft., *Wood*, 3730 !

Also in Rhodesia.

31. **C. cordata** (Bernh. ex Krauss in Flora, 1845, 81) ; an undershrub ; stems usually simple, occasionally branched, several from a woody rootstock, erect, 1–2 ft. high, somewhat angular above, terete below, even when young quite glabrous in every part ; leaves shortly petioled, when young membranous, soon firmly papery, pellucid-punctate but not warted, lower orbicular, upper ovate, shortly acuminate, base subcordate, margin slightly recurved, $\frac{3}{4}$–1$\frac{1}{4}$ in. long, $\frac{1}{2}$–1$\frac{1}{4}$ in. wide, pale green, glabrous on both sides, distinctly but not prominently reticulately veined beneath ; petiole $\frac{1}{12}$ in. long below, almost obsolete above ; flowers diœcious, white, male solitary or in pairs, female solitary ; pedicels up to $\frac{1}{4}$ in. long, male capillary, female rigid, stouter, angled ; male sepals ovate-oblong, obtuse, punctate but not warted, with a 2–4-lobate basal scale ; petals rounded-ovate, narrowed to a rather wide claw, rather shorter than the sepals, eglandular but each with about five free glands attached within their point of origin ; rudimentary ovary subcylindric, glabrous ; female sepals oblong, punctate, with a 2-lobed basal scale ; petals as in the male, eglandular ; ovary glabrous ; styles free, shortly 2-fid ; capsule $\frac{1}{5}$ in. across, subglobose, faintly warted-punctate. *Müll. Arg. in DC. Prodr.* xv. ii. 1051 ; *Pax in Engl. Pflanzenr. Euphorb. Cluyt.* 65 ; *Prain in Kew Bulletin*, 1913, 411. *C. heterophylla, Sond. in Linnæa*, xxiii. 128, and *Baill. Adansonia*, iii. 150, *both as to syn. Bernh. only* ; *Pax l.c.* 66, *as to Rehmann*, 7475, *only* ; *not of Thunb.*

EASTERN REGION : Pondoland ; Insizwa Range, *Krook*, 900 ! and without precise locality, *Bachmann*, 1112 ! Natal ; top of the Table Mountain, *Krauss*, 435 ! Durban Flats, *Gerrard*, 728 ! *Sanderson*, 716 ! *Wood* ! Pinetown, *Sanderson*, 911 ! *Schlechter* ! Friedenau, *Rudatis*, 76 ! Dumisa, 2000 ft., *Rudatis*, 401 ! Umgeni Waterfall, *Rehmann*, 7475 ! Inchanga, 2300 ft., *Engler*, 2654 ! Zululand ; Ingoma, *Gerrard*, 1160 !

32. **C. heterophylla** (Thunb. Prodr. 53) ; an undershrub ; stems usually simple, several from a woody rootstock, ascending, $\frac{1}{3}$–1 ft. high, angular, when young usually sparingly adpressed-hirsute, soon glabrous ; leaves shortly petioled, when young membranous, soon firmly papery, pellucid-punctate, when young warted, lower wideovate, upper ovate-lanceolate, acute or very rarely the lowest obtuse, base rounded or sometimes subcordate below and sometimes widecuneate above, margin slightly recurved, $\frac{2}{3}$–$\frac{3}{4}$ in. long, $\frac{1}{5}$–$\frac{1}{2}$ in. wide, pale green, very prominently reticulately veined beneath ; petiole

$\frac{1}{12}$–$\frac{1}{10}$ in. long below, almost obsolete above; flowers diœcious, white, male in 2–4-flowered fascicles, female solitary; pedicels up to $\frac{1}{4}$ in. long, male capillary, female rigid, stouter, angled; male sepals ovate-oblong, obtuse, warted-punctate, with a 3-lobate basal scale; petals rhomboid-ovate, narrowed to a rather wide claw, rather shorter than the sepals, eglandular, but each with about five free glands attached within their point of origin; rudimentary ovary subcylindric, glabrous; female sepals oblong, subacute, warted-punctate, with a 3-lobate basal scale; petals as in male, eglandular; ovary glabrous; styles free, shortly 2-fid; capsule $\frac{1}{8}$ in. across, subglobose, warted-punctate; seeds black, shining. *Willd. Sp. Pl.* iv. ii. 881; *Pers. Syn.* ii. 636; *Poir. Encyc. Suppl.* ii. 303; *Thunb. Fl. Cap. ed. Schult.* 271; *Spreng. Syst.* iii. 49; *Sond. in Linnæa,* xxiii. 128, *in small part and as to Zeyher's Borkhausen plant only, excl. var. hirsuta and syn. Bernh.; Dietr. Syn.* v. 455; *Baill. Adansonia,* iii. 150, *in part only, excl. syn. Bernh.; Müll. Arg. in DC. Prodr.* xv. ii. 1046; *Pax in Engl. Pflanzenr. Euphorb. Cluyt.* 66, *as to syn. Thunb., Willd., Müll. Arg. and Scheele, but excl. plant described; Prain in Kew Bulletin,* 1913, 411. *C. similis, Müll. Arg. l.c.; O. Kuntze, Rev. Gen. Pl.* iii. ii. 284; *Pax l.c. C. dumosa, [Harv. Mss. in sched.]* Cooper *ex Pax l.c., mainly, but excl. Bachmann,* 750. *Phyllanthus vaccinioides, Scheele in Linnæa,* xxv. 585.

COAST REGION : Uitenhage Div. ; near Uitenhage in grassy places, very common, *Thunberg* ! *Mund* ! *Boivin* ! *Drège,* 8221 ! *Prior* ! *Fraser* ! *Schlechter,* 2567 ! Borkhausen, *Zeyher* ! Van Stadens Berg, *Ecklon & Zeyher,* 34 mainly ! *Bolus,* 1665 ! *Mrs. Paterson,* 743 ! Bathurst Div. ; Trapps Valley, *Miss Daly,* 657 ! Albany Div. ; Grahamstown, *Bowie* ! *Williamson* ! *Miss Bowker* ! *Atherstone,* 92 ! Bolton ! *Prior* ! *Miss Daly,* 118 ! Stone's Hill, *Schönland,* 70 ! Fish River Heights, *Hutton* ! Fort Beaufort Div. ; Elands Berg, *Cooper,* 258 ! 259 ! 455 ! Queenstown, 4000 ft., *Galpin,* 1655 ! Cathcart Div. ; Cathcart, *Kuntze* ! East London, *Zeidler* ! *Rattray,* 688 ! *Wood,* 3354 ! Komgha Div. ; near Komgha, *Flanagan,* 815 !

EASTERN REGION : Tembuland ; Bazeia, *Baur,* 146 ! Pondoland ; without precise locality, *Bachmann,* 846 !

33. **C. daphnoides** (Lam. Encyc. ii. 54 [*Clutia*]); a shrub, erect, 4–5 ft. high, much branched; twigs spreading, sharply angular, from shortly adpressed white-pubescent to nearly or quite glabrous; leaves distinctly petioled, firmly membranous, opaque, lanceolate-oblong, obtuse, narrowed from junction of upper and middle third to the petiole, margin flat, 1–1$\frac{1}{2}$ in. long, $\frac{1}{6}$–$\frac{1}{3}$ in. wide, when young densely to sparsely white-pubescent on both surfaces, at length glabrescent; petiole $\frac{1}{8}$–$\frac{1}{4}$ in. long, puberulous; flowers diœcious, yellowish, male usually 2–5, female usually solitary; pedicels $\frac{1}{8}$–$\frac{1}{6}$ in. long, puberulous, male slender, female at length rigid; male sepals obovate-oblong, obtuse, white-pubescent externally, with a 5–7-lobate basal scale; petals rounded-spathulate, glabrous, eglandular, but each with 3–4 glands attached within their point of insertion; rudimentary ovary very short, cylindric, truncate and very slightly dilated at the tip, glabrous; female sepals as in the male, but with a 1–4-lobate basal scale; petals as in the male;

ovary glabrous; capsule glabrous, $\frac{1}{5}$ in. long, smooth ; seeds small, black, shining. *Willd. Hort. Berol.* i. 52, *t.* 52, *excl. syn. Comm., and Sp. Pl.* iv. ii. 880, *excl. syn. Comm. and syn. Thunb.* ; *Pers. Syn.* ii. 636, *excl. syn. Thunb.* ; *Ait. Hort. Kew.* ed. 2, *v.* 422 ; *Spreng. Syst.* iii. 49, *excl. syn. Thunb.* ; *Krauss in Flora,* 1845, 81; *Sond. in Linnæa,* xxiii. 126, *incl. var. incana* ; *Dietr. Syn.* v. 455 ; *Baill. Adansonia,* iii. 150 [*dapnoides*] ; *Müll. Arg. in DC. Prodr.* xv. ii. 1050, *incl. var. glabrata, but excl. var. Thunbergii and syn. Poir.* ; *Pax in Engl. Pflanzenr. Euphorb. Cluyt.* 72, *incl. both vars., but excl. syn. Poir* ; *Prain in Kew Bulletin,* 1913, 412. *C. tomentosa, Thunb. ex Schult. in Fl. Cap. ed. Schult.* 271; *not of Linn., nor of Thunb. Prodr. C. pubescens, Eckl. & Zeyh. in part, ex Sond. l.c.; not of Thunb., nor of Willd. C. pulchella, Sparrm. ex Sond. l.c.* ; *not of Linn. C. hirsuta, Pax in Ann. Naturhist. Hofmus. Wien,* xv. 49; *not of E. Meyer. C. cinerea, Burm. ex Prain in Kew Bulletin,* 1913, 412.

SOUTH AFRICA : cultivated specimens.
COAST REGION : Malmesbury Div. ; between Groene Kloof and Saldanha Bay, *Drège,* 8235 ! Ysterfontein, Darling, *Marloth,* 4040 ! Cape Div. ; near Cape Town, *Burmann* ! *Thunberg* ! *Sparrmann* ! *Sonnerat* ! *Bergius* ! *Krebs* ! *Mund & Maire* ! *Verreaux* ! *Prior* ! *Marloth,* 4415 ! Stellenbosch Div. ; Hottentots Holland Mountains, *Ecklon* ! Riversdale Div. ; mouth of the Duivenhoeks River, *Ecklon* ! near Riversdale, *Rust,* 168 ! *Schlechter,* 1813 ! Tygerfontein, 800 ft., *Galpin,* 4573 ! Corente River Farm, *Muir in Herb. Galpin,* 5333 ! Albertina, *Muir* ! Mossel Bay Div. ; Eastern side of the Gouritz River, *Burchell,* 6424 ! Mossel Bay, *MacOwan* ! George Div. ; near George, *Bowie* ! *Schimper* ! Humansdorp Div. ; Humansdorp, 300 ft., *Galpin,* 4567 ! Uitenhage Div. ; various localities, *Ecklon,* 99 ! 605 ! *Zeyher* ! *Ecklon & Zeyher,* 44 ! 45 ! *Mrs. Paterson,* 766 ! *Prior* ! *Baur,* 1012 ! *Bolus,* 1657 ! *Krauss,* 1712 ! 1713 ! Port Elizabeth Div. ; Port Elizabeth, *Drège fil* ! Bathurst Div. ; Mouth of the Great Fish River, *Burchell,* 3722 ! 3748 ! near Port Alfred, *Burchell,* 3790 ! *Galpin,* 1067 ! 3039 ! *Miss Sole,* 458 ! *Potts,* 200 ! Kowie, *Penther,* 946 ! Albany Div. ; without precise locality, *Hutton* ! *Miss Bowker* ! *Prior* ! King Williamstown Div. ; near King Williamstown, *Kennedy* ! *MacOwan,* 1340 partly ! Komgha Div. ; near Komgha, *Flanagan,* 281 partly ! British Kaffraria ; Kaboussie, *MacOwan,* 1340 partly !

34. **C. vaccinioides** (Prain in Kew Bulletin, 1913, 413) ; a shrub, much branched ; twigs prostrate, 1–2 ft. long, finely pubescent ; leaves coriaceous, sessile or very nearly so, elliptic or oblong, usually obtuse, base wide-cuneate or rounded, margin flat, not imbricate, lowest $\frac{1}{2}$ in. long, $\frac{1}{3}$ in. wide, gradually diminishing upwards, glabrous or with a few hairs along midrib above, and strongly warted, especially when young, on both surfaces, opaque, margin subhyaline ; flowers dioecious, yellowish-white with pink blotches, solitary in the leaf-axils, only male seen ; male pedicels about as long as the calyx, sparingly pubescent ; male sepals sparsely hairy, obovate, slightly retuse, $\frac{1}{6}$ in. long, with a deeply 3-lobed basal scale ; petals distinctly clawed, obcordately notched, shorter than the calyx, glabrous, each with 3 small basal scales at the very base ; rudimentary ovary very short, glabrous. *C. Thunbergii, var. vaccinioides, Pax & K. Hoffm. in Engl. Pflanzenr. Euphorb. Cluyt.* 76.

COAST REGION : Riversdale Div. ; near Riversdale, *Rust,* 619 ! 620 ! Mossel Bay Div. ; between Little Brak River and Hartenbosch, *Burchell,* 6216 !

35. **C. Thunbergii** (Sond. in. Linnæa, xxiii. 130); shrub, erect, 1–5 ft. high, much branched; twigs virgate, cylindric, puberulous or shortly adpressed white-pubescent; leaves shortly petioled, subcoriaceous, opaque, usually obovate, obtuse, narrowed from near the apex to the base, occasionally suborbicular, margin flat, $\frac{1}{3}$–$\frac{1}{2}$ in. long, $\frac{1}{8}$–$\frac{1}{5}$ in. wide, usually densely shortly white-pubescent, occasionally only puberulous, on both the surfaces; petiole $\frac{1}{12}$–$\frac{1}{10}$ in. long; flowers diœcious, white, male solitary or in pairs, female solitary, in both sexes subsessile; male sepals widely obovate-oblong, obtuse, softly shortly velvety externally, with a 3–4-lobate basal scale; petals rounded-spathulate, glabrous, eglandular, but each with about 4 glands attached within their point of insertion; rudimentary ovary very short, cylindric, truncate but not dilated at the tip, glabrous; female sepals as in the male; petals as in the male; ovary glabrous; capsule glabrous, $\frac{1}{6}$ in. long, with warted-punctate valves; seeds small, black, shining. *Baill. Adansonia*, iii. 152; *Pax in Engl. Pflanzenr. Euphorb. Cluyt.* 76, var. *canescens* only; *Prain in Kew Bulletin*, 1913, 414. *C. tomentosa, Thunb. Prodr.* 53; *E. Meyer in Drège, Zwei Pfl. Documente*, 174; *not of Linn. C. pubescens, Willd. Sp. Pl.* iv. ii. 881; *Pers. Syn.* ii. 636; *Poir. Encyc. Suppl.* ii. 303; *Spreng. Syst.* iii. 49; *Dietr. Syn.* v. 455; *not of Thunb. C. daphnoides, var. Thunbergii, Müll. Arg. in DC. Prodr.* xv. ii. 1050. *C. karreensis, Schlechter ex Pax, l.c.*

SOUTH AFRICA : without locality, *Thunberg* ! *Nelson* !
CENTRAL REGION : Beaufort West Div. ; Nieuweveld Mountains, near Beaufort West, 3000–4000 ft., *Drège*, 8230 a ! *Schlechter*, 2112 ! *Bokpoort*, 9500–4500 ft., *Drège*, 8236 b ! between Rhinosterkop and Ganzefontein, *Drège* ! Fraserburg Div. ; near Fraserburg, 4200 ft., *Bolus*, 10403 !
WESTERN REGION : Great Namaqualand ; Great Karasberg Range, Lüdhib Summit, *Pearson*, 7800 ! Little Namaqualand ; between Pedroskloof and Lelie Fontein, *Drège*, 3081 ! near Brackdam, 1800–2000 ft., *Schlechter*, 11110 ! Northeast of Stinkfontein, *Pillans*, 5687 ! near top of Rattelpoort Mountain, *Pearson*, 2965 !

36. **C. crassifolia** (Pax in Bull. Herb. Boiss. vi. 736); a shrub, erect, 4–5 ft. high, much branched; twigs virgate, quite glabrous; leaves shortly petioled, firmly coriaceous, opaque, obovate, obtuse, narrowed from near the apex to the base, margin flat, $\frac{1}{3}$ in. long, $\frac{1}{6}$–$\frac{1}{4}$ in. wide, quite glabrous on both surfaces; petiole $\frac{1}{12}$ in. long; flowers diœcious, white, male in pairs or threes; pedicels slender, half as long as calyx; male sepals rounded-oblong, obtuse, glabrous, with a 3–5-lobate basal scale; petals rounded-oblong, obtuse, spathulate, eglandular, but each with about 4 glands attached within their point of insertion; rudimentary ovary very short, discoid and dilated at the tip, glabrous. *Pax in Engl. Pflanzenr. Euphorb. Cluyt.* 71; *Prain in Kew Bulletin*, 1913, 415.

WESTERN REGION : Great Namaqualand ; Gansberg, 7000–8000 ft., *Fleck*, 465 a !

37. **C. polygonoides** (Linn. Sp. Pl. ed. ii. 1475 [*Clutia*]); an undershrub, 1–2 ft. high; stems several from a woody base, sparingly branched; twigs ascending, glabrous; leaves sessile, coriaceous,

opaque, linear-ovate or linear-elliptic, less often linear, narrowed at
the top to an obtuse tip, rounded or wide-cuneate at the base,
margin thickened, revolute, $\frac{1}{2}$–$\frac{3}{4}$ in. long, $\frac{1}{8}$–$\frac{1}{6}$ in.
wide, quite
glabrous on both surfaces, shining, usually closely imbricate, ascending, slightly convex above, flat except at the margins beneath, at
times very spreading or recurved, and then very convex above, very
concave beneath ; flowers diœcious, yellowish-white, male clustered,
female solitary, in both sexes subsessile from perulate axillary
swellings ; scales few, hyaline ; male sepals obovate, obtuse, glabrous,
with a large 5–7-lobate basal scale ; male petals rounded-ovate,
clawed, eglandular, but each with about 6 glands attached to calyx
within the point of insertion ; rudimentary ovary cylindric, glabrous ;
female sepals elliptic, larger than the male, with a large 5–7-lobate
basal scale ; petals elliptic, eglandular ; ovary glabrous ; styles
connate below, 2-fid above ; capsule globose, $\frac{1}{3}$ in. across, wrinkled,
glabrous ; seeds black, shining. *Burm. f. Prodr. Cap.* 27 *bis* [31] ;
Lam. Encyc. ii. 54 ; *Thunb. Prodr.* 53 *and Fl. Cap. ed. Schult.* 270 ;
Müll. Arg. in DC. Prodr. xv. ii. 1054, *incl. both vars.* ; *Pax in Engl.
Pflanzenr. Euphorb. Cluyt.* 78, *incl. both vars.* ; *Prain in Kew Bulletin,*
1913, 415. *C. curvata, E. Meyer in Drège, Zwei Pfl. Documente,* 174.
C. ericoides, E. Meyer, l.c. partly ; *Krauss in Flora,* 1845, 82 ; *Eckl.
& Zeyh. ex Sond. in Linnæa,* xxiii. 122 ; *not of Thunb. C. diosmoides,
Sond., l.c., incl. var. curvata* ; *Baill. Adansonia,* iii. 151, *incl. var.
curvata. C. tabularis, Eckl. Un. It.* 199 ; *Eckl. & Zeyh. ex Sond. l.c.
C. daphnoides, Eckl. & Zeyh. ex Sond. l.c.* ; *not of Lam.*

COAST REGION : Clanwilliam Div. ; Wupperthal, *Wurmberg* ! *Drège,* 8232 c !
Malmesbury Div. ; near Hopefield, *Bachmann,* 276 ! Tulbagh Div. : Saron,
Schlechter, 3 (var. *curvata*)! Worcester Div. ; Dutoits Kloof, 2000–3000 ft.,
Drège, 8233 a ! Paarl Div. ; Drakenstein Mountains, 4000–5000 ft., *Drège,* 8233 b !
Cape Div. ; Table Mountain, *Burmann* ! *Sparrmann* ! *Bergius* ! *Thunberg* !
Roxburgh ! *Burchell,* 589 ! *Drège,* 8232 a partly ! *Krebs,* 17 ! *Garnot* ! *Cooper,*
3532 ! *Ecklon & Zeyher,* 54 ! 63 ! *Ecklon,* 100 ! 115 partly ! 198 ! 199 ! 200
partly ! *Pappe,* 68 ! *Harvey* ! *Prior* ! *Bolus,* 2940 ! 4585 ! *Frazer* ! *Maire* !
Rehmann, 1395 ! *Dümmer,* 126 ! *Crosfield* ! Devil's Peak, *Harvey* ! *Wolley-Dod,*
1764 ! *Wilms,* 3620 ! near Constantia, *Ecklon* ! *Krauss,* 1715 ! Muizenberg,
Wolley-Dod, 1384 ! Stellenbosch Div. ; Stellenbosch, *Drège,* 8232 b ! Hottentots
Holland Mountains, *Ecklon & Zeyher* ! *Zeyher,* 3827 partly ! *Diels,* 1350 !
Greitjesgat, 2000–3000 ft., *Ecklon & Zeyher,* 55 partly ! Lowrys Pass, *Burchell,*
8230 ! *Penther,* 924 ! Caledon Div. ; Zwartberg, near Caledon, 1000–2000 ft.,
Ecklon & Zeyher (var. *curvata*)! Steenbraas River, *Rogers,* 11008 ! Genadendal,
3000–4000 ft., *Drège,* 8233 c (var. *curvata*)! *Roser* (var. *curvata*)! between
Palmiet River and Lowrys Pass, *Burchell,* 8174 ! Swellendam Div. ; Tradouw
Mountains, *Bowie* ! Voormans Bosch, *Ecklon & Zeyher* ! *Zeyher,* 3827 partly !
Riversdale Div. ; Garcias Pass, 1200 ft., *Burchell,* 6944 ! *Galpin,* 4571 ! Mossel
Bay Div. ; near the Gouritz River, *Ecklon & Zeyher,* 55 partly (var. *curvata*) !
Cathcart Div. ; between Windvogel Mountain and the Zwartkei River, *Ecklon &
Zeyher,* 51 partly !

XXII. CAPERONIA, St. Hil.

Flowers monœcious, rarely diœcious, dichlamydeous ; disc 0.
Male : Calyx closed in bud, splitting into 5 valvate lobes. *Petals*
5, imbricate, often with the two lowest distinctly or much smaller

than the others. *Stamens* usually 10 ; filaments connate below in a column, distinctly 2-seriate above, their free portions spreading ; anthers 2-celled ; cells dehiscing longitudinally, pendulous from the tip of the glandular connective. Rudimentary *ovary* cylindric, crowning the staminal column. *Female : Sepals* 5–6, rarely more numerous, somewhat unequal, imbricate. *Petals* 5, subequal, narrower than in the male. *Ovary* 3-celled ; ovules solitary in each cell ; styles 3, slightly connate below, ovate, deeply laciniate. *Capsule* 3-coccous ; cocci 2-valved ; valves externally covered with flattened or subulate processes mixed with or passing into gland-tipped setæ. *Seeds* nearly globose, ecarunculate ; testa minutely punctate-reticulate ; albumen fleshy ; cotyledons broad, flat.

Erect annual herbs with branching stems ; leaves alternate ; racemes axillary, with numerous male flowers above and a few basal female flowers.

DISTRIB. Species about 12, mainly in tropical South America, a few African.

1. C. Stuhlmanni (Pax in Engl. Jahrb. xix. 81) ; stems branching, rather stout, soft, hispid throughout, 2–3 ft. high ; leaves short-petioled, lanceolate or oblong-lanceolate or linear, acute, base acute, margin sharply serrate, 2–4½ in. long, ¼–1 in. wide, hispid on the nerves, especially beneath ; petiole ${}_{10}^{1}$–½ in. long, hispid ; stipules ovate-lanceolate, acuminate, or subulate, caducous ; racemes 2–2½ in. long ; rhachis and pedicels hispid ; bracts lanceolate, small ; male sepals 5, ovate, acute, hispid ; petals 5, very unequal, 3 larger spathulate-oblong, larger than the sepals, 2 very small, narrow-oblong, all clawed ; female sepals 5–6, unequal, ovate-lanceolate, acute, hispid, the 2–3 outer slightly shorter than the 3 inner ; petals 5, casually 6, oblong-lanceolate, rather shorter than the outer sepals ; ovary closely beset with narrow-subulate gland-tipped processes ; capsule muricate and setose, ¼ in. across ; seeds deep blue-grey or nearly black, spherical. *Pax in Engl. Pfl. Ost-Afr. C.* 237 ; *Prain in Dyer, Fl. Trop. Afr.* vi. i. 831.

EASTERN REGION : Delagoa Bay; Incanhini, *Schlechter*, 12039 !

Also in Tropical East Africa.

Very nearly allied to *C. palustris*, St. Hil., which is wide-spread in Tropical Africa ; the most satisfactory differential character is in the colour of the seeds, nearly black in *C. Stuhlmanni*, brown or tawny in *C. palustris*.

XXIIA. CHROZOPHORA, Neck.

The genus CHROZOPHORA, Neck., distinguished from CAPERONIA, St. Hil., by its stellate pubescence, by the absence of a rudimentary ovary in the male flower and by its 2-fid but not laciniate styles, has not yet been recorded from within our region. One member, *C. plicata*, A Juss., var. *erecta*, Prain in Kew Bulletin, 1918, 94,

closely allied to var. *obliquifolia*, Prain in Dyer, Fl. Trop. Afr.
vi. i. 835 (*C. obliquifolia*, Baill. Etude Gén. Euphorb. 322) has, however, been met with on the Limpopo north of the Zoutpansberg range and has, more recently, been communicated to Kew by the Transvaal Museum from the Limpopo at Mazambo in Gazaland, a locality very near the northern limit of our area. This species affects sand-banks in rivers and may in time be established in, if it has not already reached, the Limpopo basin south of the Tropic. From the Zambesi basin northwards to Egypt and Palestine *Chrozophora plicata* is a prostrate herb; at Mazambo, as Dr. Breijer, to whom we are indebted for this as yet most southern record of the plant, it assumes the habit of an erect low-growing shrub, 3 ft. high.

XXIII. CEPHALOCROTON, Hochst.

Flowers monœcious; petals 0.   *Male: Calyx* closed in bud, splitting into 3–4 valvate lobes. *Disc* 0. *Stamens* 6–8; filaments free, 2-seriate, inflexed above in bud, but again erect under the anthers; anthers dorsifixed, 2-celled; cells dehiscing longitudinally. Rudimentary *ovary* columnar, short, entire or 2–3-lobed. *Female: sepals* 5–6, elongated, unequal, pinnatifid. *Disc* annular. *Ovary* 3-celled; ovules solitary in each cell; styles connate below in a short column, free and multifid above. *Capsule* 3-coccous; cocci 2-valved. *Seeds* ecarunculate; albumen fleshy; cotyledons broad, flat.

Shrubs, stellately hairy; leaves alternate, 3–5-nerved at the base; stipules small, lacinulate; racemes terminal, androgynous, the male flowers aggregated in an apical peduncled subglobose head, the female flowers basal, long-pedicelled.

DISTRIB. Species about 8, all natives of Africa.

1. **C. depauperatus** (Pax & K. Hoffm. in Engl. Pflanzenr. Euphorb. Adrian. 12); a shrub, freely branching; branches and twigs clothed with harsh scattered stellate pubescence; leaves sessile or very shortly petioled, firmly membranous, spathulate-lanceolate, apex obtuse, base cuneate, margin entire, ¾–1 in. long, ⅓–½ in. wide, at first rather densely clothed with stellate pubescence, but soon becoming glabrous or with only a thin scattered stellate tomentum; stipules very small, spreading; male flower-heads almost globose, ¼ in. across, borne on a slender peduncle ¾–1¼ in. long, the flowers numerous; female flowers at base of peduncle solitary, their pedicel ⅕ in. long; male calyx-lobes ovate-triangular, acute, 1 lin. long, glabrous; stamens 5–6; rudimentary ovary columnar, dilated at the tip; female sepals densely stellate-pubescent, deeply pinnately 4–5-lacinulate on each side; ovary finely pubescent; styles very shortly connate below, much laciniate above, glabrous; capsule closely stellate-pubescent, ¼ in. across.

KALAHARI REGION: Transvaal; Koomati Poort, 1000 ft., *Schlechter*, 11779!

XXIV. ERYTHROCOCCA, Benth.

Flowers diœcious; petals 0. *Male: Calyx* closed in bud, splitting into 3–4 valvate lobes. *Stamens* 2–60, usually intermixed with small glands and sometimes surrounded by a ring of similar free or connate glands; filaments free, longer or shorter than the anthers; anthers erect, 2-celled, cells free from the base, opening by longitudinal extrorse slits. Rudimentary *ovary* 0. *Female: Calyx* 2- (less often 3–4-) partite. *Disc* usually of 2–3 free scales or lobes. *Ovary* 2–3-celled; ovules solitary in each cell; styles usually connate below; stigmas plumosely laciniate, less often smooth. *Capsule* usually 2-coccous, less often 3-coccous or by abortion 1-coccous; cocci subglobose, opening loculicidally; valves coriaceous, subpersistent. *Seeds* subglobose, covered by a thin aril; testa crustaceous; albumen fleshy; cotyledons broad, flat.

Shrubs; twigs slender; buds perulate; bark lenticelled; leaves alternate, stipulate; stipules cartilaginous, often accrescent and modified into weak thorns or spines; flowers small or very small, usually racemose; peduncles slender; pedicels capillary, articulate; male flowers usually several, female flowers usually solitary to each bract.

DISTRIB. Species about 40, all confined to Africa save one, which extends from Abyssinia to Arabia.

Leaves thinly membranous, entire or faintly crenate;
  stamens about 30 ... ... (1) **natalensis.**
Leaves firm, distinctly serrate; stamens 15–18 ... ... (2) **berberidea.**

1. **E. natalensis** (Prain in Kew Bulletin, 1911, 91); a shrub; twigs at first sparingly pubescent, soon glabrous; leaves short-petioled, thinly membranous, ovate-lanceolate, acute or obtuse, base narrow-cuneate, margin distantly glandular-crenate or subentire, 1–1½ in. long, ½–¾ in. wide, glabrous; petiole 1–1½ lin. long; stipules modified into conical thorns 1 lin. long; flowers racemose, rather small; male racemes few-flowered, their peduncle slender, glabrous, ⅓–¾ in. long; pedicels capillary, 1½–2½ lin. long, jointed near the base; bracts minute, each usually 2–3-flowered; calyx whitish, thinly membranous, deeply 4-lobed; lobes ovate, acute; stamens about 30, 10 outer, the rest central, accompanied by many ovate pilose inter-staminal glands and surrounded by a ring of 10 similar extra-staminal glands; filaments longer than the subglobose anther-cells; female flowers not seen. *Prain in Ann. Bot.* xxv. 613.

EASTERN REGION: Natal; Mount Edgecumbe, *Wood*, 1089! Mount Moreland, 500 ft., *Wood*, 1391! and without precise locality, *Gerrard*, 81!

2. **E. berberidea** (Prain in Kew Bulletin, 1911, 92); a shrub; twigs glabrous; leaves short-petioled, firmly papery, ovate-lanceolate or ovate, acute, base wide- to narrow-cuneate, margin regularly and dis-

tinctly serrate, $1\frac{1}{2}$–$2\frac{1}{4}$ in. long, $\frac{2}{3}$–$\frac{3}{4}$ in. wide, glabrous above, sparingly adpressed-hirsute on the nerves beneath ; petiole somewhat pubescent, $1\frac{1}{2}$ lin. long ; stipules modified into conical thorns 1–$1\frac{1}{2}$ lin. long ; flowers racemose, rather small : male racemes few-flowered, their peduncles slender, glabrous, very short ; pedicels capillary, over $\frac{1}{3}$ in. long, jointed near the base ; bracts minute, each several-flowered ; calyx whitish, deeply 4-lobed ; lobes ovate, subacute ; stamens 15–18, 8–10 outer, the rest central, accompanied by many ovate pilose inter-staminal glands and surrounded by a ring of 5 flattened extra-staminal glands ; female bracts each 1-flowered ; female calyx green, 3-lobed ; lobes ovate, subacute, their margins ciliate ; ovary glabrous, 3-celled ; styles 3, free to the base, laciniate throughout ; disc-scales 3, free, triangular, thin ; capsule usually 2-coccous or 1-coccous, $\frac{1}{3}$ in. wide ; seeds subspherical. *Prain in Ann. Bot.* xxv. 613.

EASTERN REGION: Natal ; near Durban, 100–200 ft., *Wood,* 7582 ! 9216 ! 9439 ! 11810 ! *Gerrard,* 549 ! Mount Edgecombe, 200 ft., *Wood,* 11593 !

XXV. MICROCOCCA, Benth.

Flowers dioecious or monœcious ; petals 0. *Male : Calyx* closed in bud, splitting into 3 valvate lobes. *Stamens* 3–30 ; filaments free, associated or not with minute inter-staminal glands ; anthers 2-celled ; cells basifixed, erect, free except at the base, dehiscing extrorsely. Rudimentary *ovary* 0. *Female : Calyx* 3–4-partite ; lobes imbricate. *Ovary* 3-celled, very rarely 4-celled ; ovules solitary in each cell ; styles 3, rarely 4, free, plumose-laciniate throughout. *Disc* composed of linear or flattened hypogynous scales alternate with the carpels. *Capsule* 3-dymous, rarely 4-dymous, breaking up both septicidally and loculicidally into 3, rarely 4, 2-valved cocci ; valves thinly crustaceous. *Seeds* globose with a thin aril and a crustaceous foveolate testa ; albumen fleshy ; cotyledons usually broad, flat.

Shrubs or herbs ; leaves alternate ; stipules very small ; racemes axillary, usually 1-sexual, occasionally androgynous ; male flowers glomerulate, occasionally with a central female flower ; rarely the racemes mainly male but with a solitary terminal female flower ; bracts small.

DISTRIB. Species 9, three confined to Africa, two to the Mascareue Islands and three in South-East Asia, with one wide-spread in all three regions.

1. **M. capensis** (Prain in Ann. Bot. xxv. 630) ; a dioecious shrub, 10–15 ft. high, rather sparingly branched ; twigs glabrous ; leaves long-petioled, membranous, elliptic-lanceolate, acute or shortly acutely acuminate, base cuneate, margin sharply serrate, 4–6 in. long, $1\frac{1}{4}$–2 in. wide, glabrous ; petiole $\frac{3}{4}$–$1\frac{1}{4}$ in. long, glabrous ; stipules lanceolate, adpressed-pubescent, small, caducous ; racemes axillary, male 3–6 in. long, female 2–4 in. long ; male flowers 6–20

to each bract; female flowers solitary to their bracts, glomerules occasionally shortly peduncled; male pedicels ⅛ in. long or less, female pedicels ⅕–¼ in. long; bracts minute, ovate-lanceolate; male calyx membranous, 3-lobed; stamens about 15; receptacular glands 0; female calyx 3-partite; lobes imbricate, their margins sparingly pilose; ovary glabrous, 3-celled; capsule thinly crustaceous. *Claoxylon capense, Baill. Étud. Gén. Euphorb.* 493, *and in Adansonia,* iii. 161; *Müll. Arg. in DC. Prodr.* xv. ii. 786; *Sim, For. Fl. Cape Col.* 317, *and For. Fl. Port. E. Afr.* 105.

EASTERN REGION: Pondoland; near the Umsikaba River, *Drège,* 4636! Isinuka, near mouth of St. John's River, 100 ft., *Flanagan,* 2502! *Bolus,* 10283! Egossa, *Sim,* 2415! and without precise locality, *Bachmann,* 805! Natal; without precise locality, *Gerrard,* 1179! Zululand; Ngoya, 1000–2000 ft., *Wylie in Herb. Wood,* 7905! *Wood,* 11570! Entumeni, 1500 ft., *Wood,* 3979! Delagoa Bay; Lourenço Marques, *Sim;* Maputa, *Sim.*

XXVI. MERCURIALIS, Linn.

Flowers diœcious; petals 0. *Male: Calyx* closed in bud, splitting into 3 valvate lobes. *Stamens* 8–20; filaments free, attached to a small receptacle; anther-cells globose or ovoid, divaricate or pendulous, ultimately ascending, opening longitudinally above. *Disc* 0. Rudimentary *ovary* 0. *Female: Calyx* 3-sect to the base. *Ovary* 2-celled, rarely 3-celled; hypogynous scales alternate with the cells, linear-subulate; ovules solitary in each cell; styles free almost to the base, short, erect or spreading, papillose, undivided. *Capsule* 2- (very rarely 3-) dymous, breaking up into 2-valved cocci; endocarp crustaceous. *Seeds* ovoid or globose, smooth or rough, testa crustaceous; albumen fleshy; cotyledons broad, flat.

Annual or perennial herbs, glabrous or pubescent; leaves opposite; often toothed, penninerved; racemes axillary, male slender, the flowers in sparse distant clusters on the distal half of the rhachis, female usually very short with 1–2 subsessile axillary flowers

DISTRIB. Species 6, all save one European; one of the European species is an introduced weed in South Africa.

1. **M. annua** (Linn. Sp. Pl. ed. i. 1035); an annual usually diœcious herb; stem ½–1½ ft. high; leaves short-petioled, membranous, opposite, ovate to ovate-lanceolate, acute, base rounded or cordate, margin crenate-serrate, 1½–2 in. long, ¾–1¼ in. wide, shortly ciliate on the margins, elsewhere glabrous on both surfaces; petiole ⅕–¼ in. long, glabrous; stipules lanceolate; male racemes up to 2 in. long; male bracts ovate, acute; female clusters sometimes with male flowers intermixed; styles simple, diverging; capsule small, tubercled, hispid, 2-coccous; seeds brown, reticulate. *Burm. f. Prodr.* 27 *bis* [31]; *Thunb. Prodr.* 78, *and Fl. Cap. ed. Schult.* 387,

as to description and partly as to specimens, but excl. loc. Outeniqua ;
Baill. *Adansonia,* iii. 158 ; *Müll. Arg. in DC. Prodr.* xv. ii. 797 ;
Prain in Ann. *Bot.* xxvii. 397.

COAST REGION : Cape Div. ; near Capetown, *Oldenland* ; *Thunberg* ! *Lehmann* !
Schlechter, 1364 ! Tulbagh Div. ; near Tulbagh, *Kässner,* 1287 !

The earliest record of this introduced species having been collected at the Cape
is that of Burmann in 1768 who had seen specimens in Oldenland's herbarium.
The next record is that by Thunberg in 1794. Müller has ignored Burmann's
statement and, owing to Thunberg having in his herbarium mixed up *Leidesia
capensis* with *M. annua,* has transferred *M. annua,* Thunb. non Linn., as a
synonym, to *L. capensis ;* Baillon's statement has also been left unnoticed by
Müller. Since 1866, when Müller wrote, Schlechter and Kässner have both
proved that *M. annua,* Linn., is actually present as a weed in cultivated ground
in South Africa. The doubt which Müller's action has thrown upon the judgment
of Burmann and Thunberg and Baillon is thereby removed.

XXVII. LEIDESIA, Müll. Arg.

Flowers monœcious ; petals 0 ; disc 0. *Male : Calyx* closed in
bud, splitting into 3 valvate lobes. *Stamens* 3–7 ; filaments short,
slender, occasionally connate below ; anther-cells globose, free from
the base, at length spreading, 2-valved. Rudimentary *ovary* 0.
Female : Calyx reduced to a single short narrow bract, or obsolete.
Ovary 2-celled ; ovules solitary in each cell ; styles 2, free, linear,
undivided. *Capsule* 2-dymous or occasionally by abortion 1-celled,
globose ; cocci 2-valved, endocarp thinly crustaceous. *Seeds* subglobose ; testa crustaceous ; albumen fleshy ; cotyledons broad,
flat.

Delicate annual branching herbs ; leaves alternate or at the branching nodes
nearly opposite, ovate or orbicular, very thinly membranous, wide-toothed or
entire ; racemes terminal or in dichasia, rhachis filiform ; male flowers minute, in
numerous fascicles towards the apex of the raceme ; female flowers few towards
the base of the rhachis, subtended by a leafy bract ; bracts and male calyx-
segments usually setose ; capsules small, hispid.

DISTRIB. Three endemic species.

Stems firm, woody at the base ; leaves narrow, thrice as
    long as broad, closely minutely crenate   ...   ... (1) **firmula.**
Stems succulent throughout ; leaves nearly as broad as
    long, wide-crenate :
    Leaves 4–7-toothed on each side   ...   . (2) **capensis.**
    Leaves 1–3-toothed on each side   ...   . (3) **obtusa.**

1. **L. firmula** (Prain in Kew Bulletin, 1912, 337) ; a rigid herb ;
stems yellowish-green, glabrous, erect, copiously intricately branched,
6 in. high ; leaves short-petioled, membranous, ovate-lanceolate or
lanceolate, acute, base narrow-cuneate, margin closely crenate, $\frac{1}{2}$–$\frac{3}{4}$ in.
long, $\frac{1}{6}$–$\frac{1}{4}$ in. wide, finely pubescent on the margins, elsewhere
glabrous on both surfaces, not puncticulate : petiole $\frac{1}{3}$–$\frac{1}{4}$ in. long,

glabrous ; stipules lanceolate ; racemes ½–¾ in. long ; male bracts ovate, acute, $\frac{1}{10}$ in. long, hispid ; male calyx 3-partite, lobes ovate, acute, sparingly hispid outside ; stamens usually 3, sometimes 4, capsule 2-coccous ; cocci hispidulous, 1 lin. long. *Prain in Ann. Bot.* xxvii. 400.

WESTERN REGION : Great Namaqualand ; Gamokab, *Schinz,* 898 ! Karukab, *Schinz,* 899 ! Groot Fontein, *Dinter,* 700 !

A very distinct species, in habit and consistence of leaves and stems bearing to the remaining members of the genus exactly the relationship which most of the species of *Adenocline* bear to *A. procumbens,* Benth. It differs most strikingly from the remaining two species in having pedicelled female flowers.

2. **L. capensis** (Müll. Arg. in DC. Prodr. xv. ii. 793, excl. syn. *Urtica capensis,* Linn. f. and Thunb.) ; a succulent herb ; stem green, glabrous, suberect or diffuse, 4–12 in. long ; leaves long-petioled, very thinly membranous, deltoid-ovate, subacute or obtuse, base truncate or shortly wide-cuneate, margin shallow-crenate, lobes 4 or more a side, ¾–2 in. long, ½–1⅓ in. wide, puncticulate, very sparingly pubescent above, glabrous or nearly so beneath ; petiole ¾–1½ in. long, glabrous ; stipules lanceolate ; racemes 1–2 in. long ; male bracts triangular-ovate, subacute, ⅛ in. long, hispid ; male calyx 3-partite, lobes ovate, acute, sparingly hispid outside ; stamens usually 6–7 ; capsule 2-coccous ; cocci hispidulous, over 1 lin. across. *Benth. in Hook. Ic. Pl.* xiii. 66, *partly and excl. t.* 1284 ; *Pax in Engl. & Prantl, Pflanzenfam.* iii. v. 50, *fig.* 31 *A, B. L. son-deriana, Müll. Arg. l.c.* 699 (*name only*). *L. procumbens, Prain in Ann. Bot.* xxvii. 400. *Mercurialis procumbens, Linn. Sp. Pl. ed.* i. 1036. *M. androgyna, Steud. Nomencl. ed.* i. 524. *M. annua, Thunb. Fl. Cap. ed. Schult.* 387, *partly (Outeniqua spec. only); not of Linn. M. tricocca, E. Meyer in Drège, Zwei Pfl. Documente,* 201 *partly (d, e, only). M. capensis, Spreng. ex Eckl. & Zeyh. in Linnæa,* xx. 213 (*name only*) ; *Sond. in Linnæa,* xxiii. 112, *partly, and excl. syn. Linn. f., and Thunb.* ; *Baill. Adansonia,* iii. 158, *partly, and excl. syn. Linn. and Lehm. Croton Ricinocarpos, Linn. Sp. Pl. ed.* ii. 1427 ; *Aubl. Hist. Pl. Guy.* ii. 883 ; *Willd. Sp. Pl.* iv. 551 ; *Geisel. Crot. Monogr.* 66 ; *Spreng. Syst.* iii. 877, *as to descript. and syn. M. androgyna, excl. syn. Boerhaave and loc. Surinam. Urtica capensis, Eckl. Un. It. n.* 814 ; *not of Linn. f. Adenocline Mercurialis, Baill. Étud. Gén. Euphorb.* 457, *partly ; not of Turcz.*

SOUTH AFRICA : without locality, *Herb. Swartz* ! *and cultivated specimens* ! COAST REGION : Cape Div. ; Table Mountain, *Ecklon,* 814 ! *Zeyher,* 3844 ! *Lehmann* ! *Masson* ! *Wright,* 426 ! Devils Mountain, 1200 ft., *Ecklon* ! *Harvey,* 504 ! *Bolus,* 2941 ! *Wilms,* 3623 ! above Overige Kloof, 2800 ft., *Schlechter,* 408 ! above Groote Schuur, *Wolley-Dod,* 607 ! George Div. ; Outeniqua Mountains, *Thunberg* ! Roode Muur, *Drège* (*M. tricocca* d) ! near George, *Burchell,* 5847 ! *Rogers* ! Uniondale Div. ; Longkloof, *Dümmer,* 1376 ! Knysna Div. ; Yzer Nek, *Burchell,* 5247 ! Karratera River, *Drège* (*M. tricocca* e) ! Albany Div. ; without precise locality, *Bowie,* 18 ! KALAHARI REGION : Transvaal ; Houtbosh, 6500 ft., *Rehmann,* 5923 ! *Schlechter,* 4427 !

EASTERN REGION : Tembuland ; between Cala and Ugie, 5000 ft., *Bolus*, 10284 !
Griqualand East ; Malowe Forest, *Tyson*, 2118 ! Natal ; Ismont, 2000 ft., *Wood*,
1867 !

This is *Mercurialis procumbens*, Linn., and *M. androgyna*, Steud. ; it is not
Urtica capensis, Linn. f. (*Acalypha decumbens*, Thunb., var. *villosa*, Müll. Arg.),
nor is it *U. capensis*, Thunb. (*Australina capensis*, Wedd., and *Droguetia Thunbergii*, N. E. Br., mixed).

3. **L. obtusa** (Müll. Arg. in DC. Prodr. xv. ii. 793) ; a succulent
herb ; stems green, glabrous, suberect or diffuse, 4–12 in. long ;
leaves long-petioled, very thinly membranous, orbicular-ovate,
obtuse, base wide-cuneate, margin crenate, lobes shallow, 3 or
fewer on each side, $\frac{1}{3}$–1 in. long, $\frac{1}{4}$–$\frac{3}{4}$ in. wide, puncticulate, very
sparingly pubescent above, glabrous or nearly so beneath ; petiole
$\frac{1}{2}$–1 in. long, glabrous, stipules lanceolate ; racemes $\frac{3}{4}$–1 in. long ;
male bracts ovate, obtuse, $\frac{1}{8}$ in. long, hispid ; male calyx 3-partite,
lobes ovate, acute, sparingly hispid outside ; stamens usually 4–5 ;
capsule 2-coccous ; cocci hispidulous, under 1 lin. long. *Prain in
Ann. Bot.* xxvii. 401. *L. capensis, Benth. in Hook. Ic. Pl. t.* 1284 ;
not of Müll. Arg. Acalypha obtusa, Thunb. Fl. Cap. ed. Schult. 546 ;
Lehm. ex Baill. Adansonia, iii. 159. *A. obtusata, Spreng. ex Steud.
Nomencl. ed.* 2, i. 10, *mainly. Mercurialis tricocca, E. Meyer in
Drège, Zwei Pfl. Documente,* 201, *partly* (a *partly,* b *partly*); *Sond.
in Linnæa,* xxiii. 111, *partly. M. capensis, Baill. Adansonia,* iii.
158, *as to syn. Lehm. only* ; *Spreng. in Herb. Berol. ex Müll. Arg. l.c.* :
not of Eckl. & Zeyh. in Linnæa, xx. 213. *Adenocline Mercurialis,
Baill. Étud. Gén. Euphorb.* 457, *partly* ; *not of Turcz.*

SOUTH AFRICA : without locality, *Thunberg* ! *Lehmann* !
COAST REGION : Cape Div. ; shore near Smitswinkel Bay, *Wolley-Dod*, 3302 !
Uitenhage Div. ; Zuurberg Range, *Drège* ! Zwartkops River, *Ecklon & Zeyher*, 35 !
near Uitenhage, *Prior* ! Port Elizabeth Div. ; Baakens River Valley, *Mrs. Paterson*,
841 ! Alexandria Div. ; Zuurberg, *Drège* (*M. tricocca* b partly) ! Bathurst Div. ;
near Barville Park, *Burchell*, 4091 ! Albany Div. ; Dassie Krantz, *Rogers*, 3961 !
CENTRAL REGION : Willowmore Div. ; Karroo between the Great Zwart Bergen
and Aasvogel Berg, 2000 ft., *Drège* (*M. tricocca* a in Herb. Kew) ! Somerset Div. ;
between the Zuurberg Range and Bruintjes Hoek, *Drège* (*M. tricocca* b partly);
woods at foot of Boschberg, 3000 ft., *MacOwan*, 1752 ! and without precise locality,
Miss Bowker !

L. obtusa, Müll. Arg., is so nearly allied to *L. capensis,* Müll. Arg., that it
may prove on closer field-study to be an ecological state of the latter species
within which it has, so far as mere description goes, been included by Sprengel,
Sonder, Baillon, Bentham and others. The two are, however, so distinct as
regards facies that *L. obtusa* has hardly ever been distributed as *L. capensis;*
the practice has been to issue it as *Adenocline procumbens,* from which it
differs widely in floral characters, but which it resembles in general
appearance.

XXVIII. SEIDELIA, Baill.

Flowers monœcious ; petals 0 ; disc 0. *Male : Calyx* closed in
bud, splitting into 3 valvate lobes. *Stamens* usually 3, less often 2,
central ; filaments shortly connate below ; anther-cells deeply

grooved and 2-globose in flower, opening apically and at length cruciately 4-valved. Rudimentary *ovary* 0. *Female : Calyx* short, deeply 3-lobed, explanate under the flower. *Ovary* 2-celled; ovules solitary in each cell; styles 2, short, slender, undivided, recurved. *Capsule* 2-dymous; cocci 2-valved, pericarp membranous. *Seeds* ovoid; testa crustaceous; albumen fleshy; cotyledons ovate, only twice as broad as the radicle.

Small annual, glabrous herbs; leaves alternate, narrow; flowers small, in glomerules or cymules at the ends of the branches or in the upper leaf-axils, shortly pedicelled; the upper densely clustered all male, those below female.

DISTRIB. Two endemic species.

Leaves ovate-oblong, margin crenate throughout, distinctly petioled... ... ... (1) **pumila.**

Leaves linear-lanceolate or linear, margin entire or 1-2-toothed on each side, sessile or nearly so ... ... (2) **Mercurialis.**

1. S. pumila (Baill. Étud. Gén. Euphorb. 466); a small annual herb; stems 2–5 in. high, erect, much intricately branched; leaves distinctly petioled, ovate-oblong, obtuse, base cuneate, margin crenate throughout, 3–5 lin. long, 2–3 lin. wide, glabrous on both surfaces; petiole 1½–2 lin. long; stipules minute, subulate; flowers axillary in androgynous sessile or very shortly peduncled glomerules; male calyx 3-partite, lobes ovate; stamens usually 3, sometimes 2–1; capsule 2-dymous; cocci 1 lin. long, smooth. *Prain in Ann Bot.* xxvii. 398. *Mercurialis pumila, Sond. in Linnæa,* xxiii. 112; *Baill. Adansonia,* iii. 160. *Tragia triandra, a pumila, Müll. Arg. in DC. Prodr.* xv. ii. 947.

COAST REGION : Uitenhage Div. ; by the Zwartkops River near Amsterdam Flats, *Zeyher,* 3843 !

2. S. Mercurialis (Baill. Étud. Gén. Euphorb. 466, t. 9, fig. 7); a small annual herb; stems 3–4 in. long, branching from the base, prostrate; leaves sessile or subsessile, linear-lanceolate or linear, apex obtuse or subacute, base narrow-cuneate, margin entire or 1–2-toothed on each side near apex, 4–8 lin. long, 1 lin. or less in width; petiole 0–⅓ lin. long; stipules minute, subulate; flowers axillary in androgynous sessile or very shortly peduncled glomerules; male calyx 3-partite; lobes ovate; stamens usually 3, sometimes 2–1; capsule 2-dymous; cocci 1 lin. long, smooth. *S. triandra, Pax in Engl. Jahrb.* x. 35, *and in Engl. & Prantl, Pflanzenfam.* iii. v. 50, *fig.* 31 *C; Prain in Ann. Bot.* xxvii. 39. *Mercurialis triandra, E. Meyer in Linnæa,* iv. 237 *and in Drège, Zwei Pfl. Documente,* 201 ; *Sond. in Linnæa,* xxiii. 113, *excl. syn. Meisn.* ; *Baill. Adansonia,* iii. 160, *excl. syn. Meisn.* *Tragia triandra, β genuina, Müll. Arg. in DC. Prodr.* xv. ii. 947.

CENTRAL REGION : Richmond Div. ; Winterveld near Limoenfontein and Groot Tafelberg, 3000–4000 ft., *Drège,* 796 ! Hanover Div. ; near Hanover, *Sim,* 13 ! KALAHARI REGION : Griqualand West ; Kimberley, *Marloth,* 869 !

XXIX. ACALYPHA, Linn.

Flowers monœcious, very rarely diœcious ; petals 0 ; disc 0. *Male :*
Calyx closed in bud, splitting into 4 valvate lobes. *Stamens* usually
8, attached to a slightly raised receptacle ; filaments free ; anther-
cells distinct, spreading, oblong or linear, usually flexuous, 2-valved.
Rudimentary *ovary* 0. *Female : Calyx* 3–4-partite ; lobes almost
free, imbricate, very small. *Ovary* 3-celled ; ovules solitary in each
cell ; styles free or connate below, laciniate or denticulate, very
rarely entire or merely 2-lobed. *Capsule* 3-celled, slightly 3-lobed,
usually small and breaking up into 2-valved cocci. *Seeds* ellipsoid
or subglobose ; testa crustaceous ; albumen fleshy ; cotyledons broad,
flat.

Herbs, shrubs or trees ; leaves alternate, 3–7-nerved from the base, the central
nerve usually also penninerved ; margin usually toothed ; petiole usually distinct ;
inflorescence various, axillary or terminal or both, androgynous or 1-sexual, when
androgynous the female flowers usually basal, rarely apical, when 1-sexual the
male in axillary spikes below the female, rarely in axillary spikelets above, more
usually in a close-set terminal spike, rarely in a loose terminal panicle ; occasionally
terminal female spikes and axillary male spikes on separate plants or on distinct
branches of same plant ; male flowers very small, glomerate in axils of small
bracts, arranged in amentaceous spikes ; female flowers solitary or paired in the
axil of a toothed or lobed bract which becomes foliaceous in fruit.

DISTRIB. Species about 300, throughout the warmer regions of both hemispheres ;
a few in extra-tropical Africa and America.

Acalypha patens, Müll. Arg. (DC. Prodr. xv. ii. 848), described from a specimen supposed to be South African, is the West Indian *A. chamædrifolia,* Müll. Arg. (l.c. 879).

Spikes normally 2-sexual ; leaves distinctly to long-petioled :
　Shrubs or small trees :
　　Spikes with a perfect terminal female flower and only
　　　male flowers below ; branches spinescent... 　　... (1) **sonderiana.**
　　Spikes with a perfect basal female flower and only male
　　　flowers above ; branches unarmed... 　　... 　　... (2) **glabrata.**
　Herbs, annual ; spikes with several female flowers below
　　and often one or more terminal abortive female
　　flowers beyond the males :
　　Spikes more or less clustered towards tips of branches 　(3) **glomerata.**
　　Spikes uniformly disposed along stems and branches :
　　　Female bracts about 20-toothed ; teeth narrow
　　　　lacinulate ... 　　... 　　... 　　... 　　... (4) **ciliata.**
　　　Female bracts 15- or fewer-toothed :
　　　　Stems simple below ; female bracts triangular
　　　　　dentate ... 　　... 　　... 　　... 　　... (5) **indica.**
　　　　Stems much branched at the base ; female bracts
　　　　　crenulate 　　... 　　... 　　... 　　... (6) **segetalis.**
Spikes normally 1-sexual :
　Leaves distinctly to long-petioled ; plants always monœ-
　　cious :
　　Female flowers in few-flowered spikelets in leaf-axils
　　　above the male spikes, very rarely aggregated in a
　　　large terminal spike ; stems slender, prostrate or
　　　climbing ; female bracts ¼ in. across or larger 　... (7) **decumbens.**

Female flowers always aggregated in a large terminal
spike :
Female bracts under ¼ in. long ; perennials :
   Stems erect ; female bracts densely stipitate-
      glandular but not setose ... ... (8) **senensis.**
   Stems prostrate or decumbent ; female bracts
      ciliate, very sparingly stipitate-glandular ... (9) **petiolaris.**
Female bracts ¼ in. long or longer ; annual ... ... (10) **Eckloni.**

Leaves sessile or subsessile ; plants normally diœcious :
   Stems procumbent :
      Leaves eglandular :
         Leaves glabrous or nearly so ... ... (11) **Zeyheri.**
         Leaves more or less pubescent ... ... (12) **peduncularis.**
      Leaves glandular :
         Leaves glabrous or nearly so ... ... (13) **glandulifolia.**
         Leaves rather densely strigose ... ... (14) **entumenica.**
   Stems erect :
      Leaves not glandular :
         Leaves with entire margins ... ... (15) **depressinervia.**
         Leaves with serrate margins :
            Leaves narrow-lanceolate... ... ... (16) **angustata.**
            Leaves ovate ... ... (17) **caperonioides.**
      Leaves more or less glandular :
         Glands on upper leaves and outside of female
            bracts sessile ... ... (18) **punctata.**
         Glands on upper leaves and outside of female
            bracts stipitate ... ... (19) **Wilmsii**

1. **A. sonderiana** (Müll. Arg. in Linnæa, xxxiv. 9) ; a small tree,
much branched ; branches rigid, leafy branches often ending in stiff
pungent spines ; bark glabrous, lenticelled ; buds perulate ; leaves
membranous, distinctly petioled, oblong-ovate or -elliptic, obtuse,
base rounded, margin bluntly rather wide-crenate, 1–2¼ in. long,
⅔–1 in. wide, glabrous on both surfaces, clustered at the apices of
short lateral twigs along the branches ; twigs densely or loosely
clothed with subimbricating rigid ovate-acute scales ; petiole ¼–¾ in.
long, sparingly pubescent ; stipules small ; inflorescence 2-sexual,
with a solitary terminal jointed female flower and with rather
distant clusters of male flowers downwards ; spikes solitary, geminate
or ternate, mixed with the leaves at the tips of the scaly twigs,
⅔–1¼ in. long ; rhachis white-puberulous ; bracts all minute ; male
several-flowered, bracts ovate, acute ; male buds glabrous, distinctly pedicelled ; female calyx 4–6-partite ; lobes ovate, acute ; ovary distinctly
3-celled, harshly puberulous ; styles 3, free to the base, laciniate ;
seeds globose. *DC. Prodr.* xv. ii. 804 ; *Prain in Kew Bulletin,* 1913,
12. *A. ? petiolaris, Sond. in Linnæa,* xxiii. 117 ; *Walp. Ann.* iii. 367.
Ricinocarpus sonderianus, O. Kuntze, Rev. Gen. Pl. ii. 618.

EASTERN REGION : Natal ; near Durban, *Gueinzius,* 11 ! 510 ! *Gerrard,* 1625 !
A very distinct species.

2 H 2

2. A. glabrata (Thunb. Prodr. 117); a shrub or small tree, 3–20 ft. high, much branched; branches flexible, bark lenticelled; buds perulate; leaves firmly membranous, distinctly petioled, ovate, acute or acuminate, base wide-cuneate or subtruncate, margin crenate or serrate except at the entire base, those produced in dry areas or seasons ½–1 in. long, ⅓–⅔ in. wide, those produced during rains 2–3 in. long, 1½–2½ in. wide, glabrous throughout or sparingly villous on the nerves beneath especially towards the base, but at length glabrous; petiole ½–1¾ in. long, channelled above and there puberulous or pubescent; sepals subulate, spreading, glabrous, persistent, 1 lin. long; inflorescence normally 2-sexual; spikes narrow, sessile, ½–1¼ in. long, with rather close-set clusters of male flowers throughout and with a solitary basal female flower, often the lower spikes with only male flowers and usually the uppermost inflorescences reduced to solitary female flowers; male bracts minute, lanceolate, glabrous, 3–8-flowered; male buds glabrous, shortly pedicelled; female bracts spathaceous, ovate, plicate, 5–7-toothed, pubescent outside, at first 1½ lin., ultimately up to 5 lin. wide; female calyx minute, 3-lobed; lobes ovate; ovary slightly 3-lobed, densely setose; styles 3, very shortly united at the base, free above, at first terete, simple, in their upper half much laciniate; seeds globose. *Thunb. Fl. Cap. ed. Schult.* 545; *Spreng. Syst.* iii. 882; *E. Meyer in Drège, Zwei Pfl. Documente,* 161; *Eckl. & Zeyh. in Linnæa,* xx. 213; *Sim in For. Fl. Cape Col.* 318, *t.* 142, *fig.* 2; *Prain in Kew Bulletin,* 1913, 12. *A. glabrata, var. genuina, Müll. Arg. in Linnæa,* xxxiv. 33, *and in DC. Prodr.* xv. ii. 857. *A. betulina, E. Meyer in Drège, Zwei Pfl. Documente,* 161, *mainly; Eckl. & Zeyh. in Linnæa,* xx. 213; *Sond. in Linnæa,* xxiii. 116; *Baill. Adansonia,* iii. 157; *not of Retz. A. betulina, var. latifolia, Sond. l.c.* 117. *A. glabrata, var. latifolia, Müll. Arg. ll.cc. Ricinocarpus glabratus, O. Kuntze, Rev. Gen. Pl.* ii. 618. *R. glabratus, α genuinus and β latifolius, O. Kuntze, l.c.* iii. ii. 291.

Var. *β*, **pilosior** (Prain in Kew Bulletin, 1913, 15); leaves more or less densely persistently pubescent or velvety beneath. *A. velutina, E. Meyer in Drège, Zwei Pfl. Documente,* 161. *A. betulina, E. Meyer, l.c.* (as to Bathurst sp. only). *A. glabrata, var. pilosa, Pax in Bull. Herb. Boiss.* vi. 733. *Ricinocarpus glabratus, α [genuinus], forma pilosior, O. Kuntze, Rev. Gen. Pl.* iii. ii. 291.

COAST REGION: George Div.; woods near George, 900 ft., *Schlechter,* 2411, partly! Uitenhage Div.; between the Kromme River and Uitenhage, *Zeyher,* 1517 a, partly! Addo, 1000–2000 ft., *Burke!* *Zeyher,* 1517 b! Zwartkops River, *Prior!* Enon, *Drège,* 2332! Port Elizabeth Div.; Krakkakamma Forest *Ecklon & Zeyher,* 72, partly! *Ecklon,* 1124! *Zeyher,* 1517 a, partly! Alexandria Div.; Bushman River, *Thunberg!* *Zeyher,* 1517 c! Bathurst Div.; near the Kowie River, *Ecklon & Zeyher,* 72, partly! Kasuga, *MacOwan,* 715! Port Alfred, 300 ft., *Schlechter,* 2692! *Potts,* 197! Fort Beaufort Div.; near Fort Beaufort, 1000–2000 ft., *Ecklon & Zeyher,* 72, partly! King Williamstown Div.; Perie Forest, *Kuntze,* 6004! East London Div.; East London, *Rattray,* 123! Komgha Div.; Kei Bridge, 560 ft., *Rogers,* 4506! Kei River near Komgha, 600 ft., *Flanagan,* 2318! British Caffraria; without precise locality, *Cooper,* 228! Var. *β*: Bathurst Div.; Fish River near Trumpeter's Drift, *Drège,* a! 4595! King Williamstown Div.; Perie Forest, *Kuntze,* 5466! Komgha Div.; Kei Bridge, 1800 ft., *Flanagan,*

1214! near Kei River, 2000 ft., *Schlechter,* 6250! Prospect Farm, 2100 ft., *Flanagan,* 409!
KALAHARI REGION : Orange River Colony ; between the Orange River and the Caledon River, *Zeyher*! Transvaal ; Zoutspansberg Range, near Goldgedacht, 3700 ft., *Schlechter,* 4602, partly! Blaauw Berg, *Schlechter*! Shilovane, *Junod* 1100! Crocodile River, *Miss Leendertz,* 716! Barberton, *Thorncroft,* 4328! Var. β : Goldgedacht, *Schlechter,* 4602, partly!
EASTERN REGION : Transkei ; Bashee River, *Drège,* b! Kentani, 1000 ft., *Miss Pegler,* 874! Pondoland ; Port St. John, Isnuka, *Galpin,* 3484! between the Umtata River and St. John's River, *Drège,* 4655! Natal ; near Durban, *Gueinzius,* 476! *Gerrard,* 82! 546! *Drège,* c! 4593! 4610! *Wood,* 1715! *Scott Elliot,* 1691! *Rehmann,* 8976! 8977! *Schlechter,* 2931! *Wilms,* 2267! Clairmont, *Engler,* 2518a! 2524! *Kuntze,* 9695! Inanda, *Wood,* 404! 430! Umgeni, *Rehmann,* 8802, partly! Friedenau, 1750 ft., *Rudatis,* 1166! Var. β : Transkei ; Bashee River, *Drège*! Kentani, *Miss Pegler,* 606! 1198! Natal ; Inanda, *Wood,* 1241! Tugela, *Gerrard,* 1623! Colenso, *Rehmann,* 7164!

A very distinct species most nearly allied to *A. fruticosa,* Forsk. (*A. betulina,* Retz.), but readily distinguished by having no glands on the leaves beneath. The variety *latifolia,* recognised by Sonder,'Müller and Kuntze, is not a valid one, but is based on specimens of twigs whose leaves have been produced in humid areas or during wet seasons. The variety termed *pilosior* by Kuntze (1893) and *pilosa* by Pax (1898) is perhaps not wholly natural, for specimens from Natal (*Wood,* 1241) as well as from the Transvaal (*Schlechter,* 4602) indicate that it too passes insensibly into true *A. glabrata.*

3. A. glomerata (Hutchinson in Kew Bulletin, 1911, 229) ; herbaceous, annual ; stems sparingly branched from the base, up to 1½ ft. high, somewhat sulcate, pubescent with short simple reflexed hairs often mixed with larger gland-tipped spreading hairs ; leaves membranous, long-petioled, ovate- or rhomboid-elliptic, obtuse or subacute, base cuneate, margin crenate, 1 2½ in. long, ½ 1½ in. wide, setulose-pubescent on the nerves and veins on both surfaces ; petiole pubescent, ½–2 in. long ; stipules subulate, ₁⁰ₜ in. long, pilose, caducous ; inflorescence 2-sexual, axillary ; spikes usually several to an axil, short, about ¾ in. long, more or less aggregated in clusters towards the ends of the branches, each with 4–10 fairly close-set female flowers below, very few male flowers above and 1–3 densely muricate abortive female flowers at the apex ; rhachis pubescent ; male bracts very minute ; male buds almost glabrous ; female bracts leafy, suborbicular or nearly reniform, margin 8–10-toothed, about ⅛ in. long, ¼ in. wide, pubescent externally, nearly glabrous within, but with several long gland-tipped hairs near the margins of the teeth on the inner and sometimes also on the outer aspect ; sepals 3, ovate-lanceolate, acute, ciliate ; ovary 3-lobed, setose-pubescent and also beset with long gland-tipped hairs on the upper two-thirds ; styles 3, free almost from the base, very slender, rather short ; seeds ovoid-ellipsoid, minutely pitted. *Hutchinson in Dyer, Fl. Trop. Afr.* vi. i. 902 ; *Prain in Kew Bulletin,* 1913, 15. *A. fimbriata, Baill. Adansonia,* i. 272, *partly* ; *not of Schumach., nor of Hochst. A. crenata, var. glandulosa, Müll. Arg. in Linnæa,* xxxiv. 43, *and in DC. Prodr.* xv. ii. 871. *Ricinocarpus crenatus, O. Kuntze, Rev. Gen. Pl.* iii. ii. 291 ; *not of O. Kuntze l.c.* ii. 617.

EASTERN REGION : Delagoa Bay ; Lourenço Marques, *Howard,* 5676!
Also in Eastern Tropical Africa from Bongo southwards.

4. **A. ciliata** (Forsk. Fl. Ægypt.-Arab. 162); herbaceous, annual; stem sparingly branched, up to $2\frac{1}{2}$ ft. high, striate, shortly retrorsely hairy; leaves membranous, long-petioled, ovate-elliptic or rhomboid-ovate, caudate-acuminate, base rounded or faintly cuneate, margin crenate-serrate, $1\frac{1}{2}$–$4\frac{1}{2}$ in. long, $\frac{3}{4}$–2 in. wide, very slightly scabrous on both surfaces and shortly pilose on the margins and main-nerves beneath; petiole 1–$3\frac{1}{2}$ in. long, sulcate, at first faintly pubescent, soon glabrous; stipules filiform, about $1\frac{1}{4}$ lin. long, slightly pubescent; inflorescence 2-sexual, axillary; spikes solitary or 2-nate, $\frac{3}{4}$–1 in. long, with about 10 female flowers below and numerous male flowers above, with usually a solitary densely muricate abortive female flower at the apex; rhachis puberulous; male bracts very minute; male buds faintly puberulous; female bracts suborbicular, $\frac{1}{6}$ in. long, if explanate $\frac{1}{4}$ in. wide, with about 20 laciniate lobules terminating as many main-nerves, sparingly patently ciliate externally and with a pubescent margin; sepals 3, ovate-lanceolate, small, membranous, ciliate; ovary 3-lobed, small, sparingly pubescent on the upper half; styles 3, free to the base, slender, laciniate; seeds ovoid, very minutely pitted. *Vahl, Symb.* i. 77, *t.* 20; *Willd. Sp. Pl.* iv. 522; *Spreng. Syst.* iii. 879; *Benth. in Hook. Niger Fl.* 504; *Baill. Adansonia*, i. 72; *Müll. Arg. in Linnæa*, xxxiv. 44, *and in DC. Prodr.* xv. ii. 873, *incl. var. trichophora, Müll. Arg.*; *Wight & Arn. in Ann. Nat. Hist.* ii. (1839) iii. *t.* 5; *Dalz. & Gibs. Bomb. Fl.* 228; *Hook. f. Fl. Brit. Ind.* v. 417; *Pax in Engl. Pfl. Ost-Afr. C.* 239; *Pax in Baum, Kunene-Samb. Exped.* 283; *Cooke, Fl. Bomb.* ii. 611; *N. E. Br. in Kew Bulletin,* 1909, 141; *Hutchinson in Dyer, Fl. Trop. Afr.* vi. i. 901; *Prain in Kew Bulletin,* 1913, 15. *A. fimbriata, Schumach. & Thonn. Beskr. Guin. Pl.* 409; *A. Rich. Tent. Fl. Abyss.* ii. 245; *Baill. Adansonia,* i. 272. *A. vahliana, Müll. Arg. in Linnæa,* xxxiv. 43, *and in DC. Prodr.* xv. ii. 873; *Oliv. in Trans. Linn. Soc.* xxix. 147, *t.* 96; *Pax in Engl. Pfl. Ost-Afr. C.* 239; *Hiern in Cat. Afr. Pl. Welw.* i. 978; *De Wild. & Durand, Contr. Fl. Congo,* i. 51; ii. 57, *and Reliq. Dewevr.* 211; *De Wild. Miss. É. Laurent,* 131, *and Études Fl. Bas- et Moyen-Congo,* ii. 284. *Ricinocarpus ciliatus and R. vahlianus, O. Kuntze, Rev. Gen. Pl.* ii. 617, 618.

KALAHARI REGION: Transvaal; Shilovane, *Junod*, 1028! 2188!

A widely spread weed in Tropical Africa, Arabia and India which just crosses the northern boundary of our area.

5. **A. indica** (Linn. Sp. Pl. ed. i. 1003); herbaceous, annual; stem simple or sparingly branched, up to 3 ft. high, somewhat sulcate, shortly pubescent; leaves membranous, long-petioled, rhomboid-ovate, ovate-oblong or ovate-lanceolate, acute or subacute, base wide-cuneate or subtruncate, margin serrate, $\frac{3}{4}$–$2\frac{1}{2}$ in. long, $\frac{1}{2}$–$1\frac{1}{2}$ in. wide, sparingly pubescent on the nerves on both surfaces, otherwise glabrous; petiole $\frac{3}{4}$–3 in. long, puberulous; stipules filiform or subulate, $\frac{1}{8}$ in. long, caducous; inflorescence 2-sexual,

axillary ; spikes solitary or geminate, $\frac{3}{4}$–$2\frac{1}{2}$ in. long, with 1–7 rather
remote female flowers below and rather few male flowers above with
usually a solitary densely muricate abortive female flower at the
apex ; rhachis puberulous ; male bracts very minute ; male buds
faintly puberulous ; female bracts 1–2-flowered, leafy, suborbicular,
obtuse, base rounded, margin triangular-dentate, $\frac{1}{6}$–$\frac{1}{4}$ in. long, if
explanate $\frac{1}{4}$–$\frac{1}{3}$ in. wide, sparingly setulose on the margin and nerves
externally, otherwise glabrous ; sepals 3, ovate-triangular, sparingly
ciliate ; ovary deeply 3-lobed, pilose with tubercle-based hairs ;
styles 3, free from the base, short, laciniate ; seeds ovoid, minutely
and closely pitted. *Willd. Sp. Pl.* iv. 523 ; *Bl. Bijdr.* 628 ; *Spreng.
Syst.* iii. 880 ; *Roxb. Fl. Ind.* iii. 675 ; *Wight, Icon. t.* 877 ; *Müll.
Arg. in DC. Prodr.* xv. ii. 868, *excl. var. abortiva* ; *Hook. f. Fl. Brit.
Ind.* v. 416 ; *Penzig in Atti Congr. Bot. Genova,* 1892, 359 ; *Pax in
Engl. Pfl. Ost-Afr. C.* 239, *and in Baum, Kunene-Samb. Exped.* 283 ;
Hiern in Cat. Afr. Welw. i. 978 ; *Cooke, Fl. Bomb.* ii. 610 ; *Durand
& De Wild. Mat. Fl. Congo,* ii. 61 ; *De Wild. & Durand, Reliq.
Dewevr.* 210 ; *N. E. Br. in Kew Bulletin,* 1909, 141 ; *Th. & Hél.
Durand, Syll. Fl. Congol.* 492 ; *Hutchinson in Dyer, Fl. Trop. Afr.*
vi. i. 903 ; *Prain in Kew Bulletin,* 1913, 15. *A. spicata, Forsk. Fl.
Ægypt.-Arab.* 161. *A. somalium, Müll. Arg. in Abhandl. Naturw.
Ver. Bremen,* vii. 27. *A. somalensis, Pax in Engl. Jahrb.* xix. 100.
Ricinocarpus indicus, O. Kuntze, Rev. Gen. Pl. ii. 618.

KALAHARI REGION : Transvaal ; Vaal River, *Burke* ! near Hammans Kraal,
4400 ft., *Schlechter,* 4185 ! Avoca, near Barberton, 1900 ft., *Galpin,* 1237 !
Komati Poort, *Kirk,* 105 ! Shilovane, *Junod,* 1321 ! 1615 !
EASTERN REGION : Natal ; Tugela, *Gerrard,* 1624 !

Also widely spread throughout Tropical Africa, the Mascarene Islands, Arabia
and South-eastern Asia.

6. **A. segetalis** (Müll. Arg. in Journ. Bot. 1864, 336) ; herbaceous, annual ; stem rather copiously branched from the base,
up to 1 ft. high, somewhat sulcate ; leaves membranous, long-petioled,
ovate, obtuse or subacute, base more or less rounded, margin crenate,
$\frac{1}{2}$–$1\frac{1}{2}$ in. long, $\frac{1}{3}$–1 in. wide, very sparingly pubescent on the nerves
beneath, otherwise glabrous ; petiole $\frac{1}{2}$–$1\frac{1}{2}$ in. long, puberulous ;
stipules subulate, $\frac{1}{8}$ in. long, caducous ; inflorescence 2-sexual,
axillary ; spikes usually solitary, short, about $\frac{3}{4}$ in. long, with 2–4
rather remote female flowers below and rather few male flowers
above, with usually a solitary densely muricate abortive female
flower at the apex ; rhachis puberulous ; male bracts very minute ;
male buds faintly puberulous ; female bracts 2-flowered, leafy, ovate,
acuminate, base subcordate, margin crenate, $\frac{1}{2}$ in. long, if explanate
$\frac{2}{3}$ in. wide, sparingly setulose on the margin and nerves externally,
sometimes with a few glandular hairs intermixed ; sepals 3, ovate,
small, sparingly ciliate ; ovary rather distinctly 3-lobed, sparingly
beset with tubercle-based hairs ; styles 3, free from the base,
slender, laciniate ; seeds ovoid, minutely and closely pitted. *Müll.*

Arg. in DC. Prodr. xv. ii. 877 ; *Hiern in Cat. Afr. Pl. Welw.* i. 979 ;
Hutchinson in Dyer, Fl. Trop. Afr. vi. i. 904 ; *Prain in Kew Bulletin*,
1913, 15.   *A. sessilis, De Wild. & Durand in Bull. Herb. Boiss.*
2me sér. i. 47 ; *De Wild. Miss. É. Laurent*, 131, *and Études Fl. Bas-*
et Moyen-Congo, ii. 284 ; *not of Poir. A. sessilis, var. brevibracteata,*
Müll. Arg. in Flora, 1864, 465. *A. sessilis, var. exserta, Müll. Arg.*
l.c.   A. gemina, var. brevibracteata, Müll. Arg. in Linnæa, xxxiv.
41, *and in DC. Prodr.* xv. ii. 866.   *A. genuina, var. exserta, Müll.*
Arg. ll.cc.   Ricinocarpus segetalis, O. Kuntze, Rev. Gen. Pl. ii. 618.

WESTERN REGION : Great Namaqualand ; Rehoboth, *Fleck*, 170 ! Little Bush-
manland ; without precise locality, *Fleck*, 472 a !
KALAHARI REGION : Transvaal ; Springbok Flats, *Sampson*, 4410 ! Shilovane,
Junod, 2346 !
EASTERN REGION : Natal ; near Colenso, 3000–4000 ft., *Wood*, 4552 ! Delagoa
Bay ; Lourenço Marques, *Quintas*, 210 ! Mabola, *Schlechter*, 11686 ! Incanhini,
Schlechter, 12043 !

Also rather widely spread in Tropical Africa.

Very, perhaps too, closely allied to *A. indica*, Linn.

7. **A. decumbens** (Thunb. Prodr. 117) ; shrubby, perennial ; stems
very slender, 6–8 ft. long, copiously branched, decumbent and
sometimes rooting at the nodes or climbing over bushes ; branches
very slender, faintly striate, glabrous ; leaves membranous, long-
petioled, ovate or ovate-lanceolate, acute, margin coarsely crenately
toothed, on the smaller twigs with the base often truncate, ¼–½ in.
long, ⅓–⅓ in. wide, on the main branches with the base distinctly
to deeply cordate, ¾–1¼ in. long, ½–1 in. wide, glabrous above, very
closely softly white- or grey-puberulous-hoary beneath ; petiole
½–¾ in. long, sparingly puberulous ; stipules linear, spreading,
glabrous, 1 lin. long ; inflorescences 1-sexual, male and female
distinct, both axillary, but the female spikes nearer the tips of
the twigs than the male and sometimes aggregated in a terminal
thyrsoid leafless spike ; male spikes solitary to their axils, very
slender, 1½–2 in. long, with a naked peduncle ¼–1 in. long ; peduncle
and rhachis finely puberulous ; bracts minute, puberulous ; buds
puberulous ; female spikes short, few-flowered, their peduncles 1–2
lin. long or shorter, with only 1–3 leafy bracts ; occasionally the
lowest with a short terminal male portion ; sometimes the leaves
towards the ends of the twigs obsolete and from 2–20 female spikes
aggregated in an ovoid mass up to 1¼ in. long and ¼ in. across ;
female bracts 1-flowered, reniform-ovate, strongly and coarsely
5–9-toothed, white- or grey-puberulous-hoary externally, glabrous
within, ¼–⅓ in. across ; calyx deeply lobed, lobes ovate-lanceolate ;
ovary 3-lobed, densely papillose ; styles 3, nearly free, densely
lacinulate : seeds subglobose. *Thunb. Fl. Cap. ed. Schult.* 545 ;
Spreng. Syst. iii. 882.   *A. cordata, Thunb. ll.cc.* ; *Spreng. l.c.* 880.
A. discolor, E. Meyer in Drège, Zwei Pfl. Documente, 161 ; *Hochst.*
apud Krauss in Flora, 1845, 84 ; *Eckl. & Zeyh. Linnæa*, xx. 213 ;
Baill. Adansonia, iii. 157 ; *Müll. Arg. in Linnæa*, xxxiv. 38.   *A.*

discolor, β *major*, *Baill. l.c.* 158. *A. capensis*, β *decumbens*, *Prain in Kew Bulletin*, 1913, 16. *A. decumbens*, β *cordata and* γ *genuina*, *Müll. Arg. in DC. Prodr.* xv. ii. 864. *Ricinocarpus decumbens*, *O. Kuntze, Rev. Gen. Pl.* ii. 617.

VAR. β, **villosa** (Müll. Arg. in DC. Prodr. xv. ii. 864); twigs, petioles and leaves beneath densely softly velvety with longish grey hairs; leaves above sparingly adpressed white-pubescent; female bracts softly strigose with spreading white hairs; leaves larger, up to 2 in. long, 1½ in. wide, more uniformly cordate at the base; male spikes as in the type. *Urtica capensis*, *Linn. f. Suppl.* 417. *U. africana*, *Linn. mss. ex Jackson in Ind. Linn. Herb.* 148. *Tragia villosa*, *Thunb. Prodr.* 14, *and in Fl. Cap. ed. Schult.* 37; *not Acalypha villosa*, *Jacq. Acalypha kraussiana*, *Buching. ex Meisn. apud Krauss in Flora*, 1845, 84; *Müll. Arg. in Linnæa*, xxxiv. 39. *A. lamiifolia*, *Scheele in Linnæa*, xxv. 587; *Baill. Adansonia*, iii. 158. *A. grandidentata*, *Müll. Arg. in DC. Prodr.* xv. ii. 823. *A. capensis*, *Prain in Kew Bulletin*, 1913, 15.

SOUTH AFRICA: without locality, *Osbeck*! *Sparrman*! *Mund & Maire*, 33! *Zeyher*! var. β: *Mund & Maire*, 659! *Drège*, 8242!

COAST REGION: Stellenbosch Div.; *Lowrys Pass*, *Schlechter*! Riversdale Div.; between Little Vet River and Garcias Pass, *Burchell*, 6925! near Riversdale, 400 ft., *Kässner*, 1164! *Schlechter*, 1961! Mossel Bay Div.; Mossel Bay, *Prior*! George Div.; Woodville, near George, *Tyson*! Knysna Div.; Gouwkamma River, *Krauss*, 1826! near Knysna, 1000 ft., *Marloth*, 2448, partly! Humansdorp Div.; Humansdorp, 300 ft., *Zeyher*, 3840! *Ecklon & Zeyher*, 73! *Galpin*, 4576! *Kennedy*! *West*, 276! Gamtoos River, *Thunberg*! *Drège*, 8240! Uitenhage Div.; Zwartkops River, *Drège*, 2313! *Verreaux*! *Ecklon*, 161! 162! *Zeyher*, 3478! 3840! *Ecklon & Zeyher*, 610! Maitland River, near the Leadmine, *Burchell*, 4404! Var. β: Riversdale Div.; banks of the Vet River near Riverdale, *Muir*, 283! George Div.; Outeniqua Mountains, *Thunberg*! Montagu Pass, *Rehmann*, 258! Knysna Div.; near Knysna, *Burchell*, 5390! 5391! 5392! *Krauss*, 1825! *Marloth*, 2448, partly! *Newdegate*, *Tyson*! Witte Drift, Plettenburgs Bay, *Pappe*! Uitenhage Div.; Van Stadensberg Range, *Zeyher*! near Uitenhage, *Prior*! Albany Div.; without precise locality, *Bowie*!

8. **A. senensis** (Klotzsch in Peters, Reise Mossamb. Bot. 96); shrubby; stems erect, woody, 1–2 ft. high, often angled or sulcate, hirsute or pubescent; leaves thinly to firmly membranous, long-petioled, ovate-lanceolate or lanceolate, sometimes linear-lanceolate, gradually acutely acuminate, base cordate, margin serrate or crenate-serrate, 1–3½ in. long, sometimes in tropical specimens up to 7 in. long, ⅓–1¼ in. wide, hirsute or pubescent on both surfaces especially on the nerves; petiole ½–1¼ in. long, pubescent and often hirsute; stipules subulate or filiform, 1½–2 lin. long, pilose; inflorescences 1-sexual, monœcious, male lateral and female terminal on the same branch; male spikes axillary, solitary, narrow-cylindric, on slender pubescent or puberulous peduncles up to 1 in. long, the spikes 1 in. (at length 2½ in.) long, dense-flowered; male bracts minute; male buds almost glabrous; female spikes at the ends of twigs, sessile, ½–1¼ in. long, ½ in. across; bracts 1–2-flowered, suborbicular or reni-form, rather deeply 7–10-toothed, under ¼ in. long, about ⅓ in. across; teeth wide-triangular, obtuse or faintly mucronate, puberulous and densely stipitate-glandular outside, but without long spreading hairs, inside glabrous; sepals 4, lanceolate, subacute, somewhat un-equal, ciliate, pubescent outside; ovary rather deeply 3–4-lobed,

pubescent and with several long-stipitate glands on the upper portion of each lobe; styles 3–4, sparingly laciniate, about ½ in. long; seeds subglobose, with a large hilum. *Müll. Arg. in DC. Prodr.* xv. ii. 845; *Pax in Engl. Pfl. Ost-Afr. C.* 239; *Hutchinson in Dyer, Fl. Trop. Afr.* vi. i. 888; *Prain in Kew Bulletin,* 1913, 18. *A. zambesica, Müll. Arg. in Flora,* 1864, 440, *and l.c.* 845; *Pax, l.c.* 239; *Gibbs in Journ. Linn. Soc.* xxxvii. 470. *A. villicaulis, Müll. Arg. l.c.* 845, *as to Meller's Manyanja plant only; Pax in Baum, Kunene-Samb. Exped.* 283; *S. Moore in Journ. Linn. Soc.* xl. 199, *partly; not of Hochst. A. villicaulis, var. minor, Müll. Arg. in Abhandl. Naturw. Ver. Bremen,* vii. 26. *A. Rehmanni, Pax in Bull. Herb. Boiss.* vi. 733. *Ricinocarpus senensis and R. zambesicus, O. Kuntze, Rev. Gen. Pl.* ii. 618.

KALAHARI REGION: Bechuanaland; Masupa River in Banquaketse Territory, *Holub*! Transvaal; near Pretoria, *Scott Elliot,* 1398! *Rehmann,* 4285! *Fehr,* 58! *Wilms,* 1321 a! *Burtt-Davy,* 695! 725! 5380! *Bolus,* 10838! *Miss Leendertz,* 56! Aapies River, *Burke*! Klippan, *Rehmann,* 5330! Houtbosch, *Rehmann,* 5915! Rustenberg, *Collins,* 70! Warmbath, *Miss Leendertz,* 1561! Waterval Onder, *Jenkins,* 6717! Lydenburg, *Wilms,* 1321, partly! Johannesberg, *Marloth,* 3830! Shilovane, *Junod,* 1039! 2178!

Also rather widely spread in eastern Tropical Africa.

9. **A. petiolaris** (Hochst. apud Krauss in Flora, 1845, 83); shrubby; stems prostrate or decumbent, less often ascending, woody, 8 in. to 2 ft. long, more or less densely pilose with long reflexed hairs; leaves thinly to firmly membranous, long-petioled, ovate or ovate-lanceolate, somewhat acutely acuminate, base deep-cordate, margin crenate or crenate-serrate, 1–3 in. long, ½–1½ in. wide, thinly pilose on both surfaces with long hairs, especially on the nerves beneath, occasionally glabrescent above; petiole ½–2 in. long, pilose with spreading or somewhat reflexed hairs; stipules subulate or filiform, 1½–2 lin. long. ciliate; inflorescences 1-sexual, monœcious, male lateral and female terminal on the same branch; male spikes axillary, solitary, narrow-cylindric, on slender peduncles beset with long-spreading hairs, up to ½ in. long, the spikes ¼ in. (at length up to ¾ in.) long, dense-flowered; male bracts minute, lanceolate; male buds almost glabrous; female spikes at the ends of twigs, sessile, solitary, oblong, up to ¾ in. long, ¼ in. across; bracts 1–2-flowered, more or less oblong, deeply 6–7-lobed, under ¼ in. across; lobes subulate or lanceolate, acute, up to 1 lin. long, setosely long-ciliate, with or without a few stalked glands on their margins towards the base, rarely with glands and hairs on the inner side; sepals 3, lanceolate, subacute; ovary 3-lobed, pilose in the upper half, with or without stipitate glands intermixed; styles 3, slender, about ¼ in. long, laciniate; seeds subglobose, with a large hilum. *Müll. Arg. in Linnæa,* xxxiv. 29, *and in DC. Prodr.* xv. ii. 847; *Prain in Kew Bulletin,* 1913, 18. *A. languida, E. Meyer, in Drège, Zwei Pfl. Documente,* 161; *Sond. in Linnæa,* xxiii. 116; *Baill. Adansonia,* iii. 157; *Müll. Arg. l.c.* 29, *and l.c.* 848. *A. brachiata, Krauss, l.c.* 83;

not of E. Meyer. *A. tenuis, Müll. Arg. l.c.* 30, *and l.c.* 848, *incl. both vars. Ricinocarpus languidus and R. petiolaris, O. Kuntze, Rev. Gen. Pl.* ii. 618.

COAST REGION : Komgha Div. ; near the Kei River mouth, *Flanagan*, 450 !
KALAHARI REGION : Transvaal ; various localities, *Wahlberg* ! *Burke* ! *Zeyher*, 1519 ! *Miss Leendertz*, 443 ! 2313 ! *Wilms*, 1321, partly ! *Schlechter*, 3897 ! *Galpin*, 513 ! 1245 ! *Bolus*, 9777 ! 9778 ! Swaziland ; low Veld near Mafutane, *Bolus*, 12294 !
EASTERN REGION : Transkei ; Bashee River, *Drège*, a ! 4594 ! Kentani, *Miss Pegler*, 870 ! Pondoland ; between Umtata River and St. Johns River. *Drège*, b ! Murchison, *Bachmann*, 791 ! Griqualand East ; Clydesdale, 2500 ft., *Tyson*, 2568 ! 2693 ! 2694 ! *and in MacOwan & Bolus, Herb. Austr.-Afr.*, 1232 ! Natal ; various localities, *Krauss*, 319 ! 367 ! *Gueinzius*, 169 ! 506 ! *Gerrard*, 518 ! 617 ! *Rehmann*, 8803 ! *Wood*, 68 ! 254 ! 1408 ! *Schlechter*, 3026 ! *Miss Franks in Herb. Wood*, 11771 ! *Rehmann*, 7795 ! 8802, partly ! *Drège*, c ! 8421 ! *Engler*, 2569 ! *Landauer*, 223 ! *Rudatis*, 1185 ! Delagoa Bay ; Ressano Garcia, *Schlechter*, 11882 ! Lourenço Marques, *Junod*, 147 !

10. **A. Eckloni** (Baill. Adansonia, iii. 158) ; herbaceous, annual ; stems sparingly branched from the base, up to 1½ ft. high, somewhat sulcate, pubescent or glabrous ; leaves membranous, long-petioled, ovate, subacute or acute, base slightly cordate or. rounded, margin crenate, ¾–1¼ in. long, ½–1 in. wide, sparingly to distinctly pubescent on the nerves on both surfaces ; petiole puberulous or pubescent, ¼–1¼ in. long ; stipules subulate, ⅛ in. long, pilose, caducous ; inflorescences 1-sexual, monœcious, male lateral and female terminal on same branch ; male spikes axillary, solitary, capitate on very slender puberulous peduncles up to ⅓ in. long, the heads subglobose, about 1 lin. across, few-flowered and very caducous ; male bracts minute, subulate ; male buds sparingly puberulous ; female spikes at the ends of twigs, solitary, wide cylindric or oblong, 1¼ in. long, ¼–¾ in. across ; bracts 1–2-flowered, leafy, wide ovate, acuminate, ¼–½ in. long, ⅓–½ in. wide, margin 5–9-lobed, lobes linear-lanceolate, ascending, pubescent externally with short gland-tipped hairs, teeth ciliate but without glandular hairs ; sepals 3, lanceolate, acute, sparingly hairy ; ovary 3-lobed, sparingly pubescent in the upper half ; styles 3, slender, free nearly to the base, undivided or very sparingly lacinulate ; seeds ovoid, smooth. *Müll. Arg. in Linnœa*, xxxiv. 30 *and in DC. Prodr.* xv. ii. 849 ; *Prain in Kew Bulletin*, 1913, 19. *A. cordata ?, E. Meyer in Drège, Zwei Pfl. Documente*, 161 ; *Baill. l.c.* 158 ; *not of Thunb. A. brachiata, E. Meyer, l.c. ; Baill. l.c. Ricinocarpus Ecklonii, O. Kuntze, Rev. Gen. Pl.* ii. 617, *and* iii. ii. 291.

COAST REGION : George Div. ; woods near George, 1000 ft., *Schlechter*, 2350 ! Uitenhage Div. ; near Uitenhage, *Burchell*, 4251 ! *Mund & Maire* ! *Verreaux* ! *Burke* ! *Prior* ! on Van Stadens Mountains, *Burchell*, 4751 ! Zwartkops River, *Drège*, 4602 ! *Ecklon*, 74, partly ! 609 ! *Zeyher*, 228 ! 3841 ! Enon, *Drège*, 2345 ! 4600 ! Bathurst Div. ; near the Kasuga River, *Prior* ! Albany Div. ; Kleinemund near Grahamstown, *MacOwan*, 1507 ! Grahamstown, *Misses Daly & Gane*, 743 ! King Williamstown Div. ; Yellowwood River, *Drège* ! King Williamstown, *Sim*, 1468 ! Perie Forest, *Kuntze* ! East London Div. ; near East London, *Galpin*, 7790 ! *Rattray*, 809 ! Komgha Div. ; near Komgha, 2000 ft., *Flanagan*, 630 !
EASTERN REGION : Transkei ; Gekwa River, *Drège*, 4601 ! Kentani, 1000 ft., *Miss Pegler*, 732 ! Pondoland ; St. Johns River ! *Drège* ! Natal ; near Durban,

Gueinzius, 8! 168! Inanda, 1800 ft., *Wood*, 3148! near Isipingo, 150 ft., *Schlechter*, 2808 !

A very distinct species, based by Baillon on a plant (*Drège*, 4600) which had been doubtfully referred by E. Meyer to *A. cordata*, Thunb., but is the same as *A. brachiata*, E. Meyer, of which no description was published. Meyer recognised two varieties within *A. brachiata* ; *a major* (*Drège*, 4602) with taller stems and almost glabrous leaves and *β minor* (*Drège*, 4601) with short stems and hirsute leaves. The more ample material since communicated shows, however, that no such subdivision is required.

11. **A. Zeyheri** (Baill. Adansonia, iii. 156); shrubby; stems slender, trailing, sparingly branched, 2–3 ft. long, bark pale, glabrous, striate ; leaves firmly chartaceous or subcoriaceous, numerous, very short-petioled, triangular-ovate, acute, base truncate or subcordate, margin distinctly serrate, 1–1¼ in. long, ⅔–¾ in. wide, quite glabrous on both surfaces or sparingly setulose on the main-nerves beneath ; petiole glabrous or sparingly setulose, 1–1½ lin. long ; stipules glabrous, herbaceous, persistent, 2 lin. long; inflorescences 1-sexual, diœcious ; male spikes axillary, solitary, peduncled ; peduncles glabrous, 1–1¼ in. long; spikes cylindric, dense-flowered, 1–1½ in. long; bracts lanceolate, sparingly pubescent, ⅛ in. long, spreading, persistent ; buds glabrous ; female spikes terminal, solitary, ½ in. (ultimately 1 in.) long; bracts 1-flowered, subsessile, leafy, wide-ovate, subacute or obtuse, base wide-cuneate, ½ in. long, ⅔ in. wide, coarsely serrate, the central tooth usually exceeding the 2–4 lateral, sparingly hispid on the nerves outside, eglandular ; sepals 3, ovate, acute, glabrous ; ovary distinctly 3-lobed, nearly glabrous ; styles 3, united in their lower fourth, ⅔ in. long, shortly laciniate upwards ; seeds subglobose. *Prain in Kew Bulletin*, 1913, 19. *A. peduncularis, var. psilogyne, Müll. Arg. in Linnæa*, xxxiv. 28, *and in DC. Prodr.* xv. ii. 846, *partly. A. Zeyheri, var. glabrata, Müll. Arg. l.c.* 29, *and l.c.* 847, *excl. syn. Sond. A. Sonderi, Gandog. in Bull. Soc. Bot. Fr.* lx. 27. *Ricinocarpus Zeyheri, O. Kuntze, Rev. Gen. Pl.* ii. 618.

COAST REGION : Uitenhage Div. ; Van Stadens Berg, *Zeyher*, 345 ! 3839 ! *Burchell*, 4726 ! *Mrs. Paterson*, 886 !

Very nearly related, as Baillon has remarked, to *A. peduncularis*, E. Meyer, but differing somewhat in habit as well as in foliage. Whether specifically distinguishable or not from *A. peduncularis*, *A. Zeyheri* seems to be a form of extremely restricted distribution.

12. **A. peduncularis** (E. Meyer ex Meisn. apud Krauss in Flora, 1845, 82) ; herbaceous ; stems slender, rather sparingly branched, prostrate, 6–18 in. long, from a stout perennial woody stock 4–8 in. long or longer and ¼–⅓ in. thick, with black rugose bark ; branches patulous, striate, rather copiously patently setulose like the herbaceous stems ; leaves membranous, numerous, very shortly petioled, ovate or ovate-lanceolate to lanceolate, acute or those lowest down obtuse, base cuneate or rounded, margin distinctly serrate, 1½–2½ in. long, ¾–1¼ in. wide, those towards the tips of the stems and branches smaller than the ones lower down, rather sparsely setulose, especially on the nerves on both surfaces and on the margin ;

petiole setulose, 1–1½ in. long ; stipules setulose, subulate, persistent, 1–1½ lin. long ; inflorescences 1-sexual, usually diœcious,
sometimes monœcious ; male spikes axillary, solitary, peduncled ;
peduncles rather copiously patently setulose, 2½–4 in. long ; spikes
cylindric, dense-flowered, slender, 1½–2 in. long ; bracts lanceolate,
setulose, ⅛–⅙ in. long, spreading, persistent ; buds setulose at the
tip ; female spikes terminal, solitary, ½ in. (ultimately 1–1½ in.)
long ; bracts 1-flowered, subsessile, leafy, wide-ovate, subacute or
obtuse, base wide-cuneate, ½–⅔ in. long, ⅔–¾ in. wide, coarsely
serrate, central tooth rather exceeding the 2–4 lateral, copiously
setulose on the margin and nerves outside and beset with glands on
the back ; sepals 3, ovate, acute, setulose ; ovary distinctly 3-lobed,
pilose and glandular in the upper half ; styles 3, united in their
lower third, 1 in. long, spirally twisted and markedly laciniate
upwards ; seeds subglobose. *E. Meyer in Drège, Zwei Pfl. Docu-*
mente, 161 ; *Eckl. & Zeyh. in Linnæa,* xx. 213 ; *Sond. in Linnæa,*
xxiii. 115, *excl. syn. Buching. and var. glabrata; Baill. Adansonia,* iii.
156, *excl. syn. Buching. ; Pax in Engl. Pfl. Ost-Afr. C.* 239, *partly ;*
Prain in Kew Bulletin, 1913, 20. *A. peduncularis, var. genuina, Müll.*
Arg. in Linnæa, xxxiv. 28, *and in DC. Prodr.* xv. ii. 846. *A. Zeyheri,*
var. pubescens, Müll. Arg. l.c. 29, *and l.c.* 847 ; *not A. Zeyheri,*
Baill. A. mentiens, Gandog. in Bull. Soc. Bot. Fr. lx. 27. *A. Dregei,*
Gandog. l.c. Ricinocarpus peduncularis, O. Kuntze, Rev. Gen. Pl.
ii. 618.

VAR. β, crassa (Müll. Arg. in Linnæa, xxxiv. 28) ; stems stouter, 4–15 in.
long, simple or subsimple ; leaves ovate, all obtuse, base always rounded, when
mature 2½–3 in. long, 1½–2 in. wide, those towards the apex of the stem much
larger than the ones lower down. *Müll. Arg. in DC. Prodr.* xv. ii. 846 ; *Prain in*
Kew Bulletin, 1913, 21. *A. crassa, Buching. ex Meisn. apud Krauss in Flora,*
1845, 83. *A. peduncularis, Sond. in Linnæa,* xxiii. 115, *as to syn. Buching. ;*
Baill. Adansonia, iii. 156, *as to syn. Buching. ; hardly of E. Meyer. A. pedun-*
cularis, var. ferox, Pax ex Prain in Kew Bulletin, 1913, 21. *A. Schlechteri,*
Gandog. in Bull. Soc. Bot. Fr. lx. 27. *Ricinocarpus peduncularis, var. ovatifolius,*
O. Kuntze, Rev. Gen. Pl. iii. ii. 292.

SOUTH AFRICA : without locality, *Masson* ! *Herb. Swartz* ! *Grondal* ! *Delalande* !
Krebs ! , Var. β : *Herb. Swartz* !
COAST REGION : Uitenhage Div. ; Bontjes River, 2000 ft., *Drège,* 2309 !
Enon, *Drège* ! near Zwartkops River, *Zeyher,* 214 ! Alexandria Div. ; Zuurberg
Range, 2000–3000 ft., *Drège* ! *Cooper,* 3580 ! Bathurst Div. ; between Blue
Krantz and Kaffir Drift, *Burchell,* 3707 ! at Rietfontein between the Kasuga
River and Port Alfred, *Burchell,* 4041 ! Albany Div. ; near Grahamstown, 1500–
2500 ft., *Ecklon,* 75, partly ! *Bolton* ! *Mrs. Barber* ! *MacOwan,* 49 ! *Galpin,* 191 !
Bolus ! *Rogers,* 174 ! 4593 ! Slang Kraal, *Burke* ! Assegai Bush and Botram,
1000–2000 ft., *Drège,* 4596 ! Howisons Poort, *Zeyher,* 3838 ! *Mrs. Hutton* ! near
Sidbury, *Burchell,* 4176 ! Stockenstrom Div. ; sources of the Kat River above
Philipston, 3000–4000 ft., *Ecklon,* 75, partly ! Peddie Div. ; Fredericksburg, *Gill* !
East London Div. ; near East London, *Rattray,* 678 ! *Wood in Herb. Galpin,*
3136 ! British Caffraria ; Hangmans Bush, *Cooper,* 147 ! 3581 !
EASTERN REGION : Transkei ; Krielis country, *Bowker,* 292 ! Kentani, 1200 ft.,
Miss Pegler, 65, partly ! Tembuland ; near Bazeia, beyond the Bashee River,
2000 ft., *Baur,* 373 ! Pondoland ; many localities, *Bachmann,* 783 ! 790 ! 792 !
793 ! 1711 ! *Beyrich,* 102 ! Natal ; near Durban, *Krauss,* 377, partly ! *Gueinzius,*
404, partly ! *Gerrard,* 619 ! *Sanderson,* 129 ! *Wood,* 1416, partly ! Inanda, 1800 ft.,
Wood, 48, partly ! Umgeni River, *Rehmann,* 8804 ! Camperdown, *Schlechter,* 3059,

partly! Var. *β* : Natal ; near Durban, *Krauss*, 377, partly! *Gueinzius*, 404, partly! *Sutherland*! *Gerrard*, 521! *Wood*, 90! Camperdown, *Gerrard*, 1166! near Clairmont, *Schlechter*, 3059, mainly! Krantz Kloof, 1800 ft., *Schlechter*, 3188! Inanda, 1800 ft., *Wood*, 640! Pietermaritzburg, *Wilms*, 2265! Riet Vlei, *Fry in Herb. Galpin*, 2722! Alexandra Distr., Dumisa, 2000 ft., *Rudatis*, 445! Highland Station, 5300 ft., *Kuntze*, 5553! Nottingham, *Buchanan*, 143! Klip River, 3500–4500 ft., *Sutherland*! and without precise locality, *Gerrard*, 373!

A well marked species the limits of which have, however, been somewhat misunderstood. The variety here recognised though, with care, readily separable, is, as Meisner pointed out when he originally described it, not really specifically distinct from *A. peduncularis*. Müller has endeavoured to distinguish between the *A. peduncularis* issued by E. Meyer and that described by Meisner ; there is, however, no justification for this action ; while it is true that *A. peduncularis*, E. Meyer, proper is rare in Natal, and that its place in that colony is largely taken by var. *crassa*, it so happens that the portion of Krauss, 377, on which in 1845 Meisner based his description of *A. peduncularis*, is not separable from the plant issued by E. Meyer in 1843 under the same name.

13. **A. glandulifolia** (Buching. ex Meisn. apud Krauss in Flora, 1845, 83) ; herbaceous ; stems slender, decumbent, 3–15 in. long, from a slender perennial woody rootstock, usually rather copiously virgately branched, striate, sparingly to copiously patently setulose ; leaves membranous, very shortly petioled, numerous, small, ovate-lanceolate to lanceolate, acute, base rounded, margin distinctly serrate, each tooth tipped by a stipitate capitate gland, $\frac{1}{2}$–$1\frac{1}{4}$ in. long, $\frac{1}{6}$–$\frac{1}{4}$ in. wide, glabrous on both surfaces or sparingly setulose on the midrib beneath ; petiole glabrous, $1\frac{1}{2}$ lin. long ; stipules minute, hyaline, membranous, caducous ; inflorescences 1-sexual, diœcious ; male spikes axillary, solitary, peduncled ; peduncles sparingly setulose, $\frac{1}{2}$ (at length 2) in. long ; spikes cylindric, dense-flowered, rather stout, $\frac{1}{3}$–$\frac{1}{2}$ in. long ; bracts linear, ciliate, $\frac{1}{8}$ in. long, spreading, persistent ; buds glabrous ; female spikes terminal, solitary, $\frac{1}{2}$ in. long ; bracts subsessile, large, foliaceous, ovate-lanceolate, acute, base rounded, $\frac{1}{2}$–$\frac{3}{4}$ in. long, $\frac{1}{3}$–$\frac{1}{2}$ in. wide, serrate with glandular margin like the leaves and in addition beset with sessile glands on the back ; sepals 3, ovate, subacute, with glandular margin ; ovary distinctly 3-lobed, closely glandular on the upper half ; styles 3, united in their lower fourth, $\frac{2}{3}$–$\frac{3}{4}$ in. long, very sparingly and shortly laciniate upwards ; seeds subglobose. *Sond. in Linnæa*, xxiii. 116 ; *Walp. Ann.* iii. 367 ; *Baill. Adansonia*, iii. 157 ; *Prain in Kew Bulletin*, 1913, 21. *A. peduncularis, var. glandulifolia, Müll. Arg. in Linnæa*, xxxiv. 28, and *in DC. Prodr.* xv. ii. 846.

EASTERN REGION : Natal ; near Durban, *Krauss*! *Gueinzius*, 170! *Gerrard*, 520! *Sutherland*! *Wood* 1416, partly! Attercliffe, 800 ft., *Sanderson*, 298! Inanda, 1800 ft., *Wood*, 48, partly! 694! Indwedwe, *Wood*, 1054! Alexandra Distr., Dumisa, 2000–2500 ft., *Rudatis*, 96! 743! 744!

A very well-marked species within which Sonder proposed to recognise two distinct forms, one with pilose nearly simple, the other with glabrous more branching stems. The material now available shows, however, that while two such forms may be distinguished, they pass into each other and cannot be treated as definite varieties.

14. A. entumenica (Prain in Kew Bulletin, 1913, 22); herbaceous; stems slender, prostrate, 1–1½ ft. long, from a perennial rootstock, simple, densely leafy from the base, striate, strigose with white patent bulbous-based hairs; leaves membranous, subsessile, very numerous, small, ovate-lanceolate to lanceolate, acute, base cuneate to rounded, margin distinctly serrate, each tooth tipped either by a stipitate capitate gland or by a stiff bulbous-based hair, ½–1 in. long, ⅕–¼ in. wide, rather densely uniformly setulose on both surfaces with bulbous-based hairs often intermixed with stipitate glands; petiole setulose, under 1 lin. long; stipules minute, hyaline-membranous, caducous; inflorescences 1-sexual, diœcious; male spikes not seen; female spikes terminal, solitary, 1–1½ in. long; bracts subsessile, large, foliaceous, wide-ovate to suborbicular, acute, base rounded, ¼–⅓ in. long, ⅓–½ in. wide, serrate like the leaves, each lobe with a stipitate apical gland and in addition setulose and rather closely covered with stipitate glands on the back; sepals 3, ovate, subacute, glandular and strigose; ovary distinctly 3-lobed, closely glandular on the upper half; styles 3, united below, upper portions not seen; seeds subglobose.

EASTERN REGION: Zululand; near Entumeni, 2000 ft., *Wood*, 3737!

A distinct species, most nearly allied to *A. glandulifolia*, Buching., but readily distinguished by its strigose leaves and its different female bracts.

15. A. depressinervia (K. Schum. in Just, Jahresber. xxvi. i. 348); herbaceous; stems slender, erect, 6–15 in. long, from a slender perennial woody rootstock, simple or very sparingly virgately branched, striate, copiously patently setulose with white bulbous-based hairs; leaves firmly membranous, numerous, small, very shortly petioled, narrow-lanceolate, acute, base rounded, margin quite entire, ¾–2 in. long, ⅙–⅕ in. wide, softly and rather copiously patently setulose with white bulbous-based hairs, nerves obscure; petiole under ½ lin. long, setulose with bulbous-based hairs; stipules minute, pilose, caducous; inflorescences 1-sexual, diœcious; male spikes axillary, solitary, peduncled; peduncles copiously patently setulose like the stems, 1½–2 in. long; spikes narrow-cylindric, dense-flowered, ⅔–1 in. long; bracts linear-lanceolate, setulose, ⅛ in. long, spreading, persistent; buds setulose at the apex; female spikes terminal, solitary, 1–1½ in. long; bracts 1-flowered, subsessile, large, foliaceous, ovate-lanceolate, acute, base rounded, ¾–1 in. long, ⅓–¾ in. wide, the lowest quite entire, those above progressively digitately 3–5- (occasionally 7-) toothed, setulose like the leaves, with white bulbous-based hairs; sepals 3, ovate, subacute, pilose; ovary distinctly 3-lobed, setulose with long white bulbous-based hairs in the upper two-thirds; styles 3, shortly united below, ½–⅔ in. long, sparingly shortly laciniate upwards; seeds subglobose. *Prain in Kew Bulletin*, 1913, 22. *A. peduncularis, var. angustata, Müll. Arg. in Linnæa*, xxxiv. 28, *and in DC. Prodr.* xv. ii. 847, *in small part, and as to Wahlberg's specimens only; not A. angustata,*

Sond. A. Schinzii, Pax in Bull. Herb. Boiss. vi. 734, *excl. var. denticulata. A. Oweniæ, Harv. ex Prain in Kew Bulletin,* 1913, 22. *Ricinocarpus depressinervius, O. Kuntze, Rev. Gen. Pl.* iii. ii. 291.

KALAHARI REGION: Orange River Colony: Besters Vlei near Witzies Hoek, 5500 ft., *Flanagan*, 1922! Harrismith, *Sankey*, 237! Basutoland; without precise locality, *Cooper*, 3577! Transvaal; Macalisberg Range, *Wahlberg*! Billys Vlei, *Mitchell*! Houtbosch, *Rehmann*, 5914! Saddleback Mountain, near Barberton, 4500–5000 ft., *Galpin*, 638! 1120! near Barberton, *Bolus*, 9776, partly! between Komati River Drift and Crocodile River, *Bolus*, 9776, partly!

EASTERN REGION: Natal; near Durban, *Miss Owen*! *Sanderson*, 508! *Wood*, 1416, partly! Inanda, *Wood*, 298! Tugela, *Gerrard*, 618! near Krantz Kloof, 1500 ft., *Schlechter*, 3186! Camperdown, *Rehmann*, 7793! Dalton, 3300 ft., *Rudatis*, 15! between Greytown and Newcastle, *Rudatis*, 2260! Mooi River, 4000–5500 ft., *Kuntze*, 5573! *Wood*, 3766!

A very distinct species, in appearance most like *A. angustata*, Sond., with which it has been confused, but readily distinguishable by its entire leaves, and by the absence of glands from the bracts and the ovary.

16. **A. angustata** (Sond. in Linnæa, xxiii. 115); herbaceous; stems slender, firm, 1–2 ft. high, from a stout woody rootstock, rather copiously virgately branched, sulcate, sparingly patently setulose with white bulbous-based hairs or at times glabrous; leaves firmly membranous, numerous, shortly petioled, linear or lanceolate, acute, base cuneate, margin distantly acutely denticulate, 1–2½ in. long, $\frac{1}{5}$–$\frac{1}{6}$ in. wide, sparingly patently pilose or nearly glabrous, nerves prominent beneath; petiole 1 lin. long, setulose or glabrous; stipules 1½ lin. long, firm, persistent; inflorescences 1-sexual, diœcious; male spikes axillary, solitary, peduncled; peduncles sparingly setulose or glabrous, 1–2 in. long; spikes narrow-cylindric, dense-flowered, $\frac{1}{2}$–$\frac{3}{4}$ in. long; bracts linear-lanceolate, sparingly setulose, $\frac{1}{8}$ in. long, spreading, persistent; buds glabrous; female spikes terminal, solitary, 1–1½ in. long; bracts 1-flowered, subsessile, large, foliaceous, lanceolate, acuminate, base rounded, $\frac{1}{2}$–$\frac{3}{4}$ in. long, $\frac{1}{4}$–$\frac{1}{2}$ in. wide, their margins sharply and strongly serrate, rather copiously glandular on the back; sepals 3, ovate, with glandular margin; ovary distinctly 3-lobed, closely glandular on the upper half; styles 3, united in their lower fourth, $\frac{2}{3}$–$\frac{3}{4}$ in. long, shortly but copiously lacinulate upwards; seeds subglobose. *Walp. Ann.* iii. 367; *Baill. Adansonia*, iii. 157; *Prain in Kew Bulletin*, 1913, 22. *A. angustata, var. glabra, Sond. l.c.* 116. *A. peduncularis, var. angustata, Müll. Arg. in Linnæa*, xxxiv. 28, *and in DC. Prodr.* xv. 847, *mainly, but excl. Wahlberg's plant. A. Schinzii, var. denticulata, Pax in Bull. Herb. Boiss.* vi. 734.

KALAHARI REGION: Transvaal; various localities, *Burke*, 349! *Zeyher*, 1518! *Nelson*, 231! *Rehmann*, 4284! *Miss Leendertz*, 363, mainly! 2582! 4821! *Jenkins*, 6735! *Schlechter*! *Wilms*, 1322! 1327! *Gilfillan in Herb. Galpin*, 6071! 6172! *Marloth*, 3669! *Rogers*, 2545! *Holub*! *Burtt-Davy*, 747! 769! 3830! 5602! 7165! 7192! 9150! *Holder*, 4548! *Bester*, 2164, partly! *McLea*!

EASTERN REGION: Natal; near Durban, *Gueinzius*, 171! *Gerrard*, 519! near Maritzburg, *Wilms*, 2262! Riet Vlei, *Fry in Herb. Galpin*, 2721!

A distinct species, most closely resembling *A. depressinervia*, K. Schum., but readily distinguishable therefrom by the glandular female bracts and ovaries and by the toothed leaves. It is most nearly allied to *A. caperonioides*, Baill., but is readily distinguished by its narrow leaves. Sonder has proposed the recognition of two varieties, distinguished by the presence or absence of pubescence, but, as Müller has already indicated, the two suggested forms pass into each other. Specimens of this species from near Pretoria (*Leendertz-Pott*, 4821) are monœcious.

17. **A. caperonioides** (Baill. Adansonia, iii. 157); herbaceous; stems slender, sparingly to copiously branched, 6–10 in. long, from a stout perennial woody rootstock 4–6 in. long or longer, $\frac{1}{4}$–$\frac{1}{3}$ in. thick with brown rugose bark ; branches erect, striate, sparingly to densely patently setulose like the stem, when young especially towards tips of twigs densely grey shaggy-pubescent ; leaves firmly papery or almost coriaceous, numerous, subsessile, ovate or cordate-ovate, acute, margin coarsely sharply serrate, $1\frac{1}{2}$–2 in. long, $\frac{2}{3}$–$\frac{3}{4}$ in. wide, sparingly setulose on the nerves above, more densely setulose and pubescent on the nerves beneath when mature, when young the leaves on both surfaces rather densely shaggy with bulbous-based hairs ; petiole setulose, under 1 lin. long ; stipules setulose, lanceolate, persistent, $1\frac{1}{2}$–2 lin. long ; inflorescence 1-sexual, diœcious ; male spikes axillary, solitary, peduncled ; peduncles sparingly to densely grey shaggy-tomentose, $1\frac{1}{2}$–2 in. long ; spikes cylindric, dense-flowered, slender, $1\frac{1}{2}$–2 in. long ; bracts lanceolate, with grey ciliate margins, $\frac{1}{4}$ in. long, spreading, persistent ; buds shaggy-setulose ; female spikes terminal, solitary, $\frac{1}{2}$ in. (ultimately $\frac{3}{4}$–1 in.) long, compact ; bracts 1-flowered, sessile, leafy, ovate lanceolate, acuminate. $\frac{1}{2}$–$\frac{2}{3}$ in. long, nearly as wide, margin strongly toothed, teeth ovate-lanceolate, acute, rather densely shaggy with bulbous-based hairs externally, but eglandular or with only a few glands intermixed ; sepals 3, ovate, acute, setulose ; ovary distinctly 3-lobed, setulose but not glandular in the upper half ; styles 3, united in their lower third, slightly or not at all laciniate above ; seeds subglobose. *Prain in Kew Bulletin*, 1913, 23. *A. peduncularis, var. glabrata, Sond. in Linnæa*, xxiii. 115. *A. Zeyheri, var. glabrata, Müll. Arg. in Linnæa*, xxxiv. 29, *and in DC. Prodr.* xv. ii. 847, *as to syn. Sond., but not as to description. A. peduncularis, var. caperonioides, Müll. Arg. in DC. Prodr.* xv. ii. 846. *A. peduncularis, Gibbs in Journ. Linn. Soc.* xxxvii. 470 ; *Hutchinson in Dyer, Fl. Trop. Afr.* vi. i. 884, *partly and as to Rhodesian plant only ; not of E. Meyer. A. transvaaliensis, Gandog. in Bull. Soc. Bot. Fr.* lx. 27.

VAR. β, **Galpini** (Prain in Kew Bulletin, 1913, 23); leaves membranous, persistently densely shaggy with long bulbous-based hairs.

KALAHARI REGION : Orange River Colony ; Parys, *Rogers*, 707! Transvaal ; various localities, *Burke*, 83! 153! *Zeyher*, 1521! *Schlechter*! *Wilms*, 1323! 1324! 1328! 1329! *Miss Leendertz*, 321! 2583! *Rogers*, 24! *Marloth*, 3866! *Engler*, 2833! *Fehr*, 57! *Rehmann*, 4553! 4554! 5485! *Bester*, 2164, partly! *Burtt-Davy*, 1988! 2305! 7171! 9211, partly! Modderfontein, *Miss Haagner*! Barberton, *Miss Thorncroft*, 30! Var. β : Transvaal ; Barberton, 4000 ft., *Galpin*, 1106!

Apparently a very distinct species with female bracts most like those of *A. punctata*, but very much firmer in consistence and hardly glandular. As regards its foliage this most resembles *A. Wilmsii*, Pax, but the absence of stipitate glands from the undersurface prevents *A. caperonioides* from being mistaken for that species which, moreover, has different female bracts. The foliage in Galpin, 1106, differs from that which is usual in the species in being thinner and more densely shaggy with long grey bulbous-based hairs. This plant may eventually prove to represent a distinct species; it agrees with *A. caperonioides* in having almost entire style-arms.

18. **A. punctata** (Meisn. apud Krauss in Flora, 1845, 83); herbaceous ; stems rather stout, rather copiously branched, 1–2 ft. high, from a stout perennial woody stock ; branches virgate, striate, pubescent to rather copiously patently setulose ; leaves membranous, numerous, very shortly petioled, ovate or ovate-lanceolate, the lowest usually obtuse, the uppermost acute, base rounded, margin sharply often coarsely serrulate, 1½–2½ in. long, 1–1½ in. wide, pubescent to strigose, especially on the nerves on both surfaces and on the margins, rarely almost glabrous, between the reticulations usually very distinctly glandular-punctate ; petiole pubescent or setulose, 1–1½ lin. long ; stipules setulose, lanceolate, persistent, 1½–2 lin. long ; inflorescence 1-sexual, diœcious ; male spikes axillary, solitary, peduncled ; peduncles puberulous to pubescent and glandular, 2½–5 in. long ; spikes cylindric, dense-flowered, rather stout, 1¼–1½ in. long ; bracts lanceolate, puberulous or pubescent, ⅓ in. long, spreading, persistent ; buds puberulous or glabrous ; female spikes terminal, solitary, sessile, 1 in. (ultimately 2 in.) long ; bracts 1-flowered, subsessile, leafy, ovate-lanceolate, acuminate, ½–¾ in. long, nearly as wide, deeply toothed, teeth lanceolate, acute, rather densely pubescent and beset with stipitate glands outside ; sepals 3, ovate, acute, glandular ; ovary distinctly 3-lobed, pubescent and glandular in the upper half ; styles 3, united in their lower third, 1 in. long, markedly laciniate upwards ; seeds subglobose. *Prain in Kew Bulletin, 1913, 23. A. peduncularis, Pax in Engl. Pfl. Ost-Afr. C. 239 partly and as to Nyasaland plant ; Hutchinson in Dyer, Fl. Trop. Afr. vi. i. 884, partly and as to Nyasaland and Gazaland specimens ; Gandog. in Bull. Soc. Bot. Fr. lx. 27 ; not of E. Meyer. A. peduncularis, var. punctata, Müll. Arg. in Linnæa, xxxiv. 28, and in DC. Prodr. xv. ii. 846. Ricinocarpus peduncularis, var. punctatus, O. Kuntze, Rev. Gen. Pl. iii. ii. 292.*

VAR. β, **longifolia** (Prain in Kew Bulletin, 1913, 24) ; stems 1–2 ft. high ; leaves lanceolate or linear-lanceolate, acute, 2–2½ in. long, ¼–½ in. wide, otherwise as in the type. *A. longifolia, E. Meyer ex Prain, l.c.*

VAR. γ, **Rogersii** (Prain, l.c. 24) ; dwarf ; stems 4–6 in. high ; leaves oblong or narrow-oblong, obtuse, ½–1 in. long, ⅛–½ in. wide, otherwise as in the type.

KALAHARI REGION : Orange River Colony ; Besters Vlei, near Witzies Hoek, 5700 ft., *Bolus*, 8251 ! and without precise locality, *Rogers* ! Transvaal ; Barberton, 2500–4000 ft., *Galpin*, 429 ! Mac Mac Creek, *Mudd* ! Var. β : Transvaal ; Witklip, 4800 ft., *Burtt-Davy*, 7264 ! Barberton, Fairview Farm, *Burtt-Davy*, 4080 ! Aapies River, *Scott Elliot*, 1449 ! *Rehmann*, 4016 ! 4283 ! *Miss Leendertz*, 1003 ! Derde Poort, *Miss Leendertz*, 363, partly ! Var. γ : Transvaal ; Waterval Boven, 4800 ft., *Rogers*, 288 ! Saddleback Mountain, 4500 ft., *Galpin*, 1121 ! Shilovane, *Junod*, 1325 ! Swaziland, *Stewart*, 8917 !

EASTERN REGION : Transkei ; Kentani, 1200 ft., *Miss Pegler*, 65, partly ! Pondo-
land ; Fort Grosvenor, *Bachmann*, 785 ! 789 ! 794 ! Griqualand East ; near
Kokstad, 3800 ft., *Tyson*, 1107, partly ! Clydesdale, 2500 ft., *Tyson*, 3107 !
Natal ; various localities, *Krauss*, 377, mainly ! *Verreaux* ! *Gueinzius* ! Bellair,
220 ft., *Schlechter*, 3105 ! *Wood*, 296 ! 697 ! 864 ! 4106 ! 6201 ! *Sanderson*, 344 !
Rudatis, 779 ! 780 ! 1201 ! *Kuntze*, 5182 ! *Sutherland* ! *Rehmann*, 6891 ! 6995 !
Zululand ; without locality, *Gerrard*, 1167 ! Var. β : Transkei ; between the
Gekua and Bashee Rivers, *Drège*, 5380 ! Tembuland ; Bazeia, 2000 ft., *Baur*,
269 ! Pondoland ; Fort Grosvenor, *Bachmann*, 788 ! Griqualand East ; near
Kokstad, 4800 ft., *Tyson*, 1107, partly ! *and in MacOwan & Bolus, Herb. Austr.-
Afr.* 1231 ! Natal ; Maritzburg, *Rehmann*, 7616 ! Camperdown, *Rehmann*, 7794 !
Dargle Road, 3000-4000 ft., *Wood*, 12605 !

Also in Gazaland and Nyasaland.

A very distinct species. The dwarf plant here treated as a variety (γ *Rogersii*)
of *A. punctata*, Meisn., may prove, when more fully known, a distinct species.

19. **A. Wilmsii** (Pax in Herb. Berol. ex Prain in Kew Bulletin,
1913, 24) ; almost shrubby ; stems firm, woody, sparingly branched,
1-2 ft. high, from a stout woody stock ; bark striate, pubescent to
rather copiously patently setulose ; leaves membranous, numerous,
very shortly petioled, ovate or ovate-lanceolate, the lowest often
obtuse, the others usually acute, base rounded or shallow-cordate,
margin shortly toothed, 2-3 in. long, 1½-2 in. wide, pubescent to
strigose (especially on the nerves) on both surfaces and with
usually numerous long stipitate glands on both surfaces on the finer
veins and distinctly glandular-punctate between the reticulations ;
petiole setulose, 1 lin. long ; stipules setulose, lanceolate, persistent,
1½ lin. long ; inflorescence 1-sexual, diœcious ; male spikes axillary,
solitary, peduncled ; peduncles pubescent to densely setulose and
glandular, 1-1½ in. long ; spikes cylindric, dense-flowered, rather
slender, 1-1¼ in. long ; bracts lanceolate, pubescent, ⅕ in. long,
spreading, persistent ; buds softly pubescent to setulose ; female
spikes terminal, solitary, sessile, 1 in. (ultimately 2¾ in.) long ; bracts
1-flowered, subsessile, leafy, wide ovate-cordate, acute, margin
toothed, ⅔-¾ in. long, 1-1¼ in. wide, teeth short, triangular,
pubescent or setulose and densely beset with long stipitate glands
outside ; sepals 3, acute, pubescent and glandular ; ovary distinctly
3-lobed, softly pubescent and glandular ; styles 3, united in their
lower fourth, ¾ in. long, markedly laciniate upwards ; seeds subglobose. *Ricinocarpus peduncularis, var. genuinus, O. Kuntze, Rev.
Gen. Pl.* iii. ii. 292 ; *not R. peduncularis, O. Kuntze, l.c.* ii. 618.
R. peduncularis, var. Radula, O. Kuntze, l.c. iii. ii. 292 ; *not
Acalypha Radula, Baker.*

KALAHARI REGION : Orange River Colony ; Harrismith, *Sankey*, 238 ! Bethlehem,
Richardson ! Transvaal ; Spitzkop, *Wilms*, 1326 ! Witklip, *Burtt-Davy*, 7260 !
Crocodile River, *Wilms*, 1330 ! Paarde Plaats, *Wilms*, 1331 ! Ermelo, *Tennant in
Herb. Burtt-Davy*, 6807 ! *Burtt-Davy*, 9390 ! *Miss Leendertz*, 2997 ! Billys Vlei,
Burtt-Davy, 9211, mainly ! Saddleback Mountain, near Barberton, 4500-4800 ft.,
Galpin, 1119 ! 1126 !

EASTERN REGION : Pondoland ; Fort Grosvenor, *Bachmann*, 787 ! Griqualand
East ; Clydesdale, 2500 ft., *Tyson*, 2602 ! 2603 ! Natal ; near Mooi River, 4000
ft., *Wood*, 4103 ! Highland Station, 5300 ft., *Kuntze*, 5655 ! Van Reenens Pass,
Kuntze, 139 !

Very nearly related to *A. punctata,* Meisn., but readily distinguished by its larger leaves beset with stalked glands springing from the finer reticulations. When the leaves are mature the pubescence and these stalked glands largely disappear. The difference thus caused has induced Kuntze to separate the two conditions, that with stalked glands on leaves as well as bracts being his var. *Radula,* that from which the stalked glands have disappeared from the leaves being his *Ricinocarpus peduncularis,* var. *genuinus,* which is therefore quite unlike the true *Acalypha peduncularis.* It seems better to treat the two forms which occur as merely conditions or states, not even as varieties; even in the adult stage *A. Wilmsii* is readily distinguished from *A. punctata* by the very different bracts.

XXX. ALCHORNEA, Sw.

Flowers diœcious, rarely monœcious; petals 0; disc usually 0. *Male: Calyx* closed in bud, splitting into 4 (rarely 3 or 2) valvate lobes. *Stamens* 8 or fewer; filaments free or nearly so; anthers dorsifixed, cells parallel or slightly divergent, free at the base, dehiscing longitudinally. Rudimentary *ovary* 0. *Female: Calyx-lobes* 3–6, usually 4, imbricate. *Ovary* 2–3-celled; ovules solitary in each cell; styles long, linear, free, usually entire. *Capsule* 2–3-coccous; cocci 2-valved; valves with a crustaceous endocarp. *Seeds* subglobose, ecarunculate; testa crustaceous; albumen fleshy; cotyledons broad, flat.

Trees or shrubs; leaves alternate, entire or toothed, often 3–5-nerved at the base and frequently glandular between the nerves at the base beneath; male flowers in axillary or lateral simple or branched spikes, several to a bract; female flowers in axillary spikes or racemes, solitary to a bract; bracts usually small or very small, sometimes long, linear.

DISTRIB. Species about 35, widely spread throughout the tropics.

Leaves narrowed to the base, penninerved throughout, not stipellate at apex of petiole; male spikes panicled; bracts minute ...   . (1) **glabrata.**

Leaves rounded and 3-nerved at the base, 2-stipellate at apex of petiole; male spikes simple; bracts long, lanceolate   ...   . (2) **Schlechteri.**

1. **A. glabrata** (Prain in Kew Bulletin, 1910, 342); a shrub or small tree up to 40 ft. high; twigs not perulate, minutely puberulous or pubescent; leaves short-petioled, membranous, oblong or oblanceolate-oblong, acute or shortly acuminate, gradually narrowed to the base, margin entire or crenately toothed, 4–6 in. long, 1–2½ in. wide, penninerved throughout, nerves rather prominent beneath, glabrous on both sides; petiole puberulous or glabrous, ⅓–⅓ in. long; stipules subulate, caducous, ¼ in. long; male spikes in lax terminal panicles up to 6 in. long, individual spikes interrupted, 2–3 in. long; rhachis slender, puberulous or nearly glabrous; flowers glomerulate, bracts and bracteoles minute; female spikes simple, 2–3 in. long, rhachis slender; flowers solitary to each bract; ovary glabrous; styles

4–5 lin. long; capsule glabrous, smooth, dark-brown, $\frac{1}{3}$ in. across; seeds globose. *Prain in Dyer, Fl. Trop. Afr.* vi. i. 916. *A. floribunda, var. glabrata, Müll. Arg. in Journ. Bot.* 1864, 336, *and in DC. Prodr.* xv. ii. 905; *Hiern in Cat. Afr. Pl. Welw.* i. 979.

EASTERN REGION: Zululand: Ongoo, *Gerrard*, 2160! Also in Tropical Africa.

2. **A. Schlechteri** (Pax in Engl. Jahrb. xliii. 221); a diœcious shrub; twigs perulate, softly sparingly pubescent; scales ovate-oblong, scarious, brown; leaves distinctly petioled, thinly membranous, oblong or obovate, acuminate, base rounded, margin crenulate-toothed, 4 in. long, 2–2$\frac{1}{4}$ in. wide, 3-nerved from the base, midrib with about 3 ascending nerves on each side above the basal pair, glabrous above, when young sparingly hirsute with long spreading hairs especially on the nerves but soon glabrous except for tufts of hairs in the angles of the main-nerves and midrib beneath, glands at the base beneath inconspicuous, the stipels at the apex of the petiole above filiform, 2 lin. long; petiole softly pubescent or puberulous, slender, $\frac{3}{4}$–1$\frac{1}{2}$ in. long, channelled above; stipules filiform, $\frac{1}{4}$–$\frac{1}{3}$ in. long, deciduous; male spikes lax, lateral, 3–4 in. long, perulate at the base like the leafy twigs but with no leaves below the flowers; rhachis softly pubescent, naked or with a few empty bracts below, above with long lanceolate bracts $\frac{1}{8}$–$\frac{1}{4}$ in. long, each subtending a cluster of 4–10 flowers; female flowers long-pedicelled, subsolitary at the end of leafy twigs, or in few-flowered racemes with each flower solitary to a bract; ovary closely and softly adpressed-pubescent; styles $\frac{1}{4}$–$\frac{1}{3}$ in. long; capsule nearly glabrous, smooth, dark-brown, $\frac{1}{4}$ in. across; seeds globose, slightly 2-sulcate on the inner face, faintly verrucose on the back.

EASTERN REGION: Delagoa Bay; Lourenço Marques, 100 ft., *Forbes*! *Schlechter*, 11530! 11531! *Junod*!

Also in Tropical Portuguese East Africa.

Nearly allied to *A. Engleri*, Pax, and *A. yambuyaensis*, De Wild., and like them a member of the section *Stipellaria*, Benth., which is perhaps best treated as a genus distinct from *Alchornea*, but which should include at least the African species of *Lepidoturus*, Baill. From the two nearly allied species, *A. Schlechteri* differs in being diœcious.

XXXI. MACARANGA, Thouars.

Flowers diœcious, rarely monœcious; petals 0; disc 0. *Male:* *Calyx* closed in bud, splitting into 3–4 valvate lobes. *Stamens* usually few, occasionally solitary, rarely numerous; filaments very short, usually free; anthers short, usually 4-celled and 4-valved, but in the only South African species almost always 2-celled and 2-valved. Rudimentary *ovary* 0. *Female:* *Calyx* truncate or toothed, ultimately wide cupular or obliquely spathaceous, rarely splitting into 2–3 lobes. *Ovary* 2–3- (rarely 4–6-) celled, sometimes by abortion 1-celled ; ovules solitary in each cell ; styles short, stout, entire, free or slightly connate at the base. *Capsule* 2- to several- (rarely 1-) coccous ; cocci 2-valved. *Seeds* globose ; testa crustaceous ; albumen fleshy ; cotyledons broad, flat.

Trees or shrubs ; leaves alternate, petioled, simple or lobed, usually 3–7-nerved from the base, often glandular-punctate beneath ; racemes or spikes lateral or forming a terminal thyrsoid panicle ; male flowers glomerulate, several to a bract ; females solitary to their bracts ; bracts entire or toothed or fimbriated.

DISTRIB. Species about 100, spread throughout the tropics of the Old World.

1. **M. capensis** (Benth. in Benth. & Hook. f. Gen. Plant. iii. 320) ; a dioecious tree 20–30 ft. high ; trunk 1–2 ft. thick ; twigs rusty-tomentose ; leaves long-petioled, coriaceous, orbicular-ovate, rather abruptly acuminate, base rounded with a minute notch at the point of junction with the petiole, or sometimes there narrowly peltate, margin entire or very shortly denticulate, 3–6 (rarely in young plants up to 12) in. long, 3–4 (rarely in young plants up to 8) in. wide, 3–5-nerved from the base, pinnately nerved with 7–9 secondary nerves on each side of the median nerve beyond the base, the nerves all raised beneath, dark green, glabrous above or when young rusty-puberulous on the nerves only, beneath rusty-puberulous, but soon glabrescent and copiously gland-dotted throughout, with the lower two-thirds of the median nerves and the lower portion of the secondary nerves rather copiously pubescent with spreading long hairs ; petiole glabrous, 2–5 in. long ; stipules oblong-lanceolate, copiously rusty-puberulous, $\frac{3}{4}$ in. long, soon deciduous ; flowers in lax axillary panicles, male up to 3 in. long, 2 in. across, female up to 2 in. long, $1\frac{1}{2}$ in. across ; bracts in both sexes lanceolate, reflexed, entire or sparingly toothed, densely rusty-pubescent ; male flowers sessile, several to each bract ; calyx 2–3-lobed, rusty-pubescent ; stamens 2–3, at first 4-celled, when fully developed 2-celled : female flowers very shortly pedicelled, solitary to their bracts ; calyx distinctly 2–3-fid ; lobes ovate, acute, rusty-pubescent ; ovary glabrous, with a resinous or waxy covering, 2-celled with 2 short free styles or as frequently by abortion 1-celled with a slightly excentric apical style and sometimes a minute rudimentary lateral style lower down ; capsule small, viscous, globose, $\frac{1}{4}$ in. across : valves coriaceous, very tardily dehiscent ; seed globose. *Sim, For. Fl. Cape Col.* 314, t. 139. *M. Bachmanni, Pax in Engl. Jahrb.* xxiii. 525. *Mappa capensis, Baill. Étud. Gén. Euphorb.* 430 (*sphalm. Adenoceras*), *and Adansonia,* iii. 155 ; *E. Meyer ex Harv. Gen. S. Afr. Pl. ed.* 2, 338. *Mallotus capensis, Müll. Arg. in Linnæa,* xxxiv. 189, *and in DC. Prodr.* xv. ii. 966. *Urticacea,* 4612, *Drège, Zwei Pfl. Documente,* 155.

EASTERN REGION : Pondoland ; between Umtentu River and Umzimkulu River, *Drège,* 4612! Port St. John, Devils Peak, 800 ft., and banks of St. Johns River, 20 ft., *Galpin,* 3444! St. Johns up to 1500 ft., *Sim,* 2417! *Bachmann,* 774!

Natal ; Dumisa, 2000 ft., *Rudatis,* 740 ! 746 ! Inanda, *Wood,* 875 ! Clairmont, *Wood,* 5526 ! Krantz Kloof, 1800 ft., *Wood,* 253 ! 544 ! Nobote River, *Gerrard,* 1 ! 1911 ! Zululand ; Ngoye, 1–2000 ft., *Wood,* 11551 ! and without precise locality, *Gueinzius* ! *Cooper,* 3142 !

XXXII. RICINUS, Linn.

Flowers monœcious ; petals 0 ; disc 0. *Male : Calyx* closed in bud, splitting into 3–5 valvate segments. *Stamens* very numerous ; filaments connate in repeatedly branching clusters ; anther-cells distinct, divaricate, distant, subglobose, dehiscing longitudinally. Rudimentary *ovary* 0. *Female : Calyx* normally spathaceous, very caducous. *Ovary* 3-celled ; ovules solitary in each cell ; styles short or long, spreading, more or less plumose, 2-fid or 2-partite, or occasionally entire. *Capsule* 3-dymous, breaking up into three 2-valved cocci. *Seed* ovoid-oblong, carunculate ; testa crustaceous ; albumen fleshy ; cotyledons broad, flat.

A tall glabrous annual, usually glaucous ; in warmer regions a perennial shrub or small tree ; leaves large, alternate, peltate, palmately 7–13-lobed, lobes serrate ; racemes more or less paniculate at the ends of the branches ; flowers rather large, the uppermost male, crowded, the lower female, shortly pedicelled ; capsules smooth or echinate.

DISTRIB. A single variable species, possibly a native of Tropical Africa, widely cultivated and often naturalised in all warm countries.

1. R. communis (Linn. Sp. Pl. ed. i. 1007) ; a small evergreen tree ; young shoots and panicles usually glaucous ; leaves large, long-petioled, membranous, palmately lobed, lobes 7–13, oblong to linear, acute or acuminate, their margin glandular-serrate, blade 1–2 ft. across, glabrous, green or reddish especially when young ; stipules green or red ; racemes stout, erect, dense or open ; male flowers about ½ in. across ; female calyx ⅓–½ in. long ; capsule 3-coccous, subglobose, ellipsoid or oblong, usually echinate, sometimes smooth or nearly so, ½–¾ in. long ; seeds with a prominent subglobose caruncle, oblong, ⅓–⅔ in. long, testa smooth, shining, mottled. *Thunb. Prodr.* 117 ; *Bot. Mag. t.* 2209 ; *Baill. Étud. Gén. Euphorb. t.* 10, 11, *figs.* 1–5, *and Adansonia,* iii. 149 ; *Bentl. & Trim. Med. Pl.* iv. *t.* 237 ; *Müll. Arg. in DC. Prodr.* xv. ii. 1017 ; *Harv. Gen. S. Afr. Pl. ed.* 2, 336 ; *Hook. f. Flora Brit. Ind.* v. 457 ; *Prain in Dyer, Fl. Trop. Afr.* vi. i. 945. *R. lividus, Jacq. Misc.* ii. 360, *and Ic.* i. *t.* 196 ; *Schkuhr, Handb. t.* 312, *fig.* 1 ; *Reichb. Ic. Bot. Exot. t.* 153 ; *Krauss in Flora,* 1845, 82. *R. communis,* δ *sanguineus, Baill. Adansonia,* i. 342. *R. communis,* δ *lividus, Müll. Arg. l.c.* 1018.

COAST AND EASTERN REGIONS : cultivated and naturalised, also cultivated in the Eastern Transvaal.

The only form, or variety, of which specimens have been sent from South Africa to Kew is var. *lividus,* Müll. Arg. (*R. lividus,* Jacq.), in which the leaves are dark red above, paler red with dark red veins beneath, red stipules, a rather dense fruiting panicle and an ellipsoid rather densely and stoutly echinate capsule : doubtless, however, other forms are also to be met with.

XXXIII. ADENOCLINE, Turcz.

Flowers diœcious, very rarely monœcious; petals 0; disc glandular. *Male: Calyx* deeply 5-lobed, open in bud; lobes imbricate. *Stamens* usually 10, sometimes fewer or more (6–12); filaments short, free, the outer alternate with the calyx-lobes, inserted round the central gland-bearing receptacle; anthers terminal, 2-celled; cells distinct, globose, divaricate, opening from the apex, and each at length 2-valved. Rudimentary *ovary* 0. *Female : Calyx* deeply 5-lobed; lobes as long as the ovary, imbricate. *Disc* of 3 broad glands alternating with carpels. *Ovary* 3-celled; ovules solitary in each cell; styles slender, 2-partite, recurved, slightly united at the base. *Capsule* 3-dymous, breaking up into 2-valved cocci. *Seeds* globose; testa thinly crustaceous; albumen fleshy; cotyledons rather narrow, flat.

Slender, erect or diffuse, usually firm, rarely succulent herbs; leaves alternate or in one species opposite, narrow or broad, entire or toothed; flowers very small, in axillary cymules often passing into a terminal raceme or panicle; male flowers usually several; female flowers few or solitary, usually forming uniparous cymes, their pedicels abruptly reflexed.

DISTRIB. Eight endemic species.

Diœcious plants; leaves herbaceous.
Leaves alternate :
  Leaves all sessile, ovate, serrate; stipules large,
    foliaceous ... ... (1) **stricta.**
  Leaves petioled below :
    Upper leaves petioled as well as the lower :
      Stipules rather large, foliaceous... ... ... (2) **ovalifolia.**
      Stipules very small ... ... (3) **humilis.**
    Upper leaves sessile or subsessile :
      Lower petioled leaves narrow-lanceolate, acute :
        Upper sessile leaves entire ... ... (4) **sessilifolia.**
        Upper sessile leaves serrate ... ... (5) **serrata.**
      Lower petioled leaves orbicular or ovate, obtuse (6) **bupleuroides.**
  Leaves opposite, petioled ... ... (7) **Mercurialis.**
Monœcious plants; leaves flaccid, below subopposite ... (8) **procumbens.**

1. **A. stricta** (Prain in Kew Bulletin, 1912, 338); a diœcious herb, woody below, strictly virgately branching upwards, 8–9 in. high; internodes ⅓–½ in. long; leaves alternate, sessile, firmly membranous, oblong to oblong-lanceolate, acute, base broad confluent with the stipules, margin sharply serrate, ⅓–½ in. long, ⅓–¼ in. wide; stipules like the leaves but slightly smaller, ⅛–¼ in. long, ⅙–⅓ in. wide; male flowers not seen; female flowers solitary, leaf-opposed, their pedicels abruptly reflexed; calyx-segments 5, their margins serrate; ovary glabrous; capsule 1½ lin. wide, 3-dymous, smooth. *Prain in Ann. Bot.* xxvii. 406.

COAST REGION: Bredasdorp Div. ; Reit Fontein Poort, 100–200 ft., *Bolus,*
8603 ! *Schlechter,* 9694 !
A remarkably distinct species.

2. **A. ovalifolia** (Turcz. in Bull. Soc. Imp. Nat. Mosc. xvi.
60) ; a
dioecious herb with numerous slender procumbent laxly branching
stems radiating from a vertical perennial woody base, which is
3–4 in. long, the herbaceous stems 1–1½ ft. long ; leaves alternate,
distinctly petioled, membranous, below ovate, acute, base rounded,
gradually narrowing upwards to ovate-lanceolate with base cuneate,
margin closely sharply serrate, ½–⅔ in. long, ¼–⅓ in. wide, glabrous
on both surfaces ; petiole ¼–⅓ in. long below, ⅛ in. long towards the
ends of the branches, glabrous ; stipules foliaceous, often rather
large and ovate-lanceolate, ¼–⅓ in. long, with margin serrate or
subentire, but usually lanceolate, ⅛–¼ in. long, with margin deeply
incised ; male flowers in axils of reduced leaves towards ends of
branches, several to an axil ; female flowers solitary, leaf-opposed,
their pedicels abruptly reflexed ; calyx-segments in both sexes 5 ;
stamens usually 10 ; ovary glabrous ; capsule 1½ lin. wide,
3-dymous, smooth ; seeds ovoid-globose, pale greyish-green. *Turcz.
in Flora,* 1844, 121, *and in Bull. Soc. Imp. Nat. Mosc.* xxv. ii. 179 ;
Prain in Ann. Bot. xxvii. 406.. *A. humilis, Baill. Étud. Gén. Euphorb.*
457 ; *not of Turcz. A. pauciflora, β ovalifolia, Müll. Arg. in DC.
Prodr.* xv. ii. 1139, δ *bupleuroides, Müll. Arg. l.c.* 1140, *partly and
as to Zeyher* 1516 *only, and* ε *serrata, Müll. Arg. l.c.* 1140. *Trian-
thema debilis, Spreng. ex Turcz. in Bull. Soc. Imp. Nat. Mosc.* xvi.
60, *and T. dubium, Spreng. ex Turcz. l.c.* xxv. ii. 179. *Mercurialis
serrata, Meisn. in Hook. Lond. Journ. Bot.* ii. 557 ; *Krauss in Flora,*
1845, 84. *M. bupleuroides ?, Kunze in Linnæa,* xx. 54, *mainly, but
excl. M. Zeyheri, Kunze ; Baill. Adansonia,* iii. 159, *partly ; not of
Meisn. Diplostylis serrata, Sond. in Linnæa,* xxiii. 114, *partly and as
to Zeyher,* 1516 *and Ecklon & Zeyher,* 39, *partly.*

VAR. β, **rotundifolia** (Prain, l.c.) ; leaves below suborbicular, subacute, base
rounded ; above ovate, acute, base rounded, otherwise as in the type. *A. pauciflora, a rotundifolia, Müll. Arg. in DC. Prodr.* xv. ii. 1139.

SOUTH AFRICA : without locality, *Masson*! *Harvey,* 626 ! *and cultivated
specimens* !
COAST REGION : Cape Div. ; Hout Bay, 0–150 ft., *Krauss,* 1190 ; *Bolus,* 7059 !
and in Herb. Norm. Austr.-Afr. 1093 ! *Schlechter,* 958 ! *Wolley-Dod,* 1652 !
Simonstown, by Noahs Ark Battery, *Wolley-Dod,* 2819, partly ! Sandhills on Cape
Flats. *Wolley-Dod,* 1882 ! Wynberg Hill, *Harvey*! Lion Mountain, *Harvey*!
Swellendam Div. ; without precise locality, *Mund (Ecklon & Zeyher,* 39, partly) !
Riversdale Div. ; Kaffir Kuils River, *Stack (Zeyher,* 1516) ! East London Div. ;
East London, *Wood,* 3137 ! Komgha Div. ; near Kei River Mouth, *Flanagan,*
842 ! Var. β : Cape Div. ; Simonstown, by Noahs Ark Battery, *Wolley-Dod,*
2819, mainly !

3. **A. humilis** (Turcz. in Bull. Soc. Imp. Nat. Mosc. xvi. 61) ; a
dioecious herb with many filiform ascending stems, rather virgately
branched, springing from a slender perennial woody base which is

1–3 in. long, the herbaceous stems ½–1 ft. long; leaves alternate, distinctly petioled, thinly membranous, below ovate or oblong or suborbicular, acute rarely obtuse, base shallow-cordate, gradually narrowed upwards to ovate-lanceolate with cuneate base, sometimes the lower leaves cuneate, very rarely the upper leaves cordate at the base, margin wide shallow-crenate rarely serrate, sometimes quite entire, ⅓–½ in. long, ⅓–½ in. wide, glabrous on both surfaces; petiole ⅕–¼ in. long below, ⅛–¼ in. long above, glabrous; stipules small, lanceolate, 1 lin. long or less, entire or faintly denticulate; male flowers in axils of uppermost leaves, several to an axil; female flowers not seen; male calyx-segments 5; stamens 6–8. *Turcz. in Flora*, 1844, 121, *and in Bull. Soc. Imp. Nat. Mosc.* xxv. ii. 179, *excl. syn. Meisn. and Sond.; Prain in Ann. Bot.* xxvii. 407. *A. pauciflora, Turcz. in Bull. Soc. Imp. Nat. Mosc.* xvi. 60, *as to the male plant with alternate leaves only. A. pauciflora, γ humilis, Müll. Arg. in DC. Prodr.* xv. ii. 1140, *excl. syn. Kunze, and θ tenella, Müll. Arg. l.c. Mercurialis tenella, Meisn. in Hook. Lond. Journ. Bot.* ii. 556; *Krauss in Flora*, 1845, 84. *M. triandra, Sond. in Linnæa,* xxiii. 113, *and Baill. Adansonia,* iii. 160, *as to syn. Meisn.; not of E. Meyer. M. bupleuroides, Baill. Adansonia,* iii. 159, *partly ; not of Meisn. M. pauciflora, Baill. l.c. partly. Euphorbiacea,* 3441, *Drège, Zwei Pfl. Documente,* 45, *partly.*

COAST REGION: Mossel Bay Div.; banks of the Great Brak River, *Burchell,* 6157! Humansdorp Div.; Zitzikamma, *Krauss,* 1911; Stockenstrom Div.; Kat Berg, *Drège,* 3441, male plant only! 8223! Queenstown Div.; Hangklip Mountain, 6000 ft., *Galpin,* 1782!

CENTRAL REGION: Graaf Reinet Div.; Cave Mountains, 4400 ft., *Bolus,* 697, partly! 4570!

Nearly allied to *A. ovalifolia,* Turcz., but readily distinguished by the very small stipules; also nearly related to *A. serrata,* Turcz., but easily distinguished by the petioled upper leaves.

4. **A. sessilifolia** (Turcz. in Bull. Soc. Imp. Nat. Mosc. xvi. 61, partly, and excl. Drège, 1867); a diœcious herb with numerous rather wiry erect virgately branching stems 1–1½ ft. high, apparently springing from a woody base, internodes 1½–2 in. long; leaves alternate, below distinctly petioled, above sessile, firmly membranous, the basal narrow-lanceolate, acute, base rounded, the upper linear-lanceolate, acute, base relatively wide confluent with the linear stipules, margin entire, involute, ¼–⅓ in. long, 1 lin. or less wide, glabrous on both surfaces; stipules laciniate at the base on the side away from the leaf; male flowers in axils of uppermost leaves, several to an axil; female flowers leaf-opposed, solitary, their pedicels abruptly reflexed; calyx-segments in both sexes 5; stamens usually 10; ovary glabrous; capsule 1½ lin. wide, 3-dymous, smooth; seeds ovoid-globose, greyish. *Turcz. in Flora,* 1844, 121, *partly, and in Bull. Soc. Imp. Nat. Mosc.* xxv. ii. 180; *Prain in Ann. Bot.* xxvii. 408. *A. sessiliflora, Baill. Étud. Gén. Euphorb.* 457, *t.* 9, *fig.* 6. *A. pauciflora, η sessilifolia, Müll. Arg. in DC. Prodr.* xv. ii. 1140. *A. pauciflora, Pax in Engl. & Prantl, Pflanzenfam.* iii. v. 49, *fig.* 30 E :

not of Turcz. Diplostylis angustifolia, Sond. in Linnæa, xxiii. 113. *Mercurialis bupleuroides, Baill. Adansonia,* iii. 159, *partly; not of Meisn.*

COAST REGION: Clanwilliam Div. ; Brakfontein, near the Olifants River, *Ecklon & Zeyher,* 40 ! near Wupperthal, 1900 ft., *Bolus,* 9090 ! Malmesbury Div. ; between Mamre and Saldanha Bay, *Drège,* 1868, mainly ! Cape Div. ; Riet Valley, Laudenbach, *Ecklon & Zeyher,* 39, partly ! Uniondale Div. ; Kamanassie Mountains, near Avontuur, *Bolus,* 2460 !

5. **A. serrata** (Turcz. in Bull. Soc. Imp. Nat. Mosc. xxv. ii. 180, excl. syns.) ; a diœcious herb with numerous filiform decumbent stems rather virgately branched, springing from a perennial woody base, the herbaceous stems 1½–2 ft. long ; leaves alternate, the lower very shortly petioled, the upper subsessile or sessile, firmly membranous, lanceolate, acute, base cuneate, margin sparingly and sharply serrate, ⅓–1 in. long, 1 lin. or under wide, glabrous on both surfaces ; petiole ½–2 lin. long ; stipules lanceolate, foliaceous, free, 1½–2 lin. long, their margin serrate ; male flowers in axils of uppermost leaves, several to an axil ; female flowers leaf-opposed, solitary, their pedicels abruptly reflexed ; calyx-segments in both sexes usually 5 ; stamens 6–10 ; ovary glabrous ; capsule 1½ lin. wide ; 3-dymous, smooth ; seeds ovoid-globose, greenish-grey. *A. sessilifolia, Turcz. in Bull. Soc. Imp. Nat. Mosc.* xvi. 61, *partly and as to Drège,* 1867 *only. A. pauciflora,* γ *humilis, Müll. Arg. in DC. Prodr.* xv. ii. 1140, *partly and to syn. Mercurialis Zeyheri only;* δ *bupleuroides, Müll. Arg. l.c. partly and as to Drège,* 1867 *only, and ζ transiens, Müll. Arg. l.c. A. Zeyheri, Prain in Ann. Bot.* xxvii. 408. *Mercurialis bupleuroides, Meisn. in Hook. Lond. Journ. Bot.* ii. 557, *partly and as to Drège,* 1867 *only ; Kunze in Linnæa,* xx. 54, *partly and as to syn. Mercurialis Zeyheri only ; Baill. Adansonia,* iii. 159, *partly. M. Zeyheri, Kunze, l.c. Diplostylis serrata, Sond. in Linnæa,* xxiii. 114, *mainly, but excl. Zeyher,* 1516, *and Eckl. & Zeyh.* 39, *from Swellendam.*

SOUTH AFRICA : without locality, *Drège,* 1867 a (Herb. Kew)! 1868, partly ! *and cultivated specimens !*
COAST REGION : Knysna Div. ; between Groene Valley and Zwart Valley, *Burchell,* 5678 ! Humansdorp Div. ; Humansdorp, under 500 ft., *Rogers,* 3082 ! Uitenhage Div. ; Sand dunes near Uitenhage, *Laidley,* 112 ! *Ecklon & Zeyher,* 39, partly ! Port Elizabeth Div. ; at Walmer and Humerood, *Mrs. Paterson,* 1030 ! Albany Div., Grahamstown, *South* ! Howisons Poort, 2000 ft., *Glass,* 191 !
CENTRAL REGION : Graaff Reinet Div. ; near Graaff Reinet, 4000 ft., *Bolus,* 697, partly !

6. **A. bupleuroides** (Prain in Ann. Bot. xxvii. 407) ; a diœcious herb with numerous erect virgately branching stems rising from a woody base ; base horizontal, 2–3 in. long ; herbaceous stems 1–1½ ft. high ; leaves alternate, firmly membranous, those nearest the base short-petioled, orbicular or obovate, obtuse, ½–1 in. long, ½ in. wide, those above sessile, ovate-lanceolate, lanceolate or linear, acute, 1–2 in. long, ⅛–⅓ in. wide, base cuneate, margin sharply serrate, glabrous on both surfaces ; petiole of lower leaves ⅓–½ in.

long, glabrous; stipules free, lanceolate, foliaceous, $1\frac{1}{2}$–2 lin. long,
laciniate near the base; male flowers in axils of uppermost leaves,
several to an axil; female flowers solitary, leaf-opposed, their pedicels
abruptly reflexed; calyx-segments in both sexes usually 5; stamens
12–20; ovary glabrous; capsule $2\frac{1}{2}$ lin. wide, 3-dymous, smooth;
seeds ovoid-globose, nearly black. *A. humilis, Turcz. in Bull. Soc.
Imp. Nat. Mosc.* xxv. ii. 179, *as to the plants of Meisner and Sonder,
not of Turcz. l.c.* xvi. 61. *A. pauciflora,* δ *bupleuroides, Müll. Arg. in
DC. Prodr.* xv. ii. 1140, *as to Krauss,* 1169 *and Gueinzius,* 172 *only.
Mercurialis bupleuroides, Meisn. in Hook. Lond. Journ. Bot.* ii. 557,
as to Krauss, 1169 *only; Krauss in Flora,* 1845, 84; *Baill. Adansonia,* iii. 159, *partly. Diplostylis bupleuroides, Sond. in Linnæa,*
xxiii. 114. *D. longifolia, Sond. Mss. in Herb. Harv.*

VAR. β, **Peglerae** (Prain in Ann. Bot. xxvii. 408); leaves all orbicular or
obovate, obtuse.

COAST REGION : Uitenhage Div. ; Elands River, *Ecklon & Zeyher,* 39, partly !
Albany Div. ; near Grahamstown, *Miss Daly,* 903 ! Featherstones Kloof, 2500 ft.,
Galpin, 255 ! East London Div. ; near Nahoon, *Rattray,* 216 ! Komgha Div. ;
near the Gwenkala River, *Flanagan,* 695 !

EASTERN REGION : Transkei ; Krielis Country, *Bowker,* 286 ! Griqualand East ;
Mount Currie, 5300 ft., *Tyson,* 1798 ! 1833 ! Pondoland ; St. Johns River,
Drège, 5173 ! Natal ; near Durban, *Krauss,* 1169 ; *Gueinzius,* 172 ! above Pinetown, 2200 ft., *Wood,* 5003 ! Clairmont, *Wood,* 10636 ! Inanda, 1800 ft., *Wood,*
253 ! Inchanga, 2000–3000 ft., *Wood,* 7184 ! Tugela, *Haygarth in Herb. Wood,*
10155 ! Ingoma, *Gerrard,* 1168 ! Nottingham, *Buchanan,* 149 ! near Krantz Kloof,
1500 ft., *Schlechter,* 3179 ! Dumisa, 2000 ft., *Rudatis,* 634 ! Zululand ; Entumeni,
2000 ft., *Wood,* 741 ! Var. β : Transkei ; Kentani, *Miss Pegler,* 871 !

7. **A. Mercurialis** (Turcz. in Bull. Soc. Imp. Nat. Mosc. xvi. 60);
a diœcious scandent herb; stems 2–5 ft. long, sparingly branched,
herbaceous, distinctly striate, jointed; leaves opposite, long-petioled,
membranous, ovate, acute or acuminate, base cuneate or truncate or
cordate, margin crenate-serrate, 1–3 in. long, $\frac{3}{4}$–2 in. wide, glabrous
on both surfaces with a few filiform glands at apex of petiole ;
petiole 1–2 in. long, glabrous : stipules minute, lacerate; flowers in
terminal simple or branched racemes; males several in the axils of
reduced leaves; females solitary, leaf-opposed, their pedicels abruptly
deflexed; calyx-segments in both sexes usually 5; stamens 10–12 ;
ovary glabrous; capsule $1\frac{1}{2}$ lin. long, 3-dymous, smooth : seeds ovoid-
globose, brown. *Turcz. in Flora,* 1844, 121, *and in Bull. Soc. Imp.
Nat. Mosc.* xxv. ii. 179. *A. pauciflora, Turcz. in Bull. Soc. Imp. Nat.
Mosc.* xvi. 60, *as to Drège,* 3441 *female, in Flora,* 1844, 121, *and
in Bull. Soc. Imp. Nat. Mosc.* xxv. ii. 179 ; *Baill. Étud. Gén.
Euphorb.* 457, *partly. A. acuta, Baill. Étud. Gén. Euphorb.* 457 ;
Müll. Arg. in DC. Prodr. xv. ii. 1141 ; *Pax in Engl. & Prantl,
Pflanzenfam.* iii. v. 49, *fig.* 30 A–D ; *Prain in Ann. Bot.* xxvii.
404. *Mercurialis caffra, Krauss in Flora,* 1845, 84 ; *Baill. Adan-
sonia,* iii. 160. *M. caffra, a brevipes, Meisn. in Hook. Lond. Journ.
Bot.* ii. 558. *M. caffra,* β *longipes, Meisn. l.c.* ; *Krauss, l.c. M.
dregeana, Meisn. l.c.* 559 ; *Buching. ex Krauss, l.c. M. subcordata,
Buching. ex Krauss, l.c. M. pauciflora, Baill. Adansonia,* iii. 159,

partly. Diplostylis caffra, Sond. in Linnæa, xxiii. 115. *Acalypha
acuta, Thunb. Fl. Cap. ed. Schult.* 546 ; *Spreng. ex Steud. Nomencl. ed.*
2, i. 9. *Euphorbiacea,* 3441, *Drège in Zwei Pfl. Documente,* 45, *mainly.*

SOUTH AFRICA : without locality, *Thunberg* ! *Masson* !
COAST REGION : Stellenbosch Div. ; Hottentots Holland, *Zeyher,* 3842, partly !
Caledon Div. ; Oaks (Farm ?), *Prior* ! Swellendam Div. ; Voormans Bosch, *Zeyher,*
3842, partly ! George Div. ; forest near George, *Prior* ! Outeniqua mountains,
Rehmann, 259 ! Woodville, *Galpin,* 4575 ! Knysna Div. ; near Gouwkamma River,
Krauss, 1192 ! Ruigte Valley, *Drège,* 2301 a ! Groene Valley, *Burchell,* 5625 !
Humansdorp Div. ; near the Kromme River, *Drège,* 2301 b ! Uitenhage Div. ; near
Van Stadens River, *Drège,* 2301 c ! Port Elizabeth Div. ; Baakens River, *Burchell,*
4339 ! *Mrs. Paterson,* 838 ! Cape Recief, *Ecklon,* 611 ! *Ecklon & Zeyher,* 37 ! Krak-
kakamma Forests, *Zeyher,* 552 ! *Ecklon,* 828 ! Alexandria Div. ; Olifants Hoek,
Ecklon & Zeyher, 36 ! Bathurst Div. ; mouth of the Great Fish River, *Burchell,* 3785 !
between Sunday River and Fish River, *Thunberg* ! near Port Alfred, *Schönland,*
788 ! 1544 ! *Schlechter,* 2732 ! Albany Div. ; near Grahamstown, *Bolton* ! *Bolus,*
2681 ! *Penther* 914 ! Howisons Poort, *Mrs. Hutton* ! *MacOwan,* 318 ! Trapps
Valley, *Miss Anstey,* 9 ! Atherstone, *Rogers,* 3301 ! and without precise locality,
Bowker ! *Williamson* ! *MacOwan,* 222 ! Fort Beaufort Div. ; near Fort Beaufort,
Ecklon & Zeyher, 38 ! Stockenstrom Div. ; Kat Berg, *Drège,* 3441, mainly !
Shaw ! near Philipton, 2000–3000 ft., *Ecklon & Zeyher* ! East London Div. ; East
London, *Rattray,* 128 ! 158 !
CENTRAL REGION : Somerset Div. ; Somerset East, *Scott Elliot,* 643 !
KALAHARI REGION : Transvaal ; Rimers Creek, near Barberton, *Thorncroft,*
250 ! 5592 ! Highland Creek, 3000 ft., *Galpin,* 841 ! Houtbosch, *Rehmann,* 5963 !
Shilovane, *Junod,* 863 !
EASTERN REGION : Transkei ; Kentani, 1500 ft., *Herb. Umtata Convent,* 495 !
Tembuland ; Bazeia, 2500 ft., *Baur,* 128 ! Perie Forest, *Schönland,* 851 ! Pondo-
land ; woods near Fort Donald, 3500 ft., *Tyson,* 1782 ! and without precise
locality, *Bachmann,* 803 ! Griqualand East ; near the River Chivenka, 4300 ft.,
Bolus, 10006 ! Insizwa Mountains ; *Krook,* 906 ! Natal ; *Nototo, Gerrard,* 32 !
Ingoma, *Gerrard,* 1175 ! woods near Byrne, 3000 ft., *Wood,* 1816 ! Zuurberg
Bush, *Wood,* 1987 ! near Van Reenen, 5200 ft., *Schlechter,* 6958 ! Umlaas River,
Krauss, 156 ! and without precise locality, *Gerrard,* 545 ! *Cooper,* 3151 !

8. **A. procumbens** (Benth. ex Pax in Engl. & Prantl, Pflanzenfam.
iii. v. 49) ; a monœcious succulent herb ; stems green, glabrous,
suberect or diffuse, 4–8 in. long ; leaves below subopposite, above
alternate, long-petioled, very thinly membranous, ovate, acute, base
somewhat cordate, margin shortly, sharply and distantly serrate,
⅓–1 in. long, ¼–¾ in. wide, glabrous on both surfaces ; petiole
¼–1¼ in. long, glabrous ; stipules lanceolate, minute ; male flowers
in sessile or short peduncled axillary glomerules ; female flowers
bract-opposed, solitary, their pedicels abruptly reflexed ; calyx-
segments in both sexes linear-lanceolate, male calyx usually 5-,
sometimes 6-, rarely 7-partite ; female calyx regularly 5-partite ;
stamens 6–8 ; ovary glabrous ; capsule 1 lin. wide, 3-dymous, smooth ;
seeds ovoid-globose, blackish. *A. Mercurialis, Baill. Étud. Gén.
Euphorb.* 457, *partly* ; *not of Turcz. A. violæfolia, Prain in Ann.
Bot.* xxvii. 403. *Acalypha obtusata, Spreng. ex Steud. Nomencl. ed.* 2,
i. 10, *partly. Mercurialis annua, Drège ex E. Meyer in Drège, Zwei
Pfl. Documente,* 201 ; *not of Linn. M. tricocca, E. Meyer, l.c., in inten-
tion but only partly as to spec. (name only)* ; *Eckl. & Zeyh. ex Krauss
in Flora,* 1845, 85 *(name only)* ; *Sond. in Linnæa,* xxiii. 111, *excl.
syn. Thunb. M. violæfolia, Kunze, Ind. Sem. Hort. Lips.* 1846, *with*

diagn. and in Linnæa, xx. 55 *and* xxiv. 162 ; *Baill. Adansonia,* iii.
159, *excl. syn. Thunb. Paradenocline violæfolia, Müll. Arg. in DC.
Prodr.* xv. ii. 793, *name only.* P. *procumbens, Müll. Arg. l.c.* 1141,
excl. syn. Mercurialis procumbens, Linn.

SOUTH AFRICA : *cultivated specimens.*
COAST REGION : Clanwilliam Div. ; Zeekoe Vley, 400 ft., *Schlechter,* 8502 !
near Zwartbosch Kraal, 400–500 ft., *Schlechter,* 5174 ! Malmesbury Div. ; Kloof
near Hopefield, *Bachmann,* 1265 ! pass near Malmesbury, 900 ft., *Schlechter,* 1603 !
Paarl Div. ; near Paarl, 1000 ft., *Drège (M. annua)* ! Cape Div. ; Green Point,
Zeyher, 3842, partly ! *Harvey* ! Waterfall on Devils Mountain, *Harvey* ! Constantia,
800–1000 ft., *Krauss,* 1821 ; Oatlands Point, *Woolley-Dod,* 2922 ! Muizenberg,
Schlechter, 1278 ! Swellendam Div. ; Kinko River near Swellendam, *Zeyher,* 3842,
partly ! Uitenhage Div. ; Addo, *Drège,* 2346 (*M. tricocca* c) ! Alexandria Div. ;
Zuurberg Range, *Drège (M. tricocca* b, partly) !
CENTRAL REGION : Willowmore Div. ; between the Zwartberg Range and
Aasvogelberg, *Drège (M. tricocca* a in Herb. Brit. Mus. and Herb. Leiden) ! Graaff
Reinet Div. ; Oudeberg Mountains, 3000 ft., and near Graaff Reinet, 2500 ft.,
Bolus, 429 !

This species, owing to the structure of its anthers, has been considered by
Müller the type of a distinct genus, *Paradenocline.* Bentham has pointed out
that there is nothing in the character of the anthers to separate *Paradenocline*
from *Adenocline.* The plant should, however, be treated as the type of a distinct
section, with androgynous in place of 1-sexual diœcious inflorescences. As regards
its facies it differs from the other species of *Adenocline* and agrees with *Leidesia
obtusa,* Müll. Arg., which it greatly resembles and for which it has often been
mistaken. There is no specimen of this plant in the Linnean Herbarium, and
there is no foundation for the assertion of Müller that this is the species described
by Linnæus as *Mercurialis procumbens.* The first description of the species,
published by Kunze in 1846, was based on specimens grown in the Leipzig Botanic
Garden. Kunze's name also perpetuates a misapprehension, for the evidence
available goes to show that this species is not the plant which was named by
Hermann (Parad. Bat., App. 10) *Mercurialis africana dicoccos folio Violæ
tricoloris.*

XXXIV. GELONIUM, Roxb.

Flowers diœcious, rarely monœcious ; *petals* 0. *Male : Sepals* 5,
rarely 6, broad, often unequal, imbricate in bud. *Receptacle* glandular or if eglandular with a small extra-staminal disc. *Stamens*
6–60 ; filaments free, filiform ; anthers dorsifixed low down, introrse,
dehiscing longitudinally. Rudimentary *ovary* 0. *Female : Sepals*
5, imbricate, narrower than in the male. *Disc* with a membranous
margin, sometimes bearing very rudimentary staminodes. *Ovary*
2–3-celled ; ovules solitary in each cell ; styles 2–3, short, spreading,
shortly 2-fid or lacerate. *Fruit* globose or 3-dymous, capsular or
drupaceous ; endocarp woody, 3- or 2-celled or casually by abortion
1-celled, indehiscent or at length opening loculicidally or breaking
up into 2–3 cocci. *Seed* without a caruncle ; testa crustaceous with
a somewhat pulpy outer coat ; albumen fleshy ; cotyledons broad, flat.

Trees or shrubs, everywhere glabrous ; leaves alternate, entire or toothed, gland-dotted, often vesicular ; stipules connate, very caducous ; flowers small, in leaf-opposed sessile or subsessile cymose glomerules ; pedicels very short.

DISTRIB. Species about 18, widespread in the tropics of the Eastern Hemisphere from East and South Africa to India, Malaya and New Guinea.

Leaves acute or acuminate, 2 in. long or longer, their
    margins sharply spinulose-serrate throughout   ... (1) **serratum.**

Leaves obtuse, usually under 2 in. long, their margins
    coarse bluntly toothed at the apex only, elsewhere
    entire ... ... (2) **africanum.**

1. **G. serratum** (Pax & K. Hoffm. in Engl. Pflanzenr. Euphorb. Gelon. 23) ; a small tree, everywhere glabrous ; leaves short-petioled, oblong or obovate-oblong, acute or acuminate, base cuneate, margin spinulose-serrate throughout, 2–3 in. long, $1\frac{1}{4}$–$1\frac{3}{4}$ in. wide, subcoriaceous, nerves rather faint, reticulate, pellucid-dotted and somewhat vesicular ; petiole over 1 lin. long ; flowers diœcious, fascicled, male few, their pedicels very short ; male sepals 1 lin. long, with ciliate margin and a dorsal gland ; stamens 8 ; receptacle glandular ; female flowers and fruit not seen. *G. adenophorum, Sim, For. Fl. Port. E. Afr.* 105 ; *not of Müll. Arg.*

EASTERN REGION : Delagoa Bay ; Umbolosi, 50 ft., *Schlechter*, 11722 ! Chope ; near Quisiqui, *Sim.*

Most nearly allied to *G. zanzibarense*, Müll. Arg., but readily distinguished by its spinulose-serrate leaves and its diœcious flowers.

2. **G. africanum** (Müll. Arg. in DC. Prodr. xv. ii. 1129), a shrub or small tree, 6–20 ft. high, everywhere glabrous ; leaves short-petioled, obovate, somewhat truncately obtuse, base cuneate, margin coarsely toothed only at the apex, $\frac{3}{4}$–$1\frac{1}{2}$ (rarely $2\frac{1}{2}$) in. long, $\frac{2}{3}$–1 (very rarely $1\frac{1}{4}$) in. wide, chartaceous, nerves prominent, reticulate, pellucid-dotted ; petiole under 1 lin. long ; flowers diœcious, fascicled, male numerous, their pedicels $1\frac{1}{2}$ lin. long ; female pedicels in fruit about $\frac{1}{4}$ in. long ; male sepals 1 lin. long ; stamens 12–14 ; female sepals unequal, the outer 3 the smallest ; hypogynous disc often with 5–6 minute subulate staminodes ; styles short, recurved, 2-fid ; capsule 3-dymous, softly coriaceous, opening loculicidally ; seed subglobose, glabrous. *Sim, For. Fl. Cape Col.* 318, *t.* 103, *fig.* 7, *and in For. Fl. Port. E. Afr.* 105 ; *Pax in Engl. Pflanzenr. Euphorb. Gelon.* 21. *Ceratophorus africanus, Sond. in Linnæa,* xxiii. 121. *Suregada Ceratophora, Baill. Adansonia,* iii. 154. *S. africana, O. Kuntze, Rev. Gen. Pl.* ii. 619.

COAST REGION : Uitenhage Div. ; Uitenhage, *Prior* ! Albany Div. ; Kowei River, below 500 ft., *Ecklon & Zeyher*, 69 ! East London Div. ; Quigney, 50 ft., *Galpin*, 3337 ! East London, *Rattray*, 671 ! Komgha Div. ; woods near Komgha, 2000 ft., *Flanagan*, 63 !

CENTRAL REGION : Somerset East Div. ; Somerset, *Miss Bowker* !

EASTERN REGION : Transkei ; Kentani, 400–1000 ft., *Miss Pegler*, 1120 ! 1259 ! 1260 ! Tembuland ; Perie Forest, 2500 ft., *Galpin*, 5911 ! Natal ; near Durban, *Gueinzius*, 104 ! *Thode in Herb. Wood*, 4724 ! Notote, *Gerrard*, 1384 ! Delagoa Bay ; Ressano Garcia, 1000 ft., *Schlechter*, 11932 ! *Sim* ; Maputa, *Sim* ; Chope, *Sim.*

XXXV. PLUKENETIA, Linn.

Flowers monœcious; petals 0; disc 0.   *Male: Calyx* 4–5-lobed; lobes valvate.   *Stamens* 8–30; filaments very short, free, inserted on a fleshy raised receptacle; anthers short, erect, their cells distinct, opening longitudinally.   Rudimentary *ovary* usually 0. *Female: Calyx* 4–5-lobed; lobes imbricate.   *Ovary* normally 4-celled; ovules solitary in each cell; styles connate in a hollow cylindric or subglobose fleshy column crowned by the short or very short free stigmatic lobes.   *Capsule* in our species depressed 4-dymous, breaking up into 2-valved cocci; cocci dorsally winged.   *Seed* globose or compressed; testa crustaceous, sometimes winged; albumen fleshy; cotyledons broad, flat.

Slender, twining, rarely erect or procumbent, herbs or undershrubs, rarely shrubs; leaves alternate, petioled, entire or toothed, 3–5-nerved from the base; flowers in axillary or leaf-opposed racemes, pedicelled, male usually minute, more or less glomerulate above, female solitary towards the base of the raceme, always few, sometimes casually absent.

DISTRIB.   Species about 12, in most hot countries; in tropical Africa four, of which the two here described extend to South Africa.

Stems erect ...   . (1) **africana**.
Stems twining   ...   . (2) **hastata**.

1. **P. africana** (Sond. in Linnæa, xxiii. 110); a branching undershrub; stems erect, 1½ ft. high, thinly pubescent; leaves short-petioled, membranous, oblong-triangular, acute, base subcordate or somewhat 3-lobed and cuneate between the small coarsely toothed lateral lobes, margin except on the basal lobes minutely and distantly toothed or subentire, 1–1½ in. long, ½ in. wide at the base, more or less pubescent especially on the nerves when young, at length nearly glabrous, nerves especially beneath purplish; petiole glabrous, ¼–⅓ in. long; stipules linear, small; racemes leaf-opposed, about 1 in. long, shortly peduncled, with 20–30 male flowers in 8–12 glomerules above and a solitary basal female flower; bracts lanceolate, small; male calyx 4-lobed; stamens about 8; filaments short; female calyx 4-lobed; ovary pilose; stigma 4-lobed; capsule 4-dymous, transversely ⅓ in., diagonally ½ in. across; cocci distinctly winged on the back; seeds complanate, under ¼ in. across, narrowly marginally winged.   *Müll. Arg. in DC. Prodr.* xv. ii. 773; *Prain in Dyer, Fl. Trop. Afr.* vi. i. 951.   *Sajorium africanum, Baill. Adansonia,* iii. 160.   *Pseudotragia Schinzii, Pax in Bull. Herb. Boiss.* 2me *sér.* viii. 635.

KALAHARI REGION: Bechuanaland; Messeringa Vlei, *Seiner*! Transvaal; Macalisberg Range, *Burke*! *Zeyher*, 1522!

Also in German South-West Africa.

2. **P. hastata** (Müll. Arg. in Flora, 1864, 469); a herb with woody base; stems up to 4–5 ft. long, slender, twining, branched, thinly pubescent; leaves short-petioled, membranous, oblong-triangular to lanceolate, base sometimes sagittate or cordate or somewhat 3-lobed and cuneate between the small coarsely toothed lateral lobes, margin except on the basal lobes minutely and distantly toothed or subentire, 1–3 in. long, ½ in. wide at the base, more or less pubescent especially on the nerves when young, at length nearly glabrous, nerves especially beneath purplish; petiole glabrous, ¼–½ in. long; stipules linear, small; racemes leaf-opposed, up to 2 in. long, distinctly peduncled, with 30–40 male flowers in 12–18 glomerules above and a solitary basal female flower which may sometimes be absent; bracts lanceolate, small; male calyx 4-lobed; stamens about 12; filaments short; female calyx 4-lobed; ovary pilose; stigma 4-lobed; capsule 4-dymous, transversely ⅓ in., diagonally ½ in. across; cocci distinctly winged at the back; seeds complanate, under ¼ in. across, narrowly marginally winged. *Müll. Arg. in DC. Prodr.* xv. ii. 772; *Prain in Dyer, Fl. Trop. Afr.* vi. i. 950. *Pseudotragia scandens, Pax in Bull. Herb. Boiss.* 2me sér. viii. 636.

KALAHARI REGION : Transvaal ; South African Gold-field, *Baines* !
EASTERN REGION : Portuguese East Africa ; Lourenço Marques, *Schlechter*, 11526 !

Also in Tropical Africa.
Very nearly allied to *P. africana*, Sond., and principally to be distinguished by its twining, not erect stems.

XXXVI. DALECHAMPIA, Linn.

Flowers monœcious, those of the two sexes together, enveloped by two showy bracts; petals 0; disc 0. *Male: Calyx* closed in bud, splitting into 4–6 valvate lobes. *Stamens* usually 20–30, rarely more or fewer; filaments united in a short or long column; anthers erect, 2-celled; cells parallel, dehiscing longitudinally. Rudimentary *ovary* 0. *Female: Calyx* usually 6-partite, lobes more or less 2-seriately imbricate, or 5-partite, quincuncial, rarely more than 6-partite; segments pinnatifid or fimbriate, less often lanceolate entire, often accrescent and coriaceous in fruit. *Ovary* 3-celled, rarely 4-celled; ovules solitary in each cell; styles united in a long column; stigma capitate, small or dilated, entire or lobulate. *Capsule* 3-dymous, rarely 4-dymous, breaking up into 2-valved cocci; endocarp crustaceous or woody. *Seeds* globose or ellipsoid, without a strophiole; albumen fleshy; cotyledons broad, flat.

Undershrubs, twining or climbing; leaves alternate, long-petioled, entire or 3–5-lobed or -partite; racemes axillary, sessile or peduncled, usually congested and subcapitate, surrounded by the 2 large veined and usually brightly coloured simple or lobed involucral bracts; flowers sessile or shortly pedicelled, the outer and lower female, usually 3 to a bract; the inner and upper 3 or more to each bract; capsules sometimes small, with rounded cocci, sometimes large with hard angular cocci.

DISTRIB.   Species about 60, spread throughout the warmer regions of both hemispheres, but most numerous in America.

Leaves divided into distinct leaflets   ...   . (1) **Galpini**.

Leaves 3–5-lobed or -partite but not divided to the base;
  Male bracts entire :
    Stigmas dilated   ...   . (2) **Kirkii**.
    Stigmas not dilated   ...   . (3) **volubilis**.
  Male bracts glandular-lacinulate   ...   . (4) **capensis**.

1. D. Galpini (Pax in Bull. Herb. Boiss. vi. 736); a prostrate undershrub ; stems slender, pubescent, spreading from a stout woody stock, 1–1½ ft. long; leaves short-petioled, 3-foliolate, segments obtuse or subacute, their margin coarsely toothed, central oblong or oblong-lanceolate, lateral unequally 2-fid, the outer segments the shorter, lobes ½–1¼ in. long, ⅓–⅔ in. wide, more or less pubescent on the nerves on both surfaces, prominently reticulately veined beneath : petiole pubescent, ⅓ in. long; stipules ovate-lanceolate, spreading, pubescent, 1¼ lin. long; heads peduncled ; peduncles slender, pubescent, ½–1½ in. long ; involucral bracts small, green, ½ in. long and broad, the upper 5-lobed, the lower 3-lobed, lobes acute, their margin entire, ciliate ; male bracts entire ; male calyx glabrous, lobes ovate, acute ; female calyx-segments lanceolate ; ovary hirsute ; style columnar, cylindric, entire, dilated at the tip ; fruit not seen.

KALAHARI REGION : Transvaal ; Banks of the Koomati River, *Bolus*, 9781 ! Berea Hill near Barberton, 2800–3000 ft., *Galpin*, 732! Queen River Valley, 2000–2500 ft., *Galpin*, 625 ! near Barberton, 2900 ft., *Thorncroft*, 73 ! Waterval River, *Wilms*, 1341, partly ! 1341 a !

2. D. Kirkii (Prain in Kew Bulletin, 1912, 363) ; a scandent undershrub ; stems slender, sparingly pubescent, twining, 2–3 ft. long ; leaves short-petioled, firmly membranous, simple, deeply 3-partite with the lateral lobes usually unequally 2-fid ; lobes lanceolate, acute, their margin sharply serrate, base cordate, 1–1½ in. long, 1½–2 in. wide, somewhat polished on both surfaces, prominently reticulate beneath, very finely pubescent on the nerves ; petiole sparingly pubescent, ⅓ in. long; stipules lanceolate, spreading, glabrous, 1¼ in. long ; heads peduncled ; peduncles sparingly pubescent, 1–2 in. long ; involucral bracts yellowish, ultimately green, ¾ in. long, 1 in. across, deep-cordate, shallowly 3-lobed, lobes ovate, acute, their margin serrate ; male bracts entire ; male calyx finely puberulous, lobes ovate, acute ; female calyx-segments linear-lanceolate, densely pinnately lacinulate on each side, lacinulæ slender, glandular, scabrous ; ovary puberulous ; style columnar, cylindric, dilated at the tip ; capsule shortly pubescent, ⅓ in. across ; cocci subglobose ; seeds globose.

KALAHARI REGION : Transvaal ; Koomati Poort, *Kirk*, 60 !

A distinct species, nearest to *D. volubilis*, E. Meyer, but with smaller leaves lobed as in *D. capensis*, Spreng. f., with which it agrees as regards style.

3. **D. volubilis** (*E.* Meyer in Drège, Zwei Pfl. Documente, 177); a scandent undershrub ; stems slender, pubescent, twining, springing from a woody base, 6–8 ft. long ; leaves distinctly petioled, firmly membranous, simple, deeply 3-partite with occasionally a pair of much smaller basal lobes superadded, lobes ovate, acute, their margin shortly serrate, base deep-cordate, $2\frac{1}{2}$–$3\frac{1}{2}$ in. long, 3–4 in. wide, somewhat polished and very minutely pubescent on both surfaces ; petiole pubescent or glabrous, 1–$2\frac{1}{2}$ in. long ; stipules lanceolate, spreading, minutely pubescent, 2 in. long ; heads peduncled ; peduncles minutely pubescent, 1–3 in. long ; involucral bracts large, yellowish-green, $1\frac{1}{2}$ in. long, $1\frac{1}{2}$–$1\frac{3}{4}$ in. wide, deep-cordate, both rather shallowly 3-lobed ; lobes acute, their margin finely toothed ; male bracts entire ; male calyx glabrous, lobes ovate, acute ; female calyx-segments linear-lanceolate, densely pinnately lacinulate on each side, lacinulæ slender, glandular, scabrous ; ovary puberulous ; style columnar, cylindric, entire, not dilated at the tip ; capsule nearly glabrous, $\frac{1}{3}$ in. across ; cocci subglobose ; seeds globose. *Baill. Étud. Gén. Euphorb.* 487, *and Adansonia,* iii. 161. *D. capensis, Sond. in. Linnæa,* xxiii. 106, *partly and as to Gueinzius,* 447, *only* ; *Müll. Arg. in DC. Prodr.* xv. ii. 1243, *partly and as to syn. Meyer only* ; *not of Spreng. f. D. natalensis, Müll. Arg. in DC. Prodr.* xv. ii. 1243.

EASTERN REGION : Natal ; Umlazi (Umlaas) River, *Drège,* 4609 ! *Krauss,* 81 ! Boroa, 150 ft., *Wood,* 565 ! 2428 ! 6186 ! near Durban, *Gueinzius,* 447 ! *M'Ken,* 558 ! *Gerrard,* 702 ! *Rehmann,* 8806 ! *Scott Elliot,* 1698 !

A very distinct but apparently a quite local species.

4. **D. capensis** (Spreng. f. Tent. Suppl. ad Syst. Veg. 18) ; a scandent undershrub ; stems slender, pubescent, twining, springing from a woody base, 10–12 ft. long ; leaves distinctly petioled, thinly membranous, simple, deeply 5-lobed with occasionally the external pair of lobes suppressed, lobes obovate, acute, their margin rather coarsely serrate, base cordate, 2–3 in. long, 3–4 in. wide, dull green, pubescent above, pubescent or villous beneath ; petiole pubescent, $\frac{3}{4}$–2 in. long ; stipules ovate-lanceolate, spreading, pubescent, 2–$2\frac{1}{2}$ lin. long ; heads peduncled ; peduncles pubescent, 1–3 in. long ; involucral bracts large, yellowish-green, $1\frac{1}{2}$ in. long, $1\frac{1}{2}$–$1\frac{3}{4}$ in. wide, shallow-cordate, deeply 3-lobed, sometimes the upper bract 5-lobed ; lobes ovate-lanceolate, acute, their margin distinctly and closely toothed ; male bracts shortly lacinulate, each lobule glandular-capitate ; male calyx glabrous, lobes ovate, acute ; female calyx-segments narrow-lanceolate, densely pinnately lacinulate on each side, lacinulæ slender, glandular, scabrous ; ovary puberulous ; style columnar, cylindric, lobulate and dilated at the tip ; capsule nearly glabrous, $\frac{1}{3}$ in. across ; cocci subglobose ; seeds globose. *Sond. in Linnæa,* xxiii. 106, *mainly* ; *Baill. Adansonia,* iii. 161 ; *Müll. Arg. in DC. Prodr.* xv. ii. 1243, *mainly but excl. syn. E. Meyer* ; *Wood, Natal Pl. t* 515,

COAST REGION: Port Elizabeth Div. ; Port Elizabeth, *Miss West*, 22 ! Uitenhage Div. ; near Uitenhage, *Burchell*, 4244 ! *Zeyher*, 3846 ! Enon, *Zeyher* ! *Drège* ! *Baur*, 1044 ! Klein Place, *Prior* ! Krakakamma, *Zeyher*, 1189 ! Tzamas, *Schlechter*, 2597 ! Alexandria Div. ; between Hoffmanns Kloof and Driefontein, *Drège* ! Bathurst Div. ; near Kaffir Drift, *Burchell*, 3862 ! between Kasuga River and Port Alfred, *Burchell*, 3969 ! Theopolis, *Bergius* ! Glenfilling, *Drège*, 2429 ! Port Alfred, *Mrs. Hutton*, 557 ! Albany Div. ; near Grahamstown, *Atherstone*, 71 ! *Bolton* ! *Miss Bowker* ! *MacOwan*, 438 ! and in *Herb. Norm. Austr.-Afr.* 511 ! Fort Beaufort Div. ; Fort Armstrong, *Fraser* ! Peddie Div. ; Fredericksburg, *Gill* ! King Williamstown Div. ; near King Williamstown, *Hutton*, 96 ! Keiskamma, *Mrs. Hutton* ! East London Div. ; Cambridge, *Miss Wormald*, 34 ! Komgha Div. ; Prospect Farm, 2100 ft., *Flanagan*, 433 ! British Kaffraria, *Cooper*, 341 !

KALAHARI REGION: Bechuanaland; Mafeking, *Herb. Marloth* ! Transvaal ; Shiluvane, *Junod*, 539 ! 2381 ! Lydenberg, *Wilms*, 1341, partly ! Warm Bath, *Burtt-Davy*, 2316 ! 5564 ! Moorddrift, *Miss Leendertz*, 2130 ! Crocodile River Drift, *Bolus*, 9780 ! Concession Creek, 2600–3000 ft., *Galpin*, 839 ! *Thorncroft*, 320 ! Middelburg, *Kässner*, 1328 ! Pretoria, *Scott Elliot*, 1424 !

EASTERN REGION : Transkei ; Kentani, *Miss Pegler*, 268 ! Pondoland ; Kletterstroom, *Bachmann*, 768 ! 769 ! 770 ! *Beyrich*, 287 ! Natal ; Colenso, *Rehmann*, 7146 ! Inchanga, *Rehmann*, 7922 ! *Herb. Marloth*, 4088 ! Inanda, *Wood*, 595 ! Avoca, 200 ft., *Schlechter*, 3015 ! Dumisa, 1600 ft., *Rudatis*, 274 ! Mariannhill Trappist Colony, *Landauer*, 233 ! Bellair. *Wood*, 11066 !

XXXVII. CTENOMERIA, Harv.

Flowers monœcious ; petals 0 ; disc 0. *Male : Calyx* 5-lobed, valvate in bud. *Stamens* 30–60 ; filaments free, capillary ; anthers dehiscing longitudinally. Rudimentary *ovary* 0. *Female : Calyx* 6-partite ; segments pectinately lobulate on each side. *Ovary* 3-celled ; ovules solitary in each cell ; styles 3, long, filiform, free, densely papillose throughout. *Capsule* 3-coccous ; cocci 2-valved ; valves crustaceous. *Seeds* globose, ecarunculate ; albumen fleshy ; cotyledons broad, flat.

Twining herbs, pubescent or rarely nearly glabrous ; leaves alternate, cordate at the base ; racemes terminal or leaf-opposed, with numerous male flowers above and 1–2 basal female flowers.

DISTRIB. Two endemic species.

Leaves unlobed or slightly 3-lobed ; lobes acute...   ... (1) **cordata.**
Leaves deeply 5-lobed ; lobes obtuse   ...   . (2) **Schlechteri.**

1. C. cordata (Harv. in Hook. Lond. Journ. Bot. i. 29) ; stems twining, 7–8 ft. long, branched, glabrous to densely pubescent, especially upwards, and armed with stinging hairs ; leaves longpetioled, membranous, triangular-ovate, acute or shortly acuminate, base rather wide-cordate, margin subentire to crenate or serrate, often slightly (rarely rather deeply) 3-lobed with lobes acute or shortly acuminate, 1¼–3 in. long, 1–2½ in. wide, glabrous to pubescent above, sparingly to rather closely pubescent beneath : petiole 1–2 in. long, glabrous to pubescent : stipules lanceolate,

reflexed, membranous ; racemes androgynous, terminal on branches
or leaf-opposed, 3-6 in. long ; peduncles naked, glabrous to pubescent,
up to 1 in. long, with many male flowers in rather remote 3-flowered
cymules or glomerules above and 1 (less often 2) basal female
flowers ; bracts lanceolate, acute ; male pedicels as long as the calyx,
female pedicels longer ; male calyx 5-sect ; lobes ovate, acute,
puberulous or pubescent outside ; stamens 40–60 ; anthers longer
than filaments, introrse ; female calyx 6-partite ; lobes pectinately
6–8-lacinulate on each side, rhachis lanceolate, somewhat accrescent
and indurated in fruit, at length ¼ in. long ; ovary setose ; styles 3,
free, ¼ in. long, densely papillose ; capsule 3-coccous, ⅓ in. across ;
cocci subglobose, sparingly to copiously setose and bristly ; seeds
globose, closely reticulate. *Baill. Adansonia*, iii. 161. *C. kraussiana*,
Hochst. ex Krauss, in Flora, 1845, 85 ; *Sond. in Linnæa*, xxiii. 110 ;
Baill. l.c. C. capensis, Harv. ex Sond. l.c. 109 ; *Prain in Journ.
Bot.* 1913, 171. *Tragia capensis, Thunb. Prodr.* 14, *and in Fl. Cap.
ed. Schult.* 37 ; *E. Meyer in Drège, Zwei Pfl. Documente*, 226, *Galye-
bosch plant only. Leptorhachis capensis, Müll. Arg., and L. capensis,
forma luxurians, Müll. Arg. in DC. Prodr* xv. ii. 926.

COAST REGION : George Div. ; near George, *Mund & Maire*, 245 ! Knysna Div. ;
woods of Knysna and Plettenbergs Bay, *Bowie* ! between Keurbooms River and
Bitou River, *Burchell*, 5292 ! Bosch River, *Drège*, 8239 ! Humansdorp Div. ; near
Humansdorp, 200 ft., *Kennedy* ! *MacOwan*, 314 ! Uitenhage Div. ; near Uitenhage,
Thunberg ! *Zeyher*, 676 ! 3845 ! by the Coega River and in the Winterhoek
Mountains, *Ecklon & Zeyher*, 71 ! Galgebosch, *Drège* ! Kleino Place, *Prior* !
Albany Div. ; Howisons Poort, near Grahamstown, *Mrs. Hutton* ! Mrs. Barber,
113 ! *Williamson* ! Komgha Div. ; near the mouth of the Kei River, *Flanagan*, 451 !

KALAHARI REGION : Transvaal ; Umvoti Creek, near Barberton, 3000 ft., *Galpin*,
1002 !

EASTERN REGION : Transkei ; Kentani, 1000 ft., *Miss Pegler*, 287 ! Natal ;
near Durban, *Gueinsius* ! *Gerrard*, 310 ! 601, partly ! *Rehmann*, 8807 ! Clairmont,
Schlechter, 2843 ! Berea, 200–300 ft., *Wood*, 6335 ! Inanda, *Wood*, 707 ! 801 !
Dumisa, 2000 ft., *Rudatis*, 799 ! near Umlaas River, *Krauss*, 186 !

Somewhat variable as regards the degree of pubescence which is copious in the
original *Tragia capensis (C. capensis*, Harv.) but almost absent in *C. cordata*,
Harv. The two are, however, connected by so many intermediates that it is not
possible to separate them as distinct varieties.

2. **C. Schlechteri** (Prain in Journ. Bot. 1913, 171) ; stems twining,
2-3 ft. long, sparingly branched, pubescent with spreading hairs ;
leaves long-petioled, membranous, palmately 5-lobed, central and
lateral lobes oblong, basal lobes suborbicular, lobes obtuse or subacute, their margin coarsely and distantly crenate or serrate, base
wide-cordate, lamina 2 in. long, as much across, central lobe 1–1¼ in.
long, ½–1 in. wide, lateral one half as long, basal short, pale green,
sparingly beset with white bristles above, densely bristly especially
on the nerves beneath ; petiole 1¼–1½ in. long, pubescent with
spreading hairs ; stipules lanceolate, reflexed, membranous ; racemes
androgynous, terminal on the branches or leaf-opposed, 2–4 in. long,
on naked patently pubescent peduncles 1½ in. long or longer, with
a few 3-flowered cymules of male flowers above, and usually 2 basal

female flowers; bracts lanceolate, acute; pedicels in both sexes long, patently hirsute; male calyx 5-sect; lobes ovate, acute, rather densely setose outside; stamens 30–40; anthers longer than the filaments, introrse; female calyx 6-partite; lobes pectinately 4–6-lacinulate on each side, rhachis lanceolate, somewhat accrescent, at length $\frac{1}{4}$ in. long, not indurated; ovary densely setose; styles 3, $\frac{1}{4}$ in. long, densely papillose; capsule 3-coccous, $\frac{1}{3}$ in. across; cocci subglobose, closely adpressed setose and bristly. *Tragia Schlechteri, Pax in Bull. Herb. Boiss.* vi. 735.

EASTERN REGION: Natal; Umkomanzi River, 3000 ft., *Schlechter,* 6701!

Nearly allied to and perhaps only an extreme form of *C. cordata,* Harv.

XXXVIII. TRAGIA, Linn.

Flowers monœcious, rarely diœcious; petals 0; disc 0 or obscure. *Male : Calyx* closed in bud, splitting into 3, rarely 4–6, valvate lobes. *Stamens* normally 3, very rarely fewer or more; filaments free; anthers 2-celled; cells parallel, dehiscing longitudinally. Rudimentary *ovary* minute or obsolete. *Female : Calyx* 6-partite; lobes more or less 2-seriately imbricate, palmately lobulate, occasionally lobes 4–5, usually accrescent and indurated in fruit. *Ovary* 3-celled; ovules solitary in each cell; styles 3, more or less distinctly connate at least at the base, entire. *Capsule* 3-dymous, breaking up into 3 2-valved cocci. *Seeds* globose, ecarunculate; testa crustaceous; albumen fleshy; cotyledons broad, flat.

Herbs, twining or suberect, usually rather copiously beset with stinging hairs on the stem, leaves and calyx; leaves alternate, stipulate; racemes leaf-opposed or terminal with many male flowers above and usually few basal female flowers; bracts small, persistent.

DISTRIB. Species about 100, spread throughout the warmer regions of both hemispheres.

Male flowers in glomerules or cymules to each bract; racemes
1-sexual :

Stems short, erect; leaves sessile ...   . (1) **Rogersii.**

Stems long, twining; leaves long-petioled   ...   ... (2) **Sonderi.**

Male flowers solitary to each bract; racemes usually
2-sexual :
Female calyx regularly 6-partite :
Stems short, erect or ascending, not twining :
Petiole 0; leaves deeply laciniately lobed ...   ... (3) **incisifolia.**

Petiole $\frac{1}{8}$–$\frac{1}{4}$ in. long; leaves shortly serrate ...   ... (4) **minor.**

Stems long, twining; petiole 1 in. long or longer :
Female calyx-segments 2–3-lobulate on each side ... (5) **Okanyua.**

Female calyx-segments 8–10-lobulate on each side... (6) **natalensis.**

Female calyx normally 3-partite but often with 1 (rarely more than 1) smaller irregular segment added :
Leaves obtuse, subsessile, $\frac{1}{2}$ in. long or less ...   ... (7) **collina.**

Leaves acute, petioled, 1 in. long or longer :
  Stems erect or ascending, not twining :
    Leaves lanceolate, with two large rounded basal
      lobes ; racemes sometimes 1-sexual ...   ... (8) **dioica.**

  Leaves triangular-ovate, not lobed   ...   ... (9) **meyeriana.**

  Stems twining, at least towards the top :
    Lobules of female calyx-segments as long as the
      width of the central rhachis :
      Leaves ovate-lanceolate, usually somewhat 3-
        lobate   ...   . (10) **rupestris.**

      Leaves triangular-ovate ...   . (11) **wahlbergiana.**

    Lobules of female calyx-segments shorter than
      the width of the central rhachis ; leaves
      triangular-ovate, rarely 3-lobate   ...   ... (12) **durbanensis.**

1. **T. Rogersii** (Prain in Kew Bulletin, 1912, 238) ; a herb with
woody base ; stems erect, tufted, simple or sparingly branched,
8–10 in. high, laxly softly pubescent without stinging hairs ;
leaves sessile, membranous, ascending, ovate-lanceolate or oblong-
lanceolate, acute, base truncate or rounded, margin minutely and
closely serrate except at the entire base, $1\frac{1}{4}$–$2\frac{1}{2}$ in. long, $\frac{1}{3}$–$\frac{1}{2}$ in. wide,
sparingly pubescent above, more densely pubescent beneath espe-
cially on the nerves ; stipules linear-lanceolate, spreading, membranous, 1 lin. long, their margin ciliolate ; racemes terminal or
leaf-opposed, 1-sexual and diœcious, male from 2–6 in. long; flowers
very many, in 3-flowered clusters above, in 3-flowered cymules
below ; bracts and bracteoles linear-lanceolate, their margin pilose ;
pedicels pubescent, shorter than the bracteoles ; male calyx usually
3-partite, but sometimes 4–5- (very rarely 6-) partite ; lobes ovate
or ovate-lanceolate, acute, sparingly puberulous within, pubescent
outside ; stamens 3, filaments longer than the anthers, rather wide-
based, incurved above ; anthers subconnivent ; female racemes
3–4-flowered ; pedicels solitary to their bracts ; female calyx 6-par-
tite, lobes rather wide-oblong, 5–6-lobulate on each side, the lobules
all ascending, narrow-lanceolate, rhachis accrescent and coriaceous,
densely pilose externally, glabrous within, in fruit $2\frac{1}{2}$ in. long ;
ovary densely setose ; styles 3, setose throughout and free almost to
the base ; capsule 3-coccous, $\frac{1}{3}$ in. across ; cocci rather closely setose,
subglobose ; seeds globose.

KALAHARI REGION : Transvaal ; Waterval Onder, *Rogers*, 2597 ! near Barberton,
3000 ft., *Thorncroft*, 624 ! 625 !

A very distinct species.

2. **T. Sonderi** (Prain in Kew Bulletin, 1912, 337) ; a herb with
woody base ; stems below erect and rigid, above twining, 3 ft. long,
sparingly branched, finely pubescent and armed with stinging
bristles ; leaves long-petioled, membranous, triangular-ovate, acute,
base cordate, margin strongly and rather coarsely serrate or
duplicate-serrate, $1\frac{1}{2}$–3 in. long, 1–$2\frac{1}{2}$ in. wide, uniformly sparingly

pubescent throughout, and setose or bristly on the nerves on both
surfaces ; petiole minutely puberulous and sparingly bristly, 1–2½ in.
long ; stipules lanceolate, reflexed, glabrous, ⅙ in. long ; racemes
lateral, leaf-opposed, 1-sexual, male 3–4 in. long, with numerous
flowers in 3-flowered cymules below, above with flowers solitary to
their bracts but with the bracteoles upraised on the pedicel ;
peduncle and rhachis finely pubescent ; male bracts linear-lanceolate,
1½ lin. long, reflexed ; male calyx 3-partite ; lobes ovate, apiculate ;
stamens 3 ; anthers subincurved ; female flowers and fruit not seen.

KALAHARI REGION : Transvaal ; Magaliesberg Range, *Zeyher* ! *Burke* ! Swaziland ; in the High Veld at Dalriach near Mbabane, 4400 ft., *Bolus*, 12290 !

Apparently most nearly allied to *T. Dinteri*, Pax, of German South-West
Africa.

3. **T. incisifolia** (Prain in Kew Bulletin, 1912, 237) ; a herb with
woody base ; stems erect, ascending or prostrate, simple or virgately
branched, 4–6 in. long, occasionally sparingly bristly, usually only
puberulous or pubescent ; leaves sessile or nearly so, membranous,
narrow-ovate or ovate-lanceolate, acute, base cuneate, margin deeply
laciniately lobed, lobes ovate-lanceolate, about 4 on each side, pointing
forward, ⅓–⅔ in. long, ¼–⅓ in. wide, sparingly setose on the nerves
above, more densely so beneath and with a solitary terminal seta at
the tip of each marginal lobe ; stipules lanceolate, reflexed, glabrous,
1 lin. long ; racemes terminal ; peduncle and rhachis puberulous or
pubescent and bristly, with numerous rather densely arranged male
flowers solitary to their bracts above and 2–3 basal female flowers ;
bracts ovate-lanceolate, entire, male 1 lin. long, female 1½ lin. long,
their margin setose ; pedicels in both sexes solitary, very short : male
calyx usually 4-partite, occasionally 5-partite ; lobes induplicate-
valvate, ovate, acute ; stamens 4–5, rarely 3 ; filaments longer than
the anthers ; female calyx 6-partite ; lobes alternately wide-ovate
and ovate-lanceolate, the wider 4–5-lobulate, the narrower 3–4-
lobulate on each side, accrescent and subcoriaceous, setose on the
back, glabrous within, each lobule tipped by a solitary seta ; ovary
densely hispid ; styles 3, free nearly to the base ; capsule 3-coccous,
over ¼ in. across ; cocci subglobose, sparingly setose ; seeds globose.

KALAHARI REGION : Transvaal ; between Koomati River Drift and Crocodile
River, *Bolus*, 9779 ! Koomati Poort, on hills at about 300 ft., *Schlechter*, 11781 !

4. **T. minor** (Sond. in Linnæa, xxiii. 108) : a herb with woody
base ; stems slender, rather rigid, erect, sparingly branched, hispidly
hairy, 6–12 in. high ; leaves short-petioled, membranous, oblong-
lanceolate to lanceolate, acute, base very shallow-cordate, margin
closely serrate, 1½ in. long, ¾ in. wide, hispid and sparingly bristly
especially on the nerves on both surfaces ; petiole rather densely
hispid, ¼–⅓ in. long ; stipules spreading, lanceolate, sparingly hispid,
1½ lin. long ; racemes terminal, with many close-set male flowers
solitary to their bracts above and 1–2 basal female flowers ; peduncle and rhachis hispid ; male bracts lanceolate, pilose, 1 lin. long ;

female bracts ovate-lanceolate, entire, pilose, 2 lin. long ; male calyx 3-partite ; lobes lanceolate, acute, nearly glabrous ; stamens 3, rather wide-based, incurved above ; female calyx 6-partite ; lobes 2-seriate, unequal, 3 wide-oblong, 3 ovate-lanceolate, pinnately 6-lobulate on each side, hispid externally, glabrous within, lobules shorter than the width of the accrescent indurated rhachis, in fruit $\frac{1}{4}$ in. long ; ovary hispid and sparingly setose ; styles 3, connate below in a distinct column, free above ; capsule 3-coccous, $\frac{1}{3}$ in. across ; cocci sparingly setose, subglobose ; seeds globose. *Baill. Adansonia,* iii. 162. *T. rupestris,* γ *minor, Müll. Arg. in DC. Prodr.* xv. ii. 940, *partly ; as to Zeyher,* 1524, *only.*

VAR. β, **longifolia** (Prain) ; leaves rather narrow-lanceolate, 1½–2 in. long, ½ in. wide, otherwise as in the type.

KALAHARI REGION : Transvaal ; Magalies River, *Burke!* *Zeyher,* 1524! Pretoria Dist., Pretoria, *Crawley,* 4144! Aapies Poort, *Rehmann,* 4121, partly! Wonderboompoort, *Rehmann,* 4555! Pinedene near Irene, *Burtt-Davy,* 2315! Crocodile River, *Wilms,* 1325! Var. β : Swaziland ; Bremmersdorp, 2200 ft., *Burtt-Davy,* 3004!

5. T. Okanyua (Pax in Bull. Herb. Boiss. vi. 735); a herb with woody base ; stems twining, 1–6 ft. high, much branched, pubescent and armed with stinging bristles ; leaves distinctly petioled, membranous, ovate, auriculate-subcordate and usually almost 3-lobate, central lobe generally elongated, acute, basal lobes rounded, sometimes triangular-ovate, with a shallow lateral sinus, base cordate, margin crenate-serrate, 1½–2½ in. long, 1¼–1½ in. wide at the base, hispid or puberulous and sparingly bristly especially on the nerves of both surfaces ; petiole 1 in. long, pubescent and sparingly bristly ; stipules broadly lanceolate, spreading, pubescent and bristly, 1½ lin. long ; racemes lateral, leaf-opposed, peduncled, 1–2 in. long ; peduncle hispid and sparingly bristly, with many male flowers above and usually 2 basal female flowers ; male bracts ovate-lanceolate, entire, about 1 lin. long, finely hispid ; female bracts 2–3-lobed, ovate, 1½–2 lin. long, hispid ; pedicels in both sexes solitary to and shorter than their bracts ; male calyx 3-partite ; lobes ovate, obtuse ; stamens 3 ; filaments longer than the anthers ; female calyx 6-partite, 2-seriate ; lobes pinnately 2–3-lacinulate on each side, rhachis accrescent, indurated in fruit, at length $\frac{1}{4}$ in. long, of one series obovate, of the other series rather narrow-lanceolate ; ovary hispid above ; styles 3, united more than half-way in a distinct basal column, free above ; capsule 3-coccous, $\frac{1}{4}$ in. across ; cocci subglobose, nearly glabrous ; seeds globose. *Prain in Dyer, Fl. Trop. Afr.* vi. i. 986. *T. angustifolia, Pax in Engl. Pfl. Ost-Afr. C.* 239, *partly, and in Baum, Kunene-Samb. Exped.* 283 ; *not of Benth. T. cordifolia, N. E. Br. in Kew Bulletin,* 1909, 141; *not of Vahl, nor of Benth. T. madandensis, S. Moore in Journ. Linn. Soc.* xl. 203.

KALAHARI REGION : Transvaal ; slopes of Masetane, *Junod,* 1051! Also widely spread in southern Tropical Africa.

6. **T. natalensis** (Sond. in Linnæa, xxiii. 107); a herb with woody base; stems twining, 6–10 ft. long, much branched, sparingly pubescent with somewhat reflexed hairs and armed with stinging bristles; leaves long-petioled, membranous, ovate or ovate-oblong, acuminate, base rounded or shallow-cordate, margin closely and sharply serrate, 2–4 in. long, 1–2 in. wide, uniformly and sparingly pubescent throughout and bristly on the nerves on both surfaces; petiole patently pubescent and bristly, 1–3 in. long; stipules linear-lanceolate, $\frac{1}{5}$ in. long, reflexed, pubescent; racemes lateral, peduncled, 1$\frac{1}{2}$ in. long; peduncle pubescent and sparingly bristly, with many male flowers above and 1–2 basal female flowers; male bracts lanceolate or ovate-lanceolate or subspathulate, acute, entire or the lowest sometimes dentate : female bracts large, subreniform or suborbicular, their margin dentate; pedicels in both sexes shorter than and solitary to the bracts; male calyx 3-partite; lobes ovate, obtuse; stamens 3; filaments as long as the anthers; female calyx 6-partite; lobes pinnately 8–10-lacinulate on each side, rhachis lanceolate, indurated in fruit, at length $\frac{1}{4}$ in. long; ovary hispid; styles 3, united almost to the apex in a slender narrow-infundibuli-form tube, the short triangular reflexed stigmas alone free; capsule 3-coccous, $\frac{1}{3}$ in. across; cocci subglobose, hispid; seeds globose. *Baill. Adansonia,* iii. 162; *Müll. Arg. in DC. Prodr.* xv. ii. 942; *Prain in Dyer, Fl. Trop. Afr.* vi. i. 974. *T. involucrata, Jacq. ex E. Meyer in Drège, Zwei Pfl. Documente,* 226 ; *Baill. Étud. Gén. Euphorb.* 461 ; *not of Linn. T. mitis,* γ *oblongifolia, Müll. Arg. in Flora,* 1864, 435, *and in DC. Prodr.* xv. ii. 942. *T. ambigua, S. Moore, and T. ambigua, var. urticans, S. Moore in Journ. Linn. Soc. Bot.* xl. 202.

EASTERN REGION: Transkei ; Kentani, Manubie. 1000 ft., *Miss Pegler,* 1257 ! Pondoland ; near St. Johns River, *Drège* ! Port St. John, 50 ft., *Galpin,* 3462 ! Natal ; Inanda, *Wood,* 741 ! Ingoma, *Gerrard,* 1164 ! and without precise locality, *Gueinzius,* 496 ! *Gerrard,* 28 !

Also in East Tropical Africa.

7. **T. collina** (Prain in Kew Bulletin, 1912, 335); a herb with woody base ; stems erect or ascending, simple or sparingly branched, never twining, 4–6 in. high, sparingly to rather copiously pubescent and sparingly armed with stinging hairs ; leaves subsessile, mem-branous, oblong, obtuse, base wide-cuneate or truncate, margin sparingly toothed except at the entire base, lateral teeth small, sometimes obsolete, the rounded apex rather strongly 3–5-toothed, $\frac{1}{3}$–$\frac{1}{2}$ in. long, $\frac{1}{4}$–$\frac{1}{3}$ in. wide, sparingly hirsute and bristly on the nerves on both surfaces ; petiole 1 lin. long or less, hirsute and bristly ; stipules spreading, lanceolate, hirsute, 1$\frac{1}{2}$ lin. long ; racemes leaf-opposed and subterminal, with many male flowers solitary to their bracts above and 1–3 basal female flowers ; peduncle and rhachis closely pubescent and sparingly bristly : male bracts lanceolate, entire, nearly glabrous, 1$\frac{1}{2}$ lin. long ; female bracts lanceolate, entire, 2 lin. long ; male calyx 3-partite ; lobes ovate,

apiculate, puberulous externally ; stamens 3 ; filaments rather longer
than the anthers, erect, thickened below ; female calyx 3-partite,
lobes suborbicular, pinnately 5–6-lobulate on each side, pubescent
externally, glabrous within, lobules lanceolate, as long as the width
of the accrescent indurated rhachis, in fruit ⅓ in. long ; ovary
densely setose ; styles 3, connate for half their length in a glabrous
column, free and recurved above ; capsule 3-coccous, ⅓ in. across ;
cocci sparingly setose, slightly angled on the back ; seeds globose.

EASTERN REGION : Natal ; near Colenso, 4500 ft., *Schlechter*, 3381 ! near
Pieters, 3000–4000 ft., *Wood*, 8881 !

A very distinct species.

8. **T. dioica** (Sond. in Linnæa, xxiii. 109) ; a herb with woody
base ; stems slender, erect or ascending, not twining, sparingly
branched or simple, 8–12 in. high, closely pubescent and sometimes
densely, sometimes very sparingly, armed with stinging bristles ;
leaves short-petioled, membranous, auriculate-subcordate and distinctly 3-lobate, central lobe lanceolate, acute or acuminate, shortly
serrate, lateral lobes rounded-oblong, deeply and coarsely toothed,
1¼–2¼ in. long, ¾–1¼ in. wide at the base, pubescent on both surfaces
and densely to sparingly bristly especially on the nerves and principally beneath ; petiole ¼–⅓ in. long, pubescent and densely to
sparingly bristly ; stipules lanceolate, sparingly pubescent, spreading or reflexed, 2 lin. long ; racemes terminal or lateral, leaf-
opposed, usually with many male flowers above and 1–2 basal
female flowers, rarely diœcious with in female plants several
female flowers ; peduncle densely pubescent and densely or
sparingly bristly ; male bracts lanceolate, 1 lin. long, pubescent ;
female bracts ovate, acute, entire, pubescent and bristly ; pedicels
in both sexes solitary to and shorter than their bracts ; male calyx
3-partite, lobes ovate, obtuse or subacute, glabrous ; stamens 3 ;
filaments dilated below, erect, longer than the anthers ; female calyx
3-partite, but with at least 1, usually 2–3, sometimes 4 smaller
additional lobes, the 3 main-lobes orbicular or oblong, pectinately
4–6-lobulate on each side, lobules lanceolate, bristly, rhachis accrescent and much indurated, at length ⅓ in. long, ¼ in. wide, additional
lobes when present very variable, sometimes many-lobulate, sometimes reduced to a single lanceolate lobule ; ovary densely hispid ;
styles 3, connate for half their length in a glabrous column, free and
recurved above ; capsule 3-coccous, ⅖ in. across ; cocci subglobose,
hispid ; seeds globose. *Baill. Adansonia*, iii. 162 ; *Prain in Dyer,
Fl. Trop. Afr.* vi. i. 993. *T. rupestris, δ lobata, Müll. Arg. in DC.
Prodr.* xv. ii. 941. *T. Schinzii, Pax in Bull. Herb. Boiss.* vi. 734.

KALAHARI REGION : Bechuanaland ; Maadji Mountain, *Burchell*, 2366 ! slopes
of Messeringa Vlei, *Seiner*, II. 284 ! Mafeking, *Duparquet*, 13 ! Transvaal ; Macalisberg Range, *Burke*, 106 ! *Zeyher*, 1523 ! Boshveld, Elands River and Drift,
Rehmann, 4979, partly ! between Elands River and Klippan, *Rehmann*, 5097 !
Pienaars River, Rooiplat, *Miss Leendertz*, 764 ! Potgieters Rust, *Miss Leendertz*,
1482 ! Moorddrift, *Miss Leendertz*, 2165 !

Also in German South-West Africa and in Rhodesia.

A very well-marked species, most nearly allied to *T. rupestris*, Sond., of which it has by Müller been treated as a variety, but better looked upon as distinct. Some of Burke's and Zeyher's original specimens are diœcious as described by Sonder; usually, however, the racemes have both male and female flowers. Within the species two very distinct forms occur; in one, represented by *Burke*, 106, *Zeyher*, 1523, *Leendertz*, 1482 and 2165, the stems are densely hispid; in the other the stems are only sparingly hispid. The two, however, do not otherwise differ and are hardly entitled to be treated as distinct varieties. The hispid form is that originally described as *T. dioica*, Sond., the less hispid form is that originally described as *T. Schinzii*, Pax.

9. **T. meyeriana** (Müll. Arg. in DC. Prodr. xv. ii. 938 as to Gueinzius' plant and excl. var. *β glabrata*); a herb with woody base; stems erect or ascending, simple or branched, never twining, 8–16 in. high, pubescent and copiously armed with stinging bristles; leaves long-petioled, membranous, triangular-ovate, acute, base rather deeply and usually widely cordate, margin closely and strongly toothed, $1\frac{1}{2}$–$2\frac{1}{4}$ in. long, $1\frac{1}{4}$–2 in. wide, above rather sparingly beset with stinging bristles, more densely bristly (especially on the nerves) beneath, very rarely almost glabrous; petiole rather densely bristly, $\frac{3}{4}$–1 in. long; stipules spreading, lanceolate or ovate-lanceolate, bristly, 2–$2\frac{1}{2}$ lin. long; racemes terminal or leaf-opposed and subterminal, with many male flowers solitary to their bracts above and 1–3 basal female flowers; peduncle and rhachis densely setose and bristly; male bracts ovate-lanceolate, entire, setose and bristly, 2 lin. long; female bracts ovate-lanceolate, with margin usually minutely toothed, bristly, 3 lin. long; male calyx 3-partite, very large for the genus; lobes ovate, acute, nearly glabrous; stamens 3, rarely 4–5; filaments longer than the anthers, incurved above; anthers subconnivent; female calyx 3-partite, with occasionally a fourth smaller lobe added; lobes wide-oblong or suborbicular, palmately 6–8-lobulate on each side, much shorter than the width of the accrescent strongly indurated rhachis, in fruit $\frac{2}{3}$ in. long; ovary puberulous and setose; styles 3, connate below in a short column, free above; capsule 3-coccous, $\frac{1}{3}$ in. across; cocci sparingly bristly, subglobose; seeds globose. *T. Bolusii, O. Kuntze, Rev. Gen. Pl.* iii. ii. 293.

COAST REGION : East London Div. ; East London, *Wood in Herb. Galpin*, 3217 ! near the mouth of the Kintza River, 50 ft., *Galpin*, 6558 ! Komgha Div. ; Prospect Farm, Komgha, *Flanagan*, 358 ! near Komgha, 2000 ft., *Flanagan*, 754 !
EASTERN REGION : Transkei ; Kentani, 1000 ft., *Miss Pegler*, 186 ! Tembuland ; Bazeia, Klipkrantz, 2000 ft., *Baur*, 458 ! near the Quinancu River, 2900 ft., *Bolus*, 10286 ! Griqualand East ; Insiswa Range, *Schlechter* ! Clydesdale, 2500 ft., *Tyson*, 1060 ! and in *MacOwan & Bolus, Herb. Austr.-Afr.* 1233 ! Natal ; near Durban, *Gueinzius* ! *Sanderson*, 302 ! Estcourt, 4000 ft., *Marshall* ! Inanda, *Wood*, 416 ! Drakensberg, Tugela River, Colenso, *Rehmann*, 7167 ! Lancaster Hill, Vryheid, *Burtt-Davy*, 11449 ! Krantz Kloof, 1500 ft., *Schlechter*, 3204 ! near Ladysmith, 3500 ft., *Wood*, 7570 ! Weenen, 3000–5000 ft., *Sutherland* !

This very distinct species is most nearly allied to *T. durbanensis*, O. Kuntze, from which it is most readily distinguished by its erect instead of climbing habit. The name *T. meyeriana*, which must be used instead of the name *T. Bolusii* proposed by Kuntze, is by no means appropriate for this erect plant of which Drège does not appear to have gathered or Meyer to have distributed specimens. It is true that in some herbaria there are specimens belonging to ' Drège 4605 ' written up by Müller as *T. meyeriana*. But these specimens do not represent the erect species collected by Gueinzius and described by Müller ; they represent the more hispid form of the climbing species described by Müller as *T. meyeriana*, β *glabrata*, afterwards described by Kuntze as *T. durbanensis*.

10. **T. rupestris** (Sond. in Linnæa, xxiii. 108) ; a herb with woody base ; stems slender, 2–3 ft. long, erect and much virgately branched below, twining upwards, pubescent and sparingly armed with stinging bristles ; leaves distinctly petioled, membranous, ovate-lanceolate, often auriculate-cordate and sometimes almost 3-lobate, acuminate, base distinctly cordate, margin closely toothed, $1\frac{1}{2}$–2 in. long, $\frac{3}{4}$–$1\frac{1}{4}$ in. wide, hirsute throughout on both surfaces and rather densely clothed with stinging bristles on the nerves ; petiole pubescent and sparingly bristly, $\frac{3}{4}$–$1\frac{1}{2}$ in. long ; stipules lanceolate, reflexed, $1\frac{1}{2}$–2 lin. long, pubescent ; racemes leaf-opposed, peduncled, with many male flowers and 1–2 basal female flowers ; peduncle and rhachis densely pubescent and sparingly bristly ; male bracts lanceolate, their margin entire, pubescent, 1 lin. long ; female bracts ovate, acute, their margin toothed, 2 lin. long ; pedicels in both sexes short, pubescent, solitary to their bracts ; male calyx 3-partite ; lobes orbicular, obtuse, sparingly pubescent ; stamens 3 ; filaments hardly longer than the anthers, incurved ; female calyx usually 3-partite with often a fourth smaller lobe added ; lobes oblong, palmately 4–5-lobulate on each side , lobules lanceolate, shorter than the width of the accrescent coriaceous rhachis, $\frac{1}{4}$ in. long ; ovary densely setose ; styles 3, connate for half their length or more in a distinct column ; capsule 3-coccous, $\frac{1}{3}$ in. across ; cocci subglobose, rather densely setose ; seeds globose. *Baill. Adansonia*, iii. 162 ; *Müll. Arg. in DC. Prodr.* xv. ii. 940.

KALAHARI REGION : Transvaal ; Mooi River, *Zeyher*, 1525 ! *Burke* ! Wonder-boom Poort, *Rehmann*, 4548 ! *Miss Leendertz*, 444 ! *Burtt-Davy*, 8058 ! Aapies Poort, *Rehmann*, 4121, partly ! hills above Aapies River, *Rehmann*, 4282 ! Pretoria Hills, 4500 ft., *Miss Leendertz*, 531 b ! Potgieters Rust, *Miss Leendertz*, 1234 ! Boshveld, Elands River and Drift, *Rehmann*, 4979, partly ! Rustenburg, 4500 ft., *Miss Nation*, 74 ! without precise locality, *Wahlberg* ! *Rogers*, 1322 !

11. **T. wahlbergiana** (Prain in Journ. Bot. 1913, 169, in obs.) ; a herb with woody base ; stems slender, prostrate or twining, 2–3 ft. long, much branched ; branches slender, puberulous and sparingly armed with stinging bristles ; leaves short-petioled, membranous, triangular-ovate, acute, base rather shallow-cordate, margin closely toothed, 1–$1\frac{1}{4}$ in. long, $\frac{1}{2}$–$\frac{2}{3}$ in. wide, with a few stinging bristles on the nerves on both surfaces, otherwise glabrous ; petiole puberulous and sparingly bristly, $\frac{1}{3}$ in. long ; stipules lanceolate, reflexed, 1–$1\frac{1}{2}$ lin. long, glabrous ; racemes terminal on the branches or leaf-opposed, peduncled, with many male flowers above and 1–2 basal female flowers ; peduncle and rhachis puberulous and very sparingly

bristly ; male bracts lanceolate, their margin entire, finely puberulous, 1 lin. long; female bracts ovate, acute, their margin toothed, 1½–2 lin. long; pedicels in both sexes very short, puberulous, solitary to their bracts; male calyx 3-partite; lobes ovate, acute, glabrous; stamens 3; filaments hardly longer than the anthers, incurved; female calyx usually 3-partite, with occasionally a fourth smaller lobe added; lobes suborbicular, palmately 4–5-lobulate on each side; lobules lanceolate, about as long as the diameter of the somewhat accrescent subcoriaceous rhachis, ¼ in. long; ovary densely setose; styles 3, very shortly connate at the base; capsule 3-coccous, ¼ in. across; cocci almost glabrous, subglobose; seeds globose, grey with bright brown blotches. *T. rupestris, β glabrata,* Sond. *in Linnæa,* xxiii. 108 ; *Müll. Arg. in DC. Prodr.* xv. ii. 940. *T. rupestris, γ minor, Müll. Arg. l.c., partly*; *as to Wahlberg's specimen only and excl. syn. T. minor,* Sond. *T. affinis, Müll. Arg. ex* Prain *in Kew Bulletin,* 1912, 334 ; *not of Robinson & Greenman.*

KALAHARI REGION : Transvaal ; Vaal River, *Zeyher,* 1526 ! *Burke !* Crocodile River, *Burke !* and without precise locality, *Wahlberg !*

Closely allied to *T. rupestris,* Sond., and treated by Sonder and Müller as only varietally distinct from that species, but easily recognised by its almost glabrous leaves with much shorter petioles and by its smaller fruits. The leaves in *T. wahlbergiana* are not auriculate-cordate as is usually the case with *T. rupestris.* More closely allied still to *T. durbanensis,* O. Kuntze, but distinguishable by its more deeply lobulate female calyx-segments.

12. **T. durbanensis** (O. Kuntze, Rev. Gen. Pl. iii. ii. 293) ; a herb with woody base ; stems slender, twining, sparingly branched, 4–8 ft. long, usually glabrous, sometimes sparingly armed with stinging bristles ; leaves long-petioled, membranous, triangular-ovate, acute, base rather deeply widely to narrowly cordate, margin closely strongly toothed, 1½–3 in. long, 1¼–2½ in. wide, glabrous or nearly so above, beneath usually very sparingly beset with stinging bristles on the nerves ; petiole sparingly bristly or glabrous, ½–1 lin. long ; stipules spreading, lanceolate or ovate-lanceolate, glabrous, 1½–2 lin. long ; racemes terminal on the branches or leaf-opposed on the stem, with many male flowers solitary to their bracts above and 1–2 basal female flowers ; peduncle and rhachis finely puberulous with sometimes a few long bristles ; male bracts ovate-lanceolate, entire, glabrous or nearly so, 1 lin. long ; female bracts ovate-lanceolate, entire, glabrous, 2 lin. long ; male calyx 3-partite ; lobes ovate, acute, glabrous ; stamens 3 ; filaments longer than the anthers, incurved ; female calyx 3-partite ; lobes suborbicular, palmately 5–6-lobulate on each side, pubescent externally, glabrous within ; lobules much shorter than the width of the accrescent indurated rhachis, in fruit ⅓ in. long ; ovary puberulous and setose ; styles 3, connate below in a short column, free above ; capsule 3-coccous, ¼ in. across ; cocci almost glabrous, subglobose ; seeds globose. *T. capensis, E. Meyer in Drège, Zwei Pfl. Documente* 226, *mainly but excl. the Galgebosch plant* ; Sond. *in Linnæa,* xxiii. 110 ; *Baill. Étud. Gén. Euphorb.* 461,

and Adansonia, iii. 162 ; *not of Thunb.　T. capensis, β, E. Meyer,
l.c.* 226. *T. meyeriana, Müll. Arg. in DC. Prodr.* xv. ii. 938, *as to
the Drège specimen only. T. meyeriana, β glabrata, Müll. Arg. l.c.*

COAST REGION : East London Div. ; East London, 50 ft., *Bolus* ! Albany Div. ;
Howisons Poort, *Williamson* ! Bathurst Div. ; near the sea, *Mrs. Barber,* 496 !
Komgha Div. ; near the mouth of the Kei River, *Flanagan,* 437 !

EASTERN REGION : Pondoland ; near the mouth of the Umsikaba River, *Drège* !
Egosa Forest, *Beyrich,* 10 ! and without precise locality, *Bachmann,* 776 ! Natal ;
near Durban, *Drège,* 4605 ! *Sanderson,* 366 ! *Kuntze* ! *Wilms,* 2273 ! *Schlechter,*
2773 ! *Rehmann,* 8805, partly ! *Wood,* 2802 ! 6343 ! Umkomaas, *Engler,* 2571 a !
Between the Rivers Umzimkulu and Umkomanzi, *Drège* ! Shafton, Howick, *Mrs.
Hutton,* 12 ! Higher Tugela, *Gerrard,* 1165 ! without precise locality, *Gerrard,*
522 ! Delagoa Bay ; Lourenço Marques, *Mrs. Howard,* 71 ! *Forbes,* 47 ! *Junod,*
198 !

Very nearly related to *T. meyeriana,* Müll. Arg., with which it has been united
by Müller, but readily distinguished by being scandent and in having male flowers
which are less than half the size. As *T. capensis,* Thunb., is not a *Tragia,* the
name *T. capensis,* E. Meyer, associated by Müller with this plant as a synonym
and employed, though without description, by Meyer, Sonder and Baillon in
previous publications, might be used to designate this species. It seems, however,
less ambiguous to employ the name suggested by Kuntze to whom we are indebted
for the first intelligible account of this species and of its ally.

XXXIX. MAPROUNEA, Aubl.

Flowers monœcious ; petals 0 ; disc 0. *Male : Calyx* shortly
2–3-lobed, slightly imbricate in bud. *Stamens* 1–3, usually 2,
exserted ; filaments united below, free above ; anthers oblong ; cells
parallel, dehiscing longitudinally. Rudimentary *ovary* 0. *Female :
Calyx* 3-lobed. *Ovary* 3-celled ; ovules solitary in each cell ; styles
connate in a short column below, above free, entire. *Capsule* subglobose or slightly trigonous, 3-coccous ; cocci 2-valved. *Seeds*
obovoid, with a large fleshy caruncle ; albumen fleshy ; cotyledons
broad, flat.

Trees or shrubs, everywhere glabrous ; leaves alternate ; spikes with many male
flowers in a dense-ovoid or subglobose head terminating specialised contracted
twigs, with usually 1–3 pedicelled female flowers at the base ; male bracts small,
imbricate, glandular at the base on each side, each 3–5-flowered ; female bracts
1-flowered.

DISTRIB. Species about five, two American and three in Tropical Africa ; one
of the latter extending to the south of the tropic.

1. **M. africana** (Müll. Arg. in DC. Prodr. xv. ii 1191, partly
and as to fruit only) ; a tree, 15–30 ft. high, everywhere glabrous ;
ultimate twigs passing into the inflorescence, up to $\frac{1}{2}$ in. long ;
leaves short-petioled, papery and at length somewhat coriaceous,
oblong or ovate-oblong, bluntly acute or acuminate, sometimes
rounded cuspidate or rounded obtuse, base rounded, margin entire,
$1\frac{1}{2}$–$2\frac{1}{4}$ in. long, $\frac{3}{4}$–$1\frac{1}{4}$ in. wide, dark shining green above, pale green

beneath, distinctly reticulate ; petiole $\frac{1}{3}$–$\frac{2}{3}$ in. long, slender ; stipules small, triangular ; spikes usually 2-sexual, male portion oblong or ovoid, rarely subglobose, $\frac{1}{4}$–$\frac{1}{2}$ in. long, dense-flowered ; female flowers 1–3, basal, long-pedicelled ; pedicels in fruit up to 1$\frac{1}{4}$ in. long ; male bracts narrow-lanceolate, acuminate from a broad base, each 3-flowered and each with two 2–3-partite basal glands ; bracts and flowers yellowish ; male calyx irregularly 2–3-lobed ; stamens 2–3, exserted ; ovary oblong ; capsule brown, slightly 3-sulcate, $\frac{1}{3}$–$\frac{2}{5}$ in. across ; seeds black, subglobose, smooth, $\frac{1}{4}$–$\frac{1}{3}$ in. long, caruncle under $\frac{1}{4}$ the size of body of seed. *Hiern in Cat. Afr. Pl. Welw.* i. 985, *partly* ; *De Wild. Miss. É. Laurent.* 141, *and Études Fl. Bas- et Moyen-Congo,* ii. 289, *partly* ; *Th. & Hél. Durand, Syll. Fl. Congol.* 499, *partly* ; *S. Moore in Journ. Linn. Soc. Bot.* xl. 204 ; *Prain in Dyer, Fl. Trop. Afr.* vi. i. 1004 ; *De Wild. Comp. Kasai,* 342. *M. obtusa, Pax in Engl. Jahrb.* xix. 116, *and in Engl. Pfl. Ost-Afr. C.* 241. *M. vaccinioides, Pax in Engl. Jahrb.* xix. 116. *M. africana, var. obtusa, Pax ex Durand & De Wild. in Bull. Soc. Bot. Belg.* xxxvii. 107, *partly* ; *Th. & Hél. Durand, Syll. Fl. Congol.* 500, *partly. M. africana, var. benguelensis, Pax & K. Hoffm. in Engl. Pflanzenr. Euphorb. Hippoman.* 180. *Excœcaria magenjensis, Sim, For. Fl. Port. E. Afr.* 104, *t.* 100 A.

EASTERN REGION : Delagoa Bay ; Maputa, *Sim*
Also widely spread in Tropical Africa.

XL. SPIROSTACHYS, Sond.

Flowers monœcious ; petals 0 ; disc 0. *Male : Calyx* usually 5-lobed, rarely 4-lobed, slightly imbricate in bud. *Stamens* 3 ; filaments united throughout in a slender tube ; anthers free, dehiscing longitudinally. Rudimentary *ovary* 0. *Female : Calyx* 5-lobed. *Ovary* 3-celled, rarely 2-celled ; ovules solitary in each cell ; styles 3, rarely 2, distinctly connate below, entire. *Capsule* 3-coccous, rarely 2-coccous ; cocci 2-valved. *Seeds* globose, ecarunculate ; albumen fleshy ; cotyledons broad, flat.

Trees or shrubs, everywhere glabrous ; leaves alternate ; spikes lateral, catkin-like, with numerous male flowers above and a few basal female flowers, or occasionally with the basal female flowers wanting.

DISTRIB. A single species rather widely spread in Africa to the south of Lat. 12° S.

1. **S. africanus** (Sond. in Linnæa, xxiii. 106) ; a tree, reaching 60 feet in height, everywhere glabrous ; twigs slender ; leaves short-petioled, alternate, firmly membranous and at length somewhat coriaceous, oblong-ovate, obtuse, base rounded, 2-glandular at the junction with the petiole, margin crenulate, about 2 in. long, 1 in. wide, medium green, paler beneath ; petiole $\frac{1}{3}$ in. long, channelled

above ; stipules minute, lanceolate, caducous ; spikes lateral, appear
ing before the leaves, sessile, densely imbricately spirally bracteate,
⅓–⅔ in. long, with many male flowers above and usually 1–3 basal
female flowers ; male bracts obtuse, eglandular, 1-flowered ; male
pedicels obsolete ; female pedicels short, in fruit reaching ⅕–¼ in. in
length ; male calyx 5- (less often 4-) sect ; lobes obovate-spathulate,
obtuse ; anthers 3, extrorse, at the apex of the staminal tube ;
female calyx 5-sect ; lobes triangular-ovate ; ovary glabrous ;
styles 3, rarely 2, connate below in a short column ; capsule
3-dymous, rarely 2-dymous, ⅓ in. across ; valves thinly crustaceous ;
seeds globose. *Walp. Ann.* iii. 360; *Pax in Engl. Pflanzenr. Euphorb.-
Hippoman.* 155 ; *Prain in Dyer, Fl. Trop. Afr.* vi. i. 1006.
*S. synandra, Pax, l.c. Stillingia africana, Baill. Étud. Gén.
Euphorb.* 522, *and in Adansonia,* iii. 163. *Excœcaria africana, Müll.
Arg. in Linnæa,* xxxii. 123, *and in DC. Prodr.* xv. ii. 1215 ; *Sim,
For. Fl. Cape Col.* 319, *t.* 144, *fig.* 2, *and For. Fl. Port. E. Afr.* 104,
t. 100 B. *E. Agallocha, Benth. in Benth. & Hook. f. Gen. Pl.* iii.
337, *partly ; not of Linn. E. synandra, Pax in Engl. Jahrb.* xliii.
223. *Maprounea africana, Müll. Arg. in DC. Prodr.* xv. ii. 1191,
mainly ; Hiern in Cat. Afr. Pl. Welw. i. 985, *mainly. Sapium
africanum, O. Kuntze, Rev. Gen. Pl.* iii. ii. 293. *Excœcariopsis
Dinteri, Pax in Engl. Jahrb.* xlv. 239. *E. synandra, Pax l.c.*

KALAHARI REGION : Transvaal ; Rustenberg, *Miss Pegler,* 974 ! near Barberton,
1350–2500 ft., *Burtt-Davy,* 2822 ! 5561 ! Sterkstroom, *Burke,* 263 ! *Zeyher,* 1528 !
Lydenberg Distr., *Burtt-Davy,* 7316 ! Pretoria Distr. ; *Reck,* 4270 !
EASTERN REGION : Natal ; Umgeni River, 100 ft., *Wood,* 3917 ! without precise
locality, *Gueinzius* ! *Gerrard,* 747 ! *Cooper,* 3494 ! Delagoa Bay ; near Lourenço
Marques, *Bolus,* 9782 ! *Sim* ; Umbelusi, *Sim.*

Also in Tropical Africa.

XLI. SAPIUM, P. Br.

Flowers monœcious, rarely diœcious ; petals 0 ; disc 0. *Male :
Calyx* usually 3-lobed, valvate, subvalvate or open in bud. *Stamens*
2–3 ; filaments free ; anthers dehiscing longitudinally. Rudimentary
ovary 0. *Female : Calyx* 2–3-lobed. *Ovary* 3-celled, rarely 2-celled ;
ovules solitary in each cell ; styles 3, rarely 2, free or slightly
connate below, entire. *Capsule* 3-coccous, rarely 2-coccous ; cocci
2-valved, rarely subindehiscent. *Seeds* globose or narrow-oblong,
ecarunculate in all S. African species ; albumen fleshy ; cotyledons
broad, flat.

Trees or shrubs, everywhere glabrous ; leaves alternate ; spikes terminal,
androgynous with numerous male flowers above and a few basal female flowers.

DISTRIB. About 100 species, widely spread in all tropical regions.

Male bracts 1-flowered ; leaf-blade shortly decurrent on
the petiole ; capsules breaking up into 3 2-valved
cocci ; valves thinly crustaceous, not horned on the
back   ...   . (1) **Simii.**

Male bracts several-flowered ; leaf-blade minutely auriculate-cordate at point of junction with the petiole :
Capsule breaking up into 3 2-valved cocci ; valves with
a thick woody endocarp and a coriaceous separable exocarp, each shortly but distinctly horned
on the back ...   . (2) **reticulatum.**

Capsule subindehiscent, 2-dymous ; valves coriaceous,
not horned on the back   ...   . (3) **mannianum.**

1. **S. Simii** (O. Kuntze, Rev. Gen. Pl. iii. ii. 293) ; a shrub, 3–5 ft.
high, everywhere glabrous ; buds perulate ; twigs slender, angular ;
leaves short-petioled, alternate, chartaceous, oblong or oblong-
lanceolate, acute or shortly acuminate, base abruptly or gradually
cuneate, margin crenate-dentate, $1\frac{3}{4}$–$3\frac{1}{2}$ in. long, 1–$1\frac{1}{4}$ in. wide ;
petiole $\frac{1}{4}$–$\frac{2}{3}$ in. long, channelled above, eglandular ; stipules small,
usually denticulate ; spikes terminal on leafy twigs, $\frac{3}{4}$–1 in. long,
slender, with many male flowers above and 2–3 basal female
flowers ; male bracts orbicular-ovate, somewhat toothed, 1-flowered,
glandular at the base ; pedicels in both sexes very short : male calyx
3-partite, lobes small, lanceolate ; stamens 3 ; female calyx 3-sect ;
lobes wide-triangular, acute, eglandular ; ovary glabrous, not horned ;
styles 3, free ; capsule globose, $\frac{1}{3}$ in. across, breaking up into
3 2-valved cocci ; valves thinly crustaceous ; seeds globose.
Excœcaria caffra, Sim, *For. Fl. Cape Col.* 319, *t.* 144, *fig.* 1. *E. Simii,*
Pax in Engl. Pflanzenr. Euphorb. Hippoman. 170.

COAST REGION : Albany Div. ; without precise locality. *Miss Bowker !* King
Williamstown Div. ; Perie Forest, 2200–2500 ft.. *Sim,* 1560 ! *Kuntze ! Galpin,*
5907 ! *Schönland,* 862 ! East London Div. ; near East London, *Thode in
Herb. Wood,* 4850 ! Needs Camp, *Galpin,* 7098 ! Komgha Div. ; near Komgha,
1800 ft., *Flanagan,* 413 ! near Kei Mouth, 100 ft., *Schlechter,* 6190 !
EASTERN REGION : Transkei ; Kentani, 1000 ft., *Miss Pegler,* 897 ! Pondoland ;
Gwenkala River, 1500 ft., *Flanagan,* 413 ! and without precise locality, *Beyrich,*
306 ! Natal ; Inanda, *Wood,* 991 ! and without precise locality, *Gerrard,* 2152 !
Zululand ; near Eshowe, *Wylie in Herb. Wood,* 7534 !

2. **S. reticulatum** (Pax in Engl. Pflanzenr. Euphorb. Hippoman.
245, fig. 46, C.D) ; a shrub, 10–18 ft. high, everywhere glabrous except
the margins of the stipules ; buds perulate ; twigs slender, spreading ;
leaves short-petioled, alternate, membranous, oblong-ovate or ovate-
lanceolate, shortly and bluntly acuminate, base rounded or wide-
cuneate, minutely auriculate-cordate at point of junction with
petiole, margin entire or serrulate, 1–3 in. long, $\frac{3}{4}$–$1\frac{1}{4}$ in. wide ;
petiole 1–2 in. long, channelled above ; stipules ovate-lanceolate,
as long as the petiole, very caducous ; spikes terminal on leafy twigs,
2–3 in. long, with many male flowers above and 1 (rarely 2) basal
female flowers ; male bracts oblong or ovate, somewhat toothed,
several-flowered, glandular at the base ; female bracts 1-flowered ;

pedicels of male flowers longer than bracts, of female flowers nearly
¼ in. long, in fruit ¾ in. long; male calyx 3-partite, lobes triangular,
toothed; stamens 3; female calyx 3-sect, wide-triangular, cordate,
acute, toothed, with either a gland or a lanceolate toothed lacinula
at each sinus; ovary glabrous, with 2 dorsal horns on each carpel;
styles 3, shortly connate at the base; capsule ovoid, acute, ½ in.
across, with a coriaceous separable somewhat wrinkled epicarp and
a thick woody endocarp, breaking up into 3 2-valved cocci; seeds
ovoid, grey, mottled with brown. *Sclerocroton integerrimus, Hochst.
in Flora,* 1845, 85; *Sond. in Linnæa,* xxiii. 107. *S. reticulatus,
Hochst. l.c. Stillingia integerrima, Baill. in Adansonia,* iii. 162.
Excœcaria hochstetteriana, Müll. Arg. in Linnæa, xxxii. 122. *E.
integerrima, Müll. Arg. in DC. Prodr.* xv. ii. 948. *E. reticulata,
Müll. Arg. l.c.* 1213; *Wood, Natal Pl. t.* 10; *Sim, For. Fl. Cape Col.*
320, *as to description mainly; and in For. Fl. Port. E. Afr.* 103,
*mainly. Tragia integerrima and T. natalensis, Hochst. ex Krauss in
Flora,* 1845, 85, *and ex Müll. Arg. in DC. Prodr.* xv. ii. 1214; *Pax
in Engl. Pflanzenr. Euphorb. Hippoman.* 245.

EASTERN REGION : Pondoland; St. Johns, *Sim,* 2422, partly! Natal; near Durban,
100 ft., *Wood,* 1012! 1033! 1417! 6529! *Rehmann,* 8984! *Wilms,* 1916! Clairmont,
Engler, 2540! Inanda, *Wood,* 92! between Pinetown and Umbilo, *Rehmann,*
8038! and without precise locality, *Gueinzius,*! *Krauss,* 351! 468! *Drège,* 4587!
Peddie! *Cooper,* 1231! *Williamson*! *Gerrard,* 83! *Gerrard & McKen,* 697!
Delagoa Bay; near Lourenço Marques, *Junod,* 68! *Sim* ; Magaia, *Schlechter,* 12046!

A very distinct species which, nevertheless, so careful an observer as Sonder in
1850 thought to be only a variety of the one next to be described. Sim, another
competent observer, has recently been led into the same misapprehension, and has
distributed under the same field-number, 2422, specimens of both these species.
Under the name *Excœcaria reticulata,* Sim based on these specimens a description
which covers both species, but has given a figure which represents only that one to
which the name *E. reticulata* does not apply. Baillon in 1863 first pointed out that
Hochstetter in 1845 had given two names, *Sclerocroton integerrimus* and *S. reticulatus,* to what are only two conditions of the same species, and that, of these two,
S. integerrimus is the name which should be used. Müller, in the same-year,
recognising this fact and, at the same time transferring the species to *Excœcaria,*
substituted for Hochstetter's two specific terms the new name *E. hochstetteriana.*
Realising the inadmissibility of the proposal, Müller in 1866 first (*DC. Prodr.* xv.
ii. 948) used the name *E. integerrima,* but later (*l.c.* 1213) altered it to *E. reticulata.*
In transferring the species from *Excœcaria* to *Sapium,* Pax has adopted the specific
name finally employed by Müller.

3. **S. mannianum** (Benth. in Benth. & Hook. f. Gen. Plant. iii.
335); a tree of moderate to considerable size, glabrous except on
the inflorescences and the scaly portion of the young twigs; buds
perulate; scales coriaceous; twigs slender, drooping; leaves short-
petioled, alternate, firmly membranous to thinly coriaceous, oblong-
ovate to elliptic-lanceolate, acute to shortly cuspidate; base cuneate
or rounded, minutely auriculate-cordate at point of junction with
the petiole, margin more or less toothed or rarely subentire, with
1-2 glands on each side along the base, 2½-6 in. long, ¾-2 in. wide;

petiole slender, 2–4 lin. long, slightly channelled above; stipules ovate-triangular, acute, usually soon deciduous; spikes terminal on leafy twigs, 2–4½ in. long, with many male flowers above and 1–3 basal female flowers; male bracts orbicular-ovate, finely toothed, several-flowered, glandular at the base; female bracts 1-flowered; pedicels of male flowers longer than the bracts; of female flowers about ½ in. long, in fruit ⅔ in. long; male calyx unequally 2–3-fid, lobes obtuse; stamens 2, less often 3; female calyx deeply 3-partite, sometimes with intercalary glandular lacinulæ in the sinuses; ovary glabrous, not horned, 2-celled, rarely 3-celled; styles 2, very rarely 3, connate below; capsule coriaceous, hardly dehiscent, usually markedly didymous, sometimes by abortion 1-coccous, ⅓ in. across; seeds sub-globose, pale brown. *Pax in Bolet. Soc. Brot.* x. 161; *Hiern in Cat. Afr. Pl. Welw.* iv. 986; *S. Moore in Journ. Linn. Soc. Bot.* xl. 204; *Prain in Dyer, Fl. Trop. Afr.* vi. i. 1016. *S. abyssinicum, Benth. l.c. S. Kerstingii, Pax in Engl. Jahrb.* xliii. 85. *S. ellipticum, Pax in Engl. Pflanzenr. Euphorb. Hippoman.* 253. *Sclerocroton ellipticus, Hochst. in Flora,* 1845, 85; *Sond. in Linnæa,* xxiii. 107; *Baill. Étud. Gén. Euphorb.* 523, *t.* 8, *fig.* 17. *Stillingia elliptica, Baill. Adansonia,* iii. 162. *Excœcaria indica, Müll. Arg. in Linnæa,* xxxii. 123, *and in DC. Prodr.* xv. ii. 1216, *partly; only the African plant. E. manniana, Müll. Arg. in Flora,* 1864, 433, *and in DC. Prodr.* xv. ii. 1217. *E. reticulata, Sim, For. Fl. Cape Col.* 320, *as to description partly and wholly as to t.* 140, *fig.* 2; *and in For. Fl. Port. E. Afr.* 103, *partly; not of Müll. Arg. Tragia elliptica, Hochst. ex Müll. Arg. in DC. Prodr.* xv. ii. 1216; *Pax in Engl. Pflanzenr. Euphorb. Hippoman.* 253.

EASTERN REGION: Pondoland; St. Johns, *Sim,* 2422, partly! Port St. John, above Tiger Flat, 100 ft., *Galpin,* 2856! Natal; Inanda, *Wood,* 954! Fairfield, *Rudatis,* 117! and without precise locality, *Gueinzius,* 175! *Krauss,* 269! *Gerrard,* 84! *Gerrard & McKen!*

Also in Tropical Africa.

ORDER CXXII. **ULMACEÆ.**

(By N. E. BROWN.)

Flowers hermaphrodite or unisexual. *Perianth* 4–5-partite, slightly imbricate, persistent. *Stamens* 4–5, free, opposite the perianth-segments, absent from the female flower; anthers dorsifixed, erect or reflexed in bud. *Ovary* free, rudimentary or abortive in the male flowers, often surrounded at the base with a ring of hairs, 1-celled; stigmas 2, terminal, subulate or stoutly filiform; ovule solitary, pendulous. *Fruit* fleshy, with a hard endocarp or "stone." *Seed* pendulous, with a thin testa, albuminous or exalbuminous; embryo curved; cotyledons flat or folded.

Trees or shrubs, spiny or unarmed ; leaves alternate, stipulate, usually 3-nerved at the base ; stipules lateral, free, deciduous ; flowers in small axillary cymes or clusters or solitary, pedicellate.

DISTRIB. Genera 9, species about 100, widely distributed, but chiefly inhabiting the warmer regions.

I. **Celtis.**—*Stipules* free. *Flowers* in small cymes or clusters or solitary, with the pedicels of the female flowers 3–9 lin. long. *Cotyledons* broad, variously folded.

II. **Trema.**—*Stipules* free. *Flowers* in small dense clusters or cymes with pedicels $\frac{1}{2}$–1 lin. long. *Cotyledons* narrow.

III. **Chætacme.**—*Stipules* connate and sheathing the buds. *Male flowers* in small cymes ; female flowers solitary ; pedicels $\frac{1}{2}$–$2\frac{1}{2}$ lin. long. *Cotyledons* unequal, the larger enveloping the smaller.

I. CELTIS, Linn.

Flowers hermaphrodite or unisexual. *Perianth-segments* 4–5, spreading, concave, green or brownish. *Stamens* 4–5, opposite the perianth-segments, absent from the female flowers. *Ovary* superior, 1-celled, rudimentary in the male flowers ; stigmas 2, terminal, sessile, filiform or subulate, very stout, widely spreading or ascending ; ovule solitary, pendulous from the top of the cell, anatropous. *Fruit* with a fleshy exocarp and hard bony endocarp. *Seed* exalbuminous ; testa thin ; cotyledons broad, variously folded.

Trees or shrubs, unarmed or spiny ; leaves alternate, stipulate ; stipules lateral, free, deciduous ; flowers in small axillary clusters or solitary, pedicellate.

DISTRIB. Species about 60, widely distributed in tropical and temperate regions.

Leaves on the flowering branches 2$\frac{1}{2}$–5 in. long, with few teeth ; fruit compressed, nearly $\frac{1}{2}$ in. long ... ... (1) **Franksiæ**.

Leaves on the flowering branches 1–3$\frac{1}{2}$ in. long, with numerous fine teeth ; fruit subglobose, less than $\frac{1}{4}$ in. in diam. ... ... (2) **rhamnifolia**.

1. **C. Franksiæ** (N. E. Br.) ; a tree 30 ft. high ; branches at first puberulous, becoming glabrous, with a grey or brown bark ; leaves alternate, puberulous on the petiole and midrib beneath, otherwise glabrous, subcoriaceous ; petiole $\frac{1}{4}$–$\frac{1}{2}$ in. long ; blade 2$\frac{1}{2}$–7 in. long, 1–3 in. broad, elliptic-oblong, acute or shortly acuminate, obliquely rounded or obliquely subcuneate at the base, entire or more or less toothed above the middle, 3-nerved at the base with 4–5 spreading primary veins on each side of the midrib ; stipules 2–3 lin. long, lanceolate, acuminate, puberulous, soon deciduous ; male and hermaphrodite flowers intermingled on axillary racemes 3–4 lin.

long; bracts $\frac{1}{2}$–$\frac{3}{4}$ lin. long, ovate, acute; pedicels of the male
flowers 1–1$\frac{1}{2}$ lin. long, of the female 4–6 lin. long, puberulous;
perianth-segments 5, spreading, 1 lin. long, oblong, obtuse, concave,
puberulous on the back; stamens not or scarcely exceeding the
perianth-segments; disc pubescent; ovary ovoid, much flattened in
young fruit, shortly 2-horned at the apex, glabrous, very rudimentary in the male flowers; styles when fully developed $\frac{1}{4}$ in. or more
long, usually twice forked, very stout, densely puberulous. *Celtis
Soyauxii, Wood in Trans. Roy. Soc. S. Afr.* iii. 56; *not of Engl.*

Eastern Region: Natal; near Durban, *Miss Franks in Herb. Wood,* 11726!

This species is allied to *C. Soyauxii,* Engl., but is distinguished by its larger
leaves, which are more distinctly toothed, different in texture and the 4–5 veins
on each side of the midrib are spreading and unite with each other in bold loops,
whilst in the true *C. Soyauxii* there are only 2–3 lateral veins, which are all very
ascending and excurrent at the margin near the apex. The fruit of *C. Franksiæ*
is also glabrous, whilst that of *C. Soyauxii* is described as thinly pubescent, but I
have not seen it, as the Angolan plant quoted under *C. Soyauxii* by Engler is
(except Welwitsch, 6285) obviously a distinct species with a totally different
venation.

2. **C. rhamnifolia** (Presl, Bot. Bemerk. 37, excl. syn.); varying from
a shrub to a tree 20–80 ft. high; branches with a brown or greyish
bark, thinly puberulous to densely pubescent on the young growth, becoming glabrous; leaves alternate, petiolate, varying from glabrous to
densely pubescent on both sides; petiole 1–3 lin. long; blade $\frac{2}{3}$–3$\frac{1}{2}$
in. long, $\frac{1}{3}$–2$\frac{1}{4}$ in. broad, ovate, acuminate, obliquely rounded or subcordate at the base, serrate with small teeth, 3-nerved at the base;
stipules 2$\frac{1}{2}$–3 lin. long, $\frac{1}{2}$–$\frac{2}{3}$ lin. broad, lanceolate or linear-lanceolate,
obtuse, brown, deciduous; flowers unisexual or hermaphrodite, soli-
tary or 2–4 in very shortly pedunculate cymes, axillary, pedicellate,
the female or hermaphrodite flowers sometimes on a distinct plant,
sometimes mingled with the males; pedicels 1–9 lin. long, slender,
puberulous; perianth-segments 4, about 1$\frac{1}{4}$ lin. long, boat-shaped,
subobtuse, very spreading, brown, with paler margins, thin,
glabrous, ciliate; stamens 4, spreading on or within the perianth-
segments and shorter than or only exceeding the latter by their
anthers, sometimes wanting in the fertile flowers; ovary shorter
than the perianth-segments, ovoid or subglobose, densely woolly;
stigmas 1–1$\frac{1}{2}$ lin. long, very stout, densely papillate-pubescent; fruit
subglobose, about 2 lin. in diam., glabrous or pubescent. *Burtt-
Davy in Transv. Agric. Journ.* iv. *t.* 111, *and* v. 433. *C. kraussiana,
Bernh. in Flora,* 1845, 87, *and in Krauss, Fl. Cap.- und- Natal.* 150;
Planch. in Ann. Sc. Nat. 3me sér. x. 295, *and in DC. Prodr.* xvii. 173;
Wood, Natal Pl. i. 25, *t.* 28; *Burtt-Davy in Transv. Agric. Journ.* v.
433; *Sim, For. Fl. Cape Col.* 306, *t.* 134 (*floral analyses very in-
accurate*). *C. opegrapha, Planch., C. vesiculosa, Hochst. ex Planch.,
and C. Burmanni, Planch. in Ann. Sc. Nat.* 3me sér. x. 294, 295, 296.
Celtis eriantha, E. Meyer in Drège, Zwei Pfl. Documente, 171, *and*

Planch. l.c. 296.—*Celtis foliis subrotundis, &c. J. Burm. Rar. Afr. Pl.*
242, *t.* 88.

SOUTH AFRICA : without locality, *Zeyher*, 314 !
COAST REGION : Port Elizabeth Div. ; Krakakamma, *Zeyher*, 570 ! 993 ! Uiten-
hage Div. ; near the Lead-mine, *Burchell*, 4498 ! Queenstown Div. ; near Queens-
town, *Cooper*, 213 ! *Galpin*, 1890 ! Komgha Div. ; near Komgha, *Schlechter*, 6160 !
CENTRAL REGION : Somerset Div. ; Bosch Berg, *Burchell*, 3146 ! by the Fish
River, *MacOwan*, 1566 ! and without precise locality, *Bowker* ! Graaff Reinet Div. ;
Oude Berg, *Drège* ; near Graff Reinet, *Drège* ! *Bolus*, 685 ! Wodehouse Div. ;
Mooi Flats, *Drège*, 8261 a ! Aliwal North Div. ; by the Orange River, *Burke* !
Albert Div. ; near Gaatje, *Drège*, 8261 b !
KALAHARI REGION : Basutoland ; Leribe, *Mrs. Dieterlen*, 193 ! Transvaal ;
hills near Aapies River, *Rehmann*, 4277 ! Bereaparle, Pretoria, *Miss Leendertz*,
259 ! 990 ! north of Klerksdorp, *Nelson*, 250 !
EASTERN REGION : Natal ; near Durban and in the Botanic Garden, *Gerrard*,
85 ! 1977 ! *Gueinzius* ! *Wood*, 1784 ! eastern side of Table Mountain, *Krauss*,
1776.

Known as White Stinkwood, Camdeboo Stinkwood and Wit-gat Boom.

Also in Tropical Africa and Arabia.

This plant is generally known as *C. kraussiana*, as Presl's earlier name was overlooked by Planchon when he monographed this genus, and has not been included
in the Index Kewensis. Dr. Rendle has shown (*Journ. Bot.* 1915, 298) that
Rhamnus celtifolius, Thunb. (*Prodr.* 44, and *Fl. Cap. ed. Schultes*, 196) is
Rhamnus prinoides, L'Hérit. Presl erroneously referred it to *Celtis rhamnifolia*,
which he founded on Drège, 8261.

II. TREMA, Lour.

Flowers unisexual. *Perianth-segments* usually 5, rarely 4, subequal or the outer slightly smaller, slightly imbricate, all concave in
the male flower, the outer two slightly concave and the others
flattened in the female flower. *Stamens* 5, rarely 4, absent from
the female flower ; filaments shorter than the perianth-segments ;
anthers erect in bud. *Ovary* sessile, erect, rudimentary or abortive
in the male, surrounded by a ring of hairs at the base in both sexes,
1-celled, with 1 pendulous ovule ; stigmas 2, terminal, not oblique,
filiform or subulate, thick. *Fruit* small, fleshy, with a thin hardened
endocarp. *Seed* with a membranous testa, albuminous ; embryo
curved, cotyledons narrow.

Trees or shrubs, without stinging hairs or spines ; leaves alternate, stipulate, often 3-nerved at the base ; stipules lateral, free, deciduous ; flowers in
small dense axillary cymes or clusters.

DISTRIB. Species about 40, widely distributed in tropical and subtropical
regions, only 1 in South Africa.

1. **T. bracteolata** (Blume, Ann. Mus. Bot. Lugd.-Bat. ii. 58) ;
a shrub or tree 8–20 ft. high ; young branches adpressed-puberulous ; leaves alternate, stipulate ; petiole 2–6 lin. long, puberulous ;
blade 1–5 in. long, ½–2 in. broad, varying from narrowly lanceolate to elongated ovate or oblong-ovate, acuminate, rounded to

cordate at the base, often more or less oblique, three-nerved at
the base, finely toothed, glabrous or thinly sprinkled with hairs
above, nearly glabrous to finely adpressed-puberulous beneath;
stipules 1–3½ lin. long, lanceolate-subulate, silky-pubescent, deciduous; flowers in small dense axillary cymes or globose clusters
¼–¾ in. in diam., which vary from shorter than to about twice as
long as the petioles, unisexual or occasionally with both sexes on
the same cyme, pubescent or puberulous; bracts ½–1¼ lin. long,
ovate, acute or tapering to a long point, pubescent; pedicels
¼–1 lin. long, jointed close under the flowers; flowers 5-partite,
green; segments of the male flower 1 lin. long, deeply boat-shaped,
obtuse, those of the female ⅔ lin. long, two of them concave, the
other three flatter and broader, elliptic, very obtuse, in both sexes
with membranous fringed margins to the inner three, the others
ciliate, glabrous; stamens 5, with stout filaments, shortly tapering
at the apex, shorter than the perianth-segments, glabrous, absent
from the female flowers; ovary compressed, elliptic, glabrous,
1-celled, with 1 pendulous ovule, columnar and abortive in the male
flowers, surrounded at the base in both sexes by a ring of hairs;
styles 2, ascending or diverging, ½ lin. long, hairy; fruit fleshy,
ellipsoid or subglobose, 1½–2 lin. in diam., black when ripe, bearing
the persistent remains of the styles. *Wood, Natal Pl. iv. t.* 356;
Burtt-Davy in Transv. Agric. Journ. v. 433; *Sim, For. Fl. Cape Col.*
305, *t.* 158, *fig.* 2 (*very inaccurate*). *T. glomerata, Blume, Ann. Mus.
Bot. Lugd.-Bat.* ii. 58. *T. guineensis, Ficalho, Pl. Uteis.* 261, *and
Priemer in Engl. Jahrb.* xvii. 426. *Sponia glomerata, Hochst., and
S. bracteolata, Hochst. in Flora,* 1845, 87, *and in Krauss, Fl. Cap.-
und- Natal.* 150–151, *and Planch. in Ann. Sc. Nat.* 3me sér. x. 321;
Rendle in Dyer, Fl. Trop. Afr. vi. ii. 11. *S. guineensis, Planch.
in DC. Prodr.* xvii. 197. *S. orientalis, β angustifolia, E. Meyer ex
Planch. l.c., under S. guineensis, Planch. Celtis guineensis, Schumach.
& Thonn. Beskr. Guin. Pl.* 160. *C. orientalis, β, E. Meyer in Drège,
Zwei Pfl. Documente,* 171.

COAST REGION: Komgha Div.; along the Kei River near Komgha, *Flanagan,*
1347!
KALAHARI REGION: Transvaal; near Barberton, *Galpin,* 848! *Thorncroft,* 3934!
EASTERN REGION: Transkei; Kentani, *Miss Pegler,* 867! Natal; Table
Mountain, *Krauss,* 41! in woods, *Krauss,* 354! between Umzimkulu River and
Umcomaas River, *Drège!* Maritzburg, *Rehmann,* 7512! Inanda, *Wood,* 625! near
Durban, *Schlechter,* 2795! Dumisa, *Rudatis,* 591! 1189! and without precise
locality, *Peddie! Gueinzius! Gerrard,* 14! *Cooper,* 1254!

Also in Tropical Africa and Arabia.

III. CHÆTACME, Planch.

Flowers unisexual. *Perianth-segments* 5, induplicate-valvate,
deeply concave with inflexed sides and all equal in the male, flattish
and one smaller than the others in the female flower. *Stamens* 5,
included in the cavity of the perianth-segments, absent from the

female flowers; anthers erect in bud. *Ovary* sessile, erect, rudimentary in the male flowers, 1-celled, with 1 pendulous ovule; styles 2,
terminal, not oblique, filiform, long and densely hairy-stigmatose.
Fruit globose, at first fleshy, becoming dry and hard, with a bony
endocarp. *Seed* globose, with a thin membranous testa, exalbuminous; embryo curved, with unequal thick fleshy cotyledons, the
larger enveloping the smaller.

Trees or shrubs, spiny, without stinging hairs; leaves alternate, stipulate;
stipules connate and sheathing the bud, deciduous as the bud expands; male
flowers in small axillary cymes or clusters, female solitary, axillary.

DISTRIB. Species 2, one in Tropical and South Africa, the other in Madagascar.

1. **C. aristata** (Planch. in Ann. Sc. Nat. 3me sér. x. 341); a shrub
or tree 9–35 ft. high, more or less armed with spines; young branches
varying from glabrous to pubescent, usually more or less zigzag;
leaves alternate, stipulate; petiole 1–4 lin. long; blade $\frac{3}{4}$–3 in. long,
$\frac{1}{3}$–1$\frac{1}{4}$ in. broad, lanceolate, oblong-lanceolate or elliptic, obtuse or
acute, tipped with a bristle 1–4 lin. long, entire or more or less
toothed, even on the same branch, glabrous on both sides or thinly
puberulous beneath; stipules connate into a sheath including the
terminal bud, 3–11 lin. long, glabrous or puberulous, brown, quickly
deciduous; spines axillary, solitary or in pairs, 2–14 lin. long;
male flowers in axillary clusters or cymes $\frac{1}{4}$–1$\frac{1}{4}$ in. in diam.;
bracts $\frac{2}{3}$–$\frac{3}{4}$ lin. long, ovate, obtuse, glabrous or pubescent, ciliate;
pedicels $\frac{1}{2}$–2$\frac{1}{2}$ lin. long, jointed close under the flower, glabrous
or puberulous; perianth-segments 1 lin. long, deeply boat-shaped
with the margins folded inwards, glabrous to pubescent, green;
stamens 5, included in the concavity of the perianth-segments;
ovary rudimentary, sometimes with 2 short style-arms at the apex,
pubescent; female flower solitary, axillary; pedicel 1–1$\frac{1}{2}$ lin. long,
pubescent (or perhaps sometimes glabrous?), with 1 minute ovate
pubescent bract; perianth-segments $\frac{3}{4}$ lin. long, $\frac{1}{2}$ lin. broad, flattish,
ovate, obtuse, puberulous; ovary erect, sessile, subglobose or ovoid,
glabrous or puberulous; styles 2, terminal, 6–8 lin. long, filiform,
diverging-erect, hairy; fruit globose, about $\frac{1}{2}$ in. in diam., at first
fleshy, flesh becoming dry and hard and adhering to the hard stone
or endocarp. *Planch. in DC. Prodr.* xvii. 210; *Sim, For. Fl. Cape
Col.* 305, *t.* 160, *fig.* 1. *C. nitida, Planch. & Harv., and C. Meyeri, Harv.
Thes. Cap.* i. 16, *t.* 25. *C. serrata, Engl. in Notizbl. Königl. Bot. Gart.
Berlin,* iii. 24; *Rendle in Dyer, Fl. Trop. Afr.* vi. ii. 14. *Celtis
appendiculata, E. Meyer, C. aristata, E. Meyer, and C. subdentata,
E. Meyer in Drège Zwei Pfl. Documente,* 171.

COAST REGION: Uitenhage Div.; forest near Van Stadens River, *Burchell,*
4666! Galgebosch and Enon, *Drège!* and without precise locality, *Zeyher,* 675!
Albany Div.; between Blue Krantz and Kowie Poort, *Burchell,* 3667! Stocken-
strom Div.; *Galpin,* 1736!
EASTERN REGION: Pondoland; St. Johns River, *Drège!* woods near coast,
Beyrich, 119, *Bachmann,* 432, 433. Natal; Inanda, *Wood,* 698! near Durban,
Wood in MacOwan & Bolus Herb. Norm. Austr.-Afr. 1366! and without precise
locality, *Gerrard,* 270! 276! 547! 1182! *Gueinzius!*
Also in Tropical Africa.

Order CXXII. A.   **MORACEÆ.**

(By N. E. Brown and J. Hutchinson.)

Flowers unisexual, in spikes or globose heads, or enclosed within a globose or pear-shaped receptacle (fig), or seated upon or immersed in a flattened receptacle. *Perianth* of 2–6 free segments or 2–6-lobed or toothed, or tubular or urceolate or absent. *Stamens* 2–6, free, opposite the perianth-segments, absent from the female flowers ; anthers inflexed or erect in bud. *Ovary* free, 1-celled, rudimentary or absent from the male flowers ; stigmas 2, equal or unequal, subulate or filiform, or style entire or bifid at the tip, filiform. *Ovule* solitary, pendulous. *Fruit* either compound and globose or oblong and formed of the enlarged fleshy perianth-segments and outer coat of the ovary, or of achenes immersed in a flattened receptacle or enclosed in a globose or pear-shaped receptacle (fig). *Seed* pendulous, with a thin testa, albuminous or exalbuminous ; embryo curved ; cotyledons flat or folded.

Trees or shrubs, with milky juice ; leaves alternate, stipulate, deciduous or evergreen ; flowers of the South African species in dense globose heads or enclosed in the receptacle (fig).

Distrib. Genera about 64, species about 950, widely distributed throughout the warmer regions.

   I. **Cardiogyne.**—Flowers in dense globose heads.

   II. **Ficus.**—Flowers completely enclosed in a globose or pear-shaped receptacle (fig).

I. **CARDIOGYNE**, Bureau.

Flowers in globose heads, unisexual, the sexes on different plants. *Perianth-segments* 4, imbricate. *Stamens* 4, opposite the perianth-segments and longer than them, inflexed in bud, absent from the female flowers. *Ovary* superior, sessile, 1-celled, with 1 pendulous ovule ; style terminal, long, filiform. *Fruit* compound, globose, pulpy, formed of the enlarged fleshy perianth-segments and bracts of the combined head of flowers, with the achenes embedded in the pulp. *Achenes* ellipsoid, notched at the apex, with a thin hard shell. *Seed* exalbuminous ; testa thin ; embryo with plicate cotyledons embracing the incurved radicle.

A climbing shrub or a bush ; leaves alternate, inconspicuously stipulate ; flowers in dense globose heads solitary or in pairs in the axils of the leaves ; fruit globose, compound, pulpy.

DISTRIB. Species 1, in Tropical Eastern South Africa.

This genus is very closely related to *Cudrania*, and only differs from it by the stamens being inflexed in bud.

1. C. africana (Bureau in DC. Prodr. xvii. 233) ; a climbing shrub or bush, armed with spines ; branches glabrous or puberulous, with pale grey bark, often ending in a spine and also with axillary spines ; leaves alternate, inconspicuously stipulate ; petiole ¼–1 in. long, slender, glabrous or puberulous ; blade 1–3 in. long, ½–1¾ in. broad, elliptic or elliptic-oblong, shortly and obtusely pointed or rounded and slightly notched at the apex, glabrous on both sides or puberulous on the midrib beneath ; stipules ¼–1 lin. long, ovate, very acuminate, quickly deciduous ; flowers mingled with bracts in dense globose pedunculate heads 3½–7 lin. in diam. ; heads solitary or two together, axillary, sessile in the female and on peduncles 2–3 lin. long in the male plant, minutely and densely velvety, greyish or whitish ; perianth-segments 4, imbricate, 1–1¼ lin. long, ⅔–1 lin. broad, obovate, broadly rounded at the apex, concave at the upper part and gibbous on the back in the male, truncate, thickened and dorsally gibbous at the apex in the female flowers ; stamens 4, twice as long as the perianth-segments, inflexed in bud ; filaments red ; anthers yellow ; ovary obovate, compressed, 1-celled, with 1 pendulous ovule ; style terminal, filiform ; female heads in fruit "the size of a walnut, uneven on the surface, yellowish cream-colour" (*Kirk*) ; achene enclosed in the perianth, 2½ lin. long, 2 lin. in diam., ellipsoid, notched at the apex, smooth, whitish-brown. *Baill. Hist. Nat. Pl. Malay. t.* 294 ; *Hook. Ic. Pl.* xxv. *t.* 2473 ; *Engl. & Prantl, Pflanzenfam.* iii. i. 76 ; *Engl. Monogr. Afr. Pflanzenfam. Moraceæ,* 5 ; *Rendle in Dyer, Fl. Trop. Afr.* vi. ii. 24. *Cudranea, Kirk in Journ. Linn. Soc. Bot.* ix. 229.

EASTERN REGION : Delagoa Bay ; on a hill near Lourenço Marques, *Bolus,* 9783 ! Also in Tropical Africa.

The fruit is edible and the wood dyes cloth yellow. In each perianth-segment there are two cavities filled with yellow granular matter.

II. FICUS, Linn.

(By J. HUTCHINSON.)

Flowers monœcious or very rarely diœcious, enclosed in a variously shaped fleshy or somewhat woody receptacle (fig). *Male* flower : perianth 2–6-lobed or -partite ; lobes or segments imbricate, usually membranous and hyaline ; stamens 1–2, or rarely 3–6, with straight

short filaments; anthers more or less oblong or ovoid, exserted or
included; rudimentary ovary 0 or very rarely present. *Female
flower* : perianth-segments often fewer and narrower than in the
male or rarely minute; ovary mostly obliquely ellipsoid or ovoid;
style almost invariably lateral, short or slender or rather long;
stigma usually oblong; ovule laterally attached and pendulous
from near the apex of the cell; achene partially enclosed within the
persistent perianth; pericarp crustaceous and dry or rarely succulent; seed pendulous, with a membranous testa; albumen often
scanty; embryo curved; cotyledons often plicate and subequal;
radicle incumbent.

Trees, shrubs or rarely climbers, with milky juice; leaves alternate or very
rarely opposite, entire, dentate, or variously lobed, very variable in shape and
venation; stipules enveloping the terminal bud, caducous at the unfolding of the
leaves or more rarely persistent; receptacles (figs) sessile or pedunculate, mostly
paired when axillary or sometimes solitary, when borne on the trunk or main
branches remote from the leaves then in leafless panicles or more usually in
fascicles, 2–3-bracteate at the base with the bracts in a whorl or more rarely
several bracts scattered on the peduncle and over the receptacle; bracts at the
ostiole (mouth) of the receptacle in several series, small, spreading horizontally
across the mouth and then visible from outside, or all descending abruptly into
the interior of the receptacle and not visible from outside, the ostiole in the latter
case being pore-like; male flowers in the African species usually very few and
near the ostiole, rarely mixed amongst the female and gall-flowers; female
flowers usually numerous and sessile; gall-flowers mostly numerous and long-
pedicellate; bracts among the flowers usually small and inconspicuous, or absent.

DISTRIB. About 700 species, spread throughout the tropics and subtropics of
ooth hemispheres, very numerous (about 180) in Tropical Africa.

Ficus Carica, Linn., the common edible fig, is much cultivated in South Africa
and is often found as a garden escape. In the Kew Herbarium there is also a
specimen of *F. retusa*, Linn., a native of India, gathered by Medley Wood (4500)
at Berea, Natal; the leaves are small, obovate, with a distinct pair of basal nerves,
and the receptacles are axillary, about the size of a small pea, sessile, glabrous,
with a few bracts overlapping the mouth of the ostiole. Dried material has also
been seen of *Ficus exasperata*, Vahl, from cultivated plants in the Natal Botanic
Garden. This species is a native of Tropical Africa, where it is very widely
spread; the leaves are so rough that they are used by the natives as a substitute
for sandpaper. Several Indian species are also grown in the Natal Botanic Garden,
mostly as ornamental shrubs or trees.

*Ostiole (mouth) of the receptacle with the bracts visible
 from the outside and spreading transversely across
 the orifice; basal bracts 3 or more :
 †Basal bracts of the receptacle arranged in a single whorl
  at the apex of the peduncle :
  Male flowers with 2 stamens; receptacles arranged
   in panicles on the main stem or branches remote
   from the leaves :
   Mature receptacles tomentose; leaves suborbicular
    or elliptic-orbicular, entire ...   . (1) **Sycomorus.**

   Mature receptacles glabrous; leaves ovate or ovate-
    elliptic, mostly repand-dentate   ...   ... (2) **capensis.**

  Male flowers with a single stamen; receptacles
   axillary :

Leaves not or only very slightly acuminate, mostly
obtuse ; young branchlets glabrous or minutely
puberulous :
Receptacles 3½–4 lin. in diam. ; leaves oblong or
oblong-elliptic, shortly acuminate ... ... (3) **Pretoriæ.**
Receptacles 4–5 lin. in diam. ; leaves ovate or
oblong-ovate, rounded or very obtusely
pointed ... ... (4) **ingens.**
Leaves acutely acuminate ; young branchlets
tomentose ... ... (5) **cordata.**
††Basal bracts somewhat scattered on the peduncle, with
some additional very small ones over the surface
of the receptacle ; leaves scabrous, often 3-toothed
at the apex ... ... (6) **capreæfolia.**
**Ostiole of the receptacle pore-like and more or less two-
lipped, with all the bracts descending abruptly into
the receptacle and not visible from the outside ;
basal bracts usually 2 :
Receptacles borne in clusters on the main branches :
Leaves oblong or lanceolate-oblong, 1¼–2¼ in.
broad, obtuse at both ends ... ... (7) **sansibarica.**
Leaves ovate or ovate-oblong, slightly cordate or
rounded at the base, caudate-acuminate, 2–4½ in.
broad ... ... (8) **polita.**
Receptacles axillary :
Leaves suborbicular or ovate-orbicular, mostly deeply
cordate or rounded at the base :
Young branchlets glabrous ; leaves sparingly and
shortly pubescent or glabrous very deeply
cordate at the base ... ... (0) **soldanella.**
Young branchlets very densely long-villous ; leaves
villous, rounded or slightly cordate at the
base ... ... (10) **Sonderi.**
Leaves elliptic, rounded or slightly cordate at the
base, about 6–15 in. long... ... ... (11) **Nekbudu.**
Leaves oblong or obovate, if slightly cordate then
small :
Receptacles sessile :
Leaves hairy and very prominently reticulate
below ... ... (12) **Stuhlmannii.**
Leaves glabrous :
Receptacles glabrous ; leaves often truncate
at the apex ; stipules sometimes subpersistent ... ... (13) **craterostoma.**
Receptacles tomentose ; leaves rounded at the
apex ; stipules caducous ... ... (14) **Petersii.**
Receptacles pedunculate :
Receptacles woolly-tomentose or pubescent ... (15) **Burkei.**
Receptacles glabrous or very minutely puberu-
lous :
Leaves with numerous flabellate nerves, 1–1½
in. long ; receptacles 2 lin. in diam., not
stipitate ... ... (16) **depauperata.**
Leaves with few or spreading lateral nerves,
if somewhat flabellate then receptacles
stipitate within the basal bracts :

Lateral nerves 8–9 on each side of the midrib ; ostiole large and umbonate ; receptacles often stipitate at the base	(17)	**natalensis.**
Lateral nerves 9–12 ; ostiole small ; receptacles not stipitate	(18)	**gurichiana.**
Lateral nerves about 5 ; venation close and delicate on the lower surface of the leaves ; ostiole small, not umbonate	(19)	**Burtt-Davyi.**

1. **F. Sycomorus** (Linn. Sp. Pl. ed. i. 1059) ; a large tree with spreading crown ; young branchlets with a circle of long slender hairs just below the node, otherwise glabrous or nearly so ; leaves suborbicular or ovate-orbicular, rounded or obtuse at the apex, cordate or rounded at the base, 2–5 in. long, $1\frac{3}{4}$–$3\frac{1}{2}$ in. broad, sub-entire or slightly undulately toothed, dull and glabrous on both surfaces or minutely puberulous below, sometimes slightly scabrous, palmately nerved at the base ; midrib prominent below, continued to the apex of the blade ; principal pair of basal nerves ascending to above the middle of the leaf-blade, with 8–10 lateral nerves on their lower sides ; remaining lateral nerves about 3 on each side of the midrib, diverging from it at an angle of about 40°, distinct on both surfaces, prominent below, bifurcate near the margin ; tertiary nerves very slender, wavy between the lateral ones ; veins very delicate and close ; petiole relatively short, $\frac{3}{4}$–$1\frac{3}{4}$ in. long, at first finely papillose and pilose, at length glabrous : stipules deciduous, those surrounding the terminal bud lanceolate, villous ; receptacles in leafless panicles produced on the main branches or on the stem, obovoid-globose, sometimes stipitate at the base, about 1 in. in diam., with a conspicuous ostiole, softly tomentose ; basal bracts 2, opposite, ovate, subcoriaceous, pubescent outside ; ostiole with numerous exserted suberect bracts ; outer bracts ovate-triangular, subacute, coriaceous, finely puberulous or glabrescent outside, inner ones spreading horizontally across the ostiole, the innermost longer and descending into the receptacle, glabrous ; male flowers sessile near the ostiole ; perianth membranous, covering the 1–3 stamens ; anther-cells free at the base ; female flowers shortly pedicellate ; style quite lateral, reddish, with an oblong yellow stigma. *Miq. in Ann. Mus. Bot. Lugd.-Bat. iii.* 282 ; *Oliv. in Trans. Linn. Soc. xxix.* 149, *t.* 99 ; *Engl. & Drude, Veget. Erde, ix. i. pt. i.* 46, *fig.* 38 ; *Mildbr. & Burret in Engl. Jahrb.* xlvi. 191, *fig.* 1, *B* ; *Hutchinson in Prain, Fl. Trop. Afr.* vi. ii. 95. *F. Sycomorus vera, Forsk. Fl. Ægypt.-Arab.* 180. *F. Chanas, Forsk. l.c.* 219. *F. integri-folia, Sim, For. Fl. Port. E. Afr.* 101, *t.* lxxxix. *F. flavidobarba, Warb., F. pallidobarba, Warb., F. ukambensis, Warb., and F. blepha-rophora, Warb. ex Mildbr. & Burret in Engl. Jahrb.* xlvi. 192, names only. *Sycomorus antiquorum, Gasp. Ricerch. Caprif. Fic.* 86 ; *Miq. in Hook. Lond. Journ. Bot.* vii. 109 *and Afr. Vijge-Boom.* 10, *t.* 1, *fig. A* ; *Kotschy, Plantæ Binderianæ,* 4. *Sycomorus rigida, Miq. in Hook. Lond. Journ. Bot.* vii. 110, *and Afr. Vijge-Boom.* 10, *t.* 1, *fig. B* ; *Kotschy, l.c.*

KALAHARI REGION : Transvaal ; Blau Berg, Zoutpansberg Range, *Burtt-Davy*,
2983 ! Ihabina, *Burtt-Davy*, 2895 a ! near Barberton, *Legat*, 1309 ! 2451 ! Bar-
berton Mountains, *Burtt-Davy*, 283 ! 323 ! 3488 ! 3503 ! 8043 ! Target Valley,
Thorncroft, 6612 ! Swaziland ; Em Babaan, *Burtt-Davy*, 2890 ! Meintjies Kop,
Pretoria, *Burtt-Davy*, 5054 !
EASTERN REGION : Natal ; without precise locality, *Gerrard*, 1632 !

Occurs also in Tropical Africa and Arabia.

2. **F. capensis** (Thunb. Diss. Fic. 13) ; a shrub or small cauli-
florous tree ; branchlets glabrous or softly pubescent, with pointed
more or less villous buds ; leaves ovate or ovate-elliptic, obtuse or
shortly acuminate, rounded, slightly cordate or shortly cuneate at
the base, 2½–9 in. long, 1½–5 in. broad or sometimes more, usually
coarsely repand and obtusely dentate, chartaceous or thinly coriaceous, glabrous or softly pubescent especially on the midrib and
lateral nerves ; midrib prominent below, gradually narrowed to
the apex of the blade ; lateral nerves usually about 6–7 on each
side of the midrib, diverging from the midrib at an angle of about
45°, arcuate, gradually fading towards the margin and branched ;
veins rather closely reticulate below ; petiole variable, up to 3 in.
long, sulcate, glabrous or pubescent ; stipules oblong-lanceolate,
acute, villous or nearly glabrous outside, caducous ; receptacles
borne in leafless simply branched panicles on the main stem or when
shrubby towards the base of the branches ; panicles up to nearly
1 ft. long, glabrous or nearly so ; peduncles ⅛–½ in. long, glabrescent ;
receptacles obovoid or obovoid-globose, subrounded or stipitate at
the base, ⅔ 1 in. long, with a usually prominent mammillate ostiole,
glabrous ; basal bracts whorled, 3, ovate-triangular, often slightly
hairy ; ostiole prominent, with numerous often hairy bracts
spreading transversely across the orifice ; male flowers subsessile ;
perianth hyaline, enveloping the 2 subsessile anthers ; female
flowers very shortly pedicellate ; perianth-segments oblong-lanceolate, acute, membranous, glabrous ; achene obliquely ellipsoid,
shining ; style laterally inserted. *Thunb. Fl. Cap. ed. Schult.* 34 ;
Mildbr. & Burret in Engl. Jahrb. xlvi. 195, *partly and excl. var.
mallatocarpa, Mildbr. & Burret, l.c.* 198 ; *Sim, For. Fl. Cape
Col.* 307, *t.* 135 ; *Hutchinson in Prain, Fl. Trop. Afr.* vi. ii. 101.
F. Lichtensteinii, Link, Enum. ii. 451 ; *Drège, Zwei Pfl. Documente,* 133. *F. Brassii, R. Br. in Trans. Hort. Soc. Lond.* v.
448. *F. thonningiana, Miq. in Ann. Mus. Bot. Lugd.-Bat.* iii. 295.
*F. capensis, var. guineensis, Miq. l.c. F. capensis, var. tri-
choneura, Warb. in Engl. Jahrb.* xx. 153, *and var. pubescens, Warb.
in De Wild. & Durand, Reliq. Dewevr.* 215. *F. Sycomorus, var.
prodigiosa, Welw. ex Hiern in Cat. Afr. Pl. Welw.* i. 1012, *and var.
alnea, Hiern, and var. polybotrya, Hiern, l.c.* 1013 *and* 1014.
F. plateiocarpa, Warb. in Engl. Jahrb. xxx. 292. *F. stellulata,
var. glabrescens, Warb. in Warb. & De Wild. Ficus Fl. Cong.* vi.
27. *F. villosipes, Warb. l.c.* 28. *F. erubescens, Warb. l.c.* 29, *t.* vi.
F. Munsæ, Warb. l.c. 29, *t.* xvii. *F. guineensis, Stapf in Johnston,
Liberia,* ii. 652. *F. kiboschensis, Warb., F. kwaiensis, Warb.*,

F. simbilensis, Warb., F. Matabelæ, Warb., F. umbonigera, Warb.,
F. oblongicarpa, Warb., F. sericeogemma, Warb., F. brachypus, Warb.,
F. grandicarpa, Warb., F. sarcipes, Warb., and F. caulocarpa, Warb.,
ex Mildb. & Burret in Engl. Jahrb. xlvi. 197–8, *names only. Sycomorus*
capensis, Miq. in Hook. Lond. Journ. Bot. vii. 113, *t.* iii. *fig. B. Syco-*
morus guineensis, Miq. in Hook. Niger Fl. 523, *and in Hook. Lond.*
Journ. Bot. vii. 112, *t.* xiv. *fig. B. S. thonningiana, Miq. Afr. Vijge*
Boom. 123, *and in Hook. Lond. Journ. Bot.* vii. 112, *t.* xiv. *A.*

COAST REGION : Knysna Div. ; near Stofpad, *Burchell,* 5294 ! Humansdorp Div. ;
Essebosch, near Zeeko River, *Thunberg* ; Uitenhage Div. ; Enon, among bushes,
Drège ! Springfields, *Paterson,* 1907 ; Albany Div. ; Blauw Krantz, in a wooded
ravine, *Burchell,* 3676 ! a large tree near the Woolwashery in Howisons Poort,
MacOwan, 2961 ! Howisons Poort, *Salisbury,* 439 ; East London Div. ; river
banks near East London, *Galpin,* 3103 ! East London, *Sim,* 2223 ! Komgha Div. ;
along streams near Komgha, *Flanagan,* 767 ; Queenstown Div. : Junction Farm,
Galpin, 8250 !
    KALAHARI REGION : Transvaal ; Houtbosch (Woodbush), *Rehmann,* 6487 ! 6489 !
Hutchins ! Magoobas Kloof, Houtboschberg Range, *Burtt-Davy,* 2595 ! 5240 !
Shewass Bush, *Legat,* 17 ! Tzaneen Estate, Zoutpansberg Range, *Burtt-Davy,*
5241 ! Medinjen Mission Station, *Burtt-Davy,* 5242 ! Target Valley, Barberton,
Thorncroft, 6616 ! Barberton district, *Burtt-Davy,* 5243 ! Crocodile River Valley,
Burtt-Davy, 7648 ! Swaziland ; Em Babaan, *Burtt-Davy,* 2828 ! 2887 ! 2890 !ı
    EASTERN REGION : Natal ; Durban, *Krauss,* 264 ! Pinetown, *Burtt-Davy,* 2387 !
Maritzburg, *Sim,* 2162 ! and without precise locality, *Gerrard.* 1183 !

Widely spread over nearly the whole of Tropical Africa ; Cape Verde Islands.

3. **F. Pretoriæ** (Burtt-Davy in Trans. Roy. Soc. S. Afr. ii. 365,
with figures) ; a large tree ; branchlets mostly short and twiggy,
glabrous and purplish when young, rarely minutely puberulous ;
leaves spreading, oblong or oblong-elliptic, rounded at the base,
mostly rather abruptly and obtusely shortly acuminate, 2–3½ in.
long, ¾–1¾ in. broad, entire, rigidly coriaceous, often somewhat
glaucous below, light-green when dry : midrib almost equally prominent on both surfaces, straw-coloured, ¾–1 lin. broad at the base,
gradually tapered to the apex of the blade : lateral nerves 10–12
on each side of the midrib, equally prominent on both surfaces,
spreading at a wide angle, rather slender, looped well within the
margin ; tertiary nerves and veins strongly reticulate on both
surfaces ; petiole ½–1 in. long, glabrous : stipules caducous, those
surrounding the terminal bud linear-lanceolate, acutely acuminate,
⅓–½ in. long, purplish, glabrous : receptacles axillary, crowded
towards the ends of the branchlets, shortly pedunculate or subsessile, glabrous, 3½–4 lin. in diam., spotted, glabrous ; peduncle
up to 1 lin. long, terete, glabrous : basal bracts 3, whorled, ovate-
orbicular, about ⅖ lin. broad, glabrous ; ostiole only slightly protruded, with 3 or 4 broad bracts visible from the outside and
spreading across the mouth, glabrous or slightly puberulous ; inner
bracts rather fleshy, curved round into the receptacle, glabrous ;
male flowers sessile ; perianth-segments 3, ovate, obtuse, slightly
fleshy, glabrous ; anther solitary, sessile ; female flowers sub-

sessile ; achene brightly shining ; style longer than the achene, slender ; gall-flowers pedicellate. *Hutchinson in Prain, Fl. Trop. Afr.* vi. ii. 116. *F. salicifolia, Balfour, Bot. of Socotra,* 282, *not of Vahl. F. salicifolia, Warb. in Engl. Pfl. Ost-Afr. C.* 162, *partly, not of Vahl ; Mildbr. & Burret in Engl. Jahrb.* xlvi. 206, *partly, not of Vahl. F. salicifolia, var. australis, Warb. in Viertelj. Naturforsch. Ges. Zürich,* li. 139.

KALAHARI REGION : Transvaal ; Wonderboom Poort, near Pretoria, *Miss Leendertz,* 150 ! *Atherstone! Galpin,* 6973 ! *Burtt-Davy,* 2645 ! 7147 ! 8066 ! 8915 ! *Schönland,* 1667 ! Wonderboom Farm, *Burtt-Davy,* 665 ! 2806 ! 2276 ! Wonderfontein, Marico District, *Burtt-Davy,* 7549 ; Kopjies near Pretoria, *Burtt-Davy,* 2750 ! Daspoort Road, *Burtt-Davy,* 2383 ! Mosilikatses Nek, *Burtt-Davy,* 7123 !

EASTERN REGION : Natal ; without precise locality, *Gerrard,* 1185 ! Portuguese East Africa ; Ressano Garcia, *Schlechter,* 11909 !

F. Pretoriæ is the famous " Wonderboom" tree of Pretoria, an account of which was given by J. Burtt-Davy in the Trans. of the Royal Soc. of South Africa for 1912. According to him " the tree spreads in a peculiar manner. Some of the branches from the centre spread out laterally in a radial direction and gradually droop towards the ground. At a distance of about 30 ft. they come in contact with it and send out roots from which new groups of stems arise. From these other branches may be given off, still in the same direction, and these coming in contact with the ground may become rooted in their turn and send up a third group of stems. . . . The whole forms a large hemispherical mass covered with evergreen leaves and small figs. Its diameter from N.N.E. to S.S.W. is 162½ ft. and from E. to W. 141½ ft. Its height . . . was 67 ft."

A photograph of the Pretoria " Wonderboom " was presented by the late Hon. C. Ellis to Sir William Thiselton-Dyer, who in turn presented it to the Royal Botanic Gardens, Kew.

The species is also found in Tropical Africa and Socotra.

4. F. ingens (Miq. in Ann. Mus. Bot. Lugd.-Bat. iii. 288) ; a tree or shrub ; young branchlets stout, with dull grey bark ; leaves ovate or oblong-ovate, rounded at the apex, deeply cordate or rounded-truncate at the base, 3¼–6 in. long, 2¼–4 in. broad, entire, chartaceous, glabrous, slightly shining above, finely warted between the veins below, finely reticulate on both surfaces, light glaucous-green when dry ; midrib flat above, prominent below, about ¾ lin. broad at the base, gradually tapered to the apex ; lateral nerves 8–9 on each side of the midrib, nearly equally prominent on both surfaces, diverging from the midrib at a very wide angle, pro-minently bifurcate ½–¾ in. from the margin ; tertiary nerves lax and reticulate, very slender like the veins ; petiole 1–1½ in. long, sulcate, glabrous ; stipules deciduous, not seen ; receptacles axillary, mostly in pairs, shortly pedunculate, globose, rounded at the base, 4–5 lin. in diam., warted when dry, glabrous or slightly pubescent ; peduncle 1–3 lin. long, or shorter, stout, very slightly puberulous ; basal bracts 2, ovate, rounded at the apex, subpersistent, coriaceous ; ostiole closed with about 3 visible imbricate bracts spreading horizontally across it, not conspicuous ; male flowers subsessile, with a solitary stamen ; female flowers sessile ; ovary smooth ; style

530   MORACEÆ (Hutchinson).   [*Ficus.*

slender; gall-flowers pedicellate. *Engl. Hochgebirgsfl. Trop. Afr.*
191; *Warb. in Engl. Pfl. Ost-Afr. C.* 161; *Hutchinson in Prain, Fl.
Trop. Afr.* vi. ii. 121. *Urostigma ingens, Miq. in Hook. Lond. Journ.
Bot.* vi. 554. *U. xanthophyllum, Miq. in Hook. l.c. U. caffrum, Miq.
Afr. Vijge-Boom.* 31. *U. xanthophyllum, var. ovato-cordatum, Sond.
in Linnæa,* xxiii. 136. *Ficus schimperiana, Hochst. ex A. Rich. Tent.
Fl. Abyss.* ii. 266, *and in Ferr. & Galin. Voy. Abyss. Atlas, t.* ii.
F. xanthophylla, Steud. ex Miq. in Hook. Lond. Journ. Bot. vi. 554.
F. caffra, Miq. Ann. Mus. Bot. Lugd.-Bat. iii. 288. *F. pondoensis,
Warb. in Vierelj. Naturforsch. Ges. Zürich,* li. 140. *F. maganjensis,
Sim, For. Fl. Port. E. Afr.* 99, *t.* xciii. *fig. B. F. cordata, Sim,
For. Fl. Cape Col.* 308, *t.* 159, *not of Thunb. F. lutea, Mildbr. &
Burret in Engl. Jahrb.* xlvi. 209, *not of Vahl.*

VAR. *β,* tomentosa (Hutchinson); mature receptacles softly tomentose.
SOUTH AFRICA: without precise locality, *Zeyher,* 1549!
COAST REGION: Albany Div.; Blauw Krantz, near Grahamstown, *Burtt-Davy,*
7814! Alicedale Poort, *Salisbury,* 442; Stutterheim Div.; Fort Cunninghame,
Sim, 2528! King Williams Town Div.; Mount Coke, *Sim,* 1523! Komgha Div.;
among rocks between Komgha and Keimouth, *Flanagan,* 850! var. *β*; Queenstown Div.; Zwartkei River Valley, Junction Farm, *Galpin,* 8173!
KALAHARI REGION: Transvaal; near Pretoria, *Rehmann,* 4434! 4435! *Miss
Leendertz,* 150! *Burtt-Davy,* 2383! 7787! 14613! Magaliesberg Range, *Burke!*
Lydenburg district, *Wilms,* 1348! between Spitzkop and Komati River, *Wilms,*
1347! Houtbosch (Woodbush), *Rehmann,* 6490! Limpopo Valley, *Hutchins,* 12!
near Mbabane, *Bolus,* 12299! Jeppes Town Ridges, near Johannesburg, *Gilfillan
in Herb. Galpin,* 6248! kopje above Geldenhuis Mine, *Ommanney,* 148! Piet
Potgieters Rust, *Miss Leendertz,* 1263! Komati Poort, *Rogers,* 2617! Majatas
Nek, *Burtt-Davy,* 142!
EASTERN REGION: Griqualand East; Mount Fletcher, *Sim,* 2407! Pondoland;
Umsibaka, *Bachmann,* 425! 429! Natal; Umgeni, *Sim,* 7126! Bushmans River,
Sim, 7124! Klipp River, *Rehmann,* 7284! Ifafa Valley, *Rudatis,* 1010! and
without precise locality, *Cooper,* 1142! *Gerrard,* 1857!
Widely spread in Tropical Africa.

5. **F. cordata** (Thunb. Diss. Fic. 8, with figure); branches somewhat elongated and straight, covered with yellowish-brown puberulous grooved bark; young lateral branchlets short and twiggy,
densely and softly tomentose; leaves ovate-elliptic or oblong-elliptic,
rounded or very slightly subcordate at the base, acutely acuminate,
1½–4½ in. long, ¾–2 in. broad, rigidly coriaceous, equally strongly
reticulate on both surfaces, glabrous; midrib continued to the apex
of the leaf-blade; lateral nerves 5–7 on each side of the midrib,
looped and much-branched well within the margin, somewhat
flexuous, equally prominent on both surfaces; petiole ½–1¼ in. long,
grooved when dry, glabrous; stipules caducous, acutely acuminate,
softly tomentose; receptacles axillary, sessile or very shortly
pedunculate, mostly in pairs, slightly depressed-globose, about ¼ in.
in diam., spotted, minutely puberulous; basal bracts rounded, hairy
outside, about one-third as long as the receptacles; ostiole with 2 or
3 overlapping rounded bracts visible from the outside; male flowers
few, near the ostiolar bracts; perianth membranous; stamen solitary;
female flowers subsessile, numerous; achene smooth. *Thunb. Fl.*

Cap. ed. Schult. 33 ; *Drège, Zwei Pfl. Documente,* 97, 147, 148 ;
Mildbr. & Burret in Engl. Jahrb. xlvi. 207 ; *Hutchinson in Prain, Fl.
Trop. Afr.* vi. ii. 119. *Urostigma Thunbergii, Miq. in Hook. Lond.
Journ. Bot.* vi. 556. *Ficus glaucophylla, Desf. Tabl.* 209, *and Cat.
Plant. Hort. Paris, ed.* 3, 346. *F. tristis, Kunth et Bouché, Ind. Sem.
Hort. Berol.* 1846, 19. *F. atrovirens, Hort. Berol. ex Mildbr. &
Burret, l.c., name only. F. cordata, var. Marlothii, Warb. in Schinz,
Viertelj. Naturforsch. Ges. Zürich,* li. 138.

SOUTH AFRICA : without precise locality, *Drège,* 4588 ! *Thom* ! *Zeyher,* 1551 !
COAST REGION : Vanrhynsdorp Div. ; Heerenlogement, *Thunberg* ; between
Ebenezer and the Gift Berg, *Drège,* a ! Attis, *Pillans,* 5480 ! foot of western slopes
of Gift Berg, *Pearson,* 5391 ! 5392 ! Clanwilliam Div. ; between Clanwilliam and
Bosch Kloof, *Drège,* 9566 ; near Brakfontein, *Ecklon & Zeyher* ! Worcester Div. ;
pass between Hex River and Outspan of Nov. 26, *Pearson,* 5264 ! TcAlee Mountains,
Pearson, 6144 !

WESTERN REGION : Great Namaqualand ; bed of Akam River, *Pearson,* 4725 !
Bushmanland ; below summit of ridge west of Aggenys, *Pearson,* 2933 ! Orange
River, between Abbasis and Ramans Drift, *Pearson,* 3103 ! Aus, *Pearson,* 4712 !
Great Karasberg ; Waterfall, Alt Ravine, *Pearson,* 8079 ! Narudas Sud, *Pearson,*
8136 ! 8162 ! 8547 ! Little Namaqualand ; between Nababeep and Modderfontein,
Bolus, 9452 ! hills near Ookiep, *Bolus,* 9453 ! Orange River, *Schlechter,* 11471 !
Koets, *Pearson,* 5729 ! base of Rattel Poort Mountains, *Pearson,* 2891 ! Plattklip,
Pearson, 3300 ! 3390 ! 3875 ! Eenriet, *Pearson,* 3695 ! Kopjes South of Kamabies,
Pearson, 3959 ; tributary of Hautams River, *Pearson,* 4891 !

KALAHARI REGION : Griqualand West ; Griquatown, *Burchell,* 1889 ! Bechuanaland ; near the sources of Kuruman River, *Burchell,* 2487/3 ! " Zululand," *Gerrard,*
1185 !

Occurs also in Tropical South-West Africa, above Brakwater pools, *Pearson,*
6090 !

6. **F. capreæfolia** (Del. in Ann. Sci. Nat. 2me. sér. xx. 94) ;
branchlets subterete, villous with whitish hairs when young, at
length becoming pubescent ; leaves often opposite or subopposite,
lanceolate or oblong-lanceolate, obtuse or subtruncate at the base,
acute or trifid at the apex, otherwise entire, 1¼–4 in. long, ½–1½ in.
broad, thinly and rigidly chartaceous, scabrous with minute tubercles
on both surfaces ; lateral nerves 5–6 on each side of the midrib, distinct on both surfaces, prominent below, arcuate, indistinctly looped
near the margin ; venation rather faint ; petiole short, 1½–4 lin.
long, pubescent or almost villous, with a pair of stipules at the base
of each ; stipules persistent, lanceolate, subacute, ¼ in. long, keeled
on the back, finely puberulous on the outside, shortly ciliate, brown
when dry ; receptacles axillary, solitary, pedunculate, subglobose,
contracted at the base, ½–¾ in. in diam., scabrid-hispidulous, without a definite whorl of bracts at the base ; peduncle ¼–½ in. long,
scabrid-hispidulous, bearing 3–4 scattered ovate obtuse coriaceous
puberulous bracts about ¾ lin. long ; ostiole broad and gaping, with
numerous imbricate bracts visible from the outside ; outer ostiolar
bracts broadly triangular, obtuse, finely puberulous on the outside,
minutely ciliate, the lower ones all descending into the receptacle,
oblong-lanceolate, obtuse, about 1¼ in. long, translucent, glabrous ;
male flowers numerous, long-pedicellate ; pedicel stout, minutely
puberulous ; perianth-segments 4–5, unequal and some connate in

the lower part, obtuse or truncate at the apex, membranous, glabrous ; stamen solitary with mostly a rudimentary female flower at
its base ; female flowers in separate receptacles, very numerous,
sessile ; perianth-segments 6, linear, acute, glabrous ; achene smooth ;
style lateral, slender, longer than the achene, with a purple stigma.
Warb. in Warb. & De Wild. Fic. Fl. Cong. 36, *t.* xxii. ; *Warb. in Engl.
Pfl. Ost-Afr. C.* 161 ; *Engl. & Drude, Veg. Erde,* ix. 1, *pt.* i. 118, *fig.*
100 ; *Mildbr. & Burret in Engl. Jahrb.* xlvi. 202 ; *Hutchinson in
Prain, Fl. Trop. Afr.* vi. ii. 107.   *F. tridentata, Fenzl in Flora,* 1844,
311, *name only.   F. antithetophylla, Steud. ex Miq. in Hook. Lond.
Journ. Bot.* vii. 236, *t.* v. B ; *A. Rich. Tent. Fl. Abyss.* ii. 272.   *F.
palustris, Sim, For. Fl. Port. E. Afr.* 99, *partly, incl. t.* xc. *fig. C.*

EASTERN REGION : Natal, *Gerrard,* 1631 !

Occurs also in Tropical Africa.

7. **F. sansibarica** (Warb. in Engl. Jahrb. xx. 171); a tree with
large spreading main branches on which are borne numerous
clusters of figs ; leaves oblong or lanceolate-oblong, obtuse at both
ends, 3½–5½ in. long, 1¼–2¼ in. broad, entire, firmly chartaceous,
finely verrucose above, otherwise dull and glabrous on both surfaces ;
midrib slightly impressed above, prominent below, about 1 lin.
broad at the base, gradually tapered at the apex of the blade ;
lateral nerves 8–9 on each side of the midrib, the lowermost pair
opposite and sharply ascending, the others diverging from the midrib at a wide angle and mostly arcuate, slender, prominent below,
looped close to the margin ; tertiary nerves very lax and branched,
distinct below ; veins closely reticulate and distinct on the lower
surface ; petiole 1¾–2¼ in. long, glabrous ; stipules deciduous ; receptacles borne in clusters on the old main branches, globose, ¾–1 in. in
diam., green when fresh, minutely puberulous, wrinkled like a plum
when dry, with a basal stipe 1½ lin. long ; basal bracts not seen ;
ostiole bilabiate, with no bracts visible from the outside, all descending into the receptacle ; male flowers with a solitary stamen.
Mildbr. & Burret in Engl. Jahrb. xlvi. 223 ; *Hutchinson in Prain,
Fl. Trop. Afr.* vi. ii. 130.   *F. Langenburgii, Warb. in Engl. Jahrb.*
xxx. 293.   *F. delagoensis, Sim, For. Fl. Port. E. Afr.* 99, *t.* xcii.
F. libertiana, Warb. ex Mildbr. & Burret, l.c., name only.

KALAHARI REGION : Transvaal, without precise locality, *Legat,* 5873 !
EASTERN REGION : Delagoa Bay, *Sim,* 5171 !

Occurs also in Eastern Tropical Africa.

8. **F. polita** (Vahl, Enum. ii. 182) ; a large tree about 50 ft.
high ; branchlets rather slender, very minutely puberulous when
young ; leaves ovate or ovate-oblong, slightly cordate or rounded at
the base, caudate-acuminate, 3–6 in. long, 2–4½ in. broad, entire,
membranous or subchartaceous, glabrous and dull on both surfaces,
5-nerved at the base ; midrib flat on the upper surface, prominent
below, about 1 lin. broad at the base, gradually tapered to the apex

of the blade ; lateral nerves about 7 on each side of the midrib,
diverging from it at an angle of about 45°, slender, prominent on
both surfaces, bifurcate about $\frac{1}{2}$–$\frac{3}{4}$ in. from the margin ; lowermost
pair and the pair above opposite, the others alternate or subalter-
nate ; veins very close and slender, forming a delicate reticulation
below ; petiole slender, up to 4$\frac{1}{2}$ in. long, finely sulcate, very minutely
puberulous ; stipules early deciduous, lanceolate-acuminate, 5 lin.
long, subcoriaceous, glabrous ; receptacles arranged in fascicles on
thick woody outgrowths from the old wood, pedunculate, subglobose, about 1$\frac{1}{2}$ in. in diam. when dry, wrinkled like a dried plum,
glabrous or very minutely and sparingly pubescent ; peduncle $\frac{1}{2}$–$\frac{3}{4}$
in. long, 1 lin. thick, glabrous ; basal bracts 3, broadly triangular ;
ostiole slightly impressed, with the bracts all descending into the
receptacle ; male flowers with 3–4 lanceolate perianth-segments and
a solitary stamen ; female flowers with 3 ovate-lanceolate obtuse
perianth-segments ; style long and slender with a short slightly
oblique stigma. *Pers. Syn.* ii. 608 ; *Roem. & Schult. Syst.* i. 499 ;
Spreng. Syst. Veg. iii. 778 ; *Mildbr. & Burret in Engl. Jahrb.* xlvi. 222 ;
Hutchinson in Prain, Fl. Trop. Afr. vi. ii. 124. *Urostigma politum,*
Miq. in Hook. Lond. Journ. Bot. vi. 553. *Ficus syringifolia, Warb. in*
Engl. Jahrb. xx. 170, *not F. syringæfolia of Kunth & Bouche.* *F.*
pseudo-elastica, Welw. ex Hiern in Cat. Afr. Pl. Welw. i. 996. *F.*
niamniamensis, Warb. in Warb. & De Wild. Fic. Fl. Congo, 14, *t.* xx.
F. umbrosa, Sim, For. Fl. Port. E. Afr. t. lxxxviii. *excl. fig. of*
receptacles. *F. barombiensis, F. stenosiphon, F. syringoides, F pachy-*
sarca, Warb. ex Mildbr. & Burret, l.c. 222–229, *names only.*

EASTERN REGION : Natal ; Durban, *Miss Franks in Herb. Wood,* 12875 ! *Burtt-
Davy,* 10418 ! and without precise locality, *Keit,* 999 !
Widely spread in Tropical Africa.

9. **F. soldanella** (Warb. in Schinz, Viertelj. Naturforsch. Ges.
Zürich, li. 136) ; a much-branched more or less globose bush ;
branchlets stout, subterete, about 4$\frac{1}{2}$ lin. in diam. near the apex,
glabrous ; leaf-scars oblong-orbicular or horse-shoe-shaped, 5–6 lin.
long, 3–4 lin. broad ; stipular scars very broad ; leaves suborbicular,
very shortly and obtusely acuminate, deeply cordate at the base,
4–6$\frac{1}{2}$ in. long, 3$\frac{1}{2}$–6$\frac{1}{2}$ in. broad, entire, chartaceous or subcoriaceous,
glabrous and dull on both surfaces or sparingly pubescent below,
9-nerved at the base ; midrib slightly raised above, prominent
below, about 1 lin. broad at the base, gradually narrowed to and
finally reaching the apex of the blade ; lateral nerves (excluding
the basal ones) 4–5 on each side, diverging from the midrib at an
angle of 45°, prominent on both surfaces, bifurcate $\frac{1}{4}$–$\frac{3}{4}$ in. from the
margin ; tertiary nerves lax and wavy, somewhat prominent below ;
veins close and distinct below ; petiole 1$\frac{1}{2}$–3$\frac{1}{2}$ in. long, narrowly
grooved on the upper surface, glabrous ; stipules caducous, not
seen ; receptacles axillary, probably solitary, sessile or nearly so,
depressed, globose, with a very slightly prominent ostiole, slightly
stipitate and slightly 7-ribbed at the base, 6–7 lin. in diam., very

minutely puberulous ; basal bracts 4, ovate, rounded at the apex,
about 1 lin. long and broad, coriaceous, reddish-brown when dry,
glabrous ; ostiolar bracts all descending vertically into the receptacle, lanceolate, subacute, ¾–1½ lin. long, reddish, with rather
narrowly membranous margins, glabrous ; male flowers very few
near the ostiolar bracts, shortly pedicellate ; perianth enclosing the
solitary subsessile anther ; female flowers sessile, scattered amongst
the gall flowers ; perianth-segments 4, linear, subacute, about 1 lin.
long, hyaline ; achene ellipsoid ; style lateral, longer than the
achene ; stigma bifid ; gall-flowers pedicellate ; pedicel stout, 1½ lin.
long, glabrous ; ovary as in female flowers, but with a larger
subsessile stigma ; scales of the receptacle subulate, acute, ¾ lin.
long, with very narrowly membranous margins. *Hutchinson in Prain,
Fl. Trop. Afr.* vi. ii. 176. *F. abutilifolia, Miq. Ann. Mus. Bot.
Lugd.-Bat.* iii. 288, *partly. F. picta, Sim, For. Fl. Port. E. Afr.* 99,
t. xciv. *fig. B. Urostigma abutilifolium, Miq. in Hook. Lond. Journ.
Bot.* vi. 551, *partly.*

SOUTH AFRICA : without precise locality, *Zeyher,* 1548 ! 1856 !
KALAHARI REGION : Transvaal ; Magaliesberg Range, *Burke,* 273 ! *Engler,*
2795 ! *Burtt-Davy,* 7161 ! Wonderboompoort, *Miss Leendertz,* 449 ! Kudus Poort,
Rehmann, 4684 ! 4686 ! Crocodile River, *Burtt-Davy,* 225 ! Singerton, near
Hector Spruit, *Burtt-Davy,* 8007 ! near Nylstroom, *Burtt-Davy,* 2362 ! Breslau,
Limpopo River, *Legat,* 5183 !

Occurs also in Tropical Africa.

10. **F. Sonderi** (Miq. in Ann. Mus. Bot. Lugd.-Bat. iii. 295) ; a
tree 20–30 ft., with a milky juice ; branchlets twiggy, very shaggy-
villous with tawny hairs when young, at length shortly pubescent ;
leaves ovate-orbicular or suborbicular, rounded at the apex, slightly
cordate at the base, 2–4 in. long, 1½–3¼ in. broad, entire, rigidly
coriaceous, long-pilose or hirsute on both surfaces chiefly on the
midrib and lateral nerves, about 5-nerved at the base ; midrib ¾ lin.
broad at the base, gradually tapered to the apex, flat above, prominent and very hairy below ; lateral nerves 4–5 on each side of
the midrib (excluding the basal nerves), diverging at an angle of
about 45°, arcuate, looped near the margin, distinct on both surfaces,
prominent below ; tertiary nerves much-branched, wavy, scarcely
visible above, prominent below ; petiole ¼–1 in. long, densely hirsute-
pilose ; stipules persistent during the flowering period, ovate-
triangular, acutely acuminate, about ¾ in. long and ½ in. broad,
reddish when dry, thinly chartaceous, villous outside, glabrous and
striate within ; receptacles crowded at the ends of the young
branchlets, axillary, geminate, sessile, subglobose, slightly umbonate,
nearly ½ in. in diam., villous ; basal bracts 2, suborbicular, about
3 lin. broad, membranous, villous with yellowish hairs outside,
glabrous within ; ostiole bilabiate, slightly projecting and a little
gaping when dry, about 1 lin. broad ; bracts all descending into
the receptacle, the middle two a little larger than the others, oblong-
lanceolate, obtuse, 1¼ lin. long, rather fleshy, glabrous, the others
subacute and more membranous ; male flowers shortly pedicellate ;

perianth enveloping the solitary subsessile anther; female flowers
subsessile; perianth-segments acutely acuminate, membranous,
glabrous; achene ellipsoid, smooth; style slender, about half as
long as the achene; stigma oblong, thick, as long as the style; gall-
flowers pedicellate, with obtuse perianth-segments and almost sessile
stigma; receptacular scales ¾ lin. long, flat, triangular, subacute.
Gibbs in Journ. Linn. Soc. xxxvii. 470; *Mildbr. & Burret in Engl.
Jahrb.* xlvi. 262; *Hutchinson in Prain, Fl. Trop. Afr.* vi. ii. 169.
Sycamorus hirsuta, Sond. in Linnæa, xxxiii. 137. *Ficus Rehmanni,
Warb. in Schinz, Viertelj. Naturforsch. Ges. Zürich,* li. 136. *F. Reh-
mannii, vars. ovatifolia and villosa, Warb. l.c. F. montana, Sim, For.
Fl. Port. E. Afr.* 101, *t.* xcv. *A. F. glumosa, Mildbr. & Burret in
Engl. Jahrb.* xlvi. 217, *partly, not of Delile. F. Engleri, Warb. ex
Mildbr. & Burret, l.c.* 219, *name only. F. Kitaba, De Wild. in Bull.
Soc. Bot. Belg.* lii. 215.

KALAHARI REGION : Transvaal ; Target Valley, Barberton, *Thorncroft,* 6611!
6614! 6615! Houtbosh, *Rehmann,* 6486! between White River and Nelspruit,
Burtt-Davy, 1515! Mooi Drift, *Leendertz,* 2117! between Nylstroom and Springbok
Flats, *Burtt-Davy,* 1726! near Nylstroom, *Burtt-Davy,* 2091! Warmbath, *Burtt-
Davy,* 2349!
EASTERN REGION : Natal; Durban, *Gueinzius,* 415! Inanda, *Wood,* 1361!
Inchanga, *Sim,* 7127! *Engler,* 2670! Camperdown, *Rehmann,* 7711! Zululand ;
Port Durnford, *Sim,* 7129! and without precise locality, *Gerrard,* 1577!

Occurs also in Rhodesia and East Africa.

11. F. Nekbudu (Warb. in Warb. & De Wild. Fic. Fl. Congo, 6,
t. iv.); a huge forest tree ; young branchlets stout, at first
adpressed-pubescent, at length glabrous ; leaves spreading horizontally, elliptic or obovate-elliptic, rounded at both ends or rarely
slightly cordate at the base, 6–15 in. long, 3½–8 in. broad, entire,
coriaceous, glabrous on both surfaces ; lateral nerves about 6 on
each side of the midrib, spreading from it at a wide angle, prominent
below, looped within the margin ; tertiary nerves rather lax, forming
with the veins a close network below ; petiole stout, 3–6 in. long,
covered with a flaky epidermis ; stipules caducous ; receptacles
axillary, sessile, paired, about ½–⅔ in. in diam., at first woolly-pilose,
at length adpressed-pilose ; basal bracts densely villous outside ;
ostiole small, pore-like, with the bracts all descending into the
receptacle ; male flowers with a single stamen. *Mildbr. & Burret in
Engl. Jahrb.* xlvi. 239; *Hutchinson in Prain, Fl. Trop. Afr.* vi. ii.
180. *F. utilis, Sim, For. Fl. Port. E. Afr.* 100, *t.* xci.

EASTERN REGION : Natal; Durban, *Burtt-Davy,* 2420! and in Herb. *Wood,*
12851! Delagoa Bay ; Lorenzo Marques, *Sim,* 6125! *Burtt-Davy,* 10556!

F. Nekbudu is grown as a street tree in Durban. It is also grown in the Royal
Botanic Garden and at the Sierres Coloniales, Brussels, where specimens were seen
in 1913. A specimen of what seems to be the same species in the Kew Herbarium
labelled "from Grahamstown" was cultivated in the Royal Botanic Gardens,
Kew, in 1876.
According to Sim the plant described by him under the name *F. utilis* occurs
throughout Portuguese East Africa ; I have compared his type specimen, 6125,
from Lorenzo Marques with the type of *F. Nekbudu* from the Congo and they
seem to be identical. Sim states that his plant is the source of all the native
cloth in the M'Chopes district.

536   MORACEÆ (Hutchinson).   [*Ficus.*

12. F. Stuhlmannii (Warb. in Engl. Jahrb. xx. 161) ; a tree about
30 ft. high ; stem about 2 ft. in thickness ; branchlets fairly stout,
ribbed, pilose with whitish or slightly fuscous hairs especially when
young ; leaves oblong, rounded at the apex, cordate at the base,
3–5½ in. long, 1¼–2¾ in. broad, entire, rigidly chartaceous or subcoriaceous, thinly pubescent especially near the midrib and with
slightly impressed venation on the upper surface, softly tomentose
or densely pubescent below ; lateral nerves 4–5 on each side of the
midrib, diverging from it at an angle of 45°, nearly straight, looped
some distance from the margin, prominent below ; tertiary nerves
few, forming with the veins a close prominent venation below ;
petiole ¾–1½ in. long, stout, grooved above, otherwise subterete,
softly pubescent ; stipules deciduous, those surrounding the terminal
bud more or less lanceolate, adpressed villous outside, glabrescent
towards the margin ; receptacles axillary, probably solitary, sessile,
globose, ½ in. in diam. or slightly more, densely tomentose ; basal
bracts connate at the base, adpressed villous outside, glabrous
within ; ostiole bilabiate, glabrous, scarcely produced, about ⅔ lin.
wide ; bracts all descending into the receptacle, the middle two
broader and more fleshy than the others, oblong, 1¼ lin. long,
glabrous, the others lanceolate, acute, 1 lin. long or less ; male
flowers shortly pedicellate ; perianth enclosing the solitary anther ;
anther-cells slightly superimposed ; female flowers sessile ; perianth-
segments 3, ovate-lanceolate, acutely acuminate, glabrous : achene
ovoid-globose, smooth ; style nearly as long as the achene, with a
rather broad flattened stigma ; gall-flowers long-pedicellate, with a
short style ; receptacular scales lanceolate, acute, membranous.
Mildbr. & Burret in Engl. Jahrb. xlvi. 220, *partly ; Hutchinson in
Prain, Fl. Trop. Afr.* vi. ii. 170. *F. howardiana, Sim, For. Fl. Port.
E. Afr.* 100, *t.* xcii. A. *F. Homblei, De Wild. in Fedde, Repert.*
xii. 195, *and in Bull. Soc. Bot. Belg.* lii. 212.

KALAHARI REGION : Transvaal ; on the Lebombo Flats, *Burtt-Davy,* 10496 !
Spelonken, Zoutpansberg, *Menne,* 3043 ! Nelspruit Hotel, Barberton, *Burtt-Davy,*
5246 !
EASTERN REGION : Lorenzo Marques, *Sim,* 6368 !
Occurs also in Tropical Africa.

13. F. craterostoma (Warb. ex Mildbr. & Burret in Engl. Jahrb.
xlvi. 247 partly) ; a shrub or small tree ; branchlets slender,
glabrous ; leaves obovate or triangular-obovate, truncate at the
apex, shortly cuneate at the base, more or less rounded in the upper
half, 1½–2½ in. long, ¾–1¾ in. broad, entire, thinly coriaceous,
glabrous ; midrib prominent and straw-coloured below, divided about
¼ in. below the apex ; lateral nerves about 7 on each side of the midrib, diverging from it at an angle of 45°, distinct on both surfaces,
rather slender, but prominent below ; veins prominent ; petiole ½–¾
in. long, smooth ; stipules persistent or subpersistent ; receptacles
axillary, sessile, subglobose, about 4 lin. in diam., glabrous ; basal
bracts puberulous outside ; ostiole large and gaping, smooth ; bracts

few, all descending into the receptacle, glabrous ; male flowers with a solitary stamen. *Hutchinson in Prain, Fl. Trop. Afr.* vi. ii. 160. *F. natalensis, Sim, For. Fl. Cape Col.* 307, *t.* 136 ? *not of Hochst.*

COAST REGION : Komgha Div. : in Gwenkala woods, near Komgha, *Flanagan,* 679 ! and in *Herb. Bolus,* 1312 !

KALAHARI REGION : Transvaal ; Houtbosch Berg, *Nelson,* 435 ! Potatabosch, *Burtt-Davy,* 1202 ! Woodbush forests, *Legat,* 6724 ! Dumisa Station, *Rudatis,* 1095 !

EASTERN REGION : Transkei ; Kentani districts, *Miss Pegler,* 1126 ! Pondoland ; Port St. John, *Sim,* 2406 ! Natal ; Durban, *Sim,* 7125 ! Umgaye, *Rudatis,* 1095 ! and without precise locality, *Gerrard,* 1184 !

Occurs also in Tropical Africa.

14. **F. Petersii** (Warb. in Engl. Jahrb. xx. 164) ; a tree up to 60 ft. high ; young branchlets crisped-pubescent, at length nearly glabrous, ribbed when dry ; leaves oblanceolate, elliptic-oblanceolate or obovate-lanceolate, narrowed to an obtuse base or rarely rounded to the base, very shortly and obtusely pointed, 1¾–3½ in. long, ¾–1½ in. broad, thinly but rather rigidly chartaceous, glabrous on both surfaces ; lateral nerves 6–8 on each side of the midrib, slightly arcuate, conspicuous on both surfaces, but more prominent below, looped within the margin ; tertiary nerves and veins lax and nearly as prominent below as the lateral nerves : petiole up to 2 in. long, slender, glabrous ; stipules caducous ; receptacles axillary, mostly paired, quite sessile, ovoid, brown, tomentose when mature, about 4½ lin. long and 3½ lin. in diam., with a conspicuous pore-like ostiole ; basal bracts small and deciduous ; bracts of the ostiole descending vertically into the receptacle, glabrous ; male flowers with a membranous glabrous perianth and a solitary stamen. *Mildbr. & Burret in Engl. Jahrb.* xlvi. 258 ; *Hutchinson in Prain, Fl. Trop. Afr.* vi. ii. 182. *F. ruficeps, Warb. in Engl. Jahrb.* xxx. 294. *F. Galpinii, Warb. in Viertelj. Naturforsch. Ges. Zürich,* li. 140. *F. Dinteri, Warb. l.c.* 141. *F. schinziana, Warb. l.c.* 143.

KALAHARI REGION : Transvaal ; between Spitzkop and the Komati River, *Wilms,* 1345 ! near Barberton, *Pole Evans,* 2962 ! *Thorncroft,* 5131 ! *Burtt-Davy,* 282 ! Rimers Creek, *Thorncroft,* 10 ! Target Valley, *Thorncroft,* 6617 ! Buffels Road, *Burtt-Davy,* 5218 ! Sandfontein, *Weber,* 6609 !

EASTERN REGION : Natal : Durban Botanic Gardens, *Burtt-Davy in Herb. Wood,* 12839 ! Albert Park, *Burtt-Davy in Herb. Wood,* 12838 ! 12849 !

Occurs also in Tropical Africa.

15. **F. Burkei** (Miq. in Ann. Mus. Bot. Lugd.-Bat. iii. 289) ; a tree ; young branchlets shortly pubescent or almost tomentose ; leaves elliptic or elliptic-oblanceolate, rounded or slightly and obtusely pointed at the apex, rounded at or slightly narrowed to an obtuse base, 1½–4½ in. long, ¾–2 in. broad, rigidly chartaceous, glabrous and dull on both surfaces, sometimes finely pustulate above ; lateral nerves 6–8 on each side of the midrib, straight or curved, looped within the margin, distinct on both surfaces, more prominent

below; tertiary nerves and veins usually nearly as prominent as the principal lateral ones, more or less straw-coloured; petiole up to 1½ in. long, glabrous; stipules caducous; receptacles axillary, paired or solitary, pedunculate, globose or ovoid-globose, 4–5 lin. long, brownish, pubescent when dry, sometimes woolly-pubescent; peduncle 1–2½ lin. long, shortly pubescent; basal bracts caducous, pubescent; ostiole pore-like, protruded, with the bracts descending vertically into the receptacle, glabrous; male flowers with a membranous glabrous perianth and a single stamen; female flower sessile; style slender; gall-flowers staked. *Mildbr. & Burret in Engl. Jahrb.* xliv. 262; *Hutchinson in Prain, Fl. Trop. Afr.* vi. ii. 202. *F. Erici-Rosenii, R. E. Fries in Wiss. Ergeb. Schwed. Rhod.-Kongo-Exped.* i. 15, *fig.* 1. *Urostigma Burkei, Miq. in Hook. Lond. Journ. Bot.* vi. 555.

KALAHARI REGION : Transvaal ; Macalisberg Range, *Burke* ! near the Wonderboom, *Burtt-Davy,* 2652 ! 7154 ! Wonderboom Poort, *Miss Leendertz,* 625 ! Kelly's Store, Buffels, *Burtt-Davy,* 2860 ! between the Oliphants and Steelpoort Rivers, *Transvaal Estate & Develop. Co.* ! Woodbush, *Hutchins* ! Modjajie Mountain, *Burtt-Davy,* 2646 ! near Nyistroom, *Burtt-Davy,* 2088 ! between Tzaneen and Medenjer, *Burtt-Davy,* 5249 !

Occurs also in South Tropical Africa.

16. **F. depauperata** (Sim, For. Fl. Port. E. Afr. 98, t. xc. fig. B.) ; branchlets twiggy, slender, very minutely puberulous ; leaves oblanceolate, rounded at the apex, gradually narrowed to the base, 1–1½ in. long, 4–7 lin. broad, entire, thinly chartaceous, glabrous and dull on both surfaces ; midrib slightly prominent on both surfaces, not continued quite to the apex of the blade ; lateral nerves about 5 on each side of the midrib, visible on the upper surface, very numerous and more or less flabellate below, prominent ; petiole slender, 2–3 lin. long, glabrous ; stipules early deciduous ; receptacles in axillary pairs, pedunculate, globose, 2 lin. in diam., laxly reticulate when dry, glabrous ; peduncle 1–1¼ lin. long, slender, softly puberulous ; basal bracts 3, persistent, broadly ovate, rounded at the apex, ⅓ lin. long, coriaceous, minutely puberulous outside ; ostiole depressed, wrinkled outside, bilabiate ; bracts all descending, the two next to the ostiole longer than the others, linear, the others subulate, glabrous ; male flowers small, sessile ; perianth-segments reddish, with membranous margins ; anther solitary, subsessile ; female flowers slightly larger than the male ; style scarcely exserted, several stigmas often cohering. *Hutchinson in Prain, Fl. Trop. Afr.* vi. ii. 204.

EASTERN REGION : Delagoa Bay ; Lorenzo Marques, *Sim,* 5031 !

Occurs also in Eastern Tropical Africa.

17. **F. natalensis** (Hochst. in Flora, 1845, 88) ; a shrub or small tree ; branchlets covered with greyish glabrous bark ; leaves oblanceolate or obovate-oblanceolate, obtusely pointed or rounded at the apex, narrowed to the base, 1¾–3¼ in. long, ½–1¼ in. broad,

entire, chartaceous, glabrous and dull on both surfaces ; midrib flat above, prominent below, gradually tapered to the apex of the blade ; lateral nerves 8–9 on each side of the midrib, diverging from it at an angle of 45° or less, looped near the margin, prominent but rather slender below ; tertiary nerves only slightly less prominent than the lateral and parallel with them ; petiole comparatively short, $\frac{1}{3}$–1 in. long, glabrous ; stipules caducous, linear-lanceolate, acuminate, about 2 lin. long, coriaceous, glabrous ; receptacles axillary, mostly in pairs, pedunculate, obovoid-globose, stipitate at the base, about 2 in. long, with a very large smooth ostiolar prominence, the rest wrinkling when dry like a dried plum ; peduncle 2–3$\frac{1}{2}$ lin. long, rather slender, glabrous ; basal bracts early caducous, leaving behind the small persistent unilateral base, glabrous ; ostiole small and pore-like ; bracts few, all descending into the receptacle, glabrous ; male flowers with a solitary stamen ; female flowers sessile, gall ones stalked. *F. columbarum, Hochst. l.c., name only.* *F. Volkensii, Warb. in Engl. Jahrb.* xx. 167 ; *Hiern in Cat. Afr. Pl. Welw.* i. 1007 ; *Mildbr. & Burret in Engl. Jahrb.* xlvi. 249 ; *Hutchinson in Prain, Fl. Trop. Afr.* vi. ii. 208. *F. Durbanii, Warb. in Viertelj. Naturforsch. Ges. Zürich,* li. 142. *F. chrysocerasus, Welw. ex Warb. in Engl. Jahrb.* xx. 167 ; *Hiern in Cat. Afr. Pl. Welw.* i. 1005. *F. Dekdekena, Mildbr. & Burret, l.c.* 255, *partly, not of A. Rich. F. natalensis, var. pedunculata, Sim, For. Fl. Port. E. Afr.* 98, *t.* xc. *fig.* A.

EASTERN REGION : Natal ; near Durban, *Krauss,* 254 ! 276 ! 288 ! *Plant* ! Albert Park, *Burtt-Davy in Herb. Wood,* 12846 ! 12847 ! 12848 ! 12878 ! Lorenzo Marques, *Sim,* 5729 !

Widely distributed in Tropical Africa.

18. **F. gurichiana** (Engl. in Engl. Jahrb. xix. 130) ; spreading habit ; branches sulcate, covered with yellowish glabrous deciduous bark ; leaves oblong or oblong-lanceolate, rounded at the apex, obtuse or rounded and sometimes almost subcordate at the base, 1$\frac{1}{4}$–4 in. long, $\frac{1}{2}$–1$\frac{1}{2}$ in. broad, entire, chartaceous, with a narrowly cartilaginous margin, finely reticulate and glabrous on both surfaces ; midrib flat above, prominent below, about $\frac{1}{2}$ lin. broad at the base, bifurcate near the apex ; lateral nerves 9–12 on each side, at an angle of about 70°, slender, looped and branched some distance from the margin, sometimes slightly impressed, but more often raised above ; petiole $\frac{1}{3}$–$\frac{3}{4}$ in. long, about $\frac{2}{3}$ lin. thick, narrowly grooved on the upper surface, glabrous, straw-coloured or reddish when dry ; stipules deciduous ; receptacles solitary or geminate, axillary, pedunculate, subglobose, about 3 lin. in diam., with a gaping slightly prominent ostiole, minutely puberulous ; peduncle 2 lin. long or less, glabrous or minutely puberulous ; basal bracts 2 (or 3 ?), ovate, obtuse, $\frac{3}{4}$ lin. long, glabrous, deciduous ; ostiolar bracts all pointing straight into the receptacle, the middle two subulate from an ovate base, fleshy, 1 lin. long, the remainder subulate or linear-subulate, $\frac{3}{4}$ lin. long, membranous ; male flowers

with 3 perianth-segments and a solitary stamen ; achene subglobose ; style nearly twice the length of the achene. *Mildbr. & Burret in Engl. Jahrb.* xlvi. 246 ; *Hutchinson in Prain, Fl. Trop. Afr.* vi. ii. 211.

COAST REGION : Clanwilliam Div. ; Annenous, *Pearson,* 5980!
WESTERN REGION : Bushmanland ; Rietfontein, *Pearson,* 3447! 3448! Little Namaqualand ; near summit of Rattel Poort Mountain, *Pearson,* 2986! south of Tweefontein, *Pearson,* 3785! south of Kamabies, *Pearson,* 3957! Modderfontein, *Pearson,* 5962! TcAlee Mountains, *Pearson,* 6165! 6148! 6135! between Nieuwe Rust and Bitterfontein, *Pearson,* 5545! and without precise locality, *Bolus,* 9454! Brakwater pools, *Pearson,* 6078!

Occurs also in Angola.

19. F. Burtt-Davyi (Hutchinson in Kew Bulletin, 1916, 232); a shrub or small tree, sometimes scrambling over rocks; branches covered with grey bark ; young branchlets minutely and softly puberulous ; leaves elliptic or oblong-elliptic, rounded at both ends or very slightly pointed at the apex, $1\frac{1}{4}$–$2\frac{1}{2}$ in. long, $\frac{1}{2}$–$1\frac{3}{4}$ in. broad, thinly chartaceous, glabrous on both surfaces ; midrib continued to the apex of the leaf-blade, fairly prominent below ; lateral nerves about 5 on each side of the midrib, much branched and reticulate well within the margin, distinct, but very slightly prominent below ; veins forming a fine close delicate network on the lower surface ; petiole about a quarter the length of the leaves, glabrous ; stipules caducous, acuminate, those towards the tips of the young shoots up to $1\frac{1}{4}$ in. long, submembranous, reddish-brown and glabrous when dry ; receptacles axillary, mostly in pairs, pedunculate, subglobose, $\frac{1}{4}$–$\frac{1}{2}$ in. in diam., minutely puberulous or almost glabrous ; peduncle 1–3 lin. long, puberulous ; basal bracts submembranous, connate at the base, finely puberulous ; ostiole gaping and pore-like, with none of the bracts visible from outside, but all descending abruptly into the receptacle ; male flowers with membranous perianths and a single stamen. *F. natalensis, Mildbr. & Burret in Engl. Jahrb.* xlvi. 255, *not of Hochst.*

COAST REGION : Riversdale Div. ; near Gauritz River Bridge, *Galpin,* 4579! Knysna Div. ; in the forest near the quarry at Knysna, *Burchell,* 5412! Uitenhage Div. ; near Enon, *Drège* a! Port Elizabeth Div. ; near the burying ground at Port Elizabeth, *Burchell,* 4306! Port Elizabeth Valley, *Paterson!* Krakakamma, *Zeyher,* 557! Bathurst Div. ; between Riet Fontein and the sea shore, *Burchell,* 4112! between Port Alfred and Kaffir Drift, *Burchell,* 3851! Queenstown Div. ; Zwart Kei River, Junction Farm, *Galpin,* 8172! Glen Grey Div. ; White Kei Falls, *Galpin,* 2507! Albany Div. ; on the rocks of Zwartwarter Poort, *Burchell,* 3411! Kowie West, *Burtt-Davy,* 7954! Howison's Poort, *Salisbury,* 440a! Alicedale Poort, *Salisbury,* 440! East London Div. ; "Cove Rock," *Galpin,* 3104! Komgha Div. ; in woods near Komgha, *Flanagan in MacOwan Herb. Austr.-Afr.* 1531!

CENTRAL REGION : Graaff Reinet Div. ; mountains near Graaff Reinet, 4200 ft., *Bolus,* 711!

EASTERN REGION : Transkei ; Kentani forests, *Miss Pegler,* 1125! 1342! Natal ; Durban, *Cooper,* 3159! *Burtt-Davy in Herb. Wood,* 12845! 12874! Maritzburg. *Sim,* 7123! Dumisa Station, *Rudatis,* 1144! and without precise locality, *Sanderson!*

This species has been confused with *F. natalensis*, Hochst., from which it may be readily distinguished by the rounded (not stipitate) base of the receptacle, the smaller ostiole, and the differently shaped leaves with their delicate reticulation.

Imperfectly known species.

20. **F. ilicina** (Sond. ex Miq. Ann. Mus. Bot. Lugd.-Bat. iii. 289); a tree; branches and branchlets thick, grooved-angular, yellowish; leaves moderately petiolate, oblong, very obtuse, sub-angular or obtuse at the base, $3\frac{1}{2}$–4 in. long, $1\frac{1}{2}$ in. broad, entire, coriaceous, pale, smooth above, reticulate below with 5–8 nerves; petiole thick, $\frac{1}{4}$–$\frac{1}{3}$ in. long; stipules ovate, acute, convolute; receptacles not known. *Urostigma ilicinum, Sond. in Linnæa,* xxiii. 136.

WESTERN REGION: Little Namaqualand; Kamiesberg, *Ecklon & Zeyher*, 4; rocky places near Kammapus, *Zeyher*, 3869.

An imperfectly known species the types of which appear to have been lost.

ORDER CXXII. B. **URTICACEÆ.**

(By N. E. Brown.)

Flowers unisexual, the sexes on the same or on different plants. *Perianth* 2–5-lobed or -partite, or urceolate or tubular and toothed at the apex, or bract-like and tubular below, obliquely open and entire or obscurely 3-lobed above, or absent from the female flowers. *Stamens* 1–5, absent or rudimentary or scale-like in the female flowers; anthers inflexed in bud. *Ovary* superior or (in *Droguetia*) adherent to or closely invested by the perianth, rudimentary, abortive or absent in the male flowers, 1-celled; style or stigma terminal, sometimes oblique; ovule solitary, basal, erect. *Fruit* a dry achene or fleshy or pulpy, naked or enclosed in or adnate to the more or less enlarged perianth. *Seed* solitary, erect, with a thin testa, albuminous or exalbuminous; embryo straight.

Annual or perennial herbs, shrubs or trees, sometimes armed with stinging hairs, with watery juice; leaves alternate or opposite, stipulate; flowers in sessile or pedunculate clusters, or heads, spikes, panicles, cymes or rarely solitary.

DISTRIB. Genera 41, species about 530, distributed throughout the temperate and tropical regions.

　　**Plants armed with stinging hairs.*

　　I. **Urtica.**—*Leaves* opposite. *Flower-spikes* two in each leaf-axil. *Stigma* a sessile mop-like tuft of hairs.

　　II. **Fleurya.**—*Leaves* alternate. *Cymes, panicles or racemosely branched peduncles* solitary in each leaf-axil. *Stigma* linear-filiform or lanceolate.

　　III. **Urera.**—*Leaves* alternate. *Cymes or panicles* solitary in each leaf-axil. *Stigma* a sessile head- or mop-like tuft of hairs.

**Plants without stinging hairs.

†Flowers all free, none enclosed in an involucre.

IV. **Pilea.**—*Leaves* opposite.   *Perianth* of 2–4 free segments in both sexes. *Stamens* 2–4, represented by scales or rudiments in the female flowers. *Stigma* a sessile head- or mop-like tuft of hairs.

V. **Pouzolzia.**—*Leaves* alternate.   *Perianth* of the male flowers 3–5-lobed, of the female flask-shaped and 2–4-toothed at the contracted mouth. *Stamens* 3–5, not represented in the female flowers.   *Stigma* filiform.

VI. **Australina.**—*Leaves* alternate.   *Perianth* of the male flowers bract-like, tubular below, obliquely open above, of the female so closely investing or adnate to the ovary that it appears absent.   *Stamen* 1.

††Several male and 1–3 female flowers enclosed in an obconic or hemispheric involucre, and some female flowers solitary in a 2-lobed or flask-shaped involucre.   Male flowers with a bract-like perianth, tubular below, obliquely open and flattish or concave above.   Stamen 1. Female flowers without a perianth.

VII. **Forskohlea.**—*Leaves* alternate.   *Bisexual involucres* conspicuously 4–7-lobed ; lobes much longer than the flowers.

VIII. **Droguetia.**—*Leaves* alternate or opposite.   *Bisexual involucres* inconspicuous, shortly toothed ; teeth much shorter than the flowers.

I. URTICA, Linn.

Flowers unisexual, both sexes on the same or on different plants. *Perianth-segments* 4, equal and concave in the male, very unequal, flat and becoming enlarged and adpressed to the fruit in the female flowers. *Stamens* 4, inflexed in bud, absent from the female flowers. *Ovary* superior, straight, erect, 1-celled, with 1 erect ovule ; in the male flowers very rudimentary and cup-shaped ; stigma sessile or rarely on a short style, formed of a small tuft of hair-like papillæ, deciduous or persistent.   *Fruit* a compressed achene, ovate, smooth or very minutely tuberculate.

Annual or perennial herbs, armed with stinging hairs ; leaves opposite, petiolate, stipulate ; flower-spikes in pairs in the axils of the leaves, unisexual or bisexual, simple or branched ; flowers in small dense clusters crowded along the spikes.

DISTRIB. Species about 40, widely distributed.

Teeth of the leaves as broad as or broader than long,
    entire ; achenes ½ lin. long, smooth, whitish-ochreous (1) **dioica.**
Teeth of the leaves longer than broad, entire or lobed :
    achenes nearly or quite 1 lin. long :
Blade of the leaf ½–2½ in. long ; achenes smooth, light
    ochraceous   ...   . (2) **urens.**
Blade of the leaf 2–8 in. long :
    Dried petioles ¼–⅜ lin. thick ; teeth entire ; achenes
        brown, dotted with dark red-brown, scarcely
        tuberculate...   ...   . (3) **Burchellii.**
    Dried petioles 1 lin. thick ; teeth usually lobed ;
        achenes brown, minutely tuberculate   ...   ... (4) **lobulata.**

1. U. dioica (Linn. Sp. Pl. ed. i. 984); perennial, unisexual or with male flowers at the lower part and female at the upper part of the same stem, or intermingled on the same spike, all parts of the plant armed with stinging hairs; stems $1\frac{1}{2}$–3 ft. high, erect, herbaceous, 4-angled; leaves opposite; petiole $\frac{1}{4}$–$1\frac{1}{2}$ in. long; blade $1\frac{1}{2}$–5 in. long, $\frac{1}{2}$–3 in. broad, ovate or elongated ovate-lanceolate, acute or acuminate, rounded to cordate at the base, serrate-dentate with teeth $\frac{3}{4}$–2 lin. long, glabrous or puberulous on one or both sides besides the stinging hairs, which are not numerous on the upper surface; flower-spikes in pairs, axillary, much shorter than the leaves, branched; sepals of the male flowers equal, $\frac{3}{4}$ lin. long, elliptic, obtuse, very minutely puberulous outside; fruiting sepals of the female flower unequal, the 2 larger $\frac{2}{3}$ lin. long, twice as long as the smaller pair, adpressed to the flattened achene and enclosing it, minutely puberulous outside; achenes $\frac{1}{2}$ lin. long, compressed, ovate, smooth, whitish-ochreous. *U. dioica, var. capensis, Wedd. Monogr. Urtic.* 78. *U. eckloniana, Blume, Mus. Bot. Lugd.-Bat.* ii. 142. *U. dioica, var. eckloniana, Wedd. in DC. Prodr.* xvi. i. 51.

COAST REGION : Albany Div. ; near the Fish River, *Miss Bowker*, 11! Queenstown Div. ; by the Zwartkei River, *Drège*, 8244! *Baur*, 972! 1137! Bongolo Poort, *Galpin*, 2038!
CENTRAL REGION : Somerset Div. ; near Somerset East, *Bowker*! Graaff Reinet Div. ; near Graaff Reinet, *Bolus*, 692! Molteno Div. ; Droughton, near Molteno, *Flanagan*, 1637! Aliwal North Div. ; by the Orange River, *Burke*!
KALAHARI REGION : Orange Free State, by the Caledon River, *Burke*, 306!
A widely distributed weed of cultivation.

2. U. urens (Linn. Sp. Pl. ed. i. 984); annual, bisexual, 10–15 in. high, all parts armed with stinging hairs; stem herbaceous, 4-angled, simple or branched ; leaves opposite; petiole $\frac{1}{4}$–$1\frac{1}{4}$ in. long, blade $\frac{1}{2}$–$2\frac{1}{2}$ in. long, $\frac{1}{3}$–$1\frac{1}{2}$ in. broad, ovate or elliptic, obtuse or acute, rounded or subcordate at the base, deeply serrate, with deltoid-lanceolate acute entire or lobed teeth $1\frac{1}{2}$–3 lin. long, glabrous on both sides, except for the stinging hairs ; stipules free, about 1 lin. long, lanceolate, acute; flower-spikes in pairs in each leaf-axil, simple, usually shorter than the petioles, bearing male and female flowers intermingled; sepals of the male flowers equal, $\frac{2}{3}$ lin. long, ovate-lanceolate, acute or subacute, very membranous ; fruiting sepals of the female flower unequal, the larger $\frac{3}{4}$–1 lin. long, broadly ovate, obtuse, enclosing the fruit and three times as long as the smaller pair, ciliate and often with a stinging hair on the back, otherwise glabrous ; achenes almost 1 lin. long, compressed, ovate, smooth, shining, dirty ochreous. *Wedd. Monogr. Urtic.* 58, t. 1, C. *figs.* 13, 14, *and in DC. Prodr.* xvi. i. 40.

COAST REGION : Cape Div. ; Rondebosch, *Milne*! Queenstown Div. ; Shiloh, *Baur*, 1138!
CENTRAL REGION : Ceres Div. ; Leeuwfontein, 2200 ft., *Pearson*, 3205!
EASTERN REGION : Natal ; near Durban, *Mrs. Stainbank in Herb. Wood*, 3145
A widely distributed weed of cultivation.

3. **U. Burchellii** (N.E. Br.); root not seen; herbaceous, apparently 2 ft. or more high, armed with stinging hairs on all parts, but very sparingly on the upper surface of the leaves and flower-spikes; leaves opposite; petiole 1–4 in. long, slender, $\frac{1}{2}$–$\frac{2}{3}$ lin. thick; blade 2–5 in. long, 1$\frac{1}{2}$–4$\frac{1}{2}$ in. broad, broadly ovate, acute, slightly notched at the broadly-rounded base, coarsely toothed; teeth 2$\frac{1}{2}$–4 lin. long, deltoid, acute, entire; upper surface glabrous or thinly sprinkled with hairs; under surface pubescent with very fine spreading hairs, besides the stings; flower-spikes in pairs, axillary, much shorter than the leaves, unbranched, unisexual or with male and female flowers intermingled; sepals of the male flowers $\frac{1}{2}$–$\frac{2}{3}$ lin. long, elliptic, obtuse, concave, all glabrous or one of them with a stinging hair on the back; fruiting sepals of the female flowers very unequal, the larger 1$\frac{1}{4}$–1$\frac{3}{4}$ lin. long and nearly 1 lin. broad, elliptic, obtuse, very scantily pubescent and armed with about 3 stinging hairs along the midrib on the back; the smaller about $\frac{1}{2}$ lin. long, oblong or ovate-oblong; achenes 1 lin. long, ovate, brown, covered with very minute dark red-brown slightly raised dots, but scarcely forming tubercles.

COAST REGION: Uitenhage Div.; without precise locality, *Zeyher*! Bathurst Div.; near Barville Park, *Burchell*, 4092!

4. **U. lobulata** (E. Meyer in Drège, Zwei Pfl. Documente, 48, 56, 58, 60, 68, and ex Blume, Mus. Bot. Lugd.-Bat. ii. 143); perennial, all parts rather densely armed with stinging hairs; stems probably 2 ft. or more high, herbaceous, 4-angled; leaves opposite; petiole 1–4$\frac{1}{2}$ in. long; blade 2–8 in. long, 2–7 in. broad, broadly ovate or reniform-ovate, acute or abruptly acuminate, more or less cordate at the base, rather deeply lobed at the margin, with the lobes 5–11 lin. long and 3–5-toothed, or occasionally nearly or quite entire, the terminal tooth being longer than the lateral teeth, both sides thickly armed with stinging hairs, puberulous beneath and thinly pubescent above or glabrous on both sides, minutely ciliate; flower-spikes in pairs in the axils of the leaves and much shorter than them, branched, with male and female flowers intermingled or the upper spikes female; sepals of the male flowers equal, $\frac{3}{4}$–1 lin. long, elliptic-oblong, obtuse, two of them with a stinging hair on the back, otherwise glabrous; fruiting sepals of the female flowers very unequal, the two larger 1–1$\frac{1}{4}$ lin. long, ovate-oblong or elliptic, obtuse, nearly three times as long as the narrow lateral sepals and enclosing the achene, armed and glabrous as in the male flowers; achenes nearly 1 lin. long, compressed, ovate, very minutely tuberculate, brown, shining. *Wedd. Monogr. Urtic.* 84. *U. lobata, E. Meyer ex Blume, Mus. Bot. Lugd.-Bat.* ii. 144. *U. Meyeri, Wedd. Monogr. Urtic.* 64, and in *DC. Prodr.* xvi. i. 44.

COAST REGION: Queenstown Div.; Table Mountain, 6000–7000 ft., *Drège*.
WESTERN REGION: Little Namaqualand; near Lily Fontein, *Drège*!
CENTRAL REGION: Graaff Reinet Div.; on the sides of Cave Mountain, near

Graaff Reinet, 4300 ft., *Bolus*, 699! near Graaff Reinet, 3000–4000 ft., *Drège*!
Sneuwberg Range, 4000–5000 ft. *Drège*! Beaufort West Div. ; between Zak River
Poort and Leeuwenfontein, 3000–4000 ft., *Drège*! Molteno Div. ; Broughton, near
Molteno, *Flanagan*, 1638!

II. FLEURYA, Gaud.

Flowers unisexual, both sexes on the same or on different plants.
Perianth-segments in the male flower 4–5, equal ; in the female
flower 4, very unequal, the upper one hooded or concave and the
lower flattened, both minute and very inconspicuous, the two
lateral much larger, flat, adpressed to the ovary and achene, becoming enlarged in fruit, thin or submembranous. *Stamens* 4–5,
absent from the female flowers. *Ovary* superior, at first nearly
straight, soon becoming oblique, 1-celled, with 1 suberect or
oblique ovule ; in the male flowers rudimentary ; stigma linear-
filiform or lanceolate. *Fruit* a compressed achene, oblique.

Annual or perennial herbs, armed with stinging hairs ; leaves alternate, petiolate,
stipulate ; stipules connate, bifid at the apex ; panicles, cymes or racemosely
branched peduncles axillary or subterminal, solitary, shorter than the leaves ;
flowers in small dense scattered clusters.

DISTRIB. Species about 13, widely distributed in tropical and subtropical
regions.

The genus *Laportea* should, in my opinion, be united with *Fleurya*. The only
distinction between them is that *Fleurya* has a tuberculate area on the sides of its
achenes, whilst in *Laportea* the achenes are smooth, but this character breaks
down, as in some species of *Fleurya* the area is smooth and in others altogether
evanescent.

Leaves with 5–8 large teeth on each side ...   . (1) **grossa**.
Leaves with 9–50 small teeth on each side :
  Flowers in small dense globular clusters ; male flowers
    sessile and 5-partite, female on narrowly winged
    pedicels scarcely enlarged in fruit ...   . (2) **mitis**.
  Flowers not in dense globular clusters ; male flowers
    on slender pedicels and 4-partite, female on broadly
    winged cuneately obcordate pedicels very much
    enlarged in fruit   ...   . (3) **alatipes**.

1. **F. grossa** (Wedd. in Ann. Sc. Nat. 4me sér. i. 183) ; a herb,
apparently 2 ft. or more high, armed with stinging hairs on the
petioles and veins on the underside of the leaves, unarmed on the
stem and inflorescence, otherwise glabrous in all parts ; leaves
alternate ; petiole ¾–3 in. long ; blade 1–5 in. long, ¾–3¾ in. broad,
ovate, acuminate, cuneately subtruncate at the base, very coarsely
toothed, green, marked with rather large white spots ; teeth 6–9 on
each side, 1–6 lin. long and as much in breadth at the base, broadly
deltoid, acute or acuminate ; peduncles solitary, axillary, about as
long as or slightly longer than the petioles, racemosely or paniculately branched ; pedicels very short, not winged, scarcely evident

in the male flowers; perianth of the male flowers of 5 equal segments nearly 1 lin. long, ovate-lanceolate, acute, concave; perianth-
segments of the female flower 4, very unequal, 1 very small and
concave, outside the others, 1 rudimentary and very narrow or
tooth-like, 2 adpressed to the ovary or achene, ½–¾ lin. long, broadly
ovate, acute, submembranous, edged with a row of very minute
sessile glands; achene 1 lin. long, obliquely orbicular-ovate, compressed, keeled, very faintly tuberculate on the flattened depressed
central area on each side, glabrous, brown, very minutely dotted
with darker brown; stigma ¼ lin. long, reflexed and closely adpressed to the achene. *Wedd. Monogr. Urtic.* 119, *t.* 1, *A, figs.* 1–3,
and in DC. Prodr. xvi. i. 76. *Urtica grossa, E. Meyer in Drège,
Zwei Pfl. Documente,* 134, 150.

SOUTH AFRICA: without locality, *Zeyher,* 3865, ex *Weddell.*
COAST REGION: Uitenhage Div.; on old trees, *Zeyher,* 565! Zuurberg Range,
Drège.
EASTERN REGION: Pondoland; near the Umtata River and between there and
St. Johns River, *Drège!* Natal; Inanda, *Wood,* 1236! Fairfield, *Rudatis,* 1335!
and without precise locality, *McKen,* 3! *Gerrard,* 710!

2. **F. mitis** (Wedd. in Ann. Sc. Nat. 4ᵐᵉ sér. i. 183); stems
herbaceous, 4–6 ft. long, erect or scrambling or prostrate and
rooting at the nodes, branching, glabrous or puberulous, varying
from nearly unarmed to thickly covered with stinging hairs; leaves
alternate, more or less armed with stinging hairs on the petioles
and veins on the under surface; petiole ½–5 in. long; blade 1–5 in.
long, ⅔–3½ in. broad, ovate, acuminate, broadly rounded or occasionally very broadly cuneate at the base, acutely toothed at the
margin, with teeth ⅔–2 lin. long, and 1–3 lin. broad, glabrous or
thinly pubescent on both surfaces; peduncles solitary, axillary,
usually branched, rarely simple, with small dense globular clusters
of flowers scattered along or at the ends of the branches, shorter or
a little longer than the petioles, glabrous or puberulous, and with or
without stinging hairs, bearing all male or all female flowers, but
often both sexes occur upon the same stem; male flowers sessile,
with 5 spreading equal segments nearly or quite 1 lin. long and
ovate-lanceolate, acute, concave; female flowers on flattened and
narrowly winged pedicels ½ lin. long, armed with stinging hairs or
unarmed, with 4 unequal segments, erect and adpressed to the
ovary, the two larger lateral, ½–⅔ lin. long, flattish, very broadly
ovate, acute or subobtuse, the other two much smaller and one of
them concave; all glabrous in both sexes, with or without stinging
hairs; achene much longer than the perianth-segments, ¾ lin. long,
compressed, obliquely orbicular-ovate, subacute, glabrous, slightly
tuberculate within a flat ovate area on each side, green or pale
brown; stigma ⅓–½ lin. long, recurved. *F. capensis, Wedd. Monogr.
Urtic.* 117, *t.* 1, *A, figs.* 7 & 8, *not of Ann. Sc. Nat.* 4ᵐᵉ *sér.* i. 183;
Wood, Natal Pl. vi. *t.* 577. *F. peduncularis, Wedd. in DC. Prodr.*
xvi. i. 75 *(incl. var. mitis). Urtica mitis, E. Meyer in Drège, Zwei*

Pfl. Documente, 127, 143. *U. peduncularis, E. Meyer, l.c.* 143. *U. mitis, Hochst., and U. ovalifolia, Buching. ex Hochst. in Flora*, 1845, 88 *and in Krauss, Beitr. Fl. Cap- und Natal.* 151.

COAST REGION: Knysna Div.; in the forest at Knysna, *Burchell*, 4544! *Krauss*, 1828, 1829! Uitenhage Div.; near Sandfontein and Matjesfontein, *Drège*! Galgebosch, *Drège*! Port Elizabeth Div.; Krakakamma, *Zeyher*, 566! Albany Div., *Cooper*, 3590! King Williamstown Div.; by the Yellowwood River, *Drège*!
CENTRAL REGION: Somerset Div.; on the Bosch Berg, *Burchell*, 3231! *Bolus*, 310.
KALAHARI REGION: Transvaal; Rimers Creek, near Barberton, *Galpin*, 1289!
EASTERN REGION: Transkei, Kentani, *Miss Pegler*, 696! Griqualand East; by the Umzimkulu River, *Tyson*, 2792! *and in MacOwan & Bolus Herb. Norm.* 768! Natal; by the Umlaas River, *Krauss*, 30! near Durban, *Rehmann*, 8808! *Gueinzius*! *Gerrard*, 277! Inanda, *Wood*, 716! 1251!

3. **F. alatipes** (N. E. Br.); a herb, 1–3 ft. high, armed with stinging hairs on all parts, unisexual or both sexes on the same plant; stem furrowed when dried, minutely puberulous with deflexed hairs under the stinging hairs at the apex only; leaves alternate, glabrous on both sides except for the stinging hairs; petioles mostly 2–$2\frac{3}{4}$ in. long, those of the two terminal leaves much shorter; blade 3–7 in. long, $1\frac{3}{4}$–$4\frac{1}{4}$ in. broad, ovate, acute or acuminate, round or notched at the 3-nerved base, acutely toothed on the margin; teeth triangular, $\frac{1}{2}$–$1\frac{1}{2}$ lin. long, $1\frac{1}{2}$–2 lin. broad, each tipped with a stinging hair; cymes or panicles solitary in the axils of the upper leaves, including the peduncle $1\frac{1}{2}$–4 in. long, 1–$2\frac{1}{2}$ in. broad, erect or ascending, branching, those of the male plant much more slender than those of the female; pedicels of the male flowers $\frac{1}{2}$–$\frac{2}{3}$ lin. long, slender, jointed just below the flower; pedicels of the female flowers at first about $\frac{1}{2}$ lin. long, becoming in fruit $1\frac{1}{4}$–2 lin. long, $\frac{1}{2}$–$1\frac{1}{4}$ lin. broad, very broadly winged, cuneately obcordate, with a deflexed flower at the notch, arranged in small compact cymules on the branches of the cyme or panicle; perianth of the male flowers of 4 equal concave ovate acute segments $\frac{2}{5}$–$\frac{2}{3}$ lin. long, glabrous; perianth of the female flowers of 4 very unequal segments, the upper concave or boat-shaped, and the lower flattened lanceolate segments rudimentary and about $\frac{1}{4}$ lin. long, ciliate, the lateral in flower scarcely $\frac{1}{2}$ lin. long, broadly ovate, acute, becoming in fruit 1–$1\frac{1}{4}$ lin. long and falcate in outline, deflexed, submembranous, green; achene abruptly deflexed, oblique and very broadly D-shaped, compressed, $1\frac{1}{4}$ lin. in diam., pitted rugulose, glabrous, light brown, with a very dark brown border; stigma $\frac{1}{2}$–$\frac{2}{3}$ lin. long, stout, densely stigmatose-hairy, orange, in fruit abruptly bent back to the margin of the achene. *Laportea alatipes, Hook. f. in Journ. Linn. Soc.* vii. 215; *Rendle in Prain, Fl. Trop. Afr.* vi. ii. 252.

KALAHARI REGION: Transvaal; forest on the Zoutpansberg Range, *Worsdell*!
EASTERN REGION: Griqualand East; Zuerberg Range, *Tyson*, 1772! Natal; in woods or bush at Enon and near Byrne, *Wood*, 1880! in bush near York, *Wood*! and without precise locality, *Cooper*, 1128!

Also in Tropical Africa.

According to Mr. Cooper, this plant in 1862 was and perhaps still is "used by the natives to punish boys." Dr. Wood remarks of his number 1880 in "Natal Plants" vi., under t. 577, that it "is much more irritating than *F. capensis* (*F. mitis*), and the natives always give it a wide berth where it is plentiful, it is much worse than the common nettle, *Urtica urens*."

III. URERA, Gaud.

Flowers unisexual, the sexes on different plants. *Perianth* 4–5- (or abnormally up to 10-) lobed or partite ; lobes or segments equal in the male, unequal in the female flowers and more or less enlarging in fruit. *Stamens* 4–5 (or in abnormal flowers 7–10), opposite the perianth-segments, absent from the female flowers. *Ovary* straight or slightly oblique, 1-celled, rudimentary in the male flowers ; ovule solitary, basal, erect ; stigma oblique, sessile, head-like or hemispherical, densely papillate-pubescent. *Fruit* a more or less compressed achene, enclosed in the enlarged perianth.

Shrubs or trees, armed with stinging hairs ; leaves alternate, petiolate, stipulate ; stipules free or connate and bifid or entire at the apex ; flowers in axillary cymes or panicles, solitary in each leaf-axil.

DISTRIB. Species about 50, in Tropical Africa, Madagascar, the Sandwich Islands and Tropical America, 2 in South Africa :

Leaves toothed ... ... (1) **tenax.**
Leaves entire ... ... (2) **Woodii.**

1. **U. tenax** (N. E. Br. in Hook. Ic. Pl. xviii. t. 1748) ; a shrub or small tree, 5–15 ft. high, armed with stinging hairs on the leaves, inflorescence and sometimes but not always on the young branches ; young main shoots or branches varying from 2 to 6 lin. thick, with a smooth reddish-brown or purplish bark ; leaves alternate, deciduous ; petioles ½–2½ in. long on the specimens seen, but possibly longer, leaving cordate scars 1½–3½ lin. in diam. after their fall ; blades 1–3½ in. long and 1–3 in. broad, very broadly ovate to suborbicular, acute or abruptly pointed, shallowly cordate at the 3-nerved base, toothed, with the teeth 1–2 lin. long and 1–3 lin. broad, triangular, acute, both surfaces (except for the stinging hairs) glabrous ; stipules spreading, 2–3 lin. long, lanceolate, acute, keeled, glabrous, brown ; flowers sometimes in laxly branching cymes or panicles, sometimes in simple raceme-like panicles and varying from ½–3 in. long, green ; male flowers usually 5-lobed, but sometimes abnormally 7–10-lobed to about three-fourths of the way down, about 1¼ lin. long and the lobes ⅔ lin. broad, elliptic or oblong, obtuse, concave ; stamens 5, or in abnormal flowers 7–10, with stout filaments ; female flowers 4-partite ; segments unequal, one minute, the others larger and ⅓–⅔ lin. long, the two lateral ovate, subacute, enlarging to 1¼ lin. long and ¾ lin. broad and becoming elliptic in fruit, the other elliptic, not enlarging ; ovary compressed, ovate,

slightly oblique; stigma terminal, oblique, consisting of a dense tuft
of short hairs, yellowish or whitish; achenes about $\frac{2}{3}$ lin. long, compressed, ovate, slightly oblique, slightly tuberculate on the sides,
pale brown. *N. E. Br. in Kew Bulletin,* 1888, 84, *with plate; Burtt-
Davy in Kew Bulletin,* 1908, 175, *and in Transv. Agric. Journ.* v. 433.

COAST REGION: Queenstown Div.; Zwartkei Valley, 3000 ft., *Galpin,* 8175!
KALAHARI REGION: Transvaal; Umvoti Creek, near Barberton, *Galpin,* 1001!
Waterval Onder, *Rogers,* 2606! Kloofs of the Magaliesberg Range and Crocodile
and Magalies Rivers, *Burtt-Davy,* 195! *Miss Nation,* 210!
EASTERN REGION: Natal; Inanda, *Wood,* 3837! Dumissa, *Rudatis,* 1069!

The bark of this plant affords a valuable fibre, of which an account is given in
the *Kew Bulletin* for 1888 as above quoted.

2. **U. Woodii** (N. E. Br. in Kew Bulletin, 1911, 96); a shrub,
only armed with a few stinging hairs on the petioles, the base of
the midrib and nerves of the leaves and sometimes on the cymes;
branches unarmed, glabrous, smooth; leaves alternate, glabrous;
petioles $\frac{1}{2}$–1 in. long; blades $2\frac{1}{4}$–$3\frac{1}{2}$ in. long, $1\frac{1}{4}$–$2\frac{1}{2}$ in. broad,
elliptic, abruptly acuminate into a subobtuse point $\frac{1}{4}$–2 in. long,
obtuse, broadly rounded, or slightly subcordate at the 3-nerved
base, entire and apparently subcoriaceous, paler beneath; stipules
connate into a tapering 2-keeled bifid body 4–5 lin. long, with 2 free
subulate points at the apex, glabrous, brown, deciduous; cymes or
panicles 1–3 in. long, axillary, unisexual and the sexes on different
plants; male flowers on pedicels $\frac{1}{4}$–$\frac{3}{4}$ lin. long, jointed close under
the flower, perianth equally 4-lobed to rather more than half way
down, $\frac{3}{4}$ lin. long, glabrous; lobes ovate, very obtuse and minutely
fringed or toothed at the apex, concave; stamens 4, longer than the
perianth; female flowers with an unequally 4-crenate or 4-lobed
shortly tubular or cup-shaped perianth $\frac{1}{3}$ lin. long, with the
lobules very minutely denticulate or fringed, glabrous; ovary compressed-ovoid, slightly longer than the perianth; stigma sessile,
oblique, large, hemispheric, densely papillate-pubescent; achenes
not seen.

EASTERN REGION: Pondoland; in a cutting to the lighthouse near Port St.
John, *Miss Pegler,* 1533! Natal; Umzinyati Falls, *Wood,* 1803! and without
precise locality, *Sanderson,* 594!

I think it probable that *Elatostemma trinerve,* Hochst. in *Flora,* 1845, 88, and
in Krauss, *Beitr. Fl. Cap und Natal,* 151, is the same as this plant, but I have
seen no specimen of it, nor does Weddell appear to have done so. It was collected
near the Umlaas River in Natal, *Krauss,* 1267.

IV. **PILEA,** Lindl.

Flowers unisexual, the sexes intermingled on the same plant, or
separate and on different plants. *Perianth* 2–4-partite; segments
equal or unequal, concave, with a gibbosity or obtuse subulate point
on the back just below the apex on one or more of them. *Stamens*

2–4, opposite the perianth-segments, rudimentary and scale-like in the female flowers. *Ovary* straight, compressed, 1-celled, rudimentary in the male flowers; stigma terminal, sessile, not oblique, formed of a dense tuft of short hairs. *Fruit* a compressed achene, scarcely or not at all oblique.

Annual or perennial herbs, often creeping, not armed with stinging hairs; leaves opposite, equal or unequal in each pair, stipulate; stipules connate, entire; flowers in axillary sessile clusters or in pedunculate heads, cymes or panicles.

DISTRIB. Species about 280, widely distributed throughout the warmer regions, but absent from Australia and only one is South Africa.

1. **P. Worsdellii** (N. E. Br.); a perennial herb up to a foot or perhaps more in height, with a creeping rhizome, glabrous, and without stinging hairs in any part; leaves opposite, stipulate; petioles ¾–2 in. long, rather slender; blades ¾–2½ in. long, ½–1¾ in. broad, ovate, acute or somewhat acuminate, rounded to subcordate at the base, rather coarsely toothed, thin; teeth ¾–2 lin. long, 1–3 lin. broad, triangular, acute, directed forwards; stipules connate into a broadly ovate obtuse 2-nerved body 3 lin. long and 2 lin. broad, membranous, brown, a pair at each node; flowers mingled with small membranous bracts in dense axillary clusters in both axils of each pair of leaves, unisexual, the sexes on different plants or intermingled in the same cluster, each pair of clusters 5–8 lin. in diam.; pedicels up to 1½ lin. long, slender, jointed close under the male flowers; perianth 3-partite and alike in both sexes, ⅔–¾ lin. long; segments equal and alike, deeply hooded-concave at the lower two-thirds, with a stout subulate blunt point on the back arising just below the apex of the concavity; stamens 3, represented by 3 minute broadly cuneate truncate rudiments in the female flowers; ovary straight, compressed, ovate, absent or very rudimentary in the male flowers; stigma sessile, not oblique, composed of a dense tuft of short hairs; achenes 1 lin. long, compressed, ovate, scarcely oblique, smooth, whitish-brown.

KALAHARI REGION: Transvaal; Houtbosch, 5500 ft., *Schlechter*, 4740! forest on the Zoutpansberg Range, *Worsdell*!

Also in Tropical Africa.

V. POUZOLZIA, Gaud.

Flowers unisexual, both sexes on the same plant. *Male flowers*: perianth 3–5-lobed; lobes abruptly pointed, valvate in bud; stamens 3–5, opposite the perianth-segments; ovary rudimentary, obovoid. *Female flowers*: perianth tubular, with an inflated basal part, flask-shaped or compressed-ovoid, minutely 2–4-toothed at the mouth; stamens or rudiments of them none; ovary ovoid, compressed, one-celled, with 1 erect ovule; stigma much exserted from the perianth, filiform, terminal, not oblique, puberulous or papillose-pubescent, deciduous. *Achene* included in the enlarged perianth and closely invested by but not adnate to it, not oblique, compressed ovoid, with a shining crustaceous pericarp.

Perennial shrubs or herbs, not armed with stinging hairs ; leaves alternate, petiolate, stipulate ; stipules free ; flowers in small dense axillary clusters, unisexual, both sexes in the same cluster, but one sex preponderating.

DISTRIB. Species about 60, in tropical and subtropical regions, mostly in the Old World, with about a dozen in Tropical America.

Leaves entire, white beneath ... ... (1) **hypoleuca.**
Leaves toothed, green on both sides ... ... (2) **procridioides.**

1. P. hypoleuca (Wedd. in DC. Prodr. xvi. i. 227) ; a shrub, with the young branches varying from very minutely puberulous to velvety-pubescent ; leaves alternate, herbaceous ; petioles 2–9 lin. long ; blades $\frac{2}{3}$–3$\frac{1}{2}$ in. long, $\frac{1}{2}$–2$\frac{1}{2}$ in. broad, ovate or orbicular-ovate, acute or shortly cuspidate-acuminate, cuneate to broadly rounded at the 3-nerved base, entire, more or less pubescent and green above, covered with a dense white felt beneath ; stipules 1$\frac{1}{2}$–3 lin. long, 1 lin. broad, tapering from the base to an acute point, brown, deciduous ; flowers in small clusters in the axils of the leaves, sessile, unisexual, both sexes present in the same axil, but with a preponderance of one sex on different plants, greenish-white, mingled with minute brown membranous bracts ; male flowers somewhat flattened and abruptly pointed in bud, 4–5-lobed to half-way down, more or less pubescent outside ; lobes $\frac{2}{3}$ lin. long, ovate-acuminate, concave, valvate in bud ; stamens 4–5, with flattened filaments longer than the perianth-segments ; ovary rudimentary, compressed obovoid, semitransparent or watery ; female flowers about $\frac{2}{3}$ lin. long, compressed-ovoid-acuminate or somewhat flask-shaped, tubular, minutely 2–3-toothed at the apex, pubescent outside ; stamens none ; ovary ovoid ; style much exserted, 1$\frac{3}{4}$–2 lin. long, filiform, minutely puberulous ; achene enclosed in the enlarged perianth, 1 lin. long, ovoid, obscurely veined, but otherwise smooth and shining, whitish. *Burtt-Davy in Kew Bulletin,* 1908, 174.

KALAHARI REGION : Transvaal ; near Moord Drift, *Schlechter,* 4315 ! Komati Poort, *Schlechter,* 11758 ! near Wonderboom, *Burtt-Davy,* 2646 ! *Miss Leendertz,* 451 ! Potgeiters Rust, *Rogers,* 2498 ! Warm Bath, *Burtt-Davy,* 2160 !
EASTERN REGION : Natal ; Umhlatuzi Valley, *Haygarth in Herb. Wood,* 11853 ! and without precise locality, *Gerrard,* 1188 !

This shrub is readily recognised by the snow-white under surface of its leaves.

2. P. procridioides (Wedd. Monogr. Urtic. 412) ; a branching shrub up to 4 ft. in height, or perhaps sometimes forming clumps of simple subherbaceous stems? branches pubescent or villous on the younger parts ; leaves alternate, herbaceous ; petioles $\frac{1}{4}$–2$\frac{3}{4}$ in. long ;

blades 1–3¾ in. long, ⅔–2 in. broad, ovate, acuminate, rounded or broadly cuneate at the base, toothed at the margin ; teeth ⅔–1 lin. long, 1½–3 lin. broad, triangular, acute or subacute ; more or less pubescent or somewhat pilose with soft hairs on both surfaces, but usually more densely so beneath ; stipules 2–4½ lin. long, broadly ovate at the base, subulate-acuminate, ciliate, and pubescent on the midrib beneath, membranous, brown, persistent ; flowers in small axillary clusters, sessile or the males very shortly pedicellate, unisexual, many male and one or a few female flowers in the same cluster, greenish-white, mingled with minute membranous brown bracts ; male flowers somewhat turbinate-subglobose and abruptly and shortly pointed in bud, 3–4-lobed to rather more than half-way down, pubescent or subpilose outside ; lobes ⅔ lin. long, ovate, abruptly and shortly subulate-pointed, deeply concave ; stamens 3–4, longer than the perianth ; rudimentary ovary obovoid, semitransparent or watery ; female flowers somewhat flask-shaped, tubular, minutely 2–3-toothed at the apex, with numerous ribs on the inflated part when in fruit, pubescent outside ; ovary compressed ovoid ; style much exserted, 2 lin. long, filiform, densely papillose-pubescent ; achene enclosed in the enlarged perianth, 1 lin. long, compressed ovoid, acute, smooth and shining, whitish or pale yellowish-white. *Wedd. in DC. Prodr.* xvi. i. 231, *excl. the American specimen quoted. Urtica?* procridioides, *E. Meyer in Drège, Zwei Pfl. Documente,* 150, 151, *name only. Margarocarpus procridioides, Wedd. in Ann. Sc. Nat.* 4*ᵐᵉ* sér. i. 204. *Bœhmeria procridioides, Blume, Mus. Bot. Lugd.-Bat.* ii. 204.

EASTERN REGION : Transkei ; forests around Kentani, *Miss Pegler,* 733 ! Pondoland ; between Umtata River and St. Johns River, *Drège* ! Zululand : Sebundini, *Haygarth in Herb. Wood,* 7910 ! and without precise locality, *Gerrard,* 1622 !

VI. AUSTRALINA, Gaud.

Flowers unisexual, the sexes intermingled on the same plant. *Male flowers* with a bract-like perianth, tubular below, open above, with the margins folded together or united in bud ; stamen 1 ; anther inflexed in bud. *Female flowers* compressed-ovoid, with the perianth so closely investing or united to the ovary that it appears absent, sometimes two flowers cohere ; ovary 1-celled, with 1 erect basal ovule ; stigma shortly filiform, more or less coiled to one side. *Achene* compressed-ovoid, keeled, with a thin crustaceous pericarp.

Annual herbs ; leaves alternate, stipulate : flowers in sessile axillary clusters, with the sexes intermingled or sometimes all the flowers of a cluster female.

DISTRIB. Species about 10, in Tropical and South Africa, Socotra, Australia and New Zealand.

Stem and branches prostrate; plant drying blackish or blackish-green:
Leaves $1\frac{1}{4}$–$3\frac{1}{2}$ lin. in diam., obtuse, scarcely toothed... (1) **paarlensis.**
Leaves 3–8 lin. in diam., pointed, distinctly toothed (2) **procumbens.**
Stem and branches erect or ascending:
Leaves entire, crenate or toothed, and the terminal tooth not or but slightly longer than broad:
Plant 4–10 in. high, branching at the base into weak flaccid stems $\frac{1}{5}$–$\frac{2}{5}$ lin. thick; leaf-blades $2\frac{1}{2}$–9 lin. long and broad, drying green ... ... (3) **capensis.**
Plant 8–18 in. high, with the main stem $\frac{3}{4}$–$1\frac{1}{4}$ lin. thick, usually drying blackish:
Upper part of stem and branches minutely adpressed-puberulous; blade of the larger leaves $\frac{1}{2}$–$\frac{3}{4}$ in. long ... ... (4) **Thunbergii.**
Upper part of stem and branches pubescent with rather coarse spreading or ascending hairs; blade of the larger leaves 1–$1\frac{3}{4}$ in. long ... (5) **lanceolata.**
Leaves very distinctly toothed and the terminal tooth usually 2–3 times as long as broad, drying green (6) **acuminata.**

1. **A. paarlensis** (N. E. Br.); a small annual prostrate herb, drying blackish; stems and branches procumbent, $\frac{1}{3}$–$\frac{1}{2}$ lin. thick, puberulous; leaves alternate, small; petioles $\frac{1}{2}$–1 lin. long; blades $1\frac{1}{4}$–$3\frac{1}{2}$ lin. long, 1–$2\frac{1}{2}$ lin. broad, elliptic or orbicular, obtuse or rounded at the apex cuneate at the base, entire or with 1–3 obscure crenations on each side, thinly pubescent; stipules $\frac{3}{4}$ lin. long, very broadly ovate, acuminate or mucronate, pubescent and ciliate, membranous; perianth of male flowers $\frac{3}{4}$ lin. long, obtuse or with an exceedingly short point; female flowers $\frac{1}{2}$–$\frac{2}{3}$ lin. long, puberulous. *Didymotoxa (Didymotoca) debilis, E. Meyer in Drège, Zwei Pfl. Documente,* 87 *and* 178, *letter* c *only.*

COAST REGION: Paarl Div.; Paarl Mountain, 1000–2000 ft., *Drège*!

2. **A. procumbens** (N. E. Br.); an annual with prostrate radiating branches, drying blackish-green; main branches about $\frac{1}{2}$ lin. thick, the others more slender, thinly pubescent with spreading hairs; leaves alternate; petioles 2–5 lin. long; blades 3–10 lin. long, 3–8 lin. broad, roundish-ovate, acute or subobtuse, very broadly or subtruncately cuneate at the base, and with 4–7 obtuse teeth or crenations on each side, thinly pubescent on both sides; stipules 1–$1\frac{1}{2}$ lin. long, very broadly ovate, acute or acuminate, membranous, glabrous, ciliate; perianth of the male flowers 1 lin. long, obtuse or acute, mucronate, ciliate on the margins and with a few hairs on the point, otherwise glabrous, often very dark purple; female flowers 1 lin. long, much compressed, with one edge very acute, solitary or two united together into one body by their more obtuse edges, pubescent (especially along the margins) with long hairs; fruit about $1\frac{1}{4}$ lin. long.

COAST REGION: Piquetberg Div.; Het Kruis, *Misses Stephens & Glover*, 8776!

3. A. capensis (Wedd. in Ann. Sc. Nat. 4me sér. i. 212); an annual herb 4–10 in. high, much branched at the base, with weak slender flaccid thinly pubescent or subglabrous stems and branches $\frac{1}{5}$–$\frac{2}{3}$ lin. thick, drying green; leaves alternate; petioles 1–12 lin. long; blades 2$\frac{1}{2}$–9 lin. long, 2–9 lin. broad, ovate or subtriangular-ovate, obtuse or rarely acute, very broadly cuneate at the base, entire or with 3–5 faint or distinct crenations or teeth on each side, glabrous or with a few hairs on both surfaces; stipules $\frac{3}{4}$–1 lin. long, broadly ovate, acute, membranous; flowers in small axillary clusters, several male surrounding 1–2 female; perianth of the male flowers, 1 lin. long, shortly tubular below, open above and obtusely angular on each side of the very obtuse mucronate apex, pubescent; female flower compressed-ovoid, $\frac{1}{2}$ lin. long, pubescent; stigma $\frac{1}{4}$ lin. long, filiform, curved; achene $\frac{2}{3}$ lin. long, ovoid, sub-acuminate, keeled on one side, dark brown. *A. integrifolia, Wedd. in Ann. Sc. Nat. 4me sér.* i. 212. *Didymotoxa debilis, E. Meyer in Drège Zwei Pfl. Documente,* 90 *and* 113. *Didymotoca debilis, E. Meyer, l.c.* 178, *letters a and b only. Didymodoxa debilis, Wedd. Monogr. Urtic.* 548, *t.* 20, B, *and in DC. Prodr.* xvi. i. 235^{61}. *D. integrifolia, Wedd. Monogr. Urtic.* 549, *and in DC. Prodr.* xvi. i. 235^{61}. *Urtica capensis as to one of the sheets marked "a" in Herb. Thunb., not as to Linn. f., nor of Thunb. Fl. Cap.;* see *N. E. Br. in Kew Bulletin,* 1913, 80. *Parietaria lanceolata, E. Meyer, a, in Drège, Zwei Pfl. Documente,* 109, *not of Thunb.*

SOUTH AFRICA : without locality, *Thunberg* !
COAST REGION : Clanwilliam Div. ; Lange Vallei, *Drège* ! Malmesbury Div. ; between Groene Kloof and Saldanha Bay, *Drège* !
WESTERN REGION : Little Namaqualand ; between Koper Berg and Kookfontein, *Drège,* 8247 !

4. A. Thunbergii (N.E. Br.) ; an erect annual herb 1–1$\frac{1}{2}$ ft. high, drying blackish ; main stems $\frac{2}{3}$–1 lin. thick, minutely puberulous ; leaves alternate ; petioles 1–5 lin. long ; blades $\frac{1}{4}$–$\frac{3}{4}$ in. long, 2–7 lin. broad, ovate, acute or obtuse, cuneate at the base, entire or with 3–4 slight crenations on each side, thinly puberulous or nearly glabrous ; stipules $\frac{3}{4}$–1 lin. long, very broadly ovate, acute or shortly mucronate, ciliate, membranous ; perianth of the male flowers $\frac{2}{3}$–$\frac{3}{4}$ lin. long, very obtuse, with or without a very short point ; female flowers $\frac{1}{2}$–$\frac{3}{4}$ lin. long ; stigma scarcely $\frac{1}{4}$ lin. long, filiform. *Urtica capensis, Thunb. Prodr.* 31, *Fl. Cap. ed. Schultes,* 155, *and as to sheet β of Herb. Thunberg ;* see *N. E. Br. in Kew Bulletin,* 1913, 80.

SOUTH AFRICA: without locality, *Thunberg* ! *Drège* !

Drège's specimen was distributed as "*Parietaria lanceolata,* Th. ? cc," but this " cc " specimen is not mentioned in *Drège, Zwei Pfl. Documente,* nor is it anywhere quoted by Weddell, although he has written the name "*Australina capensis*" upon the sheet, so that he evidently regarded it as a form of that species. I find, however, that it is most certainly distinct from the plant upon which Weddell based the name *A. capensis,* although in reality it is this species that should rightly bear that name. Drège's specimen exactly agrees with that on sheet β of *Urtica capensis* in Thunberg's Herbarium, which, as I have pointed out in the *Kew Bulletin,* is evidently the specimen from which he made his description and not from either of the sheets marked "a."

5. A. lanceolata (N. E. Br.); an annual herb 8–18 in. high, branching, often drying blackish ; main stem ¾–1¼ lin. thick, rather densely and somewhat coarsely pubescent with spreading or ascending hairs at the upper part, more thinly so below ; leaves alternate ; petioles 2–11 lin. long; blades ¼–1¾ in. long, ⅛–1⅓ in. broad, ovate, acute or subacuminate, cuneate at the base, serrate, with 5–9 teeth on each side and the terminal tooth as long as or a little longer than broad, rarely subentire or faintly crenate, thinly pubescent on both surfaces ; stipules 1–2 lin. long, very broadly ovate, acute or acuminate, coarsely pubescent and ciliate, membranous ; flower-clusters comparatively large, 1½–2½ lin. in diam. ; perianth of the male flowers ¾–1 lin. long, very obtuse, with a very short point, coarsely pubescent and ciliate; female flowers ¾ lin. long, pubescent ; stigma less than ¼ lin. long; fruit sometimes combined in pairs, varying in size from ¾ to 1¾ lin. long. *Parietaria lanceolata, Thunb. in Hoffm. Phytog. Blaetter,* i. *17, and Fl. Cap. ed. Schultes,* 155. *Urtica? radula, E. Meyer in Drège, Zwei Pfl. Documente,* 97. *Didymodoxa debilis, var. lanceolata, Wedd. in DC. Prodr.* xvi. i. 235[61].

SOUTH AFRICA : without locality, *Thunberg, Zeyher,* 1545 !
COAST REGION : Malmesbury Div. ; near Malmesbury, 200 ft., *Schlechter,* 5349 ! Cape Div. ; Muysen Berg, *Harvey* ! Hout Bay, *Wolley-Dod,* 1519 ! Farmer Pecks Valley, *Wolley-Dod,* 2818 ! Kalk Bay, 50 ft., *Bolus,* 2944 !
WESTERN REGION : Little Namaqualand ; slopes between Nababeep and Modderfontein, 3000 ft., *Bolus,* 9455 ! Van Rhynsdorp Div. ; by the Orange River, *Drège* !

I have not seen the type of *Parietaria lanceolata,* Thunb., but from the description it would appear to be identical with the subentire leaved forms of this species. The much shorter terminal point of the leaves and larger flower-clusters, as well as its different geographical range, readily distinguish this species from *A. acuminata.*

6. A. acuminata (Wedd. in Ann. Sc. Nat. 4me sér. i. 212); an annual herb 1–2 ft. high ; stem ¾–1 lin. thick at the base, branching, 4-grooved, with rounded angles, puberulous ; leaves alternate ; petioles up to 1⅓ in. long; blades ⅓–2¾ in. long, ⅛–1½ in. broad, ovate or ovate-lanceolate, acuminate, cuneate at the base, with 6–12 teeth on each side and the terminal tooth (at least of the uppermost leaves) 2–3 times as long as broad, thinly pubescent on both sides ; stipules 1¼–3 lin. long, broadly ovate, with a long awn-like point, membranous, tipped at the apex of the awn with 2–3 long hairs ; flower-clusters ¾–1½ lin. in diam. ; perianth of the male flowers ¾ lin. long, its subulate point tipped with 2–4 long hairs ; female flowers 1 lin. long, pubescent ; stigma very small, about ¼ lin. long, filiform, curved. *Parietaria cuneata, E. Meyer in Drège, Zwei Pfl. Documente, 133, 143. Urtica caffra, Thunb. Prodr. 31, and Fl. Cap. ed. Schultes,* 155 ; *see N. E. Br. in Kew Bulletin,* 1913, 80. *Fleurya capensis, Wedd. in Ann. Sc. Nat. 4me sér.* i. *183, not of Monogr. Urtic.* 117. *Didymodoxa acuminata, Wedd. Monogr. Urtic. 549. D. cuneata, Wedd. in DC. Prodr.* xvi. i. 235[61].

South Africa : without locality, *Thunberg* !
Coast Region : Uitenhage Div. ; by the river near Enon, *Drège,* b ! Stutter-
heim Div. ; forest at Fort Cunynghame, 3300 ft., *Galpin,* 2443 ! King Williamstown
Div. ; Yellowwood River, *Drège,* a ! Queenstown Div. ; mountaiu side, Bowkers
Park, 4000 ft., *Galpin,* 2605 !
Central Region : Somerset Div. ; on Bosch Berg, *Burchell,* 3144 ! 3212 !
Bolus, 1813 ! Cradock Div. ; near Cradock, *Cooper,* 3518 !
Kalahari Region : Orange River Colony ; Bloemfontein, *Rehmann,* 3835 !
Eastern Region : Natal ; Durban Flats, *Wood,* 1970 ! and without precise
locality, *Gerrard,* 265 !

Probably *Parietaria cuneata,* Eckl. & Zeyh. ex Hochst. in *Flora,* 1845, 88, and
in Krauss, *Beitr. Fl. Cap- und Natal.* 151, belongs here, and that "Eckl. & Zeyh."
is merely an error for the authority E. Meyer. It was collected near Knysna,
Krauss, 1827, but I have not seen a specimen.

VII. FORSKOHLEA, Linn.

Apparent flowers consisting of a 2–7-lobed compressed or obconic-
campanulate involucre containing one female flower or a number of
male or female flowers embedded in and clinging to densely matted
white wool. *Male flowers* surrounding the female ; perianth bract-
like, of one segment, very shortly tubular at the base (or open all
the way down ?), flattish or concave above, entire or with a short
lateral lobe below the apex on each side ; stamen one ; anther
inflexed in bud. *Female flower* consisting of a compressed-ovoid or
elliptic one-celled ovary terminating in a filiform deciduous stigma,
without perianth or stamens ; ovule solitary, erect, basal. *Achene*
compressed-ellipsoid ; with a crustaceous pericarp.

Perennial or annual herbs: leaves alternate, stipulate ; involucres sessile, in
pairs or solitary, axillary, resembling flowers.

Distrib. Species 6 in Tropical and South Africa, Atlantic Isles, Socotra and
Arabia.

Involucre 1½–4 lin. long, with broad obovate obtuse or
    subacute lobes ...   . (1) **candida.**
Involucre 4–7 lin. long, with lanceolate acute lobes   ... (2) **hereroensis.**

1. F. candida (Linn. f. Suppl. 245) ; an annual or perennial herb
½–1¼ ft. high ; stems and branches scabrous with rigid short
upcurved hairs ; leaves alternate ; petioles 2–8 lin. long; blades
¼–1¼ in. long, 2½–12 lin. broad, ovate or elliptic, obtuse or subacute,
cuneate at the base, with 3–5 teeth on each side, margins
revolute and as well as the veins beneath and the petiole scabrous
with short rigid curved hairs, upper surface rough, glabrous or
thinly pubescent, under surface thinly white-felted ; stipules ⅓–1 lin.
long, broadly ovate, mucronate, membranous, pale brown ; involucres
in pairs or solitary in each axil, sessile, obconic-campanulate, 1½–4
lin. long, 4–7-lobed to below the middle, densely covered with long

adpressed hairs on the basal part outside ; lobes obovate, often
nearly as broad as long, obtuse or subacute, glabrous, ciliate, green,
becoming whitish-brown in fruit ; flowers about 6–14 male and 3–4
female in each involucre, intermingled with clinging matted white
wool ; perianth of the male flowers entire, cuneate-obovate or
cuneate-oblanceolate, slightly hooded at the subacute apex ; stigma
$\frac{2}{3}$ lin. long. *Thunb. Prodr.* 77, *and Fl. Cap. ed. Schultes*, 385 ;
Murray, Syst. Veg. ed. xiv. 437 ; *Willd. Sp. Pl.* ii. 475 ; *Wedd.
Monogr. Urtic.* 536, *and in DC. Prodr.* xvi. i. 235[56] ; *Schinz in Bull.
Herb. Boiss.* iv. *Append.* iii. 51 (*not of Aiton*). *F. scabra, Retz. Obs.
Bot.* iii. 31.

Var. β : **viroscens** (Wedd. in DC. Prodr. xvi. i. 235ᵃᵃ); leaves green on the
under surface, otherwise as in the type, *F. viridis, E. Meyer in Drège, Zwei Pfl.
Documente,* 90, *not of Ehrenberg.*

COAST REGION : Riversdale Div. ; sandy banks of the Gouritz River, *Bowie* !
CENTRAL REGION : Calvinia Div. ; Hantam and Roggeveld Mountains, *Thunberg* !
Prince Albert Div. ; Jakhals Fontein, *Burke* ! Graaff Reinet Div. ; near the
Sundays River, *Bolus,* 474 ! Beaufort West Div. ; near Salt River, *Zeyher,* 1544 !
Var. β : Prince Albert Div. ; near the Gamka River, *Burke* !
WESTERN REGION : Great Namaqualand ; various localities, *Schinz,* 856, 857,
Steingrover, 3, 5, *Fenchel,* 113, 150, 151, *Wandres,* 23, and *Fleck,* 175, ex *Schinz* !
Little Bushmanland ; Naroep, *Schlechter* ! Van Rhynsdorp Div. ; Stinkfontein,
Schlechter, 11089 ! Var. β : Great Namaqualand ; near Warmbad, *Pearson,* 4372 !
Little Namaqualand ; Silverfontein, *Drège* ! west of Pella, *Pearson,* 3553 !

Pearson 8546 from between Klein Karas and Holoog, appears to be a hybrid
between *F. candida* and *F. hereroensis,* having the leaves and small involucre of
the former combined with the lanceolate acute involucre-lobes of the latter, and
at Kew a branch of true *F. candida* is mingled with it.

2. **F. hereroensis** (Schinz in Bull. Herb. Boiss. iv. Append. iii.
51) ; an annual or perennial herb about $\frac{1}{2}$–1 ft. high ; stem and
branches rough with rigid upcurved hairs and sometimes also
puberulous ; leaves alternate ; petioles 1–4 lin. long ; blades 3–12
lin. long, 1$\frac{1}{2}$–9 lin. broad, elliptic or lanceolate, acute or obtuse,
cuneate at the base, with 2–3 teeth on each side, margins revolute,
rough and glabrous above, white-felted on the under surface, with
short rigid curved hairs on the margins and veins beneath ; stipules
1$\frac{1}{4}$–2 lin. long, very broadly ovate, shortly mucronate, ciliate, membranous, pale brown ; involucres in pairs in each axil, sessile,
obconic-campanulate, 4–7 lin. long, usually 6-lobed to about half-
way down, densely covered with long adpressed dusky hairs on the
cup and base of the lobes outside ; lobes lanceolate, acute, ciliate
and thinly covered with adpressed short hairs on the back, green,
becoming whitish-brown in fruit ; flowers about 9–10 male and
4–5 female in each involucre, intermingled with clinging matted
white wool ; perianth of the male flowers 3-lobed, lateral lobes very
small, at the base of the terminal ovate acute middle lobe ; stigma
$\frac{3}{4}$–1 lin. long. *Rendle in Prain, Fl. Trop. Afr.* vi. ii. 300.

WESTERN REGION : Little Namaqualand ; Silverfontein, 2000–3000 ft., *Drège*
valley leading to Bethany Drift, *Pearson,* 6046 !

Also in Damaraland.

VIII. DROGUETIA, Gaud.

Apparent flowers of two kinds ; one consisting of a hemispherical shortly toothed involucre containing (and shorter than) 4–8 male and 1–3 female flowers embedded in and clinging to densely matted wool ; the other smaller, ovoid, with a flask-shaped involucre, 4-toothed at its contracted mouth, containing 1 female flower. *Male flowers* surrounding the female ; perianth bract-like, of one segment, tubular at the base, open and flattish or concave above, entire ; stamen 1 ; anther inflexed in bud. *Female flower* consisting of a compressed-ovoid or elliptic 1-celled ovary terminating in a filiform pubescent stigma, without a perianth or stamens, when 2 or more female flowers are present in a bisexual involucre they are sometimes adherent or fused into one central mass ; ovule solitary, erect, basal. *Achene* compressed-ovoid or compressed-ellipsoid, keeled, with a crustaceous pericarp.

Perennial erect or prostrate herbs ; leaves opposite or alternate, stipulate ; nvolucres axillary or in terminal spikes, sessile or subsessile, in pairs, solitary or clustered, resembling flowers.

DISTRIB. Species about 12, in South and Tropical Africa, the Mascarene Islands and India.

The involucre of this genus is difficult to distinguish from the perianths of the male flowers which project much beyond it, because it clings by means of very minute hooked hairs closely to the backs of the perianths and the wool in which they are invested, so that the perianths may easily be mistaken for lobes of the involucre.

Main stems prostrate, rooting ; leaves (including the
   petioles) ¼–1¼ in. long...   ...   . (1) **Thunbergii.**
Main stems erect, 1–3 ft. high ; leaves (including the
   petioles) ½–5 in. long :
   Leaves with 4–6 teeth on each side and the terminal
      tooth not longer than broad, under surface green... (2) **urticæfolia.**
   Leaves with 5–12 teeth on each side :
      Under surface of the leaf dull purplish ; terminal
         tooth twice or more than twice as long as
         broad   ...   . (4) **ambigua.**
      Under surface of the leaf green :
        Terminal tooth 2–3 times as long as broad   ... (3) **Woodii.**
      Terminal tooth not or scarcely longer than broad (5) **Burchellii.**

1. D. Thunbergii (N. E. Br. in Kew Bulletin, 1913, 80) ; a perennial herb, with slender creeping and rooting stems and branches ⅓–½ lin. thick, glabrous or thinly adpressed-pubescent on the same plant ; leaves opposite ; petiole 1–4 lin. long, slender ; blade 3–12 lin. long, 2–7 lin. broad, ovate, acute or acuminate ; rounded or obtusely cuneate at the base, crenate or serrate-crenate on the margins, usually thinly adpressed-pubescent above and

glabrous beneath, but sometimes also pubescent beneath even on the
same branch; stipules $\frac{1}{2}$–1 lin. long, broadly ovate, mucronate,
membranous, glabrous; involucres in each axil sometimes 1 or a
pair of tubular flask-shaped female only, 4-toothed at the apex,
sometimes also with a campanulate 6–8-lobed bisexual one mingled
with them, $\frac{3}{4}$–1$\frac{1}{4}$ lin. long, glabrous or minutely puberulous with
curved hairs outside, woolly within; male flowers in the bisexual
involucre 4–8, surrounding 1 female; perianth obliquely funnel-
shaped or campanulate at the upper part and apparently 3-lobed and
open down one side, very thin and membranous; female flowers
solitary in both the female and bisexual involucres; stigma $\frac{3}{4}$–1 lin.
long; achene about 1 lin. long, compressed-ovoid, with a keel along
one margin, slightly rugose, glabrous, blackish. *Urtica capensis as
to one of the sheets marked " a," and U. caffra sheet "a" in Herb.
Thunberg, but not as to description in Fl. Cap. ed. Schultes*, 155.

SOUTH AFRICA : without locality, *Thunberg*!
COAST REGION: Swellendam Div.; in the forest at Grootvaders Bosch,
Burchell, 7232!

As I have stated in the *Kew Bulletin*, (l.c.), Thunberg's Herbarium contains
two specimens of this plant respectively named *Urtica capensis* and *U. caffra*, but
that these specimens can in no way have been used by Thunberg when making
his descriptions of those two species. Possibly *Parietaria capensis*, Thunb. Prodr.
31, and Fl. Cap. ed. Schultes, 155, may belong here, but I have not seen
Thunberg's type.

2. D. urticifolia (Wedd. in DC. Prodr. xvi. i. 235[n], excluding
all synonyms except of E. Meyer); an erect herb, apparently 1$\frac{1}{2}$ ft.
or more high; stems obtusely 4-angled, grooved down each face,
more or less pubescent or scabrous; leaves opposite; petiole 1–10
lin. long; blade 4–18 lin. long, 3–12 lin. broad, ovate or somewhat
rhomboid-elliptic, acute or subobtuse, broadly cuneate at the base
and with 4–6 obtuse teeth on each margin, and the terminal tooth
not longer than broad, glabrous or pubescent on one or both sides;
green above and beneath; stipules $\frac{1}{2}$–1 lin. long, ovate or lanceolate, mucronate, spreading or reflexed, membranous, ciliate and
pubescent, green; involucres sometimes solitary in each axil and
female, but more usually a pair and bisexual, or sometimes bisexual
and female in the same axil, puberulous with hooked hairs; bisexual involucre about 1$\frac{1}{2}$ lin. in diam., broadly cup-shaped or hemispheric, 6–8-toothed, and the teeth with their mucronate points less
than $\frac{1}{4}$ lin. long, shorter than the flowers, green, puberulous, containing 4–8 male surrounding 2 female flowers; female involucre
ovoid or flask-shaped, minutely toothed at the contracted mouth,
puberulous, containing 1 female flower; perianth of the male
flowers bract-like, entire, concave, ovate-lanceolate, shortly mucronate-acute, much longer than the involucre, puberulous on the back
with hooked hairs, ciliate; stigma curved or coiled in a spiral;
achenes about $\frac{3}{4}$ lin. long, slightly oblique, with a stout keel along
one margin, slightly rugose, glabrous, brown. *D. urticoides, Wedd.*

in Ann. Sc. Nat. 4me sér. i. 211, *as to the South African plant only.*
Parietaria urticæfolia, E. Meyer in Drége, Zwei Pfl. Documente, 154,
not of Linn.

EASTERN REGION : Natal ; between Umtentu River and Umzimkulu River
Drège ! Durban, *Wood,* 939 !

3. D. Woodii (N. E. Br.) ; a herb, probably 3 ft. or more high,
with long slender lax branches ; stem obtusely 4-angled, grooved
down each face, slightly rough, from a somewhat harsh pubescence
of spreading hairs or sometimes the hairs are very scanty or almost
wanting, green ; leaves opposite or here and there alternate on weak
branches ; petioles of the stem-leaves 1–2 in. long, of those on the
branches 1½–6 lin. long, pubescent ; blades of the stem-leaves 2–4 in.
long, 1–2¼ in. broad, of those on the branches ½–1½ in. long and 2½–9
lin. broad, ovate, acute or acuminate, broadly cuneate at the base and
with 5–10 large obtuse teeth on each margin, with the terminal tooth
of the larger leaves 2–3 times as long as broad, thinly pubescent and
green on both sides ; stipules reflexed, 1–1¾ lin. long, lanceolate,
acuminate or subulate-pointed, membranous, ciliate ; involucres
bisexual and female intermingled or all female at each node ; bisexual
involucre 1½–1¾ lin. in diam., with about 8 subulate-pointed teeth
about ⅓ lin. long ; female involucre 1–1¼ lin. long, compressed-ovoid,
with 4 subulate-pointed teeth at the mouth, both involucres pubescent with curved hairs ; perianth of male flowers much longer than
the teeth of the involucre, concave or somewhat hooded, with an
acuminate point ; stigma densely pubescent ; achenes about ⅔ lin
long, compressed-ovoid, keeled along one margin, rugulose, blackish-
brown.

EASTERN REGION : Natal ; Inanda, *Wood,* 1243 ! and probably a specimen from
the Zuurberg Range, Natal, *Wood,* 3152, also belongs here, but has much smaller
leaves on the main stem.

4. D. ambigua (Wedd. in Ann. Sc. Nat. 4me sér. i. 211) ; a herb
apparently 1½ ft. or more high ; stems obtusely 4-angled, grooved
down each face, thinly and minutely adpressed-puberulous, smooth,
reddish when dried ; leaves opposite, or occasionally alternate on
the branches ; petioles 2–10 lin. long ; blades of those on the main
stems 1¼–2¼ in. long, ½–1¼ in. broad, of those on the branches
smaller, ovate, acuminate, broadly cuneate at the base, and with
5–12 obtuse teeth or crenations on each margin, with the terminal
tooth twice or more than twice as long as broad, thinly pubescent
and green above, glabrous or with very few hairs and dull purplish
beneath ; stipules 1½–2 lin. long, ovate, subulate-acuminate, spreading, membranous, ciliate, whitish-brown or perhaps pinkish when
alive ; involucres and flowers in arrangement and structure very like
those of *D. urticæfolia,* but the bisexual involucres are larger and up
to 2½ lin. in diam. ; achenes not seen. *Wedd. Monogr. Urtic.* 543,
partly, excl. synonyms not quoted here and Drège's plant. D. cuneata,
Buek in DC. Prodr. Index, xiv.–xvii. 122. *Parietaria cuneata, Eckl.*

& Zeyh. ex Krauss in Flora, 1845, 88, *and Beitr. Fl. Cap- und Natal.*
151. *P. capensis, Drège in Linnæa,* xx. 214, *not of Thunberg.*

SOUTH AFRICA : without locality or name of collector, a specimen named by
Weddell in *Herb. Kew*!
COAST REGION : Knysna Div. ; Blauwkrantz Pass, Zitzikama, 500 ft., *Galpin,*
4580 ! Humansdorp Div. ; between Kromme River and Uitenhage, *Zeyher,* 3866 !

5. D. Burchellii (N. E. Br.) ; a herb 1–2 ft. high ; stem subterete,
4-grooved, pubescent and slightly harsh to the touch, brownish
when dried ; leaves opposite ; petioles 1–15 lin. long ; blades ½–1¾
in. long, ¼–1 in. broad, ovate, acute, broadly cuneate at the base
and with 5–10 rather small obtuse teeth on each margin, and
the terminal tooth not or scarcely longer than broad, pubescent and
green on both sides ; stipules spreading, 1½–2½ lin. long, very
broadly ovate, with a subulate point, membranous, whitish-brown
with a dark midrib ; nodes of the short branches and of the upper
part of the stems bearing bisexual involucres, lower nodes of stems
bearing female involucres ; bisexual involucres 1½ lin. in diam., containing about 12–14 male and 3 female flowers, about 12-toothed,
and the teeth with a subulate tip ⅓–½ lin. long, ciliate with long
spreading hairs ; male perianth with the ovate tip acuminate into a
subulate ciliate point ⅓ lin. long, like the involucral teeth and
longer than them ; female involucre about ¾ lin. long ; achenes of
the bisexual involucres fused into a mass about ¾ lin. long and 1 lin.
or more broad ; achenes of the female involucres ½ lin. long, ovoid,
keeled, very minutely pitted, brown.

COAST REGION : Bathurst Div. ; near Barville Park, *Burchell,* 4084 !

ORDER CXXIII. **MYRICACEÆ.**

(By J. HUTCHINSON.)

Flowers unisexual, arranged in bracteate spikes. *Perianth* 0 ; the
female with a few hypogynous scales (*bracteoles*) around the base of
the ovary. *Male* flowers subtended by a solitary bract ; stamens
2–∞, usually 4–8 ; filaments short, more or less connate ; anthers
erect, ovoid or broadly oblong, with 2 parallel longitudinally dehiscing cells, rarely accompanied by a small subulate rudimentary
ovary. *Female* flower subtended by a solitary variously shaped
bract and with a whorl of small perianth-like hypogynous scales.
Ovary sessile, 1-celled ; style short, with two spreading or ascending
often flattened branches ; ovule 1, erect from the base of the cell,
orthotropous. *Drupe* small, globose or ovoid, usually strongly
warted, the warts often at length covered with a white waxy substance ; endocarp hard. *Seed* erect ; testa membranous ; albumen 0 ;
embryo straight, with plano-convex fleshy cotyledons and a short
radicle.

Trees or shrubs, frequently aromatic ; leaves alternate, penninerved, entire, serrate, dentate or pinnately lobed ; stipules 0 ; flowers monœcious or more usually diœcious ; male spikes axillary, usually dense-flowered, solitary (in some extra-African species fasciculate or paniculate) ; when bisexual then the male flowers below the female ; female spikes sometimes longer or shorter than the male, mostly axillary.

DISTRIB. A single genus with about 50 species, of which 15 occur within the limits of this flora, 6 in the Mascarene Islands, 11 in Tropical Africa, and the remainder are widely distributed throughout the Northern Hemisphere.

The segregation of *Myrica* into 3 genera by Chevalier (Monogr. Myricaceæ, 1901) is discussed briefly in the Flora of Tropical Africa, vol. vi. ii. p. 307.

I. MYRICA, Linn.

Characters of the Order.

Leaves sessile, broadly ovate or ovate-orbicular, as broad or nearly as broad as long, cordate at the base, undulate or repand-dentate or denticulate, rarely subentire, usually less than 1 in. long ... ... (1) cordifolia.
Leaves usually more or less petiolate, considerably longer than broad, rounded or cuneate at the base, mostly more than 1 in. long :
Rhachis of the ♂ inflorescence clearly visible between the laxly arranged flowers ; floral bracts not imbricate ; spikes slender ; branchlets glabrous ; leaves oblong-oblanceolate, entire or with a few blunt incurved teeth towards the apex ... ... (2) Burmanni.
Rhachis of the ♂ inflorescence not or only slightly visible between the flowers, the floral bracts more or less densely imbricate :
Leaves more or less rounded or obtuse at the base and subsessile, rarely very shortly cuneate, mostly entire or nearly so :
Branchlets softly pubescent or pilose with rather long hairs :
Leaves entire or sparingly and rather obscurely few-toothed :
Leaves rather acutely acuminate ... ... (3) dregeana.
Leaves rounded to an obtusely mucronate apex (4) humilis.
Leaves rather acutely crenate-serrate all round the margin ... ... (5) elliptica.
Branchlets glabrous or minutely puberulous :
Leaves very densely glandular (or when the glands have collapsed, punctate) below :
Leaves elliptic or oblong-elliptic, entire or very obscurely toothed towards the apex, very thick and rigidly coriaceous ; infructes-cences about twice as long as the leaves ... (6) kraussiana.
Leaves oblong-lanceolate, often with 3-5 rather coarse obtuse teeth on each side towards the apex, rigidly papery ; infructescences not known, but probably much shorter than the leaves ... ... (7) brevifolia.

Leaves with very few scattered glands or almost
  glabrous below :
    Branchlets quite glabrous   ...   . (8) **ovata.**

    Branchlets finely puberulous   ...   . (9) **myrtifolia.**

Leaves narrowly tapered into a distinct petiole at the
  base :
    Leaves linear or narrowly oblanceolate, entire or
      rarely with a few teeth, 4–5 times as long as
      broad :
      Branchlets greyish and minutely pubescent when
        dry   ...   . (10) **linearis.**

      Branchlets black and glabrous or nearly so when
        dry   ...   . (11) **glabrissima.**

    Leaves spathulate - oblanceolate, obscurely and
      remotely repand-dentate with 2–4 teeth on each
      side...   ...   . (12) **Zeyheri.**

Leaves more or less oblanceolate, distinctly toothed
  or incised-lobate :
    Leaves oak-like, deeply pinnately lobed with the
      lobes spreading at right angles and forming
      rather wide sinuses   ...   . (13) **quercifolia.**

    Leaves rather repand serrate   ...   . (14) **conifera.**

    Leaves closely and rather deeply crenate with
      upward pointing teeth   ...   . (15) **incisa.**

1. **M. cordifolia** (Linn. Sp. Pl. ed. i. 1025) ; a much-branched shrub
3–4 ft. high ; branches deeply leafy, pubescent or pilose, sometimes
at first quite tomentose ; leaves mostly imbricate, sessile, broadly
ovate or ovate-orbicular, cordate at the base, usually mucronate
from a rounded or obtuse apex, 2–9 lin. long, 2½–6 lin. broad, rigidly
coriaceous, usually adpressed to the branch when dry, undulate- or
repand-dentate or denticulate, rarely subentire, glabrous or sometimes glandular and slightly reticulate above, densely resinous-
glandular below, in the dry state often becoming deeply pitted below
on the collapse of the glands ; midrib usually distinct on both
surfaces, prominent below, thick at the base, gradually tapered to
the apex of the blade ; lateral nerves spreading, mostly inconspicuous ; margins cartilaginous and slightly recurved ; flowers
diœcious ; male inflorescences axillary, usually much shorter than,
rarely as long as, the leaves ; rhachis flexuous, glabrous ; bracts
solitary and below each flower, broadly ovate-triangular, obtuse,
about ½ lin. long, slightly keeled, membranous towards the margin
and shortly jagged-ciliate, otherwise glabrous ; stamens 2 ; anthers
almost sessile at the apex of a short common filament, oblong-
globose, ½ lin. long, very minutely and closely pitted ; female inflorescence axillary, solitary, about 2 lin. long ; rhachis stout, slightly
glandular, hidden by the overlapping bracts ; bracts solitary below
each flower, suborbicular, about ½ lin. broad, submembranous, shortly
ciliate, otherwise glabrous ; hypogynous scales united for half their
length, ⅓ lin. long, the free parts ovate, ciliate, fleshy, otherwise
glabrous ; ovary ovoid, covered with sessile glands ; style-branches 2,

2 o 2

subfiliform, free to near the base, $\frac{1}{2}$–$\frac{3}{4}$ lin. long, exserted from the bracts; fruits solitary or in pairs, shortly stalked, globose, about $\frac{1}{4}$ in. in diam., densely verrucose with a white waxy excretion. *Linn. Syst. Veg.* xiv. 884; *Thunb. Fl. Cap. ed. Schult.* 158; *Nouv. Duham. Traité,* ii. 193; *C. DC. in Ann. Sci. Nat.* 4*me* sér. xviii. 25, *t.* 3, *fig.* 31; *DC. Prodr.* xvi. ii. 148; *A. Cheval. Monogr. Myricac.* 168; *Marloth, Fl. S. Afr.* i. 132, *fig.* 74, A (by error named *M. quercifolia*), *and t.* 23, *fig.* B. *M. rotundifolia, Salisb. Prodr.* 396. *M. cordifolia, var. microphylla, A. Cheval. l.c.* 170.—*Alaternoides ilicis folio crasso hirsuto, Walther, Hort.* iii. *t.* 3. *Gale capensis, ilicis, etc., Petiver, Mus. Petiv.* 74. *Tithymali facie planta Æthiopica, etc., Plukenet, Almagest.* 373; *Phytog. t.* 319, *fig.* 7. *Coriotragematodendros ilicis aculeatae folio, Plukenet, Almatheum.* 65. *Myrica foliis subcordatis integris sessilibus, J. Burm. Rar. Afr. Pl.* 263, *t.* xcviii. *fig.* 3.

SOUTH AFRICA: without precise locality, *Masson*! *Thom*, 596! *Villet*! *Cunningham, Burmann, Roxburgh, Lehmann, Boivin*; *Verreaux, Vieillard & Deplanche, Zeyher,* 463! *Gillies*! *Mund & Maire*! *W. Brown*!
COAST REGION: Cape Div.; Lion's Head, *Krauss*; Cape Flats, *Zeyher*! *Rehmann,* 2050! *Pappe*! Camps Bay, *Burchell,* 894! *Alexander* (*Prior*)! *Phillips,* 237! *Worsdell*! sandy dunes near Cape Town, *Bolus,* 2939! Uitvlugt, *Wolley-Dod,* 2626! sandy flats near Simons Bay, *Frazer*! Houts Bay, *Schlechter,* 965. Caledon Div.; Genadendal, *Verreaux.* Riversdale Div.; Stille Bay, *Muir,* 176! Knysna Div.; Baak Hill, near Plettenbergs Bay, *Burchell,* 5330! Uitenhage Div.; Zwartkops and Koega Rivers, *Zeyher,* 3880! mouth of the Zwartkops River, *Zeyher,* 669! and without precise locality, *Cooper,* 1455! Bathurst Div.; near Port Alfred, *Burchell,* 3818! East India Div.; on dry downs along the sea coast at East London, *Galpin,* 5679!

This is the well-known *Wax-berry bush* of the southern coasts of the Cape; according to Marloth the plant is important in fixing the sand-dunes.

2. M. Burmanni (E. Meyer ex C. DC. in DC. Prodr. xvi. ii. 149);

branchlets numerous, crowded, fairly densely leafy, glabrous, when older marked with a few scattered brown lenticels; leaves oblong-oblanceolate, rounded to a very small tip at the apex, obtuse or a little narrowed at the base, 1–2 in. long, $\frac{1}{3}$–$\frac{3}{4}$ in. broad, rigidly coriaceous, entire or with a few blunt incurved teeth towards the apex, glabrous, punctate below from the collapsed glands; midrib slightly depressed above, prominent below; lateral nerves about 8 on each side of the midrib, visible below; petiole $\frac{3}{4}$ lin. long; flowers diœcious or monœcious; spikes unisexual or the upper part male, the lower female, very slender, axillary, solitary, from about one-third to half as long as the leaves, with the glabrous and rather stout rhachis clearly visible between the flowers; male bracts very broadly ovate, strongly concave inside, rounded at the apex, $\frac{1}{2}$ lin. long, glabrous outside, minutely ciliolate; stamens 4–5; anthers verruculose, $\frac{1}{3}$ in. long; female bracts thinner than the male; scales rounded, ciliolate; ovary small, glabrous; style-arms divergent, flattened, acute, $\frac{3}{4}$ lin. long; fruits ellipsoid-globose, about 3 lin. long, closely warted, warts waxy. *Drège, Zwei Pfl. Documente,* 106; *A. Cheval. Monogr. Myricac.* 154.

COAST REGION : Cape Div. ; Rondebosch to Hout Bay, below 1000 ft., *Drège*, a !
Caledon Div. ; mountains near the mouth of Klein River, *Zeyher*, 3875 ! Rivers-
dale Div. ; Langeberg, above Platte Kloof, *Muir in Herb. Galpin*, 5337 !
Humansdorp Div. ; Zitzikama, about 500 ft., *Galpin*, 4581 partly ! (in *Nat
Herb. Pretoria*).

According to Galpin, this species is known as "*Lottering Bush.*"

3. M. dregeana (A. Cheval. Monogr. Myricac. 155) ; young

branchlets densely leafy, softly pilose-tomentose; leaves oblong-
oblanceolate, acutely acuminate, narrowed to or somewhat rounded
at the base, 1–2½ in. long, ⅓–1 in. broad, rigidly chartaceous, glabrous
except the pubescent midrib below ; midrib prominent below ;
lateral nerves 6–9 on each side of the midrib, very slender and
slightly prominent below ; veins close but rather obscure on both
surfaces ; petiole about ½ lin. long, pubescent ; flowers diœcious ;
male spikes up to ⅖ in. long, axillary, solitary ; rhachis hidden by
the flowers, crisped-pubescent ; bracts rhomboid-ovate, 1 lin. long, a
little longer than broad, very shortly ciliolate in the upper half,
glabrous outside ; stamens 2, on very short filaments ; anthers
minutely pitted ; female flowers and fruits not known.

COAST REGION : George Div. ; by the Nuakamma River, *Burchell*, 5104 ! Uitenhage Div. ; Van Stadensberg Range, 1000–3000 ft., *Ecklon & Zeyher*, 4 ! *Burchell*,
4695 !

Chevalier quotes "Drège 4," but the specimen was undoubtedly collected by
Ecklon and Zeyher.

4. M. humilis (Cham. & Schlecht. in Linnæa, vi. 535) ; stems

erect, subsimple or with one or two branches in the upper part,
stout, about 2¼ lin. thick, clothed with scattered sessile glands and
long slender spreading hairs, more or less ridged ; leaves shortly
petiolate, fairly dense, oblong or oblong-elliptic, rounded at the base,
rounded to an obtusely mucronate apex, ¾–1¾ in. long, ¼–1 in.
broad, entire or sparingly crenate-serrate in the upper third, sparingly
pilose with long weak hairs on both sides of the midrib and around
the margin at the base, fairly densely covered with golden glands
below, the latter falling off in the fruiting stage ; midrib conspicuous
on both surfaces, more prominent below ; lateral nerves 7–10 on
each side of the midrib, diverging at a wide angle, some of them
bifurcate about half-way towards the margin, prominent below ;
veins fairly close and delicate on the lower surface ; petiole 1–1¼ lin.
long, broadened at the base, strongly wrinkled and pilose when dry ;
flowers diœcious ; male inflorescence axillary, solitary, ¾ in. long at
the time of flowering, with conspicuous spreading bracts, rather
dense-flowered ; rhachis densely pubescent ; bracts at the base of
each flower solitary, spathulate-obovate, 2 lin. long, 1¼ lin. broad,
submembranous, densely glandular outside, long-ciliate ; stamens
4–6 ; filaments connate in the lower half ; anthers large, ¾ lin. long,
sparingly pubescent ; female inflorescence much exceeding the leaves,
up to 2 in. long ; rhachis fairly slender, densely covered with sessile

glands and long scattered spreading hairs ; bracts solitary, lanceolate, about 1¾ lin. long, glandular, ciliate ; hypogynous scales 3, ovate-triangular, subacute, keeled, ⅓ lin. long, long-ciliate, coriaceous; ovary ovoid, densely warted ; styles longer than the bracts, free to the base ; fruits subglobose, 2½–3 lin. in diam., very densely warted. *C. DC. in DC. Prodr.* xvi. ii. 150 ; *Bolus & Wolley-Dod in Trans. S. Afr. Phil. Soc.* xiv. 320 ; *A. Cheval. Monogr. Myric.* 158.

SOUTH AFRICA : without precise locality, *Harvey! Desmaret! Bergius, Mund & Maire, Villet,* ♀, *Hooker,* 414 ♂ !
COAST REGION : Cape Div. ; Table Mountain, *Burchell,* 618 ! *Worsdell! Cooper,* 3484! *Galpin,* 4583, ♀! 4584! above Simons Town, *Bolus,* 4948, ♂! between Table Mountain and Devils Mountain, *Pappe!* Muizenberg, *Phillips,* 1699! Swellendam Div. ; Tradouw Mountains, *Bowie!* Zuurbraak Mountain, 3000 ft., *Galpin,* 4582! Humansdorp Div. ; Zitzikama, 500 ft., *Galpin,* 4581 partly! (in *Nat. Herb. Pretoria*)!

5. **M. elliptica** (A. Cheval. Monogr. Myricac. 166, t. viii. C, figs. 1–8 and 10); a shrub 1½ ft. high ; branches villous for some time with long weak pale yellow hairs, rather densely leafy to the base ; leaves oblong-elliptic or ovate-elliptic, rounded or truncate at the base, rounded in outline at the apex but with an acute terminal tooth ; those on the older branchlets about ¼ in. long and ⅓ in. broad, on the younger branchlets about 1 in. long and ⅔–¾ in. broad, rather acutely crenate-serrate, conspicuously nerved, rigidly chartaceous, pilose mainly on both sides of the midrib ; lateral nerves about 9 on each side of the midrib, spreading from it at a wide angle, conspicuous below ; veins not visible ; petiole 1–1½ lin. long, shaggy-pilose ; flowers (according to *Chevalier*) diœcious ; male spikes short, ovoid, 2½ lin. long ; bracts ovate, rounded at the apex, ciliolate ; stamens 4, anthers nearly sessile ; female flowers and fruits not known.

SOUTH AFRICA : without precise locality, *Burmann (Herb. Delessert).*
COAST REGION : George Div. ; Cradock Berg, near George, *Burchell,* 5903 !

6. **M. kraussiana** (Buching. ex Krauss in Flora, 1845, 89); branches fairly densely leafy, subsimple, angular or ribbed, 1½–2 lin. thick, conspicuously lenticellate in the lower part, rather densely glandular ; leaves subsessile, elliptic or oblong-elliptic, rounded at both ends, 1–1¾ in. long, ½–1 in. broad, entire or very slightly dentate towards the apex, very rigidly and thickly coriaceous, more densely glandular above than below ; midrib fairly broad and slightly prominent below ; lateral nerves 7–11 on each side of the midrib, spreading, forked towards the margin, nearly equally distinct on both surfaces ; flowers probably diœcious, neither sex seen ; male spikes (according to *Krauss*) axillary, solitary, sessile, oblong, one-third to half as long as the leaves ; bracts broadly triangular, spreading, glabrous, ciliolate ; stamens 4, anthers subsessile ; infructescences axillary, solitary, 2 in. long ; rhachis stout, about ¾ lin. thick, ribbed when dry, very densely covered with small

golden glands; bracts thick and leathery, lanceolate from a rounder base, 1–1½ lin. long, densely covered with small golden coloured glands; fruits globose, 2¾–3 lin. in diam., densely covered with warts, the warts covered with scattered glands.

COAST REGION: Cape Div. ; among rocks at the top of Steen Berg, 3000 ft., *Krauss*, 1564! Caledon Div. ; Zwart Berg, *Bowie*!

7. **M. brevifolia** (E. Meyer ex C. DC. in DC. Prodr. xvi. ii. 150); stems erect from a woody rhizome, simple, dark-purple when dry, longitudinally ribbed or angular, glandular-puberulous when young, at length glabrous; leaves fairly densely arranged, ascending, shortly petiolate, oblong lanceolate, subacute at both ends, ¾–1½ in. long, ¼–⅔ in. broad, with 3–5 rather coarse obtuse teeth on each side in the upper half, rarely a few subentire, rigidly chartaceous, shining and closely black-spotted, but at first glandular above, covered with lemon-yellow or orange glands below, the glands with interspaces about their own width; midrib prominent on both surfaces, becoming very slender towards the apex of the leaf; lateral nerves 7–10 on each side of the midrib, diverging from the midrib at a wide angle, mostly rather obscure; petiole 1 lin. long, some-times hairy; flowers diœcious; male inflorescence axillary, solitary, sessile, about one-third the length of the leaves or less, densely flowered; rhachis fairly stout, nearly glabrous, with 3–4 series of imbricate ovate-triangular obtuse bracts at the base; bracts at the base of each flower solitary, broadly obovate-spathulate, scarcely 1 lin. long and ½ lin. broad, membranous-chartaceous, glabrous except at the very shortly ciliolate apices; stamens usually 2; filaments about ⅔ lin. long, connate for two-thirds of their length; anthers ⅓ lin. long; female inflorescence solitary, axillary, about a quarter the length of the leaves; rhachis densely covered by the flowers; bracts solitary at the base of each flower, very broadly ovate, almost truncate at the top, submembranous, ¾ lin. long, glabrous except for the shortly ciliolate margin; hypogynous scales overlapping, suborbicular, ciliate, about ⅓ lin. broad; ovary glabrous; styles flattened, free to the base, 1¼ lin. long, glabrous. *Drège, Zwei Pfl. Documente*, 45 ; *A. Cheval. Monogr. Myricac.* 158.

COAST REGION: Stutterheim Div. ; Summit of Dohne Peak, 4600 ft., *Galpin*, 2458! Queenstown Div.; Winter Berg, 5000–6000 ft., *Zeyher*, 5! Stockenstrom Div. ; Kat Berg, grass fields, 4000–5000 ft., *Drège*, a!

CENTRAL REGION: Somerset Div. ; at the top of the Bosch Berg, 4500 ft., *MacOwan*, 1925!

KALAHARI REGION: Transvaal ; Belford, 6000 ft., *Worsdell*!

According to MacOwan, this *Myrica*, owing to continual burning over from grass fires, forms a creeping underground stem often an inch thick.

8. **M. ovata** (Wendl. f. in Bartl. & Wendl. Beitr. ii. 3); a glaucescent shrub about 2 ft. high; branches rather elongated, densely leafy, rather stout, ribbed, marked rather sparingly with small lenticels, dark coloured and with a glaucous "bloom," other-

wise glabrous; leaves oblong or ovate-oblong, rounded or very
slightly cordate at the base, subacute at the apex, 1–2 in. long, $\frac{1}{3}$–1
in. broad, rigidly coriaceous, entire, finely reticulate and glabrous
above, slightly glaucous below and with a few scattered golden
glands ; midrib prominent below ; lateral nerves 6–9 on each side of
the midrib, slightly prominent below, branched towards the margin ;
petiole very short, scarcely $\frac{1}{2}$ lin. long ; flowers diœcious ; male spikes
axillary, solitary, $\frac{3}{4}$–1 in. long, rather dense flowered ; rhachis
glabrous ; bracts shortly stipitate, very broadly triangular, auriculate, glabrous outside, shortly and rather closely ciliolate ; stamens
4 ; anthers with short filaments ; female spikes about $\frac{1}{3}$ in. long ;
bracts imbricate, broadly triangular, $\frac{2}{3}$ lin. long, 1 lin. broad, submembranous, glabrous outside, shortly ciliolate ; hypogynous scales
orbicular, ciliate ; ovary glabrous ; style-arms suberect, acute.
A. Cheval, Monogr. Myricac. 157.

COAST REGION : George Div. ; near the Touw River, *Burchell,* 5739 ! Knysna
Div. ; near Knysna, *Bowie! Burchell,* 5486 ! Uniondale Div. ; near the Keur-
booms River, *Burchell,* 5143 ! Humansdorp Div.; grassy hills near Humansdorp,
Kennedy ! Uitenhage Div. ; Witteklip and Vanstadens Berg, *MacOwan* !

9. **M. myrtifolia** (A. Cheval. Monogr. Myricac. 155, t. vii. fig.
A) ; older branches fairly stout, covered with greyish-brown bark ;
younger ones densely leafy, minutely puberulous ; leaves oblong or
oblong-elliptic, rather abruptly acute, rounded at the base, $\frac{3}{4}$–1$\frac{1}{2}$ in.
long, $\frac{1}{4}$–$\frac{1}{2}$ in. broad, very obscurely crenulate, coriaceous, glabrous
and finely reticulate on both surfaces ; midrib prominent below ;
lateral nerves about 5 on each side, slightly raised on the lower
surface ; petiole about $\frac{1}{2}$ lin. long, very thick ; flowers diœcious ;
male spikes axillary, solitary, one-third to a half as long as the
leaves ; rhachis angular, glabrous ; bracts stipitate, very broadly
triangular, auriculate in the middle, 1$\frac{1}{4}$ lin. long, 1$\frac{1}{2}$ lin. broad,
membranous towards the margin, shortly ciliate ; stamens 5 ; fila-
ments short ; anthers $\frac{3}{4}$ lin. long, thick ; female flowers not seen ;
fruiting spikes (according to *Chevalier*) 1–1$\frac{1}{4}$ in. long ; fruits 1$\frac{1}{2}$–2
lin. in diam., subspherical ; endocarp bony, verruculose, otherwise
glabrous.

COAST REGION : Knysna Div. ; near Knysna, *Bowie* !

10. **M. linearis** (C. DC. in DC. Prodr. xvi. ii. 154) ; branches
erect, marked with small distinct scattered lenticels, longitudinally
wrinkled and minutely pubescent, 2–2$\frac{1}{2}$ lin. thick ; branchlets
leafy, ascending, those of the male softly pubescent, the female
glabrous or pubescent ; leaves linear or narrowly linear-lanceolate,
subobtuse, narrowed at the base into a long slender petiole, 1$\frac{1}{2}$–3$\frac{1}{2}$
in. (usually about 2$\frac{3}{4}$ in.) long, 3–8 lin. broad, thinly coriaceous,
mostly entire but sometimes a few with 2–4 triangular teeth in
the upper part, sometimes with a few scattered hairs but mostly
glabrous on the upper surface, closely reticulate, with a few scattered

glands below which soon fall off; midrib fairly slender, conspicuous
on both surfaces; lateral nerves numerous, branched towards the
margin, conspicuous on both surfaces; veins forming a close distinct
network below; petiole ¼ to nearly ½ in. long, pubescent or glabrous
like the branchlets; flowers diœcious; male inflorescences axillary,
solitary, usually as long as or a little longer than the petiole when
mature; rhachis fairly stout, flexuous, glabrous; bracts solitary,
broadly obovate, obtuse, about ¾ lin. long, submembranous, shortly
ciliolate, otherwise glabrous; stamens 3–4; filaments connate in
the lower half; anthers ½ lin. long, oblong; female inflorescence
little longer or twice as long as the petiole; rhachis stout, pubescent;
bract solitary, broadly rhomboid, subacute, ¾ lin. broad, subchartaceous, ciliolate in the upper half; hypogynous scales elliptic,
fleshy, ciliate at the tips, enclosing the ovary; ovary glabrous;
styles spreading, free to the base, linear-filiform, ½ lin. long; in-
fructescence few-fruited, up to ¾ in. long; fruits sessile on the
rhachis, globose, 1 lin. in diam., strongly verrucose. *A. Cheval.
Monogr. Myricac.* 148. *M. æthiopica, Drège, Zwei Pfl. Documente*,
99; *not of Linn. M. conifera, var. subintegra, A. Cheval. l.c.*

COAST REGION: Clanwilliam Div.; near Brakfontein, *Pappe*! *Zeyher*! Clan-
william, *Zeyher*! *Schlechter*, 8026! Tulbagh Div.; Winter Hoek, *Pappe*! Paarl
Div.; between Paarl and Lady Grey Railway Bridge, below 1000 ft., *Drège*!
Stellenbosch Div.; Riverside, *Worsdell*!

11. **M. glabrissima** (A. Cheval. Monogr. Myricac. 156); a shrub
3 ft. high with glaucous foliage; branches spreading, rather densely
leafy, glabrous, marked here and there with lenticels; leaves
narrowly oblanceolate, subacute at the apex, gradually tapered into
the petiole at the base, 1–2 in. long, ⅛–⅔ in. broad, rigidly coriaceous,
entire or rarely obscurely 1–3-toothed, glabrous except for a few
scattered glands below in a young state; midrib slightly prominent
on the lower surface; lateral nerves and veins obscure; petiole 1–3
lin. long; flowers diœcious; male spikes axillary, about ½ in. long,
dense-flowered; rhachis ribbed, glabrous or nearly so; bracts very
broadly ovate-orbicular or ovate-triangular, obtuse, slightly keeled,
¾ lin. long, 1 lin. broad, glandular up the back, shortly ciliate on
the margin; stamens 4; anthers ⅔ lin. long; normal female flowers
not seen.

COAST REGION: Riversdale Div.; Langeberg Range above Platte Kloof, *Muir*,
387! George Div.; Long Kloof, above the source of Keurbooms River, in a rocky
kloof, *Burchell*, 5081! Port Elizabeth Div.; upper part of Maitland River,
Burchell, 4625!

12. **M. Zeyheri** (C. DC. in DC. Prodr. xvi. ii. 149); a densely
leafy shrub 3–4 ft. high; branches ascending, more or less distinctly
ribbed, softly pubescent; leaves spathulate-oblanceolate, gradually
narrowed at the base into a rather long petiole, with a triangular
obtuse apex, 1–1½ in. long, ¼–½ in. broad, rigidly coriaceous, obscurely

and remotely repand-dentate with 2–4 teeth on each side, minutely pubescent on the upper surface, rather densely golden-glandular below, slightly pubescent mainly on the midrib, the latter and the lateral nerves very obscure; petiole about $\frac{1}{4}$ in. long, softly pubescent; flowers diœcious; male spikes axillary, solitary, about as long as or slightly longer than the petiole, densely flowered; bracts very broadly ovate-triangular, about $1\frac{1}{4}$ lin. long, glandular outside towards the apex, slightly ciliate; stamens 4; anthers subsessile, $\frac{3}{4}$ lin. long, very minutely puberulous; female spikes about half as long as the leaves, rather slender and lax-flowered; rhachis densely hirsute with slightly reflexed hairs; bracts ovate-oblong, obtuse, rather thin, 1–1$\frac{1}{4}$ lin. long, slightly pubescent outside towards the tips; scales pubescent; ovary small, glandular; styles rather slender, about 1 lin. long, acute; fruits not seen. *A. Cheval. Monogr. Myricac.* 161.

Coast Region : Caledon Div. ; mountains near Caledon, *Zeyher*, 3878 !

13. **M. quercifolia** (Linn. Sp. Pl. ed. i. 1025) ; a low spreading bush usually about 1 ft. high; branches densely leafy, grooved, furfuraceous-pubescent and often glandular; leaves mostly imbricate, shortly petiolate, spathulate-oblanceolate in outline, gradually narrowed to the base, obtuse at the apex, mostly pinnatisect or very coarsely repand-dentate, the lobes or teeth spreading from the midrib almost at right angles, $\frac{1}{2}$–1$\frac{1}{2}$ in. long, $\frac{1}{4}$–$\frac{2}{3}$ in. broad, rigidly coriaceous, reticulate above, densely glandular and at length pitted below, usually otherwise glabrous but sometimes with a few scattered hairs on both surfaces; midrib distinct on both surfaces, prominent below; lateral nerves fairly stout and conspicuous below; flowers diœcious; male inflorescence axillary, solitary, usually about one-third the length of the leaves; rhachis densely flowered, stout, glabrous or pubescent; bract solitary below each flower, transversely elliptic or subrhomboid, about 1 lin. broad, membranous on the margin, shortly ciliate, otherwise glabrous or with a few glands outside; stamens 2–4; filaments connate in the lower half ; anthers ellipsoid-globose, $\frac{1}{2}$ lin. long, papillose; female inflorescence axillary, shorter and more slender than the males; rhachis fairly stout, rather densely glandular; bracts solitary, broadly ovate-orbicular, scarcely $\frac{3}{4}$ lin. broad, shortly ciliolate, with a few sessile glands on the back; hypogynous scales connate in the lower half, enveloping the ovary, shortly ciliate, otherwise glabrous, somewhat fleshy; ovary small, sparingly glandular; styles filiform, nearly 1 lin. long, free to the base; fruits solitary or paired, globose, 2$\frac{1}{2}$ lin. in diam., verrucose, covered with a white waxy excretion. *Willd. Enum. Pl. Hort. Berol.* ii. 1012; *Thunb. Fl. Cap. ed. Schult.* 159; *Lam. Encycl.* ii. 593; *Nouv. Duham. Traité* ii. 193; *Jacq. Fragm.* 2, *t.* i.; *DC. Prodr.* xvi. ii. 148; *Drège, Zwei Pfl. Documente*, 98, 106, 132; *A. Cheval. Monogr. Myricac.* 161; *Marloth, Fl. S. Afr. t.* 23, *fig. A*, 1–2. *M. hirsuta, Mill. Gard. Dict. ed.* viii. *n.* 6. *M. laciniata, Willd. Enum. Hort.*

Berol. 1012.   M. *ilicifolia, Burm. f. Prodr. Cap.* 31.   M. *quercifolia,*
vars. multiformis, microphylla, hirsuta, ilicifolia and *latifolia,* A.
Cheval. l.c. 163–166, *fig.* 20.—*Laurus africana minor, etc., Commelyn,*
Hort. ii. 161, *t.* 81 ; *Ray, Dendr.* 85.   *Coriotragematodendros*
africana, Botryos, etc., Plukenet, Almatheum, 65.

SOUTH AFRICA : without precise locality, *Burmann* ! *Villet* ! *Pappe* ! *Wallich* !
Roxburgh ! *Petit Thouars* ! *Bowie* ! *Lehmann* ! *Boivin* ! *Vaillant* ! *Masson* ! *Muna &*
Maire ! *Harvey* !
COAST REGION : Paarl Div. ; Achter de Paarl, *Drège,* a ! Cape Div. ; various
localities, Table Mountain, etc., *Verreaux, Ecklon,* 549 ! *Kolbe,* 2478, partly !
Pappe ! *Zeyher,* 1553 ! *Burchell,* 724 ! *Wilms,* 3633 ! 3634 ! *Drake* ! *Boivin,* 536 !
Krauss ! *Phillips,* 4295 ! *Wolley-Dod,* 2627 ! *Dümmer,* 72 ! *Worsdell* ! Caledon
Div. ; Zwartberg Range, *Zeyher,* 3879 ! George Div. ; by the Nuakamma rivulet,
Burchell, 5103 ! Uitenhage Div. ; Addo, *Drège,* c ! Vanstadens Berg, *Zeyher* !
Algoa Bay, *Zeyher,* 749 ! Humansdorp Div. ; between Tweefontein and Essenbosch,
Burchell, 4827 ! north side of Kromme River, *Burchell,* 4878 ! Albany Div. ;
hills near Grahamstown, *Bolus,* 1274 ! *Bowie,* 495 ! eastern side of Zwartwater
Poort, *Burchell,* 3433 ! Komgha Div. ; grassy slopes near Keimouth, *Flanagan,*
2582 !

14. M. conifera (Burm. f. Prodr. Cap. 31) ; a shrub or moderate-
sized tree ; branches softly and shortly tomentose, glabrous or
glandular, lenticellate, leafy nearly their whole length ; leaves
oblanceolate or linear-oblanceolate, long-attenuated into the petiole
at the acute base, subacutely triangular and mucronate at the apex,
2½–4½ (usually about 4) in. long, ½ 1 in. broad, rigidly chartaceous,
entire or rather distantly and coarsely serrate in the upper half or
two-thirds, closely reticulate and punctulate above, minutely glandular below, shortly pubescent towards the base of the midrib ;
midrib prominent on both surfaces, narrow ; lateral nerves usually
more conspicuous above than below, 12–15 on each side of the
midrib, diverging at a fairly wide angle, usually 2-furcate ; veins
scarcely conspicuous ; petiole 3–4 lin. long, puberulous ; flowers
diœcious or monœcious ; male spikes clustered or solitary, densely
flowered, more or less 1 in. long ; rhachis rather densely pubescent ;
bracts very broadly obovate, with an auricle on each side towards
the base, 1 lin. long and broad, submembranous, glandular on the
back towards the middle, ciliate ; stamens 4 or 5 ; filaments unequally connate in the lower half ; anthers rounded, ⅓ lin. long,
minutely papillose ; female inflorescence densely flowered, clustered
or solitary, ½ in. long or less ; rhachis tomentulose ; bracts very
broadly triangular-ovate, subacute, ⅔ lin. long, 1 lin. broad, thinly
chartaceous, pubescent and glandular outside ; bracteoles fleshy,
broadly obovate, ciliate, pubescent outside ; ovary pubescent ; styles
shortly connate at the base, slender, nearly 1 lin. long ; fruits
ellipsoid-globose, 2¼ lin. long, strongly warted. A. *Cheval. Monogr.*
Myricac. 144, *incl. vars. tomentosa,* A. *Cheval., banksifolia,*
A. *Cheval.,* and *glabra,* A. *Cheval. l.c.* 147–8 ; *Hutchinson in*
Prain, Fl. Trop. Afr. vi. ii. 314.   M. *æthiopica, Linn. Mant.*
Alt. 278 ; *Engl. Jahrb.* xxx. 291 ; *R. E. Fries, Rhod.-Kongo-*
Exped. 12.   M. *serrata, Lam. Encycl.* ii. 593.   M. *capensis, Hort.*

ex Steud. Nom. ed. ii. 173.   *M. banksiæfolia, Wendl. Collect.* i. 70.
M. natalensis, C. DC. in DC. Prodr. xvi. ii. 148 ; *A. Cheval. l.c.* 149.

SOUTH AFRICA : without locality ; *Brown* ! *Frazer* ! *Harvey,* 403 ! *Banks & Solander* ! *Masson* !

COAST REGION : Worcester Div. ; banks of streams in Hex River Valley, *Tyson,* 755 ! Paarl Div. ; Berg River, *Alexander* ! various localities, *Burchell,* 40 ! *Ecklon,* 42 ! *Worsdell* ! *Pappe* ! *Wallich* ! *Kolbe,* 2478, partly ! *Wolley-Dod,* 2375 ! *Bolus,* 2938 ! *Dümmer,* 1473 ! *Robertson* ! *Nelson* ! Stellenbosch Div. ; Hottentots Holland, *Pappe* ! Swellendam Div. ; mountains near Swellendam, *Bowie* ! Caledon Div. ; Houw Hoek, *Burchell,* 8109 ! *Schlechter,* 7365 ! mountains near Genadendal, *Burchell,* 8621 ! Knysna Div. ; Vlugt Valley, *Bolus,* 2468 ! Knysna, *Bowie* ! Uniondale Div. ; rocky hill near Haarlem, *Burchell,* 5026 ! Humansdorp Div. ; Diep River, *Bolus,* 2469 ! Uitenhage Div. ; various localities, *Burchell,* 4270 ! *Zeyher,* 128 ! 3876 ! *Drège* ! *Mund* ! Port Elizabeth Div. ; Krakakamma, *Mac-Owan,* 1080 ! Bathurst Div. : mouth of the Kowie River, *Atherstone* ! Albany Div. ; along the rivulet at Grahamstown, *Burchell,* 3554 ; bank of river south of Signal Hill, Grahamstown, *Galpin,* 2921 ; *Atherstone,* 18 ! *MacOwan,* 1407 ! Stutterheim Div. ; Fort Cunninghame, 3000 ft., *Sim,* 2029 !

EASTERN REGION : Transkei ; Kentani district, *Pegler,* 883 ! East Griqualand ; banks of the Umzimkulu River, *Tyson,* 2556 ; Natal ; Inanda, *Wood,* 985 ! near Durban, *Wood,* 9956 ! Friedenau Farm, *Rudatis,* 399 ! Ifafa Valley, *Rudatis,* 1422 ! and without precise locality, *Gerrard,* 103 ! 1659 ! *Gueinzius,* 55 !

KALAHARI REGION : Transvaal ; near Barberton, *Thorncroft,* 620 ! 4353 ! near Lydenburg, *Wilms,* 1353 ! Macamac, *Burtt-Davy,* 1454 !

Occurs also in Angola, East Africa, Nyasaland and Rhodesia.

15. **M. incisa** (A. Cheval. Monogr. Myricac. 150) ; older branches leafless, covered with light grey bark ; branchlets short, glandular, slightly villous with scattered hairs ; leaves lanceolate-oblong or lanceolate-linear, rounded or obtuse at the apex, long wedge-shaped at the base, 1¼–1¾ in. long, about ⅓ in. broad, deeply dentate, some-times irregularly so, with 3–5 nerves on each side of the midrib, covered with little glands on the upper surface, scarcely reticulate, reddish below with a great number of glands, furnished with a few hairs on the nerves, the midrib very conspicuous, the lateral nerves rather conspicuous ; petiole 1½–2 lin. long, puberulous ; male inflores-cences ⅓–½ in. long, with a glabrous rhachis ; bracts ciliate, devoid of glands ; stamens 4 ; filaments short ; anthers puberulous ; females not known. *Myrica foliis oblongis, opposite sinuatis, J. Burm. Rar. Afr. Pl.* 262, *t.* 98, *fig.* 1.

SOUTH AFRICA : without definite locality, *Burmann* (*Herb. Delessert*).
COAST REGION : Caledon Div. ; Zwartberg, *Worsdell* !

I have not seen the original of this species ; Chevalier says it is exactly inter-mediate between *M. quercifolia* and *M. conifera,* and suggests that it may be a hybrid from these two species. Mr. Worsdell's specimen from the Zwartberg agrees very well with the figure in Burmann ; in this figure there are infructescences, a fact not mentioned by Chevalier in his description, of which the above is a translation.

ORDER CXXIV. **BETULACEÆ.**

By S. A. SKAN.

Flowers monœcious, the male in pendulous catkins, the female in erect or pendulous catkins or in budlike heads. *Male flowers: Perianth* simple, membranous, 4- or fewer-lobed, sometimes 0. *Stamens* 2–12; filaments free, usually as well as the anthers more or less split; anther-cells 2, erect, parallel, sometimes pilose at the apex, dehiscing longitudinally. *Ovary* rudiment none. *Female flowers: Perianth* minute or sometimes nearly completely adnate to the ovary, often 0. *Ovary* inferior, crowned with the annular or toothed perianth or naked, 2-celled, often more or less laterally compressed; styles 2, free, filiform. *Ovules* 1 in each cell, pendulous, anatropous, provided with 1 integument. *Fruit* a nut or nutlet, ovoid-globose or usually laterally compressed, sometimes winged, more or less enveloped in the variously connate often much accrescent herbaceous membranous or woody bracts and bracteoles. *Seed* usually solitary by abortion, exalbuminous; testa membranous; cotyledons fleshy, smooth, rugose or ruminate; radicle short, superior.

Trees or shrubs; leaves alternate, pinnately veined, toothed, sometimes more or less lobed, rarely entire; stipules caducous; inflorescences terminal or lateral; flowers 1–3 in the axils of scale-like or sometimes leafy bracts, appearing in spring with or before the leaves or rarely in the autumn.

DISTRIB. Genera 6 and species about 80, chiefly in Europe, Temperate Asia and North America, 1 in North and South Africa and a few in Central and South America.

Betulaceæ is here defined as in Winkler's Monograph in Engler's Das Pflanzenreich. It comprises the tribes *Betuleæ* and *Coryleæ* of *Cupuliferæ* in Bentham and Hooker's Genera Plantarum.

I. ALNUS, Linn.

Male flowers usually 3 to each bract of the catkin; bracteoles 3–5, more or less united to the bract. *Perianth* sessile, deeply 4-lobed. *Stamens* 4, opposite to the perianth-lobes; filaments very short, undivided; anthers dorsifixed, ovate, not pilose at the apex. *Female flowers* usually 2 to each scale-like bract of a cylindric or ellipsoid erect spike; bracteoles 2–4, united to the base of the bract. *Perianth* 0. *Ovary* naked, sessile, compressed, 2-celled; styles 2, cylindric, short. *Ovules* solitary in each cell, pendulous; micropyle superior. *Fruiting-spikes* cone-like, ovoid or ellipsoid, with the somewhat accrescent bracts and bracteoles woody and persisting after the fall of the nutlets. *Nutlets* small, compressed, 2-winged or wingless. *Seed* 1 by abortion; cotyledons flat.

Trees or shrubs, often growing in wet places ; leaves alternate, deciduous, toothed, sometimes slightly or deeply lobed, rarely entire ; stipules enclosing the leaf in bud, caducous ; catkins or spikes 2–4 or more, usually distinctly pedunculate, racemosely arranged in terminal or axillary clusters.

DISTRIB. Species 17, chiefly in Europe, Temperate Asia and North America, 1 in North and South Africa and a few in Central and South America.

1. **A. glutinosa** (Gærtn. De Fruct. ii. 54, t. 90, fig. 2) ; usually a small tree, but sometimes reaching a height of 50–60 or even 100 ft., often a large bush ; bark in young trees smooth and greenish, afterwards brownish-black and rough ; branchlets usually glabrous, glandular ; leaves broadly obovate, suborbicular or sometimes ovate-elliptic, $1\frac{1}{2}$–4 in. long, $1\frac{1}{4}$–$3\frac{1}{4}$ in. broad, rounded, truncate or retuse at the apex, usually cuneate at the base, sometimes slightly lobed, callose-serrate, dark green, glabrous and shining above, light green, hairy along the principal veins and in their axils beneath, glutinous when young ; petiole up to $1\frac{1}{2}$ in. long, glabrous or pubescent ; stipules ovate to lanceolate, obtuse, glandular-hairy on the margin ; catkins 3–6, racemosely arranged on short terminal or subterminal branchlets ; male catkins pendulous, cylindric, $1\frac{1}{4}$–4 in. long ; female catkins erect, at first cylindric, green, up to 5 lin. long, when mature ovoid, cone-like, black, 5–9 lin. long ; nutlets obovate, about $1\frac{1}{4}$ lin. long, wingless or with a narrow coriaceous wing. *Harv. Gen. S. Afr. Pl. ed.* 2, 347 ; *Sowerby, Engl. Bot. ed. Syme,* viii. 178, *t.* 794 ; *Boiss. Fl. Orient.* iv. 1180 ; *Battand. & Trab. Fl. Algér.* 818 ; *Winkler in Engl. Pflanzenr. Betulac.* 115 ; *Elwes & Henry, Trees Gt. Brit. & Irel.* iv. 937.   *Alnus* [*sp.*], *Drège, Zwei Pfl. Documente,* 99, 101.   *Betula Alnus glutinosa, Linn. Sp. Pl. ed.* i. 983. *B. Alnus, Scop. Fl. Carn. ed.* 2, ii. 233.   *B. glutinosa, Lam. Encycl.* i. 454.   For full synonymy see *Winkler, l.c.*

COAST REGION : Paarl Div. ; by the Berg River, near Paarl, and between Paarl and French Hoek, *Drège,* 8253 ! Cape Div. ? *Banks & Solander* !

"The common Alder (*Alnus glutinosa*) is found throughout the colony, apparently wild, but whether truly so or not I cannot say " (Harvey, Gen. S. Afr. Pl. ed. 2, 347).

Also in nearly the whole of Europe, Siberia, Western Asia and North Africa. Naturalised in the North-Eastern United States of America.

ORDER CXXV.   **SALICINEÆ.**

(BY S. A. SKAN.)

Flowers diœcious, one under each bract, in cylindric catkins or more rarely in racemes, ebracteolate.   *Perianth* 0.   *Disc* of 2 gland-like scales, one posterior, the other anterior, or one only and then posterior, sometimes cup-shaped, obliquely truncate, crenate or variously lobed.   *Male flowers :* stamens 2 to many ; filaments

free or connate ; anthers ovate or oblong, affixed at the base or at
the back near the base ; cells 2, distinct, parallel, dehiscing longitudinally ; rudiment of the ovary 0. *Female flowers* : ovary sessile
or shortly stalked, 1-celled ; placentas 2–4, parietal ; style short
or 0 ; stigmas 2–4, rather thick, emarginate or 2-fid and lobed ;
ovules 2 to many, in 2 to many series, ascending, anatropous. *Capsule* ovoid or lanceolate, 2–4-valved. *Seeds* few or many, small or
minute, each with numerous long silky hairs arising from the
funicle ; testa very thin ; albumen 0 ; cotyledons plano-convex ;
radicle short, inferior.

Trees or shrubs ; leaves alternate, entire, toothed or sometimes lobed, nearly
always deciduous ; stipules free, small, scale-like and deciduous or larger, leafy,
and persistent ; flowers in catkins ; catkins axillary and sessile or terminating
short branches, appearing before or with the leaves, pendulous or erect, often
silky-villous ; bracts membranous, caducous or sometimes in the female catkins
persisting till the ripening of the fruits.

DISTRIB. Genera 2 and species about 210, widely dispersed in the Arctic,
Temperate and Tropical Regions of both hemispheres, most frequent in Europe,
Temperate Asia and North America, usually on the banks of streams or in moist
places, sometimes in Alpine Regions ; absent from Australasia and the Pacific
Islands. *Salix* alone is indigenous in South Africa, but according to Marloth
(*Fl. S. Afr.* i. 130) *Populus canescens,* Smith, and *P. nigra,* Linn., var.
pyramidalis, Spach, are well acclimatised there. *P. canescens* has been met with
in the Transvaal (Burtt-Davy, 188).

I. SALIX, Linn.

Flowers dioecious. *Disc* of very small fleshy gland-like scales.
Male flowers: *Stamens* 2–8 or sometimes up to 12 ; filaments filiform, free or rarely more or less connate ; anthers ovate, usually
small. *Female flowers* : *Ovary* sessile or shortly stalked, 1-celled ;
placentas 2 ; style often short or none ; stigmas 2, retuse or 2-fid.
Ovules usually 4–8 on each placenta, 2-seriate. *Capsule* 2-valved.
Seeds of the order.

Trees or shrubs ; leaves alternate, entire or toothed, pinnately veined, often
narrow ; stipules free, small and deciduous or larger leafy and persistent ; flowers
in catkins ; catkins usually dense and erect, axillary and sessile or terminating
short branches, appearing before or with the leaves ; bracts small, entire or rarely
toothed.

DISTRIB. Species 180–190, with the range of the order.

S. fragilis, Linn., appears to have been introduced into the Cape Peninsula. A
specimen (no. 2517) without catkins, apparently correctly named, was collected
by Wolley-Dod at Orange Kloof.

A specimen from the Cape (Ecklon ? 714 in Herb. Horn.) has been identified by
Fries (Nov. Fl. Suec. Mant. i. 77) with *S. australis,* Hildenb. and Boj. I suspect
that it is the closely allied *S. capensis.*

Leaves glabrous or if pubescent when young soon glabrescent:
Leaves up to 6 lin. broad, usually much narrower; disc-gland very small; style none or very short:
Branchlets very slender and usually short; leaves $\frac{3}{4}$–2 in. long, rarely longer ... ... (1) **capensis.**

Branchlets much stouter and usually long; leaves up to $4\frac{1}{2}$ in. long, rarely less than $2\frac{1}{2}$ in. long... (2) **Woodii.**

Leaves up to $7\frac{1}{2}$ lin. broad; disc-gland in the female flower relatively long; style prominent ... ... (3) **crateradenia.**

Leaves densely clothed on the underside with a somewhat persistent grey silky or felt-like pubescence:
Leaves narrowly lanceolate to broadly linear, up to $4\frac{1}{2}$ in. long, usually more than $2\frac{1}{2}$ in. long ... ... (2) **Woodii,** var. **Wilmsii.**

Leaves oblong-lanceolate to lanceolate, usually 1–$1\frac{1}{2}$ in. long or less, rarely up to $2\frac{1}{4}$ in. long ... ... (4) **hirsuta.**

1. **S. capensis** (Thunb. Fl. Cap. i. 139, and ed. Schult. 31); a shrub or tree up to 50 ft. high or more, very much branched; branchlets very slender, sometimes more or less pubescent when young, quite glabrous, often shining and reddish to dark brown when older; leaves narrowly lanceolate or lanceolate, rarely broadly lanceolate, usually $\frac{3}{4}$–$1\frac{1}{4}$ in., rarely up to 2 in. long, $1\frac{1}{2}$–3 lin., rarely up to 6 lin. broad, often thin, becoming more or less coriaceous, acute to acuminate at the apex, more or less cuneate at the base, entire or closely or remotely serrulate, glabrous or sometimes slightly pubescent when young, green on both sides or green above and glaucescent beneath; petiole $\frac{1}{2}$–2 lin. long; stipules minute or 0; catkins appearing with the leaves; peduncles 2–12 lin. long, bearing 1–5 shortly stalked or subsessile leaves similar to the others but usually smaller; male catkins narrowly cylindric, $\frac{1}{2}$–$1\frac{1}{2}$ in. long; bracts ovate to elliptic or suborbicular, $\frac{3}{4}$–1 lin. long, $\frac{2}{3}$–$\frac{3}{4}$ lin. broad, more or less villous inside, villous outside or glabrous except at the base, sometimes nearly quite glabrous; disc-glands fleshy, scarcely $\frac{1}{4}$ lin. long; stamens 4–8 (usually 5 or 6); filaments villous below the middle; female catkins cylindric or ovoid, $\frac{1}{3}$–1 in. long; rhachis glabrous to densely villous; bracts as in the male, soon deciduous; disc-gland cup-shaped, nearly or quite surrounding the pedicel, about $\frac{1}{4}$ lin. long; pedicel $\frac{1}{2}$–$1\frac{1}{4}$ lin. long; ovary narrowly ovoid, 1–$1\frac{3}{4}$ lin. long, glabrous; style very short or none; capsule ovoid, $1\frac{1}{2}$–$2\frac{1}{2}$ lin. long. *Fries in Öfvers. Vet.-Akad Förhandl. Stockh.* xiii. (1856) 121; *Anderss. in Vet.-Akad. Handl. Stockh.* vi. (1867) *no.* 1, 13, *t.* 1, *fig.* 11, *and in DC. Prodr.* xvi. ii. 197; *Sim, For. Fl. Cape Col.* 328, *t.* 146 (*var. a, normalis, Sim*); *Marloth, Fl. S. Afr.* i. 130, *figs.* 72–73; *Skan in Prain, Fl. Trop. Afr.* vi. ii. 319. *S. ægyptiaca, Thunb. Prodr.* 6, *not of Linn.*

VAR. β, **gariepina** (Anderss. in Vet.-Akad. Handl. Stockh. vi. (1867) no. 1, 13, and in DC. Prodr. xvi. ii. 197); diffusely branched or the ultimate branches straight, slender and often 1–2 ft. long; leaves narrowly lanceolate to linear-lanceolate, usually $1\frac{3}{4}$–$2\frac{1}{2}$ in. long, often about 3 lin. but sometimes up to $5\frac{1}{2}$ lin. broad, long-acuminate, minutely serrulate; catkins erect or suberect; capsules usually larger. *Schinz in Bull. Herb. Boiss.* iv. *Append.* iii. 50 ; *Sim, For. Fl. Cape Col.* 329. *S. gariepina, Burch. Trav.* i. 317, *t.* 6 ; *Drège, Zwei Pfl. Documente,* 217 (*S. garipina*) ; *Pappe, Silva Cap.* 30 ; *Fries in Öfvers. Vet.-Akad. Förhandl. Stockh.* xiii. (1856) 121. *S. ægyptiaca ? Drège, l.c.* 96, *not of Linn. nor of Thunb.*

VAR. γ, **mucronata** (Anderss. in Vet.-Akad. Handl. Stockh. vi. (1867) no. 1, 14, excl. forma *pubescens,* and in DC. Prodr. xvi. ii. 198) ; leaves rather thicker, oblanceolate or lanceolate, often broadest near the apex, 9–10 lin. long, 2½–3½ lin. broad in the broadest part, subacute or rounded at the apex, mucronate, pallid beneath ; catkins densely villous. *Sim, For. Fl. Cape Col.* 329. *S. mucronata, Thunb. Prodr.* 6, *and Fl. Cap. ed. Schult.* 31 ; *Fries, Nov. Fl. Suec. Mant.* i. 76, *and in Öfvers Vet.-Akad. Förhandl. Stockh.* xiii. (1856) 120.

SOUTH AFRICA : without precise locality, *Masson* ! *Thom* ! *Oldenburg,* 967 ! 1155 ! Var. β : without precise locality, *Drège,* a ! Var. γ : without precise locality, *Thunberg, Ecklon,* 1256.

COAST REGION : Clanwilliam Div., *Zeyher* ! Tulbagh Div. ; near Tulbagh, *Thunberg.* Tulbagh Waterfall, *Pappe* ! New Kloof, 500 ft., *Schlechter,* 9016 ! 9017 ; Ceres Div. ; between Ceres and Leeuwfontein, 2000 ft., *Pearson,* 3255 ! Worcester Div. ; banks of streams on mountains near Worcester, 1500 ft., *MacOwan in MacOwan & Bolus, Herb. Norm. Austr.-Afr.* 1645 ! Uitenhage Div. ; *Zeyher,* 147 ! *Tredgold,* 35 ! Enon, on the Witte River, *Gill* ! Queenstown Div. ; Shiloh, 3500 ft., *Baur,* 935 ! Zwart Kei and Klipplaat Rivers, 2200–3500 ft., *Galpin,* 7873 ! British Kaffraria, *Cooper,* 223 ! Eastern Districts, *Cooper,* 48 ! Var. β : Clanwilliam Div. ; Wupperthal, *Drège,* d, *Wurmb.* Caledon Div. ; Genadenthal and Zondereinde River, *Drège,* e. Var. γ : Paarl Div. ; damp meadows by the Berg River near Paarl, below 500 ft., *Drège,* 8251 !

CENTRAL REGION : Calvinia Div. ; Hantam, *Thunberg.* Somerset Div. ; Somerset, *Bowker* ! between the Great and Little Fish Rivers, 2500 ft., *MacOwan,* 468 ! Graaff Reinet Div. ; on the banks of the Sundays River near Graaff Reinet, 2500 ft., *Bolus,* 468 ! Aliwal North Div. ? by the Orange River, *Atherley,* 109 ! Var. β : Aliwal North Div. ; banks of the Orange River near Aliwal North, 4300 ft., *Drège,* b ! Philipstown Div. ; Orange River near Petrusville, *Burchell,* 2669 ! Prieska Div. ; banks of the Orange River, *Burchell,* 1637 !

WESTERN REGION : Var. β : Great Namaqualand ; on the Orange River, *Fleck,* 308 A. Aris Drift, *Schenck,* 235, *Pohle.* Little Namaqualand ; on the banks of the Groen River, below 1000 ft., *Drège* ! by the Orange River at Ramans Drift, 700 ft , *Pearson,* 3111 ! river bed, south of Tweefontein, 2700 ft., *Pearson,* 3759 !

KALAHARI REGION : Transvaal ; Rustenburg District, *Miss Nation,* 302 ! Var. β : banks of the Vaal River, *Nelson,* 171 !

EASTERN REGION : Tembuland ; Bazeia, 2000 ft., *Baur,* 512 ! Natal ; Coast land, *Sutherland* !

Also in Rhodesia.

Drège's specimen from the banks of the Groen River, Little Namaqualand, labelled at Kew "*Salix* an *ægyptiaca,* L. ?" is very probably the form of the variety *gariepina* distinguished by Andersson (Vet.-Akad. Handl. Stockh. vi. (1867) no. 1, 14) as *axillaris,* in which the catkins are in the axils of persistent leaves. In the material at Kew the catkins are very young. Andersson's form *puberula* (l.c.) of the same variety may be represented by Drège's specimen d from Camdeboo, Aberdeen Division. I have seen only a sterile shoot, the leaves of which are markedly different from those of any form of *S. capensis.*

2. **S. Woodii** (Seemen in Engl. Jahrb. xxi. Beibl. 53, 53) ; a shrub up to 10 ft. high ; branches more or less pubescent or sometimes densely grey-tomentose when young, soon glabrescent, red-brown when older ; leaves narrowly lanceolate to broadly linear, 1½–4½ in. long, 2–6 lin. broad, long-acuminate, narrowed at the base, closely or sometimes remotely serrulate, rarely nearly quite

entire, sometimes covered with greyish or yellowish silky hairs
when very young and not fully developed, quickly glabrescent and
often glaucescent beneath, coriaceous, rather prominently veined;
petiole 1½-5 lin. long; stipules minute, caducous; catkins appearing
with the leaves, axillary, shortly pedunculate, subtended by 1–4
small lanceolate entire leaves; male: narrowly cylindric, ¾–1½ in.
long, 2–2½ lin. broad; bracts ovate, ¾–1½ lin. long, rounded to
acute at the apex, densely villous on both sides; disc-glands fleshy,
about ¼ lin. long, the upper cylindric (sometimes wanting), the
lower broad entire or lobed; stamens 4–9; filaments villous on the
lower half; female: cylindric, rather lax, ½–1¾ in. long; bracts as
in the male; disc-gland half or completely surrounding the pedicel,
fleshy, notched or lobed; pedicel ¾ lin. long, 1–1½ lin. long in fruit;
ovary narrowly ovoid, tapering at the apex, about 1½ lin. long,
glabrous; style 0; stigma 4-lobed; capsule ovoid, 1¼–2½ lin. long.
Wood, Handb. Fl. Natal, 121. *S. natalensis, Wimm. ex Anderss. in
Vet.-Akad. Handl. Stockh.* vi. (1867) *no.* 1, 14. *S. Zeyheri, Sond. ex
Fries in Öfvers. Vet.-Akad. Förhandl. Stockh.* xiii. (1856) 121;
Anderss. in Vet.-Akad. Handl. Stockh. vi. (1867) *no.* 1, 14 (*S. Zeiheri*).

VAR. β, **Wilmsii** (Skan); leaves more persistently clothed especially on the
underside with a dense grey silky indumentum, at length glabrescent and glaucescent, quite entire or sometimes very obscurely and remotely serrulate. *S. Wilmsii,*
Seemen in Engl. Jahrb. xxvii. *Beibl.* 64, 9.

KALAHARI REGION: Orange River Colony, without precise locality, *Cooper,*
3160! Transvaal; Bruderstroom, Houtboschberg, *Nelson,* 424! Houtbosch,
*Rehmann,*6509! Magalies Berg, *Burke,* 330! *Zeyher.* 1552! along the Vaal River,
near Bloemhof, *Burtt-Davy,* 1503! Crocodile River, *Miss Leendertz,* 715! *Burke*!
Lydenberg, at 5200 ft., *Schlechter,* 3938! Komati Poort, at 1000 ft., *Schlechter,*
11847! Wonderboompoort, Pretoria, *Rehmann,* 4532! Derde Poort, near Pretoria,
Miss Leendertz, 374! Var. β: near Lydenberg, *Wilms,* 1350! 1351, 1352.
Zoutpansberg District; close to Haenertsburg, common along the Bruderstroom,
Burtt-Davy, 1266! River side, Umlomati Valley, near Barberton, 4000 ft., *Galpin,*
1278!

EASTERN REGION: Natal; Drakensberg Mountains, near Newcastle, *Wilms,*
2276! Biggarsberg Range, *Rehmann,* 7055! on the banks of the Mooi River at
6500 ft., *Schlechter,* 3341! Rovelo Hills, 7000 ft., *Sutherland*! Weenen County,
banks of the Tugela River, *Rudatis,* 1580! near Colenso, *Wood,* 4970! near
Durban, *Gueinzius,* 136.

3. **S. crateradenia** (Seemen in Engl. Jahrb. xxvii. Beibl. 64, 9);
a tree with dark red-brown branchlets which are shortly grey-
tomentose at the ends; buds also tomentose; leaves lanceolate or
oblong, up to 2¾ in. long, 7½ lin. broad, acute, narrowed to the base,
remotely serrulate, leathery, yellow-green and somewhat shining
above, grey-green beneath; midrib paler, prominent on both sides;
lateral nerves irregularly curved, prominent; petiole about ¼ in.
long, at first shortly grey-tomentose, afterwards glabrescent; male
catkins not seen; female catkins appearing with the leaves,
terminating a shortly grey-hairy branchlet 5 lin. long bearing 3
or 4 oblong leaves up to 1 in. long and 5 lin. broad, broadly
ellipsoid, up to 10 lin. long, 7½ lin. thick, loosely flowered; rhachis

densely grey-tomentose; bracts wanting; capsule ellipsoid, obtuse, at first shortly grey-hairy, afterwards glabrescent; pedicels about two-thirds as long as the capsule; style distinct; stigmas short, thick, laterally curved, emarginate; glands about a quarter as long as the pedicel which they completely surround as a little fleshy cup, higher in front, emarginate above.

KALAHARI REGION : British Bechuanaland, *Passarge,* 41 of 1896.

Native name "*Machogari.*"

4. S. hirsuta (Thunb. Prodr. 6) ; a shrub about 6 ft. high ; branches at first densely whitish-villous, later glabrescent, brown and somewhat rugose; leaves oblong-lanceolate, sometimes ovate-lanceolate or lanceolate, the largest 2¼ in. long and 1 in. broad, usually 1–1½ in. long and 5–6 lin. broad, smaller on the catkin-bearing branchlets or peduncles, mucronate, acute or acuminate at the apex, rounded to cuneate at the base, entire or sometimes remotely and obscurely serrulate, thinly covered above and densely beneath with grey silky hairs, sometimes subglabrescent above ; petiole 1½–3 lin. long ; stipules brown, membranous, obliquely ovate, ¾–1 lin. long, silky-hairy, soon deciduous ; catkins appearing with the leaves ; peduncle 4–6 lin. long, bearing 2–4 subsessile lanceolate leaves, 6–9 lin. long and 2½–3 lin. broad ; male catkins cylindric, 8–15 lin. long, 2–3 lin. broad, densely white-tomentose on rhachis and bracts ; bracts ovate to obovate, about 1 lin. long and ¾ lin. broad ; disc-glands fleshy, about ¼ in. long, irregularly lobed, crenulate or emarginate ; stamens 3–7 ; filaments villous ; female catkins (in fruit) ovoid-cylindric, 6–12 lin. long, rather dense ; disc-gland rather more than ¼ lin. long, irregularly toothed, completely surrounding the pedicel ; pedicel ½–⅔ lin. long ; capsule ovoid, 1½–2 lin. long, 1¼ lin. broad, glabrous. *Thunb. Fl. Cap. ed. Schult.* 31 ; *Willd. Sp. Pl.* iv. 695; *Fries, Nov. Fl. Suec. Mant.* i. 77, *and in Öfvers. Vet.-Akad. Förhandl. Stockh.* xiii. ; (1856) 120 ; *Trautv. Salicetum,* 17 ; *Drège, Zwei Pfl. Documente,* 217 ; *Krauss in Flora,* 1845, 88. *S. capensis, var. hirsuta, Anderss. in Vet.-Akad. Handl. Stockh.* vi. (1867) *no.* 1, 14, *and in DC. Prodr.* xvi. ii. 198 ; *Sim, For. Fl. Cape Col.* 329. *S. mucronata, Drège, l.c.* 68, 70, 109, *partly, not of Thunb.*

VAR. β, **parvifolia** (Skan) ; leaves much smaller (only 5–8 lin. long and 1½–2 lin. broad), less densely covered with white silky hairs, often soon glabrescent. *S. capensis, Thunb., var. mucronata, forma pubescens, Anderss. in Vet.-Akad. Handl. Stockh.* vi. (1867) *no.* 1, 14.

SOUTH AFRICA : without locality, *Thunberg, Pappe* ! *Masson* !
COAST REGION : Clanwilliam Div. ; Berg Valley, under 500 ft., *Drège, b* ! between Lange Valley and Oliphants River, 1000–1500 ft., *Drège, c.* Cape Div. ; on rivulets near Constantia, *Krauss.* Stellenbosch Div., *Harvey* !
CENTRAL REGION : Calvinia Div. ; between Grasberg River and Watervals River, 2500–3000 ft., *Drège, a* !
WESTERN REGION : Var. β : Little Namaqualand ; between Pedros Kloof and Lily Fontein, 3000–4000 ft., *Drège* !

Order CXXV. A. **CERATOPHYLLEÆ.**

(By S. A. Skan.)

Flowers monœcious. *Perianth* thinly herbaceous or submembranous, equal; segments 6–12, subvalvate, often toothed or lacerated at the apex. *Male:* stamens 10–20, crowded on a flat or convex torus; filaments very short; anthers linear-oblong, equalling the perianth; cells 2, linear, parallel, adnate, dehiscing longitudinally; connective produced beyond the cells into a thick coloured usually 2- or 3-toothed appendage. *Female:* ovary 1, sessile, ovoid, 1-celled; style terminal, linear-subulate, persistent; stigma unilateral; ovule 1, pendulous, anatropous. *Fruit* 1-seeded, leathery, indehiscent, ovoid or ellipsoid, somewhat compressed, tipped with the hardened style, sometimes with 2–4 spreading or reflexed spines at the base, wingless or surrounded by a narrow or broad leathery toothed wing, smooth or tuberculate. *Seed* pendulous; testa membranous; albumen 0; embryo straight, with oblong rather thick equal cotyledons; radicle very short; plumule large, many leaved.

Aquatic herbs with elongated leafy floating branches; leaves verticillate, 2-fid or dichotomously divided; segments linear or filiform, somewhat rigid, usually toothed; flowers axillary, solitary, very small, sessile, the male and female alternating at the nodes, or male at the lower nodes and female at the upper.

Distrib. *Ceratophyllum*, the only genus, includes 1 or 2 or according to some authorities about a dozen species, growing in fresh water in nearly all parts of the world.

I. **CERATOPHYLLUM**, Linn.

Characters and distribution of the order.

1. C. demersum (Linn. Sp. Pl. ed. i. 992); a glabrous perennial herb; stems much-branched, floating and submerged, 2–3 ft. sometimes up to 8 ft. long; leaves in whorls of 5–12, 1–3-times (usually twice) dichotomously divided, often about 1 in. long; segments linear to filiform, entire or more or less remotely spinulose-serrulate; perianth-segments oblong, about ½ lin. long; fruit ovoid or ellipsoid, slightly compressed, about 3 lin. long, smooth or sparingly covered with minute tubercles, narrowly or broadly winged or wingless, crowned with the long slender often curved persistent style and usually bearing 2 recurved subulate or terete spines at the base. *Smith & Sowerby, Engl. Bot.* xiv. *t.* 947; *DC. Prodr.* iii. 73; *Benth. Fl. Austral.* ii. 491; *Boiss. Fl. Orient.* iv. 1202; *Hook. f. Fl. Brit. Ind.* v. 639; *K. Schum. in Mart. Fl. Bras.* iii. iii. 746, *t.* 125; *Hiern* *in Cat. Afr. Pl. Welw.* i. 1031; *Durand, Syll. Fl. Congol.* 513; *Muschler, Man. Fl. Egypt,* i. 363; *Marloth, Fl. S. Afr.* i. 219, *fig.* 97; *Skan in Prain, Fl. Trop. Afr.* vi. ii. 326. *C. demersum, var. oxyacanthum, K. Schum. in Engl. Pfl. Ost-Afr. C.* 178. *C. oxyacanthum, Cham. in Linnæa,* iv. 504, *t.* 5, *fig.* 6 b; *Krauss in Flora,* 1844, 426. *C. vulgare, Schleid. in Linnæa,* xi. 540, *t.* 11; *Hook. Niger Fl.* 525. *Ceratophyllum sp., Wood, Handb. Fl. Natal,* 121. *Myriophyllum, Drège, Zwei Pfl. Documente,* 160. For full synonymy see *K. Schum. in Mart. Fl. Bras.* iii. iii. 746.

EASTERN REGION : Natal ; Umgeni River, *Drège,* 4457 ! *Wood,* 4000 ! on ranges 30–60 miles from the sea, 2000–3500 ft., *Sutherland* ! Umlaas (Umlazi) River, *Krauss,* 340 !

PORTUGUESE EAST AFRICA : Incanhini, *Schlechter,* 12031 !

Almost cosmopolitan.

ADDENDA AND CORRIGENDA.

THYMELÆACEÆ.

4a. Arthrosolen compactus (C. H. Wright); a densely branched undershrub; branches short, densely leafy, at first silky-pilose; leaves alternate, imbricate, oblong, obtuse, 3 lin. long, 1 lin. wide, long pilose on the margins and back, verrucose on both surfaces; flowers in terminal clusters; bracts ovate; calyx yellow, waxy, hairy outside; tube 3 lin. long, narrowly funnel-shaped above the basal constriction; lobes ovate, obtuse, the 2 outer tipped with pencils of hairs, 1¼ lin. long, 1 lin. wide; anthers oblong, obtuse; ovary ovate, compressed, ¼ lin. long; style columnar, ¾ lin. long; stigma subcapitate.

EASTERN REGION : Natal ; Nigunya, 6500 ft., *Wylie in Herb. Wood.* 10531 !

8a. Lachnæa Burchellii (Meisn. in Linnæa, xiv. 420); leaves opposite, slightly spreading, linear-lanceolate, acute, nearly flat, 3–1-nerved; heads terminal or axillary, few-flowered, sessile; bracts ovate, subacute, 1½ lin. long, 1 lin. wide ; calyx minutely puberulous outside ; scales glabrous ; flowers of the same shape and colour as in *L. diosmoides*, Meisn., very short. *L. Burchellii, var. angustifolia, Meisn. in DC. Prodr.* xiv. 577. *L. phylicoides, var. oppositifolia, Meisn. in Linnæa,* xiv. 420. *Gnidia tenuiflora, Eckl. & Zeyh. ex Meisn. in DC. Prodr.* xiv. 577.

Var. β **latifolia** (Meisn. in DC. Prodr. xiv. 577); leaves oblong or subovate-lanceolate, 3-nerved on the back, up to 4 lin. long and 1½ lin. wide ; calyx densely silky outside. *Gonophylla nana, Eckl. & Zeyh. in DC. Prodr. xiv. 577.*

COAST REGION. George Div. ; on mountains near George, *Drège*, 7370, *Ecklon & Zeyher.* Var. β, Uitenhage Div. ; Van Stadens River, *Ecklon*, 78, *Zeyher*, 3757 ! and without precise locality, *Harvey*, 823 !

This species resembles *L. striata*, Meisn., which can be readily distinguished by its glabrous calyx-tube.

10a. Lachnæa alpina (Meisn. in DC. Prodr. xiv. 578) ; an undershrub, corymbosely branched ; branches glabrous, at length rough with old leaf-scars ; leaves scattered, spathulate-linear, obtuse, flat, 1-nerved, glabrous, 4 lin. long, 1¼ lin. wide, sessile or subsessile ; head terminal, up to 20-flowered, sessile ; calyx white-silky outside, only seen young ; lobes oblong, rather longer than the tube ; scales small, filiform, slightly exserted. *Gonophylla alpina, Eckl. & Zeyh. ex Meisn. l.c.*

COAST REGION. Tulbagh Div. ; Winterhoek Mountains, *Ecklon & Zeyher*, 71 ! Allied to *L. capitata*, Meisn., which differs in having linear pungent leaves.

1a. Gnidia aberrans (C. H. Wright); a much-branched shrub; branches stout, at first pilose, soon glabrous, leaf-scars conspicuous but not very prominent; leaves alternate, approximate, ovate-lanceolate, acute, 3 lin. long, 1 lin. wide, with a few short scattered hairs above, densely silky and slightly verrucose beneath; flowers few near the apex of the branches; calyx densely silky outside, yellow (*Wylie*); ovoid base of tube 1 lin. long, narrowly funnel-shaped part 3 lin. long; lobes oblong, subacute, $1\frac{1}{4}$ lin. long, $\frac{3}{4}$ lin. wide; petals 4, fleshy, anther-like, ovate-lanceolate, $\frac{3}{4}$ lin. long; staminodes 0; anthers 4, distinctly included, oblong, obtuse; ovary compressed, $\frac{1}{2}$ lin. long; style about half as long as the calyx-tube; stigma shortly penicellate.

EASTERN REGION: Natal; Giant's Castle, 8000–10000 ft., *Wylie in Herb. Wood*. 10611.

This is an extremely anomalous plant as regards its floral structure. It differs from all the South African species of *Gnidia*, except *G. harveyana*, Meisn., and *G. anomala*, Meisn., in having only 4 perfect stamens, and from the two species mentioned in the entire absence of staminodes. The absence of the characteristic hairs around the petals excludes it from *Struthiola*, while the allied genus *Drapetes*, which does not occur in South Africa, differs in having the upper whorl of stamens perfect with the anthers exserted on long filaments.

17. Lasiosiphon Krausii (Meisn.). Add to localities :—Basutoland; Leribe, *Mrs. Dieterlen*. 7259 !

33. Gnidia Cayleyi (C. H. Wright). Page 58, line 7, for *Caley* read *Cayley*.

LORANTHACEÆ.

1. Loranthus Woodii (Schlechter & Krause). Add to Synonmy :—*Acrostachys Sandersoni, Van Tiegh. in Bull. Soc. Bot. France*, xli. 504.

SANTALACEÆ.

Page 136, line 6, for **Osyridocarpus**, read **Osyridocarpos**.
Page 136, line 8, for **Rhoiocarpus**, read **Rhoiocarpos**.
Page 145, line 10 from the bottom, *for* Kharkamo, *read* Kharkams.

24. Thesium triflorum (Thunb.). Add to Synonmy : *Osyris angustifolia, Baker in Kew Bulletin*, 1910, 238.

43a. Thesium cruciatum (A. W. Hill in Kew Bulletin, 1916, 231); a perennial subshrub, about 1 ft. high, with a thick woody rhizome; stems and branches rigid, tapering, spine-tipped, coarsely and closely longitudinally wrinkled, very minutely puberulous; leaves very small, subulate, acute or acuminate, adpressed, with brown apex, glabrous; flowers axillary, distinctly pedicellate, with the bracts crowded at the bases of the branchlets; bracts and bracteoles scale-like, triangular, minute, glabrous; perianth 1 lin.

long, externally glabrous with large ovoid external glands between
the perianth-segments ; segments ovate, subacute, erect, $\frac{1}{2}$ lin.
long, hooded, with broad undulate inflexed membranous flaps on the
margins embracing the anthers ; anthers about $\frac{1}{3}$ lin. long, exserted
from the perianth-tube; filaments $\frac{1}{4}$ lin. long ; style thick, $\frac{1}{4}$ lin.
long ; fruit immature, ribbed, glabrous.

REGION : S.W. African Protectorate ; plains south of Choaberib, *H. H. W.
Pearson*, 9447 ! sandy plains north of Areb, *H. H. W. Pearson*, 9474 !

This species should be placed next to *T. lacinulatum*, A. W. Hill, in the key of
the genus on p. 139. The chief points of difference are the glabrous flowers with
the undulate lacinulæ on the margins of the perianth-segments and the large
anthers embraced by the lacinulæ.

52. Thesium capituliflorum (Sond.). Page 168, line 13 from the
bottom, for *Harvey*, 107, read *Harvey*, 707.

EUPHORBIACEÆ.

Page 218, line 24, for **Toxicodendron**, read **Toxicodendrum**.
Page 230, lines 30 and 31, *delete* :—shrubs or bushes 2–6 ft. high.
Page 230 line 33, *after* rays 2–12 lin. long ; *add* : shrubs or
bushes 2–6 ft. high.
Page 231, for lines 1–5, substitute :—

Plants 2–6 in. high, compactly or bushily much branched ;
   involucre glabrous outside :
Branches alternate and opposite on the same plant,
   smooth when alive, not drying with a papery whitish
   wrinkled skin ; involucre 1–1$\frac{1}{4}$ lin. in diam. (& see
   65, *E. stapelioides*) ... ... (64) **gentilis**.
Branches alternate, verruculose, drying with a finely
   wrinkled whitish papery skin ; involucre 1 lin.
   in diam. ... ... (65a) **verruculosa**.

Page 232, line 26, for **angrana**, read **Angræ**.
Page 234, line 4, for **arrecta**, read **mixta**.
Page 239, line 7, for **inelegans**, read **inornata**.
Page 240, line 3 from the bottom, for **infausta**, read **falsa**.
Page 283, line 15 from the bottom, for **arrecta**, read **mixta**.

67a. Euphorbia verruculosa (N.E. Br.) ; rootstock woody, $\frac{1}{2}$ in.
or more thick, apparently extensively creeping underground, giving
off erect or ascending shoots dividing at the surface of the ground
into a compact cluster of branches 2–3 in. high, glabrous in all
parts ; branches leafless, succulent, 1–3 in. long, 3–3$\frac{1}{2}$ lin. thick,
usually bearing a compact irregular cluster of short branchlets
at the apical part ; when alive covered with alternate and more or
less spirally arranged very obtusely rounded tubercles $\frac{1}{2}$–1 lin. high,
prominent and verruculose all over ; when dried the tubercles
almost disappear and the verruculose skin shrivels and has a whitish
papillate papery appearance ; leaves practically absent, reduced to
very minute rudiments, alternate ; male involucres in small sessile

cymes or clusters at the tips of the branchlets, sessile, 1 lin. in diam. and about $\frac{3}{4}$ lin. deep, cup-shaped, with 5 or occasionally 4 transversely oblong contiguous glands about $\frac{1}{2}$ lin. in their greater diam., and with 2 or occasionally 3 broadly ovate fleshly keeled bracts at the base ; female involucres and fruit not seen.

WESTERN REGION : Great Namaqualand ; Angra Pequena, *Marloth*, 4639 !

This very distinct species is evidently allied to *E. gentilis*, *E. stapelioides*, *E. karroensis* and *E. spicata*, but differing from all in its alternate and more tuberculate branches, with a distinctly verruculose epidermis, which on dried specimens assumes a finely wrinkled whitish papery appearance, quite unlike that of any species I have seen.

Page 322, line 10 from the bottom, for **inelegans**, read **inornata**.
Page 358, lines 5 and 9, for *infausta*, read *falsa*.
Page 388, line 13, for **infausta**, read **falsa**.

SALICINEÆ.

1. Salix capensis var **mucronata** (Anderss.). In the *Journal of Ecology*, x. 70, Dr. J. Burtt-Davy regards this variety as specifically distinct from *S. capensis*, Thunb., and describes the following varieties of it :—

VAR. **integra** (Burtt-Davy, l.c. 70) ; differs from the type in the young branches being tomentose and the leaves longer and wider with entire or sparsely subserrulate margins.

CENTRAL REGION : Aberdeen Div. ; Camdeboo on the flats and at the river near Camdeboosberg, 2-3000 ft. *Drège* !

VAR. **caffra** (Burtt-Davy, l.c. 71) ; branchlets pendulous, larger than in the type ; leaves narrowly linear-lanceolate, long acuminate.

COAST REGION : Queenstown Div. ; Shiloh, 3500 ft., *Baur*, 935 ! British Kaffraria, *Cooper*, 223 ! Eastern Districts, *Cooper* 48 !

CENTRAL REGION : Graaff Reinet Div., on the banks of the Sundays River near Graaff Reinet, 2500 ft., *Bolus* 468 !

EASTERN REGION : Tembuland ; Bazeia, 2000 ft., *Baur*, 512 !

INDEX.

[SYNONYMS ARE PRINTED IN *italics*.]

2 Q

FLORA CAPENSIS.

VOL. V. SECT. 2.

(Supplement.)

FLORA CAPENSIS:

BEING A

SYSTEMATIC DESCRIPTION OF THE PLANTS

OF THE

CAPE COLONY, CAFFRARIA, AND PORT NATAL

(AND NEIGHBOURING TERRITORIES)

BY

VARIOUS BOTANISTS.

EDITED BY

SIR ARTHUR WILLIAM HILL, K.C.M.G.,
M.A., Sc.D., D.Sc., F.R.S.

DIRECTOR, ROYAL BOTANIC GARDENS, KEW,
HONORARY FELLOW OF KING'S COLLEGE, CAMBRIDGE.

*Published under the authority of the Government of
the Union of South Africa.*

VOLUME V. SECTION 2 (SUPPLEMENT).
GYMNOSPERMÆ.

L. REEVE & CO., LTD.,
PUBLISHERS TO THE HOME, COLONIAL AND INDIAN GOVERNMENTS
BANK BUILDINGS, BANK STREET, ASHFORD, KENT.
1933.

PRINTED IN GREAT BRITAIN BY
WILLIAM CLOWES AND SONS, LIMITED,
DUKE STREET, STAMFORD STREET, LONDON, S.E. 1.

PREFACE.

WHEN the *Flora Capensis* was originally planned, it was intended to include the *Gnetaceæ, Coniferæ*, and the *Cycadaceæ*. The description of *Welwitschia* was prepared for the purpose several years ago by the late Professor H. H. W. Pearson, and the Conifers were described by Dr. O. Stapf, F.R.S., late Keeper of the Herbarium and Library. Sir William Thiselton-Dyer, the Editor of the later volumes of the *Flora Capensis*, had always intended to write the account of the South African *Cycadaceæ* after his retirement, but failing health prevented him from carrying out the project, to which he had for many years devoted considerable study. Shortly before his death he handed over to me the material he had collected together and his notes, expressing the wish that I should undertake the work. While in South Africa in 1930, I was able to discuss the matter with Dr. Rattray, who has made careful studies of the South African Cycads in the field, and has grown most of them in his garden. He very kindly agreed to collaborate with Mr. J. Hutchinson, F.L.S., in the preparation of the descriptions of the Cycads for this supplemental volume. While Mr. Hutchinson is mainly responsible for the technical descriptions, Dr. Rattray's intimate knowledge of the plants, as they grow in South Africa, has added very greatly to the value of the undertaking.

The account of *Welwitschia* has been supplemented by Mr. Hutchinson to bring it up to date in the light of recent knowledge. It has been raised to family rank as distinct from *Gnetaceæ* (*Gnetum* and *Ephedra*), with which it has probably little in common apart from the Gymnospermous character. Dr. Stapf has largely re-written his descriptions of the South African *Podocarpaceæ* and *Cupressaceæ*.

The publication of this supplemental part of the *Flora Capensis* has been made possible by the generosity of the Government of the Union of South Africa, who on learning, through Kew, of the desire expressed by Botanists in South Africa for an account of the Gymnosperms, made a grant of £40 towards its publication.

The history of the inception and completion of the *Flora Capensis*, which deals with the flowering plants proper—*Angiospermæ*—was published by Sir William Thiselton-Dyer in the *Kew Bulletin*, 1925, pp. 289–293, and his valedictory preface will be found in Vol. V, Sect. 2, Part IV, of the *Flora Capensis*, written on 23rd September, 1924.

For the loan of herbarium material from South Africa we are much indebted to Dr. I. B. Pole Evans, C.M.G., Director of the Botanical Survey, to the Forestry Department in South Africa, and to the Directors of the Cape Town and Albany Museums.

ARTHUR W. HILL.

ROYAL BOTANIC GARDENS,
KEW, MARCH 25, 1933.

Addenda. **E. kosiensis.**

p. 28, 5 lines from below, *after* unbranched, *insert* :—

(see also (4) *kosiensis*).

p. 34, after notes to no. 4, *E. kosiensis* (Hutch.), *add* :—

A letter from Col. Molyneux, received after going to press, states that he has growing at the Old Fort Garden, Durban, specimens of this species 4 ft. in height, which have produced cones of a bright red colour. Probably in the wild state the stem never rises more than a few inches above the sandy soil in which it grows near the coast, but in cultivation produces a short aerial stem, as in some other species.

ORDER CXXVI. **WELWITSCHIACEÆ.**

(By H. H. W. PEARSON.)

Flowers unisexual or pseudobisexual. *Male* (pseudobisexual) *flower* : *Envelope* of 2 imbricating whorls ; outer whorl of 2 laterally placed, free, boat-shaped, keeled scales ; inner of 2 median broadly obovate or subrotund keel-less scales, connate at the base. Stamens 6, exserted ; filaments connate into a short tube at the base ; anthers somewhat 3-lobed when mature, 3-celled, dehiscing by 3 slits from the summit ; pollen ellipsoid, slightly coherent in irregular masses. Ovule solitary, terminal, erect, orthotropous, imperfect, with the single integument produced into a tubular micropyle, sharply bent near the middle, expanding at the tip into an exserted glandular-papillose stigmatiform disc. *Female flower* : *Envelope* bottle-shaped, contracted at the throat, of 2 laterally placed, connate leaves, compressed from back to front, with 2 lateral membranous wing-like expansions from the midribs. *Ovule* solitary, terminal, erect, orthotropous, perfect, with its single integument produced into a straight micropylar tube through the mouth of the envelope ; micropylar tube irregularly labiate or fimbriate, but not expanded at the apex. *Seed* flattened, closely invested by the winged envelope ; endosperm starchy, wedge-shaped below, retuse above, supporting the withered nucellar cap (*perisperm*) ; radicle erect ; cotyledons 2, rarely 3, narrow-linear ; suspensor long, coiled, persistent.

DISTRIB. Genus 1, species 1, confined to the coast region of South West Africa, from about $15\frac{1}{2}°$ S. in Angola (south of Mossamedes) to the tropic of Capricorn, in Great Namaqualand.

WELWITSCHIA,* Hook. f. (*Tumboa* Welw.).

Characters as for the family.

1. **W. Bainesii** (Carr. Conif. ed. ii. 783 (1867)) ; plant body (*hypocotyl*) woody, covered by thick corrugated cork, sometimes fused with other individuals, when injured exuding a copious gummy secretion congealing in alcohol, broadly obconic or turbinate, con-

* Name conserved according to International Rules.

cave on the top, more or less circular or elliptic in horizontal section, rising ¼–1 ft. above the ground, 1–3 ft. in diam. at the top ; epicotyl reduced to 2 leaf-bearing grooves and floriferous cushions forming a raised rim around the top of the hypocotyl interrupted at the longer diameter, and a depressed and early arrested stem apex, at length buried beneath two coalescent corky expansions overlying the concave summit of the hypocotyl and developed from the buds in the axils of the cotyledons; taproot greatly elongated, un-branched in the upper part, at length very slender, branched and brittle ; leaves 2, rarely 3, each inserted in an epicotylar leaf-groove extending round half the raised rim of the hypocotyl, oblong, entire, usually in old plants torn into few or many strap-like segments from apex to base, thick, leathery, with the main nerves parallel and distinct, growing at the base as long as the plant lives, dying at the apex, up to 4 yards long ; spikes arranged in compound dichasial cymes (rarely solitary) arising annually from pits in the floriferous cushions situated immediately above, not seldom immedi-ately beneath, each leaf ; male spike bearing 40–70 axillary flowers in 4 rows ; bracts connate, lowest pair or 2 pairs barren ; flowers concealed by the bracts until the exsertion of the anthers ; female spike bearing 40–60 flowers in 4 rows ; lowest 6–10 pairs of bracts increasing in size from below upwards, barren, the lowest 2 or 3 pairs connate. Except the micropylar tube, the naked seed com-pletely concealed by the bract at maturity. Marloth, Fl. S. Afr. i. 107, fig. 68 a and b. *Tumboa Bainesii,* Hook. f. in Gard. Chron. 1861, 1008 ; Naudin Rev. Hort. 1862, 186 ; Rendle in Cat. Afr. Pl. Welw. ii. 257 (1899) ; Engl. Pflanzenw. Afr. ii. 90–93, fig. 85 (1908). *Tumboa strobilifera,* Welw. in Gard. Chron. 1862, 71. *Welwitschia mirabilis,* Hook. f. in Gard. Chron. 1862, 71, in Trans. Linn. Soc. xxiv. 1. tt. 1–14 (1863), and in Bot. Mag. tt. 5368, 5369 (1863) ; McNab in Trans. Linn. Soc. xxviii. t. 40 (1873) ; Monteiro, Angola and River Congo, ii. t. 15 (1875) ; Schimper, Pflanzen-Geogr. 662, 664 (1898) ; Warburg, Kimena-Sambesi Exped. frontisp., p. 6 (1903) ; Karsten & Schenck, Veg.-Bild. i. t. 25 (1903) ; Pearson in Kew Bull. 1907, 347, pl. 2, figs. 3–5 ; L. Schultze, Namaland & Kalahari, t. 3 (1907) ; M. G. Sykes in Trans. Linn. Soc. Bot. ser. ii. vii. 327, t. 34–5 (anat.) ; Velenovsky, Vergl. Morphol. Pfl. iii. 775 (1910) ; iv. Suppl. t. 1 (1913) ; Coulter & Chamberlain Morph. Gymnosp. 365, 366, 374, 399 (1910) ; Church in Phil. Trans. (B) ccv. (1914), 115, with figs. ; Pearson in Prain Fl. Trop. Afr. vi. ii. 333 (1917).

SOUTH-WEST AFRICA : Namib region from north of Sandfisch Bay (23½° S.) to the northern boundary, and continued along the low coastal belt of Angola to south of Mossamedes (15½° S.).

This weird type of plant may not be so rare as has generally been supposed. According to an account in the *Diamond Fields Advertiser,* Kimberley, for April 28th, 1932, *Welwitschia* occurs in great quantity in the Kakoa-veld in North-West Damaraland ; the writer there states that he observed " an area of not less than 2,500 square miles where the gravelly surface of the ground, up hill and down dale, is covered with the plant, in some places in such profusion that it was impossible to find an opening through which to pilot the car, and we were forced to back out and make wide detours ; some plants stretched to a diameter of 16 ft." (*See also* Journ. Bot. Soc. S. Afr., pt. ʌviii, 4, 1932.)

Mr. Worsdell informs us that the plant is fairly common ("many hundreds") in the dried-up bed of a stream and on small granite hillocks about a mile to the south of Old Welwitsch Railway Station ; the crown of the stem of the largest specimen was nearly 6 ft. in diameter.

ORDER CXXVII. A. **PODOCARPACEÆ.**

(By O. STAPF.)

Diœcious, very rarely monœcious. *Male strobiles* mostly catkin-like, sometimes externally only slightly differentiated from the vegetative branches, simple or compound, terminal or axillary, solitary or fascicled, bracteate or ebracteate at the base ; fertile scales bearing basi-dorsally 2 pollen-sacs, squamiform or more or less differentiated into a claw or stalk and blade, the latter large and projecting beyond the pollen-sacs, or very much reduced, when the scales with their pollen-sacs assume the appearance of typical angiospermous stamens ; pollen grains mostly winged. *Female strobiles* usually much reduced, terminal or axillary ; lower scales barren, the upper or only the uppermost fertile, always simple, each bearing 1 (very rarely 2) ovule ; ovule usually more or less exceeding its scale, sometimes long-exserted, rarely quite enclosed. *Mature strobiles* usually little altered or the axis or also the scales becoming more or less fleshy. *Seeds* usually exserted ; seed-shell (testa) coriaceous to woody, with or without an outer covering (*epimatium*), which is either free or more or less fused with the testa, and varies from membranous to leathery or fleshy.

Shrubs or trees ; leaves usually spirally arranged, quaquaversal or dorsiventrally disposed in one plane, scale-like or linear to lanceolate, rarely ovate, always evergreen.

DISTRIB. Genera 7, with about 100 species, mostly in the tropics and the southern temperate zone.

I. **PODOCARPUS**, L'Hérit. ; Benth. et Hook. f. Gen. Pl. iii. 434.

Diœcious, very rarely monœcious. *Male strobiles* usually axillary, variously arranged, bracteate at the base, sessile or peduncled ; scales numerous, spirally arranged, imbricate, with usually broad,

triangular to ovate-rotundate, rarely lanceolate blades, and 2
relatively large dorsal pollen-sacs near the base. *Pollen* with 2,
rarely 3, wings. *Female strobiles* terminal or axillary, usually
reduced to a few sterile lower scales, which are more or less fused
with each other and with the axis and 1 or 2 terminal fertile
scales, the whole plexus often becoming ultimately fleshy (receptacle)
—rarely spike-like with few to numerous usually distant fertile
scales ; scales spirally arranged or opposite in decussate pairs, the
lower often with a foliaceous blade, the upper squamiform ; ovules
solitary, adnate to the face of the fertile scale, and usually much
exceeding it, inverted, and enclosed in a false aril (epimatium)
arising from the face of the scale and adnate to the single integument.
Seeds deciduous together with the fleshy receptacle or with the
unmodified remainder of the strobile or falling from the scales of
the persistent axis of the strobile ; seed-shell (testa) and false aril
(rarely also the fertile scale) forming a coriaceous or externally fleshy
and internally woody shell. *Embryo* axile ; cotyledons 2.

Shrubs or trees, often of great height. Leaves squamiform or linear or
lanceolate to ovate, usually spirally arranged, but placed dorsiventrally, rarely
opposite. Male strobiles solitary or clustered or disposed in compound
inflorescences, rarely apical. Seeds and receptacles, where present, greenish
or brown or sometimes vividly coloured, the seeds always conspicuously
exposed.

Distrib. About 70 species, mostly in the mountain forests of the Tropics,
a few in the Temperate Regions of the Southern Hemisphere and in Japan.

Section 1. Eupodocarpus, Engl.—Axis and scales of the female
strobiles mostly transformed during maturation into a fleshy
receptacle, or at least thickened and clavate. Inner layer of seed-
shell (testa) thin and crustaceous. Sclerenchymatic hypoderma
below the upper epidermis continuous ; stomata on the lower side
of the leaf only (at least in the South African species).

Leaves of the adult tree 1–2½ in. long, straight, shortly
    acute to almost obtuse ; receptacle distinctly
    fleshy :

    Leaves 3½–6½ lin. wide   ...   . (1) **latifolius.**

    Leaves 2 lin. wide   ...   . (2) **elongatus.**

Leaves of the adult tree 3–6 in. by 3–4 lin., tapering to
    a slender acute point ; receptacle hardly fleshy,
    though thickened and clavate   ...   . (3) **Henkelii.**

Section 2. Stachycarpus, Endl.—Axis and scales of the female
strobiles not transformed during maturation into a fleshy receptacle,
the axis ultimately dry with the scales or vestiges of the lower scales.
Inner layer of seed-shell (testa) thick and bony. Sclerenchymatic
hypoderma below the upper epidermis not continuous ; stomata
on both sides of the leaf.

Leaves of the mature tree shortly acute to almost
   obtuse, 1–2 in. by 1¼–2 lin. ; male strobiles 3–4 lin.
   long, their scales cordate-ovate, more or less obtuse (4) **falcatus.**

Leaves of the mature tree mostly long tapering to a
   sharp point, 10 lin. to 2 in. by ½–2 lin. :

   Leaves 1–2 in. by 1¼–2 lin. ; male strobiles up to
      9 lin. long, their scales ovate-triangular, acute (5) **gracilior.**

   Leaves 10 lin. by ½–¾ lin.   ...   . (6) **gracillimus.**

1. **P. latifolius** (R. Br. ex Mirb. Geogr. Conif. in Mém. Mus. xiii.
75 ; not of Wall.) ; a tree up to 100 ft. high, with a tall clean bole
on the average 2 ft. (sometimes up to 4 ft.) in diameter, and a comparatively small crown ; bark smooth, persistent ; branchlets of
the mature tree rather stunted, terete, slightly angular when young,
the decurrent leaf-bases soon becoming obliterated ; terminal
buds ovoid, up to 2 lin. long ; leaves spirally arranged, loose in the
juvenile state, crowded, particularly upwards, and then often subverticillate in the mature state, broad-linear to linear-oblong
or oblong, acute to subacute, sometimes apiculate, narrowed at
the base into a short petiole, 1–2½ in. by 3½–6¼ lin., or in the juvenile
state up to 5 in. long, straight or (the juvenile) very slightly falcate,
spreading, coriaceous, midrib slightly raised on both sides with 3
resin-ducts below the central strand ; stomata confined to the lower
side ; male strobiles solitary, cylindric, ¾–1 in. long, glaucous-
pinkish, with a few rotundate-ovate coriaceous very concave bracts
about 1 lin. long at the base ; scales imbricate, at length loose,
with a rotundate or broad-ovate fimbriate-denticulate or subdenticulate small blade up to ½ lin. long ; pollen sacs ⅔ lin. long,
conspicuous ; female strobiles borne on slender peduncles 2½–5 lin.
long, formed of 2 decussate pairs of scales fused into a fleshy
receptacle, those of one pair barren, slightly shorter, with or without
small cuspidate tips, both of the other pair or one only fertile, with
a short ovate free blade embracing the base of the ovule, the whole
receptacle 2-lobed, and if only one scale fertile more or less to very
oblique ; the receptacle when mature resembling a small dark red
cherry in colour and shape, up to ½ in. in diam. ; seeds subglobose,
3–3½ lin. in diam., dark glaucous to bluish-green or blue ; inner
layer of seed-shell thin crustaceous or almost papery, outer somewhat fleshy, very resinous, 1¼ lin. thick.   Bennett, Pl. Javan.
Rar. 40 ; C. Presl, Bot. Bemerk. 110 ; Pilger, in Engl. Pflanzenr.
Taxac. 90, 91 incl. vars. *latior* and *confertus* ; Engl. Pflanzenw.
Afr. i. 421, 422, figs. 361, 362 ; ii. 86, fig. 82 ; Marloth, Fl. S. Afr. i.
101, 102, t. 13, 17 A, and (Suppl.) Dict. Comm. Names of Pl. 101 ;
Sim, Fl. Trees & Shrubs for use in South Afr. 182, and Native Timb.
South Afr. 101, fig. 41 ; Pilger in Engl. & Prantl, Nat. Pflanzenf.
ed. 2, 247, fig. 136 ; Burtt Davy in Kew Bull. 1908, 147, Man. Flow.
Pl. Transvaal, 100 ; Levyns, Guide Fl. Cape Penins. 19.   *P. Thun-
bergii,* Hook. in Lond. Journ. Bot. i. 657, t. 22 ; Endl. Syn. Conif.
217 ; Lindl. & Gord. in Journ. Hort. Soc. v. 224 ; Pappe, Silva

Cap. 32 ; Carr. Trait. Conif. ed. i. 470 ; ed. ii. 670 ; Gord., Pinet.
ed. i. 284 ; ed. ii. 349 ; Henk. & Hochst. Syn. Nadelhölz., 398 ;
Parlat. in DC. Prodr. xvi. ii. 511 ; Fourcade, Rep. Natal Forests,
1889, 4, 121 ; Bolus & Wolley-Dod, Fl. Pl. Cape Penins. in Trans.
S. Afr. Phil. Soc. xiv. 320 ; Wood, Handb. Fl. Natal, 122 and in
Trans. S. Afr. Phil. Soc. xviii. 224 ; Agr. Journ. Cape Good Hope,
xxvi. 170, fig. opp. p. 171 ; Marloth, Kapland, 190, 191, 196, 200,
208, figs. 65, 68, 69, 73 ; Sim, Tree Plant. Natal, 236, 285, 331 ;
Bews in Ann. Natal Mus. iii. 545, 547, 548. *P. Thunbergii* var.
latifolia, Sim, Forest Fl. Cape Good Hope, 3, 332, t. 148, 149, f. 2.
P. macrophyllus, Drège, Zwei Pflanzengeogr. Doc. 123, 157, 212,
not of Don. *P. Sweetii*, C. Presl, Bot. Bemerk. 110. *Taxus latifolia*,
Thunb. Prodr. 117 ; Fl. Cap. ed. Schult. 547. *T. macrophylla*,
Banks ex Endl. Syn. Conif. 218. *Nageia latifolia*, O. Kuntze,
Rev. Gen. ii. 800 ; not of Gord.

SOUTH AFRICA : without locality, *Bowie* ! *Brand* ! *Villet* ! *Millan* ! *Mund
and Maire* !

CAPE PROVINCE : Cape Div. ; Newlands Woods on Table Mountain, *Wolley
Dod*, 2729 ! Fritz Bronn's Kraal, *Alexander-Prior* ! Orange Kloof, *Gamble*,
22002 ! Hout Bay, *Bews* ! Herb. *Sim* 19017 ! Swellendam Div. ; Groot-
vadersbosh, *Thunberg* ; near George, *Burchell*, 5843 ! forests from west of
George to the eastern boundary of the division (*Cape Forest Reports*) ;
Kaymans River Gat, in forest, *Drège* ! Knysna Div. ; all forest sections of
the division (*Cape Forest Reports*); Outeniqua Mountains, *Thunberg* ; Kaatjes
Kraal, *Burchell*, 5223 ! 5254 ! Harberville Forest, *Keit*, 524 ! Humansdorp
Div., ; forests east as far as Storms River (*Cape Forest Reports* for 1910) ;
Uitenhage Div. ; Zuurbergen (*Cape Forest Reports*) ; Albany Div. ; the wooded
kloofs near Grahamstown, *Burchell*, 3595 ! *MacOwan*, 1408 ! 1958 ! (*Herb.
Sim*) ; *Zeyher*, 3882 ! 3883 ! 3884 ! (*Herb. Sim*) ; *Atherstone*, 89 ! *Burtt
Davy*, 7816 ! King Williamstown Div. ; forests of the Perie and North Sections ;
Amatola Range (*Cape Forest Reports*); Perie forest, *Sim*, 19020 ! Stutterheim
Div. ; Koeoghe Range and Isidenge forests (*Cape Forest Reports*) ; " British
Kaffraria," *Cooper*, 1298. Fort Cunyngham, 2000–3500 ft., *Herb. Sim*,
2030 ! 2122 !

EASTERN REGION : Throughout the forests from the Great Kei River and
lower White Kei River east to the Drakensberg above Newcastle in Natal
and western Zululand (*Cape, Natal & Zululand Forests Reports*) ; Tembuland,
Engkobe, Maning, *Merwe, For. Dept. Herb.*, 2266 ! Mkonkoto forest, 3000 ft.,
Sim ! Maclear Div. ; Pot Romer Mountains, 5500 ft., *Galpin*, 6831 ! Pondo-
land ; Egosa forest, 650–780 ft., *Beyrich* ; and without precise locality,
Bachmann, 70 ! Mt. Ayliff Div. Tabun River, *Nat. Herb. Pretoria* ! Umzimkulu,
Mhlonga forest, *Kaufmann, For. Dept. Herb.*, 2168 ! 2169 ! Jutsubani forest,
Fraser, For. Dept. Herb., 2227 !

NATAL : Alfred County, near Murchison, *Wood*, 3028 ! near Durban, *Plant* !
Polela Distr., Ingwangwani, Emblazani Forest, *Household, For. Dept. Herb.*,
1957 ! Drakensberg, 3500–4500 ft., *Fourcade* ; north of Van Reenens Pass,
5000–5500 ft. (according to Bews) ; *Rehmann*, 7246 ! 7247 ! on the slopes
facing Newcastle, 5000- 6000 ft. (according to Bews). Zululand ; Odeni,
4500–5000 ft. and Ingoye forests, 1000–1500 ft. ; and without precise locality,
Gerrard, 127 ! *Sanderson* !

SWAZILAND : Forbes Reef bush, 5100, ft. *Burtt Davy*, 2748 !

ORANGE FREE STATE : Northern slopes of the Drakensberg, *Cooper*.

TRANSVAAL : Middleburg Div. : Kaapsche Hoop, *Rogers*, 21089 ! Lydenburg
Div.; forests on the eastern slope of the Drakensberg, east of Pilgrim's Rest,
Legat 2455 ! Waterberg Div. ; Nylstrom, *Col. Herb.*, 9549 ! Pietersburg Div. ;

The Downs, *Rogers*, 21910! Zoutpansberg Div.; Houtboschberg, 4750 ft., *Burtt Davy*, 1194! 2313! *Houseman, Col. Herb.* 5249! *Botha*!

This tree, commonly known as yellow-wood or Upright Yellow-wood, or Regte Geelhout, is one of the most valuable timber trees of South Africa (see Sim, l.c.). Sim distinguishes 3 varieties, namely—(a) *latifolia*, (b) *angustifolia*, and (c) *falcata*. Vars. *latifolia* and *falcata* are said to have green receptacles 2–3 lin. wide, and seeds 4 lin. in diam., with bony shells, whilst they differ from each other in having leaves 1–2 in. by 2–3 lin., and 2–5 in. by 1½–2 lin. respectively. Var. *angustifolia* on the other hand is credited with red receptacles 3–4 lin. wide and seed 5 lin. in diam., with crustaceous shells and leaves 1–2 in. by 1–1½ lin. From the fact that the author adds that var. *latifolia* is the common form throughout Cape Colony, the Transkei and Lower Pondoland, and is easily recognizable by the short wide leaves standing erect all round the rigid twigs—as seen in his plate, t. 148—it is clear that he means the common form of the tree described here; this, however, has a thin crustaceous seed-shell (testa) and receptacles which become red when ripe (see Marloth's figure in his "Flora of South Africa," t. 17 A). Var. *falcata* is said to be the common form of Upper Pondoland, Griqualand East and Natal. Of this two barren branches are figured in t. 149. The one numbered 2 is described as a robust sapling, and represents certainly the juvenile state of *P. latifolia* as understood here, whilst number 1 is doubtful. Var. *angustifolia* which the author knows only from Robertson Division, is either *P. elongatus* or an undescribed species (see next species). Pilger l.c. admits two varieties, namely, var. *latior* with leaves 2–2½ by 7½–8½ lin., from the Vogelgat Mountains in Caledon Div., *Schlechter*, 9542! and var. *confertus*, with leaves 1–1½ by 4–5½ lin. with broad rounded apiculate tips, from Table Mountain, *Bergius*; *Schlechter*, 3947. Var. *latior* is very striking on account of the unusually broad leaves, which are also somewhat thinner in texture. The specimen seen by me is a ♀ with malformed cones infested with *Corynelia uberata*, Fries, a fungus very frequently found on the South African Podocarpus. Var. *confertus*, on the other hand, comes evidently within the range of ordinary fluctuation characteristic of this species.

2. P. elongatus (L'Hérit. ex Pers. Syn. ii. 580); a tree of varying dimensions, from " small " to 80 ft. high; branchlets of the mature tree elongate, terete, grooved when young, long marked with the decurrent leaf-bases; terminal buds ovoid, small, about 1 lin. long; leaves spirally arranged, in the upper part crowded, and then often subverticillate, linear, mostly acute, rarely subobtuse, very gradually narrowed at the base into an obscure petiole, 1½–2½ in. by 2 lin., straight or slightly falcate, obliquely erect, coriaceous, glaucescent, midrib distinctly raised on the back, obscurely on the upper surface, with 3 resin-ducts below the central strand, stomata confined to the lower side in close rows; male strobiles solitary or in scanty fascicles, cylindric, slender, ⅜–1 in. long, 1½–1¾ lin. in diam., with a few rotundate-ovate coriaceous very concave bracts about 1 lin. long at the base; scales imbricate, soon very loose, with an ovate to rhombic-ovate minutely denticulate blade rather over ½ lin. long; pollen-sacs ½ lin. long; female strobiles borne on slender peduncles 2–5 lin. long and formed of a pair of scales fused with the axis into a fleshy receptacle, scales ovate, unequal, the larger fertile, embracing the base of the ovule, the whole receptacle oblique at the top, 1½ lin. in diam. (in the dry state); seeds subglobose, very slightly longer than wide, 4 lin.

in diam., dark glaucous-green; inner layer of seed-shell thinly
crustaceous. Mirbel, Geogr. Conif. in Mem. Mus. Par. xiii. 75 ;
L. C. & A. Rich. Comm. Bot. Conif. (1826) 13, t. i. fig. 2 ; Loudon,
Arb. & Frutic. Brit. iv. 2101, fig. 1997 ; Endl. Syn. Conif. 218 ;
Lindl. & Gord. in Journ. Hort. Soc. v. 224 ; Carr. Trait. Conif.
ed. i. 470 ; ed. ii. 671 (partly) ; Gord. Pinet. ed. i. 273 (partly) ;
ed. ii. 334 (partly) ; Henk. & Hochst. Syn. Nadelhölz (partly) ;
Parlat. in DC. Prodr. xvi. ii. 511 (partly ?) ; Pilger, in Engl. Pflan-
zenreich, Taxac. 89 ; and in Engl. & Prantl, Nat. Pflanzenf. ed. 2.
247 ; Edmonds & Marloth, Elem. Bot. S. Afr. fig. 267 (?) (not text) ;
Stoneman, Pl. & their Ways S. Afr. fig. 216 (?) (not text) ; Dallimore
& Jackson, Handb. Conif. 44. *P. pruinosa,* Meyer ex Endl. l.c.
Taxus elongata, Thunb. Prodr. 117 ; Fl. Cap. ed. Schult. 547 ;
Solander in Ait. Hort. Kew, ed. i. iii. 415 ; ed. ii. v. 416. *T. capensis,*
Lam. Encycl. iii. 229. *T. falcata,* Thunb. Herb. ex Juel, Plant.
Thunberg, 69.

SOUTH AFRICA : without locality, *Masson* ! *Pappe* ! *Zeyher* 3889 !
CAPE PROVINCE : Swellendam Div. ; River Zonderend, 400 ft., *Schlechter,*
5682 ! Robertson Div., on sandy islands and banks of the Breede River, *Herb.*
Forest Dept. 1247 ! Stellenbosch Div. ; without precise locality, *Miller* !
Harvey ! Malmesbury Div. ; Riebeck's Castle, *Thunberg*; Vleermuys Drift,
Thunberg. Paarl Div. ; by the Berg River near Paarl, *Drège* ! Worcester
Div. ; Dutoit's Kloof, 1000–2000 ft., *Drège* ! Piquetberg Div. ; by the Berg
River near Dooreboom, *Bachmann,* 1522, 1523 ; near Vondeling, *Bachmann,*
2211. Clanwilliam Div. ; Kradouw Krantz above the Oliphants River, among
rocks in the cliffs, facing west. *Pearson,* 5328 ! *Pillans,* 5297 ! Cedarberg
Range, Kaakadouw Kloof, 1150 ft., *Diels,* 937 !

There is a collector's note with the specimen from the Breede river to the
effect that the plant differs from *P. elongatus* in its habitat, appearance, and
in the more swollen succulent and scarlet " aril," also in the fruit-stalk having
two berries ! The specimen agrees exactly with Schlechter's from the Zonderend
river and both much resemble *P. elongatus,* except that their leaves show a
tendency to become longitudinally wrinkled, which is not observed in the
remainder of the herbarium material of *P. elongatus.* The plant is figured in
Sim, For. Fl. Cape Good Hope, on t. 149, fig. 3, as *P. Thunbergii* var. *angustifolia* (Sim, l.c. 332). It requires further investigation. The specimen collected
by Pillans at the foot of the eastern cliffs of Kradouw Krantz (no. 5297) belongs
no doubt to *P. elongatus,* although the leaves are 3½–5 lin. wide and approach
those of *P. latifolius* in shape.

In cultivation in England with Mr. Lucas at Lee, Kent, 1777 ! Dr. Salisbury,
1785 ! Dr. Vere in Kensington, 1778 !

From photographs of the Thunbergian originals of this and the following
species in the Upsala herbarium, it appears that Thunberg wrote the species
up as quoted in Juel, Plantæ Thunbergianæ, that is, *Taxus falcata* for the
western and *T. elongata* for the eastern plant, whilst the localities quoted in
Thunberg's " Flora Capensis," ed. Schult., connect the western plant with
the name *T. elongata* and the eastern with *T. falcata.* The description of the
latter and the very name *falcata* as well as its station, "in sylvis " point un-
mistakably to the name *falcata* having been intended for it, and not for the
western plant, which, moreover, was early in cultivation in England under the
name *T. elongata,* probably from seeds sent by Masson, who accompanied
Thunberg on his journey through the western region when they passed
Riebeck's Kasteel and crossed the Berg river at Vleermuys Drift in 1773 or 1774.

3. P. Henkelii (Stapf ex Dallimore & Jackson, Handb. Conif.
15, 47) ; a tall tree, usually branched from the ground unless when
standing in close associations ; bark as in *P. latifolius* [Henkel],
or in old trees coming off in sheets (*C. Ross*) ; the young branchlets
of the mature tree more or less angular, glaucous ; terminal buds
globose-ovoid, scales very broad, shortly pointed with brown
margins or the tips foliaceous ; leaves spirally arranged, loose in
the juvenile state, moderately crowded in the mature state, drooping,
linear to lanceolate-linear, long tapering to a slender acute point,
gradually narrowed at the base into a short petiole, 3–6 in. by 3–4 lin.,
straight or frequently slightly falcate, suberect or spreading, thinly
coriaceous, more or less glaucous, midrib slightly raised on both
sides with 3 resin-ducts below the central strand, stomata confined
to the lower side ; male strobiles solitary or in clusters of up to 5,
cylindric, ¾–1¼ in. long, glaucous-pinkish, with rotundate-ovate
coriaceous bracts, 1 to almost 2 lin. long at the base ; scales imbricate,
at length loose, with a broad ovate fimbriate-denticulate small
blade up to ½ lin. long ; pollen sacs ¾ lin. long, conspicuous ; female
strobiles borne on a very short peduncle and formed of a single pair
of squamiform bracts fused into a slightly clavate receptacle, one
of the bracts barren, with a slightly lower insertion, both very
broadly ovate with a short slender tip, the upper, with a narrow
membranous denticulate margin, embracing the base of the ovule,
the mature receptacle stout, clavate, hardly fleshy, greenish-
glaucous with the fertile bract patelliform, the whole up to over
2 lin. by 1½ lin., mature seeds obovoid to ellipsoid, narrowed downwards, 9–10 lin. by 7–9 lin., olive green, inner layer of seed-shell
very delicate, hardly separable, brown, the remainder of the shell
hard-leathery, and very gritty, more or less resinous, up to 1 lin.
thick. Burtt Davy, Man. Flow. Pl. Transvaal, 100, 101. *P. Thunbergii*, Burtt Davy in Transvaal Agr. Journ. 1907, 421. *P. Thunbergii* var. *falcata*, Sim, Tree Plant. Natal 236, 285, fig. 94 ; Forest
Fl. Cape Col. 332 (in part), fig. opp. p. 55, t. cxlix. fig. 1. *P. falcatus*,
Marloth, Fl. South Afr. ; Suppl. Dict. Comm. Names of Pl. 101 ;
Sim, Fl. Trees & Shrubs f. use in S. Afr. 183, Native Timb. S. Afr.
102, fig. 1 ; not of R. Br.

EASTERN REGION : " British Kaffraria," *Cooper*, 1298 ! East Griqualand,
Mt. Ayliff Div., Fort Donald, *Sim, Nat. Herb.*, 19016 ! *Cockrane* ! Gwaleni
forest, *Cockrane in Forest Dept. Herb.*, 2172 ! *Forest Dept. Herb.*, 1248 ! Pondoland : Flagstaff Div. ; Tonti Forest, Whibley, *Forest Dept. Herb.*, 2167 ! Coll.(?)
in *Forest Dept. Herb.*, 1249 !

NATAL : Polela Div. : Riverside, *Dawson* ! *Kaufmann, For. Dept. Herb.*,
2170 ! 2171 ! Ingwangwane, Xalingena Forest, *Houshold, Forest Dept. Herb.*,
1947 ! 1948 ! Emkazeni Reserve, *Houshold, Forest Dept. Herb.*, 1880 !
Pietermaritzburg Div., Zwartkop, 4000 ft., *Sim, Nat. Herb.*, 19007 ! Blink-
water, 3000 ft., *Sim, Nat. Herb.*, 19014 ! Karkloof, 3000 ft., *Sim*, 19019 ! Maritzburg, 2300 ft., *Sim, Nat. Herb.*, 19011 ! 1912 ! 1913 ! and without precise
locality, *Henkel, Forest Dept. Herb.*, 2331a & b.

SWAZILAND : Forbes Reef Bush, 5100 ft., *Burtt Davy*, 2738a !

TRANSVAAL : Barberton, *Legat*, 3467 ! Pietersburg Distr. ; Houtboschberg,
Nelson, 420 !

According to Mr. Henkel, this constitutes nearly 90% of the growing stock
in some of the Natal forests. He also states that the tree extends westwards
into the Transkei region, whilst it does not appear to cross the Tugela river
in the east. It corresponds probably to a great extent to the *Podocarpus
falcatus* (Falcate Yellow-wood) of the Natal Forestry Reports. The difference
in the structure of the seed-shell of *P. Henkelii* and *P. falcatus* is very striking.
Native name *um-Sonti* or *um-Sunti.*

4. P. falcatus (R. Br. ex Mirb. Geogr. Conif. in Mém. Mus.
Par. xiii. (1825) 75) ; a tall tree up to over 100 ft. high, with a
straight cylindrical bole about 3 ft. (sometimes 8 ft.) in diam. ;
ultimate branchlets of the mature tree stunted, crowded, terete
or more or less angular from the decurrent leaf-bases when young ;
terminal buds ovoid, up to 1 lin. long ; leaves in the juvenile tree
loose, conspicuously plagiotropic, often subopposite, linear, acute,
straight or slightly falcate, up to 3½ in. by 1½ in.–3 lin., drying
dark green or brown, moderately thick, in the mature tree crowded,
scattered, not plagiotropic, linear to lanceolate- or oblong-linear,
subacute or those towards the base of the branchlets subobtuse or
obtuse, 1–1½ in. by 1½–2 lin., straight or nearly so, firmly coriaceous,
drying dark or pale green, midrib indistinct on the upper side,
slightly raised below, with 1 resin-duct below the central strand ;
stomata on both sides ; male strobiles solitary or in subsessile
clusters of 2 or 3 or more, cylindric, 3–4 lin. long, 1 lin. thick, each
supported by broad-obovate obtuse bracts ; scales imbricate,
with a cordate-ovate subdenticulate blade, ⅔ lin. long ; pollen-
sacs almost ½ lin. long ; female strobiles (only seen in the mature
and semi-mature state) peduncled, formed of a short stipe which
does not ultimately become fleshy and 2 or 3 subcoriaceous rotundate-
ovate obtuse scales up to 1 lin. long, all deciduous except the uppermost, which supports a seed ; peduncles about 3 lin. long, marked
with the scars of early deciduous coriaceous rotundate-ovate or
sometimes leaf-like bracts ; seeds more or less globose, about 6
(rarely up to 9) lin. in diameter, glaucous-green ; inner layer of seed-
shell very hard, bony, tubercled, outer somewhat fleshy, very
resinous inwards. Endl. Syn. Conif. 219 ; Lindl. & Gord. in Journ.
Hort. Soc. v. 224 ; Carr. Trait. Conif. ed. i. 472 ; ed. ii. 670 ; Gord.
Pinet. ed. i. 286 ; ed. ii. 336 ; Henk. & Hochst. Syn. Nadelhölz. 400 ;
Parl. in DC. Prodr. xvi. ii. 511 ; Pilg. in Engl. Pflanzenreich, iv. v.
Taxac. 72 ; Cape Forest Report, 1906, plates opposite pp. iv. and 18 ;
1907, plate opposite p. 4 ; Burtt Davy in Kew Bull. 1908, 147, and
Man. Flow. Pl. & Ferns Transvaal, 101 ; Dallimore & Jackson,
Handb. Conif. ed. 2, p. 44 ; Wilson, Plant Hunt. i. 25, t. 4 and in
Journ. Arnold Arbor. ix. 145, t. 14. *P. meyeriana*, Endl. l.c. 218 ;
Parl. l.c. 512 ; Marloth, Kapland, 206. *P. elongata*, Pappe, Silva
Cap. 32 ; Carr. Trait. Conif. ed. ii. 671 (partly) ; Fourcade, Rep.
Natal For. 1889, 4 (4, 121) ; Burtt Davy in Transvaal Agric. Journ.
1907, 421 (?) ; Edmonds & Marloth, Elem. Bot. S. Afr. 172, fig. 267–1
(not fig. 267–ii, iii) ; Engl. Pflanzenw. Ost.-Afr. C. t. 1, figs. C–H ;
Sim, Tree Plant. Natal, 236, 285 ; For. Fl. Cape Col. 332, t. cl. and in

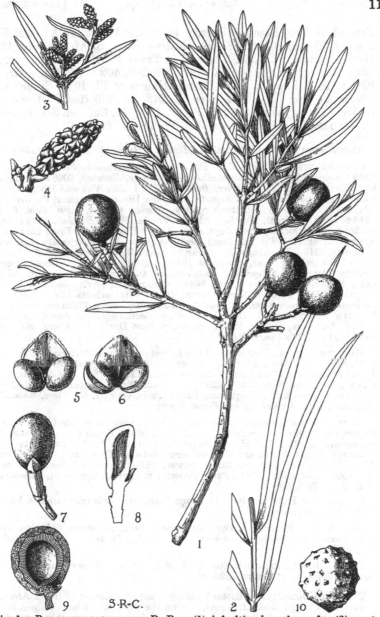

Fig. 1.—PODOCARPUS FALCATUS R. Br.—(1) A fruiting branch, × ¾; (2) part of a branch in the juvenile state (the leaves bent forward out of their natural angle 45°–50°), nat. size; (3) branchlet with ♂ strobiles, × ¾; (4) ♂ strobile, × 3; (5) scale of same from without, × 12; (6) same from within, × 10; (7) young ♀ strobile, × 2; (8) same in longitudinal section, × 2; (9) ripe fruit in longitudinal section, × 1½; (10) "stone" of a mature fruit, × 1½.

For. Fl. & For. Resourc. Portug. East Afr. 108, t. 97 A ; and Fl.
Trees & Shrubs S. Afr. 182, and Native Timb. S. Afr. 100, fig. 42 ;
Wood, Handb. Fl. Natal, 122, and in Trans. S. Afr. Phil. Soc. xviii.
224 ; Marloth, Kapland, 190, 191, 197, 208, 402 ; Fl. South Afr. i.
101, t. 18 & Suppl. ; Dict. Comm. Names of Pl. 101 ; Stoneman,
Plants & their Ways S. Afr. ed. ii. 242, fig. 215 (not 216), not of
L'Hér. *Taxus falcata,* Thunb. Prodr. 117 ; Fl. Cap. ed. Schult. 547.
T. elongata, Thunb. Herb. ex Juel, Pl. Thunb. 69.

SOUTH AFRICA : Without locality, *Masson* ! *Heward* ! *Mund & Maire* !

CAPE PROVINCE : Swellendam Div. ; Grootvadersbosch, *For. Dept. Herb.,*
1369 ! George Div. ; Sylvan Station near George, *Burchell,* 6068 ! Kaymans
Gat River, *Drège* ! Touw River, *Burchell,* 5761 ! Bier Vlei and Hooge Kraal
Forest Sections (*Cape Forest Reports*). Knysna Div. ; throughout the division
(*Cape Forest Reports*) ; Plettenberg Bay, *Bowie* ! Gowna, *Forst. Dept. Herb.*
1246 ! Kaatjes Kraal, *Burchell,* 5245 ! Vander Wats, *Burchell,* 5293 ! and
without precise locality, *For. Dept. Herb.,* 1244 ! Humansdorp Div. ; Lattering,
Storms River and Wit Els Bosch Sections (*Cape Forest Reports*) ; Hankey,
Fourcade, 3319 ! Uitenhage Div. ; Van Staadens River, *Burchell,* 4652 ! Enon,
Drège ! Zwartkops River, *Alexander Prior* ! Alexandria Div. ; without precise
locality (*Cape Forest Reports*) ; East London, *O. Kuntze* ! Albany Div. Alice-
dale, *Pulgrave, For. Dept. Herb.,* 2689 ! Queenstown Div. ; Junction Farm,
Galpin, 8179 ! Bongolo Neck, *Galpin,* 7973 ! Stutterheim Div. ; Kologha,
Isidenze and Quacu Sections (*Cape Forest Reports*). King Williamstown Div. ;
Everlyn Valley, Iselini, Donsal and Pirie Sections (*Cape Forest Reports*).
British Kaffraria, *Cooper,* 1297 ! Somerset East Div. ; Boschberg, *Burchell,*
3174 ! 3189 ! *MacOwan,* 1561 !

EASTERN REGION : Transkei and Pondoland ; Kentani, Butterworth, Tsomo,
Willowvale, Engcoba Umtatu, St. Marks, Elliot and Flagstaff Divisions (*Cape
Forest Reports*) ; and without precise locality, *Bachmann,* 74. Engcobe, Manina,
Merwe, *For. Dept. Herb.* 2268 ! 2269 ! Lusikisiki Div., Jntsubani, *Fraser,* 62 A !
For. Dept. Herb. 2226 ! Griqualand East ; Tsolo, Gumbu, Mt. Frere, Matatida
and Mt. Currie Divisions (*Cape Forest Reports*).

NATAL : From the coast to the Drakensberg above Newcastle (*Natal Forest
Report*) ; near Durban, *Sanderson,* 3015 ! *Wood,* 3005 ! Polela Div. ; Ingwangene,
Emkazeni forest, *Houshold,* For. Dept. Herb. 1949 ! 1956 ! Drakensbergen,
Rehmann, 6482 ! Zululand ; Guadeni forest between Tuzela and Insuzt rivers,
4500-5000 ft., and Ingwone forest between Mhlatuzuna and Msabazi rivers,
100-1500 ft., abundant (*Natal Forest Report*) ; Eshowe, Manga forest, Ballenden,
For. Dept. Herb. 2925 !

PORTUGUESE EAST AFRICA : Lourenço Marques ; Lebombo range in kloofs
near Estatuene, *Sim.*

TRANSVAAL : Pietersburg Div., Houtbosch Berg, *Rehmann,* 6481 (see also
the references to Mr. Rogers's specimen from " The Downs," no. 20210 ! under
P. gracilior).

BRITISH BECHUANALAND : Stellaland, in forests to the north and south of
Genesa (*Cape Forest Report,* 1895, 183).

This is the Outeniqua or Bastard Yellow-wood or Geelhout of South African
foresters. Though locally frequent, it is on the whole much less common than
the Upright or true Yellow-wood, the two Yellow woods together supplying
nearly one-half of the indigenous timber (Marloth, *Fl. South Afr.* i, 101). It
may be remarked that some of the localities quoted above for the eastern part
of the area on the authority of the Cape Forest Reports refer possibly to
P. Henkelii, with which *P. falcata* was confused for a long time.

5. P. gracilior (Pilger in Engl. Pflanzenreich, iv. v. Taxac. 71);
a tall tree up to 100 ft. or more high with a bole over 4 ft. in diam.;
ultimate branchlets of the mature tree crowded, angular from the
decurrent leaf-bases when young; terminal buds ovoid, $\frac{1}{2}$–$\frac{3}{4}$ lin.
long; leaves in the juvenile tree loose, conspicuously plagiotropic,
often subopposite, linear, long tapering to an acute point, straight
or slightly falcate, up to 4 in. by 3 lin., moderately thick, in the
mature tree crowded, scattered, not plagiotropic, linear, usually
long and gently tapering to a sharp point, rarely shortly acute,
1–2 in. by $\frac{1}{4}$–2 lin., coriaceous, drying dull green or brownish, midrib indistinct above, slightly raised below; male strobiles solitary
or in subsessile clusters of 2 or 3, supported by broad roundish
bracts, often up to 9 lin. and occasionally over 1 in. long; scales
imbricate, with a broadly ovate-triangular acute blade, $\frac{1}{2}$ lin. long;
female strobiles sessile at the end of short branchlets, carrying
reduced and often early deciduous leaves, the strobiles formed of a
short axis, 1–1$\frac{1}{2}$ lin. long, with 1–3 barren short ultimately deciduous
scales, dry and appressed or sometimes foliaceous, spreading and
recurved, the uppermost supporting an ovule; seeds ellipsoid-
globose, rounded or slightly attenuated at the base, 7–10 lin. long,
6–8 lin. across, glaucous-green to purplish-brown; inner layer
of seed-shell very hard, bony, slightly tubercled, up to 1 lin. thick,
outer usually thinner and dry or sometimes as thick as the inner
and slightly fleshy, resinous inwards. Engl. Pflanzenwelt Afr. iv.
v. 86, fig. 86, and Veget. Hara and Gallahochl. 11 (sphalm. *P.
gracilis*), Pilger in Engl. & Prantl, Nat. Pflanzenl. ed. 2, p. 245,
fig. 131); Burtt Davy, Flow. Pl. Transvaal, i. 101. *P. elongata*,
A. Rich. Tent. Fl. Abyss. ii. 278; Engl. Hochgebirgsfl. Trop. Afr.
109; Pfl. Ost.-Afr. C. 92, t. i. fig. B (not C–G); Oliv. in Journ. Linn
Soc. Bot. xxi. 404 (as forma ?); Hutchins, Forests of Kenia in Col.
Rep. Miscell. no. 41, 17; not of L'Hérit. *P. falcata*, Engl. in Hochgebirgsfl. Trop. Afr. l.c. and Veget. Usambara, 68; Pirotta in Ann.
Ist. Bot. Roma, vi. 156; not of R. Br. *Taxus elongata* Roth, in
Harris, Highl. Aeth. ii. 708; not of Ait.

TRANSVAAL : Zoutpansberg, *Houseman, Col. Herb.*, 5248! Pietersburg
Dist. ; The Downs, *Rogers*, 20210 ! (juvenile form).

Apparently throughout eastern tropical Africa as far as Abyssinia.

Podocarpus gracilior resembles *P. falcatus* so much that it has for a long
time been considered identical with it. Both exhibit a wide range of variation
in their foliage according to its age and position in the tree, the leaves becoming
smaller and more crowded as the tree reaches maturity, particularly so in the
flowering parts, and at the same time attaining more of the characters which,
as well as the floral and fruit-characters, are distinctive of the two species.
As the progress of transition from juvenile to adult stages is never quite even,
and individual trees or individual branches of one tree often lag behind or
behave precociously, barren specimens taken at random from a tree and distributed without explanatory notes are not always identifiable with certainty.
But as, moreover, the diagnostic characters even in complete specimens are
never very pronounced, *P. gracilior* and *P. falcatus* may with some show of
reason be considered as geographical subspecies or varieties of a *P. falcatus* in

a broad sense, *P. gracilior* being typical of the tropical parts of the common area and *P. falcatus* proper of extratropical South Africa with occasional linkages in one or the other direction. Houseman's specimen from Zoutpansberg (a fair-sized branch with adult foliage and a detached fruit, referred here to *P. gracilior*, agrees exactly in its foliage with typical specimens of this species from Kenya (*Hutchins*, no. 598), whilst the fruit is rather large with the outer layer of its shell too soft for typical *P. gracilior*; but it has, on the other hand, almost its counterpart in specimens of *P. falcatus* collected by Fourcade in the Humansdorp Division, Cape Province. The determination of this specimen therefore remains somewhat doubtful for the present, and the same applies to Rogers's specimen from The Downs, which is in the juvenile state.

6. **P. gracillimus** (Stapf in Prain, Fl. Trop. Afr. vi. ii. 343) ; a tree of great height (a " mammoth tree," according to Nelson), only known from a few branchlets in the adult state and some immature cones ; branchlets slender ; leaves densely crowded, sessile or subsessile, linear, about equally attenuated at both ends, or the tips more so and very acute, straight or very slightly curved, up to 10 lin. by ½–¾ lin., coriaceous, midrib obscure on both sides, particularly on the upper. Male and female strobiles unknown ; seeds (immature) globose, about 5 lin. in diam., supported by the moderately stout axis of the strobile, which is about 1½ lin. long and bears the scars of 1 or 2 barren scales, and by the fertile scale which is ovate-triangular, ½ lin. long and closely appressed to the seed. Burtt Davy, Man. Flow. Pl. Transvaal, i. 100, 101.

TRANSVAAL : Houtboschberg, *Nelson*, 423 ! *Burtt Davy*, 5083.

A very doubtful and incompletely known species. It may represent merely a state of *P. gracilior*, in which the reduction of the leaves in length and breadth characteristic of the fruiting stage has been carried to excess. Sim has already suggested this explanation. A similar state was collected by A. Whyte in the Eldama ravine, Kenya, along with typical *P. gracilior*.

ORDER CXXVII. B. **CUPRESSACEÆ**.

(By O. STAPF.)

Monœcious or occasionally diœcious. *Strobiles* small, mostly solitary and terminal. *Male strobiles :* scales opposite or in whorls of 3, squamiform and more or less shield-shaped, bearing basi-dorsally 2–6 pollen-sacs ; pollen grains without vesicular appendages. *Female strobiles :* scales few to many, all or most fertile, bearing basi-dorsally erect ovules, rarely all barren and surrounding a few terminal ovules. *Fruiting strobiles* developed as typical "cones" with enlarged leathery or woody valvate or imbricate and finally gaping scales, rarely fleshy and berry-like (galbules). *Seeds* free, rarely united in a stone, mostly with wing-like expansions of the crustaceous to woody testa ; cotyledons 2, rarely 3–6.

Shrubs or trees ; leaves spirally arranged in the juvenile state and on the long-shoots, otherwise decussate and in whorls of 3, the juvenile needle-shaped, the adult usually small, squamiform and more or less appressed, rarely both, the juvenile and the adult, needle-shaped and spreading.

DISTRIB. Genera about 15 with 125 species, of these 7 genera with over 40 species in the Southern Hemisphere ; 1 in South Africa.

I. WIDDRINGTONIA, Endl. Gen. Plant. Suppl. II. 25 (1842).

Monœcious (always ?) *Male strobiles* small, terminal, mostly on short lateral branchlets ; scales decussate, rhomboid-deltoid, produced into a beak or (upper scales) a short point or obtuse, with 2 pollen-sacs at the base. *Female strobiles* small, scattered along elongated shoots, singly or in dense clusters, rarely racemose ; scales opposite in 2 alternating pairs, divaricate at the time of pollination, then closing up, corky-coriaceous, apiculate ; ovules 3 or more at the base of each scale. *Mature strobiles* or *cones* woody, ovoid or globose, opening with 4 very thick erect valves, corresponding to the 4 scales. *Seeds* free, ovoid or trigonous, winged ; testa crustaceous. *Cotyledons* 2.

Evergreen trees ; leaves passing from a spiral arrangement in the juvenile state and in the long-shoots to a strictly decussate arrangement in the adult state, needle-shaped in the juvenile form, squamiform and tightly appressed in the adult ; cones the size of a small plum, usually clustered.

DISTRIB. Species 6 in South Africa, one of them also in Southern Tropical Africa.

Ovules 6–10 with each scale ; maturing seeds up to over 30 in a cone ; pollen-sacs protruding between the scales :

Mature cones globose-ovoid or obversely pear-shaped (when closed), up to 7 lin. in diam., smooth or blistered (but not tubercled), subapical cusps usually marked and pointed :

Cones distinctly stipitate with the stipe up to 3 lin. long, obversely pear-shaped, with a narrow top and with the cusps unequally distant from the centre of the top ; mature seeds up to 36, up to 5 lin. by 1½ lin. at the middle, wing up to 2½ lin. wide below the top ; ultimate and often also the penultimate ramifications of the branches terete, very slender, ⅓–½ lin. in diam., their scale-leaves rhombic-oblong, acute or subobtuse, 1 lin. long ... ... (1) **stipitata.**

Cones sessile or subsessile, globose-ovoid with a broad truncate top and cusps about equally distant from the centre of the top ; mature seeds up to 18, up to over 6 lin. by 2 lin. at the middle, wing up to 3½ lin. wide below the top ; ultimate and penultimate ramifications terete to subquadrangular, ½ lin. in diam., their scale-leaves rhombic, mostly ¾ lin. long (2) **Whytei.**

Mature cones (before opening) globose, up to 9 lin.
in diam., smooth, subapicular cusps (or at least
two of them distant from the top), usually short
and blunt or almost obliterated; mature
seeds up to over 20, 5–6 lin. long, including the
often oblique wings; scales of the ♂ strobile
delicate, except the sometimes subcoriaceous
acumen   ...   . (3) cupressoides.

Ovules 3–4 with each scale; maturing seeds up to 12
(rarely 14) in a cone; pollen-sacs covered by the
scales :

Mature cones not tubercled along the margins of the
valves, but more or less wrinkled all over; seeds
rather flat, 3½–4 lin. long, equally winged on
both sides, wings meeting at the apex in a notch,
up to 1 lin. wide; a shrub or small tree   ... (4) dracomontana.

Mature cones coarsely tubercled along the margins
of the valves; trees, sometimes of considerable
size :

Seeds rather flat, 4–6 lin. long, including the wings
which join over the apex and are there up to
2¼ lin. wide   ...   . (5) Schwarzii.

Seeds plump, more or less triquetrously ovoid,
4 lin. long, with a very narrow wing along 2
of the angles and over the tip   ...   ... (6) juniperoides.

1. **W. stipitata** (Stapf in Hook. Ic. Plant. t. 3126); a tree of
the habit of *W. Whytei*; juvenile state unknown; ultimate
ramifications very slender, cylindric, about ⅓ lin. in diam.; leaves
of the adult state decussate, squamiform; those of the long and
intermediate branches distant by their own length, subappressed,
lanceolate or acuminate, base adnate with parallel margins about
1 lin. long, those of the ultimate and often also of the penultimate
ramifications tightly appressed, rhombic-oblong from a cuneate
base, acute or subobtuse, 1 lin. long, rounded on the back, with the
free portion as long or nearly as long as the adnate; male strobiles
coetaneous (always?) with the mature cones, shortly cylindric,
up to 2 lin. long, sessile; scales in about 6 pairs, broadly rhomboid,
subacute to acute, ½ lin. long and wide, transversely depressed
below the middle, pollen-sacs 4, protruding between the scales;
female strobiles unknown in the flowering state; cones in loose
racemose clusters of 3–5 with a rhachis up to over 1 in. long, borne
on a stout stipe up to 3 lin. long, chestnut brown, very pruinose
towards the base, when closed inversely pear-shaped, obtuse, with
4 often pungent cusps, 10 lin. long, up to 7½ lin. in diameter below
the middle; valves more or less unequal, two ovate-oblong, up to
5½ lin. wide, two linear-oblong, up to 4 lin. wide, all usually obtuse,
smooth, their cusps unequally distant from the top; seeds up to
36, dark-brown, obovate-oblong, or oblong, with a terminal, oblong,
emarginate wing, 4½–5 lin. long, the body of the seed ovate-lanceo-

late, beaked, about 3 lin. long, 1½ lin. wide, the wing up to 2½ lin. wide below the top.

TRANSVAAL : Zoutpansberg, *Kotze* in *Forest Dept. Herb.*, 7048 ! *H. Hansen,* 7313 !

The specimens from which this species was described were taken from a tree in Mr. Hansen's garden at Piet Retief, which, according to Mr. Kotze, was obtained from the farm "Hillside," near Louis Trichardt, but specimens received from there proved to be *W. Whytei.*

2. **W. Whytei** (Rendle in Trans. Linn. Soc. ser. 2, Bot iv. 60, t. 9, figs. 6–11) ; a shrub or small tree, in the Transvaal, or up to 140 ft. high in the Tropics ; trunk up to over 5 ft. in diam., top wide, loose ; ultimate ramifications of the adult plant slender, cylindric or subquadrangular, about ⅓ lin. in diam. ; leaves of juvenile state acicular, up to 1 in. by ½–1 lin., of adult state squamiform, those of the long and intermediate branches with a lanceolate acuminate or oblong to ovate and acute somewhat spreading or appressed free blade, 2–1 lin. long, and a broad adnate base with more or less parallel margins ; those of the ultimate and sometimes also of the penultimate divisions tightly appressed (so that the contour of the branchlets is an approximately straight line or more often wavy), rhombic, about ⅓ lin. long or slightly shorter, subacute at both ends, the free and the adnate portions about equally long, slightly keeled or rounded on the back, with 1–3 slender resin-ducts, which are usually not visible externally ; male strobiles cylindric-oblong, 1½–2 lin. long, ebracteate and subsessile in the cup formed by the subtending foliage leaves ; scales in about 6 pairs, coriaceous, subpeltate, the lower deltoid, with a distinct hard beak, the upper more rounded and minutely apiculate ; pollen-sacs 4, protruding between the scales ; female strobiles in short subsessile often much reduced spikes, terminating with a vegetative bud and 2½–5 lin. long ; strobiles at the time of pollination 1½ lin. across, equalling or exceeding the subtending squamiform broad-ovate acuminate bract ; scales ovate, apiculate, face bluish-pruinose, back and margins greenish-brown ; ovules up to 5 with each scale ; cones (when closed) globose-ovoid with a broad truncate top and often pungent cusps, 9–10 lin. long, 7–8 lin. across, somewhat pruinose and resinous ; valves slightly spreading, their cusps about equally distant from the top ; seeds up to 10, obovate to obovate-oblong, up to 6 lin. by 2½ lin., dark-brown, the body ovate-lanceolate in outline, beaked, 3–3½ lin. long, the wing up to 3½ lin. wide below the top. Masters in Gard. Chron. 1894, xv. 746 ; 1894, xvi. 190, and 1905, xxxvii. 18 ; in Nature, 1894, 85 ; in Journ. Linn. Soc. Bot. xxxvii. 270 ; Whyte in Kew Bulletin, 1895, 189 ; Gard. Chron. ser. 3, xxxiii. 162 ; xxxvii. 18 ; McClounie in Kew Bulletin, 1896, 216 ; Rendle in Journ. Linn. Soc. Bot. xl. 235 ; Dallimore in Kew Bulletin, 1913, 224 ; Burkill in Johnston, Brit. Centr. Afr. 279 ; Sim, Forest Fl. Portug. East Afr. 109 ; Stapf

in Prain, Fl. Trop. Afr. vi. Sect. 2, pt. 2, 334 ; L. H. Bailey, Cult.
Evergreens, 231 ; Burtt Davy, Man. Flow. Pl. & Ferns Transvaal,
i. 102 ; Pilger in Engl. & Prantl, Pflanzenf. Ed. 2, xiii. 383 ; Dallimore & Jackson, Handb. Conif. 541 ; Chalk & Davy, Forest Trees
& Timb. Brit. Emp. i. 12 with fig. and t. opp. 18. *W. Mahonii*,
Mast. in Journ. Linn. Soc. London, Bot. xxxvii. 271 ; Sim, Forest Fl.
Portug. East Afr. 109. *Callitris Mahonii*, Engl. Pflanzenwelt Afr.
ii. 88. *C. Whytei*, Engl. l.c. 89 ; Eyles in Trans. R. Soc. South Afr.
t. 292.

TRANSVAAL : Lydenburg Distr., *Oranje, For. Dept. Herb.*, 340. Zoutpansberg
Distr. ; Blaauwberg, common in a kloof at the summit, *Houseman* ! *C. E. Gray*
in *Col. Herb.*, 4575 ! Hanglip, 3½ miles west of Louis Trichardt, *For. Dept. Herb.*,
7298 ! 5 miles west of Wylie's Poort, 4800 ft., on steep rocky slopes, *Hutchinson
& Gillet*, 4410 ! Pietersburg Distr. ; summit of the Wolkberg, near Haenertsburg, 5000-5200 ft., *Lewis* in *Col. Herb.*, 3597 ! 4310 !

Extending northwards into tropical Africa as far as Nyasaland.

3. **W. cupressoides** (Endl. Cat. Hort. Vindob. i. 209, and Syn.
Conif. 33) ; a rather compact shrub with fastigiate branches,
6–12 ft. high, rarely a tree with a trunk up to 1 ft. in diam. (according
to Fourcade), in cultivation, according to Carrière, up to almost
50 ft. high with strictly erect branches ; ultimate ramification of
the adult plant almost cylindric, the barren twigs slender, up to
½ lin. in diam. ; leaves of the juvenile state needle-shape, spreading,
up to 10 lin. by ¾ lin., glaucous below or quite green (cultivated
specimens) ; of adult state decussate, squamiform, those of the
older branches with an ovate acute usually appressed blade, the
free portion about 1 lin. long, those of the ultimate divisions tightly
appressed (so that the contour of the branchlet is approximately a
straight line or in the upper part more or less wavy), ovate-oblong
to rhombic-oblong, ⅔–¾ lin. long, subacute at both ends, but less
so at the lower, free and adnate portions about equally long,
sometimes obscurely keeled on the rounded back ; male strobiles
oblong to ovoid, about 1 lin. long, ebracteate and subsessile in the
cup formed by the subtending foliage leaves ; scales in about
6 pairs, peltate, rhombic, subacuminate, delicately scarious except the
frequently subcoriaceous acumen, the subhyaline margins minutely
denticulate ; pollen-sacs 4, protruding between the scales, female
strobiles in slender loose spikes, ½ to over 1 in. long and terminating with a vegetative bud ; strobiles at the time of pollination up
to 1½ lin. across, exceeding the subtending squamiform broad-ovate
acute bract ; scales ovate, acute or minutely apiculate, stout with
a large hump on the face, this and the base of the back bluish-
pruinose, otherwise greenish-brown ; ovules 6–7 with each scale,
bottle-shaped with 2 distinct equal wings ; cones 1–4 in a spike,
rather close, globose, up to 1 in. in diam., blackish-brown, somewhat
pruinose and resinous ; valves smooth, rarely slightly and irregularly

Fig. 2.—Widdringtonia cupressoides Endl.—(1) A cone-bearing branch,
× ⅔; (2) a branchlet of the juvenile state, × ½; (3) part of a branchlet
of the adult state, × 10; (4) a scale-leaf of same, seen from within, showing
how far it was adnate to the axis (white), × 10; (5) two ♂ strobiles, × 7;
(6) scale of a ♂ strobile, × 15; (7) same seen from within, × 15; (8) a
♀ inflorescence, × 2; (9) a ♀ strobile, ready for pollination, × 5;
(10) a scale of same with ovules at the base, × 5; (11) and (12), ovules,
× 20; (13) valve of a mature cone, showing scars of shed seeds, nat. size;
(14) and (15) seeds, nat. size.

tubercled with a short often blunt point (the morphological apex)
some distance below the top ; seeds up to 20 or more, somewhat
compressed, lanceolate to ovate-lanceolate in outline, 3½–5 lin.
long, broadly winged upwards, with the wings obovoid, emarginate,
often very oblique, and 5 to over 6 lin. long, black with a silky
lustre or the wings dark brown. Lindl. & Gord. in Journ. Hort.
Soc. V. 203 ; Knight, Syn. Conif. 13 ; Pappe, Silv. Cap. 31 ; Carr.
Trait. Conif. ed. i. 64 ; ed. ii. 518 ; Gord. Pinet. ed. i. 333 ; ed. ii.
417 ; Schlechtend. in Linnaea, xxxiii. 361, t. i. fig. 1 ; Henk. &
Hochst. Syn. Nadelhölz. 293 ; Sperk in Mém. Acad. Sc. St. Petersb.
Ser. 7, XIII. no. 6, t. 5, fig. 132-135 ; Parlat. in DC. Prodr. XVI.
ii. 443 ; Sim, For. Fl. Cape Col. 337, and Native Timb. S. Afr. 131,
t. 27 ; Bolus & Wolley Dod, Fl. Cape Penins. (in Trans. S. Afr. Phil.
Soc. XIV. 320) 1903 ; Mast. in Journ. Linn. Soc. xxviii. 270 ;
Saxton in Bot. Gaz. xlviii. 161–178, fig. 2, t. xi ; Marloth, Fl. S. Afr.
i. 101, fig. 67, a. *Cupressus juniperoides*, Linn. Sp. Pl. ed. ii, 1422 ;
Ait. Hort. Kew. ed. i, iii. 373 ; Harvey, Gen. S. Afr. Pl. ed. i. 311.
C. africana, Herm. & Oldenland ex Mill. Gard. Dict. ed. viii ; Hook,
f. in Lond. Journ. Bot. iv. 141. *Thuja cupressoides*, Linn. Mant. i.
125 ; Mant. ii. 518 ; Thunb. Prodr. 110 ; Fl. Cap. ed. Schult. 500 ;
Ait. Hort. Kew, ed. ii. v. 322 ; Willd. Sp. Pl. v. 510 ; Loudon,
Arb. Brit. iv. 2460, fig. 2316 ; Harvey, Gen. S. Afr. Pl. ed. i. 311.
T. sp. n. ?   Barrow, Travels S. Afr. i. 298. *Juniperus capensis*,
Lam. Encycl. ii. 626. *Schubertia capensis*, Spreng. Syst. iii. 890.
Pachylepsis cupressoides, Brogn. in Ann. Sc. Nat. 1ʳᵉ sér. xxx. 190 ;
Spach, Hist. Nat. Veg. xi. 346 ; Krauss in Flora, xxvii. 272 ; xxviii.
89. *Callitris cupressoides*, Schrad. in Drège, Zwei Pflanzengeogr.
Doc. 79, 115, 126, 170 ; Pappe, Fl. Cap. Med. Prodr. ed. i. 26 ; ed. ii.
37 ; Engl. Notizbl. Bot. Gart. Berlin, App. xi. 28, and Pflanzenw.
Afr. ii. 88 ; Marloth, Kapland, 116, 196, 199 ; Bolus & Wolley-Dod
in Trans. S. Afr. Phil. Soc. xiv. 320 ; Dallimore & Jackson, Handb.
Conif. 540. *Cupressus aethiopica coronata* . . . Breyne, Prodr. Fasc.
Rar. Pl. 39 ; ed. 2, p. 59. *C. nana compressis Taxi longioribus foliis
Afric.*, Pluk. Almag. Mant. 61. *C. africana lini folio*, Burmann,
Cat. Pl. Afr. Herm. 8. *Juniperus foliis frutex Afr.*, Pluk. Phyt.
t. 197, fig. 5 ; Almag. 202.

CAPE PROVINCE : Cape Div. ; Table Mountain plateau, *Wilms*, 3636 !
between Rondebosch and Wynberg, *Burchell*, 771 !   Orange Kloof, *Gamble*,
22013 ! Constantia, *Thunberg, Krause*, and without precise locality, *Kiggelaer* !
Masson ! *Harvey*, 419, *Thom*, 163 ! *O. Kuntze* ! Worcester Div. ; Dutoits Kloof,
1000-2000 ft., *Drège*. Caledon Div. ; Caledon, *Smith* ! Genadendal, 2000-
3000 ft., *Drège*. Swellendam, *Wallich, Drège*. Riversdale Div. ; Paardeberg,
Muir 5338 ! George Div. ; Cradock Berg, near George, *Burchell*, 5979 ! *Mund
(Ecklon & Zeyher* as *Thesium*, 52 partly)! Knysna Div. ; Outeniqua
Mountains, *Krauss* ! near Goukamma Rivers, *Burchell*, 5588 ! Humansdorp
Div. ; Fynbosch Hoek, *Kotze, For. Dept. Herb.* 3007 ! 3008 ! Witte Els Bosch,
900-1000 ft., *Fourcade*, 2293 ! Kromme River, *Drège* ! Storms River, *Hutchins* !
Uitenhage Div. ; Van Staadensberg Range, *Burchell*, 4688 ! Albany Div. ;
Grahamstown, *Thom* (?) 108 ! King Williamstown Div. ; Mountains, according
to *Sim*.

W. natalensis, Endl. Syn. Con. 34, which was very imperfectly described from a specimen said to have been sent by Gueinzius and Krauss from " Port Natal," is very likely *W. cupressoides.* Neither collected in the Drakensberg Mts., which were then (1839–1843) botanically quite unknown. On the other hand, both visited the area of *W. cupressoides* before they went to Natal.

Although the originals of Linnaeus' *Cupressus juniperoides*—two seedling plants—are lost, it is practically certain that they belonged to the same species as his *Thuja cupressoides,* described by him four years later. The same applies to Miller's *Cupressus africana* and Lamarck's *Juniperus capensis* and the various new combinations which rest on them. This has already been suggested by Schlechtendal, l.c., and, apart from other considerations, it is evident since the Cedarberg Mountains, the home of *Widdringtonia juniperoides,* Endl., were not known before the beginning of the 19th century, and certainly were not explored botanically until 1829. *W. cupressoides* is the Cape Cypress, Berg Cypress, or Sapree-hout of the Cape Colonists.

W. Commersonii, Endl. Syn. Conif. 34 (Syn. *Thuia quadrangularis,* Vent. in Nouv. Duham. iii. 16 ; *Pachylepis Commersonii,* Brogn. in Ann. Sc. Nat. 1ʳᵉ sér. xxx. 190), described from a specimen cultivated at Reduit in Mauritius about 1800 is possibly, as already suggested by Carrière, *W. cupressoides.*

4. W. dracomontana (Stapf ex Dallimore & Jackson Handb. Conif. 540) ;

a shrub, 8–10 ft. high, rarely a tree ; ultimate ramifications slender, about ½ lin. in diam. ; leaves of the juvenile state unknown, of the adult state decussate, squamiform, those of the older branches with an ovate subacute appressed blade, the free portion not much over 1 lin. long, those of the ultimate divisions tightly appressed with the upper part more or less bulging, so that the contour of the twig is a broken line, oblong to obovoid-oblong, hardly ever rhombic, subacute to subobtuse, with the free and adnate portions about equally long, obtusely keeled if at all, oblong, about 1 lin. long, ebracteate and subsessile in the cup formed by the subtending foliage leaves ; scales in about 6 pairs, subpeltate, rhombic-ovate, obscurely acuminate, subcoriaceous, slightly keeled upwards ; pollen-sacs 4, almost covered by the scales in the cone ; female strobiles in short, very scanty spikes, terminating with a vegetative bud ; cones at the time of pollination up to 2 lin. in diam., exceeding the subtending squamiform ovate subacute bract ; scales ovate-oblong, subobtuse, stout, particularly along the median line, but without a well-marked hump on the face, olive-green, more or less bluish-pruinose on the face when dry ; ovules 3 with each scale, bottle-shaped, with 2 obscure equal wings ; cones 1–2 from each spike, globose to ovoid-globose, smooth with a short stout point (morphological apex) from below the top, purplish-brown with a glaucous bloom, when quite mature 9–10 lin. in diam., blackish-grey ; valves wrinkled, but not or only very scantily tubercled, furrowed on the face, seed-scars obscure ; seeds up to 12 (?), somewhat compressed, oblong to lanceolate- or ovate-oblong, about 3½ by 1½ lin., equally or subequally winged on the sides, the wings meeting at the apex in a notch, about 1 lin. wide, dark brown or the nucleus black. *W. cupressoides,* Sim, Tree Plant. Natal, 234 ; For. Fl. Cape Col. 337

(the Drakensberg plant) ; Bews in Ann. Natal Mus. iii. 549 ; not of
Endl. *Callitris cupressoides*, Wood, Handb. Fl. Natal, 122, and in
Trans. S. Afr. Phil. Soc. xviii. 122, 224, not of Schrad. *C. natalensis*,
Endl. ex Fourcade, Rep. Natal For. 1889, 161, 121.

EASTERN REGION : Griqualand East ; Mt. Ayliff Distr., *Pole-Evans*, 30004 !
Transkei : Baziya Mountains, *Herb. For. Dept.*, 1375 !

NATAL : Drakensberg Range, headwaters of the Bushman's River (Langa-
libalele's location), *Fannin* ! *Sanderson*, 2011 ! Giant's Castle Game Reserve,
For. Dept. Herb. 2960 ! 2989 ! ; between Cathkin Peak and Mont aux Sources,
forming isolated woods or clumps at high altitudes, according to Fourcade ;
without precise locality, *Sim* in *Herb. For. Dept.*, 1044 !

A coloured drawing communicated by J. Sanderson, along with a fruiting
specimen, shows this species as a pyramidal tree of very regular habit with
drooping branches and twigs. Sim and Bew say it is rather a shrub than a tree,
but Sim also states that it has grown into a tree in the Pietermaritzburg Garden.

5. **W. Schwarzii** (Mast. in Journ. Linn. Soc. xxxvii. 269) ; a
tree, 50–80 ft. high, with a straight bole and pyramidal habit ;
branches ascending, ultimate ramifications of the adult plant almost
cylindric, ½ lin. in diam. ; leaves of juvenile state unknown,
of adult state decussate, squamiform, those of the older branches
with a broad-ovate, acute or subacute appressed free blade, the free
portion ¾–1 lin. long ; those of the ultimate divisions squamiform,
usually tightly appressed so that the contour is often a perfectly
straight line, or more or less wavy towards the tips of the twigs,
rhombic-oblong, ¾–1 lin. long, subacute at both ends, the free
portion much shorter than the adnate, rounded on the back or very
obscurely keeled ; male strobiles oblong, 1 lin. long ; scales in about
6 pairs, subpeltate, ovate, acute, much convex on the back, coriaceous, ½ lin. long ; pollen-sacs 4, covered by scales in the strobile ;
female strobiles unknown in the pollination state ; cones solitary
(always ?) on short clustered branches, 1½–1¾ in. long, subglobose,
not quite 1 in. in diam., grey or brownish-grey, valves more or
less spreading, upright, coarsely wrinkled and tubercled on the
concave back, the tubercles mostly along the edges, and with a
short blunt spur (the morphological apex) below the top, 10–11
lin. by 6–7 lin. ; seeds 10–14 ; body of the seed more or less obliquely
ovoid to ovoid-lanceolate, slightly compressed, 3–4 lin. long, 2-winged
upwards, including the wings 4–6 lin. by 2–3½ lin. broad-elliptic to
obovoid, emarginate, black to brownish-black with a silky lustre,
wings usually paler. Sim, For. Fl. Cape Col. 337, and Native Timb.
S. Afr. 131 (in part) ; Marloth, Fl. S. Afr. i. 101, t. 17 D. ; Dallimore
& Jackson, Handb. Conif. 541. *Callitris Schwarzii*, Marloth, in
Engl. Jahrb. xxxvi. 206, with figs. A–E ; Marloth, Kapland, 134.

CAPE PROVINCE : Willowmore Div. ; Kouga Mountains and Bavians Kloof
Mountains, near Nieuwe and Bosch Kloof farms, 2600–3900 ft., *Schwarz* !
Herb. Marloth, 3614 ! *Herb. For. Dept.* 1107 ! 1108 ! *Herb. Sim*, 2920 ! *Civil
Commissioner, Willowmore*, 1412, 1912 !

6. W. juniperoides (Endl. Syn. Conif. 32, excluding the synonymy) ; a tree, mostly 15–20 or occasionally up to over 60 ft. high, trunk up to 3 or 4 ft. in diam., branches horizontally spreading ; ultimate ramifications of the adult plant almost cylindric, $\frac{1}{2}$–$\frac{5}{8}$ lin. in diam. ; leaves of juvenile state unknown, of adult state decussate, squamiform ; those of the older branches with an ovate acutely acuminte upwards free and somewhat spreading or appressed free blade, the free portion rarely much over 1 lin. long ; those of the ultimate divisions tightly appressed, with the upper part more or less wavy, or slightly spreading so that the contour of the twigs is rarely a straight and unbroken line, rhombic-ovate to rhombic-oblong, about 1 lin. long, subacute at both ends or more obtuse at the lower, the free portion shorter than the adnate, slightly keeled on the back or rounded ; male strobiles cylindric-oblong, 1$\frac{1}{2}$–2 lin. long, ebracteate and subsessile in the cup formed by the subtending foliage leaves ; scales in about 6 decussate pairs, coriaceous, subpeltate, very broadly ovate, minutely apiculate to subobtuse ; pollensacs 4, covered by the scales in the strobile ; female strobiles in short densely scattered or crowded spikes, $\frac{1}{4}$–$\frac{1}{3}$ in. long and terminating with a vegetative bud ; strobiles at the time of pollination up to over 1 lin. across, exceeding the subtending squamiform broad-ovate acute bract ; scales ovate, subacute or obscurely apiculate, stout with a large hump on the face, olive-green ; ovules about 3 with each scale, bottle-shaped, broad, compressed, slightly unilaterally winged. Cones 1–3 in a spike, close, and if the spikes are crowded forming occasionally large compact clusters, globose, $\frac{1}{4}$–$\frac{3}{4}$ in. in diam., dark purplish-brown and usually covered with irregular roundish bosses from among which on each scale rises a stout conical pointed tubercle ; valves ultimately slightly spreading, coarsely warty or tubercled along the margins, with a stout conical often pungent mucro (the morphological apex) from below the top, and a usually striated central area ; seeds 4–8, stout more or less triquetrous-ovoid, 4 lin. long with a very narrow wing along 2 of the angles and over the tip, almost jet-black and somewhat glossy when mature or paler towards the base with a large white scar corresponding to similar scars at the base of the valves. Lindl. & Gord. in Journ. Hort. Soc. v. 203 ; Knight, Syn. Conif. 13 ; Pappe, Silva Cap. 30 ; Fl. Cap. Med. Prodr. ed. ii. 36 ; Carr. Trait. Conif. ed. i. 64 ; ed. ii. 58 ; Gordon, Pinet. ed. i. 334 ; suppl. 107 ; ed. ii. 418 ; Schlechtend. in Linnaea, xxxiii. 356, 361, tab. i. fig. 2 ; Henk. & Hochst. Syn. Nadelhölz. 292 ; Parlat. in DC. Prodr. xvi. ii. 442 ; Rehmann, Geo-bot. Verh. S. Afr. (in Bot. Centralbl. i. 1120, 1122) ; Masters in Journ. Linn. Soc. xxxiii. 268 ; Sim, Tree Plant. Natal, 234 ; For. Fl. Cape Col. 336, tab. 147 and Native Timb. S. Afr. 131, t. 27 ; Marloth, Fl. S. Afr. 101, 104, tab. 17 B, and 19 ; Dallimore & Jackson, Handb. Conif. 540 ; Wilson, Plant Hunt. t. 6. *W. Wallichii*, Endl. l.c. 34 (name) ; Lindl. & Gord. l.c. ; Carr. Trait. Conif. ed. i. 68 ; ed. ii. 62 ; Gordon, Pinet. ed. i. 335 ; ed. ii. 419 ; Schlechtend. l.c. 359 ; Henkel & Hochst. l.c. 295 ; Parlat. l.c. 433 ; Masters, l.c. 271, 273, 274.

W. Wallichiana, Gord. Pinet. suppl. 107 (name). *Callitris arborea*,
Schrad. ex Drège, Zwei Pflanzengeogr. Doc. 73 (name) ; Hutchins in
Report Conserv. For. Cape Col. 1895, 48, 49 ; in Trans. S. Afr. Phil.
Soc. xi. 62 ; in Agric. Journ. Cape Good Hope, xxvi. 661, 662 ; Storr
Lister, Rep. Chief Conserv. Fort. Cape Good Hope, figs. on p. 2.
C. stricta, Schlecht. (err. pro *C. arborea*, Schrad.) Hook. f. in Lond.
Journ. Bot. iv. 141. *C. Ecklonii*, Schrad. ex Pappe, Fl. Cap. Med.
Prodr. ed. i. 25. *C. juniperoides*, Durand & Schinz, Consp. Fl. Afr.
v. 951 ; Engler, Pflanzenw. Afr. ii. 88 ; Marloth, Kapland, 167, fig.
on p. 168. *Parolinia juniperoides*, Endl. ex Gord. Pinetum, Suppl.
107. *Pachylepis* sp., Hook. f. l.c. 142.

CAPE PROVINCE : Clanwilliam Div. ; Cedarberg Mountains, scattered singly
or in small clumps over a range of 30 miles mainly between 3000 and 6500 ft.,
Zeyher ; *Drège* ! *Wallich* ! *Leipoldt in MacOwan, Herb. Aust. Afr.* 1649 !
Budler in Herb. MacOwan, 3034 ! *Herb. For. Dept.* 1027 ! 1029 ! 1030 ! *Herb.
Sims* ! *Hutchins* !

Masters' statement that it also occurs at Swellendam is due to an old erroneous
entry in the Kew Herbarium according to which Wallich collected it in that
locality. His specimens are in fact from the Cedarberg Mts., as is shown by a
note in his own handwriting in the collections of the British Museum. Leipoldt's
specimens have unusually smooth, small and barely mature cones with small
marginal tubercles and more conspicuously, yet still narrowly, winged seeds.
The cones were collected in June when the tree was in flower, and their peculiar
condition may be due to delayed development. They were found in an
unusually low locality (2500 ft.) and are also interesting on account of the
presence of male strobiles on the female branches. The tree was certainly not
known in Linnæus' time, and probably not until Zeyher and Drège collected
it in 1829 and 1830 respectively. Linnæus' *Cupressus juniperifolia* and all the
other names connected with it have therefore to be struck out of the synonymy
of this species. A valuable timber tree (see particularly Sim and Hutchins,
ll.cc.). The Cedar-boom or Cape Cedar of the Cape Colonists.

W. equisetiformis, Mast. in Journ. Linn. Soc. Bot. xxxvii. 271, described from
specimens cultivated in the Tokai plantations near Cape Town (!) and others
(*Baur*, 1164 !) communicated to the author from the Katbergen, Stockenstrom
Division, has since been identified by the author himself as *Callitris robusta*, a
native of Australia (see *Journ. Linn. Soc. Bot.* xxvii. 332).

ORDER CXXVIII. **CYCADACEÆ.**

(By J. HUTCHINSON & G. RATTRAY.)

Male and female scales (sporophylls) arranged spirally or superimposed in cones (except in *Cycas*) ; cones diœcious. *Male cones*
solitary to several, terminal or subterminal, composed of numerous,
usually thick and fleshy or subwoody, often peltate scales bearing
on their lower surface very numerous and crowded 1-locular
sporangia, the latter often collected in small groups. *Female cones*
terminal or subterminal ; scales usually numerous, more or less

peltate, bearing 2 orthotropous inverted ovules on the lower side (or in *Cycas* several and erect in the sinuses of the segments of the leaf-like sporophyll). *Seeds* large, drupe-like, globose, ovoid or oblong, turgid or angular, the outermost layer of the integument fleshy and often coloured, middle layer crustaceous or bony ; endosperm abundant, fleshy, with one or more embryonic cavities, but with usually a solitary, slender, cylindric embryo ; cotyledons 2 ; radicle superior, attached by the spirally twisted suspensor.

Stem subterranean or above ground and attaining small tree-form, simple or sparingly branched, with a terminal tuft of leaves. Leaves spirally arranged, the spirals alternating with rows of short coriaceous prophyllary scales ; blade divided usually to the midrib into separate leaflets, the latter with or without a midrib, longitudinally nerved or rarely (*Stangeria*) pinnately nerved and the lateral nerves forked.

Species about 80, in the tropics, subtropics, and temperate regions, mainly of the Southern Hemisphere.

We are much indebted to the Director of the Botanical Survey of South Africa for the loan of specimens from the Pretoria and Durban herbaria, and for photographs of plants in their native habitats in the Transvaal ; also to the Directors of the Cape Town and Albany Museums.

An interesting collection of living South African Cycads has been brought together by Colonel Molyneux, in the Old Fort garden in Durban, which the junior author had the pleasure of visiting in August, 1930. Many plants are also grown around the Union Buildings at Pretoria, at the National Botanic Gardens, Kirstenbosch, near Cape Town, and in the Municipal Gardens, Cape Town. The majority of the species are also in cultivation in the Palm House at Kew.

I. **Stangeria**—Leaflets with a prominent midrib and spreading forked lateral nerves, the upper leaflets connate at the base and decurrent on the rhachis ; aerial stem absent.

II. **Encephalartos.**—Leaflets without a distinct midrib, the nerves parallel with the margins ; leaflets never connate at the base ; aerial stems present or absent.

I. STANGERIA, T. Moore.

Cones diœcious. *Male cones* cylindric, slender ; scales densely imbricate in many series, spirally arranged ; pollen-cells stipitate ; pollen ellipsoid. *Female cones* ovoid-ellipsoid, shorter than the male, shortly pedunculate, densely tomentose ; scales deltoid at the top with the lower side rounded, bearing at the base a pair of inverted ovules. *Seeds* broadly ellipsoid, with a dark red fleshy outer coat.

Stem subterranean, simple or branched, nude; leaves few, long-petiolate, pinnate; leaflets several pairs, opposite or subopposite, sometimes the upper ones connate at the base, entire, toothed or incised-lobate, with a conspicuous midrib and numerous spreading forked lateral nerves.

DISTRIB. Species 1, South-east Africa; coastal region from Bathurst to Zululand.

Stangeria eriopus (Nash in Journ. New York Bot. Gard. x. 164, pl. lxii (1909); stem subterranean, branched or unbranched, branches short and thick, cylindrical to obovoid, the woolly scales persistent only at the apex; leaves 1–3 to each crown; petiole as long as or longer than the blade, deeply grooved on the upper side, glabrous; leaflets 5–14 pairs, opposite or subopposite, the lower-most stalked, the upper connate and decurrent on the rhachis on the lower side, entire or serrulate to irregularly incised-lobate, elongate-lanceolate, acutely acuminate to rounded at the apex, up to 12 in. long and 2 in. broad, with numerous very closely parallel forked pinnate lateral nerves, glabrous; male cones solitary, brownish, pedunculate, cylindric, gradually tapered to the apex, 6–15 in. long, up to 3 in. in diam., with numerous spirals of closely imbricate scales; scales at first woolly, soon glabrescent, broadly triangular or rhomboid, jagged-toothed; female cones solitary, shortly pedunculate, densely tomentose, ovoid-ellipsoid, up to 7 in. long and 3½ in. in diam.; scales deltoid, with the lower side rounded; seeds broadly ellipsoid, dark red, about 1 in. long; aril very fleshy. *Lomaria coriacea*, Kunze in Linnæa xiii. 152 (1839), not of Schrad. *L. eriopus*, Kunze, l.c. xiii. 152 (1839), and xviii. 116 (1844). *Stangeria paradoxa*, T. Moore, in Hook. Journ. Bot. v. 228 (1853); Hook. Bot. Mag. t. 5121; Miq. Prodr. Cycad. 9, 18; DC. Prodr. xvi. ii. 530; Chamberlain in Bot. Gaz. lxi. 353, with figs. (1916); Pearson in Trans. S. Afr. Phil. Soc. xvi. 349, pl. viii; Schuster in Engl. Pflanzenr. Cycadac. 105, fig. 15, A–K, and t. 3 (1932). *S. schizodon*, Bull. Cat. 1872, 8. *S. paradoxa* var. *Katzeri*, Marloth Fl. S. Afr. i. 97, fig. 63, and pl. 14 (1913); Schuster, l.c. 105. *S. paradoxa* forma *schizodon*, Schuster, l.c. 105. *S. Sanderiana*, Hort. *S. Katzeri*, Regel Gartenfl. xxiii. 163, t. 798 (1874). *S. Zeyheri*, Stoneman, Plants and Their Ways in S. Afr. fig. 214 (1915). See observations by Seemann, Bot. H. M. Herald, 235 (1852-7).

GEOGRAPHICAL RANGE :—Narrow coastal strip from the Kowie River, Bathurst Div., through the native states and Natal to South Zululand; on grassy slopes, the edge of bush, or in the forest; variable according to habitat.

Bathurst Div. : between Riet River and Great Fish River, *Macowan*, 2000 ! Kentani Div. : coast belt between sand dunes, *Pegler*, 262 ! near Zolora River mouth, *Pegler*, 262 ! Manubi Forest, *Pegler*, 1247 ! Port St. Johns, *Rattray* ! *Schonland*, 3957 ! Egossa Forest, *Sim* ! Pondoland : Umteentu River, in shade of trees amongst rocks, *Burtt Davy*, 15315 ! Lusikisiki, *Doran* in Herb. S. Afr. Mus., 9412 ! without locality, *Bachmann*, 66 ! Natal : Manda, *Buchanan*, 36 ! Nahoon River, *J. M. Wood* in Herb, S. Afr. Mus. 1350 ! Dumisa, Alexandra distr., wooded slopes by the Umtwalumi River, *Rudatis*, 669 ! Pinetown, *Rehmann*, 7959 !

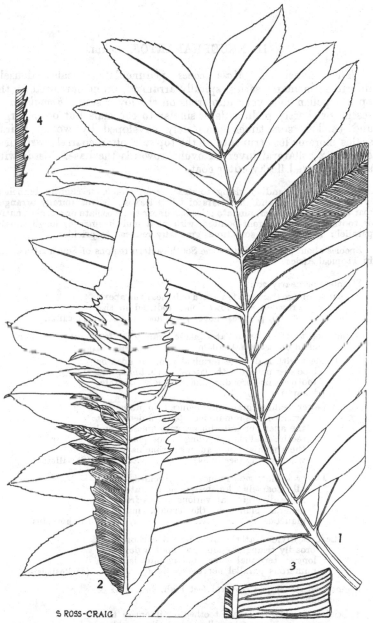

Fig. 3.—STANGERIA ERIOPUS Nash—(1) Typical leaf from open veld ; (2) leaflet from forest locality ; (3) portion of leaflet showing veins ; (4) margin of serrulate leaflet ; much reduced, except (3) and (4).

II. ENCEPHALARTOS, Lehm.

Cones diœcious. *Male cones* pedunculate ; scales densely imbricate in many series, spirally arranged, often narrowed at the apex ; pollen-cells very numerous on the lower side. *Female cones* sessile or shortly pedunculate, similar to the male but often larger and thicker, sometimes completely enveloped by woolly hairs ; scales more or less truncate at the top and often coarsely wrinkled, bearing 2 collateral inverted ovules towards the base. *Seeds* with a yellow or red fleshy outer coat.

Stems underground or rising to small trees, simple or slightly branched, covered with scales and the scars of fallen leaves ; leaves spirally arranged near the top of the stem, pinnate ; leaflets linear to lanceolate or acicular, entire or toothed, often pungent-pointed, without a midrib ; nerves longitudinally parallel ; lower leaflets sometimes gradually reduced to prickles.

Species about 20, confined to the South-eastern regions of South Africa, and in Tropical Africa.

Stemless or nearly so :
    Leaflets entire (rarely a few teeth only near the apex),
        very crowded, the lower ones reduced in size
        but not like prickles, never glaucous ...   ... (1) **caffer.**

    Leaflets toothed or lobate-spinose (rarely a few
        leaflets on each leaf entire) :
    Leaflets with a definite terminal pungent apex or
        lobe distinct from the lateral teeth, linear and
        toothed or broader and coarsely dentate-
        lobate :
      Lower leaflets gradually reduced to numerous
        prickles, the remainder often 2–3-dentate at
        the apex, broadly linear, with small lateral
        teeth ; leaflets spreading in one plane from
        the rhachis, not glaucous ; rhachis woolly
        when young (see fig. 4)   ...   . (2) **villosus.**

      Lower leaflets not reduced to prickles, nearly
        all coarsely lobate-spinose-dentate, the
        lobes diverging at various angles from the
        general plane of the leaflet, markedly
        glaucous   ...   . (3) **horridus.**

    Leaflets divided at the apex into 3-5 short very
        broadly triangular lobes, without a definitely
        longer terminal lobe, oblong-elliptic, with
        numerous parallel nerves (see fig. 5)...   ... (4) **kosiensis.**

Stems well developed, several feet high, branched or
    unbranched :
* Leaflets toothed or lobate-toothed (or if entire then
    very glaucous or broadly linear), linear-oblong
    to ovate-lanceolate ; cones not very woolly ;
    seeds red :

Leaflets glaucous, linear and tapered to the apex,
coarsely lobate-dentate or rarely entire,
markedly spinose-pointed (see fig. 6)   ...   (5) **Lehmannii.**

Leaflets green :

Leaflets coarsely lobate-dentate on the lower
margin, usually rather short in proportion
to their width, very strongly nerved with
numerous nerves (see fig. 7)   ...   ...   (6) **latifrons.**

Leaflets shortly toothed or the lower ones at
most coarsely toothed :

Mature leaflets glabrous below, with inconspicuous nerves :

Leaflets entire or toothed mainly on one
side, obliquely linear-lanceolate, one
side curved, the other nearly straight,
the lower leaflets not reduced to
prickles ...   .   (7) **longifolius.**

Leaflets equally toothed on both sides,
broadly linear, with parallel margins,
the lower not reduced to prickles   ...   (8) **Altensteinii.**

Leaflets more or less ovate-lanceolate, the
lower ones gradually reduced to
prickles (see fig. 8)   ...   .   (9) **Woodii.**

Mature leaflets pubescent below, with very
strongly marked numerous nerves, the
lower leaflets rather abruptly reduced to
prickles   ...   ... (10) **pauoidontatun.**

** Leaflets entire, fern-like, never glaucous, narrowly
linear or acicular, crowded and numerous ;
rhachis usually very woolly, especially when
young ; cones very woolly-tomentose ; seeds
yellow or dull brown :

Leaflets linear, flat, with a strong cartilaginous
but not recurved margin :

Nerves of the middle leaflets 7-9 between the
margins, very strong and rounded and completely filling the lower surface of the
leaflets   ...   . (11) **cycadifolius.**

Nerves of the middle leaflets 12-14 between the
margins, slender and somewhat obscure,
not filling the lower surface   ...   ... (12) **lanatus.**

Leaflets subacicular, with much recurved margins ;
nerves 3–4, very obscure on the lower side   ... (13) **Ghellinckii.**

1. **E. caffer** (Lehm. Pugill. vi. 14 (1834)); stem subterranean,
about 1 ft. in diam., woolly ; leaves green, up to about 14 in a crown,
up to 2 ft. long ; petiole ¼–⅓ as long as the rhachis, at first woolly
towards the base, soon nearly glabrous ; rhachis bluntly 4-ribbed,
woolly when young ; leaflets very numerous and crowded, longest
about the middle of the leaf, the lowermost becoming much reduced,

all narrowly linear-lanceolate, contracted above and decurrent at
the base, sharply spinulose-acute, entire or with 1-2 teeth near the
apex, middle leaflets 2-4 in. long, about ⅓ in. broad, pubescent when
young, about 10-nerved, with strong marginal nerves; male
cones solitary or 2-3 together, pedunculate, oblong-lanceolate,
8-12 in. long, 2-3 in. in diam.; scales rhomboid and concave at the
top, rugose; female cones pedunculate, oblong or ellipsoid, about 6 in.
long and 4 in. in diam.; scales in 4-6 spirals, broadly transversely
subrhomboid-elliptic, concave, green, with orange margins, nearly
glabrous, about 1½ in. broad; seeds broadly oblong, red, about
1¼ in. long. Miq. Monogr. Cycad. 53, partly (1842); Prodr.
Cycad. 20, partly (1861); DC. Prodr. xvi. ii. 532. *Cycas caffra,*
Thunb. in Nov. Act. Soc. Scient. Upsal. ii. 285, t. 5 (1775), as to
figure and part of description. *Zamia caffra,* Thunb. Prodr. Fl.
Cap. ii. 92 (1800); Fl. Cap. ed. Schult. 429, partly (1823). *Zamia
Cycadis,* Linn. f. Suppl. 443 (1781). *Encephalartos brachyphyllus,*
Lehm. Cat. Hort. Hamb. 1836, ex Lehm. & De Vriese Tijdschr.
Nat. Gesch. iv. 414, t. vi. and vii. (1838). *E. caffer* var. *brachy-
phyllus,* A.DC. in DC. Prodr. xvi. 2, 532 (1868).

GEOGRAPHICAL RANGE: Uitenhage, Bathurst, East London, southern part of
King Williamstown and northwards to Zululand.

Uitenhage Div.: *Tredgold,* no. 2 (Herb. Brit. Mus.)! Van Staadens, *Rattray,*
1098! between Hoffmannskloof and Driefontein, 1000–2000 ft., *Drège,* 8254!
East London Div.: rocky hill near Mt. Coke, in open ground, *Galpin,* 7839!
Kentani Div.: outside forest, rare, Sept., *Pegler,* 1124! Feb., *Pegler,*
2156! Zululand: near Ngoye, *Rattray,* 1278! without locality, *Oldenburg,*
1497!

CULTIVATED SPECIMENS: Municipal Gardens, Cape Town, *Herb. S. Afr. Mus.,*
24359! *Nat. Herb. Durban* (from Zululand), 16040! Bot. Gard. Grahamstown,
E. Tidmarsh!

According to Wylie this species is common in some parts of Zululand, where
it grows almost socially; the seeds are much sought after by baboons, who
carry them to the tops of the krantzes and thus facilitate distribution.

In his original description Thunberg confused two species, the dwarf stemless
plant which he figured and which is described above as the true *E. caffer,*
and another with a well-developed stem, *E. longifolius* Lehm.

2. **E. villosus** (Lem. Illustr. Hort. xiv. Miscell. 79 (1867)); stem
subterranean, very densely woolly-villous; leaves shining green,
usually few in a crown, slightly arcuate, up to 9 ft. or more long;
petiole, rhachis and leaflets densely woolly-villous when young,
becoming glabrous or nearly so; leaflets numerous, the lower
ones gradually and markedly reduced to prickles, the middle and
upper ones broadly linear, pungent-pointed, usually with a few small
ascending teeth and often 2-3-dentate at the apex, up to 8 in. long
and ¾ in. broad, the broadest about 25-nerved, nerves slightly
prominent below; male cones yellowish, conspicuously pedunculate,
slender, cylindric, slightly tapering to the top, about 2 ft. long

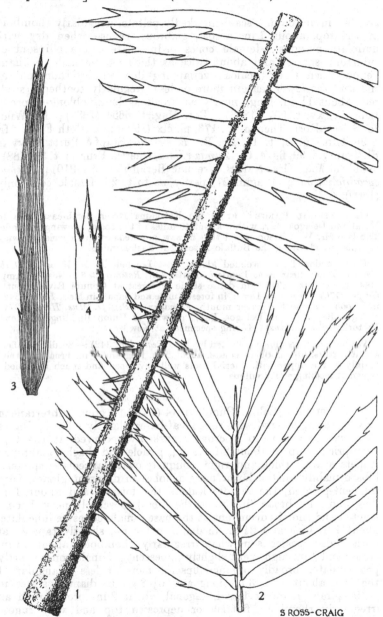

S ROSS-CRAIG

Fig. 4.—ENCEPHALARTOS VILLOSUS Lem.—(1) Basal portion of leaf showing
reduction of leaflets to prickles, typical of this species ; (2) upper portion
of leaf ; (3) single leaflet from about the middle ; (4) portion of leaflet
showing characteristic 3-pronged top ; (1), (2) and (3) much reduced.

and 3–4 in. in diam. ; scales markedly peltate, irregularly rhomboid
at the top, about 1 in. across, glabrous, rugose when dry with
undulate margins ; female cones pedunculate, more ellipsoid or
subovoid, shorter and about twice as thick as the male, brilliant
yellow when ripe ; scales overlapping downwards, rhomboid at
the top, the lower margin more or less irregularly toothed ; seeds
red, about 1¼ in. long, nearly as broad as long, oblong-ellipsoid.
Lem. l.c. xv. t. 557 (1868) ; Bot. Mag. t. 6654 (1882) ; De Wild.
Ic. Select. Hort. Thenen. iv. 173, pl. clx (1903) ; Marloth Fl. S. Afr.
i. 96, t.15, fi g. B; t. 16, fig. B.   *E. striatus,* Stapf & Burtt Davy in
Fl. Transv. i. 99, fig. 4, C.   *Zamia villosa,* Gaertn. Fruct. i. t. 3 (1788);
Willd. in Mag. Ges. Naturf. Freunde Berlin x. t. 6 (1810) ;   *Zamia
cycadifolia,* Jacq. Fragm. 27, partly, as to t. 25, female cone only
(1800).

GEOGRAPHICAL RANGE : From the Keiskama River north-eastwards to
Natal and Delagoa Bay, within about 40 miles of the coast, growing in shade.
The late Mr. W. Watson noted that he saw hundreds of this species growing
in dense woods along the Buffalo River near East London.

East London Div. : wooded kloofs, East London, *J. M. Wood in Herb.
Galpin,* 3340 ! near East London, *Smale,* 23 ! *Rattray,* 386 ! Nead's Camp,
4000 ft., *Galpin,* 3340 ! in dense shade of forest at Gonubi River mouth,
Galpin, 7767 ! Kentani Div. : in forests, male and female in May, *Pegler,* 342 !
in woods near the Kei River mouth, *Flanagan in S. Afr. Mus. Herb.,* 1374 !
Delagoa Bay : "from Delagoa Bay," growing in Union Buildings Gardens,
Pretoria (coll. *Wickins*) ! Living specimens at Kew !

We have seen seedlings collected by *Miss Pegler* (no. 342) :—Seedling leaflets
about 8 pairs, oblong-oblanceolate, about 2 in. long and ⅓ in. broad, toothed
mainly in the upper half ; aerial roots well developed and much branched ;
crown of seedling softly villous.

3. **E. horridus** (Lehm. Pugill. vi. 14 (1834)) ; stem subterranean
or very short, covered with rough leaf-bases, slightly villous ; leaves
glaucous, numerous in a crown, markedly recurved at the top,
very prickly, up to about 2 ft. long ; petiole and rhachis glabrous ;
leaflets spaced, thick and rigid, obliquely ovate-lanceolate, spinous-
lobate mainly on the lower margin, lobes in different planes, very
pungent-pointed, the middle leaflets the longest and about 4 in.
long and 2 in. broad (including the lobes), obscurely nerved, contracted and shortly decurrent at the base ; male cones pedunculate,
about 1 ft. long and 2½ in. in diam., cylin ric, slightly tapered at
both ends ; scales somewhat irregularly rhomboid, about 1½ in.
across and ¾ in. high, very slightly pubescent ; female cones shortly
pedunculate, broadly oblong-ellipsoid, more or less triangular at
the top, about 1 ft. 3 in. long and 6½–8 in. in diam. ; scales in
8–10 spirals, more or less hexagonal, about 2 in. broad, with an
irregularly rhomboid flattish or depressed top and very rugose
surface ; seeds oblong, orange-red, slightly angular, about 1½ in.
long. Miq. in Ann. Sci. Nat. Ser. II. x. 366 (1838) ; in Tijdschr. Nat.
Gesch. vi. 94, tt. 3 and 4 (1839) ; Monog. Cycad. 58 (1842), and in

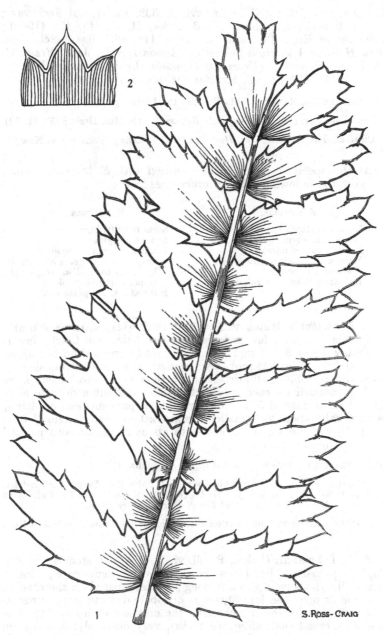

Fig. 5.—ENCEPHALARTOS KOSIENSIS Hutch.—(1) Upper half of leaf; (2) top of leaflet; reduced.

Linnæa, xvii. 726 ; DC Prodr. xvi. ii. 532, partly and excl. var. ;
Worsdell in Trans. Linn. Soc. Bot. Ser. II, v. 449, 452 (1900) ;
Schuster in Engl. Pflanzenr. Cycadac. 116, with figs., excl. vars.
Van Hallii and *latifrons* Schuster. *Zamia horrida*, Jacq. Frag. 27,
tt. 27 and 28 (1800). *Encephalartos nanus*, Lehm. Tijdschr. IV, 421,
t. 8, fig. C. (1837–8). *E. horridus* var. *nanus*, Schuster l.c.

GEOGRAPHICAL RANGE : Addo Bush, in the Uitenhage Div.

Uitenhage Div. : around Despatch, *Rattray* (S. Afr. Mus. Herb. 845 ! 1906 !).

Cultivated at Pretoria, *Nat. Herb.*, no. 8037 ! Living specimens at Kew !

As this species is liable to be confused with *E. latifrons* Lehm.,
we give the following table of differences :—

E. horridus.	E. latifrons.
No aerial stem.	Stem 8–10 ft. high.
Leaves glaucous.	Leaves not glaucous.
Leaflets rather narrow.	Leaflets distinctly broad.
Male and female cones small, about 1 ft. long.	Female cone large, about 2 ft. long, up to 60 lbs. weight, the upper part not fertile.
Habitat : Thorny scrub.	Habitat : Open grass veld.

4. **E. kosiensis** (Hutch. in Kew Bull. 1932, 512) ; stemless or nearly
so (stems at most a few inches high—*fide Aitken and Gale*) ; leaves
probably about 3 ft. long ; rhachis narrowly grooved on the upper
side, glabrous ; leaflets probably about 20 pairs, crowded and slightly
overlapping, oblong-elliptic, sessile with a very broad base, more
or less rounded on each side at the base, without a definite apex
but divided into 3–5 broadly triangular pungent-pointed lobes,
with usually about 2–4 smaller lateral teeth on each margin. 3–6 in.
long, 1½–2 in. broad, the largest with about 25 rather faint parallel
nerves ; cones not seen.

GEOGRAPHICAL RANGE : Coastal region of Zululand.

Zululand : behind sand-dune bush near Kosi Lake, East Ingwavuma district,
Aitken & Gale, 63 ; Kosi Bay, *Col. Lugge in Natal Herb.*, 16507 ! Cultivated
by *Col. G. Molyneux* at the " Old Fort," Durban, July 1930 !

Further material and information about this species is much desired at Kew.

5. **E. Lehmannii** (Lehm. Pugill. vi. 14 (1834)) ; stem up to 9 ft.
high ; persistent leaf-bases broadly ovate, acuminate ; leaves
markedly glaucous, up to 3 ft. long ; leaflets spaced on the rhachis,
the middle ones the largest, entire or lobate-dentate, very pungent-
pointed, linear or linear-lanceolate, the middle ones up to 8 in. long
and ⅔ in. broad (excluding the teeth), very obscurely or not even
visibly nerved, the teeth when present mostly on the lower margin

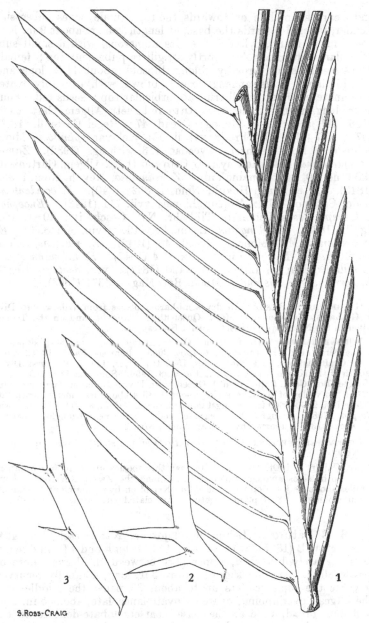

Fig. 6.—ENCEPHALARTOS LEHMANNII Lehm.—(1) Middle portion of leaf; (2) and (3) lobate leaflets; all reduced.

D 2

and near the middle or towards the top ; male cones subsessile, slender, tapered towards the base, at length yellow, about 9 in. long and 2 in. in diam., cylindric ; scales numerous, with a small suborbicular or quadrangular shortly strigillose-pubescent top ; female cones sessile, short, broadly oblong-ellipsoid, about 1½ ft. long and 1 ft. in diam. ; scales at length ruddy brown, with a broadly ovate-acuminate limb and small truncate orbicular top ; seeds red, about 2¼ in. long, including the aril.   Otto and Dietr. Allgem. Gartenzeit. 1836, 217, t. 1 ; Miq. Monogr. Cycad. 47 ; Regel Gartenfl., 1865, 197, t. 477 (♂) ; DC. Prodr. xvi. ii. 531, incl. var. *spinulosus* Miq. ; Schuster in Engl. Pflanzenr. Cycadac. 113, with figs. (1932). *Zamia Lehmanniana,* Eck. and Zeyh. ex Otto and Dietr. Allgem. Gartenzeit. 1833, no. 20, p. 158, name only. *Z. spinulosa,* Heynh. Nom. i 862 (1840). *Z. elongata,* Heynh. Nom. i. 862 (1840). *Z. occidentalis* Lodd. Cat. no. 177 ex Miq. in Linnæa xvii. 711 (1843). *Encephalartos spinulosus,* Lehm. in Tijdschr. Nat. Gesch. iv. 420, t. viii. fig. B (1838). *E. elongatus,* Lehm. l.c. 419, t. viii. A (1838). *E. mauritianus,* Miq. Monogr. Cycad. 48 (1842). *E. pungens,* Lehm. Pugill vi. 13 ; Miq. in Linnæa xix. t. 4 (1847). *E. Lehmannii* var. *spinulosus,* Miq. in Linnæa xix. 420 (1847) ; var. *dentatus* Regel. *E. horridus* var. *trispinosa,* Hook. Bot. Mag. t. 5371 (1863).

GEOGRAPHICAL RANGE : In dry semi-karoid places from Willowmore Div. to Grahamstown, and in Bedford, Queenstown, Komgha, and on the Tsoma River in the Nqamaqwe district (Native States).

Willowmore Div. : Groote River, *Gill* ! Albany Div. : northern slopes of Bothasberg, near Grahamstown, *Galpin,* 3083 ! Penrock Farm, 10–12 miles from Grahamstown, in karoid scrub, 1500 ft., *Dyer,* 1184 ! Bathurst Div. : mouth of the Kowie River, 250 ft., *Macowan,* 1959 ! Somerset Div. : Bruntjes Hoogte (Bot. Gard. Cape Town) ! Queenstown Div. : Junction Farm, *Rattray,* 1274 ! In valley of the Zwart Kei, about 5 miles above its junction with the White Kei, 2300–2600 ft., amongst dolerite rocks, *Galpin,* 8090 ! Queenstown, *Galpin,* 2708 ! Komgha Div. : exposed rocky slopes near Komgha, 2000 ft., *Flanagan,* 1373 ! Tsoma Div. : near Tsoma, *Sim,* 26 !

CULTIVATED SPECIMENS : *Nat. Herb. Pretoria,* no. 8039 ! Living specimens at Kew !

Mr. Galpin and Dr. Rattray note that this species grows in fair numbers on steep hillsides amongst doleritic rocks along the Zwart Kei River about 5 miles above its junction with the White Kei, often by suckering and seedlings forming clusters of 8–10 plants together. Associated with it was *E. cycadifolius.*

6. E. latifrons (Lehm. in Tijdschr. Nat. Gesch. iv. 424, t. ix. fig. A, B (1837–38)) ; stem up to 8 ft. high and 4 ft. in circumference, rarely branched, upper part between the leaves more or less woolly ; leaves dark green, up to 3 ft. long, markedly recurved towards the top ; leaflets up to about 33 pairs, the middle ones the largest, overlapping, ovate or ovate-lanceolate, about 5 in. long and 2 in. broad, wide at the base, coarsely lobate-dentate on the lower margin, apex and lobes pungent-pointed, very prominently nerved, the nerves numerous, pubescent below, at length nearly

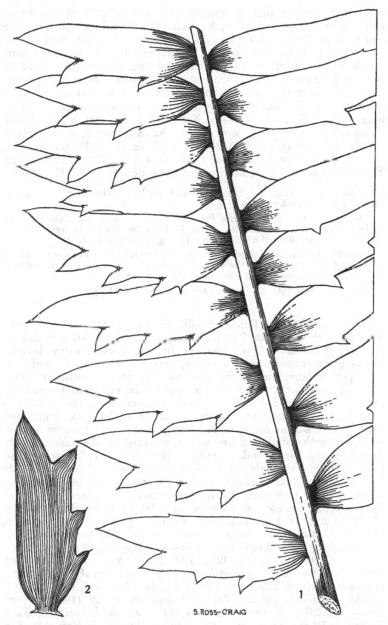

Fig. 7.—ENCEPHALARTOS LATIFRONS Lehm.—(1) Middle portion of leaf; (2) leaflet; much reduced.

glabrous; petiole with a marked yellow "collar" at the base; male cones 1–3 on a stem, with no visible peduncle, about 2 ft. long and 6 in. in diam., brownish yellow; scales much narrowed to and irregularly rhomboid at the top, not pubescent; female cones rare, up to 60 lbs. weight, sessile, broadly oblong-ellipsoid, up to 22 in. long and 10 in. in diam., with about 15 spirals of scales; scales very close, long-stipitate, umbonate and subrhomboid with a small concave, flat or beaked deeply rugose top; seeds about 2–2½ in. long, red. Lehm. in Tijdschr. Nat. Gesch. vi. 100, t. iii. (1839). *E. horridus* var. *latifrons*, Miq. in Ann. Sci. Nat. Ser. 3, x. 366, partly (1838); DC. Prodr. xvi. ii. 532; DC. Wild. Ic. Hort. Thenen. iv. 182. *E. Van Hallii*, Vriese in Tijdschr. Nat. Gesch. iv. 422, t. x. (1837–8). *E. horridus* vars. *Van Hallii* and *latifrons*, Schuster l.c. 117 (1932).

GEOGRAPHICAL RANGE: Uitenhage and Bathurst Divs., in open grass-veld and on low rocky ridges.

Uitenhage Div.: near Paarde Poort, *MacOwan*! Bathurst Div.: Trapps Valley, Clumber, 1000–1400 ft., about 10–14 miles from the sea, *Rattray*, 8439! (in S. Afr. Mus. Herb., 1100!). Living specimens at Kew!

According to Rattray the leaves persist for several years and occupy about 18 in. of the upper part of the stem; the female cones are ripe in January and the dried up male cones were then still present.

For comparison with *E. horridus* see note under that species.

7. **E. longifolius** (Lehm. Pugill. vi. 14 (1834)); stem well-developed, stout, about 12 ft. high or more, simple or rarely branched, the top dome-shaped between the bud and the mature leaves; leaves green, numerous in a crown, arcuate, recurved towards the tip; petiole fairly long, hairy when young; rhachis rounded below, with a prominent thick rib on the upper side, produced beyond the leaflets at the top; leaflets crowded, semi-erect (forming rather a wide V), lanceolate, obtuse to sharply acute at the apex, contracted and rather broad at the base, rather obscurely nerved, entire or with very few teeth usually on one side, the largest in the middle, up to 7 in. long and 1 in. broad, more rigid, thicker and darker green than in *E. Altensteinii*; male cones sessile or subsessile, oblong-lanceolate in outline, normally about 1½–2 ft. long and 6–8 in. in diam.; scales in very numerous spirals, lanceolate, acuminate with blunt hooked tops; female cones sessile, up to 90 lbs. in weight, 2 ft. long, 12–14 in. in diam.; scales in less numerous spirals than in the male, rhomboid, umbonate, very rugose; seeds red, broadly oblong, 1½–2 in. long and about 1 in. in diam. Miq. in Otto and Dietr. Allgem. Gartenzeit. 1838, 323; DC. Prodr. xvi. pt. ii. 531, incl. vars. *revolutus* Miq., *angustifolius* Miq., and *Hookeri* DC. l.c.; Schuster in Engl. Pflanzenr. Cycad. iii. figs. 3 A, 4 G, 6 A, 16 R–U (1932). *Zamia longifolia*, Jacq. Fragm. 28, t. 29 (1809); Pers. Synop. ii. 631; Spreng. Veget. iii. 908. *Zamia lanuginosa*, Jacq. Fragm. t. 30 and 31 (1809). *Cycas caffra*, Thunb. in Nov. Act. Soc. Scient. Upsal. ii. 283, as to part of descr. only. *Encephalartos caffer*,

Hook. Bot. Mag. t. 4903, not of Lehm. (1859) ; Rev. Hort. 1869, 233, fig. 56. *E. lanuginosus*, Lehm. Pugill. vi. 14 (1834). *E. Altensteinii*, Gard. Chron. Ser. iii. xl, 206, fig. 84 (1906), not of Lehm.

GEOGRAPHICAL RANGE : From Assegai Bosch in Humansdorp Div. to Van Staadens Berg, Zuurberg, in open veld.

Humansdorp Div. : Assegai Bosch, *Thunberg* ; without locality, Oldenburg 1497 (in Herb. Brit. Mus.) ! Dried cultivated specimens from various collections ! Living specimens at Kew !

Distinguishing features of this species, which was confused by Thunberg with the stemless *E. caffer*, and figured as such in the Botanical Magazine (tab. 4903), are the curled hook-like tops of the leaves, the leaflets forming a rather wide V ; they are often quite entire.

The pith is used in making Kaffir bread.

8. E. Altensteinii (Lehm. Pugill. vi. 11, t. 4, 5 (1834)) ; stem

up to about 16 ft. high and about 2 ft. in circumference ; young stem ovoid, woolly ; leaves green, numerous in crown, nearly straight, up to 5 ft. long ; rhachis soon glabrous ; leaflets spreading at a wide angle from the rhachis, moderately spaced, very numerous, the largest in the middle, the lower reduced in size but not becoming prickles, all rather broadly linear-oblong, the middle ones about 6 in. long and 1 in. broad, with about 3–5 teeth on both margins, in old plants becoming entire or nearly so ; nerves usually inconspicuous ; male cones 1 to several, shortly pedunculate, yellowish, cylindric, slightly tapered to the base, about 12–15 in. long and 4 in. in diam. ; scales beaked-acuminate, with a recurved flattish top ; female cones usually solitary, sessile, yellowish brown, broadly oblong-ellipsoid, about 1½ ft. long and 9 in. in diam. usually with about 15 spirals of scales, up to about 40 lbs. in weight ; scales rhomboid-umbonate with a truncate concave top, very rugose ; seeds red, oblong, about 2 in. long. Hort. Belge Journ. Jard. and Amat. iv. 167, pl. ix (1837) ; Miq. Monogr. Cycad. 51, incl. vars. *semidentatus* and *angustifolius* Miq. l.c. (1842) ; Miq. in Linnæa xix. 420, t. 5 (1847) ; Miq. Prodr. Cycad. 10 (1861) ; DC. Prodr. xvi. ii. 532, incl. var. *eriocephalus* Vriese Descr. Pl. Nouv. Jard. Leyd. t. 2 ; Gard. Chron. vi. 392–97, figs. 80–83 (1876) ; scr. 3, ii. 281 (1887) ; Hook. f. Bot. Mag. tt. 7162–3 (1891) ; Schuster in Engl. Pflanzenr. Cycadac. 112, with figs. (1932). *E. Marumii*, Vriese Tijdschr. Nat. Gesch. v. 188 (1838). *E. Vromii*, Malte Recherch. App. Lib. Lign. Cycad. 70, name only. *E. transvenosus*, Stapf & Burtt Davy in Fl. Transvaal i. 40, 99, fig. 4B (1926). *E. regalis* Hort. *Zamia spinosa*, Lodd. ex DC. l.c., name only. *Z. spinulosa* and *Z. spinosissima*, Hort. ex Miq. Monogr. Cycad. 51 (1842).

GEOGRAPHICAL RANGE : River valleys from the Kowie River, Bathurst Div., to Natal and the Eastern Transvaal, ascending to 3500 ft. alt. in the Amatola Mts.

Bathurst Div. : Bathurst, *Rattray in Herb. S. Afr. Mus.* 1099 ! King Williamstown, *Sim in Herb. S. Afr. Mus.* 847 ! East London Div. : Nahoon R., E. London, photo. by *Pearson in Herb. Galpin* ! East London Park, *J. Wood in Herb. Galpin* 7104 ! Komgha Div. : near **Komgha,** *Flanagan* 1372 (in Herb. S. Afr. Mus. 24360) ! Transkei Div. : Kei Road, *Rogers* 3225 ! Kentani, valleys, *Pegler,* 1116 ! Living specimens at Kew ! (I have not seen specimens from Natal and the Transvaal.—J. H.)

According to notes by Dr. Rattray, this species occurs from the Kowie River valley in the Bathurst Div. as far as Durban, and extends from the coast hills to the Amatola Mountains ; usually it is found in shady situations, and when well sheltered may reach a height of 18 ft. Plants which have not developed aerial stems are often hard to distinguish by their vegetative characters from *E. villosus,* which occurs in similar situations. In the latter species, however, the lower leaflets are much reduced and resemble prickles. Dr. Rattray also notes the interesting fact that the production of cones decreases as the southern limit of distribution is reached. In this region (Kowie Valley), in January 1908, he carefully examined more than a hundred well-developed plants and found the remains of only one male cone. Further north in the Nahoon and Gonubie valleys cones were produced in fair abundance. In Natal the species cones very freely. Baboons and monkeys play a part in distributing the seeds ; they collect and carry them to the tops of cliffs or trees, rejecting the essential hard kernel and eating only the delicate soft outer coat of the seed.

9. E. Woodii (Sander in Gard. Chron. 1908, 257, with habit figure) ; stem up to 18 ft. high, stout ; leaves green, numerous in a crown, slightly recurved, about 6 ft. long ; lower leaflets ovate, inclined to be gradually reduced to prickles, the middle ones the longest, ovate-lanceolate to broadly lanceolate, up to about 8 in. long and 1½ in. broad, the upper ones more crowded and arcuate, the lowest almost lobate-dentate on one or both margins, the uppermost leaflets often becoming entire, the apex and teeth pungent-pointed ; nerves numerous, fairly distinct ; male cones subsessile, rather slender, cylindric, up to 4 ft. long ; scales very numerous, long-pointed with a small truncate top ; female cones not seen. Prain in Kew Bull. 1914, 250, with habit fig., and 1916, 181 ; Schuster in Engl. Pflanzenr. Cycadac. 120. *E. Altensteinii* var. *bispinna,* J. M. Wood Ann. Rep. Bot. Gard. Natal 1907, 8, with fig.

GEOGRAPHICAL RANGE : Known only from Zululand.

Zululand : Ngoye, *Wylie* ! without definite locality, *Medley Wood* ! Cult. in Durban Bot. Gard. (*Natal Herb.* no. 16044 !).

This appears to be rather a distinct race or species very closely allied to *E. Altensteinii* and to *E. Hildebrandtii,* the latter from East Africa ; from the former it is distinguished by its usually much broader leaflets, which tend to become reduced to prickles towards the base, a character unknown in *E. Altensteinii,* but a prominent feature in *E. Hildebrandtii*; the latter, however, has the narrower leaflets of *E. Altensteinii.* There is a fine plant of *E. Woodii* in the Palm House at Kew.

10. E. paucidentatus (Stapf & Burtt Davy in Burtt Davy, Fl. Transvaal i. 40, 99, fig. 4, A (1926)) ; stem about 6 ft. high ; trunk about 3 ft. in diam. ; leaves green, about 8 ft. long, somewhat twisted ; rhachis shortly puberulous with crisped hairs, soon becoming glabrous or nearly so ; leaflets about 70 pairs, those towards the base spaced out and rather abruptly reduced to prickles, those higher

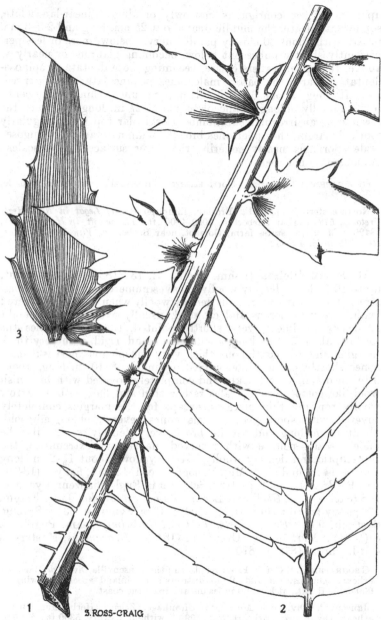

S. ROSS-CRAIG

Fig. 8.—ENCEPHALARTOS WOODII Sander.—(1) Basal portion of leaf showing
reduction of leaflets to prickles, typical of the species ; (2) upper portion
of leaf ; (3) leaflet from about the middle ; all reduced.

up more or less contiguous, narrowly or almost linear-lanceolate, somewhat falcate, the middle ones about 25 cm. long and 2·5-3 cm. broad, with about 30 closely parallel nerves below, more or less permanently shortly pubescent below, becoming glabrous or nearly so above, the upper leaflets entire, becoming more dentate to spinose-dentate towards the base ; male cones pedunculate, much curved, 1½-2 ft. long, about 6 in. in diam. ; peduncle thinly pubescent ; scales broadly oblong-oblanceolate, about 2½ in. long, glabrous, the lower ones an irregular subrhomboid-orbicular top with irregularly toothed margin, the upper ones broadly acuminate and very rugose ; male sporangia nearly covering the lower surface of the scales ; female cones not seen.

GEOGRAPHICAL RANGE : North-eastern Transvaal ; in partial shade, 3000-4000 ft.

North-eastern Transvaal : Breslau, Limpopo River, *Legat in Nat. Herb. Pretoria*, 5185 ! near Barberton, 3000-4000 ft., *Thorncroft in Herb. Rogers*, 28426 ! Moodies Estates farm, Oorskot, near Barberton, *Van Elden in Nat. Herb. Pretoria*, 10085 !

11. E. cycadifolius (Lehm. Pugill. vi. 13 (1834)) ; stem stout, up to 10 ft. high, densely woolly ; leaves numerous, up to 5 ft. long. straight or nearly so ; rhachis densely woolly when young, subterete ; leaflets very numerous and crowded, usually overlapping, straight or nearly so, linear, very shortly pointed, the middle ones the longest, about 5-6 in. long, about ⅛ in. broad, rigid, with very thick margins and 8-9 contiguous thick very conspicuous nerves ; male cones shortly pedunculate, curved, cylindric, 8-10 in. long, about 3 in. in diam., very thickly and completely covered with brownish wool like cotton wool ; scales rather thin and flat, with a narrow woolly top and thin margins, except for the margins completely covered with sporangia ; female cones several, oblong-ellipsoid, about 10 in. long and 5-6 in. broad, very densely woolly like the male ; scales peltate with recurved glabrescent extremities, the top elliptic and densely woolly ; seeds yellow, about 1½-2 in. long and 1¼-1½ in. in diam. Miq. Monogr. Cycad. 43, t. 1, fig. y-z (1842) ; DC. Prodr. xvi. ii. 531, partly ; Schuster in Engl. Pflanzenr. Cycadac. 108 (excl. syn. *E. Ghellinkii* Lem.). *Zamia cycadifolia*, Jacq. Fragm. 27, partly (as to t. 26 only *) (1800) ; Pers. Synop. ii. 631 ; Spreng. Syst. iii. 908. *Encephalartos Friderici-Guilielmi*, Lehm. Pugill. vi. 8 (1834) ; Miq. Monogr. Cycad. 44 (1842). *E. acanthus*, Masters in Gard. Chron. 1878, ii. 810.

GEOGRAPHICAL RANGE : From Kliplaat in the Jansenville Div. (*fide* Marloth). Cathcart, Queenstown and Tsolo districts ; an inland species growing at 3000-5300 ft. alt., at least 60 miles distant from the coast.

Jansenville Div. (fide *Marloth*) : Uitenhage, *Zeyher* (Herb. Mus. Brit.) ! Cathcart Div. : near Cathcart, *Sim*, 2999 ! without locality, 5300 ft., *Kuntze* !

* Jacquin's t. 25 is of a glabrous female cone, probably of *E. villosus* Lem.

Queenstown Div. : amongst rocks on mountain tops around Queenstown, 4000–4500 ft., *Galpin*, 1525 ! Summit of Mbumbula Range, Queenstown, 4700 ft., *Pearson* (photo.) ! Oxton Manor, Queenstown, 4000 ft., *Hay in Herb. Galpin*, 8411 ! King Williamstown Div. : near King Williamstown, *Sim in S. Afr. Mus. Herb.*, 1347 ! Tsolo district (*fide* Rattray).

Our knowledge of this species dates from Jacquin (Fragmenta Botanica, p. 27) in 1800. Unfortunately he confused two species. He described and figured at t. 26 a leaf of the plant which has always been known either as *E. cycadifolius* or *E. Friderici-Guilielmi.* But the female cone shown by him in t. 25 is probably that of *E. villosus* Lem., a species not recognised and described until 1867 ; this female cone is quite glabrous and the seeds are red, whereas in *E. cycadifolius* the female cone is very densely and permanently woolly, and the seeds are yellow.

12. E. lanatus (Stapf & Burtt Davy in Burtt Davy Flora Transvaal i. 40, 99, fig. 4 D (1926)) ; stems several feet high, simple ; leaves numerous in a cluster, those of the previous seasons persisting and hanging for some time ; rhachis very woolly, especially when young, subterete ; leaflets numerous, crowded, the middle and upper ones overlapping, linear, very pungent-pointed, about 4 in. long and 3½ lin. broad, glaucous and pilose above, soon glabrous, with 12–14 prominent and contiguous nerves below, margins very thick and callus-like, at first ciliate with woolly hairs ; male cones shortly pedunculate, cylindric, slightly narrowed at each end, nearly 6 in. long and 2 in. in diam., with about 12 spirals of scales ; scales narrowly rhomboid at the top and densely woolly-tomentose, about ¾ in. across ; female cones shortly pedunculate, cylindric, abruptly narrowed at each end, about 7 in. long and 3 in. in diam., with about 9 spirals of scales ; scales very broadly rhomboid, about 1 in. deep and 2 in. broad, densely but shortly tomentose except towards the margin which swells and forms a thick glabrous callus ; seeds broadly ellipsoid, about 1 in. long, probably yellow. *E. lævifolius*, Stapf & B. Davy l.c. fig. 4 e (1926).

GEOGRAPHICAL RANGE : Eastern Transvaal, from Middleburg through the Godwan River area to Barberton district, at 3000–5000 ft. altitude.

Transvaal : Middelburg district ; Mooi Kopje, *Pole Evans in Nat. Herb.*, 11427 ! Middelburg, *Merwe in Nat. Herb.*, 8041 ! Toevlugt, near Middelburg Town lands, *Weeber in Colon. Herb.*, 6471 ! Lydenburg district : Crocodile River, near Piet Schoeman, Apr. 1885, *Wilms*, 1355 ! (Herb. Mus. Brit.). Barberton district ; Godwan River, 3000–4000 ft., *Thorncroft in Herb. Rogers*, 28427 ! *Rogers*, 23689 ! Moodies, near Barberton, *Todd in Natal Herb.*, 2043 ! Edge of Black Reef quartzite escarpment, 1 mile from Berlin Forestry Station, up to 10 ft., suckering freely, *Van Nouhuys in Nat. Herb. Pretoria*, 10086 !

13. E. Ghellinckii (Lem. Illustr. Hort. xiv. Miscell. 80 (1867)) ; stems up to 6 ft. high (or more ?), in nature the lower part often resting on the ground, covered with the persistent corky leaf-bases ; leaves up to about 3 ft. long, graceful ; rhachis at first woolly-villous, with one obscure rib above and below ; leaflets numerous, widely spreading, subacicular, with strongly recurved margins, abruptly very pungent-pointed, the middle ones the longest and

about 4 in. long, woolly-tomentose when young, at length thinly
pubescent on the upper side, not visibly nerved below but with
about 4 obscure parallel nerves ; male cones sessile, slightly curved,
cylindric, about 9 in. long and 3 in. in diam., densely woolly ; scales
lax when mature, the top irregularly rhomboid, remaining densely
woolly, shortly stalked, nearly covered with microsporangia ; female
cone very shortly pedunculate, broadly oblong-ellipsoid, about 15 in.
long and 9 in. in diam., with about 9–11 spirals of very densely
woolly brownish scales ; scales remaining woolly, shortly peltate,
with recurved edges ; seeds very broadly oblong-ellipsoid, about
1¼ in. long, sordid brown or yellowish with a tinge of red towards
the base.   Lem. l.c. xv. pl. 567 (1868) ; Seward in Proc. Camb.
Phil. Soc. ix. 341, habit fig. (1898).   *Zamia Ghellinckii,* Hort. ex
Lem. l.c.   *E. cycadifolius* var. *Friderici-Guilielmi,* Schuster in Engl.
Pflanzenr. Cycadac. 109, partly (as to syn. *E. Ghellincki* Lem.).

GEOGRAPHICAL RANGE : Natal, apparently from almost sea-level to about
5000 ft. on the eastern slopes of the Drakensberg.

Natal : Umzimkulu River, *Nelson* 16 ! Umzimto district, *Wood in Natal
Herb.,* 13055 ! near the Umtwalume River, *Sanderson* 119 ! Tugela Falls,
Mont Aux sources, *Doidge* (cult. at Pretoria Nat. Herb. no. 8038) ! Tugela
Gorge, *Pulterill in Nat. Herb.,* 12446 ! *J. M. Wood,* 11865 ! In shady rocky
places, *Hutchinson,* 4536 ! Alexandra district ; stony slopes, Umgaye flats,
Dumisa, *Rudatis,* 1299 ! Without definite locality, *White* ! *Vyvyan* !

INDEX.

[Synonyms are printed in *italics*.]

Printed in the United States
By Bookmasters